MIND AS MACHINE

MIND AS MACHINE

A History of Cognitive Science

MARGARET A. BODEN

VOLUME 2

CLARENDON PRESS · OXFORD

OXFORD

UNIVERSITY PRESS

Great Clarendon Street, Oxford OX2 6DP

Oxford University Press is a department of the University of Oxford.
It furthers the University's objective of excellence in research, scholarship,
and education by publishing worldwide in

Oxford New York

Auckland Cape Town Dar es Salaam Hong Kong Karachi
Kuala Lumpur Madrid Melbourne Mexico City Nairobi
New Delhi Shanghai Taipei Toronto

With offices in

Argentina Austria Brazil Chile Czech Republic France Greece
Guatemala Hungary Italy Japan Poland Portugal Singapore
South Korea Switzerland Thailand Turkey Ukraine Vietnam

Oxford is a registered trade mark of Oxford University Press
in the UK and in certain other countries

Published in the United States
by Oxford University Press Inc., New York

British Library Cataloguing in Publication Data

Data available

Library of Congress Cataloging in Publication Data
Boden, Margaret A.
  Mind as machine : a history of cognitive science / Margaret A. Boden.
      p. cm.
  Includes bibliographical references and indexes.
  ISBN-13: 978–0–19–924144–6 (alk. paper)
  ISBN-10: 0–19–924144–9 (alk. paper)
  1. Cognitive science—History. I. Title.
  BF311.B576 2006
  153.09—dc22

                                        2006011795

Typeset by Laserwords Private Limited, Chennai, India
Printed in Great Britain
on acid-free paper by
Biddles Ltd., King's Lynn, Norfolk

ISBN 978–0–19–929238–7 (Volume 2)
ISBN 978–0–19–924144–6 (Set)

10 9 8 7 6 5 4 3

ANALYTICAL TABLE OF CONTENTS

VOLUME I

VOLUME II

WHEN GOFAI WAS NEWFAI

Difficult though it may be to believe, even one's grandparents were young once. They too knew the excitement of youth, and that first fine careless rapture. Likewise, even "good old-fashioned AI" was newfangled AI, once. And if not quite rapturous, certainly very exciting. (Sometimes careless too, as we'll see: 11.iii.a.) In short, GOFAI started life as NewFAI.

John Haugeland's GOFAI label (1985: 112–13) carries patronizing overtones absent from its alternatives: "symbolic", "classical", or "traditional" AI. It suggests a dismissive attitude all too often directed to this area of cognitive science, not to mention grandparents. My own label, NewFAI, also has a negative air. (For the record, it's my coinage, dating from the early 1980s; but others have picked it up and used it in print.) The word "newfangled" is an ancient one: I was taken aback, on a plane trip a few years ago, to encounter it in *The Canterbury Tales*. Geoffrey Chaucer used it to mean simply something *new*, but since then it has acquired a depreciatory tone. For several centuries past, according to the *OED*, it has implied *the love of novelty for its own sake or for purposes of gaudy show*. In our times, it also carries a hint of addiction to technological gizmos. All in all, then, it's hardly flattering.

My justification for using the label here is not that early AI was essentially meretricious. It wasn't. It did, however, often involve over-enthusiastic appraisals, over-optimistic predictions, premature efforts to address very complex issues, and—it must be admitted—self-deception or even dissimulation in describing what AI programs could do (see Chapter 11.iii). In brief, excitement led to overexcitement.

However, the excitement didn't spring from a shallow fascination for novelty or gadgets. It was grounded in an ambitious intellectual vision, which saw the future of theoretical psychology as the design of computational models isomorphic with (all kinds of) mental processes. Clearly expressed by Warren McCulloch and Walter Pitts in the 1940s (4.iii.e), this vision was echoed by several others in the 1950s (6.iii.c and iv.a–b). Hubert Dreyfus later traced its philosophical roots as far back as Plato (16.v.c). In its computerized incarnation, however, it was refreshingly new.

Moreover, it was seemingly endorsed by one of the first working AI programs, which found a proof for a theorem in *Principia Mathematica* more elegant than the one given by its eminent authors (6.iii.c, and i.b below). Small wonder, then, that people were excited.

A caveat, before we begin: It's not clear that all NewFAI researchers believed human thinking to be *constituted by*, or *identical with*, symbolic processes. Perhaps most did, but this was rarely made explicit. (For an early exception, see Newell and Simon 1961.) Even Alan Turing had sidestepped the issue (16.ii). "Strong" and "weak" AI (Searle 1980) hadn't yet been named, and the distinction wasn't always recognized by people at the NewFAI coalface. However, we don't need to worry about these matters until later (Chapter 16). When I speak of NewFAI/GOFAI, I'm referring to a programming methodology, not to the researcher's philosophical position. (So my usage differs from Haugeland's: in his definition, "GOFAI" includes a commitment to strong AI.)

Long though it is, this chapter would have been even longer had not much of the relevant research been discussed already. Natural language processing (NLP) was described in Chapter 9.x–xi; any NLP programs mentioned here are considered only for their relevance to AI in general. Similarly, NewFAI models of belief change and/or personality and emotion (see Chapter 7.i.a–d) are here ignored. And Allen Newell and Herbert Simon's initial work on problem solving was described in Chapter 6.iii.

The very earliest NewFAI research is described in Section i. The first AI labs and publications are noted in Section ii. Sections iii and iv, respectively, deal with the early search for generality and the later use of world knowledge in AI programs. (If asked to locate the maturing of NewFAI into GOFAI, I'd place it there.) And last, in Sections v and vi, I explain why much of the creative effort of NewFAI went into the design of new programming languages—including one which later inspired the revolutionary technology on our desks (and laps) today.

This chapter focuses on work in symbolic AI done before 1980. (In Chapter 12.vi–vii, we'll see why 1980 is a reasonable cut-off point.) As for how GOFAI fared after 1980, that story is taken up in Chapter 13. We'll see there that much of what we regard today as highly modernistic information technology was specifically inspired by visions first expressed within grandparent—or, in one case, great-grandparent—NewFAI.

10.i. Harbingers

The first paper in cognitive science, McCulloch and Pitts' 'Logical Calculus of the Ideas Immanent in Nervous Activity' (1943), was neutral as between symbolic AI, connectionism, and cybernetics. The cybernetics/symbolics schism happened later (Chapter 4.ix), largely driven by the 1950s harbingers of NewFAI.

These mid-century programs and programmatic visions implied that symbolic AI had a bright future. For some toe-in-the-water programs were capable of doing surprisingly 'intelligent' things. Moreover, the symbolic approach was better suited than the early network models to represent goal-directed action and propositional meaning. AI, it seemed, could model—maybe even match—human thought.

a. When is a program not a program?

If Gertrude Stein had said "A program is a program, is a program . . . ", she'd have been wrong. Or anyway, she'd have been wrong if she'd said this in the 1950s. For the sense in which people used the term shifted over the years.

That's not to say that there's a right sense and a wrong sense: most concepts don't have clear semantic boundaries, and this one is no exception (see 8.i.b and 9.x.d). Even today, people sometimes disagree about whether something is a program or merely an outline for a program, and about whether AI researchers should honour only the former. But the historical point, here, is that what NewFAI people were happy to count as a program changed, as implementation became more feasible.

The strong sense of *program* is a list of instructions which can—repeat, *can*—be run on some computer. (Never mind bugs: a buggy program is a program, in this sense.) But a weaker sense would cover a list of rules or instructions *written on paper and 'executed' by hand*. Such a program may not be capable of being fed to a computer. Even a *programming language* may be a purely paper-and-pencil exercise, if it hasn't yet been provided with a suitable compiler/interpreter (e.g. Newell and Simon's "Logic Language": Section v.b, below).

A yet-more-attenuated sense of the term would include a broad *design* for a program, without specification of the individual rules. In that sense, the first AI programs date back to the Second World War, when Turing—as early as 1941—wrote about how to solve problems by searching through the space of possible solutions, guided by what would later be called "heuristics". He even circulated a typescript on these matters to some of his Bletchley colleagues, including Donald Michie. Jack Copeland and Diane Proudfoot (2005: 15) describe this lost typescript as "undoubtedly the earliest paper in the field of AI", and I agree with them. However, an AI *paper* isn't the same as an AI *program*. For something to count as a program, even in the weak sense defined above, it needs to specify individual instructions, in some suitable order.

In the 1950s, weak-sense programs were often important. One example was what's *still* often referred to as "Shannon's chess program". In the late 1940s, Claude Shannon identified many of the basic problems which would have to be faced by anyone programming a computer to play chess (Shannon 1950*a*,*b*). This was highly influential work. Some ten years later, Newell and Simon acknowledged that "the framework he introduced has guided most of the subsequent analysis of the problem" (1958: 42). But their word "framework" was well chosen, for Shannon *didn't even provide a paper-and-pencil program*. What his 1950 papers provided, rather, was an abstract analysis of the task. (A paper-and-pencil chess program had already been written by Turing, but this wasn't yet well known—1947*a*: 23, 1950. It was implemented in 1951. It lost against a weak human player, was "rather aimless", and made "gross blunders"; however, it "reached the bottom rung of the human ladder"—Newell and Simon 1958: 46.)

Another example of a historically significant weak-sense program was the family role playing of Newell and Simon's Logic Theory Machine, or Logic Theorist (LT). In his autobiography, Simon recalled the historic occasion:

While awaiting completion of the computer implementation of LT, Al and I wrote out the rules for components of the program (subroutines) in English on index cards, and also made up cards for the contents of the memories (the axioms of logic). At the [CMU] building on a dark winter evening in January 1956, we assembled my wife and three children together with some graduate students. To each member of the group, we gave one of the cards, so that *each person became, in effect, a component of the LT computer program*—a subroutine that performed some special function, or a component of its memory. It was the task of each participant to execute his or her subroutine, or to provide the contents of his or her memory, whenever called by the [relevant] routine . . .

So we were able to simulate the behavior of LT with *a computer constructed of human components*. Here was nature imitating art imitating nature. The actors were no more responsible for what they were doing than the slave boy in Plato's *Meno*, but they were successful in proving the theorems given them. Our children were then nine, eleven, and thirteen. The occasion remains vivid in their memories. (Simon 1991: 206–7; italics added)

Given enough children, perhaps NewFAI could have rested content with this (weak-sense) type of program?—No. For people can't be relied on to act as though they were "a component of a computer program". Even 6-year-olds will bring their intelligence to bear, following the intention behind an instruction rather than its literal meaning and/or effortlessly (perhaps unconsciously) resolving ambiguities. I was convinced of this when watching my own small son as a LOGO novice in the mid-1970s (see Section vi.a), with Stephen Salter 'playing turtle' and helping him to get the point by deliberately misinterpreting his instruction to TURN (direction unspecified, because—so my 6-year-old son thought—'obvious').

Indeed, if people didn't do this sort of thing as a matter of course then buggy programs—ignoring syntax errors and typos—would hardly exist. Even the great Turing was a culprit, here. Newell and Simon, admitting that there's "no *a priori* objection to hand simulation of a program" but pointing out that it's usually unreliable, said:

For example, there is an error in Turing's play of his [chess] program because he—the human simulator—was unwilling to consider all the alternatives. He failed to explore the ones he "knew" would be eliminated anyway, and was wrong once. (1958: 46)

Why were they bothering to argue for the superiority of strong-sense programs over weak-sense programs as late as 1958? The reason was that most researchers still didn't have easy access to a computer, so were willing—or even forced—to rely on paper and pencil instead. Herbert Gelernter's famous geometry program, for instance, was sketched (by someone else) long before he managed to implement it (see subsection d, below). And John McCarthy started designing LISP while miles away from a suitable computer (v.c). (Many *neuroscientists* still needed to be convinced of the usefulness of functioning programs as late as 1988: Chapter 14.vi.c.)

A further reason for graduating from weak-sense to strong-sense programs was that, in the 1950s, it really wasn't clear just how much was actually achievable by a computer. It was one thing to say, with Turing, that *any* computation could in principle be computed by something like the Turing machine. But what was doable in practice might be very limited. So it was an act of creative exploration, as well as a matter of intellectual hygiene (the avoidance of self-delusion), to find out.

As soon as sufficiently powerful machines were produced by IBM (and, following McCarthy's invention of time-sharing, by DEC, the Digital Equipment Corporation), mere paper-and-pencil programs became less and less persuasive. In any case, there never would be enough "children": one very early NewFAI program contained 20,000 instructions, as we'll see.

Today, then, Stein would be applauded. A program which isn't a program in the strong sense attracts little interest from AI researchers. That's why speculative work on computational architectures, as opposed to executable specifications of psychological processes, is often undervalued (Chapters 7.i.e–f and 12.iii.d). And it's why John Holland's seminal work on genetic algorithms was ignored by many AI people for nearly twenty years, given that he'd written no GA programs himself and that no computer at the time would have had the power to run them in any case (15.vi.b).

b. The first AI program—not!

The Gertrude Stein ambiguity is one reason why identifying "the first AI program" is problematic. Another concerns what people are (or were) prepared to count as *intelligent*. Again, there's no 'right' answer: the concept of intelligence is as fuzzy-bordered as any other, and perhaps more fuzzy than most.

Received opinion holds that the first AI program was the Logic Theorist (Newell and Simon 1956*b*; Newell and Shaw 1957; Newell *et al.* 1957). This is stated baldly in many places: not just in popular accounts (e.g. Crevier 1993: 44) but in scholarly ones too (e.g. H. Gardner 1985: 145; Edwards 1996: 124; Bechtel *et al.* 1998: 53...and more). Arguably, it's also implied by the historian James Fleck (1982: 177), and even by Edward Feigenbaum (1936–) and Julian Feldman (1931–), both of whom were around at the time (1963: 108).

For all that, the oft-repeated claim that LT was the first working AI program is open to challenge—not to say false. Several programs, on both sides of the Atlantic, preceded it.

Newell and Simon began thinking about LT in autumn 1955 and, having been role-played by Simon's family in the New Year vacation, it was fully implemented some nine months later. Their first computer proof, of theorem 2.01 of *Principia*, was achieved in August 1956. And their first computer printout was triumphantly presented—right at the last minute—at the Dartmouth Summer Project in August 1956 (Chapter 6.iv.b). The triumph was deserved. This was the only computer printout to be shown at Dartmouth. And it was, *so far as most of the participants knew*, the first functioning program devoted to a task normally thought of as requiring significant intelligence. Moreover, it's clear with hindsight that it was hugely influential for the development of NewFAI as a whole (not *immediately*, however: see subsection g, below).

Newell had been inspired to start work on it by Oliver Selfridge's talk on pattern recognition, given at the RAND Corporation, Santa Monica, in November 1954 (Chapter 6.iii.b). Selfridge (1926–) had described not merely a programme for a (better) program, namely Pandemonium (Selfridge 1959), but also an already functioning, albeit primitive, system. This was a program written by Selfridge's colleague Gerald Dinneen to recognize letters of the alphabet (Dinneen 1955; cf. Selfridge 1955). It's relevant, here, that Newell later said:

Now I'd call it an artificial-intelligence program. It was not just a simple pattern-recognition device, but it actually carried out transformations and had several levels of logic to it. (interview in McCorduck 1979: 132; italics added)

What's more, the Selfridge–Dinneen program was already known about by the cognoscenti at the time of the Dartmouth event (which happened two years after that RAND seminar). Selfridge himself was one of the Dartmouth participants. However, the early alphabet-recognizer wasn't 'sexy' enough to be thought of as modelling *intelligence* or, to use RAND's preferred terminology, to be seen as a paradigm case of *complex information processing*. Pattern recognition was assumed to be relatively straightforward, at least if—as in this case—generalization wasn't involved (cf. 12.i.c).

The Logic Theorist, by contrast, was modelling one of the peaks of human thought: Bertrand Russell's logical reasoning. Accordingly, it got the historical credit for being the first AI program *even though* it owed its existence to Selfridge's ideas and Dinneen's early system.

Selfridge's Pandemonium was successfully implemented by 1958, and is described in detail in Chapter 12.ii.d. It could well have featured as an extra harbinger, here. For Selfridge's 1954 account of it not only inspired the Logic Theorist, but gave NewFAI two important concepts: parallelism and demons. Instead of writing an ordered sequence of interconnected instructions the programmer could define a set of logically independent procedures; since they didn't depend on each other, these were *in effect* parallel—even though they had to be executed one by one in a von Neumann computer. Moreover, instead of being executed immediately these procedures (demons) could *lie in wait* to be activated by a specific cue. These ideas, too, were later taken up by Newell and Simon, in defining production systems (Section v.e, below). Indeed, the demons in Pandemonium were intellectual ancestors of Simon's famous ant (7.iv.a). Considered as a harbinger, then, Pandemonium was second to none.

The Dinneen and Selfridge pattern-recognizers aren't the only rivals for the accolade of "first AI program". Arthur Samuel's (1901–90) program for playing checkers/draughts (see subsection e) is another—with an even stronger claim to the honorific label.

It was envisaged as early as 1947. At that point, it was seen as a moneymaking exercise, to raise funds for a large computer at the University of Illinois:

We would build a very small computer, and try to do something spectacular with it that would attract attention so that we would get more money. It happened the next spring there was to be a world checker champion meeting in the little neighbouring town of Kankakee, so somebody got the idea—I'm not sure it was mine, but I got the blame at least—that it would be nice to build a small computer that could play checkers. We thought checkers was probably a trivial game . . . Then, at the end of the tournament we'd challenge the world champion and beat him, you see, and that would get us a lot of attention. We were very naive. (A. Samuel, interviewed in McCorduck 1979: 148)

Naive, yes. But unlike Charles Babbage, whose similar 'moneymaking' scheme never resulted in anything (3.iv.c), Samuel did manage to make a machine play the game. His first program was implemented in 1949 at IBM (where he'd migrated from Illinois), on a '604' commercial calculating machine—and in 1951, on the brand-new IBM 701.

As for the *learning* version of the checkers program, this was up and running early in 1955 (Samuel 1959: 72). In Samuel's words:

[My program] achieved this aim—to improve its playing ability through [a] learning process involving heuristics—fairly early in its existence, certainly well before 1956 . . . Except for the fact that no publicity was made of the existence of my checker program, one could argue that a program employing learning heuristics had been "fully realized" by this time . . . My checker program was one of the first programs of any size to be run on the first experimental model of . . . the IBM 701. (interview in Copeland 1993: 253)

Despite his reference to "no publicity", Samuel had demonstrated his learning program on American TV on 24 February 1956, six months before the Dartmouth event (Samuel

1959: 72). Unlike Newell and Simon, however, he attended that meeting without bringing along printout evidence. Perhaps that's why one of Marvin Minsky's first crop of AI students, whom one might have expected to know better, would later date Samuel's implementation as "1961"—a full five years after LT (Raphael 1976: 5).

Although checkers is doubtless trounced by Russell's logic as an example of intelligence, it's not 'mindless' in the way that pattern recognition apparently is. As Samuel said:

checkers contains all of the basic characteristics of an intellectual activity in which heuristic procedures and learning processes can play a major role and in which these processes can be evaluated.

Some of these characteristics . . . are:

(1) The activity must not be determinate in the practical sense . . .
(2) A definite goal must exist . . . and at least one criterion or intermediate goal must exist which has a bearing on the achievement of the final goal . . .
(3) The rules of the activity must be definite . . .
(4) There should be a background of knowledge concerning the activity against which the learning progress can be tested.
(5) The activity should be one that is familiar to a substantial body of people . . . The ability to have the program play against human opponents (or antagonists) adds spice to the study . . . (Samuel 1959: 72–3)

In short, *organized* intelligence is needed for this game.

With respect to playing games (without learning to do better), others besides Samuel wrote programs several years before LT saw the light of day. Across the Atlantic, in May 1951, Christopher Strachey (1917–75) had coded a (buggy) heuristic draughts program for the pilot model of Turing's ACE machine at NPL (Chapter 3.v.c). Encouraged by Turing himself, he wrote an improved version for the Ferranti version of MADM. By the summer of 1952, he reported that this could "play a complete game of Draughts at reasonable speed" (Copeland and Aston n.d.). In addition, he showed that "non-mathematical" instructions outnumbered mathematical ones even in programs for doing numerical integration: in general, then, the *organization* of problem solving was the key to success (Strachey 1952). Samuel took some of his ideas from Strachey accordingly (Samuel 1959: 73).

As for chess, universally regarded as requiring intelligence (and often termed the *Drosophila* of AI, because it has attracted so much effort over the years), this too had already been programmed in Britain. Ferranti's Dietrich Prinz (1903–?)—who'd previously designed a rods-and-levers logic machine (see Preface, ii)—ran a brute-force chess program in November 1951. It played unimaginatively and slowly, and dealt only with the endgame (analysing all the several thousand possible moves in 'mate-in-two' problems).

Turing had been more ambitious. His TUROCHAMP chess program, begun (with the young Cambridge mathematician David Champernowne) in 1948 and still unfinished when he died, played a complete game—and used heuristics rather than brute force to do so (Turing 1953; Michie 1966: 37; Copeland and Aston n.d.). But Turing's ambition, here, didn't stretch to making his program known. He treated it as an entertaining, if challenging, project between friends, not as something to be touted at academic meetings. Being unknown beyond his immediate acquaintance, it couldn't have much influence even in England, still less in the USA.

One might mention even earlier cases, such as Turing's late 1940s program for writing love letters (Chapter 9.x.c). But that was just a toy, a joke. And it had no influence on the development of AI. Whereas it amused a few people, Selfridge's and Samuel's programs—and LT too, of course—excited many. Unlike them, it doesn't deserve (*sic*) to be regarded as a "discovery".

These examples bear out the general point made in Chapter 1.iii.f, that what's classed as a discovery is negotiable. At the dawn of AI, what counted as a program wasn't entirely clear, and what counted as intelligent (or even complex) wasn't clear either. So there was plenty of room for negotiation. There still is (Selfridge? Samuel? Strachey? Turing? Prinz? . . .). It's evident, nevertheless, that LT was/is widely *misperceived* as the first AI program.

c. How a program became a program

The history of Gelernter's geometry program nicely illustrates the shift from *program* (weak sense) to *program* (strong sense). It arose out of the Dartmouth conference. At that meeting, Minsky (having just seen the Logic Theory Machine) sketched a notional Euclidean "geometry machine", and got fellow organizer Nathaniel Rochester to persuade his young IBM colleague Gelernter to implement it (Chapter 6.iv.b).

The task took three years. Besides coding some 20,000 instructions, Gelernter's team at IBM had to invent a new programming language to describe geometrical 'diagrams' and operations (Gelernter 1959; Gelernter *et al.* 1960*a,b*). The operations included bisecting angles and dropping perpendiculars. This representation enabled his program to prune the search space drastically, since it tried to prove *only* those features which 'appeared' to be true in the 'diagrams'. (My scare quotes are a warning that this was *neither* computer graphics, *nor* a representation carrying the information that's available to a human being when looking at a diagram—see 13.iii.a.)

Technically, this was interesting because it showed how an internal model of a domain could be used in helping to solve problems about it. This idea, which Kenneth Craik had suggested—though in very different terms—in the early 1940s (Chapter 4.vi), would later be hugely important not only in GOFAI, but also in psychology and neuroscience (7.iv.d–e, 13.iii.a, and 14.viii).

Soon, over fifty different proofs, of up to ten steps, had resulted (Gelernter *et al.* 1960*b*: 143). But one above all attracted attention: an elegant (construction-less) demonstration of the equality of the base angles of an isosceles triangle. Gelernter himself already knew about this proof (McCorduck 1979: 188 n.). But he'd expected a different one, namely, the one used by Euclid. The unexpected proof ran as follows (see Figure 10.1):

> Consider triangles ABC and ACB.
>
> Angle BAC = angle CAB (common).
>
> AB = AC (given).
>
> Therefore the two triangles are congruent (two sides and included angle equal).
>
> Therefore angle ABC = angle ACB.
>
> QED.

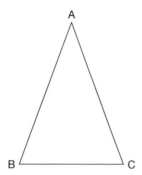

Fɪɢ. 10.1. Isosceles triangle

This proof is famous (in the history of mathematics, as in the history of NewFAI) not only because it's so elegantly simple, but also because it appears to be highly creative. A theorem concerning congruent triangles is used, which seems prima facie to be utterly inappropriate: after all, there is only the *one* triangle. Certainly, if a budding mathematician were to produce this proof spontaneously in the classroom, we'd be impressed.

(It was originally found by Pappus of Alexandria, six centuries after Euclid. But the program's computations differed significantly from those reported by Pappus himself. The reason is that, as noted above, its "diagrams" didn't carry the spatial information that real diagrams carry for human geometers. Careful comparison shows that the program was being less creative than it appeared—Boden 1990*a*: 104–10.)

What's relevant here is that Gelernter wasn't the first person to show that his program was capable of producing the proof. Indeed, he didn't even mention it in his 1959 paper. Rather, it was discovered by Minsky, using pencil and paper, long before the program was implemented.

The first public description of the incident was given by Minsky himself. This was in a 1958 forum from which, given the elegance of the proof and the identities of the participants, it was bound to spread far and wide (Minsky 1959 and Chapter 6.iv.b). And in the spreading, it was changed—by the sorts of communicative processes discussed in Chapter 8.vi. Ten years later, Seymour Papert set the record straight:

The relevant history began at a Summer Program on Artificial Intelligence held at Dartmouth in 1956. At this meeting, Marvin Minsky proposed a simple set of heuristic rules for a program [weak sense, above] to prove theorems in Euclidean geometry. *Before any program* [strong sense] *had been written* these rules were tried (by "hand simulation") on the simple theorem (sometimes called the "pons asinorum") which asserts that the base angles of an isosceles triangle are equal. *To everyone's surprise and pleasure* these rules led to an elegant proof quite different from the one normally taught in high school courses . . .

Although not entirely new, this proof is not well-known and strikes anyone with mathematical taste [in other words, not Dreyfus: Chapter 11.ii.b] as extremely elegant. So it is not surprising that *the story spread and very soon transformed itself, as stories do*, from "Minsky's rules suggested . . ." to "a program [sense ambiguous] generated a new proof" to "a computer [i.e. a strong-sense program] invented an elegant new proof". (Papert 1968: 111-6; italics added)

In the meantime, Papert continued, Gelernter—"justifiably encouraged by this little success"—used Minsky's heuristics in a program which two years later was proving theorems of much greater (i.e. ten-step) complexity. In short, the hand simulation of the Pappus proof helped spur Gelernter to convert Minsky's 'program' into a genuine, executable, program.

So who should get the credit? Who *discovered* geometry programming? Hard-headed twenty-first-century AI scientists may insist that only computer-executable (strong-sense) programs count as *programs*. If so, then the prize will be awarded to Gelernter. And indeed, he and his team did most of the hard work. But someone with historical antennae will point out that he was given the general idea, and the crucial heuristics, by Minsky—and that the most famous 'Gelernter proof' was done on the back of an envelope by Minsky too. Since *the very idea* of a geometry program was revolutionary at the time, Minsky deserves much of the credit.

d. First-footings

First-footings in the New Year are, one hopes, auspicious. The lucky piece of coal at the Dartmouth Summer Project was provided by Newell and Simon, in the form of the Logic Theory machine.

We saw in Chapter 6.iii.c why LT was so exciting for psychologists. But it galvanized budding AI researchers too, who were especially interested in *how* it worked its magic. Its programmers were well aware that this magic was limited: "LT's power as a problem solver is largely restricted to problems of a certain class", namely, proving theorems in symbolic logic (Newell *et al.* 1957: 129). A few years later, then, LT gave way to a new system: the General Problem Solver, or GPS (Newell *et al.* 1959, 1962; Newell and Simon 1961).

The first two papers on GPS combined reports of what it had actually done, when implemented, with reports of hand simulations carried out (in early 1959) by the programmers. And the latter much outweighed the former:

The program is on the RAND Johnniac. The work I am talking about here is all hand simulation. The GPS program is now being debugged, but is not solving any problems. Hence most of our information about it comes from extensive hand simulation. (Newell *et al.* 1962: 187)

Evidently, it was still acceptable to describe a program in the "weak" sense defined above. But the defensive nature of their comment shows that the GPS team themselves weren't happy with that: only strong-sense programs would really satisfy them.

What LT and—more powerfully—GPS did was to start with a symbolic represent-ation of the desired object (what would normally be termed the "goal"), and identify differences between that and the current state. The differences would then be reduced one by one, until (with luck!) all had been eliminated. This was done by applying operators to the current state, which were chosen and/or ordered by heuristics. The two key ideas, here, were drawn from previous work: difference reduction from cybernetics (Chapter 4.iii.e and vii.a) and heuristics from Simon's research on decision making in management (6.iii.a).

The term "goal" was used by the GPS team in a way not exactly equivalent to normal usage:

Each goal is a collection of information that defines what constitutes goal attainment, makes available the various kinds of information relevant to attaining the goal, and relates the information to other goals. There are three types of goals:

> Transform object A into object B,
> Reduce difference D between object A and object B,
> Apply operator Q to object A.

(Newell and Simon 1961: 284)

Strictly, then, GPS couldn't have a goal without having some fairly definite idea of how to attain it. Whether that's also true of human beings is debatable (and relevant to the distinction between a wish and an intention).

Different goal differences required different operators. If the preconditions for applying a given operator weren't satisfied, action had to be taken to establish them. For example, a logical operator considered by LT might require that the 'current state' be a conjunction, or have the term *a* before *b* rather than the other way round. If the current state didn't match this precondition, some other operator would have to be found whose actions would alter the current state to match this precondition. The preconditions of the other operator would either have to match the current state or be capable of being made to match as a result of execution of yet another operator, and so on.

In this way, sub-problems (and sub-sub-problems . . .) in LT, and sub-goals (and sub-sub-goals . . .) in GPS, were set up as means to attaining the end. If a chosen strategy couldn't be made to work, LT/GPS would backtrack to the previous choice point and try an alternative pathway through the search space. Differing search strategies could be followed in different cases. For instance, breadth-first search involved following *every* possible branch on a given level, but each one only for a few levels down before switching to the next. Depth-first search, by contrast, picked a path and followed it to the bitter end before considering the next alternative. (The combinatorial explosion, as usual, was lurking in the background. The ability to specify sub-goals on indefinitely many levels could be too much of a good thing: relentless depth-first search to an increasing degree of detail could bog the program down.)

Success wasn't guaranteed, however: this was *bounded* rationality, after all (6.iii.a). A new understanding of "algorithm" ensued as a result.

Derived from the name of the Arab mathematician al-Khwarizmi (2.i.b), the term had come to mean a formally defined (so potentially automatic) mathematical procedure guaranteed to succeed in finding the answer. And the first computer programs, designed as they were for military use, were—or aimed to be—algorithms in this sense. But because LT relied on potentially fallible heuristics to prune the search space, its programmers *denied* that it was an algorithm (Newell *et al.* 1957: 113–14, 117–18). For all that, a heuristic program *was* a formally defined, automatable, procedure—which some people continued to call an "algorithm". So while weak-sense programs were becoming strong-sense programs, strong-sense algorithms were being joined by weak-sense algorithms. For some years, while heuristic programming was still a novelty, talk of "algorithms" was sometimes ambiguous and/or misleading for this reason. Minsky, for one, complained of the "pointless argument" that arose as a result (1961*b*: 438 n. 26). Nowadays, the ambiguity has disappeared, at least for AI professionals: anything that can be run on some computer is called an algorithm, even if it implements a heuristic program.

GPS wasn't limited to logic problems, as LT had been. Even so, the name *"General Problem Solver"* was misleading. To be fair, the programmers had said this themselves:

GPS could really be better called GPSWP—sort of GPS with pretensions. By "general" we don't mean that GPS can solve all problems. "General" means that it is built in such a way that the specific content of the problem area is factored from the general problem-solving heuristics. Therefore it is capable of tackling a wide class of problems. I don't know how big the class is . . . (Newell *et al.* 1962: 189)

Any problem that could be represented in terms of goals, differences, actions, operators, and preconditions could in principle be solved by this program. The specific content was irrelevant. Moreover, some examples that human beings find tricky, such as the missionaries-and-cannibals puzzle, were solved by GPS (Chapter 6.iii.c).

However, it was the programmer who had to represent the problem in that form, which often required not only great effort but significant creativity. Arguably, that was the *real* problem solving—after which, the program itself merely 'turned the handle'. As the GPS team put it:

The real gimmick, or the good trick . . . is inventing the right spaces so that the operators we found and the differences we found made these tables of connectives small enough so that there is not too much pointless search. Most of the information we put into the system . . . was introduced by constructing the right problem spaces for the program to work in. This I think was the big selection. (Newell *et al.* 1962: 187–8)

This fact would become increasingly evident over the following decade. It was clear, for example, in Saul Amarel's (1928–2002) masterly discussion of six representations for the missionaries-and-cannibals puzzle (Amarel 1968). The most powerful of these could cope with a very large number of cannibals, arbitrarily distributed on either side of the river at the initial and/or terminal stages, using a boat whose capacity could vary during the development of the solution, and allowing for several missionary 'casualties'. (Amarel had hoped to formulate general principles whereby AI programs could create improved representations for themselves, but he had to admit defeat—Boden 1977: 334–40.)

For NewFAI, the most relevant 'functional' ideas in LT/GPS were:

* heuristic programming (NB heuristics had already been used by Samuel to evaluate alternative choices, but not to generate the search tree itself: see i.b, above),
* choice of methods (in GPS, called operators),
* similarity testing (in GPS, goal differences),
* the hierarchical sub-problem tree (in GPS, means–end analysis and planning),
* backtracking (to compare/select alternative possibilities),
* separation of the 'solving technique' from the specific content,
* the representation of the problem, or
* the search space, and
* building representations of varying structure and unbounded complexity (as opposed to fixed-size vectors).

There were technical excitements too, for LT and GPS were implemented by a revolutionary programming method, made possible by new programming languages (Section v.a, below). These provided:

* recursion,
* list processing, and
* push-down stacks.

These ideas turned out to be incalculably significant for the history of GOFAI. Over the next twenty years, the field was largely focused on theorem proving, heuristic search, planning, and—increasingly—knowledge representation. Some developments, such as flexible planning (Section iii.c), were very much in the spirit of GPS. Others, such as production systems (Section v.e) and "filtering" by constraint propagation (Section iv.b), were new approaches aimed at preserving its strengths while avoiding some of its limitations.

e. The book of Samuel

Another auspicious first-footer was the IBM researcher Samuel (see subsection b, above), later a member of Stanford's AI Department. Taking advantage of IBM's test machines at night, when his saner colleagues were safely tucked up in bed, he wrote a 6,800-instruction program for playing checkers. This was less interesting to psychologists than LT was, because he didn't claim to be modelling human intelligence. Indeed, he thought that studying the way people solve a problem gives one insight into "what the real problem is", but *not* into the brain's method of solving it (McCorduck 1979: 152). To AI people, however, his program was thrilling.

Samuel described it as playing "a fairly interesting game, even without any learning" (1959: 73). But it was the learning which really excited his contemporaries. Eight years after it was implemented, it was still "the only really successful attempt at machine learning in problem-solving" (Feigenbaum and Feldman 1963: 38). When Dreyfus first fired his cannon at AI in the mid-1960s (11.ii.a), even he saluted Samuel. This was no grudging praise: Samuel, he said, had done "very impressive work—perhaps the most impressive work in the whole of artificial intelligence" (H. L. Dreyfus 1965: 5).

In *playing* checkers, the program normally used a lookahead of three moves to evaluate all the legal alternatives, before choosing the best. (Samuel discussed a number of ways in which the lookahead might be increased, depending on the dynamics of the game: pp. 77 ff.) What counted as "the best" was decided by a method in which the program chose the move *most* likely to lead to good positions for itself and *least* likely to do so for its opponent:

It is not satisfactory to select the initial move which leads to the board position with the highest score, since to reach this position would require the cooperation of the opponent. Instead, an analysis must be made proceeding *backward* from the evaluated board positions through the "tree" of possible moves, each time with consideration of the intent of the side whose move is being examined, assuming that the opponent would always attempt to minimize the machine's score while the machine acts to maximize its score ... Carrying this "minimax" procedure back to the starting point results in the selection of a "best move". (Samuel 1959: 76)

To *improve* its play, the program used two methods: "rote learning" and "learning by generalization". In the first, it stored all the past positions, together with its initial evaluations of them. (Samuel stressed the importance of indexing, and fast sorting and searching procedures to find the record required.) If the third move in the lookahead happened to be one for which it had already calculated an evaluation, the lookahead—in effect—was increased to *six* moves. So the system pulled itself up by its own bootstraps, with the recent evaluations benefiting from a search much deeper than the official limit of three.

The second learning method was a way of improving the evaluation decision itself, by continuously adjusting the mathematical weighting of the individual test parameters according to their success in play. The program had thirty-eight parameters marking strategic features of the game (such as "threat of fork" and "center control"). Starting with only sixteen of these, it experimented to see which were the most useful, and weighted them accordingly. The least helpful ones were replaced by others drawn from the reserve list. This method was soon called *Alpha–Beta* pruning, based on the names Samuel gave to the program and its opponent (1959: 83). It gave the program an impressive strength, namely, the ability to improve its game differently in response to different people. An opponent's idiosyncratic weak spots could be discovered, and the relevant parameters weighted accordingly—even if they weren't *usually* very helpful. (The machine's play could improve even without a human opponent, if Samuel set two copies of the program 'against' each other.)

Samuel's program was a milestone for NewFAI, in three senses. First, it provided heuristic techniques which would ground further developments for years to come. To use heuristics at all in AI was original. (Turing himself still regarded machine learning as *either* random *or* systematic, by which he meant exhaustive search: A. M. Turing 1950.) In particular, Samuel's minimaxing heuristic is still crucial in machine learning.

Samuel wasn't the first to define it (John von Neumann had done that by 1928), nor to suggest using it for computerized game playing (Shannon had done that by 1950). He wasn't even the first to implement it: Strachey had done that by 1952, using a lookahead of six moves and more. But he was the first to show how minimax could be achieved by a complex evaluation function, defined in terms of various features of game strategy—"center control", for instance. (Strachey had used a simple count of 'own' and 'enemy' pieces.) He was the first to show how that function could be continuously improved. And he was one of the first to suggest that future learning programs might be able to generate their own evaluative parameters (1959: 87, 95). (Selfridge had preceded him here; soon, a new version of Pandemonium would define its own feature detectors: see 12.ii.d.)

Second, in 1962 the program became a publicist's dream overnight, when it beat Robert Nealey, a blind chess player from Stamford, Connecticut—described in the press as a former state champion. (The match had been set up at the request of the editors of *Computers and Thought*, then still in draft.) Nealey himself was impressed:

Our game...did have its points. Up to the 31st move, all of our play had been previously published except where I evaded "the book" several times in a vain effort to throw the computer's timing off. *At the 32–27 loser and onwards, all the play is original with us*, so far as I have been able to find. It is very interesting to me to note that *the computer had to make several star moves in order to win*, and that I had several opportunities to draw otherwise. That is why I kept the game going. *The machine, therefore, played a perfect ending without one misstep. In the matter of the end game, I have not had such competition from any human being since 1954, when I lost my last game.* (Feigenbaum and Feldman 1963: 104; italics added)

The publicity generated was huge. (IBM weren't happy. They'd never really favoured Samuel's program, although it was useful for testing their new computers, because "it smacked too much of machine thinking, etc., and they wanted to dispel any worry people had with machines taking over the world and all that sort of thing": Samuel, in McCorduck 1979: 151.)

Journalists announced checkers to have fallen decisively to the computer. Most AI professionals apparently believed this too, since they abandoned it for chess (e.g. Newell *et al.* 1958*b*; A. Bernstein and Roberts 1958; Bernstein *et al.* 1958)—which soon provided NewFAI with yet another highly publicized success (see Chapter 11.ii.b).

In fact, Samuel's program wasn't so good—nor checkers so trivial—as this media-friendly achievement suggested. Not only was Nealey *not* an ex-champion of Connecticut (his rating was just below the 'master' level), but he'd made a careless mistake—perhaps because he wasn't taking his opponent seriously (Schaeffer and Lake 1996). In a return match a year later, he won all ten games roundly. Samuel himself wasn't misled. He knew the program had reached only near-master level, and worked hard on an improved version (Samuel 1967). (Nearly thirty years later, he admitted in a letter that he still "had no idea how he could make his program good enough to compete with the world champion": Schaeffer and Lake 1996. Victory at Kankakee remained as elusive as ever.)

Third, and no less compelling to the public eye, the program learnt to beat Samuel himself. This was even more significant, many people felt, than LT's feat of surpassing Russell in logical elegance. The fact that Samuel was a weak player was irrelevant. The sceptic's mantra that "a program can do only what its programmer tells it to do" now had to be seen in a new light. At that time, this was IBM's mantra too (McCorduck 1979: 159). They were so worried about the IBM 704 being seen as threatening that their salesmen were instructed to quote Ada Lovelace's deflationary remark—*[The computer] has no pretentions whatever to originate anything. It can do whatever we know how to order it to perform* (Chapter 3.iv.b)—as often as possible.

Add Samuel's claim that "one can say with some certainty that it is now possible to devise learning schemes which will greatly outperform an average person" (1959: 95), and the fuse had been lit. AI was on the way up.

f. Programmatics

Just how far up was "up"? Partly because it was still difficult to communicate *anything* to a computer (Section v, below), that question was being answered less by programs than by programmatics.

Over-optimistic predictions were rife. One in particular, that a machine would be world chess champion within ten years (i.e. by 1967), would be a prime target of Dreyfus's attack a few years later (Chapter 11.ii.a). Indeed, RAND's Paul Armer reported in his upbeat 'Attitudes Toward Intelligent Machines' (1960/1963) that even *that* prediction was regarded by some as "conservative". (Armer himself didn't agree explicitly, but his preceding sentence had been "Few would have believed in 1950 that man would hit the moon with a rocket within ten years": p. 405.) This hype had such strong roots that by the mid-1960s it had abated only a little. Simon was then proclaiming, in a book on management, that "Machines will be capable, within twenty years [i.e. 1985], of any work that a man can do" (Simon 1965: 96). (We're now another forty years on . . . and still waiting.)

However, even over-optimists can sometimes be constructive. Eight of these early discussions were especially influential, and also remarkably far-sighted.

The earliest was Pitts and McCulloch's (1947) speculation about probabilistic networks and "universals" (Chapter 12.i.c). Soon after that came Shannon's (1950*a,b*) programmatic paper on chess, discussed above. Another was Selfridge's (1959) account of Pandemonium, in which (besides describing his program) he sketched future possibilities involving complex hierarchies of "demons" (12.ii.d). His vision inspired psychologists such as Jerome Bruner (6.ii.b–c), as well as AI researchers.

The fourth example of fruitful programmatics was Turing's (1950) paper in *Mind*. (His two papers of 1947 would have qualified here, had they been published; but they didn't appear until twenty years later: 1947*a,b*.) As explained in Chapter 16.ii.a, the *Mind* piece was *not* primarily about the Turing Test. That had been added as a jokey tease, which tantalized the media worldwide and diverted attention from the technical aspects. In fact, the paper was intended as an outline research programme for AI.

Turing gave only sparse hints about the results already achieved, some of which were still top-secret anyway. But what his text lacked in detail, it made up for in scope. Turing recommended AI work on intellectual activities such as chess, as well as on perception, language, and inductive learning—and pointed out that computers, like people, could make mistakes. He sketched the chess program which, when implemented a year later, would lose against a weak human opponent (Section i.a, above). And he outlined the fourfold classification of computational systems that's now normally attributed to Stephen Wolfram (15.viii.a).

Perhaps the most significant aspect of Turing's paper was that he saw AI (not yet named as such, of course) as a matter of programming computers, not building them. Quite a few people at that time were trying to produce machines capable of problem solving and/or learning, but they were doing so by means of engineering (12.ii.a–b).

However, Turing's ideas didn't immediately sweep through the scientific community: the arcane pages of *Mind* were rarely perused by them. The paper was first made more widely accessible in 1956 (see ii.b, below). Meanwhile, his equally prescient paper on 'Intelligent Machinery' (1947*b*), in which he'd declared that "intellectual activity consists mainly of various kinds of search" (of which "genetical or evolutionary" search was one possibility), remained unpublished until many years after his death. As a result, the notion that *programs*, such as the Logic Theorist, might be the royal road to AI struck many visitors to Dartmouth in the summer of 1956 as a revelation (McCarthy 1989). Even at the end of the 1950s, Frank Rosenblatt's "Perceptron" was a physical, not a virtual, machine (12.ii.e). (The capitalized word "Perceptron" names a specific piece of hardware built by Rosenblatt, whereas without capitalization it denotes a *class* of parallel-processing systems; although this class includes the Perceptron, most perceptrons are virtual rather than physical machines.)

Next, was McCarthy's 'Programs with Common Sense' (1959), which described a joint project also involving Minsky. Delivered at the seminal 1958 meeting in London (6.iv.b), this was the first published intimation of logicist AI—which would become both hugely influential and hugely controversial (Sections iii.b and e, and Chapter 13.ii).

McCarthy had been nurturing these ideas for a decade, ever since—as an undergraduate at CalTech—he'd gatecrashed the 1948 Hixon symposium and heard von Neumann's talk on automata theory (cf. 15.v.a). Today, he remembers being "turned on" to the idea of the brain as a machine by that talk, and perhaps by Wolfgang Kohler's too—although when he went back to look at their texts, he couldn't find any references to machine intelligence (personal communication). He did, however, beard von Neumann in his Princeton office, to suggest that the brain is an information-processing system *à la* Shannon. Von Neumann encouraged him to try to develop that idea. So where von Neumann's ultimate vision had been (artificial) self-reproduction, McCarthy's—from that moment on—would be (artificial) intelligence.

He soon became dissatisfied with his own equations, believing that logic must, somehow, be the way forward. One might be tempted to say that he was treading in McCulloch's footsteps. For McCulloch had used (a different type of) logic as a key idea in his 1943 paper, and had even tried to define the vocabulary of natural language in logical terms—before giving up in disgust at the difficulty of doing so (4.iii.c). But the young McCarthy didn't know about that.

Indeed, in 1955 he still knew nothing of Newell and Simon: "I had no idea that anyone was doing logic on computers" (personal communication). Meanwhile, he'd failed to persuade many people that automata theory had much to do with intelligence, and he was disappointed by the non-semantic, non-psychological, nature of the papers in the volume he'd co-edited with Shannon (6.iv.b). ("I was against behaviorism, because I used the IBM 704: lots of input/output, but also lots of internal state, which they (or anyway, Skinner) ignored": personal communication.)

So in the final section of the funding application for the Dartmouth Summer Project, he proposed "to study the relation of language to intelligence". In particular, he wanted to formulate a language that could express "conjectures" as well as proofs or instructions, and which would "contain the notions of physical object, event, etc." (McCarthy *et al.* 1955: 52–3). Now, he had his chance.

The 'Common Sense' paper described the *advice taker*, a "proposed [*sic*] program for solving problems by manipulating sentences in formal languages" (McCarthy 1959: 75). Since the inferential procedures and heuristics "will be described as much as possible [not in the particular program but] in the language itself", it followed that the *advice taker* would have a huge advantage over 1950s programs: "its behaviour will be improvable merely by making statements to it, telling it about its symbolic environment and what is wanted from it".

McCarthy's inspiration, here, was our own common sense. Most human intelligence involves natural language, where common-sense inferences are typically performed so easily that we're hardly aware that they've taken place at all. Programs, said McCarthy, should be able to do this too. Moreover, other desirable features depended on it. In particular, a program should be able to learn by taking verbal advice from a person—who needn't know anything about the structure of the program as such, and who certainly wouldn't have to *reprogram* it in order to provide an additional datum.

("Advice", here, meant telling someone a relevant fact, as opposed to instructing them on what to do.)

It followed that if a program is to be capable of learning something, it first has to be capable of being told it. How could that be achieved? The answer, McCarthy suggested, was predicate logic.

He wasn't merely saying (like the logical atomists and the young McCulloch before him: 4.iii.a–c) that natural language can be represented by some form of logic. He was saying that predicate logic could provide the common-sense inferences as well as the meaning. And the user wouldn't need to tell the computer what to do: the program would deduce not only declarative sentences, but also imperatives instructing the machine to do certain things (such as "printing sentences, moving sentences on lists, and reinitiating the basic deduction process on these lists").

As McCarthy defined his project,

A program has common sense if it automatically deduces for itself a sufficiently wide class of immediate consequences of anything it is told and what it already knows. (ibid.)

(The example he chose to discuss was deciding how to get to the airport, given that one is sitting at home with one's car parked outside.) Much of the ensuing controversy—which continues today—concerned whether a "sufficiently wide class" of consequences could, in general, be deduced by logical means.

Already, in 1958, there were sceptics. The NPL discussants included the logician Yehoshua Bar-Hillel, who'd initially been sympathetic to formal/cybernetic approaches to language but who'd recently emerged as a major critic of machine translation, or MT (Chapter 9.x.b and e). He tartly remarked that McCarthy's ideas were "half-baked", philosophically problematic, vitiated by what would later be called the frame problem (see iii.e, below), and careless of people's capacity to change their minds—not to mention the importance of *time* in deciding how to get to the airport. He ended his many criticisms by saying:

To make the argument [about getting to the airport] deductively sound, its complexity will have to be increased by many orders of magnitude. So long as this is not realized, any discussion of machines [such as that proposed] is totally pointless. The gap between Dr. McCarthy's general programme [i.e. program] ... and its execution even in such a simple case ... seems to me so enormous that much more has to be done to persuade me that *even the first step* in bridging this gap has already been taken. (in the 'Discussion' appended to McCarthy 1959; italics added)

McCarthy was unmoved. Or rather, he was moved to beef up his logic: ten years later, he admitted that the basic predicate calculus wouldn't be enough, that various modal logics would be needed too (iii.a, below). But his commitment to logicism endured—and it still does (13.i.a).

"Twin" examples of NewFAI programmatics which avoided such stringent criticism were 'The Processes of Creative Thinking' and 'Chess Playing Programs and the Problem of Complexity', both due to the LT team (Newell *et al.* 1958/1962, 1958*b*). These set out the general principles which had informed LT and would guide the writing of GPS: bounded rationality, heuristics, hierarchical structure, and so on. They caused great excitement in the small AI community and, like the other forward-looking papers mentioned here, would be cited by Minsky in his own triumph of programmatics (see below).

Soon, they'd cause excitement in other communities too. The RAND memo on creative thinking reached an interdisciplinary audience in 1962, when it appeared in a collection co-edited by the leading Gestalt psychologist Max Wertheimer (Chapter 5.ii.b). As for its twin, on chess and complexity, this was reprinted a year later in *Computers and Thought.*

Meanwhile, Simon had confidently declared in the management journal *Operations Research:* "It is not my aim to surprise or shock you . . . But the simplest way I can summarize is to say that there are now in the world machines that think, that learn *and that create*" (Simon and Newell 1958: 6; italics added). Some readers of *Operations Research* would, therefore, have been eager to see the RAND memo. But they might have been disappointed. The title mentioned "creative thinking", but the focus was on problem solving and game playing: creativity in its everyday sense was ignored. Likewise, Rochester's section on 'Originality in Machine Performance' in the proposal for the Dartmouth meeting hadn't discussed originality in the arts and sciences (McCarthy *et al.* 1955: 49). In the 1950s, that wasn't considered a fit topic for AI.

To be sure, quite a few people were working—with scant results—on "automatic programming", with the hope of avoiding some of the tedium of programming. Besides typical GOFAI programmers, these included the young Richard Laing, seeking automatic designs for cellular automata (Laing 1961*b*; cf. Chapter 15.v.b). And the 'Creative Thinking' harbinger itself predicted aesthetically valuable computer-composed music. But it said nothing specific about how this might be done. Its last-resort advice (after citing hints from George Polya: 1945), "When all else fails, try something counterintuitive", threw no light on Bach fugues, nor even on a saloon-bar pianist's mediocre efforts.

As things turned out, AI models of what's normally meant by creativity had to await the 1980s (Chapter 13.iv). Then, Simon himself would initiate work on scientific discovery (Langley *et al.* 1987). Later still, he'd consider artistic creativity too. Indeed, when invited by the *Stanford Humanities Review* to write a target article for peer commentary on the "foundational assumptions of AI", he chose instead to enter the postmodernist lions' den with a paper on literary criticism (Simon 1994*b*). But that was far in the future. In the harbinger period, it was enough if one could tackle chess or logic, or usher the missionaries and cannibals safely across the river. Creativity could wait.

g. First 'Steps'

The seventh, and most wide-ranging, example of fruitful programmatics was Minsky's 'Steps Toward Artificial Intelligence'. This paper itself developed gradually, step by step.

The earliest version (under the title 'Heuristic Aspects of the Artificial Intelligence Problem') was written in early to mid-1956, as the first draft of what would later become an MIT Technical Report. Although it was still unpublished, Minsky made the draft available informally to visitors at the Dartmouth meeting (Newell and Simon 1972: 884). It identified Minsky's major influences as Shannon, Selfridge, and Solomonoff (see 12.ii.a). Newell and Simon weren't mentioned, for the very good reason that Minsky hadn't yet heard of them.

The second draft, written after the meeting (and published by MIT's Lincoln Laboratory in December), did mention them—but only in passing (Minsky 1956*c*).

That's curious, for two reasons. First, the Logic Theorist was the only program to have been demonstrated at Dartmouth. Second, an Institute of Radio Engineers meeting held immediately afterwards (in September) had involved heated complaints backstage from Newell and Simon, who resented the fact that McCarthy—who had speculated creatively, but achieved nothing (i.e. no AI program)—had been asked to report on the Summer School. Minsky himself had been involved in the negotiations, and had agreed that this wasn't fair. The compromise reached at the IRE was that McCarthy's general report of Dartmouth was followed by Newell and Simon's detailed account of LT. But Minsky's December update of his paper still gave only scant attention to the LT team. It wasn't yet clear, at least to Minsky, that much of what they had to say might apply to intelligence in general, whether natural or artificial.

Five years later, it was. Now retitled as 'Steps . . .', and including extensive discussion of Newell and Simon's work, the paper was officially published for the first time by the IRE (Minsky 1961*b*). It's this version which was reprinted in the best-selling collection *Computers and Thought*. From that time on (i.e. 1963), it reached a wide, and ever-growing, audience.

Years later, Minsky allowed that the initial response to LT had been unfair: "like Darwin and Wallace, Darwin had done all this work and Wallace had gotten this bright idea, but they both got equal attention at the time" (in McCorduck 1979: 108). His explanation of his "perhaps surprisingly casual acceptance of the Newell–Shaw–Simon work" was that he'd sketched the Geometry Machine on the back of an envelope "in the course of an hour or so", so didn't realize just how difficult automated logic would turn out to be (McCorduck 1979: 106). Moreover, the LT team had presented their work at Dartmouth as a contribution to *psychology*, rather than to AI as such. A historical titbit, to be borne in mind when considering Minsky's later work (Chapters 7.i.e and 12.iii.d), is that he now sees the main thrust of his research and Newell-and-Simon's as reversed:

I did not realize until much later [than the Dartmouth affair] the great joke; at almost every stage, I was in various ways more concerned with human psychology, they with artificial intelligence—but neither of us would have agreed at all with that description. (Minsky, interviewed in McCorduck 1979: 107)

Even today, 'Steps' should be highly recommended to new students of AI. They'd get a sense of some of the major, still largely unsolved, problems—as well as discovering how much 'new' research is reinvention of old wheels. At the time of its first publication, it provided not just intellectual dynamite but intellectual direction too.

Minsky provided a wealth of information, as well as analysis and speculation, for 'Steps' was backed up by a comprehensive annotated bibliography, published (also in 1961) in a different journal. Given the scattering of the relevant literature, this was very useful at the time, and helped to define the scope of the recently named new field (a revised version was included as an appendix when the discursive paper was reprinted in *Computers and Thought*).

In 'Steps', Minsky reviewed the first decade of research, and prophesied future developments. Many of his prophecies have come true—not least because he forecast problems as well as achievements. He had an eagle eye for the combinatorial explosion, for example (although he didn't use that label in discussing how to avoid it). Much as a good way of gauging the success of current cognitive science is to compare it with what was envisioned in *Plans and the Structure of Behavior* (see 6.iv.c and 17.iv), so in evaluating modern AI one could do worse than consult 'Steps'.

The outstanding characteristic of 'Steps' was that it adopted a mathematical, analytic viewpoint. (Regrettably, very few NewFAI people would follow that example: 11.iii.a.) Like McCulloch and Pitts (1943) before him, Minsky was doing what McCarthy called "meta-epistemology". That is, he was trying to identify *general* representational or computational constraints on the sorts of mechanisms that are *in principle* capable of computing particular notions.

Accordingly, Minsky defined and compared various general classes of program, mentioning specific cases only as illustrations. Samuel's checkers player, for instance, wasn't considered merely as having an ingenious learning rule, but as an example of the general class in which (*a*) evaluation depends on the estimation of "imaginary" situations, or planning, (*b*) reinforcement happens "when the actual outcome resembles that which was predicted", and (*c*) there is "some way of evaluating nonterminal positions" (1961*b*: 430–1).

Minsky considered both sides of the incipient cybernetic schism (Chapter 4.ix). He himself had experience of both: he'd started as a connectionist, combining spare parts from a B24 bomber to make a pioneering learning network, but was diverted/converted to NewFAI in 1955 by his MIT colleague Solomonoff (1926–), then working on the induction of grammars (12.ii.a). So in 'Steps', he discussed adaptive "self-optimizing" systems and early connectionist machines, and also highlighted NewFAI icons such as LT/GPS and the checkers player. Significantly, he tried to analyse the weaknesses of various AI methods, as well as their potential strengths.

His section on 'Hill-Climbing', for instance, was followed by one on 'Troubles with Hill-Climbing'. This mentioned both "local peaks" (12.v–vi) and "mesas", in which "the more fundamental problem lies in finding any significant peak at all", and which are especially likely if the evaluation procedure (or "Trainer") can detect only the *solution* of a problem (1961*b*: 411, 429; cf. Minsky and Selfridge 1961).

Another knotty problem highlighted in 'Steps' was credit assignment. In a complex program, each ultimate success is "associated with a vast number of internal decisions", so we must find some way of assigning "credit for the success among the multitude of decisions" (p. 432). In LT/GPS, this was done by breaking the problem into distinct sub-goals (later, it would be dealt with also by back-propagation in PDP connectionism, and by the bucket-brigade algorithm in evolutionary programming: 12.vi.c–d and 15.vi.a).

Property-list descriptions, too, were shown to be problematic (pp. 422–5). In particular, they were problematic if the "properties" were *unary* (defined in terms of only one object) rather than *binary* or *ternary* or . . . Consider the three diagrams shown in Figure 10.2, for instance. These structures (and substructures) can be described, and therefore compared, using familiar spatial words. A computer capable of describing/comparing them, said Minsky, would need some symbolic notation that includes (1) terms for binary relations such as *inside of, to the left of*, and *above* (and, with respect to Figure 10.2 as a whole, the ternary relation *between*), and (2) a way of specifying, and shifting, the hierarchical level at which these relations were being considered. In diagram (*c*), for instance, the triangle is inside the oval, and the circle is inside the triangle. In other words, *inside of* is a recursive relation. But being recursive doesn't necessarily mean being transitive: for some purposes, it wouldn't be

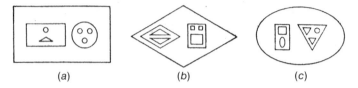

(a)   (b)   (c)

FIG. 10.2. Three diagrams for description and comparison. Reprinted with permission from
Feigenbaum and Feldman (1963: 423)

appropriate to describe the circle as being inside the oval, although for other purposes
it would. (These points would later be clarified in the ground-breaking ANALOGY
program, written by one of Minsky's students: see 13.iv.c.)

   Again, Minsky pointed out the difficulty of implementing hierarchy in neural nets
(cf. 12.viii.b and ix.a), and the need for carefully ordered training sequences "to insure
that early abstractions will provide a good foundation for later difficult problems"
(pp. 428, 434 n. 23; cf. 12.viii.c−e and x.e). And with respect to the (general) need
for recursive use of previous results, he suggested that future AI programs might
use heuristic mechanisms (as in "meiosis and crossing-over") providing for "the
segregation of groupings related to solutions of subproblems", and emulating "the
fantastic exploratory processes found perhaps only in the history of organic evolution"
(p. 434; cf. 15.vi). (John von Neumann had already said that copy errors could ground
evolution in cellular automata, but he hadn't discussed how this could be achieved in
practice: 15.v.a.)

   Minsky discussed the various types of internal "model" (e.g. "analogous, semantic,
and abstract": p. 435) which AI programs could use—and, perhaps, construct for
themselves. He showed, for example, that planning depends on simplified models of
the problem (pp. 442−3). And, sometimes citing his own mid-1950s work, he stressed
general conditions regarding what would soon be called "knowledge representation"
(KR). Minsky saw KR as "crucial":

The structure of the [representation] will have a crucial influence on the *mental world* of the
machine, for it determines *what kinds of things can be conveniently thought about.* (412−13; italics
added)

In short, he was asking what types of *virtual* world would enable machines to perform
various tasks (see Sections iii.a and v, below).

   To take just one example, what kind of "mental world" must a visual program have
if it is to be able to see *two* objects?

[The] property-list scheme is limited (for any given set of properties) in the detail of the
distinctions it can make. Its ability to deal with a compound scene containing several objects is
critically weak, and its direct extensions are unwieldy and unnatural. If a machine can recognize
a chair and a table, it surely should be able to tell us that "there is a chair and a table". (p. 422)

Besides choosing appropriate base-level property descriptors, he said, AI scientists
would have to find ways of "subdividing complex objects and describing the complex
relations between them". That quest would soon be the focus of MIT's programme of
scene analysis, carried out under his direction (Section iv.b below, and 7.v.b−d). (It was

also the key idea of Ulric Neisser's "analysis by synthesis", which eventually influenced some AI work on vision: 6.v.b.)

With hindsight, one of the most interesting aspects was Minsky's scepticism about Rosenblatt's "perceptrons", then being hyped in the press and on the airwaves (see 12.ii.e−f). This scepticism was expressed repeatedly throughout the text. For instance:

To recognize the *topological* equivalence of pairs such as those in [Figure 10.3] is likely beyond any practical kind of iterative local-improvement or hill-climbing matching procedure. (Such recognitions can be mechanized, though, by methods which follow lines, detect vertices, and build up a *description* in the form, say, of a vertex-connection table [i.e. scene analysis].) (p. 414)

How could one represent with a single prototype the class of figures which have an even number of disconnected parts? Clearly, the template system has negligible descriptive power. The property-list system frees us from some of these limitations. (p. 415)

[Rosenblatt's] nets, with their simple, randomly generated, connections can probably never achieve recognition of such patterns as "the class of figures having two separated parts", and they cannot even achieve the effect of template recognition without size and position normalization (unless sample figures have been presented previously in essentially all sizes and positions). (p. 421)

It is my impression that many workers in the area of "self-organizing" systems and "random neural nets" do not feel the urgency of this problem [of credit assignment] . . . For more complex problems, with decisions in hierarchies (rather than summed on the same level) . . . we will have to define "success" in some rich local sense. Some of the difficulty may be evaded by using carefully graded "training sequences" . . . (1961*b*: 432−3)

[The] work on "nets" is concerned with how far one can get with a small initial endowment; the work on "artificial intelligence" is concerned with using all we know to build the most powerful system that we can. It is my expectation that, in problem-solving power, the (allegedly brainlike) minimal-structure systems will *never* threaten to compete with their more deliberately designed contemporaries. (p. 446 n. 36; italics added)

And, after a careful mathematical comparison of various associative learning rules:

Incidentally, in spite of the space given here for their exposition, I am not convinced that such "incremental" or "statistical" learning schemes should play a central role in our models. They will certainly continue to appear as components of our programs [i.e. hybrid systems: 12.ix.b] but, I think, mainly by default. The more intelligent one is, the more often he should be able to learn from an experience something rather definite; e.g. to reject or accept a hypothesis, or to change a goal.

In light of the major scandal at the end of the decade, and its effects on the history of AI (Chapter 12.iii and vii.b), these early attempts on Minsky's part to counter the enthusiasm surrounding perceptrons were prophetic.

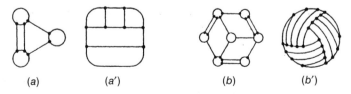

(a)   (a')   (b)   (b')

Fig. 10.3. The figures *a*, *a'* and *b*, *b'* are topologically equivalent pairs. Reprinted with permission from Minsky (1961*b*: 414).

Minsky's paper related the nascent AI to established areas of cognitive science. He cited the Gestaltist psychologists, Craik, Burrhus Skinner, Donald Hebb, Noam Chomsky, and George Miller—and Niko Tinbergen as well (4.vi and 5.ii.b–c). Twice, he praised MGP's *Plans and the Structure of Behavior* (6.iv.c). He recommended Bruner's New Look (6.ii) and Polya on heuristics (6.iii.b), and in comparing different learning strategies he mentioned mathematical psychology too (6.i.a). Neurophysiology and ethology were implicit when he defined various abstract "machines" for conditioning or noted the difficulty of "chaining" reflexes in lower animals having simpler brains (5.iii.a), and explicit when he spoke of "the bold pioneering work" of Nicholas Rashevsky (4.iii) and the "models based on brain analogies" described at the 1958 NPL gathering (6.iv.b). In addition, he remarked the recent experimental discoveries of visual feature detectors, at MIT and Harvard (14.iv.a–b).

Philosophy was included too, as when Minsky said that in thinking about "the 'mind–brain' problem" one should look to Craik and Gordon Pask, and that *freedom of will* depends on being able to distinguish our internal plan from intervention in our action (p. 447 n. 38; cf. Chapter 7.i.g). He devoted his penultimate section to 'Models of Oneself', arguing that an intelligent creature's model of the world would need to include a model of itself—which might well need to be "dual", as ours is (he'd explore this view further a few years later: Minsky 1965). The argument from Gödel's theorem (15.v. a) was based, he said, on a misinterpretation of Gödel. And the ineffability of "intelligence" and "creativity" may eventually be overcome. Programmers know, he said, that there's no mysterious "heart" in a program, only (at base) "senseless loops and sequences of trivial operations". Likewise, "[It] may be so with *man*, as with *machine*, that, when we understand finally the structure and program, the feeling of mystery (and self-approbation) will weaken" (p. 447). This was an early expression of functionalism, which Minsky's Cambridge colleague Hilary Putnam (1960) had just defined at length in an influential philosophical journal (15.iii).

In short, Minsky's attitude to these 'foreign' fields was interdisciplinarity, not condescension. He wasn't saying only that NewFAI had something to teach colleagues in other areas. It could learn from them, too.

It's relevant, here, to remark something which Minsky *didn't* say. Nowadays, he's regarded as one of the most optimistic champions of AI. He's tackling hugely difficult problems (Chapters 7.i.e and 12.iii.d), and is notorious for ebullient futuristic remarks. For instance, he once said to me that science would progress more quickly if *all* the funding money were given to AI, because AI programs will eventually surpass the intelligence of human chemists, geologists, biologists . . . and so on. That may have been tongue-in-cheek: he's nothing if not mischievous. But 'Steps' was carefully considered. Having opened by saying "we are on the threshold of an era that will be strongly influenced, and quite possibly dominated, by intelligent problem-solving machines" (p. 406), he could have closed with a vainglorious flourish. Instead, he forecast that AI programs, in the foreseeable future, would be no more than a powerful accessory to human thought:

[With the advent of time-sharing machines] we can work toward programming what will be, in effect, "thinking aids". In the years to come, we expect that these *man–machine* systems will share, and perhaps for a time be dominant, in our advance toward the development of "artificial intelligence". (1961*b*: 450; italics added)

h. The harbinger in the Bush

That final quotation from 'Steps' had an important implication: if man–machine systems were to be thinking aids for anyone other than computer scientists, they'd need to be usable by the man/woman in the street—or anyway, in the laboratory and the library. Minsky didn't mention this fact. That's hardly surprising, for AI people in the 1950s had enough to do making computers usable by their colleagues, never mind Joe Bloggs. One computer pioneer, however, had already thought about this issue before mid-century: MIT's Vannevar Bush (1890–1974).

Bush's scientific feet were planted firmly in the ground. He'd invented the (analogue) differential analyser in 1930, and was President Roosevelt's chief scientific adviser—in which role, he supervised the Manhattan Project and other military technologies during the Second World War.

Nevertheless, his speculative paper 'As We May Think' (1945, sects. 6–8) was recently described by William Gibson as "so genuinely, so embarrassingly, so rawly prescient as to make a science fiction writer squirm" (W. Gibson 2001, p. xi). This is no small accolade: it was Gibson's hugely influential novel *Neuromancer* (1984) which introduced the public to the concept, and the potential, of cyberspace. (A copy of the magazine containing Bush's paper was recently sold for $2,200: not bad, for something bought to while away a train journey—J. M. Norman 2005.)

What Gibson termed "the weird acuity of this man's imagination" was concerned with what would now be called information technology, not AI as such. Nevertheless, the paper's insights were the first explorations of issues that were later made concrete by GOFAI and which are crucial for the practice of AI today. Accordingly, it was Bush whom I had in mind when I referred, in the preamble above, to "great-grandparent" NewFAI. His work was so hugely far-sighted that it came to look even more exciting as the years passed than it had done at the time.

Bush's declared concern was to enable people to profit from, instead of being overwhelmed by, the explosion of knowledge in the previous fifty years—not least, the research done for the Second World War. Towards the end of the war, Bush—then Director of the Office of Scientific Research—was invited by the widely read *Atlantic Monthly* to suggest what scientists might do, once they had time again for non-military pursuits. His response was a visionary essay on how they could help society cope with the information explosion (Bush 1945). In fact, his vision was already some 10 years old: the magazine contribution was based on a paper he'd written *before* the war, in 1937 (Nyce and Kahn 1989, 1991).

His essay dealt with the future recording, storing, and accessing of scientific records—indeed, of texts and images in general. He forecast many of the IT applications used in offices and libraries today. (Gabriel Naudé would have been fascinated! Indeed, research libraries would be the first institutions to use computers for handling non-mathematical information.) Bush even forecast the personal computer itself. He called this the "memex", or memory extender, and described it as "a sort of mechanized private file and library".

Bush was writing science-as-speculation, but not science fiction. He took pains to relate his suggestions to specific achievements of mid-1940s science and technology—including many examples drawn from cybernetics. He anticipated add-on gizmos (automatic cameras, scanners for storing handwritten text and images, and speech-typewriters) and interface gimmicks (a push-button that would take one straight to the first page of a document's index, or a sliding "lever" that could turn the—virtual—pages of a stored document one, or ten, or twenty...at a time). As you'll know from your own experience, the modern versions of these "gizmos" and "gimmicks" aren't mere toys: to the contrary, they're near-essential, today, for many kinds of intellectual research.

Even more to the point, he also anticipated much of the internal software, or functionality. For he was envisaging a machine to aid, and even to model, creative thinking as well as mere (mere??) data retrieval. The basic idea was that the memex, and its user, could create "trails" of associations: stored memory links (aka hyperlinks) that would become increasingly idiosyncratic with personal use. Bush was deliberately contrasting *flexible associative thought* with the mid-century librarian's rigid hierarchical indexing, and imagining a machine that would fit in with it.

Part of that machine would be what we now call a search-engine. One can hardly fail to be reminded of Google, for instance, on reading Bush's description of how a medieval historian might use his own personal memex (1) to find out about the Turkish and English bows used in the Crusades, (2) to compare—and (3) to explain—their efficiency, and even (4) to speculate on why the more effective (Turkish) weapon wasn't eventually adopted by *both* sides.

The need to make digital computers human-friendly would grow inexorably as their power increased. When Bush wrote his piece, however, they were only just being invented (or perhaps one should say "reinvented", with a salute to a crowded living room in pre-war Germany: Chapter 3.v.a). Very little, as yet, could be done to make the memex a practical proposition.

Indeed, it didn't become practical in the research-lab sense until the 1960s, nor in the commercial sense until the 1980s. As remarked in Chapter 1.iii.g, then, ideas (such as hypertext) crucial to the memex were born/reborn several times within a half-century.

In 1959 Douglas Engelbart's team at the Stanford Research Institute (SRI) started trying to put Bush's vision into effect. Engelbart (1925–) explicitly acknowledged that he'd been inspired by Bush (1962: 47–69). Indeed, he'd been fired up by Bush's ideas as early as 1945:

At the end of the summer of 1945, just after the surrender of Japan, Engelbart was a twenty-year-old American naval radar technician, waiting for his ship home from the Philippines. One muggy day, he wandered into a Red Cross library that was built up on stilts, like a native hut. "It was quiet and cool and airy inside, with lots of polished bamboo and books. That was where I ran across that article by Vannevar Bush", Engelbart recalls. . . . [He] started designing computer-based problem-solving systems in 1951. (Rheingold 2000: 174–5)

Like Bush, he was thinking of a very wide range of users: "diplomats, executives, social scientists, life scientists, physical scientists, attorneys, designers". He'd also been influenced by his experiences with radar during the war, from which he concluded that a graphical display, changing in real time, might be linked to a computer—from which instructions could be fed back to alter the display.

His preliminary report on the project, prepared for the US Air Force Office of Scientific Research in October 1962, listed the team's goals partly as specific descriptions/predictions of technology (e.g. word processors), and partly as various futuristic dialogues (1962: 73–114). The latter featured an imaginary user and teacher:

[You, as a neophyte user, will be talking to] a friendly fellow (named Joe), who is a trained and experienced user of such an augmentation system within an experimental research program which is several years beyond our present stage . . .

Joe understands [that you don't know much about these matters], and explains that he will do his best to give you the valid conceptual feel that you want—trying to tread the narrow line between being too detailed and losing your over-all view and being too general and not providing you with a solid feel for what goes on. (Engelbart 1962: 73)

Those fictional dialogues with Joe are remarkable, now, for their 'modern' feel: similar interchanges go on every day in homes, computer shops, and introductory teaching labs. In addition, Engelbart foresaw graphic displays of structures which could be explored and inspected from different points of view and at different levels of detail—by an architect or a road designer, for example. (As it happened, Ivan Sutherland's *Sketchpad* was first described in the very year that Engelbart composed his report, 1962: see Chapter 13.v.c.)

At the time, however, the dialogues seemed not just fictional but science-fictional. Despite being selected as the opening paper in a forward-looking collection on information technology (Engelbart 1963), they didn't enthuse his colleagues:

Total silence from the computer science community greeted the announcement of the conceptual framework Engelbart had thought about and worked to articulate for over a decade. (Rheingold 2000: 180)

However, ARPA's Joseph Licklider had the imagination—and the intellectual fellow feeling—to get the point. He supported (generously, as ever) Engelbart's efforts to turn some of this fiction into reliably functioning reality. It took almost a decade.

At the 1968 Fall Computer Conference in San Francisco, Engelbart's team were able to demonstrate their own memex—equipped with the ancestral computer mouse. They called it NLS—or oNLine System (Nyce and Kahn 1991). This was a personalized version of human–computer "synergy", or symbiosis (Engelbart 1962: 79 ff.; cf. Licklider 1960). Three thousand computer scientists, of whom Alan Kay (1940–) was one, attended a two-hour demonstration—followed by a standing ovation. The session is now seen as "a landmark event in the history of computing" (Packer and Jordan 2001, p. xvi). Indeed, a film of the event was screened at a thirtieth-anniversary meeting at Stanford in 1998, at which the 73-year-old Engelbart received another standing ovation (Rheingold 2000: 326).

But "reliably functioning reality" in a laboratory demo isn't necessarily a reality that the man in the street can rely on. Stewart Brand, who helped with the original demonstration, soon spread the word about personal computing in his hugely influential *Whole Earth Catalog* (Chapter 1.iii.d). But it would be many years before Kay and others would be able to turn Engelbart's ideas into *commercial* realities (13.v).

A list of the pioneering features of NLS is doubly familiar. Besides reminding one of IT facilities taken for granted today, it recalls the speculations in Bush's memex paper, now 60 years old—or 70, if one considers the 1937 draft. For NLS provided the first versions of

* text linkage (not called "trails of associations", but "chains of views");
* hierarchical hypertext (the term is Ted Nelson's, coined in 1963: cf. Nelson 1965, 1974), later used as the core of the World Wide Web (Berners-Lee 1989);
* keyword search;
* whole-text retrieval by keyword;
* links between text and graphics;
* annotation;
* message passing (i.e. email);
* word-processing (copying, reordering, and adapting written text);
* collaborative working;
* multiple (non-overlapping) windows;
* the screen cursor ("bug");
* and, last but not least, that now familiar animal the computer mouse ("I don't know why we call it a mouse. It started that way and we never changed it": Engelbart n.d.).

Some of these features were already being taken up in the ARPAnet—whose very first message was sent (from UCLA) to Engelbart's lab (Packer and Jordan 2001, p. xvi). Moreover, all were routinely used by Engelbart at SRI. And they caused a sensation at the San Francisco meeting, as we've seen. But they *didn't* immediately spread into the IT community as a whole, still less to Everyman. The mouse, for instance, wouldn't start reproducing merrily—and commercially—until the 1980s.

By that time (thanks largely to Kay), the mouse, windows, icons, and text-processing had not only been much improved but had been made accessible to anyone with a PC on their desk. At long last, historians really could use their private machines to do research on the Crusades.

A final word: the memex was a harbinger of more than desktop technology. Engelbart, though apparently not Bush, saw clearly that a memex/NLS machine would *change* human minds as well as aid them. For instance, word-processing would not only aid but also affect the creative process of composing texts (1962: 74 ff.). In general, it would enable the human computer user to have "the freedom and power of disorderly [thought] processes", which in the early 1960s was impossible (p. 89).

The psychologists Bruner and Miller had recently shown that we process differently structured concepts in different ways (6.ii.b). And in his early 1960s work on cognitive technologies (6.ii.c), Bruner was pointing out that pervasive systems of representation, or "technologies"—such as language, drawing, and writing—enter into the developing mind, shaping it as well as helping it. In 1962 Engelbart—besides remarking that the precise nature of future technologies would depend on "our [psychological] understanding of the human being", including how we process distinct "concept structures" (pp. 70, 85)—was saying much the same:

In a very real sense, as represented by the steady evolution of our augmentation means [writing, printing, libraries . . .], the development of "artificial intelligence" has been going on for centuries. (Engelbart 1962: 79)

He predicted that in "augmenting" human intellect by using memex/NLS, information technology would provide the thinking processes listed above so as to enable people from all professions to solve "problems that before seemed insoluble", which may have lasted "for twenty minutes or twenty years". But increasing someone's capability "to approach a complex problem situation, to gain comprehension to suit his particular needs, and to derive solutions to problems" in this way wouldn't simply be adding more of the same, nor even providing a new box of tools. Rather, it would be changing the essence of how we think:

We do not speak of isolated clever tricks that help in particular situations. We refer to a way of life in an integrated domain where hunches, cut-and-try, intangibles, and the human "feel for a situation" usefully co-exist with powerful concepts, streamlined terminology and notation, sophisticated methods, and high-powered electronic aids. (Engelbart 1962: 1; italics added)

Forty years later, some philosophers would take the notion of cognitive technologies to heart in a highly provocative way (Chapter 16.vii.d). Following the writings of Andy Clark rather than Bruner, they argued that the mind, the self, is part *constituted* by external technologies and systems of representation. The neo-Cartesian 'mind in the head', and even the phenomenologists' 'mind in the body, ending at the skin', was dismissed. Much as anthropologists—and Wilhelm von Humboldt—had long claimed that culture and language are integral to mind (see Chapters 8 and 9.iv.a–b), so Clark was now arguing that an Alzheimer patient's notebook, or a modern person's PC, are—literally—part of their mind, or self.

The philosopher Timothy van Gelder (a sceptic with respect to GOFAI: 16.vii.c) applied this idea to Engelbart's work. In a paper proclaiming Engelbart to be Turing's equal, when judged as "the most important pioneer in the general field of computing and intelligence", van Gelder said this:

Engelbart's vision of computers augmenting human intelligence is, properly understood, *a vision of human self-transformation* through a bootstrapping process in which our current, technologically augmented intellectual capacities enable us *to refashion the spaces and practices within which we ontologically self-constitute.* Moreover, his crucial insight was that computer technology will be more profoundly and intimately connected with *the process of self-constitution through enhanced rational self-expression* than any previous technological forms. (van Gelder 2005, last page; italics added)

In other words, 'mind as machine' was being given an ontological, not merely an explanatory, air.

i. Spacewar

One harbinger seemed to many observers at the time to be trivial in the extreme. It was the first computer game, MIT's Spacewar (Levy 1994: 58–69).

It was also the first example of interactive computer graphics, where one could alter the picture on the VDU screen at will. Non-interactive graphics had been around for a decade or more. For instance, when Simon had visited RAND in 1952 he'd been met by "an eye opener. They had this marvelous device there for simulating [radar-blip] maps on old tabulating machines. Here you were, using this thing not to print out statistics, but to print out a picture, which the map was" (interview in McCorduck 1979: 125).

But you couldn't use RAND's "old card calculators" to change the picture before your very eyes.

Spacewar was seeded by Minsky's accidental discovery in about 1960 of how to make his PDP-1 display a circle. Initially, he used his Circle Algorithm to display particle interactions graphically on a cathode-ray tube screen. But the technique was soon adapted by playful MIT graduate students to allow them to pretend to chase and shoot down rockets in outer space.

The basic game was implemented by February 1962. But it was soon much improved, because the students spent most of their spare time playing it, and tinkering with it. By early 1963 (when I played with it on my first visit to MIT) Spacewar had rockets, torpedoes, explosions, sun, stars, constellations—and a joystick. You could use the sun's gravitational pull to increase your speed, provided that you didn't get too close. And you could escape into hyperspace by pushing a panic button, although you never knew just where you'd come out (if you were right next to the sun, it would gobble you up).

That wasn't the end of the line. The MIT hackers constantly advanced the program, the latest version being left in a drawer for the next person to upgrade. In fact, "the next person" might not even need to touch the drawer, for by 1963 the Spacewar code had been sent to the hackers' friends all over the USA. Eventually, the game spawned the "Intergalactic spacewar olympics" at Stanford, a heady meeting of enthusiasts reported in the magazine *Rolling Stone* and commemorated by the famous New York photographer Annie Liebowitz (Brand 1972). Minsky (1984*a*: 248) recalls that by 1965 the game had already become so addictive that daytime playing had to be banned, to leave the computer free for real research.

But how real was real? Spacewar may have seemed trivial to the outsider and trendy to Liebowitz, but over the years it would have a lot of highly practical spin-offs. For example, it soon contributed ideas to simulated displays for training pilots of aircraft and space vehicles. It was used by DEC as a diagnostic program for the PDP-1 (Levy 1994: 65). And it was a forerunner of the hugely more complex computer games we know today—indeed, of computer graphics and 'virtual reality', or VR, in general (see 13.vi).

j. The empty chair at the banquet

Computer graphics of a type hugely more sophisticated than Spacewar were already being seeded elsewhere in MIT—but the Spacewar hackers didn't know that. Indeed, even the specialists in computer graphics didn't know about it. They found out from an unscheduled talk, arranged at the last minute by Licklider, at their first official meeting in 1962 (Rheingold 2000: 149).

Ivan Sutherland, then a graduate student working in MIT's Lincoln Laboratory, had written a revolutionary computer-graphics program called Sketchpad (I. E. Sutherland 1963; Blackwell and Rodden 2003). Where Spacewar dealt with moving spots of light, Sketchpad enabled one to draw—or to instruct the computer to draw—pictures or geometrical diagrams of 2D or 3D structures. Amazingly, the diagrams could be moved, rotated, size-varied, shape-distorted, and even tidied up in various ways.

Sketchpad could well have been included here as yet another harbinger. However, it's described in Chapter 13.v.c instead. In short, there's an empty chair at our harbingers' banquet. (More accurately, there are two: remember Pandemonium.)

10.ii. Establishment

While cognitive science as a whole was being officially established by way of newly founded research groups and new publication practices (Chapter 6.iv–v), symbolic AI was coming into official existence too. But there was no rush to climb onto the bandwagon. Throughout the 1960s and early 1970s it would be studied in relatively few places, most of them in the USA—plus a handful in the UK and elsewhere (Edwards 1996, ch. 8; McCorduck 1979, chs. 5–7; Fleck 1982).

More accurately, it was being *bureaucratically recognized and officially funded* in only a few places. Interested individuals, of course, were scattered much more widely. In Bombay (Mumbai), for example, Rangaswamy Narasimhan (1926–), who'd directed the design of India's first operational digital computer, was doing early work on computer vision at the Tata Institute (Narasimhan 1964, 1966). (His interest in AI eventually led to the world-leading NCST: National Centre for Software Technology, now in a large building in Bombay's Juhu district; some twenty years ago, NCST won the international tender for the new train-scheduling program for the London Underground.)

Narasimhan wasn't the only one. Most computer science departments, in whatever country, contained one or two mavericks who spent some of their time on this new pursuit. A computer science department, however, wasn't the same thing as an AI department. And departments of electronics and/or engineering, which might also harbour a couple of AI 'oddballs', were different again. How did AI become established in its own right?

a. First labs

Several of the earliest AI workers were employees of IBM or RAND. These institutions weren't concerned with AI as an academic discipline, nor with encouraging youngsters into the field. Nevertheless, they both contributed greatly to it.

For instance, IBM supported Samuel (Section i.e, above) and RAND gave houseroom to Newell and Simon (6.iii.b–c). In addition, IBM played an important role in the Dartmouth Summer School, in the persons of Rochester, Samuel, Gelernter, and Alex Bernstein (6.iv.a). Besides IBM's regular employees, there were the occasional visitors—such as McCarthy. He'd spent the summer of 1955 at IBM Poughkeepsie, at the invitation of Rochester—who recommended the Dartmouth proposal because McCarthy "had been bending his ear about AI's potential for some time" (J. McCarthy, personal communication).

The consulting company Bolt, Beranek & Newman (BBN) was another very early AI player (Alperin *et al.* 2001: 7–37; Walden and Nickerson 2005). Temporarily lodged in two rooms inside MIT, BBN was founded in 1948 to develop the acoustic technologies initiated in the Second World War by (among others) the psychologists Licklider and John Swets—both of whom would join the company in its first ten years. (The start-up had been prompted by a vision of peace, not war: the United Nations building in New York. MIT's Richard Bolt won the architect's contract to advise on the acoustics, and decided that he needed Leo Beranek's help; and Robert Newman, then a graduate student in architecture, was soon enrolled too—Alperin *et al.* 2001: 13.) Psychology pervaded BBN from the beginning, both as informational theories of perception and attention (see 6.i and ii.a–b) and as a concern with man–machine integration. At first, that meant operator–tool environments like those studied by Kenneth Craik (4.vi) and Donald Broadbent (6.i.c). But in 1957, when digital computers came on the scene, Beranek hired Licklider specifically to move the firm nearer to them (Beranek 2005; Swets 2005). Licklider remained until 1962, and—given his key role in the history of AI (see below, and Chapter 11.i.b)—it's not surprising that BBN eventually employed several people mentioned in this book (including Oliver Selfridge, William Woods, and Ronald Brachman).

In 1966 the commercially oriented Stanford Research Institute (SRI) started an AI Center to do contract research. At first, most of its projects were government-funded, via ARPA. It soon became well known for its work on robotics and planning, and for vision research interestingly different from that done at MIT. Although these aspects are described below in separate subsections (iii.c and iv.b), and were studied separately in other labs, SRI's robot SHAKEY—so named because it wobbled—was one of the first attempts to integrate all three.

Another such attempt was the Edinburgh University robot FREDDY (Barrow and Salter 1969; Barrow and Crawford 1972; Michie 1973b). This was officially known as the Mark 1.5 project, but its more widely used name came from the acronym FREDERICK: Friendly Robot for Education, Discussion and Entertainment, the Retrieval of Information, and the Collation of Knowledge. (Edinburgh's leader, Donald Michie, was nothing if not ambitious!—see Chapter 11.iv.)

Soon afterwards, in 1970, Xerox set up their Palo Alto Research Center. (Largely because of Kay, Xerox PARC would eventually spawn a plethora of advances in human-centred AI: 13.v–vi.)

If NewFAI was to get off the ground at rapidly increasing speed, however, university-based laboratories—supporting graduate/undergraduate courses—would be needed. The first two were set up at the end of the 1950s: at CMU under Simon (6.iii.b–c), and at MIT under Minsky and McCarthy.

(In a different political climate, the honours might have gone to another Massachusetts institution instead, namely Tufts University. For a new department of "Systems Analysis" was opened there in the mid-1950s, of which Minsky was an early member. He was also one of the last members, for the department was closed by the Tufts authorities very soon after it had opened. The reason was that two of its founders, the philosopher Richard Rudner and the psychologist William Schutz, had been placed on the list of political "subversives" compiled by Senator Joseph McCarthy—Crevier 1993: 64. As a result of the witch-hunt at Tufts, Minsky left for the Lincoln lab in 1957; a year later, he transferred to MIT—initially to the maths department, but switching very soon to electrical engineering, where McCarthy was already a member.)

One might assume that MIT's first official AI project was planned at a pre-arranged meeting in someone's office, and shoehorned into official existence by several carefully crafted pages of pleas, promises, and justifications—circulated, in pristine folders, to a dozen busy committee members. Not a bit of it:

It was all very informal. One day in 1958, Minsky and I met in the corridor of Building 26, and said "Let's have an AI project! Yes, that's a good idea!" Along came Jerry Wiesner, the head of the Research Lab of Electronics [and President of MIT from 1972 to 1980], and he asked us—in the corridor—what we needed. We said: "A room, a secretary, a key-punch, and maybe two programmers?" He said, "And what about 6 grad. students?" MIT had just had a Joint Services (i.e. Army etc.) contract in electronics, related to defence, and he had all these extra students. A burnt-out computer—which was supposed to be ultra-safe!—gave us the room. (J. McCarthy, personal communication)

This was typical of Jerome Wiesner (1915–94). He was already renowned for having made the RLE, from 1952, not merely an electronics workshop but a wide-ranging interdisciplinary centre:

At one stage we had twelve different fields represented in the laboratory. The Linguistics and Psychology departments grew out of groups that were started in the lab.

[RLE's communication engineers were joined by] neurophysiologists and other biologists, linguists, economists, social scientists, and psychologists of the various persuasions... They explored each other's fields and slowly began to comprehend each other's lingo and exhibit that spirit of mental intoxication that characterizes the pursuit of an exciting idea... The two decades of RLE were like an instantaneous explosion of knowledge. (interviewed/quoted in Brand 1988: 134–5)

In short, MIT's AI Lab had an interdisciplinary home right from the start. It's little wonder that the second co-director (Papert) was a psychologist and (originally) a mathematician, not a computer scientist. And it's little wonder that MIT's Linguistics Department—headed by Chomsky (9.vi.a)—had a strongly computational slant. (Computational, but not computer *modelling*: not only did Chomsky eschew programming, but he was already distancing himself from his AI colleagues: 9.x.b.)

Generosity didn't come only from Wiesner, but—five years later—from ARPA too. Both Minsky and McCarthy had close ties with ARPA's first director, Licklider. An MIT psychologist with an early interest in computers, and in library science (Chapter 5.iv.f), Licklider had been on the Air Force's Scientific Advisory Board in the late 1950s, overseeing the ambitious SAGE project (11.i.b). As McCarthy remembers it:

Licklider said "You're spending millions of dollars, but you're not looking at the (computer) science to support it." So in 1962 they said "OK, if you'll come to Washington to be involved, we'll do it." (personal communication)

Once he was involved, Licklider supported both hardware research and software research too. And his "support" was unstinting:

[When] his office decided to support a project, that meant providing thirty or forty times the budget that the researchers were accustomed to, along with access to state-of-the-art research technology and a mandate to think big and think fast. (Rheingold 2000: 147)

For example, MIT's Project MAC received the equivalent of $25 million in mid-1990s values from ARPA between 1963 and 1970 (Edwards 1996: 269). It was directed by Edward Fredkin, a co-author—with Licklider and McCarthy—of the first paper on time-sharing (Boilen *et al.* 1963). (Later, from 1968, it would be directed by Licklider himself.) The MAC acronym was deliberately ambiguous, being variously interpreted as Multi-Access Computing (i.e. time-sharing), Man And Computers, and Machine-Aided Cognition—the last two of which recalled Licklider's "Man–Computer Symbiosis" (11.i.b).

As this catholicity implies, Project MAC covered a wide range of research, of which not all was AI—and not all of that, psychological AI. Nonetheless, NewFAI—whether psychological or technological—found shelter under the MAC umbrella. Indeed, the AI Lab set up by Minsky and McCarthy in 1959—with three whole floors of a modern building set aside for it—became a subdivision of Project MAC before graduating as a stand-alone department.

With hindsight, McCarthy sees Licklider's championship as the main reason why AI got established as early as it did, and why this happened in the USA rather than the UK—where early connectionist research, and computer science too, was at least as healthy as in America (Chapters 3.v.b–d and 12.ii.a–b). As he puts it:

The "AI establishment" owes little to the general "scientific establishment". AI would have developed much more slowly in the U.S. if we had had to persuade the general run of physicists, mathematicians, biologists, psychologists or electrical engineers on advisory committees to allow substantial NSF money to be allocated to AI research.

...AI was one of the computer science areas Licklider and his successors at DARPA consider relevant to Department of Defence problems. The scientific establishment was only minimally, if at all, consulted. In contrast European AI research long depended on crumbs left by the more established sciences. (McCarthy 1989)

(How those "crumbs" were first gleaned in the UK, and why they dried up in the early 1970s only to be restored ten years later, is described in Chapter 11.i.a, iv.a, and v.c.)

Initially, the MIT AI Lab was very much a joint enterprise between Minsky and McCarthy, who'd already worked together on organizing the Dartmouth meeting. Even in their 1955 funding application for that summer project, however, there'd been a clear difference between them. In a nutshell, McCarthy focused on logic, Minsky on robotics—including vision and anticipatory planning.

Minsky intended to study the machine learning of sensori-motor representations ("abstractions"), and problem solving by exploration of an internal model of the environment before experimental trials in the external world. (Because of this anticipatory exploration, he said, the behaviour would seem "rather clever", even "rather 'imaginative' "—McCarthy et al. 1955: 49.) McCarthy, by contrast, wanted to study "the relation between language and intelligence", in which "the trial and error processes at a higher level [than sensory data and motor activity] frequently take the form of formulating conjectures and testing them" (52–3). To do that, he'd need to construct "an artificial language" with some core strengths of English (i.e. LISP: see v.c, below). This would enable a computer to deal with conjecture and self-reference, to formulate short informal arguments, and to represent everyday notions like "physical object" and "event" (iii.e, below).

Eventually, this difference in research interests—"I never had rap sessions with Minsky" (J. McCarthy, personal communication)—led to McCarthy's decision to leave MIT in 1962 for Stanford, to establish the AI Group there a year later. (Papert took over as co-director of the MIT laboratory, and he and Minsky ran it together until 1972.) The East Coast versus West Coast rivalry that developed over the years wasn't merely about prestige or even money, but about two competing visions of what AI should aim to do. (Notoriously, papers from Stanford would be cited only rarely in work coming out of MIT.) Since AI didn't yet exist at Stanford, although computer science did, it could be moulded to suit McCarthy's own priorities.

Again, Licklider was crucial. When the Stanford AI Group was set up, "Licklider simply asked McCarthy what he wanted and then gave it to him" (Edwards 1996: 270). Later, he admitted: "It seemed obvious to me that he should have a laboratory supported by ARPA . . . So I wrote him a contract at that time." But Licklider was much more to McCarthy than an ARPA sugar-daddy. He was an intellectual partner too. As noted above, they both helped write the pioneering paper on time-sharing (Boilen *et al.* 1963). Indeed, Licklider—who put huge sums of money into the development of time-sharing as soon as he was appointed to ARPA—had been enthused by McCarthy's vision of it in the first place (see 11.i.b).

In fact, things didn't turn out quite as McCarthy had expected. For the department soon split. Largely due to Feigenbaum's presence there, a separate group—the Heuristic Programming Project—was set up in 1965, later metamorphosing into the Knowledge Systems Laboratory. McCarthy's group was now known as SAIL (the Stanford AI Laboratory). The emphases of the two research clusters were very different. McCarthy and his students were concentrating on highly abstract issues of logical reasoning and representation, based largely on "resolution" theorem proving—first formulated very soon after the founding of the department (Section iii.b). Feigenbaum's group, by contrast, were building practically useful expert systems, using facts and rules of thumb provided by domain specialists—medics, chemists, geologists, and so forth (Section iv.c).

CMU, too, benefited greatly from Licklider's largesse. But if Licklider was generous, he was also highly discriminating. When the J. Arthur Sloan Foundation provided $15 million for cognitive science in the late 1970s, they deliberately spread most of the money widely, and therefore thinly (Chapter 8.i.c). Licklider took a very different view. He felt that he knew who the top AI people were, and that between them they should get virtually all the ARPA money available. As a result, each favoured recipient got many times more than the average funding for university laboratories. That's why MIT, Stanford, and CMU quickly became *the* main university centres for AI research.

Licklider's faith in his small band of AI pioneers became doubly important in 1964, when the field suffered its first funding bombshell. A US government advisory committee blasted machine translation, declaring it well-nigh worthless—with dire consequences for MT researchers seeking to finance their work (see 9.x.e). The twenty-year shadow of ALPAC might well have been cast over other areas of AI too, whether language-based or not. And perhaps it was: the document was widely reported in the newspapers, and some university committees may have reconsidered the wisdom of embarking on AI as a result. In the few favoured labs, however, Licklider's support endured.

Meanwhile, the NewFAI gospel—in particular, its psychological version—was being spread across the seas (Chapter 6.iv.e). In 1963 Michie, enthused by his recent visit to the USA, founded a small research group with Bernard Meltzer. In 1966 this spawned Edinburgh's Department of Machine Intelligence and Perception, and (Meltzer's baby) its Department of Computational Logic. Other transatlantic visitors to MIT in this period included Ratio member Donald Mackay, and N. Stuart Sutherland—who founded Sussex University's Experimental Psychology Department in 1964 with NewFAI very much on his mind.

Visitors to the USA from even further afield included Narasimhan, who spent 1961–4 as a Visiting Scientist at the computing laboratory of the University of Illinois, Urbana (Narasimhan 2004: 250). On his return to India, he founded an AI group at the Tata Institute which soon became the National Centre for Software Technology. This was so successful that in 1986 it moved to a separate building, almost as large—if not so extraordinarily luxurious—as the Tata Institute itself.

b. The ripples spread

The AI harbingers described in Section i enthused the pioneers of cognitive science, psychologists included (Chapter 6.iv). But the NewFAI adventurers were a very small community. Moreover, they couldn't multiply the number of functioning programs overnight, since much of their effort had to go into designing new programming languages (Section v.a–e, below).

To make things worse, AI research was available only in informal laboratory research reports, or scattered across a bewildering variety of journals. Even by the end of the 1950s, then, its achievements still weren't widely known.

The possibility of NewFAI, however, was now in the air—thanks, in part, to the reprinting of Turing's daringly imaginative *Mind* paper in James Newman's *The World of Mathematics* (1956). At 2,500 pages, this boxed four-volume collection provided dozens of attractions to vie with Turing's essay—many of them already classics. But Newman's edition turned that paper into a classic too. By the time it was reprinted again in *Computers and Thought* (see below), it was already being described as "one of the best-known papers" about "the existence or nonexistence of various kinds of theoretical upper bounds on the intelligence of computing devices" (Feigenbaum and Feldman 1963: 9).

Some early programs, and their potential implications for psychology, were described in 1960, in the cognitive science 'manifesto' *Plans and the Structure of Behavior* (Chapter 6.iv.c). But the key surge of AI visibility occurred three years later. For in 1963 things changed suddenly. The very newest of NewFAI programs (the strong sense was obligatory at last: i.a, above) reached the public via Feigenbaum and Feldman's best-selling collection *Computers and Thought*.

Virtually every program later discussed by Dreyfus in his cat-among-the-pigeons RAND memo, for instance, was included there (see 11.ii.a). So were four crucial programmatic essays: 'Steps', 'Chess and Complexity', 'Attitudes Toward Intelligent Machines', and Turing's 1950 paper—the last, now placed in an appropriately *technical* context (but still mistakenly described, by the editors, as primarily concerned with the Turing Test: pp. 9–10). Admitting that AI programs were still at the low end of Armer's "continuum" of intelligence, the two young editors ended their Introduction with a rousing call to action:

What is important is that we continue to strike out in the direction of the milestone that represents the capabilities of human intelligence. Is there any reason to suppose that we shall never get there? None whatever. Not a single piece of evidence, no logical argument, no proof or theorem has ever been advanced which demonstrates an insurmountable hurdle along the continuum. (Feigenbaum and Feldman 1963: 8)

This book was hugely influential, and is still reprinted regularly. The royalties would soon be enough to support a prize for outstanding new work, awarded at the biennial International Joint Conference on Artificial Intelligence, or IJCAI. As the editors had hoped, many readers made AI their profession as a result. It would be over twenty years before another AI collection would have such an explosive effect, on readers young and old (12.vi.a).

The same year saw a collection devoted to NewFAI models of motivation and emotion (Tomkins and Messick 1963). Although this created a flurry at the time, it was grossly premature and fairly soon forgotten (see Chapter 7.i.a–c). Five years later, Minsky (1968) edited *Semantic Information Processing,* which described the NewFAI research being done by his students at MIT. And in 1975 Patrick Winston (1943–), by then running MIT's AI Lab (which he continued to do until 1997), edited *The Psychology of Computer Vision.* This book was read by many people with scant interest in computer vision, because it contained Minsky's paper on "frames"—a hugely influential discussion of schemata in general, and how they could be implemented in AI (see iii.a, below). By the mid-1970s, a fifth widely read collection on psychologically oriented AI had appeared: *Computer Models of Thought and Language* (Schank and Colby 1973).

The floodgates had opened. NewFAI work was circulating outside the laboratory walls, and attracting responses from people in various non-AI departments, including some in the social sciences and humanities. The Association for Computing Machinery's SIGART (Special Interest Group on ARTificial Intelligence) had been publishing a newsletter since the early 1960s. In 1979–80 the American Association for Artificial Intelligence (AAAI) was set up, with Newell as its first President. (For a silver-anniversary retrospective on AAAI, see Feigenbaum 2005.) Meanwhile, AISB had already got started across the pond in the mid-1960s (Chapter 6.v.c). But largely because of a bold political move in Japan, AAAI was showered with money almost from the start (see 11.v.b). Unlike most academic or professional societies, it was able to do a great deal in encouraging its area of interest.

As for regular publications devoted to the new discipline, these were now well established. By the end of the 1960s, six volumes from Michie's Machine Intelligence workshops in Edinburgh had appeared, and a seventh was already in press. The journal *Artificial Intelligence* was about to be founded (in 1970), with Edinburgh's Meltzer as Editor-in-Chief and SRI's Bertram Raphael (1936–) as Associate Editor. The *International Journal of Man–Machine Studies* had made its debut in 1969—the year which also saw the first meeting of IJCAI, held in Washington DC. As befitted the Swinging Sixties, AI was in full swing.

The philosophers helped. Following on Putnam's highly abstract 1960 statement of functionalism (in which AI wasn't mentioned), Jerry Fodor (1968) defined psychological explanation in terms of computation and AI models (16.iv.c). The psychologists helped too: Bruner's Center for Cognitive Studies encouraged psychologists to take an interest in AI—and supported AI's claim to psychological relevance (6.ii).

So far, so Sixties. The Seventies saw the students of the true pioneers setting up specialist laboratories in particular sub-fields. For example, Newell's student Feigenbaum founded the Stanford Heuristic Programming Project and the Knowledge Systems Laboratory, also at Stanford, to focus on expert systems. And the first three monographs on AI appeared in the mid-1970s. These were Raphael's highly readable trade book *The Thinking Computer: Mind Inside Matter* (1976), which—among other things—gave an insider's view on the SRI robot SHAKEY; Winston's welcome textbook *Artificial Intelligence* (1977), which explained how students could use LISP to code the programs it described; and my own *Artificial Intelligence and Natural Man* (1977), which reported not only what AI programs could do and how, but also what they couldn't (yet?) do and why—and which related AI to psychology, philosophy, and social concerns.

Thanks in part to those three books, the general interest in AI, first aroused by *Computers and Thought*, grew phenomenally. During the last years of the 1970s, students were flocking to the already established laboratories, and others were being set up elsewhere. At the same time, the journalists were getting in on the act.

Such was the interest that Feigenbaum initiated a second, more systematic, publication on AI. Whereas *Computers and Thought* had been intended to whet people's appetites, *The Handbook of Artificial Intelligence* advised them on what to swallow—and how to chew it. "Most scientists and engineers," the editors said, "though very knowledgeable about the "standard" computing methods . . . simply had never heard about symbolic computation or Artificial Intelligence" (Barr and Feigenbaum 1981: 12). The *Handbook* should function as "A self-help encyclopedia to which the aspiring practitioner could turn for explanations of fundamental ideas, descriptions of methods, and discussions of well-known programs as case studies". The three volumes appeared close on each other's heels, a total of 1,466 pages in just over a twelve-month (Barr and Feigenbaum 1981, 1982; P. R. Cohen and Feigenbaum 1982).

In brief, the ripples had spread. But it's worth noting that connectionism wasn't mentioned in the *Handbook*.

Perceptrons (which had been prominent in Feigenbaum's *Computers and Thought)* were given only two pages, and the chapter on learning opened with a curt dismissal of the approach:

[Perceptrons] failed to produce systems of any complexity or intelligence.
  Theoretical limitations were discovered that dampened the optimism of these early AI researchers (see Minsky and Papert, 1969). In the 1960s, attention moved away from learning toward knowledge-based problem solving. . . . *Those people who continued to work with adaptive systems ceased to consider themselves AI researchers*; their research branched off to become a sub-area of linear systems theory. Adaptive-systems techniques are presently applied to problems in pattern recognition and control theory. (P. R. Cohen and Feigenbaum 1982: 325–6; italics added)

The 'Vision' chapter included brief accounts of relaxation techniques (12.v.h) and Marrian low-level vision (7.v.b–d), but "parallel processing/search" had only eight index entries for all 1,466 pages.

These editorial judgements were made after consulting dozens of AI experts about the new publication (Cohen and Feigenbaum 1982, pp. xi–xii). In 1980, then, "artificial intelligence" was being perceived by most people simply as GOFAI. If connectionism was being ignored in the *Handbook of AI*, it's not too surprising that many people still use the term AI (wrongly) to cover only the symbolic variety.

c. New waves

The Journalistic ripples soon spread too. Omnibus vehicles such as *Artificial Intelligence* and the *International Journal of Man–Machine Studies* were supplemented in the 1980s by specialist journals devoted to particular sub-fields of AI.

In the 1990s, the numbers of such publications burgeoned. There are now several each in fields such as planning, problem solving, theorem proving, machine learning, computer vision, emotion modelling, legal AI, medical AI, diagrammatic reasoning, robotics, educational AI, virtual reality . . . and so on, and on. In addition, there are various journals in cognitive science, which often publish AI papers.

Most of these AI and interdisciplinary journals are complemented by their own conference series, regular workshops, and/or societies. In other words, a host of established GOFAI activities now aim to provide the cooperation and communication—including face-to-face conversations—necessary for scientific progress (2.ii.b–c).

Whether they ensure that sufficiently *widespread* communication takes place is another matter, for specialization has brought fragmentation. That's especially damaging if one's ultimate aim is to understand whole organisms, or "the whole iguana" (Dennett 1978*c*). Alan Mackworth (1945–), who entered AI at a time when virtually everyone in the field knew (not just knew *of*) everyone else, has put it like this:

Unfortunately [given the AI dream of building integrated cognitive robots], current research in AI is highly divergent with little or no overlap between specialized subfields such as computational vision, knowledge representation, robotics, and learning. Each group has its own conferences and journals, and when they do all meet at a single conference, they diverge in parallel sessions. (Sahota and Mackworth 1994: 249)

In sum: the interdisciplinarity so characteristic of NewFAI survives, but under increasing pressure.

10.iii. The Search for Generality

The hopes pushing NewFAI forward in the 1960s were focused less on technology (*pace* Licklider, Engelbart, and Bush) than on a theory of intelligence as such. (That approach was still strong in the early 1970s, leading a well-known British AI researcher to say, "I'm not interested in programs!" M. B. Clowes, personal communication.) This is one reason why so many NewFAI ideas were taken up by psychologists. For example, the various forms of "knowledge representation" discussed below were (eventually) widely employed in theories of and experiments on human memory, and AI models were continually cited in the psychology textbooks (e.g. Klatzky 1980).

For many NewFAI workers, however, "intelligence" usually meant *intelligence in general*. In McCarthy's mind, for instance, human psychology was of no special interest, "except as a clue to possible effective ways of doing tasks" (McCarthy 1989). In drafting the Dartmouth proposal, he hadn't criticized "anybody's" (i.e. the behaviourists') way of studying human behaviour: "I didn't consider it relevant". In particular, he now disclaims being part of "the cognitive revolution":

[Because I didn't consider psychology to be relevant to AI], whatever revolution there may have been around the time of the Dartmouth Project was *to get away from studying behavior* and to consider the computer as a tool for solving certain classes of problem. *Thus AI was created as a branch of computer science and not as a branch of psychology.* (McCarthy 1989)

He allows, however, that Newell and Simon ("the only participants who studied human behavior") worked "both in AI as computer science and AI as psychology".

The belief—more accurately, the often unquestioned assumption—was that a relatively small number of general mechanisms (heuristics, search techniques, learning rules, representational methods, architectures . . .) would suffice to explain intelligence in all its forms. Whatever the domain of study, from logic and chess to language and vision, the aim was to uncover that generality. Special constraints might have to be added, if one was interested in modelling human minds—limits on short-term memory capacity, for example (7.iv.b). But the underlying principles would be untouched.

The commitment to generalism was strongest at Stanford, thanks to McCarthy. In the 1970s, Minsky would give it up, now arguing that many different interacting procedures were responsible for intelligence (Minsky 1979). Indeed, he later became known as the most eminent of the "scruffies", while McCarthy was acclaimed as the high priest of the "neats" (see 13.i.c). In the early 1960s, however, generalism was dominant.

Simon was someone who did take pains to examine the generalist assumption, and sought to justify it too. For example, he argued that the core mechanism of deliberation *must* be serial, so as to handle 'what-ifs' without confusion (e.g. Simon 1967). He soon became even more influential than the building of LT/GPS had already made him. His 1962 paper on 'the Architecture of Complexity', and his 1968 Compton Lectures at MIT on *The Sciences of the Artificial*, published in 1969 and never out of print thereafter, were an inspiration (7.iv.a). Together with Minsky's 'Steps', they helped set the NewFAI agenda.

I'll mention only a few examples of NewFAI generalism here. And I'll ignore chess, which isn't my thing. It's sometimes described as the fruitfly of AI, for a great deal of technical work was done on it right from the start—and still is (for some recent examples, see Berliner and Beal 1990). That's not because people wanted an excuse to spend their professional time on their favourite Sunday afternoon pastime, but because they believed that chess involved many of the major aspects of intelligence *in general*. As the GPS triumvirate put it: "If one could devise a successful chess machine, one would seem to have penetrated to the core of human intellectual endeavor" (Newell *et al.* 1958*b*: 39).

Apparently, Everyman agreed. These efforts continually whetted the public's appetite for AI. For example, the first "respectable" chess program was featured widely in the late 1950s media, from the *Scientific American*, through the *New York Times*, to *Life* magazine (A. Bernstein and Roberts 1958). The first exhibition game between the world champion and a computer took place in August 1974 (J. E. Hayes and Levy 1976: 49; see also Good 1968). The chess master David Levy's £500 bet, in 1968, that no program would beat him within ten years caused much publicity, culminating in 1977–8 in two widely reported matches (against Northwestern's Chess 4.5 and 4.7). In both cases, Levy won—much to the satisfaction of the readers of the *New York Times,* faced with the headline 'Chess Master Shows a Computer Who's Boss'.

All those programs were 'pure' software. Deep Thought—named after the computer in *The Hitchhiker's Guide to the Galaxy*—and its successor Deep Blue, which in 1997 did beat the world champion, were very different, for they relied on special-purpose chips. (These were designed by Feng-hsiung Hsu at CMU and then IBM: Hsu 2002.) The 480-chip Deep Blue could search hundreds of millions of chess positions per second. Even so, the chips were optimizing the very general AI technique of blind search, not supporting forms of representation specific to chess—as human players do (cf. Chase and Simon 1973). This enabled the public—and the defeated champion Gary Kasparov, too—to save face: the computer might now appear to be the boss, but only because it was, at base, a bully.

For NewFAI, as remarked above, chess *as such* wasn't the point. Indeed, Minsky deliberately discouraged his MIT students from working on it (interview in McCorduck 1979: 189). Chess interested some AI researchers because it could help them to develop general AI techniques. (Today, this motivation underlies the AAAI's General Game Playing Competition, which aims at programs that can play any formally described game by using general methods of representation, reasoning, and learning: see <http://games.stanford.edu>.)

Most of the 1960s NewFAI programs, in fact, weren't designed simply to achieve a specific task. Rather, they were aimed at the general problems identified by 'Steps'—and, in the late 1960s, even by Dreyfus's *Alchemy and Artificial Intelligence* (see 11.ii.a–b). In applying (so it was widely believed) to men, mice, and Martians, they were intended as contributions to cognitive science as a whole: the study of *all possible minds*.

a. SIP spawns KR

When computers were used only to do simple mathematics, the problem of knowledge representation (KR) wasn't too pressing. Enabling a digital computer to store the fact that "$2 + 2 = 4$", and even enabling it to do the sum, was relatively painless. But things got trickier when the machines began to be used to do goal-directed problem solving, theorem proving, pattern matching, visual scene analysis, diagrammatic reasoning, and—above all—interpretation of verbal texts. Enabling the human user to communicate with the machine in English, for example, was no small feat. "Semantic information processing", or SIP, required new forms of KR, appropriate for non-numerical data and inferences.

(Strictly, people should have spoken of BR, not KR. For the term "knowledge" assumes *truth*, whereas "belief" doesn't. NewFAI writers systematically ignored the K/B distinction. Later AI researchers sometimes allowed for it, especially when discussing how to revise previously drawn conclusions now seen to be false: see McDermott 1974, and 13.i.a. However, they usually retained the term "*knowledge* representation", which had become accepted AI jargon.)

With hindsight, it's surprising how long it took people, even in scientific circles, to realize that computers might be used for SIP. Babbage arguably, and Konrad Zuse certainly, had seen the potential for non-numerical computation (see 3.iv and v.a). Zuse had even suggested using computers to do CAD/CAM carpet design, allowing for deliberate weaving errors to make the carpets appear handmade (1993: 130). And (non-interactive) computer graphics had been used since the late 1940s—by the RAND team, for instance, for radar maps (see 6.iii.b). But none of these people had tried to coax their machines to deal with the meanings carried by ordinary-language sentences.

That possibility became clearer with the theoretical marriage of Turing and the *propositional* calculus in 1943 (Chapter 4.iii.e), with the 1950s efforts in machine translation (9.x.a–d), and with the development of programming languages capable of manipulating items that looked like English words (v.c, below). By the early 1960s, non-numerical computation was no longer left to isolated visionaries. It was being explored by Ph.D. students—at CLRU, MIT, and elsewhere.

Very few people outside the MT and NewFAI communities, however, had got the message. The opening page of *Computers and Thought* bore the sub-heading *What is a Computer? Is it just a "Number Factory"?*, followed by a declaration that a computer isn't just "a high-speed number calculator" but "a general symbol-processing device" (Feigenbaum and Feldman 1963: 1). Evidently, the editors felt that the point needed to be made. And despite the enormous popularity of that collection, it still needed to be made at the end of the decade. Minsky was able to shock his intended readers in the late 1960s by calling his new book "*Semantic* Information Processing", and he had to defend the very possibility of non-formal computing yet again (Minsky 1968: 11–12).

In the general public's eyes, the use of computers to do semantic information processing was most famously illustrated by Joseph Weizenbaum's ELIZA (1966; Boden 1977: 96–7, 106–11). The program's notoriety was based on the widespread beliefs that it had passed the Turing Test, and that this is the criterion for success in AI. Both those beliefs were mistaken (see 16.ii.c).

There was a third reason, however, why the program's high status in the public's perception of NewFAI was unearned—even ironic. For ELIZA *wasn't* an exercise in AI, or anyway not in psychological AI. On the contrary, it was a successor to the very simple game-playing program described in Weizenbaum's first paper (in the newly founded commercial magazine *Datamation* in the late 1950s), significantly titled 'How to Make a Computer *Appear* Intelligent'—the italics being Weizenbaum's (interview in Crevier 1993: 133).

ELIZA was a simple pattern-matcher, with 'canned' responses triggered by keywords or phrases. Weizenbaum assembled a number of "scripts", or sets of domain-specific keywords, one of which would be used by ELIZA on any given run. The most famous of these was the script intended to simulate the type of conversation characteristic of Rogerian psychotherapy—in which the patient's remarks are continually reflected back to them by the therapist.

If Rogerian ELIZA's human interlocutor mentioned their mother or father, sister or brother, it would reply "TELL ME MORE ABOUT YOUR FAMILY". If no keyword was available, ELIZA would say "WHY DO YOU THINK THAT", or pick some previous input of the form "MY *****", and ask "DOES THAT HAVE ANYTHING TO DO WITH THE FACT THAT YOUR *****". Occasionally, this produced startlingly appropriate remarks, such as the program's "DOES THAT HAVE ANYTHING TO DO WITH THE FACT THAT YOUR BOYFRIEND MADE YOU COME HERE" in response to the keyword-free input "bullies". But that was pure chance: ELIZA had no representation of the *meaning* of "father", never mind "bullies". And the nearest it got to syntax was to respond to "I ***** you" by spitting out "WHY DO YOU ***** ME", or to turn "My *****" into "YOUR *****" (as it did in generating the BOYFRIEND response).

It was already possible to do a little better than that. In 1960 an M.Sc. student at MIT, A. V. Phillips, had used ideas from Chomsky's (1957) grammar to write a pattern-matching question-answerer that 'parsed' simple sentences into subject, verb, object, time phrase, and place phrase—everything else being ignored (Bobrow 1968: 151). And M. Ross Quillian (see below) had indicated how linguistic meaning might be represented in a program. Yet Weizenbaum didn't try to soup up ELIZA's syntax, nor its semantics either. Why not? Because he wasn't aiming to make a computer 'understand' language—i.e. represent linguistic meaning. He thought that was impossible (and would say so at length in his 1976 book). He'd written ELIZA simply to demonstrate that a programming language designed for numerical computing (i.e. FORTRAN) could be used to handle words. In short, he wasn't interested in NLP.

He wasn't interested in KR either. He chose to model Rogerian psychotherapy because it's "one of the few examples of...natural language conversation in which one of the participating pair is free to assume the pose of knowing *almost nothing* of the real world" (Weizenbaum 1966: 42; italics added). Nor was he trying to produce a prototype of something useful, even though many expert systems would later use similar pattern-matching techniques for the human–computer interface. Indeed, he was appalled when "his" ideas—as he saw it: the priority was disputed (see 7.i.a)—were appropriated by Kenneth Colby for clinical use in psychiatry.

ELIZA was irrelevant for the advance of KR, because stored English sentences were rarely suitable for representing the information being processed by a machine. They were used by Phillips's question-answerer, to be sure, but this involved mere sentence-by-sentence matching—*not* inference. The few NewFAI programs that seemed to be solving problems in English weren't actually solving problems *in English*.

So, for instance, Bert Green's (1927–) pioneering BASEBALL program conversed in English with its users, answering their questions about the statistics of baseball teams/games—such as "How many games did the Yankees play in July?" (B. F. Green *et al.* 1961). But it didn't 'think' in English. It turned the input sentences into algebraic equations, and then solved those. Its algebraic answers were transformed into English output for the user's benefit, not for the machine's. Similarly, Daniel Bobrow's (1935–) STUDENT, one of the programs highlighted in Minsky's SIP book, solved English-language problems such as: *The distance from New York to Los Angeles is 3,000 miles. If the average speed of a jet plane is 600 miles per hour, find the time it takes to travel from New York to Los Angeles by jet*, and *Mary is twice as old as Ann was when Mary was as old as Ann is now. If Mary is 24 years old, how old is Ann?* (Bobrow 1968: 147, 213). But like its predecessor BASEBALL, STUDENT would translate the problem into a set of algebraic equations, to be solved mathematically. English had nothing, essentially, to do with it.

Evidently, then, solving problems and/or answering questions posed in English didn't necessarily mean using English to do so. In fact, English wasn't a good medium for computer KR. Raphael, in his thesis on 'Semantic Information Retrieval', put it like this:

The most important prerequisite for the [machine's] ability to "understand" is a suitable internal representation, or model, for stored information. This model should be structured so that information relevant for question-answering is easily accessible. *Direct storage of English text is not suitable* since the structure of an English statement generally is not a good representation of the meaning of the statement. (Raphael 1968: 35; italics added)

If English text wasn't a good way of representing semantic information inside the computer, what was? Programs doing SIP had been written—by Samuel, for example—more than ten years before Raphael wrote those words. But those *very* early NewFAI programs hadn't aimed at representational generality, even if they'd aimed at inferential generality. Their data were expressed in a domain-specific form: for checkers, chess, or logic. If computers were to deal with a wider range of topics, some more neutral KR would be needed. Raphael, again: "models which are direct representations of certain kinds of relational information usually are unsuited for use with other relations. [AI needs] a model which can represent semantic content for a wide variety of subject areas" (ibid.). Likewise Minsky, announcing the key claim of his SIP book: "the route toward generality [in AI] must lie partly in . . . the representation of more and better kinds of knowledge" (1968: 13).

The most influential KR-for-SIP at that time, besides logic, was the method of semantic networks developed by Quillian (1931–). These were implemented in the mid- to late 1960s, when Quillian was working at Bolt, Beranek & Newman, or BBN (Quillian 1967, 1968, 1969). But they'd been envisaged when he was a graduate student of Simon's at Carnegie Mellon (Quillian 1961, 1962).

Quillian had come into AI after being trained in sociology. His AI aims were to enable automatic word disambiguation, comparison between word meanings, interpretation of anaphora, and associative inference. But his prime concern, doubtless encouraged by Simon, was to cast light on human memory and language understanding:

The central question asked in this research has been: What constitutes a reasonable view of how semantic information is organized within a person's memory? In other words: What sort of representational format can permit the "meanings" of words to be stored, so that humanlike use of these meanings [by computers] is possible? (Quillian 1968: 227)

Indeed, over half of his Ph.D. thesis of 1966 would be based on experiments recording the thinking aloud of a woman interpreting language. A summary of the thesis appeared in the psychologists' journal *Behavioral Science* (Quillian 1967), and the writers he cited there included Frederic Bartlett on memory (5.ii.b), Miller on the "Magical Number Seven" (6.i.b), Bruner and Earl Hunt on concept learning (6.ii.b−c and v.a, and iii.d below), and Walter Reitman on thinking (6.v.b and 7.i.b). In addition, he drew on an up-to-date review of what psychologists knew about associative meaning (Deese 1962). And some years later, he did a systematic series of psychological experiments on memory and language (A. M. Collins and Quillian 1972). In short, he was "disposed to consider this model a psychological theory" (1967: 429).

The core insight—that word-ambiguity might be dealt with by looking for intersections of meanings—had been stated in the 1950s, by CLRU's Margaret Masterman and Robert Richens (see Preface, ii, and 9.x.d). And Quillian presented his earliest ideas at a 1961 colloquium in King's College, Cambridge, where the CLRU team reported on the latest version of their computerized thesaurus, or "semantic interlingua". But whereas their focus was solely technological (machine translation), his was primarily psychological. Moreover, his methodology was very different from the analyses/programs used at CLRU.

Specifically, Quillian pioneered localist connectionism (12.ii.i). (David Rumelhart, normally thought of as a guru of PDP connectionism, later recalled being "inspired" by Quillian to begin his own work on psychological simulation—J. A. Anderson and Rosenfeld 1998: 272.) A semantic network was composed of nodes and links between nodes. The nodes stood for concepts, objects, and events. The links represented relationships of various types, such as *superordinate, subordinate, isa, has-part* (and, if one wished, *aunt of, married to* . . . and so on). Such a network could represent both general concepts and individual instances: both cats and Tibbles, the latter linked to the former by an *isa* (IS-A) link.

On Quillian's view, both human long-term memory and the knowledge (data, meanings) possessed by a programmed network are implemented as nodes and links. Or rather, "denotative, factual information" is stored in this way. People's "plans for doing things", their "feelings about words", and their knowledge of "the conditional probabilities of word sequences" were specifically excluded (1967: 410). All of those were being explored by psychologists trying to capture the nature of meaning. Piaget, for instance, had discussed action plans in terms of sensori-motor schemas (5.ii.c); word sequence probabilities had been stressed by the informational psychologists (6.i.a and 9.x.b); and the feelings associated with words had been intriguingly highlighted by Charles Osgood's group in Illinois (Osgood *et al.* 1957). A complete theory of semantic memory, said Quillian, would have to include all those aspects too.

In a Quillian network, the inferences justified by the program's knowledge were drawn by a mechanism of "spreading activation". Whenever a node was activated, the activation would automatically spread along the links that connected it to other nodes—thus forming inferential pathways of various kinds. This simple arrangement met Raphael's demand: that the information required to generate the output from a KR be "easily accessible". (That wasn't true for logicist KR: there were no stored relevance links between propositions in the predicate calculus, only general rules of inference which could link them *if they could be found.*)

For example, consider property inheritance in semantic networks. If the class name *bird* was linked by *has* to *wings*, and if *robin* was linked as a *subordinate* to *bird*, then robins could easily (*sic*) be inferred to have wings too; and if Fred was linked by *isa* to robin, then Fred's wings could be easily inferred as well. Similarly, spreading activation enabled the network to make explicit comparisons and contrasts between the meanings of any pair of words, and to interpret ambiguous words according to the linguistic context—where the key factor in disambiguation was the *intersection* of activated pathways. Figure 10.4 shows two different pathways associating *plant* with *live*, and Figure 10.5 shows an associative structure leading to the intersection node (*sad*) for *cry* and *comfort*.

Quillian tested his networks on the English text used in the thinking-aloud experiment mentioned above. There were seven sentences, which between them contained nineteen ambiguous words. (This was pushing at the technological limits: because the program itself used up most of the computer's memory, there was room for no more than twenty definitions of English words—Crevier 1993: 82.) His first program interpreted twelve of them correctly. (It couldn't decide on four, so they remained ambiguous; and it got three wrong.) An improved version, developed in concert with Bobrow, dealt also with ambiguous prepositions. For instance, *I threw the man in the ring* was correctly interpreted, in context, as *While in the ring I threw the man*, or as *I threw the man who was in the ring*, or as *I threw the man into the ring* (Quillian 1968: 261–2).

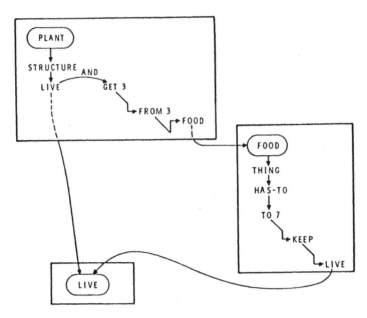

FIG. 10.4. Two paths direct from "Plant" to "Live". Reprinted with permission from Quillian (1968: 250)

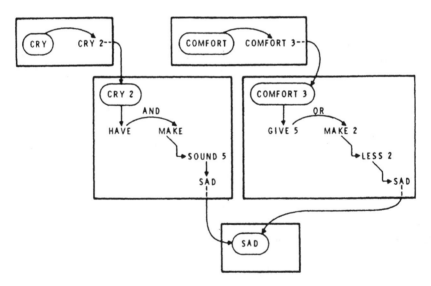

FIG. 10.5. A path from "Cry" and a path from "Comfort" which reach the same intersection node, "Sad". Reprinted with permission from Quillian (1968: 250)

In the late 1960s, semantic networks were seen as a liberation. Many people (in AI, psychology, and linguistics) adopted, or adapted, them. They weren't used only for language. For instance, they were applied to vision as well as concepts in John R. Anderson's ACT system (7.iv.c).

Granted, there were hazards lying in wait for the unwary. Various difficulties in using/interpreting semantic nets were detailed in 1975 by Quillian's BBN colleague William Woods (see Chapter 9.xi.e). Later still, some of these were elaborated by James Allen and Alan Frisch (1982), and by Ronald Brachman—another BBN colleague at the time (1977, 1979, 1983). But the difficulties didn't kill the interest. Indeed, nearly twenty years later the editors of a comprehensive volume on KR judged them to be "arguably the most popular of all techniques employed to represent knowledge in AI systems" (Brachman and Levesque 1995, p. xviii).

If Quillian's work inspired a huge amount of research over the years, so did McCarthy's harbinger project, logicism. Further claimants to the KR crown appeared as AI matured, some coming from other disciplines within cognitive science (social psychologists' *scripts* and linguists' *semantic primitives*, for instance: 7.i.c; 9.viii.c). By the mid-1980s, there was a thirty-page bibliography on the topic (Brachman and Levesque 1995: 335–65). In short, KR was a leading concern—many would say *the* leading concern—of symbolic AI.

One shouldn't conclude that people were faced with an *embarras de richesse*: a choice between myriad utterly different forms of KR. For the *richesse* consisted largely in superficial, not fundamental, differences.

This was made clear in a hugely influential paper by Minsky, circulated in draft (as was his wont) for some years before appearing as an MIT memo in 1974, and an official publication in 1975. 'A Framework for Representing Knowledge' unified the work of a number of scientists both within and outside AI. These people had said essentially similar things about how concepts work. The core idea was that concepts are structured, and that this structure—by storing co-relevant things together and by indicating just how they're related—enables sensible inference.

For example, Bartlett had posited structured "schemas" in the 1930s, and Hebb outlined hierarchical "cell-assemblies" in the 1940s (5.ii.b and iv.c). Quillian's semantic networks represented meaning structures too, although they could also depict merely contingent associations. AI vision researchers in the 1960s had defined structured object models for interpreting images (see iv.b, below). In the early 1970s, Robert Abelson had added "scripts" and "themes" defined as interpersonal concepts, and then (with Roger Schank) "scripts" understood as stereotyped ways of behaving (7.i.c). Meanwhile, Michael Arbib was describing sensori-motor schemas as well as conceptual ones (and he'd later attribute them to robots as well as to many animal species: 14.vii.b–c).

Minsky referred to some of that literature in trying to unify the superficially disparate work on KR "in Artificial Intelligence and Psychology" (1975: 211). That phrase "and Psychology" was crucial: "I draw no boundary", he said, "between a theory of human thinking and a scheme for making an intelligent machine." The paper's key claim was that common sense depends on structured "'chunks' of reasoning, language, memory, and perception". This claim was explicitly opposed both to behaviourist psychology (specifically, the first two tenets: see 5.i.a) and to AI logicism, which—Minsky said—tried "to represent knowledge as collections of separate, simple fragments".

Minsky added recommendation to unification. And what he recommended was the KR notion then favoured at MIT: the "frame". Frames were hierarchical structures ("networks of nodes and relations") representing complex objects. Each frame had fixed attributes at the higher levels, and many terminal "slots" for properties that could be filled in various ways. They occurred as "systems" of frames—for instance, the many images of a "scene from different viewpoints", or the different situations brought about by "actions, cause–effect relations, or changes in conceptual viewpoint". Since the various frames within a frame system share the same terminals, it's possible to "coordinate information gathered from different viewpoints"—in other words, to make inferences drawing on distinct sources/types of knowledge. (Later, these ideas would spawn object-oriented programming languages, in which frames were in effect treated as primitives: see v.c, below.)

Examples of frames included the concept of "arch" employed by Winston's learning program (subsection d, below), or the everyday concept of a room. A room typically has one door (occasionally, more), one or more windows, a roof, and a cuboidal shape; it normally contains furniture (distinct frames define tables, chairs, etc.); and it's usually contained in a larger building, such as a house.

However, "typically", "normally", and "usually" are weasel words. To be ultra-safe, the *number* in the *door* slot for *room* (for instance) could simply be left open, to be filled in by the programmer for each specific room considered. But that would be highly time-consuming; and it would also mean that a vision system looking for an unknown room or door, or trying to interpret an unfamiliar image which happened to depict a room, couldn't use one-door-ness as a cue. Alternatively, the number-of-doors in a *room* frame could be given a "default" value *(one)*, which would stand unless overridden by the programmer or, as a result of inferences, by the program.

In practice, said Minsky, much of the knowledge possessed by a frame-using program would be stored as default values. This was economical and efficient, but it could lead to problems. Consider exceptions, for instance. Cats have tails . . . but Manx cats don't. How could an AI representation—whether a hierarchical frame, or a set of logicist axioms, or a semantic network, or a Schank–Abelson script . . . —deal with that fact? It would be counter-productive simply to omit the tail slot from a cat frame; or to remove the axiom *If x is a cat then x has a tail* from a theorem-prover; or to delete the link between *cat* and *tail* in a semantic net; or to drop the rule, in a stereotyped script, that friendly visitors would stroke the tail of the family's cat. After all, the inference that a particular cat has a tail is far more often right than wrong. Minsky suggested various ways in which something could be assumed by an AI program unless it was specifically contradicted. Indeed, "default reasoning" soon became a key issue in AI (13.i.a).

Not for the first time, then, Minsky was surveying the land in preparation for core debates in AI. But whereas 'Steps' had (unavoidably) been programmatic, 'Framework' discussed many programming details. Its influence was due as much to the authoritative exposition of the MIT approach as to the general observations it contained.

Despite his efforts at unification, significant differences remained. Indeed, Newell (1982: 92) complained of "a veritable jungle of opinions" about KR, adding that "There is no consensus on any question of substance." Today, over twenty years later, KR is still a major concern. The early 1990s saw a 408-page special volume of the journal

Artificial Intelligence, and a further update appeared very recently (Brachman *et al.* 1991; Brachman and Levesque 2004). Evidently, KR and SIP were prizes which would be very hard-won.

b. A resolution to do better

Theorem proving had competed with checkers and chess as the first topic for real (i.e. implemented) programs. The fact that LT had found one proof more elegant than Russell's in *Principia Mathematica* had been hugely encouraging (Chapter 6.iii.c). And some, including the LT authors themselves, had been further encouraged by believing that they were modelling the thought processes of human logicians.

However, the mathematical logician Hao Wang (1921–95), then at Oxford but about to return to Harvard, soon criticized LT in pretty strong terms (H. Wang 1960). Not only could *Principia* proofs be generated more efficiently by other mechanized methods, but these methods—unlike LT—would eventually find *any* provable proposition within the relevant set. He concluded that LT wasn't a good ideal for mechanized theorem proving. More generally, he rejected human simulation as an approach to automated reasoning: it was logic that was important, not human thinking (cf. MacKenzie 1995).

Minsky disagreed, arguing in 'Steps' that whereas the *Principia* problems were simple enough to yield to exhaustive procedures, more complex problems wouldn't be. In particular, such procedures couldn't help with "the fundamental heuristic problem of *when to decide to give up on a line of attack*" (Minsky 1961*b*: 437).

In terms of Gottlob Frege's distinction between logical norms and psychological laws (2.ix.b), Minsky was in effect sitting on the fence. Unlike Wang, he wasn't restricting himself to pure logic—for he was considering the untidy practicalities of real-time computation. In terms of a distinction drawn 100 years later—and mentioned above, in contrasting McCarthy and Minsky—Wang was a neat whereas Minsky was a scruffy (see 13.i.c). On the other hand, he wasn't insisting that those practicalities be drawn solely, or even primarily, from human psychology: he was interested in the practical constraints on *any* complex computational system. He was happy to learn lessons from human psychology, when it seemed helpful. But unlike Newell and Simon (whom Frege would have accused of coming close to "psychologism"), he didn't believe that automatic theorem proving should necessarily mimic human thought.

The divide between theorem proving as logic and theorem proving as psychology was a special case of the divide between technological and psychological AI. And it endured, as this area of NewFAI advanced (MacKenzie 1995). Whichever side of this fence one favoured, the mere fact of such disagreement shows that while LT had kicked off the ball, it hadn't established the rules.

Broadly speaking, three different approaches to theorem proving developed, each of which is still with us today: logical/axiomatic, heuristic, and model-based. These were exemplified respectively by the resolution theorem-prover, the Logic Theorist, and Gelernter's geometry program.

Resolution theorem proving was driven by the same aim as Wang's: to discover a mechanized procedure that would be *guaranteed* to find a proof. It was defined in the early 1960s by the British logician J. Alan Robinson (1930–) at Rice University (J. A. Robinson 1963, 1965), and implemented by others soon afterwards (Wos *et al.* 1964,

1965). The immediate result, for Robinson himself, was a "spectacular rise to stardom in AI" (E. A. Feigenbaum, interviewed in McCorduck 1979: 219).

The reason why Robinson's work had such an impact was that it could deal with a relatively complicated form of logic. With respect to Russell's *propositional* calculus (the focus of McCulloch and Pitts: see 4.iii.c), there was no difficulty in principle. All proofs could be found by constructing truth tables. To be sure, the size of the search increased exponentially as the number of variables grew: if establishing *if (p and q) then (a or b)* was practically feasible, proving *if (p and q and r and s . . . and z) then (a or b or c . . . or n)* might not be. However, there was nothing intellectually deep about the methods, and almost any programmer could write programs for applying them. Russell's *predicate* calculus (the focus of McCarthy: see i.f, above) was a very different kettle of fish. It included existential and universal quantifiers of varying scope, and terms denoting relations of various degrees (binary, ternary . . . *n*-ary). Accordingly, it required more sophisticated methods of problem solving.

That's where resolution theorem proving came in, for it was a method for dealing with (first-order) predicate logic. Robinson wasn't the first to try to define some such method. The logician Jacques Herbrand, as long ago as 1930, had defined—though not implemented, of course—a basic algorithm for proof procedures (sometimes called "refutation procedures") which tried to establish a theorem by showing that its negation is inconsistent. In other words, the basic rationale was the assumption that not-not-X implies X.

That seems like common sense, although it was disputed by "constructivist" mathematicians (they claimed that X had to be proved directly, not inferred from a double negative: P. M. Williams, personal communication). However, non-constructivists were happy with it. Accordingly, after computers had come on the scene a number of people defined mechanizable versions of Herbrand's algorithm. One of these people was Robinson's ex-tutor Putnam (M. Davis and Putnam 1959/1960). (In his eyes, this was *logic*, not psychology; that's why he didn't mention it in his paper on functionalist philosophy of mind, written at the same time: 16.iii.b.)

Soon after Putnam's logic paper appeared, Robinson tried to implement the method defined in it—and discovered that it was highly inefficient. But he was also trying to implement another logician's suggestion at much the same time, and he found that a combination of ideas drawn from both papers was superior to either (MacKenzie 1995). That method was resolution, which—besides its relative efficiency—could deduce *new* statements from existing knowledge. That is, they could make information explicit which had previously been implicit and unrecognized. The other neo-Herbrand techniques, including Putnam's, couldn't do that: they produced only trivial consequences, by instantiating variables in logical expressions (Wos and Veroff 1987: 895).

In one respect, Robinson's theorem-prover was elegantly simple. For it used only one essential rule of inference. (The other rules, added over the years, provided the guidance/restriction strategies.) It always tried to infer a *contradiction* between the desired conclusion and the premises. This may sound paradoxical, but it's a form of argument (*reductio ad absurdum*) that's common even in everyday chit-chat. If someone makes a statement, and their opponent can show that it contradicts their premises and/or common knowledge, then (if the premises are true) it must be false. So we often suppose—'for the sake of argument', as we may say—that something is the case, *precisely in order to show that it isn't*. If we can't find any contradiction between X and the shared premisses, then the disputed statement can stand.

A resolution theorem-prover aiming to prove X, then, would first try to show that X is impossible. And it would do *that* by seeking some contradiction between X and the premisses. If it found such a contradiction, then not-X (the negation of X) would have been proved. So in trying to find a plan enabling a robot (such as SHAKEY) to do something, for instance, the program would start by assuming that no such plan exists.

As usual, it wasn't all plain sailing. Robinson's initial version was very inefficient, tending to get bogged down in masses of valid-but-boring conclusions. These were drawn because they *could* be drawn, not because (in respect of the task) they *should* be drawn. In other words, it suffered from the "combinatorial explosion", a form of information overload that bedevilled many NewFAI programs (and which often bedevils human brains too). In general, the difficulty was that the computer might try to consider and/or infer too many items to make problem solving practically possible. Sir James Lighthill, in his savage attack on AI some ten years later, would write as though the combinatorial explosion was an unseen and insuperable obstacle (11.iv.a). It may have been insuperable: we'll come to that. But it certainly wasn't unseen. It was a bugbear familiar to all computer scientists, and had been mentioned many times in 'Steps' (though not under that name). And it could be ameliorated, if not wholly overcome.

At first, people assumed that better hardware (faster processing and larger memories) would enable resolution theorem-provers to scale up beyond the initial barrier of a few dozen items. But it soon became clear that better software would be needed too. Accordingly, several people tried to improve on the basic resolution algorithm. They included Robinson himself (1968), his collaborator Larry Wos (in several papers), David Luckham (1967), and Cordell Green (1969)—who applied it to question answering. These people (and their successors in later decades) defined increasingly efficient "strategies" for guiding and restricting the many inferences made by a resolution theorem-prover, so that the system knew where (not) to go and, as Minsky had demanded, when to give up.

Robinson's approach caused huge excitement in the early 1960s. One didn't have to be a logician to find it intriguing. McCarthy's harbinger paper, after all, had suggested that predicate logic could represent common-sense knowledge (and could be used to give advice while leaving the core program untouched)—so the new technique might be useful in any problem domain. Moreover, the resolution theorem-prover did what the *advice taker* had been proposed to do: it drew new conclusions. Tell it A (given that it already knew B, C, and D), and it might be able to infer E. A wide variety of resolution-based NewFAI programs were soon developed. In question-answerers, for example, the program's world knowledge was expressed as axioms, and questions were presented to it as theorems to be proved (C. C. Green 1969). Logic even entered robotics, in the STRIPS resolution system that planned actions for the SRI robot (see subsection c, below).

There was a time bomb hidden here, however. It sputtered into life about twenty years later, when a new programming language with resolution built into it was widely used in practical applications—namely, expert systems (v.f below, and 11.v.a). Non-logician users sometimes assumed that if their program *failed* to find a contradiction, then not-not-X hadn't been established so not-X had been proved. This so-called "negation by failure" was a mistake. Failing to find a contradiction meant nothing, because predicate calculus isn't what logicians term "decidable" (cf. 4.i.c).

In real life too, there's all the difference in the world between proving that something is false (real negation) and failing to prove that it's true (negation by failure). Just compare being able to prove that your partner isn't having an affair with not being able to prove that he/she is. The reason, of course, is that there are many potentially relevant pieces of evidence (premises) of which you may be unaware, and which you have no realistic hope of discovering/deducing. (Even the private detective may fail to find all of them.) Indeed, you may even have difficulty in recognizing *just which* matters are "potentially relevant" (Chapters 8.vi.c and 7.iii.d).

Problems in very well-understood areas of science are less open to threat on this account. For one can hope (*sic*) to have identified all the relevant knowledge, and one can hope (again, *sic*) that the situation is relatively 'closed' against unpredictable complications. But even there, it's not obvious that the problems are in principle decidable. If they're not, then assuming that not-X had been proved merely because of the failure to find a contradiction isn't logically justified. In other words, proof of non-existence is very much more difficult than proof of existence.

So why wasn't resolution laughed out of court? Well, partly because expert logicians—the Putnams and Robinsons of this world—wouldn't be tempted to make that mistake. Moreover, in a logical system that's technically *complete* the drawback outlined above doesn't apply. All the required premises and/or axioms are (by definition) given, and the rules of inference are capable of leading to every possible true conclusion. It follows (in principle) that if they *fail* to draw a particular conclusion then it's not there to be drawn, so cannot be true. And Robinson showed that resolution theorem proving is complete in this technical sense. Provided that the premises suffice to imply the conclusion, it can *in principle* be found by the resolution method.

Even so, *in practice* it was often difficult, or impossible, to work through the deduction far enough to complete the proof. (Remember Minsky's point about knowing when to give up.) Robinson's star status waned, accordingly. Feigenbaum remembers:

[Here] he was, [a logician] propelled to the front ranks [of AI], and suddenly felt heavy obligation to extract the AI researchers from the pit into which they were falling, the pit of the combinatorial explosion. He understood this, but was really helpless to do anything about it since he was a logician who invented a method, not an AI researcher interested in formalizing the world's knowledge. Finally he gave up, decided he was really sorry he'd got people into this trap, but he couldn't do anything about it. As AI moved away from the Resolution Method, he moved back to logic and resigned his position on the editorial board of the AI Journal, and retreated from the whole scene. (interview in McCorduck 1979: 219–20)

c. Planning progresses

AI planning sprang from Newell and Simon's mid-1950s work on LT and GPS (see Sections i.b and d above, and Chapter 6.iii.c). That work had been a revelation—indeed, a revolution.

Nevertheless, GPS had many weaknesses. One was mentioned in i.d, above: its tendency to get bogged down by forming sub-goals on too many levels. Another was its inability to outline a plan at an overall, abstract, level, leaving the details to be decided later (on execution). This made it especially ill-suited for use in robotics, where the real-world situation may change while the program is running. By the same token, it wasn't amenable to interrupts during processing. Nor could it learn from its mistakes: an error recovered by backtracking on Monday would be repeated relentlessly on Tuesday. (Some capability for learning was soon provided, however: Newell *et al.* 1960.) Certain errors, such as undoing a previously achieved (and essential) sub-goal, weren't even identifiable by the system.

In the 1960s and early 1970s, these weaknesses—and others—were overcome as means–end analysis was developed into planning programs of increasing power. Many of these were written for the Stanford robot, SHAKEY (and others for Edinburgh's FREDDY).

The STRIPS planner (an acronym for STanford Research Institute Problem Solver) used predicate calculus theorem proving to plan SHAKEY's movements in much the same way as GPS had done for abstract problems (Fikes and Nilsson 1971; Fikes *et al.* 1972*a*,*b*). It relied on a novel form of KR: a "triangle table" storing operators, their preconditions, and the newly true facts resulting from their execution: see Figure 10.6. All other facts were assumed to remain unchanged (an assumption which raised the spectre of the frame problem: see iii.e, below).

Triangle tables enabled STRIPS/SHAKEY to act, and also to learn to do so better. The key point was that they identified whole sets/subsets of actions, for achieving particular goals/sub-goals. One triangle table could be inserted (as a chunk) into another one, so representing several levels of plan complexity without having to reconsider each individual operator. Moreover, STRIPS could generalize its plans. For example, having worked out *how to get from Room 7 to Room 3 via Room 6, in order to open Window 2 (using Box 1 fetched from Room 4)*, it could express a plan schema representing *how to get from any room to any other, via some other if necessary, to open any window, using any box found in any place.* (In essence, this was done by replacing logical constants with variables.)

Since it identified the expected result of a given action, or set of actions, a triangle table could be used to monitor success—and to plan subsequent actions accordingly. It could even be (intelligently) broken up, the relevant portions being incorporated into new plans. And these, in turn, could be inserted into others.

However, STRIPS (in its first incarnation) was limited in a number of ways. As a result, successive improvements were developed in the early 1970s (see Boden 1977: 280–7, 357–70).

For instance, STRIPS (like GPS) couldn't produce an overall strategic plan of the problem before the detailed solution was started. For very simple problems, this didn't matter. But if there were many different operators, each with many preconditions, planning would be scuppered by the combinatorial explosion. What was needed was a further level (or levels) of abstraction, allowing a solution to be stated in very general terms. These would later be transformed into lower representational levels, each more detailed than the one above it—with the last having direct control of the robot's movements. (Similarly, one may plan a journey in terms of the towns to be visited, before worrying about the order in which to visit them—still less, just how to get from this town to that one.)

Initially True
Preconditions (ITPs)

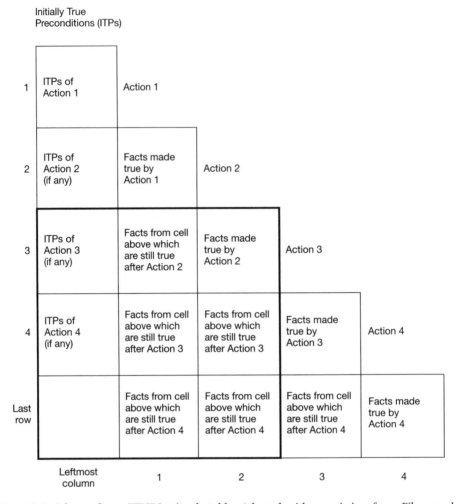

FIG. 10.6. Schema for a STRIPS triangle table. Adapted with permission from Fikes *et al.* (1972*a*: 259)

Accordingly, Earl Sacerdoti (1948–) developed ABSTRIPS, which solved its problems on a hierarchy of representational levels (Sacerdoti 1974). The first was the most abstract. Details were considered only when a successful plan in a higher-level problem space gave strong evidence of their importance. The crux of ABSTRIPS was that the lists of preconditions for the various operators were ordered for priority, so that the most critical preconditions were considered first. For example, if SHAKEY's goal was to push a box from one room into another, ABSTRIPS enabled it to get to the former room first (assuming that it had started out from yet another room), and then to open the relevant door (assuming that it was shut) before positioning itself by the box so as to be able to push it.

Like GPS's difference orderings, the criticality levels of the preconditions for each operator were provided in the database, not worked out by the program itself. ABSTRIPS could pass easily from one plan level to another, mapping more abstract onto less abstract representations, because the abstraction involved was merely a question of *ignoring* details in an operator's definition (details which, if necessary, could be taken up later). If the content of the definition had differed according to plan level, it would have been more difficult for the program to pass from one level to another.

Similarly, if—when considering the details—ABSTRIPS discovered that the sub-plan wasn't executable in the current circumstances, it could return smoothly to the *more* abstract level, much as GPS had backed up to previous choice points on failing to achieve a sub-goal. As Sacerdoti pointed out, this was especially important for robotics, since the real world can't in general be expected to stand still. A state of affairs that pertained at the outset might have changed by the time it became relevant.

In fact, the Stanford team deliberately teased/tested SHAKEY by occasionally moving a block to a different position. The robot wasn't capable of noticing this being done. But it might later realize that something was amiss when it came to execute a plan that assumed that the block was in its *original* position. In that case, SHAKEY would recompute the plan. (As the situated roboticists would later point out, this was very different from being in constant interaction with a dynamically changing world: e.g. R. A. Brooks 1991*b*, n. 1.)

While Sacerdoti was developing ABSTRIPS at Stanford, Gerald Sussman (1947–) at MIT was working on HACKER (Sussman 1975; Boden 1977: 286–97). This was a planner with a difference, for it monitored its own performance and corrected its mistakes. It did this by exploiting what Papert was already referring to as a "powerful idea": *bugs* (see vi.a, below). As Sussman put it:

I believe that effective problem solving depends as much on how well one understands one's errors as on how carefully and knowledgeably one makes one's initial choices at decision points. The key to understanding one's errors is in understanding how one's intentions and purpose relate to his plans and actions . . . [So one needs a] teleological commentary about how the subparts [relate to the overall goals]. [And one needs] knowledge about how to trace out bugs and about the kinds of bugs that might be met in applying a given kind of plausible plan. (Sussman 1974: 236)

He identified five main types of bug: PCB, PM, PCBG, SCB, and DCB (that is: Prerequisite-Conflict-Brothers; Prerequisite-Missing; Prerequisite-Clobbers-Brother-Goal; Strategy-Clobbers-Brother; and Direct-Conflict-Brother). These concerned inter-actions between plan components, manifesting themselves (for instance) as unsatisfied prerequisites, unnecessary double moves, or failure to protect a condition that must continue to exist until a specific point in the plan. An unsatisfied-prerequisite bug differed from a prerequisite-missing one. The former was a hitch discovered only on actually trying to execute the plan; the latter would be discovered by a critical analysis of the plan's structure before any attempt was made to execute it. (Compare encountering roadworks on your way to the station with planning to reach it by a mistaken route.)

While engaged in planning, HACKER employed several CRITICS, self-monitoring procedures designed to spot/fix particular bugs. For instance, one CRITIC looked out for cases where the program, having achieved an essential sub-goal, later undid it in order to achieve another one. If such a case was found, the program would devise a "patch" for the original plan. (Besides spotting and fixing bugs, HACKER learnt to avoid them in future: see subsection d, below.)

Sussman's work helped Sacerdoti to make a further advance in planning. Part-inspired by HACKER, he abandoned ABSTRIPS for NOAH: Nets of Action Hierarchies (1975*a,b*). ABSTRIPS, like HACKER, had been a linear planner. Both programs assumed that sub-goals are additive, so could be considered independently—even if they were best considered in a particular order. However, sub-goals are sometimes non-additive: they interact, so that the achievement of sub-goal A may have to be deliberately undone before it's possible to achieve sub-goal B. This would happen, for example, if HACKER were asked to solve the problem shown in Figure 10.7 (Sacerdoti 1975*b*: 105–9).

ABSTRIPS could sometimes use backtracking to reach a clumsy solution to non-additive problems. But it couldn't achieve the optimal solution, because it couldn't take sub-goal interaction into account. NOAH, by contrast, could. It did this by ignoring temporal order in its initial plans. Sub-goals were represented merely as logical conjuncts to be achieved in parallel. When the plan was elaborated, on successively detailed levels, NOAH would scan for potential interactions between sub-goals and specify temporal order whenever this was needed to avoid them. The result would be a partially ordered plan, constructed so as to minimize backtracking in execution.

NOAH's secret lay in its CRITICS, inspired by HACKER's subroutines of the same name but constructive rather than destructive in nature. For these Stanford-based CRITICS added constraints to a partially specified plan, instead of rejecting incorrect assumptions that shouldn't have been made in the first place.

There were five general-purpose CRITICS to oversee the elaboration of NOAH's plans. They watched out for ways of resolving potential conflicts between sub-goals; of specifying existing objects for use rather than leaving their identity vague; of eliminating redundant preconditions; of resolving "double crosses" in which each of two conjunctive purposes denies a precondition for the other; and of optimizing disjuncts so that a choice between alternative sub-goals could be predetermined or postponed, whichever was the more sensible.

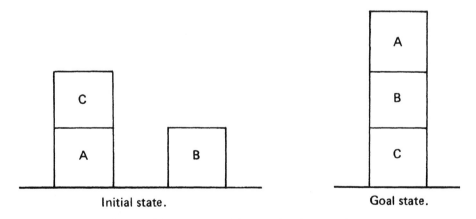

FIG. 10.7. A non-additive problem. Reprinted from Boden (1977: 360)

In addition, NOAH could be provided with task-specific CRITICS. This was done for Stanford's CBC project (Computer Based Consultant), an interactive program designed to give on-the-job advice on how to assemble a machine to novice mechanics having varied levels of expertise (Hart 1975; Sacerdoti 1975*b*: 90–109, 117–20). In order to give advice that was appropriate for the particular user, NOAH used the specific queries posed by the human mechanics as cues directing it to answer at one hierarchical plan level rather than another. So one mechanic might be told simply to "Replace the pump", whereas a less experienced person might be instructed to "Remove the 4 mounting bolts at the base of the pump using a $\frac{3}{8}$-inch open-end wrench." Similarly, the program could ask helpful questions when the user got into trouble, because it had some idea of just where in the overall task the human's difficulty might lie. In short, CBC was an early forerunner of today's "Steve", a VR system that enables mechanical advice to be given by a humanoid animation, bodily movements—and voice recognition—included (Chapter 13.vi.b).

NOAH managed to solve a problem which bemuses many humans, and which early 1970s theorem-provers had been quite unable to handle. Michie had recently identified the "keys and boxes" puzzle (too complex to state here) as a benchmark problem for AI, but NOAH worked it out in only twenty-one steps (Sacerdoti 1975*b*: 70; Michie 1974).

Similar work on non-linear plans was now being done in Michie's Edinburgh laboratory, by Austin Tate (1975). But problems remained. For instance, neither Sacerdoti's NOAH nor Tate's INTERPLAN could cope reliably with cases where each of two conjunctive goals denies a precondition for the other. (The "double cross" CRITIC could sometimes diagnose and resolve the trouble, but in other cases this wasn't possible.) As Sacerdoti put it, "the system must be creative and propose additional steps that will allow the two purposes to be achieved at the same time" (1975*b*: 42).

That sort of creativity was provided by BUILD, a program written by Scott Fahlman (1948–)—who, by the way, also invented the "smiley face" emoticon (Fahlman 1974). BUILD represented plan structures in such a way that it could solve the problems shown in Figures 10.8 and 10.9.

Both of these see-saw tasks require an additional manipulation early in the plan, which is later undone by the builder. Without going into the details of how BUILD worked (cf. Boden 1977: 366–70), let's just note that it 'knew' enough about weights and levers to foresee unwanted effects (i.e. the collapse of the structure being assembled),

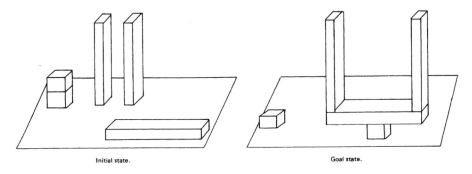

Initial state.   Goal state.

FIG. 10.8. Blocks-world problem (only one hand; no sliding). Reprinted from Boden (1977: 364)

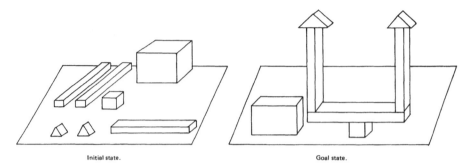

Initial state.   Goal state.

FIG. 10.9. Blocks-world problem (only one hand; no sliding). Reprinted from Boden (1977: 365)

and could forestall them—by adding steps into the plan—without introducing further contradictions. (Well, that's what Fahlman claimed, but it's not quite true: the cube used as a scaffold in Figure 10.8 can be removed at the end only by sliding, which is forbidden. When someone pointed this out on a sultry summer's day in 1978, during a talk I was giving at RAND, there was an appalled silence in the room—until a man at the back shouted "It's an ice cube!")

Ice cubes aside, it's clear that by the mid-1970s planning structures of some complexity were being handled in AI. Backtracking, generalization, abstraction, contingency planning, diagnostic critics, anticipatory corrections, on-the-spot adjustments, mapping of the current context of action, memory of past contexts . . . all these were now possible. Some programs (such as BUILD) could recognize that a current plan threatens to become overly complicated, and could interrupt it to switch to an alternative—with the option of resuming it *at that very point* if it turned out later that the first plan had been better after all. (This flexibility was due to new programming languages: Section v.d.)

By the mid-1970s, then, a number of planning techniques were available. General-purpose methods were sometimes supplemented by special-purpose data and tricks (NOAH's $\frac{3}{8}$-inch open-end wrenches, for instance). But the teleological structure of plans *in general* was the focus of all this GPS-inspired NewFAI work. And Newell and Simon's production systems provided further general features, such as tolerance of interrupts (7.iv.b and Section v.e below).

The generality was limited, however. In particular, there was no good way of modelling systems having several different, and potentially conflicting, goals. Research on multi-purpose planning had been done in the 1960s: Walter Reitman's Argus (7.i.b), and the Mars robot controller designed at MIT by William Kilmer and McCulloch. But the problems involved were by then insuperable. It was difficult enough, as we've seen, to model plans for achieving a single goal. Not until the end of the century would the Kilmer–McCulloch controller become a practical proposition (14.v.a), or multi-motive scheduling begin to be understood (7.i.e–f).

Moreover, the 1970s theoretical advances didn't prevent there being major problems in practice. The Stanford robot, for instance, spent an inordinately long time deliberating about what it should do—not because of any Hamlet-like indecision on its part, but because of plan-hierarchy complexity. (The combinatorial explosion, again.) Even when, thanks to Sacerdoti, the robot was able to recover from unexpected events, this wasn't the work of a moment. So in a constantly changing real-world environment SHAKEY, like all other NewFAI robots, was useless.

That's largely why many roboticists in the 1980s would reject planning *as such*, relying on insect-inspired reflexes instead (Chapter 13.iii.b). Nevertheless, SHAKEY had prompted a great deal of useful work—and not only in AI. The neuroscientist Arbib, for instance, was inspired by SHAKEY to start on a programme of research on visuo-motor control in frogs, which led eventually to an ambitious theory of human thought—even including religion (see 14.vii.b).

d. Early learning

During the early to mid-1950s, most AI research on learning was connectionist. Even Minsky had tinkered a connectionist learner in 1951, and devoted his Ph.D. to explaining why it worked (12.ii.a). In symbolic AI, learning machines were thin on the ground. By the mid-1950s there was only one of any real interest: Samuel's checkers player.

In the late 1950s, however, a number of people started to work in this area. One was Amarel (1962), who wrote a heuristic program which, in a sense, learnt *how to learn*: its task was to induce routines equivalent to the sixteen basic operations of the propositional calculus, but once a particular operator had been learnt it could help in learning others. (Much the same would later apply to STRIPS, as remarked above.) His draft paper was listed in Minsky's 'Steps' bibliography, but (perhaps because it wasn't officially published until 1962) it wasn't included in *Computers and Thought*. Three other early 1960s papers, however, were included—all intended as general theories, and all largely based on psychological ideas.

The first was a pattern-recognizer, an adaptation of Pandemonium (Uhr and Vossler 1961*a*/1963). It learnt its own "feature detectors", instead of having to be supplied with them. (The authors compared it with David Hubel and Torstein Wiesel's contemporary work on vision: Chapter 14.iv.b.) The paper in Feigenbaum and Feldman's collection described how the program learnt to recognize written letters and doodles (with respect to the latter, it surpassed human beings). But it had also learnt to deal with diagrams of speech acoustics, which aren't distinguishable by humans unless they're expert in the field (Uhr and Vossler 1961*b*). Its generality lay in the fact that a pixel is a pixel . . . that was all the program needed to know.

The second was Feigenbaum's (1961) EPAM: Elementary Perceiver and Memorizer, briefly mentioned in Chapter 6.iii.c. This had started out as an attempt to simulate aspects of decision making in business (Newquist 1994: 81). But it moved away from business towards psychology, and from decision making towards memory. The published paper described it as "an attempt to state quite precisely a parsimonious and plausible mechanism sufficient to account for the rote learning of nonsense syllables".

The program used its "elementary information processes" to build, and when necessary to revise, treelike "discrimination nets" representing the various syllable pairs. The details of the nets, and of the processes using them, were decided partly by general principles and partly by human psychology. For instance, *any* learning system needs to store more information about a stimulus if it has to reproduce it, as opposed to merely recognizing it. But only systems—such as EPAM—simulating human psychology need to notice the order of letters in syllables, since our tendency to be more influenced by end letters than by middle letters, and by first letters than by last letters, is presumably not a cognitive universal.

EPAM was more valuable for psychology than for AI. Feigenbaum (with Simon) had already used it to simulate some classic paired-associate experiments, when it had shown stimulus generalization and various types of interference (where later learning seems to undo earlier learning) even though these hadn't been specifically anticipated. And they soon applied it to explore the difficulties involved in learning serial lists where the same item occurs twice (Feigenbaum and Simon 1962). Even today, improved versions are still being reported in the psychological journals (e.g. Feigenbaum and Simon 1984; Gobet 1998).

The last of the *Computers and Thought* trio was valuable for psychologists and AI alike (Hunt and Hovland 1961; see also Hovland and Hunt 1960). Indeed, its ideas would eventually bear huge fruit in technological AI (13.iii.e). It was a model of strategies for learning from examples, based largely on Bruner's psychological work of the late 1950s (6.ii.b–c).

The senior author was Yale's Carl Hovland, whose mid-century analyses of the information available in concepts had inspired Bruner to work on that topic in the first place. The junior author was his student Hunt (1933–). Nowadays, their approach is normally attributed to Hunt, for Hovland died (in 1961) soon after their initial program was completed, and the work was taken forward by the younger man—who published an entire book on it just before *Computers and Thought* appeared (Hunt 1962).

Following Bruner, Hunt and Hovland defined strategies for learning both conjunctive and disjunctive concepts, given both positive and negative examples. And (again, like Bruner) they considered not only concepts whose instances share common properties, such as redness or triangularity, but also concepts defined by "common relationships"—such as having *a large figure on top and a small one below*, irrespective of the shape of either figure.

Broadly speaking, they modelled the inductive strategies informally described in *A Study of Thinking*—which they thought of as MGP's TOTE units (Hunt 1962: 184–5). In addition, they leant on the computer simulation of human memory that Hunt had presented for his Ph.D. in 1960, which featured—for instance—the limited size of STM. That was significant. For they were doing psychology, not technological AI:

We have attempted to write a computer program which, when given as input coded representations of the stimuli, will give as output coded responses that can be used to predict the responses of a human subject. *Accurate prediction of the responses, not the development of a good hypothesis developer*, nor, solely, the reproduction of previously obtained protocols, is our goal. (Hunt and Hovland 1961: 146; italics added)

So, for instance, their model favoured "positive focusing", in which hypotheses are developed from positive instances rather than negative ones—even when this isn't the most efficient strategy. Why? Because BGA's experiments had found that human subjects favour positive focusing.

Hunt's (1962) book discussed a number of strategies—some implemented, others not. The way in which they were described was less 'logical', and more 'psychological', than in the Hunt–Hovland paper. Consider "conditional focusing", for example—a procedure for learning disjunctive concepts. In the *Computers and Thought* paper, this had been expressed as a recursive function operating on sets, an idea taken from McCarthy (1960). But Hunt described it as a method for building decision trees (see Figure 10.10). These two approaches were equivalent: logic had been re-expressed, not dropped. Indeed, Hunt still declared that "Concepts are essentially definitions in symbolic logic" (p. 8).

We needn't follow through this strategy in detail. But notice that step (3) involves a simple frequency measure, seeking the description *most commonly* applicable to positive instances. Later versions of this general inductive approach would employ increasingly sophisticated frequency measures (see 13.iii.e). Notice, too, that different random choices at step (3)—where no single description is "the most common"—would result in different decision trees for one and the same concept. Suppose, for example, that the training set was as follows (Hunt 1962: 234):

Positive instances	Negative instances
Large, black circles	Large, black triangles
Large, white circles	Large, white triangles
Small, black triangles	Small, black circles
Small, white triangles	Small, white circles

Two decision trees derived from those data are shown in Figure 10.11.

As Hunt pointed out, conditional focusing *as shown in Figure 10.10* was "an algorithm for producing answers", which "[clearly] will not do as a simulation" (p. 235). The Hunt–Hovland paper had already suggested limiting the number of recursions permitted, so restricting the maximum length of a path through the decision tree. Now, Hunt had done further experiments on human subjects, learning concepts of various kinds. As a result, he suggested changes to make the machine strategies more human-like—which sometimes, though not always, made them more efficient too.

For instance, a "positive focusing" modification could give quick solutions for conjunctive concepts (see Figure 10.12). This new routine was inserted between steps (0) and (2) of the procedure shown in Figure 10.10. It would seem at first sight, therefore, that conditional focusing had been made more complex. But that wasn't so. For the calculations shown in Figure 10.12 take less time (especially in list-processing languages) than those of step (2) in Figure 10.10.

Various other modifications were suggested as well. In short, this work was not only an achievement, but also an intriguing promise of further achievements. That promise would be fulfilled, for Hunt and Hovland's research led eventually to the ID3 algorithm (13.iii.e). (It was part-inspiration also for the NewFAI program that discovered geometric analogies, using the "common relationships" mentioned by H&H and by Minsky in 'Steps'—T. G. Evans 1968: 283, and 13.iv.c.) However, that didn't happen immediately.

Despite being honoured by inclusion in *Computers and Thought*, these early learners didn't set GOFAI learning afire at the time. In part, that was because they were so heavily biased towards psychology. Besides, NewFAI was concerned with other things: planning, theorem proving, and especially KR. McCarthy had said in his harbinger paper

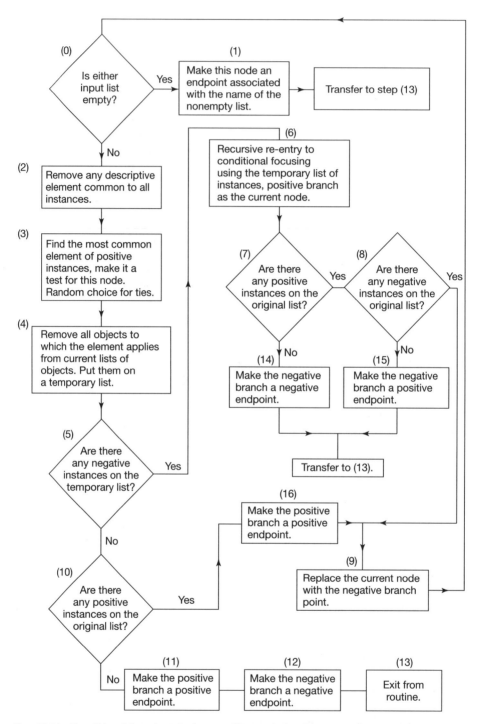

FIG. 10.10. Conditional focusing. Redrawn with permission from Hunt (1962: 232)

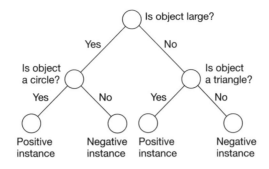

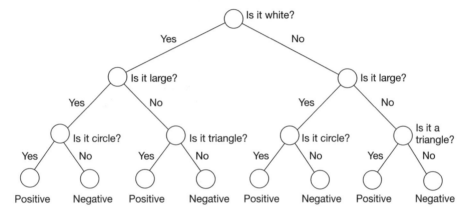

FIG. 10.11. Alternative decision trees defining the same concept. Redrawn with permission from Hunt (1962: 234, 235)

that if a machine were to learn something, it would first have to be able to represent that something—and the 1960s, as we've seen, were largely devoted to finding ways of representing things. Occasionally, creative efforts in KR were aimed at automatic learning: Michie's "Adaptive Graph Traverser", for instance, which extended Samuel's work from games to problem solving (Michie and Ross 1969). By and large, however, learning was put on the back burner.

In the early 1970s, a few pots were brought to the front of the stove. The first was Winston's concept-learner (1970a), the second Sussman's HACKER (1973/1975), and the third McDermott's TOPLE (1974). This trio addressed many interesting aspects of intelligence, for the programs learnt (respectively) by comparing examples, by doing, and by being told. They created some local warmth, especially at MIT. But they barely raised the temperature in the kitchen, as we'll see.

By the time Winston's program was officially published (in 1975), it was already well known. It had been developed during the late 1960s, presented to MIT for a Ph.D. in 1970, and circulated unofficially from then on—amid no little excitement. It was the first NewFAI learner since Samuel's to be widely lauded within the field (although I felt at the time, and still do, that its title to fame was largely due to the efficiency of the MIT publicity machine).

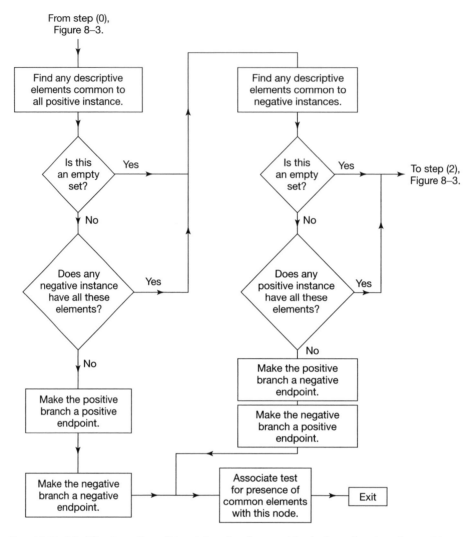

From step (0),
Figure 8–3.

Find any descriptive
elements common to
all positive instance.

Is this
an empty
set?   Yes

No

Does any
negative instance
have all these
elements?   Yes

No

Make the positive
branch a positive
endpoint.

Make the negative
branch a negative
endpoint.

Find any descriptive
elements common to
negative instances.

Is this
an empty
set?   Yes   To step (2),
                       Figure 8–3.

No

Does any
positive instance
have all these
elements?   Yes

No

Make the positive
branch a negative
endpoint.

Make the negative
branch a positive
endpoint.

Associate test
for presence of
common elements
with this node.

Exit

FIG. 10.12. Modification of conditional focusing for a rapid solution of conjunctive problems.
Redrawn with permission from Hunt (1962: 236)

Winston, like Bruner and Hunt before him, assumed that concepts can be defined
by lists of necessary and sufficient conditions (but cf. 8.i.b, 9.x.d, and 12.v–vi and
x.b). His program learnt what counts as an *arch*, or an *arcade*, in the blocks world. On
being presented with inputs labelled as examples or counter-examples (including near-
misses), it generated an articulated description of each one: a hierarchical semantic
network. Gradually, it modified the current candidate to construct a description
matching *only* the examples.

That's easily said. In fact, the comparison and modification techniques were a
significant KR achievement at the time (see Boden 1977: 252–67). Quillian had shown
how to build semantic networks, and how to use them for simple inferences; but

he'd said nothing about how to compare them. Winston's new network-matching techniques were designed to do just that. Indeed, they—not the program's power as a learner—were the main reason for people's interest.

An arch, for instance, consists of two vertical blocks with another block on top of them—not lying on the floor beside them. The shape of the top block is irrelevant: it may be a bulky cuboid, or a plate, wedge, or pyramid. Moreover, there must be a significant gap between the two vertical blocks: if they abut, it's not an arch but a near-miss. These facts wouldn't be evident on seeing just one arch. They'd become clear only after many different inputs—including some non-arches as well as various types of arch. The program's final concept is shown in Figure 10.13.

Much as Bruner and Hunt had found that the order of presentation matters when people learn concepts, so it mattered for Winston's program. And he had to avoid introducing several differences at once—hence the importance of the near-miss. But the inputs, and the list of candidate properties, were carefully tailored to suit the blocks world. Because this was so narrowly defined, Winston *couldn't* present the program with irrelevant information. For the same reason, it didn't have to decide which ones were *likely* to be salient. Real-life learning is different. It's possible only if one assumes that most possible descriptions are irrelevant, and (usually) that some relevant descriptions may be missing—or even false. Such issues were ignored by Winston. (And, to be fair, by most other AI workers until many years later: see 7.vi.d–f, 12.viii.c–e, and 13.iii.e.) Moreover, structural comparisons of complex semantic networks were difficult

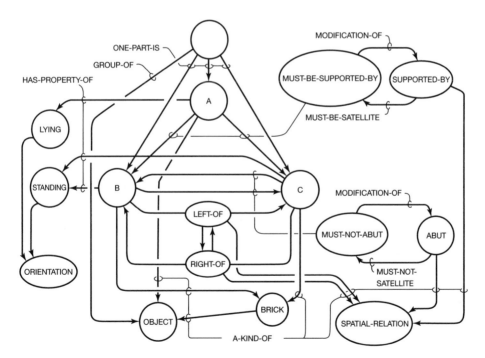

FIG. 10.13. Semantic network representing the concept of arch. Redrawn with permission from Winston (1975: 198)

to program, and to generalize (much more difficult than the property-counting done by Hunt's strategies). For both these reasons, Winston's program didn't hold much promise of practical applications.

Another issue ignored by Winston, and by most of his contemporaries, was *just why* certain programs did or didn't manage to learn. This question was an aspect of meta-epistemology, which most NewFAI scientists ignored—Minsky being an honourable exception (see i.g, above). It was highlighted in the mid-1970s, when researchers at Edinburgh not only proved that Winston's program had unsuspected limitations but also explained its successes in a principled way (Richard M. Young *et al.* 1977).

Their own concept-learning algorithm was a generalization of a focusing strategy earlier identified by Bruner. It could deal with hierarchical concepts (such as arch or arcade), and dealt uniformly with descriptive features (such as predicates, relations, and multi-valued dimensions) which Winston had treated as special cases. It worked irrespective of the order of presentation. And it explained the logical role of near-misses, which Winston had recognized intuitively. The key idea was to represent a candidate hypothesis by *two* nodes in the semantic network, one describing the specific instances already known to fall under the concept, the other holding the most general description known to be possible, given the instance encountered so far. So the two nodes set boundaries (identifying sufficient and necessary conditions) to the search-space within which the concept must lie. An efficient learning strategy would shift these boundaries until they met.

This Edinburgh algorithm was later used as the logical core of the widely used method of "version spaces" (T. M. Mitchell 1979; Mitchell *et al.* 1983). Its importance here, however, is as an early attempt to meet Drew McDermott's criticism of NewFAI in general: that it failed to analyse why programs were, or weren't, successful (see 11.iii.a).

Soon after Winston's 1970 memo weighed down the postmen's mailbags, his MIT colleague Sussman circulated a memo of his own (1973/1975). This described HACKER, a planning program engaged on a very different task: learning by doing. It wasn't the only NewFAI system that learnt by doing. Another was the Edinburgh pole-balancer (Michie and Chambers 1968), and yet another the STRIPS learner developed for the SHAKEY robot (Fikes *et al.* 1972*a,b*; Boden 1977: 280–6). But HACKER exhibited a capacity for deliberate self-criticism. Among other things, this suggested how Piagetian error-led constructive learning might be possible (see 5.ii.c, and Section vi.a below).

Much as Winston's program looked out for specific types of discrepancy, which it tried to reduce by relevant amendments to its conceptual models, so HACKER looked for discrepancies of particular kinds. As noted in Section iii.c above, it corrected its own mistakes by using a "teleological commentary" on its performance, couched in terms of various types of bug. But it didn't have to correct the same mistake repeatedly, because it learnt from its errors.

HACKER's learning took two forms. On the one hand, it stored its corrective "patches" in a central LIBRARY, indexing them according to the bugs they'd been written to correct. On the other hand, a patch devised in particular circumstances could be generalized to cover a general class of cases. (That had also been true of STRIPS, as we've seen.)

For instance, consider Figure 10.14. HACKER solved (*a*) instantly, using the primitive action for moving blocks. But (*b*) caused trouble, because the most nearly relevant primitive action required that the block to be picked up have no other block on top of it. However, HACKER classified this correctly as a prerequisite-missing bug, and patched the plan accordingly: A was moved off B, and then B could be put on C. This patch was stored in the LIBRARY, and brought out for problem (*c*)—which was therefore solved *immediately*, despite having the very same structure as (*b*). Moreover, problem (*d*) was also solved immediately, because the patch generated in response to (*b*) was sufficiently general to direct these steps (resulting in Figure 10.14.*e*):

> Wants to put A on B
>> Notices C and D on A
>>> Puts C on TABLE
>>> Wants to put D on TABLE
>>>> Notices E on D
>>>>> Puts E on TABLE
>>>> Puts D on TABLE
> Puts A on B.

HACKER's learning-by-practice wasn't just a matter of repetition, where doing something several times makes it more likely to be done in future. (Think connectionism.) Rather, the program had learnt something about the teleological structure of what was being done, and used this to do something else—something better—next time. So when it used the primitive action to solve problem (*a*), it made a note that this was being done *in order to* get A onto B. Similarly, its comments on the patch noted that putting A on the table was done purely *in order to* clear B, *in preparation for* moving it onto C. In short, Sussman was following Papert's lead in proclaiming the "virtuous" nature of bugs (Sussman 1974; cf. vi.a, below).

As for learning by being told (McCarthy's dream for the Advice Taker), McDermott's TOPLE was intended to show that this is a highly active form of cognition. Besides interpreting the meaning of the input sentences, one has to make judgements on their truth. Indeed, the interpretation itself may depend on judgements about truth, as Winograd's SHRDLU had recently shown (see the discussion of *Put the blue pyramid on the block in the box*, in Chapter 9.xi.b). Those judgements are grounded in the cognitive system's world model.

So TOPLE didn't just add newly input sentences ("Advice"?) to a list, but interpreted them in the light of what it already knew—or believed—about the world being described. Ambiguities were resolved, and anaphora understood, by comparing the input sentence with the program's current world model. The model itself could be amended, if it conflicted with the facts provided in the input. For each input, a "tree of hypothetical worlds" would be constructed, capturing every possible interpretation. (That was done by using the CONNIVER programming language, co-developed by McDermott himself: v.d, below.) This tree would then be pruned, leaving only the "plausible" possibilities—i.e. those consistent with the world model. As McDermott put it:

TOPLE is basically a very skeptical program. It wants to resist any change to the data base . . . so it forces itself to believe what it is told as cheaply as possible . . . [It] knows that some interpretations of the situations that it is told about are unlikely. It resists having to believe in such interpretations at all if more plausible interpretations of what it hears are available; and, if it must accept them, it demands some kind of compensating belief. (1974: 57)

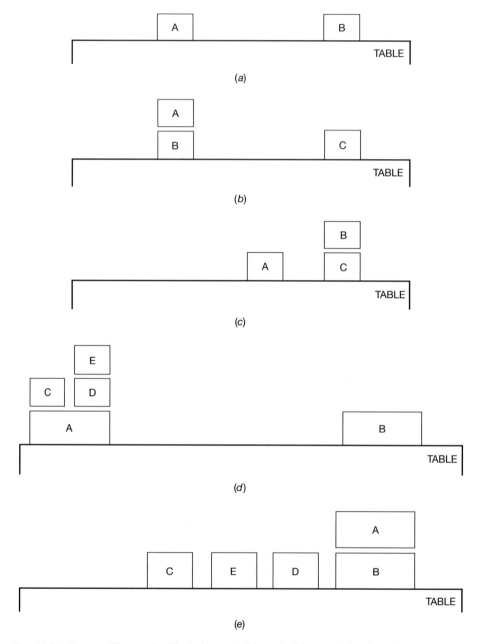

FIG. 10.14. Four problems solved by HACKER. Adapted with permission from Sussman (1975: 9–11) (*a*) Problem: Starting from this situation, put A on B. (*b*) Problem: Starting from this situation, put B on C. (*c*) Problem: Starting from this situation, put C on A. (*d*) Problem: Starting from this situation, put A on B. (*e*) HACKER's solution to problem (*d*).

The world model employed by TOPLE was a logical one, for McDermott was then still wedded to logicism (see 13.ii.a). But it was open to others to suggest non-logical versions, and the psychologist Philip Johnson-Laird soon did so (7.iv.d–e). He discussed the role of spatial mental models (and sentence–model comparisons, and even possible-worlds ontology) in reasoning and language understanding—not least, in learning by being told.

Despite these attempts to advance GOFAI work on learning, and despite the praise handed out to Winston's program (and, to a lesser extent, HACKER), the topic was still largely neglected. Ryszard Michalski, a key player in the revival of AI learning in the early 1980s, remembers that it was seen in the 1970s as "a 'bad' area to do research in (the dominant AI research was then on problem solving)" (R. S. Michalski, personal communication). What eventually brought it back to life was an exciting development of Hunt's work. But that's a story for a later chapter (13.iii.e).

e. 'Some Philosophical Problems'

In 1969 McCarthy and Patrick Hayes (1944–), then a Ph.D. student in Meltzer's Metamathematics Unit at Edinburgh, gave a paper at the fourth Machine Intelligence Workshop which very soon became a classic. Indeed, drafts had already been circulating for some time, and Hayes had been asked to give seminars on it in the months preceding the Workshop: "People were kind of primed for something important, and this was generally reckoned to be John's magnum opus in some sense" (P. J. Hayes, personal communication).

"John's" *magnum opus*, you'll notice. Why not "Pat's", too? Well, McCarthy was one of the three biggest names in NewFAI (alongside Minsky and Newell-and-Simon). Hayes was a mere research student. In fact, when he first impressed McCarthy with his ideas, he hadn't even achieved that far from exalted status: he'd just started a diploma in Machine Intelligence.

That was in 1966, when McCarthy came to Edinburgh for the second MI Workshop. Being too lowly a creature to be allowed to attend the sessions, Hayes begged Rod Burstall to arrange an audience with the great man. For McCarthy's grapevine-circulated memo on 'Situations, Actions, and Causal Laws' (1963) had excited him "hugely", as being "a natural way to put [my hero] Carnap into a computer". Burstall came up trumps, and scheduled the meeting:

Rod warned me that John was a very busy man, I had 30 minutes at most, and not to make him cross. So I turned up at the appointed time . . . and I was so nervous that I had written out my questions on a piece of paper, in case I forgot them. So John snarled "Come in!" rather grumpily, obviously irritated at having to talk to this kid as a favor to Rod, and I sat down and started asking my questions. John tossed the first couple of answers off in a few words but then he started to get thoughtful, and by about the 5th question he figured I was reading from a list, and he suddenly stood, reached over the (tiny) desk and grabbed the list out of my hand, "Lemme see that!" And then he sat in total silence for several minutes reading it, and I sat in utter terror that I had committed some cardinal sin of intellectual discourse by having a list written out and would be banished from academic society for ever. And then he looked up and said "How would you like to come to California?" And I was so flabbergasted that I couldn't speak. And then I said "When?" and he said "Maybe in the summer" and I said "Wow! Great" and so on and then my 30 minutes was up and I didn't see him again . . . (Hayes, personal communication)

After several months of total silence, during which Hayes put it out of his mind as a dream, a telegram arrived summoning him and his wife and baby son to visit California—all expenses paid. That was in the spring of 1967, and Hayes remembers:

While [I was] there we wrote the main draft of the paper. In fact, John had already written a lot of it, and my role was to be a kind of in-house critic/questioner, like a court jester, for most of the stuff; but I did write the section on situation calculus and modal logic (Section 4 of the paper) which, apparently, introduced the actual term "situation calculus" into the language though at the time this didn't seem like a big deal. (personal communication)

Their declared intention was to discuss some familiar philosophical problems *from the standpoint of* AI—and this, they did. But they also identified a huge philosophical problem *concerning* AI. Specifically, they defined the frame problem.

The roboticist Lynn Andrea Stein (1965–), some twenty years later, said that "a definition of the 'frame problem' is harder to come by than the Holy Grail" (L. A. Stein 1990). I'll offer one, nevertheless—or rather, two. To see why a dual definition is needed, one must first consider just how McCarthy and Hayes introduced the problem in their 1969 paper.

They didn't do this in a defeatist spirit. On the contrary, they located the frame problem in the context of a constructive discussion of how new forms of logic might be used for AI purposes. (Thirty-five years later, McCarthy's home page on the Web described it tersely as "the basic paper on situation calculus".) In other words, the generalities being sought by NewFAI were set alongside, and to a large extent identified with, the generalities sought by philosophical logic. And the McCarthyite faith in logic was undimmed: in their eyes, the frame problem was potentially solvable.

They weren't afraid of reaching for the skies. The philosophical chestnuts addressed in their paper were:

* causation;
* the nature and origins of knowledge;
* purpose, action, and ability;
* counterfactual conditionals;
* self-knowledge;
* and free will/determinism.

In addition, they discussed various modal (i.e. non-truth-functional) logics. These had been developed by philosophers in order to express certain highly general aspects of the world clearly, and to enable inferences about them to be deduced accordingly. The subject matter, ranging over metaphysics and mind, included:

* possibility and necessity;
* probability;
* ontology;
* knowledge and belief;
* tensed statements and time;
* decision and purposive action;
* commands;
* obligation and permission;

* interrogatives;
* and the conditions necessary for communication.

So McCarthy and Hayes were optimistic, not to say hubristic. Although modal logics were a twentieth-century invention (mostly pioneered after mid-century), the problems they were focused on were much older. As for the "chestnuts", these had puzzled philosophers for hundreds of years. Questions about the basic ontology of the universe (states, events, properties, changes, actions . . . ?), for example, had kept generations of grey-beard metaphysicians busy. Now, here were two NewFAI researchers, one of them a mere pipsqueak, brashly claiming to be able to illuminate them—even, *mirabile dictu,* to solve some of them. Causation and counterfactuals, for instance, could—so they said—be explicated by considering systems of finite automata interacting according to deterministic rules (McCarthy and Hayes 1969: 470–7, 479–80).

This was strong meat. As remarked in Chapter 16, philosophical problems don't get solved in a hurry. So their paper was nothing if not ambitious. Admittedly, their hubris had limits. For they were also claiming that AI "needs" philosophy. Even though most philosophy was declared by them to be "irrelevant", NewFAI scientists could learn a lot from some philosophers.

They weren't the only people who thought that philosophy and NewFAI needed each other. Others were saying that, too (Minsky 1965; Boden 1965, 1970, 1972; Newell 1973*b*). By the end of the 1970s this claim was even more common (e.g. Newell and Simon 1976). It was argued at length—and with no little passion—by Aaron Sloman, who declared that "within a few years philosophers . . . will be professionally incompetent if they are not well-informed about these developments" in AI (Sloman 1978, p. xiii). Moreover, some of the early cyberneticians—especially Craik (1943) and McCulloch (1948, 1961*a*; Pitts and McCulloch 1947)—had been equally bold in their claims to be able to help solve long-standing philosophical problems (see Chapters 4.iii.a–c and vi.b, and 12.i.c). So the coupling of philosophy with computers wasn't new.

However, McCarthy and Hayes were looking to philosophy in a more focused way. Their approach was distinctive in its heavy reliance on modal logics. Most of these had been developed very recently, in the 1960s, by philosophers such as Arthur Prior, Jaakko Hintikka, Georg von Wright, Saul Kripke, and Nicholas Rescher.

In recalling the many ancient philosophical disputes, McCarthy and Hayes reiterated the claim made in McCarthy's harbinger papers (1959, 1963), that natural language and common-sense knowledge could be represented in predicate calculus terms. And they promised that (their improved version of) resolution theorem proving could be used to explore the implications of all the highly problematic concepts listed above—and even to prove the correctness of complex action strategies. It followed that any philosophy of knowledge (epistemology) which could not, at least in principle, help AI "to construct a computer program to seek knowledge in accordance with it, must be rejected as too vague" (1969: 467).

For most AI scientists, however, what was especially interesting was the paper's relevance not for philosophy, but for the practice of AI. And there, despite the optimism, lurked a huge problem—one which would dominate the research of both men, and many of their AI colleagues, for decades. This was what they called "the frame problem" (p. 487).

McCarthy and Hayes introduced the frame problem by reference to their attempt to write an AI program that knew something about human communication: specifically, telephone conversations. When representing one person's ability to get into a conversation with another, they'd been "obliged to add the hypothesis that if a person has a telephone he still has it after looking up a number in the telephone book", and also to assume if someone looks for a number they'll know it, and that if they dial it they'll speak to their friend (pp. 485, 487, 489). Clearly, this spelled trouble.

Indeed, it spelled potentially endless trouble. It may not actually be the case that if Peter looks up John's number in the book, he will know it, or that if he dials the number he'll soon be talking to John. As the authors pointed out (on p. 489):

1. The page with John's number may be torn out.
2. Peter may be blind.
3. Someone may have deliberately inked out John's number.
4. The telephone company may have made the entry incorrectly.
5. John may have got the telephone only recently.
6. The phone system may be out of order.
7. John may be incapacitated suddenly.

Of course, they said, we could add terms contradicting any one, or even all, of these exceptions—"But we can think of as many additional difficulties as we wish, so it is impractical to exclude each difficulty separately."

They suggested three methods of escape, the first of which was to use the notion of a state vector, or "frame", associated with a theorem-prover. Here, various aspects of the (formally described) situation are listed, and the effect of any action is defined by stating which situational aspects are changed—all the others being presumed unchanged (p. 487). Earlier in the paper, for example, they'd mentioned a simple action sequence enabling a monkey to reach some bananas suspended from the ceiling: *Move the box under the bananas; climb onto the box; and reach for the bananas* (p. 481). A clear, though usually unstated, assumption of this strategy was that moving the box doesn't move the bananas. (That's usually true, of course. But others soon pointed out that it's not true in the situation depicted in Figure 10.15.)

Second, they said, modal terms like "normally" or "probably" could be introduced into the logic. This would act as a default assumption, yielding the expected conclusion unless there were a specific statement saying, for instance, that the phone system wasn't working (p. 489). (They rejected the idea that each sentence should be given its own probability measure, because in most cases we wouldn't know how to do that, and because it's not always clear how to match probabilities to people's subjective conviction: p. 490. Although they didn't say so here, this was a dig at fuzzy logic: Zadeh 1965—see 13.i.a.)

And third, they suggested using Rescher's (1964) recently published philosophical work on hypothetical reasoning and counterfactuals. His logic, they said, might enable a program to search for, and sensibly fix, inconsistencies arising from its (deliberately false) assumption that *nothing* would change as a result of a given action (p. 499).

None of these suggestions was spelt out in detail in their paper. And they made this telling confession:

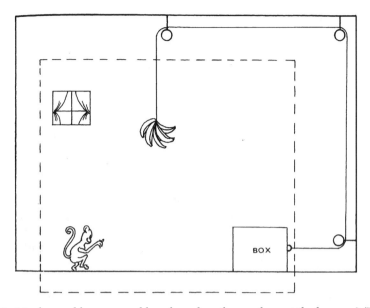

FIG. 10.15. Monkey and bananas problem: how does the monkey get the bananas? (The usual approach to this problem assumes, though doesn't necessarily explicitly state, that the relevant 'world' is that shown inside the dotted-line frame. In other words, nothing exists outside this frame which causes significant changes in it on moving the box.) Reprinted from Boden (1977: 387)

We hereby warn the reader, if it is not already clear to him, that these ideas are very tentative and may prove useless, especially in their present form. However, *the problem they are intended to deal with, namely the impossibility of naming every conceivable thing that may go wrong, is an important one for artificial intelligence, and some formalism has to be developed to deal with it.* (p. 490; italics added)

While they were writing these words, the first of their three escape routes was already being developed by some of McCarthy's Stanford colleagues. On the one hand, by Green (who described it at the same Edinburgh workshop: C. C. Green 1969); on the other, by the programmers of STRIPS, the triangle-table planner used by SHAKEY (Fikes and Nilsson 1971; Fikes *et al.* 1972*a*, *b*). And as soon as the Edinburgh paper was off the press, people started writing about "the frame problem" in AI (Raphael 1971; Sandewall 1972).

Roboticists in particular, who couldn't avoid real-world exigencies even in highly artificial environments, were well aware of "the impossibility of naming every conceivable thing that may go wrong". But common-sense reasoning was threatened too, as McCarthy and Hayes' telephone example had shown. The impossibility of giving cut-and-dried definitions of concepts, or natural-language words, were further sources of this general difficulty.

In fact, their find-the-phone-number example had identified *two* problems, not one. Because both of these were introduced in relation to the same example, they're often conflated—even confused. Both are called "the frame problem", although it's usually clear from the context which one the speaker has in mind.

The first was the problem of knowing just which aspects of a situation would be changed by a particular action, and which would not. For instance, whether it's true that "if a person has a telephone he still has it after looking up a number in the telephone book", or that moving the box doesn't move the bananas. That is "the frame problem" strictly so called, because that is the one addressed by the first of their three escape routes, which is where the word "frame" entered the discussion.

The second was the problem of reasoning with incomplete knowledge—where one doesn't know, for instance, whether the page has been torn out of the phone book, or whether the phone system is out of order. This problem (often called the frame problem, but sometimes the qualification problem: see 13.i.a) is a feature both of our inevitable ignorance about the facts of the real world, and also of the irredeemable vagueness of ordinary-language concepts. It's normally dealt with by some form of default reasoning—their second escape route. But of course the default doesn't always apply.

In general, the first version of the frame problem was relevant for robot planning, and the second for common-sense reasoning. Given the importance of both of these (especially the second) for AI in general, it's hardly surprising that the frame problem (dual sense) has been on the lips of AI researchers, not to mention AI's opponents, ever since it was first named (e.g. Raphael 1976: 146–52; Pylyshyn 1987; Dennett 1984*b*; Ford and Hayes 1991; Ford and Pylyshyn 1996; Sperber and Wilson 1996; Shanahan 1997)—nor that it's mentioned in various other chapters in this book (e.g. 7.iii.d, 8.i.b, 9.d–f).

McCarthy and Hayes themselves were no exception. They spent many years trying to defuse the problem, working respectively on non-monotonic reasoning and naive physics (Chapter 13.i.a–b).

(Today, Hayes feels that they were over-sanguine in the 1960s and 1970s, and that the frame problem is probably unsolvable: see 13.ii.b. Certainly, the claim—see Shanahan 1997—that it has already been solved is over-optimistic for the general case.)

The statement of the frame problem, and the suggestions regarding a situation calculus, were the most influential features of 'Some Philosophical Problems'. But a third was sometimes picked up too. This was McCarthy and Hayes' distinction between three sorts of representational "adequacy", whether in minds or in machines: metaphysical, epistemological, and heuristic. (They might have embarked on a philosophical discussion of reductionism here, but they didn't; Fodor 1968 had just used broadly comparable ideas to do so.)

A representation of a problem is metaphysically adequate "if the world could have that form without contradicting the facts of the aspect of reality that interests us" (p. 469). Such representations were mainly useful, they said, for constructing general theories. For example:

1. The representation of the world as a collection of particles interacting through forces between each pair of particles.
2. Representation of the world as a giant quantum-mechanical wave function.
3. Representation as a system of interacting discrete automata [a system which they went on to explore in their paper]. (p. 469)

A representation is epistemologically adequate "if it can be used *practically* to express the facts that one actually has about the aspect of the world [being considered]" (italics added). The language of particles-and-forces, they said, can't be used to express the facts that "dogs chase cats" or "John's telephone number is 321–7580", but ordinary language can. On the other hand, English can't express the information processing involved in recognizing a particular face. (Nor, one might add, can it express the NLP rules involved in interpreting Chinese—despite John Searle's famous claim to the contrary: see 16.v.c.)

With respect to philosophically problematic concepts such as cause, ability, and knowledge, a main aim of their paper was to find a way of representing/expressing these which would be epistemologically adequate for a computer—or, indeed, for a metaphysician. In that case, they would *no longer be* philosophically problematic.

Finally, a representation is heuristically adequate "if the reasoning processes actually gone through in solving a problem are expressible in the language". Only in very simple cases is the representation which is epistemologically adequate *also* heuristically adequate. They said virtually nothing more about this level of adequacy, but it underlay the already recognized need for high-level programming languages, and would soon aid the development of AI planning.

There may be more than three representational levels, for heuristic adequacy can exist on multiple levels of abstraction. High-level programming languages are needed because people can't think fruitfully in machine code when they're trying to solve relatively large-scale problems—in other words, machine code isn't heuristically adequate for such problems (Section v, below). So solutions found in the PLANNER programming language must be translated into LISP, and then into machine code, before they can actually be executed.

We've seen, for instance, that neither GPS nor early STRIPS could produce an overall strategic plan before starting on the detailed solution. In effect, they made no distinction between heuristic and epistemological adequacy. As problem complexity increased, they were soon scuppered by the combinatorial explosion. But that distinction was recognized in Sacerdoti's ABSTRIPS. This could ignore the low-level details while formulating the strategic plan, and also (during execution) translate that plan into detailed instructions about what needed to be done. Whereas STRIPS couldn't see the wood for the trees, ABSTRIPS could represent a hierarchy of woods, in which copses and trees of decreasing size could be successively specified in heuristically adequate terms, until an epistemologically adequate level was reached.

In sum, the logicist case was argued more explicitly, and (thanks to philosophical logic) more persuasively, in 'Some Philosophical Problems' than in McCarthy's harbinger papers. But the ever-hungry worm nestling at the heart of logicism had been made clearly visible.

10.iv. The Need for Knowledge

Minsky had said (in 'Steps') that work on AI "is concerned with using all we know to build the most powerful system that we can" (1961*b*: 446). But much of the newest NewFAI, as we've seen, was ignoring much of what we know. In other words, it was seeking *general* theories of intelligence. By the 1970s, however, Minsky's remark was being interpreted more seriously. For generality wasn't working as well as had been hoped. Increasingly, the 1970s saw a flight from domain-independent models. (This didn't apply to Minsky himself, who complained about this trend in his high-visibility Turing Award lecture: Minsky 1969.)

As Winograd put it (in March 1974), there had been a growing split between people "making programs do more and more intelligent things" and people studying "general issues of representation and problem-solving". But from now on, things would be different:

Today I see a movement towards a middle ground. The theorem-proving craze is slowing down. People are aware that very general systems are not going to be the basis of practical programs, and people who have been doing specialized programs are asking what these programs have to offer which can be brought to bear on more general problems. (Winograd 1974: 92)

a. A triumph, and a threefold challenge

Winograd was a mere graduate student in September 1970, and a neophyte postdoc in September 1971. Nevertheless, he was invited to give a talk at the second IJCAI Conference in London that autumn. This was no mere parallel session, fighting for attention among half a dozen others. On the contrary, it was the inaugural *Computers and Thought* Lecture, supported by the royalties donated by Feigenbaum and Feldman (ii.b, above). Those of us in the UK who were planning to attend the meeting were enthused by the rumours of his participation, for his work—available as a Technical Report from MIT's AI Lab since February of that year—was already a glowing beacon for the cognitive science community.

Why all the fuss? Well, Winograd's SHRDLU—written as his Ph.D. thesis of 1971 and officially published, to huge acclaim, a year later—was a triumph of NLP (9.xi.b). It was also a triumph of AI programming. But Winograd had done more than write an amazing program: he'd challenged three basic assumptions of the conventional NewFAI approach.

First, he abandoned generality and relied instead on detailed domain knowledge. Indeed, he was one of the strongest voices arguing for this move. Since he wasn't merely arguing but also achieving, and that in a spectacular fashion, his words carried weight. In interpreting the input sentences, SHRDLU relied heavily on its detailed knowledge of syntax. Its unprecedented ability to parse sentences as complex as *How many eggs would you have been going to use in the cake if you hadn't learned your mother's recipe was wrong?* was the result. And Winograd drew a general moral: whatever the problem domain happened to be, successful AI required that the relevant domain knowledge be provided to the program.

He wasn't the only one in the early 1970s to be saying this. Within NLP, HEARSAY combined syntactic, semantic, pragmatic, lexical, phonemic, and phonetic knowledge—all simultaneously available via the blackboard architecture (Reddy *et al.* 1973; Newell *et al.* 1973). Outside NLP, work in vision and expert systems was similarly relying on specific knowledge of the domains concerned (see subsections b–c, below). And Joel Moses (of MIT) had seen the writing on the wall for NewFAI generalism as early as 1967:

The word you look for and you hardly ever see in the early AI literature is the word knowledge. They didn't believe you have to know anything, you could always rework it all ... In fact 1967 is the turning point in my mind when there was enough feeling that the old ideas of general principles had to go [which is why, for instance, Danny Bobrow ignored GPS to work on STUDENT] ... I came up with an argument for what I call the primacy of expertise, and at the time I called the other guys the generalists. I was antigeneralist for many years. (interviewed in McCorduck 1979: 228–30)

But Winograd was especially influential. Indeed, Moses went on to say that "it took some difficult doing" for knowledge-based AI to be recognized as necessary, and "I think what finally broke [the generalists'] position was Winograd" (McCorduck 1979: 229).

The second major challenge to NewFAI posed by Winograd was his criticism of "declarative" programming and his urging of "procedural" programming instead. The logicist dream was that intelligence could be modelled as theorem proving carried out propositions in predicate calculus. Such propositions are declarative: they state *It is the case that* ... Even those NewFAI workers who favoured non-logical representations (such as semantic networks) thought of programs, in essence, as starting from facts and inferring other facts from them. (For instance, the *isa* network links between *cat/mammal/animal* meant that *A cat is a mammal* ... etc.) So what AI programmers needed to do was to declare (*sic*) the facts relevant to the domain concerned, and then let some general-purpose inference engine—a theorem-prover, or a network-rambler—take over. In short, providing domain knowledge to a program meant telling it *what is the case* rather than *what should be done*.

Winograd disagreed. His definitions of words weren't 'factual' statements or logical axioms expressing the word's meaning (see 9.xi.b). Nor were they semantic networks, whose pattern of nodes and links defined the meaning. Rather, they were mini-programs ("procedures"), which would be run when the word was encountered. On encountering the definite article *the*, for example, SHRDLU would immediately start looking for one and only one thing which fitted the following description, e.g. *pyramid*. Similarly, SHRDLU's definition of *and* was a mini-program that interrupted the normal parsing sequence to tell the system to "look for another one [i.e. a syntactic category] like the one you just found" (Winograd 1973: 179). This enabled it to cope sensibly—and relatively quickly—with both *The giraffe ate the apples and peaches* and *The giraffe ate the apples and drank the vodka*. Sometimes, the definition of a word would include a program for examining *previous* sentences—as for the word "one" in *Pick up the green one*. (As Winograd pointed out, all this was possible largely because he was using a version of the new PLANNER programming language: v.d, below.)

A wide range of work in AI and computational psychology was profoundly affected by Winograd's recommendation of procedural programming (see 7.ii.d and iv.c). Throughout the 1970s, as we've seen (iii.a, above), knowledge representation was a key focus of GOFAI research. The already familiar question of whether predicate logic could represent all of human knowledge was now joined by the broader question of whether *any* declarative representation could do so. For a while, there were heated discussions between declarative and procedural camps. The former presented the latter as tinkerers, content to cobble programs together without worrying about logical consistency. At the same time, logicist approaches (and logic-based languages) were viewed with suspicion, not to say disdain, by the proceduralists. (Often, they drew an analogy with Gilbert Ryle's distinction between knowing *how* and knowing *that*: pulling such a famous philosopher on board seemed to raise their spirits significantly—see 16.i.c.)

By the late 1970s, however, most people had realized that the divide between the two programming styles wasn't absolute. After all, even LISP instructions (procedures) could be used as data by other instructions: see v.c, below. And Newell and Simon's productions, despite being based on a system of *logic*, were rules which *made things happen*. So whether one chose to regard a particular representation of knowledge as "procedural" or "declarative" was largely a matter of emphasis, or point of view.

Winograd himself eventually admitted that AI work should combine the two types of representation, so as to get the best of both worlds (1975). Indeed, a prime aim of the new programming language (KRL) he developed with Bobrow was to do just this (Bobrow and Winograd 1977). KRL made quite a stir at the time. It was given pride of place as the very first paper to be published in *Cognitive Science*, and was later described as "one of the more ambitious efforts in the history of AI representation frameworks" and "the high point of a certain style of Knowledge Representation" (Brachman and Levesque 1995: 263). (Not everyone was impressed, for it was also described as "a splendid edifice [constituting] a castle in the air": McDermott 1978.)

KRL tried to get "the best of both worlds", because the two approaches had complementary advantages and drawbacks. It's easier to add new knowledge to a declarative representation; for although simply *adding* an instruction may (sometimes) be just as easy, its implicit effects are less clear. In general, declarative languages are easier for people to understand, whereas procedurally embedded knowledge may be highly opaque to the human user. But adding knowledge isn't enough: one must also specify how it can be used. Often, it's easier to embed knowledge implicitly inside a procedure (which will actually get the relevant task done) than to provide it as a theoretical background from which the necessary procedures must be inferred. Imagine trying to give a theoretical justification of all the heuristics used in AI programs, for instance: even deducing *Protect your queen* wouldn't be trivial. For effective action, a set of domain-specific procedures may be more economical than long lists of facts to be manipulated by general deductive procedures. In short, the *advice* "If you know Turing's human, assume that he's fallible" may be more helpful than the *data* "Turing is human, and all humans are fallible".

All of that became evident, however, largely as a result of the discussion prompted by Winograd's unflinching support for procedural programming in his early work. The familiar Popperian point applied: your scientific theory doesn't have to be right, but it should lead to challenges and tests that show whether it may be right, and—if not—just what theory might be better.

The last of Winograd's three challenges was directed against hierarchical programming: "heterarchical" programming was recommended instead. In a hierarchical program, one part—the master program, or central executive—has overall control and the others are subordinate to it, as mere subroutines in the service of its goals. Hierarchy had been crucial to GPS, and even to Pandemonium (whose demons could communicate only upwards, not laterally or downwards: 11.ii.d). Moreover, Simon (1962, 1969) had declared hierarchy to be necessary for both artificial and biological intelligence, despite certain disadvantages pointed out by his colleague Newell (1962). Winograd didn't deny its importance. But he did deny that inflexibly top-down control was the best way of exploiting it. In place of rigid hierarchy, he said, what AI needed was heterarchy.

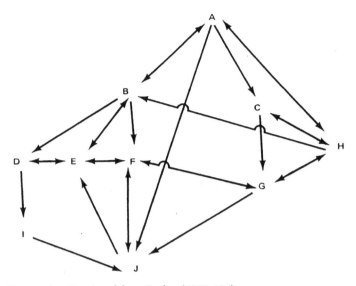

Fɪɢ. 10.16. Heterarchy. Reprinted from Boden (1977: 126)

In a heterarchical program, control is more equally distributed throughout the system, and internal communication between the various subroutines is increased. The component programs can call upon each other upwards, downwards, or sideways (see Figure 10.16). Moreover, they can do this at various points in their (potentially independent) functioning. SHRDLU, for example, could do it many times in parsing a single sentence (see 9.xi.b).

The added flexibility which heterarchy provided to AI problem solving was avidly welcomed for work on many domains. Fahlman, for instance, identified heterarchy as the key source of BUILD's ability to plan blocks-world structures. And an example applied to vision, for use by the MIT robot (for which SHRDLU had been written in the first place: see 9.xi.b), is shown in Figure 10.17.

The various parts of a heterarchical program were *thought of* very differently from the subroutines in GPS. Winograd's colleague Winston, for instance, put it like this:

Communication among these modules should be more colorful than mere flow of data and command. It should include what in human discourse would be called advice, suggestions, remarks, complaints, criticism, questions, answers, lies [better: approximations], and conjectures ... Note particularly that ... programs normally thought to be low level may very well employ other programs considered high level ... [In computer vision, for instance] line-finders that work with intensity points are low level but may certainly on occasion call a stability tester that works with relatively high-level object models. (Winston 1972: 444)

In short, a heterarchical program was conceptualized as a community of autonomous *agents*—although that terminology didn't become widespread in AI until the 1980s (see Chapter 13.iii.d).

The term "heterarchy" wasn't coined by Winograd. It had already been used by neurophysiologists such as McCulloch (1947). And they'd borrowed it from the political philosophers. Today, it's typically used—by sociologists, political scientists,

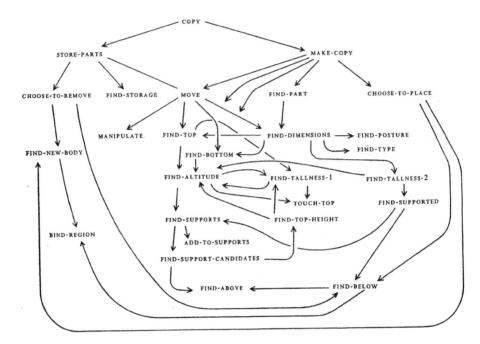

Fɪɢ. 10.17. Heterarchical organization of the MIT robot's vision system. Reprinted with permission from Meltzer and Michie (1972: 456)

and management gurus—to describe human organizations (e.g. Stark 1999). Within AI, it has fallen out of use, being replaced by "agents" or "distributed AI".

These terminological points are relevant because they highlight the human and political aspects of the concept—which were never far from the surface, even in technical treatises on computer programming. Autonomous agents deserve respect. Someone who's allowed (see the quotation above) to advise, suggest, criticize, and complain . . . is an equal, not a mere lackey. And so it was with the components in heterarchical programs. Winston, again: "the modules interact not like a master and slaves but more like a community of experts" (1972: 443). The new programming style was repeatedly described in terms of "committees" of "experts", bound by mutual respect and judicious negotiation rather than autocratic power relations.

In other words, socio-political metaphors were commonly used by AI scientists to recommend heterarchy over hierarchy—at the height of the Cold War, with resonant implications (see 1.iii.b–d). (Twenty years later, when "multi-agent systems" became all the rage, their popularity would be partly fuelled by similar political considerations, even though the Cold War had thawed by then: 13.iii.d.)

(As for the wunderkind Winograd, what happened to him? It's an interesting story, and not one which AI people like to hear. For in their view, their hero betrayed them: see Chapter 11.ii.g.)

b. Clearer vision

Vision is the prime example that shows why AI is no longer defined as Minsky originally defined it: "the science of making machines do things that would require intelligence if done by men" (Minsky 1968, p. v). We don't normally think of vision as requiring intelligence: after all, everyone who isn't blind can do it—and so can squirrels. Accordingly, the initial NewFAI attitude to vision was that it must be pretty simple. As late as 1966, Minsky asked a bright first-year undergraduate, namely Sussman, to spend the summer linking a camera to a computer and getting the computer to describe what it saw. This wasn't a joke: both Minsky and Sussman expected the project to succeed (Crevier 1993: 88).

In the very earliest days of NewFAI, "vision" had meant pattern recognition, modelled by highly general techniques (see Uhr and Vossler 1961*b*). This might conceivably have been enough for the bookreader-for-the-blind envisaged by Selfridge and McCulloch. But it wasn't enough to capture vision as described by the "New Look" psychology of Bruner and Richard Gregory, which posited top-down influences from perceptual hypotheses about the real world. Indeed, Gregory (1967) was arguing that visual computers would suffer from illusions just as humans do, and for similar reasons (6.ii.e).

The implication was that pattern recognition should make way for scene analysis, where the program's task was to interpret visual input *as depicting objects in real-world scenes*. (Most of the early scene-analysis programs are detailed in Boden 1977, chs. 8–9.) This, in turn, implied that the image should be articulated into distinct parts—which pattern-recognizers couldn't do. So some NewFAI pioneers started thinking about the *structure* of images. Narasimhan in Bombay, and (following Narasimhan's lead) Max Clowes in Oxford/Canberra, offered hierarchical "picture grammars" inspired by Chomsky's syntax (Narasimhan 1964, 1966; Clowes 1967, 1969). And Clowes, whose earliest work had been in technological character recognition (Clowes and Parks 1961), now drew inspiration also from psychology and neurophysiology: specifically, from Miller's 'Magical Number Seven' and retinal/cortical feature-detectors (Clowes 1967: 181, 195–6; see 6.i.b and 14.iv.a–b).

In particular, pattern recognition wasn't enough for robotics. An AI robot—as opposed to a fixed-routine *Unimate* bolted to a factory bench—needs to know where something is in 3D space; what size and shape it is; what the currently invisible parts of it are like (e.g. the occluded corners of the wedge numbered "11" and "12" in Figure 10.18); what the orientations of the various surfaces are; and how to distinguish one 3D something from another one in the first place. Accordingly, the people associated with the MIT robot, Stanford's SHAKEY, and Edinburgh's FREDDY tried to write programs enabling a computer to interpret 2D images in terms of 3D scenes.

Those last nine words were easy for me to write, and for you to understand: today, the distinction between (retinal) "image" and "scene" is clear. In the 1960s, it wasn't. With the 20:20 vision of hindsight, it's amazing how blind early NewFAI was to it. When AI workers first tried to write 2D-to-3D programs, they often made mistakes precisely because they hadn't taken this distinction properly on board. In other words, they hadn't realized the need to include *knowledge of the image-forming process* in their programs.

The image-forming process held to be most relevant at that time was projective geometry. 2D-to-3D algorithms based on physical optics would be written later, in the mid- to late 1970s (14.v.f and 7.v.b–d). But before then, optics was used only in 2D-to-2D line-finders (Roberts 1965; Shirai 1973) and region-finders (Tenenbaum *et al.* 1974; Tenenbaum and Weyl 1975). Computer vision used hand-drawn line drawings, or line drawings generated from grey-scale camera input by line-finders (Winston 1975).

Because the geometrical knowledge embodied in the first scene-analysis programs wasn't made fully explicit, they succeeded for largely mysterious reasons. Likewise, they suffered apparently inexplicable failures.

The prime example of that type of blindness was due to Adolfo Guzman (1943–), a Mexican graduate student at MIT (today, working at Mexico's National Polytechnic Institute). On one level, Guzman's SEE program (1967, 1968, 1969) was a triumph. For it could interpret a complex image as representing many separate physical objects: eleven, in Figure 10.18.

It did this by first finding the *vertices* (classified into nine basic types), and then using them to guide linkages between line-bounded *regions*. A link between two regions assigned them to one and the same physical thing. Linked regions were usually neighbours, but not always (see Figure 10.19). So seven non-adjacent regions in Figure 10.18 (labelled 3, 21, 22, 23, 24, 28, and 29) were all assigned to "one" object, as were regions 1, 2, and 33.

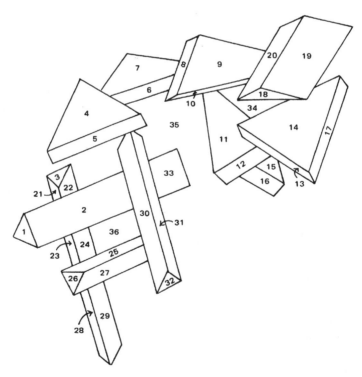

Fɪɢ. 10.18. Guzman's "HARD". All bodies were correctly identified by SEE, even though one (regions 6:7) has no "useful" visible vertices. The background (34:35:36) also was correctly found. Reprinted with permission from Guzman (1969: 273)

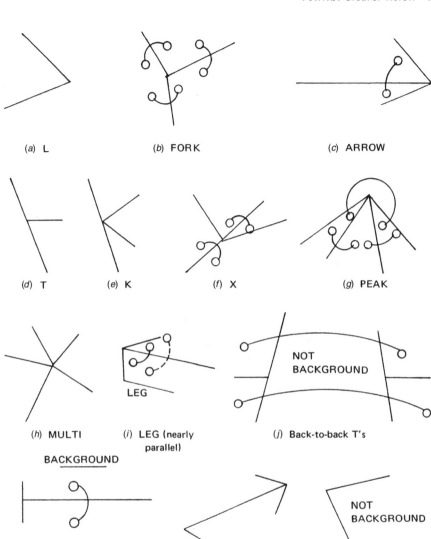

(a) L   (b) FORK   (c) ARROW

(d) T   (e) K   (f) X   (g) PEAK

(h) MULTI   (i) LEG (nearly parallel)   (j) Back-to-back T's

BACKGROUND

(k) 3-parallel T on background   (l) L inhibits ARROW-link   (m) Use of L's in finding background (See caption)

FIG. 10.19. Guzman's vertices and their associated links. Items (b–c), (f–g), and (i–k) show strong region-links. The absence of links in items (a), (d–e), and (h) means that these vertices provide no reliable cues about physicality. Item (i) shows a weak link, placed on ARROW by an adjacent LEG. Items (j–k) show strong links placed on complex patterns. Item (l) shows link-inhibition by an L-vertex (assumes convexity). And (m) shows the convexity assumption in the use of Ls to identify the background. Adapted with permission from Guzman (1969: 258; 1968: 87)

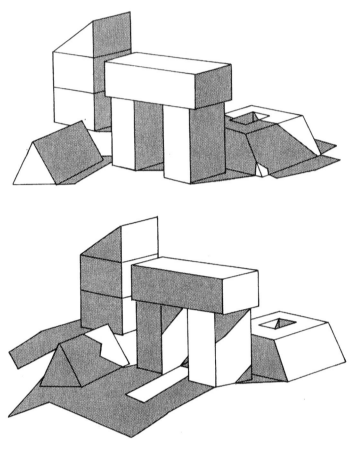

Fɪɢ. 10.20. Two differently shadowed images of the same scene. Adapted with permission from Waltz (1975: 20)

It turned out, much to Guzman's puzzlement, that there were some things which SEE *couldn't* see. Holes, for instance. Even without the shadows (which SEE couldn't cope with), Guzman's system wouldn't have recognized the object on the far right of Figure 10.20 as a single thing.

His colleague Winston (1970*b*) was as puzzled by this as Guzman, and tried to extend SEE so as to cope with holes—without success. Nor could SEE have recognized images of impossible objects as impossible (see Figure 10.21). Moreover, even when it did see things correctly Guzman didn't really understand why. He didn't know *why* a fork normally allows one to place region-links across all three lines, nor *why* the few link-inhibition rules were helpful yet not foolproof: see items (*b*) and (*l*) in Figure 10.19. Even more to the point, he didn't ask.

Admittedly, another NewFAI scientist had already produced a program that could articulate an image in terms of distinct 3D objects. Larry Roberts (1937–) had cut his teeth at MIT by trying to speed up the perceptron (J. A. Anderson and Rosenfeld 1998: 100). But he'd turned to scene analysis for his doctoral thesis, written in 1961 but

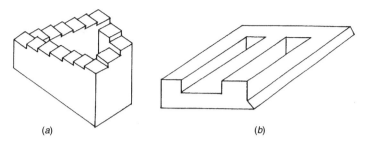

FIG. 10.21. Two impossible objects. Item (*a*) adapted with permission from Meltzer and Michie (1971: 296); item (*b*) reprinted with permission from Clowes (1971: 105)

not officially published until later (Roberts 1963, 1965). Indeed, it was his early 1960s research which initiated the polyhedral blocks world. (He initiated something else, too: appointed head of the information-processing group at ARPA in 1966, he was largely responsible for the creation of the ARPAnet.)

In some ways, Roberts's program was more powerful than SEE. Besides picking out individual objects it could also identify them as cuboids, or wedges, or . . . which SEE couldn't do. In addition, it could discover precise sizes, locations, and surface orientations (by number crunching guided by information about the picture plane). And it could accept camera input. The hope was that some future version, besides being used in robotics, would be able to interpret aerial photographs—like those which had shown the Soviet missiles on Cuba just as Roberts was embarking on his doctorate. But line drawings were what the program was really about, since the grey-scale image was immediately converted into a drawing. (The line-finder looked for edge fragments in four orientations, inspired by David Hubel and Torstein Wiesel's recent discovery of cortical feature-detectors: 14.iv.b.)

However, this system was less general than Guzman's. It relied on internal geometrical models of three types (cuboids, wedges, and hexagonal prisms), each carrying rules about how to interpret specific 2D cues in 3D terms. Anything that couldn't be analysed by applying those rules was, in effect, invisible. In dealing with the line drawing, the program's first move was to find the lines bounding separate *regions*. Using Roberts's list of "approved polygons" (polygons that could depict a face of one of the three basic polyhedra), it then assigned the regions to distinct physical objects. It could also cope with "compound" polyhedra, some of whose faces wouldn't be approved polygons (see Figure 10.22). At base, however, it had to know precisely what it was looking for in order to find it.

Both these scene-analysis programs had a strong affinity with the New Look. For they both went "beyond the information given" (Bruner 1957*a*), by using top-down perceptual hypotheses about the objects they expected to find. But only Roberts identified the theoretical source of those hypotheses, namely projective geometry. Guzman, by contrast, based his interpretative heuristics in intuition. He didn't ask *why* they worked, *why* a corner of a certain type would appear in the drawings as a vertex of a certain type.

Indeed, he didn't distinguish clearly between "corner" and "vertex" in the first place (and sometimes used these terms interchangeably). The former was a denizen of the

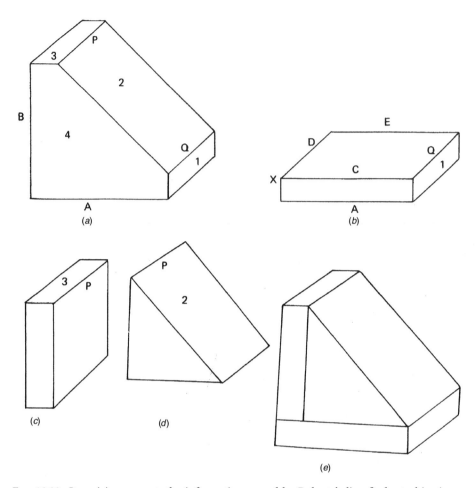

Fɪɢ. 10.22. Item (*a*) represents the information passed by Roberts's line finder to his picture interpreter, which searched for significant picture fragments (suggesting "approved polygons"). It found the polygon bounding region 1 with line A connected to it, and constructed and analysed out solid (*b*) accordingly. Similarly, (*c*) and then (*d*) were found. No picture lines remained unaccounted for. The program's final reconstruction of the target compound object is shown in (*e*). Reprinted with permission from Roberts (1965: 182)

scene, the latter of the image. But that was never clearly stated, because Guzman didn't explicitly consider (as Roberts had done) the geometry of the 3D-to-2D projection involved. Instead, he implicitly embedded some of his unorganized spatial intuitions in the region-linking rules. SEE's successes were due to the nuggets of geometrical gold buried inside those intuitions, and to the (atheoretical) improvements that resulted from trial and error with the program.

In the 1960s, it wasn't only Guzman who confused image and scene. Most NewFAI people did. That's why Winston's (1970*b*) treatise on holes added piecemeal procedural patches to SEE, instead of asking systematic questions about 3D/2D representation. It was left to two other people to do that.

One was Clowes (1933–81), by then at the University of Sussex. The other was the mathematician David Huffman (1925–99), who'd recently left MIT to found the Computer Science Department at UC Santa Cruz. Both Clowes (1971) and Huffman (1971), independently, suggested labelling each line according to its *physically possible* interpretations, and then finding the set of *mutually consistent* interpretations across the image as a whole. Clowes recommended breadth-first search to do this, Huffman depth-first search—but their core ideas were so similar that people soon spoke of "Clowes–Huffman labelling". Huffman's labels were the ones most widely borrowed (for his improved version, see Huffman 1977*a,b*). However, Clowes's paper was the more interesting for AI, because he reported a scene-analysis program showing how the shared theoretical insights could actually be used.

Clowes and Huffman pointed out that there are four possible interpretations of any line in pictures of polyhedra: either a convex or a concave edge with both associated body surfaces visible, or a "hiding" edge that obscures a surface to one or other side of it. Figure 10.23 shows these four edges, labelled respectively by plus and minus signs and arrows (the visible surface lies to the right of an ant crawling in the direction shown by the arrow). However, no line can be labelled in different ways at either end, because no (non-curved) edge can be convex at one point and concave at another, nor hiding an adjacent surface at one end but not at the other. This allows for tests of mutual consistency between line labels, given that vertices can be sensibly interpreted in only a few ways.

Clowes had originally been puzzled about why SEE worked at all, not only why it sometimes didn't (personal communication). That puzzlement was now solved. He and Huffman had distinguished clearly between image and scene, and analysed the possible relations between them in an orthodox 3D/2D projection. They'd shown why Guzman's "arrow" was so useful: the shaft *must* show an edge, convex or concave, with both adjacent surfaces visible and both belonging to the same object. And they'd shown why his "Ls" were so unhelpful: there are six possible line–label combinations for an L, so Ls have no reliable region-linking rules—see item (*a*) of Figure 10.19. They'd explained why SEE had been defeated by holes, and even by concavities, and they'd offered ways of recognizing these physical features. And both had shown that their methods (for Clowes, a program; for Huffman, a paper-and-pencil algorithm) could find *all possible* picture labellings—which meant that they could also identify *impossible* labellings, and even locate the impossibilities at specific picture points. (Guzman had instructed SEE

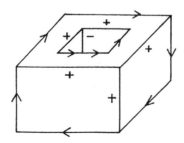

FIG. 10.23. The four physically possible line labels (see text). Reprinted with permission from Huffman (1971: 305)

to reject "illegal" pictures, but his criteria of illegality—e.g. lines not forming part of the boundary of a closed region—weren't properly understood. An illegal picture isn't at all the same thing as an impossible object.)

Two important additions to the domain knowledge used in scene analysis followed in the mid-1970s. One was Mackworth's (1973) gradient space. This was an extension of projective geometry that enabled vision programs to interpret some images that had previously defeated them: namely, drawings of polyhedra with skewed surfaces (see Figure 10.24). Also, by including explicit constraints relating the *relative slopes* of surfaces to their visibility or invisibility, Mackworth explained why Clowes–Huffman labelling worked. This is historically interesting not least because David Marr, despite his scathing criticisms of scene analysis, would later borrow gradient space for use in his own work (Marr 1982: 17, 240–3; see Chapter 7.v.b–d).

The other addition to visual knowledge was David Waltz's (1975) treatment of shadows and cracks. Building on Clowes–Huffman and Mackworth, he allowed lines to represent the boundaries of shadows as well as of objects, and to depict several types of "crack" between adjacent, but separable, objects (e.g. the lines separating the three stacked objects on the left of Figure 10.20). His program could interpret images such as those in Figure 10.20, and could also recognize that they represent one and the same scene.

Even before Waltz's thesis was officially published, further domain knowledge had been added by others. For instance, Kenneth Turner (1974), of the FREDDY team in Edinburgh, extended Waltz's program to include certain classes of curve: see Figure 10.25. (Turner's program started out with a TV image, which it converted to a line drawing, and used hierarchical visual models of the objects that FREDDY might be expected to encounter: see Figure 10.26.)

However, Waltz had done much more than add extra domain knowledge. He'd provided a general method—for coping with multiple constraints—which could be applied in all areas of AI, from resolution theorem proving to connectionist pattern recognition. So today, what people remember best about his early 1970s work isn't the shadows, but the filtering algorithm.

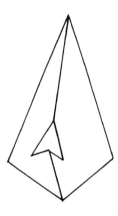

Fɪɢ. 10.24. An object with a skewed surface. Reprinted with permission from Mackworth (1973: 134)

FIG. 10.25. Image of scene with curved (and shadowed) objects. Reprinted with permission from K. J. Turner (1974: 246)

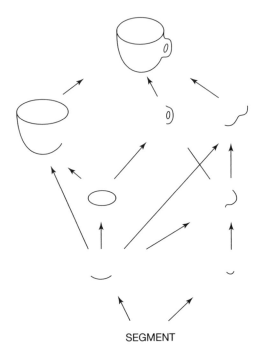

SEGMENT

FIG. 10.26. A hierarchical representation of a cup. Redrawn with permission from K. J. Turner (1974: 244)

Waltz-filtering was a technique developed to overcome the combinatorial explosion when there were many possible interpretations of a single cue (in this case, of a single line). Because Waltz allowed for shadows and cracks, his program had eleven different line labels, as compared with the Clowes–Huffman four. This multiplied the possible vertex interpretations horrendously.

There were 500 sensible interpretations of FORK and T, and 100 each for MULTI, X, and K; ARROW and L both had about 75. Only PEAK was relatively manageable, with merely 10. (Things would have been even worse if Waltz hadn't used his *physical knowledge* about shadows, cracks, and separable edges to exclude impossible interpretations: there were no fewer than 6 million combinations for the "relatively manageable" PEAK.) This meant that provisionally assigning all possible labels to every line (as in the Clowes–Huffman approach), or even to every vertex, wasn't feasible.

Instead, his program chose a pair of neighbouring vertices at random, and provisionally assigned all—and only—the *mutually consistent* line labels to them. Then, it crawled along to the next vertex, and assigned *only those labels which were consistent with the ones already assigned*. To his astonishment, this simple iterative procedure, when applied to simple scenes, found a unique label for every line. That was also true for many complicated scenes. Even when it wasn't, the numbers of "possible" labels remaining were so small that the familiar Clowes–Huffman approach (comparing labels across the whole image) could be used to disambiguate them.

The overall interpretation was speeded up still further when Waltz decided to make the initial choice of vertex only *semi*-random. In doing that, he relied—again—on specific domain knowledge. Given that some vertices have fewer sensible interpretations than others, Waltz ensured that two relatively unprolific vertices were "randomly" chosen. Also, reminding his readers that one usually attacks a jigsaw puzzle by finding the edge-pieces first, he made the program pick vertices that probably contained *object-boundary* lines. (So PEAKS were shunned at this stage, because only two of the ten PEAK interpretations make sense on the scene–background boundary.)

His AI colleagues were quick to get the point:

Waltz' work on understanding scenes surprised everyone. Previously it was believed that only a program with a complicated control structure and lots of explicit reasoning power could hope to analyze complicated scenes. Now we know that understanding the constraints the real world imposes at junctions is enough to make things much simpler . . . It is just a matter of executing a very simple constraint-dependent iterative process that successively throws away incompatible line arrangement combinations. (Winston 1977: 227)

Waltz-filtering, like gradient space, was another legacy of scene analysis which would be utilized by Marr, despite his harsh words about GOFAI computer vision. It was used by many others too, for it wasn't specific to vision. Any domain involving multiple constraints, and where some local interpretations were mutually inconsistent, could be tackled more efficiently by means of it.

There was a major problem, however. Namely, a single mistake in the Waltz filter's depth-first search might cut off the very branch of the search tree on which the correct solution lay. Geoffrey Hinton (1976, 1977) referred to this as "computational gangrene". As we'll see in Chapter 12.v.h, he formulated an improved constraint-filtering strategy, called "relaxation" (the term was borrowed from other work on scene analysis: Rosenfeld *et al.* 1976). Relaxation found the best overall interpretation *even though* this might contain some minor inconsistencies. In other words, the gangrene was prevented.

Initially, Hinton's ideas were developed in the context of a GOFAI vision system, the Sussex POPEYE project (see below). Soon, however, he (and others) would apply them in PDP models. Today, multiple constraint satisfaction by relaxation is a key pillar of connectionism. But the pillar's base, and even its name, lies in good old-fashioned scene analysis.

Good old-fashioned scene analysis was more varied than I've suggested so far. For instance, the SRI approach differed from MIT's: Jay Tenenbaum's group, which included the young Harry Barrow (1943–), started from photos rather than drawings and from regions rather than lines (Tenenbaum *et al.* 1974; Tenenbaum and Weyl 1975; Tenenbaum and Barrow 1976). Their program, with helpful input from the programmer, wasn't restricted to the blocks world but learnt to recognize real-world objects such as telephones. There were some other (non-interactive) early 1970s programs that started from regions rather than lines (Brice and Fennema 1970; Barrow and Popplestone 1971; Yakimovsky and Feldman 1973). But one of the most interesting examples of 1970s computer vision, albeit not widely noticed at the time, differed even more from MIT's blocks-world paradigm.

This was the POPEYE project, directed by Sloman at the University of Sussex (Sloman 1978, ch. 9). POPEYE modelled the interpretation not of fully connected drawings of perfect polyhedra (or even polyhedra-with-shadows), but of highly ambiguous, noisy, input—with both missing and spurious parts: see Figure 10.27. And it simulated the *complexity* of perception to an extent that was unusual at the time.

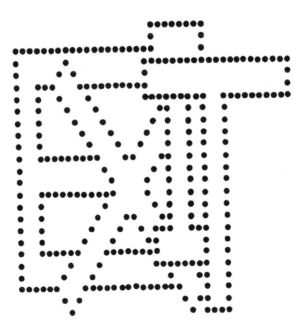

FIG. 10.27. Noisy visual input. Reprinted with permission from Sloman (1978: 219)

POPEYE's complexity had several roots. Many different sorts of background knowledge were combined in the one program. In addition, processing could occur concurrently in different domains, determining which sub-processes would dominate the scarce computational resources. This was different from the heterarchy so popular in the early to mid-1970s. In SHRDLU, for example, there was only one locus of control at any moment; and control was transferred to process X by an explicit call from process Y—as when the syntactic procedure for interpreting the word *the* passed control to a perceptual subroutine, to see whether there was *one and only one* thing fitting the predicate concerned. Each knowledge domain in POPEYE had its own priorities for finding/processing information, and these priorities could change suddenly as a result of new information arriving unexpectedly. Complexity (and flexibility) was increased also by the fact that some of the internal representations constructed by POPEYE were temporary, rather than provisional. (Something provisional may become permanent, but something temporary should not.) At the point when its funding was cut, the POPEYE program was about to be turned into a hybrid system (12.ix.b). A neural net was being designed (by Hinton) to replace the GOFAI spell-checker included in POPEYE. The idea was that the network would be trained on a collection of known words, and would then suggest the most likely word when presented with a "half-baked" letter sequence (A. Sloman, personal communication).

Considered as a practical visual system for a robot, POPEYE wasn't at all impressive. But Sloman wasn't trying to advance robotics, least of all robotics confined to toy polyhedral worlds. Rather, he was trying to advance Immanuel Kant's argument that the mind must provide some prior knowledge for even the "simplest" perceptions to be possible. Kant had talked about the mind's comparing representations, combining and separating them . . . and so on, but "What Kant failed to do was describe such processes in detail" (Sloman 1978: 230).

Whereas for Kant the principles of organization were very general, and innate (see Chapters 2.vi.a and 9.ii.c), for Sloman they also included highly specific examples. Citing recent work on how we avoid word-by-word parsing of commonly heard phrases and on how we learn visual "phrases" too (Becker 1973, 1975), and referring also to Bartlett (see 5.ii.b), he pointed out that visual schemata can aid recognition enormously.

For instance, the familiar upper-case sign "EXIT" helps us—and POPEYE—to recognize the four letters in Figure 10.27. Likewise, a head, or leg, in a photo of a human being not only speeds up our interpretation of that picture-part but also guides our recognition of other parts. In both cases, the computational complexity has a chicken-and-egg aspect: if one recognizes a particular set of dots in Figure 10.27 as colinear, that can help one to recognize an "E"; but if one has already recognized "EXIT", one will be much more likely to recognize *those very dots* as colinear (Sloman 1978: 228–32). POPEYE was an attempt to move towards a realistic degree of visual complexity. To do that, it needed to be fairly complex itself:

Our program uses knowledge about many different kinds of objects and relationships, and runs several different sorts of processes in parallel, so that "high-level" processes and (relatively) "low-level" processes can help one another resolve ambiguities and reduce the amount of searching for consistent interpretations. It is also possible to suspend processes which are no longer useful, for example low-level analysis processes, looking for evidence of lines, may be terminated prematurely if some higher-level process has decided that enough has been learnt about the image to generate a useful interpretation. This corresponds to the fact that we may recognize a whole (e.g. a word) without taking in all of its parts. (Sloman 1978: 229)

One reason why POPEYE had relatively little influence on AI vision research is that, in effect, it was killed in about 1978. By that time, Marr's exclusively bottom-up approach to vision had become popular with the referees advising the grant-awarding bodies, and POPEYE was marked down accordingly. Sloman's account of what happened is given in the historical note added to the online version of his 1978 book. (One of the many factors he mentions is the widespread move from AI languages such as LISP or POP-11 to more general, more efficient, languages such as Pascal or C/C++. These make it very difficult to express "complex operations involving structural descriptions, pattern matching and searching", and to permit "task-specific syntactic extensions ... which allow the features of different problems to be expressed in different formalisms within the same larger program"—see Section v, below.)

However, POPEYE wasn't wholly forgotten. Quite apart from Sloman's own later work, it was cited by the neuroscientists who recently caused a sensation by positing two visual pathways in the brain—one for perception, the other for action (Goodale and Milner 1992; Milner and Goodale 1993: cf. Chapter 14.x.c). The dorsal pathway apparently locates an object in space relative to the viewer, who can then grasp it; the ventral pathway may (this point is contested) locate it relative to other objects, and enables the viewer to recognize it. The evidence lies partly in brain-scanning experiments with normal people, and partly in clinical cases: damage to these brain areas leads to visual ataxia and visual agnosia, respectively. For example, one patient can recognize an envelope but is unable to post it through a slot, whereas another can post it efficiently but can't say what it is.

These 1990s researchers, too, asked a host of questions about how the different pathways can be integrated in various circumstances. In Sloman's view (personal communication), even they didn't really get the point—which was that there are many visual pathways, not just two. Of course, he was talking about computation, not neuroscience: there may or may not be distinct neuranatomical pathways related to distinct types of function. (In low-level vision, it appears that there are: 14.i.c.) But the moral of POPEYE, as of Sloman's more recent visual work (1989), was that *many* types of background knowledge play their parts *concurrently* in perception.

In a sense, POPEYE wasn't really about vision. Rather, it was an exercise in architecture building. For Sloman saw computer vision as a way of keying into the computational architecture of the mind as a whole. (More accurately: as a way of keying into the space of all possible minds—1978, ch. 6.) Over the years, he would focus increasingly on the control structure of the entire mind, and—after a long period of relative neglect—his work on that topic is now attracting significant attention (Chapter 7.i.f).

A final word: GOFAI vision and robotics were typically pursued in terms of the Cartesian sensori-motor sandwich (see 2.iii.a). MIT's early Mars robot was a rare exception—and even there, the sandwich was ignored rather than challenged (13.iv.a). The slices of bread, one white (the sensory input) and the other brown (the motor output), enclosed the meat at the centre: vertices, line labels, models of "EXIT" ... in other words, mental representations linked by theorem proving, planning, and other forms of inference. Marrian research, which eclipsed scene analysis in the 1980s

(7.v.b–d), fed on the same sandwich, for Marr too treated perception and motor action as separate aspects of intelligence.

This didn't sit well, for instance, with the finding that only *self-activated* locomotion helps kittens to develop visual discrimination (Held and Hein 1963), for that fact suggested a much more intimate relation between sight and movement. And it ran counter to the approach of the two mid-century Williams, namely Ross Ashby and Grey Walter (4.viii). But they were (temporarily) out of fashion. Although computer vision had become increasingly enactive, utilizing the changes in input caused by the movements of the system/camera itself (including 'foveal' focusing and continual 'saccades'), internal representations—and the action–perception split—had endured. Not until the "situated" and "dynamical" roboticists of the late 1980s (13.iii.b–c) would AI seriously question the sensori-motor sandwich.

c. Expert Systems

Expert systems showed the 1970s "need for knowledge" especially clearly. For these AI programs embodied highly specialist knowledge gleaned from human professionals. The first examples were written in the late 1960s, and more appeared in the 1970s (Michie 1979). The target areas were very strictly confined, and were originally drawn from science, medicine, or science-related businesses such as computer manufacture or oil prospecting. (Soon afterwards, law was tackled too: see 13.ii.c.)

Developed for *practical* use, the early expert systems were then the prime example of technological AI. As remarked in Chapter 1.ii.b however, technological AI is often based in psychological insights. Expert systems were intended to capture the knowledge and reasoning processes of human experts. In the 1970s, moreover, the venture capitalists hadn't yet moved in on the field. Lacking strong commercial pressures, expert-system builders still aimed, in part, to investigate intelligence as such. So at the close of the decade the co-directors of MIT's AI Lab wrote that "*most* of the people doing artificial intelligence", besides trying to make computers more useful, hoped to formulate theories applicable to "any intelligent information processor, whether biological or solid state" (Winston and Brady 1979, p. ix; italics added).

Throughout the 1970s, AI scientists developed what Feigenbaum (1977) dubbed knowledge engineering, defined as "The art of designing and building expert systems and other knowledge-based programs". This included the "art" of coaxing the real experts to make their human expertise explicit, so that the programmer could program it. As soon became clear, much of the expert's knowledge wasn't available as facts/theories in the domain textbooks. Rather, it consisted in informal heuristics developed over the years, rarely verbalized and almost never communicated. All agreed that making this knowledge explicit wasn't going to be easy—and some critics doubted whether it's even possible (see 13.ii.b).

In general, knowledge engineering required special methods of questioning (analogous to the "protocol analysis" of the Gestalt psychologists and LT/GPS authors), developed by trial and error over the early years (R. Davis 1976/1982). Sometimes, the questioning would be done by human beings. But sometimes, it was done by a program: an expert system whose expertise lay in knowing *how* another expert system (e.g. MYCIN) represents its domain knowledge, and *how* it reasons about it. Using this "meta-knowledge", the automatic knowledge engineer would be able, for instance, to criticize the format of a new item of knowledge added by the non-specialist user—and might also suggest a way out of the difficulty.

AI was notionally aimed at expert systems right from the start. Licklider (1988) had contributed to AI so generously, both intellectually and (via ARPA) financially, because he hoped for military and civilian applications—many of which would assist human experts, or occasionally replace them (see 11.i.b). NewFAI's search for generality, however, implied that expert systems weren't its prime concern. So even MIT's Project MAC, an ambitious (ARPA-backed) 1960s programme of mathematical research, was generally regarded more as NewFAI's close cousin than as its twin. But as NewFAI matured into GOFAI, AI projects focused increasingly on specific practical domains.

That focus wasn't due merely to people wanting to be useful and/or rich. Some were disillusioned with the failure of generalist programs to progress beyond the first, exhilarating, days. And some felt that the intellectual challenge would be greater as a result.

Feigenbaum, for instance, embarked on AI as blue-sky research:

At that time, [as a student] in the fifties, I didn't think of practical applications at all. I was intrigued by the vision of a highly intelligent, maybe super-intelligent, artifact. (interviewed in Shasha and Lazere 1995: 213)

Some ten or fifteen years later, when visiting Newell and Simon at CMU in the late 1960s, he still had that vision. But now, he put it like this:

You people are working on toy problems. Chess and logic are toy problems. If you solve them, you'll have solved toy problems. And that's all you'll have done. Get out into the real world and solve real-world problems. (Feigenbaum and McCorduck 1983: 62)

Naturally, Feigenbaum felt that this finger of scorn couldn't be pointed at himself. For he'd arrived to give his talk at CMU carrying the defensive shield of DENDRAL on his arm.

DENDRAL (the name came from the Greek for *tree*) was the first expert system. It was developed at Stanford by Feigenbaum with Bruce Buchanan (1940–), Georgia Sutherland, and the Nobel geneticist Joshua Lederberg (1925–). Soon after the start, they were joined by the chemist Carl Djerassi, who acted as the domain specialist. It was begun in the mid-1960s, as a result of conversations between Feigenbaum and Lederberg.

Lederberg had been sympathetic to AI for some years (Lederberg 1987: 17–19). He'd been captured by Minsky's infectious enthusiasm, had read his 'Steps Toward Artificial Intelligence' on its official publication in 1961, and had been impressed when Minsky showed him the interactive possibilities in Spacewar. He'd even tried to persuade Minsky to move to Stanford (the Medical School), in McCarthy's wake. He could use a computer, and had done some computer-based medical work—on genetic epidemiology. However, that was merely computerized card shuffling. (Often, using appallingly low-quality data—which implied, for instance, that some mothers give birth at three-month intervals.) He was already convinced that "computers [and especially interactive computers] were going to change the whole style of scientific investigation". Specifically, by 1963 he was hoping to use computers to study chemical structures. But he realized full well that "This was not going to happen with card deck data entry."

In short, when Feigenbaum first met Lederberg—introduced by Karl Pribram in 1963, when *Computers and Thought* was hot off the press (Lederberg treasures his signed copy)—and communicated his youthful hopes to the older man, he was pushing at an open door. The graduate student Feigenbaum wanted a tractable AI problem to work on, dealing with what he called "empirical induction in science" (Feigenbaum and Watson 1965). The Nobel prizewinner Lederberg had a suitable domain already in mind. He'd even sketched a structural taxonomy of organic molecules, and had outlined some problem-solving procedures. But he had "no idea how one would go about translating these structural representations into a computer program" (1987: 22). In 1965, when Feigenbaum arrived at Stanford, they agreed to collaborate on a specific chemical problem chosen by Lederberg—and the result was DENDRAL (so called, because it generated *trees* of candidate structures).

The DENDRAL program was completed by the end of the decade, when it was presented alongside "Some Philosophical Problems" at the fourth Machine Intelligence Workshop (Buchanan *et al.* 1969*a*). More accurately, the *first* completed program was presented there. Increasingly powerful versions soon appeared, and were described at the next three Edinburgh Workshops (Buchanan *et al.* 1969*b*, 1972; Feigenbaum *et al.* 1971; Buchanan and Sridharan 1973). The most interesting improvement was meta-DENDRAL, which was under development from 1970 to 1976 (Buchanan and Mitchell 1978). Besides learning domain-specific rules inductively, this could propose historically new rules itself, which could be included alongside the ones it had been given (so it was one of the few early "creative" programs—see Boden 1977: 327–32). The final version of DENDRAL, and a retrospective view, appeared in 1980 (Lindsay *et al.* 1980; see also Barr and Feigenbaum 1982, chs. vii–viii).

DENDRAL's expertise lay in a particular corner of organic chemistry, concerning the steroids used in contraceptive pills—at that time, still a novelty. Analytical chemists often discover the structure of an unknown compound by using mass spectroscopy, in which an electron beam breaks the molecules into fragments. Having identified the fragments (by their spectrographs), chemical theory suggests how they could fit together in a whole molecule. DENDRAL was designed to do the same sort of thing, by combining chemical knowledge with AI techniques of heuristic search. A "toy" problem, this was not. Nevertheless, its performance compared well with that of human chemists.

Chemical theory played a large part in DENDRAL's success. The program could formulate probable hypotheses about an unknown compound's molecular structure (on the basis of its spectrograph), and then test these hypotheses by way of further predictions. Besides analysing specific molecules, it came up with inductive hypotheses based on the data (e.g. "IF the graph of the molecule contains the estrogen skeleton, THEN break the intramolecular bonds between nodes 13–17 and 14–15"). It didn't stop there: it selected the most likely hypotheses, by reference partly to chemical considerations (such as the number of bonds assumed broken) and partly to logical features such as simplicity, uniqueness, and evidential strength. Moreover, DENDRAL provided, for the first time, a complete list of the set of possible isomers of a given empirical formula within several families, including amines and thioethers.

As for meta-DENDRAL, this discovered mass spectrum fragmentation rules for several types of compound (such as aromatic acids) which hadn't previously been recognized by expert chemists. It did this by first seeking regular patterns in the experimental data, and then applying known concepts to see whether it could find any explanation of them. So it would ask, in effect, "What's so special about nodes 13–17 and 14–15? Why do the intramolecular bonds break there, rather than elsewhere?" Its answer would be couched in terms of some smaller, more general (sub-molecular) structure in the immediate environment of the broken bonds. The latest version could even accept mixed data drawn from several different molecular structures, and separate out the different sub-groups while finding a characteristic explanation for each.

A final advantage, for chemistry, of the DENDRAL research was that the process of extracting the human expert's knowledge led to some new insights. For example, one of the non-chemists involved remarked that it seemed to him that amines are something like ethers. That analogy hadn't ever occurred to the chemists themselves, but they agreed that there was an interesting similarity. This was a special case of the general point—which was becoming more evident with every NewFAI year that passed—that trying to express a theory (in chemistry, or psychology . . . whatever) clearly enough for it to be programmed can lead to new insights, both before and after the computer is plugged into the wall. That general point was illustrated also by the new psychological insights about expert reasoning *as such* which sprang from these exceptionally intensive interviews.

In short, DENDRAL was fulfilling Lovelace's prophesy that some future Analytical Engine might help scientists to "adequately express the great facts of the natural world" (3.iv.b). It was doing this, at the level of the virtual machine, not by what she'd called "mathematical" (algebraic) computation, but by logical-symbolic inference. However, Ada's mentor Babbage had believed that logic and algebra were essentially one. So one can see NewFAI research on KR as an attempt to justify his hunch *in practice.* DENDRAL had apparently done that—and had inched chemical science forward in the process.

Close on the heels of DENDRAL came MYCIN, also from the Stanford stable (Shortliffe *et al.* 1973, 1975; Buchanan and Shortliffe 1984; Cendrowska and Bramer 1984). The core programmers were Buchanan and Edward Shortliffe, who soon qualified in medicine. (One co-author, Stanley Cohen, later developed the basic technology for recombinant DNA.) A production system of 450 rules, MYCIN simulated a medical consultant specializing in the diagnosis and treatment of infectious diseases. Clinicians reported that it performed much better than junior doctors, and as well as some experts—a report that was later confirmed by blind testing (Yu *et al.* 1979).

MYCIN was much more than a table of bugs-and-drugs, for it employed then advanced AI techniques. It engaged in question-and-answer conversations, lasting twenty minutes on average, with doctors needing specialist help. The NLP was of little interest, but it was backed by complex domain reasoning. The physician would ask MYCIN for advice on the identification of micro-organisms, and on the prescription and dosage of antibiotic drugs. That advice would be based on blood tests and histology, and on the patient's symptoms and past medical history. If the relevant evidence hadn't been provided, MYCIN could ask the doctor for it. In addition, MYCIN could explain its advice, at the appropriate level of detail.

Whereas DENDRAL had had a scientific theory (as well as experts' reported intuitions) to go on, MYCIN didn't. Why? Because chemistry does, and this area of medicine doesn't. MYCIN was wholly dependent on rules (heuristics) gleaned from clinical textbooks and from the experts consulted by the knowledge engineers. Also, and again unlike DENDRAL, it had to express various degrees of uncertainty—in a way that reflected how doctors assess the impact of evidence on their diagnosis. So it could present *several* hypotheses to the user, prioritizing them according to the various confidence levels—which, in turn, could be explained.

In brief, this was a user-friendly interface designed for people knowing nothing about computers (cf. 13.v). It conversed in English. It asked questions when it needed to, and in so doing it often reminded the user about things as yet undone. It offered helpful advice, and explanations too (so it could be used as a tool for teaching, as well as clinical practice). It gave estimates of how reliable its advice was likely to be. And it even explained just why it couldn't be sure. Indeed, if providing incomplete medical data counts as "putting the question in wrong", then MYCIN (after requesting the missing data) could do what Babbage's lady visitor had required of the tiny Engine in his salon: have "the answer come out right" nevertheless (3.i.a).

Both DENDRAL and MYCIN were production systems (v.e, below). Feigenbaum and Buchanan had been inspired to use that methodology by their ex-colleague Newell, who'd visited Stanford to lecture about it in 1967. One of the advantages, for expert-system builders, was that each production rule was stated independently. So one didn't need to rewrite the whole program in order to add an *extra* item of knowledge gleaned from the human expert, including his/her explanation of the program's past failures. (This didn't mean, of course, that its interactions with existing rules could be foreseen: the combinatorial explosion was always lying in wait.)

Feigenbaum and Buchanan tried to extract the logical skeleton of the production systems they'd pioneered, so that others could use it to write expert systems in indefinitely many domains. In doing that, they developed EMYCIN: the first expert-systems "shell".

The "E" stood for Empty, for a shell was a content-free inference engine. (The term "inference engine" was coined by Randall Davis, in homage to Babbage—interview in Crevier 1993: 157.) It specified, for example, the basic blackboard architecture, and the method of conflict resolution to be used when more than one production was satisfied. Also, it provided methods for "forward-chaining" and "backward-chaining" inference. The former was used bottom-up to generate conclusions from data, the latter top-down to find evidence for a supposition (and to explain an item of advice by recapitulating the previously fired rules). Several expert-systems shells were being marketed by the close of the 1970s, for use in building new systems for commercial clients.

By the early 1980s, then, expert systems had become a recognized field within AI. At that point, the field was given an added boost by Japan's Fifth Generation project (Chapter 11.v). This prompted several popular publications from AI leaders, such as Feigenbaum's *The Fifth Generation* (1983) and Winston's *The AI Business* (1984). There was a burst of technical discussion, too (e.g. Stefik *et al.* 1982; R. Davis and Lenat 1982; F. Hayes-Roth *et al.* 1983; Merry 1985; Waterman 1985). Optimism reigned.

But there were problems lurking in the undergrowth. Some expert knowledge was highly elusive. And some of that tacit knowledge might elude capture for ever. By 1990, various insightful critiques had highlighted the difficulties, and the need to keep human beings somewhere inside the loop (see Chapter 13.ii.b–c).

10.v. Talking to the Computer

Abstract proofs about universal Turing machines (4.i.c) and universal programming languages are all very well. But in practice, as George Orwell might have put it, some languages are more universal than others. Indeed, a list of epigrams on programming, written by CMU's Alan Perlis (who led the team that designed ALGOL in the late 1950s), includes this salutary warning: *"Beware the Turing tar-pit in which everything is possible but nothing of interest is easy"* (Perlis 1982).

When the first computers were built, it was difficult even for experts to tell them to do anything interesting. For one had to use the machine code (or some very closely associated assembly language). That involved long sequences of 0s and 1s, which human beings find almost unintelligible. So J. Clifford Shaw, the person primarily responsible (aided by Newell) for the *programming* of the Logic Theorist and GPS, remembers:

It was very painful [in 1954] to try to program anything, to make progress towards a chess-learning machine, because we didn't have an adequate language for communicating. [Newell and I] had done a number of programs in what was essentially machine language . . . but it was far too low-level a language to begin to specify the chess-playing program. *As programmers, we had a creative task each time with trying to invent a representation in the machine corresponding to what we were communicating fairly loosely in English.* (Shaw, interviewed in McCorduck 1979: 141; italics added)

For NewFAI to be practicable, then, it would have to escape the shackles of the binary code. To do that, it relied on three Ls: List processing, Logic, and LOGO. Each of these enabled people to avoid thinking of computers only as shufflers of 0 and 1, even though that's what they are at base.

List processing was the first of the three to rule the NewFAI roost, in the mid-1950s. Logic programming became prominent twenty years later. As for LOGO, this was less widely used by AI researchers. But it was much more interesting to other people. It wasn't merely a programming language, but embodied a specific psychology of thinking, learning, and teaching.

a. Psychology outlaws binary

There's a reason why binary code is near-unintelligible: namely, the "magical number seven" (Chapter 6.i.b). It's because of that limitation on human memory that computer science as we know it, NewFAI included, couldn't have got off the ground without programming languages.

When the psychologist Miller first identified the "magical number" he pointed out that it can be cheated by chunking. He also pointed out that since there can be no more than about seven chunks, we regularly compose larger chunks out of smaller chunks . . . on indefinitely many levels. Programming languages are chunkers: one term in the high-level language is defined by many in the machine code. That's why programmers can write complex software using these languages which they couldn't have hoped to write without them.

The need for programming languages had already become clear in practice even as Miller was writing his paper (published in 1956). And his colleague Licklider was well aware of this. In his famous call to arms on enabling "Man–Machine Symbiosis"

(11.i.b), he listed them as an essential prerequisite for that happy state of affairs (1960: 4). The lack of powerful programming languages, he said, was "the most serious obstacle to true symbiosis".

Why was Licklider so concerned? After all, by the mid- to late 1950s a few such languages were already available. The most widely used were IBM's FORTRAN (the acronym came from FORmula TRANslation); the algebra-like ALGOL, designed in Germany with some input from Zuse; and COBOL, created by Grace Hopper (who'd also written the first compiler, in 1952).

However, these weren't much use to NewFAI. Programming languages are like traps, designed to catch different animals in different ways: they store and process information differently, and are best suited to different computational tasks. FORTRAN and ALGOL had been custom-made (for scientists and engineers) to represent mathematical formulae or number arrays, not to model hierarchical problem solving or associative thought. (Weizenbaum used FORTRAN to program ELIZA, as we've seen: but he didn't try to *do* anything with it, besides superficial template matching.) Similarly, COBOL was designed largely to help businessmen and accountants to do their sums. But AI was focused on symbol processing *in general*, not just numbers. (Zuse's Plankalkul was non-numerical, but was still unknown: see subsection f.)

Since AI required new types of task to be accomplished, the AI community had to create new ways of telling computers what to do. That explains Licklider's concern, for he'd been closely involved with the NewFAI pioneers right from the start. It also explains why Minsky could justify the "slow" rate of progress in 1950s AI by saying that "Much of [our] time has been spent on the development of programming languages and systems suitable for the symbol-manipulation processes involved" (1961c: 215).

The design of such languages depended on three things: (1) how people find it easiest to think; (2) what NewFAI folk wanted to think about; and (3) how the new language could be implemented.

The last question was crucial. Someone can define a formal language and then use it 'intuitively'. That's what Bertrand Russell did when he proved theorems in his propositional and predicate calculi (Chapter 4.iii.c), and what Chomsky did when he applied his transformational grammar to English sentences (Chapter 9.vi). But to enable a machine to use the language, someone has to show how to represent the newly defined chunks in the computer's memory, and how to translate them into machine code so that the program can actually be run.

The same applies to the development of user-friendly computer interfaces (Section i.h above and Chapter 13.v). As Engelbart put it, in discussing the desirability of word-processing:

[The] internal structure [of symbols in the computer for words, phrases, sentences, paragraphs, cutting, pasting, copying, etc.] may have a form that is nearly incomprehensible to the direct inspection of a human (except in minute chunks).

But let the human specify to the instrument his particular conceptual need of the moment, relative to this internal image. Without disrupting its own internal reference structure in the slightest, the computer will effectively stretch, bend, fold, extract, and cut as it may need in order to assemble an internal structure that is its response, structured in its own internal way . . . [It] portrays to the human via its display [i.e. interface] a symbol structure designed for *his* quick and accurate perception and comprehension of the conceptual matter pertinent to his internally composed substructure. (Engelbart 1962: 87)

These psychological facts of life affected the development of NewFAI very deeply. It required special programming languages right from the start, and couldn't have matured into GOFAI without continual (and still continuing) improvements in *how* to tell a computer what to do. Deciding *what* to tell it to do, which is what people normally focus on when they think about AI, was therefore just one side of the coin. NewFAI researchers had to originate both the head and the tail.

b. Entering the lists

When Newell died in 1992, the *AI Magazine* printed three full-page pictures of him, including one on the front cover. In addition, they ran a twenty-six-page obituary, a memoir by Simon, and a previously unpublished speech of Newell's—championing AI as a "fairy-story" full of "enchantment"—that he'd given nearly twenty years before his death (Laird and Rosenbloom 1992; Simon 1992; Newell 1976/1992). Fully one-third of the Winter number was given over to him.—Why?

For the obituarists, his long-time SOAR collaborators John Laird and Paul Rosenbloom, his research as a cognitive scientist was paramount (see Chapters 6.iii.b–c and 7.iv.b). The nature of mind, they said, was "the ultimate question", which they hoped wouldn't be lost in AI's rush to money-spinning commercial applications (p. 41).

Newell's own theory of mind was the paradigm of GOFAI, even including the claim of strong AI (Chapter 16.v.c and ix.b). His SOAR colleagues made that very clear. They didn't mention the fact that he'd encouraged other cognitive science approaches, too. The connectionist Hinton, recalling Newell's offering him a job at CMU, has said:

[Newell] was very eclectic. He was very broadminded. He realized that sooner or later there was going to be a connection between what went on in the mind and what went on in the brain... [He] was in favor of having people do all sorts of things [at CMU], so he was basically in favor of having someone who worked on neural nets there. He could see it coming back into fashion...

I was very impressed by the fact that Newell was open to getting somebody in an area that he didn't believe in. It's very rare to see that in academics. (J. A. Anderson and Rosenfeld 1998: 375)

But if the obituarists didn't praise Newell's eclecticism, they did laud his work on programming languages—which enabled him, and many others, to ask that "ultimate question" in computational terms. By inventing new programming languages, Newell—with Simon—revolutionized the potential of AI. One might even say that they *made AI possible.* For in the mid-1950s, they implemented the first list-processing language.

I say "they", but the younger man was the senior in terms of computing expertise. Most of the technical details were worked out by Newell, although the overall rationale was developed jointly. (The same was true of their second revolution in programming, namely, production systems: subsection e, below.) Presumably that's why Newell, not Simon, was elected as AAAI's first President. (His vision for AAAI was recently reprinted in their twenty-fifth anniversary number, and extensively quoted in the opening paper: Newell 1980/2005; Reddy 2005: 9.)

List processing enabled computers to handle problems of greater hierarchical complexity than had been possible before. And it enabled them to be more than mere number-crunchers. That is, it supported what Minsky (1968) termed "semantic" information processing, concerning matters normally expressed not in numerals but in logical symbols or even natural language.

In principle, of course, list-processing languages weren't necessary. Any universal programming language, in theory, can support any computation (Chapter 4.i.d). Indeed, Colby later switched from a Newell–Simon list processor to ALGOL (designed to represent numerical arrays) when writing the later versions of his neurotic program (7.i.a). What's possible in principle, however (the Turing tar-pit), may be very different from what's feasible in practice. And in practice, list processing was a huge advance.

Its ability to represent hierarchy, a prime feature of human thought and action, made it both easier to work with during the programming process and more apt for modelling thought. The second point was especially important to Newell and Simon, who were psychologists before they were computer scientists (Chapter 6.iii). When they first designed new languages for AI, they weren't doing an abstract exercise in computer science. Rather, they were trying to get the Logic Theorist off the ground (Newell and Simon 1956b; Newell and Shaw 1957).

Their initial attempt at going beyond machine code was the "Logic Language", which they used for paper-and-pencil simulations (Newell and Simon 1956a). They soon buckled down to implementation, however, with the help of a RAND colleague: the professional computer scientist Shaw. Through the late 1950s, they pioneered a six-sibling family of Information Processing Languages, or IPLs (Newell and Shaw 1957; Shaw et al. 1958; Newell 1960; Newell et al. 1964).

The first sibling wasn't actually implemented. For IPL-I was none other than the Logic Language, its name being "a label we put on retroactively" (Shaw, in McCorduck 1979: 142).

IPL-II, however, was implemented—and was powerful enough to be used for the Logic Theorist. It had no compiler, and had to be translated into machine code by hand. (This was a labour-intensive exercise: Newell and Simon worked out the relevant binary numbers simultaneously, saying them out loud to ensure that they agreed.) IPL-III, Newell said many years later, was "the best one of the bunch", because it had "no syntactic structure at all" (Laird and Rosenbloom 1992: 23). But it had to be abandoned, because it "was so space intensive that you simply couldn't run it on the machines"—which, in those days, were still tiny. IPL-V was more tractable, and was used for writing GPS.

In these IPL programs, each component ("statement") was a list structure. In general, a list needn't be a list of single items: it may be a list *of lists*. Just as a shopping list can be hierarchically organized (Grocer: sugar, tea, biscuits; Greengrocer: lettuce, carrots, courgettes; Draper: ribbons . . .), so could an IPL routine. So an IPL program wasn't best thought of as a sequence of instructions all conceptualized at the same level. Some instructions were simple (compare: "Get money from bank machine"), others more complex (compare: "Go to the grocer—and while you're there, buy sugar, tea, and biscuits and order some coffee beans").

Since *everything* was a list, statements could be treated either as data (information, facts . . .) to be noted, copied, or even changed, or as routines to be executed. So routines could be altered during execution, by the same processes that altered lists of data. These processes included substituting one symbol for another, adding an item to the end of a list, and inserting one list into another.

The last facility was necessary for solving problems whose goal/sub-goal complexity wasn't known beforehand. Think of going downstairs to fetch some milk from the fridge: it may turn out that you first have to search for the key of the kitchen door. Even in theorem proving, which was Newell and Simon's concern at this time, one doesn't usually know how many levels of means–end analysis will be necessary.

This required creative thinking about basic aspects of computer science. I implied, above, that if one could implement lists then hierarchy (lists of lists) would come for free. But this wasn't going to happen by magic: AI wasn't such a "fairy-tale" as that. How, then, were lists to be implemented?

For instance, how could a lower-level list be contained within a higher-level one? To do this (which was needed for means–end analysis, as we've just seen), memory access and allocation had to be handled in a new way. Specific sections of the memory couldn't be pre-assigned to this or that function, as was usual, because the programmer couldn't foresee how much memory would be needed when the program was run, nor just which items would need to be associated with (or included within) each other.

Newell and Simon's solution was to label each item with the memory address of the next item on the list—where 'adjacent' items needn't be close together in the physical memory store. The listed items could then be accessed in the right order *despite* being scattered across the memory as a whole. Functional ordering was liberated from physical ordering. Similarly, list insertion could be effected not by 'pushing other items aside' (compare making room in the crockery cupboard for a new set of bowls) but by changing a couple of memory addresses. As Simon put it later:

[The] entire memory could be organized like a long string of beads, but with the individual beads of the string stored in arbitrary locations. "Nextness" was not determined by physical propinquity but by an address, or pointer, stored with each item, showing where the associated item was located. Then a bead [or a sub-string of beads] could be added to a string or omitted from a string *simply by changing a pair of addresses*, without disturbing the rest of the memory. (Simon 1991: 212; italics added)

Another fundamental contribution was the 'push-down stack'. The IPLs were recursive languages, meaning that one procedure could be nested inside another. To achieve that, Newell and Simon implemented the push-down stack. (An earlier form of push-down stack had been suggested, but not implemented, by Turing in the ACE report: Chapter 3.v.c.) This is analogous to a stack of unpaid bills placed on a spike, to be dealt with one by one and removed accordingly. If instruction A includes the instruction to do B, then A is temporarily stored ('pushed down') in such a way that it can be refound ('popped up'), and control passed back to it, as soon as B has been completed. This is possible even if instruction A is nested inside *itself*: the push-down stack ensures that the computer won't 'lose its way' by forgetting which level should be in control at any given stage.

That's how GPS was able to represent, and be efficiently controlled by, means–end hierarchies with goals and sub-goals on various levels. And it's how computers were able to accept self-reflexive procedures such as this (expressed here in English):

```
CLIMBSTAIRS:
    push right toe forwards
    IF toe meets obstacle
        THEN raise right foot onto next step
            AND raise left foot onto same step
        CLIMBSTAIRS
    ELSE stop.
```

In such cases, *repetition* (of reaching the next step) was implemented as *recursion* (of CLIMBSTAIRS).

A third innovation concerned memory loss. Earlier computer programming had followed the principle that when one routine hands over control to another it instantly forgets everything it had needed to know about the local context while it was working. After all, why waste memory space? However, instant memory loss on the relinquishing of control would make recursive routines impossible. One clever feature of IPL-V was that the programmer didn't need to state explicitly, every time a different goal took over the control, that the relevant context should be remembered (for push-down) or recalled (for pop-up). Nor did he/she need to state that when a given routine had been completed its contextual information could be deleted to save space, or that if it was abandoned then *everything* currently in the push-down stack could be removed simultaneously. The programming language took care of all that automatically.

As it turned out, IPL-V was soon dropped like a hot potato (see below). So why did Newell's obituarists make such a song and dance about it? The reason was that Newell and Simon hadn't merely invented a new type of programming language. They'd also developed a new approach to programming *in general*—and here, Simon was at least as important as the more technically skilled Newell.

Simon summarized his general ideas about hierarchical complexity soon after the birth of IPL-V (Simon 1962). He expanded them as the Karl Compton Lectures on 'The Sciences of the Artificial' in 1968—since when, they were revised and reprinted many times (Simon 1969). But the seeds had been evident in the first IPL-V manual, Newell's (1960) RAND memo on the language.

That memo was widely read within the NewFAI community—indeed, it helped to *create* that community. After circulating within the core of insiders, it was put out by a trade publisher in 1961; and a second edition appeared a few years later (Newell *et al.* 1964). By that time, IPL-V had been well-nigh abandoned. So the interest wasn't primarily in that particular language. Rather, it was in the manual's general advice on programming: e.g. recommending top-down design, short subroutines, and the avoidance of GOTO instructions. This was valuable to all NewFAIers, irrespective of their choice of programming language. As Newell's obituarists put it, he and Simon had introduced "*a design philosophy for programming* that years later would be reinvented independently as structured programming" (Laird and Rosenbloom 1992: 24; italics added).

In short, IPL-V was an important contribution to computer science, to software engineering, and (via GPS) to cognitive science too. That's partly why the *AI Magazine* honoured Newell so generously (see also subsection e, below).

c. LISPing in 'English'

Newell and Simon weren't the only ones to program in IPL-V. Colby, for instance, did so too, when writing the first version of his neurotic program (7.i.a). Nevertheless, it wasn't easy to use.

It was slow, because it had no compiler to translate instructions *directly* into machine code (it depended on an interpreter). It forced the programmer to keep track of some boring 'housekeeping' details, such as declaring whenever an unwanted list could be deleted to make room for a new one. And it was defined at a very low level, more like an assembly language (which is just one step up from the machine code) than what people think of today as a programming language. It therefore looked more like algebra than English—and not pretty algebra, at that. Indeed, Sloman (personal communication) describes it as "a horrible inelegant mess, [which] had no hope of surviving".

It didn't follow that the IPL-V *manual* was destined for the dustbin, as we've seen. But despite their respect for the programming strategy spelt out there, most computer scientists were loath to use IPL-V as their programming medium. Several other list-processing languages were soon designed accordingly.

MIT's Bobrow and Raphael compared some of them (Bobrow and Raphael 1964), as J. Michael Foster of Aberdeen University did a few years later (1967, ch. 7). Their differing advantages and disadvantages included speed, computational potential, and ease of use. For example, were the instructions interpreted or compiled, or both? Did the user have to oversee the 'garbage collection' (tidying up the used/unused memory locations)? Could lists be represented in only one fixed way? Was sharing of sub-lists allowed? And what about circular lists? These questions guided the very early AI researchers' choice of language, given that some form of list processing was desired.

So IPL-V soon had rivals. Gelernter, for instance, developed FLPL to write his geometry program (Gelernter *et al.* 1960*a*; see Chapter 6.iv.b). And Victor Yngve's (1958) COMIT was designed to deal with words, for the purposes of machine translation. COMIT was more easily intelligible than either IPL-V or FLPL—indeed, even beginners (such as myself: see Preface, ii) could use it. This was due not least to its relatively user-friendly programming manual, which stood out from the crowd in those days of aggressively 'macho' technicality (Yngve 1962*a*,*b*). (In fact, COMIT remained in use outside AI well into the 1970s: Yngve 1972.)

Within AI, the rival which won out was LISP—an acronym for LISt-Processing language. This was developed by McCarthy, and was one of his most important contributions to the field.

He'd already declared the need for a new, and English-like, language in his funding application for the 1956 Dartmouth Project (see i.h, above). Following discussions there with Newell and Simon, he started work on LISP that autumn (Chapter 6.iv.b). (He recalls: "I wasn't much influenced by Dartmouth—unless, perhaps, Newell and Simon pushed me into logic as a way of expressing information": personal communication.)

McCarthy was then teaching at Dartmouth, which didn't yet have an IBM 704 machine, so he was confined to using pencil and paper (Gabriel 1987: 520). When he went to MIT, he "concentrated on doing LISP" (J. McCarthy, personal communication). He wrote his first LISP program, for differentiating algebraic expressions, while visiting

IBM in summer 1958: it had if–then conditionals, and recursive use of conditionals (personal communication).

He outlined his ideas for the language publicly two years later, both in the very first MIT AI Lab memo and at Albert Uttley's November 1958 NPL meeting in London (McCarthy 1958, 1959). More detailed descriptions soon followed (McCarthy 1960; McCarthy *et al.* 1962). (The implementation details, and many subsequent improvements made by McCarthy and others, are described in Gabriel 1987.)

Introducing the famous CONS and CDR, McCarthy described a new way of implementing lists. This owed something to IPL-V and its recent cousins, but went far beyond them too. LISP quickly became the natural language of choice for AI (Bobrow 1972). Three decades later, when NewFAI had long given way to GOFAI, it was still being described as "the most widely used language in AI programming" (Gabriel 1987: 527).

The first version of LISP had several technical strengths. It was up to two orders of magnitude faster than IPL-V (and less demanding of memory space), because it had a compiler as well as an interpreter. And it provided automatic garbage collection. Besides tidying the memory store, this kept track of the partial results of various computations without any need for human monitoring.

LISP made the activity of programming easier, too. Because it offered built-in arith-metical functions, the programmer who happened to need them didn't have to define them anew. And it enabled one to write conditional (*if–then* . . .) instructions easily, as well as straightforward imperatives (*do this*). This was due to its intellectual origins in logic: specifically, Alonzo Church's lambda calculus (Chapter 4.i.c) and Gottlob Frege's higher-order functions (2.ix.b). (The first of these was later commemorated in the name of the hugely popular VR environment LambdaMOO: see 13.vi.d.)

Since functions could be arguments in other functions, recursion could be made explicit in the program, instead of being handled purely behind the scenes by the push-down stack. (Strictly, a LISP program wasn't a sequence of *instructions* but a set of *function calls*: one function call could include arguments returned as values from other function calls, and so on.) Like the IPLs, then, LISP allowed the definition of a term to include that term, so that processes could invoke themselves: CLIMBSTAIRS, all over again.

With respect to *what one could talk about* when talking to the computer, however, the most important innovation was its syntax. Recursion was represented more 'naturally' in LISP than in IPL-V, by nesting one bracketed expression inside another. Compare: *(The cat (I saw on Saturday night (when I came to supper)) sat on the mat)*. And there were no arcane symbols. The simplest LISP expressions consisted of up to thirty capital letters and digits, with commas and brackets as well: e.g. ((ABCDEF,239),(MNO,835)) or ((FATHER,MAGGIE),(HUSBAND,DOROTHY)). This syntax carried a liberat-ing—but potentially deceptive—advantage, and an annoying disadvantage too.

The disadvantage was that the more complex LISP expressions could have many, many nested brackets. This meant that structures of significant hierarchical complexity could be represented. So far, so good. But typing errors were common. One of SHRDLU's sentence parsings, written in a LISP-based language, ends with a sequence of eleven brackets, thus:))))))))))) (Winograd 1972: 176). These are difficult to count accurately, even if one uses a pencil tip to do so. (Try it!) Today, the computer editor could count the brackets for you. But in those days, the brackets had to be matched by hand. As a result, bracket bugs were common in LISP programs. (And the name was sometimes glossed as "Lots of Inane Stupid Parentheses".)

The advantage was that LISP allowed the user to employ symbols visually identical with English words—as in the (FATHER,MAGGIE) example, above. Indeed, Mc-Carthy's (1959) introductory paper had stressed this fact. LISP's origins were in logic, as we've seen. But McCarthy wanted to have his cake and eat it, by combining the deductive power of logic with the representational power of natural language. So in his talk at Uttley's meeting, he'd proposed the Advice Taker: a (notional) program that would accept statements in English and draw inferences from them (Section i.f, above).

The option of using English terms made it much easier to program complicated procedures in LISP than in IPL-V. For each step or sub-procedure could be named and/or expressed by symbols that looked like semantically appropriate, and therefore easily remembered, English words. In terms of ease of use, then, LISP was a liberation.

For this reason (plus the speed, arithmetic, garbage collection, and conditionals), AI professionals in the USA overwhelmingly opted for McCarthy's language. IPL-V was dropped, as were its other rivals. LISP ruled supreme.

(UK researchers used LISP too, in the early days. By the late 1960s, however, the Edinburgh-based POP2 was widely preferred, and remained dominant there—spreading to other European centres—for many years: Burstall and Popple-stone 1966. Indeed, its successor POP-11 is still used by some AI researchers east of the Atlantic. Unlike LISP, which was designed for offline use, POP2 was a pioneering online computer language. That is, it was intended for use "by a person communicating directly with a computer *via* a typewriter"—Popplestone 1966: 185. It combined facilities offered by ALGOL and LISP: arrays, as well as lists. And its syntax avoided the plethora of brackets that plagued LISP programmers, making bugs due to typing mistakes less likely.)

However, there was a sting in the tail. LISP's major advantage—that it appeared to be using English words—was a mixed blessing. Too often, programmers unthinkingly assumed that the wordlike symbols in the program meant much the same as the corresponding words in natural language. As a result, they deceived themselves and others about the power and psychological relevance of their research.

They'd be rapped sharply over the knuckles for this by an AI colleague in 1974 (11.iii.a). But despite the knuckle rapping, the temptation to over-interpret procedure names and data labels persisted. Twenty years later, leading AI researchers were still being accused of believing that their programs were doing something (e.g. finding/recognizing analogies) in a 'human' way largely because this was implied by the mnemonic labels they'd chosen (Hofstadter and FARG 1995: 275–91).

Another way of putting this is to say that LISP made it easy for AI workers to fall into the trap described by Searle (16.v.c). Even though (*pace* Searle) a program *is not* all syntax and no semantics (see 16.ix), it can be so regarded for certain purposes. That is, it can be seen as an uninterpreted formal system. Purely formal systems can sometimes be mapped onto structural relations between meaningful concepts. For instance, a simple letter-shuffling or necklace-building game may map onto the rules of arithmetic—something one realizes spontaneously, as one plays the game (Hofstadter 1979: 46–54; Boden 1990a: 42–6). Where AI programs are concerned, there will always be *some* degree of match to the domain being modelled. The more insightful and/or competent the programmer, the closer the mapping will be. So people, programmers included, will be tempted to understand LISP's 'English' symbols as though they were the equivalent English words.

In general, this will be a mistake. Setting aside the philosophical question of whether there can be any real understanding there (Chapter 16), a program's quasi-understanding will inevitably be less rich than the human's. For in practice, not all of the relevant semantic associations will have been provided. Even arithmetical calculations by a computer can fail to map onto the human equivalent (13.ii.b). And natural-language words are interpreted by us in terms of types of relevance undreamt of by NewFAI (7.iii.d and 9.x).

This wasn't clear to AI workers in 1960, despite Bar-Hillel's recent critiques of MT (9.x.e) and Ludwig Wittgenstein's just-published *Philosophical Investigations* (9.x.d). Nor was it clear to 1960s cognitive scientists in general. Their common assumption, which guided Bruner's work on concept formation for instance, was that words could be defined by necessary and sufficient conditions. Indeed, the programme of logical atomism was still close to many people's hearts, from McCulloch (4.iii.c) to McCarthy himself. Small wonder, then, that 'English' was commonly assumed to be English.

In sum, what looks like gobbledygook isn't likely to be misread as good plain sense, but what *looks like* good plain sense is easier to write than gobbledygook is. If IPL-V hadn't given way to LISP, there'd have been less misunderstanding of AI. On the other hand, there'd have been much less AI to be misunderstood.

d. Virtual cascades

Even LISP had its weaknesses. There were certain things that late 1960s AI programmers would have liked to be able to do which simply couldn't be done, or which could be done only with great difficulty. So several people at MIT started asking themselves whether they could design better languages.

They might have started from scratch: *Forget LISP! Let's invent a fundamentally new language!* After all, that's what Robin Popplestone (1938–2004) had done when he designed POP. And it's what Kay at Xerox PARC did in 1971–2 when he designed SMALLTALK, the first object-oriented programming language (A. C. Kay 1993).

Object-oriented languages are 'natural' in the sense that they allow the programmer to think in terms of objects, or hierarchical concepts (think *frames*), without always having to worry about *just how* those objects can/will interact. In effect, the object (as defined in the programming language) already 'knows' how it behaves, and/or how it relates to other objects. Often, it can 'tell' the user what it can do, and ask just which feature the user wants to focus on. (That's partly why object-oriented languages are useful for designing computerized *agents*: see 13.iii.d.) The reason is that object-oriented languages have an "inheritance" mechanism, whereby properties already defined with respect to one concept are automatically inherited by any concept defined as a subclass of it. (Compare: if I tell one of my grandsons that a lynx is a type of cat, he immediately knows a great deal about lynxes because he already knows a great deal about cats.) Java is a well-known modern example, but there are others besides. (SMALLTALK itself is still used occasionally: Shasha and Lazere 1995: 49.) Licklider had remarked the need for object-oriented languages in his 1960 paper, pointing out that computers at that time had only two elementary symbols (0 and 1), and "no inherent appreciation either of unitary objects or of coherent actions".

But although Kay was using SMALLTALK for his own research (see 13.v.d), it was too far ahead of its time to be effectively implementable. (Today's Java is partly inspired by SMALLTALK.) At MIT, moreover, LISP ruled the roost. It was so well established there that the natural option was to use it as the basis for designing more powerful languages.

It was already known that one programming language could be defined in terms of another. (Gelernter's FLPL, for instance, had been a list-processing language embedded in FORTRAN—hence the "F" in the acronym.) In other words, there could be a *cascade* of computational environments.

Just as computer scientists weren't restricted to working in machine code, so they weren't restricted to working in a 'base-level' programming language such as IPL-V or LISP. The facilities made available by a programming language on level n could be used to define new facilities in language $n + 1$, which in turn could support the computational environment of language $n + 2$ and so on. In the terminology used in Chapters 7.i.f and 16.ix.a, each language provides a particular *virtual machine*, and one virtual machine may be implemented in another . . . until we reach the machine code—which alone can make something happen in the electronic entrails.

In such cases, successive levels of internal translation have to be done by the computer. Because this costs time and memory space, multilevel languages are inefficient. Eventually, new languages and/or hardware would be developed in which functions previously considered as high-level (needing one or more layers of compilation/interpretation) were directly implemented in the hardware. For instance, in the late 1970s MIT developed the LISP Machine, in which basic LISP functions were embodied in the machine code (Bawden *et al.* 1977), and also the Connection Machine (Hillis 1985), in which parallelism—instead of being simulated by a von Neumann computer, as is still usual today—was built in. (Neither was a commercial success: 11.v.a.) But that was for the future. Meanwhile, NewFAI was still struggling to find better ways of telling computers what to do.

At MIT, several virtual cascades were defined, of which the two best known were PLANNER and CONNIVER. Indeed, the former became world-famous as a result of Winograd's thesis, which electrified the AI community—and cognitive scientists in general—early in the 1970s.

Recall, for instance, Winograd's opening remark to SHRDLU: "Pick up a big red block" (iv.a, above). This wasn't a straightforward task, because the green block had to be cleared away first—which involved noticing it, picking it up, finding empty space for it on the table, and putting it down. But Winograd didn't have to say any of that: he left SHRDLU to work it out for itself. If he'd been using LISP, he couldn't have written such a vaguely specified instruction. He'd have had to say not only what he wanted done, but precisely how it was to be achieved.

Instead of LISP, Winograd had used (a cut-down version of) Carl Hewitt's new language, PLANNER (Hewitt 1969, 1972; Sussman *et al.* 1970). PLANNER was the first "goal-directed" language, in which one could specify a goal in general terms without having to identify all the particular objects and operations involved in attaining it.

This was possible because procedures were indexed not by some individual descriptor or name, but by general patterns specifying the type of data on which (or towards which) they were supposed to work. So procedures could be called indirectly, using (unbound) "open variables" instead of (bound) "closed variables".

To see the difference, compare asking a friend to bring you a copy of today's newspaper with asking them to walk to the third counter on the right in the newsagent's shop on Middle Street with the purple awning, riffle through the newspapers until they find *The Independent*, and then buy it, handing over the money to the shop assistant and bringing back the change. The first version of your request is much easier to give. But you have to assume that your friend has enough nous to be able to execute it. They need to know already, for example, that if one wants to buy a newspaper the best place to do so is a newsagent's, that there's a newsagent nearby (on Middle Street or elsewhere), and that all the newspapers will be laid out together.

So it was with PLANNER. Winograd's instruction "Pick up a big red block" activated the PICKUP function by name. But the PLANNER procedure defining this function included within itself the general information: "If you wish to pick up X (whatever it is), and there is (any) Y on top of that X, then you should call CLEARTOP to get rid of Y before you try to move X". In this example, the first occurrence of X is open, as is the first occurrence of Y, but the subsequent occurrences of both are closed. When PICKUP was called, *the program itself* could be relied on to fill in the requisite values of the variables. In LISP, by contrast, every procedure call had to include only closed variables (otherwise, the LISP compiler would spit it out, with an error message asking the programmer to bind the open variable).

As this simple example illustrates, a 'single' PLANNER procedure could include calls to other procedures within it (these calls might be recursive, hierarchical, or heterarchical: Section iv.a, above). More complex PLANNER procedures could include heuristic advice about which other procedures *might* be relevant, and which of these should be tried first. When no heuristic advice was specifically included, the PLANNER procedure would search the database for *any* item whose goal matched the desired pattern, trying each one in turn.

All of this involved hidden processing. A PLANNER program would *automatically* remember what was the last decision, and which alternatives had remained at the time. If the current attempt failed, it would backtrack to the last decision point and try another alternative—or, if none remained, it would back up to the next-highest level in the search tree and try any alternative remaining there. This processing strategy was implicit in the programming language, whereas a LISP programmer (or in the case of GPS, an IPL-V programmer) would have had to specify it explicitly. Since such details were hidden, PLANNER programs were relatively 'chunky' and so relatively easy to understand. The "magical number seven" had been cheated, again.

By the same token, however, the hidden details could include unsuspected inefficiencies and dead ends. Consider the housekeeping involved in the automatic backtracking. PLANNER handled its push-down stacks in such a way that the details of the local environment at each previous decision point were saved, to be reinstated when failure caused the program to pop up to some higher level. However (as in the earlier list-processing languages), abandonment of a goal on level n would automatically delete the contextual information at level n. Human thinking isn't like that: when someone decides to return to a previous decision point so as to restart the search from there, they'll usually want to remember *why* the previous attempt failed. For if they don't, they might try a second alternative (and a third, and . . .) that's doomed to failure for precisely the same reason. Also, they might want to remember *how the world changed* while they were making the unsuccessful attempt. For if they can do that, they may even be able try the same alternative again—this time, making sure that a specific troublesome change is either prevented or immediately remedied.

To avoid PLANNER's stupidity in situations of failure, Hewitt's MIT colleagues Sussman and McDermott designed yet another LISP-based language, called CONNIVER (Sussman and McDermott 1972a,b). A failed CONNIVER procedure could not only tell its higher-level module why it failed, but also pass on information about the sequence of world changes it encountered on the way. This was a prime reason why HACKER (written in CONNIVER) could learn from its mistakes, whereas SHRDLU couldn't (see iii.d, above).

A prime reason why CONNIVER programs could do things which no PLANNER program could do was the language's economical (and automatic) storage of contextual information. A series (or tree) of local contexts was stored in such a way that the *shared* information didn't have to be repeated. After the first context frame, any information that was explicitly coded represented only environmental *change*. So a robot wouldn't have to reiterate the fact that a square box *stays square* no matter how many times it's moved. Or rather, it wouldn't have to do so if the programmer had had the foresight to build that proviso in. (This was one way of trying to avoid, or anyway lessen, the notorious frame problem.)

Another strength automatically provided by CONNIVER was the ability to choose *the best* of a set of alternative procedures, instead of (as in PLANNER) making an unordered list of candidates and trying them out one by one. To be sure, some PLANNER programs could make such choices too—but that ability had to be specifically written in by the programmer. In CONNIVER, it was tacitly provided by the language itself.

PLANNER and CONNIVER were, in effect, LISP with knobs on. Other AI languages were developed later which weren't based on LISP. But the point of general importance here is that a new programming language, even if it's defined in terms of a previous one, can hugely increase the ease of programming—and the power of the ensuing computation.

e. NewFAI in parallel

Having pioneered list processing, Newell and Simon could have been forgiven for resting on their laurels—at least so far as the technological aspects of AI were concerned. But they didn't.

In the mid-1960s, they (again, mostly Newell) developed yet another way of talking to computers: by using production systems, or PSs. Strictly, these were a type of computational architecture rather than a specific programming language. Like semantic networks, or ATNs (9.xi.b), they could be defined in many different languages (see below). In other words, they were a new class of virtual machine for AI's armoury—providing a form of *parallel* processing, of a richness far surpassing Pandemonium (Chapter 11.ii.d).

Thanks to the ever-present grapevine, NewFAI's inner circle knew about this development from the start. In 1967 Newell's lectures on PSs at Stanford had inspired Buchanan and Feigenbaum (who'd collaborated with Newell on IPL-V at RAND, ten years earlier) to use them as the basis of DENDRAL, officially reported in 1969. But it wasn't until the early 1970s that Newell and Simon themselves published on their new approach (Newell 1972, 1973a; Newell and Simon 1972).

Through the rest of the 1970s, that approach was widely adopted within both technological and psychological AI. Encouraged by DENDRAL and its cousin MYCIN, PS-based expert systems burgeoned. And psychological modelling flourished too, following Newell and Simon's massive tome on *Human Problem Solving* (Chapter 7.iv).

Like LISP, PSs grew out of logic. In 1947 the Polish-born logician Emil Post (1897–1954) had defined "rewrite production systems" as a way of representing recursion. As formally defined by Post (1943), a production system was a set of logically independent condition–action rules, or "productions". As implemented by Newell and Simon, a production system was a set of *if–then* pairs: *if* the condition is satisfied, *then* the action is executed. In effect, then, the whole system consisted of demons (or of ants: 7.iv.a–b). This was a huge liberation from the recursive programming languages, in which each procedure had to be explicitly invoked by some other procedure. Productions, by contrast, were triggered by the current state of affairs.

Post's logic had already entered cognitive science before Newell and Simon took an interest in it, for Chomsky (1957) had based the rewrite rules of his phrase-structure grammar on it (Chapter 9.vi.c). It had also affected *computer* science, where it was being used in the early 1960s to design compilers. Indeed, Simon first heard about production systems from a young computer scientist called Bob Floyd, who was working in a Boston software company before Simon gave him a job at CMU (Crevier 1993: 150). But it hadn't yet spread to AI. Although Chomsky had done important theoretical work in computer science (including a proof that push-down stacks can handle context-free languages: 9.vi.a), he wasn't interested in AI—not even computational linguistics. His only venture into this area was in collaboration with a psychologist colleague (G. A. Miller and Chomsky 1963: 464–82). Newell and Simon, by contrast, had been committed to computer modelling since the mid-1950s (6.iii.b). When they discovered Post's logic, they showed how it could be implemented for AI purposes—and how it could be *put to work*.

Since sets of IF–THEN rules can in principle be used to represent any computation, PSs were in effect a universal (very high-level) programming language. Indeed, Newell's student Michael Rychener (1976) reimplemented many classic NewFAI systems as PSs. However, they're usually thought of not as a programming language but as a general type of program. For PSs are a *class* of control architectures, which have been implemented in a variety of (lower-level) programming languages.

Some of these languages were designed with PSs specifically in mind. One of the first widely used examples was PSG. This was the final version of a series of possible languages outlined by Newell (1973a) to implement different control systems for PSs—"different", for instance, in the ways in which conflict resolution and/or memory matching and/or backtracking were handled.

PSG, in turn, gave rise to successive versions of OPS, or Official Production-System language (Forgy and McDermott 1977; Forgy 1981, 1984). These were all developed at Newell and Simon's home base in Pittsburgh, and partly by Newell himself. One of the first commercially successful expert systems, John McDermott's VAX configurer (11.v.b), was written in OPS4 and OPS5.

The main aims behind OPS were to support Newell and Simon's claims that PSs could function in complex environments (which provide many interruptions), and could learn general lessons from experience while so doing (Forgy and McDermott 1977: 933). In other words, OPS was a departure from the early expert-systems approach, in which a PS would be tailored for a single task: specialized, uninterrupted, and fixed. (Likewise, Newell and Simon's PS models had been a departure from GPS—which also was specialized, uninterrupted, and fixed.)

In general, PSs used a "blackboard" architecture. Although production systems, like GPS, were sequential (only one rule could be executed at a time), they weren't pre-ordered instruction lists: first do *this*, then do *that* . . . Instead, all the currently active conditions were made simultaneously visible to the whole system, by being placed on a central blackboard. The PS progenitors, Newell and Simon, allowed no more than nine symbols on the blackboard, so as to match the "magical number seven" of human psychology (Chapter 7.iv.b). But not all PS designers limited their systems in that way. Blackboard architectures were pioneered in the early 1970s HEARSAY speech system, a project led by Newell himself. The blackboard was helpful here because it enabled the program to integrate evidence drawn from widely different areas—namely, phonetics, syntax, and semantics (Chapter 9.xi.g). (Eventually, blackboard PSs were used as the basis of most expert systems.)

In essence, if a production found its condition on the blackboard, then it would fire automatically. Because that was so, and because (as Post had specified) the individual productions were logically independent, a PS was in effect a parallel-processing system. Like Selfridge's (1959) Pandemonium (11.ii.d), the virtual machine was parallelist even though the basic implementation wasn't. And since hundreds, or even thousands, of productions could be involved, most PSs were immeasurably richer than Pandemonium.

In practice, however, this richness presented a problem. For two (or perhaps many more) productions might find their conditions satisfied simultaneously. So some way—preferably sensible, not random—had to be found of enabling the system to choose *just one* of these for execution. Various conflict-resolution methods were used. For instance, the unfired productions might remain visible, as in SOAR, or they might be repressed—as in the earlier PSs (see Chapter 7.iii.b).

In other words, one of the major architectural dimensions on which PSs could differ was their method of dealing with simultaneously satisfied productions. Another was their method of backtracking, and yet another was memory matching (Newell 1973*a*). Memory matching was problematic, for instance, because there might be tens of thousands of individual rules in a single system: how can the blackboard be efficiently inspected, given this fact? Partly because of differences in memory matching, forward and backward chaining (iv.c, above) could be done in various ways.

In sum, production systems offered a rich store of virtual machines implemented in other virtual machines (including languages specially designed for PSs) . . . and differing among themselves not only in low-level detail but also on several broadly defined computational dimensions. The potential for both psychological and technological AI had been hugely increased. Those twenty-six pages of obituary were well deserved.

f. It's only logical!

LISP was designed so that the programmer could tell the computer what to do, when. So the general form of a programmed procedure was *do this, then do that, followed by the other* . . . Even the 'knobby' LISPs, such as PLANNER and CONNIVER, required the programmer to lay down a sequence of goal-oriented actions, although the details would be decided by the program itself at run time.

Given the sequential architecture of the von Neumann machine, this made sense. But AI workers sometimes wanted to shrug off the sequential straitjacket. This was so, for example, if their programs had to deal with situations arising at unexpected times. Indeed, it would in general be nice—so they thought—not to have to trouble themselves with issues of temporal order. How much simpler if the machine could be given the facts and then left to do what it had to do, in whatever order it pleased.

As we've seen, one way of avoiding specifications of task order was to provide anticipatory demons, which would lie in wait until aroused by their input cue. Another was to use a 'parallelist' PS. And a third was logic programming.

(A word of clarification: programming *in* logic—i.e. logic programming—must be distinguished from programming *inspired by* logic, and from programming a computer to *do* logic. Moreover, one logic must be distinguished from another. Three have given rise to hugely influential AI languages: Russell's predicate logic, Post's production systems, and Church's lambda calculus. However, when people mention "logic" in the context of programming languages, or "logicism" in cognitive science, they usually mean predicate logic—perhaps supplemented by modal logics of various kinds: see 13.i.a–b.)

Logic programming was PROgramming in LOGic—hence the acronym of the best-known language of this type, PROLOG. But PROLOG wasn't the only example, nor even the first. Zuse had designed a logic-based programming language, the Plankalkul, as early as 1946. However, this wasn't published until 1972 (and translated in 1976). It didn't catch on: despite being better than its rivals in some ways, they were already well established (Bauer and Wossner 1972). Adopting a new programming language is almost as difficult as abandoning the QWERTY keyboard. It does happen, but only if the advantages are significant.

In Anglo-American computing, theoretical work on 'declarative' programming languages had started in the early 1960s (e.g. Laski and Buxton 1962), and the turn of the decade saw the first fully implemented example. This was ABSET (with its compiler ABSYS), named after ABerdeen and SET theory (Foster and Elcock 1969; Elcock *et al.* 1971).

The authors' motivation was clear. In a preliminary paper tellingly subtitled 'Programs Written Without Specifying Unnecessary Order', Foster had said:

The basic elements in this system are not instructions to do something, as are the statements of ALGOL, but assertions about the data, such as $a = b$ or $a + b = c$. The evaluation of a program of assertions *is not* the obeying in specified sequence of a set of instructions, but *an attempt to find data which satisfies the assertions.* (Foster 1968: 387; italics added)

At that time, his "Assertions" language (like Ted Elcock's work on 'Descriptions': 1968) was only partly implemented. But within a year or so the ABSET/ABSYS version had been completed. Again, the aim was to avoid the sequential straitjacket:

The overall design aim of ABSET was to devise an interactive programming language in which it is possible, at will, to take or defer decisions about a program: we therefore require that *decisions which are logically separable can indeed be taken separately and in any order*. (Elcock *et al.* 1971: 467; italics added)

We have concentrated on more primitive ideas [including notions taken from set theory], *the distinction between an ordering of decisions and an order of evaluation*, and the manipulation of partly-evaluated program. (pp. 467–8; italics added)

Two themes run through [our work]: that it should be clear what a program says, and that *the language should not force the programmer to commit himself to decisions he would prefer to postpone*. (p. 469; italics added)

Like set theory in mathematics, ABSET/ABSYS was focused on the primitive logical concepts underlying more familiar notions—here, not *numbers* but *list processing, matrix multiplication, text processing,* and so on. That is, it was intended as a general programming language. By 1980, however, it—and respected logic-based rivals such as SYNICS, named for SYNtax and semantICS (Edmonds and Guest 1977*a,b*; Edmonds 1981: 406–17)—had been overtaken by PROLOG.

In a retrospective review, two of the Aberdeen authors showed that "anything expressible in PROLOG was capable of straightforward translation into a subset of ABSYS" (Elcock and Gray 1988: 1; see also Elcock 1988). Indeed, they quoted Robert Kowalski's (1988) admission that "ABSYS, a declarative programming language . . . anticipated a number of PROLOG features"—including resolution, unification, backtracking, and the notorious 'negation by failure' (see iii.b, above).

Kowalski's word, here, was significant. For it was he who'd made PROLOG work. Or rather, it was he who'd made PROLOG work best. PROLOG was a language based on Russell's predicate calculus, and using an efficient form of resolution defined by Kowalski in 1971 (Kowalski and Kuehner 1971). It was first conceived—also in 1971—by Alain Colmerauer (1941–) and Philippe Roussel at the University of Aix–Marseille, for use in NLP. It was implemented by them in ALGOL a year later (Colmerauer and Roussel 1993). Kowalski's more efficient implementation was done at Edinburgh (until 1975) and then at London's Imperial College (Kowalski 1979).

One of the advantages of PROLOG was that it provided "for free" much of the inference machinery which had to be specifically programmed in LISP. By the mid-1980s, it was being widely employed in Europe (van Caneghem and Warren 1986), and introductory accounts had been written for AI students (Clocksin and Mellish 1981; Clocksin 1984). Not the least spur to its success was its use by the Japanese in their Fifth Generation project (11.v.a).

Like PSs, a PROLOG program—according to the blurb—would provide the machine with a set of items, and leave it to cope with them in whatever order it pleased. Unlike PSs, the items in question weren't expressed as conditionals (*if–then* rules) but as declaratives (assertions). The program would use general methods, such as logical *unification* and *resolution* (J. A. Robinson 1965), to draw whatever inferences it could from the items available.

PROLOG attracted a lot of attention in its early days partly because it was seen as a protagonist in the then raging 'procedural/declarative' controversy (iv.a, above). Logic, as such, says nothing about the *order* in which assertions are to be tested (or, in a Post production system, the order in which rewrite rules are to be applied). That's decided by the human logician. Indeed, logically independent assertions (such as the various conjuncts within one long conjunction) could in principle be considered simultaneously. That's why many PROLOG programmers insisted that PROLOG, as an implementation of logic, provided declarative program specifications—not procedural programs. (Hence my reference to "the blurb".)

It gradually became clear, however, that it had to be thought of in both these ways. In trivially simple cases, to be sure, a PROLOG program could be seen as purely declarative. When there are only a few assertions to be tested, the order of testing won't matter. But a large PROLOG program (like a large PS) will include many individual items—and may require long chains of inference too. Random testing could take for ever and a day. Some ordering preferences must be followed—which is to say that some *procedural* information must be provided in the program. Eventually, as people realized the futility of this dispute, the language was judged more by a 'horses for courses' criterion. (Logic programming was *combined* with list processing in POPLOG, a programming environment widely used in the UK and Europe which integrated PROLOG with POP2: Hardy 1984.)

With respect to horses for courses, there was an embarrassment lurking in the undergrowth. PROLOG was logically elegant—indeed, logically reliable. But whether it was *practically* reliable was less clear. For as we saw in Section iii.b, resolution theorem-provers can't distinguish between *proving that X is false* (i.e. genuine negation), and *failing to prove that X is true* (i.e. negation by failure). The same applied, then, to any program written in PROLOG. However, there's a huge difference between these two types of negation (if you doubt this, read X as *my partner is having an affair*). If the program is doing theorem proving in a closed logical system, this may not matter. But the case is different for reasoning in 'open' systems, where one can't assume that all the relevant information is present. The use of PROLOG for real-world expert systems was therefore problematic.

This was pointed out very early on (K. L. Clark 1978). Even so, the Japanese adopted PROLOG enthusiastically in their Fifth Generation programme a few years later. Their assumption, apparently, was that every essential feature of the relevant specialist domain could be adequately represented in a PROLOG-based expert system—"adequately" for technological purposes, if not for Marcel Proust or for jealous lovers. (The adequacy of PROLOG for various *legal* purposes was discussed at length by Kowalski himself: Chapter 13.ii.c.) Soon, various people tried to provide 'real' negation, combining goal-directed user questioning with an underlying three-valued logic: *true, false,* and *don't know* (Aida *et al.* 1983; Edmonds 1986).

When Kowalski first implemented PROLOG, he used von Neumann machines. But the virtual parallelism of PROLOG—i.e. the logical independence of individual assertions—was in principle well suited to hardware parallelism. So when the Japanese committed themselves to PROLOG, they committed themselves also to developing a PROLOG machine, where the inferential methods of unification and resolution weren't programmed in, but implemented in the hardware. (In effect, logic was the *machine code* of the computer.) Today, PROLOG is still used—with or without PROLOG machines—by many people in AI.

10.vi. Child's Play

LOGO wasn't just yet-another-programming-language: it was more like a philosophy of life. And it had a fascinating life story of its own. It started in the 1960s, in a form so simple that it was easy for AI researchers to ignore. Around 1990 it metamorphosed into something radically different, and very exciting.

From the start, it brought Piagetian psychology into the world of programming, and (soon afterwards) programming into the world of child education. Eventually, it even helped to bring computing to Everyman, in personal computers like the one you may have sitting on your desk at home.

a. The power of bugs

In 1967, while his MIT colleague Hewitt was struggling to sophisticate LISP as PLANNER, Papert (1928–) was defining a new language, LOGO. This was specifically intended for use by children as young as 4 or 5 (Papert 1972, 1973). As such, LOGO needed to be both simple and intuitively accessible.

Accordingly, the virtual world it defined contained a "turtle" moving about on a CDU screen, according to the child's instructions. These were expressed in familiar terms such as FORWARD, BACK, STOP, TURN, RIGHT, LEFT, 5, 10, PEN UP, and PEN DOWN. And the human–computer interface had to be designed anew: instructions weren't given by typing, but by button pressing (how would a 4-year-old manage to type "FORWARD"?).

Very soon, Papert took a leaf out of W. Grey Walter's book (Chapter 4.viii.a–b) and implemented LOGO as a language for controlling robot turtles moving around the floor. (The first turtles were made in the AI Lab's workshop, but they were commercially manufactured from 1977, by Terrapin Software.)

Whether in the virtual (CDU) or the real world, LOGO turtles carried a "pen" which they could lower so as to leave a trace on the screen/paper as they moved. The possibility arose, then, of the child's defining routines and subroutines for drawing a SQUARE, a ROOF, a HOUSE, a MAN, a FACE, and so on.

Usually, the first attempts wouldn't work (see Figure 10.28). Perhaps the child, trying to draw a square, would tell the turtle to TURN. But should it turn LEFT or RIGHT? Often, the child wouldn't say, assuming that this was obvious (especially if the first turn had already been made). And how far should it turn? In a program for drawing a square, "TURN 90" was needed; for starting to add a roof, some other angle would be required. And PEN-UP would be needed if, having drawn the house walls, the child now wanted to add a door or a window. Otherwise, an unwanted line would mar the drawing.

All pretty trivial . . . ? For Papert, absolutely not. The object of this playful exercise was education, not meaningless play—and education not in drawing houses but in self-critical, creative thinking *in general*. As he put it, he was concerned less with mathematics than with *mathetics*: "the set of guiding principles that govern learning" (Papert 1980: 52). In other words, he was doing cognitive science, not technological AI.

His five years with Jean Piaget in Geneva had convinced him that learning comes about by *construction*, not *instruction*. That is, the child's interactions with objects in

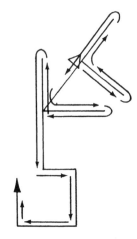

(a) Intended MAN

(b) Picture drawn by buggy MAN-program. (The small triangle shows starting position of the turtle.)

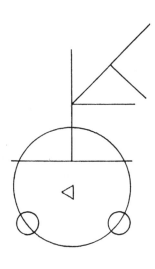

(c) Intended FACEMAN

(d) Picture drawn by buggy FACEMAN-program.

FIG. 10.28. Drawings done by children's LOGO programs. Reprinted with permission from Goldstein (1975: 250)

the world (and virtual worlds) enable him/her to construct problem-solving schemata, which become increasingly well adapted to the task as a result of active reconstruction prompted by errors (Chapter 5.ii.c). So Papert relied on the child's own exploratory activities to foster progress through constructive (*sic*) self-criticism, perhaps aided by a sensitive teacher.

A key idea was "powerful ideas": concepts that would help the child (and his/her teacher) to understand what's involved in thinking. Perhaps the most powerful idea of all was *bugs*.

Bugs are errors. But they're errors of a very special kind. Because they occur in programs, they're in principle both recognizable (as Lovelace had realized: see 3.iii.b) and correctable. In addition, they're displayed by the (non-judgemental) computer, rather than being pointed out by a (possibly scornful) human teacher. As illustrated in Figure 10.28, Papert enabled children to use their own buggy programs to gain insight into their thinking processes, and thereby to improve them.

One mid-1970s MIT project aimed to devise a computer program to play the role of the teacher. So Ira Goldstein's MYCROFT used general ideas about debugging, like those already being used in HACKER, to evaluate and correct *drawings done by computers programmed to draw by children* (Goldstein 1975). Given a mismatch between the intended and actual drawings, MYCROFT would inspect the child's program to locate the bugs. But it wouldn't merely spit out a corrected program. (The child probably wouldn't learn much if it did.) Rather, it would analyse the faults in relation to specific bugs, which the child might learn how to correct—and how to avoid—in future. For instance, if one starts to draw a man (as in Figure 10.28*b*) by drawing the two legs splayed out from the starting point, one will have to insert an extra orienting step before starting on the body. Otherwise, as in this buggy picture, the body will be aligned with the second leg instead of being at an angle to it.

Reorientation bugs *in general* involve the failure to restore some previous state of affairs so as to continue with the next main step of the procedure. They occur in all sorts of contexts, not just in drawing. Papert's hope was that LOGO-based experience, whether with MYCROFT or with a human teacher, would help children to recognize—and avoid—common bugs in many different domains, as well as increasing their self-confidence by 're-educating' their attitude to error.

His own writings, and those of his adherents, abounded with anecdotal evidence that this was so (e.g. Papert 1980; Howe *et al.* 1980; Weir 1987). As a result, LOGO in some form (including LEGO/Logo: see below) was being used "in about one-third of the elementary schools in the United States" by the mid-1990s (Resnick 1994: 26).

However, some careful 1980s attempts to assess the educational value of LOGO programming were ambiguous. The effects were there, to be sure—but they were neither so strong nor so generalizable as Papert had hoped (Pea and Kurland 1984; Kurland *et al.* 1986; Pea *et al.* 1987).

Today's verdict, after many empirical studies (focused on other programming languages as well as on LOGO), is that positive effects depend on the extent to which the instructions, and the teachers' comments, stress *metacognition*, or thinking about thinking. Merely teaching children to program, without such metacognitive encouragement, doesn't necessarily help them to develop transferable thinking skills.

The Harvard educational psychologist David Perkins, who'd been familiar with Papert's work from the start, reviewed the literature and concluded:

Originally, one might [like Papert, and others too] have thought of programming as a kind of cognitive playground; mere engagement in the activity of itself would exercise the mind as real playgrounds exercise young bodies, without any need for *instruction finely tuned* to provoke such consequences. Unfortunately, the research argues against such a vision.

Instead . . . certain inconvenient conditions must be met. [These include huge time commitments for practice, and significant programming expertise on the part of the teacher.] Taken together, these points suggest that the *straightforward* teaching of elementary programming is not a very good way to foster cognitive skills. (Salomon and Perkins 1987: 163–4; italics added)

Some twenty years later, his judgement remains unchanged (D. N. Perkins, personal communication).

b. Complication and distribution

Even in the 1970s, LOGO wasn't restricted to drawing-with-turtles. For instance, Harold Abelson and Andrea diSessa (1980) showed how LOGO could be used to teach children about other types of mathematics, such as spherical geometry and algebraic vectors—and the laws of physics, too. Part of the interest here was that a law of physics could be deliberately varied, or omitted, and the result would be visible on the CDU screen.

In today's terminology, the children were constructing a virtual reality whose behaviour one could actually *see*. But because the computer graphics used by Abelson and diSessa were so minimalist, these imaginary physical worlds couldn't begin to compare with what we now think of as VR (13.vi).

The 1980s saw the addition of list processing to LOGO. This suited it, for instance, for helping children to learn writing skills (Friendly 1988). Also in the 1980s, Papert added *real* bells and whistles. In other words, he helped the LEGO toy company to design special sensors and actuators to be connected to LEGO components (Resnick 1994: 24–31). The sensors could respond to touch, infra-red, and light; and the actuators included gears, pulleys, wheels, motors, and light switches. Using this LEGO/Logo construction kit, children could build/program their own machines. These ranged from toy cars through ovens to robots—with Valentino Braitenberg's lifelike "vehicles" in between (15.vii.a).

Still more was to come. In the 1990s, LOGO was hugely improved/expanded so as to be suitable for 'real' AI programming (B. Harvey 1997). In addition, a parallelist version called StarLogo was developed towards the end of the century (Resnick 1994: 31–47; Colella *et al.* 2001).

StarLogo provided a high-level language whereby *thousands* of turtles (agents) could be simulated simultaneously. Each one had its own "program", in the form of a set of rules about what to do in various situations. These rule-sets could be identical in every agent, or there might be several different rule-sets distributed (equally or unequally) across the population. Think of a beehive: there's only one queen, but there are many workers and drones. Queen, worker, and drone follow very different rules, live very different lives. And the entomologist can study the overall organization within the beehive, as well as studying the behaviour of individual bees.

In a StarLogo program, each little local environment ("patch"), too, could execute StarLogo commands. So it could grow more 'food' in certain circumstances, for instance, or release 'chemicals' towards neighbour patches at various rates. As a result, the user could model *interactions with and within* the environment, as opposed to *agents' actions on* a passive environment.

In short, StarLogo gave people a way of studying *distributed* systems. These are made up of many autonomous agents—which may be very simple, or relatively mindlike (see 13.iii.d–e and 15.viii–ix). They're all around us: not just beehives, but traffic jams too. But being ubiquitous doesn't make them easy to understand.

StarLogo might even be called the first postmodernist programming language. It's not that postmodernists are enamoured of traffic jams, or less anxious than the rest of us when near a beehive. But, as we saw in Chapter 1.iii.d, they eventually progressed beyond their counter-cultural predecessors of the 1960s–1970s (whose rejection of computer-grounded psychologies was total) by warming to certain types of computational psychology and AI/A-Life: namely, those focused on distributed systems. Much as heterarchy was more politically congenial than hierarchy (see iv.a, above), so cooperation between autonomous individuals was more acceptable—and also more veridical—than rigidly other-directed behaviour.

c. Pointers to the future

The children hadn't dropped out of the picture. Far from it.

On the one hand, they could use StarLogo too. And in playing (*sic*) with it, they could, for the first time, get an intuitive sense of the behaviour of distributed systems.

On the other hand, Papert's passion for child-friendly computing had helped lead to technologies without which many children's lives today—and probably yours, too—would be very different. Namely, user-friendly interfaces and virtual reality. In other words, Papert's work was crucial in turning the futuristic visions of Bush and Engelbart into everyday realities.

But that's a story for a later chapter (13.v–vi).

11

OF BOMBS AND BOMBSHELLS

NewFAI didn't excite only its practitioners: it electrified plenty of other people as well.

Some were cognitive scientists: psychologists, linguists, and philosophers enthused by NewFAI's promise of turning *mind-as-machine* from a metaphysical slogan into a scientific research programme. Their excitement was raised almost to fever pitch by the consciousness-raising meetings of the late 1950s (see Chapter 6.iv). But some were vociferous opponents, fiercely scornful of any such idea—and worried about AI's technological potential, to boot.

This wasn't a private fight: virtually everyone could join in. In other words, the critiques came largely from journalists, social commentators, and the general public. As we'll see, their passionately held views had important (and usually damaging) effects on the intellectual advancement of AI and related disciplines.

Early connectionism, too, met opposition—but this came only from AI insiders. It sometimes aroused passions, to be sure. Indeed, a major scandal ensued within the field when Marvin Minsky's mid-1950s criticisms in 'Steps . . .' were brought to a head in 1968 (Chapters 10.i.g and 12.iii). But outsiders knew little or nothing of that. Their sights were set on the symbolic footpath, not the cybernetic one (1.ii.a). In essence, that was because it was symbolic AI which claimed to represent specific propositional content, and rational thought (4.v.d and 10.iii.a).

(In the 1980s, when connectionist AI became more widely known, some of the public opposition to NewFAI was generalized to cover that as well. Only "some": many members of the public were seduced by the connectionist propaganda machine—see Chapter 12.vi–x.)

The opposition to AI, and to cognitive science in general, was both theoretical and political—sometimes, closely interrelated. Mostly, it came from critics proud to be outsiders, and anxious to distance themselves from computational ideas in any way possible. But basic criticism occasionally came from insiders as well—arousing no little glee in the outsiders, of course.

Between the mid-1960s and the mid-1980s, a number of fundamental challenges arose which affected symbolic AI's practices and public relations—and which spilled over to some extent onto cognitive science as a whole. Each of them, to a greater or lesser degree, illustrates the claim made in Chapter 1.iii.a, that the Legend of wholly disinterested science is a myth. We can think of these challenges in terms of (literal) bombs and (metaphorical) bombshells.

"Bombs", because the US military's view of the psychology of decision making and rationality (Chapter 6.iii.a–b), and especially its sponsorship of AI technology for weapons and strategic planning, troubled many people. As the bombs got nastier, the opposition got more pointed. In the 1980s, when the bombs had become very nasty indeed, and the governmental decisions about their use increasingly irrational (by common-sense standards), the AI researchers themselves made concerted efforts to counter these lunacies. The bombs are discussed in Section i.

As for "bombshells", by my count there have been nine of those. For a criticism to qualify as a bombshell, it didn't have to be correct: some were, some weren't. But it did have to be so widely accepted by the general public and/or those holding the purse-strings for research funding that the field was undermined, or at best temporarily held back. The one exception to this was the early 1980s bombshell from Japan, a challenge which invigorated AI instead of damaging it (Section v, below).

The earliest bombshell of all has been described already: a US government committee report of 1960 that paralysed the funding of machine translation (MT) in the West, recovery coming only years later (see 9.x.f). (AI *as a whole* wasn't severely damaged by that particular blast, because the pioneering AI labs at MIT, Stanford, and Carnegie Mellon weren't doing MT: it's no accident that none of the "harbingers" of Chapter 10.i discussed it.) The latest, namely "situationism" AI, will be discussed in Chapters 13.iii.b–e and 15.vii–viii and xi. (The dust from that one is still settling.) So that leaves seven to be covered here.

Three philosophical bombshells, only one of which was detonated by a philosopher, are highlighted in Section ii, below. They were aimed at symbolic AI and cognitive science. Their original target was NewFAI-based work, but each was—and still is—directed also at later research.

Next, in Section iii, we'll consider an admonishment demanding better intellectual hygiene in AI—again, initially aimed at NewFAI but still sometimes needed today. Sections iv and v outline the effects, for both GOFAI and connectionist work, of an explosive charge from 1970s England, and of another (ten years later) from Japan: namely, the Lighthill Report and the Fifth Generation project.

Finally, Section vi identifies the early 1980s development that devastated some parts of the field (at least in the public eye) while clearing obstacles from others. That ultra-brief section heralds Chapter 12, in which this parcel of connectionist dynamite (whose fuse had been smouldering since mid-century) is described at length.

11.i. Military Matters

Mathematical research has long been exploited for military purposes. Even Archimedes was persuaded by King Hieron of Syracuse to put his beloved geometry to practical use (Chapter 2.i.a). That's why he designed the ship-destroying crane, and other weapons aimed against the besieging Romans in 212 BC. Nearly 2,000 years later, much the same thing was happening. But now, as well as geometry, there was AI.

AI's precursor, cybernetics, had already been used for urgent military purposes in the Second World War (4.vii.a). The techniques made available by AI as such could serve an even wider range of military purposes, involving reasoning as well as missile tracking.

This section outlines how the military helped give rise to NewFAI, and how the ups and downs of defence spending were reflected in changes in support for AI in general. Connectionism wasn't immune: as we'll see in Chapter 12.vii.b, DARPA's mid-1980s decision to fund it was driven by their hope for "next-generation, fire-and-forget, autonomous weapons" (DARPA 1988, p. xxv).

(Now, in the post-9/11 world, the military are expecting even more from AI. Reporting a significant rise in technical papers submitted to the 2005 'Innovative Applications of AI' (IAAI) conference, a journalist noted that "Various strains of AI are viewed as critical to homeland security, antiterrorism, and emergency response. Governments are allocating huge amounts of funding for AI research"—Hedberg 2005: 12. And in Europe the journal *IEEE Intelligent Systems* devoted a special issue to 'Artificial Intelligence and Homeland Security' in autumn 2005. One ludicrous result of this situation was a widely cast military request for predictive models of the emotions of terrorists: see Chapter 7.i.f.)

Cognitive science was often affected too. By the early to mid-1960s, for instance, even 'straight' linguistics and cognitive psychology were being generously funded in the USA for defence-related reasons, including the national loss of face caused by the 184 pound Sputnik in 1957, and the Soviets' space dog Laika, sent into orbit in Sputnik-II a few months later (6.iv.f; see also Koerner 1989). Of course, Sputnik didn't represent only loss of face. Potentially, it presented a military threat (so prompting the formation of ARPA, as we'll see): spies, and worse, in the sky.

As we saw in Chapter 2.ii.c, science depends on wealthy philanthropists of some kind. Today, AI is supported by many different institutions. These include academic grant-givers such as the USA's National Science Foundation and the UK's various research councils, and major charities like the Rockefeller and Nuffield foundations. But such is the wide range of AI applications that funds are available also from many governmental departments, and from private businesses—some of which have in-house AI teams.

Military and space-related sources (such as RAND, NASA, and Lockheed) are of course included. But war-related funding was even more important half a century ago. Indeed, *all* post-war scientific research was heavily in debt to the military. In 1947–50 the USA's Office of Naval Research alone was funding half of all the sponsored science being done at MIT (Mirowski 2002: 200). Ten years later, NASA too was a prime sponsor. Formed by President Eisenhower in July 1958 as the Aeronautics *and Space* Administration (replacing NACA: the National Advisory Committee on Aviation), this was a speedy—not to say panicked—response to the two Soviet Sputniks of a few months earlier. Where NACA's budget had been a mere $117 million, for 8,000 people, NASA soon received $6,000 million (for a staff of 34,000). Inevitably, some of that munificence would reach the universities—perhaps not for research in medieval poetry, but certainly for engineering and computing, and for cognitive psychology too.

This fact not only made AI possible, but to some extent affected the topics it chose to study. Only *to some extent*: even Noam Chomsky, no friend of the military, has allowed that "When MIT was funded maybe 90 per cent by the military it had no constraints on what it should do. As it has moved from the Pentagon to corporate funding there are more and more constraints" (quoted in Swain 1999: 28–9). (After 1973, this liberal attitude on the uses of military funding was tightened up: see below.)

In addition, AI's dependence on military funding, and the military support for computational psychology too, influenced public attitudes towards both fields.—Rarely, of course, for the better.

a. Nurtured in war

Warfare was part of the context even when NewFAI was merely a twinkle in Alan Turing's eye. Computer technology as such, and cybernetic control-systems too, were first developed for military ends (Chapters 3.v and 4.vii.a).

The Harvard Mark 1, for example, was largely funded by the US Navy. Howard Aiken was a commander in the US Naval Reserve, and Grace Hopper a lieutenant—later, an admiral. And the ENIAC was built for the US Army, via their Ballistic Research Lab in Maryland. (Unusually, the military resources were limited, for resistors were very hard to find; the ENIAC was ready on time only because someone discovered that one of the manufacturers had bins full of rejected samples, rejected not because they were physically faulty but because the colour markings were wrong—M. V. Wilkes 1982: 54.)

(Today, things haven't changed as much as one might think. "The most powerful publicly admitted [*sic*] computer in the world" in 1997 was mainly used for "*in silico* nuclear weapons testing"—J. M. Taylor 2001: 168. As for software, a single project was said by DARPA's director to have paid back "the entire investment that ARPA had made in artificial intelligence since the beginning"—namely, the conversion of an ARPA-sponsored program written for factory scheduling into a logistics aid for moving soldiers, tanks, etc. in the first Gulf War—Shasha and Lazere 1995: 220. And one doesn't have to be employed by the military to join in. In 2003–4 DARPA announced a cash award of $1 million for the first team whose autonomous vehicle could cross Death Valley, from Los Angeles to Las Vegas, within a specified time—DTI 2004: 34. If the prize is won by a gaggle of pacifist garage tinkerers, so be it: DARPA's intent is to use whatever ingenuity they can find, to accelerate a technology that could be used for military purposes.)

The one exception to the military dominance in early computing was the series of computers designed in pre-war Germany by Konrad Zuse (Chapter 3.v.a). By 1941, these included an operational version of the Z3: in effect, a stored-program von Neumann machine, capable of floating-point arithmetic—and potentially of 'semantic' computing too.

Fortunately, the German military (although it had partly funded Zuse's research) turned out to be interested in neither. Soon after being conscripted into the army in 1941, Zuse asked for leave to work on his machine (which could be used to calculate matrices predicting wing flutter in aircraft flying at different speeds). His application was supported by a letter from a respected professor. In his autobiography, he recalled the result:

The Battalion commander, a major, summoned me, first informed me that as a completely new soldier I had no right to take leave anyway, and continued "What does it mean here [in the letter] when it says that your machine has applications in aircraft construction? The German *Luftwaffe* is top-notch, what needs to be calculated there?"—What was I to say to this? How could I have responded to this? Permission denied. (Zuse 1993: 56–7)

Zuse's cryptographic skills went to waste too: his work on coding wasn't needed, he was informed, because the Germans had the Enigma machine. (If he'd been told that Turing was already breaking the Enigma code, he'd have been as surprised as the major—not least, because he hadn't yet heard of Turing.) Instead, he was put to work on weapons development: the successors to the flying bombs that ravaged London in the last years of the war.

It was said later that when Adolf Hitler was told of Zuse's invention "he replied that he didn't need any computing machine, he had the courage of his soldiers" (Zuse 1993: 81). Zuse never learnt the truth about that titbit: "Fact or fiction?—It was a time of rumors." But it's clear that computing wasn't high on the German military's agenda.

Things were very different across the Channel, and across the Atlantic too. After the Second World War ended, the urgency had gone. And so had much of the money—but by no means all.

Whereas US defence spending after 1918 dropped to a minimum, in the post-1945 period it dropped only to just under half (see the graph in Molina 1989: 11). Money was still needed, the powers-that-be decided, to deal with the military challenges posed by the Cold War: e.g. the Berlin blockade of 1948, the Chinese revolution of 1949, the first Soviet A-bomb test (also 1949), and the Korean war in 1950–1. That war caused defence spending to peak again, and although it did sometimes decrease it never returned to the low post-1945 level.

During the 1950s, AI was getting off the ground. War-related funding was crucial (Edwards 1996, chs. 2 and 8; Mirowski 2002: 161–90). Some of the money was spent on obviously military applications. As early as 1953, for example, the US military asked a budding NewFAI scientist (Alex Bernstein) to try to simulate Washington DC's missile air defence system (McCorduck 1979: 154). But much of it was spread more widely.

For instance, mid-1950s MT research in the UK, USA, and Soviet Union was being funded by military bodies. The US Navy, as we've seen (Chapter 9.x.a), was supporting British MT work focused on such blamelessly pacific sentences as *agricola incurvo terram dimovit aratro*. That may seem odd—but if one could translate sentences about farmers and ploughs, one could probably translate sentences about admirals and battleships as well. And while the Navy was sponsoring descriptions of the ploughing farmer, and Frank Rosenblatt's perceptrons too (12.ii.f), one of its sister services was developing "techniques of air warfare" at the RAND Corporation.

RAND—the acronym stands for Research ANd Defence—had been founded in 1946 by the US Air Force (with Douglas Aircraft, who dropped out two years later). The aim was to apply game theory and systems analysis (developed during the Second World War) to weapons design and to nuclear strategy. Throughout the 1950s, its computation and software development centre was one of the largest in the world. For computers were *essential* to RAND's purposes: planning for nuclear warfare could be done only by simulation. In addition, RAND employed psychologists and other social scientists to improve the efficiency of the human beings in warfare's technological loops. So Allen Newell and Herbert Simon were (separately) hired to focus on air defence simulations and game-theoretic/bounded-rationality models of warfare, respectively (Chapter 6.iii).

At the end of the 1950s Paul Armer, then heading RAND's Computer Science Division, wrote an influential survey-cum-prospectus of AI (Armer 1960/1963). This was published not by an academic journal, but by the Wright Air Development Center

(Wright, as in Wilbur and Orville). That wasn't surprising, given RAND's Air Force roots. A full third of the paper was devoted to Russian attitudes to AI (these were divided, as they were in the West), and to the Russian research that was going on in the area.

Clearly, and unsurprisingly (given the global reach of the Cold War mentality), there was a good deal of that. One Soviet scientist had remarked (speaking in English) that simulating the brain with a computer was "*the* number one problem"—and, said Armer, the emphasis was clear in his intonation (p. 402). The Russians were already discussing "chess playing by machines, and the deciphering of ancient Mayan manuscripts" (p. 404). Trivialities, perhaps, like the ploughing farmer. But Armer made it very clear that he agreed with the President of General Dynamics Corporation that "if the area has real military or psychological value to them, they'll put massive concentration on it" (p. 404).

In that, of course, the Soviets weren't alone. Defence spending in the USA tracked the national politicians' view of military threat. After Korea there were several more peaks:

* one in the post-Sputnik and Cuban missile era,
* one linked to Vietnam (where US involvement lasted from 1965 to 1973),
* and one—in 1992, still rising—in connection with President Reagan's military programme of the early 1980s (see Figure 11.1).

Some of this money went on building more bombs. But much of the R & D tranche was spent on the development of information technology—primarily hardware (including $3 million from DARPA for the ill-fated Connection Machine: Hillis 1985), but also

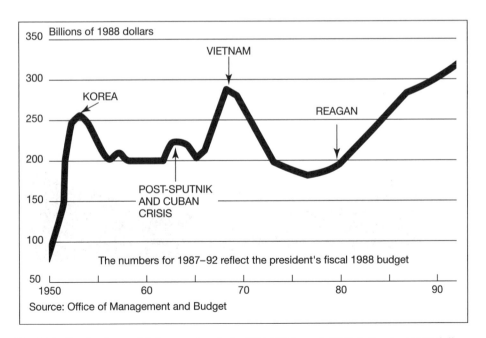

FIG. 11.1. Total volume of defence outlays in the USA (1950–early 1990s). Constant 1988 dollars. Redrawn with permission from Molina (1989: 17)

software. Indeed, creating new sorts of bomb required a huge investment in new computers, and in new ways of using them. (For some fascinating, if frightening, details, see MacKenzie 1991*a,b.*) In effect, the microelectronics industry we take for granted now was developed by the military. It was RAND's 1950s efforts in computer building, for instance, which led IBM—despite Thomas Watson's early scepticism (2.iv.c)—to decide to manufacture digital computers (Edwards 1996: 122).

The vicissitudes of the Department of Defense's (DOD's) budget, and especially the early 1980s funding rise (subsection c, below), also influenced the development of AI. One change—not in the amount of money, but in its favoured recipients—was due to the Mansfield Amendment of 1970: ARPA's annual budget of $26 million for information-processing research was now directed only to *military* applications (Chapter 6.iv.f). But various other changes affected the *amount* of money available. The course of US government funding for mathematics and computer science (i.e. excluding the bombs, as such) is shown in Figure 11.2.

b. Licklider as a military man

In the 1950s and early 1960s, military aims—and therefore funding—were already suspect to some people. Probably most, however, considered them respectable—after all, military efforts had recently saved the world from Nazi/Imperial-Japanese domination.

AI benefited hugely from that fact, not least because Joseph Licklider was in the second group:

I had this little picture in my mind of how we were going to get people and computers really thinking together. [And I thought that] "command and control essentially depends on interactive

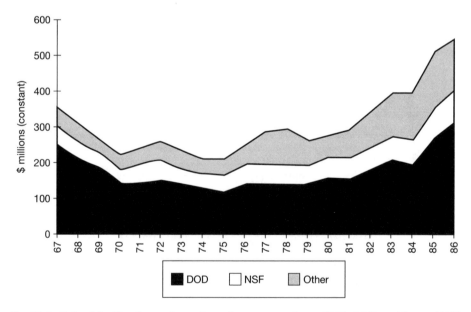

Fig. 11.2. Federal funding for mathematics and computer science, 1967–86. Data: Flamm (1987: 46). Redrawn with permission from Edwards (1996: 283)

computing [i.e. time-sharing] and there isn't any interactive computing so the military really needs this". *I was one of the few people who, I think, had this positive feeling toward the military. It wasn't just to fund our stuff, but they really needed it and they were the good guys.* (Edwards 1996: 267; italics added)

Today, Licklider is widely remembered as the prime sponsor of the technological advances in computer communication that led from the pioneering ARPAnet (first suggested in 1967) to the Internet (Waldrop 2001). Besides providing the money, he'd also provided the vision—in his influential essay on 'The Computer as a Communications Device' (Licklider and Taylor 1968). But what's more relevant here is that he had a crucial formative influence on AI in its earliest days. So his attitude to the military mattered.

ARPA had appointed him the first director of their information-processing section in 1962 (see Chapter 5.iv.f) largely because of his visionary 'Man–Computer Symbiosis'—which cited several early AI programs, both symbolic and connectionist (Licklider 1960). One historian has said that this paper "rapidly achieved the kind of status as a unifying reference point in computer science (and especially in AI) that *Plans and the Structure of Behavior* [see 6.iv.c], published in the same year, would attain in psychology" (Edwards 1996: 266). It wasn't a detailed review, orienting future AI research. That place was held by Minsky's 'Steps'. Rather, it was a vision of some potential *applications* of AI.

The 'Symbiosis' paper forecast the use of interactive (time-sharing) AI systems as assistants to human beings in many situations—explicitly including military command and control, where real-time functioning is often crucial. The idea of time-sharing had come from John McCarthy. Indeed, Licklider confessed to having undergone "a religious conversion to interactive computing" as a result of McCarthy's lectures at MIT (Edwards 1996: 264).

He'd been influenced also by McCarthy's harbinger paper of 1959 (see 10.i.f). Although the computers would provide speed and accuracy and the humans flexibility and common sense, he insisted that the contributions of the two sides couldn't easily be teased apart (hence, "symbiosis"). For instance, he said, theorem-proving programs could learn from experience (as McCarthy had suggested), and the SAGE program could suggest courses of action which the human being may not have considered.

SAGE was an ambitious *military* program, whose interface had been designed by Licklider himself (Edwards 1996, ch. 3). The interface was almost the first in which information was shown on a VDU; and different graphic displays could be brought onto the screen by touching it with a lightpen. Licklider's experience as a psychologist of perception was being put to good practical use.

SAGE was the first large computerized system for "C3": command, control, and communications. The "SA" in the acronym stood for Semi-Automatic, marking the man–machine symbiosis he'd recommended. (The "GE" stood for Ground Environment.) The machines, in this case, included a line of radar stations in Canada—and used as much power as a small city (J. McCarthy, personal communication). In other words, SAGE was the 1950s predecessor of the fully automatic "Star Wars" of the 1980s (see subsection c, below).

When Licklider joined ARPA, it was only 4 years old. President Eisenhower had set it up early in 1958, in a near-panic response to the Soviet Union's Sputnik (launched just a few months earlier, in October 1957). Some scientists feared that this political panic would sideline non-military research, or even squeeze it out of consideration—and funding—altogether. For instance, the Stanford biologist Joshua Lederberg recalls:

Starting with the observation of Sputnik . . . I had set out [in late 1957–early 1958] to assure that fundamental biological science was properly represented in the programs of space research that were just emerging. *The danger was that scientific interests would be totally submerged by the international military and propaganda competition.* They have never gained first priority; they might have been totally excluded. (Lederberg 1987: 17; italics added)

Given Licklider's background (not to mention the benefit of hindsight), it's not too surprising that he also appreciated "fundamental" scientific interests. And where Licklider led, ARPA (and ARPA's money) followed.

ARPA's newness was one reason why Licklider was able to have such a seminal effect on the organization's policy. Another is that, as a member of MIT and as one of the Macy participants, he was already close to many other cyberneticists and budding cognitive scientists. Besides Donald Hebb, whose *magnum opus* he'd commented on in draft (Chapter 5.iv.f), these included Norbert Wiener, Jerome Lettvin, Warren McCulloch, George Miller, McCarthy, and Minsky (and many others too—Edwards 1996: 263–4). Indeed, his long-standing links to Minsky would be wryly remarked by many people when DARPA funds for connectionism dried up in the late 1960s (Chapter 12.iii.e).

On his appointment to ARPA, Licklider funded not only time-sharing (which he'd helped design: 10.i.h) but also AI—including psychological AI (Licklider 1988). He'd started out as a psychologist himself, working alongside Miller on speech and psychoacoustics; and he helped to design the HEARSAY speech-processing project (Newell *et al.* 1973). Indeed, he'd set up MIT's first Psychology programme in the late 1940s. If that hadn't been so, early AI would probably have been very much more "technological" in aim than in fact it was.

He soon sponsored MIT's ambitious Project MAC, launched in 1963. The acronym was spelt out sometimes as Man And Computer, sometimes as Machine-Aided Cognition, and sometimes as Multi-Access Computing—i.e. time-sharing. Project MAC's funding was exceptionally generous: $2 million as the start-up sum, and $3 million for each of the following years. The young AI Lab at MIT benefited to the tune of nearly $1 million. (All in early 1960s values.)

In addition, Licklider funded the first four US graduate programmes in AI—at MIT, CMU, Stanford, and SRI. For the next fifteen years, he continued to provide large amounts of money to these four groups, his attitude being that ARPA money should go to the people he knew to be excellent scientists (see 10.ii.a). Unlike the more 'academic' National Science Foundation (NSF), ARPA didn't have to function via peer review. In effect, this was an old-boy network from his days at MIT and Bolt, Beranek & Newman: the favoured few—notably, Minsky and McCarthy (and Newell and Simon)—were given the money and left to get on with it.

If Licklider retained his faith in the military as "the good guys", many members of the general public didn't. The military's respectability was soon undermined, in their eyes. This was one aspect of the counter-culture that emerged in the late 1960s, and which was strengthened by the psychological aftermath of warfare and US defeat in Vietnam (1.iii.c).

Although the US politicians stopped well short of wearing flowers in their hair, they did listen. In the mid- to late 1970s, there was less money available for R & D in America. MIT, for example, suffered financial difficulties across the whole institution as a result (see 14.vi.a).

To be sure, this situation was largely due to OPEC's sudden tripling of the oil price in 1973. Further price-hikes followed: in a ten-month period in 1974, the price of a barrel of oil rose 228 per cent (Boal *et al.* 2005: 14). But in addition, a greater proportion of the R & D tranche was socially directed. For a "new set of demands and values" was in play, in reaction against those which had previously driven technological development (Molina 1989: 24). An important part of the counter-cultural movement was:

[a] critique of the social consequences of unfettered technological development, ranging from the environmental damage caused by the side-effects of modern science-based production processes to the use of sophisticated electronics in the war in Vietnam. (Dickson 1984: 30)

As a result, defence expenditure declined (as a percentage of GNP, it dropped from 2.9 to 2.3—Molina 1989: 24). This meant, among other things, that there was less money available for AI research. A very cold wind was blowing: ARPA's cutbacks to AI funding in 1974 had several "two-million-dollar-a-year contracts cut to almost nothing" (Hans Moravec, interviewed in Crevier 1993: 117). And ARPA cut back its support of *basic* research still further, accordingly (*THES* 1975).

But that state of affairs didn't last. By the late 1970s, the "new set of values and demands" had been sidelined by the economic crisis that followed the recent oil shocks. One historian of technology says:

[The counter-cultural] demands for greater social responsibility and accountability in the development of science and technology were superseded. Such demands could not provide an impetus similar to war or international competition—either politico-military or economic. This period of socially-aware demands was a time of little growth for the R&D system. As soon as the spectre of the Vietnam defeat began to fade, replaced by alarm at the economic crisis, war and international competition began to reassert themselves and, with them, the same dominant social interests [namely, those of the scientific, corporate, and military elites]. (Molina 1989: 26; italics added)

By the early 1980s, then, American R & D in computing was rising steeply again, largely due to increased defence spending (see subsection c).

The AI community benefited, not least because part of the DOD's agenda was to advance US information technology *in general* so as to meet the Fifth Generation challenge (Molina 1989: 76–7; see v.a–b, below). The politicians were persuaded to authorize the vast sums involved partly because of fears that the USA's economy would be outstripped by Japan in the coming information age (Roland and Shiman 2002). The $4 million provided to NASA in 1985 for space-related AI (which they'd been supporting at a lower scale since 1982) was justified not only by national pride with respect to the space race, but also by the expectation of technological spin-offs. The law passed by Congress to authorize this stated that "the development of [advanced automation and robotics] shall be estimated to cost *no less than 10 per centum of the total space station cost*" (Montemerlo 1992: 51; italics added).

By the late 1980s, the annual budget for NASA's AI research programme had risen to $13 million, and it stayed at this level until at least 1992. But not all of this new money had come in response to the Fifth Generation. Nor was it attributable to Licklider's efforts. Rather, it was the result of a highly controversial political decision.

As such, it wasn't welcomed by the AI community as a whole—as we'll now see.

c. Star Wars and AI qualms

Some members of the AI community had been wary of accepting military money from the start. Wiener had stopped all his military work in 1946, at the end of the Second World War (so Theodore Roszak's impassioned attacks on him had been partly misguided: see 1.iii.c–d). Unlike Licklider, they didn't see the military in uncomplicated terms as "the good guys".

Vietnam provided further ethical complications. And the burgeoning cruise-missile technology added yet more. Benjamin Kuipers (1949–), for instance, had been funded as a graduate student by NSF and "didn't think much about military funding and AI research at that time" (Kuipers 2004). On graduating in 1974, he accepted one year's funding from DARPA before changing from AI vision to AI in medicine: for DARPA, the only agency willing to fund his vision work, wanted to use it for developing better-targeted cruise missiles.

Kuipers (a conscientious objector during the Vietnam war: see ii.e, below) may have been more scrupulous than most. But ten years later, the ambivalence had spread—and deepened. The AI profession of the mid-1980s was highly uneasy about the project then gobbling up most of the generous AI funding. This was the "Star Wars" programme, or Strategic Defense Initiative.

Star Wars was announced on national TV by President Reagan himself, in March 1983. Immediately, DARPA was allocated $600 million—later rising to $1 billion—over five years to develop the computing required (Roland and Shiman 2002). (The similarly named Strategic *Computing* Initiative had already been planned by DARPA, as its response to Japan's Fifth Generation project. In the event, SCI was swallowed up by SDI: Edwards 1996: 298–9.)

Most of this huge sum of money, DARPA decided, would be devoted to five specific aims, each involving AI at its core (Edwards 1996: 296). These were an intelligent image-processing system, for battlefield reconnaissance; two AI battlefield management systems; an autonomous land vehicle for use by the US Army; and an intelligent fighter pilot's assistant, which would possess 10,000 words of English and be able to understand the pilot's speech even in a noisy jet cockpit.

You may smile, realizing the many impracticalities being blithely ignored. But in the geopolitical circumstances of that time, it wasn't funny. The least funny aspect of all was that man–computer *symbiosis* had virtually disappeared. The goal of Star Wars (or so people were told: see below) was to provide an AI shield to defend the USA—if necessary, by triggering machine-made attacks. So Star Wars computers would not only detect an imminent Soviet attack but would also deploy an *automatic* nuclear response. (Hence the DARPA director's assurance to a US Senator that "we might have the technology so he couldn't make a mistake", where "he" was the President of the United States, faced with the awesome decisions required in a nuclear confrontation—Edwards 1996: 71.)

This removal of the human being from the loop caused explicit opposition from AI scientists around the world. And from many others too—including Hollywood directors, literary theorists, sociologists, and philosophers. For instance, the hugely successful 1984 film *The Terminator* played on people's fears of failures in early-warning systems, as well as their distrust of AI technology in general (Edwards 1996: 22–6). And Donna Haraway's monitory *Cyborg Manifesto* was written at much the same time (Chapter 1.i.b).

But one didn't have to be a counter-culturalist to be sceptical. The historian Paul Edwards states that Reagan "did not bother to consult Pentagon analysts before announcing the plan", but that "had he done so, *most would have refused to support it*" (1996: 289; italics added).

Computer scientists, of course, were in a special position to understand the technical problems. They knew only too well that a program consisting of 10 million lines of code (which was what the Star Wars directorate was proposing) couldn't, in practice, be bug-free.

What's more, they knew that even a bug-free program won't always do *precisely* what it was intended/expected to do. All the standard programming languages were ambiguous. This applied even to the supposedly mission-critical Ada (named after Ada Lovelace), which the DOD in 1975 had insisted be used for any real-world military application. That is, a given instruction in Ada (or any other source code) could be translated in different ways by different compilers when it was run. And *concurrent* execution was often ambiguous too, in that the programming language left it open for the compiler to choose which instruction should be followed first (E. Robinson, personal communication). In time-critical real-world applications, this might matter.

Later, a very carefully defined subset of Ada, called SPARK, would avoid ambiguity entirely (Chapman *et al.* 1994*a*,*b*; Chapman 2001). In the 1980s, however, it was rife. Admittedly, this rarely (if ever) caused problems in practice. But many felt that launching nuclear weapons wasn't a context in which to take that for granted.

Over and above those highly general points, AI professionals had by then learnt some lessons. The days of the "ten years" hype were (mostly) over. They knew that many of the AI problems assumed by Star Wars protagonists to be solvable were intractable in practice, and perhaps even in principle. These included real-world examples subject to the frame problem, and the (proposed) automatic programming of complex and bug-free code. Moreover, they'd learnt that AI successes could often be derived only by heuristic programming, not by success-guaranteed algorithms. Where an inappropriate decision could have such dire effects, the essentially fallible nature of heuristics was more than a little troubling.

As if all that weren't enough, several failures in early-warning systems *had already happened*. In one case, an unusual (so hard-to-spot) fault in a 46 cent computer chip had caused the North Atlantic Defence (NORAD) computer to report a Soviet attack, and 100 B-52 bombers had been readied for take-off (Edwards 1996: 285). In another, a major alert had been caused by someone's forgetting about leap years when programming the calendar, needed for predicting when and where the rising moon would appear in the sky (H. Thompson 1984). If even something so straightforward as the calendar could be misprogrammed, how much greater was the potential for mistake lurking in the arcane science and mathematics built into nuclear warning devices. In one eighteen-month period (after NORAD installed new computers), there'd been no fewer than 147 false alerts—four of which had moved the US defensive forces one step closer to a nuclear response.

The public concern was so great that the US House of Representatives held two days of hearings in May 1981 on "Failures of the NORAD Attack Warning System". These hearings "revealed a long and mostly secret history of spectacular failures in the computerized . . . Ballistic Missile Early Warning System" (Edwards 1996: 284).

All this explains why, when US Vice-President George Bush (senior) visited England in 1985, he was greeted by a media campaign criticizing the technical assumptions implicit in the Star Wars project. That campaign included a letter signed by seventy British computer professionals, including some mentioned elsewhere in this book. At the same time, the British Computer Society organized several meetings to look into the military implications of computing.

Bush didn't have to go abroad to encounter such resistance. There was plenty of it at home:

A pledge not to take SDIO [Star Wars] funding, in order not to support *the impossible fantasy* of effective nuclear defense, circulated widely on [US] university campuses. Thousands of scientists signed. (Edwards 1996: 289; italics added)

On both sides of the Atlantic, the recently founded CPSR (Computer Professionals for Social Responsibility) was one of many sources of professional criticism. CPSR gave strong backing, and publicity, to the computer scientist David Parnas (1941–). Besides running several pieces by/about him in their own newsletter, they sponsored a volume on *Computers in Battle: Will They Work?* (Bellin and Chapman 1987), in which one chapter was written by Parnas.

Parnas had been invited to give evidence on this question to the US Senate. In December 1985 he assured the senators that Star Wars *couldn't* be made to work. Moreover, he said, since mistakes were so likely and their effects so horrendous, it shouldn't even be attempted.

His evidence was especially weighty because he'd been deeply involved in defence-sponsored projects. He'd done eight years' study of real-time software to be used in military aircraft, and from 1979 to 1982 he was head of software engineering research for the US Navy. So far was he from having any *political* axe to grind in these matters, that in June 1985 he'd accepted the US government's invitation to join their top advisory group on the computing aspects of Star Wars (the SDI Committee on Computing in Support of Battle Management).

Soon after the committee's first meeting, however, he'd resigned. He published his letter of resignation (with eight brief explanatory appendices), which later formed the basis of his Senate testimony (Parnas 1985a,b, 1986, 1987; cf. D. B. Davis 1985). So Parnas's protest wasn't a principled anti-militaristic, or even anti-nuclear, one. Rather, it was pragmatic: war is sometimes necessary, and so are nuclear weapons—but *this* military strategy was just too dangerous to use.

(This all seems so obvious. So why didn't Reagan listen? Why didn't he halt Star Wars forthwith? Indeed, why had he suggested it in the first place?—One common answer is that its being "an impossible fantasy" didn't matter, because the primary aims weren't military at all. Rather, they were political and ideological: Reagan intended to bankrupt the Soviet Union, which would be bound to try to respond in kind even though it couldn't afford to do so. Whether it was his original intention or not, that's what happened. With the fall of the Berlin Wall in 1989, and the break-up of the USSR soon afterwards, the political "need" for Star Wars disappeared. The fear lessened—and, as one might expect, the Federal money available for AI lessened too.)

d. *Les mains sales?*

Given the insuperable technical difficulties detailed by Parnas and others, and the huge risks to humanity implied by the nuclear dimension, it was easy for AI scientists to disapprove of Star Wars. As in the case of Parnas himself, one didn't have to hold an especially subtle and/or left-wing moral–political position to want to have nothing to do with the project. And it was relatively easy—at least for those employed by universities, and/or those outside the USA—to refuse to accept grants for work that was directly, explicitly, related to Star Wars. But deciding on one's position with respect to military work *in general* was more tricky (Ladd 1987).

Some individuals accepted money only for military research that was relatively benign in intent. After all, autonomous vehicles could prevent human soldiers from getting killed, as bomb-disposal robots do. (Several decades later, the title chosen for the 2006 International Conference on Automation and Robotics, which dealt with "demining, search/rescue missions, homeland security … etc.", was 'Humanitarian Robotics'.) Some, such as Kuipers (a Quaker), refused to work on any military projects. However, others objected that for non-pacifists to do that was as squeamishly hypocritical as for meat-eaters to refuse to work in abattoirs. The British AI scientist Yorick Wilks declared that if our soldiers had to risk their lives on our behalf then he, safe at home in his laboratory, was glad to support them. And to refuse *any* military funding was to be cut off from the most lucrative grants (so one could employ only graduate-student assistants, not research fellows: Kuipers 2004). In short, AI people faced dilemmas like those facing the hero of Jean-Paul Sartre's *Les Mains Sales*.

One could be ambivalent about this, even in the midst of a terrible war. Zuse, for instance, had had mixed feelings about his work on flying bombs:

Technically speaking, our assignments were very interesting. The fact that the sophisticated weapons we were working on in the final analysis served war and death, was easily forgotten. And yet we all would have preferred working on civilian projects. (Zuse 1993: 60)

(Tom Lehrer might have taken that claim with a pinch of salt, for when Zuse's workplace was bombed in 1945 he spent a few weeks working with the Lehrer-immortalized rocket scientist Werner von Braun.)

AI researchers varied widely in their preparedness to accept/seek military funding. In Edinburgh, for instance, Bernard Meltzer was wholly against accepting grants from military sources whereas Donald Michie baulked only at doing *classified* research. MIT in the 1960s adopted a policy of doing no classified research on campus, although its special laboratories, such as the Draper and Lincoln labs, did do military work. (It wouldn't need a cynic to say that that's why they were founded in the first place: the Lincoln lab was set up in 1953–4 to design SAGE—Rheingold 2000: 142–3.) McCarthy and Edward Feigenbaum, by contrast, had no problem with accepting money for classified purposes (Fleck 1982: 206).

Feigenbaum even argued—both in a best-selling book and in evidence to a Congressional hearing—that the Fifth Generation project threatened the USA's military superiority, which the government should take urgent steps to preserve (Feigenbaum and McCorduck 1983: 215–20). He became a national leader in this area of applied AI (and today, specifically mentions it on his web site). He was also more up front than most in acknowledging the military's role in getting AI off the ground:

When no corporation or foundation chose to take AI seriously, or could afford to, the Advance [*sic*] Research Projects Agency (ARPA) of the Department of Defense supported it through two decades of absolutely vital but highly risky research. Since the Pentagon is often perceived as the national villain, especially by intellectuals, it's a pleasure to report that in one enlightened corner of it, human beings were betting taxpayers' money on projects that would have major benefits for the whole human race. (Feigenbaum and McCorduck 1983: 215)

And in urging an immediate US response to the Fifth Generation project, he said, "It is essential that the newest technological developments be made available to the Defense Department . . . [which] needs the ability to shape technology to conform to its needs in military systems" (p. 217). Perhaps it's not too surprising, then, that he eventually became chief scientist for the US Air Force.

His Stanford colleague Terry Winograd (who'd founded CPSR with Brian Cantwell Smith) had a very different attitude—and avoided military funding as a result. He couldn't deny that military research may have beneficial spin-offs. (In applying for government funds, DARPA had made a point of stressing the likely commercial benefits from Star Wars research.) But he was very uneasy at the extent to which *everyone* receiving DARPA money—including about 85 per cent of the AI research done at the four major centres (MIT, CMU, SRI, and Stanford)—was now fairly directly implicated in military research (Winograd 1991). For DARPA had announced in 1983 that even basic research now had to be related to one of the five military aims specified above. (The mid-1970s policy shift mentioned below had merely required that it be relevant to military use in general.)

AI researchers' consciences were troubled also by the fact that even seemingly innocent AI work could be used for military purposes. That's why ARPA, which wasn't engaged in the disinterested pursuit of truth, had been prepared to fund it in the first place. Their mandate had specified the responsibility to fund research which the DOD wouldn't otherwise have supported "because the feasibility or military values of the new capabilities were not apparent at the beginning" (Edwards 1996: 260).

Consider, for instance, NLP research on anaphora, such as the little—but decidedly tricky—word *it*. MIT's Eugene Charniak (1972, 1973, 1974) wrote a program to analyse stories about children's birthday parties. One sentence was *He will make you take it back*. How could this be interpreted? Suppose the gift-giver had just said *I put the toy train on the table*: for sure, the table (mentioned nearer to *it* than the train was) would not be what the birthday boy would want his guest to take back. But how was the program to know that? When ARPA turned more mission-oriented in the relatively cash-strapped mid-1970s, eschewing basic research in favour of projects with military relevance (*THES* 1975), some NLP researchers encountered funding difficulties for the first time. A friend told me that he'd solved the problem by resubmitting his proposal, now using examples like *The battleship was spotted North-East of the rock. The gunner aimed at it*. This man was no more concerned with battleships than Charniak had been with toy trains: both were interested only in *it*. However, anyone writing an NLP-based battle management system would need to be able to cope with *it*.—Morally, then, there were no easy answers.

Specifically military research proceeded, regardless. By the mid-1980s, an impressive—or depressing, depending on your point of view—roster of (non-secret) applications had been developed. Some of these had advanced AI *as such*. That's not surprising, for real-world problems in general provide a useful discipline to the programmer. And military applications had some specially challenging features:

Military operations . . . possess significant characteristics that have not always been prominent in other AI application domains. One such characteristic is the time-critical nature of tactical decision making—the need for appropriate, real-time response to dynamic situations. (J. Franklin *et al.* 1987: 605)

Typically, the "dynamic situations" weren't merely rapid (like the movement of a guided missile), but also highly complex. That is, they involved battles, not just single weapons. As a result, much military AI focused on increasing the efficiency of search and pruning techniques, and of *distributed* reasoning under uncertainty (see 13.iii.d)—where the data were not only uncertain, but also vast in quantity, multi-sourced, and used for different purposes by different types of personnel (J. Franklin *et al.* 1987: 606).

Among the many programs developed in this general area were one (called BATTLE) for deciding what weapons to send where, before and during a battle, and another (ANALYST) for integrating sensory information betraying the current location of enemy units (Slagle and Hamburger 1985; Bonasso 1984). Others were aimed at building (for example) "smart" bombs, mine-laying robots, and (non-intelligent) radio sensors. The latter were disguised as twigs, jungle plants, and animal droppings; and they were used for detecting movement, body warmth, the noise of truck engines, or even the scent of human urine (Edwards 1996: 3).

Clearly, then, the computer scientists of the 1980s weren't short of ingenuity. But political critics already saw them as morally compromised. Some of Chomsky's fiercest polemic, for instance, had been directed against the use of high-tech weapons on an undeveloped country halfway across the world (Chapter 9.vii.a). In his eyes, indeed, DARPA/RAND were doubly guilty. Not only had they invented these fearsome devices, but their game-theoretic strategists had advised that it was "rational" to use them (cf. 6.iii.a).

Today, twenty years later, robot planes and tanks are only too familiar. Indeed, the US Congress has approved a programme to make fully one-third of all US ground attack vehicles and deep-strike aircraft unmanned by 2010 (Almond 2005).

As for robot soldiers (i.e. robots used instead of soldiers), they're normally justified in terms of lessening the mortality rate of human soldiers—which, no doubt, they do. But one especially cynical (or especially honest?) military expert, the Director of America's "Global Security" research body, was recently quoted as saying this:

[Robots] will kill without pity or remorse. Historically, only about 1% of soldiers in real war are like that. They are sociopaths who these days are used as snipers. Half of the rest just spray bullets around and the other half never fire at all. (John Pike, quoted in Almond 2005: 21)

Given the cold bellicosity behind those words, it would be comforting to think that this newspaper quotation was as inaccurate as many of them are. But I wouldn't bet on it.

In sum, Archimedes and King Hieron live on: the link between AI and the military is as strong as ever. And still, expectations of AI are sometimes so unrealistic as to be risible. A US Army general in 1970, on being shown SHAKEY, asked whether a bayonet could be mounted on it—apparently assuming that it could skip across grassy meadows and crawl through ditches and tunnels . . . whereas in fact it could barely stand up (B. Raphael, personal communication). Three decades later, a US-government employee asked an AI researcher whether he could model the emotions of terrorists—something that's way beyond the state of the art (although the other AI workers he'd approached hadn't admitted this: see Chapter 7.i.f). Any opportunity to put AI to martial use, however fanciful, is eagerly explored. *Plus ça change, plus c'est la même chose.*

11.ii. Critics and Calumnies

Alongside the bombs, there were the bombshells. Between the mid-1960s and mid-1970s three waves of intellectual criticism washed over NewFAI, and the cognitive science projects associated with it. These were wide-ranging and widely disseminated critiques, whose influence is still strong today (see Chapters 13.ii and 16.vii).

The first, and most famous, came from an outsider: the phenomenological philosopher Hubert Dreyfus (1965, 1972). The next best known came from a neighbour: the computer scientist Joseph Weizenbaum (1976). And the third came from a highly respected AI insider (see Section iii, below).

Dreyfus's charge was twofold: that GOFAI wasn't succeeding, and that it never could. Weizenbaum's was that even if it could appear to succeed (which he, too, doubted), it shouldn't be used—at least, not in 'human' contexts.

Criticism, and the debate it engenders, isn't always good-tempered. These two critics repeatedly hurled calumnies at their opponents, who were tempted to answer in kind. The more passionate the writing, the less effective it was in changing the minds of the enemy, i.e. NewFAI-influenced cognitive scientists. Points of criticism that were both just and important received less attention than they deserved.

Since many supporters of Dreyfus and/or Weizenbaum were members of the general public already disposed to dismiss AI, these heartfelt criticisms spread like wildfire. The 1970s critique of GOFAI was no mere intellectual exercise, confined to the academy: it was a passion-ridden cultural phenomenon too.

a. The outsider

Hubert Dreyfus's first extended critique of AI bore the RAND Corporation imprimatur, and he was employed by MIT at the time, a formative point in his career (he was born in 1929). Nevertheless, he was an outsider.

He was a professional philosopher (with a first degree in physics), not an AI researcher. In the mid-1960s he was teaching the subject at MIT, but he soon moved to Berkeley's Philosophy Department (alongside John Searle). Even more to the point, his philosophy—a form of phenomenology taken from Martin Heidegger, Maurice Merleau-Ponty, the later Ludwig Wittgenstein, and Michael Polanyi (see 16.vi–vii)—was deeply antipathetic to GOFAI.

The RAND connection was due to his brother Stuart, a mathematician working there (and at MIT) at the time. He'd persuaded Armer, then head of RAND's Computer Science research, to invite Hubert to visit their Santa Monica centre for two weeks in the summer of 1964 (S. E. Dreyfus, personal communication). This sojourn on the Pacific coast would enable him to observe the AI work being done there by Newell, with a regular visitor from the other side of the continent—namely, Simon (Chapter 6.iii.b–c). So far as the siblings were concerned, he was going there to meet the enemy at close range.

It wasn't a friendly meeting, and it didn't lead to a truce. On the contrary, after his fortnight sojourn at RAND Hubert wrote a stinging ninety-page critique of GOFAI, based on a talk he'd given there in August 1964. Provocatively, he called it *Alchemy and Artificial Intelligence* (H. L. Dreyfus 1965).

Alchemy argued two different, though closely related, points. On the one hand, it declared the overall project to be in principle impossible for philosophical reasons—in essence, because the "higher" forms of intelligence are *necessarily* derived from "lower" forms concerned with bodily action (this claim is discussed in Chapter 16.vii.a). On the other hand, it mocked the performance of the programs that had actually appeared thus far.

Dreyfus accused NewFAI of four general performance failings, each missing out some "essential" aspect of human intelligence. These were reliance on the fringe of consciousness; discrimination between the essential and the accidental; tolerance of ambiguity; and perspicuous grouping. The lack of those aspects, he said, resulted in the two glaring weaknesses of NewFAI: crudity and brittleness.

Some types of intelligence, he allowed, were in principle programmable: "associationist" and "simple formal" thinking. But "complex formal" and "non-formal" intelligence weren't. The latter types included chess and "intuitive" theorem proving, and riddles and translation. Whatever successes NewFAI had achieved so far, and might achieve in future, lay within the first two classes of problem. These could be tackled by formalized "counting out" and "trial and error". Heuristic search wasn't a general answer, because AI heuristics were derived by human insight—and (in people) heuristics often need to be applied insightfully, too.

Moreover, he said, intelligence *did not* constitute a "continuum", as assumed by Armer and by Feigenbaum and his co-editor Julian Feldman (see the quotes given in 10.ii.b). On the contrary, there were fundamental qualitative differences between the four classes. In any event, it was likely that "the body plays a crucial role in making possible intelligent behavior" (H. L. Dreyfus 1965: 59). If so, then computers simply couldn't qualify.

One of the specific criticisms he made was that GPS relied crucially on its programmers' human ability to distinguish between the essential and the accidental, in choosing how to represent the problem. Another pointed out that pattern recognition programs belied the fact that human concepts are networks of Wittgensteinian family resemblances, not definable by necessary and sufficient conditions. Yet another rejected Rosenblatt's vision of a mechanical secretary: no one knew how to begin making such a device, Dreyfus said, because acoustically identical speech signals are heard as different phonemes depending on the "global" factor of expected meaning (but see 9.xi.g). Some others were directed at MT in general, Yehoshua Bar-Hillel being repeatedly quoted.

In essence, these criticisms weren't new. Quite apart from the by-then-familiar complaints about MT (9.x.d–f), the relevant features of human thinking had been stressed also by the Gestalt psychologists. Merleau-Ponty (one of Dreyfus's three major influences) had borrowed extensively from the Gestaltists, and had also highlighted the importance of having a *body* (16.vii.a; cf. H. L. Dreyfus 1967).

These aspects of thought had even been noted by the AI researchers themselves. Newell and Simon owed much to the Gestaltists (6.iii.c). As for the frame problem, which exemplified the 'essential/accidental' difficulty, this had been pointed out by NewFAI people in the 1950s and named by them in the 1960s. Moreover, ambiguity was being confronted (in as yet unpublished work) by Yorick Wilks (9.x.d), and perspicuous grouping by Adolfo Guzman (10.iv.b).

But although Dreyfus failed to acknowledge the extent to which NewFAI had already *recognized* such issues, his main claim was correct: attempts to *deal with* them, when they weren't simply ignored, had been crude. This, he said, wasn't a question of AI's being a very young science. Rather, the crudity was inevitable, because the last two of his four "classes" of intelligence were immune to AI methods. Famously, he compared NewFAI claims about progress to a man climbing a tree in order to reach the moon (1965: 17, 86). The climber would move a tad nearer to his goal, to be sure. But clambering up even the Californian redwoods wouldn't ever get him there.

Besides criticizing specific programs, Dreyfus scornfully quoted various wildly optimistic predictions made by NewFAI enthusiasts. Chief among these was Simon's forecast, in a talk given in 1957, that a computer would beat the world champion at chess within ten years (Simon and Newell 1958: 7). The prospects of that happening were already dim when he wrote *Alchemy*:

As described in their classic paper, [Newell, Shaw, and Simon (1958*b*) admitted that] their program was "not yet fully debugged" . . .

In fact, in its few recorded games, the NSS program played poor but legal chess, and in its last official bout (October 1960) was beaten in 35 moves by a ten-year-old novice (xxxi). Fact, however, had ceased to be relevant. [Their] claims concerning their still bugged program had launched the chess machine into the realm of scientific mythology. (H. L. Dreyfus 1965: 6)

The notorious chess forecast was one of many examples of hype in AI (both symbolic and connectionist) that were used by opponents to damage the field. Simon himself produced another gem a few years later: the hubristic "ten years" had been doubled to a more modest twenty, but the substance of the prediction had multiplied a thousandfold—"machines will be capable, within twenty years, of doing any work that a man can do" (Simon 1965: 96). Even if "any work" actually meant only office work (he was writing about his old love, management), this was optimism gone wild.

Dreyfus wasn't the first opponent to mock such claims made by AI researchers. Bar-Hillel, for instance, had preceded him (9.x.b. and e). Nor was he the last: Sir James Lighthill (iv.a, below) would target several of them.

There were plenty of less highly publicized criticisms over the years, of course. There's some evidence, for instance, that DARPA killed their programme on speech understanding in the mid-1970s not simply because the allotted five years had ended, but because they felt let down—even deceived—by the fact that the grammar used when speaking to the system had to be very highly constrained. I'm tempted to say that this should have been obvious to them right from the start (although SHRLDU may have led them to expect better), except that the research director himself has protested,

But nobody had ever said anything about not constraining the grammar! Nobody even understood that that was a parameter that could be adjusted at the beginning of the project! (Raj Reddy, interviewed in Crevier 1993: 116)

That example reminds us that there are two types of hype. In honest hype, the optimists themselves are misled. In irresponsible or dishonest hype, they know that they're exaggerating (see iii.b, below). Dreyfus's arrows of contempt were aimed primarily at the first.

b. Scandal

Alchemy was no dry intellectual critique, but a passionate attack on a project perceived as the enemy. And it engendered responses in kind.

Dreyfus's mockery of the 1957 chess hype soon came back to plague him. In 1966, only a year after *Alchemy*, he himself was beaten by Richard Greenblatt's MacHack program in a match organized by Seymour Papert. (Greenblatt was still an MIT undergraduate when he wrote the program—in response to Dreyfus's earlier remarks.) The match was reported in the *New Yorker*, and in less exalted publications too.

Dreyfus's defence—that he hadn't claimed that no program would ever play a competent game, and that he was a "rank amateur" anyway—was true enough (1972, pp. xxxii–xxiii, 223 n. 45). (Legends are long-lived: he's still falsely accused of making "the blunt assertion that no computer program would be able to play a good enough game of chess to beat a ten-year-old"—Levy 1994: 89.) But he *had* said that any program play would be inhuman and mechanical. And Simon—not a disinterested observer, of course—described the 1966 game as "wonderful" and "a cliffhanger" (McCorduck 1979: 199).

So, in the eyes of the NewFAI community, Dreyfus had egg on his face. The incident delighted them, and prompted gleeful comments and witticisms in several issues of the SIGART newsletter.

Their satisfaction wasn't drily professional/technical, but all too human. Most of SIGART's readers loathed Dreyfus. That's hardly surprising, for the feeling was mutual. His contempt for their activities was unplumbed. Even his choice of title was insulting, intended (as his text made clear) as *Alchemy alias Artificial Intelligence*. Significantly, he had nothing good to say about alchemy. Years later, he'd mellowed enough to agree with Winograd's (1977) remark that "it was the practical experience and curiosity of the alchemists which provided the wealth of data from which a scientific theory of chemistry could be developed" (H. L. Dreyfus 1979: 2). His RAND fireball, however, *didn't* allow this. According to what he said there, chemistry had been substituted for alchemy—not developed from it.

The way in which Dreyfus had presented his arguments was at least as important in triggering the ensuing scandal (not too strong a word) as were the arguments themselves. This cat wasn't merely stalking the pigeons, but teasing and sneering at them unmercifully. The sharpness of his tongue is illustrated by this remark:

This output of confusion [i.e. various NewFAI predictions about chess machines] makes one think of the French mythical beast which is supposed to secrete the fog necessary for its own respiration. (1965: 8)

NewFAI researchers, and people doing "cognitive simulation", were described (in Dreyfus's words) as credulous, smug, unscientific, casual, dogmatic, and naive. They were accused (again, Dreyfus's words) of self-delusion, methodological confusion, and absurdity; of spreading intellectual smog; and of preferring adventure to patience.

Newell and Simon (especially Simon) got the worst of it. Besides the epithets just quoted, they were said to show "either a will to obscure the issues or a total misunderstanding" of the Gestaltists' position, and to have "surreptitiously" introduced their own insight into their programs. And Simon's hugely over-optimistic chess prediction of 1957 was mocked over and over again. There were only two people for whom Dreyfus had a good word to say: "Only Shannon", he declared, "seems to be aware of the true dimensions of the problem" of ambiguity in language and perception (p. 75). And only Donald MacKay, he said, had seen that *analogue* processing, and/or "wet" engineering, might be needed for successful AI (cf. 4.v.b and 12.ix.b).

The mockery—and the philosophy too—alienated Armer, already an influential champion of AI (see 10.i.f). (Years later, he said that if he'd known about the brothers' 1962 attack on AI he wouldn't have invited Hubert to RAND in the first place—McCorduck 1979: 194.)

He wanted to suppress Dreyfus's paper, and had "a big squabble" with some RAND colleagues who liked it (interview in McCorduck 1979: 195). It's pretty certain that they weren't paid-up phenomenologists. So perhaps they liked it largely because they were 'pure' computer scientists, who saw AI as a betrayal of the proper standards of theoretical rigour (see iii.a, below). Or perhaps they were merely speaking out for academic freedom: let a hundred flowers bloom, and so on. (Unlikely: academic freedom wasn't high on RAND's agenda, and many of its reports were classified.) In any event, Armer faced a mini-rebellion in the RAND ranks.

After about nine months of this—while rumours about the draft spread from the Pacific to the East Coast, and across the Atlantic too (I remember being all agog to see it)—Armer allowed it to come out as a RAND memo in December 1965. This mimeographed paper was eventually followed by a properly printed version, in 1967. A few changes had been made, and some abusive remarks removed.

But by that time, of course, the first—venomous—version had been widely circulated. Indeed, it had already achieved a mention in the *New Yorker's* Talk of the Town column of 11 June 1966, which repeated Dreyfus's scornful remarks on the weakness of computer chess. Naturally, the AI community weren't pleased. Simon, the chief target of attack, was even more incensed by Armer's decision to publish than by the journalist's high-profile follow-up. RAND's imprimatur, he said, gave Dreyfus a credibility which was "really false pretences" (McCorduck 1979: 194).

Alchemy is usually thought of as Dreyfus's first public assault on the field. In fact, the earliest was an attack—co-authored with his brother—prompted by a 1961 meeting convened by MIT's School of Industrial Management to celebrate the Institute's Centennial Year. Among the interdisciplinary delights in that conference were an opening speech on scientists and policy making by C. P. Snow, and a paper on digital libraries that would have made Gabriel Naudé's hair stand on end (Kemeny *et al.* 1962). But NewFAI was featured too.

Besides a paper on EPAM by Simon and Feigenbaum, there was a general talk on AI given by John Pierce (of ALPAC fame: 9.x.e). This was followed by invited comments from Claude Shannon and Walter Rosenblith (introduced by Vannevar Bush), and then a General Discussion. Not liking what they heard, the two brothers wrote a note that was eventually published as part of the official record (H. L. Dreyfus and Dreyfus 1962). Ink was already being mixed with pepper: AI wasn't merely opposed, but scorned. (What later became the tree-to-moon argument appeared there as mountain-to-moon.)

However, an editor's footnote (on p. 321) admitted that their remarks "were not made at the Pierce session itself, but were submitted in writing a short time thereafter". Simon was enraged by that too, as he admitted later to Pamela McCorduck:

It was not a discussion that took place during the meeting, it was an afterthought they had. And it was a nasty little diatribe about this preposterous stuff that was being peddled. (interview in McCorduck 1979: 193)

His rage would increase on publication of *Alchemy*, wherein the pepper had grown even hotter.

The brothers Dreyfus weren't the first to attack NewFAI, and the nascent cognitive science associated with it, in strongly emotional terms. As we saw in Chapter 9.x.f, Mortimer Taube (1910–65) had done this a few years before, in his book *Computers and Common Sense: The Myth of Thinking Machines* (1961). And the ratio of abuse to factual/analytic discussion had been even higher there. The psychologist Walter Reitman had commented:

This book is the work of an angry man.... [Taube] concludes that an uninformed, science-worshipping public is being deceived, hoodwinked, and bilked of millions of dollars by electrical engineers and computer enthusiasts. (Reitman 1962: 718)

Reitman wasn't imagining things, for Taube's disdain near-scorched the pages. His remark that Turing himself displayed "the tendency of computer experts to be pontifical about subjects in which they have no competence" was a relatively mild example (Taube 1961: 51). He'd accused MT researchers of "writing science-fiction to titillate the public and to make an easy dollar or a synthetic reputation". And he'd specifically said that "the nation's monetary resources" were being wasted when used to support AI research.

Armer had bridled then, too—not least, because Taube's book led to a "[negative] climate of research in the field" (Armer 1960/1963: 389). Nor was he the only one. Besides the hostile review penned by Reitman, another attack was mounted by Richard Laing (1962), then a member of Arthur Burks's team at the University of Michigan and doing work on neural networks and automata theory (1961*a,b*). (Like several others in

Burks's group, he was later recognized as an early worker in A-Life: see Chapter 15.v.b, and Richard Laing 1975, 1977, 1989.)

Stuart Dreyfus (1962), by contrast, had approved of the book. (Simon "blew up at me in his office" at RAND, as a result: S. Dreyfus, personal communication.) In a review written for a friend's literary magazine, he did distance himself from it in some degree, saying that "not all of Dr. Taube's arguments are convincing" and allowing that "some innocent researchers are doubtless encompassed by some of Taube's sweeping condemnations". However, he described chess-playing machines as "a haven for frauds for centuries". (True, but hardly tactful.) And he continued:

[The] state of the artificial intelligence art has been grossly exaggerated. It is further clear that the natural desire of the bored public for believable science fiction, of the newspapers for sales, of computer companies for publicity, and of hard-working scientists for adulation contribute to this systematic delusion. (S. E. Dreyfus 1962: 54)

He poured scorn on the "multitudinous, disorganized, helter-skelter projects" of AI, and on LT/GPS (and Simon's chess prediction) in particular. And he described Taube's volume as the first book on AI childcare, remarking: "Taube may not be a wise or sympathetic parent, but at least he is willing to stand up to the pampered child."

One sentence was prophetic. He didn't have a "non-existence proof" of intelligent machines, but: "Conceivably, someday, a philosophic study of positive attributes of intelligence might [provide one]." In short, his and (still more) his brother's major intellectual project was already being flagged.

If even Stuart Dreyfus remarked on Taube's sweeping condemnations, and his lack of wisdom and sympathy, it's not surprising that the book cut little ice with the AI community. Despite his international eminence, Taube's over-the-top rhetoric led most of them to overlook his criticisms. The brothers Dreyfus each had a quiverful of insults too, as we've seen. Naturally, these were resented by NewFAI folk:

[Dreyfus's attack on AI contains] a set of almost defamatory personal allegations. (Papert 1968: 0–8)

Dreyfus' mission does not end with showing that people associated with Artificial Intelligence are wrong or even foolish. He is called to expose them as obscurantists and liars. (III–1)

The infamous RAND memo, then, expressed—and aroused—so much antagonism that it wasn't taken as seriously by NewFAI people as it might have been. (And, at that time, its author had no international eminence to command respect.) AI-leaning philosophers, too, were largely unimpressed. That's evident in Daniel Dennett's "little scalpel job" (personal communication) of 1968 (his first publication), and in my own critiques as well (Boden 1972: 146–50, 222–3; 1977: 435–41).

The situation wasn't helped by the fact that Dreyfus had made many mistakes, and even some disingenuous remarks—and omissions. The errors were ruefully acknowledged even by Weizenbaum, who shared many of Dreyfus's worries about the AI project: "It was terribly incompetent. [Dreyfus] knew so little about computers, and made so many mistakes!" (interview in Crevier 1993: 123). But the disingenuousness was even worse than the errors, and both were noted by Papert (1968), in a paper mischievously called 'The Artificial Intelligence of Hubert L. Dreyfus: A Budget of Fallacies'.

Papert being the AI star that he was, and Dreyfus already a notorious intellectual gadfly, this was a high-visibility contest. And likely to be a fierce one, too. Papert's flair for combativeness was evident from the tone of the already well-known draft of *Perceptrons* (see 12.iii.b). He would pull no punches.

And so it turned out. Papert accused Dreyfus of "questionable honesty". He showed, for example, that Dreyfus had deliberately omitted a brief sentence from a passage by W. Ross Ashby (about Herbert Gelernter's program: 10.i.c) which would have made nonsense of his ill-informed criticism (Papert 1968: III-8–9). Similarly, he'd stopped short of quoting a sentence from Wiener which confessed to a huge weakness of chess programs (III-4). And he'd claimed that Simon said in 1962 that his 1957 chess prediction was "almost realized", whereas Simon had said no such thing.

As well as bad faith, there was a failure to do his homework:

The most astonishing feature of Dreyfus' texts [the two versions of *Alchemy*] is their bibliography. His references to experiments on Artificial Intelligence are almost entirely confined to early work that has filtered into anthologies of "classical papers" [i.e. *Computers and Thought*]. (Papert 1968: 0–11)

One of the missing items was Guzman's work, which though still unpublished was in the air, and being done by "people who live within ten miles of him" (II-9). Dreyfus's ignorance of the literature was compounded by his ignorance of what the literature was trying to achieve. On that topic, *A Budget of Fallacies* outdid even *Alchemy* in its level of calumny. For instance:

[His] discussion is irresponsible. His facts are almost always wrong; his insight into programming is so poor that he classifies as impossible programs a beginner could write . . . (Papert 1968: 0–2)

[His] comments on actual projects [show] that he systematically misunderstands their purpose, their methods, and their difficulties. The reason is simple. He knows nothing about the technical issues and barely understands the language used. In addition he is sufficiently suggestible to take people as meaning what he thinks they must. (0–7–8)

[He ludicrously alleges that Ashby misled people.] That "naive" readers might be misled is undeniable. But scientific writing is not addressed to the ignorant and the naive (even if they are professors of philosophy). (III-9)

(*Ouch!*)

If *Alchemy* was as bad as this, why bother with it? Surely Papert, as one of the half-dozen top people in AI at the time, had better things to do? He answered:

I have been told that it is a waste of time [to rebut *Alchemy* in detail]. I have been told that only a pedant would object to the technical nonsense that pervades every paragraph of Dreyfus' papers about Artificial Intelligence since his real purpose is to provide insight into the rich subtlety of human intelligence. I have been told that his arguments must be read as literary conceits with rich "humanistic" content.

I think it does matter. I sympathize with "humanists" who fear that technical developments threaten our social structure, our traditional image of ourselves and our cultural values. But there is a vastly greater danger in abandoning the tradition of intellectually responsible and informed enquiry . . . The steady encroachment of the computer must be *faced*. It is cowardice to respond by filling "humanities" departments with "phenomenologists" who assure us that the computer [cannot encroach further] into areas of activity they regard as "uniquely human".

   Our culture is indeed in a desperately critical condition if its values must be defended by allowing muddled thinking to depose academic integrity. (Papert 1968: 0–2–3)

The protest about "filling" humanities departments with phenomenologists was aimed at the postmodern turn in Western society: *les événements* were already spreading far beyond Paris (cf. 1.iii.c–d, 6.i.d, and 8.ii.b–c). As we'll see below (subsection e), had Dreyfus written his RAND squib a decade earlier Papert might not have been so concerned to demolish it.

   Papert could have given a more literal answer to the question "Why bother?" He could have said, "I wrote it because Armer asked me to." As it turned out, Armer was disappointed: RAND's lawyers forbade publication for fear of slander suits (McCorduck 1979: 196). MIT's lawyers were less nervous, or perhaps they weren't consulted. Papert's paper, after circulating widely in draft, eventually appeared (still unfinished) as a Project-MAC Report.

c. After Alchemy

RAND Corporation memos aren't found in airport bookstores—although *Alchemy*, despite Armer's attempt at censorship, was the most widely read RAND memo of all time. But Dreyfus soon started work on a book-length statement. On its publication in 1972, it became a best-seller overnight. Many members of the general public got their first extended vision of AI from Dreyfus's book.

   What Computers Can't Do (1972) contained much more philosophical discussion than *Alchemy*, and was graced by a Preface written by Harvard's Anthony Oettinger (1929–), a well-known critic of MT. But the technical aspects were virtually unchanged. Indeed, most of the text of *Alchemy* was reproduced verbatim. (This explains why Dreyfus remarked that "the funny thing is, there was practically no additional response when the book was published"—McCorduck 1979: 195.) Some extra examples were added, to be sure; for instance, he pointed out (contra McCarthy) that being "at home" isn't simply a matter of being in a particular physical location but of owning (or renting . . .) the real estate in question—which can't be defined in logicist terms (1972: 150). But McCarthy's logicism wasn't new.

   In the book's discussion of 'Ten Years of Research in Artificial Intelligence (1957–1967) [*sic*]', the AI programs that had been written since the original RAND squib were almost all ignored—even those trying to address the very problems which Dreyfus had highlighted in the mid-1960s. Apart from McCarthy and Newell–Simon, Dreyfus focused only on MIT. (Even so, Winograd, whose work had electrified the AI grapevine and was about to electrify the wider world, wasn't mentioned. This may have been because SHRDLU wasn't featured in Minsky's *Semantic Information Processing*, from which Dreyfus drew heavily.) AI work had been going on elsewhere too—but from his new pages, you'd never think so. As Bruce Buchanan put it:

One can only speculate why the author fails to acknowledge recent AI work. To this reviewer, and other persons doing AI research, programs developed in the last five years seem to outperform programs written in the tool-building period of 1957–1967 . . . One would hope that a criticism of a growing discipline would mention work in the most recent one-third of the years of work. (Buchanan 1972/1973)

The anti-GOFAI rhetoric, too, was as virulent as before. The ever-courteous Buchanan chided Dreyfus gently, but firmly:

It is lamentable that the critique of AI in this book has taken the form of a popular-press attack on AI work. The author's phrases are damning but his arguments are not convincing. As the author mentions, the popular press has often given over-enthusiastic impressions of AI work; this book is written in the same vein, but with a negative sign. (Buchanan 1973)

A few leading GOFAI workers admitted that he'd said some things worth saying, and the GOFAI psychologist Zenon Pylyshyn (1974), while rebutting him at length, did so too. For example, Wilks (1976) allowed that "the naive, and from this distance in time, absurd over-optimism of many in early AI is now sad reading", and that—as Dreyfus had insisted—what Wittgenstein called the human "form of life" must underlie/inform natural language. He did say: "Dreyfus's empirical arguments are not sound, they disprove nothing." But he added:

Whether or not its survey of research is fair, or its arguments are sound, the book has done a great deal of good in a field whose lack of self-generated criticism is scandalous. Dreyfus should review the field every five years and change his empirical arguments each time—AI workers should then be grateful and ask no more of him, for he would be doing them a considerable service. (Wilks 1976: 184)

In effect, Wilks got his wish. For Dreyfus's critique would run and run. Quite apart from other books and papers attacking AI (e.g. H. L. Dreyfus and Dreyfus 1986, 1988), *the* book turned out to have many lives. (It's still being reprinted roughly once a year.)

Dreyfus's book lived on in other outsiders' writings, too. Most of these authors were in his home discipline, philosophy (Chapter 16.vii). But a few had studied the psychology of human skills, and/or had experience—if only as observers—of GOFAI in practice.

One person inspired by him was Harry Collins, whose late-century critique of expert systems will be outlined in Chapter 13.ii.b. Another, who was initially inspired by Heidegger but who later took Dreyfus's work willingly on board, was David Sudnow (1978/2001). His subtle descriptions of the phenomenology of learning a motor skill, namely piano playing, were mentioned in Chapter 7.vi.h. In general, the descriptions of skills offered by the Dreyfus brothers and their followers often provided interesting data. What was missing, and what's required by cognitive science, was an *explanation* of those data. (In the last quarter-century, some such explanations were attempted: see 14.viii.b, 15.vii–viii, and 16.vii.)

An unrepentant second edition of *What Computers Can't Do* appeared in 1979, consisting of the first version unchanged but with an added Introduction, in which some post-1967 AI work featured. Winograd—who hadn't yet come out as a lapsed GOFAIer (see subsection g, below)—was now discussed, with respect to both SHRDLU and KRL (pp. 5–15, 48–55). Scene analysis and robotics made an appearance (pp. 15–25). And frames, conceptual-dependency theory, and KRL were explored in relation to early to mid-1970s research on knowledge representation (pp. 27–55)—see Chapters 9.xi.d and 10.iii. But the verdict was unchanged: all were declared valueless.

(Dreyfus, like Licklider and Bar-Hillel, had predicted from the beginning that a "symbiosis" of man and computer might achieve more than unaided human intelligence can—1965: 83. Now, he admitted that specialist applications such as DENDRAL were useful, but insisted that these had nothing to do with general intelligence—1979: 5.)

Dreyfus's ink was still peppery. Citing an MIT report that had described research on flexible motor action by robots as "somewhat dormant", he tartly remarked, "[We] can only take 'dormant' as a polite synonym for stagnant or even comatose" (1979: 25). And having praised Winograd for being "admirably cautious" in his claims about modelling language as a whole, he commented drily:

[Everyone] interested in *the philosophical project of cognitive science* will be watching to see if Winograd and company can produce a moodless, disembodied, concernless, already adult surrogate for our slowly acquired situated understanding. (1979: 55; italics added)

In brief, cognitive science was chasing a philosophical mirage. GOFAI programs were "not at all promising as contributions to psychology" (p. 18).

This new edition opened with a mini-history of the field, which boiled down to *I told you so!* During the previous decade, he said, his critique "has been more or less acknowledged", and "the wishful rhetoric characteristic of the field has been recognized and ridiculed by AI workers themselves [specifically, McDermott: see Section iii below]" (Dreyfus 1979: 1). In sum:

Almost everyone now agrees that representing and organizing commonsense knowledge is incredibly difficult, and that facing up to this problem constitutes the moment of truth for AI. (1979: 3)

d. Dreyfus and connectionism

If 1979 was "the moment of truth" for GOFAI, it was also the moment for the first multidisciplinary meeting on distributed connectionism (Chapter 12.v.b). If Dreyfus knew anything of the work reported there, he didn't mention it in his new edition.

Indeed, in a Panel Discussion held in New York five years later, he declared:

The ability that people have . . . to have the right thing pop into their head . . . requires having images and having memories since it involves seeing the current situation as resembling earlier situations, where resembling is a tricky notion because resembling doesn't mean identical with respect to any particular features, *which is the way machines always have to analyze resemblance,* but simply overall similarity. (Pagels *et al.* 1984: 341; italics added)

He was right about the "trickiness" of the notion of resemblance. But if he'd known anything about PDP connectionism at that time (which in all conscience he should have done), he would never have said that machines "always" have to analyse it as sharing identical features.

Two years after that, it became abundantly clear, even to outside observers, that GOFAI was no longer the only AI game in town (12.vi.a–c). The American Academy of Arts and Sciences therefore organized a wide-ranging debate on AI. The brothers Dreyfus contributed a paper giving guarded approval to connectionism, and saying of GOFAI: "The rationalist tradition [in philosophy] had finally been put to an empirical test, and it had failed" (H. L. Dreyfus and Dreyfus 1988: 34; cf. Dreyfus 1972: 215). The editor of the trade book that carried this debate evidently agreed with them, for it bore the tendentious subtitle *False Starts, Real Foundations* (Graubard 1988).

Accordingly, Dreyfus started work on another 'revision' of his book, defiantly called *What Computers STILL Can't Do* (1992). Like the earlier editions, this attracted a great deal of attention. The journal *Artificial Intelligence* devoted almost an entire number to it, including reviews by McCarthy, John Haugeland, and Collins (among others), and Dreyfus's reply (Stefik and Smoliar 1996).

Dreyfus's new Introduction provided some predictable mockery of Douglas Lenat's CYC project (1992, pp. xvi–xxx; see Chapter 13.i.iii), and an equally predictable salutation to the now reconstructed Winograd and to Philip Agre and Lucy Chapman (pp. xxxi–xxxiii: see 9.xi.b and 13.iii.b). Mainly, however, it focused on PDP connectionism. Again, there came the *I told you so*:

In retrospect, the stages of my critique of attempts to use computers as physical symbol systems [which a connectionist system is not] to simulate intelligence now fell into place. My early appeal to holism, my concern with commonsense understanding as know-how, Stuart's phenomenology of everyday skills, and the capacities of simulated neural networks all added up to a coherent position—one that predicted and explained why GOFAI research should degenerate [*sic*] just as it had. (1992, p. xv)

("Degenerate", because GOFAI was said to be "a paradigm case of what philosophers of science call a degenerating research program": p. ix.)

But Dreyfus's new essay was less a matter of *Three cheers for connectionism!* than of *So much the worse for AI!* AI of some (non-GOFAI) sort, he said, might not be impossible: "no one has been able to come up with such a negative proof" (p. ix). But the prospects didn't look rosy.

Dreyfus admitted that neural networks were superior to GOFAI in some ways (1992, pp. xiv–xxxvi; cf. H. L. Dreyfus and Dreyfus 1988). For instance, they allowed for family resemblances in concepts (12.x.b). They showed that skilled behaviour needn't involve abstraction of a *theory* of the domain (12.vi.e). And they demonstrated that learning could take place without all the relevant 'beliefs' being explicitly represented beforehand (12.vi.c).

But, he continued, "The commonsense-knowledge problem resurfaces in this work and threatens its progress just as it did work in GOFAI" (1992, p. xxxvi). So neural networks can make inappropriate generalizations, such as learning to recognize the presence/absence of shadows instead of tanks; and they'd probably "stupidly" fail to learn *our* generalizations, or to adopt our priorities, such as symmetry (cf. Chapter 8.iv.b). Perhaps only a neural net with cell numbers, architecture, and initial connectivities identical to the human brain could ever model our intelligence.

As for learning, supervised PDP learning is dependent on the human trainer. Even unsupervised (reinforcement) learning, if the network were to be intelligent like us, would require—he said—that it have needs and perspectives like ours, and a comparable sense of relevance. And, of course, artificial neural networks lack bodies, just as GOFAI programs (and robots) do.

In short, he wasn't hopeful: "it looks likely that the neglected and then revived connectionist approach is merely getting its deserved chance to fail" (p. xxxviii). Even technological AI, if aimed at implementing *general* intelligence, was futile. As for psychological AI, this was a waste of time.

e. The neighbour

Weizenbaum (1923–) was a neighbour not only in the sense that he was located in MIT's Computer Science Department, but also in the sense that his ELIZA program was regarded by the public as an example of NewFAI. That was a mistake, for ELIZA hadn't been an effort in AI (see 10.iii.a). However, this wasn't widely known. The publicity afforded to his mid-1970s attack on AI was therefore all the greater.

Weizenbaum's book *Computer Power and Human Reason: From Judgment to Calculation* (1976) was memorable for three things. It provided perhaps the best introductory account of a Turing machine (cited in Chapter 4.i.b). It gave a superbly funny description of the priorities and social (or antisocial) habits of a new subspecies of *Homo sapiens*, the dedicated "hacker" (pp. 115–24). (The hacker lifestyle had already been described in *Rolling Stone*, but with more appreciation and less humour: Brand 1972.) And it argued at length that GOFAI could never replace human judgement—and that even if it could, it shouldn't.

Weizenbaum didn't discuss actual AI programs in any detail, wanting "to avoid the unnecessary, interminable, and ultimately sterile exercise of making a catalogue of what computers will and will not be able to do, either here and now or ever". He complained that Dreyfus was too concerned with "the technical question of what computers can and cannot do" (1976: 12). This, he said, wasn't the key issue:

[If] computers could imitate man in every respect—which in fact they cannot—even then it would be appropriate, nay, urgent, to examine the computer in the light of man's perennial need to find his place in the world. (Weizenbaum 1976: 12)

As for that "which in fact they cannot", this was a matter of principle. Man-as-machine was an absurdity:

Whether or not [the programme of AI] can be realized depends on whether man really is merely a species of the genus "information-processing system" or whether he is more than that. I shall argue that an entirely too simplistic notion of intelligence has dominated both popular and scientific thought, and that this notion is, in part, responsible for permitting artificial intelligence's perverse grand fantasy to grow. I shall argue that an organism is defined, in large part, by the problems it faces. Man faces problems no machine could possibly be made to face. Man is not a machine. (p. 203)

His insistence that AI systems, even if they could be hugely improved, should not be substituted for human judgement was especially fierce with respect to contexts involving interpersonal respect or emotions. In such cases, the very idea of replacing human judgement with computer calculation was "obscene": merely contemplating such projects "ought to give rise to feelings of disgust in every civilized person" (1976: 268). So Weizenbaum denounced his rival Kenneth Colby's plans to use ELIZA-like programs for interviewing mental patients (see 7.i.a).

In part, he was (understandably) worried by the crudity of the NLP involved in Colby's programs, and the consequent crudity of the computer's performance. In part, however, he was concerned that the patients' naivety would lead them to believe that Colby's computer was really intelligent. As evidence for this concern, he reported that his secretary had once asked him to leave the room when she was interacting with ELIZA. In truth, this was no cause for alarm: if *you* were playing around with ELIZA and had decided to answer its quasi-questions honestly, would you want your boss reading over your shoulder? Nevertheless, it was possible that some programs might fool some people into attributing genuine intelligence to them.

Weizenbaum also poured scorn on the idea that AI programs might be used in the law courts, to aid the judges in their deliberations. He'd already crossed swords about this with McCarthy, for his views were already familiar to "the artificial intelligent-sia"—both from face-to-face meetings and from the pages of *Science* (Weizenbaum 1972). McCarthy hadn't shared his horror at the prospect of AI-in-the-law-courts. Weizenbaum recalled:

As Professor John McCarthy once put it to me during a debate [IJCAI-1973, at Stanford: Kuipers *et al.* 1976: 13], "What do judges know that we cannot tell a computer?" His answer to the question . . . is, of course, "Nothing." And it is, as he then argued, perfectly appropriate for artificial intelligence to strive to build machines for making judicial decisions. (Weizenbaum 1976: 207)

Weizenbaum gave an answer to McCarthy's question, but was horrified that it should have been asked at all:

What could be more obvious than the fact that, whatever intelligence a computer can muster . . . it must always and necessarily be absolutely alien to any and all authentic human concerns? *The very asking of the question, "What does a judge (or a psychiatrist [see 7.i.a]) know that we cannot tell a computer?" is a monstrous obscenity.* That it has to be put into print at all, even for the purpose of exposing its morbidity, is a sign of the madness of our times. (pp. 226–7; italics added)

(No one, then, was seriously suggesting that AI programs might be seated on the bench: McCarthy had merely been responding to a rhetorical question from Weizen-baum—see below. But people were already working on expert systems representing relatively clear-cut laws, as we'll see in Chapter 13.ii.c–d. And it was the philosoph-ically sophisticated Buchanan who'd first suggested that this be done: Buchanan and Headrick 1970.)

In short, expert systems were (potentially) all very well for dealing with geology for oil prospecting, or medical diagnosis of bodily diseases. At a pinch, they might even be used as aids, though not as substitutes, in searching legal databases for precedents. But where decisions about human lives were concerned, AI programs should be forever eschewed.

Various NewFAI leaders would make fierce ripostes. That's not to say that they rejected every one of Weizenbaum's warnings. But not all were willing to spend time in planning how to pre-empt them. Take McCarthy, for instance. His answers to *What worries about computers are warranted?* showed that he thought there were some (Kuipers *et al.* 1976: 10). But he hadn't lost much sleep in worrying. He hadn't thought seriously, for example, about *What do judges know that we cannot tell a computer?* He didn't even remember saying it, although he was happy to reply to it (in one brief sentence) as a "rhetorical" question. His reply was typically recalcitrant: "I'll stand on that if we make it 'eventually tell' and especially if we require that it be something that one human can reliably teach another" (p. 9). In other words, he didn't take Weizenbaum's question about AI judges as a *practical* issue, but as an attack on the philosophical principle (which he'd been among the first to adopt) that all human knowledge can be formally represented by AI.

Nor had McCarthy thought seriously about any other social implications of the widespread use of AI. Indeed, he'd said in the early 1970s that it wasn't worth speculating about these ahead of time. Once the technology was actually being used in society, it would be clearer what the real problems were. On those grounds, he'd opposed Donald Michie's setting up a meeting of a few carefully selected participants to discuss these issues (The Serbelloni File 1972, letter from J.M. to D.M.M.).

Michie's meeting did eventually take place, at the Rockefeller Foundation's Villa Serbelloni in Italy. That fact alone indicated that some NewFAI people disagreed with McCarthy on this matter. (My own view was that it was irresponsible *not* to consider the possible social implications, despite the unavoidable uncertainties at that early stage: Boden 1977, ch. 16.) They felt that Weizenbaum had been right to raise these concerns, even if he'd done so in an unnecessarily emotive, near-hysterical, fashion. One didn't have to go all the way with him to be worried about such things.

In particular, one didn't have to share his socio-political views. But as we'll now see, many did. Some of the feeling in his favour, on the part of professionals and public alike, was grounded in the cultural context of the time.

f. A sign of the times

Both Dreyfus and Weizenbaum were important voices in the counter-culture (Chapter 1.iii.c–d). But the counter-culture itself had already prepared the general public to pay attention to them.

Let's consider Dreyfus first. His book sold well, in all three versions, not only because he had some important things to say, nor only because many people agreed with him. In addition, they *wanted* to agree with him, and empathized with the virulence of his attack. Indeed, that's largely why Papert had taken such pains to rebut *Alchemy*. Had the RAND memo been published only ten years earlier, when the most powerful McCarthy in the land was Senator Joseph, he'd probably have agreed with his friends that to criticize it in detail was "a waste of time" (see above). In the late 1960s, however, it wasn't.

The first edition of Dreyfus's book fitted well with the emerging counter-culture. One aspect of that cultural emergence was that Dreyfus's background position, namely neo-Kantian phenomenology, was enjoying a renaissance in the Anglophone world. By the time that *What Computers Can't Do* appeared, even Hilary Putnam, the founder of functionalism, was already starting across the philosophical divide (Chapter 16.vi).

The philosophical movement gained strength—or anyway, numbers. By 1990, several prominent cognitive scientists (including Agre and Chapman, and Brian Cantwell Smith) were regularly quoting Merleau-Ponty and, especially, Heidegger. The MIT roboticist Rodney Brooks even felt the need to say of his own seminal paper in the technical journal *Artificial Intelligence* that "It isn't German Philosophy," though he did admit to "certain similarities" (1991*a*, sect. 7.5; cf. M. W. Wheeler 2005, ch. 1). According to that stream of philosophy, as we saw in Chapter 2.vi, *no* naturalistic account of human thought is possible, not even a neuroscientific one: AI is just the worst of a bad bunch.

Another counter-cultural aspect was that AI was closely associated with military-industrial concerns. Dreyfus himself didn't stress this. Nevertheless, most young readers in the 1970s were very happy to see AI denounced.

Unreconstructed aficionados of the old philosophical/political order, of course, weren't. Feigenbaum, for instance, had little patience for Dreyfus:

What artificial intelligence needs is a *good* Dreyfus. The conceptual problems in AI are really rough, and a guy like that could be an enormous help . . . We do have problems, and they could be illuminated by a first-class philosopher. But Dreyfus bludgeons us over the head with stuff he's misunderstood and is obsolete anyway [see above] . . . And what does he offer us instead? Phenomenology! That ball of fluff! That cotton candy! (interview in McCorduck 1979: 197; italics added)

(Buchanan, whose Ph.D. had been in philosophy, was less dismissive of phenomenology: "If there is any reason to read the book at all, it is to become acquainted with this current view of man and the world which is different from the traditional scientific view"—1973: 21.)

The "cotton candy" is considered at length in Chapter 16.vi–viii and ix.d–e, together with attempts, even on the part of the Dreyfus brothers (1990), to reconcile phenomenology and cognitive science—but not GOFAI. Here, what's relevant is that outsiders in general were readier to consume it in the 1970s than they would have been at any previous time.

That's largely why the young Dreyfus, who didn't have international (or even national) eminence when he wrote *Alchemy*, very soon achieved it.

Similar remarks apply to the historical context of Weizenbaum's book, published just four years after Dreyfus's. Indeed, he himself saw it in this light (1976: 10 ff.). So he thanked the counter-cultural figures Lewis Mumford, Steven Marcus, and Chomsky: not just for their constructive advice on the drafts, but for their encouragement when he "despaired" at others' having said what he had to say more eloquently:

[As] Lewis Mumford often remarked, it sometimes matters that a member of the scientific establishment say some things that humanists have been shouting for ages. (Weizenbaum 1976, p. x)

Specifically, he was firing a salvo in the "science wars" that caused havoc in social psychology and, especially, anthropology (Chapters 1.iii.b–d, 6.i.d, and 8.ii.b–c). Having said that "This book is only nominally about computers" (p. ix), he declared "scientism" to be the true enemy—of which psychological AI was a special case (pp. 222–3).

It's hardly surprising, then, that his views were widely welcomed by the general public. Many of them were already highly sceptical about (and frightened by) AI, so were delighted to receive support from someone in the know. Dreyfus, after all, had been a mere philosopher: Weizenbaum, by contrast, knew which side of a computer was which. Many also cheered his bitter critique of the use of computer programs—often incomprehensible to the user—for destructive military purposes in Vietnam and elsewhere (1976: 236–43).

A proviso: with respect to Weizenbaum's strictures on computer judges, the "general public" didn't include young black males in inner-city USA (Turkle 1995: 292). Their view of judges was very different from that of white middle-class students. Whereas the latter usually agreed with Weizenbaum (saying, for instance: "Judges weigh precedents and the specifics of the case. A computer could never boil this down to an algorithm"), their black contemporaries said things like:

This is a pretty good idea. He [the computer judge] is not going to see a black face and assume guilt. He is not going to see a black face and give a harsher sentence.

[From another young man:] I know that if I ever had to go before some judge, there is a good chance that he is going to see my face and he is going to think "nigger".

The computer judge would have a set of rules. He would apply the rules. It might be safer. (both speakers quoted in Turkle 1995: 292–3)

Those interviews were done in 1983. By 1990, PDP had come on the scene. An MIT student (race unspecified, but the context suggests that he was black) now pointed out:

If you were training a [neural net] computer to be a judge and it looked at who had been found guilty and how they were sentenced, it would "learn" that minority people commit more crimes and get harsher sentences. The computer would build a model that minorities were more likely to be guilty, and it would, like the human judges, start to be harder on them. The computer judge would carry all of the terrible prejudices of the real ones . . . Of course, you would never be able to find a "rule" within the system that said you should discriminate against minorities. The computer would just do it because in the past, people had done it. (quoted in Turkle 1995: 293)

In short, a PDP judge would have racist flaws similar to those of the NewFAI expert system designed to select candidates for a London medical school (see 13.ii.c).

Weizenbaum hadn't simply attacked specific applications of AI, such as military or legal programs. AI in general, he said, depended on a "corruption" of language whereby intentional terms such as *problem*, *understanding*, and *knowledge* were tacitly redefined to fit the mechanistic mould. Minsky, Newell, and Simon were specifically named. But this wasn't a matter restricted to the laboratory:

[The] computer, as presently used by the technological elite, is . . . an instrument pressed into the service of rationalizing, supporting, and sustaining the most conservative, indeed, reactionary, ideological components of the current *Zeitgeist*. (p. 250)

Such remarks went down well on campuses across the world, and with sensation-hungry journalists. Within the profession, however, they were to some extent counter-productive. They antagonized the AI cognoscenti, many of whom were themselves out of sympathy with US policy, above all in Vietnam. His MIT colleague Kuipers, for instance, complained:

I had some strong reactions to Joe Weizenbaum's book *Computer Power and Human Reason*. The book mentions some important concerns which are obscured by harsh and sometimes shrill accusations against the Artificial Intelligence research community. On the whole, it seems to me that the personal attacks distract and mislead the reader from more valuable abstract points . . . *I see [many of his ideas] as being quite current in the AI community, so I was quite puzzled by Weizenbaum's vehement attacks on us for not sharing them.* (Kuipers et al. 1976: 4; italics added)

Kuipers may have been more annoyed than most, being a conscientious objector who did alternative service (before going to graduate school) instead of submitting to the

Vietnam draft. But he certainly wasn't the only NewFAI researcher to be offended by Weizenbaum's accusations of lack of moral/political integrity.

Nevertheless, the book couldn't be ignored by the NewFAI community. It was the subject of many debates, both formal and informal. McCarthy, for one, saw it as a threat to AI research in general, being all of a piece with the counter-culture of the time:

This moralistic and incoherent book uses computer science and technology as an illustration to support the view promoted by Lewis Mumford, Theodore Roszak, and Jacques Ellul, that science has led to an immoral view of man and the world. I am frightened by its arguments that certain research should not be done if it is based on or might result in an "obscene" picture of the world and man. Worse yet, the book's notion of "obscenity" is vague enough to admit arbitrary interpretations by activist bureaucrats. (Kuipers *et al.* 1976: 5; cf. p. 9)

McCarthy wasn't one of those who, like Kuipers, sympathized with Weizenbaum's politics. He complained that "Certain political and social views are taken for granted," accused Weizenbaum of "political malice" and "new left sloganeering", and—by implication—objected to his description of US policy in Vietnam as "murderous" (Kuipers *et al.* 1976: 6, 7, 9).

But a significant number of his colleagues did share Weizenbaum's concerns. Partly due to the efforts of Winograd, a new society and newsletter were founded in Palo Alto in 1981: Computer Professionals for Social Responsibility, or CPSR (see <http:www.cpsr.org>). Offshoots soon sprang up at other US locations, at the universities of Edinburgh and Sussex, and in France, Germany, and Italy.

Within a few years, CPSR was lobbying against the technical inanities of the USA's Star Wars programme (see i.c, above). One prime concern was the issue already mentioned by Weizenbaum: programs so complex that they're incomprehensible to the layman (including military personnel) and sometimes even to computer professionals. A new journal, *Computing and Society*, was founded in 1988 by CPSR's executive director, Gary Chapman, to draw attention to these matters. Across the ocean, several attempts were made in the 1980s to warn the public *and AI professionals themselves* about the risks of using certain types of AI system (H. Thompson 1984, 1985; Whitby 1988; Council for Science and Society 1989). Sometimes, these warnings were placed in the context of a specific, though unorthodox, philosophy of science—as we'll now see.

g. The unkindest cut of all

Another bombshell of this general type was due, believe it or not, to Winograd. In the mid-1980s, he publicly accepted the arguments already put forward by Dreyfus and Weizenbaum, and added those of the Chilean biologist Humberto Maturana.

All three of these men were seen as highly maverick by GOFAI folk, and were mostly ignored by them accordingly. (Maturana was rarely even read, partly because he wrote in such an opaque manner; his work became popular only much later: see 15.viii.b.) Winograd, by contrast, couldn't be ignored.

What caused most consternation in GOFAI circles wasn't the force of his newly borrowed arguments. Rather, it was his offering the enemy the glamour and authority of his hugely famous name. For this was the man who, only a couple of months after getting his Ph.D., had been invited to give the inaugural *Computers and Thought* Lecture at IJCAI-1971. His youthful achievement in NLP, and his influence on 1970s symbolic AI in general (and on cognitive science too), had been nothing short of phenomenal (see 9.xi.b and 10.iv.a).

A person of lesser intellectual honesty could have rested comfortably on those early laurels for a lifetime. Through the 1970s, however, he'd become increasingly sensitive to the difficulties inherent in NLP—already emphasized by Bar-Hillel and Dreyfus. And he said as much in print, writing alongside Minsky and Papert (and others) in an official national report:

> The AI programs of the late sixties and early seventies are much too literal. . . . This gives them a "brittle" character, able to deal well with tightly specified areas of meaning in an artificially formal conversation. They are correspondingly weak in dealing with natural utterances, full of bits and fragments, continual (unnoticed) metaphor, and reference to much less easily formalizable areas of knowledge. (Winograd 1976: 17)

By the early 1980s, he'd long abandoned SHRDLU, and put his major efforts into founding CPSR instead. He was disenchanted also with his programming language KRL, which was "collapsing perhaps under the weight of its own features" (Brachman and Levesque 1995: 263). His long-heralded treatise on *Language as a Cognitive Process* had a promising title, from the point of view of orthodox cognitive science. But it never progressed beyond the first instalment, on syntax (Winograd 1983).

Semantics, he now felt, was very much more difficult. He'd hinted as much in a talk he gave at IJCAI-1973 in Los Angeles. But there, he was saying in effect, "When I wrote SHRDLU, I hadn't really thought much about semantics. I'll have to do so"—and he praised Roger Schank, as someone who was already attempting the taxing task of bringing the communication of meanings into AI (see 9.xi.d). At that point, he hoped to emulate, even surpass, Schank's efforts. But by the mid-1980s, when his *Syntax* book appeared, he'd given up. He'd decided that the task wasn't just taxing, but impossible. And he soon said so, at length, in a book co-authored with Fernando Flores (Winograd and Flores 1986).

Flores had been the key player in leading Winograd to change tack. It was he who'd introduced Winograd to his compatriot Maturana's writings—and to Maturana himself, who later gave them feedback on the draft of their book (Winograd and Flores 1986, p. xiv). Flores didn't go in for nitty-gritty linguistic theory, still less the highly abstract semantics of Richard Montague, then gaining popularity on another part of the Stanford campus (9.ix.c). Indeed, he was less interested in language as such (i.e. sentences) than in the forms of human communication it made possible.

Winograd was drawn to him by political sympathies as well as intellectual ones. Deeply influenced (as a student of engineering) by Stafford Beer's cybernetic work in the 1960s, Flores had been largely responsible for inviting Beer to design the Cybersyn network for the newly elected Allende government in Chile (Chapter 4.v.e). He'd been the minister responsible for nationalizing Chile's industry, and after Colonel Pinochet's CIA-led coup in 1973 he spent three years in prison. He was released in 1976, thanks to the good offices of the San Francisco chapter of Amnesty International—which brought him to Stanford, three years after Winograd's arrival there.

The two men's shared intellectual sympathies eventually went much further than a dissatisfaction with formalist NLP. For, to his previous colleagues' bewilderment,

Winograd followed Flores (and Dreyfus) and turned to hermeneutics and phenomenology (9.xi.b and 16.vii), and to autopoietic biology (15.viii.b and 16.x.c). The co-authored
book of 1986 bitterly disappointed most of his long-time admirers as a result.

Even Winograd himself hadn't foreseen this volte-face. When he went to Stanford in
1973, he'd known of Dreyfus only through the scandal described in Section ii.a–c above.
He says now that he "wasn't pre-inclined" to him (personal communication). I take
this remark to be code for something very much stronger: benignly non-competitive
though he was, Winograd could hardly have escaped infection by the virulent hostility
to Dreyfus then current at MIT. But his inclinations changed. For Dreyfus turned out
to be one of a small group, also including Searle and Daniel Bobrow, attending "a
very interesting series of informal private lunch seminars". (If Winograd had adopted
the macho 1980s motto "Lunch is for wimps", maybe things might have turned out
differently!)

One person whose life was changed by the Winograd–Flores book, which he read
while it was still in preprint form, was Randall Beer. Then, he was a graduate student in
GOFAI, happily writing expert systems in LISP. Now, he's a renowned A-Life roboticist
and dynamical systems theorist (15.vii.c and viii.d).

For Beer, Winograd's new book was an epiphany. This wasn't because of its views
on language as such, but because it introduced him to the biological ideas of Maturana
and Francisco Varela (M&V). Their theory of autopoiesis (15.viii.b and 16.x.c) had
persuaded Winograd and Flores to think of 'Language as a Biological Phenomenon'
(pp. 38–53). On reading their book, and then M&V's (Maturana and Varela 1972/1980),
Beer was convinced that cognition couldn't be understood except as a property of life,
and that both these concepts had to be theorized in dynamical, embodied, terms. He
now sees his own work as "an attempt to concretely express, illustrate and apply some
of the insights that [M&V's ideas on autopoiesis] bring to understanding biological
behavior and cognition" (R. D. Beer 2004: 310).

But if the new publication was a welcome wake-up call for this young AI student, for
most of its older AI readers it was a very nasty shock. Like M&V themselves, who avoided
all talk of "representations" and even of "feature detectors" (15.vii.b), Winograd was
rejecting the computational approach wholesale. Considered as a philosophical position,
functionalism and "strong AI" were being declared intellectually bankrupt (16.iii–iv,
v.c, and ix.b). Considered as a practical proposition, GOFAI-based NLP was being
dismissed as a waste of time.

In the eyes of the GOFAI community, this was betrayal. Winograd had started as a
heretic and rapidly became a high priest of the orthodoxy. Now, he was an apostate. That
in itself was enough to qualify as a bombshell. And GOFAI's philosophical enemies,
naturally, made the most of the ammunition.

11.iii. A Plea for Intellectual Hygiene

Neither Dreyfus nor Weizenbaum, nor even the newly reconstructed Winograd, had
been asking for AI to be done better. They didn't really want it to be done at all.

Admittedly, they were happy for AI to be applied for highly limited technological
purposes. But they didn't want it to be used as the intellectual core of cognitive science.

Indeed, they didn't want cognitive science: *mind as machine* (they believed) was absurd, and *mind as formalist/symbolic machine* was anathema.

Our next bombshell, by contrast, was specifically asking for AI to be done better (Mc-Dermott 1976). The person who dropped it in the mid-1970s believed that human intelligence could be explained by symbolic AI in principle, and that logicist programs might model much of it in practice. (He later changed his mind about logicism: see 13.ii.a.) But there was no chance of that, he said, if AI research carried on in the way it had started.

He accused the early AI programmers of being scientifically ill-disciplined and often self-deluding. To turn over-excitable NewFAI into respectable GOFAI, they'd need to mend their ways.

a. The insider

Drew McDermott (1949–) was a well-respected AI researcher at Yale (previously at MIT) when he first threw a cat among the AI pigeons. He was also an accomplished philosophical logician, with a good grasp of work in epistemology and metaphysics. He'd won his GOFAI spurs by helping develop CONNIVER (Chapter 10.v.c), and by modelling some of the automatic inferences made when we hear familiar words (10.iii.d). So he was 'one of the boys'.

However, while the AI community was still basking in the reflected glory of SHRDLU, he wrote a squib about the field for SIGART, provocatively called 'Artificial Intelligence Meets Natural Stupidity' (1976). This was a critique (even, as he put it, "self-ridicule"), not an attack. But it was so cutting, and so well aimed, that when Haugeland later included it in his collection *Mind Design* (1981) it was sandwiched between David Marr's (1977) strongly anti-GOFAI 'Personal View' and Dreyfus's (1981) characteristically dismissive 'AI at an Impasse'.

McDermott charged his NewFAI colleagues—and, disarmingly, himself—with systematic self-deception and obfuscation, various types of sloppy thinking, and lack of persistence in improving first-effort programs. In short: he diagnosed an overall lack of scientific self-discipline, which compared unfavourably with common practice in other areas of computer science.

It was no accident, he said, that many professional computer scientists looked down their noses at AI. It wasn't simply that, as *mathematicians*, they preferred proven theorems to unreliable heuristics. In addition, they deplored the lack of intellectual discipline that was all too common in the NewFAI days. (Resentment at the glamour and media publicity attending AI probably played a part too: I certainly got that impression in several conversations with computer scientists in the 1970s.) "If we are to retain any credibility", he declared, "this should stop" (1976: 143).

Today, 'pure' computer scientists still look down their noses at AI (see 13.vii). But, partly because AI now is technically so much deeper and more rigorous than it was then, McDermott sees this as a primarily methodological difference:

AI people [today] are willing to try an algorithm and test it empirically even though theorists have proven it won't work. I've had more than one theorist tell me that the value of AI is in its practitioners' willingness to try things. When they work better than expected, often lots of interesting theoretical issues are created as a byproduct, and the theorists move in.

I'm not saying that AI shows the theory was *wrong*. [We know, for a problem that's NP-complete, that] there's no hope of finding an algorithm that solves every instance of the problem. The question is whether there are interesting clusters of instances where better algorithms can be found, and there's no way to know without empirical exploration. (McDermott, personal communication, 2004)

NewFAI workers, said McDermott in his bombshell paper, weren't being irresponsible at random. A systematic source of self-deception was their common habit (made possible by LISP: see 10.v.c) of using natural-language words to name various aspects of programs.

These "wishful mnemonics", he said, included the widespread use of "UNDER-STAND" or "GOAL" to refer to procedures and data structures. In more traditional computer science, there was no misunderstanding; indeed, "structured programming" used terms such as GOAL in a liberating way. In AI, however, these apparently harmless words often seduced the programmer (and third parties) into thinking that *real* goals, if only of a very simple kind, were being modelled. If the GOAL procedure had been called "G0034" instead, any such thought would have to be proven, not airily assumed.

The self-deception arose even during the process of programming: "When you [i.e. the programmer] say (GOAL...), you can just feel the enormous power at your fingertips. It is, of course, an illusion" (p. 145). Indeed, he said, one of the advantages of CONNIVER over PLANNER was that it was *harder* to program in CONNIVER, because FETCH & TRY NEXT was used instead of GOAL, METHOD instead of THEOREM, and ADD instead of ASSERT (144–5). (But he too, he confessed, had been guilty of obfuscation: CONNIVER's so-called "CONTEXTS" were no such thing.) The rot had also affected AI thinking about *deduction*. The meaning of this familiar, and apparently unambiguous, word had been unwittingly changed (by AI work on resolution theorem proving: 10.iii.b) to become "something narrow, technical, and not a little sordid" (p. 145).

Self-deception was rampant in the names given to subroutines. Among the countless examples McDermott could have cited were Colby's FINDANALOG (7.i.a), and Robert Abelson's REINTERPRET FINAL GOAL, ACCIDENTAL BY-PRODUCT, and FIND THE PRIME MOVER (all procedures for "rationalization": 7.i.c). As for the names of entire programs, the famed GPS, with its spurious promise of being a *general problem-solver*, would have been better called LFGNS—"Local-Feature-Guided Network Searcher". And Ross Quillian's (1969) 'Teachable Language Comprehender' would have caused less excitement had it been—honestly—named 'The Teachable Language Node Net Intersection Finder'.

Indeed, whole sub-fields had been optimistically misnamed, with unfortunate effects:

And think about this: if "mechanical translation" had been called "word-by-word text manipulation", the people doing it might still be getting government money. (p. 148)

Here, McDermott was referring to the disastrous effects of the ALPAC bombshell of 1964 (Chapter 9.x.e), and to the general disenchantment with machine translation seeded in 1960 by another insider, Bar-Hillel (9.x.b and e). He might have added that a similar reproof could have been made—and, already, often was—against McCarthy's name for the field as a whole: artificial *intelligence*. Some AI sympathizers had foreseen this from the start, and had recommended other names accordingly (6.iv.b).

Quillian (1961, 1968) had pioneered another near-ubiquitous example of self-deception, namely, work on semantic networks (10.iii.a). These, said McDermott, were neither semantic nor networks (p. 160). For instance, the apparently simple IS-A link (aka *isa*) was fraught with misinterpretation—usually, over-interpretation. McDermott's worries about IS-A (146 ff.), and about the handling of "a" and "the" (152–3), were broadly similar to those of William Woods (see 9.xi.e). In a nutshell, highly limited—though precisely defined—program procedures were being *interpreted* in a much vaguer, and richer, way.

The sloppy thinking connected with the IS-A link was largely due, McDermott argued, to the lazy assumption that natural language provided both problems and solutions. In other words, language-like internal representations in the computer were felt to be intrinsically superior to more abstract ones. (This was implicit in Quillian's little note of 1961, for example.) To the contrary, any really illuminating (and workable) representation of a familiar concept or an English sentence would be "a directly useful internal representation, probably as remote as possible from being 'English-like' " (p. 150).

Moreover, natural language was being underestimated. Not only is it often/usually used for purposes other than transmitting information, but its meaning can't be easily corralled. For instance, the common NewFAI assumption that words such as *the*, *a*, *all*, *or*, and *and* are expressible in something like predicate calculus terms was an illusion. Russell and the logical atomists had believed it, to be sure (Chapter 4.iii.c). But their philosophical critics had been legion. Even the exercise of "examining two pages of a book for ten-year olds, translating the story as you go into an internal [programmed] representation" (p. 153), said McDermott, would show the task to be horrendously difficult. (Here, he was thinking of Eugene Charniak's work: Charniak 1972, 1973, 1974.)

In general, he added, people should distinguish clearly between what their program actually did and what some other version of it might do in the future. As it was, "Performance and promise run together like the colors of a sunset" (p. 157). What the program *didn't* do should also be made clear:

AI as a field is starving for a few carefully documented failures. Anyone can think of several theses [including some highly publicized ones] that could be improved stylistically and substantively by being rephrased as reports on failures. I can learn more by just being told why a technique won't work than by being made to read between the lines. (p. 159)

(It's worth remarking, by the way, that one of the many commendable—and often commended—aspects of Winograd's Ph.D./book had been his unusual honesty in pointing out the *limitations* of his program.)

Last but not least, McDermott identified a "common idiocy" in AI research: the abandoned-program syndrome. Far too often, people supposed "that having identified the shortcomings of Version I of a program is equivalent to having written Version II" (p. 156). (Again, he identified himself as one of the culprits: McDermott 1974.) There was some sociological excuse for this, in that the culture of awarding Ph.D.s in AI prioritized theoretical comment over effective programming, and sought new work on new problems rather than improvements on research already done (whether by Self or Other). But scientific excuse, there was none. What true scientist would happily report only his initial, half-baked, ideas, without carrying them forward in later work?

Colleagues, too, should be able to carry them forward. But if AI was to benefit from normal scientific cooperation (Chapter 2.ii.b–c), programs should be properly debugged so that other people could run them, test them, and develop them. Although McDermott didn't say so (because he didn't know it at the time: personal communication), Winograd was at fault here. The man–machine conversation he'd reported was the only one of that length ever produced (even so, it may have been 'cobbled together' rather than being one continuous interchange), and it turned out that SHRDLU couldn't be effectively run by walk-up users because of the bugs remaining in it (Winograd, personal communication—see 9.xi.b). McDermott admitted, however, that a first-effort program may be very useful—and surprising—in showing what can be done, and what with persistence *might* be achieved.

This was strong stuff: McDermott didn't mince his words ("stupidity", "idiocy" . . .). But he didn't suffer the calumnies that the AI profession was directing against Dreyfus and Weizenbaum. Not only was he an insider, and a self-confessed culprit to boot, but he was undeniably right. His arrows had hit the mark. Moreover, he wasn't saying (as they were) that AI was a worthless enterprise. Even logicist AI was a respectable project (so he believed at that time: but see 13.ii.a).

b. Natural Stupidity survives

Some others, too, were complaining that AI's "lack of self-generated criticism is scandalous" (Wilks 1976: 184). Their remarks, and McDermott's, shamed some of their colleagues into being more self-critical. Even so, complaints about "natural stupidity" still had to be made in the 1980s . . . and occasionally have to be made now.

(To be fair, natural stupidity wasn't monopolized by GOFAI. It was a feature of some *connectionist* AI as well. "Why think when you can simulate?", as James Anderson put it—see Chapter 12.ii.a.)

A decade after McDermott's squib, two similar complaints were made on either side of the Atlantic. One came from McCarthy, then President of the AAAI (McCarthy 1984).

His brief President's Message (constantly threatening, he said, to turn into a paper) argued that we need better standards for AI work. Not that AI people were any less motivated than other scientists to do good work: they weren't "fooling" the funders, or the public. But in any science evaluative standards have to develop, and this takes time. In AI, he said, they "are not in very good shape" (p. 7). He identified seven principles that should help AI to progress faster, some of which (such as repeatability) had been anticipated by McDermott. And, by the way, he pointed out that although the Turing Test can be an interesting "challenge" for AI research, it *isn't* a "scientific criterion" of AI value (8)—see Chapter 16.ii.

The other complaint published in that year came from Edinburgh's Alan Bundy (1947–). He criticized GOFAI work as—still—insufficiently analytical and (therefore) non-cumulative (Bundy 1984; Bundy *et al.* 1985). Although learning was the particular domain he considered, his conclusion was directed at AI in general. A more scientific, not to say mathematical, approach was required if the field was to progress.

With respect to learning, Bundy showed that a powerful "focusing" algorithm defined by some of his Edinburgh colleagues in the year following McDermott's 'Natural Stupidity' tirade subsumes many AI programs modelling superficially different types of learning in apparently different ways (discrimination, generalization, version-spaces . . .). That is, the principled core of these seemingly diverse programs was the same. Moreover, it helped explain the differences in the ways in which they modify their rules on the basis of positive or negative information, and in learning conjunctive or disjunctive concepts.

The explanation of *why* these programs worked, as opposed to mere reports *that* they worked, lay in such theoretical analysis. In Chomsky's language (which Bundy borrowed), AI research shouldn't merely describe a program's performance, but should identify its competence as well (Chapter 7.iii.a).

By that time, Marr too had asked for more systematic theoretical analysis in AI (D. C. Marr 1977). Indeed, he'd gone further than either Bundy or McDermott, claiming that AI programming is a waste of time if the abstractly definable task underlying the domain concerned hasn't been identified (7.iii.b).

Minsky and Papert, of course, had provided an abstract computational analysis of one type of AI—namely, simple perceptrons—twenty years earlier (12.iii.a–c). Indeed, ten years earlier still Minsky had shown (in 'Steps': 10.i.g) that analytical questions and comparisons were crucial. But the point hadn't sunk in. For far too long, the Minsky–Papert exercise remained the Lone Ranger of *theoretical* AI, or what McCarthy called meta-epistemology. If McDermott's advice had been truly taken to heart when he wrote his paper, companions might have turned up sooner.

As it is, most of McDermott's complaints still ring true of much AI work. That's clear from these two late-century remarks—one astringent, the other defensive:

We need to discipline ourselves so that our tests are both proper in themselves and connectible, through common data or strategies, with those done by others. Because if we do not seek the standards [that] doing information science should imply we will be open, correctly, to the claim that we're just inventing copy for the salesmen. (Sparck Jones 1988: 26)

Computer science is really like physics in 1740. People are so happy for any results, that more than haphazard repeatability and duplication of other people's results are too much to ask. It's still at the stage where people are discovering rather than colonizing. (D. B. Lenat, interviewed in Shasha and Lazere 1995: 229)

It's clear, too, from the fact that AI hype still occurs.

For instance, the official web site for RoboCup, the ongoing competition in robot soccer, opens by declaring this as the aim:

By the year 2050, [to] develop a team of fully autonomous humanoid robots that can win against the human world soccer champion team. (<http://www.robocup.org>, accessed 9 Apr. 2004)

In 1999, soon after RoboCup was launched—at the IJCAI-1997 meeting in Nagoya, Japan—with ninety AI teams competing (Asada *et al.* 1999, 2000), one of its initiators quoted those words and commented:

The current technology is nowhere near the accomplishment of that goal, as most robots use wheels instead of legs, and have serious difficulties seeing a ball and other robots. Nevertheless, we believe that the goal can be accomplished by 2050. (Kitano 1999: 189)

One of the main issues, he said (p. 190), was collaboration: How can robots learn to collaborate? How could teamwork emerge? How can we program it? Do we even need to program it? (One of the papers representing the then state of the art described an entirely behavior-based approach: Werger 1999, and cf. 13.iii.d–e.)

The competition's spokesman Mats Wiklund apparently hadn't read the first sentence of that last quotation. For he was reported by Reuters in 1999 as predicting man-sized soccer automata by 2002, "although we don't expect the robots to stand a chance against humans at that stage". If Reuter's quotation was accurate, Wiklund's words "at that stage" were more lunatic than cautious.

A few years later, more than 300 research groups around the world were involved with RoboCup; and the 2003 competition in Padua registered 243 teams and 1,244 participants. By that time, much of the more successful (i.e. least embarrassingly unsuccessful) efforts were going into small puppy robots, not large two-legged ones. (Most of these were AIBO robots bought off the shelf from Sony: see <http://www.eu.aibo.com>. The challenge was in the programming, not the engineering.) And the dreaded offside rule was still nowhere to be seen. But the overall aim hadn't changed.

There's no doubt that something of interest will ensue (has already ensued) from this exercise, which involves real-time world-modelling and real-time decisions about cooperation and strategy. That's argued persuasively in a recent paper co-authored by the eminently sane Alan Mackworth (Sahota and Mackworth 1994). It was he who first recommended soccer as a fruitful domain for AI. Although he's now given up on soccer (because it doesn't allow for long-term planning), he's still working with ball-kicking robots relying on recognition and guidance—i.e. not on lasers or sonar (13.iii.c). Indeed, he suggests replacing "GOFAI" with "GOFAIR", where the "R" stands for *situated* or constraint-based Robotics (Mackworth 1993, 2003). (Unlike the more radical situated roboticists described in Chapter 13.iii.b, however, he doesn't deny representations or planning: GOFAI is part of GOFAIR.) But he bemoans the lack of scientific discipline in most of the RoboCup work:

Our original robot soccer-players were the world's first . . . We did enter the simulation track [of RoboCup] three times. We were focussed more on the science than performance (at least, that was our excuse)—we won the odd game, but never a competition. I tried as a member of the Board to get people to accept a standard hardware platform but failed. So much of the effort went into hacking a better kicker etc., not into smart controllers and planners. Although it has had a tremendous stimulating effect on students and good PR, not a lot of great new science has come out of it. That disappointed me since I started it all. But in retrospect soccer has lots of great features as a test bed but it lacks longer term time horizons and hence planning and the like. (A. K. Mackworth, personal communication, 2004)

Most RoboCup competitors are less judicious than Mackworth. There are occasional signs of caution in the PR material: the report on the latest bout of the competition, for instance, opens by saying, "We do not promise that this [officially stated] goal will be reached by 2050" (Pagello *et al.* 2004: 81). Nevertheless, the mind boggles. The defence given to me in 2004 by one of RoboCup's winning team leaders (tact forbids . . .) that "It's not a prediction, it's a challenge. No one takes it seriously!" was disingenuous. For this admission came only after repeated needling from me. At first (and second, and third . . .), this person had insisted that the declared goal was achievable. I was repeatedly given the PR party line, obviously compiled for communication to journalists, with no attempt to engage with the scientific questions I'd raised.

In short, RoboCup isn't unambiguously "good PR". It may well have brought more students into AI, as Mackworth says. It's even been named by Paul Cohen (also an eminently sane individual) as one of the "new challenges" to replace the Turing Test as a measure of advance in AI (P. R. Cohen 2005: 63–4, 67). And it undoubtedly gets AI onto our TV screens. But as *the very first item* on the RoboCup web site, the ludicrous prediction cited above is potentially damaging. For such attention-seeking irresponsibility can lead people to devalue AI as a whole—as McDermott pointed out over thirty years ago.

McDermott, here, should have the last word. In a recent interview (in the ACM's first electronic publication), he was asked how he feels about being an AI researcher today:

I'm thrilled with my station in life now. I think AI has matured a lot and has developed into a much more hard-edged discipline. I'm glad just to be a part of that. (McDermott 2001*a*)

Even so, he said in the same interview that youngsters entering AI should beware of the superficially "exciting" projects, and question their supposedly technical vocabulary so as not to fall into "fantasy":

If somebody starts talking about meta-rules and you're thinking "Wow, this is really neat, this is like something that is self-conscious, it's able to think about its own structure, etc.", think again. I cringe when I hear people talking about meta-knowledge representation of something. It seems to me there are still a lot of people in AI who are living in a fantasy world, and it ultimately hurts the field. (2001*a*)

In sum, his rules of intellectual hygiene for AI are still valid—and sometimes still needed.

11.iv. Lighthill's Report

As René Descartes foresaw long ago, progress in scientific research requires not only intellectual creativity but also financial support (2.ii.b–c; cf. 6.iv.f). So the historian must sometimes consider money: where it came from, why it occasionally dried up, and what—sometimes—made it start flowing again.

GOFAI suffered four funding bombshells. These were especially damaging, not least to overall morale, when they were grounded not in fiscal/military policy in general (as recounted in Section i) but in the funders' disaffection with AI in particular. That applied to the first and second financial bombshells: two government-initiated reports which, in essence, declared AI to be a waste of time—and money.

The first was the ALPAC Report of 1960 (9.x.f), the second the Lighthill Report some ten years later. The former was launched in the USA, and virtually stopped MT funding on both sides of the Atlantic. The latter originated in the UK, and caused significant damage there—with shock waves spreading across the Pond.

It's worth remarking that the explosive force of the second of these bombshells was largely due to the personality of one man. His personality, you'll notice: not his research. As such, it's a telling objection to the Legend of wholly disinterested science (Chapter 1.iii.b).

a. A badly guided missile

The Lighthill Report was twenty-one pages of dynamite, sent from on high to wreak havoc on the ground. It caused a recurring nightmare for AI in Britain, a bad dream that was eventually ended by dawn trumpets from Japan. It even engendered a few restless nights in the USA. This extraordinary episode wouldn't have happened but for the personal characteristics of a single individual: namely, Michie.

Michie was the founder of symbolic AI in Britain (see 6.iv.e). He was clever, literate, charming, self-confident, persistent, energetic, entrepreneurial . . . and each of these, in spades. In a word, charismatic. But he was also, for such a suave man, surprisingly brash in his claims about his own, and AI's, achievements and potential.

Before turning to AI, he was already well known—and, significantly, well connected—as a geneticist (a protégé of Conrad Waddington at Edinburgh: see 15.iii.b). In the 1950s, with his first wife, Anne McLaren (now Dame Anne, FRS), he'd published research that helped lay the groundwork for today's reproductive technology—and later earned him a Pioneer Award from the International Embryo Transfer Society. So although AI became his passion, it wasn't his first scientific success.

As for his achievement in AI, this was significant. Indeed, he was recognized by IJCAI's Reward for Research Excellence in 2001. For example:

* His pole-balancer, written on his MENACE-inspired visit to Stanford in 1961 (see 6.iv.e), was the first reinforcement-learning program (Michie and Chambers 1968). It wasn't a simulation, but controlled a real pole on a real cart.

* The pioneering Graph Traverser (Doran and Michie 1966; Doran 1968*b*; Michie and Ross 1969) provided ideas that are now standard in heuristic search, and it still lives on as the core of a widely used planning program developed by Austin Tate (Michie 2002: 4).

* His later work (with Ivan Bratko) on chess endgames was an important contribution to chess programming—and to our knowledge of chess (e.g. Bratko *et al.* 1978; Bratko and Michie 1980).

* And his recent StatLog project, which ranges from statistics through tree building to dynamical systems, has been described as "by far the most exhaustive investigation into the comparative performance of learning algorithms" (Russell and Norvig 2003: 675; cf. Michie *et al.* 1994).

For present purposes, however, what's most relevant is his sociological influence as a research leader and highly visible AI guru. In those roles, he was hugely constructive, on the world scene as well as nationally. But he also sowed some destructive seeds.

His Department of Machine Intelligence and Perception was the first in the world to be dedicated to AI. (For the story of how it came to be founded, see Chapter 6.iv.e.) Among the AI leaders who came from that stable were Patrick Hayes, Robert Kowalski, Aaron Sloman, Michael Brady, Rod Burstall, Robin Popplestone, James Doran, Bundy, and Tate. In addition, many US scientists spent brief visits there—and some recorded their gratitude for Michie's guidance and suggestions (e.g. Quinlan 1983: 481). His Edinburgh colleague Meltzer was the first editor of the *Artificial Intelligence* journal (founded in 1970). And Michie's international Machine Intelligence meetings and publications, initiated in the mid-1960s, were the first regular series in the field.

All that was highly commendable—and hugely influential. So much so, that when Donald Broadbent and others in 1970 lobbied to get some much-needed government funding to support the maverick AI being done by Gordon Pask's team in Richmond, they lost out to Michie's team in Edinburgh (Mallen 2005: 87). But there was trouble in the offing.

By the time of the break-up of the Edinburgh triumvirate in 1970 (Chapter 6.iv.e), two of Michie's colleagues, Meltzer and Christopher Longuet-Higgins, had become his bitter enemies (P. J. Hayes 1973: 36–7, 43–4; Howe 1994). Indeed, Meltzer (1971) had recently regaled their international peer group, the readers of the *SIGART Newsletter*, with a squib containing some caustic remarks about the Edinburgh robot FREDDY. (His subtitle, 'Bury the Old War-Horse!', referred to the Turing Test, wrongly seen by the public as a criterion of success in AI—but one might have been forgiven for interpreting the title in another way!) Since he was the first Editor of the *Artificial Intelligence* journal, Meltzer's criticism of his colleague's pet project carried weight as well as venom.

The discord had spread way beyond Edinburgh. Michie's tireless lobbying for and publicizing of AI had offended many people. In part, they were annoyed by (even jealous of?) the huge public attention his department attracted: so many TV crews and other visitors made their way there that such visits eventually had to be restricted. In part, too, they were upset by his unrealistic optimism. Domestic robots, according to him, were just around the corner. (That's not quite fair: sometimes, he was more realistic—e.g. Michie 1968*b*.)

Worse, some doubted his reports of the present almost as much as his hubristic hopes for the future. They suspected that his glowing descriptions of FREDDY were, to put it delicately, inaccurate (see below).

The squabbling in his department was the most important of three factors which prompted the Science Research Council (SRC) to commission Sir James Lighthill (1924–98) to write a "dispassionate" report on AI research in the UK. They could hardly have chosen anyone more high-profile. For Lighthill, a world leader in hydrodynamics, was the holder of the Lucasian Chair of Applied Mathematics at Cambridge—previously held by Isaac Newton and Charles Babbage (and today by Stephen Hawking).

The second factor behind the decision was a recent SRC review of computer science. This had found a troubling division in professionals' attitudes towards the centrality/marginality of AI (SRC 1972: 19). And the third was Michie's recent (failed) application for a seven-year grant, which had included a request for a PDP-10, a machine larger than any the SRC had ever provided for research in any discipline (Michie 1972).

The Report, submitted in September 1972, had been intended as an internal document for the SRC's eyes only. But it was so controversial that the Council decided (uniquely, so far as I can discover) to publish it—and with four commentaries alongside it (SRC 1973).

The original version had contained corrosive personal comments about Michie, who'd been *very* generously funded by SRC. One member of the relevant SRC committee (who prefers to remain anonymous) recalls, for instance, that they were shocked to discover that Michie's widely shown film of FREDDY was speeded up by sixteen times, and that the assembly being filmed wasn't fully autonomous but was being controlled by someone at a keyboard. The defence, he tells me, was (1) that the measurements being used by the human controller were correct, so they *would* work in an autonomous robot; and (2) that the slowness was due to the human's being unable to tell the arm/gripper to move *up 7, left 21, down 3*, but having to type in each step incrementally: *UUUUUUULLLLLLLLLLLLLLLLLLLLLLLLDDD*. That defence might have been grudgingly accepted if these shortcomings had been mentioned up front, but they weren't.

In short, Michie hadn't delivered what he'd promised, nor even what he'd announced. To be fair, he wasn't the only person playing this sort of game. Hans Moravec, a graduate student at Stanford when grant reports/applications were being prepared for the SHAKEY robot, remembers a similar deception:

An entire run of SHAKEY could involve the robot going into a room, finding a block, being asked to move the block over to the top of a platform, pushing a wedge against the platform, rolling up the ramp [formed by the wedge], and pushing the block up. SHAKEY never did that as one complete sequence. It did it in several independent attempts, which each had a high probability of failure. *You were able to put together a movie that had all the pieces in it*, but it was really flaky. (interviewed in Crevier 1993: 115; italics added)

Moravec added: "Not that it fooled anybody: the DARPA people who were reading those reports had been students in AI a few years before!" So perhaps Michie's 16-accelerated film might have been viewed more indulgently had it not been for the composition of the committee. My informant tells me that the panel included highly influential theoretical computer scientists who would have had scant time for AI, which is largely experimental, even if it had been done by the Angel Gabriel.

The Lighthill Report fizzed with bubbles of contempt. The fiercest criticisms were excised before publication, although they'd already spread on the grapevine to some extent. However, everyone in the AI community was able to decode the expurgated text, considered as a *roman-á-clef*.

What caused the major scandal (as opposed to the insider gossip) was Lighthill's dismissive judgement of AI in general. Or rather, *what he took to be* AI in general (see subsection b).

He divided AI into three categories. Category A was Advanced Automation for specific purposes—including expert systems, speech recognition, and MT. Category C was Computer-based CNS research—including models of cerebellar function, associative memory, language, and vision. These two categories, he said, had only "a minor degree of overlap of interest" (Lighthill 1973: 6). They were closer to already established (code: respectable) disciplines—computer science and control engineering for A, neurophysiology and psychology for C—than to each other. The third was category B: the Bridge Activity of Building Robots, where "aims and objectives are much harder to discern but which leans heavily on ideas from both A and C and conversely seeks to influence them" (pp. 2, 17). It sought to "mimic" hand–eye coordination, visual scene analysis, use of natural language, and common-sense problem solving (p. 7).

Although the Report started by including all three categories within AI, it went on to treat category B alone as "AI" properly so-called. Broadly speaking, Lighthill approved of work in categories A and C. Indeed, in 1973 he told MIT that Stephen Grossberg was doing, as Grossberg has put it, "exactly what AI should have done" (Chapter 14.vi.a). "Bridging" projects, however, were lambasted.

The only AI person for whom Lighthill had a good word to say was Winograd—who'd clearly charmed him, as he charmed everyone else, with his refreshing modesty. But even Winograd's work, which he classified as C, was used (because of its reliance on detailed world knowledge) as a stick to beat category B (pp. 16–17).

Lighthill's tendentious definition of category B could be decoded even by outsiders as *What's valuable in B has almost all been pinched from A or C, neither of which have benefited from B—and they never will.* Insiders could just as effortlessly decode it as *The work done at Edinburgh that was directed by Michie, rather than by his colleagues Longuet-Higgins, Meltzer, and Richard Gregory.* And just in case someone hadn't already worked it out, there came a passage which—in the British context—pointed the finger straight at Michie:

[Disappointments about category B arose because] claims and predictions regarding the potential results of AI research had been publicized which went even farther than the expectations of the majority of workers in the field, whose embarrassments [*sic*] have been added to by the lamentable failure of such inflated predictions. (p. 8)

So Lighthill said, for instance:

There is . . . a widespread feeling that progress in this bridge category B has been even more disappointing [than in A and C], *both as regards the work actually done and as regards the establishment of good reasons for doing such work* and thus for creating any unified discipline spanning categories A and C. (p. 3; italics added)

[The] whole case for the existence of a continuous, coherent field of Artificial Intelligence research (AI) depends critically on whether between categories A and C *there exists* a significant category of research that may be described as a "Bridge" category, B, as well as on the strength of the case for *any* researches in that category. (p. 6; italics added)

And again, there came a remark that pointed straight to Michie:

Here, letter B stands not only for "Bridge activity", but also for *the basic component* of that activity: Building Robots. [This is] seen as an *essential* Bridge Activity justified primarily by what it can feed into the work of categories A and C, and by the links that it creates between them. (p. 7; italics added)

The motivations of Robot Builders were sneered at, being said to include "to minister to the public's general *penchant* for robots [from medieval fantasy to science fiction]", and perhaps "to compensate for the lack of female capability of giving birth to children" (p. 7). (One of Edinburgh's small team of roboticists, Pat Ambler, was a woman: perhaps, to this very Establishment male, she was invisible?)

He listed the "past disappointments" in all three categories, and especially in B, with some relish. The vast amounts of money spent on MT and general speech recognition had been "wholly wasted". (This, remember, was advice to a funder.)

Lighthill allowed that B-research on high-level programming languages had been useful. (These included Edinburgh's POP, used for most AI work in Britain for many years: 10.v.c.) He also allowed that computational modelling had encouraged "a new set of attitudes to psychological problems", by distinguishing "possible candidates for consideration and theories that simply cannot be made to work" (p. 12). But he classified this research (including SHRDLU, and early connectionist AI in general) as category C—i.e. not AI at all.

And his diagnosis? Here he pulled rank, using masterly understatement. "As a mathematician", he said blandly—his readers would know that he held Sir Isaac Newton's Chair—he was inclined "to single out one rather general cause" (p. 9). Namely, "failure to recognize the implications of the 'combinatorial explosion' ". (That was unfair: as we saw in Chapter 10.iii, the combinatorial explosion had been recognized since the late 1950s. Waltz-filtering, for instance, was developed to avoid it.)

The rubble created by this missile was widespread. Summaries and titbits from the Report appeared in the media: *Sun* readers might not be assailed by it, to be sure, but opinion-formers taking *The Times* or other broadsheets were. Moreover, Lighthill's eminence (besides holding Newton's Chair, he'd amassed twenty-four honorary doctorates by the time he died—Mialet 2003: 443) ensured that his judgements would be taken as authoritative by the general public, and by the scientific community at large. They'd even be *welcomed*, since—as explained in Section ii.f, above—most people were suspicious and/or fearful of AI in the first place. (A close parallel was Sir Roger Penrose's ill-informed but widely popular attack on AI, around 1990: Chapter 14.x.d.)

In short, the picture looked grim.

b. Clearing up the rubble

Such was the concern at Lighthill's conclusions that, besides publishing the Report (largely at Michie's insistence, I'm told), the SRC Chairman Brian Flowers announced in the Preface that "The Council would welcome readers' comments on the importance of artificial intelligence research, and the extent of the support the Council should plan to give to it." As a start, the SRC publication itself included four sections of commentary written by people with differing views.

The most important objection, which was made by many people, was that Lighthill's (regrettably vague) A–B–C classification, and his tendentious definition of category B, betrayed fundamental misunderstanding of the field. As the psychologist Stuart Sutherland put it:

Area B [best interpreted not as Building Robots but as "Basic" research] has clearly defined objectives of its own. Its aim is to investigate the *possible* mechanisms that can give rise to intelligent behaviour, to characterize these mechanisms formally, and to elucidate general principles underlying intelligent behaviour. These seem to be valid scientific aims and are clearly different from those of work of types A and C. (N. S. Sutherland 1973: 22)

Similarly McCarthy:

If we take [his] categorization seriously, then most AI researchers lose intellectual contact with Lighthill immediately, because his three categories have no place for what is or should be our main scientific activity—*studying the structure of information and the structure of problem solving processes independently of applications and independently of its realization in animals or humans.* (McCarthy 1974: 317)

Sutherland, Michie, McCarthy, and Hayes all pointed out that many examples of AI work couldn't be readily fitted into *only one* of Lighthill's three categories. All agreed that B-research didn't cover only robotics. (Lighthill's use of "Robot" was unclear in any case, for he sometimes included chess-playing programs—P. J. Hayes 1973: 39–40.) And Hayes remarked that the A–B–C classes "fit altogether too neatly" with the "idiosyncratic" views of three mutually hostile personalities at Edinburgh: Meltzer, Michie, and Longuet-Higgins (P. J. Hayes 1973: 43). Their interests, as well as their personalities, differed. (Meltzer was a theoretical computer scientist, and the first Editor of *Artificial Intelligence*; Longuet-Higgins, who'd won his FRS as a chemist, had already done seminal research on associative memories, language, and the perception of music: see Chapter 12.v.c and Longuet-Higgins 1962, 1972, 1976, 1979; Longuet-Higgins and Steedman 1971.) (An intriguing snippet: Lighthill and Longuet-Higgins had been separated by only one year at Winchester public school, where they used to work on maths conundrums together—H.C.L.-H., personal communication.)

Definitions can always be nit-picked, of course. But as Sutherland said (1973: 23), definitions *matter* when what's being discussed is which areas should be supported by a funding council, and which should not.

Lighthill's misunderstanding wasn't too surprising, for his "independence" was close to ignorance. He'd spent only two months investigating AI, and hadn't contacted some whom one would have expected to be approached. His fifty-one names included many non-AI people, and were mostly British—although McCarthy, Minsky, Winograd, David Hubel, Bertram Raphael, and Ira Pohl were also listed. Indeed, it appeared that he'd spent most of his time talking to people at Edinburgh (Hayes 1973: 37). That was too bad. Even Lucasian professors need to do their homework.

The main set of Comments published alongside the Report was written by Sutherland, Professor of Experimental Psychology at Sussex. This was a masterly demolition of Lighthill's paper, explaining why psychological AI is the intellectual core of cognitive science.

The other comments came from Longuet-Higgins (by then, also at Sussex), Roger Needham, and Michie. Longuet-Higgins (1973) praised Lighthill's paper as "shrewd", "penetrating", and "comprehensive", and (predictably) endorsed his positive valuation of category C. But, he said, Lighthill had been too concerned with the brain's hardware, as opposed to its software. Even Marr's abstract models of the neocortex (14.v.c–e) would "for some time to come" be less valuable for the cognitive sciences (*sic*: see 1.ii.a) than non-neurological AI theories of cognition would be.

The most like-minded response came from Needham, a leading computer scientist at Cambridge (see Preface, ii). But even he admitted that the A–B–C classification was "contentious" (SRC 1973: 32).

The commentary didn't stop there. The UK's/Europe's AI Society ran two reviews in their newsletter. One was a delicious two-page spoof supposedly written by "Sir Grogram Darkvale FRS" (the ogre's name having been turned upside down), but actually authored by Sussex's Max Clowes (Anon. 1973). The other was a more serious piece, by McCarthy's collaborator Hayes—who defined AI as "the development of a systematic theory of intellectual processes, wherever they may be found" (1973: 40).

Meanwhile, across the seas, McCarthy (1974) reviewed Lighthill for the *Artificial Intelligence* journal. He allowed that AI faced great difficulties, and had been "only moderately successful" (p. 322). Nevertheless, he concluded that "Lighthill has had his shot at AI and missed" (p. 321). This was hardly surprising, since the missile's goal had been so badly defined.

(In private, of course, the sparks flew even higher. Hayes was heard to remark that accusing roboticists of having repressed maternal instincts was like accusing hydro-dynamicists of suffering from premature ejaculation.—Lighthill, you'll remember, was a hydrodynamicist.)

Certainly, Sutherland, Hayes, and McCarthy were all *partis pris*. Even so, their reviews would astonish the innocent reader. Rarely can such an important piece of advice on scientific funding, and by such a hugely eminent author, have been so shoddy. (An Edinburgh friend, who shall be nameless, told me that someone he knew phoned Flowers to say: "You idiot! If you wanted to bury Michie, why didn't you pick someone who'd do the job properly?")

Michie, of course, was the *parti* most *pris* of all. He didn't lie down quietly. Indeed, it was largely due to him that the Report had been made public in the first place. His official reply filled eight pages in the SRC's Lighthill booklet. This concentrated on the substantive issues, and on the implications for funding needs in the UK (for a PDP-10, for instance, and an interface to the ARPAnet). To reach a much wider readership, he also wrote four papers timed to coincide, more or less, with the publication of the Report. One was for the popular *Computer Weekly*, for whom he was soon writing a regular column ("Michie's Privateview") in which AI was often lauded (Michie 1973*a*). Another, comparing the vast sums spent on largely irrelevant nuclear physics with the much smaller amounts needed for practically useful AI, appeared in the equally popular *New Scientist* (1973*d*). A third, specifically defending Edinburgh's AI work, graced the pages of the business-oriented *Management Informatics* (1973*b*). And the last, aimed at his scientific peers, was for *Nature* (1973*c*).

He didn't stop there. Within a twelvemonth, he republished some of his less technical articles (written from 1961 onwards), aimed not at AI specialists but at administrators and scientists in general (Michie 1974). Other trade books on AI would eventually follow (Michie 1982; J. E. Hayes and Michie 1983; Michie and Johnston 1984). True to form, these pulled no punches: the question of the potential of AI "is believed by some to be the most consequential ever posed" (J. E. Hayes and Michie 1983, p. ix).

But the damage was done: journalists, businessmen, and even scientists were more aware of Lighthill's scandalous conclusions than of the detailed rebuttals. The Lighthill Report thus undermined the morale of AI researchers in the UK. Not only did they fear the SRC funding stop which Lighthill had recommended, but their scientific reputation in the eyes of outsiders (including alternative funders) had been lowered.

Among those who left for the USA as a result were Hayes and Brady, both then at Essex. (After helping to run the MIT AI Lab for some years, Brady eventually returned; he's now "Sir Michael" and Professor of Information Engineering at Oxford.) They didn't escape entirely, however, for the psychological shock waves were felt across the seas.

This is evident from contemporary issues of the *SIGART Newsletter*. Indeed, one of the best-attended sessions at IJCAI-1973, held in August at Stanford, was a hot-off-the-press video of the Royal Institution's (July) debate between Lighthill, Michie, Gregory, and McCarthy. (I remember the atmosphere's being more like a pantomime than an academic occasion: the dramatis personae drew repeated hisses and cheers from the Stanford audience.) And although various US colleagues tell me that they can't recall research funding being affected, there was a cutback in US money for robotics (Fleck

1982: 192). In addition, as noted above, ARPA started favouring mission-directed research. The historian Fleck grants that this happened in a social context of *general* cutbacks for 'irrelevant' scientific research (see i.b and c, above), but he believes that Lighthill was an additional factor (ibid.).

Whether the Report hugely delayed the advance of AI in the UK is controversial. The administrative leader of the Alvey Programme (see v.c, below) stated that Lighthill "dealt a heavy blow to AI research in Britain from which, in 1981, it had yet to recover" (B. Oakley and Owen 1989: 15). So did the post-Alvey evaluation commissioned by the government:

Prior to Alvey [i.e. in the early 1980s] the UK AI community was small and fragmented. Government support for AI was minimal. The Lighthill Report of 1973 *had resulted in a significant reduction in AI funding and led to a brain drain of many key figures from the UK.* By 1980 skilled personnel were concentrated in a handful of academic centres. (Guy *et al.* 1991: 16; italics added)

The SRC itself had said much the same, in 1979:

The Panel [on Proposed New Initiatives in Computing and Computer Applications] *has no doubt* that the reluctance of the present community to take up the challenge (of industrial robots research) is due at least in part to *the general discouragement of Artificial Intelligence* which took place in this country several years ago and that it is now up to SRC to take steps to remedy the situation. (quoted in Fleck 1982: 215 n. 87; italics added)

But Sloman disagrees:

In the UK, it certainly did not nearly (even temporarily) kill off AI, as some people often say. The main effect was to facilitate the spread of funds for AI to other parts of the UK, so that it became less heavily concentrated in Edinburgh. (personal communication)

Although the funding for Edinburgh became less generous, they did get the PDP-10 and ARPAnet access that Michie had asked for. And other centres, Sussex included, were given access to it for SRC-funded AI research. So work didn't grind to a halt.

It's true that AI funding in 1973–83 didn't reach the level that had been recommended in the earlier SRC review (SRC 1972). But science funding in general during that period, in the UK as in the USA, fell as a result of the OPEC-initiated oil crisis. Much of the UK's 1970s AI was funded not as "Computer Science" but via the new SRC Panel on Cognitive Science. This, as it happens, was set up largely as a result of the anti-Lighthill arguments from Longuet-Higgins and Sutherland.

As for Michie, he was officially sidelined. A new Department of Artificial Intelligence (headed by Meltzer) was formed in 1974, which took over most of the resources, including the robotics equipment, built up under his leadership. He was shunted off into an independent Machine Intelligence Research Unit, and forbidden to work on robotics (Fleck 1982: 189).

But he was irrepressible. Through the rest of the 1970s he accepted many visiting appointments abroad—mostly in Virginia (with his ex-Bletchley colleague Jack Good) and Illinois, but also Canada and the USSR. At home, he spent much of his time in publicizing expert systems. Besides his contributions to the media, he set up a British Computer Society special interest group on expert systems, which ran its own conferences and newsletter.

By the early 1980s, he'd even achieved some distance from the crisis. He now explained the Lighthill incident as a "mishap of scientific politics" due to all-too-human frailties:

Officialdom has subsequently indicated eagerness to repair the damage. Bodies such as the Science Research Council may find it hard to accept some of the remedies required—notably *the return from abroad and re-habilitation of some of those whose work was pilloried.*

At the time, I felt amazement. Ecclesiastes, that incomparable analyst of the dark side, has warned about this feeling: "If you see in a province . . . justice and right violently taken away, do not be amazed at the matter; for the high official is watched by a higher, and there are higher ones over them . . .".

A number of officials attempted at the time to palliate my wrath by explaining in just such terms some of the actions which their jobs had obliged them to take. I concluded that nothing but ignorance at the top could be the cause of the abuses, and that no good could be accomplished until the minds of men were better informed [and so I wrote many "popular" articles about the achievements and promise of AI]. (Michie 1982: 247; italics added)

How had he regained his urbanity? The "re-habilitation" had been made possible when the UK's AI nightmare was abruptly ended by loud noises from Japan. As we'll now see, this clarion call was shamelessly economic/political, not Legend-arily scientific.

11.v. The Fifth Generation

The third funding bombshell was constructive rather than destructive, rescuing non-military AI from the doldrums—and, to some extent, from other scientists' scepticism. It can be called a "bombshell" nevertheless, because its advent was unexpected and its immediate results explosive. For it came in the form of a challenge from Japan, one so frightening that it reinvigorated official sponsorship of GOFAI (and relevant aspects of cognitive psychology) on both sides of the Atlantic.

a. A warning shot from Japan

In October 1981 the Japanese used AI to fire a warning shot over the bows of the Western economies. It had huge repercussions, for it led to an injection of governmental and industrial funds on a huge scale. It was this which revivified the field in the UK, and which took it up to a new level of public interest in the USA and elsewhere. AI was suddenly ubiquitous: it was featured at length in *Fortune, Forbes,* and *Time* magazine (where a computer was chosen as "man of the year"), and even made the front cover of *Newsweek.* By 1985, the biennial IJCAI was attracting some 6,000 participants (twelve years earlier, there'd been only about 250).

Had the Japanese made some great intellectual advance? An AI equivalent of the genetic code, perhaps? No. What caused the excitement, and the frissons of fear, was Japan's announcement in October 1981 of a ten-year national plan for developing "Fifth Generation" computers.

This came to most people, including journalists, as a bolt from the blue. The worldwide AI grapevine, however, had had fair warning. The project's overall director, Tohru Moto-Oka (1929–85), later recalled:

[Before the conference] there had been workshops and so forth to hear the opinions of international scholars; through [Japan's IT association] there were discussions with research institutions of the principal countries of Europe and America, and some information was obtained from a survey visit to America.

In addition, there were invitations to participate from [Japan] to American and European governments, and because the project had already become well known abroad, more than eighty overseas participants attended the Fifth Generation Computer Conference.

Before this conference I spent a month … touring America, Germany, France, and Great Britain, in preliminary discussions with participants [including Colmerauer and Feigenbaum]. (Moto-Oka and Kitsuregawa 1985: 6)

What caused the huge public—and political—excitement was less the ambitious technological project than Japan's economic motivation for launching it. They weren't seeking to appropriate just another market, to be added to cameras and video equipment. Their aim, baldly stated to the Western politicians and industrialists they invited to the "Announcement" meeting in Tokyo, was world domination in information technology (Moto-Oka 1982). Pre-eminence here, they argued, meant economic pre-eminence *tout court*.

Indeed, as the opening speech from Moto-Oka made clear, for Japan it meant economic survival:

Japan … cannot attain self-sufficiency in food [or energy]. On the other hand, we have one precious asset, that is, our human resources. Japan's plentiful labor force is characterized by a high degree of education, diligence, and high quality. It is desirable to utilize this advantage *to cultivate information itself as a new resource comparable to food and energy*, and to emphasize the development of information-related, knowledge-intensive industries which will make possible the processing and managing of information at will. (quoted in Feigenbaum and McCorduck 1983: 12; italics added)

Similarly, an official of the Ministry of International Trade and Industry, the body overseeing the Fifth Generation programme in Japan, told an American journalist:

Until recently we chased foreign technology, but this time we'll pioneer a second computer revolution. If we don't, we won't survive. (Feigenbaum and McCorduck 1983: 135)

In other words, industrial economies needed AI—and Japan, lacking rich agricultural or oil/mineral resources, needed it most of all.

Fifth Generation computers were thus defined not (like the first four generations) by their hardware components, but as computers designed to be suitable for supporting AI applications (Moto-Oka and Kitsuregawa 1985). This included not only parallel computers but also powerful machines dedicated to a specific AI language, such as LISP or—especially—the recently developed PROLOG. (For several years, PROLOG's implementer Kowalski virtually lived on the plane from England to Japan.)

As it turned out, LISP machines didn't catch on widely—not even as exotic workstations for the laboratory bench. Designed at MIT's AI Lab in the mid- to late 1970s (Greenblatt 1974; Bawden *et al.* 1977; Greenblatt *et al.* 1984), they were eventually launched—amid much excitement, fanned by the goings-on in Japan—by two companies set up by MIT personnel (Symbolics and LMI). But they were very expensive, and—one Symbolics employee believes—suffered guilt by association when the over-inflated early 1980s expectations for AI failed to be satisfied (Withington 1991). The same was true of the massively parallel, and massively expensive, Connection Machine—also designed at the MIT AI Lab (Hillis 1985).

But that was for the future. Meanwhile, there was the breathtaking announcement in Tokyo. This was astonishing in terms of both the scale and the nature of the Japanese commitment.

The Fifth Generation plan was to be jointly funded by the country's government and electronics industry, to a minimum of $810 million—some observers were already speaking of billions. (As it turned out, the Japanese government provided less money than any of the eight companies was devoting to its own R & D programme, and vastly less than the $1 billion per year that IBM was spending on its research—Newquist 1994: 212.)

What's more, the project had been allowed to overturn the traditional Japanese attitude to seniority. The director of Tokyo's newly formed AI institute (ICOT) was a surprisingly young man, in his early middle age. He insisted that ICOT be staffed only by *very* young computer scientists, under 30 years old, whose training was up to date and whose creativity was still comparatively high. Each of the Japanese companies involved would be expected to provide four or five of their best young researchers, free of charge to ICOT.

Not the least surprising aspect of the Fifth Generation project was that it was officially announced to an *international* audience. The declared reason for this was that, in this very expensive R & D area, competitors in the capitalist marketplace needed to *cooperate*.

Such cooperation would soon be initiated in Texas. August 1982 saw the foundation of MCC (the Microelectronics and Computer Technology Corporation) in Austin. The AI section was headed by W. W. (Woody) Bledsoe (1921–95), and the overall director was Vice-Admiral Inman—previously Deputy Director of the CIA, but described by Douglas Lenat as "the most charismatic person I'd ever met" (Shasha and Lazere 1995: 234). (In 1993 Inman would be nominated as Secretary of Defense by President Clinton; a few weeks later, however, he decided against accepting the post, which was given to William J. Perry instead.) The good impression was evidently mutual: MCC would soon provide the tens of millions of dollars required for Lenat's CYC project (see 13.i.c).

b. Self-defence in the USA

Capitalist cooperation—and in Texas, of all places!—was just one example of the huge effect this announcement had on AI research funding in the USA, and in the UK and Europe too. The 6,000 participants of IJCAI-1985 included venture capitalists as well as ordinary businessmen, all prepared to provide money for AI research.

The money reached both 'pure' and 'applied' AI workers. Quite apart from the funds coming directly or indirectly from the US government, the newly founded AAAI benefited hugely. Feigenbaum, having been a core AI researcher for many years and having also got wind of the incipient challenge from Japan, was a leader in establishing AAAI in 1980 (Feigenbaum 2005; Reddy 2005). Unlike its long-time UK/European predecessor, AISB (see 6.v.c), it found itself showered with money. Hayes, a refugee from the Lighthill debacle across the Atlantic (who'd be elected President of AAAI a few years later), was impressed—and not a little envious:

I remember having a kind of distant amazement that the USA could find so much money in such a short time. AI societies in Europe were run on shoestrings, almost entirely by volunteer academics, but AAAI had piles of cash from day one (well, actually maybe something like day three) thanks to the AI explosion and the trade fair at the early meetings. Nothing like that had happened outside the USA. (Hayes, interviewed in 2005—Reddy 2005: 11)

Even inside the USA, however, nothing like that would have happened if it hadn't been for the Fifth Generation. In short, AAAI blossomed not just because the USA was rich, but because it was frightened.

Two AI pioneers, Feigenbaum and Buchanan, founded Teknowledge, the first commercial expert-systems company in the USA. (A much smaller enterprise had been started by Michie in the UK, but Teknowledge was on an altogether more ambitious scale.) They had no trouble finding eager recruits:

[We] hired as many smart people as we could afford. At one point, I had taken a leave from Stanford to manage the company and a friend at IBM Watson labs complained to me that IBM couldn't hire enough talented AI scientists because Teknowledge was hiring them all. (Buchanan 2003: 3)

The task of this new company was to write expert systems to commission and/or to develop off-the-shelf shells for an indefinitely wide range of clients. At much the same time, Feigenbaum also co-founded a specialist medical company called IntelliGenetics. This was based on Stanford's MOLGEN expert system, which advised biochemical researchers on how to clone specific DNA sequences. Other software, and hardware, AI companies soon followed them.

(Few of these companies prospered to the expected degree. The author of one history of these events puts this down to a number of reasons, not least the "egos, frailties, and foibles" of scientific researchers with no head for business—Newquist 1994, p. xiii.)

Some commentators compared the Fifth Generation's influence in stimulating scientific research in the USA to that of the Soviet Sputnik in 1956. (In part, this was because the USA—unlike Japan—saw the Fifth Generation largely in terms of its military significance.)

Hard-headed businessmen and ambitious politicians, naturally, had scant interest in psychological AI: technological AI was what they cared about. Nevertheless, AI in general shared the benefits: not only would more young people need to be trained in this new technology, but basic AI research would be needed too.

Moreover, Japan's stated goals—which included MT, speech processing, intelligent assistants, advanced problem solving, and robotics—would require advances in psychological AI, even if the human example was to be dropped as soon as it became inconvenient. By the same token, they'd require advances in other areas of cognitive science, especially psychology.

AI, at last, had been shown to be *useful*. And it was being widely *said* to be useful. Feigenbaum wrote a trade book, which sold over 200,000 copies in the English-language edition (plus several translations), outlining 'Japan's Computer Challenge to the World' and arguing that hugely increased funding for AI was needed if the USA were to retain its economic success (Feigenbaum and McCorduck 1983).

His AI colleagues weren't greatly impressed, although they'd be among the beneficiaries if his wake-up call succeeded. For even if one ignored the vulgarity of the tone (which wasn't easy to do), many of the judgements were questionable. However, that didn't worry most of the readers, who didn't know what questions were relevant.

The book was hugely influential—if not within the field, then outside it. It swiftly propelled AI onto the TV screens, and Feigenbaum onto the chat shows. Quite apart from his (well-deserved) status as a pioneer of expert systems, he had special access to the Japanese developments. For his wife, the computer scientist Penny Nii, was Japanese herself. Five years later, they'd cooperate in a book brashly proclaiming the business successes recently achieved by AI technology (Feigenbaum *et al.* 1988).

One can't say that the hype had ceased: far from it. (McDermott's lessons hadn't been fully taken on board.) Both of Feigenbaum's volumes, not to mention the journalists' offshoots, needed to be taken with a pinch of salt.

So did Winston's *The AI Business*, which appeared hot on the heels of Feigenbaum's *The Fifth Generation* (Wilson and Prendergast 1984). The book's nicely ambiguous title should have warned readers that it was a piece of advertising, not a mere report of scientific advance. Besides straightforward accounts of existing expert systems (including one for configuring VAX machines uniquely to suit clients' needs, which had already saved its parent company DEC a fortune: Kraft 1984), this collection contained its fair share of shaky predictions. It even offered a tendentious editorial defence of past NewFAI claims, such as Simon's notorious vision of a chess program beating the world champion by the mid-1960s (see Section vii.a–b). The people making those predictions, said Winston, were conscientious scientists "simply trying to fulfill their public duty" by preparing people for something that seemed quite plausible at the time (Wilson and Prendergast 1984: 3).

Even NASA, whom one might have expected to be more hard-headed, fell foul of the demon of hype. The manager of NASA's AI research admitted as much, a few years later:

[We held] a week-long workshop in April 1985 in which each NASA center got to describe the AI and robotics projects that it had under way. By the end of the second day, the total number of expert systems *that had been described as having been developed* was about 100. However, *none of these expert systems ever became operational.*

NASA's AI research program *was not immune to overselling and overpredicting.* The objectives stated [in an official "white paper" of 1986] for the program during its inception [were hugely ambitious]. These goals were to be achieved by 1995! (Montemerlo 1992: 51–2; italics added)

The hype of the early 1980s (especially at NASA) was largely due to inexperience. As work on automation and expert systems progressed, people's expectations matured and their predictions became more realistic. By the mid- to late 1980s, NASA was turning "from revolution to evolution", and devising many useful systems—partly by picking the "easy" problems first, and partly by including in-the-loop human control (Montemerlo 1992: 53–7). (An insightful defence of the need for humans to remain in the loop was published in 1990 by Dreyfus's disciple Collins: 13.ii.b.)

NASA wasn't alone. The VAX configurer (XCON) and its successor XSEL (J. P. McDermott 1980, 1982) was providing a crucial service to DEC's clients *and* saving the company over $40 million a year (Russell and Norvig 2003: 24). From 300 rules (and very low accuracy) in 1979, it had grown to more than 3,000—and could now reliably configure ten different DEC computers, not just one (Bachant and McDermott 1984).

XCON/XSEL was the top success story. But there were now many other proven money-savers. At last, AI was successful enough, and visible enough, to attract huge financial support from both government and industry. The AI industry in the USA burgeoned from a few million dollars in 1980 to billions of dollars in 1988.

It wasn't all good news: as usual, the hype became counter-productive. The boomerang effect peaked around 1990. Misled by the optimism, not to say disingenuousness, of AI's vociferous champions, the industrialists expected too much—and met with inevitable disappointments. The period of 1980s largesse was followed by a brief "AI Winter", in which "many [AI] companies suffered as they failed to deliver on extravagant promises" (Russell and Norvig 2003: 24). Eventually, the promises grew less extravagant—and were very often delivered.

(Today, AI is—in practical terms—even more successful. But it's much less visible. Both GOFAI programs and neural networks are hidden inside a myriad of everyday products and services, from cars to refrigerators, call centres to credit checks. That's partly why, when people mistakenly say that GOFAI has "failed" as a technology, they're often believed: see 13.vii.b.)

For all the obvious cultural reasons, the main effect in the West of the ICOT initiative was seen in the USA. Indeed, Feigenbaum's first book relentlessly exploited American fears that US economic and political interests, and ascendancy, were threatened.

Describing the Fifth Generation as "comparable in human intellectual history to the invention of the printing press, with the certainty of making even greater changes in the life of the mind than books did", he chided counter-cultural American intellectuals for seeing the computerization of university campuses not as an opportunity but as "the new barbarism" (p. 210). "Most self-styled intellectuals", he complained in disgust, "don't even recognize what's happening" (p. 211). As for more explicitly political comments, he said:

The Japanese plan [for AI applications] is bold and dramatically forward-looking. It is unlikely to be completely successful in the ten-year period. But to view it therefore as "a lot of smoke", as some American industry leaders have done, is a serious mistake . . .

We now regret our complacency in other technologies [small cars, videos, computer chips]. Are we about to blow it again? The consequences of complacency . . . will be devastating to the economic health of our most important industry. The Japanese could thereby become the dominant industrial power in the world. (p. 2)

We believe that Americans should mount a large-scale concentrated project of our own; that not only is it in the national interest to do so, but it is essential to the national [military] defense. (p. 216)

A superiority in knowledge technology provides whoever holds it with . . . an unequivocal advantage—whether we are speaking of personal power, national economics, or warfare.

The Japanese understand this perfectly . . . Other nations recognize the soundness of the Japanese strategy—and, of course, its inevitability. In response to the farsighted Japanese, ambitious national plans are being drawn up in many places. But the United States, which ought to lead in such plans, trails along in disarrayed and diffuse indecision. (p. 239)

He got his wish. The US government, via ARPA in the Department of Defense, committed $1,000 million over five years—twice as much as ARPA's total expenditure on computing over the previous twenty years.

c. Lighthill laid to rest

In the UK, the response to the Japanese initiative was less up front in tone. But it happened. However, it couldn't happen as quickly as it did across the Atlantic. Quite apart from cultural differences in attitudes to innovation and entrepreneurship, the UK was still carrying the baggage of the Lighthill Report.

If this had been an embarrassment before, it was doubly so now. Perhaps triply. For Michie, as he often said (e.g. Michie and Johnston 1984: 38), had written the very first commercially marketed expert-system shell, back in the 1960s. (Called AL/X, Ray Reiter wrote his Illinois MS thesis on it in 1981—and see Harmon and King 1985: 109.) The prospect of hearing "I told you so!" from him was not one his former detractors found agreeable.

Accordingly, yet another report was commissioned by the SERC—the new name for the SRC (Alvey Committee 1982). This one, too, was entrusted to an eminent mathematician: Sir Peter Swinnerton-Dyer—although the criteria he was given ignored scientific importance, being based only on how information technology could aid UK industry in general (Alvey Committee 1982: 5–6; cf. Cornwall-Jones 1990). And this time, the intended verdict was delivered: a green light for AI.

Given the hangover from Lighthill, however, the term "artificial intelligence" was declared inadmissible (a further reason why, in the UK at least, AI has low visibility today). The powers-that-be insisted on "Knowledge-Based Systems" instead. (So when Sussex received new money for "technology transfer", in the form of thirty guaranteed M.Sc. studentships a year, we had to change the name of the already existing course: AI became IKBS.)

The key strategy group involved, which included Michie, had no illusions about this. A couple of months before the Japanese announcement, they'd met to discuss the future of AI in Britain:

"A bunch of us had felt for quite a time that we really ought to try to do something about rehabilitating AI in the U.K. after Lighthill." . . . [The aim was to set up a special SERC programme in AI.] "We were all sitting around a table wondering what to call this new area", says John Taylor [a big wheel in the Admiralty's Weapons Establishment, and in the computing section of SERC]. "Should we call it artificial intelligence? *We didn't want to call it artificial intelligence because of all the Lighthill connotations*, and we didn't want to call it expert systems, and we came up with this awful phrase 'intelligent knowledge-based systems' or IKBS." (B. Oakley and Owen 1989: 15; italics added)

For public consumption, Lighthill discreetly wasn't mentioned. (The official evaluation of the Alvey project would be less coy—Guy *et al.* 1991: 15–16.) The explanation given for the change in terminology was that "AI" was overambitious and unnecessarily provocative, whereas "IKBS" was not.

As a psychological observation about people's prejudices and mental associations, this was doubtless correct. But a moment's clear thought would have shown that the term "knowledge" was no less philosophically problematic than "intelligence"—which anyway hadn't been dropped, but merely transformed into an adjective. Some people, accordingly, preferred the term "inference computing" (B. Oakley and Owen 1989: 270). But it didn't catch on. (Feigenbaum now favoured KIPS: knowledge information processing systems, of which expert systems were a special case.)

After the Japanese fired their public warning shot two months later, this budding SERC activity was merged into the national Alvey Programme, which spent £350 million over five years on advanced information technology (B. Oakley and Owen 1989: 294). Business and war were both represented: besides the SERC, the money came from the Department of Trade and Industry and the Ministry of Defence.

Of the four main areas of IT involved, one was AI (the others were man–machine interfaces, VLSI, and software engineering). It was agreed at governmental level that £21 million should go via the SERC to universities for AI and computer science (half each)—plus more money to work done, and 50 per cent funded, by industry (Alvey Committee 1982: 2). This would be partly for immediate/short-term demonstrations of IKBS systems with clearly commercial uses, and partly for longer-term, including highly general, AI research (Alvey Committee 1982: 17). (This was the first time that the SERC, supposedly an autonomous body, had been instructed by government to set aside a certain sum of money for university research in a particular area.)

In addition, the UK was a contributor to the European ESPRIT Programme, which had a budget of $650 million from governments and another $650 million from industry. Other European countries, too, supported their own AI research more generously—in some cases ensuring that most of the increase went on basic, not applied, research (Dickson 1986).

Looking back on the five-year Alvey Programme, its director, Brian Oakley, asked 'Was It All Worthwhile?' (B. Oakley and Owen 1989: 265–94). His answer wasn't an unqualified "Yes". For the hoped-for cooperation between industry and academia, always a problematic area in Britain, had been less fruitful than expected. In some cases—the work on speech technology, for example (a dozen projects, the largest of which was at Edinburgh)—the major industrial funder had pulled out after deciding that progress wasn't fast enough (p. 291).

Nevertheless, AI applications—in connectionist pattern recognition as well as GOFAI-based data handling and expert systems—had burgeoned. So had work on human–computer interfaces (13.v): "for many years to come human intervention, and so the significance of the human interface, will remain essential for most applications of AI" (B. Oakley and Owen 1989: 271). Hardware issues had progressed, and software engineering had benefited too. Market awareness had been hugely raised, which had been one of the aims of Alvey from the start. "It remains only a matter of time", said Oakley, "before inference-computing applications far exceed today's so-called conventional uses of computers" (p. 284).

Finally, the need for interdisciplinarity had been underlined. The Alvey directorate, Oakley reported, had exerted "a bit of pressure" to foster cooperation in the face of "the rigid departmental structure of universities" (p. 292).

Despite the call for interdisciplinarity, Oakley—and Alvey—was primarily concerned with AI as an economic force, not as basic and/or psychologically oriented research. The same was true of the effort in Japan, and of the responses in Europe and the USA.

Indeed, Minsky complained that the success of the current expert systems "bodes ill for making further progress", because commercial companies weren't prepared to look ahead—to incorporate learning, for instance. In particular, he said:

There is no significant increase in the number of people working on the ideas that we will want to use in ten years. The number of people doing *basic* research in Artificial Intelligence is probably under one hundred people and maybe under fifty. (Minsky 1984a: 245; italics added)

[Large companies who boast about their AI groups are loath to provide small amounts of money to support a few students.] They do not seem to understand where the ideas came from and where the new ones will come from in another decade. (p. 251)

For all that, the effects on AI in general had been broadly beneficial, especially in the UK where (thanks to Lighthill) Establishment scepticism had reigned. By the late 1980s, Oakley saw a change in the "fashion" for public support of research in the West, with "a renewed emphasis on basic research in academia and a weakening of academic/industry ties" (B. Oakley and Owen 1989: 292–3). Lighthill had been laid to rest at last.

11.vi. The Kraken Wakes

John Wyndham's sci-fi novel *The Kraken Wakes* narrates the havoc caused when a sea-monster, asleep in the depths of the ocean for many years, suddenly becomes active again. From GOFAI's point of view, the Kraken was connectionism. This (eighth) bombshell comprised an intellectual watershed for AI, as well as a financial one.

a. Small fry and sleeping draughts

Connectionist AI was spawned in the 1940s (Chapter 12.i). And some of the small fry had swiftly grown appreciably larger. So in 1958 Rosenblatt published his seminal paper on "Perceptrons"—in the very same volume of *Psychological Review* which, a few months earlier, had carried the GPS authors' 'Elements of a Theory of Human Problem-Solving' (Newell *et al.* 1958a).

Minsky and Papert, however, were (at that time) less eclectic than the editor of *Psychological Review*. From 1956 (i.e. 'Steps') onwards, they expressed fundamental doubts about the value of perceptrons. The small fry swam on, regardless—until the MIT duo administered a powerful sleeping draught in the late 1960s (12.iii).

But the ensuing slumber was a doze, not a coma: important connectionist research was still going on beneath the waves. In 1979 connectionism raised its nose above the waters, at an interdisciplinary meeting in La Jolla (12.v.b), and the whole of the Kraken's head became visible soon afterwards (12.vii.a).

b. Competition

Worse was to come. The monster leapt high out of the ocean in 1986, on publication of the two-volume connectionist 'bible' (12.vi).

Almost immediately, the USA's main research sponsor had second thoughts about its generous funding of GOFAI. Late in 1987, DARPA sponsored a detailed two-month study to reconsider their policy (12.vii.b). The result was a shift in their priorities. To be sure, symbolic AI wasn't put in the dire position that connectionist AI had been in before. (Since the late 1960s, ARPA/DARPA had favoured the former almost to the exclusion of the latter—not least, because of Licklider's closeness to Minsky.) But the limited research money now had to be shared.

Perhaps even more to the point, the flow of research students into GOFAI slowed. That was no accident. The connectionist bible had been written, priced, and marketed as a deliberate attempt to induce youngsters onto the connectionist pathway through the AI field (12.vi.a).

Outsider interest shifted as well. Many psychologists and (especially) philosophers transferred their concern from GOFAI to connectionism, which promised a more plausible account of concepts (12.x). Symbolic AI was commonly said to have failed. As remarked above (i.d), one high-status collection on the symbolic/connectionist AI debate was subtitled *False Starts, Real Foundations* (Graubard 1988)—hardly a neutral way of putting it.

Coincidentally, Haugeland's patronizing definition of GOFAI dates from this general period. His critique was driven by phenomenology, not connectionism—which wasn't even discussed (the bible was still in press when his book appeared). But his readers would be even more ready to accept his criticisms in the pro-connectionist climate.

The journalists, too, swam with this tide. They saw only the connectionist Kraken, dismissing the still-advancing GOFAI as a wounded minnow floundering in the shallows. Symbolic AI was no longer glamorous, no longer sexy. And cognitive science in general, if it was to retain respect and gain new adherents, had to follow in the Kraken's wake.

The history of this intellectual watershed, and of GOFAI philosophers' attempts to meet it in a principled fashion (12.x.d), is detailed next, in Chapter 12. Then, in Chapter 13, we'll see how symbolic AI fared in the aftermath.

CONNECTIONISM: ITS BIRTH AND RENAISSANCE

"Only connect", said E. M. Forster as the motto of *Howards End*. Connectionists say this too. But whereas the novelist was talking about people, they're talking about neurones—or, more accurately, abstract computational units inspired by neurones. This chapter explains the "connect", and casts doubt on the "only".

If connectionism reminds one of *Howards End*, it also reminds one of *The Sleeping Beauty*. For the infant field, spawned in the 1940s by cybernetics, went into hibernation in the late 1960s. Ironically, the sleeping draught had been forced down the Beauty's throat in a no-holds-barred attack by one of her earliest suitors—an extraordinary story, told in Section iii.

Her heart was still ticking over during dormancy, however, and some important thinking was done during the twenty-year sleep. She stirred visibly in 1979, aroused by an interdisciplinary get-together on the Californian coast. The return to full activity happened, amid a blaze of publicity, in the mid-1980s. She rose from her bed and danced around the world, posing for the newspapers and entrancing almost all who met her.

The awakening of connectionism invigorated many philosophers and psychologists, but it constituted a bombshell from the point of view of symbolic AI (see 11.vi). Indeed, the few who remained unentranced by the merrily dancing Beauty were long-time devotees of GOFAI (x.d, below). But a bombshell, despite delivering a very nasty shock, need not end one's life. It may leave certain strengths unharmed—and even unsurpassed. The sections below (viii–ix) which cast doubt on the "only" show why GOFAI didn't leave the floor. It would continue steadily on its way during the next quarter-century (a story told in Chapter 13), achieving results which connectionism couldn't—and still can't—match.

That human minds *do have* rich associative powers had been obvious for centuries. And two early twentieth-century writers had provided a host of intriguing examples illustrating the subtleties involved. One was Sigmund Freud, whose work on dreams and slips of the tongue suggested countless strings of surprising, but psychologically possible, associations of ideas (see Chapter 5.ii.a). The other was the literary scholar John Livingston Lowes (1930), a specialist on Samuel Taylor Coleridge's poetry. In a masterly detective story hunting out the sources of the imagery in *The Ancient Mariner* and

Kubla Khan, he gave highly detailed, and intuitively convincing, hypotheses about just where Coleridge's ideas had come from. But in answer to the question *How?*, all he could say was to repeat Coleridge's own metaphor of "the hooks and eyes of memory" (cf. Boden 1990*a*, ch. 6).

In other words, by 1940 *the fact that* our memory is richly associative was clear, but *how it manages* to associate wasn't. Indeed, things had been made worse by Karl Lashley's recent failure to find localized memories (5.iv.a). As he put it, "I sometimes feel, on reviewing the evidence . . . that learning just is not possible. *It is difficult to conceive of a mechanism* which can satisfy the conditions set for it" (Lashley 1950: 477–8; italics added).

The connectionists tried to answer that *How?* question, in scientific rather than merely intuitive terms. They hoped to explain not only the maverick associations found in poetry or dreams, but also everyday perception, language, learning, and rational thought. And they showed how, in principle, a memory could be distributed over a whole network of cells, rather than being stored in one place.

Connectionists thought of the hooks and eyes of memory in two different, but closely related, ways. On the one hand, they wanted to define idealized computational networks, made up of many simple interconnected units, which could in principle underlie this or that psychological phenomenon. On the other, they wanted to identify the computations actually carried out in the brain, and to understand how the relevant brain mechanisms implement them.

The first is the aim of connectionist AI/psychology, the second of computational neuroscience. The two disciplines are discussed here and in Chapter 14, respectively. Some people's research was guided by both aims, so is mentioned in both chapters. Walter Pitts and Warren McCulloch in 1947, for example, defined two novel types of information processing *and* related them to specific neural structures. And Albert Uttley, in the early 1950s, focused on real neurones as well as artificial ones—and turned towards neuroscience even more strongly in his mature work.

Connectionism seeks to explain (and to model) psychological capacities that intrigue almost everyone: perception, memory, analogy, creativity, language, learning, and development. It puts computational flesh onto the associative bones sketched by Freud and Livingstone Lowes, suggesting how haunting images can arise—whether neurotic or poetic. And it carries a strong whiff of the brain. So far, so good.

At base, however, it's the study of two highly abstract topics: the configuration (anatomy) of neural nets and the functioning (physiology) of learning rules. Inevitably, then, much of this chapter concerns the mathematics of various learning rules and nets. I'll spare you (and myself) the equations, for this isn't a practitioners' manual. Even so, some paragraphs may be tough going for some readers. The overall story, however, should be readily intelligible—and exciting, too. For connectionism has aroused heady hopes, passionate controversies, huge media interest, and intense philosophical debate.

The first two sections below focus on the opening quarter-century of connectionism: first, the initial ideas, whose implementation—if any—could only be highly schematic; second, some pioneering computer models. Section iii deals with a notorious 1960s critique of some of that early work, which is widely (though perhaps unjustly) accused of having set the field back for almost twenty years. What was actually going on in those years of hibernation is described in Sections iv and v.

The heyday of parallel distributed processing (PDP)—which is what most people today take "connectionism" to mean—is discussed in Sections vi and vii, as is the explosion of the "bombshell". Sections viii and ix consider how PDP research later tried (and failed) to recover some of the lost strengths of GOFAI. Finally, Section x explains why many philosophers were excited by PDP, and how some contributed to its development.

12.i. Lighting the Fuse

Connectionism's fuse was laid down in the mid-eighteenth century, and patted into the ground more firmly several times over the next 200 years. It wasn't until the 1940s, however, that it began to smoulder. That happened when McCulloch and Pitts (1943) defined idealized neural nets based on logic (see 4.iii–iv).

During the 1930s, neural feedback loops, or reverberating circuits, had often been posited to explain psychopathology of various kinds (4.iii and v). But by the early 1940s, McCulloch was thinking of them mostly as the brain's *normal* activity. As for how they might be acquired, he and Pitts suggested in their 'Logical Calculus' paper (1943) that learning could—in principle—be modelled by cyclic nets.

Even then, there was no intellectual explosion—only a dull glow. Initially (as we saw in Chapter 4.iii.f), their paper faced hostility or indifference from most psychologists. One reason was its neurological implausibility. Another was its rebarbative abstractness. And a third was its blue-sky character: at that time, its ideas could be implemented only in a very primitive fashion (with soldering irons to the fore). The dull glow wouldn't brighten until computers came on the scene.

a. A long gestation

Connectionism was conceived almost three centuries ago (in the 1740s), but had a long gestation. In its embryonic form, it was merely biologized introspection: David Hartley's view that thinking is grounded in associative mechanisms in the brain (2.x.a). These, he said presciently, make two things possible. First, they link different concepts so that the one comes to mind when one thinks of the other. Second, they enable a mere fragment to recall a whole stimulus. (Hum the first three chords of Beethoven's Fifth Symphony, or of Three Blind Mice—and see whether your neighbours can stop themselves from thinking of the fourth.)

Hartley was unusual in thinking about brain mechanisms at all. His contemporary David Hume, an immeasurably greater philosopher, theorized about mental associations but not about any detailed goings-on in the brain (2.x.a). His supposed Newtonian "attractions" were even more mysterious than Hartley's hypothetical "vibrations".

A hundred years later, Hartley's hints were echoed by the father-and-son philosophers James and John Stuart Mill (2.x.a). They, too, supported their mentalistic focus on the association of ideas with speculations about the brain. They argued that learning could be due to an increased likelihood of brain activity at certain points, as a result of simultaneous activity earlier. But *just what* this activity was, and *just how* its probability could be modified, remained mysterious.

In the late nineteenth century, some then unorthodox biologists ventured that the mechanism might involve links between separate units, or brain cells—as opposed to a continuous network, or reticulum (2.viii.c). And in 1890, thanks to William James's widely read *Principles of Psychology*, that general idea became familiar in the new discipline of psychology.

Yet familiarity didn't bring clarity, still less proof. For nearly 200 years, Hartley's hunch remained no more than intriguing.

By the 1940s, however, it had begun to look highly plausible. The neurone theory had been vindicated. "Animal spirits" and "vibrations" had given way to action potentials. And various synaptic properties—facilitation, inhibition, thresholds, and refractory periods—had recently been established (see Chapter 2.viii).

These ideas led McCulloch and Pitts to describe psychology as the study of precisely defined neural nets (4.iii.e). Lashley and Donald Hebb proposed that psychologists should also think about the activity of less precisely defined *populations* of neurones (5.iv). And in 1949, Hebb put the central idea into a more tractable form, suggesting that a "connectionist" (his word) learning rule for modifying synapses is the neural mechanism underlying memory and conceptual thought (5.iv.b–e).

By that time, too, ideas about computation, and about "self-organization" (as cybernetic control) in animals *and* machines, had arisen (Chapter 4). Since it was now conceivable that some artefacts might work in a way somehow similar to the brain, connectionism was often expressed in computational terms—drawing sometimes on logic, sometimes on statistics.

Some of these early systems managed to show how a whole group of (artificial) units could store a 'single' representation—and access it, as Hartley had guessed, when given only a partial cue. And the 1950s saw the development of models that could learn concepts as well as associating them. Typically, these used some version of Hebb's 'ft/wt' (fire together, wire together) rule.

Connectionism was no longer just a twinkle in Hartley's eye. The baby was now well and truly born.

b. Turing and connectionism

The prototype of the first modern computer, MADM, wasn't completed until five years after the 'Logical Calculus' paper. And the von Neumann computer, whose design rested in part on McCulloch and Pitts' 1943 theory, became available later still, in the early 1950s (Chapter 3.v.e). Only then could functioning models of the mind—whether symbolic or connectionist—be based on their ideas.

Before that time, various people had predicted that the advent of digital computers would make exploratory modelling possible. They were mostly engineers, or psychologists with a background in physics or engineering.

Belmont Farley and Wesley Clark, engineers at MIT's Lincoln Laboratory, had even sketched potential simulations of trial and error learning (Farley and Clark 1954). Describing "self-organizing networks" based on Turing machines, they distinguished "conditioning" and "serial learning" in terms of short-performance and long-performance units, respectively. But these were merely suggestions: in the early 1950s, functioning models were still thin on the ground.

One of the very first was due to Alan Turing. He'd already implemented a learning network by 1947, on an embryonic version of MADM. He reported observing "the sequence of externally visible actions for some thousands of moments" (Turing 1947*b*: 20). Most of his connectionist experiments, however, had been pencil-and-paper simulations.

It may seem strange to refer to Turing and connectionism in the same breath. For he's usually thought of as the guru of symbolic AI, thanks to the Turing machine (4.i) and the Turing Test (16.ii.c)—not to mention his early efforts in programming (3.v.b). It's widely assumed (by non-specialists) that he said nothing about connectionism. But that's not so. His oft-repeated aim "to build a brain", which dated at least from 1944 (Hodges 1983: 290), *isn't* relevant here, for it didn't necessarily imply a brainlike (micro-)anatomy. But in a speculative report written in September 1947, he discussed a range of computational networks consisting of standard binary units, each one linked to (zero or more) others (Turing 1947*b*).

Such networks could be specifically designed to perform a pre-set task, as McCulloch and Pitts had remarked. Alternatively, they could be "unorganized". In the latter case, their links were randomly assigned. This, said Turing, is "about the simplest model of a nervous system with an arrangement of neurons [such as the cortex] whose function is largely indeterminate" (1947*b*: 10, 16).

For learning to occur in an initially unorganized network sounded, on first hearing, like magic. But he suggested that some training procedure, analogous to the education of a child, might be imposed which would gradually alter the linkages. A random network could be converted into one that behaves in systematically adaptive ways, he said, if the positive and negative influences on the links ("pleasure and pain") were set appropriately. His observations of "thousands of moments" of computer activity were focused on a pioneering model of this type.

Turing's remarks here were, as ever, intriguing. But they had no influence on the field. The paper was an internal report for NPL (the UK's National Physical Laboratory), and remained unpublished until 1969.

Certainly, he sometimes mentioned these ideas to compatriots and foreign visitors. For instance, at a seminar held by the Manchester philosophers in 1949, at which luminaries from various disciplines were present (see 16.ii.a), he speculated about "a machine using neuron-models" (Manchester Philosophy Seminar 1949). But just how this might be achieved wasn't spelled out. Even when lecturing to the London Mathematical Society in the year of the NPL report, he spoke only about ACE (3.v.b), without mentioning the possibility of more brainlike machines (Turing 1947*a*).

He himself failed to develop these ideas further because, by the late 1940s, his own interests were turning from learning to embryology. He was still concerned with the gradual organization of an initially unorganized substrate. But he was now asking how it might happen *without* the "external interference" of a teacher (A. M. Turing 1952: see 15.iv.a). In other words, he'd turned to self-organization.

c. 'How We Know Universals'

While Turing was still writing his NPL report, a very different suggestion about brainlike machines was being published on the other side of the Atlantic. This was contained in an essay on 'How We Know Universals: The Perception of Auditory and Visual Forms' (Pitts and McCulloch 1947), and it did make connectionism's fuse glow brighter.

The new paper by Pitts and McCulloch was rooted in cybernetics. One of its sources was a statistical theory of heart fibrillation, in terms of the spread of impulses within a network of excitable elements. This had been developed in the mid-1940s by Norbert Wiener and Arturo Rosenblueth, and Pitts had recently worked on it.

Pitts and McCulloch had now switched from logic to statistics, from single units to collectivities, and from purity to noise. Although they didn't use the modern vocabulary, it's clear that their 1947 paper was describing the brain in terms of what's now called distributed, cooperative, computing (Chapter 14.ii.b and iii.a).

Whereas their 1943 networks hadn't been intended as a biologically plausible theory, their 1947 networks were. For they now proposed neurological hypotheses to match specific neural nets (see 14.ii.b). One of the three reasons (mentioned above) for the former indifference to their work had therefore vanished—and indeed, their second paper got a warmer welcome than the first (Bishop 1946). Another obstacle would soon vanish too, when computer technology became available.

The two authors didn't disclaim their earlier, logic-based, work. On the contrary, they saw the new approach as an "extension" of it, and as "a systematic development of the conception of reverberating neuronal chains". They still insisted that "any theoretically conceivable net" based on neuronal properties demonstrates that the computation concerned could be physically effected. But they were now ready to be more realistic about what those (anatomical and physiological) properties are.

They allowed that one can't assume that the brain is made up of such neat and tidy connections as they had previously discussed. Nor can one assume that actual neurones have such constant, reliable, thresholds. Plausible neural nets therefore have to be error-tolerant:

It is wise to construct . . . these nets so that their principal function is little perturbed by small perturbations in excitation, threshold, or details of connection within the same neighborhood. Genes can only predetermine statistical order, and original chaos must reign over nets that learn, for learning builds new order according to a law of use. (Pitts and McCulloch 1947: 46)

Moreover, perception itself is error-tolerant, in the sense that it involves the recognition of categories—what philosophers call "universals"—whose particular instances may differ in many details. Each cat is somehow unlike every other, but they're all recognized as cats. A theory of perception should therefore explain how a physical mechanism such as the brain (or, by implication, a computer) could do this.

What they sought, then, was "general methods for designing nervous nets which recognize figures in such a way as to produce the same output for every input belonging to the figure" (p. 47). And "output", here, covered both *classification* (*recognition*) and *action* (*movement*). For instance, we'll see below that Pitts and McCulloch asked how the eye muscles can be instructed to lock one's gaze on the centre of attention.

Instead of strings of precisely positioned logic gates, they now considered parallel-processing, statistical, devices. These were first defined as abstract mathematical systems, and then related to possible implementations in the brain.

"Parallel", here, meant logical independence rather than precise temporal simultaneity. Time (all but ignored in the 1943 paper) was assumed to be an important factor.

But the real-time details were deliberately ignored. For instance, their system might compute a moving average over the preceding five synaptic delays, but these were unrealistically assumed to be equal and constant (p. 47). As for "statistical", this meant not probability theory but some sort of averaging, or assimilation to a norm.

In general, they argued, a universal is recognized by identifying an invariant under some group of mathematical transformations. They defined two types of mechanism capable, in principle, of doing this.

In the first, an input organ (such as the retina or cochlea) made up of many individual receptor units is "scanned" to find the average of all the independent measurements, or mini-decisions. This could enable the system to recognize (for example) shapes of any size, or chords of any pitch. Their mathematical analysis highlighted a general principle they called *the exchangeability of time and space*. This said that the delay in scanning any stimulus dimension corresponds to the number of distinct places on that dimension. In other words, the fewer discriminable points, the less time needed for discrimination—and vice versa.

In the second, a group of distinct perceptual inputs is "reduced" to a standard form. The "uniform principle of design" here is to select some *one* of the many appearances (transformations) of the category as the norm, and then to define parameters enabling one to locate all the others with reference to it. If any of the transformational equivalents is perceived, then the category will be recognized as such. All triangles, for instance, are topologically equivalent and have straight sides. (Compare John Locke's puzzlement over how a triangle can be thought of as "neither oblique nor rectangle, neither equilateral, equicrural, nor scalenon; but all and none of these at once"—see 2.x.a.)

The example they gave was the gaze reflex, in which the eye centres on some stimulus in the visual field. The norm, here, was defined as the "center of gravity of the distribution of brightness". They showed how a neural mechanism could compute this centre of gravity, move the eyes towards it, and then—through negative feedback—keep them there. The mechanism they suggested, based partly on work done on cats by the neuro-anatomist Julia Apter (1946), involved two point-to-point mappings: one from retinal cells to the superior colliculus, and one from the colliculus to the eye muscles.

This paper aimed for a high standard of theoretical rigour. Its approach was "basically simple and completely general, because any object, or universal, is an invariant under some groups of transformations" (W. S. McCulloch 1961b: 10). (Identifying all the relevant transformations was, of course, the $64,000 question.) It even included proofs that a certain net was *the simplest one possible* for effecting a certain computation. But it tried not to substitute abstraction for practicality. Some conceivable systems (designs for brains or computers) were defined only to be rejected, because of the combinatorial explosion.

However, it dealt with neither learning nor randomness. Its topic was how we know (recognize, represent) universals, not how we learn them. The "small perturbations" in thresholds and connectivities weren't supposed to result from the processing history of the system, but were taken as given. (They corresponded to the spontaneous neural firings discovered in the 1930s, and caused by chemical changes at the synapse: see 2.viii.f.) Moreover, they had to be small if the mathematics was to fit: massive changes or randomness would significantly alter the computed average and/or distance from the norm.

Randomness was widely recognized by cyberneticians as an important challenge, given the behaviourist anti-nativism prevailing at the time (see 5.i.a and 7.vi). If there was little or no prior structure to the newborn's cerebral cortex, how did interaction with the environment manage to produce it? Or if there *was* significant prior structure at birth, could this be ascribed to universal mechanisms of self-organization working in the womb, rather than to individual heredity? (see Chapter 14.viii.a). The encephalographer William Ross Ashby, for example, was working on self-organization in random systems, and would soon unveil his adaptive Homeostat (4.vii.c–d).

Likewise, McCulloch and (especially) Pitts were already trying to generalize their ideas to random nets. But they'd achieved suggestive analogies rather than definitive theorems.

So in a lecture given in the previous year (but not published until 1952), McCulloch had summarized their approach in terms of each neurone's being surrounded by "a nest of surfaces such that the chance of connection with our neuron is the same for all neurons on one and the same surface" (W. S. McCulloch 1946: 270). If only the connections are important, the nest will be spherical. If the directions and/or positions also matter, it will be egg-shaped or somehow lopsided.

In the same lecture, he'd compared learning to the physics of magnetizing a bar of steel. The myriad tiny magnets, initially positioned at random, end up (because of many local interactions) all pointing in the same direction—and, having reached equilibrium, they stay there. After spelling out the analogy between the various stages of magnetization and the formation of synapses, he said:

> It is not too much to hope that with these things in common the mathematics for one may be shaped to fit the other. If it will serve, then we may someday state how random nets may learn by taking on this or that local structure . . . *This may be done in some few years.* (W. S. McCulloch 1946: 273; italics added)

Those years turned out to be many, not few. Pitts had done some work on the mathematics of the three-dimensional "nests" mentioned by McCulloch. But he didn't persist, and never published his results. McCulloch (1961*b*: 110) later said this was "because we could not make the necessary measurements". But part of the problem was the dreadful mental decline that struck Pitts in 1952 (see Chapters 4.iii.d and 14.v.b). He burnt his manuscript on 3D nets, and no trace was ever found—despite major efforts by his friend Jerome Lettvin in searching for it.

Pitts described his new ideas in public only once, at the second Macy conference on cybernetics (see 4.v.b). That lecture wasn't published, either. However, as McCulloch remarked later, "[It] was enough to start John von Neumann on a new tack" (1961*b*: 110).

d. From logic to thermodynamics

This new tack—on which McCulloch, too, was working—was a search for a probabilistic logic (von Neumann 1956). That's one in which the functions, not just the arguments, are more or less probable.

Von Neumann did sketch a probabilistic network of formal neurones. But he wasn't satisfied with it, because the units were far more reliable, and far simpler, than real neurones. So instead of focusing on that network, he generalized his research to cover a diversity of "complicated automata"—now called cellular automata (15.v).

The result, in effect, was invisibility. These highly abstract ideas reached few mathematicians or computer scientists, and even fewer neurophysiologists. Von Neumann did publish a lecture he gave at the Hixon Symposium in 1948 (von Neumann 1951). But otherwise, his work—including five lectures given at the University of Illinois in 1949—remained unavailable until the mid-1960s, well after his death (15.v.b). Again, light was being hidden under bushels.

If von Neumann's work on cellular automata didn't influence the nascent connectionist community, his Silliman Lectures did (von Neumann 1958). These were intended for delivery at Yale, but were still unfinished at his death. They'd grown, at McCulloch's instigation, out of a talk von Neumann gave to the American Psychiatric Association in 1955.

There, he'd stressed the computational complexity of individual neurones as well as of collectivities, and argued that mere redundancy (large numbers of units) couldn't account for the brain's power and reliability. In the Silliman Lectures, too, he argued that brains ("natural automata") are so different from digital computers that computational neuroscience can't be grounded in logic.

The differences he mentioned included:

* the statistical (imprecise, yet reliable) character of the nervous message;
* its implementation as periodic pulse-trains;
* its anatomical restriction (often) to only a few synaptic steps;
* the continuous (analogue) properties of the synapse;
* the rich connections of individual neurones;
* the huge numbers of neurones;
* their functioning largely in parallel;
* and the possibility that neural thresholds and/or connections may change over time.

All these, he said, implied a need for a very different approach:

[Whatever] language the central nervous system is using, it is characterized by less logical and arithmetical depth than what we are normally used to . . . [The] language here involved may well correspond to a short code [i.e. a basic machine language], rather than to a complete code [i.e. an assembly code or programming language]; when we talk mathematics, we may be discussing a *secondary* language, built on the *primary* language truly used by the central nervous system . . . [Whatever] the system is, it cannot fail to differ considerably from what we consciously and explicitly consider as mathematics. (von Neumann 1958: 81–2)

As for what the new approach should be, however, this unfinished text didn't say.

To be sure, there'd been a hint already. For in the 1948 lecture, von Neumann had suggested that "thermodynamics, primarily in the form it was received from Boltzmann" might be a good analogy for cerebral processing. (Compare McCulloch's earlier remarks on "magnetization", above.) But this idea couldn't yet be explored by computer modelling. The special-purpose differential analysers he'd used since 1940 weren't suitable, and the infant von Neumann machine wasn't sufficiently powerful.

It wasn't even obvious that a more powerful version could fit the bill, until McCulloch proved (in 1959) that a digital computer could implement a probabilistic logic (W. S. McCulloch 1961*b*: 16). But this was a point of principle, not practice. Only very much later would "thermodynamic" connectionism be feasible (see Sections v.e–f and vi.b, below). Meanwhile, von Neumann's fellow Hungarian Andras Pellionisz was inspired by his book to develop a "geometrical" form of connectionism (see 14.viii.b). Tucked away in Budapest, however, his ideas weren't taken up.

Von Neumann's scepticism about logic-based approaches had some influence, nevertheless—not least, because it was seen as a recantation by the master, the man whose eponymous machine had made GOFAI possible. Right from the start, a close friend recalled later, "one of [von Neumann's] motives for pressing for the development of electronic computers was his fascination with the working of the nervous system and the organization of the brain itself" (Ulam 1989: 19). Indeed, he'd altered his own computer design to incorporate the McCulloch–Pitts neurone (see Chapter 4.iv). The fact that he now went to such pains to stress the differences between brains and computers attracted attention accordingly.

Many readers of his posthumous book inferred that GOFAI (still NewFAI at the time) was doomed to failure. They assumed that *brainlike* computer models were needed instead. But just what might that mean?

12.ii. Infant Implementations

Several early cyberneticians had already tried to build brainlike models, some of which were theoretically interesting (see Chapter 4.vi–vii). But they'd had to rely on their own engineering skills, working with the proverbial biscuit tins and string to build special-purpose connectionist *hardware*.

This might involve a monstrous apparatus, like the analogue associative-memory machine constructed at University College London in the early 1950s by Wilfred Taylor (1956, 1959). This machine could recognize patterns, such as alphabetic letters, written in varying tones of black and white, and presented in varying orientations. In effect, it learnt its own feature detectors (see 14.iii). And it worked fast: it could find the pattern edges (high-intensity gradients) in less than a microsecond.

The apparatus had a nine-cell artificial retina. This involved lateral inhibition between neighbouring input units, and a "maximum amplitude filter" that passed on—to the sixteen-cell associative layer—*only* the strongest signal within a given area, provided that it was larger than the others by some minimal amount. (Later, Taylor applied these ideas to the brain, suggesting that specific cortical neurones and connections act as maximum-amplitude filters: W. K. Taylor 1964.) The light intensities were measured discontinuously, as ten distinct levels. Patterns were coded in terms of the most useful sensory input features, such as vertical, horizontal, and oblique lines. These features weren't built in, but were identified (from around 500 possibilities) by means of lateral inhibition.

By the mid-1950s, however, general-purpose computers were becoming available in the leading research centres. These were huge monsters too, of course. But each one could be used for many *different* purposes. This meant that (small) connectionist models could now be implemented more conveniently, by being simulated in digital computers—as Turing (1947*b*) had foreseen.

Not everyone switched from analogue to digital. Taylor, for instance, continued to develop his hardware model through the early 1960s, because the data couldn't be fed into it quickly enough in simulated form. But many others did take advantage of the new technology. The scene was now set for an explosion of work in both symbolic *and* connectionist AI.

a. B24 bricolage

While MADM and EDSAC (see 3.v.b) were still being put through their pioneering paces, Marvin Minsky, then an undergraduate at MIT, was puzzling about trial-and-error learning. He was largely inspired by McCulloch's ideas and by Hebb's then recent book, both of which he'd learnt about from his teacher George Miller.

On graduating in 1951, and before starting work on his Ph.D., he turned his puzzling into practice. With a $2,000 grant from the US Navy (officially, for Miller), he built a machine that simulated four rats in a maze learning to avoid each other:

In the summer of 1951 Dean Edmonds [a physicist] and I went up to Harvard and built our machine. It had three hundred tubes and a lot of motors. It needed some automatic electric clutches, which we machined ourselves. The memory of the machine was stored in the positions of its control knobs, 40 of them, and when the machine was learning, it used the clutches to adjust its own knobs. We used a surplus gyropilot from a B24 bomber to move the clutches. (J. Bernstein 1981a: 69)

Each of the forty "neurons" had a probability of firing, implemented by the machine's potentiometer. The reward/punishment clutch was fired by a gas tube. And when the reward/punishment button was pushed, the shaft would turn clockwise or anti-clockwise, thus increasing/decreasing the probability of passing a signal next time.

(Where is this swords-to-ploughshares machine now? Well, after the 1956 meeting McCarthy's mentor at Dartmouth, John Kemeny, said that he wanted to have it there. Minsky and McCarthy got it up to Hanover, in full working order. However, it was never used for anything else. "Maybe it's still there, in some basement": J. McCarthy, personal communication.)

The contraption's success—up to a point, it worked—had much the same effect on Minsky that NewFAI was to have on many people a few years later:

We sort of quit science for a while to watch the machine. We were amazed that it could have several activities going on at once in this little nervous system. Because of the random wiring it had a sort of fail safe characteristic. If one of the neurons wasn't working, it wouldn't make much difference and with nearly three hundred tubes, and the thousands of connections we had soldered there would usually be something wrong somewhere . . . I don't think we ever debugged our machine completely, but that didn't matter. By having this crazy random design it was almost sure to work no matter how you built it. (J. Bernstein 1981a: 69; italics added)

However, he soon abandoned this happy-go-lucky approach. Not only was he unable to interest his friend Burrhus Skinner in the project (he'd been trying to implement Skinner's ideas—Crevier 1993: 35), but his mathematical instincts overpowered his engineering ones.

The switch away from gizmo-gazing was triggered by his meeting someone else who'd worked on learning. Specifically, someone studying the induction of Miller's artificial grammars (Chapter 9.v.d and x.b):

[In around 1955] I met a young man named Ray Solomonoff who was working on an abstract theory of deductive inference . . . He had worked on a learning machine . . . that was pretty formal. I was so impressed I decided this was much more productive than the neural net system, in which you built a piece of hardware and hoped it would do the right thing. With [this new] approach, *you tried to make theories of what kind of inferences you wanted to make, and then asked "How would I make a machine do exactly that?"* (Minsky, interview in Crevier 1993: 37; italics added)

In other words, bricolage was no longer enough. Nor were intriguing but ill-understood 'successes'.

Nor, significantly, was it enough to try "to understand how the brain works" as opposed to "understanding what it does" (interview in McCorduck 1979: 84). Ray Solomonoff, in short, had made him realize that McCulloch-and-Pitts wasn't sufficient. He also needed *a better theory of the task*.

A few years later, in his seminal 'Steps' paper (Chapter 10.i.g), Minsky would apply this insight right across the AI board. But in this early case, it was being applied to learning. To be sure, his Ph.D. of 1954, part-supervised by von Neumann, described similarly tinkered machines—constructed from "SNARCs": Stochastic Neural-Analog Reinforcement Calculators. But it discussed some of their limitations, as well as their potential (cf. also Minsky 1956a). For example, he speculated about how one network could be enabled to control another (a prime focus of research today: Section ix.a, below).

Paul Werbos's diagnosis, years later, was that Minsky—like many others "in the learning-business"—simply "didn't understand numerical analysis [or] the concept of numerical efficiency" (J. A. Anderson and Rosenfeld 1998: 344). However that may be, by the mid-1960s Minsky favoured careful mathematical analysis of the computational power of AI models in general. He dropped connectionist research for a while, disappointed by his own pioneering machine. Indeed, his abstract analysis of early connectionism's limitations would lead to a major scandal in the field (see Section iii).

It wasn't only Minsky who'd matured by the 1960s. Throughout the previous decade, some other connectionists had progressed from isolated tinkering towards systematic analytical research. (Some, but not all: "Why think when you can simulate?" was still a common attitude—J. A. Anderson and Rosenfeld 1998: 245.)

These influential early modellers included the cyberneticians and psychophysiologists mentioned in Chapters 4.v–vi and 14.i–ii. Others, discussed below, were Uttley, Raymond Beurle, Oliver Selfridge, Frank Rosenblatt, and Bernard Widrow.

b. Self-organizing networks

One swallow doesn't make a summer, and half a dozen neurones don't make a network. It was accepted in the 1950s that McCulloch–Pitts neurones could implement logic gates, and many strings and simple loops for performing and/or learning specific tasks were modelled accordingly. But other questions concerned the properties of *whole groups* of neurones, as involved in the reverberating circuits posited by neurophysiologists—and by Hebb (1949).

Turing had recently described embryogenesis in terms of interacting waves (Chapter 15.iv.a), and others now used similar ideas to explain *neuropsychological* self-organization. One such approach was pioneered in 1954 by Beurle at Imperial College, London, and developed by Clark and Farley at MIT (Beurle 1956, 1959; Clark and Farley 1955). They studied the dynamical properties of two-dimensional arrays of randomly connected cells, or neurones. Those properties were conceptualized as waves of activation: building, spreading, persisting, dying—and sometimes meeting and interacting.

The randomness was important not just as a mathematical challenge but as an attempt to match the biological facts. The increasingly common comparisons between computers and organisms, said Beurle, were "interesting when made in relation to abstract fundamental concepts, but less productive when details are considered" (Beurle 1956: 83). For whereas the components of computers are "connected to some exact specification", many parts of the cortex show a "very large random factor in the interconnexions between neurons". In short:

The aim of this paper [is] to show that some of the basic forms of behaviour of living organisms can be simulated by a mass of simple units *without the necessity of postulating a large degree of specific organization* of these units. (Beurle 1956: 83; italics added)

Among those "basic forms of behaviour", said Beurle, were chains of conditioned responses and memory in general (1956: 81–3).

Beurle made constant references to "servo-control", but his 1956 paper was a theoretical one. In the implementations of his wave-interaction theories, B24 bombers were left far behind. In some of Farley and Clark's later work, a model network was built out of 1,296 analogue units, with as many as thirty connections each (Farley and Clark 1961).

By that time, even larger networks were being *simulated*, not built with soldering irons, by Richard Laing at Michigan—up to 10,000 neurones, with up to 200 synapses per neurone (Laing 1961a). But Beurle and Clark-and-Farley were earlier. Moreover, MIT—Clark and Farley's home base—had much more visibility in the AI community than Michigan did (see Chapter 15.v.b). Light under bushels, yet again.

That's not to say, however, that the Imperial and MIT hardware implementations were so primitive as to be a waste of time. Far from it. This work remained influential long after it was done, as we'll see.

The Beurle-inspired networks were guided, to some extent, by mid-century neuro-physiology (2.viii.d–e). For instance, although the connections were random, the probability of two units' being connected decreased with their distance from each other. This was based on (admittedly shaky) evidence about the distribution of cell bodies and axons in the cat's visual cortex.

One physiologically unrealistic assumption made by Beurle was that there was a fixed synaptic delay, and a fixed refractory period—after which a cell would suddenly recover its excitability. These timings were important, because the behaviour of the whole network would depend on the proportion of currently excitable cells. Some of his followers, such as Farley and Clark, tried to keep closer to the biology. That is, they allowed for exponential recovery, and for variations in the thresholds and connectivities involved. Their waves could be visually displayed, and even 'frozen' for inspection (Farley and Clark 1961).

Beurle claimed that if several cells at the centre were simultaneously stimulated, the activation would spread out over a proportion of the net. If the stimulus was weak, the wave would eventually extinguish. But a strong stimulus would generate a wave that spreads out over the whole sheet: "[The wave] will increase in amplitude until it 'saturates,' when it uses all the cells in the medium through which it passes, and the amplitude cannot increase further" (Beurle 1956: 63). By contrast, stimulation of a single cell wouldn't spread.

In certain circumstances, a wave of excitation could leave a process of self-excitation behind it:

Fully saturated waves may follow each other at intervals equal to the period taken by the cells to recover sensitivity completely, and unsaturated waves may follow even closer. This makes it possible for a wave to pass through a region, and return again to the same region some time later when the majority of the cells has recovered. This would allow a relatively local circulation of activity which might be of some importance. (Beurle 1956: 65)

If two waves of excitation were generated independently, specific types of interference pattern would result when they met.

In some of Beurle's computational experiments, neighbouring units mutually inhibited one another. In others, a unit's sensitivity was slightly increased with each activation, so that the collectivity learned to respond differently as time passed. These results were used to ground hypotheses about the difference, and the interaction, between short-term and long-term memory.

Beurle's hypotheses were not merely qualitative, but quantitative too: he offered a mathematical theory describing the origin of wave-interference patterns. It turned out later that some of his mathematics was mistaken. But many of his ideas remained interesting, especially his demonstration that a *part* of a wave could regenerate the rest.

The neurophysiologist Jack Cowan (1933–) came across Beurle when he was at Imperial College in 1956. He now says that "a lot of the basic ideas about the dynamical properties of networks of neurons are sitting in Raymond Beurle's work", and "I found what Beurle had done was really interesting mathematically," despite the mistakes (J. A. Anderson and Rosenfeld 1998: 105).

In the early 1960s, this part-to-whole result would be cited in developing holographic theories of associative memory (see Section v.c). And when Cowan took over Nicholas Rashevsky's then flagging group at that time, he started by trying to develop Beurle's work—hiring the young chemist Hugh Wilson to help him (alongside Stuart Kauffman, "right out of medical school": see Chapter 15.viii.b).

In the early 1970s, Cowan and Wilson formulated an influential theory of neural dynamics that included Beurle's system as a special case. Later still, Cowan realized that they'd created the neural analogue of Turing's (1952) work on morphogenesis, which was a *universal* model of pattern development (J. A. Anderson and Rosenfeld 1998: 111, 118). Small wonder, then, that he regards Beurle's research as mathematically interesting.

c. Connections with the Ratio Club

Uttley (christened Albert, but nearly always called Pete) had outlined an associative memory even before Beurle did. Indeed, Beurle explicitly acknowledged Uttley's help, in the 1956 paper. Uttley grounded his pioneering models in current ideas about the brain—and in McCulloch and Pitts' neuronal logic (Sholl and Uttley 1953). And he tried very much harder than Beurle to match the biological details (see Chapter 14.ii.b).

Uttley was one of the seventeen founder members of the Ratio Club (see Chapter 4.viii). Indeed, it was he who'd chosen the name—because its etymology covered reasoning, relations, and numbers. Among his fellow founders were Ashby, Horace Barlow, Patrick Merton, William Grey Walter, and Donald MacKay. Other members included the visual psychologist William Rushton, the neurophysiologist John Pringle, and the AI pioneer Jack Good (Husbands forthcoming). (At least two of these men are widely regarded as being unlucky not to have won a Nobel Prize: namely, Barlow and Rushton.)

The Ratio members had got used to cross-disciplinary research during the Second World War, but Uttley was even more interdisciplinary than most. Originally a mathematician, he was involved in some early developments of computer technology. During the war, and afterwards at the Royal Radar Establishment, he developed analogue devices for target-tracking. He designed an early digital computer called TREAC (short for Telecommunications Research Establishment Automatic Computer) in the 1950s. This was the UK's first parallel processor. He also designed a dynamic computer memory, using feedback implemented by mercury delay lines—an idea anticipated by Turing (see 3.v.c). His prime interest (and graduate training), however, was in psychology. He left RRE for NPL in the mid-1950s, where he founded the "Autonomics" division to work on learning in brains *and* machines.

Combining mathematical expertise with a healthy respect for experimental data (on both brain and behaviour), his early ideas attracted various people's attention. Computer modellers were interested of course, but so were psychologists and, especially, neurophysiologists. His Ratio colleagues Barlow and Merton, for instance, were strongly influenced by him: the former for his application of information theory to neurones; the latter for his work on the automatic aiming of gun turrets in aircraft, which informed Merton's servo-theory of muscle control (H. B. Barlow, personal communication).

Uttley's reputation was international, too. He was hailed as the founder of "a revolutionary mechanical logic" by a French cybernetician (de Latil 1953: 280). And one of his papers was included by Claude Shannon and John McCarthy in their seminal collection on *Automata Studies* (1956).

In addition, Uttley co-organized the select London (NPL) seminar at which Selfridge and Rosenblatt first presented Pandemonium and perceptrons, respectively (see below), and Barlow first announced his coding theory of perception (14.ii.a). This NPL meeting, like the Dartmouth Project just before it, was an important event in the history of cognitive science (see 6.iv.b). Its participants included many people mentioned in this book.

For all the respect in which he was held, Uttley's ideas weren't always explained clearly. Barlow (personal communication) recalls that "he was not a great communicator", and remembers him "trying, and failing, to get across the idea of what he called 'unitary coding' at the Ratio Club". (Barlow now thinks this may have been a form of "sparse" coding: see 14.x.e.) Nevertheless, Uttley undoubtedly helped foster the nascent interest in cognitive science.

In his early days at NPL, Uttley developed a range of "conditional probability machines" (Uttley 1956, 1959*a,b*). These were intended to simulate various neural mechanisms underlying the conditioned reflex (see Chapter 2.viii.b) and innate releasing mechanisms (see Chapter 5.ii.c). One design, implemented by hydraulics as well as electronics, enabled a computer to count coincidences, and then to estimate the probabilities that two distinct inputs would occur together. He saw such probability estimations as the essential ground of all learning (14.ii.b).

Other models were focused on positive and negative conditioning, neural inhibition, memory decay, and time of onset. Uttley showed that if the temporal order of two inputs of equal strength were recorded, twelve different patterns could result—and could be detected automatically. In general, he explained how a single neural circuit used for inductive learning could distinguish several classes and subclasses, depending on the connectivities and thresholds involved.

These mid-century models are still recognized as important early studies of "reward-modified learning" (Minsky and Papert 1988: 282), and of the role of redundancy and probability estimations (Barlow 2001*b*, acknowledgements). Uttley's mature work, by contrast, is rarely cited today (see Section iv.a, below).

Quite apart from his specific models, his general approach—and his hugely exciting 1956 meeting—inspired a number of people in the 1950s to take computer models of intelligence seriously. As Richard Gregory (personal communication) has put it: "The idea that mind is due to physical mechanisms was still quite alien to almost everyone. Uttley saw very clearly that this was the way to go, so he shone out as a pioneer."

d. Pandemonium

One of Uttley's near-contemporaries on the other side of the Atlantic was Selfridge. (Yes, it was his family—his grandfather—who had founded the famous Oxford Street store.)

In today's publication-obsessed research evaluations, he would hardly figure. (His eight first-author entries in this book's bibliography span forty-seven years.) So much the worse for bibliometrics as a measure of intellectual value. For Selfridge was a hugely important figure in the history of cognitive science. His ideas, in conversation as much as publication (and in the circulation of unpublished drafts), would influence not only connectionism but several other areas besides (see Chapters 6.ii.c and iii.b, 10.i.b, 13.iii.d, and 14.ii.a).

As a young man at MIT, Selfridge was at the heart of the cybernetic community. He was close to both McCulloch and Wiener (who acknowledged his help in the second edition of *Cybernetics*). And his near friends included his contemporaries Jerome Lettvin and Pitts, who were his room-mates for a while. By 1947, he was revising Wiener's mathematical theory of heart flutter (with Pitts working alongside, on fibrillation), and already knew of the ideas on stimulus generalization in Pitts and McCulloch's universals paper.

Ten years later, he was publishing on pattern recognition and learning (Selfridge 1955, 1956)—and enthusing Allen Newell and Herbert Simon to embark on simulating thought (see 6.iii.b and 10.i.b). In addition, he was acting as the patron for Minsky's SNARC machine.

But above all, he was implementing his own "paradigm for learning": a program called Pandemonium, named after John Milton's vision of a demon-packed hell (Selfridge 1959). As remarked in Chapter 10.i.b, this could well have been included in the list of AI harbingers there. Its influence was incalculable.

McCulloch described the Pandemonium paper (written in July 1958) as "the outstanding American contribution" to Uttley's seminal NPL symposium on 'The Mechanization of Thought Processes' (W. S. McCulloch 1961*b*: 224). But that's not to say that he was surprised by it.

The fundamental idea—that perception involves a hierarchy of specialized feature-detectors, or "demons"—had been much discussed between himself, Wiener, and Selfridge, and he'd already applied it in designing a tonal 'reading machine' for the blind. It was familiar also to Minsky, Lettvin, and Ulrich Neisser (all acknowledged in an endnote). Indeed, Selfridge had outlined the core ideas several years earlier, at a meeting on learning attended also by Clark, Farley, Newell, McCulloch, Miller, and Selfridge's Lincoln colleague Gerald Dinneen (Selfridge 1955, 1956; cf. Dinneen 1955).

Pandemonium was part-program, part-programmatic. Selfridge had implemented a simple version on an IBM computer, and would describe its progress in the pages of *Scientific American* a few years later (Selfridge and Neisser 1960). The main excitement, however, lay in his vision of its *potential*. This vision included speculative glimpses of mindlike software "agents", capable of cooperation and communication not only with each other but with a human user too (13.iii.d–e).

The program was a NewFAI achievement. (That's why it could have been listed as a NewFAI harbinger.) However, it was conceptualized as a multilevel parallel-processing network (see Figure 12.1). In spirit, then, it was a connectionist system. It used a *localist* representation, in which one unit corresponded to one feature, or concept. From the bottom up, Selfridge distinguished four levels: "data" demons (cf. a cochlea or retina), "computational" demons, "cognitive" demons, and one or more "decision" demons. In effect, the higher-level demons were grandmother cells (14.x.e), each one looking out for a particular feature, or set of features, in the level below.

If a demon found what it was looking for, it "shrieked" (as Selfridge put it) to the demon/demons on the level above. When the input demons responded to specific sensory data, such as sound or light, the computational demons measured their activity—and some computed *compound* features (such as equality, greater than, less than, maximum, and average). The cognitive demons looked out for more complex patterns, such as the letters A or B. Their decisions were continuously graded: the more confident the demon was, the louder it shrieked. Since the image was simultaneously on view to all the cognitive demons, they all shrieked in parallel—hence, pandemonium. The top-level demon identified which one of these was the loudest, and its output represented the network's final decision about what the input pattern was.

Pandemonium could have been programmed to discriminate clearly defined patterns, or classes. But the aim was for it to *learn* to recognize new ones, and to do so even if they *could not* be precisely defined beforehand. Selfridge assumed—as Jerome Bruner

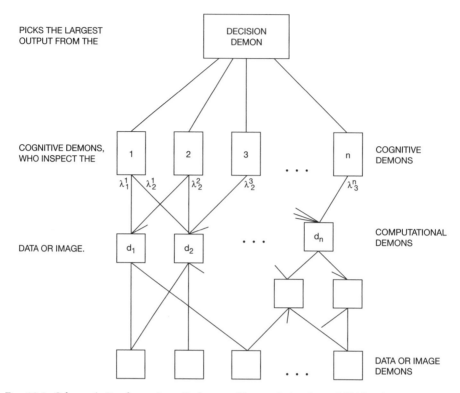

PICKS THE LARGEST
OUTPUT FROM THE

DECISION
DEMON

COGNITIVE DEMONS,
WHO INSPECT THE

COGNITIVE
DEMONS

DATA OR IMAGE.

COMPUTATIONAL
DEMONS

DATA OR IMAGE
DEMONS

Fɪɢ. 12.1. Schematic Pandemonium. Redrawn with permission from Selfridge (1959: 514)

was doing, too (6.ii.b)—that people's classifications (concepts) approximate logical functions of the features concerned. He designed the data demons and computational demons accordingly, and Pandemonium's task was to learn (via the cognitive demons) just which features are involved in this or that class.

It started out with a set of cognitive demons already defined in plausible terms:

For the [cognitive level], we collect a large number of possibly useful functions, eliminating a priori only those which could not conceivably be relevant, and make a *reasonable* selection of the others, being bound by economy and space. We then guess *reasonable* weights for them. *The behavior at this point may even be acceptably good*, but usually it must be improved by means of [several kinds of adaptive changes]. (Selfridge 1959: 516; italics added)

In other words, each cognitive demon computed a weighted sum of the outputs of all the computational demons, but the weights for individual features differed. In recognizing an A or an H, for instance, slanted and vertical lines were weighted differently.

The system was shown examples of various patterns, labelled as such, and had to guess the label for a newly presented example. It was then told whether its guess was right or wrong.

It learnt by hill climbing, a technique introduced in Arthur Samuel's checkers-player (10.i.e). So it continually computed the "score" of its overall performance. On each trial, the weights of some relatively highly weighted feature were slightly altered, in random directions, and the new score was compared with the previous one. This process was repeated until the overall score was maximized—or rather, until every such change reduced it. Selfridge realized that the system might get trapped on a local maximum, or "false peak", but could suggest no way of avoiding this. He did point out, however, that different learning rules would suit statistically different problem landscapes.

The network's method for learning was similar to what Selfridge saw as the basic principle of adaptive movement (see Section iv.b). But there was a crucial difference, which made him distinguish "supervised" and "unsupervised" learning. In these two types of learning the success is monitored, respectively, by the human trainer or by the system itself.

This was an important distinction, and (with the caveat mentioned below) a lasting one.

Today, the three major computational paradigms for learning are *supervised, unsupervised*, and *reinforcement* learning. (And there's some evidence that different brain structures and neurochemicals, such as dopamine, may be involved in the different types: e.g. Schultz *et al.* 1997; Kitazawa *et al.* 1998; Doya 2000.) Each type may result in sequences of input–output pairs (compare chains of conditioned reflexes: 5.iii.a). But this is achieved in different ways:

* In the clearest cases of supervised learning, the human programmer "trains" the system by defining a set of desired outcomes for a range of inputs, and providing continual feedback to the system about whether it has achieved them. When it fails to do so, it has made an "error"—and error messages of various sorts are crucial in the learning rules concerned. In other cases, the target is specified by the system itself, based on sensory signals and higher-level goals (Wolpert *et al.* 2001: 489). Examples include the "emulator systems" discussed in Chapter 14.vii.c, in which the brain predicts the sensory consequences of current motor activity and adjusts the relevant motor control if those predictions aren't borne out.

* In unsupervised learning, there are no desired outcomes or error messages—nor any reinforcers either (i.e. punishments or rewards). Rather, learning depends merely—as Hebb had suggested—on simultaneous activations of the pre-synaptic and post-synaptic neurones. Put crudely, cells that fire together, wire together (ft/wt). As we saw in Chapter 5.iv.b, Hebb himself gave two versions of the ft/wt rule; since then, there have been many further variations. If this seems to favour sequences built up by mere happenstance, perhaps it does: "The main problem with purely unsupervised learning is that *there is no guarantee that the representations learned will be useful for decision making and control*" (Wolpert *et al.* 2001: 490; italics added).

* In reinforcement learning, no desired outputs are specified beforehand. However, each actual output receives either reward or punishment from the environment. This may be the biological environment, or a human "teacher". Sometimes, whole sequences of individual outputs are strongly reinforced at their endpoint, in which case a problem of "credit assignment" arises: which of the individual outputs were most responsible for the happy ending? (Credit assignment is a common problem for non-connectionist AI too: see Chapter 15.vi.a.)

Returning to Selfridge, the promised caveat is that, according to the modern terminology, he sometimes conflated both supervised and unsupervised learning with reinforcement learning. He pointed out that animals, unlike Pandemonium, possess intrinsic measures of success, or innate reinforcers. (Even John Watson had allowed that a young baby will be frightened by loud noises: 5.i.a.) Moreover, he said, reinforcement tells the learner whether it has done well or badly, without providing detailed information on *just which* aspects of its decision were correct/mistaken (i.e. without what was later termed credit assignment).

Since Pandemonium couldn't judge whether one decision was "better" than another, it needed to be told. That is, it was restricted to supervised learning. Unsupervised learning might be possible, said Selfridge, but only if it rode on the back of supervised learning. For instance, if Pandemonium were to discover that the classes were very clearly divided, it might then reject classifications based on more ambiguous evidence.

Selfridge made three intriguing suggestions about how the functions, as opposed to the weights, of the computational and cognitive demons might be adaptively altered. This would enable the network to define its own operators.

(Pandemonium couldn't do this, but a daughter program did so a few years later: Uhr and Vossler 1961*a*/1963. That system initially learnt to recognize alphanumeric characters, and arbitrary doodles, by spotting—defining—their characteristic features for itself. But it was soon used also to learn to recognize spectrograms of spoken digits. These are complicated visual diagrams of acoustic output; non-experts can't distinguish them, but the program learnt to do so, even when the word was spoken by different people: Uhr and Vossler 1961*b*.)

His first suggestion, here, was that a relatively useless demon might be removed, or disconnected. Second, a new demon might be produced by "conjugating" two sub-demons, so that it coded the simultaneous presence of *two* specific features, or *one of them only*, or *one or both*, or *neither*, etc. (He was being influenced here not only by McCulloch and Pitts' 1943 paper, but also by Bruner's work on concept learning: 6.ii.b.) And third, a single demon might be randomly "mutated", and the best survivor retained.

(Simultaneous mutations were specifically excluded, because Selfridge didn't know how to attribute any ensuing improvement to *this* new demon rather than *that* one. In other words, the credit assignment problem again. We'll see in Chapter 15.vi.a that this nut was eventually cracked by John Holland—whose early modelling had located hidden contradictions within Hebb's verbal theory: Rochester *et al.* 1956.)

The implemented Pandemonium program was the first step in the construction of a machine to convert hand-keyed Morse code into a typewritten message. Its task was to learn to discriminate dots and dashes. This is more tricky than one might think: although dashes are in theory exactly three times as long as dots, in practice their length varies. Accordingly, there were only two cognitive demons: one for *dot*, one for *dash*.

But an automated Morse typewriter would need to do more than that: it would have to group the dots and dashes in appropriate ways. So Selfridge envisaged future programs having extra levels of cognitive demons, for computing "symbols" (e.g. *dot–dot–dot*), letters (e.g. *S*), and words (e.g. *SOS*). Beyond that, syntactic and semantic interpretation would require yet more levels.

The implication was that networks with many hierarchical levels exist in adult human brains, and might one day be modelled in computers. Perhaps all these levels could be learnt/constructed from scratch (by iterated conjugation, for instance), or perhaps they're largely built in—by evolution or by the programmer. Selfridge didn't address that question in this paper. But he soon would, as we'll see.

e. The perceptron

Exciting—and influential—though it was, Pandemonium was overshadowed by the near-simultaneous *perceptron*. Perceptrons were described in a leading psychological journal in 1958, shortly before Selfridge gave his NPL talk. (They were also described at the NPL meeting, so reached a widely interdisciplinary group: Rosenblatt 1959.)

The perceptron paper was a distillation of several years' work at Cornell, where the psychologist Rosenblatt had borrowed the Aeronautical Laboratory's IBM-704 computer to do his modelling. This arrangement had started as an inconvenience, necessary because his own department didn't have a suitable machine along the corridor. But it eventually helped spread his ideas. For while psychologists read about them in the *Psychological Review*, physicists—thanks to his borrowed colleagues—encountered them in *Reviews of Modern Physics* (H. D. Block 1962; Block *et al.* 1962).

It was Rosenblatt, even more than Selfridge, who fanned the smouldering fuse of connectionism into a shining flame. A perceptron's task is the same as Pandemonium's: to learn to discriminate input patterns. Indeed, Rosenblatt's diagram of a simple perceptron was highly similar to Selfridge's of Pandemonium (see Figures 12.1 and 12.2). But Rosenblatt's work was both more ambitious and more systematic than Selfridge's.

It was more ambitious, in that Rosenblatt assimilated learning to fundamental self-organization. Self-organization is the emergence of order from a relatively unordered state (see Chapter 15), so *all* learning exemplifies it, in a sense. But this terminology is especially appropriate when the starting point is highly unordered, and Rosenblatt focused on initially random systems. That is, he didn't start out—as Selfridge had done—with networks provided with "reasonable", or "appropriate", features and weights.

Selfridge soon argued that this focus was *too* ambitious. Random networks, he said, are useful only for small local tasks, such as correlating or classifying inputs (Minsky and Selfridge 1961). As we'll see in Section iii, his co-author here would eventually publish a mighty roar of disapproval at Rosenblatt's random-network approach. But that was for the future. Meanwhile, this new work was very well received.

It was more systematic than Pandemonium, in that Rosenblatt compared many different types of perceptron, involving a variety of learning rules. Some of these

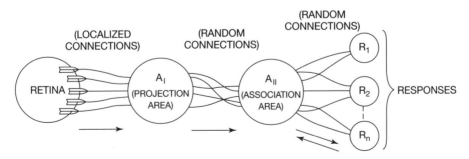

FIG. 12.2. Organization of a perceptron. Adapted with permission from Rosenblatt (1958: 389)

were defined in his paper for *Psychological Review*, others in his eagerly awaited book, previously available only as a lab report (Rosenblatt 1962). So although, as he admitted, "the mechanism for pattern generalisation proposed by Clark and Farley is essentially identical to that found in simple perceptrons" (1962: 24), he took their approach further.

Rosenblatt's ideas were expressed—and compared—in mathematical terms. However, the behaviour of the more complex perceptrons was described qualitatively, not numerically. In many cases, he backed up his theoretical discussion by precise measurements on experimental simulations.

His book included a speculative chapter on "back-coupled" networks. These contained at least one level of what are now called hidden units, plus "back-connections" whereby error messages could be passed backwards from level to level. Such error messages, he said, would enable "layers of units which are relatively remote from the sensory end of the perceptron [to] modify the activity of layers which are relatively close to the sensory end" (1962: 471).

He didn't give an algorithm by means of which this could be done. Nine years later, in 1971, Werbos did so—but no one realized this at the time. The first *recognized* algorithm for back propagation had to await the mid-1980s (see Section vi.c–d, below). Back in the early 1960s, however, Rosenblatt had explored how various ways of doing it—by magic, so to speak—would modulate the overall behaviour of the network.

For example, he argued that a particular back-coupled network illustrated "the simplest conditions under which 'selective attention' might be said to occur in a perceptron" (Rosenblatt 1962: 479). He outlined an experiment in which this back-connected perceptron would learn to distinguish not only triangles and squares and/or the top and bottom halves of the field, but triangle-in-the-top and square-in-the-bottom. It would be able to pick out (pay attention to) one of these stimuli at a time:

[The] system should give a consistent description of one of the two stimuli, in terms of shape and location, and ignore the other stimulus; it will not name the shape of one and the position of the other, even though both shapes and both positions are simultaneously present. (p. 478)

Like Uttley, Rosenblatt was aiming for biological realism—hence the title of his book: *Principles of Neurodynamics: Perceptrons and the Theory of Brain Mechanisms*. But he did this in much more general terms than Uttley. A perceptron, he said, is:

a hypothetical nervous system, or machine . . . designed to illustrate some of the fundamental properties of intelligent systems in general, without becoming too deeply enmeshed in the special, and frequently unknown, conditions which hold for particular biological organisms. (Rosenblatt 1958: 387)

So Pitts and McCulloch's 1947 attempt to model detailed neuro-anatomy (see 14.ii.b) wasn't an exemplar for Rosenblatt.

Nor was their 1943 paper. Quite the contrary. Rashevsky's followers in general, he complained, had produced "a profusion of brain models which amount simply to logical contrivances for performing particular algorithms . . . in response to sequences of stimuli" (1958: 387). Admittedly, they'd been more concerned with possibility proofs than with brains as such. But they assumed that it would re-quire "only a refinement or modification of existing [logical] principles to understand the working of a more realistic nervous system" (p. 388). (This criticism was un-fair. Certainly, the fashion at the time was logicist; but Rashevsky's biophysics group had long suggested using differential equations to describe neural networks: see Chapter 4.iii.c.)

Rosenblatt took a fundamentally different approach: no "idealized wiring diagrams" for him. Indeed, he was closer to the cybernetic tradition. Declaring Ashby (4.vii.c–d) and von Neumann as his intellectual ancestors, he aimed to show how an initially random network could organize itself.

(As for how it could identify the *relevant* stimuli in the first place, he argued that perceptual similarities aren't objective features, like McCulloch and Pitts' geometrical properties. Rather, they depend on the nature and history of the perceptual system itself: cf. Chapters 15.vii.c and 16.vii.c. In a sense then, the fundamental input sensitivities aren't random at all.)

The theoretical language needed, Rosenblatt said, wasn't Boolean logic but—as the cyberneticians had suggested—probability theory. Some network-modellers had already recognized this, as we've seen. However, "it is frequently hard to assess whether or not the systems that they describe could actually work in a realistic nervous system, and what the necessary and sufficient conditions might be" (p. 388). This two-pronged criticism was a clue to his own approach. He used both experimentation and mathematical analysis to study a wide range of machines.

In the simplest cases, a perceptron is a parallel-processing system comprising an input layer of sensory units (a "retina" of S-points), an output layer of response cells (R-units), and a single layer of associator neurones (A-units) between them. Once a classification has been learned, some individual R-unit fires if and only if the relevant pattern is input. In the ultra-minimalist version, there is only one R-unit, hence only one possible classification.

Each A-unit receives excitatory connections from several S-units, and each R-unit from several A-units. Between S-units and A-units, there are only feed-forward connections. But (usually) each R-cell has inhibitory feedback links to all the A-cells that *don't* excite it. This helps make the responses mutually exclusive: if this R-cell fires, then that one doesn't.

The units are all-or-none McCulloch–Pitts neurones: if the (fixed) threshold is reached, the cell fires. However, the activity of an A-unit has some graded (and alterable) "value", which makes it more or less effective in its influence on the relevant R-cell. The information in the system is thus stored, or represented, as a set of connections of varying strength.

Learning takes place by some form of negative feedback, which Rosenblatt termed "reinforcement". He usually ignored positive reinforcement: learning in simple per-ceptrons follows error, not success. (This fitted some theories of learning, such as Jean Piaget's, but was controversial nonetheless: see Chapter 7.vi.h.)

The paper described more complex perceptrons, too. And the complexities involved were many and various. I'll list some of them here (without discussing them), so as to convey a sense of the visionary scope of Rosenblatt's thinking at the time—and so as to allow specialist readers to compare his visions with today's achievements. For in effect, he was outlining a long-term research programme for connectionism:

* A network, he said, might contain a larger number of cells,
* and/or several associative layers.
* There might be lateral (inhibitory) connections between A-units.
* Or there could be R-units coding "discriminating features" instead of mutually exclusive responses (so seven R-units could represent 100 patterns).
* Another possibility was to have mini-networks of A-units, specialized to detect discriminating features.
* Further complications included changeable thresholds in the A-units,
* and continuously graded activity in the S-units, proportional to stimulus intensity.

That wasn't all:

* Complex perceptrons might employ a (probably more realistic) form of feedback, in which each R-unit excites its own A-units.
* There might be positive feedback (increase in connection strengths) as well as negative feedback.
* And the training might "force" some specific initial connections, the rest of the system organizing itself around them.
* Also, there might be a "projection" layer placed before the first associative layer.
* There might be systematic connections between S-units and the projection layer, to reflect "focal points" on the retina.
* And the system could be sensitive to temporal, as well as spatial, order.

The crucial point, irrespective of the complexity of the network, was that learning involves some change in the A-units or in their connections. For instance, an A-unit might output a weaker impulse; or there might be a change in the probability that its output will actually reach the relevant cell in the next layer.

Rosenblatt pointed out that there are various ways in which successful connections might be encouraged, and several different criteria for judging success. Usually, he relied on error correction. A connection would be strengthened if the R-unit failed to respond when it should have done, and weakened if it fired when it shouldn't have done. But various types of learning rule could achieve this:

In the alpha system, an active cell [A-unit] simply gains an increment of value for every impulse, and holds this gain indefinitely. In the beta system, each source-set [all the cells that feed forward to a particular R-unit] is allowed a certain constant rate of gain, the increments being apportioned among the cells of the source-set in proportion to their activity. In the gamma system, active cells gain at the expense of the inactive cells of their source-set, so that the total value of a source-set is always constant. (1958: 392)

Different systems could be specified by substituting different numbers for the variables concerned.

When Rosenblatt did this, and measured the learning curves observed in his computer implementations, he found systematic differences across learning rules. What's more, these differences could be predicted from six quantities: the number of excitatory/inhibitory connections per A-unit, the threshold of an A-unit, the proportion of R-units to which an A-unit is connected, and the number of A-units/R-units in the system.

Connectionism, it seemed, had come a very long way from playing around with bits of abandoned B24 bombers. The B24 bricoleur himself would soon declare this apparent progress an illusion (see iii, below). But meanwhile, in describing his achievements and predicting their future, Rosenblatt didn't hold back. He even designed a secret weapon (not kept secret for long): a mathematical proof that perceptrons had astonishing potential.

f. Excitement, and overexcitement

Rosenblatt made bold claims for perceptrons. "*Each of [the six] parameters*", he pointed out, "*is a clearly defined physical variable, which is measurable in its own right, independently of the behavioral and perceptual phenomena which we are trying to predict*" (1958: 406). It followed, he said, that his system far surpassed behaviourist learning theories—including Clark Hull's, which he saw as the best of a bad bunch (Chapter 5.iii.b).

His theory, or so he claimed, was superior in respect of parsimony, verifiability, generality, and explanatory power. It enabled precise quantitative predictions of learning curves from neurological data, and vice versa. It could be used both to analyse the behaviour of biological organisms, and to guide computer modelling ("the *synthesis* of behaving systems, to meet special requirements"). Hebb had hinted at the organic substrate underlying behaviour, and at a plausible bridge between biophysics and psychology (5.iv.b–d). But, Rosenblatt declared, "[my] theory represents the first actual completion of such a bridge" (p. 407).

This was exhilarating stuff: Edward Tolman's schematic sowbug had been well and truly outdone (see 5.iii.c). To cap it all, Rosenblatt soon (in 1959–60) proved the "perceptron convergence theorem".

The convergence theorem applied only to simple perceptrons, with only a single layer of units between the input and output layers. It stated that a particular learning rule is *guaranteed* to find the correct set of connection values for any pattern that is in principle learnable by a perceptron. That is, a perceptron using this rule can learn to do anything that it's possible to program it to do.

The relevant rule was a form of error correction, in which the A-units' values are slightly raised or lowered whenever their R-unit fires incorrectly. It's true that, eventually, it would always converge on a solution. But that wasn't entirely clear at the time. For Rosenblatt's proof was very demanding, and so too was the simpler version produced by a physicist colleague (H. D. Block 1962, sect. 9b). Ironically, the theorem would later be proved more elegantly by Rosenblatt's fiercest critics (see Section iii.b).

No less exciting than the mathematics and the experiments was the long list of things that perceptrons—considered as a general class—could apparently do. Their capabilities, Rosenblatt reported, had already been *seen* to cover the following:

* both spontaneous and trained pattern recognition;
* tolerance of changes in size or rotation;
* recognition of examples never yet encountered;

* responsiveness to patterns in any sensory modality (or combination thereof), and in both space and time;
* sensitivity to visual contours;
* selective attention and recall;
* distributed memory (stored not, as in the localist Pandemonium, in a single high-level unit but in the collective activity of many units);
* damage tolerance;
* associative learning;
* trial-and-error learning;
* and learning of ordered sequences of response.

Was there anything perceptrons *couldn't* do? Yes, said Rosenblatt:

The question may well be raised at this point of where the perceptron's capabilities actually stop . . . [Is it] capable, without further modification in principle, of such higher order functions as are involved in human speech, communication, and thinking? Actually, the limit of the perceptron's capabilities seems to lie in the area of relative judgment, and the abstraction of relationships. In its "symbolic behavior", the perceptron shows some striking similarities to [certain] brain-damaged patients . . . As soon as the response calls for a relationship among stimuli (such as "Name the object left of the square") . . . the problem generally becomes excessively difficult for the perceptron . . . Some system, more advanced in principle than the perceptron, seems to be required at this point. (1958: 404–5)

In his book a few years later, and especially in the final chapter, Rosenblatt admitted further limitations. For instance, a single-layer perceptron can't deal with multiple simultaneous inputs, such as a circle next to a square (although a multi-layer version could: see above). And no perceptron of the types he had considered was capable of using a "temporary memory", which is needed for many tasks (1962: 577).

Despite these caveats, however, many people were deeply impressed by Rosenblatt's claims. (Perhaps some who read the paper didn't read the book?) His ideas were soon widely applied (e.g. David and Selfridge 1962).

The flame he had lit ignited the journalists, too. Rosenblatt received "wall-to-wall media coverage" (J. A. Anderson and Rosenfeld 1998: 304). The media interest was caused partly by the 'brainlike' nature of perceptrons, and partly by his special-purpose hardware, the "Mark 1" perceptron (H. D. Block 1962, sect. 8).

This machine—which learnt, for instance, to recognize letters from the alphabet—used photoelectric cells for the retina, and (like Minsky's earlier gizmo) motor-driven potentiometers for varying the weights. It had a 20×20 input array, 512 associator units, and eight binary response units, with each S-unit having up to forty connections to the A-units. The eight R-units communicated and competed with each other, eventually agreeing on a (winner-take-all) single response. Rosenblatt found, as Minsky had done before him, that his hardware needn't be perfect. Change a few connections, knock out some units, or use somewhat unreliable components, and the Mark I might still give an acceptable result.

He himself was ambivalent about the publicity. Obviously, it had its advantages. But it usually involved a fundamental misunderstanding of what he was up to.

For the journalists, he was building a machine. It might be only the first in a long line of improved machines, but the object of the exercise was to engineer a gizmo. For

Rosenblatt, of course, that wasn't the point at all. In explaining why he'd coined the term "neurodynamics" (and used it in his book title), he said:

The term "perceptron", originally intended as a generic name for a variety of theoretical nerve nets, has an unfortunate tendency to suggest a specific piece of hardware, and it is only with difficulty that its well-meaning popularizers can be persuaded to suppress their natural urge to capitalize the initial "P". On being asked, "How is Perceptron performing today?" I am often tempted to respond, "Very well, thank you, and how are Neutron and Electron behaving?" (Rosenblatt 1962, p. v)

He regretted the scornful scepticism aroused among other scientists by the media attention. The Office of Naval Research, who had funded his work (cf. 11.a), probably regarded it less as an exercise in neurodynamics than as a prelude to militarily useful pattern-recognizers. But scientists in general, reading the press reports, might think his aims were far more grandiose than either:

[In] the first public announcement of the program in 1958 by the popular press, [they] fell to the task with all the exuberance and sense of discretion of a pack of happy bloodhounds. Such headlines as "Frankenstein Monster Designed by Navy Robot That Thinks" (Tulsa, Oklahoma *Times*) were hardly designed to inspire scientific confidence. (Rosenblatt 1962, p. v)

He was wise to be wary. The brouhaha helped trigger a hugely damaging attack from two critics at MIT (Section iii, below). In effect, Rosenblatt's research programme was killed off by it—or anyway, put into lengthy hibernation.

That wouldn't be the last time when "scientific confidence" in an entire area of research was undermined by overenthusiastic reports composed by, or even *for*, the press. Given the public's (and other scientists') tendency to see computer modelling as hubris, this is an occupational hazard of AI in general. A telling example of what can happen if one is too ready to boast to the press (and to everyone else) was described in Chapter 11.iv.

g. Enter the Adaline

Rosenblatt wasn't alone at that time in being fêted by the newspapers for building an intriguing adaptive machine. Widrow's Adaline (ADAptive LINear Element) and Madaline (a group of Many Adalines) also caused much media excitement.

One widely publicized example of Widrow's pattern-learning hardware, which he called "Knobby Adaline", was an electrical circuit with sixteen input switches and a manual threshold control. Another worked by electrolysis. The connections were made of copper-plated pencil leads, whose resistance varied with the thickness of copper gradually laid down on the graphite core.

You may be reminded of Gordon Pask's electrochemical "concept machine" mentioned in Chapter 4.v.e—which, like Pandemonium, had recently been described at Uttley's 1958 conference. This is no accident. The cybernetic movement had experimented with many adaptive machines based on physical (as opposed to programmed) functions, even including chemistry. And Widrow, as an electrical engineer trained at MIT in the years around 1950, was firmly rooted in the cybernetic tradition.

His special interest was digital filtering, wherein a signal is strengthened by compensating for the noise. Originally a topic for engineers, signal detection theory had been applied to psychophysics by David Green and John Swets at the famous 1956 symposium at MIT (see Chapter 6.iv.b). Widrow, however, was especially enthused by his experience of the 1956 summer school at Dartmouth.

Widrow questioned Wiener's (then orthodox) theory that one has to know the statistics of the signal in order to design the best filter. Widrow wanted an adaptive filter, which would optimize itself *no matter what* the relation of signal to noise.

Wiener had used "mean square error" to measure a filter's success. (*Mean*, because it's the average error that matters; *square*, because squaring a number always gives a positive number, so the direction of error can be ignored.) Widrow had to find a way of minimizing this error measurement automatically. This would require some form of supervised learning, wherein the machine could gauge the difference between actual and desired outputs. In other words, it would require something broadly similar to a perceptron. (This isn't a remark about Widrow's thought processes, for he didn't know about Rosenblatt's work until it hit the newspapers.)

Widrow's first solution resembled the perceptron convergence procedure, for it involved repeatedly making small changes in the filter's parameters and measuring the resulting error each time. Those changes which were found to give the greatest average improvement were then followed until improvement ceased. However, this method of gradient search by "steepest descent" required many successive measurements and calculations. In 1959 (having just moved from MIT to Stanford), Widrow and Marcian (Ted) Hoff together discovered a much more efficient way of finding the error gradient: the "LMS", or least mean square, algorithm (Widrow and Hoff 1960) (see Section v.a, below).

Within only half an hour, they had implemented the LMS algorithm on the department's analogue computer. The next day, they cobbled together an adaptive digital circuit with a 4×4 array of binary input switches—which they used as a demonstration model to amaze their colleagues, not to mention the press. "You'd be surprised how many different geometric patterns you can make with a little four by four array of switches," Widrow said later (J. A. Anderson and Rosenfeld 1998: 54).

That was the first hardware Adaline. The pencil-lead version—as it happens, the beginning of liquid-state electronics—was devised and commercially marketed soon afterwards. It was much more efficient than its predecessor, and also than the Mark I perceptron. Indeed, Widrow regarded the latter as "a disaster", because of its unwieldiness, slowness, and unreliability (J. A. Anderson and Rosenfeld 1998: 59).

One obvious commercial application of Adalines was in adaptive telephone modems, to filter out the noise in the telephone line. But *noise* is a generic concept, so these "adaptive neurons" were potentially useful for many different purposes, including some of interest to cognitive science. Clearly, they might be valuable in technological AI. Whether biological organisms can compensate for noise by using some equivalent of LMS was less clear, not least because LMS involved supervised learning. But the question seemed worth pursuing.

Where adaptive networks were concerned, then, at the turn of the 1960s all systems were *Go!* The interest aroused was phenomenal, and "probably sucked away 10 or 15 years of all the smartest people in cybernetics" (Minsky 1984*b*: 122).

Admittedly, as Widrow recalled many years later, the publicity hype that he and Rosenblatt received "infuriated" many of their colleagues (J. A. Anderson and Rosenfeld

1998: 65). But this annoyance could be dismissed, at least by enthusiasts and outsiders, as being driven by professional jealousy rather than reason. By the end of the 1960s, over 100 research groups in AI and psychology were focusing on perceptrons and similar systems.

Yet only ten years after that, a science journalist wrote: "[The perceptron] had fundamental limitations and has, in recent years, pretty much been abandoned" (J. Bernstein 1981*b*: 100).

"*Abandoned? Surely not! (Whatever happened to the shining flame?) And if so, why?*"—The "Surely not!" will be addressed in Section iv, where we'll see that the flame had been damped, but by no means extinguished. First, let's discuss the "Why?".

12.iii. Attack Without Apology

The journalist just quoted had interviewed Minsky at length. It was he who reported Minsky's story about being beguiled by his B24 bricolage before returning to "science". This probably explains his dismissive remark about connectionist research.

For in so far as perceptrons were "abandoned"—which they weren't, not entirely (see Sections iv–v)—it was largely Minsky's doing. If an outsider got the impression that connectionism was finished, talking to Minsky could have been responsible. This presumably explains also why the science journalist Pamela McCorduck, who knew Minsky personally but wasn't an expert in the field, wrote in her late 1970s history of AI that connectionism had "died" (McCorduck 1979: 47).

The extent of Minsky's responsibility for the Sleeping Beauty's going into hibernation is arguable, as we'll see. But that he had something to do with it is certain. He was one of the devilish duo—for so the connectionist community viewed them—who published a radical attack on Rosenblatt's methodology (Minsky and Papert 1969).

a. The devilish duo

That Minsky jeopardized network research may seem ironic. For he'd built one of the very first connectionist learning machines (see Section ii.a). In fact, Minsky's approach to the two main types of AI swung like a pendulum.

His initial work on learning was connectionist in spirit, as we've seen. But his early paper on neural networks was logicist, not probabilistic. Then, partly due to the influence of John McCarthy, he turned to GOFAI (property-list) methods, making them the focus of research in the AI Lab they co-founded at MIT (Chapter 10.ii.a).

Soon afterwards, he embarked on his first principled critiques of connectionism. These were early exercises in meta-epistemology (10.i.g). For example, he and Selfridge pointed out some intrinsic limitations of the work on perceptrons outlined above (Minsky and Selfridge 1961). And in his seminal paper 'Steps Towards Artificial Intelligence' (Minsky 1961*b*), he explicitly favoured symbolic AI above adaptive networks for modelling learning (10.i.g). In the late 1960s, he even described perceptrons as "sterile" (Minsky and Papert 1969: 232).

By the 1980s, however, his interest had swung back towards parallelism. His "K-lines" account of memory explained it as the partial reactivation (by single units) of *sets* of simple units, comparable to cell assemblies, working in parallel (Minsky 1980; 1985, ch. 8).

But he hadn't abandoned GOFAI (see subsection d, below). On the contrary, his intellectual pendulum had come to rest in an attempted integration of symbolic processes with parallel architecture (Minsky 1985). This hybrid position had been anticipated, in outline, a quarter-century before, in the 'Steps' paper.

Minsky was a hugely influential figure in early AI, as we saw in Chapter 10. He was respected by the wider computing community too. Not only was he the first AI person to receive the ACM's high-profile Turing Award, but his three predecessors were the pioneering 'pure' computer scientists Alan Perlis, Maurice Wilkes (3.v.b), and Richard Hammond. So when he attacked perceptrons, people sat up and listened. It's no wonder that he was so deeply resented by the connectionists: for them, it was a public relations disaster.

Minsky's partner in the devilish duo was Seymour Papert, a South African who'd left home in the 1950s to study mathematics at the University of Cambridge. They were brought together by McCulloch in autumn 1963, at a party in his house. (I was there—sitting on the carpet with Papert!) Papert had come straight from the airport, on his arrival from Geneva where he'd been working with Piaget for five years.

As a Piagetian, Papert saw learning not as the gradual self-organization of an initially random system, but as a process structured by a succession of internal schemata (see Chapters 5.ii.c and 10.v.f). So although he was already thinking about thinking in computational terms (he'd been commuting to London to use the NPL computer, no machine being available to him in Switzerland—Crevier 1993: 86), he was doing so in a way that was at odds with the connectionist approach. As for Minsky, he was already directing the NewFAI research dominant at MIT.

Besides these shared grounds for scepticism about perceptrons, the two men had already ventured into the relevant mathematical theory:

We had both been interested in the perceptron since its announcement by Rosenblatt in 1957. In fact we had both presented papers related to its "learning" aspect at a symposium on information theory in London in 1960. (Minsky and Papert 1969: 239)

Eighteen months later, with McCulloch's encouragement, they were together at MIT—where they started a "serious" collaboration.

b. The opening salvo

Their joint critique was highly abstract. They didn't build and test actual computer models, still less "quit science for a while to watch the machine". They focused instead on the underlying principles, to determine the scope and limits of the methodology's computational power.

This wasn't a new departure: Minsky's 'Steps' had been written in that way too. In other words, they were doing what Drew McDermott and Alan Bundy had criticized NewFAI workers for not doing (11.iii)—and what Rosenblatt himself had been doing when he formulated the convergence theorem.

It's clear, however, that they were far from disinterested:

Our first formal presentation of the principal results in [our] book was at an American Mathematical Society symposium on Mathematical aspects of Computer Science in April 1966...We were pleased and encouraged by the enthusiastic reception by many colleagues at the A.M.S. meeting and no less so by the doleful reception of a similar presentation at a Bionics meeting. However, we were now involved in establishing at MIT an artificial intelligence laboratory largely devoted to real "seeing machines", and gave no attention to perceptrons until we were jolted by attending an IEEE Workshop on Pattern Recognition in Puerto Rico early in 1967.

Appalled at the persistent influence of perceptrons (and similar ways of thinking) on *practical pattern recognition*, we determined to set out our work as a book. (1969: 242)

Even this passage gives little hint of the passion that had perfused the early drafts. According to Robert Hecht-Nielsen (1947–), these dripped with "vitriol" and amounted to "an unseemly personal attack" on Rosenblatt (J. A. Anderson and Rosenfeld 1998: 305).

Competition for money was certainly one factor. The DOD "was seriously thinking about funding Rosenblatt's work on the Perceptron" (Newquist 1994: 72). This alarmed Minsky and Papert, because every ARPA dollar devoted to Cornell would be one less for MIT. In the early 1960s that might not have mattered too much, given the depth of ARPA's coffers at that time. But "rumors swept through the [AI] research community around 1968 and 1969 that ARPA was going to cut back on its seemingly unlimited largesse" (Newquist 1994: 139). That may be why, having circulated the book in draft for some years, they now decided to publish it more widely.

There may well have been personal jealousies too. Minsky would understandably be annoyed by Rosenblatt's superman image in the media, especially the mistaken assumption that the Mark I was the first learning network ever built. Possibly, there was some long-standing rivalry: Rosenblatt and Minsky had been fellow pupils at high school in the Bronx. And it probably didn't help that, in 1957, Marshall Yovits of the Office of Naval Research—a friend of Minsky's, who had provided much of the early funding for AI—"had publicized Rosenblatt's work and made a big splash about it" (J. A. Anderson and Rosenfeld 1998: 99).

However that may be, and even though the later printings (after Rosenblatt's death) were dedicated to Rosenblatt's memory, it's pretty clear that "the whole thing was intended, from the outset, as a book-length damnation of Rosenblatt's work and...neural network research in general" (Hecht-Nielsen again: ibid.). Eventually, Minsky and Papert followed "the strong and wise advice of colleagues", and expunged almost all of the vitriol.

Those wise colleagues weren't trying to protect Rosenblatt. On the contrary, they wanted to ensure that the damning critique was all the more effective for being seen as objective. For Minsky and Papert weren't the only ones at MIT to doubt Rosenblatt. Cowan remembers attending a "terrible" lecture he gave there in 1958, at which the audience "went after him and really attacked him". "Almost everyone" at MIT, says Cowan, was very sceptical: "by and large, it was clear that the perceptron wasn't doing the things that Frank claimed it could do" (J. A. Anderson and Rosenfeld 1998: 99–100).

The scepticism was experienced with passion. Many of these critics were "infuriated" by Rosenblatt's media popularity partly because they feared that the irresponsible claims involved might cause a backlash against neural networks in general.

Their fears were justified. The VLSI engineer Carver Mead (1934–) would later attribute the "twenty-year famine" in neural networks to the "overhype" about perceptrons (J. A. Anderson and Rosenfeld 1998: 141). Even if Mead was exaggerating, there was much truth in his charge (see subsection e, below). And ARPA (soon to be renamed DARPA: see 5.iv.f), who had funded early 1960s neural-network research but withdrew their support because of the criticisms emanating from MIT, repeatedly warned against similar "hype" when they considered funding connectionism again (see Section vii.b, below).

In the final version, then, Minsky and Papert's critique had lost the invective that had peppered the drafts. It was expressed almost entirely in the ascetic terms of computational logic—or, as they put it, computational geometry.

So they defined various types of simple perceptron: all with a single associative layer, and all locally connected, with no feedback loops. And they compared their computational properties. For instance, they considered the effect of altering the number of S-units leading into an A-unit, and/or the size of the A-unit's receptive field. Mostly, they considered the behaviour of a single (output-layer) neurone, as opposed to the joint activity of several such neurones.

They saw their work as relevant not only to assessing the potential for practical applications, but also to theorizing about the brain. For if perceptrons can't do some of the things that brains do, then some neural processes must be different. Primarily, however, it was an exercise in the general theory of computation: a study of self-organizing machines that make decisions by weighing many independent items of evidence. As such, they suggested, it might throw light not only on the brain but even on "how the genetic program computes organisms" (Minsky and Papert 1969: 1).

The interest and optimism aroused by Rosenblatt's work had been due in part to his convergence theorem (see ii.f, above). Minsky and Papert agreed that it was "seductive" (1969: 14), even "amazing" (1988, p. xi). But they cautioned against taking it at face value.

A theorem, of course, is a theorem—and they even produced new (more elegant) proofs of their own (1969, ch. 11). They argued, however, that certain relevant questions had rarely, if ever, been addressed. These included:

* Is the perceptron an efficient form of memory?
* Does the learning time become too long to be practical even when [learning] is possible in principle?
* How do the convergence results compare to those obtained by more deliberate design methods?
* What is the perceptron's relation to other computational devices? (1969: 243–4)

The last of these questions became "more and more important" in their eyes. It prompted comparisons, for instance, with Ashby's Homeostat (Chapter 4.vii.c–d) and with GOFAI's techniques for problem solving and learning. Minsky and Papert detailed the threat of combinatorial explosion, arguing that scaling up simple perceptrons to deal with higher-order problems wouldn't be feasible.

As they pointed out, the existing successful perceptrons relied on operators specially designed for the task—such as Selfridge's low-level demons, or feature-detectors. This, they said, was no accident. In general, "significant learning at a significant rate presupposes some significant prior structure" (p. 16). This was a version of the lesson already learnt the hard way in GOFAI, that "general" problem-solvers are highly limited in power (see Chapter 10.iii.c and iv). In sum, perceptrons could be useful only for simple, pre-specified, problems.

One might think that the connectionists would have been grateful. After all, GOFAI was thriving despite the threat of the combinatorial explosion, and despite the fact that digital computers are only *approximations* to universal Turing machines. To warn against over-optimism about perceptrons isn't necessarily to dismiss them as a waste of time. Rosenblatt himself had suggested important limitations, as we've seen.

But Minsky and Papert went further. They concluded that this type of research could lead only to a dead end, since even idealized perceptrons aren't equivalent to Turing machines. In other words, the combinatorial explosion was only part of the difficulty. What, above all, put the sceptical cat among the connectionist pigeons was their proof that (simple) perceptrons cannot do certain things which, intuitively, one might have expected them to be able to do.

One intractable problem was the computation of parity. Although a "diameter-limited" perceptron (where each operator can inspect only a small part of the retina, or stimulus, at one time) can tell whether any particular part of an image contains a dot, it can't tell whether the number of dots overall is odd or even. Indeed, Minsky (with Selfridge) had already noted this point in 1960, and repeated it in 'Steps'. Similarly, McCulloch had already noted that a perceptron can't compute *exclusive-or* (so some of the disjunctive concepts studied by Bruner—see 6.ii.b—were out of reach).

Now, Minsky and Papert elaborated on these limitations, and identified more. In general, they showed that perceptrons can discriminate only between classes that are "linearly separable". (Suppose two classes are each represented by a set of points in some mathematical space: if a line can be drawn between the two sets, the classes are linearly separable.)

Most of the specific examples they discussed were clearly relevant to pattern recognition. For instance, they proved that although simple diameter-limited perceptrons can distinguish concave and convex lines, they can't compute spatial connectedness (pp. 12–13, 70–95). They can't say, for example, whether a scribble consists of one continuous line or several. (People sometimes fail too: the two doodles printed on the cover of their book were visually equivalent, but tracing-by-finger shows that one is connected and the other isn't.) Nor can such perceptrons recognize connectivities in line drawings of 3D solids, as the current GOFAI scene-analysis programs could (10.iv.b).

To many people, this was highly counter-intuitive. If perceptrons can cope with convexity, then surely they should be able to deal with connectedness too? Besides referring to existing scene-analysis programs (pp. 232–9), Minsky and Papert gave principled arguments to show that serial, not parallel, computation was required for recognizing connectedness (and parity). And they pointed out that, since all the biological receptor cells identified thus far were diameter-limited (see 14.iii.a–b), an animal needs "more than neurosynaptic 'summation'" to perceive connectedness.

Allen Newell (1969) hailed their critique as "a great book", not least because it furthered "the appropriate shaping of computer science into a disciplined field of enquiry". Many GOFAI scientists welcomed it for less laudable reasons, seeing it as a powerful weapon in a turf war for funding and prestige. But if people in symbolic AI welcomed it, connectionists understandably didn't.

Psychologists who had been excited by perceptrons saw the book as "a major disaster" (R. L. Gregory, personal communication). The network modellers themselves countered it by saying that their research had only just begun, and that Minsky and Papert had considered only very simple systems. Rosenblatt's co-worker Herbert Block (1970), for instance, argued that multi-layer perceptrons would be able to do everything they had said to be impossible.

The critical duo had anticipated this riposte. On their view, more complex, many-levelled, networks would offer no significant advance. They admitted that this was merely a hunch, meriting further research:

The perceptron has shown itself worthy of study despite (and even because of!) its severe limitations. It has many features to attract attention: its linearity; its intriguing learning theorem; its clear paradigmatic simplicity as a kind of parallel computation. There is no reason to suppose that any of these virtues carry over to the many-layered version. Nevertheless, we consider it to be an important research problem to elucidate (or reject) our intuitive judgment that the extension is sterile. (Minsky and Papert 1969: 231–2)

They even allowed that "some powerful convergence theorem" might be discovered for describing learning in multi-layered machines. Clearly, however, they weren't holding their breath.

c. Intransigence

Twenty years later, it would seem—to many—that their judgement had been mistaken. The late 1980s saw an explosion of interest in connectionism, both from AI researchers and in the media. Besides countless intriguing applications, the results included parity recognition (Rumelhart *et al.* 1986*a*), a "powerful convergence theorem" for multilevel networks, and a multilevel learning rule of great generality (see Section vi.b–c). It was even proved that a three-layer network can in principle solve *any* problem (Hornik *et al.* 1989).

On top of all that, DARPA started funding neural networks again, explicitly admitting that their previous disdain had been "undeserved" (DARPA 1988: 23). In short, many people felt that Minsky and Papert had ended up with egg on their faces.

Not so!, they replied. For their judgement remained unchanged. Minsky made this clear in his contributions to the DARPA study, and also more publicly in a second edition of *Perceptrons*—the text unchanged, but with a new Prologue and Epilogue. He and Papert were unrepentant and unbowed:

[We] were tempted to "bring [our] theories up to date". But . . . we found that little of significance had changed since 1969 . . . Our position remains what it was when we wrote the book. We believe this realm of work to be immensely important and rich, but we expect its growth to require a degree of critical analysis that its more romantic advocates have always been reluctant to pursue—perhaps because the spirit of connectionism seems itself to go somewhat against the grain of analytic rigour. (Minsky and Papert 1988, p. vii)

Their main complaint concerned that central topic of GOFAI research, the representation of knowledge (see 10.iii.a):

What we discovered was that the traditional analysis of learning machines—and of perceptrons in particular—had looked in the wrong direction. Most theorists had tried to focus only on the mathematical structure of what was common to all learning, and the theories to which this had led were too general and too weak to explain which patterns perceptrons could recognize. As our analysis... shows, this actually had nothing to do with learning at all; it had to do with the relationships between the perceptron's architecture and the characters of the problems that were being presented to it. The trouble appeared when perceptrons had no way to represent the knowledge required for solving certain problems. The moral was that one simply cannot learn enough by studying learning by itself; one also has to understand the nature of what one wants to learn. No machine can learn to recognize X unless it possesses, at least potentially, some scheme for *representing* X. (Minsky and Papert 1988, p. xii)

That final sentence, of course, was a near-verbatim rendering of McCarthy's statement about learning in the early days of GOFAI (10.i.f), and a restatement of Minsky's concern with knowledge representation in 'Steps' (10.i.g).

They did make one concession. They answered the question "How much, then, can we expect from connectionist systems?" by saying "Much more than the above remarks might suggest, since reflective thought is the lesser part of what our minds do" (1988: 280). They even said: "We see no reason to choose sides... [And] we expect the future of network-based learning machines to be rich beyond imagining" (p. xiv).

But they were far from endorsing the purist type of self-organization envisioned by Rosenblatt and most, though not all, PDP workers (see Sections vi–viii, below). High-level intelligence, they said, does not and cannot arise from utter randomness, nor from a wholly non-sequential system:

Most probably, the human brain is, in the main, composed of large numbers of relatively small distributed systems, arranged by embryology into a complex society that is controlled in part (but only in part) by serial, symbolic systems that are added later. (p. xiv)

d. The hybrid society of mind

Innocent readers might have been bemused by the passage just quoted, which was given no explanation. But it tacitly reflected a wide range of work done by cognitive scientists in the twenty years after the first edition. This included research on nativism and modularity (see Chapter 7.vi). Above all, however, it reflected Minsky's ideas on *The Society of Mind* (Minsky 1985). Strictly, one should rather say "Minsky and Papert's ideas", for they'd cooperated very closely in developing them.

The *Society* book had appeared (from a trade publisher, not an academic press) only recently, almost simultaneously with the PDP bible (see Section vi.a). But most cognitive scientists already had some idea of its main themes. Minsky had been mentioning them in conversation since the early 1970s, had given many people drafts from the beginning, and had published a "taster" ten years before (Minsky 1979). Indeed, it "was already cited in detail in a dozen books that came out before it did" (Brand 1988: 102). What they didn't expect was the rhetorical simplicity and structure of his text.

As for simplicity, the book was deceptively lucid. Minsky later said that during the ten years of drafting it, he'd ruthlessly removed jargon: "Whenever someone asked me what a term meant, I explained what it meant and took it out" (interview in Brand 1988: 102).

But the real surprise was the textual structure. The book consisted of 270 half-page to two-page snippets, each discussing a particular psychological and/or computational topic. These ranged from grasping blocks, through recognizing chairs, to emotion, jokes, self, and consciousness; and (as AI topics) from means–end analysis, through frames, demons, and defaults, to reinforcement in neural nets. The relevance of one snippet to another was largely tacit, the overall picture emerging only when the links had been activated in the reader's mind.

This apparent disorganization was deliberate. (It was widely rumoured that Minsky had wanted to present it in a loose-leaf folder, so that the pages could be endlessly rearranged, but that the publisher had refused; however, none of that is true: Minsky, personal communication.) The unusual structure of the book reflected Minsky's unorthodox ideas about the structure of the mind:

My explanations rarely go in neat, straight lines from start to end. I wish I could have lined them up so that you could climb straight to the top, by mental stair-steps, one by one. Instead, they're tied in tangled webs.

Perhaps the fault is actually mine . . . But I'm inclined to lay the blame upon the nature of the mind: much of the power seems to stem from just the messy ways its agents cross-connect. (1985: 17)

The questions he was addressing, and the general tenor of his reply, were broadly similar to those addressed in Aaron Sloman's theory of motivation and intelligence (see Chapter 7.i.e–f). For Minsky, too, was concerned with the mind as a whole, not just cognition—still less, a specific subset of cognitive abilities. And he too was concerned with architecture, not algorithms.

To be sure, when he could gesture towards an existing or plausible AI algorithm, he did so. ("Gesture", because he gave no references within the text: he left his specialist readers to decode his remarks using their previous knowledge, and provided a glossary to supplement the common sense of his general readers.) But the crux of his discussion was how the algorithms would fit together.

The fit he described was far from snug. Instead of likening the mind to a carefully designed top-down program, he compared it to a loosely organized collective, or society. A society is a distributed system made up of many different agents (see 13.iii.d). Each one has different skills, knowledge, and interests. And all of them do their own thing in parallel, though some may be dormant while others are active.

So it was for Minsky's theory of mind. Various types of communication between agents allowed for the emergence of ordered sequences, hierarchies, conflicts, negotiations, dominance, and cooperation. For example, different instincts or motives would compete for control of the hands, or of attention. Some "supervisor" agents would be able to prioritize one goal over another. But many motives—to eat or sleep, for instance—would spontaneously build up in strength until they broke through such top-down controls.

The GOFAI research of the past three decades, including work on hierarchy, planning, and learning, was a major source of ideas for Minsky's book. His Acknowledgment mentioned many well-known names in symbolic AI (1985: 322–5), and diagrams based on Patrick Winston's blocks-world arches adorned the chapter on learning (see 10.iii.d). But GOFAI concepts such as demons, production systems, blackboard architectures, and heterarchy (all parallelist in intent) were stronger influences than single-minded top-down programs such as GPS.

Connectionist AI was a strong influence too. So were Minsky's years of friendship with (among others) Pask, Hebb, Ashby, von Neumann, and especially McCulloch and Selfridge—all acknowledged in the text. And he leant heavily on his own theory of K-lines: a discussion of associationist memory (both semantic and episodic), and of the hardware likely to be best suited to it (Minsky 1980; cf. 1985, ch. 8).

Despite his long-standing infamy in connectionist circles, the K-line theory had already been greeted as "an important advance" by the organizers of the meeting that heralded the connectionist revival (see Section v.b, below). As they had put it:

The central idea of the model, that partial mental states are re-created by activating particular agents that designate them, is an interesting intermediate position with respect to the issue of local versus distributed representations . . .

The real value of Minsky's model will only be known when the model is specified in sufficient detail for it to be simulated, but the general approach of trying to implement sophisticated computational processes in parallel neuron-like hardware seems extremely promising. (Hinton and Anderson 1981: 23)

Another intellectual source (via the book's effective co-author, Papert) was Piagetian psychology. This highlighted the importance of developing new ways of organizing existing knowledge, as opposed to acquiring utterly new knowledge (see Chapter 5.ii.c and Section viii.c–d, below). Yet another was Sigmund Freud's psychodynamics. Unlike Kenneth Colby, Minsky didn't present a 'Freudian' program (7.i.a). But he used many Freudian ideas in his discussions of intelligence, motivation, emotion, humour, and self.

Not least, the book drew on evolutionary biology, with its emphasis on opportunistic bricolage rather than anticipatory design. As Minsky put it, the "messy ways [the] agents cross-connect [are] only what we must expect from evolution's countless tricks" (1985: 17). This supported his claim that the mind is a more or less well-integrated collection of separately identifiable systems (cell assemblies, networks, programs, procedures . . .) on many different levels.

That claim wasn't new. William McDougall (1923, 1926) had described the mind as "a colony of monads" (mindlike subsystems) interlinked by functional relations broadly similar to those posited for Minsky's "society" (see Chapter 5.ii.a). More to the point (since McDougall was by then almost forgotten), Hebb had stated this view clearly in 1949, and Herbert Simon had recommended hierarchy as a general principle of complex "design" (including evolution) some years later (1962). In addition, it had recently been stressed in various areas of cognitive science, for instance by David Marr and—as "mental modularity"—by Noam Chomsky and Jerry Fodor (Chapters 7.vi.d–e, 9.vii, and 16.iv.c–d, respectively).

But Minsky was especially interested in subsystems (modules) that were *not*, or not wholly, innate and *not* informationally encapsulated in Fodor's sense. How could they be learnt, and how could they come to interact and control each other? How could modules provided "by embryology" be honed by experience? In particular, how could cultural practices and beliefs (such as religion: Chapter 8.vi) develop, and enter into motivation and emotion as well as problem solving?

Minsky seemed to share Fodor's view that there was no hope of a detailed predictive science of higher mental processes (7.iii.d). But (like Sloman) he was aiming for a scientific understanding of the sort of computational architecture that can make such phenomena possible.

(Minsky's next book, *The Emotion Machine*, would develop these ideas further: see Chapter 7.i.e. As I write these words, it's not yet published. Or rather, it's not published *in print*. Draft chapters have been available for some time, however, on Minsky's web site: Minsky in preparation. Instead of dog-eared mimeographs or fiddly microfilms of "faint purple typewriting with handwritten annotations"—see Chapter 9.vi.a and ix.e—we have clearly legible and easily searchable text. And comments are welcomed, to be sent to Minsky's email address. In short, a good example of Solomon's House in cyberspace: see Chapter 2.ii.b. As for programs grounded in *The Emotion Machine*, he says: "It has almost enough (hidden) technical detail to program it. In fact we already have an early running version of its architectural structure, in which two robots cooperate to solve problems in a simple world, by communicating with some natural language"—personal communication, December 2004.)

The long-awaited *Society* book was rich in insightful comments, questions, and suggestions. But its hybrid spirit, sharing both GOFAI and connectionist insights, didn't fit either orthodoxy. (The need for hybrid systems had been hinted at in 'Steps', as we saw in Chapter 10.i.g; now, it was being made fully explicit.)

Even more to the point, it offered no mathematical analyses, no new techniques for instant application, and no new programs. As such, it was more likely to appeal to non-practitioners than to hands-on AI technologists. Daniel Dennett, for example, was strongly influenced by it (Chapter 16.iv.a).

However, Minsky being Minsky, the book couldn't be ignored by the AI community. It was given eighty pages of reviews/reply in the *Artificial Intelligence* journal (Stefik and Smoliar 1991).

One AI reviewer felt it "unfortunate that cognitive scientists have, for the most part, reacted to Minsky's book as though it were light reading or a minor conversation piece, to be relegated to a coffee table" (Stefik and Smoliar 1991: 321). But another, while allowing that the book was "fascinating and provocative", was less sympathetic:

How, *exactly*, is this mediation [between agents in conflict] to be accomplished? Without an answer to this, the concept of an agent is an idea with no practical value; it is so general that it might encompass anything. The science, the engineering, the *value* of ideas lie in specifications and the associated falsifiable claims that Minsky chooses not to make.

It is all well and good to invent interesting theories and speculate about their consequences, but such an approach almost inevitably leads to an examination of the strengths of these theories and not of their weaknesses—and Minsky's inability to choose between conflicting agents is indeed a weakness of his theory. A formal approach would have forced him to focus on this weakness until some resolution was obtained; his informality allows him to deal with it by sticking his head in the sand . . . All told, *The Society of Mind* is as Minsky describes it: an adventure in speculation, and not substantive science. It is well worth reading, but only in that light. (Stefik and Smoliar 1991: 338–9)

That was the polite way of putting it. I myself heard many exasperated complaints, even contemptuous dismissals. These were voiced by AI workers (including callow

MIT graduate students, hardly fit to lick his boots) more frustrated by Minsky's lack of technical detail than inspired by his subtlety and vision. His notorious disinclination to programming, and the lack of programs to illustrate his architectural ideas, didn't help (Chapter 10.i.a).

The complaints were understandable. But the dismissals were undeserved. The Popperian appeal (above) to "falsifiable claims" appeared to forget that Karl Popper himself had defended speculative metaphysics as a fruitful predecessor of "substantive science" (K. R. Popper 1935). In short, *Society* was setting an agenda for future AI much as 'Steps' had done over twenty years before (Chapter 10.i.f), albeit now in an even more informal way. The immediate difficulty was that neither symbolic nor connectionist AI was ready to get very far in working on it.

Minsky and Papert, clearly, were using connectionist ideas in their society-theory of mind. The "coffee table" book of 1985 emphasized K-lines, associative memory, and distributed systems. Why, then, were they so unrelenting in the (near-simultaneous) second edition of their notorious critique of Rosenblatt?

The reason was that, even by the late 1980s, most of the new perceptrons were conceptualized as initially unstructured self-organizing systems, doing only one thing—recognizing or expressing only one pattern—at a time (see Sections v–vi). The importance of domain-specific multilevel structure, whether emergent or inbuilt, was being acknowledged by some of the field's leaders. But it was a goal for the future rather than an achievement of the past (see Sections viii–ix). In respect of the *intellectual substance* of their earlier attack on connectionism, then, Minsky and Papert offered no apology.

e. Were they to blame?

As for their *historical influence* on the field, they insisted that—despite the demono-logy—they had nothing to apologize for. Yes, interest in connectionism had plummeted. But they weren't to blame. (*"Not us, guv!"*)

Many readers were surprised by this. For it was—and still is—commonly believed that their 1969 book called a halt to connectionist research, which (with a few honourable exceptions) lay latent until the PDP-led renaissance of the late 1980s. Indeed, some commentators thought that such research had ceased completely.

For instance, a history of AI published in 1979 declared that GOFAI now dominated cognitive psychology, while "the attempt to imitate cell behavior [had] produced only trivial results, *then withered and died*" (McCorduck 1979: 47; italics added). Similarly, the Dreyfus brothers have said that "Everyone who knows the history of the field will be able to point to the proximal cause" of GOFAI's becoming "the only game in town": namely, the mid-1960s circulation of drafts of *Perceptrons* (H. L. Dreyfus and Dreyfus 1988: 21).

This descent into the "wilderness years", it's often added, was due not least to Minsky's unparalleled access to key personnel in the governmental funding bodies. Above all, he was close to Joseph Licklider (5.iv.f, 10.ii.a, and 11.i.b), whose good offices he'd procured for the type of AI being done in his laboratory at MIT. Hecht-Nielsen, for example, says:

Minsky and Papert's book, *Perceptrons*, worked. The field of neural networks was discredited and destroyed. The book and the associated conference presentations created a new conventional wisdom at DARPA and almost all other research sponsorship organizations that some MIT professors have proven mathematically that neural networks cannot ever do anything interesting. The chilling effect of this episode on neural network research lasted almost twenty years. (J. A. Anderson and Rosenfeld 1998: 305)

Certainly, most of the excitement, and the money, throughout the 1970s and early 1980s was in GOFAI rather than connectionism. There's even some suggestion that Rosenblatt himself could no longer get his papers published in the technical journals (Newquist 1994: 74). DARPA later admitted publicly that they had "largely abandoned" neural networks, seeing symbolic AI as "apparently more promising" (DARPA 1988: 23).

Nevertheless, the role of Minsky and Papert's critique is disputed. Widrow's comment, for instance, is very different from Hecht-Nielsen's:

[They] were able to prove that [the perceptron] could do practically nothing. Long, long, long before that book, I was already successfully adapting Madaline, which is a whole bunch of neural elements. All this worry and agony over the limitations of linear separability, which is the main theme of the book, was long overcome.

We had already stopped working on neural nets. As far as I knew, there wasn't anybody working in neural nets when that book came out. I couldn't understand what the point of it was, why the hell they did it. But I know how long it takes to write a book. I figured that they must have gotten inspired to write that book really early on to squelch the field, to do what they could to stick pins in the balloon. But by the time the book came out, the field was already gone. There was just about nobody doing it. (J. A. Anderson and Rosenfeld 1998: 60)

Their own assessment of their book's historical influence on the AI community explicitly rejected the received view:

How did the scientists involved . . . react to [our analysis]? One popular version is that the publication of our book so discouraged research on learning in network machines that a promising line of research was interrupted. Our version is that progress had already come to a virtual halt because of the lack of adequate basic theories, and the lessons in this book provided the field [i.e. AI in general] with new momentum—albeit, paradoxically, by redirecting its immediate concerns. (Minsky and Papert 1988, p. xii)

They (or anyway, Papert) admitted to a degree of "hostility" in their critique:

Did Minsky and I try to kill connectionism . . . ? Yes, there was *some* hostility in the energy behind the research reported in *Perceptrons*. . . . [Part] of our drive came, as we quite plainly acknowledged in our book, from the fact that funding and research energy were being dissipated on what still appear to me (since the story of new, powerful network mechanisms is seriously exaggerated) to be misleading attempts to use connectionist methods in practical applications. (Papert 1988: 4–5)

But the funding collapse that followed on the book was blamed by Papert not on their own quest for money but on the schismatic nature of 1960s AI (see Chapter 4.ix) and the "universalist" tendencies of their AI colleagues:

The desire for universality . . . was nurtured by the most mundane material circumstances of funding. By 1969, the date of the publication of *Perceptrons*, AI was not operating in an ivory-tower vacuum. Money was at stake. . . . [That's largely why our book was misinterpreted as claiming] that neural nets were universally bad. (p. 7)

Even more important, he said, was the computational weakness of the connectionist approach: "We did not think of our work as killing Snow White; we saw it as a way to understand her" (pp. 7–8). Moreover, he and Minsky pointed out that work on neural networks had become "virtually dormant" twice: once in the late 1950s, and again in the late 1960s (Minsky and Papert 1988, p. xi). The implication was that their 1969 attack wasn't to blame. The work had "stalled" because of its inherent limitations.

There's some evidence for their view in Widrow's comment, above, and there's more in Sections iv–v, below. Even so, their disclaimer is only a half-truth.

Drafts of their work had been circulating for at least eight years before the book appeared. What's more, they'd attacked perceptrons at several high-visibility meetings from the late 1950s on, such as the Chicago conference on self-organizing systems in 1962 (organized by Yovits: Yovits *et al.* 1962), and the Wiener Memorial meeting (in Genoa) in 1965. They'd also shared these critical ideas with the funding agencies: not just Yovits at the Office of Naval Research, but their contacts at ARPA too (J. A. Anderson and Rosenfeld 1998: 109). Those "contacts", of course, included Minsky's long-time friend Licklider, the supremo in charge of AI funding at ARPA/DARPA (Chapter 5.iv.f).

And people had listened. Cowan, who attended the Chicago meeting, says: "There's no question that after '62 there was a quiet period in the field" (J. A. Anderson and Rosenfeld 1998: 108). In short, if many researchers in 1969 were already disillusioned, the book's authors—if not the book itself—were largely responsible.

As for what they meant by giving "new momentum" to the field of AI, this involved a swing towards the central problem then being addressed by GOFAI:

It seems to us that the effect of *Perceptrons* was not simply to interrupt a healthy line of research. That redirection of concern was no arbitrary diversion; it was a necessary interlude. To make further progress, connectionists would have to take time off and develop adequate ideas about the representation of knowledge. (Minsky and Papert 1988, p. xiii)

"In any case", they added, "the 1970s became the golden age of a new field of research into the representation of knowledge [see Chapter 10.iii.a]. And it was not only connectionist learning that was placed on hold; it also happened to research on learning in [symbolic AI]."

Their comment on symbolic AI was justified. Although Winston's model of learning had some influence, most of his GOFAI colleagues were working on the representation of knowledge for problem solving, planning, perception, and parsing (see Chapters 10 and 9.x–xi). Indeed, Minsky later admitted that "In all modesty, we were unduly influential" in dissuading GOFAI people from working on learning (Minsky 1984b: 122).

But whether connectionist learning really was "placed on hold" is questionable, as we'll now see.

12.iv. Lamps Invisible

The 1970s and early 1980s were the time of the Sleeping Beauty's sleep. Indeed, they're often called the Dark Ages of connectionism. But Cowan, for instance, has protested:

"I don't think it was the dark ages at all. There was a lot going on" (J. A. Anderson and Rosenfeld 1998: 110).

That's true. If the funding famine caused by the Minsky–Papert critique prevented connectionism from running riot, it didn't stop it in its tracks entirely. Nor did it prevent important neuroscientific work on connectionism, which didn't depend only on DARPA for financial support (see 14.v.a–b and x.a).

Most of the 1950s pioneers continued working throughout the 1970s. (Rosenblatt didn't: he died in a boating accident in 1971.) Some younger folk focused on networks as associative memories. (These generate pattern B on being shown pattern A; they include "content-addressable" memory, where A is a fragment and B the whole pattern.) And Marr and others, by the mid-1970s, diverted interest in AI vision towards connectionist models (see Chapter 7.v.b–d).

In short, by the time connectionism became publicly visible again in the mid-1980s, it rested on three decades of continuous research (J. A. Anderson and Rosenfeld 1998; Widrow 1990). The many enthusiastic outsiders who regarded it as "new" were thus sadly—or rather, happily—misguided.

Some of what went on in those "Dark Age" years excited fellow hibernators at the time, and those responsible would eventually bask in the limelight of the connectionist renaissance (see Section v). But some, discussed here, received scant attention even from like-minded colleagues. Uttley and Selfridge continued working but would never play a leading role again; and the pioneering John Andreae was unappreciated by his peers. (Werbos wasn't appreciated either: his contribution, an important learning rule, was recognized only in the post-hibernation period—see Section vi.d, below.)

a. Relegation to the background

Uttley, in the 1970s, was no longer regarded as an important force in cognitive science. On leaving NPL in 1966, he'd joined N. Stuart Sutherland's psychology group at the University of Sussex, where he implemented a large number of neural networks and moved increasingly towards computational neuroscience (see 14.ii.b). But, for reasons explained below, his ideas now fell on relatively stony ground.

By the standards of the time, Uttley's computer models were more complex than most. His numeral-learner, for example, contained over 200 units ("informons"), each with up to fifty variable inputs. And these might represent many more actual neurones: Uttley allowed that whenever he spoke of a "neurone", one could probably substitute "neurone pool". (Some of the more well-known connectionists said much the same thing.)

His mature theory, which described synapses in informational terms, was summarized in the *Journal of Theoretical Biology* (1975) and detailed in a large book soon afterwards (1979). One of the central ideas was that a single neurone can become a "classifier", not through external reinforcement or training but by "subset reinforcement". That is, the neurone calculates the input correlations, and disconnects itself from the uncorrelated inputs—thereafter responding only to the relevant subset. "Calculates" is the right word, here, for Uttley gave mathematical equations defining which probabilities were supposedly assessed, and how. (He didn't treat *all* learning as unsupervised: the neurones between cortex and muscle, he said, do need a teacher. In other words, they need conditioning.)

Uttley's work was nothing if not ambitious. For instance, he sketched networks explaining why Bach's music is perceived differently today from how it was appreciated in his own time. These ideas, in turn, rested on his more detailed account of how successive layers of neurones in auditory cortex may spontaneously (i.e. with attention, but without training) discover deeper and deeper relationships between musical notes. These were assumed to be hierarchical classifications, defined across time and in harmonic space.

He admitted that his approach threw little light on language. But he claimed that it helped us understand [take a deep breath, here] perception, motor skills, generalization, hierarchical concepts, memory, imagination and hallucination, music, art and craft, and purposive behaviour in general—even including birdwatching and religious belief. (He was a keen birdwatcher, and a committed Christian.)

Given these wide-ranging claims, one might have expected Uttley's *magnum opus* of 1979 to make a splash in the cognitive science community. However, it didn't.

Besides his failings as a communicator in face-to-face conversations (see Section ii.c), and the fact that he "was not someone who went out of his way to interact with other people" (A. Baddeley, personal communication), his highly abstract writing style didn't help. Even his shorter book, written a few years later to show "the enquiring layman" how adaptive neural networks can have mindlike properties, was beyond the capacity of most laymen, enquiring or otherwise (Uttley 1982). Valentino Braitenberg's (1984) contemporaneous account of mindlike "Vehicles" grounded in interacting reflexes was far more accessible (14.vi.a and 15.vii.a).

The main problem, however, was the change in popularity of his theoretical approach. Information-theoretic psychology and psychophysiology had fallen out of favour since around 1960. Even within behind-the-scenes connectionism, Uttley's circuit-based approach had been pushed aside by studies of random networks, and of single impulses rather than spiking frequencies. The focus was on trainer-led (supervised) rather than spontaneous learning, and conditioning had been replaced by pattern recognition as the centre of attention. (Interest in models of conditioning would revive in the late 1980s.) Indeed, the first—and highly influential—PDP conference was held in the very year that Uttley's technical book was published (see Section v.b).

Moreover, network models *in general* were still out of fashion in 1979. Most of Uttley's Sussex colleagues in cognitive science (myself included) were more interested in GOFAI approaches, and were concentrating on vision and knowledge representation rather than learning. This applied even to Geoffrey Hinton (1947–), who in the early to mid-1970s at Sussex was still using GOFAI methods (see below). The general feeling was shared by Uttley's postdoc William Phillips, hired to work with him at Sussex for four years on a reading aid for the blind (which wasn't successful):

I didn't study Uttley's theories very closely, even though I had to help him give a presentation on the "Informon" to the Royal Society. At that time I was very much under the spell of Max Clowes's arguments for the necessity of a syntactic approach to cognition in general, including image processing and scene understanding. As Uttley's work did not seem relevant to those issues it didn't seem worth while spending much time on it. (W. A. Phillips, personal communication)

Ironically, Phillips—currently Professor of Psychology at Stirling's Centre for Cognitive and Computational Neuroscience—now favours information-theoretic and other essentially statistical perspectives on cognition. So he now appreciates Uttley's work better than he did then. But it had no overt influence on his development.

In sum, various factors combined to make Uttley's book invisible—even at very close range.

b. Run and twiddle

Selfridge, too, now received less attention than he deserved—especially from AI scientists. To be sure, he was still a respected voice within the core AI community. (He'd left academia for Bolt, Beranek & Newman but retained his links with his fellow pioneers.) And cognitive psychologists valued his earlier research. Pandemonium was highlighted in two early textbooks (Neisser 1967; P. H. Lindsay and Norman 1972), and fetchingly illustrated in one of them (see Figures 6.5 and 6.6). But his own theory of learning was so different from the then current GOFAI paradigm that most AI people ignored it. (Minsky didn't, but he was always a maverick.)

Even the embattled connectionists paid little or no attention. This was partly because Selfridge didn't make his new work easily available. A few brief accounts appeared in out-of-the-way publications, and drafts of his future book were circulated to friends—including myself (Selfridge 1981). But it was never officially published.

Focusing on *the whole animal* rather than isolated systems such as vision, Selfridge argued that the fundamental mechanism of all adaptive movement is a form of unsupervised trial-and-error learning. He called this "RT", or "run and twiddle", identifying the core principle as "if things are getting better, don't change what you're doing" (Selfridge 1984: 23). Even *E. coli* can engage in a random walk, follow a positive value gradient, and then stop if things don't get any better. In other words, he said, even simple creatures—unlike man-made perceptrons (which he criticized for this reason)—"want" something.

What *E. coli* wants is an optimal concentration of certain chemicals, which can be achieved by simple chemotaxis. Animals with more complex wants, or purposes, need more complex mechanisms to guide and structure their behaviour. That is, they need hierarchies of adaptive control loops—but how can such systems, and their development, be conceptualized? As he admitted, it was hard to know how to pose the relevant questions, never mind answer them.

(His answers today are posed in terms of "EAMs": Elementary Adaptive Mechanisms: Selfridge and Feurzeig 2002. Now based at MIT's Media Lab, Selfridge is working on DARPA's "ABC" programme. The abbreviation stands for Agent-Based Computing, an area currently of great interest to GOFAI too—see Chapter 13.iii.d–e.)

Another reason for the lack of interest in Selfridge's 1970s ideas was that unsupervised learning had become unfashionable. Thanks to perceptrons, Adalines, and—ironically—Pandemonium, most connectionism in the 1970s involved a human trainer. Those few people who were still modelling reinforcement were swimming against the current.

c. Reinforcement and purpose

One of the people still modelling reinforcement was Andreae (1927–), an Englishman based since 1988 at the University of Canterbury, New Zealand. He'd started in the early 1960s, with a model of noise-tolerant trial-and-error learning called STeLLA (Andreae 1963, 1969a; Gaines and Andreae 1966). (The name, and its typography, indicated Andreae's former post: Standard Telecommunications Laboratories Learning Automaton.)

STeLLA was interesting not least because it was an attempt to embrace both sides of the symbolic–subsymbolic and virtual–interactive dichotomies. That wasn't always recognized by Andreae's contemporaries, however. For example, Edinburgh's James Doran (1940–), who'd already simulated a "pleasure-seeking" rat robot inspired by the Grey Walter tortoises (Doran 1968a), said this:

[It] is far from clear how such network systems can be persuaded to yield really complex behaviour. The results of the work [by Minsky and Papert: 1969] on perceptrons and other such self-organizing systems do not encourage optimism. (Doran 1969: 524)

Andreae described STeLLA as constructing, and being guided by, an internal plan of action and model of the environment. These were implemented as matrices coding probabilistic transitions between various pattern–action pairs. He soon realized, however, that probabilistic transitions weren't going to be enough. His associative robots needed to be able to learn fixed sequences too. They couldn't be programmed (for instance) to recite *one, two, three, four, five* . . . flawlessly. But they should be able to learn to do so: even a 99 per cent probability of coming up, next, with *six* wouldn't be acceptable. So he redesigned STeLLA in the early 1970s, to produce a fundamentally different machine: PURR-PUSS (Andreae 1977, 1987; Andreae and Cleary 1976).

The acronym stood for Purposeful Unprimed Real-world Robot with Predictors Using Short Segments. The "PURR" coded Andreae's guiding aims: that the robot should set its own goals; learn patterns and behaviours that he hadn't anticipated (even though some "innate" biasing might sometimes be needed); and have real-world capacities comparable with ours, such as motor control, 3D vision, hearing, and even language. The "PUSS" coded the complex associations that were crucial in the robot's processing and performance.

For example, this new system was able to construct temporary memories on various hierarchical levels. These were broadly equivalent to the pushdown stacks and "contexts" used in GOFAI planning (Chapter 10.iii.c and v.b–d), in the sense that they enabled the system to monitor its own performance, keeping track of where it was in a plan structure. Instead of being stored as symbolic expressions, they were stored as associations with environmental changes (compare: tying a knot in one's handkerchief) or, more often, trivial bodily actions (compare: raising an eyebrow). Fixed sequences, by contrast, require long-term memories—but these too were stored as complex associations constructed by the system.

One of Andreae's papers (1969a) was chosen as the first item in the newly founded *International Journal of Man–Machine Studies*, whose readership included psychologists, and he contributed to an interdisciplinary encyclopedia co-edited by a linguist (Andreae 1969b). More recently, he has been cited in historical remarks in the reinforcement-learning literature (Sutton and Barto 1998: 19, 84, 109).

Nevertheless, and although he was respected by control engineers, he had relatively little influence on cognitive scientists.

One reason was place of publication. Most of his papers appeared in journals aimed at engineers. And the forty technical reports on PURR-PUSS that were lodged in leading libraries, including the Library of Congress, were easy to find only if one already knew about them. (Modulo information overload, the Web should decrease this sort of invisibility.)

A closely related reason was timing. Andreae's earliest work appeared when people excited by connectionism were looking in Rosenblatt's direction. By the time he described his second system, connectionism in general was out of fashion. One reviewer of his 1977 book (Schubert 1978), for example, wheeled out John McCarthy's GOFAI objection that one must understand a cognitive skill well enough to program it before one can enable a computer to learn it (Chapter 10.i.f). Andreae's position, by contrast, was that complex associative robots probably won't be fully intelligible, even to their designers. (The same may apply to detailed neuroscientific models: see Chapter 14.v.d.)

Only very recently, thanks to the renewal of interest in this area, has his later research been made widely available—although it remains to be seen how many people will read it (Andreae 1998). Previously, it was neglected: "*Unfortunately*, [Andreae's] pioneering research was not well known, and did not greatly impact subsequent reinforcement learning research" (Sutton and Barto 1998: 19; italics added).

12.v. Behind the Scenes

Although no behind-the-scenes connectionist was attracting much attention in the cognitive science community, some were making significant theoretical advances nonetheless.

Very few were talking to each other: their shared project didn't become truly communal until the end of the 1970s (see subsection b below, and Section vii). Meanwhile, they were working in very different disciplines, and on rather different problems.

a. Left alone to get on with it

If Andreae was unfortunate, Widrow—in his own estimation—wasn't. Not only did he (unlike Andreae) receive enormous attention in his early career, but he managed to escape the near-inevitable consequences.

On his view, the anti-perceptron backlash largely passed him by simply because he'd called his networks by a different name:

I looked at [Minsky and Papert's] book, and I saw that they'd done some serious work here, and there was some good mathematics in this book, but I said "My God, what a hatchet job." I was so relieved that they called this thing the perceptron rather than the Adaline because actually what they were mostly talking about was the Adaline, not the perceptron. (J. A. Anderson and Rosenfeld 1998: 60)

Another reason for Widrow's lucky escape, perhaps, was that—unlike Rosenblatt— he hadn't provocatively presented his Madalines as models of human psychology.

He had, however, suggested in the paper on the LMS algorithm (Widrow and Hoff 1960) that AI could be hugely advanced by using "computers built of adaptive neurons". This suggestion may have looked like an afterthought to the contemporary readers of that paper, but AI was already close to Widrow's heart. He'd attended the Dartmouth Summer Project in 1956 (Chapter 6.iv.b), and decided then and there "to dedicate the rest of my life to that subject" (J. A. Anderson and Rosenfeld 1998: 49). Adaline-based computing, he believed, could do things that NewFAI couldn't:

[Samuels's checkers-player] would be considerably more powerful if it were possible to extract from the memory previous situations that are *similar* and not necessarily *identical* to the current situation. Far less experience and storage would be needed to adapt to a given level of competence. (Widrow and Hoff 1960: 133)

In general, he said, AI needed not "rote memory" but "recall-by-association parallel-access memory systems"—which could be implemented in "Memistors". These were Adaline-based devices, providing a simple form of content-addressable memory.

A third reason why Widrow was less affected by the backlash was that, as a control engineer, he could demonstrate practical applications that undeniably worked.—But *how* did they work? The answer lay only partly in the special-purpose hardware. Even more important was the mathematics. If the journalists had been turned on by the switches and the pencil leads, the scientists were more excited by the learning rule:

[The key to the LMS algorithm] was to get the gradient not by taking many samples of data and measuring mean square error over a long period of time. The idea was to be able to get the gradient from a single value of error—a single number, square it, and say that's the mean square error. Then when you work out the gradient of that error with respect to the weights, it's really simple. You get an algebraic expression and you realize that you don't have to square anything; you don't have to average anything to get mean square error. You don't have to differentiate to get the gradient. You get all this directly in one step. Not only that, but you get all components of the gradient simultaneously instead of having to make measurements to get one gradient component at a time. *The power of that, compared to the earlier method* [i.e. steepest descent], *is just fantastic.* (J. A. Anderson and Rosenfeld 1998: 53; italics added)

(*Plus ça change*... It became clear years later that the "fantastic" Widrow–Hoff adaptation rule was essentially the same as Hull's early 1940s equation for calculating habit strength: Sutton and Barto 1981; see 5.iii.b.)

Widrow wasn't the only one to address his work mathematically. In general, the behind-the-scenes research of the 1970s–1980s which—unlike that discussed in Section iv—*did* eventually receive recognition was mathematical in spirit. It aimed to find more powerful training procedures, especially for multi-layer nets (which Rosenblatt had found intractable).

The problem with multi-layer nets was that they contained "hidden" units. These weren't directly connected to *both* input and output units, and might even be directly connected to *neither*. And that fact carried unavoidable problems with it:

* How could an individual hidden unit be reached by a learning rule (which monitors the output, given a certain input)?
* Could one define rules that would enable multi-layer networks to adapt efficiently?
* Was there any multi-layer equivalent of the single-layer convergence theorem?

* What were the limits, if any, on the patterns or structures that specific kinds of rule and/or network could learn?
* How few hidden units would suffice for any given task? (Too many, and the system would in effect be a look-up table.)
* What could be said about the potential storage capacity of associative memories with different numbers of cells?
* And what were the functional merits of distributed memories as compared with localist networks, or with GOFAI?

Answers to these abstract questions were sought by a wide variety of people. Some were control engineers, seeking a multi-layer equivalent of the LMS algorithm. Some were computer engineers, keen to implement their answers in VLSI chips. Others were electrical engineers or physicists, interested in dynamical models of associative memory. Yet others were computer scientists already working in AI.

And many were mathematically minded psychologists, usually with experience of GOFAI modelling and/or interests in neuroscience. A few of them tried to model the nervous system in some detail (see 14.iv–v).

b. A problem shared . . . ?

One shouldn't assume that all these people communicated with each other, or even knew of each other's existence. During the Dark Ages most of them were working in isolation, separated not by monastery walls but by disciplinary boundaries. This is why priority disputes in this area are so common. Indeed, a recent collection of interviews abounds with coded and not-so-coded priority claims, backed up by hard-luck stories about lack of recognition in the wilderness years (J. A. Anderson and Rosenfeld 1998).

To be sure, GOFAI-based cognitive scientists already had a thriving interdisciplinary community (see Chapters 6–10), and many influential connectionists had strong links to that. Some had even developed GOFAI-inspired psychological theories.

For instance, David Rumelhart at UC San Diego had developed formal models of story plots, or "story grammars" (Rumelhart 1975). And he had used both ATNs and ideas from HEARSAY to model reading errors (see Chapters 10.v.e, and 9.xi.b). Indeed, HEARSAY was later acknowledged by the leaders of PDP as having "played a prominent role in the development of our thinking" (Rumelhart, McClelland, *et al.* 1986: 43). Rumelhart had also been excited by NewFAI's "semantic networks" (Chapter 10.iii.a), which had helped lead him and others to work on associative memories (Findler 1979).

Similarly, Christopher Longuet-Higgins had produced GOFAI models of musical perception (see Chapter 2.iv.c). Moreover, there was a small but growing community of connectionist workers on vision, including Marr, Tomaso Poggio, Stephen Grossberg, and Hinton. Although they were wary of GOFAI approaches, most of them had started from that base; and a few, such as Hinton, had worked on GOFAI projects for some years (see below, and Chapters 10.iv.b, 14.ii–iii).

However, specialists in AI and psychology, even if they communicated with each other, had little or no contact with engineers and physicists. If "A problem shared is a problem halved", there were no helpful interdisciplinary half-measures available.

This situation would begin to change in the summer of 1979, when many connectionists (including supporters of *non*-distributed memories) met for the first time. The occasion was a meeting on 'Parallel Models of Associative Memory'. It was hosted in the Cognitive Science Center in La Jolla, by Rumelhart and Donald Norman—key members of the PDP group (see below). (Ironically, 1979 was the year in which connectionism was announced to have finally "withered and died": Section iii.e, above.)

This was one of the many projects funded around 1980 by the Sloan Foundation, who were then switching support from neuroscience to cognitive science (see Chapter 8.i.c). The organizers were the psychologist Hinton and the neurophysiologist James A. Anderson (1940–), at Brown. Hinton was then based in the Cambridge University research unit that had housed Kenneth Craik, but he'd visited UCSD's Cognitive Science group and was about to return there. As for Anderson, he'd been at UCLA in the early 1970s but was now based at Brown.

(NB. One shouldn't confuse James A. Anderson with John R. Anderson, whose ACT* system was discussed in 7.iv.c. It's easy to do so, however, because they both did important work on associative memory. Indeed, Hinton and J.A.A. described J.R.A.'s ideas on this topic as "seminal"—Hinton and Anderson 1981: 15. In this chapter, "Anderson" refers to J.A.A. unless otherwise noted.)

Like Christopher Langton's "Call for Papers" for the seminal conference on A-Life (15.ix.a), Hinton and Anderson's invitation had been sent to a very varied group:

We deliberately chose people from fields as diverse as neurophysiology, cognitive psychology, artificial intelligence, mathematics, and electrical engineering. *Most of the participants had not met each other previously and in some cases were unaware of each other's work.* Yet as things progressed, it became clear that there were large areas of agreement, as well as strong disagreement about details. (Hinton and Anderson 1981, p. vii; italics added)

The new-found community, however, wasn't destined for sweetness and light. Within a few years, the "strong disagreement about details" would help engender what Bart Kosko later described as "many feuding factions", some of whose leaders "literally hated each other's guts" (J. A. Anderson and Rosenfeld 1998: 402–3). Human beings being what they are, this lack of collegiality was largely due to jealous resentments caused by the *success* of the field, with its attendant publicity and commercial promise (see Section vii).

However, all that was yet to come. At the end of the 1960s, success seemed very far off. Quite apart from the Minsky–Papert fusillade, there were many practical difficulties.

c. How large is your memory?

One difficulty concerned multiple storage in associative memories. Both Rosenblatt and Widrow had found that their machines could store more than one pattern at a time—but the more patterns involved, the less efficient the storage: interference effects came to the fore.

This made sense, given that any one pattern was distributed across many different units, and that each unit might contribute to several patterns. Indeed, it raised the spectre of a rapid descent into chaos as the number of stored patterns increased. What was needed, beyond these empirical observations, was some *principled* (mathematical) estimate of the potential storage capacity of connectionist systems, and of the efficiency of distributed memories.

Such questions had been asked before, in relation to punch card systems as well as brains. Turing himself had been asked by John Z. Young "whether one can make an estimate of how much information can be stored in a brain with a given number of nerve cells", and had sketched the bare bones of a reply (S. S. Turing 1959: 145 ff.).

Moreover, Peter Greene, in the Chicago group founded by Rashevsky, had recently developed that earlier work. He'd shown how random superimposed ("hash") coding could be used both to store and to recall associative memories (Greene 1965). (Cowan regards this as "virtually identical with David Marr's cerebellum model, except it's not applied to the cerebellum"—J. A. Anderson and Rosenfeld 1998: 110; see Chapter 14.iii.d.)

But Greene's paper wasn't widely read. The first to reach an appreciable audience for these matters were Longuet-Higgins and his students David Willshaw (1945–) and Peter Buneman, at Edinburgh University (before Longuet-Higgins's move to Sussex in 1974). In the late 1960s, Longuet-Higgins outlined a quasi-holographic model of memory (1968*a,b*), and then collaborated with his two students on a non-holographic model (Willshaw *et al.* 1969).

He presented some of these ideas informally at one of the renowned Serbelloni meetings on theoretical biology run by his Edinburgh colleague Conrad Waddington (15.v.b). On that occasion, Cowan made him "absolutely furious" by saying they were "virtually identical with Steinbuch's learning matrix" in the late 1950s (J. A. Anderson and Rosenfeld 1998: 110). Longuet-Higgins in "absolutely furious" mode was, as all his friends know, fearsome to behold. But this storm in a teacup didn't matter. For the main audience wasn't the select group overlooking Lake Como, but the wide readership of the *Proceedings of the Royal Society*, Series B, and the even wider readership of *Nature*.

Later recalling the critical reception of his 1968 *Nature* paper, Longuet-Higgins said it "attracted altogether too much attention when it first appeared in print". But he had a ready defence:

I should have known better than to speculate so shamelessly about the neural mechanism of temporal memory; the idea of a neuron's "learning" a particular frequency has scarcely a leg to stand on. But the discussion the paper provoked—in both physical and physiological circles—testified to the fact that ideas, however far-fetched, are the life blood of scientific research, and that a demonstration of how something could possibly happen (in this case, cued temporal recall) may be almost as welcome as the discovery of how it actually does happen. (Longuet-Higgins 1987: 367)

Quite apart from any second thoughts about that paper, he had other reasons for moving on to a non-holographic theory. Several people in the 1960s had likened the brain to a hologram, a newly invented device for storing optical images. Patterns in a hologram are not stored at specific points but are distributed across the whole system. They can be reconstructed from a damaged hologram or from a small part of it, so the device acts as a content-addressable memory.

The physicist P. J. van Heerden, when he originated the brain–hologram analogy in 1963, had compared holograms to Beurle's models of neural nets, which were inspired by current ideas about the cortex (see Section ii.b). And the neuroscientist Karl Pribram, co-author of *Plans and the Structure of Behavior* (6.iv.c), had likened them to brains in some detail (Pribram 1969). Later, in a memorial volume for his ex-teacher Lashley, he said:

The properties of holograms are so similar to the elusive properties that Lashley sought in brain tissue to explain perceptual imaging and engram encoding, that the holographic process must be seriously considered as an explanatory device. (Pribram 1982: 176)

Willshaw and Longuet-Higgins, however, were sceptical.

Although the analogy was suggestive, they said, it breaks down on closer examination. They had recently generalized the hologram to deal with sound, in the "holophone" (Willshaw and Longuet-Higgins 1969a). But their main aim hadn't been to widen the technology (from optics to acoustics). Rather, they wanted to estimate the computational power of holograms *considered as a general class of device*. Since temporal patterns are one-dimensional, the holophone's mathematics would be relatively tractable.

Now, they argued that some crucial physical properties of holograms can't be attributed to brains (Willshaw *et al.* 1969). Moreover, the quality of the regenerated patterns is *in principle* unnaturally low: they'd wryly noted the "rather distressing noise characteristics" of the holophone (Willshaw and Longuet-Higgins 1969a: 356). As for computer modelling, holographic systems would be excessively slow in simulation: the holophone needs N^3 separate multiplications for a signal of length N. This was a theoretical point, but had been borne out by experimental simulations.

In short, holographic models of memory, although admittedly suggestive, didn't fit the biological facts. (Some neuroscientists, including Pribram, retained faith in them nonetheless: Pribram *et al.* 1974.) And they weren't computationally tractable either.

Instead of holographs, Willshaw and Longuet-Higgins offered a simple matrix schema of an associative network having the relevant functional properties (Willshaw *et al.* 1969). This schema was both easier to simulate and more biologically plausible. They predicted that it "may well find application in computing technology, especially when parallel computation techniques become generally available" (Willshaw and Longuet-Higgins 1969b: 351). And although their approach was primarily "a mathematical investigation", not an empirical hypothesis, they outlined how it might apply to actual synapses (Willshaw *et al.* 1969: 962; cf. Willshaw 1981).

Their matrix model could store more information than a hologram of comparable size, and could reconstruct patterns more faithfully. Indeed, "in optimal conditions [the network] has a capacity which is not far from the maximum permitted by information theory" (Willshaw *et al.* 1969: 960). Specifically, the informational capacity was 69 per cent of the theoretical maximum, and proportional to the number of possible connections.

This was encouraging: it meant that interference between superposed representations was less of a threat than had been feared. Admittedly, simulation had shown that "[even] in the absence of damage there is a fairly sharp limit to the number of associations that can be stored; if this number is exceeded the performance of the system degenerates rapidly" (Willshaw and Longuet-Higgins 1969b: 358). Nevertheless, the high potential efficiency (within that limit) of distributed systems was a strong point. The more patterns a memory model can store, the better.

Or so it seemed. At the turn of the next century, the ex-Ratio Club neurophysiologist Barlow would show that this assumption was mistaken (Barlow 2001*a*: 603–4). He'd already suggested as much, many years earlier (Barlow 1961). Now, he provided mathematical arguments to prove that *learning*, as opposed to mere *storage*, requires "sparse" representations—which use many more units to store each pattern (see 14.x.e).

At the turn of the 1970s, however, Willshaw and Longuet-Higgins's analysis seemed to be very good news.

d. Disillusion on distribution

The bad news was that learning in multi-layer nets was still intractable. When Willshaw and Longuet-Higgins published their work (and Minsky and Papert their broadside), disillusionment had begun to set in. Indeed, Widrow later said: "[The] field was already gone. There was just about nobody doing it" (J. A. Anderson and Rosenfeld 1998: 60)—see Section iii.e.

Widrow himself had given up in disgust. He'd returned to his early interests: "We stopped doing neural nets because we'd hit a brick wall trying to adapt multilayer nets. On the other hand, in adaptive filtering and adaptive signal processing, we were making great strides" (J. A. Anderson and Rosenfeld 1998: 61).

Some connectionists avoided these problems by focusing on localist memories, instead of distributed ones. In a localist system, a concept or pattern is represented by a single unit. An early 1980s example, published in a leading psychological journal (largely because it was closely compared with experiments on human subjects), was a model of word recognition broadly similar to the localist Pandemonium (McClelland and Rumelhart 1981; Rumelhart and McClelland 1982). In this model, letter strokes, letters, and (finally) words were recognized by three separate layers.

Localist networks have certain processing advantages, and were defended by Julian Feldman—co-editor of *Computers and Thought*—in the core journal *Cognitive Science* (J. A. Feldman and Ballard 1982). Moreover, the 1970s saw increasing experimental evidence for "grandmother cells" in the brain, and Barlow (1972) had proposed an influential theory that emphasized them (Chapter 14.x.e). So even if the cerebral cortex was a distributed system, it might have some important localist features.

Nonetheless, distributed systems seemed more interesting to most people. The authors of the word recognition network, for instance, were already moving firmly in that direction (see Section vi).

Gradually, the disillusionment lessened, as light became visible at the end of the distributed tunnel. The exciting new results included linear associative memories (in 1972), Hopfield nets (in 1982), the Boltzmann machine (in 1985), and above all back propagation (in 1986). Widrow, again: "We would have given our eye teeth to come up with something like backprop" (J. A. Anderson and Rosenfeld 1998: 60).

These ideas, all discussed below, provided complex networks with increasing computational power and efficiency. They also—up to a point—offered increasing biological plausibility.

(Only "up to a point". For one thing, back propagation is unrealistic: brain synapses don't transmit backwards. For another, connectionist theories and artificial networks were/are drastic oversimplifications of real neurones: see 14.i.)

e. Linear associative memories

Linear associative memories were independently defined in 1972 by James Anderson of UCLA (who had developed a simpler memory model in the late 1960s), and by Teuvo Kohonen (1934–) of the Helsinki University of Technology. Both men had long-standing interdisciplinary interests, but they came into cognitive science from diametrically opposite directions.

Anderson was an experimental neuroscientist whose teenage passions had been amateur radio and science fiction, whereas Kohonen was an electronic engineer whose "special hobby" at high school had been psychology—not least, Gestalt psychology (J. A. Anderson and Rosenfeld 1998: 242, 147). Whereas Anderson studied memory in a neuroscientific context, even developing a detailed "cortical model" of its implementation in the brain (J. A. Anderson 1972: 190 ff.), Kohonen entered from the engineer's corner. He was intrigued by various recent analyses of holographic memory, and by some suggested alternatives—including that of Willshaw and Longuet-Higgins (Kohonen 1972: 174).

Coming from distinct professional communities, Anderson and Kohonen knew nothing of each other's work for several years (although both attended the La Jolla meeting in 1979). Indeed, Kohonen had often found his professional background to be an obstacle in biological circles.

To be fair, even his fellow engineers were initially sceptical. He originally became interested in neural networks in 1962, and in the late 1960s submitted his first two papers on the topic to two very different journals: *Nature* and the *IEEE Transactions on Computers*. Both submissions "came back like a boomerang", because his views were considered so "weird" by the editors and referees (J. A. Anderson and Rosenfeld 1998: 150).

His theory of associative memory was eventually published in 1972. But it wasn't widely accepted until the 1980s. Part of the problem was that although he was "desperately trying to relate [his] networks to biology", he presented himself as "an inventor" (p. 160). He attended some neurophysiological meetings in the 1970s, but found it "very difficult to get an engineer's point of view accepted" (p. 155).

The engineer's point of view is more mathematical than the biologist's, even though neurophysiology had been moving in this direction ever since Rashevsky (see Chapter 4.iii.c). Anderson himself, whose own paper appeared in the successor to Rashevsky's journal, has said:

It is interesting to see how differently the two authors develop the same basic idea: Kohonen is concerned with the mathematical structure, while Anderson is concerned with physiological plausibility ... Kohonen does not make the neural analogies as explicitly as does Anderson, but the mathematics is identical. (J. A. Anderson and Rosenfeld 1988: 171)

The mathematics was not only identical, but also simple (linear): at base, multiplication and addition. These new systems—which Anderson called "interactive" memories and Kohonen "correlation matrix" memories—were in effect complex perceptrons, in which a *group* of *analogue* (not McCulloch–Pitts) units is connected to another such *group*. The crucial biological analogies were thus neuronal firing frequency, which *varies* with the strength of the stimulus, and cell assemblies (5.iv.c). Grandmother cells were tacitly denied, or anyway ignored (see Chapter 14.x.e).

The point to grasp, for those not mathematically inclined, is that these systems boiled down to *arithmetic*. Each cell group was represented by a vector (a matrix of many individual values) defined over its component units. And its activity level was expressed by adding the activity levels (summing the weights) of all the units. As for the synaptic weight changes that effected Hebbian learning, these were proportional to the product reached by multiplying the pre- and post-synaptic activity levels.

At the 1979 La Jolla meeting, Kohonen wowed the audience by presenting a "spectacular" example of content-addressable memory (Hinton and Anderson 1981: 9). To near-universal amazement, an entire face was reconstructed by inputting either a mere fragment, or a smudged version, of the image. Moreover, this was "unsupervised" learning, done without any feedback from the programmer.

Kohonen stressed that 100 different faces had 100 *distinct* representations in the network. But these were distributed ("holological") matrices, not localized units. More important, since even Simon had allowed that nothing in GOFAI-based psychology required that memories be stored at a specific point, the computational operations were *structured interactions* correlating these collective states (Kohonen *et al.* 1981: 107).

At one level, simple correlation matrix memories could be seen as "the extreme and purest cases of S–R mapping", since there is either no feedback whatever from response to stimulus, or (in certain recurrent systems) feedback involving *all* the stimuli (Kohonen *et al.* 1981: 112). At another level, they could be seen as models of real neural structures. Indeed, Kohonen made various comparisons between 3D matrices and laminar cortex, and between his auto-associative memories and synaptic learning.

Similarly, he suggested that his account of topographical mapping (Kohonen 1982) explained why there are various types of *interneurone* in the brain (see Chapter 14.x.a). (This work on topological mapping was akin to that of Willshaw and Christoph von der Malsburg, who were cited several times in Kohonen's paper: see 1.iii.h and 14.v.b.)

f. The physicists have their say

Ten years later, in 1982, a very different model of associative memory burst onto the scene: Hopfield nets. These are named after the Caltech physicist John Hopfield (1933–), who intended them as "a form of general (and error-correcting) content-addressable memory". They'd actually been described many times by others, earlier. But for various reasons, explored below, it was Hopfield who got the credit.

Hopfield nets were globally connected, in that every unit was linked to every other. So there weren't any separate layers: each unit acted as an input, output, and processing unit.

To be set up for an application, the inputs would be coded as (for example) sensory feature detectors, while the final output was the relevant concept, or pattern. The units were McCulloch–Pitts neurones. The (initially random) connection weights were continuous, and symmetrical (each unit would excite its 'pair' to the same extent). At each change point—which were random, not time-stepped—the unit would sum the weights of its synaptic inputs, and then 'fire' only if its threshold had been reached.

In physical systems made of many simple elements, said Hopfield, stable collective phenomena (such as vortex patterns in fluids) emerge spontaneously. The explanation is dynamical:

The equations of motion of the system describe a flow in state space. Various classes of flow patterns are possible, but the systems of use for memory particularly include those that flow toward locally stable points from anywhere within regions around those points. (Hopfield 1982: 460)

In other words, the system undergoes myriad local changes until it settles into equilibrium. (Compare Steve Wolfram's four dynamical patterns in cellular automata, only one of which is suitable for computation: see 15.viii.a.) Memory storage is analogous to the stability of the relevant points (attractors). Error correction and generalization correspond to the fact that one and the same attractor will be reached from anywhere within the relevant region. And content addressability is comparable to the reconstruction of the attractor state from an approximation to it.

As for how this happens, in physics the settling-down process involves energy minimization. So Hopfield defined an algorithm, based on the theory of spin glasses, that enabled a simulated network to reach equilibrium in a comparable way. This is reminiscent of von Neumann's hunch that the logic of the brain might be closer to thermodynamics than to formal logic (see Section i.c).

The global "energy" of the network as a whole was defined by summing across all units. And the effect of (binary) state changes in the thresholded units was either to decrease this global measure or to leave it unaffected. When there were no more unit changes (i.e. when the network had settled down), this measure would have reached a minimum.

Having defined this class of networks, Hopfield ran a large number of simulations varying the numbers of units and the starting points (i.e. the binary values assigned to the units before the system was allowed to run). He asked (for instance):

* how many memories a network of a given size could store before producing retrieval errors, or even being totally overwhelmed;
* how many units needed to be set "correctly" at the start in order for the *closest* attractor to be reached;
* and how dissimilar two inputs must be if they were to be stored as distinct memories (distinct attractors).

Rosenblatt, too, had combined theoretical analysis with systematic simulation experiments (see Section ii.e–f). But Hopfield saw his networks as an advance on perceptrons in three ways:

* They involved backward as well as forward connections (this was implied by the symmetry condition);
* they focused on abstract emergent properties, as opposed to the specific results of encountering a particular pattern in the real world;
* and they didn't require synchronous change.

Similarly, the brain involves backwards connections (though not two-way synapses), and it doesn't work in time steps. However, the brain isn't made of McCulloch–Pitts neurones. Two years later, Hopfield generalized his results to cover more biologically realistic units. The paper's title said it all: 'Neurons with Graded Response have Collective Computational Properties Like Those of Two-State Neurons' (1984).

Hopfield's 1982 paper contained two exciting claims. One—echoing the comparable memory-network research described above—was that "the ability of large collections of neurons to perform 'computational' tasks may be in part a spontaneous collective consequence of having a large number of interacting simple neurons" (Hopfield 1982: 460). That is, intelligence could arise without the need for highly specific circuitry to be designed by evolution—or by technologists. The other was more novel, and more provocative: that "the model could be readily implemented by integrated circuits hardware" (p. 460).

It's not surprising that many people were enthused. For there was promise here for psychologists and technologists alike. Whereas Kohonen's first paper on neural nets had come back "like a boomerang", Hopfield's work was immediately acclaimed—and was soon featured in the *Los Angeles Times*.

When, only six years later, DARPA considered funding neural networks again, they would identify it as a watershed:

From the mid-1950s to the mid-1970s, a wide variety of associative memory models were studied . . . Widespread interest in such models seems to have waned by 1975. Although many application areas were explored, these early memory models had little technological impact. The available technology favoured the use of [the digital memories used in GOFAI] . . .

Hopfield rekindled widespread interest in associative memory when he proposed his 1982 energy minimization model based on the outer product storage rule. *While outer product memories had been studied extensively, Hopfield's version captured attention* because of its timeliness from a technological standpoint and the theoretical appeal of energy minimization. (DARPA 1988: 81; italics added)

Anderson agrees: "As far as public visibility goes, the modern era in neural networks dates from the publication of this paper" (J. A. Anderson and Rosenfeld 1988: 457).

The public, here, included not only the *Los Angeles Times* but also many of Hopfield's professional colleagues—a fact tartly remarked by the neurophysiologist Cowan:

The theoretical physics community is like a swarm of locusts. There are far, far more theorists around than there are problems. There are only two or three problems, and whatever problem gets hot, a lot of them swarm onto it. That's what happened with the Hopfield paper. (J. A. Anderson and Rosenfeld 1998: 113)

Undeniably, however, the "locusts" helped advance the field. (And some stayed within it, rather than flying off to pastures new.) They had exceptionally keen mathematical skills, and a pre-existing body of theory (the physics of spin glasses) that could be applied in analysing these networks. The increase in mathematical expertise would be a lasting benefit, even if more neuroscientific data were also needed (see Chapter 14).

As for Hopfield himself, his careful analysis of the computational properties of 'his' networks was highly influential. It was more easily intelligible, for example, than Grossberg's many earlier remarks on such topics (see below, and 1.iii.h). It helped, of course, that Hopfield's paper had been published in the highly respected *Proceedings of the National Academy of Sciences*. So had some of Grossberg's (e.g. Grossberg 1971). But, characteristically, they'd covered a host of neuropsychological details that many readers would find distracting.

In addition, Hopfield himself was no shrinking violet. He "travelled all over the country talking about his results", even reaching itinerant visitors by speaking in small rooms located in major airports (Barrow 1989: 14). In other words, he was deliberately taking his ideas into the scientific marketplace (1.iii.h). As we'll see below, this was just one reason why many of his colleagues became deeply irritated by him.

But "the public" also included the many people in business, the military, and academia who hoped to gain from putting associative memories onto commercially marketable VLSI computer chips. Indeed, Mead—at the California Institute of Technology—achieved this within a few years.

By 1990, Mead had even designed silicon chips replicating a 2,500-cone human retina (Sivilotti *et al.* 1987; Mead and Mahowald 1991). These multi-layer networks—the first examples of what he termed "neuromorphic engineering"—provided not just photoreception but also various types of low-level visual processing, including automatic light adaptation. He also produced a VLSI cochlea, and a sound localizer based on the auditory brainstem (Mead 1989). Had Hopfield published the same paper ten years earlier, when VLSI was still confined to the research lab, it wouldn't have made such a splash.

All this is *not* to say that Hopfield was the first to have the core idea of energy minimization. A number of researchers interviewed by Anderson and Edward Rosenfeld (1988) for their fascinating "oral history of neural networks" insisted—with some force, not to say venom—that several people had already had much the same idea. Some had even explicitly mentioned spin glasses.

At least eight such individuals were named in these interviews, including two as far back as the 1950s. And two interviewees picked out Shun-Ichi Amari (1936–), of the University of Tokyo, as the real pioneer.

Amari had been modelling associative networks since the late 1960s. In 1977 he defined "learning based on potential" in terms of a mathematical function identical to Hopfield's (Amari 1977: 275). Although he gave only fourteen lines to this topic, he situated it within an elegant (and equally condensed) mathematical discussion of various learning rules, of the differing effects of noise, of two new "convergence theorems", and of recurrent networks in which the output is fed back as the input. Hopfield cited Amari (alongside Kohonen, Willshaw and Longuet-Higgins, Anderson, and Marr—but not Grossberg) as a precursor. But whereas Amari had given only fourteen lines to the idea, Hopfield discussed it at greater length.

Hecht-Nielsen, in these same interviews, mistakenly accused Hopfield of stealing Amari's idea *without* citing him (J. A. Anderson and Rosenfeld 1998: 301). This mistake, I assume, was an overflow from his general irritation at the attention, indeed the publicity, that Hopfield received. Hecht-Nielsen's lasting irritation, and the venom of the other interviewees' remarks, was due in part to Hopfield's relentless self-promotion, and in part to his repeated failures to cite other people's work, as a result of which he got the credit for priority.

A glaring example was his paper comparing networks composed of binary (McCulloch–Pitts) and continuously graded (sigmoid) neurones. The opening page announced "new" computational results, "unexpectedly" discovered (Hopfield and Tank 1986: 265). This appeared in the high-prestige journal *Science*, whose referees evidently didn't know that these results weren't new at all. But others did, and wrote in to complain vociferously. Most unusually, the editor agreed to publish several Technical Comments pointing out not only that Hopfield and Tank's contribution wasn't original but also that others had already taken them further.

One brief comment ended by saying drily that "The application of neural network theory to technology would be expedited by further consideration of known results," and gave a footnote citing several reviews of "early [*sic*] and recent" relevant work (Carpenter *et al.* 1987: 1227). That was for public consumption. What was said in private on the grapevine, which was part-surfacing in those oral-history interviews, was much less tactful. (So much, again, for the Legend: 1.iii.b.)

g. The power of respectability

The main reason why Hopfield's 1982 paper had attracted so much more attention than its predecessors was his reputation *as a physicist*. As Anderson has put it:

John Hopfield is a distinguished physicist. When he talks, people listen. Theory in his hands becomes *respectable*. Neural networks became instantly legitimate, whereas before, most developments in networks had been the province of somewhat suspect psychologists and neurobiologists, or by those removed from the hot centres of scientific activity. (J. A. Anderson and Rosenfeld 1988: 457; italics added)

What can happen when someone isn't perceived as "respectable" is indicated by Werbos:

So he [Grossberg: see Chapter 14.v.a] handed me his papers. I can honestly say, based on those papers—this was really early, like '71,'72—I know that he was talking about what we now call Hopfield nets because that's what was in those papers. He was having trouble getting it published. It may be that what had happened with Grossberg is in part what happened with me [with respect to backprop: see Section vi.d]—namely, we had the exact same idea in the exact same form [as someone else did later], but people weren't willing to publish it. He had to dance it through and change it and modify it and screw it up before people would allow it to get through the system. *And then after the screwed-up version got through, then people would allow the full form of it to come in from places that they trusted. We were not people they trusted, neither Steve nor I.* (J. A. Anderson and Rosenfeld 1998: 343; italics added)

Grossberg first wrote about these ideas—his "Additive Model"—as a freshman (at Dartmouth) in 1957, and had published "at least fifty papers on it" by the time the current label was coined. (One of those "fifty" papers was a review article, which one might expect to have been widely read: Grossberg 1974.) Naturally, he resents Hopfield's high visibility: "I don't believe that this model should be named after Hopfield. He simply didn't invent it. I did it when it was really a radical thing to do" (J. A. Anderson and Rosenfeld 1998: 172, 176).

One of the problems with priority claims is that if an idea is very new, it may not be intelligible to other people even if they do come across it. That's especially so if it's presented alongside many other novel ideas, and a range of highly disparate empirical data that the writer is seeking to explain. As remarked above, Hopfield's abstract, data-free, work was more influential than Grossberg's "equivalent" for these very reasons.

Werbos apparently understood Grossberg's ideas on first reading. But Rumelhart has confessed that he (as a research student in mathematical psychology), *and* his more senior colleagues at Stanford (including William Estes), didn't:

[In about 1966, Grossberg] sent them a copy of his big, fat dissertation, which had all of these differential equations and everything, and this big cover letter in which he explained how he had solved all the problems in psychology. Of course, the people at the Institute for Mathematical Studies in Social Sciences were a little perplexed about what this could all be about. I remember *we spent a great deal of time and effort trying to figure it out and failed completely, I should say, to understand* what it was that Grossberg had actually done. In the end, I guess I gave up on it. (J. A. Anderson and Rosenfeld 1998: 271; italics added)

As we'll see in Chapter 14.v.a, the people at Stanford were sufficiently impressed to invite Grossberg there as a graduate student. Nevertheless, even first-rank mathematical psychologists were "perplexed".

I shan't attempt a definitive priority judgement here (Grossberg? Amari? Hopfield? . . .). As remarked in Chapter 1.iii.h, identifying "the same" idea, even in mathematical contexts, is often a highly subtle matter. *The very same equation* may be presented in two different ways, so that the 'second' presenter is understandably, and perhaps even justifiably, regarded as the 'first'. I suggested, above, that this applies in the case of Hopfield and Amari, respectively. It may well apply in the case of Hopfield and Grossberg too, but mathematical skills way beyond my own would be needed to say this with any confidence. It's undoubtedly true, however, that Hopfield was heard immediately whereas Grossberg, a quarter-century before, wasn't.

Whomever one credits with having had the idea of energy minimization first, the great advantage of Hopfield's presentation—and of his respectability—was that it attracted the "locusts". Their analytical skills helped people to understand the fundamental principles of the networks they discussed.

However, there were two problems, one sociological and one theoretical. First, most physicists knew nothing about psychology (and many couldn't have cared less). Occasionally, they looked at neuroscience: Hopfield himself had regularly attended neurobiological conferences and Princeton seminars in the 1970s. But Kohonen was exceptional in choosing psychology as a "hobby". So they couldn't simulate the experimental data, nor identify theoretically significant psychological problems on which to test out their ideas. It wouldn't have occurred to them, for example, to model the formation of the past tense (Section vi.e, below), or the visual perception of colour and form (see 14.v.d).

Second, Hopfield faced the difficulty that Selfridge had faced with Pandemonium (see ii.d, above): the danger of getting trapped in a local minimum. (*Minimum*, this time: in hill climbing the relevant quantity is increased, whereas in energy minimization it is decreased.) As remarked above, his algorithm never allowed the overall energy measure to rise, no matter what changes took place in the units. There was therefore no way in which the system could escape from a local minimum, even if a much deeper "valley" existed nearby in the state space. For Hopfield's immediate purposes (namely, storing memories), this didn't matter: all that was required was that *some* attractor be found. But if, for any reason, one wanted to locate the attractor with the lowest minimum of all, Hopfield's method would usually miss it.

As we'll see in Section vi, both these problems would be solved a few years later by Hinton and his colleagues in the PDP group—to some extent inspired by Grossberg's recent papers (especially 1976a,b, and 1980).

Between them, the members of the PDP group had many years of experience in AI modelling and/or computational psychology, drawing on both GOFAI and connectionism. For example, in the early to mid-1970s Norman had co-authored the textbook that used Pandemonium as a theoretical motif (see Figures 6.5 and 6.6), Rumelhart had developed computational theories of reading errors and story grammars, and the two had co-edited a collection on computational models (D. A. Norman and Rumelhart 1975).

Hinton's experience was especially relevant, however. For even before joining the PDP group he'd devoted several years to Hopfield's core question—and had offered a similar answer.

h. Hinton relaxes

The "core question" was how a computational system making many different local decisions could reach an internally coherent (and stable) state. Hinton, being a psychologist not a physicist, thought of it in terms of a specific example: how can a vision program simultaneously satisfy many different constraints when interpreting a line drawing? Where Hopfield would minimize overall "energy", Hinton had minimized overall "inconsistency".

Hinton first worked on this issue in the early to mid-1970s, as a student of Longuet-Higgins (in Edinburgh and then Sussex) and a colleague of Sloman and Max Clowes, both AI vision experts in the Cognitive Studies Programme at Sussex. Having been told about neural networks by his schoolfriend Inman Harvey (now a key researcher in A-Life: see 15.vi.c), he got interested in distributed memories. That's why he sought out Longuet-Higgins (who was by then working on GOFAI models of music and language). While still at Edinburgh, he decided to address some of the basic issues in the context of vision. Later, he worked as a Research Fellow on Sloman's POPEYE vision project at Sussex for a year or two (see 10.iv.b).

He started from an idea drawn from scene labelling, a method of interpreting line drawings pioneered by Clowes: the Waltz filter. This was a form of depth-first search that involved the sequential propagation of constraints (10.iv.b). However, there was a danger of what I remember Hinton calling "computational gangrene". A contradiction at any point would lead to the permanent abandonment of search below that point—so if one of the labels happened to be mistaken, the solution would be blocked. In short, noise was disastrous.

But noise is also ubiquitous. In particular, it's inescapable in grey-scale images: a light-intensity gradient in the image may or may not represent an edge in the scene. David Waltz had already added new labels for line drawings, representing shadows for instance. And Marr (1975a,b) had recently suggested a set of constraints (in effect, labels) involved in interpreting grey-scale images, corresponding to edges, textures, reflectances, and orientations (Chapter 7.v.b–d). Hinton wanted to find some way of maximizing the mutual consistency of any such set of low-level labels.

He was aiming for optimal, though not necessarily perfect, constraint satisfaction. This required parallel processing and alterable decisions, guided by a measure of inconsistency that was minimized as the system moved towards the optimum interpretation.

The first published version of such a method was called "relaxation" (Rosenfeld *et al.* 1976), and Hinton used this term in his own work (1976, 1977). His contour-finding program provisionally assigned continuous numerical values (between 0 and 1) to hypotheses about various areas in the image: that this is an edge part, for instance. If it is an edge part, then colinear edge parts are to be expected nearby. They aren't guaranteed, however. The edge part may be the endpoint of a sharp corner, or its continuation may be prevented by some other object obstructing the view—or it may not correspond to a real edge at all.

Whereas the first relaxation technique had allowed individual hypotheses to interact (support or inhibit) directly, Hinton's didn't. Instead, the logical relations *between* hypotheses were also expressed as numbers, which acted as feedback loops guiding the local decisions. The overall measure of logical consistency could be calculated, and was used to achieve the optimal overall value in the circumstances. At that point, each hypothesis would be announced "true" or "false", and contradictions (by now, minimized) would be ignored.

In other words, whereas Waltz had dealt only with *strong* constraints (which must be satisfied), Hinton focused on *weak* ones (which need be satisfied only so far as possible). That was a significant advance.

But this method, like Hopfield's, was threatened by extremity traps. And now it mattered, since the aim was to find the best *global* interpretation. Suppose the image were ambiguous—not in the sense that two alternative interpretations are equally consistent (as in the vase/faces of Figure 5.1), but in the sense that one of two plausible interpretations is better than the other. For instance, consider a photo that could be of either Jill or Mary: which is the more likely? If Hinton's system happened to reach the less consistent solution first, it would never find the other one. (Eventually, Hinton would show how to escape such traps: see Section vi.b.)

One counter-intuitive finding of Hinton's early work on vision had nothing to do with relaxation labelling. This was his proof that the most efficient way of coding for features (to recognize shapes, for example) is to use a few very coarsely tuned units, not a large number of finely tuned ones (Hinton 1981; cf. Hinton *et al.* 1986: 90–4). Overlapping groups of coarsely tuned receptive units, within specified size constraints, can compute spatial properties with specifiable gains in efficiency. So the fact that biological neurones don't respond to a perfectly defined narrow class of stimuli may be a strength, not a weakness.

As he pointed out, this is a mathematical result. As such, it applies to distributed processing in general, not just to vision.

i. Passing frustrations

If the main problems facing Dark Ages connectionism were mathematical, there were technical obstacles too. In a nutshell, the computers weren't up to it.

In the 1970s, computers were still relatively limited. Procedures requiring many cycles of simultaneous calculations weren't feasible. As a result, some good theoretical ideas couldn't be implemented by AI modellers, or only in a drastically cut-down form. And some were even ignored by them.

This applied to Holland's work on evolutionary computing, for instance. Although he'd long solved the problem that had defeated Selfridge, namely how to deal with *multiple* mutations, his theory couldn't be implemented and wasn't yet widely known (see Chapter 15.vi.a–b). Similarly, it was clear from the early 1970s that linear associative memories were promising. But they couldn't yet be used to deal with complex patterns, because large vectors need very many multiplications and additions. Connectionism couldn't flourish without much-increased computer power. This finally became available in the 1980s.

Implementation aside, however, in the wilderness years plenty had gone on behind the scenes. By the early 1980s, the actors were waiting in the wings—and their voices were beginning to be heard, at least by members of the AI community.

For example, Douglas Hofstadter (personal communication) gave a well-attended seminar at MIT in 1983–4, in which he covered a wide variety of parallelist models. These were HEARSAY, PDP (including a recent simulation of typing errors: Rumelhart and Norman 1982), simulated annealing, sparse distributed memories, and his own computer models of cognition—which included a network of nodes with spreading activation (see Section x.a, below). And Hinton's work was exciting many members of AISB in the United Kingdom, not just his erstwhile colleagues in Sussex.

Outside certain sub-groups of the cognitive science community, however, even PDP was still unknown. Moreover, connectionism in general was widely thought to have been *proven* useless (see Section iii). The money and public attention were directed elsewhere. AI was constantly featured in the media, thanks to Japan's Fifth Generation project (11.v). But it was expert systems and chess programs that were discussed, not connectionist AI. Similarly, most philosophers discussing AI, whether pro or con, considered only symbolic computation (see Chapters 11.ii.a–d and 16.iii–vi).

Hubert Dreyfus was no exception. In 1979, the very same year as the La Jolla conference, the revised edition of his book *What Computers Can't Do* contemptuously declared:

Since those pursuing this course [the study of self-organizing systems], sometimes called cybernetics, have produced no interesting results—although their spokesman, Frank Rosenblatt, has produced some of the most fantastic promises and claims—they will not be dealt with here. (H. L. Dreyfus 1979: 130)

To make it worse, he was so confident in this dismissal that he even quoted his own arch-enemy, Minsky, in support.

The frustrated neural networkers could be forgiven for resenting this cultural invisibility—and they did. Passions ran deep, and sometimes surfaced in surprising ways. In 1981 for instance, in an otherwise highly convivial twenty-person weekend on adaptive systems organized by Selfridge, a leading cybernetician (never mind who . . .) expressed the most extraordinary personal hostility at my merely *mentioning* symbolic AI.

By the end of the 1980s, however, the shoe would be on the other foot.

12.vi. Centre-Stage

Connectionism exploded into the public consciousness for the third time in 1986. Granted, Douglas Hofstadter's hugely popular writings had already alerted his readers to the ideas of "subcognitive" parallel processing and "active" memory (see Section x.a, below). But there'd been no accessible account of *just how* this could be implemented. When one finally appeared, the fuss was even greater than had attended Rosenblatt and Widrow, or Hopfield.

It reached across the general public, from professional philosophers to the tabloid-reader on the Clapham omnibus. In the late 1980s it seemed that one could hardly turn on the television without seeing some reference to the magical new machines.

"Magical", in the popular view, because they had no clearly identifiable memory traces, because they could learn without being explicitly taught, and because some could start out from a merely random state.—And impressive, too. For these network machines really did seem to be a sort of "artificial brain". In fact, this 1950s phrase was now widely resurrected.

The type of connectionism concerned was the newly named "parallel distributed processing", or PDP for short. Localist and circuit-based connectionism were largely ignored. Many people didn't even know of their existence.

Moreover, the subtext—not to say the propaganda—changed. Previously, the focus of connectionism had been wholly positive: to discover the nature of associative memory. Now, it was largely negative: to counter GOFAI's influence on theoretical psychology and the philosophy of mind. In this context, the central message was that effortless intuitive thinking is *not*, as GOFAI had assumed, just like conscious inference but without the consciousness.

The downplaying of GOFAI was one reason for PDP's popularity with the public. At last, it seemed (but see Sections viii–ix, below), the 'dehumanizing' image of mind-as-von-Neumann-computer had got its come-uppance.

The leading connectionists themselves were less thoroughly dismissive of symbolic AI:

It would be wrong to view distributed representations as an *alternative* to [GOFAI] representational schemes like semantic networks or production systems . . . It is more fruitful to view them as one way of implementing these more abstract schemes in parallel networks, but with one proviso: Distributed representations give rise to some powerful and unexpected emergent properties. These properties can therefore be taken as primitives when working in a more abstract formalism. [Examples are] content-addressable memory, automatic generalization, and the selection of the rule that best fits the current situation. (Rumelhart, McClelland, *et al.* 1986: 78)

But those wise words were largely ignored. Most outsiders, and many insiders too, evidently felt the humbling of GOFAI to be far too satisfying to pay attention to them. (Some exceptions are discussed in Sections viii–ix, and especially ix.b.)

a. The bible in two volumes

The new name, and the sudden surge of interest, arose with the publication of the two-volume PDP bible in the autumn of 1986 (Rumelhart, McClelland, *et al.* 1986; McClelland, Rumelhart, *et al.* 1986). This was PDP's manifesto, a clarion call to join the new movement.

Unlike the cognitive science manifesto (*Plans and the Structure of Behavior*: see 6.iv.c), its hand-waving was relatively restrained. Most pages reported work that had actually been done—some (in volume I) defining general methods, some (in volume II) applying those methods to specific topics. This made the argument more persuasive than an informal account would have been, but it also made for fairly dry reading. Fortunately, many readers had been prepared—and hugely excited—already. A Pulitzer-prizewinning book published seven years earlier had introduced the *idea* of distributed processing by a host of memorable, and intuitively intelligible, examples (see x.a, below). Now, the bible was adding mechanism to idea.

Like Norman and Rumelhart's GOFAI collection *Explorations in Cognition* (1975) and Minsky's *Semantic Information Processing* (1968), the PDP bible was a collection of papers largely written by research students or only-just-postdocs. If that sounds boring, it wasn't: these two volumes were even more influential than their GOFAI predecessors.

The editors were Rumelhart and James, or Jay, McClelland (1948–) at UC San Diego. (Almost immediately after the bible was published, Rumelhart moved to Stanford, where he worked from 1987 to 1998.) Hinton had originally been an editor too, but withdrew to concentrate on the Boltzmann machine. He hadn't been a mere dogsbody editor: Terrence Sejnowski described him, years later, as "the seed that led to PDP" (J. A. Anderson and Rosenfeld 1998: 323).

Further members of "the PDP Research Group" were officially based both in and outside San Diego. They included (among others) Norman, Sejnowski, Ronald Williams, Paul Smolensky, Jeffrey Elman, Michael Jordan, and Francis Crick.

Crick's name implies an interest in biology, and the PDP bible did contain several neuroscientific papers—one of which stressed the very *un-neural* character of artificial networks (Crick and Asanuma 1986) (see Chapter 14.i). The book even declared that the PDP group accepted "the brain metaphor" in place of "the computer metaphor". But although the (limited) biological plausibility "definitely enhanced" their approach, the neuroscience was strictly secondary:

We are, after all, cognitive scientists, and PDP models appeal to us for psychological and computational reasons. They hold out the hope of offering *computationally sufficient* and *psychologically accurate* mechanistic accounts of the phenomena of human cognition which have eluded successful explication in *conventional computational formalisms* [i.e. GOFAI]; and they have radically altered the way we think about the time-course of processing, *the nature of representation*, and the *mechanisms of learning*. (Rumelhart, McClelland, *et al.* 1986: 11; italics added)

The italicized phrases are the keys to why other cognitive scientists were so interested. *Computational sufficiency* seemed to be provided by the Boltzmann machine and back propagation (see below). These alleviated two of the major mathematical difficulties of the Dark Ages.

Psychological accuracy was the aim of PDP models of experimental data, some of which had great theoretical interest. For instance, the past-tense learner described below challenged Chomsky's nativism (Chapters 7.vi.a and 9.vii). More broadly, the PDP perceptrons—unlike Rosenblatt's simple versions—allowed a role for success. There was "extensive experimental evidence" suggesting that "learning when the organism is correct usually appears to be more important than learning when a mistake is made" (Hinton and Anderson 1981: 15).

The criticism of *conventional computational formalisms* cast doubt not only on the psychological relevance of GOFAI, but also on philosophies of mind that shared the same conceptual roots (see 16.iii–iv). The suggestion that *representation* could be thought of in a new way was exciting not only for psychologists but also for philosophers (Sections viii–x, below). And the notion—not new, but resurrected—that machines could *learn* without being explicitly taught added the promise of commercial applications.

Having these intellectual delights to offer, plus the increased computer power of the 1980s that made persuasive demonstrations possible, the PDP group had been hopeful about prompting interest in connectionism. However, they realized that this would need some special effort on their part.

Their report on the 1979 conference had kicked off with the warning: "This is a difficult book" (Hinton and Anderson 1981: 1). It was indeed—not least because state vectors specifying holistic activity patterns were a mathematical nightmare for many people. They were unfamiliar to non-physicists, and very different from GOFAI instructions in LISP's 'English' (10.v.c). Now, they decided to provide not only a research report but a tutorial too.

The PDP bible was intended to fulfil both functions. Because they especially wanted to get the youngsters interested, they wrote it in an unusually accessible way. Moreover, they reduced the cost by doing all the typesetting and proof-reading themselves, and even paying for the preparation of the camera-ready copy. They hoped to have well-thumbed, coffee-stained, volumes on students' desks—not pristine pages sitting on library shelves.

But even they were amazed by the scale and immediacy of their success. There were so many advance orders for the PDP books that the second printing was started even before the first had been released. Forty thousand copies were sold in the first seven years.

The scale of the advance orders suggests that the rumour-mill in the final years of the Dark Ages had primed many people to be receptive. Sejnowski recalls that even a few months before publication "very few people knew what was happening by word of mouth" (J. A. Anderson and Rosenfeld 1998: 324). Nevertheless, enthusiastic students "from everywhere" turned up at the pre-publication Connectionist Summer School he and Hinton organized at CMU. Some may have been drawn in by Hinton's easy-to-read description of Boltzmann machines in the popular magazine *Byte*, in 1985. And some may even have found out about it when "somebody at MIT put out a spoof [announcement] advertising a connectionist cooking summer school" (J. A. Anderson and Rosenfeld 1998: 324).

Primed or not, what readers found on opening the books was a fascinating mix of abstract theory and psychologically significant models. As Rumelhart recalled later, the former had had a long gestation:

[In the late 1970s, Jay McClelland and I developed our interactive model of word recognition that] turned out to very much like these models that settle into stable states. Indeed, it was those features that we eventually worked out. But what I remember are hours and hours and hours of tinkering on the computer. *We sat down and did all this in the computer and built these computer models, and we just didn't understand them. We didn't understand why they worked or why they didn't work or what was critical about them. We had no important theory yet.* We struggled and struggled and began to get insights into the nature of these interacting systems and became totally obsessed by them. (J. A. Anderson and Rosenfeld 1998: 275; italics added)

The "important theory" they'd been hoping for was eventually achieved. It included, for instance:

* new methods of analysing hidden units (see below);
* Hinton's proof about the efficiency of coarsely tuned units (Section iv.h);
* McClelland's recasting of GOFAI's HEARSAY architecture in PDP terms (Section iv.b);
* and Smolensky's harmony theory.

But the most exciting theoretical advances, and the ones that received the most attention, were

* the Boltzmann machine, and
* back propagation.

b. Bowled over by Boltzmann

Boltzmann machines were announced informally in 1984, in a Carnegie Mellon Technical Report. This was written by Hinton (then at CMU) and the CMU computer scientist David Ackley, together with Sejnowski—an ex-physicist turned neurobiologist, whom Hinton had met at the 1979 La Jolla function.

Sejnowski (1947–) was first drawn to neurobiology by Charles Gross's research on feature-detectors (14.iii.b), and had been won over (in 1978) by studying electroreception in fish:

[Seeing the molecular receptors with the electron microscope] changed my life. After that I wasn't interested anymore in just abstract understanding of the brain. I really wanted to understand how it was made. I was committed to the idea that you had to understand the actual substance that the brain was made from if you're going to understand how it works. (J. A. Anderson and Rosenfeld 1998: 321)

By the mid-1990s, his prime interest would be "using techniques and tools from network modelling to understand different pieces of the brain" (J. A. Anderson and Rosenfeld 1998: 329). In the early 1980s, however, he was still doing 'abstract' connectionism. There was no neuroscience in NETtalk, for instance (see subsection f, below)—nor in Boltzmann machines.

The title of the CMU report described these machines as "constraint satisfaction networks that learn", so highlighting the topic that Hinton had been researching for a decade (see Section v.h, above). But three new features had been added. These were energy minimization, simulated annealing, and learning.

The first of these was borrowed from Hopfield. In the summer of 1982, Hinton and Sejnowski—already cooperating on relaxation models of vision—attended a small meeting of connectionists at the University of Rochester. There, they heard Hopfield give a talk on his about-to-be-published work. As Hinton remembers it,

As soon as Hopfield gave that talk, I realized that what you wanted to use these nets for was to solve constraint-satisfaction problems. You wanted to let the energy function correspond to the objective function you were minimizing. I'm not sure Hopfield had really understood that. He was using these nets for memories at that point, 1982. (J. A. Anderson and Rosenfeld 1998: 372)

In other words, "energy" could be made to stand for "inconsistency". And Sejnowski recalls:

Both Geoff and I within seconds realized that this was the convergence proof we needed to show that our constraint-satisfaction scheme in vision could actually be implemented with hardware. That's how the Boltzmann machine was born. (J. A. Anderson and Rosenfeld 1998: 322)

But translating relaxation into energy minimization wouldn't be enough: the problem of local minima would remain. It was Sejnowski who saw the way out.

He recalled some recent work by Scott Kirkpatrick (at IBM) on optimization by simulated annealing (Kirkpatrick *et al.* 1983). As long ago as 1953, physical chemists had used the Boltzmann equations of statistical mechanics in a method (the Metropolis algorithm) that modelled collections of atoms in equilibrium at various temperatures. Kirkpatrick had recently broadened the notion of equilibrium to cover optimization in general, including computer circuitry design and the travelling salesman problem (i.e. *How can one find the shortest route for visiting several scattered towns?*). Moreover, he had added simulated annealing (see below) to find the equilibrium more quickly.

Now, Sejnowski suggested doing the same thing. The Boltzmann machine "was the result of a week of solid thinking about what would happen if we had a noisy Hopfield network" (J. A. Anderson and Rosenfeld 1998: 322).

Annealing is a technique used by metallurgists to cool metals evenly. The quickest way is to start off at a high temperature, and cool the metal down gradually. Analogously, in Hinton and Sejnowski's networks equilibrium was reached by allowing the changes in the first few cycles to be relatively large ('noisy'), becoming smaller as time went on (Ackley *et al.* 1985; Hinton and Sejnowski 1986). This made it unlikely that the system would be trapped in a sub-optimal attractor.

Hopfield would have left it at that. Indeed, Hopfield would have spared the words and gone to town on the equations. To him, and to his fellow "locusts", the mathematics would be obvious. Kirkpatrick had offered more explanatory text than Hopfield. But he too had addressed his paper to people familiar with the mathematics: namely, computer scientists and engineers. The PDP group were aiming for a much wider audience and, thanks largely to their rhetorical skills, they reached it.

For example, in the PDP book (though not in their paper for fellow professionals, in *Cognitive Science*) Hinton and Sejnowski took pains to describe simulated annealing in intuitive terms. They imagined a ball-bearing on a two-dimensional landscape, and asked how one should shake the whole system to get it to lie still at the bottom of the deepest valley (see Figure 12.3).

Violent shaking, they pointed out, wouldn't do. For the ball-bearing would constantly jump from one valley to another—often, upwards. Gentle shaking would do the trick—but it might take a very long time. "A good compromise", they said, "is to start by shaking hard and gradually shake more and more gently" (Hinton and Sejnowski 1986: 288). With that strategy, the ball-bearing would probably find the bottom point and stay there.

Admittedly, in a landscape with hills and valleys at sufficiently awkward heights and locations, it might not. But they remarked that interesting problems have many

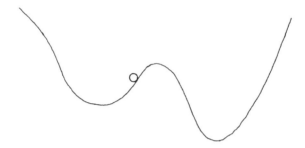

FIG. 12.3. Ball-bearing on an (energy-)landscape. Reprinted with permission from Rumelhart and McClelland (1986: 287)

dimensions, which implies many more potential pathways between one valley and another—especially when one shakes hard. (They predicted that simulated annealing would work well only for high-dimensional state-spaces.)

They spelt these things out because their intended audience was unfamiliar with neural networks, and perhaps even with physics. Hopfield's fellow locusts were very welcome to read the books too, but they weren't the primary target. The PDP bible was an educational enterprise aimed at cognitive scientists in general—and, as such, it worked. It even reached the more intrepid members of press and public.

The Boltzmann machine was—dare one say it?—a *general* problem-solver. Given suitable input units, the system could discover the appropriate consistency relations for itself. Or they could be provided beforehand, in the connection weights.

For example, the authors pointed out that one could use a Boltzmann machine for visual interpretation (Hinton's long-standing interest). The input units would fire when the relevant features (light-intensity gradients, texture cues, etc.) were present, and the weights would ensure that optimal equilibrium corresponded to the correct interpretation (cf. Ballard *et al.* 1983).

In principle, any 'environment' of interacting constraints—visual, linguistic, logical, financial . . . —could be implemented. So this approach could be applied to many issues in theoretical psychology. What's more, it could be applied to many practical problems in industry and commerce.

These networks were multi-layer perceptrons, whose units were not deterministic (as in Hopfield nets) but stochastic. The relative probabilities of the possible states of the system, and of the activity of the hidden units in different contexts, were defined by Boltzmann equations. And where Hopfield nets had modelled memory but not learning, the Boltzmann machine modelled both. The network was said to have learnt a pattern when the probabilities in the net matched those in the environment. In reaching this match, the system would in effect have identified (mimicked) the higher-order structure of the environment—which need not have been known beforehand by the programmer.

The new learning rule which brought about this match was simple (though slow), but far from obvious. It estimated and compared the probabilities of co-excitation of unit pairs in two equilibrated situations: with the environmental input present, and in the free-running state. ("Having the input present" meant clamping the input and output units into the appropriate states, before allowing equilibration across the hidden units.) Using these statistics, it gradually altered the connection weights so that the output layer would spontaneously adopt the relevant activity pattern even without any input, and complete the pattern if only part was provided.

The equilibria were reached by simulated annealing. In principle, the learning rule could have involved annealing too, but Hinton and Sejnowski couldn't implement that. However, the noise in the probability measures sometimes enabled it to escape from a local minimum anyway.

The Boltzmann learning algorithm provided what Minsky and Papert had asked for in 1969: a powerful convergence theorem *guaranteeing* successful learning in multi-layer nets (see Section iii.b). This was the aspect of their work of which Hinton and Sejnowski were most proud:

Perhaps the most interesting aspect of the Boltzmann Machine formulation is that it leads to a *domain-independent* learning algorithm that modifies the connection strengths between units in such a way that the whole network develops *an internal model which captures the underlying structure of its environment* ... [When a network with hidden units] does the wrong thing it appears to be impossible to decide which of the many connection strengths is at fault. *This "credit-assignment" problem was what led to the demise of perceptrons* [Minsky and Papert 1969; Rosenblatt 1962]. The perceptron convergence theorem ... could not be generalized to networks of [decision] units when the task did not directly specify how to use all the units in the network.

This version of the credit-assignment problem can be solved within the Boltzmann Machine formulation. (Ackley *et al.* 1985: 641; italics added)

The underlying reason for this guarantee was that the Boltzmann equations in physics guarantee thermal equilibrium. Some day, sun and moon and everything else will have the same temperature—but there's a very long time to wait. Analogously, this new convergence theorem (like Rosenblatt's) showed what's inevitable in principle, not what's feasible in practice. Even if one could be sure that one was dealing with a problem for which a satisfactory set of weights in principle exists, one couldn't be sure of finding them.

In practice, the learning rule was very slow, and useless for large networks. That's why, despite the huge interest aroused on publication, relatively few people besides the authors themselves actually used Boltzmann machines for psychological research (but see Plaut and Shallice 1993). As Hinton and Sejnowski had pointed out, even small networks could be problematic:

[One] difficulty is that there is nothing to prevent the learning algorithm from generating very large weights which create such high energy barriers that the network cannot reach equilibrium in the allotted time. Once this happens, the statistics that are collected will not be the equilibrium statistics required for [the relevant equation] to hold and so all bets are off. We have observed this happening for a number of different networks. They start off learning quite well and then the weights become too large and the network "goes sour"—its performance deteriorates dramatically. (Hinton and Sejnowski 1986: 298)

There were other difficulties, too—not least, that more training would make the network "go sour", its performance getting worse and worse.

In brief: Boltzmann machines were mathematical magic. But they cast their spells unreliably in the real computational world.

c. Backprop hits the headlines

A more trusty piece of mathematical magic announced in the PDP volumes (and simultaneously in the high-prestige *Nature*) was back propagation (Rumelhart *et al.* 1986*a,b*). This was the idea which Widrow "would have given [his] eye teeth to come up with" (see v.d, above).

Backprop, too, solved the credit assignment problem for multi-layer networks, and although it was still slow it was much faster than the Boltzmann machine. (In the ensuing years, a lot of effort was put into speeding it up.) Although it couldn't be *proved* to converge, it was certainly applicable to a wide range of problems. At the end of the 1980s, it was judged to be "the most effective current learning algorithm for complex, multilayer systems" (J. A. Anderson and Rosenfeld 1988: 674).

Back propagation was a generalization of the long-familiar LMS algorithm, or "delta rule" (see ii.g, above). But it corrected errors on several levels, instead of just one. It did this by tracing responsibility back from the output layer into the hidden layers, identifying the individual units that needed to be adapted *even though* many units had been acting simultaneously. It was thus analogous to Holland's contemporaneous bucket-brigade algorithm (Chapter 15.vi.a).

Backprop requires the programmer to know the precise state of the output layer when the network is giving the right answer. A unit-by-unit comparison is then made between this exemplary output and the output actually obtained from the network being trained. Any difference between an output unit's activity in the two cases counts as an error. (Unlike the Boltzmann machine, the units here were deterministic.)

So far, so familiar. But the (current or desired) state of the hidden units isn't known, for they can't be directly inspected. The backprop algorithm therefore *assumes* that the error in an output unit is due to error(s) in the hidden units connected to it.

It attributes a specific amount of error to each hidden unit, depending on the connection weight between it and the output unit. Blame is shared between all the hidden units connected to the mistaken output unit. If a hidden unit is linked to several output units, its mini-blames are summed.

Proportional weight changes are then made to the connections between the hidden layer and the preceding layer. That layer may be another (and another...) stratum of hidden units. But ultimately it will be the input layer, and the weight changes will stop. This process is iterated until the discrepancies at the output layer disappear.

The resulting overall state of the hidden units was taken to be the "internal representation" of the input. Such representations were necessary, the authors said, whenever "the similarity structure of the input and output patterns are very different" (Rumelhart *et al.* 1986*a*: 318). "A classic example of this case", they pointed out, is the *exclusive-or* (XOR) problem, which single-layer perceptrons had been notoriously unable to solve (see Section iii.b). Pandemonium could solve it, because (as pointed out in Section ii.d) it could embody demons looking out for just that logical feature. But there had been no learning rule—no convergence theorem—that *guaranteed* success in this general class. Backprop, by contrast, did.

Backprop aroused enormous interest. The need for it had long been recognized:

* Rosenblatt himself had speculated about "back-coupled" perceptrons twenty years earlier, but he hadn't defined an algorithm for computing and communicating the error messages (see Section ii.e, above).
* His arch-critic Minsky, in his hugely influential AI prospectus of the same period (1961*b*), had named credit assignment as one of the key problems that would have to be solved (10.i.g).
* And Widrow had "hit a brick wall trying to adapt multilayer nets" in the early 1960s, largely because he couldn't solve it.

Now, someone had mastered it at last.

Those words "at last", however, are highly misleading. The adulation—not to mention Widrow's eye teeth—might have been offered long before. For backprop wasn't strictly new. Quite apart from Holland's analogous work, it had been independently discovered several times already.

d. Backprop anticipated

Even within the PDP group, backprop had been several years a-brewing. Rumelhart had prepared a talk on it for a meeting in 1983, but chose "at the last minute" to give another talk instead (J. A. Anderson and Rosenfeld 1998: 279).

A full year later, he mentioned the idea to Sejnowski. He "immediately understood" and was the first to use it for a large task (in NETtalk: see below). Rumelhart himself had attempted only toy problems.

When Rumelhart first mentioned it to Hinton, by contrast, Hinton declared that it could never work. Nevertheless, and despairing of getting the Boltzmann machine to work efficiently, a few months later Hinton thought he might as well try backprop. His research students were so sceptical that none was willing to program it, even though he assured them that "It would only take a day or so" (J. A. Anderson and Rosenfeld 1998: 377). So he (with Williams) implemented it himself, and then spent some months improving it. Their description of the 'new' technique was added as an extra chapter just as the PDP volumes were going to press. This was a full three years after Rumelhart's phantom talk.

Even more to the point, there were others outside the PDP group who had come up with much the same idea.

The PDP authors themselves cited two "similar" learning rules. These had been published very recently (in 1985) by David Parker and Yann le Cun. Indeed, in 1982 Parker had tried to get it patented by Stanford, but was rejected because they thought it had no commercial value (S. Grossberg, personal communication).

Moreover, several people had defined it long before that. Grossberg had done so, as we saw in Chapter 1.iii.h (and his then current version allowed for many brain/behaviour facts ignored by the PDP group—Grossberg 1987*a*: 232–40). Several non-biologists had done so too. Its mathematical core (the chain rule of differentiation) had been discussed in the 1960s by control theorists (Bryson and Ho 1969), and by Amari also (1967). And, even more significant, it had been *implemented* in the early 1970s, as a routine statistical technique for automatic calculation.

The person who did that, and who (so far as is known) was the true originator of backprop *as a computerized algorithm*, was Werbos (1947–). And he did it for the most surprising of reasons.

Werbos was a highly interdisciplinary animal: a mathematician interested in the intersection of psychology, economics, and political science. While a new Ph.D. student at Harvard in 1970 or so, he had suggested using "this little method for adapting multilayer perceptrons". But he didn't want to use it for modelling supervised learning. Rather, he wanted to translate Sigmund Freud's ideas on psychic energy into mathematics.

Werbos's overall aim had been to model intelligence. His elders and betters weren't impressed:

I had an adaptive critic backpropagated to a critic, the whole thing, in '71 or '72. I actually mailed out copies of that paper to some people out at Stanford and certain people around Harvard.

 The thesis committee said, "We were skeptical before [i.e. when he had mentioned the mathematics but not the AI], but this is just unacceptable. This is crazy, this is megalomaniac, this is nutzoid." (Werbos, interviewed in J. A. Anderson and Rosenfeld 1998: 342)

Even allowing for the verbal embroidery, Werbos's frustration (remember "We were not people they trusted, neither Steve nor I": v.g, above) is understandable. Not until he'd presented his ideas as statistics rather than neural networks, and as applied political science rather than AI, did Harvard (in 1974) agree to pass the thesis. The blandly respectable title was *Beyond Regression: New Tools for Prediction and Analysis in the Behavioral Sciences*—but Werbos used the ideas to make some highly influential predictions about the conduct of the Vietnam war (J. A. Anderson and Rosenfeld 1998: 339–40, 349).

 Meanwhile, finding himself penniless, Werbos had got a full-time job on a joint MIT–Harvard computing project. This sustained him while he was doing the Ph.D. He generalized his algorithm to deal with time-varying processes, and put it into the "Time Series Processor" as part of MIT's standard software. Besides being included in the official report to the military funding agency, a description of the algorithm (which he called dynamic feedback) sat on many people's desks: "It was part of the computer manual, so the first publication of backpropagation was in a computer manual from MIT for a working command for people to use in statistical analysis. I was a second author of that manual" (J. A. Anderson and Rosenfeld 1998: 348).

 Computer manuals, of course, aren't high-profile publications. DARPA reports are. But—most unusually—Werbos's mid-1970s research report to DARPA containing details of his method was never published, nor even officially listed. The reason was that it also contained politically embarrassing results (J. A. Anderson and Rosenfeld 1998: 351).

 Moreover, Werbos's 1978 paper (in an IEEE journal) announcing that he'd defined a powerful method of "dynamic feedback" was printed without the Appendix, because of length limits. Without the details, why should readers get excited, or even believe him?

 Another publication, in 1982, even added discussions of neural nets and various parts of the brain. It provided diagrams of multi-layer perceptrons, doing reinforcement learning by back propagation. But although this paper got Werbos some job offers, backprop was still left on the shelf. His new employers tried to get MIT's engineers to implement it as a useful analytical tool: they wouldn't.

So much for the vicissitudes of Werbos's dynamic feedback. The sources for Parker and le Cun were similarly inaccessible. They were (respectively) a Technical Report of MIT's Center for Computational Research in Economics and Management Science, and a low-profile journal and edited collection.

Rumelhart and colleagues, by contrast, published simultaneously in the hugely influential journal *Nature* and (at much greater length) in a book written in a highly accessible style, with a provocative overall theme. That book was an instant best-seller, as we've seen. And everyone who was a scientific anyone read *Nature*. Their version of back propagation became famous virtually overnight. Until Werbos was remembered/discovered by the neural network community some years later, the PDP group were mistakenly given the credit for defining and/or implementing the "first" backprop algorithm.

e. Wonders of the past tense

There was another reason, besides the elegant mathematics, why the PDP approach became famous overnight. Namely, the 'bible' described simulations of a number of important psychological phenomena. In other words, it seemed to be describing the real world, not just doing exercises in fancy mathematics. And the "real world", here, was the world of human beings—not spin glasses.

Perhaps the most widely read chapter in this category was Rumelhart and McClelland's (1986). It described a Kohonen associative memory (with 460 input units and 460 output units), equipped with Rosenblatt's perceptron convergence procedure. This network modelled how a child learns the past tense of English verbs—which most of the bible readers, of course, would have heard in their own infant children or siblings.

Outsiders (and locusts) might have asked "Who cares? What's important about that?" But cognitive scientists did care, because this model presented a fundamental challenge to two closely related orthodoxies: nativism and rule realism. (By rule realism, I mean commitment to the psychological reality of processing rules, of which grammatical rules are a special case.)

Both of these orthodoxies were prominent in Chomsky's linguistics and Fodor's philosophy of mind (see 9.vii and 16.iv.c–d). And the second was a linchpin of Allen Newell and Simon's psychology (6.iii, 7.iv.b, and 16.ix.b). These four men had been hugely influential in the formation of the field. So it's not surprising that nativism and rule realism were accepted, even assumed, by most cognitive scientists at the time.

Rumelhart and McClelland questioned all that:

We propose an alternative to explicit inaccessible rules. We suggest that lawful behavior and judgments may be produced by a mechanism in which there is no explicit representation of the rule. Instead, we suggest that the mechanisms that process language and make judgments of grammaticality are constructed in such a way that *their performance is characterizable by rules, but that the rules themselves are not written in explicit form anywhere in the mechanism.* (Rumelhart and McClelland 1986: 217; italics added)

The italicized claim had been made years earlier by Dreyfus, as we saw in Chapter 11.ii.a. But whereas he couldn't say anything positive about what such mechanisms might be like, beyond a vague reference to "bodily skills", Rumelhart and McClelland could.

They even claimed to have demonstrated success. If they were to be believed, Chomsky, Fodor, Newell, Simon, and almost everyone else in cognitive science were barking up the wrong tree.

Rumelhart and McClelland started from the fact remarked in Chapter 7.vi.a, that young children go through three stages in forming the past tense. First, a few past-tense verbs, and no mistakes; then, more past tenses, plus frequent over-regularization (*he goed* or *he wented* instead of *he went*); finally, lots of past-tense verbs, without mistakes. This development is gradual, and also includes more detailed changes—in the relative frequencies of *he goed* and *he wented*, for instance.

Their PDP model apparently produced the very same sort of behaviour. ("Apparently", because this claim was later challenged: see Section x.d, below.) But it was neither provided with, nor did it develop for itself, any explicit rules about past-tense formation. The patterns in its behaviour were based in memories (sets of connection weights) implicit in the network as a whole—and which reflected the statistical structure of its environment.

(Sets of connection weights were often referred to as "memories", largely in order to stress the radical difference between PDP and GOFAI representations. However, that could be misleading, because the word normally suggests something that is closely related to the inputs presented—which the weight sets produced by PDP learning rules are not. In some connectionist systems, inputs—sentences, for instance—are stored verbatim, and processing is done on the collection of examples when the system is presented with a query.)

Here, the "environment" was the language that the system—child or computer—experienced as input. The input to the computer wasn't continuous speech. Rather, it was (phonological representations of) a set of stem/past-tense pairs.

Rumelhart and McClelland claimed that the number of recognizable past-tense verbs heard by the infant is at first so small that there's little scope for spotting statistical regularities. Moreover, many everyday (i.e. frequent) words (*go, come, give, take, get . . .*) are irregulars, so the very early corpus includes a relatively large proportion of past-irregulars, such as *he went*. So the first input they provided consisted of only ten verbs: eight irregular, two regular; later, another 410 would be added. Present–past pairs (for instance, *go–went*) are evidently learnt by rote, as one-offs. As more speech is heard, however, the regular verbs massively outnumber the irregular ones.

In other words, the addition of *-ed* (and other past-tense rules) is increasingly prominent in the input statistics. The PDP model—and, its authors suggested, the child—picked up this statistical fact, and reflected it in its own performance. Only much later, after numerous repetitions of *went* had made it a statistically recognizable exception, did the network generate it reliably again.

The past-tense learner was theoretical dynamite:

* It cast doubt on nativism and modularity, for it suggested that a domain-general learning rule—namely, Rosenblatt's perceptron convergence procedure—can suffice for the acquisition of grammar.
* And if grammar, why not other cognitive structures too? (Twenty years later, McClelland would argue much the same point in respect of semantic categories such as *animal* and *plant*, which are often described as innate: see Chapter 8.i.d.)

* It undermined belief in the psychological reality of explicitly represented processing rules,
* and of explicit (symbolic) representations.
* And it threatened the then popular forms of philosophical functionalism (Chapter 16.iii–v).

Little wonder, then, that people flocked to read Rumelhart and McClelland's chapter.

f. Escaping from the black box

Very soon after the past-tense learner, other 'sexy' psychological simulations were described, and hit the media in their turn. Among the most visible—or perhaps one should say "audible"—was NETtalk (Sejnowski and Rosenberg 1986, 1987).

This network, which used phonetic representations that could be fed into a speech synthesizer, gradually learnt to pronounce English words. Listening to its increasingly human-like output, one was reminded irresistibly, and eerily, of an infant passing from babbling to speech. In short, it was guaranteed to *épater les bourgeois*—and in 1986 it did so, on the *Today* show.

But it captured the attention of cognitive scientists, too. For it addressed a methodological problem that faced PDP in general: namely, how to analyse the behaviour of a complex network, instead of treating it as an unintelligible black box. (Compare Minsky's confession "we sort of quit science for a while to watch the machine", and Rumelhart's: "[We'd] built these computer models, and we just didn't understand them . . . why they worked or why they didn't work or what was critical about them.")

Of course, one could always specify the learning rule being used. But what, exactly, had been learnt? Or, to put it another way, what exactly were the hidden units doing? How, for example, was the network representing the hard 'c' in 'cat' and the soft 'c' in 'city'?

NETtalk was developed at Johns Hopkins University in 1985–7, by Sejnowski and the psychologist Charles Rosenberg. Using backprop, it learnt the correct pronunciation of printed English words (from a 20,000-word dictionary) after only three months of work on their part. Indeed, to their amazement it took only one day to master the very first (100-word) version of the task. Three months later, it could learn a 2,000-word corpus in the same time. But just how it was doing it, and just how one might find out, were mysteries.

The inputs weren't letters or sounds, but phonetic representations called Wickelgren features. These are trios coding the target phoneme and its immediate predecessor and successor. So NETtalk didn't have to learn to distinguish the many English vowel sounds, for instance. Much of the power of the system was due to this choice of representation.

That fact illustrated an important point about PDP applications as a class. Just as GOFAI's "General" Problem Solver relied on the programmer to choose and articulate the problem, so a "general" connectionist learning rule works on a representation selected by the designer. One can't simply take a problem and set the learning rule to run on it, any more than one can state the missionaries and cannibals puzzle in English and expect GPS to solve it. This had been true ever since Rosenblatt's first perceptron, but—perhaps partly because "knowledge representation" was a prime focus of the rival, GOFAI—it wasn't much discussed.

Though small by today's standards, the network was very large for its time. It had 309 units, 18,629 connections, and about 30,000 weights. There were 26 output units, 80 hidden units, and 203 input units. Given this complexity, explaining its final performance—as opposed to stating the learning rule—wasn't easy. It took the designers not three months, but three years.

In the end, Sejnowski and Rosenberg recommended three different methods for analysing NETtalk's behaviour in terms of what the hidden units were doing. These analyses were based on network pathology, records of unit activation, and cluster analysis.

Network pathology involved systematic "lesioning" of the network, either before or after it had learnt its task. The authors suggested that this method could be used to throw light not only on artificial networks but on clinical neurology too. For instance, NETtalk might help in studying acquired dyslexias.

Soon afterwards, the British clinician Timothy Shallice (one of the dyslexia experts they'd cited) worked with Hinton on a connectionist model of reading, parts of which they deliberately disabled in a variety of ways. Forty stimulus words, from five semantic classes, activated twenty-eight (position-specific) letter-detecting units; in addition, sixty-eight semantic units represented features such as size, texture, typical location, use, and so on. On average, each input word was supposed to activate fifteen semantic units, with much overlap between words of similar meaning (such as *cot* and *bed*) and less between semantically different words (such as *cot* and *cat*). Hinton and Shallice found that the system exhibited a range of "dyslexias" that are qualitatively similar to those seen in human patients. Various neuroscientific hypotheses were suggested, accordingly (Hinton and Shallice 1989/1991).

Unit activation, they suggested, was related to neuroscience too. After learning, any one of NETtalk's hidden units seemed to be responding to several different input characters, so had no obvious coding or single interpretation. A few years later, follow-up studies found that some units responded more strongly to certain phonetic classes (vowels, for example), but that strict localization didn't develop (J. A. Anderson and Rosenfeld 1988: 662).

As for cluster analysis, this identified the associations most often used by the system in partitioning the "space" of its performance. It turned out—to the authors' surprise—that the network had developed an implicit hierarchy of (nearly eighty) distinctions like those explicitly defined in phonetics: vowels and consonants, voiced and unvoiced consonants, and so on. A GOFAI speech system might have employed an explicit representation of this phonetic structure, but NETtalk didn't. Rather, it implicitly reflected the statistical structure of its environment (English words, each one input as a series of trio-phonetic representations). And this, as we've seen, is what the past-tense learner had done, too.

Statisticians (such as Werbos) might have accused them of reinventing the wheel, for cluster analysis is a common method in statistics. Indeed, it became increasingly clear through the 1990s that many 'insights' of connectionism were differently named versions of statistical techniques. (And of behaviourist equations too: we saw in Section v.a that the Widrow–Hoff algorithm was equivalent to Hull's "habit strength".)

This was pointed out by Donald Michie, a long-time worker in AI learning—mostly of the GOFAI variety (Michie *et al.* 1994). It was also remarked by an outsider, Warren Sarle of the SAS Institute, North Carolina (Sarle 1995).

SAS produces software for statistics, artificial neural networks, data mining, etc. It's not surprising, then, that Sarle remarked that networks learning from noisy data are doing data analysis—which is what statistics is all about. Neural net researchers concentrate on the algorithms, statisticians on their results. And neural networkers tend to ignore the distributional assumptions they have made, whereas statisticians explore their consequences. Nevertheless, there are many close parallels.

Sarle listed over seventy near-equivalences between connectionist and statistical concepts. Let's pick out just two examples. First, feed-forward nets with one hidden layer are closely related to what statisticians call projection pursuit regression. And second, probabilistic neural nets are identical to what's called kernel discriminant analysis. In general, he said,

many results from the statistical theory of nonlinear models apply directly to feedforward nets, and the methods that are commonly used for fitting nonlinear models . . . can be used to train feedforward nets. (Sarle 1995)

Sarle's analysis confirms Gerd Gigerenzer's (1994) view that "new ideas" in science commonly arise from pre-existing tools, and that experimental psychology in general has long exploited statistics in this way. It also confirms the central claim of this section and the preceding one: that connectionism needed mathematical advances to progress. The growth in computer power helped. Indeed, it was essential. But it would have been useless without the mathematical *ideas*.

12.vii. The Worm Turns

The explosive response to the PDP bible, and to the first summer schools at CMU and San Diego, revived the flagging fortunes of connectionism. It wasn't just intellectual appreciation that was involved, but financial support too.

a. Joyful jamborees

Much as a family gathering can be all the more joyous if it brings together people long parted, so an intellectual meeting will be all the more appreciated if it unites people from many different directions. If some or all of these can be cast in the role of the prodigal son, so much the better for fostering satisfaction all round.

The first connectionist conferences are remembered with great pleasure by many of the contributors to the oral history of connectionism (J. A. Anderson and Rosenfeld 1998). They played an important role, not only in raising the confidence of the field but in showing that it *was* a field—and a large one—in the first place.

Connectionism had always included two streams. There were those (like Rosenblatt and the PDP group) who leant towards psychology and/or neurobiology. And there were those (like Widrow and Hopfield) who came from engineering, physics, computing, and/or mathematics. In the 1970s and early 1980s, as we've seen, the two streams rarely mixed.

In the late 1980s, however, several international conference series were initiated which deliberately involved both. The first connectionist jamboree had been in 1979 at La Jolla (see Section v.b, above). But that was a mere taster. These new initiatives were more meaty.

The first "fairly large" neural network meetings were held in Santa Barbara in 1985, and at Salt Lake City's Snowbird Conference Center in 1986 (J. A. Anderson and Rosenfeld 1998: 300). The Snowbird meeting, organized by researchers at AT&T Bell Labs, is fondly recalled by several of the contributors to the *Talking Nets* oral history (a fascinating set of reminiscences and hindsighted judgements, arising in conversations between leading connectionists: J. A. Anderson and Rosenfeld 1998).

Primarily, it's remembered for the exciting mingling of people and sharing of ideas. Indeed, it later gave rise to the NIPS (Neural Information Processing Systems) series of meetings, now regarded by many as the premier scientific conference on neural networks. But it's remembered also, at least by the electrical engineer Robert Hecht-Nielsen, for the organizers' bizarre ideas on how to win friends and influence people:

They had created this obnoxious conference registration form, where you wrote a little paragraph justifying why you should be invited to this meeting, and then underneath there were two boxes that they could check. One was "accept" and one was "reject". And as I understand the numbers, there were approximately five hundred of these forms returned, and something on the order of 125 of them were accepted. The others got them back saying "rejected". To say the least, there were a lot of angry, disenfranchised people. (J. A. Anderson and Rosenfeld 1988: 300)

Quite apart from his disgust at the crass way in which the Snowbird selection had been carried out, Hecht-Nielsen regretted that it had been done at all. Given the "sudden enthusiasm" for the topic, it was clear to him—and to his friend Bart Kosko—that "having a major meeting with unrestricted attendance was an absolute prerequisite for further growth and progress in the neural net field".

Accordingly, the two electrical engineers organized another, larger, meeting in 1987: the first IEEE International Conference on Neural Networks, for which Grossberg was the general chairman. This was intended as an educational exercise, rather than a straightforwardly scientific one. For the newly visible field was riven by priority disputes, and by worrying examples of ignorance and resentment that threatened its future. As Hecht-Nielsen later put it:

There was already [in 1987] a very large clash under way between people who had been in the field for a long time and who had really prepared the foundation in the field and had all the good ideas, and new people who had come along and really didn't add much to the field, if anything, except perhaps an infectious enthusiasm and a following... There was this body of individuals who really had created the field and really had all the original ideas at that point, who didn't even exist as far as maybe a majority of the new people interested in the field were concerned. (J. A. Anderson and Rosenfeld 1998: 301)

One of the main aims of the meeting, then, would be "to firmly and unequivocally establish the history of the field in people's minds in a more correct sequence of events"

(p. 301). This meant inviting as many as possible of the half-forgotten originators, as well as the newly fashionable ones. But, as the co-organizer Kosko remembered:

By the fall of 1986 we . . . went around to the many feuding factions in neural networks. The feuds were really beginning to heat up now. The PDP books were out. There were different camps, and we felt the only thing to do was have a level playing field . . . Many of the leaders of the neural field [whom we invited] literally hated each other's guts . . . (J. A. Anderson and Rosenfeld 1998: 403)

Given this unhappy background, Kosko and Hecht-Nielsen were far from confident that all the invitees would accept (they didn't), or that many other people would bother to turn up. But they were willing to put their money where their mouths were. Together with the local branch of the IEEE, they took on a million-dollar risk to underwrite the conference. (Sheraton hotels don't come cheap.) In the event, however, 2,000 people signed up.

Almost as many attended the summer 1987 meeting on neural networks at San Diego. A few months before that, Grossberg, Kohonen, and Amari had co-founded the International Neural Network Society (INNS), with Grossberg as its first President. They also invited papers for a new journal, *Neural Networks* (Grossberg was Editor-in-Chief). INNS held its first conference in Boston in 1988, co-sponsored by six IEEE societies. As for IEEE itself, it ran a 750-person scientific meeting (on 'Neural Information Processing Systems—Natural and Synthetic') in Denver in November 1987; and the second IEEE 'Neural Networks' conference—no longer a source of financial nail biting—took place in San Diego in July 1988.

Very soon (in 1989), these two series of meetings (INNS and IEEE) would unite as the International *Joint* Conference on Neural Networks. There were over 1,800 participants, including many from industry, at the initial assembly. The atmosphere of the first IJCNN reminded one participant of the movie mythology of *Star Wars*:

There was very much a sense of excitement, almost euphoria. The Neural Network Jedi had returned and a rebellion was in progress. Bernie Widrow, one of the pioneers of the field, was perfectly cast in the role of Obi-wan Kenobi. The Evil Empire of AI, which had wrongfully suppressed and oppressed the innocent, was already declared to be dead. A new millennium was about to dawn. (Barrow 1989: 6)

One didn't need to be in Hollywood's home nation to share the sense of excitement. Other meetings were springing up all around in 1986–7, in the UK/Europe and Japan as well as in the USA.

The PDP group, in particular, put a lot of effort into stirring the pot. They instituted a series of highly successful summer schools, to introduce both students and fellow cognitive scientists to hands-on experience of connectionist computing. These were equivalent in intent to RAND's summer 1958 "Research Institute" and the other GOFAI summer schools (see Chapter 6.v.a). Just as in RAND-1958, a number of already established cognitive scientists welcomed the opportunity to be trained in the new methodology—and the new way of thinking about mind and psychology.

In addition, there were various attempts at commercial application. As Mead said later, these were interesting because they involved "actually looking at what [neural networks] do rather than making up stories about what they might do" (J. A. Anderson and Rosenfeld 1998: 142).

b. DARPA thinks again

At much the same time as these sudden and worldwide developments, DARPA began to have second thoughts about their current funding—or rather, no-funding—policy. ("No-funding" isn't quite fair: they'd supported a few selected projects from 1982 onwards, and had funded Hecht-Nielsen in building neurocomputers.)

The key trigger of their rethinking was the series of lectures given by Grossberg at MIT's Lincoln Laboratory in 1986–7, videos of which were circulated in the lab and reached the Director. He found them impressive, but there was no unanimity: some scientists and DARPA offices still held to the Minsky–Papert critique. He therefore persuaded DARPA to set up an urgent five-month enquiry into neural networks (DARPA 1988). Grossberg (personal communication) was invited to head it, but had to decline because of his other commitments. (As we've seen, besides doing his own research he was then the founding President of INNS, and editing the new journal *Neural Networks*.) Widrow was appointed as Director instead.

The enquiry was kicked off, in October 1987, by a two-day symposium at the Lincoln Laboratory. The task of the symposium was to outline the study's central themes. The fourteen speakers included several cognitive scientists and neurophysiologists. These were Feldman, Rumelhart, and Sejnowski; Grossberg and Gail Carpenter, both of whom spoke on vision; and David Hubel, Nobel prizewinner for research on visual cortex (see 14.iii.b).

The team who undertook the subsequent work, and eventually wrote the 650-page Report, were drawn mainly from computer science and engineering (in industry as well as academia). They included Hopfield and Hecht-Nielsen, as well as Widrow. But there were cognitive scientists, too: Grossberg, Carpenter, Selfridge, Waltz, Feldman, Williams, Rumelhart, and Richard Sutton.

In addition, the six Panels consulted many individual researchers, some of whom were invited to give a talk to the official DARPA team. Not the least of these was Minsky, who soon used his response as the core of his intransigent comments in the second edition of *Perceptrons* (Minsky and Papert 1988) (By this time, of course, he was both criticizing connectionism and using it in his own work: see Section iii.c–d, above.)

The reason behind this particular choice of speakers and Panel members was simple: DARPA's concerns were military. Neuroscience and psychology were included not for their own sake, but because they might help in producing useful working systems—where the military decided what was "useful". (There was little new in that. AI had been largely funded by military money from its inception: see Chapters 4.vi.a, 11.i, and 9.x.a.)

The Report baldly stated that any money that might be forthcoming would support "advancement of technologies critical to the development of next-generation, fire-and-forget, autonomous weapons" (DARPA 1988, p. xxv). Hence the emphasis on vision—for instance, Hubel's involvement. But any research topic might be relevant, if it advanced the understanding of connectionist systems in general. So, for instance, Sejnowski was invited to speak on 'Analysing the Hidden Units in Multilayered Neural Networks', one of the methodological issues raised by NETtalk.

As Hecht-Nielsen's remarks (above) clearly show, there were social sensitivities to be respected here. Widrow, as Study Director, noted that "there are many different and even conflicting views of what neural networks are, what they can do, and how they should be implemented", and disarmingly pointed out that "representatives of every school of thought in neural networks were consulted" (DARPA 1988, pp. xxix, xxx).

There was also the delicate matter of why DARPA had seemingly been so blind, in their near-refusal to fund connectionist AI for the last twenty years. Largely, no doubt, because—like Yovits—they'd been persuaded by Minsky in the 1960s (see Section iii.e, above). But there was another reason too, which was mentioned repeatedly in the Report: counter-productive hype on the part of connectionists themselves.

As Widrow put it in his introduction,

Among scientists, the presence of hype and extravagant claims casts a dark shadow and makes the work controversial in scientific circles and amongst the world at large. (DARPA 1988, p. xxix)

The study specifically asked whether connectionism was "just a repackaging of old ideas and promises", as "some" (notably, Minsky and Papert) were claiming, and whether it was more "hyperbole" than achievement (DARPA 1988, pp. xxvii, xxix). Connectionist hyperbole, both past and present, was sternly noted. Media publicity and consorting with venture capitalists were decried too. Widrow tacitly defended DARPA's long-standing absence by complaining that "involvement is difficult . . . in the presence of hype" (p. xxv).

For all their disapproval of hype—plenty of which was now being spread by the PDP groupies, if not by the PDP group themselves—the scientific Panels didn't react as Sir James Lighthill had reacted to GOFAI sixteen years before (see Chapter 11.iv). As a result of their detailed assessment, DARPA decided to initiate "a major new program in neural networks beginning in 1989" (p. xxv).

As though this revenge wasn't sweet enough, Minsky and Papert were given a coded rebuke: "Neural network research is not new—it is, rather, newly revived from an obscurity and even disrepute which is now understood to have been undeserved" (DARPA 1988: 23). What Mead had called the "twenty-year famine" was over.

12.viii. A la recherche . . .

In the late 1980s, work on the new perceptrons ballooned. Much was largely 'more of the same', in the form of an ever-lengthening list of commercial applications driven by the venture capital remarked on by DARPA.

Examples ran all the way from mineral prospecting in the field to banking in the City of London or Wall Street. (Physics graduates were snapped up by the money men, to use 'thermodynamic' networks for financial predictions of various kinds.) These applications exploited PDP's power of pattern recognition, which Mead soon affirmed in uncompromising terms:

[Recognition problems are] an area where it's very hard to do anything without a neural net. The AI people tried for years. With a neural network, you can do better in a week than people have done fooling around with AI programs for years . . . With real-world data, where you don't know what the information is, neural network paradigms can pull out that information and make it useful. That is very, very clear. It wasn't clear five years ago [in the late 1980s]. Then it was a gleam in all of our eyes, but at that time it wasn't clear to anybody who was objective about it that it was going to be better than sitting down with a smart guy and writing a program. Now it's clear that there's no contest at all. That's been a big change. (J. A. Anderson and Rosenfeld 1998: 141)

It was also clear—within the field, if not in its press releases—that "Every silver lining has a cloud, and PDP is no exception" (A. J. Clark 1989: 127). Specifically, there's more to intelligence than pattern recognition. It followed that PDP wasn't the answer to every maiden's prayer—nor to every businessman's, either.

Relying not on madeleines but on arguments, some critics exhorted the new enthusiasts to remember, and try to recover, some key abilities of symbolic AI. For various things that had been easily achieved by GOFAI—notably, serial order and hierarchical planning—weren't yet possible using PDP.

Hierarchy couldn't be ignored, for it wasn't merely (merely?) a characteristic of intellectual thought. As Lashley had pointed out, it's a feature of skilled behaviour in general:

There is a series of hierarchies of organization; the order of vocal movements in pronouncing the word, the order of words in the sentence, the order of sentences in the paragraph, the rational order of paragraphs in a discourse. *Not only speech, but all skilled acts* seem to involve the same problems of serial ordering, even down to the temporal coordination of muscular contractions in such a movement as reaching and grasping. (Lashley 1951a: 121–2; italics added)

Propositional reasoning was impossible too, but the new connectionist modellers didn't even try to recapture that. (As you'll remember, the superiority of symbolic AI for representing propositions was one reason why it overtook cybernetics/connectionism in the 1950s: see 4.ix.b.) Even today, it's still an unconquered obstacle (but see Section x.e, below).

It wasn't clear that these things ever would be achievable by connectionist means. Perhaps only GOFAI could do certain sorts of problem solving, and/or perhaps the new networks would have to simulate GOFAI systems if they were to do so too. If one wants accurate reasoning or precise rule following, the famed noise tolerance of PDP may even be a disadvantage: two plus two really does equal four—not "probably 4", "maybe 3", or even 4.0001. Even "binding" different predicates (e.g. colour and shape) together to apply to *one-and-the-same* object—child's play in GOFAI—is problematic in connectionism.

These doubts were similar, at base, to Walter Reitman's doubts about the capacities of purely "Hebbian" programs (see Chapter 7.i.b). Now, some thirty years later, they engendered a new generation of PDP systems, doing things the ancestor networks couldn't do.

Admittedly, a few connectionists had already addressed these problems. Andreae had been doing so in working systems since the early 1970s (see Section iv.c, above). Minsky, in his "society" theory, had been attempting this too (iii.d, above). And John R. Anderson had combined localist networks with GOFAI insights about goal-directed thinking (7.iv.c). But the high-visibility (i.e. PDP) connectionist modellers had focused primarily on pattern recognition, not on hierarchy or sequence. Now, they made those the aim of their research.

(Meanwhile, two rather different things were going on. From around 1990, some young members of the recently named A-Life community developed a new form of neural net, which worked in continuous time and could simulate smoothly changing dynamical systems: see Chapter 15.viii.d. In addition, the now unfashionable GOFAI workers carried on, pushing their achievements still further: see Chapter 13.)

a. Emulating the ancestors

The first of these descendant networks had been conceived in 'biblical' times. For the PDP group themselves shared some of Minsky and Papert's continuing doubts about perceptrons. Indeed, they agreed with Minsky's hunch in his 'Steps' paper of 1961, that some *combination* of neural networks and GOFAI would be needed (see Section iii.b).

Even in their biblical volumes they admitted that, in order to do many kinds of problem solving, a PDP network would have to emulate a von Neumann machine. And Hinton, for example, had already tried to enable a PDP system to perform logical inference (Touretsky and Hinton 1985), and to implement a production system (see 10.v.e) (Touretsky and Hinton 1988).

Unlike some of their followers, then, the PDP group didn't claim that GOFAI had been a "false start" (Graubard 1988) or that it had "failed" (H. L. Dreyfus and Dreyfus 1988: 34; Dreyfus 1992, p. ix). Still less did they see "fooling around with AI programs" as a waste of time. On the contrary:

[The] idea of distributed representations is consistent with some of the major insights from the field of artificial intelligence concerning the importance of structure in representations and processes. (Hinton *et al.* 1986: 104)

[Mental] processes that take longer [than 0.5 seconds], we believe, have a serial component and can more readily be described in terms of sequential information-processing models. For these processes, a process description such as a production [see Chapter 7.iv.b] would, we imagine, provide a useful approximation. (Rumelhart *et al.* 1986c: 56)

The mental processes in question were characteristic of human thinking—the main focus of traditional AI (and of many group members' early research). Rosenblatt himself had surmised that perceptrons couldn't embody "such higher order functions as are involved in human speech, communication, and thinking" (see ii.f, above). Similarly, the PDP group admitted almost thirty years later that:

The PDP system is fine for perception and motor control, fine for categorization. It is possibly exactly the sort of system required for all of our automatic, subconscious reasoning. But [we] think that more is required—either more levels of PDP structures or other kinds of systems—to handle the problems of conscious, deliberate thought, planning, and problem solving . . . People interpret the world rapidly, effortlessly. But the development of new ideas, or evaluation of current thoughts proceeds slowly, serially, deliberately. People do seem to have at least two modes of operation, one rapid, efficient, subconscious, the other slow, serial, and conscious. (D. A. Norman 1986b: 541–2)

The *microstructure* of the serial mode, they believed, is connectionist. This provides welcome emergent properties such as content addressability and graceful degradation, which sequential processing can exploit though not explain. Nevertheless, they recognized that PDP work would have to build on the "major insights" gained by years of GOFAI research.

Even in 1990, they weren't expecting any quick success. Hinton, by then based at the University of Toronto, edited a special issue of *Artificial Intelligence* on how connectionists might add "symbol processing" to pattern recognition. But the lost strengths of GOFAI weren't going to be resurrected overnight:

[In future, if current learning] techniques can be applied in networks with greater representational abilities, we may see artificial neural networks that can do much more than just classify patterns. But for now, the problem is to devise effective ways of representing complex structures in connectionist networks without sacrificing the ability to learn the representations. *My own view is that connectionists are still a very long way from solving this problem* . . . (Hinton 1990: 3; italics added)

As for just how one can make a PDP network pretend to be a serial machine, the PDP books offered a number of ideas. For example, one chapter suggested that sequential thinking might be modelled by pattern matching in *multi-level* networks (Rumelhart *et al.* 1986c).

It described a simple computer model that used two mutually influential self-equilibrating networks to produce successive moves in a game of noughts and crosses (tic-tac-toe). The relaxation constraints of both networks embodied the rules and strategy of the game, but one network represented "player's move" and the other "opponent's move". The perceptual (input) model of each opposing move could drive equilibration in the player network, so as to produce an appropriate counter-move. In that case, the seriality would depend on the environment. But if the perceptual models were stored, or internalized, they could then be used by the player network to plan (and evaluate) hypothetical moves—in other words, to deliberate.

The ensuing years, to the end of the century, saw important advances such as recurrent nets, developmental minimalism, representational trajectories, input histories, multi-networks, hybrid systems, and constructive networks.

Each of these approaches lessened the distance between PDP and traditional AI. And each engaged with fundamental issues in computational psychology and philosophy of mind—not least, hierarchical structure and nativism. The first four are closely linked, and are described here; the last three will be discussed in Section ix.

b. Recurrent nets

Lashley had pointed out long before that serially ordered behaviour, such as speech, requires anticipatory (planning) mechanisms that define the 'place' of a behavioural unit in the overall sequence—typically, by imposing a hierarchical structure on it (see Chapter 5.iv.a). Planning had been a prime focus of GOFAI, and of GOFAI-based psychology. But whereas list processing had been ideally suited to represent both sequence and hierarchy (10.v.b–c), it wasn't obvious how—some doubters even said "whether"—a PDP system could do so.

Jordan (1956–) took up the challenge, and confronted the doubters with a recurrent network that could plan (Jordan 1986/1989). The general idea of recurrent networks, in which the output from some later layer is fed back as input to some earlier layer, wasn't new. It had been stressed by the cyberneticians, in their focus on "circular causation" and self-stimulating circuits (see Chapter 4.iii.c and v). But the early network-modellers had avoided feedback loops, because of the complex mathematics involved (the input to a given level changes continually). Now, Jordan provided a functioning model that could be used to generate ordered sequences of behaviour.

Constant feedback from output units to input units enabled Jordan's net to keep track of where it was, with respect to the sequence concerned. For instance, if a quickstep were represented as *slow, slow, quick, quick, slow* the net would know whether it had just output a *slow*, and which one of the three possible *slows* this had been. This information could affect the next signal from the input layer to the hidden layer, which in turn would determine the next output step.

So far, so rigid: a particular starting state would always trigger the same sequence. So Jordan added "plan units" to the input layer, which sent their own signals on to the hidden layer. Whether a hidden unit fired depended on the input it received from *both* the ordinary input units (receiving continual feedback) *and* the plan units. Hence: different plans, different behaviour—even though the perceptual input at the start might be identical.

As described so far, this required the researcher to specify all the unit thresholds and weights involved. In the toy networks (some less than a dozen units) described by Jordan, this was feasible. But pre-setting a network to execute 'at will' the plan of either 'Three Blind Mice' or 'The Death of Cock Robin', or any one of dozens of nursery rhymes (as humans can do) would be an impossible task. Clearly, what was needed was for the network to learn how to produce those sequences for it-self. In other words, it had to learn to predict the next item in the behavioural sequence—and to reflect, in its own behaviour, any hierarchical structure in the input domain.

The pioneer in making this possible was Elman (1948–), who was working on speech processing (1990, 1993). This had been a key topic for connectionism since its beginnings. McCulloch, Pitts, and Selfridge had been involved in MIT's early project of developing an automatic speech-typewriter; and Rosenblatt, Longuet-Higgins, and Willshaw had all considered temporal input patterns, as we've seen.

Classical AI had modelled speech processing too, notably by HEARSAY in the early 1970s (see 9.xi.g and 10.v.e). In principle, this GOFAI model was a parallel-processing system, using multiple constraint satisfaction to interpret its auditory input. Over a decade later, and partly inspired by HEARSAY, Elman's PDP networks learnt to do much the same thing.

Elman's recurrent networks differed from Jordan's in using constant feedback from the hidden units (not the output units) to a special group of "context" units (which replaced Jordan's predetermined plan units). But the general principle was similar. An Elman network kept track of what the immediately preceding state of the hidden units had been. Indeed, it implicitly stored information about *all* its earlier states, because—given that the context units sent signals back to the hidden units—the current activity of a hidden unit would indirectly reflect its previous state, which in turn would have reflected *its* previous state . . . and so on. The context units (unlike Jordan's plan units) could learn, for the signals they sent to the hidden units were automatically adjusted by back propagation at each cycle.

These networks learnt to predict temporal dependencies, including dependencies within dependencies. For example, they learnt to predict phonemes so as to pick out individual words from continuous auditory input (Elman 1990). Sometimes, they'd get it wrong: *anelephant* would be segmented as *a nelephant*—a mistake that children often make too.

Given strings of individual words (Elman 1993), they distinguished verbs from nouns, from adjectives . . . and even animate from inanimate objects. And they learnt to reflect nested grammatical hierarchies. A plural noun at the beginning of an output sentence, for instance, would be matched by a plural verb later, even if there was an embedded relative clause in between; and a *subject-noun* would always be followed by some *verb*, and sometimes preceded by *adjectives*. (Strictly, this is an exaggeration: the longer the sentence, the less reliable the network's performance.)

All this was achieved in virtue of the distributional statistics of the various words. The Bloomfieldians, then, would have been happy (see Chapter 9.v). But whereas they had measured (timeless) coexistence within a whole corpus, Elman's networks were assaying the temporal statistics of sentences presented to them word by word and one by one.

The grammatical structures weren't represented explicitly, just as the rules of the past tense hadn't been stored *as such* in the past-tense learner. Network analysis showed, however, that the activities of the hidden units differed according to the grammatical role of a given word: e.g. *cat*, used as subject or as object.

In other words, Elman's networks posed yet another challenge to Chomsky's nativism. As already remarked, they were closer to the structuralist linguists who had preceded him. They had no inbuilt English grammar, nor any domain-specific Language Acquisition Device. Where SHRDLU and LUNAR (see Chapter 9.xi.b) had been provided with procedural definitions of English syntax, these PDP systems seemed to be able to pick it up for themselves by relying on domain-general principles of learning.

This was a "challenge" rather than a refutation, not least because backprop isn't biologically plausible. Synapses don't run backwards, nor are brains provided with error corrections by some outside supervisor. Moreover, Elman didn't ask whether there are types of 'grammar' that can be learnt by unstructured PDP systems but *can't* be learnt spontaneously by humans. If there are—which later turned out to be so (N. V. Smith 1999: 134–5)—then his results didn't disprove Chomsky's claim that humans depend on innate linguistic structure.

Nevertheless, his work revived questions about the development of language which Chomskyans—which is to say, most cognitive scientists at that time—had thought already settled.

c. Start simple, develop complex

Elman's work also raised other questions about children's language learning, which led to a general idea that one might call developmental minimalism. This sees development as passing from simple types of representation to more complex ones, where each of the higher levels is constructed 'on the shoulders' of the one below. (As remarked in Chapter 7.iv.h, it's a special case of the boundedness of rationality being a Good Thing.)

For example, Elman (1993) soon found that his networks failed to learn satisfactorily if the initial input set contained complex structures. By experimenting, he discovered that if the first inputs were all simple, more complex sentences could be learned later. It was as though the basic syntax was distinguished first, initiating a series of stepping stones leading to the full grammar.

This result fitted well with the view of some psycholinguists at the time, that parents use a special dialect ('motherese') when talking to infants (see Chapter 9.vii.c). The syntactic simplicity of motherese, it was said, enables the child to acquire grammar. However, those claims were by now controversial; and Elman, in particular, wasn't convinced. One problem was that even if parents do use motherese, children overhear adult conversation too—so why aren't they confused by that? Is the mere lack of active engagement enough to protect them?

Accordingly, Elman changed his strategy. Instead of minimizing the network's input (environment), he now minimalized its memory. That is, he added a procedure that automatically 'emptied' the context units after every third or fourth input word, by resetting their activity to zero. So whereas these units had originally been designed to save traces of *all* of the network's previous activity (see above), they now held the impress of only the previous one to four words.

Clearly, the revised network had a good chance of learning *John punched Jack*, but not *John punched Jack hard on the nose*. And so it turned out: the new network could learn only sentences within the memory span of the context units. Once that had been achieved, however, Elman gradually increased the memory span, and found that this enabled the system to learn increasingly complex sentences. This strategy wasn't wholly 'artificial': evidence already existed suggesting that limits on human memory span affect the complexity of parsing, and explain 'garden-path' sentences (Chapter 7.ii.b).

The moral Elman drew was that if a recurrent network—whether artificial or biological—is learning something from scratch then a weaker, smaller-span, memory isn't a flaw (as he had earlier assumed). On the contrary, it's an advantage. Hence the subtitle of Elman's 1993 paper: 'The Importance of Starting Small'.

Babies and infants, he assumed (he wasn't an unreconstructed nativist: see 7.vi.g), learn virtually everything from scratch. Suppose there were some maturational process—myelinization of nerve fibres, perhaps?—that gradually extends the child's memory span. If so, it could afford domain-general assistance, and would result in abilities developing from simpler to more complex forms.

Developmental minimalism had started (for Elman) as an ad hoc computational trick, designed to avoid disappointing empirical results. But it was soon recognized as relevant to learning and development in general. The central principle was the need for repeated recoding, or representational trajectories.

d. Pathways for representation

The term "representational trajectories" was due to Andy Clark, a philosopher at the University of Sussex (1993, ch. 7). But the basic idea had been stated in the 1970s by Annette Karmiloff-Smith, then working with Piaget in Geneva, and formed the core of her theory of "representational redescription", or RR (1992). As we saw in Chapter 7.vi.h, RR theory posited the spontaneous development of increasingly powerful and flexible representations in the infant's mind.

Now, in the 1990s, she and Clark related it to the constraints faced by *any* learning system (A. J. Clark and Karmiloff-Smith 1993; Clark 1993). And soon afterwards, Clark and Chris Thornton (an AI colleague at Sussex) outlined a method whereby a PDP system could use representational trajectories to learn increasingly complex structures (A. J. Clark and Thornton 1997; Thornton 2000).

Their key insight was that the new perceptrons were limited in principle much as the old ones had been. Instead of linear separability, the focus now was on statistical regularity. Some regularities in the input are so clear that even an 'uninformed' learner can spot them. (Hence Mead's triumphalist remark quoted in the preamble of this section.) Others are more subtle, more elusive. In these ("Type-2") cases, the learner needs to know something about what regularities to look for if he/she/it is to have a realistic chance of finding them.

In effect, this was a generalization of Chomsky's anti-empiricist argument about the need for some innate grammar to support language learning (Chapter 9.vii). But unlike Chomsky, Clark and Thornton allowed that a suitably 'informed' learner may *either* be built/evolved to possess such guidance from the start, *or* may somehow learn, or construct, it for itself. What is needed in either case is a recoding of the input data, to make the marginal regularities stand out. A sequence of such recodings is a representational trajectory.

One example is Elman's research on increasingly complex grammars; indeed, Clark and Thornton described their work as an extension of Elman's.

* Another is the sequence of redescriptions posited by Karmiloff-Smith, whereby an automatic skill—such as drawing, or speech—develops into a flexible, generalized, and consciously self-reflective activity (see Chapter 7.vi.h).

* Yet another is the series of visual computations involved in constructing the Marrian Primal Sketch (Chapter 7.v.b–d).

* And yet another is the history of scientific theories and analogies used in developing current optics (P. M. Churchland 1995: 271–86) or astronomy (Thornton 2000, ch. 3). If Aristotle, or even Isaac Newton, were to come back today, we couldn't teach them modern physics by simply *telling* them: they'd need a sequence of conceptual bootstraps.

In general:

[A] wide variety of superficially distinct ploys and mechanisms can be fruitfully understood in these terms. Such ploys and mechanisms range from simple evolved filters and feature detectors all the way to complex cases involving the use and reuse of acquired knowledge. The goal, in every case, is to systematically reconfigure a body of input data so that computationally primitive learning routines can find some target mapping—that is, to trade representation against computation. (A. J. Clark and Thornton 1997: 57)

The "ploys" used by *Homo sapiens* include a host of examples based in language, writing, technology, and cultural institutions. These are all "adaptations enabling this representation/computation trade-off to be pursued on an even grander scale" (p. 57), and are essential for, even intrinsic to, distinctively human thinking. So far, so psychological—but Clark would soon use this point to ground an unorthodox philosophy of the self (see Chapter 16.vii.d).

Non-human animals can't generate these language-based "cognitive technologies" (Clark's term, borrowed from Bruner: 6.iii.c). But even they needn't possess inbuilt modules suited to every structure they can learn, provided that they can recode the current information. If we want to understand why different species have different learning limits, we must ask how their representational trajectories differ, and why.

At the end of the century, Thornton told a fairy story about a Machine That Can Learn Anything—let's call it MTCLA (Thornton 2000, ch. 1). According to its designer, and to "no fewer than three TV stations", this machine never failed to make the correct association (to pick up the regularity concerned) in the majority of cases, *no matter what* problem it was presented with. What's more, the designer wasn't lying. Nonetheless, it turned out as the story proceeded that MTCLA was a confidence trick—or rather, three.

* First, its achievements were too few to be useful. A success rate of 50.01 per cent would satisfy the "majority of cases" criterion, but for any practical purpose it needs to be much higher. (Whether 100 per cent reliability is required, or even possible, is a question discussed by philosophers of science as well as by stockbrokers: Thornton 2000, ch. 9.)

* Second, the users of the fabled machine had to specify its learning tasks in a highly restricted way. In short, the power lay in the input representation. (Compare the so-called *General* Problem Solver: see Chapter 6.iii.c.) If no suitable input could be found, the machine wouldn't actually *fail*, because it couldn't be presented with the problem in the first place.

* And third, every problem that could be presented was trivial: a simple statistical trick sufficed to learn the regularity.

In fact, this wasn't a mere fairy story. The networks of Rosenblatt, Widrow, Hopfield—and now, PDP—had involved comparable hype, and comparable con tricks. The moral of Thornton's mocking tale was that, even after forty years of connectionist research, the core of (non-trivial) learning had hardly been touched.

If we really want a machine that can learn anything, he said, we must provide a domain-general method whereby the learner can produce useful truths from what looks, initially, like "trash". It looks like trash because of what Clark had called "snowblindness" (1993: 149), in which the system is (at first) so overwhelmed by detail that it can see neither the high-level regularities nor even the lower-level ones. Building on their joint 1997 paper, Thornton described an incremental process of recoding ("recursive relational learning") that provides a series of "scaffolds" for learning.

On this account, learning is continuous with creativity:

Without the ability [for] relational learning, an agent is . . . effectively *trapped* within its immediate sensory space. Its only recourse is the establishment of simple partitions within that space. But [with] relational learning, the agent acquires the ability to escape its immediate embodiment—to interact with properties that are not explicitly manifest in the world to which it has direct sensory access. (Thornton 2000: 181)

As the number of recoding levels rises, the highest ones are "substantially the creative artefacts of the learner's *own* processing . . . [i.e.] their properties are [less] severely constrained by the source data" (p. 193).

Thornton didn't claim that his method could learn literally *anything*, nor that the human brain can do so. There may be inescapable constraints on the recodings that are possible in human thought. It's sometimes suggested, for example, that we won't ever understand the relation between brain and consciousness, much as dogs won't ever learn arithmetic (see Chapter 14.x.d). But if we want to understand systems that even *approximate* an MTLCA, we must allow for representational trajectories.

e. The importance of input history

As for just what these trajectories will be, it became increasingly clear that the input history—the temporal sequence of the trash—matters. For instance, how (and whether) a PDP system learns to emulate rule-governed syntactic processing depends on its training history.

Elman's decision to present simple sentences before complex ones was an early illustration of this point. But in the early 1990s, the effects of more subtle variations of the training set were explored by the Oxford developmental psychologist Kim Plunkett, with Virginia Marchman and Chris Sinha (Plunkett and Marchman 1991, 1993; Plunkett and Sinha 1992). Their test bed was an improved version of the past-tense learner described in Section vi.e (they'd added a hidden layer, and backprop).

Instead of artificially 'splitting' the input, they presented a fixed vocabulary set in varying quantities and proportions, and in varying orders. They found that both gradual and sudden (stagelike) changes in performance are driven by internal reorganizations that depend crucially on the input history. For instance, a training sequence can sometimes inhibit later learning that would have been possible if a different input sequence had been used.

In doing this study, they had to analyse the network's performance in greater detail than Rumelhart and McClelland had done. Beyond considering the overall statistics, they tracked the learning of individual verbs, and of distinct sub-groups of regulars and irregulars. They found many differences in the pattern of errors and error recoveries concerned, often including "regressions" in which old errors returned and/or new ones appeared as learning progressed. Another interesting result was contamination by irregulars, as certain regular verbs were apparently assimilated to subclasses of irregulars.

In all these cases (and others), developmental linguists had recently observed similar error patterns in children. In both humans and PDP systems, then, the learning of the past tense is a more complex—and contingency-driven—phenomenon than had previously been suspected.

But why stop at the past tense? Plunkett found that the effects of different input histories were often predictable, given the general properties of the PDP system concerned (the learning rule being used, and the constraints necessarily involved in storing many distributed patterns in a single network). The moral, then, was that *in general* input histories matter. One might say this insight wasn't new: as we've seen, Hopfield had asked how many units should be set correctly at the start to give optimal results. But no one had studied the effects of *extended* training histories in such detail.

12.ix. Still Searching

The four advances considered so far—recurrent nets, developmental minimalism, representational trajectories, and input histories—were closely linked, as we've seen. The other three were more diverse.

Research on multi-networks, hybrid systems, and constructive networks used very different methods to increase the power of connectionist systems. One even risked betraying the faith by resurrecting GOFAI, though admittedly only in combination with networks. By the turn of the twenty-first century, however, these too still left much to be done.

a. Assemblies of cell assemblies

An early instance of a multi-network was described in the PDP books, namely, the noughts-and-crosses player mentioned in Section viii.a. An even earlier example had been outlined by McCulloch in the late 1960s, in his controller for a Mars robot (Chapter 14.iv.a). Others were sketched or implemented later.

For example, Clark and Karmiloff-Smith (1993) suggested that a new network might come to represent the inner structure of an older one in a more "symbolic" way as a result of skeletonization (cf. Chapter 7.g). This technique copies an initial net into another one, and then prunes the connections that correspond to inessential, relatively inactive, connections in the first (Mozer and Smolensky 1989).

But two-network systems, such as these, are unrealistic. Hebb had argued long ago that adult human brains must involve numerous cell assemblies; and if Minsky and Papert's "society of mind" theory was on the right lines, then complex intelligence requires many interacting networks (iii.d, above). The PDP group agreed. As Norman put it in their manifesto:

[People] can do multiple activities at the same time, some of them quite unrelated to one another. So, in my opinion, a PDP model of the entire human information processing system is going to require multiple units. That is, the complete model requires that the brain consists of several (tens? hundreds? thousands?) of independent PDP-like systems, each of which can only settle into a single state at a time. (Norman 1986*b*: 542–3)

Moreover, each of these "independent PDP-like systems" would very likely be a modular hierarchy itself.

This claim may have seemed strange to readers over-impressed by the connectionists' constant references to representations being "distributed over the *whole* network". But the PDP group's earlier work in GOFAI had already alerted them to the need for hierarchical structure. What's more, backprop had been designed with this in mind:

One of the best and commonest ways of fighting complexity is to introduce a modular, hierarchical structure in which different modules are only loosely coupled [here, Hinton cited Simon's *Sciences of the Artificial* (see Chapter 7.iv.a)] ... Self-supervised back-propagation was originally designed to allow efficient bottom-up learning in domains where there is hierarchical modular structure ... It is possible to learn codes for all the lowest-level modules in parallel. Once this has been done, the network can learn codes at the next level up the hierarchy. The time taken to learn the whole hierarchical structure (given parallel hardware) is just proportional to the depth of the tree ... [although] it is helpful to allow top-down influences from more abstract representations to less abstract ones, and [this has been done in] a working simulation. (Hinton 1989: 228)

Hinton's recommendation of "top-down influences from more abstract representations" was yet another example of GOFAI footsteps preserved in the connectionist sand.

If PDP could provide massive modularity in principle, however, it didn't follow that it could do so in practice. Hinton's reference (above) to a "working simulation" didn't mean that large multi-network systems were practically feasible by 1990, for they weren't. But it was already clear that one of the most important aims of PDP research into the next century would—or anyway, should—be to master this problem.

That qualification ("or anyway, should") is necessary because there was surprisingly little work on this topic in the 1990s. To be sure, the anthropologist David Hutchins (1995) modelled the interacting roles within a ship's crew (a part-hierarchical assembly) by way of *networks of networks* (see Chapter 8.iii.b). And some researchers, including Jordan and Hinton, soon developed pattern recognition systems within which distinct sub-networks emerged that were sensitive to different aspects of the input—men's, women's, or children's voices, for example, or a particular vowel (Jacobs *et al.* 1991).

This was done by extending the 'winner-takes-all' approach from single units to groups of simultaneously active units. (In a winner-takes-all network, the learning rule doesn't increase the weights on *all* the units, but only on the most highly activated one.) A higher-level 'referee' sub-network was backprop-trained alongside the emerging experts, to decide which mini-network should be the winner when more than one was activated by a certain pattern.

In effect, this was a self-organizing PDP system that learnt to mimic a simple version of Pandemonium. But a multilevel Pandemonium, or the interacting "multiple unit" hierarchies mentioned by Norman, would be harder to implement—never mind to learn.

Much as the early computational psychologists had abandoned multi-motive systems, because getting a program to pursue one goal was difficult enough (see Chapter 7.i.b), so PDP workers found that getting a network to achieve a single task was more challenging than they had expected. This is just what Thornton's analysis would predict: many interesting tasks are Type-2 problems, requiring one or more levels of recoding—which 1990s networks couldn't deliver. Accordingly, hardly anyone tried to address the sort of multi-network complexity that Norman had in mind. (An interesting exception, based partly on his own psychological theory, is described below.)

Many focused instead on continuing the mathematical turn described in Sections iv–v, seemingly forgetting that the original purpose of perceptrons had been psychological. As Thornton put it in 2001 (personal communication):

My perception is that connectionists are currently executing a kind of mass retreat into mathematics, perhaps in a subconscious attempt to deny operational reality. Certainly the kind of psychologically-aware approach pioneered by Geoffrey Hinton [in his models of semantic memory] doesn't get much of a look-in these days. (Thornton 2001, personal communication)

To be sure, the 1990s had seen significant work on 'real' learning, some of it deeply informed by psychological and/or neuroscientific data (O'Reilly and Munakata 2000). But, as yet, there was little overlap between the biologically oriented connectionists and their more analytically oriented cousins. The conferences described in Section vii.a (above) were driven by the hope that empirical and analytic workers would collaborate, the distinction between them becoming much less clear. That hope hadn't been fully satisfied—which is why most of the research described below is more mathematical than psychological, and more psychological than neurological.

b. Hands across the divide

The assumption—shared by the PDP group and society-Minsky—that much human thinking requires some *combination* of connectionist and GOFAI methods led to a new research field: hybrid systems. This term is sometimes used to mean programs/robots that combine situated and deliberative procedures (see 7.iv.b and 13.iii.c). More often, however, it's used to mean systems combining connectionist and symbolic architectures.

The general idea of combining two very different types of computation in one machine wasn't new. Indeed, it dated back to the earliest days. MacKay, with McCulloch's encouragement, had suggested developing part-analogue, part-digital computers in the late 1940s (MacKay 1949/1959). By the mid-1960s, he was even saying:

We on the circuit side had better be very cautious before we insist that the kind of information processing that a brain does can be replicated in a realizable circuit. *Some kind of 'wet' engineering may turn out to be inevitable.* (Mackay 1965: 329; italics added)

Minsky had said in his widely read paper 'Steps Toward Artificial Intelligence' (available in draft from 1956 on, and published in 1961) that hybrid architectures would probably be needed (Chapter 10.i.g). A few years after that, Reitman (1965) had built Argus as an exercise in combining sequential and parallel processing (see 7.i.b). Since then, however, hardly any work had been done on it.

Some commentators in the mid-1990s expressed impatience that this combination hadn't already been achieved:

The whole field has been hovering around this issue, pendulum-like, for over three decades and it is time that these two approaches are reconciled, so that we can begin to take advantage of and build on the strengths of both. (Honavar and Uhr 1994, p. xiii)

After all, the two types of AI were in principle equivalent, since both were using general-purpose computational systems. In practice, however, they were different. The *degree* of difference was a matter of taste, not to say self-regard or professional pride (I was reminded of the horse in *The Wizard of Oz*, whose colour changed with the lighting as it trotted along: Boden 1991). But there were differences nonetheless, and it wasn't clear how they could be surmounted. (For the record, it still isn't.)

The organizers of a 1995 conference on the topic located the core problem in the ontologies we use to describe the world. As they put it, we must somehow "reconcile the static ontologies of standard knowledge representation with a continuously changing world described using the ontologies of physics" (Hallam 1995, p. v). In other words, we must find a way to translate effectively between von Neumann's "primary" and "secondary" languages (see Section i.c). What is needed, they said, is "a theory able to treat continuous and discrete processes and their interactions in a uniform way"—and that was still distant.

Perhaps, other authors added, it was even impossible:

There is no clear method for ensuring the preservation of a symbolic algorithm in a dynamical equation that does not itself constitute an algorithm; nor one for ensuring the preservation of a symbolic semantics (or function) in a connectionist account of the discipline of real world entities... There is no possibility of an exact mapping between the connectionist and the symbolic. (Franks and Cooper 1995: 69)

This question couldn't be decided, they argued, until we have a better way of individuating algorithms. The current definition relies on the intuitive notion of a "step", which might be interpreted in differing ways. This raises questions about all computer models, hybrid or not, and "the issue remains to be solved by the cognitive sciences" (p. 70).

Meanwhile, a useful distinction was made between three types of hybridization—only one of which reflects hybrid psychological theories (Franks and Cooper 1995). In the first, two *physically* distinct types of computer architecture are used to compute different functions. For instance, a rule-based expert system implemented in a von Neumann machine might use neural networks on VLSI chips to recognize some rule-triggering patterns. This may be fine for engineering purposes, but it doesn't reflect the physical construction of the brain.

In the second type, the hybridness is viewer-dependent, since it rests on how one chooses to *describe* the system. For instance, the past-tense network can be described either as a PDP model or *as if* it were a rule-learner. The psychological theory being modelled, however, was purely connectionist. Only the third type is "truly" hybrid, in that the system's behaviour is generated by both connectionist and symbolic functions, *and* by theoretically significant causal relations between them.

The PDP bible itself had suggested that, in doing paper-and-pencil arithmetic, one combines perceptual pattern matching with rule-governed mathematical reasoning. Most PDP modelling wasn't driven by psychological theories that featured strong interactions between symbolic and connectionist processing. But Hinton (1988/1990) made an early attempt, by combining PDP with localist connectionism in order to represent part–whole hierarchies—such as family trees, for instance.

The localist network was a reduced version of the distributed one, wherein the 'relevant' units could be identified and accessed more directly. In other words, "the very same object can be represented in different ways depending on the focus of attention" (Hinton 1988/1990: 65). Putting the same point the other way around, he said:

[Some] patterns of activity in [the network] need to exhibit the double life that is characteristic of symbols. The patterns must allow remote access to fuller representations, but so long as the patterns are also reduced descriptions this remote access need only be used very occasionally (e.g. a few times per second in a person). Most of the processing can be done by parallel constraint satisfaction on the patterns themselves. (p. 49)

And rationality, as opposed to intuitive thinking, was shown *not* to be a matter of seriality. Its defining characteristic, rather, was that "the way in which entities in the domain are mapped into the hardware *changes* during the course of the inference" (p. 50; italics added). It followed that the crucial GOFAI assumption—that fast intuitive inference is similar in form to conscious inference, but without the consciousness—was judged by Hinton to be "a major psychological error" (p. 72).

Hinton's 1988/1990 approach was soon developed by others (J. B. Pollack 1990; Chalmers 1990; Plate 1995; Sperduti 1992; Sperduti and Starita 1994; Goller and Kuchler 1996; Goller 1999). So his *distributed reduced descriptors* later gave way to *recursive auto-associative memory* (RAAM), *holographic reduced representations* (HRRs), *labelling recursive auto-associative memories* (LRAAM), and *folding architecture networks* (FANs). In general, these new acronyms named methods by which vectors representing the individual components of a compositional structure were combined into a single vector standing for the whole structure. However, although they were more powerful than Hinton's family-tree program, they still couldn't deal with large-scale tasks: only small examples were tractable.

Hierarchy is important not only in linguistic behaviour, such as family trees or syntactic parsing. As Lashley pointed out (5.iv.a), it's an observable feature of motor skills in general. And if Simon was right in his *Sciences of the Artificial*, it's also near-inevitable, for reasons of processing economy. Indeed, recent neurological work shows that the spinal cord itself (not just the brain) stores a number of basic motor functions as *patterns* of individual muscle activations, which mini-hierarchies are combined in various ways to produce larger-scale action sequences (Mussa-Ivaldi 1999; Tresch *et al.* 1999). It follows that weakness in handling hierarchy is a severe drawback for any computational methodology that hopes to simulate even non-intellectual behaviour.

By the mid-1990s, there were several part-PDP working models that took psychological hybridization seriously. (The long-established family of ACT architectures outlined in Chapter 7.iv.c were hybrid in spirit, but didn't come from the PDP stable.)

One of these learnt to cope with simple arithmetic—specifically, the multiplication tables (J. A. Anderson *et al.* 1990*b*). The number-names were stored symbolically, but the numbers were also represented (according to their size) on a "sensory" topographic map. The justification for this was the fact that many (most?) humans seem to represent numbers spatially, especially when making judgements about relative size.

"Coping with" arithmetic, in this case, didn't mean conquering it. Like Dr Johnson's dog walking on two legs, the wonder was not that the system did its task well but that it did it at all. For it made many mistakes: 30 per cent errors in multiplying two integers. However, most of these were "intuitively reasonable", and matched a variety of error-types observed in human beings. Tongue-in-cheek, the authors declared: "It may seem remarkable to spend hours of supercomputer time to obtain the wrong answers to simple arithmetic, but that is cognitive science."

Another PDP hybrid was implemented by the neuropsychologist Shallice (now at UCL's Institute for Cognitive Neuroscience) and other London-based colleagues (Cooper *et al.* 1995). It was based on the theory of action that Shallice had developed in the late 1970s with Norman, a PDP group member who had worked for many years on the psychology of error (D. A. Norman and Shallice 1980/1986).

Whereas Norman had focused on everyday cases, such as typing errors (Rumelhart and Norman 1982), Shallice added clinical examples. Besides his interest in dyslexia (which he and his colleagues continued to model: Plaut and Shallice 1993; Plaut *et al.* 1996; Plaut 1999), Shallice was an expert in clinical ataxias, or disorders of action.

For instance, brain-damaged patients may apparently forget that the letter should be put in the envelope before the sticky flap is licked, or frequently get into bed on going upstairs to change their clothes, or pick up the kettle when they mean to pick up the teapot. Similar mistakes—errors of *order*, *capture*, and *object substitution*, respectively—occur occasionally in all of us (and Norman had asked "Why?"). But they're much more common in certain clinical syndromes. Moreover, brain-damaged patients can't easily recognize or correct them, whereas normal people can.

In 1977–8, on first reading about GOFAI planning, Shallice told me of his hopes that this approach might help him analyse such errors. But he soon decided that connectionist ideas were needed too. He and Norman argued in 1980 that over-learned action is generated by one or both of two types of control. Errors, of differing classes, occur when one or both of these controls breaks down at specific points.

The first control mechanism ("contention scheduling") is automatic. It involves competition between various interacting and hierarchically organized action schemata. Control goes to the one whose activation passes some threshold.

The second mechanism ("executive control") is attentional—that is, conscious. Because this involves deliberate supervision and modulation of the operation of the first, it's sometimes called the supervisory attentional system. (An action schema is defined by a goal, and a partially ordered set of sub-goals; each schema controls a partially ordered sequence of actions.) It's required in five types of situation:

* planning or decision making;
* especially difficult actions;
* repairing mistakes;
* inhibiting a strong already learnt response;
* and learning new actions.

The original theory had been purely verbal: the techniques available in 1980 couldn't produce a computer model. Fifteen years later, Shallice's team built a model that simulated both normal error-free behaviour and, with different parameter settings, patterns of error seen in various clinical syndromes, including Parkinson's disease.

Their prime example was coffee making. This wasn't a trivial, spur-of-the-moment, choice. They focused on coffee making partly because it's a familiar multilevel action schema. But they picked it also because it had been closely studied in neurological patients. In short, there was already a detailed 'library' of errors in coffee making, observed in both normal and brain-damaged people.

This study, which turned out to be very influential, was part of a more general programme of research in computational psychology (Cooper *et al.* 1996). One of the core collaborators was John Fox, of the Imperial Fund for Cancer Research—a long-standing leader of expert-systems work in Great Britain. That was no accident, for the team saw both GOFAI and PDP as essential:

The implementation is hybrid in that it contains components which carry out continuous-valued operations using a broadly interactive activation approach [based on the late 1970s work of Grossberg, McClelland, and Rumelhart] and other components which carry out discrete symbolic operations. (Cooper *et al.* 1995: 28)

They pointed out that this connectionist–symbolic mix wasn't a mere implementation detail, or engineering 'fix', but crucial to the psychological theory being modelled. The discrete, serial, symbolic aspect concerned the *selection* of this or that action schema or goal (sub-schema, sub-goal . . .) at a given point in the action sequence. The connectionist aspects regulated the *levels of activation* of the schemata in varying circumstances.

To some extent, the activation levels depended on the simulation of neuroscientific factors, such as lateral inhibition and dopamine concentration. Future research, the team said, would consider a range of neurotransmitters; differing reaction times for different senses; finer-grained object representations; and the attentional level of control (the 1995 program concerned only the automatic level). Each of these additions should result in a more discriminating and more realistic model of errors.

For instance, suppose that the sugar sachet mentioned in the plan schema were represented not only (as in the 1995 model) by *is-sachet* and *contains sugar*, but also by other perceptual and semantic features. In that case, one would expect more "substitution" errors, in which one object is picked up instead of another which shares some features and/or associations with it.

Besides applying to everyday errors and to debilitating psychopathology, the Norman–Shallice theory has been applied to hypnosis (see Chapter 7.i.g). With the help of philosophical work defining various levels of "explicit" and "implicit" representation, hypnosis can be explained as executive control *without* the highest level of explicitness—at which the person not only has an intention but knows that, and can report that, they have it (Dienes and Perner forthcoming). The intention can be monitored and flexibly executed by the hypnotized person. But these processes go on without conscious attention.

In sum, there's a growing appreciation that we need both GOFAI and PDP to model the mind. Michael Arbib, when he was interviewed for the "oral history" of connectionism, put it like this:

I don't believe that neural nets are a magic panacea. I think there's still a place for knowing how to add numbers exactly, rather than using a neural net. *I think in the future we'll see hybrid systems* where we have something like schema theory [see Chapter 14.vi.c] or modular design to understand how to take a complex problem, break it into pieces, and then for *some* of those pieces find that neural nets will do the best job. (J. A. Anderson and Rosenfeld 1998: 236; italics added)

c. Constructive networks

Quite soon, connectionists began to experiment with "constructive" networks. In these, the number of hidden units didn't have to be decided by the designer. For hidden units could be deleted or added during the running of the system. (A wide range of examples are compared in Fiesler 1994.)

Self-deleting networks were defined by Rumelhart as early as 1987. Starting with many hidden units, these were selectively pruned—weights were permanently reduced to zero—so long as this reduced the overall error. This approach was sometimes compared to 'neural selection' in the brain, in which profuse neuronal connections are progressively pruned as a result of experience (Chapter 14.x.d). In essence, however, it was a slight variation of the normal PDP network: some weights were eliminated once and for all, not merely (temporarily?) reduced.

The complementary type of constructive network, in which hidden units are added, is more interesting. One influential example was developed by Fahlman (Fahlman and Lebiere 1990). His "cascade correlation" networks started out with no hidden units at all: the input units were directly connected to the output units. But they added new ones incrementally until the error couldn't be further reduced. Each newly added unit would be randomly connected to the input layer, and would receive input also from the previously added hidden unit—hence the name, *cascade* correlation.

This network construction was a cycle of two phases. Even before the first hidden unit was added, some learning would take place, during which the (direct) input–output weights were adjusted in the usual way. Cascade learning would start when these weight changes had settled to equilibrium. A number of potential hidden units would be provisionally added, and their performance monitored. The one found to be most efficient in reducing error would then be permanently included. After another 'normal' learning phase in which the whole network (now carrying a new hidden unit) adjusted its weights, the cascading would kick in again—and so on, until no current provisional unit was able to improve performance.

Cascade correlation enabled networks to emulate certain aspects of GOFAI programs. For example, we saw in Chapter 7.vi.g that Piaget's developmental theory of seriation—the ability to produce rank order, as in building a staircase out of blocks—posited a final stage in which the child apparently plans his/her actions as a rationally ordered sequence. We saw, too, that his three-stage theory was first simulated by GOFAI programs (e.g. Richard M. Young 1976). This is perhaps just what one would expect, given the GOFAI-ish nature of the final behaviour. But we also noted that a connectionist model of RR for seriation was achieved twenty years later (Shultz 1991; Shultz *et al.* 1994, 1995). This was done by using cascade correlation. Two separate output layers were required: one to decide which block to move, the other to decide where to put it.

One might object, however, that this seriation network involved cheating. For the error assessment didn't use a perceptual test, but relied on Piaget's formal–operational rule for building a staircase: *Move the smallest block that is out of place into its correct place.* In other words, it's not clear that the connectionist system was generating GOFAI-type performance 'spontaneously'.

d. What had been achieved?

By the end of the century, as we've seen, the lost strengths of symbolic AI hadn't been fully recovered. Research, not to say *Recherche*, remained necessary.

In practice, many people decided to *combine* rather than to *recover*. As Arbib said in 1993:

There is so much fast-computing hardware around that a lot of ideas which were purely theoretical in the '70s are now eminently practical in the '90s ... A lot of work now has to be either very focused, on a very specific application, or you have to develop a hybrid system where you apply some standard signal-processing and some standard AI expert systems stuff, using a neural net for a module or two. I think the day of the magic single neural network has gone. (J. A. Anderson and Rosenfeld 1998: 235)

But there were now many more ways to skin a connectionist cat (and yet more arcane statistical mathematics used to do so). To mention just a few, most of which have been discussed above:

 * Some networks used basic computing elements (as identified by the mathematics) that were implemented not as single units but as groups of interconnected units, or "neuron pools" (Amari 1977).

* Besides a variety of learning rules that adjusted weights on existing connections, there were others that added new links and pruned old ones.

* Some learning happened across generations instead of in individuals, using the newly popular technique of genetic algorithms (Chapter 15.iv); an example involving a PDP group member, namely Elman was (Nolfi *et al.* 1994).

* Some connectionist work led back to Hopfield rather than PDP: 'attractor' neural networks were completely connected nets, understood as dynamical systems converging towards various attractors (global equilibrium states) as a result of different inputs, or 'perturbations'.

* The puzzle of how to enable ANNs to deal with hierarchy was addressed by methods in which a network could contain representations of a whole as well as representations of its parts—although, as remarked above, only small-scale tasks were tractable.

* Some people, such as Shallice, developed truly hybrid systems, in which neural networks and GOFAI components were combined.

* Judea Pearl (1988) described how to use (non-recurrent) Bayesian networks to do approximate inference, or reasoning, not mere pattern recognition.

* And, building on Pearl's work, methods for "loopy belief propagation" (LBP) enabled recurrent networks containing many loops, of different sizes, to converge instead of running endlessly (McEliece *et al.* 1998; Murphy *et al.* 1999; Yedidia *et al.* 2000). LBP is now being used for various purposes in computer vision (e.g. Coughlan and Ferreira 2002; Sigal *et al.* 2003; Isard and MacCormick in preparation), and applied also to make inferences of a type that would previously have been handled by a GOFAI expert system—medical diagnosis, for example.

Moreover (as we'll see in Chapter 14), the distinction between neural networks as an AI technology and as a way of modelling the brain was becoming less clear. For example, attractor networks were studied in relation to Walter Freeman's dynamical theory of the neurophysiology of perception (see Chapter 14.ix.b). And in 1998, Hinton was appointed head of the new Gatsby Computational Neuroscience Unit, at University College London.

Hinton's reputation, already high in the 1980s, had soared still higher: Cowan said the best advice he could give to a young researcher would be to "try to work with Geoff Hinton" (J. A. Anderson and Rosenfeld 1998: 124). At the outset of the new century, he received the first Rumelhart Prize for "contributions to the formal analysis of human cognition". This wasn't, you'll notice, a prize confined to connectionists. Accordingly, it was awarded at the 2001 (Edinburgh) meeting of the Cognitive Science Society.

More generally, connectionists moved towards techniques inspired by neuroscience (see Chapter 14). Some new learning rules, for instance, took account of spiking frequencies and synchronies, while others modelled the effects of widely diffusing chemicals in the brain. One, based on the functioning of the cerebellum, was "two to four orders of magnitude faster" than backprop (McKenna 1994: 79, 82).

The influences went both ways. A persuasive computational argument for loc-alism—recently swamped by the D in PDP (*pace* Hinton's localist–PDP mix: see above)—implied that neuroscientists should *expect* to find grandmother cells as well as cell assemblies. The localist resurrection (and grandmother's revenge), however, had to await the new millennium (Chapter 14.x.e). In the last twenty years of the old one, unalloyed PDP was the most high-profile form of connectionism.

12.x. Philosophers Connect

The high visibility of PDP, described in Sections 8.vii–viii, was largely due to its *philosophical* interest. At an informal level, this was true of its appeal to journalists and the general public. But the professional philosophers were involved too. PDP was widely perceived by them as a more human alternative to classical functionalism (Chapter 16.iii–v), and also as a counter to Chomsky's nativist philosophy of mind (9.vii). Most of the interest came from analytic philosophers already sympathetic to cognitive science—although, as remarked in Chapter 1.iii.d, some postmodernists saw connectionism as buttressing their own, very different, approach (Globus 1992; Canfield 1993; Wilson 1998). But the GOFAI classicists fought back, as we'll see.

The widespread debate involved not only card-carrying philosophers, but also leading model-builders on both 'sides' of AI.

a. A Pulitzer prelude

Many philosophers first became aware of (broadly) connectionist ideas in 1979–80, as a result of Hofstadter's book *Gödel, Escher, Bach: An Eternal Golden Braid* (1979). This extraordinary intellectual and rhetorical feast drew many people to cognitive science, including what would later be called A-Life, for the first time. In addition, it inspired some already within the field to think about computation in a fundamentally new way.

One example of the latter was Mitchel Resnick, whose work on StarLogo was the result (Resnick 1994, p. xvii, and Chapter 12.iii.d). Another was Randall Beer (personal communication), then a graduate student writing GOFAI programs. His vision of how one might approach AI problems was irreversibly broadened by reading the book (see 15.vii.c and viii.c).

Begun in the early 1970s, the manuscript was essentially complete when Hofstadter joined Indiana University in 1977, in his first 'real' job. (He remained at Indiana thereafter, apart from interludes at MIT in 1983–4 and at the University of Michigan from 1984 to 1988.)

Being utterly impossible to pigeonhole, and very long to boot, it didn't easily find a publisher. This was so even though Hofstadter himself would do the highly complex, and potentially expensive, 'typesetting'. Fortunately, Martin Kessler of Basic Books was prepared to fight for it, provided he could find a non-US partner to share the risk. As the co-founder of the British publishers (Harvester Press), I had to fight for it too. On being told (by the other co-founder) that it was "totally unintelligible", I said it would either "fall deadborn from the press" like Hume's *Treatise* or become a cult book—but that we shouldn't be publishing at all if we weren't willing to accept such a hugely original, and insightful, manuscript.

A cult book, indeed, it turned out to be. And the cult was worldwide: *GEB* was translated many times. It soon won a Pulitzer Prize, and is still much admired. In 1999 the *New Scientist* magazine asked the mathematician John Casti to invite several people to choose a science book from the last quarter-century to take to a desert island; three, including Casti and myself, chose this one. (I had to make a second choice, as a result. I picked Andrew Hodges' biography of Turing: a rich book for British social history, not just for science/mathematics.)

Hofstadter explored, and insightfully interrelated, many superficially diverse topics—enough for many years on a desert island. These included the art of fugue, visual metamorphoses, translation, analogy, creativity, paradoxes, Gödel's theorem, GOFAI, *Alice's Adventures in Wonderland*, ants and bees, life, DNA . . . and connectionism.

Connectionism was presented, here, less as an AI modelling technique than as a guiding idea expressed through metaphor and analogy. Ant colonies were as relevant as brains, life as mind, memory as mathematics. The fundamental notion concerned association and interaction within a highly fluid parallel-processing system, an active memory from which higher-level properties of adaptation and intelligence emerge.

Surprising as it may seem, Hofstadter hadn't been turned in this direction by the connectionist work discussed earlier in this chapter, but by the pioneering HEARSAY program:

Perhaps the deepest influence on me personally . . . was the Hearsay II speech-understanding system, which was developed in the mid-1970's by a team headed up by Raj Reddy, and including Victor Lesser and Lee Erman, among others. Hearsay II was a highly parallel system (it actually ran on truly parallel hardware), and it was the original implementation of the famous "blackboard architecture". I still think, personally, that the ideas in the Hearsay project are every bit as deep and as important as those in the PDP volumes. (personal communication)

As this recollection shows, HEARSAY II wasn't a typical GOFAI program. Nor was the first version of HEARSAY, in so far as it was parallelist in spirit (Newell *et al.* 1973; Reddy *et al.* 1973; Reddy and Newell 1974). Nevertheless, the original research was done within the GOFAI stable, and headed by one of GOFAI's high priests: namely, Newell. This, then, is yet another reminder that the two types of AI aren't so wholly opposed as they're often assumed to be.

Soon afterwards, besides co-editing a volume on "self and soul" with a leading philosopher of mind (Hofstadter and Dennett 1981), Hofstadter wrote several pieces on analogy and creativity in the *Scientific American* (many of these are reprinted in Hofstadter 1985*a*). He'd been invited in 1981 to alternate with Martin Gardner in a regular column, and ran it alone for a while on Gardner's retirement in 1982. And in 1983 he wrote a paper for an interdisciplinary volume which, when it was reprinted—with a new Post Scriptum—alongside some of his *Scientific American* pieces two years later, was also widely influential. The message of 'Waking Up from the Boolean Dream' was this:

[Until] AI has been stood on its head and is 100 percent bottom-up, it won't achieve the same type of intelligence as humans have. To be sure, when that type of architecture exists, there will still be high-level, global, cognitive events—but they will be epiphenomenal, like those in a brain. They will not in themselves be computational. Rather, they will be constituted out of, and driven by, many smaller computational events, rather than the reverse. In other words, *subcognition at the bottom will drive cognition at the top*. And, perhaps most importantly, the activities that take place at that cognitive top level will neither have been written nor anticipated by any programmer. This is the essence of what I call *statistically emergent mentality*. (Hofstadter 1983/1985: 285/654; italics in 1985 original)

However, although this paper was more explicit in comparing traditional and connectionist AI than *GEB* had been, it was still rhetorical rather than technical—deliberately so, for Hofstadter was addressing a mixed, non-specialist, audience. *Precisely how* a parallel-processing system could implement concepts—and model their role in perception, memory, analogy, and thought—wasn't explained.

In the Post Scriptum of 1985, Hofstadter mentioned the PDP group, whom he'd recently visited in San Diego. He hailed their work with "delight", as "a hotbed of subversive PDP activity" (Hofstadter 1983/1985: 654). But he gave no details. He also dropped hints, in the *Scientific American* as in the Post Scriptum, about his own computer model of concepts and creative analogy (13.iv.c)—but again, no details.

He'd given seminars on his model at MIT in 1984, and a detailed lab report was made available to the cognoscenti a few years later (Hofstadter *et al.* 1987). But official publication was long delayed (Hofstadter and Mitchell 1993/1995). Indeed, actual implementation—as opposed to mouth-watering promises—had to wait until the end of the century (McGraw 1995; Rehling 2001).

In the event, then, this intriguing but difficult work made much less of a splash than *GEB*. It didn't even enthuse the professional community, in part because it didn't exemplify the now all-conquering PDP approach. Hofstadter's model of analogy was parallel and distributed, but only partially non-symbolic. So it was connectionist only in the broad sense.

Indeed, he never considered himself part of the "connectionist" movement, and avoided their conferences (personal communication). His deepest influences from cognitive science were the HEARSAY "blackboard" model and Rumelhart and Norman's work on errors (see Chapter 10.v.e and Section ix.b above, respectively).

In 1986, of course, the *Precisely how?* question was given an initial answer in the PDP manifesto (see Section vi.a). By that time, *Gödel, Escher, Bach* and 'Waking Up from the Boolean Dream' were already part of the general intellectual background. Without such hugely popular heralds, the two technical volumes from San Diego, reader-friendly though they were, might not have found such a wide and immediate welcome.

b. Connectionist concepts

The main philosophical interest of Hofstadter's approach had been his insistence that *concepts* are active, interactive, fuzzily defined (holistic), and constantly changing. This was clearly opposed to the GOFAI picture, in which "a passive memory [consists of] data-structures [that] simply wait around to be inspected or manipulated" (Hinton and Anderson 1981: 11).

Ludwig Wittgenstein had argued for a broadly similar view, rejecting the logicism of his own—and McCulloch's—younger days (Chapters 9.x.d and 4.iii.c). For him, and for his many followers, concepts marked overlapping "family resemblances" rather than lists of necessary and sufficient conditions.

In addition, philosophers in touch with cognitive psychology knew of the experimental evidence that concepts function as "prototypes" with an ill-defined penumbra of similar cases (see 8.i.b). (Few realized, however, that James Anderson, for instance, had been relating this evidence to connectionism since the mid-1970s.) So Hofstadter's ideas had fallen on well-prepared philosophical ground.

The PDP volumes endorsed such views. In general, their authors made a point not only of describing their models clearly (ball-bearings in valleys) but also of indicating their philosophical implications. In particular, they repeatedly argued that concepts aren't cut-and-dried, but inherently fuzzy. This rhetorical strategy helps explain why their books were so crucial to the culturally prominent connectionist renaissance described in Sections vi–vii. But the philosophy would have been less persuasive without the technical details.

These showed, at last, how various puzzling phenomena could happen:

* how a single concept could be distributed over an entire network;
* how several different concepts could be stored simultaneously;
* how multiple constraint satisfaction could tolerate unclarities, and even contra-dictions;
* how a single unit could contribute to the representation of many concepts, and
* how it could 'mean' different things in different contexts;
* how 'one and the same' concept used in two different contexts could involve different sets of units;
* how two different inputs could be classified under the same concept;
* how semantically similar concepts could be represented by similar activity patterns;
* and how a fragmentary memory could excite a whole concept.

PDP systems, then, seemed to do 'naturally' many things which brains can do—and which might underlie the conceptual abilities long highlighted by philosophers unsympathetic to GOFAI.

Strictly, the phrase "at last" (above) is unwarranted, for others had shown these things before, as we saw in Sections ii and iv–v. But very few philosophers knew that. In England, some of those at Sussex did, largely because of Longuet-Higgins's and Hinton's presence there (I myself, for example, had been writing about connectionism since the mid-1970s). But, on both sides of the Atlantic, philosophers who weren't on the AI grapevine didn't.

This applied even to Paul Churchland. He had argued since the early 1970s that the philosophy of mind, and of science, should be based on brain dynamics (16.iv.e). But while still at the University of Manitoba, and even for some time at Stanford (where he went in the early 1980s), he was unaware of connectionist AI:

At the time [when I wrote my 1979 book] I would have guessed that a new paradigm [based on the brain] was at least twenty-five years away, and probably more like fifty.

In this I was wrong, for in fact it already existed and had existed, at least in stick-figure form, since the late fifties. By 1959 F. Rosenblatt had developed the Perceptron paradigm of vector-to-vector transformations in a parallel network of neuronlike processing units . . . Unfortunately, that paradigm did not catch on, and for two decades it was almost forgotten . . . [and] faded to invisibility. (P. M. Churchland 1989, p. xiii)

Churchland first "stumbled across" ideas about massive parallelism and vector transformations in 1983, when reading about the cerebellum (see Chapter 14.viii.b–c). He used these neuroscientific ideas to buttress his long-standing philosophical views, in essays written early in 1984 and published two years later—one of them in *Mind*

(P. M. Churchland 1986*a,b*). Thanks to *Mind*'s high profile, that item was widely read by philosophers. But whether it convinced many of them is another question. Its account of the processes underlying conceptual thought (and sensori-motor integration) was highly speculative, and there was no mention of existence proofs from computer modelling. It was the near-simultaneous publication of the PDP bible which *proved* to Churchland's *Mind* readers that vector-transforming parallel systems really could do interesting things. In short, the PDP volumes were a revelation to most philosophers, even including those already committed to a science-oriented approach.

Functionalism seemed to have widened its grasp, to include mental—and computational—phenomena very different from the formalist Turing tables highlighted in its initial definition by Hilary Putnam (16.iii.b). The PDP group encouraged such views, saying that:

[PDP models] in no way can be interpreted as growing from our metaphor of the modern computer . . . [and involve] a new form of computation, one clearly based upon principles that have heretofore not had any counterpart in computers. (Norman 1986*b*: 534)

Anti-logicist doctrines that had previously seemed vague now appeared to have clear empirical support, vindicating those Wittgensteinians—notably Dreyfus—who had argued for years that GOFAI misdescribed concepts (see 11.ii.a).

Indeed, Dreyfus (and his brother) soon hailed connectionism as a philosophical advance, because it allowed sub-conceptual units and didn't presuppose that there must be an explicit theory of every domain (H. L. Dreyfus and Dreyfus 1988). Their paper was one of a set of 'reviews' written for the *Artificial Intelligence* journal, but reprinted under the tendentious title *The Artificial Intelligence Debate: False Starts, Real Foundations* (Graubard 1988). This new-found sympathy for connectionism belied Dreyfus's 1979 opinion that it had produced "no interesting results" (cf. Section iv.i).

From the late 1980s onwards, then, many people saw PDP as offering a scientifically respectable philosophy of mind that made room for mental subtleties previously ignored. However, there were dissenters.

Dreyfus himself rejected connectionism, despite seeing it as less implausible than GOFAI, saying that "building an interactive net sufficiently similar to the one our brain has evolved may be just too hard" (H. L. Dreyfus and Dreyfus 1988).

Indeed, neo-Kantians in general held that a "scientifically respectable philosophy of mind" is a chimera, since there can be no naturalistic account of meaning or normative rationality (Chapter 16.vi–viii). They saw even neuroscience as philosophically irrelevant, never mind GOFAI or PDP. And John Searle, who disagreed with them on the latter point, was no more convinced by connectionist functionalism than he had been by the GOFAI variety (see 16.v.c).

Even within the functionalist camp, PDP didn't sweep the board entirely. For the leading GOFAI computationalists resisted this newly fashionable view. The central point of contention was the relation between connectionist and symbolic processing.

c. The proper treatment of connectionism?

According to Paul Smolensky (1955–), a member of the PDP group, "the proper treatment of connectionism" was to regard it as concerned with *subsymbolic*, or *subconceptual*, processing. That is: "the units do not have the same semantics as words of natural language" (Smolensky 1988: 6), and do not correspond to the categories we use "to consciously conceptualize the task domain" (p. 5). This conscious conceptualization includes not only everyday examples, but also the verbal protocols observed in problem-solving experiments (see Chapters 6.iii.b–c and 7.iv.b).

Typically, a PDP unit represents some tiny detail or microfeature, whose individual significance may be very difficult to express in non-technical terms and quite impossible to access introspectively. In a system-modelling stereopsis, for example, each unit codes a comparison between the light falling on corresponding points on the two retinae (see Chapter 7.v.d and 14.iv.f). To be sure, the system as a whole computes the *depth* of the object being looked at. But no single unit does so. Similarly, NETtalk units represent not whole words, but Wickelgren features—and very few people have heard of those.

Even when some units do code for familiar features, as in Hinton's (1988/1990) model of Jets and Sharks (coded as married, single, divorced . . .), they don't do so in a "symbolic" way. For—as Smolensky pointed out—a PDP unit's meaning may vary according to context (i.e. the simultaneous activity of other units), whereas the meaning of GOFAI symbols is fixed.

In sum, when people introspect they aren't aware of the computations actually involved: "serial, symbolic descriptions of cognitive processing are *approximate* descriptions of the higher level properties of connectionist computation" (Smolensky 1987a: 103; italics added). So folk psychology, though not wholly illusory, isn't the true picture of thought. For Smolensky, as for Hofstadter, concepts and cognition properly so called exist only as emergent properties of processes occurring at the subsymbolic level. For certain purposes, that processing can be ignored and the symbolic level taken for granted. But even symbolic thinking is, at base, a PDP phenomenon—and this must be remembered, if human thinking is to be understood.

Two philosophers who agreed with this wholeheartedly were Paul and Patricia Churchland (1942– and 1943–). They had discovered PDP on moving to San Diego in late 1984—by which time the neuroscientifically oriented *Mind* paper was already "in the system". Unsurprisingly, given their earlier philosophical views, they welcomed it with alacrity.

Those earlier views had included a concern with neuroscience. Indeed, Patricia Churchland was already moving towards philosophically informed neuroscience, as opposed to neuroscientifically informed philosophy (P. S. Churchland 1986). She soon collaborated with the neurophysiologist Christof Koch and with Sejnowski, of NETtalk fame, in defining "computational neuroscience" as a distinct research area (Sejnowski *et al.* 1988: see Chapter 14, preamble). Later, she and Sejnowski would plan neuroscientific experiments together, and co-author a textbook (P. S. Churchland and Sejnowski 1992). Paul Churchland, however—while keeping abreast with neuroscience—stayed closer to 'pure' philosophy.

As we'll see in Chapter 16.iv.e, he was already well known as a proponent of eliminative materialism. This doctrine holds that the mental states named by everyday psychological language have no scientific basis, and do not really exist: beliefs, desires, hopes, regrets . . . all these are as illusory as the medieval humours, or witches (P. M. Churchland 1979, 1981). On encountering PDP, he immediately appropriated it to reinforce his long-standing eliminativism, and to develop his early ideas about the philosophy of science (1989). Later, he would apply PDP modelling and neuroscience to moral and political philosophy too (1995).

Much as Hofstadter had spoken of perception, explanation, and analogy in terms of fluid and inter-associated concepts, so Paul Churchland gave a "non-sentential" account of science—and of thought in general. He'd already argued that perception is fundamentally imbued by scientific theories, and that cognitive representations aren't language-like (Chapter 16.iv.e). Now, he said that the classical view of science—theories-as-sentences, related by deductive rules of explanation—should be replaced by a PDP-based account in which explanation boiled down to analogy.

He saw this as a truer picture of scientific knowledge and explanation, and also as a way of avoiding certain notorious difficulties that attend the classical view. Citing PDP models such as NETtalk, he argued that concepts could be represented as prototype-plus-penumbra within a (tacitly) hierarchical activation space. And he related this account to various aspects of scientific observation, learning, skilled experimentation, and reasoning—including Kuhnian paradigm shifts, and the stubborn opposition that typically greets them (1989: 191 ff.).

Anticipating the objection that this was psychology, not philosophy, Churchland said that philosophers must respect the facts about how scientists think—not least, because only thus would certain seeming difficulties (about the role of simplicity in science, for instance) be resolved. Against Popper and the neo-Kantians alike, he argued that *even if* rationality and epistemology are irreducibly normative notions, facts about how concepts are implemented in the brain can be philosophically relevant, since normative claims have empirical presuppositions which may in fact be false (1989: 195–6).

Churchland insisted that when the system is learning some generalization, it is "*theorizing* at the level of the hidden units, exploring the space of possible activation vectors, in hopes of finding some partition or set of partitions on it that the output layer can then exploit . . ." (1989: 179). He even said:

[It] is clear that no cognitive activity whatever takes place in the absence of vectors being processed by some specific configuration of weights. That is, no cognitive activity whatever takes place in the absence of some theory or other.

This perspective bids us see even the simplest of animals and the youngest of infants as possessing theories . . . [Their] theories are just a good deal simpler than ours, in the case of animals. And their theories are much less coherent, less organized, and less informed than ours, in the case of human infants. Which is to say, they have yet to achieve points in overall weight space that partition their activation-vector spaces into useful and well-structured subdivisions. But insofar as there is cognitive activity at all, it exploits whatever theory the creature embodies, however useless or incoherent it might be. (pp. 188–9)

Many philosophers, not least the neo-Kantians (Chapter 16.vi–viii), would predictably disagree. But computational philosophers had been accustomed to the assimilation of perception to scientific theorizing ever since the "New Look" pioneered by Bruner and Gregory (see Chapter 6.ii). So one might have expected them to be sympathetic.

Some were. Paul Thagard (1988, 1989, 1990), for example, was already developing a philosophy of science similar to Churchland's, and implementing various computer models accordingly (Holland *et al.* 1986; Thagard *et al.* 1988; Holyoak and Thagard 1989). Indeed, he used these empirical studies to challenge the Fregean orthodoxy

(2.ix.b) about the relation between logic and psychology, and spoke of "revising logical principles in the light of empirical psychological findings" (Thagard 1982).

Others, however, were not.

d. The old ways defended

Fodor in particular was not. Although strongly influenced by Bruner in the 1950s, by the early 1980s he was arguing for "mental modularity" (Chapters 7.iii.d and vi.d, and 16.iv.d). This led him to reject Churchland's account of perception as theorizing (Fodor 1984).

Perception (Fodor argued), being modular, is *not* theory-laden through and through. Admittedly, evolution may be said to have built certain assumptions, even "theories", about the world into our perceptual apparatus. But these aren't changeable: our fundamental perceptual processing isn't cognitively penetrable by the theories of science. So the sorts of perceptual shift envisaged by Churchland even in the 1970s, and by Thomas Kuhn before that, are impossible. (For Churchland's reply, see his 1989.)

Fodor had other reasons, too, for rejecting connectionism. In a widely read critique co-authored with the psychologist Zenon Pylyshyn, he described PDP as old wine in new bottles—and fatally adulterated, at that (Fodor and Pylyshyn 1988; cf. Pylyshyn 1984).

The old wine, here, was associationism—which had been laid down in philosophy's cellar three centuries ago (see Chapter 2.x.a). Expressing a "gnawing sense of *déjà vu*", he claimed that Hebb and Hull had been "conclusively" refuted twenty or thirty years before, and that PDP had merely added frills to their ideas. Indeed, this seemingly modern high-tech approach threatened to lead us back over 200 years, to "a psychology not readily distinguishable from the worst of Hume and Berkeley" (Fodor and Pylyshyn 1988: 49, 64).

More specifically, Fodor (contra Churchland) accepted the psychological reality of belief, desire, and other folk-psychological categories—which he glossed, significantly, as "*propositional* attitudes" (Chapter 16.iv.c–d). As such, they involve sentence-like mental representations, composed of concepts and processed by unconscious formal rules. These rules aren't mere emergent patterns of behaviour, but have psychological reality in their own right. Similarly, concepts aren't inherently fuzzy patterns of activation or prototypes-plus-penumbra, but semantically interpretable formal symbols, whose meaning—unlike that of Smolensky's "subsymbolic" units—is fixed irrespective of context (see also Fodor 1998a: 22).

These disagreements were closely linked to another. Fodor (and Pylyshyn, who regarded even visual imagery as "propositional": Chapter 7.v.a) insisted that PDP, in principle, couldn't model sequence and hierarchy in general, or language and conceptual thought in particular. A holistic pattern of activation, they argued, doesn't have the sort of internal structure that's required by the compositionality, productivity, and systematicity of language.

Specifically, they said, the holism of PDP can't allow that the meaning of a sentence depends recursively on the meaning of its component parts, enabling the production of indefinitely many new sentences. Nor can it admit that a concept has the same meaning irrespective of context. And nor can it satisfy the generality constraint (G. Evans 1982), that one and the same concept can be used in indefinitely many systematically related thoughts. For instance, someone able to think *John loves the girl* must also be able to think *The girl loves John*; and someone who thinks that roses are red and violets blue must also be able to think that roses are blue and violets red. A PDP network, said Fodor and Pylyshyn, might represent any of these thoughts without being able to represent its permutational partner, because it treats them as holistic patterns—not as *thoughts*, at all.

Lastly, Fodor—like many others influenced by Chomsky—was committed to nativism (see Chapters 7.vi.d–e and 16.iv.c–d). And nativism had been challenged by PDP. The challenge based on representational trajectories was still in the future (see Section viii.d, and subsection e below). But the past-tense learner had already thrown the connectionist cat among the nativist pigeons. This network was taken by PDP enthusiasts to prove not only that explicit rules aren't needed for using grammar, but also that an inborn Language Acquisition Device isn't needed for learning it. *Pace* Chomsky, mere statistical analysis of the input seemed to do the trick.

The most influential Chomskyan reply, which Fodor in general endorsed, came from linguists Steven Pinker and Alan Prince (1988). They crawled over the past-tense learner with a fine toothcomb, devoting 120 pages to their critique. The connectionists had claimed that the network made the same errors as children do (errors which Chomskyans regarded as strong evidence for nativism). But Pinker and Prince showed that the network's errors were *not* precisely the same as children's, and they argued that some of the errors made only by children can be explained by GOFAI but not by PDP.

It turned out later that the psychological data on which both they and Rumelhart and McClelland relied were faulty. For example, psycholinguists had reported, or anyway implied, that—for a while—children always over-regularize all irregular verbs. But they don't: they over-regularize only 5–10 per cent of irregulars, and correct uses co-occur with the incorrect ones. Moreover, psychologists hadn't reported any *irregularizations* of *regular* verbs—which, indeed, a Chomskyan would never expect. However, they do happen. Both these unexpected phenomena, and many others, fell out 'for free' from the network statistics explored in the early 1990s by Plunkett and Marchman (see Section viii.e).

Some of Pinker and Prince's criticisms of the past-tense learner could be applied to PDP in general, at least as it existed around 1986. For instance, the training set had contained no sentences (made up of nouns, adjectives, and variously tensed verbs), but only associative pairs: the *stem/past tense* of 420 verbs. Moreover, these pairs were input in two stages, in the first of which only ten (including eight irregulars) were used. In effect, then, the network had been given grammatical structure for free—or anyway, pointed so firmly at it that the learning required was trivial (compare Thornton's MTCLA: Section viii.d, above). Real parents, by contrast, don't tiptoe around their children mouthing only ten verbs.

But if the past-tense learner was in this sense too weak, in another it was too strong. Pinker and Prince complained that a connectionist learning rule could in principle learn *any* linguistic regularity, whereas children *do not* (because natural languages share certain structures, and lack others) and *cannot* (because some pre-existing bias is needed to enable interesting structure to be picked out). Even non-Chomskyans had to admit that if people find it difficult to learn a grammatical structure (mirror-image reversal of word strings, for example), then a PDP simulation should do so too. What's more, said Pinker and Prince, all the microfeatures were treated equally: the statistical learning picked out the most frequent patterns, not the most important ones. Top-down influences could counter this—but these go against the spirit of connectionism.

The question arose, then, as to what connectionism *could* offer to cognitive science. Fodor and Pylyshyn's answer, in effect, was "Not much". They saw it as a theory not of cognition as such, but of its implementation. All very well, perhaps, for neuroscience. But, given the functionalist assumption of multiple realizability (Chapter 16.iii), facts about implementation have no relevance for philosophy, or even for cognitive psychology properly so called. Newell and Simon agreed. Connectionism deals with the microstructure of (the implementation of) thought, but nothing below 100 milliseconds of brain activity, on their view, is significant in the study of cognition (Newell 1980, 1990).

In short, these computationalists still insisted that language and thought must be explained in formalist terms. The philosophy of mind needs the language of thought (alias LOT: Chapter 16.iv.c) and/or Physical Symbol Systems (16.ix.b), not connectionism. As Fodor put it, both then and later, a GOFAI-based psychology is "the only [theory of cognition] we've got that's worth the bother of a serious discussion" (2000*b*: 1).

Worth the bother or not, serious discussions of PDP soon abounded. The disputes outlined above triggered a huge, and still growing, philosophical literature. Paul Churchland's work had been provocative from the start, and it continued to raise philosophical hackles. As for Fodor and Pylyshyn's critique, Smolensky published a counter-blast even before it was officially published—to which Fodor soon replied (Smolensky 1987*b*; Fodor and McLaughlin 1990). Other professional philosophers soon joined in.

e. Microcognition and representational change

The most important newcomer to the debate was Clark. Besides bringing a taste of PDP connectionism to a general philosophical audience in an Aristotelian Society paper (1990), he provided countless details in a highly influential book, *Microcognition* (1989). This described PDP technology in plain English, and related it to many issues in the philosophy of mind and language—including the disagreements mentioned above.

For instance, Clark countered Pinker and Prince by pointing out (among other things) that "importance" could be coded by assigning higher weights to some microfeatures than to others, and that Karmiloff-Smith had already outlined how GOFAI-like processing might develop from a PDP base (see Section viii.d).

He also rejected Churchland's eliminativism. Individual thoughts (beliefs, hopes, intentions . . .), he said, could conceivably map onto identifiable computational states in the brain, even though these states aren't sentence-like. And even if they don't, it doesn't follow that they aren't "real". The reason is that the categories of folk psychology are holistic interpretations of behaviour, logically independent of the underlying causal mechanisms (cf. 16.iv.a–b). Ironically, Clark's anti-eliminativism was directed also against the arch-realist Fodor:

Fodor's approach is dangerous. If one accepts the bogus challenge to produce syntactic brain analogues to linguistic ascriptions of belief contents, he opens the Pandora's box of eliminative materialism. For if such analogues are not found, he must conclude that there are no beliefs and desires. The mere possibility of such a conclusion is surely an effective *reductio ad absurdum* of any theory that gives it house space. (A. J. Clark 1989: 160)

As for Fodor and Pylyshyn's arguments about systematicity, Clark wasn't convinced by these either—though he admitted that productivity was more problematic than sequential order (1989, ch. 8; 1991*b*). His main point was that compositionality needn't be built into the basic architecture, but could emerge from it.

The core idea, here, was mental modelling (see Chapters 4.vi, 7.iv.d, and 14.vii–viii), together with the PDP group's suggestion that our reasoning capacity may result from "our ability to create artefacts—that is, our ability to create physical representations that we can manipulate in simple ways to get answers to very difficult and abstract problems" (Rumelhart *et al.* 1986*c*: 44; cf. A. J. Clark 1989: 223). Given that words are external (publicly accessible) things, which can be modelled in the mind/brain just as physical objects can, Clark saw the compositionality of thought as derived from that of language. That is, the brain somehow creates a *virtual* machine with formalist properties—and conceptual thought is a property of this machine. It follows, said Clark, that GOFAI-explanations of thought and language aren't mere *approximations* to finely detailed subsymbolic accounts, as Smolensky had claimed, but are *true* (or in some cases *false*) descriptions of the relevant virtual machine. (The implication was that only humans have language-like internal states; for Fodor, by contrast, even animals—if they have representations at all—possess some innate language of thought: see 16.iv.c.)

As we saw in Section ix.b, Hinton had recently suggested a model that tallied with Clark's approach (Hinton 1988/1990; cf. A. J. Clark 1991*b*: 215). Each concept, Hinton had said, might have *two* representations: a localist (single-feature) "reduced description", and a distributed "expanded description" involving many microfeatures. The first might be manipulable in a GOFAI-like way, even though the second isn't. But the reduced description isn't an empty syntactic token, as a GOFAI symbol is (cf. Searle 1980), because it is linked to the expanded one—whose microfeatures can be accessed if the concept's *meaning* needs to play a role. As Clark put it: "Classical representations, minus such a sub-structure, threaten to be brittle and contentless shells" (1991*b*: 217). In sum, "even when a classical virtual machine is somehow implicated in our processing, its operation may be deeply and inextricably interwoven with the operation of various connectionist machines" (1991*b*: 215).

Clearly, then, Clark—despite being the leading philosopher of connectionism—was never committed to *pure* PDP, the view that "Every cognitive achievement is psychologically explicable by a model that can be described using only the apparatus of PDP" (1989: 128). This position, he said, was just as implausible as its GOFAI-equivalent. Even in his first book, he argued for mental "multiplicity":

I am advocating that cognitive science is an investigation of a mind composed of many interrelating virtual machines with correct psychological models at each level and further accounts required for the relations between such levels. Only recognition of this multiplicity of mind, I suspect, will save cognitive science from a costly holy war between the proponents of PDP and the advocates of more conventional approaches. (A. J. Clark 1989: 141)

Whether any of these virtual machines is as closely GOFAI-like as Fodor claimed was another matter. But this, for Clark (as for Hinton), was an empirical question. As for the answer, he was agnostic: he closed his second book by saying "No one knows" (1993: 227).

If Clark's second book still didn't answer the question about the relevance/irrelevance of GOFAI, it did offer some important advances. Earlier, he'd described symbolic reasoning as "merely ingenious icing on the computational cake" (1989: 135). Now, he asked how that icing could have been concocted in the first place. In other words, his interests had shifted from established (static) conceptual abilities to representational change. He now tried to show *how* "mental multiplicity" could develop from a PDP base. And his answer described the scaffolding of learning by various recodings of existing representations (1993, esp. chs. 7–8)—see Section viii.d.

As we've seen, he'd recently collaborated on this issue with a leading *developmental* psychologist, Karmiloff-Smith. She was one of a group of developmentalists who were reinterpreting the idea of innateness (Elman *et al.* 1996). Specifically, they'd revived Piaget's biological concept of *epigenesis*: a series of self-organizing interactions between pre-existing mechanisms and the environment (both in and outside the organism), by which the embryo/infant bootstraps itself into maturity (Chapters 7.vi.g–i, 14.ix.c, and 16.vii–viii). Unlike Piaget, this group had the advantage of PDP techniques and developmental neuroscience. Like Clark, then, they asked how a connectionist system—whether natural or artificial—could undergo representational changes in epigenetic development.

In other words, "Is it innate?" was no longer being treated as a *Yes/No* question (cf. 7.vi.g–i and 9.vii.c–d). The new "Minimal Nativism", as Clark termed it (1993, ch. 9), drew the sting of the nativism debate, for it stressed both inbuilt and bootstrapped processing biases ("knowledge"), as well as environmental interactions.

By the 1990s, there was a wide range of empirical research (not all done from a computational perspective) suggesting what these biases and interactions might be. Some philosophers interested in mental development called on cognitive neuroscience (see Chapter 14.xi.b). Others stressed the mother–child interactions that naturally scaffold the development of pointing, turn-taking, and language, and/or culture-specific scaffolding in the form of language and artefacts—what Bruner (6.iii.c) had called cognitive technology (E. L. Hutchins 1995—cf. 8.iii; Hendriks-Jansen 1996; A. J. Clark 2001, ch. 8). By the end of the century, then, research on (lifelong) conceptual *development* was influencing the philosophy of mind and cognitive science.

f. Non-conceptual content

Interest in development was evident also in philosophical work on the relation between concepts and non-conceptual content. One might say that these correspond to Hinton's reduced and expanded descriptions respectively, but that would be misleading: the distinction arose within 'pure' philosophy, and—as we'll see—was linked to connectionism later.

Non-conceptual content is meaning (intentional content) carried by sensory-perceptual experience and/or action that's untouched by language. One familiar example is the content *bug* that's supposedly available to the frog in virtue of the 'bug-detector' cells in its retina (and the adaptive links to the muscles of its limbs and tongue): Chapter 14.iii.a.

The word "supposedly" is needed here because the notion of non-conceptual content is controversial. René Descartes had seen no need for an equivalent (and is widely reviled for callousness as a result: Chapter 2.ii.e). Even today, some philosophers argue that there can be no such thing: in their view, the very idea of non-conceptual content, or purely sensory experience, is incoherent (Chapter 16.viii.b). Many others disagree. Indeed, the suggestion that there is some middle ground between adult human experience and the mental vacuum of insensate beings has a strong intuitive appeal—not just to pet-owners and zoo-keepers, but to people wanting to understand the *human* mind.

The Oxford philosopher Gareth Evans (1946–80), for instance, argued that non-conceptual content is the base from which true concepts develop (G. Evans 1982). He recognized that it's difficult to pin it down. Does the frog really employ the content *bug*—or, rather, *bug-over-there* or *edible-object-to-upper-right*, or *black spot there*, or ... ? The answer, he said, lies in the frog's sensori-motor skills. For to have access to non-conceptual content of a certain sort just is to have bodily skills of a certain kind. The skills aren't mere evidence for (symptoms of) the existence of the content, but are criterial of it. As for concepts, Evans defined these in terms of the generality constraint (see above). The frog lacks the concept of "bug", for it can't respond to bugs in many different ways, nor (use language to) think of them in many different contexts. In other words, it can't conceive of bugs *objectively*—which is to say that it can't conceive of them at all.

But the same applies to babies: as Piaget had pointed out long before, being able to shake and suck a rattle, or to pick it up and drop it, isn't the same as being able to think of rattles in indefinitely many ways. Evans, then, was less concerned to give dumb animals their due than to explain the origin of concepts in humans, whether in evolution or infancy.

Largely inspired by Evans's intellectual legacy (he died in 1980, aged only 34, and his book was published posthumously), Adrian Cussins (1960–) drew on connectionist ideas in trying to show how this conceptual development could take place (Cussins 1990). While still a Research Fellow (at Oxford and Stanford) in the late 1980s, he produced the C3-theory of cognition, so called because it dealt with the 'Connectionist Construction of Concepts'. Almost all of the C3 paper dealt with *just what counts* as content, and *just what counts* as a concept. Having discussed these two questions for sixty pages, he closed with a brief section arguing that connectionism might show how, in practice, concepts emerge from a non-conceptual base.

Theories based on GOFAI (such as Fodor's), he said, don't explain the origin of conceptual content but instead take it for granted. In this, he was echoing Searle, for whom GOFAI was "all syntax and no semantics" (Chapter 16.v.c)—and, for that matter, Fodor himself (in his discussion of methodological solipsism: 16.iv.d). Both Searle and Fodor had argued that GOFAI can't explain how concepts in someone's mind can refer to objective facts or events. Cussins suggested that his theory can do so, for it shows how there can be "organisms *in* the world which are capable of thinking *about* the world" (Cussins 1990: 368).

Following Evans, he took non-conceptual content to be the representation of the world that is implied by the possession of language-free bodily skills. These include visual navigation, knowing the position of one's limbs, and catching a bug with one's tongue. Our friend the frog has such skills—but no concepts. We need to explain both how non-conceptual content can arise and how concepts can progressively be constructed from it. Concepts and objectivity (the mind–world distinction) are essentially connected, and to explain objectivity we must show how conceptual representations can have semantic properties such as reference, truth, and falsity. As Cussins put it:

We need to understand how there can be a principled, if not sharp, distinction between creatures like paramoecia and creatures like us, between infantile and adult cognition, perhaps also between normal and demented cognition. With a fair grasp of the principle of this distinction, we can *then* address ... how best to model the computational processes that can transform a creature from lying on one side of the distinction to lying on the other side. (Cussins 1990: 414; italics added)

His answer (again, following Evans) was that "contents are non-conceptual in virtue of being perspective-dependent" (p. 425), and that concepts can be constructed from a non-conceptual base by learning increasingly perspective-independent abilities. In humans, this process culminates in the "emergence" of objectivity and the mind–world distinction. At that point, the person satisfies not only the generality constraint, but also objectivity conditions such as referring to specific objects in the world, assessing truth-values, making valid inferences, and—in general—making judgements and perceptual discriminations that are independent of their own perspective.

So far, so philosophical: this was a theory of C2 (the Construction of Concepts), not C3. Connectionism now entered the picture, albeit briefly. Cussins commended Smolensky's account of the "proper treatment" of connectionism, and rebutted Fodor and Pylyshyn's claim that connectionists can't explain systematicity. And he suggested that this methodology might be able to show how concepts arise in practice (pp. 429–37).

However, he didn't discuss any specific models illustrating the movement towards objectivity that he had in mind. He might have cited Marr's theory of vision (it was mentioned in a footnote, but in a different context). For this had indicated how a connectionist system might construct decreasingly perspective-bound representations (see Chapter 7.v.b–d). From the retinal image, through the Primal Sketch, to the $2\frac{1}{2}$D sketch and beyond, Marr had pictured a subjective-to-objective progression satisfying some of the criteria noted by Evans and Cussins.

He might have cited Hinton too, for Hinton had already outlined how a visual network could construct increasingly viewpoint-independent representations of arbitrary 3D shapes (Hinton 1981; cf. Boden 1988: 80–6). ("Outlined", because he'd modelled his abstract theory only in toy systems.) Indeed, Hinton would have been a better example for Cussins's purposes than Marr, because his objective representations didn't depend, as Marr's did, on top-down influences from ready-made concepts (object models).

An even better example was soon provided by Ronald Chrisley (1965–), a Sussex philosopher with hands-on programming skills who developed Cussins's suggestion about a connectionist base for objectivity (Chrisley 1990, 1993; Chrisley and Holland 1995). He designed a recurrent PDP network called CNM (Connectionist Navigational Map), which would enable a robot to learn to navigate (to reach "home"), gradually decreasing its perspective-dependence as learning proceeded.

A series of experiments showed how various types of spatial generalization were, or were not, generated (in the hidden units) by distinct representational codes. For example, how could CNM (first) learn to identify specific locations, and (then) to relate them to each other more or less systematically? (This wasn't straightforward, for the robot's "sensations" and "actions" weren't associated one-to-one: the same action sometimes yielded different sensations, and different actions might yield the same sensation.)

CNM could be seen as an exercise in robotics. Chrisley compared it also to empirical work on learning (artificial) grammars, on spatial maps in the hippocampus, and on the development of object permanence in children. But his primary aim was philosophical: to clarify what sorts of representational trajectory might count as "increasing objectivity".

Cussins and Chrisley were both firmly rooted in analytic philosophy (they were fellow graduate students at Oxford in the 1980s). But their work on objectivity exemplified the growing stress on *embodiment* that characterized cognitive science in the 1990s, and which was often expressed in terms of the very different phenomenological tradition (see Chapters 15.vii and 16.vi–viii). Indeed, Cussins—and Clark (1997, 2005), too—moved increasingly towards that approach (Cussins 1992).

By the end of the century, then, Dreyfus was only one of many proclaiming the importance of bodily skills. Even Evans had highlighted such skills, in his account of non-conceptual content. But neither Dreyfus nor Evans had discussed connectionist implementations of intelligence, and Cussins had done so only in the sketchiest terms. Chrisley, by contrast, tried to show how specific PDP networks might be used to illuminate the grounding of objective thought in the dynamic interplay of bodily action and perception.

Although Cussins was exploring a core assumption of connectionism—that concepts can emerge from a subsymbolic base—he had fewer readers than Fodor, Churchland, or Clark. He was a philosopher's philosopher, whose highly abstract and closely argued work bristled with technical terms. Indeed, his C3 paper had been rejected as "too philosophical" by the editor of the interdisciplinary *Behavioral and Brain Sciences*, himself a philosopher—namely, Stevan Harnad (Cussins, personal communication).

He influenced the field, nevertheless. Paul Churchland was so impressed by the C3 paper that, shortly after it appeared, he invited Cussins to join him at San Diego. *Mind* published his second paper on content and embodiment soon afterwards (Cussins 1992). Clark referred to C3-theory in developing his own account of representational change (A. J. Clark 1993: 73–6). Chrisley spelt out the third 'C' in C3 in some detail, as we've seen. And Brian Smith (1996) was strongly influenced by Cussins's publications on embodiment, and (even more) by his ongoing research on content, objectivity, and ontology. As we'll see in Chapter 16.ix.e, Smith was attempting a major upheaval in the philosophy of mind and computation, in which insights of the analytical and phenomenological schools would be combined.

g. An eye to the future?

I said above (viii, preamble) that the late-century connectionists didn't try to tackle propositional reasoning. However, Clark (2005, in preparation) has very recently outlined a research programme aimed at doing so—and at taking *embodiment* seriously in the process. As a philosopher, he won't be doing hands-on modelling. But he will be taking modelling techniques into account—and empirical psychology and neuroscience too.

He sees language as a form of cognitive technology, whose origin and use can be understood as an aspect of embodied cognition (Clark 2003a; and 16.vii.d). Its epistemic artefacts aren't axes, pencils, or computers but publicly audible/visible words—and internal representations of them, and of the motor actions we've learnt to associate with them. At base, these are forms of perceptual imagery, not language-specific types of representation.

Clark isn't the only philosopher, today, to stress the embodied, perceptual, aspects of language. A similar idea imbues Jesse Prinz's radically empiricist philosophy of mind (Prinz 2002). And cognitive linguists such as Larry Barsalou and Rolf Zwaan have recently developed "situation semantics", which holds that words don't *have* meaning so much as *prompt* it (Barsalou 1999*a*,*b*; Zwaan 1999; Feldman and Narayanan 2004).

The abstract aridities of Montagovian semantics are here left far behind (see 9.ix.c). For on this view, language works by eliciting specific perceptual and motor imagery in the hearer/speaker, who thus *simulates* the relevant semantic content. Situational linguists can call on psychological studies of visual imagery (e.g. Stanfield and Zwaan 2001), and on fMRI (brain scanning: see 14.x.b) data showing that words elicit activity in the neurones that are normally active when the person senses, or enacts, the relevant thing or action (e.g. Tettamanti *et al.* 2005). They can also call on seventy years' work on schema theory, thanks to Frederic Bartlett and his many successors (see 5.ii.b, 7.i.c, and 14.vi.c, and Fauconnier and Turner 2002).

The idea that the concepts named by words may be based in concepts derived from the body isn't new to cognitive science. For instance, George Lakoff and Mark Johnson suggested a quarter-century ago that abstract concepts, such as *justice*, may be metaphorical extensions of body-grounded concepts, such as *balance* (Lakoff and Johnson 1980; M. D. Johnson 1987); and some even earlier claims about the bodily origins of language, including syntactic structure, are cited in Boden (1981*a*). What's new is the greater attention to psychological *processes*, that is, the suggestions about *how* language works within our minds.

Applying this approach to the language of *belief*, *desire*, and *intention* (the core vocabulary of Theory of Mind), it may seem that "simulation theory" has won out over "theory theory", at last. However, this depends—as we saw in Chapter 7.vi.f—on *just what* sub-personal cognitive machinery is involved. And that's not easily discovered. The current empirical evidence for simulation semantics has resulted from presenting either single words or isolated sentences describing simple actions. But fMRI can't distinguish *Man bites dog* from *Dog bites man*, nor reflect *reasoned series of propositions*. One of Clark's aims is to grasp those nettles.

He reminds us that a biological niche containing self-constructed nests, or webs, enables birds or spiders to do things they couldn't do without them. Similarly, a cultural niche containing axes and pottery (and shipboard instruments: 8.iii.a) enables us to do things, and to think things, that would otherwise be impossible. In addition to specific artefacts such as these, the spatial organization of objects can be used as a memory-prompt enabling us to perform potentially difficult tasks with relative ease. Consider, for example, the physical arrangement of the instruments aboard ship (8.iii.a), or of unfilled glasses on a bartender's counter (K. Beach 1988); and remember the many times you've positioned the salt and pepper pots to make a *non-spatial* point.

But language, given what linguists call "the arbitrariness of the sign" (9.iv.c), is even more generally useful than is arrangement-in-space. Words can stand in for non-spatial and/or abstract items and relations, from sameness/difference to justice. It thus provides a "superniche", in which we can represent *all* external epistemic tools. Uniquely "poised between the inner and the outer, the public and the private", it acts to "stabilize, anchor, and scaffold individual thought and interpersonal coordination" and helps to create "our selves" (A. J. Clark 2004: 725).

This explains the special power of language. Words lead us to do things which we wouldn't, couldn't, have done otherwise. They do that by selectively *directing our attention* to other items in our minds—including other words, and representations of bodily actions. As Clark puts it, "Words aren't meanings, but clues to meanings." (Here, he's quoting Elman, who in turn was quoting a remark made in conversation by Rumelhart many years ago—Elman, personal communication, and 2004: 301; for Rumelhart's explanation of it, see Rumelhart 1979: 85.)

Associative processes can link internal representations directly with each other, without any current engagement with the external environment. For instance, we can remember something we did, said, or thought yesterday. In general, we can monitor and control our own thoughts (beliefs, intentions, moral principles . . .) by using language to focus our attention on them—an ability that makes human 'freedom' possible (7.i.g). In short, our high-level knowing can be 'decoupled' from our sensori-motor behaviour, despite being mediated by representations that are grounded in embodied perception and bodily action. It's this, Clark argues, which causes the undeniable discontinuity between *Homo sapiens* and other species, despite the continuity of the basic psychological mechanisms (i.e. perception, association, and bodily skills).

Clark mentions many examples of empirical work which support his view. For instance, chimps trained to discriminate between sameness and difference pairs (e.g. cup–cup and cup–shoe) by picking out a red or blue tag are able later to learn to recognize higher-order sameness and difference—as in cup–cup and shoe–shoe, or cup–cup and cup–shoe (Thompson *et al.* 1997). Chimps who'd learnt the lower-level discrimination but *without* using a separable token to mark it couldn't learn to recognize the higher-level property. The experimenters suggested that what the tag-trained chimps were actually doing was to compare *a memory of a red tag* with *a memory of a blue tag*—which is a lower-order problem. (So Thomas Evans's task for his ANALOGY problem, one might say, was to provide his NewFAI program with the relevant tags: 13.iv.c.)

Another example cited by Clark is work done by the Paris-based group led by cognitive neuroscientist Stanislas Dehaene. This shows that arithmetical reasoning involves both verbal and visuo-spatial representations (Dehaene 1997). Experiments on animals and infants indicate that a primitive visuo-spatial sense of number has evolved independently of language, but can be supplemented by language in mature human beings (Spelke 1994; Dehaene *et al.* 1998a). Prior to linguistic development, there seem to be two core visuo-spatial "codes" for counting: one exact (for numbers up to between three and five), and one approximate (for larger quantities). Exact calculation of larger numbers requires language (Pica *et al.* 2004).

Experiments with bilingual subjects display systematic differences between arithmetical reasoning that's exact or merely approximate, and which relies on over-learned associations in one language or the other (Dehaene *et al.* 1999). These differences concern not only the time taken to complete the task, but also the locus of fMRI-scanned brain activations. And they appear to depend on the extent to which the reasoner employs *words functioning as perceptual cues*. For exact calculations, the solution is faster if the sums are presented in the language in which they were taught—and the brain's language area is activated. For approximate calculations, the language-of-teaching is irrelevant, and a different (parietal) brain area is activated.

Clark sees these results as showing that what we normally regard as high-level reasoning may be grounded in visual perception, but improved by words—which are *essential* for certain reasoning tasks. Even for those tasks, however, words function by means of learnt associations to language-specific perceptual (auditory and visual) imagery.

All very interesting . . . not least, because it can be seen as a twenty-first-century version of Hume's talk of "impressions" and "ideas" (2.x.a). (It's also an updated version of the Whorfian hypothesis, at least where vocabulary is concerned: see 9.iv.c.) But there are two major, and familiar, problems: hamsters and hierarchy.

If language and reasoning rest on perception and association, or in other words what Elizabeth Bates called "sundry old parts" (see 7.vi.c), why can't hamsters learn English? Why have all the attempts to teach language to non-human animals foundered, just as Descartes predicted (2.iii.c and 7.vi.c)? We've seen that Clark rejects Chomsky's nativist answer, and Fodor's too. He also rejects Kim Sterelny's (2003) recent suggestion that the brains of newborn humans today are representationally and computationally deeply different from those of early hominids, due to some combination of genetic changes and the effects of our unique developmental cocoon. He admits (personal communication) that there must be *some* inherited difference between animals' brains and ours, but has nothing convincing, nor even intriguingly plausible, to say about just what it might be.

As for hierarchy, the rock on which Burrhus Skinner's theory of language foundered (9.vii.b) and—as we've seen—one still unconquered by PDP, Clark ignores it. His project focuses on individual words and over-learned phrases, not whole newly generated sentences. Apart from saying (correctly) that a representation of structure X doesn't have to be X-structured, he has nothing to add about hierarchy.

So why mention his work here? In particular, why mention a largely speculative project, still (in February 2005) in the process of grant application? Why not confine the discussion to already peer-reviewed, or anyway already drafted, research reports? This book, after all, is history—not futurology.

The reason is that Clark's chosen problem is highly important, with significant implications for every discipline within cognitive science. Maybe he won't achieve the breakthrough that he's hoping for. But he'll probably come up with something of interest—a better understanding of attention, for instance. Psychologists have done a huge amount of work on this, including much which today would be described as the study of consciousness (6.i.a–c and 14.ix.a), but the concept is still far from clear. Again, his work may help to fill some of the gaps in situation semantics: the currently popular "blending theory", for instance, abounds with hand-waving about *just what* computations might be involved, and *just how* they can be triggered (Fauconnier and Turner 2002). Other goodies may ensue too—though we can't be sure of that now (see 17.i). Science-as-she-is-done, and philosophy-in-the-making too, isn't always a matter of dotting the *i*s and crossing the *t*s. It also involves speculation, hunches, boldness . . . in a word, risk.

12.xi. Pointing to the Neighbours

Connectionist AI/psychology had two next-door neighbours, and another directly opposite. The next-door neighbours were experimental psychology and neuroscience. The one facing across the street was symbolic AI.

These fellow residents spoke to each other less than one might think. As we've seen, a large part of connectionism's project was *negative*: to deny GOFAI's claims to ground a science of the mind. So cosy chats on the doorstep weren't very likely there. But even neighbour neuroscience was largely neglected by the connectionists I've discussed in this chapter. Indeed, the neglect ran both ways: for many years, most neuroscientists were territorially suspicious of computer models brought onto their patch (see 14.iv.d and v.c).

There were some exceptions. Clark, for instance, clearly realized that if his philosophy of mind/language is to be vindicated, it must be consistent with the data of experimental psychology and neuroscience. And the Churchlands drew even more inspiration from neuroscience. However, very few philosophers of connectionism paid much attention to either. Indeed, many of the *hands-on* connectionists, especially those coming from physics or engineering, near-ignored these experimental sciences too. Even if they considered psychology carefully, which they didn't always do, they sidelined neuroscience. (Again, there were exceptions: some individuals, such as Sejnowski and J. R. Anderson, came into the field from neurophysiology.)

Their networks were inspired by the brain, to be sure. But they focused on mathematical analyses of general computational properties: pattern recognition, distributed representation, associative memory, learning, development, and the low-level grounding of high-level thought. They took on board some very general features of biological neurones, but didn't pay attention to detailed aspects of the nervous system. Their work wasn't much help, for instance, if one wanted to know *just how, in fact,* a frog manages to catch a fly with its tongue, or *just how, in fact,* someone recognizes their grandmother.

Potentially, the connectionism and neuroscience neighbours had much to teach each other. Both were concerned with the probabilistic properties of cell networks (cf. 14.ii.b); and hypotheses about what *computations* (what information processing) the nervous system is performing could guide neuroscientific questions about the bodily mechanisms involved. In some cases, the party wall between the next-door residences was pretty thin. In other words, the distinction between connectionism and computational neuroscience was fuzzy.

With advances in neuroscience, it's getting fuzzier. One indication of that is that a recent textbook written by two students of McClelland is significantly more neural/biological than the first PDP volumes were (O'Reilly and Munakata 2000). Indeed, it's boldly subtitled *Understanding the Mind by Simulating the Brain*. Published at the turn of the century, it presages a new phase of research for the new millennium. To be sure, associative memory isn't being explained directly in terms of ion channels.

But the brain—and even ion channels—is being given more attention than before. (In part, of course, this is a result of neuroscientists' having learnt more about it.)

Twentieth-century computer models that tried to cast specific spotlights on the brain will be considered in Chapter 14. Before talking to the next-door neighbour, however, let's look at what happened to the neighbour across the street.

The spectacular rise of PDP didn't cause GOFAI to collapse. If Fodor and Pylyshyn hadn't given up on the symbolic approach, neither had their GOFAI colleagues. And much of what they were doing had implications stretching beyond technology, into cognitive science.

13

SWIMMING ALONGSIDE THE KRAKEN

The long-submerged connectionist Kraken identified in Chapter 11.vi surfaced in 1979, and leapt into the air a few years later (see 12.v–vi). There's no denying that it posed a threat to symbolic AI. It commanded philosophical interest, stole public sympathy, and—after DARPA's U-turn in 1986 (12.vii.b)—competed for funds. But it didn't chase GOFAI from the seas.

While the monster frolicked in the limelight, symbolic AI carried on swimming steadily. It's been doing that for twenty-five years now, during which time a mini-Kraken has risen to visibility—situated, or embodied, AI. (That too, the ninth bombshell, was spawned around mid-century: Chapter 4.viii.)

This chapter describes some work from GOFAI's post-Kraken period. I've ignored natural language processing as such, because various examples of post-Kraken work in NLP were described in Chapter 9. However, it's worth mentioning that current AI models of planning and multi-agent cooperation, including virtual-reality set-ups in which a VR agent interacts with a human being (Section vi.b, below), owe much to earlier NLP research on the recognition of intentions and beliefs (see 9.xi.f).

Some of the chosen themes—VR, for instance—have a decidedly modern feel. But every one was presaged in the very earliest days of NewFAI. In short, the harbingers (10.i) were coming home to roost.

Not all, however, found a comfortable place to perch. Logicism's efforts to do so are described in Section i. Some post-Kraken criticisms of logicism, and of the explosion of expert systems dating from the early 1980s, feature in Section ii.

Next, I focus on five topics which despite being heralded in the harbinger papers attracted scant attention until around 1980. These are (in Section iii) diagrammatic reasoning; the situationist attack on planning; agents and distributed cognition; inductive learning; and (in Section iv) creativity. GOFAI's contribution to human–computer interfaces and VR is outlined in Sections v and vi, which also explore some potential psychological effects of VR.

Finally, I give a historical perspective on two common beliefs: that AI isn't a discipline so isn't worthy of respect, and that GOFAI has failed.

13.i. Later Logicism

Post-Kraken work on classical AI themes included the use of logic to represent knowledge. This research programme had been announced in 1959 by John McCarthy, and was given a huge boost ten years later by McCarthy and Patrick Hayes (10.i.f and iii.e). Since the late 1970s there have been many developments.

Some of these, as McCarthy and Hayes predicted, concern classical metaphysical questions. For instance, Judea Pearl's long-standing logicist research has focused on causation and counterfactuals, on probability, and on the nature of scientific explanation (Pearl 1988, 2000; Halpern and Pearl 2005*a*,*b*). It's hardly surprising that it has drawn the attention of professional philosophers. For causation was grist to the metaphysician's mill even in Aristotle's time; counterfactuals have vexed logicians ever since the Middle Ages; probability has perplexed philosophers for 200 years; and scientific explanation was much discussed in the twentieth century (e.g. Popper 1935; Salmon 1989; Lipton 1991).

Overall, GOFAI's logic-based advances have included work on at least three fronts. Namely, non-monotonic logic, naive physics, and an AI encyclopedia (not to be confused with an encyclopedia of AI). But whether these three (described below) should really be called "advances", as opposed to *continued research activity*, is highly controversial. For it's still not universally agreed within AI that logic is a suitable medium for KR.

It still has champions around the world. For instance, Germany's Wolfgang Bibel, asked to comment on "AI's greatest trends and controversies" for the millennial issue of a leading computer journal, insisted that "the language as well as the inferential machinery of logic [is] fundamental for the endeavor of realizing an artificial intelligence" (Hearst and Hirsh 2000: 8). But a recent President of AAAI devoted much of his Presidential Address to casting grave doubts on this (Waltz 1999). In short, by no means all of today's AI is imbued by the McCarthyite passion for logic.

a. Less monotony

"Monotony" was lessened by post-Kraken GOFAI in order to deal with the (second sense of the) frame problem—namely, situations involving incomplete knowledge (see 10.iii.e). Symbolic systems now have a flexibility way beyond what was achievable in 1980.

In ordinary (monotonic) logic, any conclusion that follows from a set of assumptions, or premises, also follows from any *larger* set of assumptions. If the original assumptions were true, then the conclusion is true, no matter what other true premises are added. In non-monotonic reasoning, by contrast, the entailment of the conclusion can in some circumstances be annulled.

Consider the common-sense decision to abandon one's plan to take the train, and to take the bus instead—because on reaching the station one discovers that the trains aren't running today.

Using a monotonic logic, one would have to avoid inconsistency by *revising* one's initial assumptions ("Drat! I thought the trains were running today, but they aren't. Delete that premise") and then replanning the journey. (So-called *truth-maintenance* techniques use monotonic logic, and attempt to restore consistency whenever new information comes in which contradicts some previous assumption.) Using a non-monotonic logic, one could *add* the new premiss "No trains today", but allow it to override the older premisses so as to *avoid* drawing the conclusion that "I shall take the train today".

That's an illustration of what's sometimes called the "qualification" problem: the fact that reasoning from incomplete knowledge is unavoidably risky. If new information comes in, one may have to change one's mind in some way. Various attempts to compensate for incomplete knowledge had been discussed in the 1970s (e.g. A. M. Collins *et al.* 1975; Norman and Bobrow 1975). Among them was GOFAI work which, following McCarthy and Hayes (1969), aimed to do this without abandoning representation-by-logic.

Ways of doing that had begun to poke their heads above the water in the 1970s (Winograd 1980*b*). For instance, Uppsala's Erik Sandewall (1945–) extended predicate calculus logic by defining the *Unless* operator, which prevented the system from making inferences it would otherwise have drawn (Sandewall 1972; cf. Raphael 1971). If told both *A* and *A Unless B implies C*, Sandewall's program would set up a sub-goal to assess the truth of *B*, before deciding whether to infer *C*. (This could happen on several goal levels, as *Unless* was recursive.) Similarly, the default slot-fillers in AI 'frames' led programs to assign default values *only if* those slots hadn't already been given some other value (Hewitt 1969; Minsky 1975). And Earl Sacerdoti's ABSTRIPS prevented failures by postponing on-the-spot executive planning until the last minute (10.iii.c).

In all those cases, disaster was pre-empted by *avoiding or delaying* decisions. In "fuzzy logic" too, contradictions were pre-empted—in this case, by using probabilistic reasoning (Zadeh 1965, 1972, 1975; for an 'update', see Bouchon-Meunier *et al.* 1995). Since there was no firm commitment to the truth of the conclusion, there was no contradiction if it turned out to be false.

In non-monotonic reasoning, by contrast, decisions/conclusions were actually made—only to be withdrawn later. Work of this type began in the mid-1970s, and a small workshop on the topic was held at Stanford in 1978. By that time, several ways of enabling a logic-based program to 'change its mind' were being suggested.

McCarthy himself developed "circumscription" as a possible solution (1977, 1980*a,b*, 1986). The key idea was that the program should assume, at any given time, that it knows all the relevant facts. Consider the missionaries and cannibals problem, for example:

Imagine giving someone the problem, and after he puzzles for a while, he suggests going upstream half a mile and crossing on a bridge. "What bridge", you say, "No bridge is mentioned in the statement of the problem." And this dunce replies, "Well, they don't say there isn't a bridge." You look at the English and even at the translation of the English into first order logic, and you must admit that "they don't say" there is no bridge. So you modify the problem to exclude bridges and pose it again, and the dunce proposes a helicopter, and after you exclude that, he proposes a winged horse or that the others hang on to the outside of the boat while two row. (McCarthy 1980*a*: 30)

Clearly, given the inventiveness of "this dunce", no additions to the axioms of the problem could ever defeat him.

The only strategy, said McCarthy, is to assume that all the relevant facts are known—so if bridges aren't mentioned they're not needed. In *puzzles*, like the missionaries and cannibals brain-teaser, that's actually so. But in real-life problems, *that all the relevant facts are known* is an assumption, not a datum. It may turn out to be false. If so, the problem-solver must be prepared to reconsider some of the previous conclusions. (Fuzzy logic wouldn't help, he said. For not only can we not assign meaningful probabilities to bridges or to Pegasus, but we don't even think about them. Nor should we: when would such thinking ever stop?)

As McCarthy pointed out, circumscription wasn't an example of non-monotonic *logic*, but of non-monotonic *reasoning* that augmented ordinary first-order logic. (So was Jon Doyle's "truth maintenance system": Doyle 1979. This stored the reasons for its conclusions, and used backward chaining to revise its beliefs if a reason statement turned out later to be untrue.)

McCarthy advised against altering the logic itself, because it's difficult to ensure that all such modifications are compatible (1980*a*: 37). Non-monotonic logics were suggested nevertheless, by Drew McDermott and Ray Reiter (McDermott and Doyle 1980; McDermott 1982*a*,*b*; Reiter 1980). Reiter's "default reasoning", for example, was extended by a complete proof theory for a large class of common defaults, which could be presented to a resolution theorem-prover.

It wasn't always easy to compare the various forms of non-monotonic reasoning. Reiter, for instance, said that a particular result "suggests some deeper relationship between closed world default theories and [McCarthy's] method of circumscription although I have been unable to discover just what this might be" (1980: 129). McCarthy, besides offering technical comparisons between his own approach and Reiter's (McCarthy 1980*b*), argued that default reasoning in general is less useful than circumscription, because it's less generalizable (1980*a*: 37). For example, a block x that isn't explicitly stated to be on block y is taken (by default) not to be on y; similarly, for block z; but there's no way of inferring what circumscription allows one to assume: that there are *no* (unmentioned) blocks on y.

In 1980 non-monotonic reasoning was the subject of a special double issue of the *Artificial Intelligence* journal. This heralded an explosion of work in the area. Some of it was soon to be featured in the widely read compendium of 1985 that also included Brian Funt's and Aaron Sloman's thoughts on analogical representation (see iii.a, below).

Many people—even McDermott himself (ii.a, below)—pointed out that none of these attempts to formalize non-monotonic reasoning could solve the frame problem in the general case. (Hayes agrees.) But not all reasoning concerns the general case. Just as expert systems can be useful if they're restricted to a certain topic, so programs can approximate common sense where the types of rethinking that may be needed can be specified beforehand. Accordingly, research on non-monotonic reasoning (and logics) continued despite the critics' strictures. And it still does (cf. Gärdenfors 1992; Besnard 1989; Antoniou and Williams 1997). Indeed, just as this book was going to press McDermott received the AAAI Classic Paper Award for an influential paper on the topic (Hanks and McDermott 1986), and for his work on "causal and temporal reasoning in a wide variety of fields" (Leake 2005: 3).

As for the frame problem, it's still exercising many billions of brain cells. Hayes himself co-edited a collection of sixteen papers on it in the early 1990s, and also took time out to write a summary for Stevan Harnad's online journal *Psycoloquy* (Ford and Hayes 1991; P. J. Hayes 1992). Unsolvable though it may be, the problem won't lie down.

b. More naivety

Just as non-monotonic reasoning was presaged in 'Some Philosophical Problems', so was Hayes' work on naive physics (1979, 1985a,b). Hoping to rescue AI from the "toy problems" so characteristic of NewFAI, his goal was to capture "a sizeable portion of common-sense knowledge about the everyday physical world: about objects, shape, space, movement, substances (solids and liquids), time, etc." (P. J. Hayes 1979: 242).

These highly abstract questions were relevant to a wide range of practical problems. Many robots then being envisaged would need some mastery of the answers; and some expert systems, as his past collaborator was saying (McCarthy 1983), would need it as well.

Hayes relied on his own intuitions about what naive physics is. He didn't run any experiments, such as those done by the psychologist Michael McCloskey (1983a,b). McCloskey asked, for example, how people envisage the trajectory of a stone (held at shoulder height) dropped by someone walking fast. He found that his subjects' expectations were usually mistaken. Most said the stone would fall straight down (although a few thought it would move backwards, to land behind the point of its release); even Ph.D. students of physics often failed to give the right answer—a parabola. In other words:

[our misconceptions] appear to be grounded in a systematic, intuitive theory of motion that is inconsistent with the fundamental principles of Newtonian mechanics [but] resembles a theory of mechanics that was held by philosophers in the three centuries before Newton. (McCloskey 1983b)

McCloskey suggested that these false beliefs arise from specific visual illusions (to which the medievals were also subject), a perceptual base which makes them highly resistant to modification by reason ("many [students of physics] emerge with their intuitive impetus theories largely intact"). However, *everyone* in McCloskey's experiments assumed that the stone would fall to the ground *somehow*. Indeed, he didn't bother to ask whether or not they formed that expectation: it was too obvious to mention.

That shows why Hayes didn't run experiments. What he wanted to make explicit was *that the stone would fall*. Also, he wanted to spell out just what that means: that it would follow a smooth, roughly vertical, downward path ("parabola or straight line?" wasn't relevant), ending on the ground roughly where it was dropped ("ahead, behind, or underneath?" wasn't relevant). Likewise, he wanted to say just what it is to *drop* something. (If a piece of fluff falls from my coat, have I dropped it? If I release a bird from my hand and it flies away, have I dropped it? If I'm lying face down on the floor clutching a pencil and I unclench my fingers, have I dropped it? . . .)

His two 'Naive Physics Manifestos' were a valiant attempt to make explicit a myriad of things which even McCarthy's "inventive dunce" wouldn't have thought of mentioning. These taken-for-granted assumptions and inferences pervade our everyday action, thinking, and language. They're just as crucial for subtle NLP as for successful manipulation and flexible problem solving. (Indeed, Roger Schank's "conceptual dependencies" had been an attempt to capture some of what Hayes was trying to express more thoroughly: 9.xi.d.)

Hayes asked, for instance, what happens when liquid pours out of a tilted glass. Or, to put it another way, just what is it for liquid to pour out of a container? What generally causes it? And what generally results from it? What's the difference in meaning between *pour*, *flow*, *spill*, *drop*, and the like? How might a robot be enabled to see that a container was just about to spill its contents? And how could it select a specific action to prevent the spillage, or to pre-empt a particular unwanted result of it?

As Hayes pointed out, in our everyday thinking we don't get our answers to such questions from theoretical physics. Even when someone happens to know the physics, they don't usually rely on it (a fact which McCloskey's experiments confirmed). What we rely on is naive (intuitive) "qualitative" reasoning. What we need for AI purposes, then, is a formal representation of that. And predicate calculus, according to Hayes, should be a suitable formalism.

Naive physics, alias qualitative reasoning, became an important focus of post-Kraken research. The journal *Artificial Intelligence* published a special volume in 1980, and two brief retrospectives in the early 1990s (Kuipers 1993*a*,*b*). Two influential collections appeared within six years of each other (Bobrow 1984; Weld and de Kleer 1990), and work in this area has continued ever since (Kuipers 2001).

One question that received attention was whether one should assume, following Hayes, that naive physics is *consistent*. Andrea DiSessa (1988) said it wasn't. He posited a number of "phenomenological primitives" (not theoretical principles), such as an intuitive notion of *impetus* as an unanalysed combination of force and motion. His logically fragmented view of intuitive physics allowed different, and potentially conflicting, primitives to be activated in different situations. It was similar in spirit, then, to contemporary theories of "simple heuristics" in psychology (Chapter 7.iv.g).

Another general question, which became increasingly insistent over the 1990s, was whether a purely intellectualist approach, consistent or otherwise, can capture our everyday physical knowledge. Sceptics argued that it's our *embodied experience* of solids and liquids which gives us access to their properties. For these critics, it wasn't only naive physics which was under attack. The whole of "disembodied" GOFAI, including plan-driven robotics, was put in doubt (Chapters 15.viii and 16.vii, and iii.b–d below).

Tacit conceptual structures exist for naive psychology too (7.vi.f), and for animals, plants, and time (8.i.b and d). Their core aspects are universal, not culture-specific. And they underlie thoughts that go way beyond common sense: notably, myth and religion (8.vi). In short, all human thinking involves intuitive reasoning—what Hayes called naivety.

c. The AI en-CYC-lopedia

If all human thinking involves intuitive common sense, it follows that an encyclopedia can't be understood without the reader's relying on it. But if the "reader" is an AI program, that common sense has to be explicitly supplied. This fact (or fancy?) informed the mid-1980s hope of building an AI encyclopedia: an organized body of knowledge to be consulted not by people, but by other programs.

The AI encyclopedia was Douglas Lenat's (1950–) CYC project. With the enthusiastic backing of Edward Feigenbaum, by then the public champion of expert systems, CYC was launched in late 1984 at the newly founded Microelectronics and Computer Technology Corporation (MCC) in Texas, and officially announced in the business-oriented *AI Magazine* (Lenat *et al.* 1986).

Why would the business community be interested? Well, Lenat had found in his survey of expert systems (Davis and Lenat 1982) that time was often wasted by one AI team's effortfully representing knowledge already expressed by others. Moreover, naive physics was usually ignored, since the introspection encouraged by the interview techniques of knowledge engineering wasn't capable of capturing it. A central store of common-sense knowledge, accessible by many different expert systems, would thus be extremely useful. It was the hope of attaining this practical goal which led to CYC's hugely generous funding: $50 million, enough for "two person-centuries" of work, to be spread over ten years (Lenat and Feigenbaum 1991). (Later, the funding was upped to reach six person-centuries.)

The key idea was that concepts/facts would be represented in organized frames (10.iii.a), with just a million frames sufficing to store all (non-idiosyncratic, non-arcane) human knowledge. Lenat and Feigenbaum (1991: 210) allowed that their "million" was speculative, but pointed out that two AI leaders (Marvin Minsky and Alan Kay) and a group working on electronic dictionaries had all made comparable estimates. (These included the "back-of-the-envelope calculations" mentioned in Chapter 1.iii.g.) In other words, instead of building common sense into the logic, as default reasoning, it was to be expressed within the widely shared knowledge base (Lenat and Feigenbaum 1991).

Once the encyclopedia had reached a certain size, said Lenat, it would (like McCarthy's notional Advice Taker) be able to learn new things not by hand coding, but simply by being told. And the telling would involve not only semantic inheritance, but analogy and questioning too.

For instance, CYC might be told that *a tiger is a big, fierce, cat with black and orange stripes, living in the jungle*. The system's previous knowledge of *cats* would be wheeled in, to be modified only in respect of size, fierceness, stripes, and habitat. That a tiger is a carnivorous furry mammal, with four legs and a heart, susceptible to hunger and to gravity . . . wouldn't have to be stated anew. All that information would already be included in the 'cat' frame, and automatically inherited by the 'tiger' frame. If CYC chose to ask the human being what specific type of meat tigers eat, it could do so—and slot the answer into the relevant frame accordingly.

Moreover, jungles would have been represented as *large, hot, rainy, forests with abundant flora and fauna*. So CYC would know, without needing to be specifically told, that tigers like hot, rainy, places. Given appropriate inferential mechanisms to access such knowledge when needed, an expert system for zoo-keepers would be able to advise that the tigers shouldn't be housed on the artificial ice-floe with the polar bears.

All this, however, was easier said than done. To make CYC work would require powerful AI techniques for analogical thinking (see iv.c, below). Significant effort—and, ideally, psychological research—would have to go into choosing the central case of each class, what Eleanor Rosch had called the prototype (8.i.b). Otherwise, why not define a cat as a sort of tiger? A variety of inference mechanisms would need to be provided. Eventually, CYC ended up with thirty: some already familiar—such as inheritance, demons, and if–then rules—and others not, such as following metaphorically sensible links.

The database would inevitably contain global inconsistencies, so the CYC team would have to provide non-monotonic reasoning. They'd have to provide modal operators too, including some dealing with the non-truth-functional logic of intentions, knowledge, and belief. (For instance, if Mary knows that Jack lives at No. 5, it *doesn't* follow that she knows that Poppy's uncle lives at No. 5—even if Jack is, in fact, Poppy's uncle.) To do all that, they'd have to create "a powerful constraint language which is essentially predicate calculus" (Lenat and Feigenbaum 1991: 211–12).

The amount, and range, of knowledge to be entered into the system by Lenat's thirty staff members was daunting. It concerned "thousands of everyday activities like shopping, football, visiting doctors, and so on" (Brand 1995: 236). So Schank–Abelson scripts (7.i.c) had found a new home, expressed in CYC's new language. Last but not least, the tacit knowledge that Hubert Dreyfus had said protected humans from the frame problem was to be explicitly incorporated in CYC. That meant that all of Hayes' questions would have to be answered. CYC would need mastery of naive physics—and naive psychology and folk biology too.

Whether that "mastery" was to be conceived in logical or pragmatic terms, however, was controversial. Lenat himself leant towards the second option, and favoured it even more strongly after some years of experience in building CYC. In the early 1990s, he said this:

We avoided the bottomless pits that we might have fallen into by basically taking an engineering point of view rather than a scientific point of view. Instead of looking for one elegant solution, for example, to represent time and handle all the cases, look instead for a set of solutions, even if all those together just cover the common cases. (interviewed in Brand 1995: 237)

The philosopher Arthur Prior (1914–69), who'd pioneered work on temporal logic, would have been turning in his grave. And the logicist proponents of "neat" AI were uneasy too. But the a-theoretical "scruffies" were more accepting. Their question, in general, was not *Is it logically consistent?* but *Does it work?*

The *neat/scruffy* distinction between thinking styles is an old one, although those labels didn't come into common use in AI until about 1980. Then, they spread throughout cognitive science as a rampant meme (8.v.c). (They originated with Schank, according to his closest colleague: R. P. Abelson 1981*b*.)

The distinction is often seen as separating those driven by respectable intellectual standards from those too lazy to be bothered. That "good guys, bad guys" dichotomy informed William Woods's published critique of Schank's NLP, for example (9.xi.e), not to mention the scathing remarks he made face to face on conference panels, and in private. But some self-confessed scruffies offered strong arguments in defence of their position.

Minsky (1994), for instance, gave various reasons for believing that in *any* highly intelligent system, "consistency and effectiveness may well be incompatible". And forty years earlier, in 'Steps', he'd already argued that various types of non-logical computation are in practice essential even for proving theorems in formal logic (see 10.iii.b). As co-author of the fiercely analytical *Perceptrons*, he couldn't be accused of being intellectually lazy or scared by logic. Neither could Alan Bundy, who outlined how to combine the strengths of the two methodologies in undergraduate teaching without being misled by "slogans", or trapped into narrow AI "apprenticeships" (1981).

As an indication of CYC's *effectiveness*, Lenat—in the early 1990s—mentioned that when asked *Show me an image of shirtless young men in good physical condition*, the system had delivered a picture captioned "Pablo Morales winning the men's 1992

Olympics 100-meter butterfly" (Brand 1995: 240). The caption was crucial here, for CYC had no visual ability. But how had that query been linked to that caption?

The annotated transcript showed that a variety of common-sense knowledge had been involved. Some items were logicist axioms, such as *If X won event Y, then X participated in event Y*. Some were property-inheritance frames, implying for instance that *All Olympic men's events are instances of men's sports competitions*. Some were frames including common-sense default assumptions, such as *Participants in sports competitions are young* and *Participants in sports competitions have an athletic physical build*. Some were axioms grounded in common-sense knowledge, such as *Men who are swimming wear swim trunks*. And some were McCarthyite circumscription assumptions, such as *If CYC can't infer that a person is wearing X, assume he isn't wearing X*. So although CYC couldn't *see* that Morales was athletic, or shirtless, it found his picture in response to the user's request nevertheless.

One may wonder just how far this proffered example was a set-up. As in Lenat's Automated Mathematician (AM) program (Section iv.c, below), cynics might suspect that certain heuristics or frame-slot fillers had been provided precisely *in order that* specific 'discoveries'—such as this one—would be made. Such doubts could be allayed by inspecting CYC's data and finding that it included very much more than sport and swimming trunks, and by testing it with unexpected queries about a wide range of topics.

Lenat's original plan had been for CYC to achieve self-learning by 1994, and "completion" by 2000. Two mid-term reports—a book, and a brief résumé for the AI professionals—appeared in the early 1990s (Lenat and Guha 1990; Lenat and Feigenbaum 1991, esp. 210–28). MCC technical reports gave further details, some too recent to be mentioned in Lenat's book: e.g. further aspects of CYC's ontology of causality; early work on its taxonomy of types of change, such as transfer, destruction, etc.; and suggestions on how it might communicate with programs having simpler ontologies—Guha and Lenat 1989. (Another brief paper aimed at his peers came off the press more recently: Lenat 1995.)

CYC was proudly announced to be "still on schedule", even though "almost ten times as many" frames were required as had been expected. The team still hoped to "finish" by 1994, where this meant:

[reaching the point] where it will be more cost-effective to continue building CYC's knowledge base by having it read online material, and ask questions about it, than to continue the sort of manual "brain-surgery" approach we are currently employing. (Lenat and Feigenbaum 1991: 212)

As it turned out, by the end of the century CYC had a core ontology of 6,000 concepts with 60,000 facts—allowing the system access to about a million facts in all (S. Russell and Norvig 2003: 363–4). These included: trees are usually found in the open air; a person, or a robot, should hold glasses of liquid the right way up; and (a particular favourite of mine) when people die they stop buying things. Countless other examples were reported by CYC's programmers, but those three items alone indicate the range of applications they were hoping to aid and abet.

By that time, several other attempts at designing and filling large common-sense knowledge bases were under way (S. Russell and Norvig 2003: 364). And a conference series on 'Formal Ontology in Information Systems' had been established, plus various web sites describing examples of ontology research around the world (e.g. <http://ksl-web.stanford.edu/kst/ontology-sources.html>).

In part, this activity was driven by a shared vision of the so-called Semantic Web, an extension of the World Wide Web foreseen by its 1989–90 inventor Tim Berners-Lee (2000: 21, 30, ch. 13). This would be "a web of data that can be processed directly or indirectly by machines" (2000: 177). It would allow not only keyword search—by Google, for example—but also automatic communication between programs, eventually by means of the "intelligent agents" mentioned in iii.d, below. For computers on the Semantic Web will achieve "at first the ability to describe, then to infer, and then to reason" (Berners-Lee 2000: 184).

Although web pages contain much which can't be understood/processed by machines, there is "a vast amount of data in them, such as stock quotes and many parts of online catalogues, with well-defined semantics" (Berners-Lee 2000: 180). If the well-defined semantics can be standardized, and the ill-defined semantics upgraded, the possibilities for automatic data search will be hugely augmented. So "the desperate need for the Semantic Web" (ibid.), given the explosion of information present but not practically accessible on today's web pages, is behind much of the current work on KR ontologies.

It's difficult even to merge very simple databases (Doan and Halevy 2005). For instance, if one database on the Web uses three descriptors for houses (*location, price* ($), *agent id*) and the other uses four (*area, list price, agent address, agent name*), how can the merging system match these schemata so as to get a sensible result? It's even more difficult to enable knowledge sharing that will support/involve simple reasoning. Consider, for example, an automated version of searching the Yellow Pages, so as to match *the books I want to buy* with *the books available from various sources* (Burstein and McDermott 2005).

Even that task is relatively cut-and-dried. The sharing of general knowledge is more taxing still. If the basic ontology of concepts like *space, move, time, cause,* and *animal* (and perhaps *goal, intention, purpose,* and *choice*) differs from program to program, then knowledge sharing will be faulty or impossible. Similarly, if one wants robots (e.g. Mars rovers or autonomous vehicles) to be able to communicate with each other then they must have a common ontology. Otherwise, a suggestion from robot A that robot B follow plan X in order to achieve situation Y couldn't be properly assimilated by the intended recipient.

This partly explains why, by the end of the century, the construction of "formal encyclopedias of knowledge and methods and techniques for tailoring them to particular ends", ultimately covering "all human knowledge and methods", was being listed as one of the long-term goals of AI (Doyle and Dean 1997: 98). In short, the CYC team's claims, in a discussion of the foundations of AI (Kirsh 1991*a*), that something like their approach was essential for the advance of AI *in general* had been widely accepted (Lenat and Feigenbaum 1991).

Widely . . . but by no means universally, as we'll see (ii.a–b, below). In any case, these were promissory notes rather than reports of actual achievements. With respect to the latter, what verdict should one reach on CYC's contribution?

If the money men's hopes of providing expert systems with full common sense hadn't been realized, it didn't follow *even from the point of view of the neats* that nothing of interest had been learnt. CYC had taught AI workers a great deal about the problems, and some of the tentative solutions, involved in ontology building. And it had contributed also to AI thinking about KR-by-frames, non-monotonic inference, analogical reasoning, and prototype-based thought.

Something had been learnt, too, about the topics that concern metaphysicians. As was already evident from 'Some Philosophical Problems', AI logicists got some of their ideas from previous work done by philosophers. Where Warren McCulloch had been inspired by logical atomism, the later logicists drew on Prior's (1967) temporal logic; on Willard Quine's (1953*a*) account of natural kinds and Donald Davidson's (1980) of events; on Saul Kripke's modal logic and ontology of possible worlds (1963, 1980); and on Richard Montague's model-theoretic semantics (9.ix.c) . . . to name just a few. Two AI scientists who've worked in this area (and on CYC) for many years have concluded that

The discipline imposed in AI by the need for one's theories to "work" has led to more rapid and deeper progress than was the case when these problems were the exclusive domain of philosophy (although it has at times also led to the repeated reinvention of the wheel). (S. Russell and Norvig 2003: 363)

Whether these results were worth the initial $50 million, never mind the larger sums provided later, is controversial. Minsky, for one, wanted an even larger investment:

[One] can imagine a system that acquires most of its knowledge by experience. [However, in 1994] we still don't have suitable learning techniques. Lenat and I agree in the view that in order to learn as a person does, one will need to begin with a considerable body of built-in knowledge about a variety of effective ways to learn. The problem is we have not done enough research yet to know how to do this . . .

It seems to me there is a tragic aspect to the present situation. Over the last 15 years, almost all theoretical effort has gone into seeking alternatives where none exist. Consequently, all those years have slipped by, with no project other than CYC under way. The result is, we're all waiting to see how Lenat's work comes out, while doing nothing to help or compete . . .

I find it heartbreaking there still are not a dozen other such projects in the world, while there are thousands each of attempts committed to logical deduction systems, situated action and autonomous robot schemes, feed-forward neural networks, and rule-based expert systems. Each of these has particular virtues, but none of them show much promise of making inroads into that problem of scaling up the exploitation of large accumulations of knowledge. (Minsky and Riecken 1994: 27–8)

Those words "tragic" and "heartbreaking", unusual as they are in scientific discourse, show that Minsky firmly believes a much-improved CYC to be possible—if only we try hard enough, and dig deep enough into our communal pockets. Indeed, his sci-fi novel *The Turing Option* features a futuristic CYC much as Arthur C. Clarke's *2001* featured the futuristic HAL (A. C. Clarke 1968; H. Harrison and Minsky 1992). (For the record, Minsky had been a technical consultant for the film *2001*; since Clarke's book was based on the Clarke–Kubrick screenplay, one could even say that Minsky was part-responsible for that novel, too.)

Lenat may not see CYC as a close rival to HAL, but he does hope that a future version will (for instance) help Alzheimer's patients to remember things they used to know, or prevent someone who's just broken their leg from receiving online advertisements for running shoes (Brand 1995: 239, 242). What's more, he thinks it's already good enough to be useful. In 1994 he founded a company called CyCorp, whose officially declared aim is "to create the world's first true artificial intelligence, having both common sense and the ability to reason with it". He launched an open-source version (OpenCyc) early in the new century (later than originally planned, as the expected release date of April 2001 was delayed). Clearly, then, he expected many people to find it practically serviceable. (A wide range of potential applications are listed on the web site: <http://www.cyc.com>.)

13.ii. Choppy Waters

If GOFAI was still swimming, it was doing so in choppy waters. The late-century critics of logicism and applied AI were provoked, and emboldened, by the huge attention being aroused by CYC, and by the explosion of expert systems triggered by the Fifth Generation project. To some extent, their complaints had been prefigured in NewFAI days, by Dreyfus and Joseph Weizenbaum (Chapter 11.b). But there was now much more work to attack, and the new attacks were more detailed accordingly.

Predictably, perhaps, McCarthy was recalcitrant. He still insists that "the best hope for human-level AI is logical AI, based on the formalizing of commonsense knowledge and reasoning in mathematical logic" (2005: 39). But not all his AI colleagues agree. Even his co-author Hayes has had second thoughts, as we'll see.

Some post-Kraken criticisms came from social scientists who shared Dreyfus's philosophical sympathies. The British sociologist of science Harry Collins (1943–) was a prime case in point. Others came from legal researchers sympathetic to Weizenbaum, such as Philip Leith (1954–). But some came from within the AI community itself.

a. Apostasy

Logicism's opponents didn't need to send in any Trojan horses. For this long-standing area of technical research was being attacked from within.

One such assault was mounted by the computer scientist Brian Cantwell Smith. He wrote a bitingly critical paper on CYC for the multiple review in *Artificial Intelligence* (B. C. Smith 1991). Although CYC was at the bull's-eye, logicism in general was targeted in the surround.

However, Smith had always been a maverick. His radical views on the semantics of programming languages had received the accolade of being included in the major collection on KR (B. C. Smith 1985). But his further work on computation was a very different kettle of fish. Although it wasn't yet officially published (that would happen in 1996), it had been circulating in draft for some years. It was brilliant, or crazy . . . or maybe both. What it *wasn't*, was orthodox. Indeed, it was deeply counter-intuitive in a host of ways (see 16.ix.e). It was relatively easy, then, for logicists to ignore Smith's critique of CYC. If not a true outsider, he wasn't a typical insider either. His views could be more readily dismissed as a result.

More worrying, from their point of view, was the fact that the arch-logicist Hayes was growing more pessimistic about the logicist programme as a whole:

[With respect to] the extent to which formal logical ontologies can be said to adequately capture human intuitive knowledge . . . I've become very cynical as I grow older. For example, remember the Frame Problem? That was a warning flag: we should have thought about that a lot harder, and seen it as a critique of our inadequate ideas, rather than trying to solve it with a technical hack. (personal communication 2004)

His new-found "cynicism" covered more than the frame problem. It also applied to formalizations of common-sense notions such as knowing what one is doing; having a sense of oneself; understanding the instruction "Proceed with caution"; and making, keeping, and breaking promises. How could a program, or a robot, have such abilities?—"We don't have any idea of how to begin answering questions like this" (personal communication). He didn't (and doesn't) think that AI is impossible. But the problems of representation were more difficult than most logicists believe.

Accordingly, he now had doubts about the extent of CYC's relevance to AI. It was generally accepted that CYC would need mastery of naive physics and naive psychology. On Hayes' view, its mastery of such matters wouldn't need to match ours. It wouldn't need to know (for instance) how a robot could obey instructions to be "careful", or how it could make or break a promise. So much the better for Lenat: if CYC didn't need to know such things, he didn't need to worry about them. But so much the worse for the prospect of CYC's realizing the goal of logicist AI as a whole: "Even if CYC had succeeded beyond Doug's wildest dreams, it wouldn't be making a jot of progress towards answering questions like these" (Hayes, personal communication, 2004).

One might object that CYC *would* have to understand promises, if it were to include entries about contracts or international treaties, and carefulness if it were to include items about legal negligence. Lenat could point out, in reply, that it had never been intended as an encyclopedia of politics or law. But some logic-based expert systems had been—and are now—intended to deal with legal matters such as these (see subsection c, below). So Hayes' late-century views on the difficulty of formulating naive physics and, especially, psychology were very uncomfortable for committed logicists to hear.

At least Hayes stopped short of accusing logicism of *impossibility*. But to the consternation of many in the AI community, another leading figure in the development of logicism had now abandoned it entirely. This was genuine apostasy, and as such highly disturbing.

McDermott, who'd tried so hard to define a non-monotonic logic, was now describing the solution of the frame problem as impossible in principle. In a paper cheekily called 'A Critique of Pure Reason' (1987), first given at a small AI meeting at the end of 1985, he argued—against his own previous convictions—that the dream of representing everyday knowledge by predicate calculus (even with modal knobs on) is unattainable.

This 'Critique' was unexpected, and all the more shocking for that. For he'd only recently published several logicist papers on cognitive science, and an AI textbook recommending the McCarthy–Hayes approach to beginners (McDermott 1978, 1982a, 1985; Charniak and McDermott 1985). Now, he undermined it. His new paper criticized some well-known cases, offered a general diagnosis of the "meagre results" attained so far, and rebutted six familiar logicist "defences" (including reliance on non-monotonic logics).

So, for instance, he faulted Hayes' work on naive physics and Stanley Rosenschein's (1981) on planning. These were among the best exemplars of logicism but, McDermott argued, they'd failed. As one illustration, Hayes' (1985b) account of liquid pouring into a container such as an open bath or a closed tank included "[what] seems like a beautiful pair of arguments". On closer inspection, however, Hayes had deduced a *reductio ad absurdum* so as to refute an assumption, but had then inferred that an assumption not thus refuted was therefore *proved*.

As for the most highly visible, not to say notorious, example of logicism, namely CYC, this—said McDermott—was doomed to failure. It didn't follow that nothing of interest to AI (and psychology, and even philosophy) would be found out by Lenat's project. As he'd remarked in his 'Natural Stupidity' paper (1976), failures can be instructive. But Lenat was hoping for more.

For all that, McDermott remained committed to GOFAI in general. He *wasn't* emulating Dreyfus's attacks on the formalization of common-sense knowledge, for his faith in formalizability was intact. His doubts concerned axiomatizing knowledge, not programming it.

The crucial logicist fallacy, he said, is to confuse deduction with computation, so to assume that all thinking is essentially deductive merely because—as he still believed—it's computational. (In 'Natural Stupidity' he'd already accused AI of misusing the term *deduction*: 11.iii.a.) This explained the paradox of an expert philosophical logician such as Hayes making the apparently elementary mistake noted above. The various efforts by AI logicists, himself included, to define non-monotonic logics had failed to deal with everyday temporality, for instance (McDermott 1982*a*). Yet we, as normal human beings, deal with it successfully as a matter of routine.

The inferences actually made by people, or by programs, may of course be formally describable even if they're not strictly deductive. Indeed, McDermott still retained his faith that this is so. However, since axiomatic analysis is restricted to deductive domains, non-deductive AI programs can be *scientifically* understood only by means of a general theory of non-deductive inference. This, McDermott pointed out, had long been sought by philosophers but not found.

Admitting the troubling possibility that much of AI may be dead-end research "buoyed by simple ignorance of the past failures of philosophers", McDermott still hoped nevertheless that its concepts and techniques will help us discover the general theories which traditional epistemology has not. Until then, on his view, there can be no comprehensive science of intelligence.

Some people assumed that this meant that AI is mere pragmatism, a technology with no scientific pretensions and, in particular, no implications for cognitive science. In other words, all one could say in 'justification' of a non-deductive AI program was *Look, Ma: it works!* But that wasn't—and isn't—McDermott's view.

He's still convinced that AI can help us understand intelligence as a *computational* phenomenon, even if it isn't an axiomatic/deductive one. So in his recent book *Mind and Mechanism* (2001*b*), he argues that programs based on probability theory or heuristics aren't deductive. They can, however, be rationally justified, in that they prevent consideration of irrelevant possibilities, and/or prune these possibilities more efficiently than other programs do. Perhaps the human mind does the same sort of thing.

b. Can the fox catch the rabbit?

A rabbit may elude the fox because although the fox sees it, and wants it, he can't run fast enough to catch it. Or the fox may not realize that the rabbit's there, so doesn't even try to catch it. Both sorts of elusiveness found analogues in logicism, and in expert systems too. And both led to attacks on late-century GOFAI.

Consider the observant but slow-running fox, first. The reasoning used by NewFAI's DENDRAL and MYCIN had been restricted to theoretical chemistry and IF–THEN logic. By the end of the 1970s, however, some people were trying to include reasoning of other kinds—causal inferences, for example. So an expert system designed for cardiology included an early causal model of how the several components of the heart lead to distinctive ECG, or electrocardiogram, traces (Bratko and Mulec 1981; cf. Bratko *et al.* 1988). Such research was driven by McCarthy's harbinger-hope of giving AI programs common sense.

That was a tall order. Obviously, satisfying McCarthy's hopes would be a useful achievement. (The rabbit was clearly visible.) But even if it was in principle possible to formalize common-sense notions such as *cause*, which in itself was controversial, it certainly wasn't going to be easy. (The fox would have to run very, very fast.) A great deal of AI work in the final decades of the century was aimed at that goal, as we've seen.

And besides, there was the (second) question of the visibility of the rabbit. In other words, there was the question whether professional expertise is describable, even in *words*, in its entirety. A fundamental assumption of the expert-systems enterprise was that it is. (The rabbit could be seen if the fox looked hard enough—and if seen, it could eventually be caught.) However, that assumption had been roundly criticized well before expert systems came on the scene.

Sometimes, it was criticized merely by a telling example. For instance, as early as 1950 the computer scientist Lord Bowden had said:

a machine is unlikely to be able to answer satisfactorily such a question as this. "If a man of twenty can gather ten pounds of blackberries in a day and a girl of eighteen can gather nine, how many will they gather if they go out together?" (Bowden 1972, p. vi)

As he went on to observe, "it is problems like this which dominate the thinking of ordinary human beings".

Sometimes, however, the assumption was dismissed more systematically, in terms of a specific epistemological theory. The chemist–philosopher Michael Polanyi (1891–1976) is a good example here. He had long stressed the importance of "tacit" knowledge in science (1958, 1966, 1967). He allowed that tacit knowledge, once identified, can be made explicit. But he held that some (unspecified) tacit knowledge will always remain:

Tacit knowing is the fundamental power of the mind which creates explicit knowing, lends meaning to it and controls its uses. Formalization of tacit knowing immensely expands the powers of the mind, by creating a machinery of precise thought, but it also opens up new paths to intuition. (Polanyi 1966: 18)

Polanyi had used his critique to oppose philosophies of science, such as Karl Popper's, which took scientific knowledge to consist only in (explicit) theories, hypotheses, and empirical data. But it was later applied by his followers to expert systems, too.

One of the younger generation of scholars deeply influenced by Polanyi was Dreyfus. Another was Collins—who was led to consider AI largely by Dreyfus's critique of it.

Dreyfus had pointed out, for example, that if you tell an automated travel agent that you'd like a flight to San Francisco leaving "a little earlier" than its first suggestion of 6.30, you don't want it to come back to you with a flight leaving at 6.29 (Pagels *et al.* 1984: 340). Moreover, adding a rule saying *"Earlier" means at least fifteen minutes earlier, and at most two hours earlier* would often help—but not always. As Dreyfus put it, this expert system "doesn't contain any knowledge of human temporality—what spans of time are important to human beings".

Towards the end of the century, Collins explored that line of thought in writing at length about expert systems (1989, 1990; Collins and Kusch 1998). His prime motive was to attack cognitive science, to persuade his readers that *mind-as-machine* is a philosophical absurdity. That's not our main interest here (but see Chapter 16). Rather, our question is what worried him about post-Kraken expert-systems *technology*, and what AI could learn from him about its own practices.

What it *could* learn, not what it *did* learn. Partly because they associated him with Dreyfus, most AI workers didn't bother to read Collins, or didn't take him seriously if they did. He certainly wasn't a friendly voice. Nevertheless, he said some things worth hearing.

In comparing rule-following expert systems with human experts, including scientists, Collins showed that our knowledge is indeed largely tacit. So far, so familiar—from the point of view of the AI practitioner. For by the time Collins's book appeared, knowledge engineers—whether they'd read Polanyi or not—had long recognized that professional skills are largely unstated. Indeed, the interview techniques they'd established had been developed precisely in order to uncover normally hidden inferences and assumptions, including informal heuristics that wouldn't be found in the textbooks (see 10.iv.c).

But Collins had two more arrows in his quiver. One was that scientific understanding is largely a matter of practical, bodily, skills. As such, these skills can't be fully described in verbal/logical terms. Again, this was familiar territory to readers of Polanyi. Another sociologist of science, Donald Mackenzie, would soon make the same point—in a reassuring, not apocalyptic, spirit—with respect to nuclear-weapons technology (see 4.ix.a). But Collins made it with specific reference to expert systems.

He drew much fascinating evidence from his closely detailed case study of laboratory work in crystal growing, and of an attempt to write an expert system for this domain (1990, chs. 10–12). This evidence cast doubt on whether AI programs could be relied on to interpret new experimental data. "DENDRAL had done pretty well", an AI defender might say. But Collins would have replied that the novel spectrographic data were perceptually monitored by a human being, to check for noise, before being passed on to the program (1990: 128–32). Without such monitoring, DENDRAL could easily have been misled (cf. the discussion of BACON in H. M. Collins 1989).

In addition (the last arrow left in the quiver), Collins argued that *using* an AI system requires one to draw on background skills and assumptions that aren't normally made explicit. Indeed, he said, the same applies with respect to using a hand-held calculator, or even adding on one's fingers.

For example, in counting on one's fingers one must be able to recognize a finger, to remember when a finger has been used, and so on (Collins 1990: 48). Any or all of these skills might be lost after neurological damage. And in using a calculator to work out someone's height in centimetres, one must know what type of answer counts as "correct" (pp. 53–8). To say that a man who is 5 foot 9 inches tall has a height of 175.26 centimetres, because the calculator says so, is *not* to use the machine properly.

In that case, of course, the calculator does at least *do the arithmetic* correctly. But Collins described several intriguing examples where the computer gets the arithmetic wrong (pp. 66–70). For instance, 7 divided by 11, multiplied by 11, came out on one machine as 6.9999996. On another, the same sum resulted in 7; but *the very same machine, set to work more "accurately"* also delivered the answer 6.99999988079071. In such cases, an arithmetically sensible human being is needed to retrieve the situation. If they have some glimmering of understanding of the machine concerned, so much the better—but this isn't strictly necessary.

Moreover, using numbers, never mind using computers, is a community-specific practice. It's governed by norms of correctness that one learns—or not—from one's culture. These norms don't merely influence the numerical exactitude we deem suitable in describing someone's height, but also determine what doing arithmetic *is*. (The same applies to mathematical proof, and to its automation: Mackenzie 1995.) "In lands remote from Western culture", said Collins, "the calculator is no more an arithmetician than two sticks rubbed together are a slide rule" (p. 71).

Similarly, one must have some sense of the purposes and limits of an expert system to be able to use it properly. In other words, an expert system *purely by itself* doesn't, and can't, supply reliable answers. It can do so only when used by a person. And that person, besides having common sense and a vast store of worldly knowledge, is a member of a community, whose norms of "correctness" matter:

The intelligent computer is meant to counterfeit the performance of a whole human being within a human group, not a human being's brain. An artificial intelligence is a *"social prosthesis"*. In the Turing Test the computer takes part in a little social interaction. Again, when we build an expert system it is meant to fit into a social organism where a human fitted before. An ideal expert system would replace an expert, possibly making him or her redundant. It would fit where a real expert once fitted without anyone noticing much difference in the way the corresponding *social group* functions. (1990: 14–15)

Although he didn't say so here, it had been the express intention of many expert-systems builders that their programs be consulted *in lieu of* human experts. The resulting consultation might not happen "without anyone noticing much difference". But in the absence of the AI system, it might be that no consultation would happen at all. (Human experts are expensive, and geographically focused too.)

Collins didn't dismiss the AI enthusiast's claim that expert systems can be useful, even *very* useful. By 1990, after all, their usefulness had been demonstrated over and over again. And Polanyi himself had said that formalization "immensely expands the powers of the mind". For many practical purposes, then, Collins's argument could be ignored. But it came into its own in contexts where expert systems were being considered for purely automatic use. If Collins was right, the human being should never be taken out of the loop.

During the 1970s, very few people had been considering doing that. Even Kenneth Colby's plans for computer psychotherapy, already denounced as "obscene" by Weizenbaum, hadn't suggested that the psychiatrist be sidelined, nor (yet) that patients self-administer the computer therapy in their own homes (see Chapter 7.i.a). At that time, expert systems clearly weren't good enough to be left to run the show by themselves.

In the 1980s, however, expert systems entered deeply into the business world—and the military world too. Many in the AI community found this deeply worrisome, and some passionately resisted the US government's "Star Wars" plan for using automatic AI experts in defence (see 11.i.c). Most of that resistance drew its power from their expert realization that the *technical* correctness of huge AI programs couldn't be guaranteed. But some of it was based on considerations similar to those which Collins would publish a few years later.

For instance, Henry Thompson (an NLP specialist at the University of Edinburgh) told the salutary tale of a nuclear near-disaster which had been averted only because common sense—and a sense of human community—had led someone to doubt what the computer was saying (H. Thompson 1984). A nuclear red alert in the USA had been caused by an unknown object on the horizon. The reason why this frightening episode didn't escalate further was that someone ruminated that the Soviets hadn't been making especially threatening remarks recently. The norms of political behaviour, even during the Cold War, therefore made it highly unlikely that this mysterious object was a Soviet attack. And the same norms deemed it inadmissible to launch defensive nuclear weapons on the basis of such weak—i.e. *politically* implausible—evidence. Accordingly, the computer was overridden. (The unknown object eventually turned out to be the rising moon: 11.i.c.)

That's not to say that whenever someone uses their common sense to contradict the computer, they're right. Consider another salutary tale, the story of the 1979 Three Mile Island accident in Pennsylvania (Reason 1990, ch. 7). There, a nuclear meltdown very nearly happened (and all children and pregnant women were evacuated, by order of the Governor) when someone sensibly decided to ignore the computer's advice—"sensibly", because the advice would have been disastrous *in any circumstances other than the highly improbable situation which actually obtained.*

That particular decision might have been prevented by a better human–computer interface. For it happened largely because the operator was overwhelmed by myriad dials changing simultaneously: nearly 200 alarms went off within the first few minutes (Michie and Johnston 1984: 57). After the official inquiry into this crisis, the instrumentation in some manned control rooms was simplified accordingly. But there's no way of avoiding such mistakes entirely.

Nevertheless, if the person in Thompson's moonrise story, with their community-sensitive political antennae, had been taken out of the loop then he might not have been there to tell the tale, and you might not have been here to read it. Many less dramatic examples were discussed in warnings given by other AI-friendly writers besides Thompson. For instance, people working in or sympathetic with the field of expert systems gave various warnings about their use in psychotherapy (7.i.a), law (see below), medical diagnosis (Dennett 1986), and other socially relevant areas (Boden 1977, ch. 15; Council for Science and Society 1989, chs. 3–5).

Some Western AI scientists advised using the Japanese term "job assistant systems" instead, because it made the responsibility of the human user, and the (partial) inferiority of the program, more evident (Council for Science and Society 1989: 5). Some warned their fellow AI professionals not to mislead the public by hyping their programs, and asked that they try to prevent the marketers from doing so too—backed up, preferably, by an official "Code of Conduct" for the profession (Bundy and Clutterbuck 1985; Whitby 1988). And some even included specific warnings in their programs and/or the accompanying literature, to ensure that users didn't ascribe real intelligence or, worse, responsibility to the machine (Sieghart and Dawson 1987).

In sum, Collins's hostility to AI in general and expert systems in particular didn't prevent him from having some highly pertinent things to say. Whether many AI researchers (as opposed to Collins's philosophical sympathizers) actually bothered to read them is another matter.

c. Matters-in-law

As a special case of the type of worry raised by Collins, and one which raises many issues within other areas of cognitive science, let's consider expert systems in law. After outlining (in this subsection) how these have been developed over the years, we'll look (in the next) at some of the "human" problems regarding their use today.

When Herbert Simon modelled Supreme Court decisions as wiring diagrams in the 1940s (6.iii.a), he was doing this *post hoc*: there was no suggestion that the electronic circuits might be used by the judges themselves—still less, that they might take over. And when Lucien Mehl, at the famous NPL meeting in 1958 (6.iv.b), envisaged using automated logic for legal inference he was scolded in the discussion by Yehoshua Bar-Hillel for being "very premature, to put it mildly" (Mehl 1959: 783).

Premature, it certainly was. But in principle, not totally unrealistic. In 1976, when Weizenbaum published his impassioned critique of "AI judges" (11.ii.d), a few people were already starting to speculate about how legal reasoning might be modelled by AI.

Bruce Buchanan (with Thomas Headrick of Lawrence University) had written a speculative paper on this topic for the *Stanford Law Review* some six years earlier (Buchanan and Headrick 1970). And two years after that, in April 1972, the Stanford Law School had held a 'Workshop in Computer Applications to Legal Research and Analysis'. Most of the twenty-eight participants were lawyers (including one from the early-established Institute of Computing and Law in Toronto). But they also included Feigenbaum, Buchanan, and Stanford's L. Thorne McCarty (1944–).

Whereas Feigenbaum and Buchanan (and Headrick) talked about AI and expert systems in general, McCarty (1973) reported on the "pilot" for TAXMAN. For despite his modest disclaimers "I don't yet have any results . . . [only] tentative and incomplete [remarks]" and "Most of what I describe has not yet been implemented" (pp. 1, 4), he'd already started programming the first expert system in law.

TAXMAN's area of expertise was the taxation of corporate reorganizations. Its aim wasn't to do arithmetic ("You owe $350,000 in tax") but to identify tax-free, or relatively tax-free, reorganizations. In addition it should argue for *and also against* the case that a particular reorganization would be free of taxes. The data—including rules defining legal concepts such as *stock, voting stock, equity, control, acquisition . . .*—were represented in a semantic net, modifiable at any time. (Interaction with users, McCarty intended, would improve the system continually.) Every assertion was linked to an additional data structure giving its legal justification—and also "some indication of how it can be subsequently attacked" (p. 6).

McCarty saw TAXMAN as having wide potential. Part of his reason for choosing corporate taxation as the topic was that

the law here has an interesting history: it shows a complex interplay between broad common-law judicial concepts and specific statutory declarations, and this developmental process is one aspect of the law that I want the TAXMAN program eventually to focus upon, however difficult that may be. (McCarty 1973: 2)

For example, the crucial concept of "business purpose" was probably *not* capturable by a set of legal rules, but was grounded in "a complex set of arguments based on a vast collection of knowledge about [corporations] and the purposes of the reorganization laws" (p. 10). It remained to be seen how far this could be captured by AI heuristics. If case law were ever to be included, this would require programs for dealing with analogies/disanalogies between cases: a problem "more important and interesting than any I have talked about so far", and "the really challenging part of the research" (pp. 12, 14).

But even in the case of failure, he said, everyone would gain. On the one hand, legal analysis would be as helpful for advancing AI in general as chess, blocks world, or even mass spectrometry (i.e. Feigenbaum and Buchanan's DENDRAL: 10.iv.c). On the other, the attempt to write legal AI programs would provide insights into jurisprudence and the nature of legal reasoning.

McCarty was right (Rissland *et al.* 2003). However, he was a lone voice at that time. Buchanan was interested, as we've seen. But his energies were focused more on chemistry (DENDRAL) than law. It would be some years before any serious attempts, other than McCarty's (1977, 1980), were made to design legal reasoners as opposed to speculating about them.

The most influential of these was due to the PROLOG-implementer Robert Kowalski. In the mid-1980s, he and his student Marek Sergot would be invited to participate in an IJCAI Panel Discussion on the legal implications of AI (Boden *et al.* 1985). Whereas some panellists focused on questions about legal responsibility for the use/non-use of expert systems in general, Kowalski and Sergot discussed computer representation *of the law itself*. Today, their influence is still strong: many current legal AI systems are based on their pioneering example (and Sergot has been a President of the International AI and Law Association).

Their first exercise in this area was a PROLOG version of the British Nationality Act (BNA) of 1981 (Kowalski and Sergot 1985; Sergot *et al.* 1986; Kowalski 1992). Soon afterwards, they produced comparable programs representing legislation in other areas: immigration, government grants to industry, pensions, and taxes; and sick-pay regulations (Sergot *et al.* 1986: 385).

They'd chosen to model the BNA for three reasons. Two are unsurprising: the BNA was an example of statute law, which is "basically definitional in nature" and so less difficult to handle than case law; and, being then so recent, it was still relatively free of "the complicating influence of case law" (Sergot *et al.* 1986: 370).

The third reason was less obvious—and less open to counter-cultural scorn. For they chose this piece of legislation partly because it was highly controversial. The civil rights movement in the UK was up in arms, because the recent Act had introduced several new classes of British citizenship. Instead of all citizens being equal, some were now more equal than others. "We hoped", said Kowalski, "that formalization of the various definitions might illuminate some of the issues causing the controversy" (ibid.).

His hope was reasonable. For even if—as critics of logicism would have said—the attempt to corral legal concepts of nationality within strict PROLOG definitions was doomed to failure, it might help clarify the concepts concerned. Indeed, trying to write laws in the form of expert systems often exposes unrecognized gaps or ambiguities, just as it does for psychological theories.

By the mid-1980s, for instance, one US state legislature was already drafting its regulations in programmed form for this reason (Council for Science and Society 1989: 23). Similarly, a program produced at much the same time to mimic the initial selection of applicants for St George's Hospital Medical School in London was exposing previously unsuspected racism in the human selectors' procedures (CRE 1988). The racism may have been (to borrow Weizenbaum's word) obscene, but the whistle-blowing program wasn't.

Kowalski and Sergot's early systems were never used as a matter of routine. They weren't installed in lawyers' (or civil servants') offices and let loose on legal/administrative decision making. But besides any help they may have provided to legislators in drafting, they did have a strong influence on many advice-giving systems—such as an adviser on social security entitlements (Browne and Taylor 1989).

Today, BNA-type programs are being used by companies and governments in the UK, Europe, Australia, and USA (M. Sergot, personal communication). These deal, for instance, with building regulations, safety rules, and tax laws. The human–computer interfaces are hugely better now than in the 1980s, but the logical core is broadly similar.

So Kowalski's vision has been vindicated? Well, yes and no. For Kowalski's main interest was/is in cognitive science, not technological AI. (The same applies to McCarty.) His main aim wasn't to produce AI programs for lawyers to use. Rather, it was, and still is, to understand and help to improve everyday thinking in general—even including conflict resolution where, as in the Palestine–Israel dispute, the two sides have logically incompatible goals (Kowalski 2001, 2003, in preparation; Kowalski and Toni 1996). He sees legal thought as a helpful halfway house between logic and full informality. The hardy perennial of default reasoning, for example, is somewhat clearer in legal/administrative contexts than in idle gossip about the day's latest news (Bondarenko *et al.* 1997).

However, technological AI has benefited too, for logic programming as such has been affected. Kowalski (2002) has recently suggested several extensions based on types of language and argument needed in legal and everyday contexts. One of these adds "ordinary negation" to "negation by failure" (see 10.iii.b and v.f).

Whereas Kowalski and Sergot came to AI-and-law from their expertise in GOFAI, others came to it from law and jurisprudence. Leith, now Professor of Law at Queen's University, Belfast, is a well-known supporter of the use of computers in law, but a long-standing critic of AI models of legal reasoning. He was among the first to criticize logicist programs such as Kowalski's (Leith 1986*a*).

His core complaint was, and remains, that law is not—or not usually—a matter of cut-and-dried rules, but involves social negotiations of various kinds on the part of the barristers and judges who interpret the law:

It seems to me to be all very well to draw up a collection of rules from legislation; but, as lawyers all know intimately, a piece of legislation is but one thing in the legal world. (Leith 1986a: 551)

In principle, of course, certain interpretations can be favoured in an AI system. But if that's done, then other possibilities are tacitly excluded.

Leith follows up Kowalski's admission that common law is especially difficult to model in AI terms, and that a concern for precedents can complicate even statutory law. Precedents in law provide the human paradigm of "case-based reasoning", a technique whose AI versions are widely used in many different contexts (Kolodner 1992, 1993; Leake 1996). For example, it was incorporated in the MINSTREL storytelling program, enabling it to generate the concept of *suicide* from that of *killing* (S. R. Turner 1994; see iv.c below).

In case-based reasoning, a previous problem/solution is identified, somehow (methods vary), as being similar to the current problem, and is then used as a template for solving it. Various aspects of the previous case's structure, with modifications as necessary, are mapped onto the current situation. This isn't straightforward, for both the identification and the modification involve analogical reasoning.

Modelling case-based reasoning in law, as McCarty had forecast in the early 1970s, would require a very powerful theory of analogy (Rissland 1985). But analogy isn't a simple matter—which is why the initial judgement of one person (lawyer) can be altered by reasoned negotiation with another. Common sense, not to mention cognitive scientists' work on relevance (7.iii.d), suggests that programs will interpret the law in ways much more crude, and much less just, than humans do. That applies even to legal expert systems with a case-based component (e.g. Rissland and Skalak 1991; Skalak and Rissland 1992; O'Callaghan 2003; O'Callaghan *et al.* 2003*a,b*).

More recently, Leith (1992) has pointed out that law-in-use typically involves spoken argument, not only in court but also in preparation (identifying the "facts" and the core issues, and even negotiating with clients about their instructions). This adds further cognitive dimensions, over and above those relied on by readers of law books. And it's not only computer scientists who forget this. Academic lawyers forget it too:

The history of academic thinking about law and legal analysis in the 20th century has been the history of the textual nature of law. This most solid strand of legal thought has almost completely dominated and blotted out all other views; it is the view that all that lawyers need to know about law sits on the bookshelves of law. (Leith 1992: 227)

Legal *practitioners*, said Leith, know better. But they don't write the law books. (They may, of course, write novels and screenplays, in which the narrative can turn on just what's said in the witness box, even on *just how* it's said; remarks in letters can have similar effects on juries: e.g. the insurance assessor's "You must be stupid, stupid, stupid" in John Grisham's *The Rainmaker*.) Deep analysis of rules is all very fine for the legal academic, Leith admitted, but "if we want to develop and encourage gifted lawyers, then [we should focus on] the communicative skill of rhetoric—as the essence of advocacy" (p. 233). This realization, he continued, "takes us far away from current research strategies in AI and law" (p. 234).

He was right: AI modelling of 'law as she is spoke' is a tall order. This is evident, for instance, from Bernard Jackson's (1995) subtle exploration of the numerous interpretative complexities. And although some current AI work does focus on the rhetorical structure of legal arguments, it's still relatively crude (e.g. Ashley 1990; C. Reed 1997; Reed and Norman 2004). Grisham needn't fear that an AI-advised rival will replace him on the best-seller lists.

Law, logic, analogy, even rhetoric . . . it's clear that this is a wide-ranging area. Both professional lawyers and AI researchers take part in the ongoing International Conference on AI and Law (the ninth in the series was held in Edinburgh in 2003). Similarly, both contribute to journals such as *Artificial Intelligence and Law* and the *International Journal of Law and Information Technology*. And very recently, in November 2003, the core *Artificial Intelligence* journal published a special issue on AI and law, with contributions from both sides of this disciplinary fence. In addition, the more general research is sometimes reported also at cognitive-science meetings (e.g. Ashley and Keefer 1996).

The technical discussions range from dedicated models of specific legislation to wide-ranging work on the psychology of case-based reasoning and argument structures. In short, insights from AI, law, and psychology contribute to what is an essentially interdisciplinary enterprise.

d. Judgements about judges

Not all of the discussions on AI and law are "technical", however. Some are ethical. Indeed, we've seen that some law programs, such as Kowalski's version of the British Nationality Act, were written partly in order to clarify morally troubling aspects of the law.

But that was clarification, not application. Now that various legal programs are actually being used, and more are in the Establishment-approved pipeline, arguments about their appropriateness abound (e.g. Susskind 1987, 1993, 2000; Leith 1986*a*,*b*, 1988, 1992; Whitby 1996*b*: 44–63). The disputants include lawyers, as well as AI and cognitive-science researchers.

Many of these disputes concern the reliability of expert-systems representations of statute and/or common law, including computational and jurisprudential questions about the relation between law and logic. Others spill over into by-now-familiar questions about how to model the flexibility of human thought. For instance, the psychologist Jerome Bruner has recently discussed the combination of system and messiness in legal reasoning; he sees this as being grounded in case-based reasoning, but stresses the *narrative* aspect of the "cases" concerned (Bruner 2002; Shore 2004: 148–9). Yet other disputes express worries about situations in which the human being might cede responsibility for judgment to the program.

In general, the concerns differ according to just who is (or will be) using the expert systems, and why.

Most law-related programs today are intended for in-house use in lawyers' offices, or by people giving advice to citizens on entitlement to benefits and other regulatory matters. Examples include the latest version of TAXMAN, and the regulatory programs modelled on the work of Sergot and Kowalski. "Merely information retrieval", you may say. But even these programs can raise Collins's worry about removing human discretionary judgement from the loop (Browne and Taylor 1989).

Others are intended for use in the law courts, to aid decisions on the verdict (in non-jury trials) or on sentencing by judges or magistrates. These are even more problematic.

For they revive the McCarthy–Weizenbaum question (11.ii.d): whether computers might ever replace judges—and if so, whether this is desirable.

Just in case you think that it so obviously *isn't* desirable that it's simply not worth discussing, you should remember the remarks of the several young black Americans quoted in Chapter 11.ii.f. For instance:

I know that if I ever had to go before some judge, there is a good chance that he is going to see my face and he is going to think "nigger".

The computer judge would have a set of rules. He would apply the rules. It might be safer. (quoted in Turkle 1995: 292–3)

The question, then, is whether such fairness could be achieved (whether "the rules" could be adequately expressed), and if so whether this would do significant harm to the human dimensions of the law courts.

Such topics are no longer maverick, and many people today share some of Weizenbaum's concerns. For instance, in 1989 the UK's Council for Science and Society, an opinion-forming group with close links to both Houses of Parliament (and other sectors of the Establishment), predicted that an expert system might in future be used to help a Crown Court judge pass sentence. So the judge might say:

Norman Stanley Fletcher, you have pleaded guilty in this court to a series of offences. In the light of your previous record, my normal inclination would be to put you on probation. However, I have consulted DISPOSAL, the expert system [in the laptop] I have on the Bench with me. Having entered all the relevant particulars about you, it tells me that the prospects of your responding favourably to probation are only in the range of 14 to 17 per cent, and that the statistical significance of this result is very high. It also appears that the prospects of success of any other non-custodial sentence that I might consider would be even lower. Accordingly, I have no option but to sentence you to a term of imprisonment of seven years . . . (Council for Science and Society 1989: 26)

Whether a successful appeal in such a case could be based on the judge's admission that he overruled his own judgment because of "what his microcomputer 'told him' ", the Report continued, "is an intriguing question".

That wasn't a joke—even though Norman Stanley Fletcher was the lead character in a hugely popular TV sitcom set in one of Her Majesty's prisons. (You may also feel that it's no joke.) This particular section of the CSS Report was drafted by the prominent British lawyer Paul Sieghart, a few months before his unexpected death in 1988. Sieghart was the founder of CSS, and also of an influential group (Justice) for considering tricky juridical/constitutional questions, especially in the area of human rights. Several important law reforms have been instigated by him, and he's still honoured by an annual Memorial Lecture that's been given by some of the country's most renowned judges and legislators. In short, he was a provocative thinker whose provocations often ended up being accepted as common sense. Moreover, he can't be written off as someone with no human sensitivities. Sieghart may have been smiling when he wrote those words, but they were seriously meant.

Comparable worries have been addressed by Richard Susskind (1961–). He's the Gresham Professor of Law and Visiting Professor at Strathclyde University's Centre for Law, Computers, and Technology. But he's no cloistered academic. (Is anyone, nowadays?) Besides his academic posts, he's the official adviser on IT to Lord Woolf, Lord Chief Justice of England and Wales. As such, his opinions are more influential than most.

Susskind has long held that "there are no theoretical obstacles, from the point of view of jurisprudence, to the development of rule-based expert systems in law of limited scope [such as the Scottish law of divorce]" (1987, p. vii). Accordingly, he's broadly in sympathy with the BNA-based approach. Indeed, he's helped to develop several similar implementations (e.g. Capper and Susskind 1988).

With respect to the Weizenbaum–McCarthy debate, he said this (in the late 1980s):

A detailed study of this matter is sorely needed both to dispel the profusion of misconceptions and to assure the public that while computers will no doubt provide invaluable assistance to the judiciary in the future, it is neither possible *now* (or in the conceivable future) nor desirable *ever* (as long as we accept the values of Western liberal democracy) for computers to assume the judicial function. (Susskind 1987: 249)

As for what one might expect a "detailed study" to show, Susskind was very clear about the difficulties involved:

Moreover, any legal theorist will recognize that construction of an expert system in law that could solve *all hard cases* would require . . . the development (explicitly or implicitly) of theories of legal knowledge, legal science, individuation, structure, legal systems, logic and the law, and of (judicial) legal reasoning, all of a far greater degree of richness, sophistication, and complexity than those that have been generated by the most adept jurisprudents in the past. (p. 250; italics added)

In other words, *Don't hold your breath!* Without using inflammatory terms such as "obscene", Susskind was echoing Weizenbaum's plea for human judges.

He's repeated this plea recently, by reissuing an early paper (1986/2000) in which he expressed his "horror" at McCarthy's reply to Weizenbaum, pointing out that among the many capacities of judges which computers don't possess are "moral, religious, social, sexual, and political preferences" and "creativity . . . intuition, commonsense, and general interest in our world that we, as human beings, expect not only of one another as citizens but also of judges acting in their official role". (Consider the scorn meted out to the judge—perhaps apocryphal, or joking—who asked "Who are the Beatles?")

The 1987 quotation, above, mentions "hard" cases. But even "clear" cases may not be so clear as is sometimes assumed. Susskind, again:

In any clear case, *according to positivists*, the judge is committed—perhaps by the value of legality—to making one particular legal decision (or else, we might say, the case is not clear). If the judge wishes to abide by the principle of legality, or for some other reason decides to remain faithful to the law, he has no choice but to make that decision, even if he finds it morally unacceptable . . . As a matter of logic, then, there is no need for judgment in clear cases; there is nothing required other than a conclusion/decision reached through the inexorable application of deductive inference procedures . . . (1986/2000: 286; italics added)

The positivist separation between law and morals, Susskind argues, is mistaken. However, many arguments intended to refute it do no such thing (pp. 282–92). In particular, arguments assuming that positivism, mistakenly, "tends to see judges as computers; as, at least in clear cases, simply certifying what in fact the law is, without necessary moral commitment" (p. 282) are often flawed. They fall foul of logical non sequiturs, to the author's failure to be "a natural lawyer" (p. 291), and—significantly—to ignorance of AI in general.

In sum, Susskind (like Collins) sees many problems regarding tacit judgements lying in wait here. Nevertheless, his views on jurisprudence don't outlaw AI from legal offices, nor even—as quasi-intelligent aide-memoires—from the judge's bench. The relatively clear cases, at least, are potential grist to the computational mill. Like Sieghart, Susskind is an influential voice in the British legal profession. So we can reasonably expect to see AI entering the law courts. Whether it will always be reasonably used there, and/or reasonably regarded by the lay members of the court, remains to be seen.

13.iii. Advance and Attack

If logicism was a key research focus right from the beginning of AI, many other themes mentioned by the NewFAI pioneers were not. They couldn't be fruitfully studied until other matters (e.g. KR) had been advanced. In the closing quarter of the twentieth century, however, they came into their own.

The result was both advance and attack. Some late-century advances delivered on promissory notes issued in the 1950s. Others went so far beyond the early efforts that they engendered a very different (i.e. non-psychological) research spirit. One development, rooted in old ideas, attacked the most cherished of NewFAI assumptions: this mini-Kraken, namely situated AI, is regarded as a 'hot' topic today. Similarly, "distributed" cognition, in which *multiple* agents (human and/or artificial) collaborate, is now "one of the main directions for AI research in coming years" (Doyle and Dean 1997: 95; cf. Grosz 1996).

a. Gelernter revivified

Herbert Gelernter's late 1950s implementation of Minsky's back-of-the-envelope sketch (10.i.c) turned out to be much more than a nine days' wonder. For the general idea of using *a model, or schema, of the problem* (of which a diagram is a special case) had a lasting effect on techniques for controlling search. Moreover, diagrammatic reasoning itself attracted increasing interest in the last two decades of the century.

For instance, in 1980 Funt described an ingenious way of implementing diagrams-as-analogue-spatial-representations (as opposed to Gelernter's diagrams-as-symbolic-line/angle-specifications). He used a simulated parallel processing 'retina' (a 2D visual array) made of many individual cells, wherein image transformations could be defined spatially rather than symbolically. One application modelled the image-rotation ex-periments that had caused such excitement in experimental psychology in the 1970s (Chapter 7.v.a); not only could the system decide whether one image was indeed a rotation of another, but the time it took to do so was proportional to the degree of rotation involved (Funt 1983). Another, even more relevant here, was the WHISPER program (Funt 1980).

WHISPER solved problems about stability and "chain reactions" in the blocks world. But unlike Scott Fahlmann's BUILD (10.iii.c), it didn't use abstract reasoning about

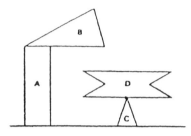

FIG. 13.1. Is block B stable? If not, what will happen when it falls? Adapted with permission from Funt (1980: 214)

physics to do so. Instead, it used qualitative knowledge about physics, and ways of continuously transforming one retinal image into another, to infer what would happen in certain situations. For instance, it could 'see'—or anyway, 'imagine'—that object B in Figure 13.1 is unstable, and that on falling it will hit object D, which will then tip over until the lowest point rests on the floor to the left of object C (Funt 1980, esp. 213–29).

WHISPER's method of representation was very different from most AI programs, and Funt's 1980 paper was chosen as one of those reprinted in the influential *Readings on Knowledge Representation* (Brachman and Levesque 1985). Given that this collection offered a thirty-page bibliography of relevant publications which hadn't been included, that was no small accolade.

Alongside it was a theoretical paper on analogical representation by Sloman (1975), whose earlier work on this theme had inspired Funt to build WHISPER in the first place (Sloman 1971; cf. 1978, ch. 7). As the editors of the *Readings* pointed out (p. 431), Sloman was one of the few people to have thought deeply about the difference between analogical and formal representations—a topic that was causing huge controversy in psychology at that time (7.v.a).

A main aim of Sloman's 1971 paper had been to clarify what's going on in "intuitive" reasoning. Since WHISPER's inferences were based on its imaginative transformations on the retina, they were intuitive rather than deductive. Indeed, this is true of diagrammatic reasoning in general. But how could logicians or mathematicians be confident that intuitive reasoning is valid?

Over the past quarter-century, Alan Bundy's research group at Edinburgh have addressed that question, seeking to justify intuitive reasoning in deductive terms. Specifically, they've worked on mathematical induction, in domains ranging from program synthesis, through arithmetic, to geometry (e.g. Bundy 1988; Bundy *et al.* 1990). In mathematical induction, of which diagrammatic reasoning is a subclass, specific results (achieved by intuitive "eureka steps") are used to justify general conclusions.

The Edinburgh group's general "theory of diagrams" enabled them to show how a diagrammatic proof can be generated, then schematized, and finally verified (Jamnik *et al.* 1998; Jamnik 2001). More recently, their interactive program "Dr. Doodle" has helped students to reason more easily about a mathematical domain that's usually represented algebraically (Winterstein *et al.* 2004). (One shouldn't assume that diagrams or visual images always aid problem solving; sometimes, they block it—Boden 1990*a*: 100–10.)

Many others, too, took up Gelernter's baton. In the 1980s, the journal *Artificial Intelligence* ran a 400-page special number on geometric reasoning (Kapur and Mundy 1988). By that time, too, researchers were able to knit new AI garments by wielding it alongside Ivan Sutherland's baton of computer graphics (v.c, below). Sutherland's pioneering ideas of the early 1960s had now been properly implemented, so that diagrams of three-dimensional structures could be generated, moved, rotated (with automatic deletion of hidden lines if necessary), and transformed. Plane geometry doesn't need such changes, but some diagrammatic thinking does.

In brief, diagrams have come in from the cold. Even McCarthy now allows (or "I would, if I had a definite idea of what a diagram is": personal communication) that they may be a more efficient form of representation for some purposes than logic is—although where logic can be used in a running program it's preferable, and it should in any case be used to specify what a program does. Today, there's a biennial interdisciplinary conference called 'Diagrams' (devoted to "the theory and application of diagrams in any scientific field of enquiry"), besides sessions and workshops scattered across various other meetings.

The 'Diagrams' series attracts people from AI and cognitive psychology, through education and human–computer interaction, all the way to architecture and carto-graphy. So besides being highly interdisciplinary, it involves both psychological and technological AI.

b. Planning attacked—

The NewFAI researchers, one might say, had been committed Marxists. For Karl Marx had stressed the role of anticipatory planning in intelligent human labour:

[Man] sets in motion the natural forces which belong to his own body, his arms, legs, head and hands, in order to appropriate the materials of nature in a form adapted to his own need . . . [So do animals—but human labour is different.] A spider constructs operations which resemble those of the weaver, and a bee would put many a human architect to shame by the construction of its honeycomb cells. But what distinguishes the worst architect from the best of bees is that the architect builds the cell in his mind before he constructs it in wax . . . Man not only effects a change of form in the materials of nature; he also realizes his own purpose in those materials. (Marx 1890: 283–4)

Simon, Fahlman, Sacerdoti, Allen Newell, Gerald Sussman, Richard Fikes, Nils Nilsson . . . all had agreed. Indeed, it would be difficult to name any NewFAI sci-entist, or any psychologist following MGP's manifesto (6.iv.c), who didn't. Many NLP researchers in the 1980s, for example, studied ways in which communicative (and other) plans can be recognized and/or put into effect (Cohen *et al.* 1990). And mainstream robotics continued to rely on planning (Brady 1985).

In the mid-1980s, that comfortable consensus was challenged. The attack came from several different research areas: behaviour-based robotics, interactionist AI, and situated automata theory.

The common theme uniting these "situationist" approaches was their focus on "[systems] having an intelligent ongoing interaction with environments that are dynamic and imperfectly predictable" (S. J. Rosenschein and Kaelbling 1995: 149). They emphasized embodiment, and rejected planning (and mental representations in general). In effect, they advised workers in AI/robotics to emulate Marx's spiders and to forget his architect. (It was even suggested—by David Cliff: 15.vii.b—that "AI" should be read not as "artificial intelligence" but as "artificial insects".)

Their work was soon dubbed nouvelle AI (which also covered animate vision, in which the creature's own movements provide crucial information: 15.vii.b). It sparked methodological changes in robotics research and philosophical changes in cognitive science. Eventually, it excited members of the general public too.

Situated automata theory was the first to make its mark. It was originated in the early 1980s by the computational linguist Stanley Rosenschein at Stanford's CSLI, and developed with Leslie Kaelbling at Brown (S. J. Rosenschein 1981, 1985; Rosenschein and Kaelbling 1986; Kaelbling 1988). The theory was used to design robots—the first, disarmingly named Flakey. But Flakey and SHAKEY, despite the rhyming similarity, were very different:

[Situated agents] are very difficult to program because of their close interaction with the environment in which they are situated. Specifications of correctness for situated agents amount to specifications of their interactions with the environment: what action should the agent take when the environment is in a particular configuration? . . .

The emphasis on an agent's connection to its environment is *an important change* from that of traditional theories of representation and control. [We] use high-level symbolic languages to describe the informational content of agents *without* requiring the symbolic structures to be implemented in the agent. (S. J. Rosenschein and Kaelbling 1995: 149–50; italics added)

Rosenschein used a logical specification of the task, and a model-theoretic semantics of motor action, to design robots driven by hardware circuits rather than programs. The circuits ensured that a specific action would take place in a specific situation, and the logic ensured that the overall task would be achieved. Flakey's designers later rejected the "folk wisdom" that representation and situated automata were incompatible (1995: 150). Nevertheless, it was clear that they understood *representation* in a way far removed from the GOFAI norm.

Their approach had some influence within AI, and was later used to design mobile robots for exploration and factory use (Kaelbling and Rosenschein 1990; S. J. Rosenschein and Kaelbling 1995). But it was too austere to excite outsiders. The most widely prominent situationists came from the other two groups mentioned above.

The most well known were the situated roboticists, hard-headed engineers building mobile (and media-friendly) gizmos to perform practical tasks. Some of these gizmos were dubbed "epigenetic robots", because their control systems and behaviour *developed* according to the features of their physical and social environment (see 15.viii.a). But intellectual waves were made also by the interactionists: maverick computer scientists who hoped to develop Heideggerian philosophy into a new way of doing AI. (Increasingly, some people fitted both descriptions: M. W. Wheeler 2005.)

The situated roboticists made four positive claims, and a negative one. Positively, they said:

1. that "intelligent" robots could be engineered rather than programmed, and would be faster, more reliable, and more ecologically valid as a result;

2. that modelling/understanding human intelligence will be possible, if at all, only if we have first modelled/understood simpler animals, such as insects;
3. that AI needs to model *complete* sensori-motor systems (what had been termed "the whole iguana": Dennett 1978c), as opposed to specialist AI sub-areas; and
4. that biology should be respected, not ignored.

The last claim reflected the fact that late-century neuro-ethologists were describing reflex mechanisms showing how Marx's spiders construct their delicate "operations" (Barth 2002), and how various other animals manage to live their lives.

In new-style robotics labs around the world, cockroaches and crickets inspired artificial creatures with ne'er a plan in sight (see Chapter 15.vi.c, vii.b–c, and viii.c). These were far less shaky than SHAKEY, being capable (as ever: up to a point) of walking on rough ground and climbing over obstacles, and of adaptive response in real time. By the end of the century, they would be used for applications ranging from Mars rovers to hugely popular children's toys (R. A. Brooks 1999). Some jokers, as we've seen, remarked that "AI" now stood for "artificial insects".

The negative claim was that intelligence doesn't require planning—indeed, that it doesn't need internal representations of any kind. (That was the extreme version. Sometimes, it was stated more ambiguously: the key situationist paper remarked that "much" human-level activity needs no representations, tacitly implying that logical/linguistic thought might do so—R. A. Brooks 1991a: 148–9.)

This claim was guaranteed to raise the hackles of functionalists (16.iii–iv)—which is to say, all traditional cognitive scientists. By contrast, it delighted people who were already hostile to cognitive science, and especially to GOFAI. Unsurprisingly, then, it spread apace. The seductive combination of telegenic robots and fierce attacks on symbolic AI attracted significant public attention—and sympathy.

Often, these new roboticists used a catchy phrase to express their view: *the cheapest store of information about the real world is the real world*. In fact, this remark was so catchy that they'd apparently caught it unawares from their predecessors. For that very phrase, and near equivalents, had been used long before by the now-scorned NewFAI folk: for instance, Feigenbaum (1969: 1011), Donald Michie and Richard Gregory (Michie 1970: 76), and even MGP themselves (G. A. Miller *et al.* 1960: 78). However, that didn't prevent its being true. GOFAI robots, such as SHAKEY, did indeed take an unconscionable time to update their world models, and to decide what to do next. Besides the huge computational expense, this put them at the mercy of events even in a sluggishly changing world.

The most high-profile iconoclast was MIT's Rodney Brooks (1954–), with Case Western's Randall Beer (1961–) as runner-up in the visibility stakes. Brooks was known to roboticists by the mid-1980s (R. A. Brooks 1986, 1987). But he opened the 1990s with a more generally noticeable splash, publishing two highly provocative papers previously available (as MIT memos) only to the cognoscenti. In them, he made the four positive claims distinguished above and drew the obvious implication: that symbolic AI was a waste of time—at least if one's "dreams" were for psychological understanding, rather than techno-dollars: "traditional Artificial Intelligence offers solutions to intelligence which bear *almost no resemblance at all* to how biological systems work" (1991b: 1; italics added). And he defended the negative claim too:

[In building several autonomous robots we] have reached an unexpected conclusion (C) and have a rather radical hypothesis (H):

(C) When we examine very simple level intelligence we find that explicit representations and models of the world simply get in the way. It turns out to be better to let the world itself serve as its own model.

(H) Representation is the wrong unit of abstraction in building the bulkiest parts of intelligent systems. (1991*a*: 139–40)

In a sense, this was a step back in time. For the anti-mentalist (fifth and sixth) tenets of behaviourism were being revived, even though the underlying physical/neurophysiological causes were now being considered (5.i.a). GOFAI-based computational psychology, by implication, was on the wrong track.

Despite having his home in an AI lab, Brooks's work was widely regarded by the mid-1990s as an example of A-Life, not of AI. Indeed, he often published in A-Life sources (e.g. Brooks 1992), and he co-edited some A-Life volumes (Brooks and Maes 1994; Steels and Brooks 1995). (That's why I've chosen to describe his—and Beer's—robots in the A-Life chapter, rather than this one.) Moreover, his negative attitude to GOFAI was shared by A-Lifers in general. Accordingly, the media-hyped success of situated robotics didn't persuade the general public that AI had advanced, but that it had failed (see vii.b, below).

The negative claim was argued also by the interactionist Philip Agre (1960–), initially at MIT and then at UCSD. He did much of his early work in cooperation with David Chapman at MIT, and the anthropologist (and ethnomethodologist) Lucy Suchman at Xerox PARC, who'd arrived at a similar position independently (Suchman 1987; Agre 1990).

In the 1990s, Agre would focus on human–computer interaction (Agre and Schuler 1997; Agre and Rotenberg 1997), and on human–human interaction too (his advice on the social conventions for Internet communication became a widely cited item on the Net: Agre 1998–2003). He even developed a "participatory" account of computation *as such* (see 16.ix.d). But the seeds had been sown in his interactionist research of the mid- to late 1980s.

In his MIT thesis written at that time, Agre had attacked planning even more explicitly than Brooks was doing. He said that, of all the non-classical AI work being done at the time, Brooks's was "by far the most similar project to my own" (Agre 1988: 247). But whereas Brooks started from robots and insects, Agre started from (largely introspective) human psychology and philosophy—especially Martin Heidegger. (Believe it or not, his thesis was supervised by none other than Michael Brady—not someone whom one readily associates with Heidegger.)

Calling for "an inversion of values" in both cognitive science and technological AI, he declared:

Faced with an empirical phenomenon to explain, our first explanatory recourse should be to dynamics, not to machinery. Faced with a technical problem to solve, our engineering should begin with dynamics, not with machinery. (1988: 27)

Dynamics, he said, "concerns the interactions between an individual (robot, ant, cat, or person) and the world" (p. 22).

The "ant", here, was Simon's ant, whose path across the beach was continuously determined by the sand grains it encountered (Chapter 7.iv.a). And the "person" was any one of us, engaged in our everyday activities. Most of these, said Agre, are routine, involving little conscious thought and no anticipatory planning.

Like the anthropologist Anthony Wallace (Chapter 8.i.a), Agre picked his daily journey to work as a prime example. But his approach was very different. Whereas Wallace had analysed "driving to work" in terms of Plans and TOTE Units, Agre described "walking from his home to the subway" as a routine consisting of a number of major steps, either run off automatically or triggered by interaction with (aka perception of) the environment (1988: 39–51). Even automatic actions, however, would be executed slightly differently each time, depending on the specific situation (the height of the kerb, the location of other pedestrians, and so on).

Agre's descriptions of routine activity were heavily influenced by the transactional psychologist Eric Berne (1964, 1972), the ethnomethodologist Harold Garfinkel (1967), and the anthropologist Pierre Bourdieu (1977). But he went beyond these speculative theorists by implementing his ideas in an AI program, called Pengi (Agre and Chapman 1987, 1991; Agre 1988: 189–233).

This was a LISP Machine reimplementation of a commercial video game called Pengo in which humans used a joystick and a button to control a cartoon penguin. The Pengi penguin controlled itself—or rather, it was controlled by the events happening in its simulated world. The penguin had a goal: to evade simulated killer bees, while trying to zap them by kicking virtual ice cubes at them. But it didn't Plan how to achieve that goal, nor represent it internally as a goal-state. Rather, it engaged in "an improvisatory interaction with continually evolving surroundings" (Agre 1988: 189). Instead of "elaborate world models", it used temporary "deictic" (situation-bound) representations—*the-ice-cube-I-am-kicking, the-direction-I-am-headed-in, the-bee-I-am-attacking, the-bee-on-the-other-side-of-this-ice-cube-next-to-me*—to organize its interactions with its environment from moment to moment (pp. 190–5).

Agre reported that the program played the game about as well as he did, although less well than Chapman. And he claimed that it played in much the same way as humans do—indeed, in much the same way as people engage in any routine activity. Planning, on his view, is neither necessary nor even possible. He remarked (as Karl Lashley had done long before him: 5.iv.a) that some skilled motor action is so fast that it simply can't be controlled by deliberate plans and decisions. Pengo experts, for instance, can push the right buttons amazingly quickly when responding to what's on the screen.

More accurately, he thought that Planning-with-a-capital-P (i.e. as envisaged by NewFAI and by MGP) is impossible. People do plan and make plans. But a plan, he said, is very different from a Plan:

No plan could ever be so exhaustive that you could mechanically "execute" it. Carrying out a plan requires continual improvisation, interpretation, and fine judgement—especially about whether to revise or abandon the plan in mid-course . . . Real plans can be concise compared to Plans because they can rely on many aspects of how, where, and by whom they'll be used. (Agre 1988: 40–1)

ABSTRIPS, of course, had been an attempt to capture this run-time flexibility (10.iii.c). And as Agre pointed out, even "the very earliest definitions of Planning", by MGP and the LT–GPS team, had emphasized that the lower, tactical levels "can often be left unelaborated until it comes time to execute them" (1988: 235). But the elaboration, when at last it did come, was itself envisaged by GOFAI as a Plan.

Agre went even further: mental representations, the theoretical core of functionalist cognitive science, were denied. Deictic representations were admitted, but "objective" world models weren't. Mentalism was rejected:

Mentalism refers to any psychology or philosophy organized around metaphors of Inside and Outside . . . Above all, mentalism leads one to make theories that posit objects and processes residing entirely within the head. (1988: 19)

Mentalism provides a simple formula that provides plausible answers for all questions: put it in the head. If agents need to think about the world, put analogs of the world in the head. If agents need to act in situations, put datastructures called "situations" in the head . . . The tacit policy of mentalism, in short, is to reproduce the entire world inside the head: a "world model". (p. 20)

In arguing this position, Agre relied heavily on neo-Kantian philosophers such as Heidegger, Dreyfus, and Richard Rorty (Agre 1988: 36–7; see 16.vi–viii). But he was influenced also by cybernetics (4.v–viii) and what was about to be named A-Life (15.vii–x). That's clear from this list of "interactionist words" chosen to describe his own account: *interaction, conversation, involvement, participation, servocontrol, metabolism, regulation, cooperation, improvisation, turn-taking, symbiosis, routine*, and *management* (1988: 20).

"Mentalism and interactionism", he continued, "are incompatible . . . [Each] offers its own distinctive way of approaching every phenomenon of human existence." He made a seeming concession to mentalism: "Sure, perhaps some of these structures *are* entirely inside of agents' heads, but that's just an unusual special case with no particular privilege" (p. 20). However, whatever structures there might be inside our heads were assumed to be very different from the GOFAI norm. (A plan, remember, was not a Plan.)

Nouvelle AI quickly became a fashion, not to say a fad. It was reviving themes that had been neglected by most computationalists: notably, ecological validity (7.iv.g and v.e–f), animal intelligence (2.ii.d–e and 15.vii), and embodiment (16.vii). One of the people hugely influenced by its stress on reactivity and on dynamics was the young Beer (personal communication), who abandoned GOFAI partly as a result of Pengi (see 15.vii.c and viii.d).

Especially intriguing, to many observers, were the philosophical implications. The Cartesian ego engaged purely in thought (2.iii.a), and the sensori-motor sandwich (10.iv.b), were shunned—to be replaced by an alternative that had been waiting in the wings since the eighteenth century (2.vi).

Ironically, then, Brooks's (and Beer's) anti-mentalist emphasis on nuts-and-bolts engineering encouraged a renewal of interest in the type of philosophy favoured by the counter-culture. Heidegger and Maurice Merleau-Ponty, long wheeled in by Dreyfus to criticize GOFAI, were now names for cognitive scientists themselves to conjure with (M. W. Wheeler 1996, 2005). Even Brooks referred to Heidegger: he said of his own approach "It isn't German Philosophy", despite "certain similarities" to Agre's Heidegger-inspired research (R. A. Brooks 1991*a*: 155). Clearly, and whatever his private opinion (if any) about German philosophy, he realized that his name was being spoken in the same breath as Heidegger's by many of his fans.

c. —and defended

Not everyone, however, jumped on the situationist bandwagon. One who resisted it was the philosopher David Kirsh (1950–). He'd been at MIT for five years (before leaving for San Diego in 1991), so was familiar with Brooks's MIT memo on 'Intelligence Without Representation' (1987). Indeed, he organized a small Workshop on 'The Foundations of AI' at MIT's Endicott House, at which Brooks was an invited speaker.

When Brooks's MIT memo was officially published, in a special volume of *Artificial Intelligence* based on the Workshop, Kirsh's reply appeared alongside it (Brooks 1991*a*; Kirsh 1991*b*). Although specifically directed at Brooks, much of what he said ran counter to Agre as well.

Concentrating on the situationists' negative claim, Kirsh argued that any behaviour that involves concepts needs internal representations. He allowed, as Brooks had done too (1991*a*: 149), that "representation" is a weasel word, carrying multiple ambiguities. But *conceptual* representations were needed for a wide range of behaviours.

Full-blooded concepts, he said, enable a creature to recognize perceptual invariance, to reify and combine invariances (in referring predicates to names, for instance, or in drawing inferences), and to reidentify individuals over time.

They also allow creatures to engage in anticipatory self-control (i.e. planning), and to negotiate between, not just to schedule, potentially conflicting desires. (That robots might use reflex mechanisms to schedule conflicting motives had been pointed out twenty years earlier by the MIT team asked to design a Mars robot: see 14.v.a.) Moreover, they enable one to think counterfactually, to use the cognitive technology of language to create new abilities (6.ii.c), and—by teaching these abilities to others—to make cultural evolution possible (8.v–vi). Human adults possess all these capacities, chimps most of them, dogs some of them, and newborn babies hardly any (see 7.vi).

Kirsh wasn't arguing that non-human animals must possess symbolic, compositional, representations. He allowed that non-linguistic concepts and/or meanings may be implemented in other ways (see 12.x.f, 14.viii, and 16.iv.c). He even welcomed situated robotics, for having extended the domain of concept-free behaviour and for indicating the wide range of mechanisms that may be termed representations.

But he insisted that logic, language, and thoughtful human action all require symbolic computation. These may be, as Agre had put it, a "special case" of intelligent behaviour. But they're there. And they're significant. In sum, GOFAI might not be needed when studying cockroaches, but it was essential for modelling important aspects of human minds.

Simon, the high priest of GOFAI, also fought back against the situationist tide. As Brooks himself pointed out (1991*a*: 149), Simon had long allowed that intelligent systems are largely directed by the environment (7.iv.a). He'd already reformulated planning, now seeing it as a complex web of reflexes, or productions (7.iv.b). However, he thought of these as mental representations, some containing abstract variables to be instantiated before the rule could be run. (That's why Brooks, besides denying that his work was German philosophy, denied also that it was "production rules", and why Agre rejected production rules, ACT*, and SOAR—R. A. Brooks 1991*a*: 155; Agre 1988: 240–4.)

In a paper written with the psychologist Alonso Vera, Simon expressed the essence of situationism thus:

In its extreme form, the SA [situated action] view argues that there is no need to include internalized world models in the equation. Such internal states, some proponents of this view have said, have no causal effect on behavioral output. (Vera and Simon 1993: 12)

Reiterating his Physical Symbol System approach (16.ix.b), he argued that "no one has described a system capable of intelligent action that does not employ at least rudimentary representations" (p. 38). He insisted that the situationists themselves used representations in their models. And, doffing his hat to his famous ant, he pointed out that GOFAI systems could be highly responsive to environmental control.

The GOFAI–situationist dispute aroused great interest. Kirsh's critical paper, for example, was soon assigned in AI seminars at MIT and elsewhere (personal communication). The special number of *Cognitive Science* carrying the Vera and Simon paper presented Replies by Agre (1993), Suchman (1993), and others. Agre allowed that *Pengi* had "internal state" (e.g. deictic representations), but disputed Simon's assumption that *any* internal state counts as a cognitive representation. And he claimed that Simon had missed the crucial point: "The critical issue is whether one's categories locate things in agents and worlds separately or in the relationship between them" (1993: 68–9).

The debaters were soon accused of having "shed more heat than light . . . with neither side yielding enough ground to reach a point of productive dialogue" (M. D. Byrne 1995: 118). But dialogue, productive or otherwise, continued. Two consecutive volumes of *Artificial Intelligence* were devoted to interactionist AI in the mid-1990s, and prompted so much attention that they were soon republished as a book (Agre and Rosenschein 1995). And the 1991 papers by Brooks and Kirsh were reprinted in various collections on the philosophy of AI and cognitive science (e.g. Boden 1996; Haugeland 1997; Chrisley 2000, vol. iii).

Today, the Brooks–Kirsh battle is still undecided, although it's less fiercely oppositional, on both sides, than it was initially (Chrisley 2000: iii. 6–7). And Kirsh is getting strong support from some philosophical colleagues. For instance, Richard Samuels (forthcoming) has written a robust defence of the "Standard Model" in cognitive science. Unfashionably, in these largely situationist times, he argues that symbolic representations are needed to support means–end planning and much of the flexibility of human behaviour. (Often—shock! horror!—he even cites SOAR: see 7.iv.b.)

AI professionals are usually more eclectic than philosophers, even if they happen to have been philosophically trained. So McDermott included "dynamic planning" (which he attributed to the ideas of Agre, Chapman, Rosenschein, and Kaelbling) as a subclass of planning when he co-edited a special volume of *Artificial Intelligence* (McDermott and Hendler 1995: 6). And although some roboticists today rely only on situationist techniques, others also employ symbolic (deliberative) planning. Indeed, a recent volume supposedly—according to its title—focused on situated, behaviour-based, robotics devoted an entire chapter to the topic of how to combine reactive and deliberative mechanisms (Arkin 1998, ch. 6).

Probably, the future lies in systems combining the two approaches. For many such systems already exist. By the late 1980s Flakey was being driven by a combination of an improved STRIPS-based, world-modelling, planner and a more reactive, situationist, system (D. E. Wilkins 1988, ch. 12). And even before that, the mind had been informally described by Minsky as a combination not only of symbolic and connectionist representations, but also of top-down GOFAI and largely autonomous, bottom-up, subsystems (Chapter 12.iii.d).

Symbolism and situationism needn't be clumsily bolted together, as they are in many current "hybrid" systems. Alan Mackworth has recently put it like this:

The methodologies [for building integrated perceptual agents] are evolving dialectically. The symbolic methods of Good Old-Fashioned Artificial Intelligence and Robotics [GOFAIR] constitute the original thesis. The antithesis is reactive Insect AI. *The emerging synthesis*, Situated Agents, needs formal rigor and practical tools. A robot is a hybrid intelligent dynamical system, consisting of a controller coupled to its plant . . . Even though a robotic system is, generally, a hybrid system, its [overall control should be] unitary. Most other robot design methodologies use *hybrid models of hybrid systems*, awkwardly combining off-line computational models of high-level perception, reasoning and planning with on-line models of low-level sensing and control. We have developed a [unitary] testbed for multiple, visually-controlled, cheap robot vehicles performing a variety of tasks, including playing soccer. (Mackworth 1998; italics added)

Mackworth, then, has abandoned the sensori-motor sandwich—but without abandoning planning. Instead, his "Constraint Net" model is based on, and verified by, a formal specification of a *symmetrical coupling* between robot and environment (Sahota and Mackworth 1994). He's no longer entering his ball-kicking robots in the RoboCup competition that he helped initiate (with Hiroaki Kitano) in the late 1990s (see 11.iii.b). But his constraint-satisfying robots are a significant progression from both SHAKEY and Flakey, and also from Brooks's early menagerie of Allen, Herbert, Genghis, Attila, and Hannibal.

As for what Mackworth calls "the original thesis", this hasn't stood still: post-Kraken advances in symbolic planning have been significant (McDermott and Hendler 1995; S. Russell and Norvig 2003, chs. 11–12). They're still continuing. The 1995 edition of a widely used AI textbook described algorithms generating plans with dozens of steps, but the second edition only eight years later presented algorithms that "scale up to tens of thousands of steps" (S. Russell and Norvig 2003, p. viii).

One powerful GOFAI planner, for instance, controls the operations of a spacecraft, given high-level goals specified from Space Control on earth (Jonsson *et al.* 2000). It generates structured plans and also monitors their execution, diagnosing and (often) fixing problems as they occur. It's not HAL, to be sure (A. C. Clarke 1968). But, in its own (relatively predictable) domain, it's highly effective.

Up to a point, less predictable domains can be handled too, for some recent planning systems include ways of reasoning under uncertainty. In short, planning is here to stay. But whether it can survive as a *general-purpose* topic, as the NewFAI researchers had intended, is unclear. McDermott isn't optimistic:

[Possibly, architectures and/or formalisms for general-purpose planning may be developed.] This would allow a recoupling of theory and practice, and could lead in exciting new directions.

However, we're pessimistically forced to conclude that another solution is somewhat more likely. It may be inevitable for the field of planning to split into even smaller subfields, each with its own domain of interest (manufacturing, deliberation scheduling, logistics planning, etc.). After all, there may not be much in common between designing the behavior of a robot and designing the behavior of a military logistics system. It would be a pity to see this happen, but if what is gained is a set of elegant and powerful theories coupled with useful implementations replacing the current elegant but weak theories coupled with toy systems, then maybe it will be worth it. (McDermott and Hendler 1995: 13)

d. Agents and distributed cognition

Both Rosenschein and Agre, in the quotations given above, referred to situated *agents*. Indeed, situationism helped spark a flurry of research on "intelligent agents", or "autonomous agents". The *Artificial Intelligence* journal devoted over 800 pages to the topic, which were also published as a stand-alone book (Agre and Rosenschein 1995); and the interest continued (Wooldridge and Jennings 1995; Huhns and Singh 1997; Wooldridge 2001). Likewise, situationism encouraged work on the closely associated topic of "distributed" cognition (Bond and Gasser 1988; Demazeau and Muller 1990; G. Weiss 1999).

In distributed cognition, the overall task is achieved by numerous agents acting in concert, no one of them having access to all the relevant information or skills. The individuals may be more or less complex, in psychological terms. But they can achieve the task only by (knowingly or unknowingly) cooperating. In short, the interactions *between the many agents* are at least as important as the nature of each agent as an individual.

Examples found in the social (*sic*) insects include termite nests and ant trails, in which the phenomenon (nest or trail) emerges from the behaviour of many individual agents. Ants, of course, *are not* mindlike, and the cause of ant trails is very simple: each ant automatically drops chemicals ("pheromones") as it walks, and automatically follows pheromones dropped by others (see Chapter 15.x.a). A human example, discussed in Chapter 8.iii, is the navigational knowledge aboard a modern ship, which is distributed across many crew members. (The "ultimate" distributed workforce, it's been suggested, would be the world community of seafarers, linked by the Internet to their employers, unions, weathermen, and families: P. Collins and Hogg 2004.)

By century's end, it had become clear that the interactions *between agents and their environment* were crucial too. The situationists had stressed embodiment and environmental triggering, of course. But they'd seen "agent" and "environment" as clearly distinct, despite their causal interactions. In the late 1990s, that view was challenged by some specialists in HCI (human–computer interaction: see Section v, below).

Combining Bruner's work on cognitive technologies (6.ii.c), Edwin Hutchins's on environmentally embodied knowledge (such as shipboard instrumentation: 8.iii), and Clark's on the extended self (16.vii.d), they defined "distributed cognition" as a new theoretical approach in HCI. "One major benefit", they said, "is the explication of the complex interdependencies between people, artifacts and technological systems *that can often be overlooked when using traditional theories* of cognition" (Y. A. Rogers forthcoming; italics added). Blurring the conceptual boundaries between self and environment, individual and culture, helped them to analyse collaborative workplaces in general and computer interfaces in particular.

In the 1980s, however, those developments lay in the future. Meanwhile, distributed cognition was understood in the "traditional", more individualistic, way. Indeed, the term was used interchangeably with "multi-agent systems"—which inevitably raised the question "Just what is an agent?"

The answer wasn't clear. The term "agent" had officially entered AI long before, in Oliver Selfridge's Pandemonium paper (having originated in discussions with McCulloch). His then futuristic term had had a double meaning: (1) a self-directed task-achieving (i.e. mindlike) system, or demon; and (2) a demon performing some relatively menial (but purposeful) task on behalf of human users, who would otherwise have to do it themselves. For the 1980s situationists there was a third meaning: (3) a very simple demon, which might be program code or a situated robot.

(For the record, Selfridge himself is still working on agents, within DARPA's agent-based computing programme: see Chapter 12.iv.b. He's trying to build adaptive purposive hierarchies made up of lower-level purposive mechanisms. These multi-loop-controlled agents are much more complex than those favoured by most "situationists" today.)

These three demon-types, all implemented in the 1980s (or earlier: see Doran 1968*a*), were very different. For instance, mindlike agents had goals of their own, and ways of deciding how to achieve them, while simple demons didn't. And whereas most agents were largely dormant, waiting to be triggered by their environment, some mindlike agents were continuously active. Minsky's *Society of Mind*, widely circulated in draft from the late 1970s, had referred repeatedly to "agents" of various kinds. By an agent, Minsky meant

a machine that accomplishes something, without your needing to know how it works. You call it an agent when you want to treat it as a black box . . . When I call my travel agent Roy, I'm less concerned with how Roy works than with having him do what I need him to do. (Minsky and Riecken 1994: 24)

It followed that one shouldn't expect to define an agent in terms of how it worked, because the whole point (according to Minsky) was that one didn't need to know this.

Even now, AI workers debate the difference between an agent and *just any* AI program (S. Franklin and Graesser 1997). They also ask whether "multi-agent systems" are better described more anthropomorphically, in terms of the interactions within "societies of agents" (Huhns and Stephens 1999, sect. 2.4). And even now, they disagree about just what an agent is:

[There] is no universally accepted definition of the term agent, and indeed there is a good deal of ongoing debate and controversy on this very subject. Essentially, while there is a general consensus that autonomy is central to the notion of agency, there is little agreement beyond this. (Wooldridge 1999: 28)

These differences of definition reflect differences in methodology. While some post-Kraken programmers were developing societies of mini-minds, each with internal state representing their own and even others' goals, others were modelling agents of a much more minimalist kind. So Agre, in his introduction to the special double volume of *Artificial Intelligence* devoted to agent research, pointed out that it was "inspired by a great diversity of formalisms and architectures" (Agre 1995: 1).

Let's consider the minimalists first. These were the situationists in general, including some people more closely associated today with A-Life than with AI (because distributed systems are often thought of as "self-organizing" systems: Chapter 15). As we've seen, they described intelligent behaviour—whether in individuals or in groups—as emerging from a collection of independent reflexes, not directed by top-down control or abstract plans.

The most minimal examples of all were cellular automata in which the rules were (*a*) very simple and (*b*) identical for each agent, or cell (15.v). Work on cellular automata—together with Douglas Hofstadter's discussions of ant colonies in *Gödel, Escher, Bach* (12.x.a)—inspired Mitchel Resnick's development of Seymour Papert's LOGO language into the massively parallel StarLogo (Resnick 1994, pp. xvii, 31, 59).

Each agent in a StarLogo system was typically very simple, and similar to all the others. (Multi-agent systems composed of mindlike agents, by contrast, would have fewer—and more diverse—components.) As we saw in Chapter 10.vi.b, StarLogo enabled programmers to simulate not only interactions between situated agents, and agents' actions on the environment, but also the actions/interactions of the many different parts of the environment itself. Its potential was huge. Resnick's agents could represent turtles, or termites (ants), or cars and lorries in traffic jams . . . indeed, indefinitely many classes/colonies of interacting individuals.

Besides being used for professional AI research, StarLogo made distributed computation/cognition accessible to the general public. In two trade books, one of them specifically aimed at teachers, Resnick explained how StarLogo could be used to build systems simulating hundreds, even thousands, of individual agents (Resnick 1994; Colella *et al.* 2001). And he made the software available on the Web (<http://www.media.mit.edu/~starlogo/>). Using StarLogo, or the broadly similar SimAnt software, "thousands of . . . households [began to] play with ants on their computer screens". Ants, indeed, were the flavour of the day: the leading ethology volume on these little insects became "a sort of cult classic, attracting attention far outside the [biological] community" (Resnick 1994: 59).

Resnick hoped both to teach StarLogo users and to learn from them (1994: 5). The former, in helping them to understand decentralized systems in general. The latter, in discovering the *"folk* systems science" (comparable to naive physics) used when people think intuitively about distributed systems. Much as naive physics contains many mistaken assumptions, so—he expected—naive-systems science does too. Indeed, the "centralized mindset" of modernism had specifically inhibited a correct understanding of distributed systems. The illusory notions it had encouraged included the Freudian ego, creationism in biology, conspiracy theories in politics, and authoritarian views of family and pedagogy (1994: 119–23; 129). Work in distributed AI, of which StarLogo was a special case, could help to change that:

When something happens, [most people] assume that one individual agent must be responsible.
   But the centralized mindset is neither unchanging nor unchangeable. As decentralized ideas infiltrate the culture—through new technologies, new organizational structures, new scientific ideas—people will undoubtedly begin to think in new ways. People will become familiar with new models and new metaphors of decentralization. They will begin to see the world through new eyes. They will gradually recraft and expand their ways of thinking about causality. (Resnick 1994: 130)

   Resnick didn't claim that *all* cognition is distributed:

In economics, an unyielding commitment to decentralized, laissez-faire strategies can be just as debilitating as an unyielding commitment to centralized planning. So too with thinking: an unyielding decentralized mindset is no better than a centralized one. Many phenomena in the world *do* have centralized explanations. As [we] construct theories about the world, [we] should be able to draw on both centralized and decentralized ideas. (Resnick 1994: 148)

In this, he was following Papert himself. For Papert, in close cooperation with Minsky, had described the mind as a *hybrid* society of agents (12.iii.d). That is, there were GOFAI-ish agents as well as situationist ones. Indeed, there were connectionist agents too, so the society of mind was hybrid in both senses of the term (see 12.ix.b).

During the 1980s–1990s, research into GOFAI agents went on alongside work on the minimalist variety. Indeed, "alongside" sometimes meant *within the same family*: Stanley Rosenschein's brother Jeffrey outlined how to design "rational agents" that would use game theory to make "deals"—language better suited to GOFAI than to Brooks (J. S. Rosenschein and Genesereth 1985; cf. Rosenschein and Zlotkin 1994). Broadly, a mindlike agent was a robot or chunk of software (sometimes called a softbot) with three characteristics. Its action was autonomously controlled by its own internal processes, and by cues in its real/virtual world; the action was relatively complex, or purposive; and its own goals and self-generated sub-goals might, if the agent so decided, take priority over the goals of other agents or even of the system as a whole.

Sometimes, agents were *defined* as having "mental state, consisting of components such as beliefs and commitments" (Shoham and Tennenholtz 1995: 231). And sometimes, their mental state was analogous to the Theory of Mind, or ToM (7.vi). In such cases, the agent could represent not only other agents' abilities, beliefs, and intentions but also the social relations between self and other: "In order to cooperate effectively with its peers, an agent must *represent* any social structures in which it plays a part, and *reason* with these representations" (d'Inverno *et al.* 1997: 600).

These agents used their knowledge of others' capacities and intentions to generate shared plans and/or to decide whether to assist in another's plan. That decision, in turn, could rest either on monitoring the other's performance or on anticipating future performance or intentions. As for how they acquired their model of their peers' intentions, they normally used inductive techniques of "plan recognition" based on their observation of one or more interactions with the fellow agent concerned. In many cases, their "peers" were human beings, monitored and/or anticipated in much the same way.

One must hope that their ToM is realistic, for the applications include NASA's "VISTA" ground-control support system for the space shuttle, which went live in 1992 (Horvitz and Barry 1995). This program monitors the status of the five major engine subsystems, deciding *how to present the data on the screen being watched by the human operators*.

As Eric Horvitz (personal communication) remarks, VISTA was "perhaps the first exploration of the use of two rich probabilistic models running side by side". One of these is of the shuttle itself, based on a stream of telemetry; the other concerns "the beliefs of a user of a particular level of expertise, based on the displayed information". The inferences of both models are combined to make decisions about the psychological costs and benefits of revealing or highlighting different information on the screen. Besides avoiding information overload, VISTA aims to capture/shift the user's attention when something significant happens. In short, it tries—in real time—to pre-empt the problems facing machine operators that were studied long ago by Kenneth Craik (4.vi.c) and Donald Broadbent (6.i.c), and (in respect of short-term memory) by George Miller too (6.i.b).

Selfridge's second meaning (above) became especially relevant in the late 1980s. For by that time, many autonomous software systems were being built into computer interfaces, designed to handle administrative tasks of information retrieval/filtering on behalf of the human user (see v.c–d, below). The variety of these unseen agents increased with the years. When Berners-Lee's vision of the Semantic Web finally comes about, it will be thanks not only to shared ontologies but also to reliable agents working behind the scenes on behalf of the human users.

As Kay put it, these 1980s agents were "computer processes that act as guide, as coach, and as amanuensis" (1989: 130). For instance, they might devise timetables (respecting the entries already in the user's diary), book hotel rooms, or arrange for flights and car hire—perhaps without bothering the user, or perhaps making suggestions for human ratification (D. A. Norman 1994). Some could even learn how to do better in future by inferring, or being told, why their suggestion was rejected. One such agent could learn up to twenty new timetabling rules each night, by applying a version of ID3 to the information gleaned from the user on the previous day (T. M. Mitchell *et al.* 1994: 87).

By the end of the century, intelligent/autonomous agents had caught the imagination of many AI scientists. The ever-growing body of research had led (for instance) to international conferences—and journals—on 'Multi-Agent Systems', 'Autonomous Agents', and 'Adaptive Systems', and to specialist sessions in the IJCAI meetings (e.g. M. E. Pollack 1997: 578–654). Distributed AI was being used in industry as well as in interfaces (Parunak 1999; Alonso 2002). Moreover, affordable robots were now available off the shelf, so that people could study not only single home-made robots (such as FREDDY, SHAKEY, Flakey, or Cog) but also groups of interacting robots (Arkin 1998, ch. 9).

The general public was pricking up its ears, too. "Agent" had even become a buzzword in the marketplace. The Apple Newton, for example, was proudly advertised as "agent software", and General Magic trumpeted their new "messaging agents" (Riecken 1994: 18).

Agents had first achieved media popularity through Minsky's 1985 trade book *The Society of Mind*. But one didn't need to read the book to hear of the idea. Quite apart from reports by journalists, Minsky himself had been spreading the gospel among the business community for some years. At fundraising meetings in MIT's new Media Lab, he (and Kay, too) had sung the praises of agents as a way not only of evolving minds but also of developing useful and user-friendly computer technology (Section v, below).

As for distributed cognition, this idea had entered the public consciousness in 1979 through Hofstadter's extraordinary best-seller *Gödel, Escher, Bach* (Chapter 12.x.a). It had been reinforced in the mid- to late 1980s by the PDP 'bible' (12.vi.a). And—as remarked above—self-organization in "ant colonies" had become so evocative that a specialist tome on ants had become a cult book.

The popularity of both these notions was partly due to the widespread counter-cultural disaffection with centralized control and affection for the concrete and specific (see 1.iii.c–d). Indeed, Resnick said as much in lauding the fall of the Soviet Union *and* the decentralization of IBM in December 1991: "Throughout the world", he remarked, "there is an unprecedented shift toward decentralization" (Resnick 1994: 7; cf. Toulmin

1999; Turkle and Papert 1990). Systems where many autonomous individuals 'did their own thing' were felt to be even more culturally/politically appealing than committees, which in turn had been seen as more attractive than autocratic hierarchy (10.iv.a).—So much, yet again, for the Legend.

e. Social interaction and agents

Modelling collections of mindlike agents raised analogues of problems in social management that people face every day. In the absence of a top-down executive control, questions arose about how the knowledge possessed by the various components of the distributed AI system is integrated within the system as a whole. How could robots, or softbots in a virtual world, cooperate so as to achieve a goal which none could achieve alone?

The "social" behaviour of agents thus became an issue: the challenge was "to guarantee the successful coexistence of multiple program and programmers . . . [where the] agents' actions, and perhaps also goals, may be either facilitated or hindered by those of others" (Shoham and Tennenholtz 1995: 231). Increasingly, then, designers of distributed AI focused on the *interactions* between agents.

Social behaviour, in this context, included various types of communication: not just informing and questioning, but negotiating, bargaining, and voting too. The simple demons in Pandemonium had 'voted' via the loudness of their shouts (12.ii.d). But some late-century agents had to 'think' much harder before deciding which way to vote: for instance, by comparing their existing goals with the probable results of what they were being asked to do by some other agent. Negotiation and bargaining were even more complex. (In 1997 a special issue of the *Artificial Intelligence* journal was devoted to these "economic" aspects of distributed AI: Boutilier *et al.* 1997; see also J. S. Rosenschein and Genesereth 1985; Sandholm 1999.)

Various types of negotiation and bargaining were modelled, Rosenschein's "deals" being among the first. In systems based on *contract nets* (R. G. Smith 1980), for instance, the agents made "bids" that described how their abilities (including their current beliefs) were relevant to solving the overall problem. A central "manager" then decided which "contracts" to assign to each agent. One recent example of this approach was developed for NASA, to aid astronauts on long-term missions (Naghshineh-pour *et al.* 1999). The NASA contract net schedules the production and transportation of food—planting, monitoring, harvesting, recycling, and processing—in outer space.

As that example indicates, some modern agents can negotiate about several different goals, with incomplete information, and differing time deadlines (Fatima *et al.* 2004). An individual agent may even have deadlines, known only to the agent itself; if so, these deadlines affect its bargaining about which tasks it will volunteer and which it will accept (Sandholm and Vulkan 1999).

Besides communication, there was cooperation: *when? how? with whom? why?* In other words, post-Kraken agent modellers had to think about the issues raised many years earlier by the psychologist Robert Abelson, regarding the identification of plan junctures at which one particular agent might be able to help another (Chapter 7.i.c).

That's not to say that they used Abelson's work, for they didn't. But a late-century abstract analysis of "cooperation structures" dealt with the questions he'd raised, and related them to specific difficulties in the programming of distributed AI—how to determine, for instance, whether cooperation is even in principle possible, given some group of agents and a particular agent's goal (d'Inverno *et al.* 1997).

Similarly, Abelson's questions arose when people discussed how to generate collaborative plans (Grosz and Kraus 1996, 1999). And although most AI researchers ignored the fact (which Abelson had highlighted) that an agent may want to aid *or obstruct* another agent's plans, a few didn't. One team, for instance, said: "[Unlike most of those working on cooperation, we do not] assume benevolent agents" (d'Inverno *et al.* 1977: 605)—and their abstract model of cooperative structures specifically allowed for indifference or malevolence in some agents. Again, a general ontology for describing social interactions between AI agents explicitly allowed for "selfish" as well as "collaborative" behaviour (Castelfranchi 1998).

A central feature of 'mindlike' distributed AI was the communication of one agent's internal state (its beliefs, goals, and willingness to help) to others, and/or the communication of non-shared knowledge about the environment (e.g. the constraints on or goals of the problem as a whole). The same had been true of previous efforts to model cooperation—for instance, the two robots discussing how to open a bolted door (see Chapter 9.xi.f). Indeed, people usually take it for granted that communication, and a fortiori cooperative communication, *must* have evolved so that information could be transmitted to creatures currently ignorant of it.

Researchers in distributed AI typically did so too—and so did most A-Lifers. Even minimalist examples like ant trails, formed by the pheromones dropped by individual ants (15.x.a), could be shoehorned into this assumption. And A-Life models of the adaptive evolution of language usually relied on it heavily (e.g. Steels 1998*a,b,c*; Steels and Kaplan 2001; Steels and Belpaeme forthcoming). Some writers made it absolutely explicit: "A first prerequisite for communication is that some organisms have access to information (knowledge) that others do not, for if they all have access to the same information, no communication is necessary" (MacLennan and Burghardt 1994: 165).

However, some turn-of-the-century work, broadly inspired by Francisco Varela's work on autopoiesis (15.viii.b and 16.x.c), challenged it. The Sussex A-Life researcher Ezequiel Di Paolo (1970–) showed that communication, and cooperation, can in principle evolve *without* information being transmitted from one agent who has it, to another, who doesn't (Di Paolo 1999, ch. 8; 1998).

He built an intellectually rich series of evolutionary computer models, which evolved various types of social coordination between (simulated) robots. Systematic experimental variations discovered which parameters, under which conditions, were most important, and suggested implications for biological theories of communication. In some versions of the model, information wasn't equally available to all, and (sometimes) cooperative action–response strategies evolved so that the robots could benefit from each others' knowledge. In other versions, all agents possessed the same information—but even then, cooperative communication sometimes developed.

Each robot needed to replenish its energy regularly by accessing food. If it failed, it would die. And only if its energy level was above a certain threshold would it be able to reproduce. The food was finite, so the more one robot consumed the less there was for the others. In that sense, there was a conflict of interest: prima facie, not a situation in which one would expect cooperation.

In the informationally asymmetric versions of the model, more energy was made available to those robots which happened to signal the presence and type of found food (as opposed to either 'cheating' or remaining silent) and/or which happened to approach suitable food when perceiving the signal. However, no robot 'knew' that a particular action was a signal. Indeed, an action's communicative function *as* a signal, denoting a certain type of food and prompting neighbour robots to approach it, evolved as an aspect of the increasing coordination between the robots.

In the 'universal information' version, fitness didn't depend on single actions by individuals (emitting, or responding to, a signal). Rather, it depended on *a sequence of alternating actions by both agents. For example, imagine* that you and I are both gazing hungrily at a tall cherry tree. You must say "bing", then I must say "bong", then you say "tring", and I say "trong"—for *only then* will any cherries drop from the tree. Or rather, *only if you happen to say* "bing" . . . etc. will the cherries fall: whether you're aware that you "must" say this to get that result is immaterial. An apple tree would require a different four-signal sequence (perhaps tring, trong, bing, bong—with me to start, this time) before either of us could sink our teeth into the fruit.

Similarly, energy in Di Paolo's model was released "partially depending on the actions being correct at the required steps of the sequence" (Di Paolo 1999: 163). At first, there was no "sequence", only random actions. But an accidental 'correct' action would provide some energy, and a correct four-action sequence would provide even more. Ideally, the robots would select their next action depending on the food type and the previous actions—all perceptible to both robots. And evolution, due to differential reproduction, often resulted in a final generation that coordinated their actions perfectly. (This was true *even though* one and the same robot might have to perform a given action in first place or in third place, depending on the food type: apples or cherries, as it were.)

Di Paolo's robots couldn't have 'learned' to produce the right actions as lone individuals. They could achieve high energy levels only by acting cooperatively in locating the food. This type of cooperation, however, was very different from that studied by Abelson, or by AI scientists designing cooperating mindlike agents. There were no plans, no intentions, no symbolic references to food . . . no internal state at all. Accordingly, Di Paolo argued that this work cast serious doubts on orthodox cognitive science (1999, ch. 10).

One implication, he said, was that an individual's successful performance needn't imply a corresponding competence at the individual level (Chapter 7.iii.a). An agent who can get the cherries by cooperating with another agent may have *no* individual competence for getting cherries. Similarly, a robot that produces rhythmic behaviour when interacting in a particular acoustic world may show no hint of rhythm when acting on its own (Di Paolo 1999, ch. 9; 2000*a*).

Another implication—highly subversive, from the point of view of orthodox cognitive science—was that AI work on "social" agents wasn't truly social. AI agents were conceptualized as *individuals*, who happened to be interacting. Certainly, things occurred when they interacted which wouldn't have occurred otherwise. (Remove the diary-checker from the multi-agent system responsible for personal timetabling, and the human user wouldn't be happy.) But there was little or no notion that social behaviour may depend largely on *the dynamical properties of the interaction as such.*

For purely technological purposes, this might not matter. But for explaining human social behaviour (7.i.c), Di Paolo felt that it did. Even situated AI, although superior (in his view) to classical GOFAI, hadn't gone far enough:

To take situatedness seriously sometimes implies renouncing the simplicity of explanatory monism and embracing the complexity of the multiple concurrent and interdependent factors that form a historical process. This may not always be an easy task. However, once a situated process is properly understood many issues that are initially mysterious can be explained in a natural manner. Thus, cooperative coordination in games with conflict of interest can be explained by a combination of Darwinian selection and ecological situatedness [as outlined above]. (1999: 190; italics added)

From this viewpoint, my claim (in Chapter 1.ii) that cognitive science has studied social matters as well as individual cognition is misleading. To be sure, an adequate cognitive science would have to encompass social interactions, and some cognitive scientists have *attempted* to do so. If Di Paolo is right, however, they've gone about it in a fundamentally wrong-headed way. Even the respected *social* psychologist Abelson is individualistic at heart.

This dispute is an old one, although Di Paolo has put it in a new, and scientifically intriguing, way. In a nutshell, the question is whether individual selves constitute society or whether they're constituted by it. (Note: constituted by it, not just influenced by it.) Opposing answers have been defended by philosophers and social scientists for at least a century (Hollis 1977).

f. Technology swamps psychology

Inductive learning is a prime example of psychological AI's spawning, and eventually being overtaken by, technological AI.

There's no doubt that the psychologists got there first, for they were already studying concept learning by mid-century (see 6.ii.b–c and 10.iii.d). Informational analyses were published by Carl Hovland in 1952, computational strategies by Bruner in 1956, and computer models by Hovland and Earl Hunt in 1960–2.

Through the 1960s, Hunt continued his research on concept learning. No longer working with Hovland, who'd died in 1961, he now cooperated with Janet Marin and Philip Stone. They defined a new method for inductive classification: the Concept Learning System, or CLS rule (Hunt *et al.* 1966). Having only five steps, it was much simpler than the strategies discussed in Hunt's 1962 book (see Figures 10.10 and 10.12). But like them, it was a recursive algorithm which, given a training set of examples, learnt a top-down decision tree for classifying them.

CLS started by asking whether either *all* or *none* of the input cases were labelled as being examples of one concept. If so, it would stop. If not, it would divide the set into subsets and ask the same question again. At each hierarchical level, it would choose a property with more than one value (including present/absent), and partition the set accordingly. So flowers might be sorted into subsets by scarlet, blue, yellow, or white (think poppies, bluebells, daffodils, and snowdrops).

One weak spot in this procedure, at least in the early version, was that the property used to partition the set was chosen at random. This could result in a very inefficient search. So the programmers made it possible for the human user to choose the property to be considered. In other versions, simple frequency counts (successors of step 3 in Figure 10.10) would select the most useful property for partitioning the set. A property possessed by only 1 per cent of the examples, for instance, isn't a good (i.e. computationally efficient) one with which to start—even if it's a *sufficient* condition of the relevant concept.

Another weak spot was that concepts were (still) assumed to have necessary and sufficient conditions: as though all bluebells were blue, and all daffodils yellow. In fact, some bluebells are white, or pink. But most are indeed blue, which is why blueness is a very useful indicator of bluebells. CLS, on being presented with a pink bluebell, wouldn't treat it as an exception to be tolerated. Rather, it would reject blueness as a criterion of bluebells—thus throwing out the baby with the bathwater.

CLS could learn from its mistakes. If a new example (a pink bluebell, perhaps) didn't fit its current model, it would reconstruct the entire decision tree so as to include it. Because the combinatorial explosion raised its ugly head as the number of past classifications grew, Hunt's team eventually amended CLS so that it considered only a random subset of its 'memories' while revising the decision tree.

Not everyone studying GOFAI induction in the 1960s was doing similar things. For instance, other methods were devised by several members of the Edinburgh AI group, its leader Donald Michie among them. They occasionally cited Bruner (e.g. Popplestone 1969: 203), and/or defined strategies which were generalizations of his (e.g. Richard M. Young *et al.* 1977). But their work was logical, not psychological. Much of it was an attempt to widen theorem proving to cover induction too (Popplestone 1969; G. D. Plotkin 1969, 1971).

In the 1970s, as remarked in Chapter 10.iii.d, learning was seen—in the USA, at least—as "a 'bad' area to do research in" (R. S. Michalski, personal communication). Nevertheless, this period saw several advances. In the mid-1970s, meta-DENDRAL induced previously unknown rules about the points at which certain chemical molecules are likely to split (Chapter 10.iv.c). Simon's psychology student Pat Langley (1953–) started work on BACONian induction at much the same time (iv.c, below). And Ryszard Michalski used a multi-valued version of predicate logic to define a new learning algorithm, AQVAL (Michalski 1973; Michalski and Larson 1978).

When it made a mistake, AQVAL reconsidered only those previous classifications which were most relevant—not a random subset, as with CLS. And it reconstructed only that part of the decision tree which had made the mistake, not the whole tree. In those ways, it was an improvement on CLS. But that hadn't been Michalski's aim in designing it. He'd first modelled learning (of handwritten characters from several alphabets, including Roman, Greek, and Cyrillic) in 1966, in Poland. When he started work on AQVAL, soon after his arrival in the USA in 1971, he knew nothing of Hunt's research (personal communication).

Suddenly, around 1980, induction turned from being a "bad" area to a popular one. The first international workshop on machine learning was held (at CMU) in 1980, and three issues of the *International Journal of Policy Analysis and Information Systems* were given over to it. A few months later, in spring 1981, a special number of the *SIGART Newsletter* followed suit. In 1983 an authoritative collection of papers appeared, which was very widely read (Michalski *et al.* 1983). The Japanese Fifth Generation project was spreading hopes of providing expert systems with learning capabilities. The mid-1980s saw Bundy's analytical unification and critique of superficially diverse learning programs (11.iii.b). And by 1986, the new journal *Machine Learning* had been established, with Langley as its founding editor.

The methods used included connectionism as well as GOFAI, and genetic algorithms as well as fixed programs (J. H. Holland *et al.* 1986; Goldberg 1987). By the late 1980s, some machine learning was taking account of underlying causal mechanisms (Bratko *et al.* 1988, 1989). And the field was influencing the philosophy of science (Thagard 1988, 1989, 1990; P. M. Churchland 1989).

Not all of this post-Kraken literature concerned Hunt's theme (and ours): what one might term "property-list" induction. But much of it did.

Why the sudden blossoming of interest? After all, people had been working on GOFAI induction for some years. They included Michie, Robin Popplestone, and Gordon Plotkin in Edinburgh; Ivan Bratko in Prague; Langley and Simon at CMU; Michalski at the University of Illinois; and Thomas Mitchell, also at Illinois—who'd cut his AI teeth (at Stanford) on the learning aspect of meta-DENDRAL, but who became well known for his method of "version spaces" (T. M. Mitchell 1979; Mitchell *et al.* 1983). One name, however, stood out above all: J. Ross Quinlan (1943–), ensconced at the RAND Corporation by 1983.

It was Quinlan's work, from the late 1970s on, which tipped the popularity scales in favour of machine learning. Specifically, he devised the ID3 algorithm (Quinlan 1979). This was instigated in the context of chess endgames, thanks to a "challenging" question posed by Michie: could one tell, from the state of the board alone, whether a certain endgame was likely to be lost within a fixed number of moves? (Quinlan 1986: 84; cf. Quinlan 1983: 481). But Quinlan soon moved beyond chess, generalizing his algorithm a few years later (1983, 1986).

ID3 revolutionized machine learning, setting the GOFAI agenda for many years to come. It was the first inductive program specifically designed to handle "large masses of low-grade data" (1983: 463). And it was mathematically guaranteed to find the most efficient classification method for a given domain, provided that it was given a representative sample of examples (i.e. all types of example included, and the rarities suitably rare). Potentially, the commercial applications were legion.

One might argue that the *theoretical* interest of Quinlan's work on machine learning hadn't increased—had perhaps even decreased—with ID3. For his initial research, presented at the very first IJCAI meeting in 1969, had been different. Like Arthur Samuel's game-player, Sussman's HACKER, and Michie's pole-balancer (Michie and Chambers 1968), his first learning program had used adaptive self-monitoring to improve its own problem-solving ability (Quinlan 1969). By contrast, ID3 discovered already structured concepts. That's a perfectly respectable type of learning—but it's not the only one, nor the deepest.

It was, however, the one with the greatest technological promise. But if ID3 pointed towards technology, it had started out in psychology. For Quinlan had been deeply influenced by Hunt. He'd even co-published with him, on problem solving (Hunt and Quinlan 1968). That had been during his time as the first Ph.D. student in computer science at the University of Washington, where Hunt had a home in the Psychology and the Computer Science departments (E. B. Hunt, personal communication). On turning to learning as his prime topic, he benefited from Hunt's ideas. Later, he described ID3 as "a relative" and "a descendant" of the CLS classification rule (1979: 171; 1983: 465). Indeed, he acknowledged CLS as "the patriarch" of the large family of post-Kraken inductive programs (1986: 84).

In outline, ID3 was given a training set of a large number of instances of various concepts: chess endgames, perhaps, or the nineteen common diseases that affect soya beans (see below). Each instance was described in terms of a set of properties (the position of king or rook, for example), and the program learnt rules for defining each concept. It did this by a hierarchical series of binary partitions of the training space. When the space could be divided no further, the program would be tested on sets of new examples (instances and non-instances).

So far, so CLS. But ID3 was hugely more efficient than its psychological precursor. Its advantages were:

* the much greater size of the training set,
* a larger number of attributes used in defining concepts,
* greater complexity of the concepts (measured by the number of nodes on the decision tree),
* much greater speed and computational efficiency,
* the ability to accept classifications accounting only for *most* of the data, and
* the possibility of discovering some classification-relevant attributes automatically (i.e. local *patterns* of attributes, as in chess positions—1983: 477–81).

ID3, said Quinlan, was "five times as fast as the best alternative method that I could devise" (p. 463). And it surpassed all the other inductive algorithms favoured at the time, such as Michalski's AQVAL or Mitchell's version spaces. Version spaces had been designed to deal with large data sets too, by setting boundaries of maximal and minimal generality to the concepts being developed. But as Quinlan pointed out, when the number of examples was *very* large, Mitchell's method (which required the system to remember all the still-possible maximal/minimal rules) would become unmanageable. Instead, he suggested a simple iterative strategy:

** select at random a subset of the given instances (called the *window*)
** *repeat*
  * form a rule to explain the current window
  * find the exceptions to this rule in the remaining instances
  * form a new window from the current window and the exceptions to the rule generated from it *until* there are no exceptions to the rule. (Quinlan 1983: 469)

(For generating *approximate* classification rules, that final "until" had to be relaxed—1983: 474–7.)

Experimenting with two different ways of forming the window, Quinlan used this approach to classify a data set of nearly 2,000 objects, involving fourteen attributes, and requiring a decision tree with no fewer than forty-eight nodes (Quinlan 1979). The results were staggering.

Irrespective of the window-forming method used, he found that *only four iterations* were normally needed to find a correct decision tree. The final window could contain only a small fraction of the 2,000 objects, and the size of the initial window didn't matter much. Moreover, the time required increased only linearly as the task difficulty rose (measured by the number of objects, attributes, and decision tree nodes). So whereas Hunt had been restricted to very simple concepts, like the experimental examples in Bruner's *A Study of Thinking*, Quinlan's method could be used to deal with much more interesting cases.

Consider soya bean diseases, for instance—no mere triviality in America's Midwest, since soya beans are Illinois's main commercial product. The nineteen common diseases are recognized by thirty-five descriptors, covering various types of leaf spots (with/without haloes, or water-soaked margins, or . . .) and holes, seed shrivelling, and information about season and rainfall.

There's no simple one-to-one mapping of symptom to disease. Rather, there's a complex pattern of symptoms for each disease, complicated by the fact that not all the symptoms need to be present in a particular case. (Pink bluebells, again.) So even experienced soya bean farmers can't always be sure what's ailing their plants. Illinois has long provided a number of Agricultural Extensions Offices which they can phone for advice. In especially difficult cases they can also arrange for microscopical tests to be done, at the farmer's expense, by the university.

Michalski and Richard Chilausky, both working—appropriately enough—at the University of Illinois, decided to use ID3 to write an expert system that might save time and money for all concerned (Michalski and Chilausky 1980). With the help of the textbooks, augmented by forty-five hours of consultation with a local plant pathologist, they identified the probably relevant descriptors and provided them to ID3. In addition, they designed a descriptive questionnaire to be filled in by the farmers whose crops were afflicted—and the farmers' answers were then used as input to the expert system (see Figure 13.2).

It turned out that, for fifteen diseases, their ID3-based program outperformed the world expert—who'd written the textbook on which the initial data classification had been based. Having been trained on 307 different cases, it was tested on 376 new ones. (The tree was automatically converted into a coherent set of production rules, implying that IF such-and-such symptoms were present, THEN such-and-such a disease was present.) It got only two diagnoses wrong, whereas humans following the textbook rules failed in 17 per cent of cases. Indeed, the computer-generated rules were accepted for daily use by soya bean farmers and pathologists alike (Michie and Johnston 1984: 111).

Even though all the data in ID3's training set (regarding soya beans or anything else) were presented simultaneously, the algorithm made sensible decisions about which attributes to consider first. The window was chosen at random, as we've seen. But the item "form a rule to explain the current window" involved frequency counts that improved on those used by CLS. In Quinlan's words:

The whole skill in this style of induction lies in selecting a useful attribute to test for a given collection of objects so that the final tree is in some sense minimal. *Hunt's work* used a lookahead scheme driven by a system of measurement and misclassification costs in an attempt to get minimal-cost trees. ID3 uses *an information-theoretic approach* aimed at minimizing the expected number of tests to classify an object. (Quinlan 1983: 466–7; italics added)

That was the thin end of the technological wedge. Whereas Hunt had started from experimental data, common sense, and simple logic to define thinking strategies that

Environmental descriptors
 Time of occurence = July
 Plant stand = normal
 Precipitation = above normal
 Temperature = normal
 Occurence of hail = no
 Number of years crop repeated = 4
 Damaged area = whole fields

Plant global descriptors
 Severity = potentially severe
 Seed treatment = none
 Seed germination = less than 80%
 Plant height = normal

Plant local descriptors
 Condition of leaves = abnormal
  Leafspots–halos = without yellow halos
  Leafspots–margin = without watersoaked margin
  Leafspot size = greater than $\frac{1}{8}$"
  Leaf shredding or shot holding = present
  Leaf malformation = absent
  Leaf mildew growth = absent
 Condition of stem = abnormal
  Presence of lodging = no
  Stem cankers = above the second node
  Canker lesion color = brown
  Fruiting bodies on stem = present
  External decay = absent
  Mycelium on stem = absent
  Internal discoloration of stem = none
  Sclerotia–internal or external = absent
 Conditions of fruit–pods = normal
  Fruit sports = absent
 Condition of seed = normal
  Mould growth = absent
  Seed discoloration = absent
  Seed size = normal
  Seed shrivelling = absent
 Condition of roots = normal

Diagnosis:
    Diaporthe stem canker( ) *Charcoal rot*( ) *Rhizoctonia
    root rot*( ) *Phytophthora root rot*( ) *Brown stem root rot*( )
    Powdery mildew( ) *Downy mildew*( ) *Brown spot*(x)
    Bacterial blight( ) *Bacterial pustule*( ) *Purpose seed stain*( )
    Anthracnose( ) *Phyllosticta leaf spot*( ) *Alternaria leaf
    spot*( ) *Frog eye leaf spot*( )

Fig. 13.2. A questionnaire completed by a soybean farmer, used as input to an ID3-based expert system, with the program's diagnosis (Brown spot) shown underneath. Reprinted with permission from Michalski and Chilauski (1980: 138)

might actually go on in human heads, Quinlan was using highly abstract formulae drawn from information theory. In effect, ID3 was doing entropy calculations in order to choose which attribute to consider next (1983: 467 ff., 475 ff.). These calculations were justified *not* by psychological theory (even though some psychologists were starting to use ideas about entropy in their models: 12.vi.b), but purely by mathematical efficiency.

As the post-Kraken period progressed, the psychological roots of automatic induction grew ever more obscure. For Quinlan (e.g. 1988, 1993) and others used increasingly arcane statistical measures to enable more efficient machine learning. Much as connectionist learning, during these years, grew closer to mathematical statistics *as such* (Chapter 12.vi.f), so most of the GOFAI versions did too.

This was pointed out in the early 1990s by Michie (Michie *et al.* 1994). Indeed, most of the papers given at his 1992 Machine Intelligence workshop, which was devoted to the topic of learning, exemplified the mathematical trend (Furukawa *et al.* 1994). Despite a handful of people-centred presentations, including one by a leading developmental psychologist based in Edinburgh (C. B. Trevarthen 1994), human beings weren't much considered. Like most other 1990s work on automatic induction, this was *machine* learning: technological, not psychological, AI. (The exceptions included Michalski's theories of plausible inference and human learning; based on a model of dynamic links between conceptual hierarchies, these were developed through the 1980s with the psychologist Allan Collins: Collins and Michalski 1989; Hieb and Michalski 1993.)

Some even began to claim that this research wasn't AI at all (see vii.b, below). Statisticians and computer scientists didn't want to be labelled as doing AI. This attitude was—is—due partly to intellectual territoriality ("Get off my patch!"), and partly to a wish to avoid diversionary philosophical challenges. Hard-headed technologists and mathematicians didn't want to be saddled with provocative psychological terms such as *intelligence*, *knowledge*, or even *learning*.

As a result, the most successful type of machine learning is near-invisible today, because it's blandly named "data mining". (Two of the field's pioneers now use both names in the same breath—and the same book title: Michalski *et al.* 1998.)

By "successful", I really do mean successful. As one commentator has put it, the AI equivalent of Columbus' looking for India and finding the resource-rich America is looking for machine learning and finding data mining (Whitby 2004). Huge sums of money are now spent on partitioning enormous (and noisy) data sets into useful categories. For example, advertising agencies are commonly asked to find detailed descriptions of the people likely to be interested in buying such-and-such a product, or voting for such-and-such a policy. It's a rare client, however, who thinks of this as machine learning—still less, as AI. Yet without AI, there'd be no data mining.

The *psychological* origins of automatic induction are less visible still. (Statistics rules!) For practical purposes, that doesn't matter. For historical purposes, however, machine learning's debt to experimental psychology should be recognized.

13.iv. Explaining the Ineffable

'Explaining the Ineffable' was the title—and 'Intuition, Insight, and Inspiration' the key words in the subtitle—of the paper which Simon, near the end of the century (Simon

1995*b*), said that AAAI Fellows should read if they doubted that AI is a *science* (see vii.a, below). Clearly, he felt that AI had gone some way towards fulfilling the promise of his 1958 RAND harbinger (10.i.f). But that promise hadn't been taken up explicitly until the 1980s.

a. Creativity ignored

Creativity was an obvious challenge for AI right from the start—or even before it. In the imaginary conversation that launched the Turing Test, Turing had depicted a computer as interpreting/defending a sonnet, tacitly implying that a computer might compose poetry too. However, once people started writing AI programs, this particular challenge was parked on the sidelines.

When GOFAI was still NewFAI, creativity was a no-go area. To be sure, Donald Mackay (1951) had described a probabilistic system which he said would show "originality" in a minimal sense—but that had been an aside, not the main object of the exercise. Similarly, problem-solving programs were occasionally described as creative. And Simon, in 1957, had mentioned the "theory of creativity" he and Newell were developing—and had praised the pioneering work of Lejaren Hiller and Leonard Isaacson, whose computer-generated Illiac suite (a string quartet, first performed in 1957) was, in his estimation, "not trivial and uninteresting" (McCorduck 1979: 188). Nevertheless, creativity in the layman's sense (i.e. art, music, and science—and maybe jokes) was all but ignored by AI professionals.

Outside the field, this was less true. Zuse's vision of automatic carpet design, with deliberate weaving-errors to add 'authenticity', wasn't yet in the public domain. But visual computer art was getting started in Europe and the USA by 1963, and interactive computer art had already begun in the 1950s (see Section vi.c, below).

In the 1950s, too, Hiller—a professional chemist, but also a Master of Music—had initiated the Illiac suite (the score is given at pp. 182–97 of Hiller and Isaacson 1959). The first three movements were generated from rules defining various musical styles (sixteenth-century counterpoint, twelve-tone music, and a range of dynamics and rhythms), sometimes combined with tone-pairs chosen by chance; the fourth movement was based not on familiar styles but on Markow chains (Hiller and Isaacson 1959). Systematic experiments in computer music, including instrumentation, were done at IRCAM in Paris from the early 1960s. And a competition for computer-composed music was organized at the 1968 International Federation for Information Processing meeting in Edinburgh (and led to the founding of the Computer Arts Society).

In general, however, these projects involved the artistic avant-garde, with a few art-oriented scientists—not the leaders of GOFAI. Their references to "creativity", "invention", and "discovery" in the Dartmouth proposal weren't being echoed in AI research (McCarthy *et al.* 1955: 45, 49 ff.). Even creative problem solving was rarely described as such, despite the LT team's use of the term in their late 1950s call to arms.

That's why my AI colleagues were bemused when, in the early 1970s, I told those who asked that I'd decided to include a whole chapter on the topic in my book on the field. Several protested: "But there isn't any work on creativity!"

In a sense, they were right. Admittedly, the 1960s had produced an intriguing model of analogy (see below), and meta-DENDRAL had touched on creativity in chemistry.

Moreover, much NewFAI work was an essential preliminary for tackling creativity as such. For instance, NLP researchers had asked how memory structures are used in understanding metaphysics (Wilks 1972: see 9.x.d), metaphor (Ortony 1979), or stories (Charniak 1972, 1973, 1974; Schank and the Yale AI Project 1975; Rieger 1975*a,b*). One brave soul was generating story-appropriate syntax (Davey 1978: see 9.xi.c), and another studying rhetorical style (Eisenstadt 1976).

But apart from a handful of painfully crude "poets", "story writers", or "novelists", there were no models of what's normally regarded as creativity (Masterman and McKinnon Wood 1968; Masterman 1971; Meehan 1975, 1981; Klein *et al.* 1973). The renowned geometry program was only an *apparent* exception (see 10.i.c and Boden 1990*a*: 104–10).

This explains why John Haugeland (1978, sect. 7), when he questioned the plausibility of cognitivism, raised general doubts about GOFAI analyses of "human insight" but didn't explicitly criticize any specific models of it. His first worry was that GOFAI systems "preclude any radically new way of understanding things; all new developments would have to be specializations of the antecedent general conditions". But Galileo, Kepler, and Newton, he said, invented "a totally new way of talking about what happens" in the physical world, and "a new way of rendering it intelligible". Even to learn to understand the new theory would be beyond a medieval-physics GOFAI system, "unless it had it latently 'built-in' all along". (He may have chosen this particular example because he'd heard of the about-to-be-published BACON suite on the grapevine: see below.) The underlying difficulty, he suggested (following Dreyfus: 11.ii.a), was that "understanding pertains not primarily to symbols or rules for manipulating them, but [to] the world and to living in it".

Before the 1980s, then, creativity was a dirty word—or anyway, such a huge challenge that most programmers shied away from it. The most important exception, and still one of the most impressive, was due to someone from *outside* the NewFAI community.

b. Help from outside

Harold Cohen (1928–) was a highly acclaimed abstract artist of early 1960s London. (Those words "highly acclaimed" weren't empty: a few years ago, a major exhibition on 'The 1960s' was held at London's Barbican. Even though this was focused on 1960s culture as a whole, not just on the visual arts, two of Cohen's early paintings were included.) He turned towards programmed art in 1968. He spent two years in Stanford as a visiting scholar with Feigenbaum, in 1973–5, where he not only found out about AI but also learnt to program.

In the four decades that followed, while at the University of California (San Diego), he continuously improved his drawing and colouring program, called AARON (H. Cohen 1979, 1981, 1995, 2002; McCorduck 1991; Boden 2004: 150–66, 314–15). Successive versions of AARON were demonstrated at exhibitions in major galleries (and at Science Centres) around the world, and received huge publicity in the media. (An early version is now available as shareware.)

Unlike his 1960s contemporaries Roy Ascott and Ernest Edmonds (vi.c, below), Cohen wasn't using computer technology to found a new artistic genre. Rather, he was investigating the nature of representation. To some extent, he was trying to achieve a better understanding of his own creative processes. But, over the years, he came to see his early attempts to model human thought as misdirected:

The common, unquestioned bias—which I had shared—towards a human model of cognition proved to be an insurmountable obstacle. It was only after I began to see how fundamentally different an artificial intelligence is from a human intelligence that I was able to make headway.... The difference is becoming increasingly clear now, as I work to make AARON continuously aware of the state of a developing image, as a determinant to how to proceed. How does a machine evaluate pizazz? (personal communication, July 2005)

Cohen asked how he made introspectively "unequivocal" and "unarbitrary" decisions about line, shading, and colour. (From the mid-1990s his main focus was on colour.) And he studied how these things were perceived—by him and others—as representing neighbouring and overlapping surfaces and solid objects. Moving towards increasingly 'three-dimensional' representations, he explored how his/AARON's internal models of foliage and landscape, and especially of the human body, could be used to generate novel artworks. (He was adamant that they *were* artworks, although some philosophers argue that no computer-generated artefact could properly be classified as art: O'Hear 1995.)

Two examples, dating from either side of 1990, are shown in Figures 13.3 and 13.4. Notice that the second, drawn by the later version of the program, has more 3D depth than the first. (Which, in turn, has more than the drawing by AARON's early 1980s 'Acrobat and Balls' period—see Boden 1990*a*/2004, frontispiece.) In particular,

FIG. 13.3. An example of AARON's 'jungle' period, late 1980s; the drawing was done by the program, but the colouring was done by hand. Untitled, 1988, oil on canvas (painted by Harold Cohen), 54″ × 77″; Robert Hendel collection. Reproduced with permission of the artist

Fɪɢ. 13.4. An example of early 1990s AARON; the drawing was done by the program, but the colouring was done by hand. *San Francisco People* (1991), oil on canvas (painted by Harold Cohen), 60″ × 84″, collection of the artist. Reproduced by his permission

'Jungle-AARON' didn't have enough real 3D data about the body to be able to draw a human figure with its arms overlapping its own body. His early 1990s city-people could have waved, if he had wanted them to; but his late 1980s jungle-dwellers couldn't have crossed their arms instead of waving.

By 1995, Cohen had at last produced a painting-machine-based version of AARON which could not only draw acceptably but also colour to his satisfaction—using water-based dyes and five brushes of varying sizes. However, his satisfaction was still limited.

By the summer of 2002, he'd made another breakthrough: a digital AARON-as-colourer, whose images could be printed at any size. The program was regularly left to run by itself overnight, offering up about sixty new works for inspection in the morning. Even the also-rans were acceptable: one gallery curator who exhibited it told Cohen that "he hasn't seen AARON make a bad one since it started several weeks ago" (Cohen, personal communication). One might even say that AARON has now surpassed Cohen as a colour artist, much as Samuel's program surpassed Samuel as a checkers (draughts) player nearly half a century before. For Cohen regards this latest incarnation of AARON as "a world-class colorist" whereas he himself is merely "a first rate colorist" (personal communication).

A comparably spectacular, though later, AI artist was "Emmy"—originally EMI: Experiments in Musical Intelligence. This program was written in 1981 by the composer

David Cope (1941–), at the University of California, Santa Cruz. (It wasn't easy for other people to run, or experiment with, the early version. For the now familiar MIDI, or Musical Instrument Digital Interface, which defines musical notes in a way that all computers can use, wasn't yet available. Indeed, its inventor, Dave Smith, first had the idea in that very year, and didn't announce the first specification until August 1983: Moynihan 2003. Today, Emmy's successors are based on MIDI, so can be run on any PC equipped with run-of-the-mill musical technology.)

This wasn't the first attempt to formalize musical creativity (and nor was the Hiller and Isaacson effort). A system of rules for do-it-yourself hymn composition was penned in the early eleventh century by Guido d'Arezzo, who also invented the basis of tonic solfa and of today's musical notation (A. Gartland-Jones, personal). But Cope, almost 1,000 years later, managed to turn "formalize" into "implement". Moreover, his program could compose pieces much more complex than hymns, whether within a general musical style (e.g. baroque fugue) or emulating a specific composer (e.g. Antonio Vivaldi or J. S. Bach). It could even mix styles or musicians, such as Thai–jazz or Bach–Joplin, much as the Swingle Singers do.

Emmy gained a wide audience, though not as wide as AARON's. Part of its notoriety was spread by scandalized gossip: Cope has remarked that "There doesn't seem to be a single group of people that the program doesn't annoy in some way" (Cope 2001: 92). But as well as relying on word of mouth, people could read about it, and examine some Emmy scores, in Cope's four books (1991, 2000, 2001, 2006). Enthusiasts could even try it out for themselves, following his technical advice, by using one of the cut-down versions (ALICE: ALgorithmically Integrated Composing Environment, and SARA: Simple Analytic Recombinant Algorithm) provided on CDs packaged inside his books.

They could listen to Emmy's compositions, too. Several stand-alone CDs were released (by Centaur Records, Baton Rouge) in the late 1990s. In addition, several live concerts of Emmy's music were staged to public audiences. (These featured human instrumentalists playing Emmy's scores, because the program didn't represent expressive performance: it laid down what notes to play, not how to play them.)

However, the concerts were mostly arranged by Cope's friends: "Since 1980, I have made extraordinary attempts to have [Emmy's] works performed. Unfortunately, my successes have been few. Performers rarely consider these works seriously" (Cope 2006: 362). The problem, said Cope, was that they (like most people) regarded Emmy's music as computer "output", whereas he had always thought of it rather as *music*. Moreover, being "output" it was infinitely extensible, a fact—he found—that made people devalue it.

In 2004 he took the drastic decision to destroy Emmy's historical database: there will be no more "Bach" fugues from the program (2006: 364). Emmy's "farewell gift" to the historical-music world was a fifty-page score for a new symphonic movement in the style of Beethoven, which required "several months of data gathering and development as well as several generations of corrections and flawed output" (2006: 366, 399–451). From now on, Emmy—or rather, Emmy's much-improved successor—will be composing in Cope's style, as "Emily Howell" (p. 374).

Hofstadter, a fine amateur musician, found Emmy impressive despite—or rather, because of—his initial confidence that "little of interest could come of [its GOFAI] architecture". On reading Cope's 1991 book, he got a shock:

I noticed in its pages an Emmy mazurka supposedly in the Chopin style, and this really drew my attention because, having revered Chopin my whole life long, I felt certain that no one could pull the wool over my eyes in this department. Moreover, I knew all fifty or sixty of the Chopin mazurkas very well, having played them dozens of times on the piano and heard them even more often on recordings. So I went straight to my own piano and sight-read through the Emmy mazurka—once, twice, three times, and more—each time with mounting confusion and surprise. Though I felt there were a few little glitches here and there, I was impressed, for the piece seemed to *express* something . . . [It] did not seem in any way plagiarized. It was *new*, it was unmistakably *Chopin-like* in spirit, and it was *not emotionally empty*. I was truly shaken. How could emotional music be coming out of a program that had never heard a note, never lived a moment of life, never had any emotions whatsoever?

[. . . Emmy was threatening] my oldest and most deeply cherished beliefs about . . . music being the ultimate inner sanctum of the human spirit, the last thing that would tumble in AI's headlong rush toward thought, insight, and creativity. (Hofstadter 2001*a*: 38–9)

Hofstadter was allowing himself to be *over*-impressed, here. For Frederic Bartlett's "effort after meaning" (5.ii.b) imbues our perception of music as well as of visual patterns and words. The human performer projects emotion into the score-defined notes, much as human readers project meaning into computer-generated haikus (9.x.c). So, given that "Chopin-like" scores had been produced, it wasn't surprising that Hofstadter interpreted them expressively.

What was surprising was the Chopin-like musicality of the compositions. Simon's 1957 prediction that a computer would write aesthetically valuable music within ten years had failed, and had been mocked accordingly (H. L. Dreyfus 1965: 3). But Cope had now achieved this, though fourteen years late.

Emmy's basic method was described by Cope as "recombinatory", and summarized by Hofstadter as "(1) chop up; (2) reassemble" (p. 44). In fact, Emmy was exploring generative structures as well as recombining motifs. It showed both combinational and exploratory creativity—but not, as Hofstadter (2001*b*) was quick to point out, transformational creativity (Boden 1990*a*/2004). A 'new' style could appear only as a result of mixing two or more existing styles.

The program's database was a set of 'signatures' (note patterns of up to ten melodic notes) exemplifying melody, harmony, metre, and ornament, all selected by Cope as being characteristic of the composer concerned. Emmy applied statistical techniques to identify the core features of these snippets, and then—guided by general musicological principles—used them to generate new structures. (Some results worked less well than others—e.g. Cope 2001: 182–3, 385–90.)

Strictly, Emmy wasn't an exercise in "explaining" the ineffable. For Cope's motivation differed from Simon's (and Hofstadter's: see below). His aim wasn't to understand creative thought, but to generate musical structures like those produced by human composers. Initially, he'd intended EMI to produce new music in *his* style, but soon realized that he was "too close to [his] own music to define its style in meaningful ways", so switched to the well-studied classical composers instead (2001: 93). (A quarter-century later, having destroyed the historical database, he switched back to computer compositions in his own style—2005: 372 ff. and pt. III *passim*.)

In short, he was modelling music, not mind. (The task wouldn't necessarily have been easier had he known the psychological details; for instance, limits on short-term memory rule out the use of powerful generative grammars for jazz improvisation: Johnson-Laird 1989/1993.)

Nevertheless, the implication of Cope's writings was that all composers follow some stylistic rules, or algorithms. Sonata Form, for example, was supposed to be a formal structure rigidly adhered to by everyone composing in that style. This assumption has been questioned. It's known that Haydn, Mozart, and Beethoven (for instance) worked diligently through the exercises given in various musical textbooks. But whether they stuck rigidly to those rules in their more original, creative, music is quite another matter. Peter Copley and Drew Gartland-Jones (2005) argue that they did not.

On their view, the formal rules of style are extracted and agreed *post hoc*, and are then followed to the letter only by musical students and mediocrities. Once sonata form, or any other musical structure, has been explicitly stated, it tends to lose its creative potential. (This explains the "paradox" of sonatas in the romantic period being *far less* free than in the classical times: p. 229.) But the rules are flexible enough to evolve in use:

> It would be tempting to view this process as *generally agreed* forms, evolving. But . . . [it] would be more useful to see the general acceptances of common practice as *present to provide a frame for musical differences, changes, developments etc.* (Footnote: This is the basis for Boden's transformational creativity). If there is a constraint it comes organically from within a complex network of practitioners rather than [a] set of stated constraints that are accepted until someone decides they need breaking. This is a complex mechanism indeed, and even if we argue that algorithms might explain certain emergent patterns, it is not at all sure that such patterns stem from a simple, if enormously lengthy, set of rules. (Copley and Gartland-Jones 2005: 229; final italics added)

Cope's Emmy, they admit, can indeed compose many acceptable pieces. But even if it could generate fully "convincing" examples, "without a model of how [the abstracted rules] change we are capturing an incomplete snapshot of musical practice" (p. 229). In other words, Cope is modelling musical creations—not yet musical creativity.

c. In focus at last

It's no accident that those two highly successful programs, AARON and Emmy, were written by non-AI professionals. For they depended on a reliable sense of how to generate and appreciate structures within the conceptual space concerned (Boden 1990*a*/2004, chs. 3–4). In general, a plausible computer artist stands in need of an expert in art. Even if (like Hiller) the person isn't a professional artist, they need (again, like Hiller) to be a very highly knowledgeable amateur.

Often, artist and programmer are different people (e.g. William Latham and Stephen Todd, respectively: Todd and Latham 1992). But some artists are sufficiently computer-literate to design their own systems. Edmonds was a professional computer scientist in the 1960s, as well as being an influential artist. (Although he's always used his programs to help him understand creative thinking in general, he was less concerned than Cohen or Cope to talk about *the program as such*: if his viewers needed to realize that there was a program involved, they didn't need to think about just how it worked: see vi.c, below.) Today, forty years later, many young artists are computer-literate even if they're not computing professionals.

If artist programs need experts in art, much the same is true of programs focused on literature, maths, and science. A competent programmer is the sine qua non, but domain expertise is needed too. Since most AI researchers were reasonably proficient in those areas, one and the same person could be both programmer and expert. That's why AI work on creativity, once it got started, usually focused on them (rather than on music or, still less, visual art).

The everyday skill of analogy, which features in both literature and science, had been modelled as early as 1963 by Thomas Evans (1934–) at MIT. His program was a huge advance (T. G. Evans 1968). Implementing the ideas briefly intimated by Minsky in 'Steps' (see Figure 10.3), it not only discovered analogies of varying strength but also identified the best.

It could do this because it described the analogies on hierarchical levels of varying generality. Using geometrical diagrams like those featured in IQ tests (see Figure 13.5), Evans's program achieved a success rate comparable to that of a 15-year-old child. But it wasn't followed up. (This was an example of the lack of direction in AI research that so infuriated McDermott: 11.iii.a.)

By the early 1980s, analogy had returned as a research topic in AI. For instance, an international workshop held in 1983 included five papers explicitly devoted to it, plus several more that could be seen as relevant (Michalski 1983, esp. 2–40).

Evidently, the AI scientists concerned didn't read Fodor's *Modularity of Mind* (1983), or anyway didn't accept its pessimism about explaining the higher mental processes. Fodor despaired of any attempt to understand analogy in scientific terms. Despite its undeniable importance, he said,

nobody knows anything about how it works; not even in the dim, in-a-glass-darkly sort of way in which there are some ideas about how [scientific] confirmation works. (Fodor 1983: 107)

And, according to him, they never would: "Fodor's First Law of the Nonexistence of Cognitive Science [states that] the more global ... a cognitive process is, the less

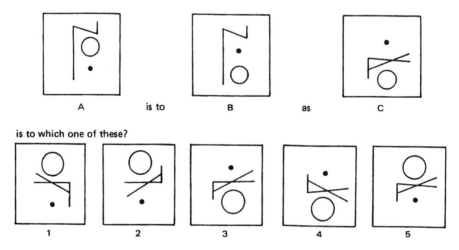

FIG. 13.5. Analogy problem tackled by Evans's program. Reprinted with permission from (Minsky 1968: 332)

anybody understands it." (As we saw in Chapter 7.vi.h, he was right in arguing that we'll never be able to predict/explain every case of analogy in detail, but wrong in concluding that nothing of scientific interest can therefore be said about it.)

Fodor's scepticism notwithstanding, several GOFAI scientists in the 1980s tried to model how conceptual analogies are generated, interpreted, and used (Gentner 1983; Holyoak and Thagard 1989; Thagard *et al.* 1988; Gentner *et al.* 1997). Like Newell and Simon before them, they sometimes went back to the Gestalt psychologists' 1930s work on problem solving (Chapter 5.ii.b). For instance, they asked (taking Karl Duncker's example) how the notion of a besieging army could help in discovering how to irradiate a tumour without killing the surrounding tissues.

In general, they wanted to know just how an analogous idea could be identified, and fruitfully mapped onto the problem at hand. The answer usually given was to show how distinct conceptual structures could be compared, and—if necessary—adapted so as to match each other more closely. Sometimes, the problem-solver's goal was allowed to influence the comparison. But the basic process was like that used long before by Patrick Winston's concept-learner (10.iii.d): comparing abstractly defined, and pre-assigned, structures.

A blistering critique of these "inflexible" and "semantically empty" approaches was mounted by Hofstadter (Hofstadter and FARG 1995: 55–193). He complained that they were less interesting than Evans's work of twenty years before, and radically unlike human thinking. (Although both these charges were fair, his stinging critique wasn't entirely justified; for 'structuralist' replies, see Forbus *et al.* 1998; Gentner *et al.* 1997.)

But Hofstadter didn't share Fodor's gloom about the impossibility of *any* scientific understanding of analogy. To the contrary, he'd already spent many years studying it, using a basically connectionist approach. He'd been thinking—and writing—along these lines since the early 1970s (Chapter 12.x.a).

By the mid-1980s, he and his student Melanie Mitchell had implemented the Copycat program (Hofstadter 1985*a*, chs. 13 and 24, 2002; M. Mitchell 1990/1993; Hofstadter and Mitchell 1993/1995; Hofstadter and FARG 1995, chs. 5–7). This was described in several seminars at MIT in 1984, although without attracting much attention at the time (personal communication).

Intended as a simulation of human thinking, it modelled the fluid perception of analogies between letter strings. So it would respond to questions like these: "If *abc* goes to *pqr*, what does *efg* go to?", or (much trickier) "If *abc* goes to *abd*, what does *xyz* go to?" (Lacking a 'circular' alphabet, it couldn't map *xyz* onto *xya*; instead, it suggested *xyd*, *xyzz*, *xyy* . . . and the especially elegant *wyz*.) Its descriptors were features such as *leftmost, rightmost, middle, successor, same, group*, and *alphabetic predecessor/successor*. Like Evans's program, Copycat could generate a range of analogies, and compare their strength. Unlike Evans's program, it was probabilistic rather than deterministic, and could be 'primed' to favour comparisons of one type rather than another.

Whereas Copycat worked on letter strings, Hofstadter's Letter Spirit project focused on complex visual analogies. Specifically, it concerned the letter-likenesses and letter-contrasts involved in distinct alphabetic fonts. An *a* must be recognizable as an *a*, no matter what the font; but seeing other letters in the same font may help one to realize that it is indeed an *a*. At the same time, all twenty-six letters within any one font must share certain broad similarities, all being members of *that* font (see Figures 13.6 and 13.7).

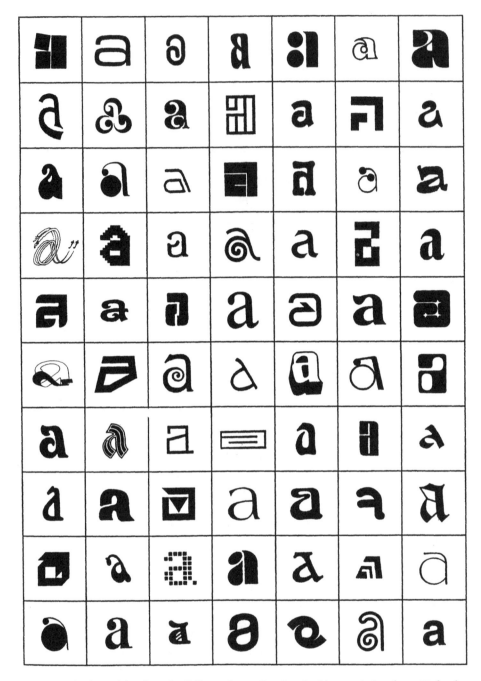

Fɪɢ. 13.6. The letter 'a' written in different fonts. Reprinted with permission from Hofstadter and FARG (1995: 413)

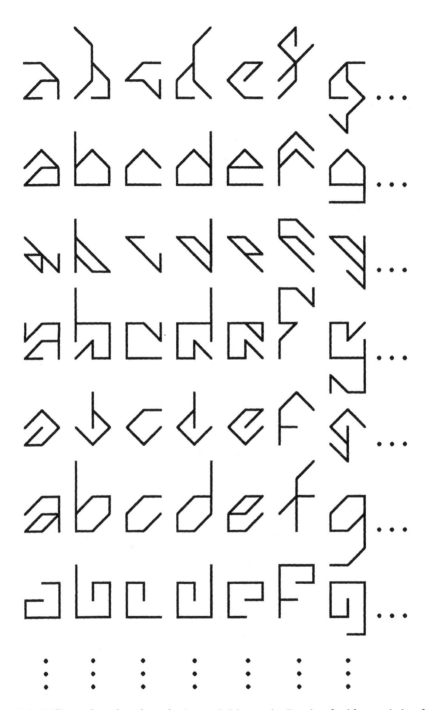

FIG. 13.7. Different fonts based on the Letter Spirit matrix. Reprinted with permission from Hofstadter and FARG (1995: 418)

Hofstadter's early writings on Letter Spirit had outlined a host of intriguing problems involved in interpreting and designing fonts (e.g. Hofstadter 1985*b*). And in the mid-1990s he described a program capable of recognizing letters in a variety of styles (Hofstadter and FARG 1995: 407–96; McGraw 1995; Hofstadter and McGraw 1995). By the new millennium, the design aspect of Letter Spirit had been part-implemented too (Rehling 2001, 2002). The new program could design an entire alphabet, given five 'seed' letters (*b, c, e, f, g*) as a guide. Today, the group's aim is to generate an alphabet from only a single seed.

Letter Spirit is the most ambitious AI analogy project, and in my view the most interesting. Whether it can readily be applied to other domains, however, is unclear. Even the much simpler Copycat would be difficult to generalize. However, if what one is interested in is how it's possible for human beings to engage in subtle and systematic analogical thinking, then it's a significant contribution. It's no simple matter to generate or appreciate an alphabetic font. So much so, indeed, that hardly anyone other than Hofstadter would have dreamt that anything cogent could be said about the computational processes that may be involved.

Turning from analogy to story writing, the best AI story-writer was authored not by a novelist or literary critic but by a computer scientist, now a professional games designer: Scott Turner (1994). (He's not to be confused with Mark Turner, a teacher of English at the University of Maryland who turned to cognitive science to illuminate the interpretation and creation of literature: M. Turner 1991; Fauconnier and Turner 2002.)

This program didn't produce high-quality literature. Its plots were simplistic tales about knights, princesses, and dragons, and its use of English left a great deal to be desired. But it had three interesting features.

First, it used case-based reasoning (ii.c, above) to create new story plots on the basis of old ones. For example, it generated a concept/episode of *suicide* (derived from *killing*) when the story's plot couldn't be furthered by third-party fighting. Second, it relied on the latest version of the Yale analysis of motivational and planning schemata to decide what might plausibly be done, and how (cf. 7.i.c and 9.xi.d). And third, Turner had realized—what previous AI programmers had not (e.g. Meehan 1975)—that a story needs not only goals and plans for each character involved in the plot, but also *rhetorical* goals and plans for the storyteller. Accordingly, a character's goals were sometimes rejected by the program, or their expression suppressed in the final narrative, for reasons of story interest or consistency. (For a sustained critique of Turner's approach, see Bringsjord and Ferrucci 2000.)

By the mid-1990s, AI had even made good on Babbage's suggestion that puns follow "principles" including double meanings and similar pronunciation of differently spelt words (Chapter 3.iv.a). Kim Binsted, at the University of Edinburgh, wrote a program—JAPE—that originated punning riddles fitting nine familiar templates, such as *What do you get when you cross an X with a Y?, What kind of X has Y?, What kind of X can Y?*, and *What's the difference between an X and a Y?* (Binsted 1996; Binsted and Ritchie 1997; Binsted *et al.* 1997; Ritchie 2003*a*). JAPE used a semantic network of over 30,000 items, marked for syllables, spelling, sound, and syntax as well as for semantics and synonymy. The program would consult the templates (no simple matter) to generate results including these: *What do you call a depressed train? A low-comotive; What do you call a strange market? A bizarre bazaar;* and *What kind of murderer has* *fibre? A cereal killer.* Babbage's "triple pun" (cane/Cain, a bell/a belle/Abel) couldn't have been created by JAPE, but its puns were no more "detestable" than most. (In fact, it's the most successful of today's AI jokers—Ritchie 2001; 2003*b*, ch. 10.)

As for creative mathematics, great excitement (within AI, if not outside it) was caused in the late 1970s by Lenat's program AM, or Automated Mathematician (Lenat 1977). This was his doctoral thesis at Stanford. Starting from a few simple concepts of set theory, it used 300 heuristics for modifying concepts, and criteria of mathematical "interestingness" (occasionally supplemented by specific nudges from the programmer) to generate many concepts of number theory. These included *integer*, *addition, multiplication, square root*, and *prime number*. It even generated a historically novel concept, which was later proved as a (minor) theorem, concerning *maximally divisible numbers*—a class which Lenat himself hadn't heard of. Later, he recalled how he found out about them:

> [I wondered whether any mathematician had thought of such a thing. Polya seemed to be the only one who knew.] He said, "This looks very much like something the student of a friend of mine once did." Polya was about 92 at the time. It turned out the friend was [Godfrey] Hardy and the student was Ramanujan. (D. B. Lenat, interviewed in Shasha and Lazere 1995: 231)

Coming up with something that had been discovered by such a giant as Srinivasa Ramanujan (1887–1920), people felt, was no mean feat.

Whether the excitement was fully justified was another matter. Critics pointed out that Lenat hadn't made clear just how the interesting concepts were generated, and suggested that a heuristic that was crucial for generating the notion of primes had been included, whether consciously or not, so as to make this discovery possible (Ritchie and Hanna 1984). What's more, they said, it may have been used only the once. (No detailed trace of the program was available.)

Lenat replied that AM's heuristics were fairly general ones, not special-purpose tricks, and that (on average) each heuristic contributed to two dozen different discoveries and each discovery involved two dozen heuristics (Lenat and Seely Brown 1984). He admitted, however, that writing AM in LISP had given him a tacit advantage. For minor changes to LISP syntax were relatively likely to result in expressions that were mathematically interpretable.

A few years later, yet more excitement was caused by Lenat's EURISKO, which—satisfying McDermott's plea for program development (11.iii.a)—incorporated heuristics for modifying *heuristics* (Lenat 1983). Besides being used to help plan experiments in genetic engineering, it came up with one idea (a battle-fleet design) that won a war-game competition against human players, and another (a VLSI chip design) that won a US patent—which are awarded only for ideas not "obvious to a person skilled in the art". Lenat himself then switched to research on CYC. But his AM program led others to focus on how mathematical "interestingness" could be used in automating creative mathematics (for a review, see Colton *et al.* 2000).

What of Simon himself? His late-century programming efforts were devoted to explaining creativity in science, although he gestured towards the humanities from time to time (Simon 1994*b*). With Langley and others at CMU, he wrote a suite of increasingly powerful programs intended to model the thought processes of creative scientists such as Francis Bacon, Joseph Black, Johann Glauber, Georg Stahl, and John

Dalton. These were initiated in the late 1970s, and continually improved thereafter (Langley 1978, 1979, 1981; Langley *et al.* 1981, 1987).

These inductive systems generated many crucial scientific principles—quantitative, qualitative, and componential. They came up with Archimedes' principle of volume measurement by liquid displacement, the very origin of *Eureka!* And they rediscovered Ohm's law of electrical resistance, Snell's law of refraction, Black's law of the conservation of heat, Boyle's law relating the pressure/volume of a gas, Galileo's law of uniform acceleration, and Kepler's third law of planetary motion. Occasionally (e.g. with Snell's law), they used a symmetry heuristic to choose between more and less elegant, though mathematically equivalent, expressions. Some could produce hypotheses to explain the observed data patterns, whether mathematical (e.g. Black's law) or qualitative (the chemical patterns observed by Glauber, and componentially explained by Stahl and Dalton). And the later versions could use the real (i.e. messy, imperfect) historical data, not just data doctored to make the sums come out exactly right.

Simon wanted to *explain* the ineffable, not just—as in meta-DENDRAL, or Emmy—to mimic it. So his group aimed to keep faith with human psychology.

Their programs, accordingly, respected the details recorded in the laboratory note-books of the scientists concerned. For example, they used the same data that the long-dead authors had used. (Or rather, they used the same verbal/mathematical data: they couldn't accept visual/auditory/haptic input from the real world, nor recognize similarities between sensory patterns—cf. Chapter 12.v–vi.) And as well as generating the same scientific laws, they tried to match the heuristics that had been used by the human scientists, and the temporal order of their hunches and discoveries—and mistakes. Later work by the CMU group focused, for instance, on the general principles of how to suggest and plan experiments (Kulkarni and Simon 1988), and on the use of diagrams in scientific discovery (Cheng and Simon 1995).

At one level, the BACON program and its siblings (BLACK, GLAUBER, STAHL, and DALTON) were highly impressive. They offered successful models of induction, and illuminated certain aspects of how many human scientists go about their work. (Despite the common propaganda, some science is *not* concerned with numbers, nor with componential structure: Chapter 7.iii.d.) But the creativity involved was exploratory rather than transformational.

In other words, these programs were spoon-fed with the relevant questions, even though they found the answers for themselves. What their human namesakes had done was different, for they'd viewed the data in new ways. Indeed, they'd treated new features as data. To identify *mathematical* patterns in the visual input from earth or sky had been a hugely creative act when Galileo Galilei first did it. To look for numerical constants was another, and to seek simple linear relationships *before* ratios or products (search priorities that were built into the heuristics used by the BACON suite), yet another. The CMU programs were deliberately provided with the ways of reasoning which Bacon, Glauber, Stahl, and Dalton had pioneered for themselves. They were roundly criticized by Hofstadter as a result (Hofstadter and FARG 1995: 177–9; see also H. M. Collins 1989).

(This, of course, recalls Haugeland's principled objection to GOFAI-based models of "insight". It also relates to the worries about "open-ended" evolution mentioned below and discussed in Chapter 15.vi.d. We'll see there that *totally new types* of sensor have been evolved in artificial systems, but only by unexpectedly taking advantage of the contingencies of the physical environment and/or hardware. Compare these 'biological' examples with Haugeland's psychological claim that understanding pertains to "the world and to living in it".)

In short, Saul Amarel's (1968) hope that an automated system might come up with a radically new problem representation remained unfulfilled. And not only Amarel's, for Simon himself had expressed much the same hope in the 1950s. The LT team's harbinger memo had listed four characteristics of "creative" thought, of which the last was this: "The problem as initially posed was vague and ill-defined, so that part of the task [is] to formulate the problem itself."

That's *still* beyond the state of the art. Langley (1998) recently reviewed seven cases of AI-aided discovery that were sufficiently novel, and interesting, to be published in the relevant scientific journals. He pointed out that, in every case, the programmers had been crucial in formulating the problem and/or manipulating the data and/or interpreting the results. So Ada Lovelace's futuristic vision of science-by-machine has been realized: many new answers have been found automatically, and some new questions too (e.g. new experiments). But *fundamental reformulations of old problems*, still less *radically new* ones, have not. (As we saw in Chapter 3.iv.b, Lovelace wouldn't have been at all surprised by that. On the contrary, it's just what she expected.)

In one sense, it's what Simon himself expected, too. For machines, at present, lie outside the cooperative loop (Chapter 2.ii.b–c). Simon described scientific discovery as a matter of "social psychology, or even sociology" (1997a: 171). His main reason was that open scientific publication provides a "blackboard" that hugely extends the individual scientist's memory (p. 172). At present, machine discovery systems "are still relatively marginal participants in the social system of science". We might be able to hook them up to data-mining programs, to reduce their reliance on human beings for providing their data and problems. However, "Even this is a far cry from giving machines access to the papers, written in a combination of natural, formal and diagrammatic language, that constitute a large part of the blackboard contents" (p. 173).

As for negotiations about the value of supposed "discoveries" (Chapter 1.iii.f and Boden 1997), Simon's view was that "the machine (augmented now by the computer) has already, for perhaps a hundred years, been a member of the society of negotiators" (Simon 1997b: 226). (Here, he cited the prescient Henry Adams, who'd been so deeply troubled by his visit to the dynamo hall in an industrial exhibition: H. Adams 1900.) Future AI programs, said Simon, might persuade us to value their discoveries above our own previous judgements (1997b: 225). He added that this was already happening in the area of mathematical proof (but that's not straightforward: see MacKenzie 1995).

Disputes about what counts as a discovery are especially likely in cases of transformational creativity, in which previously valued criteria are challenged. This type of creativity *was* eventually modelled, up to a point, by evolutionary programs. Some were focused on art (Sims 1991; Todd and Latham 1992), some on music (Gartland-Jones and Copley 2003; P. W. Hodgson 2002, 2005), and some on science or engineering (Goldberg 1987; Sims 1994; Ijspeert *et al.* 1997).

To some extent, genetic algorithms (Chapter 15.vi) could transform the conceptual space being explored by the program. For example, Paul Hodgson, an accomplished jazz saxophonist, wrote several programs designed to improvise Charlie Parker-style jazz in real time. The first two, IMPROVISER and VIRTUAL BIRD, used brief melodic motifs as primitives (these weren't statistically culled, as in Cope's work, but were based on a systematic theoretical analysis of music: Narmour 1989, 1992). VIRTUAL BIRD played well enough, early in the new century, that the world-famous Courtney Pine was willing to perform alongside it. Another, an evolutionary version called EARLY BIRD, used only dyadic (two-note) primitives (P. W. Hodgson 2005). It explored 'Bird-space' even more adventurously than its predecessor, partly because of the transformations it generated and partly because its primitives were less highly structured. Even so, it wasn't 'transformational' in the sense of generating a recognizably different musical style. Hodgson felt that this would be possible, but only if he himself added a great deal more musical information to constrain the changes allowed (personal communication).

That's hardly surprising, for in all such evolutionary programs the fitness function was due to the human being, whether built into the program or provided interactively. Peter Cariani (1997) has criticized current GAs accordingly, arguing that they're incapable of the sort of "open-ended" creativity seen in biological evolution. We'll see in Chapter 15.vi.d, however, that wholly new sensory organs *have* been evolved in artificial systems, by a combination of GAs and environmental/hardware contingencies. Perhaps future research on psychological creativity will take a leaf out of this biological book?

Biological evolution is not only open-ended, but unpredictable. And most creative ideas are unpredictable, too. There are various reasons why that's so (Boden 1990*a*/2004, ch. 9; see also Chapter 17, below). But that doesn't mean that AI creativity researchers were wasting their time. One may be able to explain something—to show *how it is possible*—without also being able to predict it (7.iii.d). Much as a theoretical psychology could never predict every passing fancy or every suicidal thought within Jo Bloggs's mind, so it could never predict every creative idea. It might be able to say a great deal, however, about the general types of ideas that were likely or unlikely, and why. That, you'll remember, had been Cohen's aim when he embarked on AARON in the first place; and it was Simon's aim, too.

By the turn of the millennium, then, AI research on creativity had at last become respectable. Several books on the topic had appeared, written or edited by long-standing members of the community (Michie and Johnston 1984; Boden 1990*a*; Shrager and Langley 1990; Partridge and Rowe 1994; Hofstadter and FARG 1995). The interest was spreading way beyond a few enthusiasts. The *Stanford Humanities Review* published two book-length special numbers on AI and creativity, especially in relation to literature (Guzeldere and Franchi 1994; Franchi and Guzeldere 1995).

The respectability suddenly snowballed into a range of professional meetings. IJCAI-1997 commissioned a keynote presentation on the topic (Boden 1998*a*). And a flurry of creativity conferences and workshops were mounted by AI and A-Life researchers. These included a continuing series on 'Discovery Science', which complemented the 'Creativity and Cognition' meetings on computer art that had been organized in the UK for many years past by Edmonds (vi.c, below).

If the ineffable hadn't yet been fully explained, *that it was truly ineffable* was now highly doubtful.

13.v. Outreach to Everyman

You may remember Babbage's lady visitor, who asked: "Now, Mr. Babbage, there is only one thing that I want to know. If you put the question in wrong, will the answer come out right?" (Chapter 3.i.a). Harriet Martineau was outraged, seeing this as a waste of the great man's time.

In 1960 many might have shared her outrage, especially if they'd never read Vannevar Bush (10.i.h). After all, most people hadn't ever touched a computer, and it wasn't at all clear that they ever would. The easy assumption, at that time, was that anyone who used a computer would be a computer expert so the issue of "putting the question in wrong" simply wouldn't arise.

What a mistake that was! Even by then it was clear that programming languages were needed, both to prevent *the computer experts themselves* from putting the questions in wrong and to enable them to ask the right questions in the first place (10.v.a–c). And error-detecting compilers were soon devised to alert the computer buffs to some of their mistakes. (In the mid-1970s, IBM's PL/1 tried to help programmers to remove bugs: Des Watson, personal communication. But even today, 'silent' automatic error correction is used only for very simple/common cases; instead, diagnostic error messages help the human to correct what's been put in wrong.)

Above all, it was a mistake to assume that *ordinary people*—from doctors and lawyers to the man on the Clapham omnibus—wouldn't be interacting with computers. Indeed, a few privileged doctors were interacting fruitfully with MYCIN by about 1970. But the man on the Clapham omnibus would have to wait until the mid-1980s, when user-friendly home computers became available.

Today, computer scientists take it for granted that users will often be ignorant, careless, impatient . . . in a word, human. So anticipating their mistakes, and enabling human–computer interaction (HCI) to be so intuitively natural that mistakes are minimized, is a thriving research area (D. A. Norman and Draper 1986; Sharples 1996; Rheingold 2000). The softbots, or intelligent agents, mentioned in Section iii.d are an example: human users who can rely on a softbot's doing something not only save time and effort, but avoid the mistakes they might make if they tried to do it for themselves.

This section (and the next) describes how GOFAI helped to bring user-friendly computing about. As we'll see, GOFAI itself benefited from the exercise. In making Bush's "memex" a reality, AI scientists would devise ways of interacting with computers without which today's AI simply couldn't exist. Computer graphics and object-oriented programming languages, for instance, were initially developed to help fulfil Bush's prophetic vision.

a. Papert and the media lab

In a sense, "user-friendly home computers" had been available ever since the late 1960s. For Papert had provided a programming language, LOGO, that was intelligible to very young children (10.vi). He'd designed homely forms of implementation too: push-buttons instead of keyboards, for use by clumsy-fingered infants, and line-drawing "turtles" moving on the CDU screen or on the floor. Much of LOGO's success lay in the fact that it resulted in (real or virtual) drawings, rather than text. For visual imagery is more basic, more intuitive, than writing.

Someone might object that LOGO turtles weren't really computers, just programmable toys. But that would be too quick, and too dismissive. For Papert had always aimed to make computing in general more accessible to non-specialists. What's more, his vision and example played a large part in inspiring others to develop the home computers we know today.

Papert's role in encouraging user-friendly computing grew even more prominent in the 1980s than it had been before. Besides developing StarLogo, which made distributed computing possible even for children, he intensified his activities in public education. For instance, in 1985 he started an educational/experimental programme in Boston's James E. Hennigan School, offering the LOGO experience to children drawn from a deprived area of the city (Brand 1988, ch. 7).

Still more to the point, he moved sideways within MIT: having begun in the young AI Lab, in 1986 he joined Nicholas Negroponte's (1943–) newly founded Media Lab. In other words, he was a key player in the Media Lab's activities from the start. Indeed, he'd been a player even *before* the start. For the Media Lab's real beginnings were twenty years earlier: it was the successor of Negroponte's late 1960s Architecture Machine Group, or AMG (Brand 1988: 137–54). Papert had been closely involved with AMG. (So had his close collaborator Minsky, who is now the Toshiba Professor of Media Arts and Sciences, as well as being an MIT Professor of Electrical Engineering and Computer Science.)

The Media Lab would turn out to be crucial in developing various futuristic technologies, especially VR. Whereas the AI Lab itself had grown more technological and less psychological (and so less to Papert's taste), the Media Lab took the two dimensions of AI equally seriously. That's because their prime focus was on human–computer *interaction*. (A secondary focus was on what's now called A-Life—see Chapter 15, and Brand 1988, ch. 6. Selfridge, currently working on adaptive agents, is now yet another of the Media Lab's pioneer stars: Selfridge and Feurzeig 2002.)

The AMG had pioneered 'exploratory' computer graphics of the type foreseen by Douglas Engelbart (10.i.h). They made sure that even architectural designers—and even architects' clients—with no prime interest in computers would be able to use their software. Besides experimenting with the computer-aided design (CAD) of individual buildings, they also explored Negroponte's (1970, 1975) interests in cityscapes. In general, they aimed to achieve Joseph Licklider's (1960) "symbiosis" between human and computerized visual designers, whether in architecture or beyond.

For example, AMG produced the first detailed VR simulation of an actual town (Aspen, Colorado) in 1978–9. Its users could explore the simulated streets as they wished, see the buildings from varied perspectives, and even enter some of them—and all this in any one of the four seasons.

The "Aspen Movie-Map" had been funded by DARPA, because of the Pentagon's interest in the use by Israeli commandos of a (physical) mock-up of Entebbe airport for practising the freeing of the hostages held in 1973 (Brand 1988: 141). Not quite everyone was impressed: the Movie-Map was one of the visionary research projects ridiculed by Senator William Proxmire (see 6.iv.f). But as so often happened, the Senator's scorned targets eventually had the last laugh—and everyone playing video games today should be laughing with them. (Although the US Army's current VR training environment and children's game are no laughing matter: see <http://www.ict.usc.edu> and <http://www.americasarmy.com>.)

Also funded by DARPA, and also a seed of today's technology for VR 'telepresence', was the Talking Heads project (Brand 1988: 91–3). This was intended to enable the US's five top leaders to run the country together while sheltering from a nuclear attack in five different locations. Among other things, it initiated attempts to model realistic lip–sound synchronization for speech (9.xi.g).

The Media Lab, then, was highly interdisciplinary. Architects mingled with biologists, psychologists, linguists, acousticians, musicians, and educationists—and everyone mingled with AI scientists. Indeed, as befits genuine interdisciplinarity, much of the mingling went on *inside* individuals' heads.

One of the key interdisciplinary heads in this enterprise was Papert. On joining the Media Lab, he set up the Epistemology and Learning Research Group (Harel and Papert 1991). This was no ivory-tower containment. Besides many educational consultancies, he founded a successful commercial venture (LOGO Systems Inc.) in 1981, and the non-profit Logo Foundation ten years later (<http://el.media.mit.edu/logo-foundation>). These provided a wide range of software, books, and advice on educational methods inspired by Piagetian ideas in general and LOGO in particular.

Papert is well known today for various reasons. His fame within the AI community—and with most cognitive scientists—rests primarily in his cooperation with Minsky on *Perceptrons* and *The Society of Mind* (Chapters 12.iii.d and 7.i.e), and secondarily in his trenchant rebuttal of Dreyfus (11.ii.b). In addition, these people remember his pioneering work on LOGO.

The general public, and educational psychologists, know of him through LOGO and also through his wider interests in pedagogy. He's been a highly visible gadfly on the back of the educational establishment for nearly forty years. Like the counter-cultural Ivan Illich (1971), but with a much stronger theoretical base, he was a fierce critic of orthodox practices in the 1970s. Many years later, he's still pursuing radical reform of the school system (Papert 1993).

There's a third reason why Papert *should* be remembered. It's unknown to most people, even among his professional peers. But it's the one which is especially important here. Namely: his AI work has influenced every computerized desktop, and almost every multimedia application, of the twenty-first century.

For it was Papert's Piagetian approach, and his child-friendly LOGO language and display, which inspired Kay to design computer interfaces for use by Everyman (see subsection d, below). Without those, today's world of widespread personal computing would be impossible. And he encouraged Negroponte to try to simulate architectural spaces (in AMG), and to set up the Media Lab to explore the possibilities of realistic simulation in general (including A-Life). That too was an exercise in the design of novel, and relatively intuitive, interfaces. It was guided, according to Negroponte's initial proposal, by studies in "epistemology [and] experimental psychology" (Brand 1988: 11)—code words meaning Papert.

Minsky and Jerome Wiesner were also hugely important influences. Indeed, the well-connected Wiesner—just retired, in 1980, as President of MIT and previously science adviser to the Kennedy and Johnson administrations—was able to be even more supportive of the Media Lab than he had been of the infant AI Lab, a quarter-century before (10.ii.a). He and Negroponte "toured and lectured and demoed and bargained for seven years, and raised the requisite millions" (Brand 1988: 11; cf. 131–7).

In short, Papert helped to develop virtual *machines* into virtual *reality*. His help was largely indirect, but crucial nevertheless. The user-friendly LOGO was eventually surpassed by SMALLTALK, the push-button boxes by menus and icons. And the on-screen LOGO turtles made way for on-screen Aspen...and for the host of VR applications we know today.

b. The H in HCI

HCI is a special case of what Donald Norman (1986*a*) called "cognitive engineering": the area of "applied cognitive science" which deals with how we use tools—from bath taps to computers. Since *Home sapiens* is above all a tool-using species, cognitive engineering has ramifications in many areas. Indeed, if—with Bruner (and Engelbart: 10.i.h)—one regards even language and drawing as cognitive *technologies*, then cognitive engineering is potentially relevant to virtually every aspect of human life.

As Norman pointed out, any adequate study of tool use is going to need not only an analysis of task complexity, but also "a theory of action". It's no accident, then, that he's helped to develop a general psychological theory of action. Besides influencing HCI (and tool design in general: D. A. Norman 1988), this is now influential in clinical neuroscience (Chapters 12.ix.b and 14.x.b). It's also no accident that two early papers on the general principles of HCI were largely due to him (D. A. Norman 1986*a*; W. J. Hutchins *et al.* 1986).

HCI work falls broadly into two groups. On the one hand, it involves writing user-friendly programs *to be run on a computer*. These may concern education and professional training, for instance—a huge area today, with several dedicated conferences/ journals. On the other hand, it involves designing user-friendly interfaces *to be part of the computer itself*. As this section will indicate, these concerns may overlap. A successful interactive VR program, for example, needs to be easily usable by people who aren't AI experts.

One aspect of a successful interactive program is that it makes the computer, and the program running it, invisible. That is, it gives users

the qualitative feeling that we are *directly* engaged with control of the objects [of interest to us]—not with the program, not with the computer, but with the semantic objects of our goals and intentions. (E. L. Hutchins *et al.* 1986: 95; italics added)

The writer using a word processor, for example, has the illusion that they're dealing directly with words, lines, or paragraphs—none of which exist as such inside the physical machine. Rather, they are structures in the virtual machine. Likewise, someone playing a computer game feels that they're zapping the goblin itself, or negotiating a real ski jump. It should be no surprise, then, that early HCI involved the development of 'direct' image manipulation on the one hand, and of object-oriented programming on the other.

The invisibility of the tool, to the tool-user, is a very old idea. René Descartes pointed out that the blind man's stick is, in effect, an extension of his body. More recently, phenomenologists such as Merleau-Ponty have written at length on this theme; and David Sudnow (1978/2001) has described the simultaneous growth of power and invisibility in learning to play the piano. So HCI is a special case of the study of "in-dwelling" in one's tools.

HCI has both technological and psychological aims, for it seeks to expand *the mind* as well as the computer. (This, of course, is just what Engelbart predicted nearly half a century ago.) It's also an aspect of cognitive science. To write easily usable software, one has to know something about how the "H" in HCI functions—and how that mental functioning enables, or prevents, the I between the H and the C. Kay, perhaps the most famous of all interface designers, put it like this:

[The] actual dawn of user interface design first happened when computer designers finally noticed, not just that end users had functioning minds, but that *a better understanding of how those minds worked* would completely shift the paradigm of interaction. (A. C. Kay 1990: 123; italics added)

Kay dated this realization as happening "to many computerists in the late sixties". And, as noted above, it was the guiding principle of MIT's Media Lab. However, both Bush and Gordon Pask had achieved it much earlier. Bush, indeed, was the first person to realize the need for HCI (Bush 1945, sects. 6–8). His memex was intended to achieve a man–machine fit by exploiting associative processing—and visual imagery—in both (10.i.h).

The memex was purely imaginary. Twenty years later, Pask looked more closely at *just how* people make what Bush had called associations. And he used what he found to pioneer human-friendly AI teaching machines. Pask's mid-1960s machines were designed for use by learners with three fundamentally different cognitive styles (Chapter 4.v.e). If he didn't allow for the questions being put in "wrong", he did allow for them being put in three contrasting ways. However, his ideas didn't catch on. In part, this was because they'd need to be reapplied in each subject domain. For Pask's HCI concern wasn't how to improve the computer itself, but how to write helpful educational programs for it.

Papert, too, was developing teaching machines in the mid-1960s. He wasn't allowing for human users' individual differences, as Pask was. But he was allowing for differences in age: his machines were designed to be accessible even to very young children, as we've seen.

Partly because of the child-friendly interface, Papert's ideas did catch on. During the 1970s, LOGO systems spread like wildfire (or like successful memes: Chapter 8.v.c). By the end of the decade, others were applying his Piagetian ideas about self-critical learning (as opposed to the LOGO language) in AI teaching machines. So they were used, for instance, to improve children's arithmetic, or university students' programming skills (Brown and Burton 1978; O'Shea and Young 1978; Brown and VanLehn 1980; Burton 1982; Sleeman and Brown 1982; O'Shea and Self 1984).

Babbage's lady visitor certainly wouldn't have outraged Papert, for his key point was that "putting the question in wrong" can be highly instructive. In other words, *bug* is a "powerful idea", and bug correction helps one think at a meta-level about one's own thinking (see 10.vi). But if that aspect of his psychological theory was widely taken up by others, his turtles weren't. They were restricted to simple drawing and/or locomotion.

What was required to go further was an interface that could be used for many different purposes. The 'universal' digital computer needed a 'universal' interface.

c. Good ideas in hibernation

The most important interface designers, besides Engelbart, were Sutherland and Kay. (For others, see: Laurel 1990; Rheingold 2000.) They wouldn't have been outraged by Babbage's visitor either. But besides hoping to minimize the non-specialist user's mistakes, they aimed to enable users—including AI scientists—to do things 'naturally' on computers which couldn't be done at all before.

To do that, they designed fundamentally new kinds of interface, for inclusion within the computer itself (i.e. the second type of HCI distinguished above). But their work took many years to be put into effect, because their ideas when first mooted were far ahead of their time.

Sutherland (1938–) was a wunderkind. While at high school in the early 1950s, he not only programmed simple computer calculators but also—with his elder brother Bert, and under the direction of Edmund Berkeley (1909–88)—built an electro-mechanical "squirrel" called Squee (D. G. Bobrow, personal communication).

Berkeley was an apt teacher for two machine-minded youngsters. He'd already designed a very early computer for personal use, called "Simon" (Berkeley 1949, 1956). It had 129 relays and a five-hole paper-tape feed. Given numbers of up to 255 binary digits, with coded instructions for up to nine different arithmetical operations, it displayed its answers in lights. The prototype was completed in 1950, and by the end of the decade he'd sold over 400 plans or kits for the machine (J. M. Norman 2004: 73).

In addition, he'd written the first popular book on electronic computers: *Giant Brains or Machines that Think* (Berkeley 1949). After describing various existing (and planned) machines, such as the Mark 1 and the ENIAC, he devoted the final chapters to speculations on how computers would affect society. In a sense, that was what it was all about. Indeed, the book's original title, cited in the contract he signed in 1946, had been *Machines to Help Us Think* (J. M. Norman 2004: 73). In other words, he was closer to Bush, Licklider, and Engelbart than to Minsky, McCarthy, and Newell-and-Simon. Possibly, the young Sutherland may have absorbed some of Berkeley's interest in the practical uses of computers.

For the time being, however, he and his brother were concentrating on their robot squirrel. Squee, which made the front cover of Berkeley's *Radio Electronics* magazine and was also featured in *Life* (19 March 1956), looked like a slightly more complex Grey Walter tortoise. It had two phototubes ("eyes") at the top of the steering post, two switches for sensing contact, and a scoop ("hands") at the front; guided by light, it searched for a "nut" (a torchlit tennis ball), which it picked up and took back to its "nest" (Berkeley 1956). Having exhibited this delight across the USA (New York, Pittsburgh, and Minneapolis) by 1956, Sutherland published a description soon afterwards (Sutherland *et al.* 1958).

Like the tortoises (4.viii.a–b), Squee was meant to draw a general moral—one which applied to human intelligence too. And many readers were persuaded. When he was still only 20 (in March 1959), Sutherland won a prize from the American Institute of Electrical Engineers for a paper on 'Parallels—Men and Machines'.

In later life, he deepened this early interest in mechanical creatures, writing about walking robots for the *Scientific American* (Raibert and Sutherland 1983) and patenting a robot arm in 1990. And his IT skills were recognized when, still only 26 years old (in 1964), he followed Licklider as head of ARPA's office for information processing technology.

But his fame arose from two achievements of his early to mid-twenties that had nothing to do with squirrels, artificial or otherwise, and which were decades before their time in terms of technology. For that reason, they had to go into hibernation until it was technically possible to put them into effect. These were his Ph.D. thesis on Sketchpad (1963), which initiated interactive computer graphics; and his 1965 vision of virtual reality, or VR.

Given that Sketchpad was described as early as 1962, why didn't I include Sutherland as a "harbinger" in Chapter 10.i? Well, that research—like Kay's, which was based on it (from 1966 on)—couldn't come to fruition until the late 1980s. Indeed, Kay said in 1989 that user-interface design was *still* hardly accepted as "a real subject" (A. C. Kay 1989: 123). Similarly, Sutherland's mid-1960s ideas on VR couldn't be fulfilled until the 1990s. Because both men had to wait twenty years or more for computer technology to catch up with them, their legacy to AI has a very modern, post-Kraken, feel.

Their CVs feel pretty modern too. Kay's includes appointments at Xerox, Apple, Disney, and Hewlett-Packard. As for Sutherland, his first job was as ARPA's Director of Information Processing in 1962–4 (supporting Project MAC, for instance). He's now Vice-President at Sun Microsystems. And, having co-founded a company doing VR simulations for aerospace and defence applications, he's now a billionaire.

Beginning life as a Ph.D. thesis, Sketchpad eventually became "one of the most influential computer programs ever written by an individual" (according to the citation for Sutherland's 1988 Turing Award). The thesis itself was published only as an MIT Technical Report, accessible to relatively few people until it was made available on the Web (Blackwell and Rodden 2003). But it was summarized in a conference publication (I. E. Sutherland 1963).

As it turned out, that was enough. For the details were far less important than the overall picture. Considered as a program, Sketchpad was highly limited. (Hardly surprising, given its date.) It couldn't even be shared, for it could be run only on the TX-2 at MIT's Lincoln Laboratory. But considered as a vision, it was dynamite.

For one thing, it was the first program which enabled the user to change the content of a computer's memory interactively, without reprogramming. (The SAGE operators in the 1950s had used lightpens to alter graphic displays by touching the screen; but they were changing *which stored memories were shown*, not *the memories that were stored*; see 11.i.) For another, it provided a suite of facilities for drawing and transforming pictures systematically. It wasn't the first graphics program: Spacewar, for example, had preceded it (10.i.i). But it was the first where the user could manipulate images *as such*.

The human could produce multiple instances of various image schemas ("master drawings"), and alter them at will in several ways. For example, the user could change the size of an existing drawing. Or if one had drawn a line on the screen that wasn't quite vertical, one could tell Sketchpad to make it vertical, maintaining the continuity of the whole object as it did so. Similarly, a carelessly drawn corner could be automatically transformed into a 'proper' corner. Physics could play a role too. So a truss in an engineering drawing, described as having a particular mass, could be made to bend as if it were governed by Newton's laws—that is, 'realistically'. In addition, Sketchpad allowed individual instances to be moved around the screen (with a lightpen) as whole objects, rather than remapping a collection of individual pixels.

Changing the master drawing would immediately change all its daughter images in the relevant way: so adding a third ear to a rabbit schema would magically add an extra ear to all the "copied" rabbits. What's more, images could be superimposed: "You could make a picture of a rabbit and a picture of a rocket, and then put little rabbits all over a large rocket. Or, little rockets all over a large rabbit" (Nelson 1977).

The details would disappear from view if the picture were made small enough, but would reappear when it was enlarged again. And the superpositions were recursive: you could have rabbits on rockets, on rabbits, on rockets. . . . Describing Sketchpad in a chapter with the title 'The Most Important Computer Program Ever Written', Ted Nelson (1977) admitted that "the rabbits and rockets are a frivolous example". But he pointed out many "obvious" applications: "blueprints, or electronic diagrams, or all the other areas where large and precise drafting is needed. . . . [A] new way of working and seeing was possible."

As for the still-untapped potential, Sutherland suggested (for instance) that facial characteristics in a drawing could in future be altered by the user. What's more,

if the almost identical but slightly different frames that are required for making a motion picture cartoon could be produced semi-automatically, the entire Sketchpad system could justify itself economically in another way. (Blackwell and Rodden 2003: 4–5)

Yes indeed!

d. The human face of the interface

Sutherland had intended his program to be used by artists and engineering draughts-men, as a tool for CAD. In other words, 'interface' HCI had begun. But it would be Kay who saw how the ideas in Sketchpad could be exploited in designing interfaces for an even wider range of users.

Kay first encountered Sketchpad in 1966, as a very new graduate student at the University of Utah. This was one of the first few locations to be put on the ARPAnet, and it was to be Sutherland's home for many years. As Kay remembered it:

At Utah before you got a desk you got a stack of manuscripts and you had to read the stack. It described Sketchpad. Basically, you had to understand that before you were a real person at Utah.

They also had a tradition there that the latest graduate student got the latest dirty task to do. Mine happened to be on my desk—a pile of tapes and a note which said "This is the Algol for the UNIVAC 108; if it doesn't work, make it work." It turned out to be the first Simula. (interview in Shasha and Lazere 1995: 42)

SIMULA, which had been developed in Norway just one year earlier, is now seen as a primitive object-oriented programming language. The first person to see it that way was Kay—whose "dirty task" would change not only his life, but (eventually) millions of other lives too. Like Sketchpad, SIMULA distinguished general classes from specific instances, which shared core properties with the master. Kay, who'd studied biology (and maths) as an undergraduate, felt that he was encountering something vaguely familiar in both cases: "The big flash was to see this as biological cells" (interview in Shasha and Lazere 1995: 43).

If the images/objects in Sketchpad and SIMULA were truly analogous to living cells, they'd have autonomy (as whole-object individuals), property inheritance, communication (passing and receiving messages), and the capacity to congregate in many-levelled hierarchies. Kay, then, sensed the possibility of a powerful object-oriented programming language (his term). Six years later, he'd designed one: SMALLTALK (Chapter 10.v.d). (And by 1991, an article about object-oriented programming, optimistically called 'Software Made Simple', was emblazoned on the cover of *Business Week* magazine—Resnick 1994: 42.)

But that was for the future. Meanwhile, on leaving Utah in the late 1960s, Kay—enthused by Licklider's concern for man–machine symbiosis (a near-synonym for HCI)—helped to design the ARPAnet. In 1970 he moved to Stanford's AI Lab, and in 1972 to Xerox PARC, where he headed the interdisciplinary Learning Research Group.

During the late 1960s, he worked on the FLEX machine, intended for use by doctors, lawyers, and the like (Shasha and Lazere 1995: 43–4). This experience—at a time when most people had hardly heard of computers, never mind seen one—led Kay to speculate about future interfaces which Everyman would find easy to use.

To address the H in HCI, Kay drew on Marshall McLuhan's *Understanding Media* (1964) and also on computationally informed theoretical psychology. As for the first, he recalled later:

Though much of what McLuhan wrote was obscure and arguable, the sum total to me was a shock that reverberates even now. The computer is a medium! I had always thought of it as a tool, perhaps a vehicle—a much weaker conception. What McLuhan was saying [when he talked about the printing press and television having changed the thought patterns of those who learned to read, or view TV] is that if the personal computer is a truly new medium then *the very use of it would actually change the thought patterns* of an entire civilization. (A. C. Kay 1989: 124; italics added)

(Bush had said much the same about the memex: remember that medieval historian, researching the Crusades? 10.i.h.) Kay was struck, too, by McLuhan's idea that "anyone who wishes to receive a message embedded in a medium *must first have internalized the medium* so it can be 'subtracted' out to leave the message behind" (p. 124; italics added).

Clearly, the closer the medium was to the already existing mind, the more chance that it would be "internalized" effectively. Hence Kay's interest in the psychology of thought, creativity, and learning. This interest was deepened by his conversations with Papert from the late 1960s on, including his visit to "one of the earliest LOGO tests within a school" (A. C. Kay 1989: 125). Indeed, he was "greatly encouraged" by Papert's work (A. C. Kay and Goldberg 1977: 171), and "*possessed* by the analogy between print literacy and LOGO" (A. C. Kay 1989: 125; italics added).

On reading "many" books in theoretical psychology, he soon turned (at Papert's suggestion) to Jean Piaget. Piaget had pointed out that the child's ways of thinking, as compared with the adult's, aren't just 'less of the same' but *different*. Clearly, this had implications for how one should best design interfaces for use by children. But Piaget's writing was too like McLuhan's—intriguing, but obscure—to be really useful. What was useful was Bruner's new interpretation of Piaget's results, especially as described in his book *Towards a Theory of Instruction* (Bruner 1966*b*).

In particular, Kay was inspired by Bruner's notion of cognitive technologies (Chapter 6.ii.c). These were different systems of mental representation: enactive, iconic, and verbal–symbolic. Kay saw them as "multiple separate mentalities with very different characteristics" (1989: 126), and aimed to exploit all three. In his words:

> Now, if we agree with the evidence that the human cognitive facilities are made up of a *doing* mentality, an *image* mentality, and a *symbolic* mentality, then *any user interface we construct should at least cater to the mechanisms that seem to be there*. But how? One approach is to realize that no single mentality offers a complete answer to the entire range of thinking and problem solving. *User interface designs should integrate them* at least as well as Bruner did in his spiral curriculum ideas. (1989: 127; last two italics added)

Accordingly, he invented—initially, for FLEX—HCI facilities including a computer mouse (enactive); screen icons (visual); and overlapping windows of text (verbal–symbolic).

As is usual (see 1.iii.g), his highly original ideas didn't come out of thin air. Early versions of Kay's interface items had been available since the 1960s (10.i.h). But Kay's were more powerful, and easier to use—and, as we've seen, more directly connected with current psychological theory. They were also better grounded empirically:

> [It] took our group [at Xerox PARC] about five years and experiments with hundreds of users to come up with the first practical design that was in accord with Bruner's model and really worked. (A. C. Kay 1989: 128–9)

For example, his windows *overlapped* because there was psychological evidence—from a best-selling book on tennis coaching, believe it or not (Gallwey 1974, ch. 7)—that people think/learn better if they're concentrating on only one thing at a time. (Later, and probably thanks to Kay, the tennis author—Tim Gallwey—would become a consultant at the Media Lab—Brand 1988: 100.) So having one window almost completely obscuring another was a Good Thing, provided that the other/s could be made the centre of attention at the click of a mouse. (Gallwey's book had included advice on how to learn to focus on one aspect of performance at a time; where Kay generalized this to computer interfaces, Gallwey later published best-sellers on other sports, and on "work" . . . etc.)

These technical advances were made possible by means of object-oriented programming. Today, this is hugely important across GOFAI as a whole. But it was developed by Kay to build flexible interfaces.

The SMALLTALK programming language (completed in 1972) was also driven by psychological criteria. It matched people's natural tendency to think in terms of concepts, and to 'place' a concept in a new context without worrying about just how it will be able to fit in. This made it easier to design computer interfaces where the user simply has to press a button or click on an icon, instead of knowing—and telling the machine—just what computations have to be performed. It also made it easier to include graphics in the interface, which—following Bruner and Papert—Kay saw as especially helpful to children.

He didn't stop there. He predicted personal computers much more convenient than the 350 lb FLEX machine: "This note speculates about the emergence of personal, *portable* information manipulators . . . " (A. C. Kay 1972; italics added). He went on to design the Dynabook, a notebook-sized gadget "as small and portable as possible" combining pictures, animation, music, speech, and text (words and musical notation). The Dynabook was intended as "a dynamic medium for creative thought" of many different kinds, in which—for instance—maths could become "a living language in which children could cause exciting things to happen" (A. C. Kay and Goldberg 1977: 170, 177; see also Kay 1977, 1989; Ryan 1991).

Although this multimedia system was never fully built, its core ideas were implemented in the Xerox Alto computer by 1974 and it was the ancestor of today's laptops and palmtops. And as if all that weren't enough, he also instigated desktop publishing (thanks to multi-font word-processing), and helped develop the Ethernet, laser printing, and client–server architectures.

I say "he", but of course others at Xerox PARC were involved too. One of these was Kay's assistant Adele Goldberg. Another was David Canfield Smith, who coined the term "icons", and whose 1975 Stanford thesis—based on SMALLTALK—eventually fed into the Xerox Star machine (D. C. Smith 1977). This programming environment enabled the user to select, and hop about among, different visual presentations of the interface by choosing from a menu. In Smith's mind, his Pygmalion program was not only assisting creative thought, but—in a sense—modelling it too. The same had been true, of course, of Bush's "associative" memex.

Kay's success in HCI eventually helped foster the success of Apple Computers—and, indirectly, of Microsoft too. Xerox PARC had refused to manufacture machines or to market software based on his work, rejecting his visionary prediction of huge technological applications. They wanted to develop the paperless office, not home computers for Everyman. As for the multitude of visitors to PARC, most realized that Kay's team were making "spectacular advances", but "very few understood either in depth or in scope the applications of what they were seeing" (L. Tesler, interviewed in Brand 1988: 173). Steve Jobs, by contrast, saw the point:

When I went to PARC [in November 1979], I thought it would be an interesting afternoon, but I had no concept of what I'd see. Larry Tesler was my guide for an hour, to show me around. My mind was just totally blown. The minute I saw an Alto (PARC's personal computer) and the mouse and the multiple fonts, I *knew* that we had to have it. I came back to Apple a raving maniac about this stuff and I grabbed a bunch of people . . . and dragged them over there. (Steve Jobs, interviewed in Brand 1988: 172–3)

After this "mind-blowing" visit to Kay's lab, Jobs tried to buy the SMALLTALK software, but Xerox refused to sell it—despite being a minor shareholder in Apple (Shasha and Lazere 1995: 49). Apple used ideas inspired by Kay in developing the AppleMac (and appointed Kay as an Apple Fellow in 1984). The rest, as they say, is history. (A chequered history, to be sure: because of IBM PC marketing patterns, Microsoft's "Windows" software is much more widespread than the AppleMac version, on which it's based.)

Over the years, Kay has become increasingly interested in children's education— partly as a result of a 1968 visit to Papert's lab (Shasha and Lazere 1995: 45, 47 ff.).

Following Papert's example, he regularly tested the accessibility of his HCI ideas by trying them out on children, both at a Palo Alto junior school and—via Papert—at Boston's Hennigan School. A second reason for working with children was that, being used to having coloured paints and musical instruments, they "really needed" computing facilities that were pushing the state of the art in the 1970s (A. C. Kay and Goldberg 1977: 171).

Having worked for Apple in the mid-1980s, and for Disney Imagineering in the 1990s, Kay is now a Hewlett-Packard Fellow. He was—and still is—an important presence/visitor for the Media Lab, where he's advised on projects ranging from Hollywood animation to A-Life (Brand 1988, ch. 6). And he's involved in various educational ventures, such as the non-profit Viewpoints Research Institute (founded in 2001). But despite his journey through the highest hills of high tech, he retains his interest in psychology. Indeed, he still calls Bruner on the phone about once a fortnight (Shore 2004: 87).

Bruner's ideas on cognitive technology are enthusing a new generation of HCI enthusiasts, too. For instance, Mike Scaife (who died suddenly in 2003) and Yvonne Rogers at Sussex have instituted a research programme called Interact. Explicitly inspired by Bruner (Y. A. Rogers *et al.* 2002*b*), this is centred on children's (often collaborative) use of external representations of various kinds: words, diagrams, pictures, toys . . . and so on.

These include 'tangibles', which are physical objects (bricks, balls, puppets, clothing . . .) electronically augmented so as to trigger events in other electronic artefacts. For instance, the children's play activities with a tangible interface can cause changes in a screen interface. The VDU may show a virtual animal, which engages the children's attention: as they do this or that with their electronically enhanced toys, the screen animal shows this or that aspect of its form or behaviour (Price *et al.* 2003). And those changes are recorded by the computer, to be used later by the children for discussion and self-reflection of many kinds.

The Interact programme is essentially interdisciplinary, for it has two closely related aims. One is to devise new technologies—in this case, for use by young children. The team's new designs go way beyond Papert's turtles, for they involve recent advances such as "pervasive" computing (i.e. the use of tangibles) and multimedia VR. (For examples, see Scaife and Rogers 2001; Rogers and Muller forthcoming; Marshall *et al.* 2004.) The second aim is to augment, as well as to exploit, our understanding of how children play, learn, collaborate, and represent (e.g. Scaife and Rogers 1996; Scaife 2005; Rogers *et al.* 2002*a*).

Of the many advances in these areas of psychology since Bruner's seminal 1960s work, some were initiated in the 1970s by Scaife himself (see 6.ii.c). Moreover, Rogers—whose expertise is strongly technological—was initially trained as a psychologist, as were several others in their group. In short, Interact is just one example illustrating the fact that in HCI research, psychology and GOFAI (and sometimes connectionism too) walk hand in hand.

Increasingly, they walk unnoticed—indeed, invisibly. In a futuristic article in the *Scientific American*, Xerox PARC's chief scientist Marc Weiser (1952–99) predicted the advent of what he called "ubiquitous" computing, or ubicom (Weiser 1991). This, he said, would be the "third wave in computing", following on time-shared mainframes and the personal computers foreseen long ago by Bush. What even Bush hadn't foreseen was the use of miniaturization and telecommunications (e.g. wireless technology, offering local links and/or entry onto the Internet) to embed tiny adaptive computers into almost every niche in our environment: offices, houses, cookers, cars, toys... even jewellery and clothes. The human–computer interface, traditionally understood, has largely disappeared: as Weiser (1994) put it, "the world is not a desktop".

There's now an international conference series on ubiquitous computing (Ubicomp), dedicated to putting his vision into effect. IBM preferred their own terminology: pervasive computing, or "percom". But whether driven by Xerox or by IBM, or by the many start-up companies at the turn of the millennium, the third wave of computing is gaining strength. The closer it gets to human lives, the more important the psychological issues become. And as we've seen, much of it was seeded by psychology in the first place.

13.vi. Virtual Reality

Virtual reality is the epitome of current high tech in IT. But it's not of interest only to 'techies'. It raises some deep philosophical issues (see 16.vii.d and viii.c), many intriguing psychological questions (e.g. 14.viii.b and 15.xi.a), and various social-psychological worries too (subsections d–e, below). Nor is it really new. The idea, and even the pioneering technology, dates back to the late 1950s.

One might argue that the general idea dates back even further than that, if one counts E. M. Forster's dark sci-fi story 'The Machine Stops' (1909). The characters are a mother, Vashti, and her son Kuno. Vashti, like most of the civilized human race, lives underground in a wholly machine-tended capsule. She communicates with her son (and "several thousand" others: "in certain directions human intercourse had advanced enormously") only by a videophone, never face to face. The videophone—"rightly", she thinks—doesn't "transmit nuances of expression": it conveys instead "a general idea of people ... good enough for all practical purposes". Kuno prefers the despised real world, and eventually stops the machine—to his mother's desperation, who prays for it to restart some day.

Kuno's (Forster's) horror at the human isolation and machine dependence suffered/enjoyed by Vashti parallels some of today's worries about computer companions and avatars (see below). But even Forster's imagination didn't stretch to wholly machine-generated images, still less to the ambition and detail of VR as we know it a century later. (Nevertheless, a quotation from his story opens a recent paper on the philosophical implications of VR: H. L. Dreyfus 2000.)

The term "virtual reality" was coined in 1989 by Jaron Lanier, to denote "three-dimensional realities implemented with stereo viewing goggles and reality gloves" (M. W. Krueger 1991, p. xiii). Lanier, a computer scientist and musician, had already founded the first commercial VR company (in the early 1980s). So he wasn't announcing a new vision. On the contrary, his new label (like the term "artificial life": 15.x.a) was intended to pull together pre-existing work of diverse kinds.

VR has since been redefined in several ways (Steuer 1992). These vary on both technological and philosophical dimensions. We don't need to consider those variations here, but it's worth noting that "VR" means different things to different writers.

It also means different things to different writers in the sense that it's ethically approved by some and roundly condemned by others. By the mid-1980s, various commentators were raising worries about VR.

Some were sociopolitical. Donna Haraway (1944–) was especially influential, here (Haraway 1986/1991). (For recent examples, concerning multi-user VR and the BodyNet, see Rheingold 2002.) I'll ignore those, however. There will be plenty of psychological worries (and some philosophical implications) to discuss. That's only to be expected, given that VR is such an intimate form of cognitive engineering.

a. Intimations of VR

Among the pre-existing work gathered under Lanier's label was Sutherland's. Indeed, Sutherland's second claim to fame is as the prophet of VR.

Besides being its prophet, he was an early pioneer. His revolutionary VR headset, announced in the 1960s and completed by 1970, superimposed a transparent TV image over the real-world perception (I. E. Sutherland 1968). (Previous efforts had *substituted* a TV display for the real image: Comeau and Bryan 1961.) It provided a full-colour, three-dimensional, visual display that filled the user's entire field of view. (Moreover, Sketchpad-derived techniques enabled the wearer to manipulate the virtual images with various input devices.) The experience of 'total visual immersion' was new—and, even though it lacked stereovision, remarkably compelling.

Filmgoers, of course, had long got drawn into Bambi's hand-drawn virtual world. But while watching him they could see the walls, and the seats and people and popcorn too. The conventional movie screen took up a mere 5 per cent of the spectator's field of vision. Even Cinerama and Cinemascope, still novelties at the time, didn't get anywhere near 100 per cent visual immersion (Heilig 1955: 283–4).

One of the first cinematographers to be enthused by that "100 per cent" idea was Hollywood's Morton Heilig (1955). His description of 'The Cinema of the Future' saw the viewers near-inhabiting the film, being wholly immersed in a virtual world even more all-encompassing than Cinerama. Besides all-round visual experience and stereo sound, it would offer taste, touch, and smell—each sense to "dominate the scene in roughly the same proportion we found them to have in man: sight, 70%; sound, 20%; smell, 5%; touch, 4%; and taste, 1%" (Heilig 1955: 292). (Unappreciated in Hollywood, Heilig moved to Mexico, where he invented two early virtual reality devices that later inspired more powerful versions from others—Packer and Jordan 2001: 220.)

However it was Sutherland's prophecies, not his technical achievements, which caused the stir in the 1960s. For what he had in mind was a very high degree of reality attribution. Visual immersion was just part of it.

Even before announcing the VR goggles, he'd already urged his NewFAI colleagues to produce computer-generated pictures (and sounds, and forces) that would be experienced by the viewer as 'real'—and which might represent unreal "Wonderland" worlds too. So in his speech to a large IFIPS meeting in 1965, he'd said:

We live in a physical world whose properties we have come to know well through long familiarity . . . A display connected to a digital computer gives us a chance to gain familiarity with concepts not realizable in the physical world [such as non-realistic geometries, or negative mass]. *It is a looking-glass into a mathematical wonderland.* (I. E. Sutherland 1965: 506; italics added)

To be convincingly realistic, he continued, the display "should serve as many senses as possible". Excellent sound generation was already achievable, although speech generation wasn't (nor would it be, until much later: see 9.xi.g). And computer-generated "kinesthetic displays" were well within reach, by extending the controls of flight simulators—which gave trainee pilots of the 1960s "the feel of a real airplane".

Besides providing joysticks with force-feedback capability (already available), future VR computers—he said—would "easily sense the positions of almost any of our body muscles". Because they'd be able to monitor not only hands and arms but eye movements too, we might use "a language of glances" to control a computer: "For instance, imagine a triangle [in the VR display] so built that whichever corner of it you look at becomes rounded."

The futuristic speech ended with a flourish:

The ultimate display would, of course, be a room within which the computer could control the existence of matter. A chair displayed in such a room would be good enough to sit in. Handcuffs displayed in such a room would be confining, and a bullet displayed in such a room would be fatal. *With appropriate programming such a display could literally be the Wonderland into which Alice walked.* (p. 508; italics added)

One of the people in the audience for Sutherland's lecture was Frederick Brooks (1931–), who'd headed the team designing IBM's System/360 computers in the late 1950s. For him, it was an epiphany:

[Sutherland] said, "Think of the screen as a window into a virtual world. The task of computer graphics research is to make the picture in the window look real, sound real, interact real, feel real." And I said, "That's where it's at." I've been working on that challenge ever since. (interview in Shasha and Lazere 1995: 170)

Brooks wasn't the only one. And once the technology had caught up with Sutherland's vision, VR found many industrial applications.

Vibrating seats, fan-generated winds, and chemical 'smell banks' arrived relatively early. By the late 1980s, thanks largely to work done at NASA-Ames and the Media Lab, and funded by Hollywood as well as by government grants, VR had added stereovision, headphones, microphones, gesture interaction, and fibre-optic data gloves that could measure the ever-changing bend of each finger joint and the distance between the fingers (Fisher 1990; Brand 1988). (An early pressure-sensitive glove, the "Teletact", was invented by Robert Stone and Jim Hennequin, but wasn't taken up by other VR workers: R. Stone, personal communication.)

Brooks ended his reminiscence by saying "Any year we'll get there." For already "by 1994", on his view, "one could honestly say that VR 'almost works' " (Brooks 1999: 16). This means that "any year" is roughly *now*. Presumably, he's disappointed. For despite all the hype, a near-fully realistic experience is only slightly nearer.

Complete sensory immersion still isn't possible. (Even if it were, it's not clear that it could ever be phenomenologically equivalent to having a bodily presence in a real situation—H. L. Dreyfus 2001: 59–72; Dreyfus 2003.) In research laboratories today, and even in artists' studios, VR goggles and gloves—and body suits—can be fed with signals from afar, or from simulated worlds instead of the real one. And entire virtual environments can be built, obviating the need for goggles or gloves. But the VR dreamt of by Sutherland, where the simulated world is wholly imaginary and yet utterly convincing, is still a very long way off.

Indeed, some multimedia experts are already claiming that VR, as originally under-stood, has not only "failed" to live up to its promises but has no chance of doing so in the future. More success, in Stone's (2005) view, has been achieved within what's called "serious gaming"—whether intended purely for entertainment or for training purposes. This is partly a matter of accessibility and affordability: unlike VR installations confined to specialist high-tech laboratories, these systems are already available to, and easily usable by, Everyman. But it's also a matter of verisimilitude in the characters and avatars depicted. As of January 2006, perhaps the best example of 'realistic' human movements and facial expressions in such a game is the "G-man" character in Half Life 2, which has "40 facial muscles and a skin texture that beggars belief" (R. J. Stone, personal communication). Indeed, this technology has inspired an eerily plausible simulation of the court room scene in the Jack Nicholson film *A Few Good Men*—called, as you might expect, *A Few Good G-Men* (see <http://www.machinima.com/films.php?id=1154>).

Partly because of the huge difficulties—and expense—involved (how many dollars per simulated muscle?), by the late 1990s much of the research emphasis had switched from the interface itself to the human's *interaction* with it. The 'reality' was largely projected onto the interface by that interaction (effort after meaning, again).

Accordingly, some current definitions of VR don't specify particular technologies (e.g. gloves, graphics, or headsets) but focus rather on the interactive experience. For instance, when the UK's Department of Trade and Industry launched their VR Awareness Initiative in 1996, they adopted Stone's definition: "Virtual Reality refers to a suite of technologies supporting intuitive, real-time interaction with multi-dimensional databases" (R. J. Stone, personal communication).

The omission of any reference to immersion, as envisaged by Sutherland and Brooks, was deliberate. For excessive hype had led to widespread scepticism. (The same old AI story!: see 11.ii–iii and 12.ii.f.)

b. VR as a practical aid

To illustrate the (limited) degree to which VR has already "got there", let's consider four practical applications: one for information access, one for doing surgery, one for surgical training, and one for training in hands-on mechanics. (Then, in subsection c, we'll look briefly at VR in contemporary art and entertainment.)

The first is an example of "wearable computing", intended for (literally) everyday use. This is a "Personal Information Architecture" gadget (for belt, briefcase, or handbag) called BodyNet (because the components communicate by low-powered radio waves extending no more than six feet from the body). The aim is to combine a person's radio, TV, video/CD-player, laptop, diary, and mobile phone. Like a waterproof wristwatch, this is "an intimate interface: always there, always on" (Shivers 1993: 7).

BodyNet is based on "magic goggles": clear-glass goggles which, besides enabling the wearer to see the real world, present small inset colour displays to each eye. These may carry information about the current state of the stock market, or they may present a TV newscast or chat show, or the wholly imaginary VR world of a computer game. Miniature earphones and microphone enable the wearer to hear and deliver speech within the virtual world. The focus of the (stereoscopic) goggle display was initially identical, irrespective of where the person was looking. But by the mid-1990s the system could monitor the wearer's eye movements to discover what they were looking at. If this is something in the virtual world, the next image sequence and/or audio input could be generated accordingly.

Wearing/carrying BodyNet, one could be walking through a simulated VR fantasy while also walking down the High Street. For the virtual images accompany the person's real perception rather than replacing it. Partial immersion, such as this, raises interesting HCI questions about how people manage to focus on physical and virtual worlds simultaneously, and how they can be induced by the VR engineer to attend to what's 'relevant' (Ohta and Tamura 1999). In other words, psychology and computing in a close embrace, again.

The second practical application concerns 'distance surgery', still in the experimental stage. The thinking and planning is done by the human surgeon. But he/she is immersed in a virtual world. This represents the real world at a distance, where the real patient is really located. In other words, this is a form of *telepresence*, as opposed to a wholly fictitious VR construction.

The surgeon receives camera input and force-feedback (haptic) data from the real-world site, thanks to VR goggles and gloves, and sends body-movement signals to the robot 'surgeon' at the other end of the line. If miniature robots are involved, the surgeon's movements are scaled down during transmission. In short, distance surgery combines sensory (visual, touch, and kinaesthetic) VR with sophisticated robotics.

One leading centre for this research is Oussama Khatib's laboratory at Stanford. Khatib, currently co-editor of the *Robotics Review*, directs a wide range of research in robotics. This includes "human-centred" robotics, in which the robot interacts and/or cooperates with a human being in real time (Khatib *et al.* 1999, 2002).

Human-centred robotics in general raises a number of problems which traditional robotics didn't have to face. Some concern the 'psychological' interlock between the person and the (largely) autonomous artefact: just how can their plans/perceptions be integrated? A mismatch between surgeon and robot could have horrendous consequences. Others concern their physical interlock: if they're sharing the same space, the robot must (usually) have no sharp edges, no sudden movements, and no excessive forces (Zinn *et al.* 2002).

Excessive force can be avoided, whether the robot is human-centred or not, by deliberately limiting its mechanical strength and/or by building in haptic sensors. These provide force feedback to control the power of its autonomously generated movements. In other words, the robot 'knows' by touch not only *that there's something there* but also *how resistant it is*, and therefore *how strongly the robot should push at it*. (Compare the feel of flesh with that of bone; and imagine pressing on a cushion, a balloon, or a bottle of wine.) In addition, 'kinaesthetic' feedback tells the robot how much force it needs to exert to lift the object, and/or carry it.

Yet other new problems concern the VR aspects—if any. In distance surgery, the surgeon doesn't share the robot's physical space but the patient does. There's an exception to the "no sharp edges" rule, since the robot manipulates a scalpel. It's especially important, then, to ensure that haptic information is conveyed to the surgeon effectively. If the robot responds autonomously to its own sensors, it must do so immediately and delicately. (In many distance-robotics applications the haptics don't need to be quite so delicate.)

One might argue that, in a very real—i.e. phenomenological—sense, the surgeon *does* share the robot's space. For as the surgeon becomes more familiar with the robotic tool, the tool (much like a blind man's cane) comes to be experienced as part of his/her own body. To put it another way, the boundary of his/her body seems to extend to the region of the robot. This is a special case of the sort of action-led adaptability in body image that was first studied by George Stratton over 100 years ago (see Chapter 14.viii.b). (Many other Strattonesque reorientations based on VR and/or telepresence are described in A. J. Clark 2003*a*.)

In principle, the same equipment can be used to convey 'sensory' information from a *virtual* world. The better the VR design, the greater the person's sense of immersion in a "genuine" reality will be. (I can vouch for that, having experienced Khatib's experimental visual/haptic system when visiting his lab in autumn 2003.) In one sense, this isn't new: as Sutherland pointed out in his 1965 talk, flight simulators have existed for years. But they've got steadily more convincing.

Distance surgery isn't the only use of VR by surgeons: our third example concerns the use of 3D simulations of bodily organs for surgical *training*. For instance, consider a VR brain being developed at the University of Nottingham under the direction of the neurosurgeon Michael Vloeberghs (Wang *et al.* forthcoming).

Not only does this VR brain have the appropriate 3D anatomy, but it's realistically squashed/deformed (in real time) when the trainee's virtualized surgical tool touches it. As well as *looking* different, it *feels* different too. Given that brains are very soft (compared with hearts, for example), the visual/haptic information at issue here is highly complex. For instance, the "touching" may be prodding, pinching, or cutting—each of which has its own distinctive material and perceptual effects.

The mathematical techniques being used by the Nottingham team have already been applied to other parts of the body too. And organs that aren't surrounded by a bony skull may be easier to simulate. For in listing their VR tasks in order of difficulty, these authors identify the deformations caused by "the contact of the tissue with itself and with its surroundings" as the most challenging of all (Wang *et al.* forthcoming, sect. 3).

The last illustration of practical VR is a recently developed system for training mechanical skills. A number of "embodied conversational agents" are already being developed for HCI-based training in a wide range of domains (Andre 1999; Cassell *et al.* 2000). That word "embodied" is misleading for these agents aren't robots. Rather, they're VR creatures existing in cyberspace—and interacting there with real people.

One of a variety of AI training systems developed by Jeff Rickel and Lewis Johnson, at the University of Southern California's School of Engineering, is called Steve (Rickel and Johnson 1998, 1999, 2000). The acronym stands for SOAR Training Expert for Virtual Environments. As that "S" indicates, this computer tutor is grounded in GOFAI work. It builds not only on SOAR (which itself is highly complex: Chapter 7.iv.b), but also on GOFAI agents, teaching machines, computer graphics, and NLP (e.g. B. G. Deutsch 1974; Grosz 1977; Grosz and Sidner 1979; Rickel *et al.* 2002). And, true to the interdisciplinary nature of HCI in general, it also uses psychological research—such as Paul Ekman's work on the facial expression of emotion (cf. Chapter 7.i.d).

Steve's designers describe him (*sic*) as "a new breed of computer tutor", since he's "a human-like agent that can interact with students in a virtual world to help them learn". (One of the first of the old breed of computer tutor was the programmed mechanic CBC, mentioned in Chapter 10.iii.c.)

The human pupil, in order to share the VR world with Steve, wears VR-display goggles and VR gloves, and carries a 3D mouse. And Steve—visually presented to the learner as a 3D humanoid (disturbingly, lacking hips and legs)—is able to watch and monitor the pupil's actions, and to give spoken advice, accompanied by gestures, when difficulties arise. (The "-oid" in "humanoid" is only too apt, for Steve also lacks realistic skin texture and facial expressions: he's hugely inferior in these respects to the Half Life 2 example mentioned above.)

Despite the lack of nether regions, Steve is a 'believable' agent. This is largely because he/it integrates perception, cognition, and motor control:

The perceptual module monitors the state of the virtual world, maintains a coherent repres-entation of it, and provides this information to the cognition and motor control modules. The cognition module interprets its perceptual input, chooses appropriate goals, constructs and executes plans to achieve those goals, and sends out motor commands. The motor control module implements these motor commands, controlling Steve's voice, locomotion, gaze, and gestures, and allowing Steve to manipulate objects in the virtual world. (Rickel and Johnson 1999: 343)

Specifically, what Steve teaches his students is how to perform *physical* tasks, where bodily demonstrations, as well as verbal instructions, may be helpful. Examples mentioned by his designers include the operation and maintenance of complex—and often dangerous—equipment in ships, factories, power stations, and the like. (In principle, other physical tasks, from cooking to surgery, might be demonstrated by Steve too. But this would require detailed simulation of the relevant physical objects in the VR world.)

Steve's tasks can involve more than spanners and knobs. They have a social dimension as well:

In addition to training students on individual tasks, he can also help them learn to perform multi-person team tasks: he can serve as a tutor for a student learning a particular role in a team, and *he can play the role of a teammate when a human teammate is unavailable.* (Rickel and Johnson 1998: 30; italics added)

(In subsection d, we'll consider some worrying implications of the use of the words *teammate* and *he* in that quotation.)

c. VR in art and play

Sutherland's talk of Wonderland had carried associations of play, and of art. And indeed, artists were inspired by AI/VR to develop new artistic genres (Brand 1988; Stiles and Selz 1996; Packer and Jordan 2001).

Some of these could be seen as versions of performance art, in that the prime interest wasn't in the sensory qualities of the product (images, music...) but in the mental processes and physical skills that produced it. Arguably, *all* art, even including the *Mona Lisa*, is performance art in that sense (D. Davies 2004, esp. chs. 7–10). If so, then the new VR art wasn't as marginal as a conventional aesthetics would imply. In any event, the art world of the late twentieth century was greatly influenced by experiments in AI and VR (Stiles and Selz 1996: 384–498; Packer and Jordan 2001).

Sometimes, computers were used merely as "rather complicated tools, extending the range of painting and sculpture, performed music, or published literature". But in the more interesting cases, they made possible "*a whole new field of creative endeavor* that is as radically unlike each of those established genres as they are unlike each other" (Ascott 1990: 245; italics added). A key feature of *interactive* computer art was that the viewer wasn't just a viewer, but a playful participant—even a co-creator.

Computer art began in the mid-1960s (Klutsch 2005; Nake 2005). One of the first recognized computer scientists to support this new activity was Berkeley, who'd encouraged the young Sutherland brothers in their Squee project shortly before (and who'd written the first popular book on electronic computers: Berkeley 1949). The February 1963 issue of Berkeley's magazine, recently renamed *Computers and Automation*, announced a competition for "examples of visual creativity in which a computer plays a dominant role".

At much the same time, the first examples were already being produced. Some in New York, by Michael Noll (aided by the cognitive psychologist Bela Julesz), and some in Stuttgart, by Frieder Nake (1938–). Nake even used a new flatbed drawing machine constructed by Konrad Zuse himself (Nake 2005: 55; cf. Chapter 3.v.a).

As for interactive art, the first highly visible instance was *SAM* (Sound Activated Mobile), built by the cybernetic sculptor Edward Ihnatowicz (1926–88). Exhibited to public delight at the 1968 exhibition in London (and ending up in the Exploratorium in San Francisco), this looked like a large fibre-glass flower on a flexible stalk, with four petals and four stamens. But the stamens were radio sensors. They picked up noises made by the viewers, and caused the hydraulic pistons in the stalk to move the flower left or right, up or down, towards the source of the sound (Zivanovic 2005: 103).

That was the principle, too, of Ihnatowicz's even more famous interactive sculpture, *The Senster*. This eerie creature, built in the laboratories of University College London, had been commissioned by Philips, Eindhoven, and was exhibited by them from 1970. Looking like a 15 foot tall Meccano giraffe, it was equipped with sound and movement sensors attached to hydraulic motors that moved its head and neck towards the person making the sound/movement. It was even more compelling than Grey Walter's Festival of Britain tortoises (4.viii.a), because it seemed to engage with *you*, as an individual. In addition, it seemed to have feelings of its own: it would shy away from loud noises, and if the noise became overwhelming it would raise its head and "disdainfully ignore further sounds until the volume subsided" (Zivanovic 2005: 104–5).

(Philips eventually dismantled it, without informing Ihnatowicz, because they had problems in keeping it working properly. The visitors to their site, many of whom had come there only to view *The Senster*, had been complaining that it was malfunctioning—and Philips had had enough. Today, the reassembled and slowly rusting mechanical structure sits, *sans* electronics, on the premises of an engineering firm who'd done some work on it in its heyday.)

Other original applications were due to the Greek Australian performance artist Stelarc (1946–). His exceptionally imaginative performances included experiments linking/integrating his own muscles with robots, and with random or human-originated messages from the Internet (Stelarc 1986, 1994, 2002*a,b*; M. Smith 2005). They weren't mere physical exploits to be gawped at, but activities that raised questions about the limits of body and self (see Chapter 16.vii.d). Stelarc's artistic journey wasn't initiated by VR. He'd started out, in the late 1960s, with non-electronic man–machine linkages: meat hooks inserted into his skin, and attached to pulleys (Paffrath 1984). But as soon as it was possible to exploit robotics/telepresence/VR, he began to do so.

Most artists, however, were engaged in less intimate interactions, and with VR rather than robotics. Moreover, they weren't necessarily playing with different *realities* so much as a different *medium*: the computer-generated sensory image. That image might be shown on the machine's VDU, or projected onto a wall, or (occasionally) transferred onto canvas. And it might or might not be presented as an alternative world.

One of the first to produce "cybernetic" artworks, and explicitly to define a new aesthetic relating to them, was the British painter Roy Ascott (1934–), later described by the art historian Frank Popper as "the outstanding artist in the field of telematics" (Popper 1993: 124). While working at the Ealing School of Art in the 1960s, he helped initiate an artistic revolution. This was done partly through his own work, and partly by inviting like-minded pioneers such as Pask to lecture to his students.

His first interactive art didn't involve computers, but consisted of canvases with items/images on them that could be continually moved around by the viewer. So the "viewer" of the resulting collages was their *maker* too. This aspect was retained, and strengthened, in the computerized art that followed. The value of AI-based interactive art, for Ascott, was its ability to engage the viewer–participant *as creator* (Ascott 1964, 1966/1967; cf. 2003). By the same token, he held that its aesthetic crux was the nature of the interaction itself, not the (visual and/or musical) end result.

Inevitably, the authority of the single artistic signature was undermined, for crucial creative choices were now being made by the viewer. A quarter-century later, with the advent of the Internet and the Web, authorship of the "telematic" artwork might be very widely dispersed:

[The] status of the art object changes. The culturally dominant objet d'art as the sole focus (the uncommon carrier of uncommon content) is replaced by the interface. Instead of the artwork as a window into a composed, resolved, and ordered reality, we have at the interface a doorway to undecidability, a dataspace of semantic and material potentiality. (Ascott 1990: 237)

Telematic culture means, in short, that we do not think, see, or feel in isolation. Creativity is shared, authorship is distributed, but not in a way that denies the individual her authenticity or power of self-creation, as rather crude models of collectivity might have done in the past. On the contrary, telematic culture amplifies the individual's capacity for creative thought and action . . . *Networking supports endless redescription and recontextualization such that no language or visual code is final and no reality is ultimate.* (Ascott 1990: 238; italics added)

In brief: Ascott's telematic art is VR as a buttress of postmodernism. That particular philosophical interpretation is relatively recent. But right from the start Ascott was always more interested in the conceptual implications of cybernetic technology than in the technology itself. That's why, when the art critic Jasia Reichardt organized the now famous Cybernetic Serendipity exhibition in 1968, she didn't include him (J. Reichardt, personal communication).

Another early AI art experimenter was the artist–computer-scientist Edmonds (1942–), of the University of Loughborough—now, director of the Creativity and Cognition group at Sydney's University of Technology. In contrast with Ascott, he *was* primarily concerned with how the new technology could be used to create novel forms of art.

His inkjet print *Nineteen* (a static array of twenty coloured oblong patches on a near-white background) was first exhibited in 1968–9 and, having been accidentally destroyed, was reconstructed for a historical exhibition in the late 1990s. It was generated by a FORTRAN program designed to find the best of all possible placements of the twenty patches (each of which had already been designed by Edmonds), subject to a few constraints—e.g. that two particular pieces shouldn't be on the same row or column (because this would dominate the canvas).

Edmonds's aesthetic aim, here, wasn't merely to produce a visually interesting composition, but to focus the viewer's mind on the structural possibilities:

It was intended that the assembly in the array be such that a feeling of finality be avoided. The variety of possible relationships between the elements is left for the viewer to sense without actually being able to move them about as he would in an assemblage. (Cornock and Edmonds 1970/1973)

In other words, Edmonds had used the computer "as a problem solver", to tackle an aesthetic problem which was "very hard to solve by hand or analysis" (Edmonds 2002). (In fact, in the three hours allocated, the computer didn't manage to find a solution satisfying all of Edmonds's constraints. But it got very near to doing so, and Edmonds himself was able to modify its solution slightly so as to get what he felt was the best arrangement. His Notes on how he did this were sometimes exhibited alongside the piece.)

His work in later years showed how to use computers as essential aspects of new creative styles, as well as aids in stylistic exploration. For example, he exhibited the world's first generative time-based artwork: *Fragments 1984/5* (a special case of what he named Video Constructs). Here, viewers could sense "the variety of possible relationships" by watching an ever-changing series of abstract compositions generated by a PROLOG program. *Time* was part of the artwork, as well as geometry and colour: variations in the order of images, and in the pace of change, provided different aesthetic effects. Even before Video Constructs, Edmonds had pioneered *interactive* art (e.g. Edmonds and Lee 1974). His novel HCI designs often involved threefold dependencies between action, graphics, and music. By the mid-1980s, he was developing complex VR technology too, for exploring 3D (or 4D, if one includes time) instead of 2D, as before (Wernald and Edmonds 1985).

In all these artistic endeavours, the underlying generative structure, not the individual images/sounds, was the focus of interest. So Edmonds opened a retrospective statement of his general approach by quoting Paul Cézanne's remark that "The technique of any art consists of a language and a logic", and Kasimir Malevich's that art-making involves "a law for the constructional inter-relationships of forms" (Edmonds 2003). (He also cited serialist composers, such as Pierre Boulez.) His very earliest art had been figurative. However, he was already tending towards abstractionism when computers came on the scene, and this technology led him (enabled him) to commit even more fully to abstract, formal, work.

The development of interactive art in the UK was strongly influenced, from the 1960s on, by Edmonds and Ascott. (Strictly, they weren't the ultimate pioneers, for Pask had placed his adaptive "Musicolour" in UK dance halls in 1953: see Preface, ii, and 4.v.e.) Across the Atlantic, the Bell Labs engineer Billy Kluver (1927–) cooperated in the 1960s with several New York artists—Jasper Johns, Andy Warhol, Robert Rauschenberg, and the composer John Cage—to produce electronic art of various kinds (Kluver 1966). By the end of the decade, the avant-garde interest was enough to prompt a seminal international meeting on "cybernetic" art at London's Institute of Contemporary Arts (J. Reichardt 1968). (Avant-garde, but not counter-culture: as remarked in Chapter 1.iii.d, this August 1968 exhibition couldn't have happened in Paris, given the political *événements* in July.)

In the 1970s the interest grew, and widened (Brown and Lambert in preparation). For example, the University of Wisconsin's Myron Krueger (1942–), trained as a computer scientist, was inspired by Cage's work on randomness to design interactive environments as artworks. In some of these, images of people located in different places were projected into a single virtual space—enabling the real individuals to experience the telepresence of their fellows, and to communicate accordingly. Like Ascott and Edmonds, he defined a new aesthetic, suggesting how "responsive environments" were best appreciated (M. W. Krueger 1977). (Krueger envisioned his interactive system spreading way beyond the art world, for use in education, psychology, and psychotherapy; the examples of surgery and training, above, wouldn't have surprised him.)

Today, computer-generated and/or interactive art is still cutting-edge, if no longer avant-garde (e.g. Krueger 1991; Candy and Edmonds 2002; Ascott 2003; Whitelaw 2004). It exploits VR, multimedia, the Internet, the Web, and A-Life evolutionary programming—all of which are dependent on AI research.

Both Ascott and Krueger had explicitly defined their new aesthetic in terms of the gallery visitor's *interaction* with the electronic object/environment. The beauty or interest of the product (the visual display, the music . . .) was much less important to them than the nature of the interaction. (This left open many questions about just what sort of interaction was to be valued, and why: Boden forthcoming.) At the turn of the century, a similar concern was taking over in VR research more generally—for education, certainly, but also for entertainment.

Now, in 2006, entertainment rules the VR roost. There are a number of massively multi-player online simulation games, such as Everquest. These may have hundreds of thousands, even millions, of human players, all continually building and extending online virtual worlds, and online relationships for play, trading, or debate. While most are fantasy worlds, in which players build imaginary cities and forge imaginary alliances, some are based on actual places and/or events—such as the Battle of Britain. The computer game based on this Second World War conflict provides detailed simulations of English airfields, towns, and cities (Bradbury 2003). The players create (or re-create) air crews, and engage in virtual bombing missions or Spitfire–Messerschmitt dogfights that may mimic specific events from August 1940. Clearly, AI/VR has come a long way since Spacewar (see 10.i.i).

Occasionally, the human–VR interaction spills intriguingly from simulation to reality in unprecedented ways:

Some people are making their (real) living from playing virtual online games—by building up powerful characters or buildings, amassing virtual wealth, and then selling it for real money through eBay. (M. Sharples, personal communication)

And not only eBay: several dedicated web sites specialize in auctioning virtual characters and attributes for decidedly *non-virtual* money. Large sums of money may be spent on a single item, whether a carefully crafted VR person or a VR magic sword, and the overall spend is huge. Indeed, one US economist has estimated that the VR gaming world has a GDP per capita roughly the same as Namibia's (Mueller 2004: 13).

'Crazy', perhaps—but innocent enough. Indeed, less harmful than the quest for the black tulip in seventeenth-century Holland, in which livelihoods were lost as well as fortunes made. But whether all merging of the virtual and the real is as benign as this is questionable, as we'll now see.

d. Computerized companions

VR applications prompt the human subject to have 'reality experiences' to differing degrees. But they all foster experience as opposed to imagination. In other words, the willing suspension of disbelief with which one approaches films and theatre (and novels) differs radically from the *spontaneous experience* of reality that's brought about by successful VR.

For example, the world of a 1990s MUD (Multi-User Dungeon) chat room makes possible a set of shared experiences which simply weren't available before (Curtis 1992; Turkle 1995: 180–209, 248 ff.). It may generate a very strong sense of reality, leading MIT's Sherry Turkle (1948–) to say: "Our experiences [in cyberspace] are serious play. We belittle them at our risk" (Turkle 1995: 269). Some philosophers have even claimed that virtual worlds have *a new kind of reality*, rather than being simulations of something else (Graham 1999: 158–66; see also Haraway 1986/1991 and Turkle 1995).

These facts underlie some common misgivings about VR. For instance, the quotation above, referring to the virtual mechanic Steve as "he" and as a "teammate", raises some tricky questions. So besides asking whether VR (as Brooks predicted) has "got there", one might also ask *whether we'd really want it to*.

That question was addressed in the 1980s–1990s by many people. Some were professional VR technologists, such as Pavel Curtis (1960–) of Xerox PARC, creator of the popular MUD LambdaMOO (Lambda from LISP, MOO for MUD, Object-Oriented). LambdaMOO is an ever-growing multi-user virtual environment, with thousands of different rooms and outside spaces, described in words by its players (see Rex 2005). Besides creating LambdaMOO (in 1990), Curtis described his experience of maintaining it, which had shown him how people were using—and misusing—it (Curtis 1992). Some were social scientists: the sociologist Turkle (1995), for instance, or the social psychologist Neil Frude (1983). And some, such as Ascott (1990, 2003) and Haraway (1986/1991), were artists, littérateurs, or philosophers. The items represented by their various worry beads ranged widely over human experience, both actual and potential.

Concerns were raised, for instance, by VR applications that weren't just intellectually engaging, but emotionally engaging too. One example was Creatures, a computer game or (better) a computer world based on evolutionary programming (Grand *et al.* 1997; Grand and Cliff 1998; cf. Chapters 15.vi and 16.x.b).

Creatures caused considerable interest and experimentation on the Internet when it was commercially released in 1996–7, and swiftly led to several hundreds of dedicated web sites—and to improved versions in subsequent years (Cliff and Grand 1999). Indeed, it has aroused attention from cultural commentators, who see it as a paradigm of the technology of autonomy that underpins "the posthuman cyborg" (see Kember 2003: 83–109, and Chapter 1.iii.d). Here, however, what's relevant is that it turned out to be surprisingly addictive in a personal/emotional sense.

The user could hatch, teach, and evolve up to ten virtual animals ("norns") at a time. Each norn had an individual (genetically specified) outward appearance, a neural-network "brain", a particular temperament, and an underlying metabolism. And each learnt to behave in different ways, depending on its environmental history—including the situations devised, and the rewards and punishments administered, by the user. The properties that could be selected for breeding the next generation included temperament and bodily (i.e. computer-graphic) appearance.

The effect on users was compelling. Some youngsters became so emotionally attached to their carefully nurtured creatures that they would grieve if one was wiped from the disk—even though new norns could be hatched on command (M. Sharples, personal communication). Frustration and annoyance would have been appropriate, since the individual norns had been evolved and taught through many hours of highly concentrated interaction with the system. But grief, however fleeting, is another matter. It showed that the norns were being experienced as analogous to real people, or anyway pets (see 7.i.f).

Similarly emotion-ridden effects were often seen in the craze for Tamagotchis in the mid-1990s. (The Japanese word means 'egg-friends'.) These virtual beasties were even less real-looking than norns, having a near-minimal interface (only 10 × 10 pixels). The attraction, and the addiction, was in the need (if the beast was not to "die") for the user to nurture, feed, and even entertain it—all in real time. Norns and Tamagotchis were thus heralds of the end-of-century switch of VR emphasis from *interface* to *interaction*.

If even pixie-like norns and 100-pixel Tamagotchis can engage some people's emotions to this extent, what of more humanoid creatures? Pask's CASTE teaching machine had led some early 1960s users to experience "a sense of participating in a competition (some say a conversation) with a not dissimilar entity" (see 4.v.e). And Rodney Brooks's Cog robot could induce even a *technically knowledgeable and philosophically hostile* visitor to engage with it "as though in the presence of another being" (see 15.viii.a).

But those examples were merely tasters. Cog's long-eyelashed stablemate Kismet, built and nurtured (*sic*) by Brooks's students Cynthia Breazeal and Brian Scassellati, apparently expressed emotions as well as eliciting them (Breazeal and Scassellati 1999, 2000, 2002*a,b*). Like its predecessor *The Senster*, it seemed not only to pay attention but also to have its attention captured appropriately by the human's actions. Visitors to the MIT lab found it much more effective in eliciting emotional reactions than *The Senster*, and even more seductive than Cog (Fox Keller forthcoming). That's hardly surprising, given that its HCI was deliberately modelled on that of human infants/mothers (e.g. C. Trevarthen 1993).

Indeed, psychological research figures increasingly prominently in today's efforts to produce humanoid robots, and even "robot companions". For example, there have been systematic studies of how children's expectations of and interactions with robots compare with those of adults (e.g. Dautenhahn 1999, 2002; S. Woods *et al.* 2004, 2005).

Physical robots, such as Kismet, are limited by their creators' skills as mechanical engineers. With the potential of today's VR, the human's experience of "presence" can be made yet stronger.

Many years ago, the social psychologist Frude (1983) predicted personal GOFAI companions, in the form of talking sex dolls or screen displays. These would go far beyond Descartes's ill-fated (and probably apocryphal) automaton Francine (2.iii.f). They'd engage in intelligent and sympathetic conversations, speaking in friendly/sexy synthesized voices. And they'd learn what interested their human partner—so they wouldn't bore my son by talking about jewellery, or me by talking about cars. Moreover, they'd cater for a wider range of "interests" than one might think: cuddly/furry robots "would extend the range of potential activities and permutations to present a whole new realm of challenges to the sexually inventive" (Frude 1983: 174).

Frude's prediction hasn't come to pass. However, plenty of AI/VR work has been commenced which might help in moving towards it. (Some, indeed, is already being exploited for that purpose: see below.) For instance:

* The synthesized voices could now be given appropriate local accents (Chapter 9.xi.g).
* They could be given emotional intonations, too (Chan 1990; Cassell *et al.* 1994). In principle, they might one day include 'realistic' emotional features like those in John Clippinger's 'neurotic' NLP program (7.ii.c).
* The lip–sound synchronization for VR speech can be based on photographs of the lip movements of a specific person (9.xi.g).
* VR systems may also be able to 'lip-read' to aid their understanding (only sixteen lip positions are needed to represent English speech: Lucena *et al.* 2002).
* Animators are applying David Perrett's data on computer-simulated facial attractiveness (see 14.iv.d), and in principle could adjust their VR faces to exploit the female user's time of the month (Penton-Voak *et al.* 1999).
* Work's going on around the world on enabling AI systems to recognize people's facial expressions of emotion. (Steve is one example, mentioned above; others include robots developed by Fumio Hara's group at the Tokyo University of Science.)
* If the human user were to wear suitable VR gloves (as Steve's interlocutors already do), the AI companion could also use electrical signals from the skin to gauge the person's emotional state.
* Conceivably, the system might even be able to tell when the human is telling lies. For besides acting—through the gloves—as a conventional lie-detector, it might pick up facial cues to lying (7.i.d gives references to Ekman's work on lying).
* The companion's own 'face' might express emotions too, even if it were a doll (as opposed to a VDU screen). Roboticists in Japan and the USA have been modelling our detailed facial musculature, to enable their robot's rubberized 'face' to change its shape

according to one basic emotion or another—e.g. Hara's group, again. (Advances in materials science might provide more realistic 'skin' and 'flesh'.)

    * Rod Brooks's Kismet robot—carrying floppy pink ears, huge blue eyes, and ultra-long eyelashes—engages people's interaction (and "caring" responses) despite its limited range of behaviour (15.viii.a);

    * mutual gaze and joint attention are so convincingly modelled in Brooks's humanoid robots that even people who are in the know seem to sense an animate "presence" (15.viii.a);

    * Negroponte's Media Lab have been aiming at highly personalized conversational computers since the 1970s, their ultimate test case being the intriguing all-anaphors interchange quoted in 9.xi.g.

    * Finally, teledildonics (the mind boggles!) already offers a wide range of sex at a distance, sometimes involving wearable VR gear such as the "cybersex suit" or the "Virtual Sex Machine" (*Teledildonics* 2003). Future advances in haptics will probably be adapted for these purposes.

Frude saw convivial computerized companions as inevitable (1983: 58). The AI technology will be developed, and the businessmen (and the pornographers: p. 174) will then take over. The huge consumer demand, he said, will be driven by the universal human tendency to animism (which can now be understood as drawing on the concepts and inferential processes of Theory of Mind), in which intentional agency is ascribed to non-human beings (see 7.vi.f and 8.vi.d).

One must agree about the animism, and about the businessmen: just think of Tamagotchis, again. Although it's relevant to note, here, that robots are already being used for interacting with people who *lack* the normal tendency to animism. As we saw in Chapter 7.vi.f, autistic children lack Theory of Mind. Some people working in social robotics have tried to use robots to help mediate social interaction *between the autistic child and other humans* (Robins *et al.* 2005; cf. Robins *et al.* 2004). They've had some success, but they've also noticed that the robots can reinforce the autistic's tendency to fall into repetitive stereotyped behaviour. The use of non-human, and therefore non-threatening, robots in teaching autistic children isn't new: see the LOGO-based approach described in Weir and Emanuel (1976). What Weizenbaum would think about these projects probably doesn't bear repeating in polite company; but, as with Colby's automated psychiatric interviews, perhaps the key question is whether or not the child is actually helped (cf. 7.i.a).

But if one agrees with Frude about the animism, one may disagree about the full-blown technological prediction. Intellectually stimulating, and even emotionally engaging, robots and/or (Interneted) VR interfaces, yes. And pseudo-satisfying 'sexual' encounters, too. After all, both of these exist already. But the arguments elsewhere in this book suggest that the fully "intimate" machines envisaged by Frude are unattainable. NLP will never be up to it (Chapter 9.x–xi). Nor will our computational theories of motivation and emotion (7.i.e–f), nor of reasoning, relevance, and common sense (7.iv and iii.d, 10.i.f and iii.e, and i–ii, above).

Some AI professionals, however, might take Frude's side here. Attendees at the IEEE's latest international "Ro-man" symposium, whose theme was 'Getting to know Socially Intelligent Robots', probably would (see <http://ro-man2006.feis.herts.ac.uk>). And when I asked one of Steve's authors whether he shared my scepticism about this, he didn't say that I was obviously wrong—as some AI enthusiasts would have done. More cautiously, he said:

Well, ever is a long time. I'm leery of speculating about what will ever happen in the future, not without some convincing argument that such a computerized companion is altogether impossible.

I see rapid advances these days on computational models of emotion. I am not sure how far progress will go, but I don't see obvious barriers at the present time.

And corpus-based NLP techniques are also making rapid advances. They make broad-based question–answer systems much more feasible than was the case in the past.

Still, there are many barriers to creating a general conversational companion. Even if there is progress in some areas the overall goal still seems very far off. (W. L. Johnson, personal communication; quoted by permission)

As for whether such technical advances would be *ethically* welcome, Frude sat on the fence. He granted "the danger that contact with an artificial intimate [*sic*] might prove so fascinating and so satisfying that natural inter-human friendships would be overshadowed" (p. 174). But he pointed out that old people in Western cultures are often neglected, and suggested that—in the absence of real human interactions—this futuristic HCI would be better than nothing: "Such a machine promises freedom from loneliness and boredom and could end for ever the blight of social isolation with all its attendant evils."

You may find this suggestion chilling, a prime case of what Weizenbaum called "obscenity": substituting computers for humans in quintessentially personal contexts (7.i.a and 11.ii.d). If so, I agree with you. Nevertheless, at least one AI colleague—whom I shan't name—disagrees. Moreover, pushed by the demographics of an ageing population, Japan's Ministry of International Trade and Industry has recently launched a major research programme into humanoid robots, intended as helpers *and companions* for the elderly (Fox Keller forthcoming, n. 18). Presumably, these politicians have no ethical quibbles either.

Chilling or not, we don't need to worry too much about computerized companions—at least, not yet. VR avatars are another matter. People not at all worried by Frude's intimate machines, because they doubt that they'll ever exist, may already be very worried by these.

e. Psychology and avatars

In Hinduism, an avatar is an animal incarnation or representative of a god. In VR, it's a screen name/image that represents a particular human user (Damer 1998; Wilcox 1998). When someone uses the same avatar over a period of time, it grows a database for future use. (Steve isn't an avatar: he/it doesn't represent a specific person, merely a generalized human tutor.)

In general, an avatar prompts the psychological processes that enable us to interpret other people's behaviour. In other words, it taps into the user's Theory of Mind—as screen-borne or robotic companions would, too (7.vi.f).

Today's avatars include printed names/initials or abstract icons; 2D cartoons or photos of humans or animals; and 3D images of a person with face, body, and fingers—and bodily movements as well. The range of humanoid avatars is huge, not least because a company in Iceland sells software enabling you to design your own (Wilcox 1998, ch. 8). Already, several recent simulation games, such as The Simms, involve the DIY design of avatars.

Tomorrow's screen stand-ins will be even more realistic. Naturalistic gestures and facial expressions (remember Steve), and suitable voices, will be designable (Cassell *et al*. 1994). The more detailed the avatar's face, the more important that its lip movements match its voice plausibly. So special synchronizing algorithms will doubtless be employed too (Waters and Levergood 1995; Lucena *et al*. 2002).

An avatar converses either by writing text messages or by audible/visible speech. And it may walk to different locations in the virtual world displayed, taking up different points of view (affording different information) accordingly. It may soon be able to recognize emotions in the interlocutor's voice or written text, and *automatically* simulate them in its own facial expression (Olveres *et al*. 1998). (It's already possible for the avatar-holder to key certain facial expressions in by hand.) As ever in HCI, for the avatar to be realistic the psychologists must previously have found out just how *people* behave in comparable contexts (Cassell *et al*. 1994, again).

We don't have to look to the future to be worried by avatars, for they've already aroused social and psychological anxieties. The prime reason is that they enable a person to hide their identity, including their age and gender, and even to pretend that they're someone else. (Significantly, the chapter on avatar design that reports on the Icelandic company mentioned above is subtitled 'Who Would You Like To Be Today?'—Wilcox 1998: 187.)

Now, in the new millennium, several million people may be continually presenting (misrepresenting?) themselves as avatars within a single MUD. It's worth noting that these MUDs are often called MMORPGs: Massively Multi-Player Online *Role-Playing Games*.

The social psychologist Irving Goffman (1959) long ago pointed out the potential for deception and confabulation in role playing, and in the everyday "presentation of self". But VR is Goffman in spades. So one MUD enthusiast, for instance, said to Turkle:

You can be whoever you want to be. You can completely redefine yourself if you want. You can be the opposite sex. You can be more talkative. You can be less talkative. Whatever. You can just be whoever you want, really, whoever you have the capacity to be. You don't have to worry about the slots other people put you in as much. It's easier to change the way people perceive you, because all they've got is what you show them. They don't look at your body and make assumptions. They don't hear your accent and make assumptions. All they see is your words. (quoted in Turkle 1995: 184–5)

This threatens not only to deceive other people, but also to confuse—and perhaps to destabilize—oneself. Just what might follow from a new realization of "whoever you have the capacity to be"?

Sometimes, the illusory avatars are harmless enough. For one thing, people only rarely adopt "a coherent character with features distinct from their real-life personalities" (Curtis 1992: 327). For another, their MUD avatars—according to Curtis—are driven by wish-fulfilment of a fairly innocent kind: "I cannot count the number of 'mysterious but unmistakably powerful' figures I have seen wandering about in LambdaMOO." The virtual world, he said, seemed to be enabling people to emulate various attractive characters from fiction. (How many Mr Darcys, I wonder? And how many James Bonds?)

But less attractive characters could be emulated too. This was illustrated by another of Turkle's interviewees, a male student whose avatars are often highly violent. He acknowledged that the violence is "something in me", and defended himself thus: "but quite frankly I'd rather rape on MUDs where no harm is done" (p. 185).

Well, perhaps. No woman is harmed (yet??). But the user himself may be harmed by repeatedly acting out such behaviour. "Perhaps", and "may be", because this is of course an empirical question. Given the ambiguous results of the massive research done on the psychological effects of watching violence on TV, it's a question that's not likely to be easily answered.

Again, a MUD user can sometimes take control of *someone else's avatar*, forcing it to do things which the genuine owner wouldn't want it to do. This allows for an especially 'realistic' version of virtual rape (Turkle 1995: 250–4). Here, the perpetrator doesn't merely act out the rape of a specially created (wholly imaginary) VR character, but forces cooperation (and 'signed' messages expressing satisfaction, and pleading that he should continue . . .) from a pre-existing avatar victim. Not surprisingly, the hidden humans violated in this way—who may or may not be female—don't like it. The common defence is: "This is a GAME, nothing more. [People need] to chill out and stop being so serious. MUDs are supposed to be fun, not uptight" (Turkle 1995: 253). I'm not persuaded.

In short, it's not at all clear that "no harm is done" by engaging in these unsavoury activities. But neither is it clear that simply *avoiding* unsavoury activities would neutralize all the potential problems.

Specifically, a number of VR critics have argued that merely adopting different personae at different times, whether unsavoury or not, may unsettle someone's psychological equilibrium. As Dreyfus put it, quoting Søren Kierkegaard, "the self requires not 'variableness and brilliancy' but 'firmness, balance, and steadiness' " (H. L. Dreyfus 2001: 82). You may feel that cognitive scientists shouldn't be troubled by this. For cognitive science sees the "self" as a narrative construction rather than an encapsulated essence (Chapters 7.i.e–f and 16.vii.d). But as a virtual machine (*sic*) that underpins the purposive coherence of the person's life, it had better have "firmness, balance, and steadiness". VR experiments in "multiple personality" are all very well, but in some cases they're playing with fire (Turkle 1995: 206 ff.; cf. 7.h).

Postmodernists, however, may not agree. For example, Kenneth Gergen, one of the first counter-culturalists to attack 'scientific' social psychology in the 1970s (see 6.i.d), has recently welcomed the relativizing effects of what McLuhan (1964: 3) called the global village:

> [We] exist in a state of continuous construction and reconstruction; it is a world where anything goes that can be negotiated. Each reality of self gives way to *reflexive questioning, irony, and ultimately the playful probing of yet another reality*. The center fails to hold. (Gergen 1991: 6; italics added)

He was talking primarily about the multiplication of real (i.e. geographical/cultural) environments in our experience, but he later included VR too (Gergen 1994).

The italicized phrases, above, express attitudes typical of VR users. So it's not surprising that other postmodernists, who already favoured the deconstruction of the self for independent philosophical reasons, welcomed the variation of self afforded by avatars in cyberspace. The leading proponent of that view was Haraway (1986/1991). But the influential artist Ascott, too, promulgated the notion of "Cyberself":

We are each made up of many selves: de-centred, distributed, and constructively schizophrenic. We are the embodiment of technoetic relativity ... [where a "technoetic" aesthetics focuses not on the surface image of the world but on] creative consciousness and artificial life. (Ascott 1998/2003: 375, 381)

And Turkle remarked approvingly that

The rethinking of human ... identity is not taking place just among philosophers [she meant postmodernists, but some philosophers of cognitive science are included too: Chapter 16.vii.d] but "on the ground," through a philosophy in everyday life that is in some measure both proved and carried by the computer presence. (Turkle 1995: 26)

Raising the question whether MUDs offer "psychotherapy" or "addiction", Turkle cited cases—including herself—where someone's experience of acting within a particular (real or VR) communicative world helped them achieve a more satisfactory personal balance (1995: 196–209, 262–3). And she endorsed postmodern relativism, "humor" and "irony" about these matters: "We do not feel compelled to rank or judge the elements of our multiplicity. We do not feel impelled to exclude what does not fit" (p. 262).

She also pointed out, however, that avatar experimentation has its limits. For instance, a man pretending to be a woman on the Internet *must* miss out on many aspects of the experience of being a woman, from the fear of pregnancy to worries about how much make-up to wear to a job interview (p. 238). And despite her postmodernist sympathies, she said: "Virtual environments are valuable as places where we can acknowledge our inner diversity. *But we still want an authentic experience of self*" (p. 254; italics added). She saw VR as offering us "a multiple but integrated identity whose flexibility, resilience, and capacity for joy comes from having access to our many selves" (p. 268). However, she added that "if we have lost reality in the process, we shall have struck a poor bargain".

In a nutshell, nothing here is straightforward. The evidence shows that avatars can be psychologically liberating as well as oppressive. They enable social experimentation and self-exploration that may be advantageous, up to a point. Possibly, they may even display 'the real person' for almost the first time. That was suggested by Turkle's interviewee Thomas, who told her: "MUDS make me more what I really am. Off the MUD, I am not as much me" (p. 240). (Of course, this isn't conclusive: he may have seen himself as much closer to Darcy or Bond than he really is.) It was suggested by a Texan muscular dystrophy sufferer featured in the Alter Ego photographic exhibition in London of MUD/MMORPG players and their online avatars (Mueller 2004: 13). And it was suggested, too, by William Horwood's (1987) haunting novel *Skallagrigg*.

That word "haunting" reminds us that novels, plays, films ... provide forms of VR which may seem very real indeed. Kenneth Walton (1990) has argued that such arts depend on "make-believe"—in which we experience what it would *really* be like to be shipwrecked on a desert island, or to consider killing one's stepfather the king, but without experiencing the nastier aspects of those realities. (It's not clear that he's right:

if you know you can't drown, where's the "real" horror of shipwreck?—Graham 1999: 155–8; cf. Turkle 1995, ch. 9.) Walton's view implies that VR is just 'more of the same'.

If so, we should worry about it no more—though also no less—than we do about the violence in Quentin Tarantino's movies, or the pornography filmed on the other side of those Hollywood hills. There's a difference, however. VR worlds/avatars can offer an experience of 'immersion' stronger than that involved in reading a book or watching a film, and can even be individually crafted. As a result, they may be both more haunting and, in their practical consequences, more nasty.

Over and above the personal practicalities that have been the focus of this subsection, and the one before it, philosophical issues concerning self and embodiment are highlighted by VR, and especially by telepresence. These issues were raised long ago, for instance in a hugely intriguing thought experiment described by Daniel Dennett (1978e). The core question concerns the location, and/or the boundaries of the self. Does it stop at the skull? Or at the skin? Or does it encompass the cultural artefacts without which we couldn't function as we do—and wouldn't be the people that we are? If the latter, as argued for instance by Andy Clark (1997), then VR technology adds further opportunities, further complications . . . and further questions (A. J. Clark 2003a: 90–8). (The philosophy of self and embodiment is discussed in Chapter 16.vii.d; and the VR ideas that inform *The Matrix Trilogy* are related to philosophical arguments about realism in Chapter 16.viii.c.)

Much more could be said about this new cognitive technology, the realization of Bush's and Engelbart's dreams. There's already a huge literature on the virtues, and the dangers, of VR and the Internet. (Extensive bibliographies are given in Turkle 1995, Shields 1996, C. H. Gray 1995, and Haraway 1986/1991; for a strongly hostile approach, see the references cited in Kroker and Kroker 1997.) However, much of it is either technically ill-informed or philosophically naive—and some is both. My advice is to read Turkle's fascinating interviews and subtle psychological insights, and to look out for an informative and thought-provoking book being written by Yorick Wilks: a long-time aficionado of NLP (see 9.x.d), originally trained in philosophy, and currently a Fellow at the Oxford Internet Institute (Y. A. Wilks in preparation).

13.vii. Coda

Two hostile questions, one old-and-continuing, the other more recent, are often raised about AI. The first is whether it's a 'proper' discipline. In other words, is there a well-defined subject matter which AI is *about*? The second is whether AI—in particular, GOFAI—has failed.

Many critics answer "No" to the first and "Yes" to the second. I'll argue that the "No" is justified, up to a point, but that—again, up to a point—the "Yes" isn't.

a. Is AI a discipline?

In 1968 Michie and his co-editor Ella Dale opened *Machine Intelligence 2* with these questions:

What is Machine Intelligence? Is it a theory, a discipline, an engineering objective? Or is it a pretentious name for the more peculiar parts of computer science? (Dale and Michie 1968, p. ix)

They weren't asking purely on their own account. Indeed, they answered immediately that

In our view the matter is quite clear. Since the objective is to bring into existence an intelligent machine, those engaged in the attempt must look on themselves as engineers. (ibid.)

Rather, they were reflecting the puzzlement, not to say scepticism, being expressed by others.

There was a lot of that about. Even Lord Bowden, the leading computer scientist who agreed to write the Preface to one of Michie's *Machine Intelligence* volumes, asked, "is this enterprise of yours part of the main stream development of computers, or is it merely an interesting, but irrelevant, side line?" (Bowden 1972, p. v). Michie may have regretted inviting him to contribute the Preface, for he went on to say:

How far can computers be used to solve the apparently insoluble intellectual problems of the day? I have probably misunderstood your work, but I have been disheartened by some of the research in which your members are engaged. It is pressing against the limits of knowledge and this is splendid, but are you right to be so worried about problems which appear to me to be semantically significant rather than technically important, and philosophically interesting rather than economically useful? (Bowden 1972, p. viii)

As for using AI to help us understand the workings of the human mind/brain, such as the recognition of gestalts, he despaired: "Heaven knows these problems defy solution—why therefore should one expect that they can be studied or solved or practised by computers?" (p. ix).

Bowden wasn't complaining that NewFAI was doing its job badly, rather that it had chosen the wrong job to do. Psychology (or the imitation thereof), he believed, was beyond the remit of AI as an intellectual discipline. But many other computer scientists at that time were asking not so much whether AI was a discipline as whether AI workers were disciplined. That is, the doubts about hype and lack of self-criticism which Drew McDermott would later discuss (see 11.iii.a), and which would soon land Michie himself in very hot water (11.iv.a), were already bubbling beneath the surface.

Those doubts were still bubbling in the late 1970s, when a historian of AI was told by orthodox computer scientists that it's "a 'freaky', rather dubious fringe activity" (Fleck 1982: 207). And they're still bubbling now.

In 1996, for instance, the AAAI Fellows had an email debate about whether AI is "engineering" or "science". Some voted for the former, as Michie had done long ago. Simon's contribution was brief, and to the point. He referred his fellow Fellows to two very recent papers, including a "case-study" on creativity (Simon 1995*b,c*), and to his classic *Sciences of the Artificial* (1969; see especially chs. 1, 3, and 4). He insisted that—alongside its engineering aspects—"artificial intelligence is science". And he added testily, "I do think that members of the community who wish to discuss these matters have some responsibility for familiarity with the serious literature on the topic" (AAAI Fellows' email list, 22 May 1996).

Two years later, the editors of the 100th volume of *Artificial Intelligence* exulted that it had been firmly established as "a discipline" (Bobrow and Brady 1998*a*). It was now respected even by engineers, they said—adding that Sir James Lighthill had got his come-uppance at last (cf. 11.iv).

But this upbeat editorial carried a whiff of Hamlet's "Methinks the lady doth protest too much". Other remarks in their essay bemoaned the specialist splintering that was preventing NLP or vision research, for example, from being submitted to the core journal. This splintering soon brought complaints also from two presidents of the core society, AAAI. Ronald Brachman's Presidential Address in 2004 complained about the centrifugal tendencies of AI, and his successor, Mackworth, declared that "The way to overcome these centrifugal tendencies is to develop better theories of cognitive architecture and to work in integrated applications [i.e. GOFAIR: see Section iii.c above, and 11.iii.b]"—adding that AAAI was "needed more than ever", being "the premier venue for bringing together the subdisciplines" (Hedberg 2005: 12). (However, he's optimistic: "There was a lot of evidence [at the AAAI and IAAI meetings in 2005] that AI is in a healthy state, [with many] exciting new ideas and applications. Our field is reaching out to many related disciplines in a new way. That to me is a sign of maturity: we are no longer insecure as a discipline"—Hedberg 2005: 15.)

Besides their complaints about subdisciplinary splintering (and the profusion of subdisciplinary journals), the *Artificial Intelligence* journal editors even complained that too many papers were now purely mathematical, rather than implementational. In short, they were implying that some of their own colleagues didn't have a good sense of AI *as a discipline*. (Soon afterwards, they underlined the wide scope of current AI by reprinting the fourteen invited lectures from IJCAI-1997—ranging from Bayesian analysis and non-monotonic logic, through NLP, planning, and visual identification, to social robotics and creativity: Bobrow and Brady 1998*b*.)

Further evidence of continued bubbling comes from the delightful Father Hacker, who's still a regular contributor to the *AISB Newsletter/Quarterly*. A key item of advice in his recent 'Guide for the Young AI-Researcher' is this:

The basis for your reputation is a notable piece of science or engineering, with which you are associated worldwide. But in this environmentally aware world, recycling will avoid unnecessary effort on your part. The recent history of AI is replete with the revival of old techniques (neural nets, genetic algorithms), rejected as unfeasibly inefficient in the Mid-Twentieth Century [see Chapters 12.iii and 15.vi.b], but given new life by Moore's Law [i.e. the inexorable growth of affordable computing power] in the Twenty First. A bit of historical research, plus some judicious renaming to place the old ideas in a modern context, will pay dividends. (*AISB Quarterly* 2003: 12)

If today's AI researchers aren't deliberately stealing and renaming old ideas, as Father Hacker advises, they are sometimes reinventing the wheel—and lack of historical knowledge doesn't help. For instance, 1980s–1990s work in machine learning often replayed insights available in traditional statistics (see 12.vi.f, and iii.f above); and the NewFAI researchers Michie and Gregory both said that the best model of the real world is the real world itself, long before the situated roboticists did (iii.b, above).

A related failing is AI's vulnerability to fashion—and to the uncollegial scorn that comes with it. Dissension, if not active scorn, has characterized AI almost since the beginning.

(AI isn't alone. Professional psychology, too, has long been subject to fashion and unpleasantness: see 7.vii.b and e. Anthropology has suffered dreadfully from professional dissent: see 8.i.d and ii.a–c. As for philosophy, that's not all sweetness and light either. Just look at Bertrand Russell's comments on Ludwig Wittgenstein's mature work, or at some of the epithets hurled at Heidegger by Anglophone philosophers: see Chapter 16.vii.a.)

Much of the rivalry within AI was due to the explanatory/methodological schism that developed in the 1950s (4.ix). The late 1960s scandal over perceptrons was just one high-visibility example (Chapter 12.iii).

One might have expected the dissension to have settled down by the mid-1980s. But when AI was already some 30 years old, Newell identified no fewer than thirty-six issues on which AI researchers had been divided, and in many cases still were (Newell 1983: 191). Worse, the division was often unscientific, not to say ungentlemanly. Newell lamented the "sloganeering character" of these disagreements (p. 189). And he complained bitterly about the disputatious tone of an anti-GOFAI paper published in the same volume, "written to accomplish science entirely by means of commentary" (p. 292).

Today, things are scarcely much better. As remarked in Chapter 4.ix.a, a recent AISB editor complained: "The lack of tolerance [between different research programmes in AI] is rarely positive, often absurd, and sometimes fanatical" (Whitby 2002b). The charge of *being ill-disciplined*, then, still has some force.

But the scepticism that Michie was seeking to counter may have been caused also by AI's use of two very different methodologies, symbolic and connectionist. Lee Cronbach, had he seen Michie's questions, might have replied that AI wasn't a "discipline" for that reason (see Chapter 7, preamble). Were he alive today, Cronbach would have been even more puzzled, for AI now involves evolutionary computing too—not to mention the so-called nouvelle AI of situated robotics and enactive perception (15.vii–viii and 7.v.e–f). Even though all of these are in principle compatible, individual AI scientists tend to choose one rather than another.

There was another worry, too. That reference to "the more peculiar parts of computer science" could be interpreted as referring to peculiar *topics*, not just peculiar (even ill-disciplined) *methods*. If AI was truly a discipline, what was it studying which wasn't better studied by some other part of computer science?

Nearly twenty years after Dale and Michie, on the occasion of the *Artificial Intelligence* journal's fifteenth birthday, a different editorial duo addressed much the same doubts. Bobrow and Hayes (1985) invited about twenty people to answer ten questions about the achievements of AI, including both GOFAI and connectionism. One of their ten queries asked whether it was a single discipline or just a miscellaneous collection of ideas about non-numerical computing. The unspoken implication was that these ideas were perhaps not only disparate, but largely half-baked and/or "pretentious" too.

In answer to the editors' questions, a number of leading AI experts said it *wasn't* a single discipline. For example:

AI, broadly conceived, is just too large to be a single discipline—mainly because intelligent perception and behavior touch so many aspects of computer science, control theory, and signal processing theory. (Nils Nilsson, quoted in Bobrow and Hayes 1985: 376)

If AI is a single discipline, it is more like Biology than Neuro-biology. It is inherently multi-disciplinary, and the more AI is applied the more disciplines will be involved. (Aaron Sloman, quoted in Bobrow and Hayes 1985: 377)

At the same time, some of the very same people said that there *was* a core discipline there. But this was identified in very different ways, sometimes as "declarative" GOFAI, sometimes much more broadly—as the intellectual core of cognitive science in general:

[That] part of AI based on reasoning with declaratively represented knowledge (e.g. logical formulas) with semantic attachment to specialized procedures and data structures is a coherent field. I think it will inherit the name AI and continue to develop as the core discipline underlying intelligent mechanisms. (Nilsson, quoted in Bobrow and Hayes 1985: 377)

There is a central coherent core of AI/Cognitive Science [*sic*] which is the systematic study of actual and possible intelligent systems. This seeks general principles, not just successful designs. (Sloman, quoted in Bobrow and Hayes 1985: 377)

Very few people, Sloman added, were working on that—"partly because it is the hardest part of AI, and has least short-term pay-off". (His own work, however, had already been focused on it for some years: Chapters 7.i.f and 16.ix.c.)

My own answer said, in part:

AI suffers from some of the same problems as philosophy: it tackles the unanswered, or even the unanswerable, questions. As soon as it manages to find a fruitful way to answer one of them, the question gets hived off as a specialist sub-field. The special sciences started emerging from philosophy in the Renaissance. Specialist areas of study have been emerging from AI for only thirty years. But already we have distinctive sub-fields—such as pattern-recognition, image-processing, and rule-based systems. Their executors often speak of AI (if they speak of it at all) as something not quite respectable, something nasty in the woodshed that was once glimpsed but is better forgotten. What should not be forgotten is that the respectable topics were excluded from what people are prepared to call "AI" as soon as they became "respectable". (M. A. Boden, quoted in Bobrow and Hayes 1985: 376)

Today, some twenty years later still, I stand firm by that reply—except that I should have made it clearer that the "sub-fields" (and the discoveries) are often attributed to *computer science*, or to *software engineering*, not to AI itself. Time sharing, for instance, was developed by McCarthy, and so were fast-prototyping and data mining. But they're not usually thought of as having hailed from AI.

A proliferation of specialist journals has appeared since *Artificial Intelligence* first came off the press. New ones are founded with relentless regularity, accompanied by specialist (and sub-specialist . . .) conferences whose papers, and even whose CFPs (Calls for Papers), are nigh-unintelligible to all but dedicated experts. Computational methods are there being perfected, details added to answers, new questions raised—and, often, new applications made possible. It ought to be acknowledged, however, that the pioneering methods, questions, and answers came largely from AI.

Those questions for which reliable methods and answers haven't yet been found still lie within AI itself. Since they aren't yet well defined, it follows that—in a sense—AI *isn't* a proper discipline. But it doesn't follow that it's not worth pursuing. The very existence of so many specialist offshoots, in areas which initially would have been regarded as "peculiar", "pretentious", or even "freaky", shows the contrary.

Some philosophers, of course, argue that the central goals and assumptions of AI, and especially of strong AI (16.v.c), are fundamentally incoherent. If so, then strong AI certainly isn't a proper discipline—much as fairyology isn't, and couldn't be. On their view, it's *philosophically* pretentious, since it pretends to possess an explanatory potential which, in principle, it lacks. Whether that's true is explored throughout Chapter 16. But even if it is, it doesn't apply to most AI work—for most AI researchers avoid making claims about explaining or constituting "real" intelligence.

b. Has GOFAI failed?

The second hostile question often heard today is sometimes expressed in terms of "AI", but is often intended primarily for GOFAI. The charge is that AI has *failed*.

   This charge was made by Dreyfus (hardly a disinterested witness), in his 1998 Houston University lecture on 'Why Symbolic AI Failed'. And a few years earlier, he'd said:

Like the dissolution of the Soviet Union, the speed of *collapse* of the GOFAI research program has taken everyone, even those of us who expected it to happen sooner or later, by surprise. (H. L. Dreyfus 1992, p. xiv; italics added)

The rationalist tradition [in philosophy] had finally been put to an empirical test, and it had *failed*. (H. L. Dreyfus and Dreyfus 1988: 34; italics added)

The connectionist Robert Hecht-Nielsen might not agree with Dreyfus about the philosophical "why", but he does agree with him about the practical verdict:

[As] it turned out, this approach [GOFAI] *never amounted to much*—a realization which it took two decades and billions of dollars to establish and which did yield *a handful of valuable accidental discoveries*. (in J. A. Anderson and Rosenfeld 1998: 303; italics added)

A negative verdict is strongly implied, too, by Brooks—another non-disinterested witness—in his 1991 paper 'Intelligence Without Representation', which argued that GOFAI was/is fundamentally wrong-headed (Section iii.b, above, and Chapter 15.viii.a). Indeed, given the huge media publicity which Brooks's work attracted, his habit of using the term Artificial Intelligence to cover *only* GOFAI was partly responsible for the general belief today that AI has failed (see 1.iii.g).

   Two of those three men have a numerous following, among the public as well as in cognitive science itself. The feminist Sarah Kember, for instance, cites Dreyfus with approval (though adds that he's "exceeded by Donna Haraway's formulation of situated knowledge"—Kember 2003: 9). She has sympathy for Brooks's approach, even though she accuses him of working "without any real regard to human implications, social and scientific applications or philosophy" (p. 69). Indeed, she devotes many pages of cultural commentary to celebrating the successes and potential of A-Life. In her mind, however, "AI" (i.e. GOFAI) is different: her opening page proclaims "the failure of AI as a project" as an apparently self-evident proposition.

   So an opinion poll on the matter of AI's failure would have depressing results for GOFAI pollsters. However, people—even *most* people—can be wrong. Are they wrong, in this case?

   If the measure of failure is whether AI has lived up to its most extravagant promises, then they're right: it's failed. The "world chess-champion by 1967" didn't happen (although a world-beater had emerged by 1997). And much of the connectionist hype that irritated Bernard Widrow (12.vii.b) hasn't been fulfilled either.

   To be sure, if something hasn't happened yet it may happen tomorrow, or the day after. Undated promises must be treated with care. However, my own view—some AI colleagues would disagree—is that some promises won't be fulfilled for hundreds of years, if ever. Perfect, or anyway human-level excellent, translation is one example (9.iv.b and x.e).

Further 'failures' are doubtless in store—especially if we listen to the futurologists. One such, British Telecom's Ian Pearson, recently hit the headlines (see BBC Online, 18 January 2002: Sci Tech News). He predicted that by 2010 the first robot will have passed its GCSE exams, which British children take at age 16; A Levels (age 18) will follow a few years later, and the robot will be ready for its degree a few years after that.

This fantasy makes RoboCup's predictions of a world soccer champion by 2050 seem cautious by comparison (11.iii.b). The AISB journal editor judged it extraordinary that a professional within a high-tech company like BT should believe such nonsense, and I agree with him:

It could be that BT has something very special in a secret lab somewhere. Then again, it could be that Mr. Pearson has based this prediction on observation of tealeaves rather than of robots. (Whitby 2002*a*)

(In Britain, reading the tea leaves left in someone's cup is the equivalent of Roman augurers' reading chicken bones. Unfortunately, this charming practice is dying out, because of the widespread use of teabags.)

But fulfilment of hyped-up promises isn't a reasonable criterion of success-or-failure. That's especially true, given that many (most?) people in the field didn't subscribe to them in the first place.

By the same token, 'passing the Turing Test' has never been an appropriate criterion of success in AI (see Chapter 16.ii.c). So the fact that—with a handful of (arguable) exceptions—it hasn't been passed is neither here nor there. Nevertheless, many people still assume that it's highly relevant. Dreyfus's 'Failure' lecture, for example, opened with a reference to Turing's remarks about human–computer indistinguishability by 2000.

Rather, we should ask whether AI—symbolic, connectionist, evolutionary, or situationist—has made significant progress towards the goals expressed in less careless moments. In Naudé's terminology (see Preface, preamble), we should credit AI researchers with a concern not for Magick but for Mathematicks, and ask how effective their Mathematicks has been.

To answer that question, we must recall the distinction between technological and psychological AI. The goal of the former was/is to build useful computer systems—doing or, as Minsky forecast in 'Steps' (10.i.g), *assisting with* tasks which humans want done. (One can't say "tasks which humans would otherwise have done", for some tasks can be performed only by means of huge computer power.) The goal of the latter is to develop explanatory theories of mind, and perhaps also (according to strong AI) to build computer systems that are genuinely intelligent in themselves.

As regards technological AI, the response to "Has AI failed?" must be a resounding "No!" Even though this book virtually ignores applications, a few are briefly mentioned. Examples include:

* software for translation (Chapter 9.ix.f) and speech generation (9.xi.g);
* computer vision techniques (7.v.c) and 'retinal' VLSI chips (12.v.f);

* expert systems in many different areas (10.iv.c and ii.b–c, above);
* Hollywood animation (15.ii.a and x.a);
* interface design (v, above);
* virtual reality systems (vi, above);
* computer-assisted and/or interactive art (iv and vi.c, above);
* robotic surgery (vi.b, above);
* and, regrettably, battlefield management and missile guidance (3.v.a, 4.vii.a, and 11.i).

A volume even longer than this one could be written about technological AI as a whole. Besides the many high-profile examples, it would have to mention the host of *invisible* AI applications, of which the general public aren't even aware. Sometimes they're invisible because the businesses that commissioned/built them regard them as trade secrets. More often, they're invisible because miniaturization, and sometimes wireless communication too, have made them so. Weiser, in his *Scientific American* article of 1991, spoke of "calm technology, when technology recedes into the background of our lives".

Even before the arrival of literally "ubiquitous" computing, AI already pervades our lives to an unrecognized degree. Everyone in industrial societies is surrounded by it: in the street, the office, the supermarket, the factory, the bank, the car, the airport, the home . . . and so on, ad nauseam. Try to launder money for the Mob, for example, and you may be caught out not by a detective but by a machine (Senator *et al.* 1995). Some of these AI systems are neural networks, like those used (unseen by their customers) by financial institutions of various kinds. Others are GOFAI programs, like those used (again, unseen) by insurance salesmen in telephone call centres.

Even when the public are aware that such programs are being used, they may not think of them as *examples of AI*:

Knowledge-based systems have permeated almost all areas of modern life. Indeed this area of AI has been so successful that many people no longer associate it with AI. They simply see advice-giving as yet another thing that computers can do. That is, I suppose, the ultimate success for any branch of science. (Whitby 2004: 36)

We've seen, for instance, that the longed-for Semantic Web—typically regarded as a futuristic project *for computer science*—will require advances in ontology research (Section i.c, above). But work on computerized ontologies was pioneered by NewFAI, both for AI planning and for NLP. Indeed, this was a key theme of the seminal paper on 'Some Philosophical Problems' (Chapter 10.iii.e).

Besides specific application programs (many of which outperform human beings in their particular specialized task), a host of widely used computer technologies, not normally thought of as AI, actually originated in AI research:

Work in AI has pioneered many ideas that have made their way back to mainstream computer science, including time sharing, interactive interpreters, personal computers with windows and mice, rapid development environments, the linked list data type, automatic storage management, and key concepts of symbolic, functional, dynamic, and object-oriented programming. (S. Russell and Norvig 2003: 15)

AI technologies underlie many Internet tools, such as search engines, recommender systems, and Web site construction systems. (S. Russell and Norvig 2003: 27)

[There are many] examples of AI technology that are now so pervasive that few people even realize they involve AI concepts. These include spreadsheets, speech recognizers, digital maps, and even the World Wide Web, which is just a large semantic network. (Pat Langley, email to AAAI Fellows, 14 August 1998; quoted by permission)

Video games and animated movies are using techniques developed in robotics. For example, path planning algorithms are used to move soldiers around in video games. [Added May 2004: Bodily control and perception algorithms are also being used to produce plausible characters in visual reality.] Machine vision techniques are also in use to search and index images in large databases such as the world wide web. (Matt Mason, email to AAAI Fellows, 17 August 1998; quoted by permission)

[User-friendly AI-based systems include] programs which can detect the presence of known or new viruses in computer programs; checkout scanners which can identify fruit and vegetables through the use of scent sensors; car navigation systems which can guide a driver to a desired destination; password authentication systems employing biometric typing information; ATM eyeprint machines for identity verification; and molecular breath analyzers which are capable of diagnosing lung cancer, stomach ulcers and other diseases. (Lofti Zadeh, email to AAAI Fellows, 17 August 1998; quoted by permission)

As two of those quotations point out, many of these things are thought of as "mainstream" computer science, not as AI at all. (This is yet another illustration of AI's successes being relabelled as belonging to some *other* discipline.) Sometimes, such relabellings are truly bizarre: an AI colleague of mine was recently astonished, and outraged, to be told by a professional computer scientist that "NLP isn't AI, it's computer science" (Sharon Wood, personal communication).

Sometimes, to be sure, there are disasters. The newspapers abound with horror stories about hugely expensive government software failing to do what it was intended to do. Nevertheless, the notion that AI, or even GOFAI, has failed as a *technological* enterprise is absurd. So also—IMHO, as they say on the emails—is Hecht-Nielsen's reference to "a handful of valuable accidental [GOFAI] discoveries".

More to the point, for our purposes, is the question whether *psychological* AI has failed. This can be posed with respect to both weak and strong AI. The success/failure of strong AI will be discussed in Chapter 16, so let's concentrate on weak AI here.

People often say that whereas connectionist AI has succeeded in helping theoretical psychologists, GOFAI hasn't. At best, they'll allow that GOFAI was historically signific- ant, in encouraging psychologists of the 1960s–1970s to think computationally and to build computer models (Chapters 6.iii–iv and 7). Considered as psychology, they say, GOFAI has failed.

The Dreyfus brothers, for instance, declared in 1978 that "classical, symbol-based, AI appears more and more to be a perfect example of what Imre Lakatos [1970] has called a degenerating research-programme". And fourteen years later, Hubert Dreyfus repeated that charge almost word for word (1992, p. ix).

It's true that some psychological topics haven't been, and perhaps never can be, adequately addressed by hands-on GOFAI modelling or even by detailed computational theorizing. The identification of relevance (7.iii.d and 8.vi.c) is an example. Such topics are intractable because of the frame problem (10.iii.e and i–ii, above); the fuzziness of concepts (8.i.b, 9.x.d, and 12.x); cultural variations in language and custom (8.i and 9.iv.b); the highly idiosyncratic experience and world-knowledge of human individuals; and, not least, human *embodiment* (16.vii, and iii.b–c above).

Nevertheless, GOFAI modelling is still being used—sometimes alone, sometimes in hybrid systems—to develop theories of hypnosis (1.i.g); of absent-mindedness and pathological disturbances of everyday action (12.ix.b); of artistic and scientific creativity (iv, above); and of anxiety (7.i.f). And GOFAI ideas are helping us to explore the mental architecture of the mind as a whole (7.i.e–f and 12.iii.d). The people engaged in this work include clinical neurologists and psychologists as well as AI scientists, so it can't be dismissed as merely a fantasy of the nerds, nor even as pure (impractical) theory.

Indeed, we saw in Section iii.b and 12.viii–ix (and shall see again in 14.ix.b) that symbolic AI's methodological rivals—connectionism, situated AI/A-Life, and dynamical systems—can't cope with some phenomena that are crucial features of human psychology. Means–end planning, deliberate reasoning, and flexible thought using objective concepts are the prime examples. Philosophers such as Fodor, Kirsh, and more recently Samuels, despite the disagreements between them (over modularity for instance: 7.vi.d–e), have spelt out some of the reasons. Plenty of puzzles remain: SOAR, or the GOFAI aspects of ACT* (7.iv.b–c), haven't solved all the problems. But the point is that, *in principle*, some sort of GOFAI approach will be needed (as well as connectionist and situationist ideas) to understand the human mind. Clearly, then, the charge that GOFAI is "a degenerative research programme" is mistaken.

Blay Whitby has put it in a nutshell:

A myth has developed that AI has failed as a research programme. This myth is prevalent *both inside and outside* AI and related scientific enterprises. In fact AI is a remarkably successful research programme which has delivered not only scientific insight but a great deal of useful technology. (Whitby 2004: 1; italics added)

He points out that GOFAI and connectionism alike have been—and are—useful, in both academia and the marketplace. Indeed, one very widely used commercial application, the UK's prizewinning Clementine data-mining system, relies on an amalgam of both methodologies—with evolutionary computing (15.vi) thrown in for good measure (Khabaza and Shearer 1995).

In sum, GOFAI hasn't "failed", and it hasn't ceased either. Today's AI champions can still produce defensive rhetoric that makes one wince:

Researchers in AI have traditionally met problems of formulation joyfully, courageously, and proudly, accepting the severe risks ensuing from such exploratory work ... (Doyle and Dean 1997: 88)

(Entire boxfuls of medals couldn't do justice to such heroes!) But the same writers aren't going over the top when they say:

As a field, AI embarks on the next fifty years excited about the prospects for progress, eager to work with other disciplines, and confident of its contributions, relevance, and centrality to computing research. (p. 99)

As Mark Twain might have said, the rumours of its death are greatly exaggerated.

14

FROM NEUROPHYSIOLOGY TO COMPUTATIONAL NEUROSCIENCE

In his peppery *Dictionary of Psychology* (1995), N. Stuart Sutherland said that cognitive science and neuroscience are "almost mutually exclusive" (see Chapter 1.ii.c). This was more than a little naughty. As an expert in vision and a friend and admirer of David Marr, and as a colleague of Albert Uttley, Sutherland knew very well how necessary that "almost" was. They'd long pointed out that a theory of the brain needs a theory of what it is that the brain is doing.

Alan Turing had said this long ago (in 1951), in a letter to the anatomist J. Z. Young:

I am afraid I am very far from the stage where I feel inclined to start asking any anatomical questions. According to my notions of how to set about it, that will not occur until quite a late stage when I have a fairly definite theory about *how* things are done. (quoted in S. S. Turing 1959: 146; italics added)

That "how", of course, was a computational how. And the principle hadn't changed since 1951. By the time Sutherland made his remark, many psychologists had said the same thing (e.g. Mehler *et al.* 1984), and so had Chomsky, in 1965 (Chapter 7.iii.a). What's more, many neuroscientists agreed. Michael Gazzaniga (1939–), for instance, had recently insisted that "neuroscience needed cognitive science . . . to attack the central integrative questions of mind–brain research" (1988: 241).

This is a very general point. Even in GOFAI, as we saw in Chapter 10.v, what's usually of interest is the *virtual* machine defined by the programming language and/or the program, not the manufactured tin can which—in *physical* terms—is doing the work.

Sutherland's claim (in the same dictionary entry) that cognitive science concerns the brain's *software* and neuroscience its *hardware* also needed qualification. He himself had long rejected Hubert Dreyfus's view that "we must leave the physical world to the physicists and neurophysiologists", to the exclusion of AI (see Chapter 16.vii.a). On the contrary, he'd said, we need high-level (software) concepts to understand brains as well as computers:

Understanding the workings of the brain also involves us in developing the appropriate concepts to summarise blocks of similar operations conducted by it and an appropriate language in which such concepts can be embedded and manipulated . . . e.g. negative feedback, receptive field, iconic store, and so on . . . (N. S. Sutherland 1974: 265)

In short, it's not true that computation is irrelevant if we have the neuroscience (a claim made also by Dreyfus's Berkeley colleague John Searle—1992, ch. 10). For, so Sutherland was saying, the one is integral to the other.

If neuroscience looks at the brain (and the rest of the nervous system) and asks "What does this bit do?", *computational* neuroscience asks "How does it manage to do it?" And that "How?" is computational. As John Mayhew, another vision expert (and a follower of Marr: see Chapter 7.v.e), has put it:

Finding a cell that recognizes one's grandmother does not tell you very much more than you started with; after all, you know you can recognize your grandmother. What is needed is an answer to how you, or a cell, *or anything at all*, does it. The discovery of the cell tells one what does it, but not how it can be done. (Mayhew 1983: 214; italics added)

Even if the detailed wiring diagram is available, as it was for the cerebellum by the early 1960s (see Section iv.c), *what the wires are doing* may be unknown. The key questions concern what information is received and/or passed on by the cell or cell group, and how it's computed by them. Put another way, they concern "how electrical and chemical signals are used in the brain to represent and process information" (C. Koch and Segev 1989: 1).

The differences between the systems discussed in Chapter 12 and those described here are outlined in Section ii. Then, we'll look at some early examples of neuroscientific work driven by computational questions, though rarely aimed at implementation. Section iv explains how computational ideas helped inspire the discovery of feature-detectors. Section v outlines early models of specific parts of the brain (such as the reticular formation and the cerebellum), while Section vi considers some especially 'realistic' system-level simulations. The whole-animal approach of computational neuro-ethology is considered in Section vii.

Various examples and definitions of "representation" are discussed in Section viii, followed (in Section ix) by some common sources of scepticism about AI neuroscience. Lastly, in Sections x and xi, we'll ask whether cognitive neuroscience has moved us—or ever could move us—closer to an understanding of consciousness.

Before embarking on the substantive issues, however, let's look very briefly at the history of two labels.

14.i. Notes on Nomenclature

In one sense, the story told in this chapter is anachronistic. For I often use vocabulary that's relatively recent to describe work done many decades ago. That's deliberate, for I'm seeking to show the links, indeed the continuity, between that older work and what went on later.

Specifically, if you go back to read the neuroscientific literature published before the early 1960s, you won't find the word "neuroscience". And you won't find the phrase "computational neuroscience" until a quarter-century after that. (As for "cognitive neuroscience", those words first left the presses after the one and before the other: see Section x.b, below.)

a. The naming of neuroscience

Neurophysiology became computational before it became (i.e. before it was called) an aspect of neuroscience. When I was a medical student at Cambridge in the 1950s, some of our supervisors (notably Horace Barlow) were already starting to talk about neurones in informational terms. But the word "neuroscience" was never used. We spoke rather of neurophysiology and neuro-anatomy (hardly, yet, of neurochemistry) ... and of clinical neurology, psychopathology, and psychiatry. Moreover, we thought of these as largely *distinct*, in practice if not in principle.

The new term was coined at MIT in the early 1960s by the neurochemist Francis Schmitt. (He also coined the term "information substances" twenty years later, to cover "a variety of transmitters, hormones, factors, and protein ligands": S. Beer 1999. That is, even at the level of the neurochemicals themselves, *messages* were deemed to be as important as molecules.)

Schmitt needed the new name for a nationwide project specifically aimed at uniting several sociologically distinct camps. He was bringing together at least half a dozen experimental scientific disciplines (ranging from molecules to behaviour), and clinical expertise too. As he recalls:

> [Now] in the 1990s such a program seems entirely reasonable, *but in the 1950s and 1960s it was not so*. The boundaries between the various biomedical disciplines ... were clearly defined, particularly for didactic purposes in medical school curricula. Professionals in individual fields looked somewhat askance at experts in other fields who presumed to be knowledgeable also in theirs, as would certainly be the case in the new field I then had in mind. (Schmitt 1992: 1; italics added)

He took care to involve the US funding agencies right from the start—a wise move, given their crucial role as René Descartes's wealthy philanthropists enabling research to be done (2.ii.b–c). Soon, the new label gained international respectability, when the editor of *Nature* invited him to write a piece explaining just what "neuroscience" was (Schmitt 1992: 10).

The date of this multidisciplinary foundation was perhaps no accident, being determined not only on intellectual grounds (e.g. the recent advances in biochemistry and molecular genetics) but also by sociological factors. We saw in Chapter 1.iii.c–d that the general cultural challenge to formalist modernism in the 1960s involved an increased respect for concrete, situated, studies, including what had previously been sidelined as mere "applied" sciences (Toulmin 1999: 165). So perhaps Schmitt had sensed that the "pure" scientists were now more willing to listen to what the medics had to say, and to think of *their own* problems in practical, clinically oriented, terms.

The Society for Neuroscience was founded a few years later, when (according to its web site) "neuroscience barely existed as a separate discipline". Today, the SFN still makes a point of describing itself as a group of "basic scientists *and physicians*". Now, SFN has over 36,000 members. But when it began in 1970 there were only 200. And it was almost ten years after that before cognitive psychologists, as opposed to brain scientists, started to take *clinical* data really seriously (see Section xi.b, below).

Here as elsewhere, McCulloch had been an ambiguous figure. On the one hand, he was (for cognitive science) *the* seminal formalist. On the other hand, he'd always valued practical/therapeutic relevance as much as rigour, and many-methods rather than just one (see Chapter 4.ii–v). Some of the more orthodox, more modernist, scientists had been wary of him accordingly.

Perhaps encouraged by *Nature*'s interest, the new word caught on. The journal *Neuroscience* was founded in the mid-1970s. Twenty years later Sutherland (1995), acerbic as ever, defined the term as "a fashionable catch-all phrase that includes all disciplines that directly study the nervous system . . . and all that attempt to relate behaviour to the nervous system (e.g. physiological psychology)".

He explicitly contrasted it with cognitive science: "an equally fashionable term, comprising a set of disciplines that barely overlap with neuroscience". But this was misleading. The overlap, though admittedly limited (and viewed with suspicion by some: see Section viii.d, below), was by then significant. Indeed, it had been growing steadily for fifty years.

b. The computational species

Computational neuroscience, focusing on information processing, was pioneered at mid-century, as we'll see. But the name dates only from the early to mid-1980s.

Its first high-visibility usage was in a paper in *Science*, written by three leading cognitive scientists—drawn from neuroscience, connectionism, and philosophy (Sejnowski *et al.* 1988). They used it to cover any research on the brain and nervous system which asked computational questions and/or used computational modelling.

For example, it would cover research that asked *just how* cells in a monkey's temporal cortex could detect—i.e. compute—another monkey's face looking in a particular direction (see Section iv.d, below). Merely finding those cells (by experiments with micro-electrodes), without ever considering such questions, would be interesting. But, as Mayhew would be quick to point out, it wouldn't count as *computational* neuroscience.

But there are stricter definitions. Three years earlier, in 1985, Eric Schwartz wrote to a number of authors inviting them "to contribute to a book whose purpose was to define the term *computational neuroscience*" (E. L. Schwartz 1990, p. ix). The result was a 1987 symposium in California, followed by the promised book in 1990.

In explaining why he'd bothered to do this, Schwartz reported "violent turbulence over the past four decades" in the fortunes, and even the name, of the area he'd been working in. All the names he listed (*cybernetics, neural networks, brain theory, artificial intelligence, neural modelling*), he said, had been "abused, and none felt untainted for the present purpose" (p. ix). His preference, *computational neuroscience*, was described as "that area of overlap between neuroscience and computer science which required sufficient specialized expertise to justify a new subdiscipline". More particularly:

computational neuroscience, then, is the problem area in which *difficult algorithmic or implementational questions are intimately related to the data of the nervous system*. The interplay of neural data and of computation and applied mathematics define the scope of this term. (E. L. Schwartz 1990, p. x)

Since Schwartz's definition specifically mentions difficult applied mathematics, it doesn't cover all of the work discussed in this chapter. Of course, what's difficult for one person may not be difficult for another. Nevertheless, the seminal research described in Sections iii–v below, even including Marr's early models of the cerebellum and hippocampus, doesn't fall under Schwartz's rubric. Nor does the early work in computational neuro-ethology (Section vii).

To refuse to regard any of that research as computational neuroscience, however, would be to lose sight of two key points. First, it would jettison Mayhew's important distinction between what-questions and how-questions. Second, it would obscure the historical connections between the relatively straightforward discussions of the 1950s–1960s and the (often fearsome) mathematics of neuroscientific research in the 1980s–1990s.

Accordingly, I'll rely on the more catholic definition given above. What's more, I'll interpret it so as to include approaches inspired by cybernetics and/or information theory (cf. 1.ii.a).

This chapter, then, narrates how purely biological neurophysiology and neuro-anatomy were joined by computational neuroscience. The plot of the story is concerned less with the specific discoveries that were made, fascinating though these are, than with the changes in *the sorts of question being asked*. Over the past sixty years, the central nervous system came to be appreciated not only as a biological organ but also as an informational/computational machine. Besides enriching neuroscience as such, this change made possible its growing rapprochement with psychology. Without it, neurophysiology would have remained outside cognitive science.

14.ii. Very Non-Neural Nets

The computational properties of neurones and cell assemblies were studied in very general terms by the connectionist AI/psychology described in Chapter 12. From the neuroscientist's point of view, that research had four major deficiencies: too neat, too simple, too few—and too dry.

a. Too neat

Connectionist AI tried to be neat, whereas the brain is likely to be messy. Connectionist learning rules, as we saw in Chapter 12, were valued for their increasing mathematical power and elegance. But this is largely irrelevant, from the neuroscientist's perspective.

Francis Crick (1916–2004) put it like this:

[I] suspect that within most [connectionist] modellers a frustrated mathematician is trying to unfold his wings. It is not enough to make something that works. How much better if it can be shown to embody some powerful general principle for handling information, expressible in a deep mathematical form, if only to give an air of intellectual responsibility to an otherwise rather low-brow enterprise? (Crick 1989a: 132)

All very well, he continued, but evolution is a tinkerer, employing a ragbag of "slick tricks" rather than purist mathematical principles. And those tricks are what biologists want to identify:

Why not look inside the brain, both to get new ideas and to test existing ones? The usual answer given by psychologists is that the details of the brain are so horrendously complicated that no good will come of cramming one's head with that sort of information. To which the obvious reply is, "If it's as complicated as that, how do you hope to unscramble its workings by a purely black-box approach, by merely looking at its inputs and outputs?" (Crick 1989a: 132)

Crick himself was oversimplifying. Many neuroscientists value the mathematical elegance of computer models, if only in providing 'existence' proofs. Sometimes, they show that something broadly similar to a real neural network, *but even simpler*, can do very surprising things. For instance, an utterly random network—which the embryo brain is not—can self-organize into structures very like those found in real brains (see Sections vi.b, ix.a, and ix.c).

One can't deny, however, that over-fierce mathematics may put neuroscientists off. The history of the field would have been rather different if that weren't so (see Sections v.d, vi.a, viii.b, and ix.b). Nor can one deny that the more 'lifelike' a neural model gets, the less formally intelligible it may be. Large 'evolutionary' networks, for example, may embody clear rules (even Type-I theories *à la* Marr: see Chapter 7.iii.b), but their final state may be nearer to a Type-II theory—which is almost to say, no *theory* at all (see Section ix.d).

But Crick was right about the slick tricks. Various ecologically relevant feature-detectors found in crabs, frogs, and mammals are mentioned in Section iv, and other special-purpose mechanisms (in frogs and insects) are described in Section vii and Chapter 15.vii. Significantly, Walter Pitts—a mathematician—was *disappointed*, even *a little appalled*, by his seminal co-discovery of bug-detectors, because they weren't formally tractable (see Section iv.a).

b. Too simple

The second deficiency of connectionism was its unrealistic 'neurophysiology'. A connectionist unit—of whatever type—is a mythical beast, as elusive in the biological world as the gryphon. The computational power of a real neurone is closer to that of a complex artificial network, or even a small computer, than to a single network-element.

Real neurones aren't (or aren't just) components of logic gates: one cell may have thousands of possible outputs. As Crick (1989a) has remarked, there may be many glutamate receptors on a single neurone, whose number and locations are surely relevant but which are simply ignored by connectionists. And it was discovered in the mid-1980s that dendrites, of which a neurone can have many thousands, aren't passive conduits but dynamical transmitters (with varying conductance). It's even been suggested, on the basis of a simulation (McKenna 1994: 83), that *an individual dendrite* may be able to compute exclusive-or—which Frank Rosenblatt's perceptron couldn't do (Chapter 12.iii.b).

What counts as a 'plausible' model, of course, changes as our knowledge grows. Implausibility can vanish overnight. For instance, multiplication rules were included in artificial networks (12.v.e) to provide mathematical neatness, certainly not biological credibility. But in the mid-1990s, "multiplication synapses" were discovered in the brain (Elman *et al.* 1996: 105).

In addition, *time* matters in real brains. For example, there's a wide range of 'windows' (measured in milliseconds) during which two inputs can have a joint effect; and spiking frequencies may be more relevant than single spikes. Yet connectionist networks are typically time-less. (Some exceptions are discussed in Section ix.g.)

Furthermore, the most widely used connectionist learning rule—back propagation—is unrealistic. Synapses don't transmit in two directions.

c. Too few

The third drawback of connectionist AI concerned its less than generous 'anatomy'. Judged by the numbers of units, artificial neural networks were tiny by biological standards.

True, some biological species have only 302 neurones (see v.b, below). And some have specific networks with only thirty neurones or fewer, whose connections and patterns of excitability can be individually modelled. For instance, there are alternative simulations of the network that enables a sea slug (*Aplysia*) to swim away from danger (Getting 1989; Kleinfeld and Sompolinsky 1989).

But even insect brains are enormously larger than that (M. O'Shea, personal communication). The bee brain is estimated to have about 1 million neurones (plus perhaps 100 million glial cells). And some molluscs, namely octopodes, have insect-like numbers of neurones: hundreds of thousands, if not millions. This is due to their very advanced visual systems: lower molluscs, such as sea slugs, have only about 20,000–30,000. As for the human cerebral cortex, this has about 20,000 million neurones (i.e. 20 US billions), and the number of dendrites and synapses is orders of magnitude greater. The largest of today's artificial networks is puny, indeed microscopic, by comparison.

Admittedly, network models soon started growing. In the late 1960s, for example, a Hungarian researcher simulating the cerebellum achieved a huge jump in scale: from a network of 1,296 units to one of over 64,000 (see Section viii.b). By the mid-1970s, he was coping with more than 1,700,000 components. But even that number, impressive though it is, was niggardly alongside the cerebellum itself.

Moreover, these numbers *matter*. It's not good enough to say there are "many" neurones, each one with "many" synapses. To understand the functioning of a particular part of the brain, one often needs to know *just how many* (see Section v.c).

Even more to the point, the brain isn't a formless mass of identical cells. So building a super-duper-computer that matched the huge *numbers* wouldn't help much. The networks described in Chapter 12 were very rarely assigned to specific parts of the brain. The assumption seemed to be that a neurone is a neurone, is a neurone . . . But there are many (perhaps twenty-five) distinct types of neurone, with different locations and physiological—and computational—properties.

(The even more numerous glial cells, which outnumber the neurones by about 100:1, seem to have mainly structural/guiding and nutritive functions, not informational ones. For instance, some of them form the myelin that coats mature axons. However, there's recent evidence that they can modulate the responses of the neurones, so in that sense they play a computational role: M. O'Shea, personal communication. For more on neuromodulation, see Section ix.f, below.)

This anatomical diversity of the neuronal cells was already being noted at the beginning of the century (2.viii.c). Santiago Ramón y Cajal identified five types of cerebellar neurone (the Purkinje, basket, stellate, Golgi, and granule cells), and distinguished some of their interconnections (the mossy, climbing, and parallel fibres)—see Figures 2.1–2.3. That was the beginning of the anatomical work used to ground the modelling of the cerebellum seventy years later (Section v.c, below). Ramón y Cajal also showed that the retina contains several types of unit: rods, cones, bipolars, horizontals, amacrines, and ganglion cells (whose axons form the optic nerve).

Later, John Eccles and others combined microscopy with single-unit recording, making the anatomical complexity even clearer (2.viii.e). Under the microscope, the cortex has six layers. And single-cell recordings showed functional distinctions, too. Visual cortex, for instance, contains over fifty distinct areas or neurone groups, simultaneously processing many different features, such as colour, lightness, texture, depth, and shape (Zeki 1993). Today's functional map of the striate cortex is more complex than that of the London Underground.

Most AI connectionists ignored such facts. Even Carver Mead's VLSI chips (12.v.f) only *approximated* the real retina, cochlea, and cortex. So when connectionists boasted that their models were more biologically plausible than GOFAI, they should rather have said they were *less implausible*.

To be fair, some did. Warren McCulloch and Pitts themselves, in 1943, had admitted that their formal units and "temporal" expressions, or TPEs, were idealizations (4.iii.e). This was confirmed in the 1950s by work on stochastic firing, grounded in continuous synaptic variation that's independent of the input from other neurones. By the 1980s, the idealization had become even more evident (Perkel 1988). Indeed, the PDP bible included a chapter on how PDP units differ from real neurones (Crick and Asanuma 1986).

But most of the connectionists featured in Chapter 12 paid little or no attention to the brain *as such*. Here, we'll consider research using computational theories and/or computer modelling *to cast light on the nervous system itself*.

d. Too dry

Computational neuroscience aims to match the details of the software to specific aspects of the biological hardware—or, as it's sometimes called, the wetware. And "wetware" is often interpreted as *wireware*. Certainly, the computational neuroscientists discussed in this chapter concentrated on the detailed anatomical connections within the nervous system. They weren't concerned with chemical/biophysical properties. The "wetness" of real neurones was ignored: linkage was all. In that sense, they're 'connectionists' too.

Strictly, however, wetware includes more than wireware. For neuroscientists study not only the networks of neurones but also what they are made of, how they actually work. The Hodgkin–Huxley equations for the nervous impulse were an early example (2.viii.e).

Accordingly, some computational neuroscience looks *below* the neuronal level. (For some examples, see C. Koch and Segev 1989, chs. 2–5; Feng 2004, chs. 2–7.) This is a relatively recent development: wireware models came first, as we'll see.

By the 1990s, however, a few researchers were simulating the computational functions of membranes and dendritic spines, and the action of the calcium ion and neurotransmitters (e.g. C. Koch 1990, 1999; Gazzaniga 1995, pt. i). A specific Hebbian learning rule was modelled in terms of biophysically realistic data on the strength, and temporal properties, of synaptic currents and membrane potentials (Rao and Sejnowski 2001). And some computer models included the effects of neuromodulators, which can temporarily change a neurone's properties (Philippides *et al.* 1998).

Two interesting wetware examples are described in Section ix.f, below. One concerns the computer simulation of neurochemical modulation; the other seeks to produce non-linear computation by harnessing real chemicals and/or neurones to electronic circuits. With those exceptions, I'll consider only models of the wireware: theories at the level of neurones or large neurone groups. This is partly because wetware models are state of the art rather than historical, and partly because they are less relevant for cognitive science in general than wireware models are.

I'll also ignore 'neuro-informatics'. This deals with the wireware too. It applies computerized statistical analysis to huge data-banks of results, so as to discover possible structure–function relations in the brain. It's increasingly necessary, given the flood of neurological data. Over 14,000 reports of connections in the rat's brain, in over 900 separate papers, appeared between 1980 and 2000, since when there will have been many more (M. P. Young 2000: 3, 56). And then there's cats, and monkeys . . . Neuro-informatics may help inspire computational theories, by leading people to ask what the anatomical data *mean* in functional terms. But it's still in its infancy. The theories I'll discuss here weren't based on statistical data analysis, but on individual experiments and observations.

An objection often heard from the more traditional neuroscientists is that computer models are (to put it politely) of dubious value, because they hugely oversimplify the neurophysiology *even if* they attempt to take it on board. Indeed, we'll see that it took many years before such models were seriously regarded by professional brain scientists (v.d and vi.c, below). However, two leading computationalists, namely Christoph Koch and Idan Segev, have given this scepticism a suitable reply:

One frequently hears the criticism that simplifying brain models lack this or the other physiological feature and are thus unbiological and irrelevant for understanding the nervous system. . . . This argument, taken to its extreme, implies that we will not be able to understand the brain until we have simulated it at the detailed biophysical level! . . . [But] such a simulation with a vast number of parameters will be as poorly understood as the brain itself. Furthermore, if modeling is a strategy developed precisely because the brain's complexity is an obstacle to understanding the principles governing brain function, then an argument insisting on complete biological fidelity is self-defeating. (C. Koch and Segev 1989: 2)

It's only recently that books have started to appear which try to take both low-level neurophysiology and high-level psychology seriously. Among the first, in the 1980s, were two by Stephen Grossberg, whose pioneering work is described in Section v. And in the 1990s, Arnold Trehub of UMass Amherst published a monograph describing computational theories (and models) of *planning* and *narrative comprehension* (Trehub 1991).

Trehub's theories of these higher mental processes are based on Hebbian learning constrained by detailed neural structures, and even by neurochemicals. The biophysical facts involved concern the mechanisms of long-term/short-term memory (e.g. ATF and DTF: axon and dendrite transfer factors), and the key neural structures are the "synaptic matrix" and the "retinoid". Trehub's claim is that these wetware matters mustn't be ignored, because they enable us to rule out a wide range of alternative cognitive hypotheses.

As for textbooks integrating connectionist modelling with real neuroscience, the first example came off the press at the turn of the century. Randall O'Reilly and Yuko Munakata, both ex-students of James McClelland (and now based at the University of Colorado), hoped to follow in their tutor's footsteps in more ways than one. For in their Preface, they said:

Our objective in writing this text was to replicate the scope (and excitement) of the original PDP volumes in a more modern, integrated, and unified manner that more tightly related biology and cognition . . . (O'Reilly and Munakata 2000, p. xxv)

They didn't entirely succeed, for the "excitement" that had been aroused by the original PDP volumes had been so great that probably no one could have matched it (see 12.vi.a). For one thing, the new textbook wasn't aimed at the still-uncommitted student, as the PDP bible was. (Still less would it enthuse the journalists and the general public, as its predecessor did.) For another, O'Reilly and Munakata were doing something which very obviously needed to be done, whereas the PDP volumes had struck most readers as a bolt from the blue.

Nevertheless, their book was important, for it did—as they hoped—relate biology and cognition much more tightly together. O'Reilly and Munakata, and many of the computational researchers discussed by them, showed a healthy respect for the wetware. Neuroscientific findings, which in any case had ballooned since the mid-1980s, were taken on board wherever possible.

Their hope of replicating the "scope" of the PDP books was also fulfilled. For their discussions (and models) ranged psychologically from language, memory, and attention to specific learning rules, and biologically from large-scale brain anatomy to axons, ion channels, and membrane potentials. In short, their volume was not merely a textbook, but an interdisciplinary tour de force.

One way in which their text surpassed the first edition of the PDP equivalent was that they made available (on the Web) a powerful software tool with a detailed manual, and gave tutorial appendices in the printed text. In addition, they discussed many examples of psychological simulations developed by using this system.

They called it "Leabra": Local, Error-driven and Associative, Biologically Realistic Algorithm. The phrase "Error-driven and Associative" was code for the fact that Leabra combined both reinforcement and Hebbian "ft/wt" learning.

Like Allen Newell's SOAR and John R. Anderson's ACT* (see 7.iv.b–c), this was intended as a "unified" model of cognition. They shared Newell's view that a coherent unified model is much more difficult to produce, and therefore much more informative, than specialized theories of specific phenomena (p. 11). All areas of experimental/clinical psychology were supposed to be covered, in an integrated fashion.

Leabra drew on core theoretical insights from connectionist AI. But relevant aspects of neuroscience were included too—and given greater prominence. For example, the activation function controlling the spiking of the simulated neurones was only "occasionally" drawn from mathematical connectionism (p. 42). Usually, it was based on facts about the biological machinery for producing a spike (the sodium pump: see 2.viii.e). These included detailed data on ion channels, membrane potentials, conductance, leakages, and other electrical properties of nerve cells (pp. 32–48).

One might think that *from the psychologist's point of view* this was over-egging the pudding: surely such details can't be relevant for understanding high-level phenomena such as reading, word meaning, or memory?

O'Reilly and Munakata readily admitted that "higher levels of analysis . . . provide a more useful language for describing cognition" (p. 40). They pointed out, however, that the basic equations governing the activity of Leabra when it was simulating reading, or memory, or . . . weren't (except occasionally) picked out of an abstract connectionist mathematics. Rather, they were painstakingly drawn from detailed biophysical data. This was true, for instance, of the equation used in Leabra for integrating all the inputs into a neurone (see their explanation of equation 2.8 on pp. 37 ff.).

They drew the line at applying this equation "at every point along the dendrites and cell body of the neuron, along with additional equations that specify how the membrane potential spreads along neighbouring points of the neuron" (p. 38). They had no wish "to implement hundreds or thousands of equations to implement a single neuron", so used an approximating equation instead. But, characteristically, they provided references to other books which did explain how to implement such detailed single-neurone simulations (including the delightfully named *Book of GENESIS*: J. M. Bower and Beeman 1994).

In general, the psychological models developed by O'Reilly and Munakata would have been different had the neuroscientific data been different. Their discussion of dyslexia, for instance, built not only on previous connectionist work (e.g. Plaut and Shallice 1993; Plaut *et al.* 1996: see Chapter 12.ix.b), but also on recent clinical and neurological information (2000: 331–41).

As our knowledge of the brain advances (thanks to brain scanning, for example), future psychological models—they believe—will, or anyway should, be different again. They see their book as "a 'first draft' of a coherent framework for computational cognitive neuroscience" (p. 11). In short, the functionalist philosophers' "multiple realizability" is now under strong attack (16.iii). (It's under attack not only in 'human' neuroscience, but in neuro-ethology, too; indeed, the electric fish studied long ago by Lord Cavendish have prompted some strong arguments against it: Keeley 2000*a*.)

Even so, the computational level of theorizing is still crucial. Future researchers who follow O'Reilly and Munakata's lead—or who use Michael Arbib's NSL language to model macroscopic brain structures (see Section vii.c, below)—will *not* be fulfilling Wittgensteinian philosophers' hopes for "the disappearance of psychology as a discipline distinct from neurology" (Rorty 1979: 121). Or rather, they'll be doing so only if one interprets that word "distinct" to mean something like "paying no attention whatsoever to . . .". Indeed, the main theme of this chapter is that neurology itself has become more computational, more psychological.

This fact, combined with the demise of multiple realizability in its strong form, makes the demarcation between "psychology" and "neurology" more fuzzy than it used to be. But that's not to report, nor even to predict, the "disappearance" of psychology.

14.iii. In the Beginning

Detailed computational questions first arose in neurophysiology at around mid-century. McCulloch and Pitts, and Kenneth Craik too, had considered such questions in general terms in the 1940s. What was different in the early 1950s was that some people started using ideas from information theory and computing to help them study the neurophysiological nitty-gritty.

a. Computational questions

Even in the late 1940s and 1950s, when 'neuroscience' was still unheard of, people had already started to ask computational questions about specific aspects of the nervous system.

Their answers were rarely put in terms of implemented models, because computers were still primitive. But the notion that the brain (with the sense organs) is an information-processing machine, whether digital or analogue, was being taken seriously by a diverse group of people:

* by neurophysiologists such as McCulloch, Barlow, and Jerome Lettvin;
* by cyberneticians such as William Ross Ashby;
* by physiological psychologists such as Uttley;
* by psychologists with a strong engineering bent, such as Craik and Richard Gregory;
* and by computer scientists such as Turing and Oliver Selfridge.

Moreover, these people were now talking to each other. This was new. Before the Second World War, and excepting the leaders of the cybernetics community discussed in Chapter 4 (including Nicholas Rashevsky's group for mathematical biophysics), there hadn't been much interchange between psychologists and physiologists.

Admittedly, Charles Sherrington himself had joined the British Psychological Society on its foundation, and contributed a paper (on binocular flicker) to the first issue of its journal in 1904. But this was more an expression of hope and encouragement than a mark of central research activity. Behaviourism treated the brain as a black box, and Gestalt psychologists had little constructive to say about it. Freudians and Piagetians, even less. Only William McDougall had made a concerted effort to bring the two disciplines together, and he didn't get very far.

What brought brain and behaviour—and computers—closer together was the challenge set by the war (see 4.vi–viii and 11.i). Craik and Frederic Bartlett, for example, were closely involved (even before war was officially declared) in applying psychology to various tasks undertaken by servicemen. And most members of the interdisciplinary Ratio Club of the 1950s (4.viii and 12.ii.c) had been involved in the war effort as very young men, cooperating across disciplines to study not only radar, sonar, and radio, but also a wide range of machines *and* their human operators.

Turing's war work had nothing to do with physiology, or psychology either (3.v.d). But we saw in Chapter 12.i.b that he was thinking about neural networks by the mid-1940s and gave a brief description of "cortex" in his NPL report of 1947. However, his account of the brain, and of the "pleasure–pain systems" that might "organize" the cortex, was brief—and very general:

Many parts of a man's brain are definite nerve circuits required for quite definite purposes. Examples of these are the "centres" which control respiration, sneezing, following moving objects with the eyes, etc.: all the reflexes proper (not "conditioned") are due to the activities of these definite structures in the brain. Likewise the apparatus for the more elementary analysis of shapes and sounds probably comes into this category. But the more intellectual activities of the brain [such as speaking English or French] are too varied to be managed on this basis . . . We believe then that there are large parts of the brain, chiefly in the cortex, whose function is largely indeterminate. (A. M. Turing 1947*b*: 16)

Turing's notional neuroscience had little or no influence. The 1947 report remained unpublished until 1969, and Turing himself soon turned from cortex to embryology (Chapter 15.iv). When invited to give a talk to the Ratio Club he spoke first on AI and later on morphogenesis, not on brains (H. B. Barlow, personal communication).

While Turing was writing his NPL report, Pitts and McCulloch (1947) were thinking about the brain in much more detail. Their earlier, 'logical', paper of 1943 is accepted by neuroscientists today as "one of the great boosters of modern brain science" (Braitenberg 1984: 108). It had been based on the recent discoveries about synaptic activity, which would soon be confirmed by work on cell membranes and neurotransmitters (Hodgkin and Huxley 1952; Eccles 1953, 1964).

But it had ignored both the 'noise' in the nervous system and its neuro-anatomy. The computational/statistical aspects of their 1947 theory dealt with the noise, and were outlined in Chapter 12.i.b. Here, our interest is in what they said about the anatomy.

The first point to note is that they said anything about it at all. Their opening sentences put the goal and justification of computational neuroscience in a nutshell:

To demonstrate existential consequences of known characters of neurons, any theoretically conceivable net embodying the possibility will serve. It is equally legitimate to have every net accompanied by anatomical directions as to where to record the action of its supposed components, for experiment will serve to eliminate those which do not fit the facts. (Pitts and McCulloch 1947: 46)

In other words, abstract networks such as those discussed in Chapter 12 are worth discussing, but one can also try to map them onto real brains—in which case one is generating empirical hypotheses that can help brain science to advance.

Ashby, to be sure, had said the same sort of thing (4.vii.c–d). But Pitts and McCulloch were more specific than Ashby in their suggestions about *just which* cells might be doing *just what*. In their words: "We endeavor particularly to find those [net designs] which fit the histology and physiology of the actual structure" (p. 47).

They dismissed Gestalt theories of brain–world isomorphism, seeing them as hand-waving. The Gestaltists relied on some mysterious pre-established harmony (which they *did not* explain in evolutionary terms), and didn't seriously consider how the brain can identify and represent percepts/concepts. That was what Pitts and McCulloch wanted to explain.

They rejected what was later termed the grandmother cell theory, because (for example) a square could conceivably be represented by spatio-temporal averaging over a "mosaic" of cells:

[For this reason, we disagree with] the neurologists of the school of Hughlings Jackson, who must have it fed to some specialized neuron whose business is, say, the reading of squares. That language in which information is communicated to the homunculus who sits always beyond any incomplete analysis of sensory mechanisms and before any analysis of motor ones neither needs to be nor is apt to be built on the plan of those languages men use toward one another. (p. 56)

In short, *one concept, one cell* is both unnecessary and implausible (see Section ix.e).

Their hypothetical representation of a square, they said, might be thought to arise in a brain area one level up from Area 17. But in fact, it didn't:

A square in the visual field, as it moved in and out in successive constrictions and dilatations in Area 17, would trace out four spokes radiating from a common center *upon the recipient mosaic*. This four-spoked form, not at all like a square, would then be the size-invariant figure of square. *In fact, Area 18 does not act like this*, for during stimulation of a single spot in the parastriate cortex, human patients report perceiving complete and well-defined objects, but without definite size or position, much as in ordinary visual mental imagery. *This is why we have situated the [computational] mechanism . . . in Area 17, instead of later in the visual system.* (pp. 55–6; italics added)

Other neural hypotheses attributed temporal "scanning"—supposedly involving the alpha waves—to a specific brain area (the stripe of Gennari), and even to particular types of neurone within that layer. Similarly, they described the gaze reflex as a "servo-mechanism" (*sic*) wherein the superior colliculus "computes by double integration the lateral and vertical coordinates of the 'center of gravity of the distribution of brightness'", and sends impulses to the eye muscles so that the eyes turn towards the centre of gravity, gradually slow down, and stop when it's fixated. Claiming "considerable support for this conjecture in the profuse anatomical and physiological literature", they related the computations supposedly involved (expressed as mathematical equations) to particular laminae within the colliculus.

The issue, here, is not whether their suggestions were correct. Indeed, their more specific hypotheses were highly provisional: "It is evident that many details of [the] hypothetical nets of this paper might be chosen in several ways with equal reason; we have only taken the most likely in the light of present knowledge" (p. 54). But the general potential was clear:

We have focussed our attention on particular hypothetical mechanisms in order to reach explicit notions about them which guide *both histological studies and experiment*. If mistaken, they still present the possible kinds of hypothetical mechanisms and the general character of circuits which recognize universals, and give practical methods for their design. (p. 65; italics added)

And, quite apart from the contingent neuro-anatomy, they claimed to have established a mathematical principle covering the design of neural networks in general:

These procedures are a systematic development of the conception of reverberating neuronal chains, which themselves, in preserving the sequence of events while forgetting their time of happening, are abstracted universals of a kind. Our circuits extend the abstraction to a wide range of properties. By systematic use of the principle of the exchangeability of time and space [see Chapter 12.i.c], we have enlarged the realm enormously. (p. 65)

By implication, biologists would do well to remember that natural selection might have discovered this principle too, exploiting it in various ways within real nervous systems.

As for whether 'the facts' confirmed Pitts and McCulloch's theories, the jury is still out. Intracerebral EEG scanning is no longer accepted, but some of the processes they ascribed to visual cortex may actually go on there.

(In the Introduction to his *Cybernetics*, Norbert Wiener reported, with no little satisfaction, that the leading neurophysiologist Gerhardt von Bonin, on catching sight of the Pitts–McCulloch paper lying on his desk, had immediately asked him: "Is this a diagram of the fourth layer of the visual cortex of the brain?"—Wiener 1948: 22. In truth, it was a diagram of the abstract architecture of their computational theory, not of the brain as such. But the similarities were hardly surprising, and Wiener's satisfaction scantly justified, since their theory had been inspired by the neuro-anatomy in the first place.)

Current biologically informed models of "how we see universals" are hugely more subtle and complex (see Sections vi–vii, below). Nevertheless, Pitts and McCulloch had made the first serious attempt to locate specific computational functions in particular cell groups in the brain. As Christopher Longuet-Higgins said in a different context (12.v.c), "a demonstration of how something could possibly happen . . . may be almost as welcome as the discovery of how it actually does happen".

McCulloch, neuroscientist though he was, would have agreed. According to his close colleague Lettvin, he himself had little desire to test his new theories by the traditional methods. To be sure, he and Lettvin would co-author an experimental report ten years later that ruled out one of the mechanisms he'd posited in 1947 (Lettvin *et al.* 1959: 253). But McCulloch's general attitude to experimental neuroscience had turned to disappointment. He now preferred an unconfirmed but mathematically coherent theory to a ragbag of empirical facts:

[By the mid-1950s] he was dedicated to knowing how the brain works in the way that the creator of any machine knows its workings. The key to such knowledge is not to analyze observation but to create a model and then compare it with observation by mapping. But the poiesis must come first, and McCulloch would rather have failed in trying to create a brain than to have succeeded in describing an existing one more fully. (J. Y. Lettvin, in Dupuy 2000: 137)

Another person newly critical of the traditional methods of neuroscience was Gregory, who'd worked on the development of radar in the war. At Uttley's NPL meeting on 'The Mechanization of Thought Processes', and in a discussion in Cambridge a few years later, he caused no little consternation by giving an engineer's perspective on brain ablation studies (Gregory 1959, 1961).

His critique was based in his understanding of complex systems—such as a radio set:

Suppose that when [a] condenser breaks down, the set emits howls. Do we argue that the normal function of the condenser is to inhibit howling? Surely not. The condenser's abnormally low resistance has changed the system as a whole, and the system may exhibit new behaviour, in this case howling. (Gregory 1959: 678)

What's needed instead is a theoretical model of how the brain is working (and what it's doing), in terms of which the data can be interpreted. This applied also, he said, to experiments based on electrical stimulation; but ablation studies had led to especially shaky inferences.

Gregory was—and still is—sometimes accused of having "made clear [that] neuroanatomical investigations would have to be abandoned in favour of the electrical models pioneered by Craik and Ashby" (Hayward 2001*b*: 304). This is nonsense. He didn't rule out neuro-anatomical investigations, nor even ablation experiments. Rather, he used a cybernetic argument to warn against simplistic interpretations of them.

Even in the twenty-first century, such warnings are sometimes necessary. Several have been issued recently, applying to "double dissociations" in clinical cases, to animal experiments, and to studies of deficits in vision, face recognition, and dyslexia (M. P. Young *et al.* 2000; Shallice 1988*a*, chs. 10–11; McClelland 2000, pp. xxi–xxii). McClelland, for instance, puts it like this:

[An] explicit computational perspective often leads to new ways of understanding observed [neuropsychological] phenomena that are apparently not always accessible to *those who seek to identify subsystems without giving detailed consideration to the mechanisms involved.* . . . [Several examples illustrate that] the inference from data to the modular architecture of the mind is not at all straightforward, and that explicit computational models can provide alternatives to what in some cases appears to be *a fairly simplistic reification of task or item differences into cognitive modules,* and in other cases manifests as *a reification of types of errors (semantic, visual) into lesion [sites].* (McClelland 2000, pp. xxi, xxii; italics added)

So Gregory shouldn't be accused either of saying that brain ablation is a waste of time, or of issuing unnecessary warnings about reification. As a historical point, however, his strictures may have helped fuel the resentment of those brain scientists who later criticized the hype attending AI models in general, and simplistic 'brainlike' networks in particular (see Section vi.c).

b. Computations in the brain

By the early 1950s, and partly because of McCulloch and Pitts' two papers, more neurophysiologists were thinking about neural functions in informational/computational terms. One of these was Uttley (who *did* combine them with traditional experimental methods, as we'll see).

As explained in Chapter 12.ii.c, he was already using Claude Shannon's theory of information to describe neuronal function, and to design computer models. His "informon theory", developed through the 1960s and 1970s, represented synapses as calculating and communicating information—and also made various suggestions about the neurophysiology involved (Uttley 1975, 1979).

Uttley defined synaptic conductivity in terms of average impulse (spiking) frequencies, changes in which were the physical basis of learning. And he argued that synapses must contain three distinct counters, probably implemented as three neurochemicals of varying concentrations, in order to estimate the probability of co-occurrence of pre- and post-synaptic impulses. It is because these counters have to sample over time, and are imperfect ("leaky") anyway, that learning takes many minutes or even hours. (Long-term memory, he suggested, requires a second mechanism, involving some near-irreversible chemical process that fixes the short-term memory.)

He argued, too, that negative feedback from inhibitory neurones is needed to stabilize conductivities. Synapses are essentially variable, nevertheless, so learning is complemented by forgetting (extinction):

The reinforcing signal to an informon corresponds to the unconditioned stimulus which reinforces a conditioned stimulus in animal learning experiments; without such a reinforcing signal synaptic conductivity decays; correspondingly an animal forgets what it has learned. (Uttley 1982: 30)

Remembering, on this theory, isn't threatened only by confusion, where one and the same neurone group is asked to store too many different representations (cf. Chapter 12.v.c–f). Even a single memory would decay in the absence of continual reinforcement.

Uttley actually designed a dynamic computer memory based on this principle. A photocell looked at a cathode-ray tube screen displaying the output, and its signal went through a mercury delay line to the CRT input. This feedback loop continually refreshed the CRT image, maintaining the dots on the screen (R. L. Gregory, personal communication).

He even tried to provide new data, collaborating on various experimental studies of synaptic conductivity. Some of these were done with Benedict Delisle Burns, who'd argued (in the 1950s) that some synapses must have more complex properties than current models of learning allowed (Delisle Burns 1958: 96 ff.). Part of Delisle Burns's reasoning was based on the recent discovery of noise in the nervous system, which moved neurophysiology from deterministic ('telephone exchange') to stochastic theories (see 2.viii.f).

Uttley's work reflected this new approach. He tried to explain why (as he put it) "the brain ticks by itself", and to clarify the functional implications. And he offered accounts—and simulations—of the action of specific cells, such as the basket and granule cells of the hippocampus and the orientation detectors in visual cortex (discovered in the late 1950s: see Section iv.b). For instance, he modelled the newborn kitten's (fairly poor) 'innate' orientation detectors, and the learning processes that improve these networks as the kitten encounters straight lines and boundaries. (As for whether these detectors really were *innate*, see Section x.a and c, below.)

In short, Uttley made a serious attempt to take account of the neurophysiological data, in considering the different *types* of synapse that might exist in the brain. His early work was described as "an invaluable contribution to the study of learning in animals" (Delisle Burns 1968: 151).

One young neurophysiologist strongly influenced in the 1950s by Uttley's use of information theory was his fellow Ratio member Barlow (1921–), a great-grandson of Charles Darwin. Indeed, Barlow now says, "I was very fortunate to have belonged to [the Ratio Club], for when I come to think about it I can see that much of what I've been interested in since has stemmed from those evenings" (personal communication). Others included Patrick Merton, also a Ratio member (see Chapter 12.ii.c), and Jack Cowan. Cowan went from Imperial College to spend a year with Uttley at NPL before moving to Chicago to direct Rashevsky's group (12.ii.b).

Barlow was then a researcher at the University of Cambridge (and a Royal Society Research Professor there, years later). He first outlined his hugely influential "coding" theory of perception at the same two meetings where Gregory had ridiculed howl-inhibitors (Barlow 1959, 1961).

He cited Craik (see Chapter 4.vi), saying that the brain must build a "model" of the external world, which can be used to guide appropriate behaviour (1959: 542). He wanted to show what such an internal model must be like, and how it might actually be built. So, borrowing Shannon's idea of optimal coding, he asked how a machine might be designed to reduce redundancy when recording sensory information. In other words, how can a machine—or a nervous system—find simplicity in complexity?

Low-level perceptual systems (such as the retina), said Barlow, appear to possess *built-in* mechanisms for doing this with respect to types of redundancy that are always present in the environment. But high-level perception—of one's grandmother, for instance (though Barlow didn't mention that example)—requires that the nervous system be able to *learn* to reduce the relevant redundancies. This, he argued, must involve changes in nerve signalling rather than neurone numbers:

> In the nervous system the number of nerve fibres available for a specific task must, to a large extent, be determined genetically. One may expect evolutionary adaptation to have performed part of the [communication] engineer's job in selecting codes for the sensory signals, but such inherited codes obviously cannot be adapted to the redundancy of sensory input which is peculiar to each individual. Now although the number of nerve cells available is probably determined genetically, the number of impulses in the nerve cells is not, and some of the advantages of optimal coding would apply if the incoming information were coded—not onto the smallest possible number of nerve fibres each working at its optimal mean frequency—but into the smallest possible number of impulses in a relatively fixed number of nerves. This type of coding can be epitomized as *economy of impulses*. (Barlow 1959: 550)

Barlow was more interested in how perceptual redundancies, or patterns, could be recognized than in how they might be learnt. So he offered no learning rules. Nor did he claim that optimal coding had been 'proved' to happen. But he did give neurophysiological evidence for it, including experiments on adaptation and lateral inhibition—in vision, hearing, and touch. These showed that spiking frequencies can depend on specific neuronal mechanisms—such as the ON–OFF/movement receptors he himself had recently found in the frog's retina (Barlow 1953).

(A few years later, he would discover *directional* detectors in the rabbit's retina—Barlow *et al*. 1964; Barlow and Levick 1965. Again, he asked just what information was being computed, and how. And he showed how inhibition of one cell by another functions in the discrimination of direction of motion.)

As for intelligence, a word that "was added to the title in an incautious moment", Barlow saw this as largely continuous with perception. Concepts are relatively high-level input redundancies, and reasoning is based in redundancies too:

> [When] one considers the two main operations required for optimal coding there is a striking parallel with the two types of reasoning which underlie intelligence.
>
> The outputs of a code can be thought of as logical statements about the input, and, if the code is reversible [in a sense explained in his paper], these logical statements, taken together, are sufficient to determine the exact input. Forming these statements and ensuring that they fulfil this condition are straightforward problems of deductive logic . . . [As for inductive reasoning, this depends] on counting frequencies of occurrence of events. Having been presented with 1000

white swans and no black ones, the relevant parts of a code would say "henceforth regard all swans as white unless told otherwise". Inductively one would say "All swans are white." *The tools of logical reasoning appear to be the same as those needed for optimal coding,* so perhaps they can be regarded as the verbal expression of rules for handling facts which our nervous system uses *constantly and automatically* to reduce the barrage of sensory impulses to useable proportions. (1959: 555; italics added)

The emphasis in Barlow's coding theory, as in Uttley's work, was on the brain's measurement and comparison of the *probabilities* of stimuli. This approach, Barlow has recently recalled, fell into the background for many years while people focused instead on how the brain generates *transformations* of physical stimuli (Barlow 2001*b*). But the original emphasis, he now thinks, was sound. Although earlier theories of perception were "wrong in over-emphasizing the role of compressive coding and economy in neuron numbers", they were "right in drawing attention to the importance of redundancy" (see Section ix.e, below).

In short, "Think probabilities":

In any [perceptual] representation probabilities are key elements, because they are the fuel for accurate decision making. *So the take-home message for the neuroscientist should be: "Think probabilities*: What probabilities are needed? How are they represented? How are they estimated? How are they modified? How are they transmitted to other places in the brain? And how are they combined for making the moderately rational decisions that we observe brains making[?]" (Barlow 2001*b*: 251; italics added)

(As we saw in Chapter 12.vi.f and ix.d, "thinking probabilities" is just what the mathematical connectionists had been doing.) Barlow still recognizes his fifty-year debt to Uttley, whose name is one of nine mentioned in the Acknowledgments of his 2001*b* paper.

c. Formal synapses

One of Barlow's colleagues at Cambridge—and at the Ratio Club—was the physiologist Giles Brindley, who worked there until moving to London in 1968. In the 1960s, Brindley developed a formal theory of "modifiable" synapses. And Mayhew would have been happy, for Brindley was asking *just how* learning can be achieved by the brain.

Specifically, he was asking—as Uttley (also a Ratio member) had done before him—(1) what different types of synapse might exist, and (2) how their computational properties compared.

Brindley started from Donald Hebb's (1949) hypothesis about learning (5.iv.b). But Hebb's ideas had been expressed verbally, not mathematically, and Brindley (1967) pointed out that they could cover many distinct types of synaptic change. For instance, the neurones involved might be excitatory or inhibitory, and the modification might affect the pre-synaptic and/or post-synaptic cells. So several "Hebbian" learning rules were conceivable. Indeed, four had been described already (though only one had been defined formally).

Brindley suggested that different types of observable behaviour—such as classical and operant conditioning—might involve different learning rules. And if forgetting was to be explained, then extinction rules were needed too. Rashevsky's journal had reported a model of Pavlovian conditioning as early as 1950, but it hadn't allowed for extinction.

By exploring the possible combinations, Brindley defined ten types of synaptic modification. He showed that their logical input–output relations fell into three classes: A, B, and C. (The crucial question was whether the set-theoretic intersection of the two inputs was equal to one of them, or zero, or neither.) Six of the ten synapses were class A, three B, and one C. And these classes were systematically related:

[Any] two members of the same class can, with the aid of non-remembering elements that perform simple logical operations [i.e. McCulloch–Pitts neurones combined as *and*, *or*, and *and-not* logic gates], replace one another in any net; but a member of class A cannot replace one of class B or C. A member of class B can replace one of class A or C only if non-logical elements, for example noise generators, are included in the net. (Brindley 1967: 361)

In effect, Brindley was giving a computational justification for the newly discovered noise in the nervous system (Delisle Burns 1968).

A mathematical Appendix described a wider range of modifiable synapses. These involved multiple inputs, closed (re-excitatory) loops, and temporal summation. They, too, could be formally classified in terms of their input–output relations.

Using this classification, Brindley designed (pencil-and-paper) models supporting different types of learning—such as classical and operant conditioning, both with and without extinction. His network for *classical conditioning with extinction*, for instance, used two modifiable synapses, the first of class B and the second of class A (p. 367). This was no accident, for he proved that if extinction was to be modelled at all then "classical conditioning *requires* modifiable synapses of classes A and B or of class C, and operant conditioning of class B or C" (p. 361; italics added). It followed that if Hebbian synapses are indeed involved in storing learnt information, then some of the nervous system's modifiable synapses *must* be of class B or C.

One methodological moral was that computational arguments based on the behaviour of the whole nervous system can't tell us just which synapses are present. For there are "a number of quite different but roughly equally plausible kinds of modifiable synapses that will do the same things (i.e. perform the same computations) in networks of nearly the same complexity" (p. 365).

But such arguments can suggest hypotheses about which *classes* of synapse are present, in what proportions. They prove, for example, that it is "uneconomical to use class B synapses to produce a class-A input–output relation, and infinitely uneconomical to do the reverse". And since the nervous system as a whole can exhibit class B input–output relations, it must (as noted above) involve synapses of class B or C.

In a follow-up paper, Brindley (1969) discussed the construction of much larger networks, in which only a few simple details need to be accurately specified and which can learn indefinitely many different things. (All previous models had been devoted to a single learning task.) And the *numbers* of neurones now entered the picture:

The number of cells required for performing tasks of the kind considered [e.g. verbal rote learning] as well as the human brain can perform them, is only a small fraction of the number of cells in the brain.

[The] models proposed [here] are likely to be the most economical possible for their tasks, components and constructional constraints, and . . . any others that approach them in economy must share with them certain observable features, in particular an abundance of cells with many independent inputs and low thresholds. (Brindley 1969: 173)

Various experimentally testable predictions about synapses and cell numbers could be generated accordingly. In other words, this computational approach not only 'made sense' of the brain, but promised to help illuminate its detailed structure.

Among those whom Brindley acknowledged at the end of his 1969 paper were Barlow (already widely respected) and two unknowns: his students Stephen Blomfield and Marr. As we'll see in Section v.c, Marr was soon to publish the first formal theory of the brain *as such*. Eventually, because of his later research (on vision), he would become a household name for cognitive scientists in general. Brindley, by contrast, remained largely unknown.

But his work had been crucial. It confirmed what McCulloch, Uttley, Barlow, and Gregory had already argued: that *traditional neuro-anatomy must be combined with computational analysis*, if the nervous system is to be understood.

14.iv. A Fistful of Feature-Detectors

If Barlow's theory of perceptual coding was important for the rise of computational neuroscience, so was his practical work. Indeed, some of his early experiments led indirectly to perhaps the most famous paper (and certainly the best title) in the field.

Neurophysiologists had known since the late 1930s, thanks to Keffer Hartline at Rockefeller University, that some retinal cells in frogs aren't mere passive receptors of points of light. Rather, they respond to simple patterns of illumination: ON, OFF, or ON–OFF (Hartline 1938). In the early 1950s, Barlow reported in the *Journal of Physiology* that such cells are more sensitive to edges (light contours) than to unstructured light, and that the ON–OFF variety are especially sensitive to movement (Barlow 1953). Half-jokingly, he even called these cells "bug detectors".

In 1959 that term would start to ring still-resounding bells.

a. Bug-detectors

Most people today associate bug-detectors not with Barlow, but with the four authors of 'What the Frog's Eye Tells the Frog's Brain': Lettvin, Humberto Maturana, McCulloch, and Pitts (1959).

McCulloch had been a founder of the cybernetics movement (4.v–vi), and the 'Frog's Eye' work was done at the hub of cybernetics: MIT. So it's prima facie not surprising that although three of the four authors were neuroscientists, the paper appeared—like many 'cognitive' pieces at that time—in a journal officially aimed not at biologists but at radio engineers. However, this didn't happen by choice. Lettvin remarked in a lecture, years later, that they'd tried to get the paper published in several biological journals but were "laughed off the stage". The reason was that they saw coding not as all-or-none but as implemented by variable *bursts* of nerve signals (John Collier, personal communication).

The paper's message was as memorable as its title. For it showed that the frog's eye has a great deal to tell its brain, over and above the presence of light—or even of Barlow's edges and movement.

As Lettvin's group put it: "the eye speaks to the brain in a language already highly organized and interpreted, instead of transmitting some more or less accurate copy of the distribution of light on the receptors" (Lettvin *et al.* 1959: 251). Specifically, they had found single cells in the frog's retina that were sensitive to one of four different light patterns:

(1) local sharp edges and contrast; (2) the curvature [within specific limits] of edge of a dark object; (3) the movement of edges; and (4) the local dimmings produced by movement or rapid general darkening. (p. 253)

These results were observed even when the stimulus, instead of being artificially isolated (as in Barlow's experiments), was surrounded by a *naturalistic* image of the countryside as presented to a frog.

The 'Frog's Eye' authors also found that the four cell-types mapped onto four distinct cell sheets in the frog's brain (the optic tectum, or colliculus). And these were precisely co-registered: neighbouring points in the retina were connected to neighbouring points in sheet 1 . . . and sheet 4. What's more, the nerve fibres would grow back to the right place if they were cut. The idea of *some sort of* point-to-point mapping had been suggested by Julia Apter, and by Pitts and McCulloch in their 1947 paper (see 12.i.c). Now, there was detailed experimental evidence. (The explanation for *how* this mapping is established was then unknown: see ix.a, below.) And there were various subtleties. For instance, some class 3 cells fired only if the movement was in a certain, very broadly defined, direction.

In fact, there were some subtleties so surprising that they weren't even mentioned in the published paper. Lettvin recalls:

What we did not report, and what to me is still [in 1994] the most astonishing thing about the bug detectors, is the following property . . . You bring one spot in, move it into the [receptive] field, and as long as you move it around, wonderful response.

You bring in two spots; if they're rigidly coupled in their motion by a fixed distance between them and are moved around, you get a good response almost as if they are only one spot.

You bring three rigidly coupled spots in, and it doesn't matter their size or their disposition or their distances from each other: move them around as a rigidly coupled triad and there's no response at all.

. . . If you have a white background and two black spots, OK; three spots, forget it. You connect any two spots of the three by a barely visible black line, and all of a sudden it's a two-spot system. It becomes visible. Or else you move any one spot with respect to the other two, now there's a response. (J. A. Anderson and Rosenfeld 1998: 18)

How, Lettvin asked himself, could a mere retina possibly compute such distinctions? He still has no answer—so still hasn't officially published the finding. In his judgement, then, this discovery wasn't really a discovery (see Chapter 1.iii.f). Putting it in Mayhew's terms, he'd discovered the cells, and what they were computing—but not *how* they were computing it.

What's more, it seemed that the eye might be telling the brain very interesting things—interesting, that is, *to the frog*. These concerned the approach of predators (casting a shadow over the scene: see class 4 above), and the whereabouts of food:

The operations thus have much more the flavor of perception than of sensation if that distinction has any meaning now. That is to say that the language in which they are best described is the language of complex abstractions from the visual image. We have been tempted, for example, to call the convexity detectors "bug perceivers". Such a fiber responds best when a dark object, smaller than a receptive field, enters that field, stops, and moves about intermittently thereafter. The response is not affected if the lighting changes or if the background (say a picture of grass and flowers) is moving, and is not there if only the background, moving or still, is in the field. Could one better describe a system for detecting an accessible bug? (Lettvin *et al.* 1959: 253–4)

Indeed, the retinal cells seemed to guarantee the frog a healthy diet of fresh food. They responded only to stimuli that were moving, so—in its normal environment—probably *alive*. This tallied with previous observations of the whole animal's behaviour:

The frog does not seem to see or, at any rate, is not concerned with the detail of stationary parts of the world around him. He will starve to death surrounded by food if it is not moving. His choice of food is determined only by size and movement. He will leap to capture any object the size of an insect or worm providing it moves like one. He can be fooled easily not only by a bit of dangled meat but by any moving small object. (p. 231)

Remarkably, the radius of curvature that strongly triggered the convexity detectors corresponded to bug size. In short, the frog's eye was seeing only what the frog needed to know.

Later, Lettvin would complain about the use of such "anthropomorphic" language even in talk about human brains, adding that "people have screwed up the frog because they're taking bachtriomorphic views of the frog. If you really want to understand the frog, you must learn to be objective and scientific" (J. A. Anderson and Rosenfeld 1998: 344). The 'Frog's Eye' paper itself, however, had initiated this perspective.

The 'Newtonian' assumption inherited from the associationists, and dating back to John Locke's distinction between "sensation" and "reflection" (Chapter 2.x.a), had been that the sense organs—and by extension, the peripheral parts of the brain—are both passive and atomistic. Sensitivity to points of light, *yes*; to complex patterns of light, *no*. To perceive (infer, compute) a light–dark gradient with a specific radius of curvature, it had been tacitly assumed, was a matter for the cerebral cortex—present only in mammals and birds—and must involve a large number of cells. Now, that assumption was shattered.

But the surprise, though great, wasn't unqualified. Recent work in both neurophysiology and cybernetics/AI had implied that some such results might be found. Barlow's experiments a few years before had modified the assumption already. Indeed, others besides the MIT team were looking for feature-detectors of some sort (see below). And Pitts and McCulloch (1947) themselves had posited several layers of visual computation in the brain, going from simple features to increasingly complex concepts.

(Admittedly, they'd assumed that the features would be relatively simple, and formally tractable. Pitts was therefore, so Lettvin recalls, "a little appalled by the results, which were very different from what he expected", and "while he accepted the work enthusiastically, at the same time it disappointed him"—J. A. Anderson and Rosenfeld 1998: 10.)

Moreover, many cyberneticians—though not the orthodox neurophysiologists—had encouraged attention to what *the animal* needs to know, given what it wants to do. In particular, Lettvin's mentor McCulloch had taught him to think of sensory neurophysiology as "experimental epistemology". (I remember being startled to find these words written on the door, on my first visit to Lettvin's lab.) He meant that one should ask not merely how perception tells animals about the world, but also how it makes appropriate action possible (Chapter 4.iii.b).

A visual neurophysiologist who became close to the MIT team a few years later was told by Lettvin that "Pitts and McCulloch actually were on the eye/brain paper as a friendly courtesy. They had no role in the planning or execution or anything of the eye/brain thing" (C. G. Gross, personal communication). However, McCulloch's lack of hands-on contributions doesn't rule out his having had a background influence on Lettvin's thinking.

Similarly, Selfridge had argued—on computational grounds, and partly influenced by Pitts and McCulloch—that complex perceptions *must* be based on many distinct feature-detectors (or "data demons"), where the "features" have survival value for the animal. And he'd illustrated part of what he meant by the Pandemonium program. (Only "part", because the program didn't need to survive, so didn't *want* anything: see Chapter 12.ii.d and iv.b.)

Lettvin was a very close friend of Selfridge, and of McCulloch and Pitts. Their mutual friendship had led to the 'Frog's Eye' work in the first place. The paper described itself as "an outgrowth" of the 1947 study of universals, which—in effect—had postulated layers of feature-detectors. In addition, it carried this acknowledgement: "We are particularly grateful to O. G. Selfridge, whose experiments with mechanical recognizers of pattern helped drive us to this work and whose criticism in part shaped its course" (Lettvin *et al.* 1959: 253, 254). In short, although the authors were doing "wet" neuroscience, not describing a computer model, they were plainly inspired by computational ideas.

What actually happened, according to Maturana (personal communication), began with Maturana's accidental discovery (in November 1958) of movement-sensitive cells in the frog's retina. He found these interesting because they confirmed his hunch that such cells might exist, a hunch based on purely anatomical grounds (microscopic evidence of asymmetries in the connection patterns of some of the retinal cells). But Lettvin's familiarity with Selfridge's ideas enabled him to see the wider importance of Maturana's finding. He immediately switched the laboratory's main research programme to a hunt for other specialized, and ecologically significant, cells—and they found them.

A further surprise was in store. Within a few years, Maturana rejected the computational interpretation of these experiments. He developed their whole-animal emphasis into a neo-Kantian philosophy of biology, in which the concepts of computation and information had no place (see 15.vii.b and 16.x.c). Orthodox neuroscientists, however, knew little or nothing of this—and cared less. Even with Maturana's blessing, neo-Kantianism was out of fashion.

b. And more, and more . . .

The whole-animal emphasis was unusual in the 1950s: most people looking for feature-detectors at the time weren't guided by it. David Hubel and Torstein Wiesel, for instance, were not.

They'd arrived at Stephen Kuffler's lab at Johns Hopkins (where Barlow was also working) in 1958, and were searching not in the frog's eye but in the cat's brain. Their discovery of feature-detectors in the striate cortex in effect confirmed Turing's hunch, that "the apparatus for the more elementary analysis of shapes and sounds probably comes into this category [i.e. definite structures in the brain]".

Their results, which revolutionized the study of visual cortex, were even more surprising than the MIT group's were. In the late 1950s they identified brain cells responsive to very highly specific patterns on the retina: lines/edges of many different orientations (Hubel and Wiesel 1959). More accurately, they found cells that responded to small line *segments*. Anything we'd normally regard as a 'line' would be represented by a *set* of 'line-detectors'. (Twenty years on, Hubel confessed that how the fragmentary information from such cells is assembled to build a percept of a line "is still a complete mystery" Hubel 1982: 519.)

Later (at Harvard), they would find even more types of feature-detector, at various levels in visual cortex. (Right from the start, they produced putative circuit diagrams for their various cell-types.) The levels seemed to be functional as well as anatomical, for some neurones, the "hypercomplex" cells, were fired by particular *combinations* of lower-level features. Moreover, the orientation detectors were arranged in regular columns, perpendicular to the brain's surface: same orientation, same column; and cells responding to similar orientations inhabited neighbouring columns (Hubel and Wiesel 1962, 1968, 1974).

They also found that many cells in visual cortex were binocular, whereas those in the first way-stage (the LGN, or lateral geniculate nucleus) were 'wired' to only one eye. The binocular cells were evidently wired with remarkable precision, each having two receptive fields closely similar in size, complexity, orientation, and position. However, most of them responded more strongly to one eye than to the other: Hubel and Wiesel called this effect "ocular dominance".

Their initial discovery had amazed them as much as anyone else, for they hadn't been expecting "features" quite like that. A year after receiving the Nobel Prize for their work, Hubel recalled:

We had been doing experiments for about a month . . . and were not getting very far; the cells [in the cat's cortex] simply would not respond to our [two stimulus-classes, namely] spots and annuli. One day we made an especially stable recording . . . The cell in question lasted 9 hours, and by the end we had a very different feeling about what the cortex might be doing. For 3 or 4 hours we got absolutely nowhere. Then gradually we began to elicit some vague and inconsistent responses by stimulating somewhere in the midperiphery of the retina. We were inserting the glass slide with its black spot into the slot of the ophthalmoscope when suddenly over the audiomonitor the cell went off like a machine gun. After some fussing and fiddling we found out what was happening. The response had nothing to do with the black dot. As the glass slide was inserted its edge was casting onto the retina a faint but sharp shadow, a straight dark line on a light background. That was what the cell wanted, and it wanted it, moreover, in just one narrow range of orientations.

This was unheard of. *It is hard, now, to think back and realize just how free we were from any idea of what cortical cells might be doing in an animal's daily life* …

It took us months to convince ourselves that we were not at the mercy of some optical artefact … We did not want to make fools of ourselves so early in our careers. (Hubel 1982: 516; italics added)

Orientation detectors, bizarre though they seemed, were just the beginning. Hubel and Wiesel themselves soon discovered detectors for directional movement, for instance. And then, *mirabile dictu*, other researchers found cells higher up in the visual system (in inferotemporal cortex) that were responsive to hands, or to faces. These discoveries were amazing—indeed, apparently incredible (see below). But they weren't totally unexpected, and nor were they accidental.

The Warsaw neurophysiologist Jerzy Konorski (1903–73) had been the first to suggest, in 1960, that highly complex features such as these (which he called "unitary perceptions") might be coded by single neurones, or "gnostic units". In his book a few years later, he'd argued that these units might be grouped in separate places in the inferior temporal cortex. What's more, he speculated that there were different units coding for faces, limbs, animals, handwritten letters, or emotional expressions, and so on (Konorski 1967).

Konorski's ideas were still maverick in the early 1960s, when "most visual psychologists had never heard of this area [inferotemporal cortex]" and thought of the striate cortex alone as the ground of vision (C. G. Gross 2002: 86). But they hadn't come out of nowhere. He'd been inspired not only by Hubel and Wiesel's findings but also by Karl Pribram's work on the visual effects of lesioning monkey temporal cortex, and by the descriptions of highly specific clinical agnosias then coming from Luria in the Soviet Union (Pribram and Mishkin 1955; Luria 1968).

Maverick or not, Konorski was largely right: we can now see that he "anticipated subsequent discoveries to an amazing degree" (C. G. Gross 2002: 85). Even so, and despite a highly complimentary review in *Science* (C. G. Gross 1968), Konorski's book was largely ignored by visual psychologists and neurophysiologists:

For at least the next decade virtually all the many citations to the book were to the parts concerned with learning rather than with perception; learning theory [aka behaviourism] still dominated American psychology. (C. G. Gross 2002: 87)

One American who didn't ignore the parts concerned with perception (and who wrote the glowing *Science* review) was Charles Gross (1936–). While still based at Cambridge University as a graduate student (see Preface, ii), he'd visited Konorski's Warsaw lab in 1961 to learn about his strange new ideas.

Gross was as excited as anyone else by Hubel and Wiesel's recent discoveries. And he knew (from lesion studies like Pribram's, being carried out at Cambridge by his supervisor Larry Weiskrantz) that inferotemporal (as well as striate) cortex was crucial for vision (personal communication). So on moving to MIT a few years later, he decided to try to find further feature-detectors and/or gnostic units in that part of the brain—which he did.

But he could hardly believe his own eyes. He found a cell fitting one of Konorski's speculations: well, not for a limb—but for a hand. When Gross first drafted his research paper, he "did not have the nerve" to mention the cell that coded for the monkey's hand (C. G. Gross 1998: 199). He was as wary as any seventeenth-century gentleman of witnessing to the astonishing (see Chapter 2.ii.b). Like Hubel and Wiesel before him, he didn't want his career to be over before it started because of being seen as either sensationalist or gullible. A senior colleague—Hans-Lukas Teuber, a member of the cybernetics group—encouraged him to be brave. As a result, the hand-detector was duly announced (C. G. Gross *et al.* 1969).

However, it evidently beggared belief. People didn't even *try* to replicate it. No one else reported any follow-up experiments for over a decade. And Gross himself, "perhaps because of the general skepticism", waited just as long before fully describing the face-detectors (Bruce *et al.* 1981). Yet they, too, had shown up in his initial experiments.

By then, the scepticism had lessened. A flood of papers soon appeared on visual (and auditory) single-cell detectors in various parts of the monkey's cortex. They included many from David Perrett's lab at the University of St Andrews, Scotland (Perrett *et al.* 1982, 1985, 1990, 1992).

By the mid-1990s, cells had been found which discriminated (for instance) hands; faces; facial expressions; humans walking towards or away from the monkey subject; and anyone—human, monkey, or robot—picking up a raisin or performing some other action salient to monkeys. In addition, 'bimodal' detectors had been discovered, which responded to *visual and tactile* stimuli from the same body area, such as the arm and the space near to it (Graziano and Gross 1993).

Even the humble (humble?) retina had offered up further riches. Today, neuro-scientists can distinguish "over 70 retinal neurons, well-differentiated in terms of morphology, connections, and transmitters and therefore presumably function too" (C. G. Gross, personal communication).

As that word "presumably" implies, not all of these retinal cells can be classi-fied—yet—as detectors, for we don't know what they detect. But the fact remains that neural detectors of many different kinds, including the recently discovered mirror neurones (Section vii.c, below), are still being found in the nervous system.

c. But how?

As Mayhew's remark reminds us, however, the discovery of a single-cell detector doesn't tell one *how* that detection is achieved.

Of course, systematic experiments were done right from the start to identify the relevant stimulus classes. Lettvin's team, for example, studied many different stimuli, including diverse radii of curvature and various types of (smooth/jerky, fast/slow) movement. Similarly, Gross used a range of stimulus classes in testing the monkey's hand-detector (see Figure 14.1). Such experiments indicated which aspects of the stimulus are important. But they didn't show how they are computed. (Even now, the precise circuitry isn't known.)

By the mid-1980s, the biological experiments were being compared with specific computational theories. So, for example, when the St Andrews group discovered a wide variety of face-detector cells in the brain, they discussed them in the light of Marr's theory of vision (Chapter 7.v.b–d), Marvin Minsky's ideas about visual frames (10.iii.a), and Robert Baron's (1981) computational model of face recognition.

FIG. 14.1. Examples of shapes used to stimulate an inferior temporal cortex neuron "apparently having very complex trigger features. The stimuli are arranged from left to right in order of increasing ability to drive the neuron from none (1) or little (2 and 3) to maximum (6) . . . The use of [these] stimuli was begun one day when, having failed to drive a unit with any light stimulus, we waved a hand at the stimulus screen and elicited a very vigorous response from the previously unresponsive neuron . . . We then spent the next 12 hrs testing various paper cutouts in an attempt to find the trigger feature for this unit. When the entire set of stimuli used were ranked according to the strength of the response that they produced, we could not find a simple physical dimension that correlated with this rank order. However, the rank order of adequate stimuli did correlate with similarity (for us) to the shadow of a monkey hand" (C. G. Gross *et al.* 1972). Reprinted with permission from C. G. Gross (1998: 199)

Baron, a neurophysiologist at the University of Iowa, had worked on formal models of the nervous system since the early 1970s. In his account of face recognition, he'd posited various viewer-centred templates, such as distinct views of the head as it would appear at different stages of rotation. A face can be recognized (by brain or computer), he said, only if such templates are present, and "In every case, recognition is determined by correlating the processed input patterns against all stored templates" (Baron 1981: 145).

Before template comparison could get started, the system had to fixate on the relevant face features (if possible, the eyes first), and to standardize the image size. Since the relevant features were fixed (eyes, nose, chin, etc.), the program's task wasn't to distinguish between faces and other objects, but between this face and that one: Janet or Jill. Each face was represented in memory by up to five features, with up to four distinctly different templates for that feature—hence up to twenty templates in all. And, in accordance with experimental data on human eye movements, recognition was sometimes "Gestalt" (immediate), sometimes sequential (feature by feature).

Baron suggested that the various types of abnormal face recognition observed in clinical patients, such as prosopagnosia (where even one's spouse may not be recognized), are caused by breakdowns at different points within the system (pp. 171–5). He assigned specific computational roles to cells in the retina, lateral geniculate bodies, primary visual cortex, and cortex.

When discussing the types of feature-detector involved in locating a face in the first place, he said:

It is perhaps no accident that the eye responds so readily to patterns having a dark center and light surround. The iris of the eye, eye-ball, eye socket, eyebrow, face, and hairline form a series of light and dark concentric circular regions that can easily be detected in the visual field. (p. 168 n.)

As this remark implies, he generalized his theory to visual recognition in general. And he stressed that in outlining the brain's "logical architecture", one must "carefully distinguish between the information and control processes involved" (p. 138). In short, Baron was addressing Mayhew's question about *how* a given cell computes what it does.

d. Monkey business

Perrett's group in Scotland, part-inspired by Gross and Baron, made some remarkable discoveries (Perrett *et al.* 1985). Although some cells were responsive to faces as such, others responded only to faces in particular orientations: full-face, profile, looking upwards, or looking downwards. Moreover, many of these were sensitive also to the direction of gaze, coding full-face-with-eye-contact, full-face-with-averted-gaze, or profile-with-averted-gaze.

The team suggested that these are monkeys' business. In other words, they are important social signals for monkeys, as they are for humans.

(With respect to humans, these researchers would later do hugely influential work on universal/cultural facial recognition, including a study of the attractiveness of facial symmetry—a topic discussed in Chapter 8.iv.b: Perrett *et al.* 1999; Penton-Voak *et al.* 1999; Keysers and Perrett 2004. The results of their experiments in presenting subjects with systematically varied computer simulations have been widely applied in Hollywood animation, and for VR faces too (see 13.vi): Burt and Perrett 1995; Calder *et al.* 2000.)

They observed, in separate experiments, that young monkeys "give appeasement gestures at an increased frequency for faces making eye contact compared with faces where the eyes are averted" (Perrett *et al.* 1985: 309). But, for social purposes, it presumably doesn't matter (to monkeys) whether the eyes are averted by 20, 50, 60, or 70 degrees. And indeed, the experimenters *didn't* discover a continuous range of eye-aversion detectors, analogous to Hubel and Wiesel's line-orientation cells:

It is important to note that in the present study no cells were found that were maximally active to the face positioned at some intermediate angle between face and profile. *Only a small number of distinct views of the head seem to be receiving independent analysis in [this part of the brain]* but for each view there is considerable generalization over size and isomorphic rotation. (p. 315; italics added)

In general, they said, one might expect to find cells sensitive to features relevant to the animal's lifestyle:

To avoid jumping into the lion's mouth one has to know which direction to jump. The importance of such information might have encouraged the evolution of neural mechanisms capable of direct and independent computation of the orientation of the head and body with respect to the observer and hence a viewer-centred analysis of these objects. (p. 315)

One must remember, here, that Perrett was studying monkeys. Most animals wouldn't be able to see *the lion*, even if they were able to avoid its mouth. In other words, very few animals—only birds and mammals—can do pattern recognition, or see whole objects such as lions or faces. All the others use "slick tricks" instead, including many species-specific feature-detectors.

For a crab, for example, if something is above the horizon it's dangerous and if it's below the horizon it's a crab. Or rather, the action that the visual stimulus evokes in the crab—either freezing/escape or claw-waving—is appropriate as a response to seagulls or crabs, respectively. The size and shape of the object involved don't matter, because they can't be perceived.

Moreover, the crab doesn't need any "neural 'software' image" to code the horizon. Instead, it's coded in such a way that "an object will be treated as a threat *if it appears on a particular part of the retina*, and thus the behaviour can be 'hard-wired' " (Layne *et al.* 1997: 52; italics added). That is, the crab has an appropriate slick trick. (Compare the crickets discussed in Chapter 15.vii.c.)

In relation to Marr's theory of object recognition, Perrett's group pointed out that whereas some cells were responding to faces as objects independent of the viewer, others were coding viewer-centred information. Because the oriented-face-detectors weren't representing ("making explicit") all the information available in the $2\frac{1}{2}$D sketch, their functioning was described as "a stage beyond two-and-a-half-dimensional sketches" (Perrett *et al.* 1985: 315).

Independent analysis of distinct views of the same object was necessary, they said. (*Computationally* necessary, that is.) But only relatively few distinct views may be needed (here they cited Minsky 1975). And, as remarked above, they'd found only very few views of eye gaze to be independently coded. Accordingly, they criticized Baron's model for requiring a distinct template for every 20 degrees of head rotation.

Their general conclusion was that "the recognition of one type of object may proceed via the independent high level analysis of several restricted views of the object (viewer-centred descriptions)" (p. 293). This, they said, challenged Marr's theory, which did *not* predict cells sensitive to certain views of faces. Partly, this was because viewer-centred representations were supposed to exist only at the level of the $2\frac{1}{2}$D sketch. Also, Marr had favoured distributed representations, considered as emergent properties of large groups of neurones (Chapter 12.v–vi). But Perrett's group had seemingly discovered paradigm cases of grandmother cells (individual neurones coding crucial information).

For our purposes, the important point is not whether Perrett's specific claims were right or wrong. What's important is their general nature. In a word, they were *computational* claims.

Such analysis simply couldn't have appeared in the early 1960s, when feature-detectors were both new and few. At that time, Marr and Minsky hadn't published on vision. There were no computer models of face recognition. And Barlow's theory of grandmother cells hadn't been formulated (see Section ix.e).

To be sure, by 1961 there was already a computer model of feature detection that was seen, by some, as biologically relevant. I'm thinking of Leonard Uhr and Charles Vossler's version of Pandemonium, which learnt its own basic operators (Chapter 12.ii.d). Citing Hubel and Wiesel, Uhr and Vossler described their work as embodying "relatively weak, [biologically] plausible, and 'natural-looking' assumptions" (1961*a*/1963: 265). The young Marr apparently agreed, for he would soon describe the model's operators as near-equivalents to what he termed "codons" in the brain (D. C. Marr 1969: 469)—see Section v.c, below.

Most neuroscientists at that time, however, would have disagreed. The general view—based on studies of newborn kittens—was that feature-detectors, and their anatomical organization within the brain, aren't learnt but innate. They might be fairly crude in the newborn animal, to be perfected in the critical period; and they might be lost through lack of use. (There was evidence for both of these claims.) But in essence, they seemed to be built in.

MIT's experimental epistemologists, following McCulloch, had expressed this in explicitly Kantian terms (Chapter 12.ii.c). Describing the organization of eye–brain connections, they said:

The way that the retina projects to the tectum [to the four co-registered sheets: see above] suggests a nineteenth-century view of visual space . . . By transforming the image from a space of simple discrete points to a congruent space where each equivalent point is described by the intersection of particular qualities in its neighborhood . . . every point is seen in definite contexts. *The character of these contexts, generally built in, is the physiological synthetic 'a priori.'* [. . . If] there is any randomness in the connection of this [retina–tectum] system, it must be at a very fine level indeed. (Lettvin *et al.* 1959: 252–3; italics added)

Such views fitted well with Barlow's seminal arguments about low-level coding mechanisms (see Section iii.b), if not with behaviourist psychology. And they would endure for some years. For they cohered with the newly fashionable nativism of the late 1960s (Chapter 7.vi.d–e), including the growing evidence for a host of anatomical specializations related to language (see Chapter 9.vii and Lenneberg 1967).

In sum, feature-detectors were unexpected examples of what Turing had called "definite structures" in the cortex. They were initially assumed—by Hubel and Wiesel (1962) as well as the MIT Kantians—to be genetically determined.

By the mid-1970s, this assumption would be overturned. For computer models would show that the structure of visual cortex might arise spontaneously (see Sections vi.b and ix.a).

14.v. Modelling the Brain

The Pitts–McCulloch paper of 1947 had presented a computational theory of (part of) the nervous system, but not a computer model. Twenty years later, McCulloch would help to design such a model. It was still impossible in the 1960s, however, to simulate the brain in any detail.

Nevertheless, this decade saw the beginnings of four ambitious—and interestingly different—research programmes in computational neuroscience. Each eventually involved computer simulation, and each grew closer to the biological data over the years. They focused on brain modelling (discussed here and in Sections vii and viii.b); neuro-ethology (Section vii); and psychological realism (Section vi).

They were pioneered by Marr, Arbib, Andras Pellionisz, and Grossberg. All four men were highly interdisciplinary. The first three turned to neuroscience (and in Arbib's and Marr's case, to psychology too) after being trained in mathematics and/or physics and engineering. Grossberg's initial training was in psychology and, despite his voracious reading in neurophysiology and mathematics, he always regarded himself (and still does) as a theoretical psychologist (personal communication).

In addition, the equally interdisciplinary McCulloch initiated a brain model in the 1960s that was perhaps the first computer simulation to be constrained by specific details of neuro-anatomy and neurophysiology. After lying largely forgotten for many years, it has very recently been resuscitated. Let's start with that.

a. The Mars robot

One of the earliest—and most perceptive—working simulations of a specific brain region was produced by an engineer, a neuroscientist, and a programmer: William Kilmer, McCulloch (the senior partner), and Jay Blum (Kilmer *et al.* 1969).

This program—named RETIC—was a model of the reticular formation (RTF), a structure in the vertebrate brainstem that was attracting much interest at the time. The received wisdom was that the RTF switched the animal between sleep and waking. But some had suggested that it might also switch between different modes of activity, or instinctive behaviours.

Under the microscope, it appeared to be a stack of tissue slices (likened to poker chips: cf. Ito 1984). There were visible interconnections within each slice and also between slices—and sensori-motor connections, too. But what the microscope couldn't show was *just what* pattern of connections would be capable of activity switching of the type proposed. That was the biological question driving the project.

There was a technological aspect too, for the program was intended as a controller for a Mars robot. Indeed, it was an early example of situated robotics (see Chapter 15.vii.a), although it was a simulation rather than an actual robot. The situatedness was a technological necessity, because of physics. Because radio signals between Earth and Mars travel slowly, neither new sensory information (picked up by the robot) nor new goals (suggested by the humans back home) can be swiftly communicated between the two planets. So a Mars robot must be largely autonomous, switching automatically from one activity to another as the local circumstances require.

GOFAI planning (then in its infancy: see Chapter 10.iii.b), used for NewFAI robots such as SHAKEY, was eschewed partly because it was too slow. In addition, however, the anatomical evidence didn't favour a central controller.

Many vertebrate species have an RTF but no cortex, which suggests that their behavioural 'decisions' aren't made top-down. For the same reason, a hierarchical program like Pandemonium, with its autocratic master demon, seemed biologically implausible. Rather, the lower-level units should somehow decide the outcome themselves, "shouting" not to some higher-level demon but to each other.

What was needed, then, was distributed, cooperative, computation. McCulloch had already described this in his 1947 paper (12.i.c). But he called it "redundancy of potential command". This terminology—and the idea itself—was a legacy from McCulloch's Navy days (4.iii.a). He'd learnt then that, in battle conditions, control can shift continually from one ship to another, according to the information they're receiving. Even the Admiral's flagship isn't all-powerful—nor, unfortunately, ever-present (Arbib 2000: 194).

The Mars robot was designed accordingly. Each RTF slice was modelled by a small network, or module, which estimated how likely (how appropriate) a particular type of activity was, given the current sensory input. These within-slice estimations were tentative, in that they could be influenced by between-slice connections. The final selection was distributed over the whole system. That is, the overall network would identify the maximum inter-module consensus, so choosing the most appropriate activity as its motor output.

In principle, one might have expected occasional confusion and/or paralysis, with no *one* activity being a clear winner. (Remember Buridan's ass: placed exactly halfway between two identical bundles of hay, and lacking 'freewill', the creature would supposedly starve to death.) In fact, this very rarely happened. Irrespective of the input, RETIC almost always converged to a majority-based choice of a single mode of action.

McCulloch's team (let's call them KMB) were making a crucial assumption, here. Namely, that distinct bodily activities have distinct neural mechanisms driving them, plus a 'switch' to activate one rather than another.

Peter Greene at Chicago, building on the ideas of Beurle and Belmont Farley (Chapter 12.ii.b), had already shown that in principle this needn't be so (P. H. Greene 1962). He'd sketched a hypothetical horse whose gaits—walking, trotting, cantering, galloping—were all generated by a single system of four interconnected 'neurones'. Gait selection involved tipping *the whole system* into one of four modes, each with its "natural" or "resonant" frequency. If the brain was anything like this mathematical model, he said, single-unit analysis and/or theories of separate neural circuitry could never provide the true explanation.

In modern parlance, Greene was talking about attractors in dynamical systems (see Section ix.b, below). But for most neuroscientists at that time, McCulloch included, specific neural circuitry was more attractive than attractors.

Attractive or not, RETIC was only very superficially related to the brain. This was partly due to the fact that it was a functioning program, designed when computing power was still very limited.

The early model has recently been updated by Kilmer (1997), one of the original authors. Incorporating a heterarchical 'committee' of up to thirty interconnected modules, the new network converges to a single output (action pattern) in less than thirty cycles of computation. It's still described by Kilmer as being inspired by the brain (the RTF). But it's more strongly oriented to practical robotics than to neuroscience.

Some other models of action selection are grounded in hypotheses about different areas of the brain. For instance, in a simulation built at the University of Sheffield, specific behavioural gating is done by mechanisms based on the anatomy of the basal ganglia (e.g. Prescott *et al.* 1999; Gurney *et al.* 1998). The RTF wasn't represented in the model. At that stage, Tony Prescott (personal communication) allowed that the RTF affects the general arousal state (sleep/wakefulness), but believed that its sensory innervation is too primitive to change the organism appropriately from one behaviour to another.

Besides their basal-ganglia simulation, Prescott's team have built a robot embodying the same principles. It switches cleanly between actions of wall following, search, food pickup, corner finding, and food depositing (Gonzalez *et al.* 2000). The overall behaviour appears coherent nevertheless—indeed, this integration was a prime constraint guiding the research (Prescott 2001). It's achieved partly because competing (potential) actions don't distort the chosen action, and also because the winning action persists even when the strength of the salient input falls somewhat below the level that was required to initiate it.

At least, that's true when the simulated dopamine level is "normal". But varying the dopamine level results in behavioural changes analogous to those seen in humans. Low dopamine leads to difficulty in approaching and picking up "food" objects, and/or in slower movements (compare Parkinson's disease). High dopamine may lead to inappropriate behaviour, such as repeated lifting/lowering of the robot's arm, caused by two action systems being selected simultaneously (compare Tourette's syndrome).

Yet more recently, Prescott's team have returned to the 1960s RTF Mars robot (Humphries *et al.* forthcoming). They now acknowledge that the basal ganglia can't be the only action selection system operating in vertebrates. For both helpless neonates and decerebrate cats "have a limited behavioral repertoire that can be expressed in the absence of basal ganglia (in [helpless] neonates [the basal ganglia are] not connected; in decerebrates [they] have been lesioned)" (sect. 2.2). For instance, these animals can switch between grooming, feeding, locomoting, escaping, and self-defence. It follows that "some neural structure *within the intact brainstem* must also be capable of functioning as a limited action selection system" (italics added). And the prime candidate for that structure, as McCulloch suggested long ago, is the RTF.

The Sheffield group tested the KMB model (and Kilmer's updated version of it) in three ways:

* by simulating it,
* by implementing it as a control architecture for a robot,
* and by comparing its performance with various other controllers.

They found that the original KMB version often worked, but not always. For some input sets, it didn't converge in the simulation (although it did when embodied as a robot). And in the simple task they set to it, it did no better *for the robot* than a random-choice controller, and less well than a winner-takes-all controller. In other words, the sequence of action selections didn't reliably lead to behaviour enabling the robot to "survive". However, when they "fixed" the KMB model by preventing random reassignments of connections, performance improved dramatically.

In addition, they used GAs (genetic algorithms; see 15.vi) to test various controllers of the same general type but which *weren't* constrained by the facts of neuro-anatomy. Among the huge number of configurations available in KMB space, they found many that could function as robot controllers. Some of these outperformed the winner-takes-all strategy that had beaten KMB's original design. Moreover, some could function even with very noisy input sets—alias relatively primitive sensors.

Their paper ended with some speculations about how the RTF and basal ganglia may interact in vertebrate action selection (sect. 6.2), and how KMB models might be useful—even advantageous—in robotics (sect. 6.3).

In a nutshell, then, the Sheffield team had been able "to fulfill McCulloch's original hope for the model for the first time" (sect. 2.3).

b. The musician in the spare room

While the early RTF program was being designed in one Cambridge, a much more ambitious formal theory of the brain was being developed in the other—but without a computer program.

The location was a spare room provided by Crick and Sydney Brenner at the Molecular Biology Unit, sponsored by the Medical Research Council in 1963 (the successor of Max Perutz's 1950s Cavendish Laboratory group). Crick, with James Watson and Brenner, had recently discovered the nature of the genetic code—and before that, in 1953, the structure of DNA. As for Brenner, he was initiating a long-term study on the development of every single cell (under the electron microscope) of the nervous system of the 1 mm-long nematode worm *C. elegans* (Brenner 2001: 125–36). With only a few hundred cells (it turned out to be 302), always interconnected in the same way,

C. elegans was destined to become hugely important for neurobiology and genetics. (Both Crick and Brenner received Nobel Prizes: in 1962 and 2002, respectively.)

The focus of the new theory that's of interest here, however, wasn't biochemistry or genetics. Rather, it was computation.

The occupant of the spare room was Marr, who'd been given a job in the lab "simply because he was working on something interesting" (Brenner 2001: 136). (Those were the days!) He turned out to be a wizard in dealing with the Unit's first computer—acquired only after much wrangling with the MRC funders, who didn't see the point of biologists having their own machine (Brenner 2001: 140–1). But that was more of a hobby than a serious research interest. Although Marr—persuaded by Brenner, persuaded in turn by his childhood friend Seymour Papert—saw the scientific potential in computing, the Unit's hard-won machine wasn't suitable for the ambitious work he wanted to do. Specifically, his ambition was to understand the brain.

After completing a degree in mathematics at Cambridge, Marr had done a year's intensive study of neuroscience and biochemistry—when he wasn't playing the clarinet, at which he was a master. He'd been selected for the National Youth Orchestra, had played various types of music at Rugby public school, and had been a member of a flourishing jazz group, the J. L. Dixie 7. His musical interests were an added bond with his neuroscience supervisor, Brindley—who is a respected amateur composer, some of whose works have been performed by the London Sinfonietta (P. Husbands, personal communication).

But music wasn't Marr's only passion. Like McCulloch and the 12-year-old Pitts before him (4.iii), Marr had been seduced in his youth by Bertrand Russell and Alfred North Whitehead's *Principia Mathematica*. One of his schoolfellows, also a Dixie player, remembers him—as a teenager in the early 1960s—reading this with "fascination and vast admiration", alongside the work of William Grey Walter on robots and John Dunne on time (P. M. Williams, personal communication). By the late 1960s, Marr's intellectual gurus also included Hebb—and therefore McCulloch too (see Chapter 5.iv.b). It was their concern with network-logic and learning which inspired his postgraduate studies.

His dissertation for his Trinity College Fellowship of 1968 had considered associative memory in the brain. It asked not only (qualitatively) how this is possible, but (quantitatively) how large the memory storage could be, given a certain network size. Others, notably David Willshaw and Longuet-Higgins, were addressing similar topics at that time (see Chapter 12.v.c). But Marr combined his abstract mathematical argument with a profusion of neuroscientific detail.

Soon, he converted his Trinity dissertation into three closely related theoretical papers (Marr 1969, 1970, 1971), plus a fourth co-authored with Blomfield (Blomfield and Marr 1970). All four appeared in leading scientific journals. One of these—the *Journal of Physiology*—had never published a non-experimental text before. In short, Marr was engaged in a fundamentally new *type* of neuroscientific research.

His four papers discussed the cerebellum, cerebrum (neocortex), and hippocampus (archicortex). They aimed to explain how the cerebellum learns and perfects movement in bodily skills, and to show that the neocortex and hippocampus are the sites of long-term and temporary memory respectively, with many memory-relevant inter-connections. And they did this in terms of specific hypotheses about *what, where,* and—Mayhew's question (directly inherited from Marr)—*how*. These hypotheses covered many different things:

* input–output relations;
* neural connectivities, numbers, and activity levels;
* and the ways in which different classes of synapse,
* assigned to specific types of neurone,
* can be modified during learning
* and later used for retrieval.

In other words, Marr was following in Brindley's footsteps. Indeed, Brindley was acknowledged in each of Marr's papers. And "Brindley synapses" were repeatedly mentioned in the text—for instance, in the discussion of (Marr's equivalent of) feature-detectors (D. C. Marr 1970: 205–6, 215; 1971: 32–3, 78–9). But besides his classification of learning rules, Brindley had also inspired Marr to think about the cerebellum.

c. Secrets of the cerebellum

The cerebellum doesn't initiate skilled (for Marr: "voluntary") movements: that's done by motor cortex. Rather, it coordinates and controls them. More accurately, it coordinates skilled, learnt, or voluntary movements. (Reflex movements, including integrated patterns involving all four limbs, are controlled by the spinal cord and/or brain stem; they can occur even in decerebrate animals: Chapter 2.viii.d.)

In 1964 Brindley had published a brief paper on how the cerebellum uses sensory information. He'd suggested that the task of the cerebellum is to learn motor skills in such a way that when a simple or incomplete message (compare: an intention to pick up a glass, without knowing exactly how) is sent to it from the cerebral cortex, the required movement is performed automatically. That is, the cerebellum functions as an associative memory.

Marr agreed (1969: 438). But his discussion was very much fuller than Brindley's, and included a host of predictions about synaptic functions, connections, and numbers.

Even that fact, however, was partly due to his supervisor. Many years later, Brindley modestly said he'd given Marr no more than "a few" ideas—plus "one or two pep talks". The purpose of those pep talks was "to persuade him that unless his work led to experimentally testable predictions whose prior probability was neither almost zero nor almost unity, no experimenter would read his work" (Vaina 1991: 308).

Marr's view of what the cerebellum was doing reflected the several different 'languages' found even in the early digital computers. For instance, he said:

Where it is possible to *translate* the combined activity of many cerebral fibres rather simply in *directives*, doing so in the cerebellum would free the cerebrum from an essentially tedious task. In these circumstances, the cerebellum becomes rather more than a slave which copies things originally organized by the cerebrum: it becomes an organ in which the cerebrum can set up *a sophisticated and interpretive buffer language* between itself and muscle. . . . The automatic cerebellar *translation* into movements or gestures will reflect in a concrete way what may in the cerebrum be diffuse and *specifically unformulated*, while the analysis leading to that diffuse and unformulated state can proceed *in its appropriate language*. (Marr 1969: 468; italics added)

So, where Craik had talked of "translation" from sensory to cerebral to motor languages (Chapter 4.vi.b), Marr implied several (many?) cerebral languages suited to modelling different things. One brain level may pass messages to another, which translates them into some other internal language ... and ultimately into muscle movements.

Besides this general notion of computer 'languages', Marr would employ many specific computational arguments. His thinking started, however, from the neuroscientific data.

This is clear from the rhetorical structure of his first paper. Its opening sentence was "The cortex of the vertebrate cerebellum has a simple and extremely regular fine structure."

He cited the recent (1967) book by Eccles and two younger colleagues (one of whom was a Hungarian expert on control-engineering), with the telling title *The Cerebellum as a Neuronal Machine* (Eccles *et al.* 1967). (These authors had suggested that learning takes place in the cerebellum, perhaps by the growth of dendritic spines on the Purkinje cells.) And he immediately went on to summarize what was known of the (cat's) neuro-anatomy:

The axons of the Purkinje cells form the only output from the cortex of the cerebellum; and these cells are driven by two essentially different kinds of input, one direct, the other indirect. The first is the climbing fibre input, and the second the mossy fibres, *whose influence on the Purkinje cells may be complicated.*

The inferior olive is the only known source of climbing fibres ... Each [olivary cell] sends out an axon which terminates in one climbing fibre on just one Purkinje cell; there are very few exceptions. The climbing fibre completely dominates the dendritic tree of the Purkinje cell, and its action has been shown to be powerfully excitatory ... (Marr 1969: 438–9; italics added)

(It's now known that there are more than a "very few" exceptions: although each Purkinje cell has only one climbing fibre, each climbing fibre connects with ten to fifteen Purkinje cells—Rolls and Treves 1998: 191.)

Bearing these biological constraints in mind, he then outlined his solution. What the cerebellum was doing, and *how*, could be understood in terms of associative memory. And, unlike the writers discussed in Chapter 12, Marr ascribed this memory to distinct structures in the brain:

[I suggest] that each olivary cell corresponds to a "piece of output" [an *elemental movement*] which it is necessary to have under control during movements.

... Every action therefore has a defining representation as a sequence of firing patterns in the olive.

The final assumption ... is that the nervous system has a way of converting the (inhibitory) output of a Purkinje cell into an instruction which provokes the precise movement to which its uniquely related olivary cell responds.

[I shall argue that] the Purkinje cell can learn all the "situations" in which the olive cell movement is required, and later, when such a situation occurs again, can implement the movement itself ... [So] the cerebellum could learn to carry out any previously rehearsed action which the cerebrum chose to initiate, for as that action progressed, the context for the next part of it would form, would be recognized by the appropriate Purkinje cells, and these would turn on the next set of muscles, allowing further development of the action. In this way, each muscle would be turned on and off at the correct moment, and the action would be automatically performed. (Marr 1969: 439)

As for the mossy fibres ("whose influence on the Purkinje cells may be complicated"), they provide the information that defines the "context" for each Purkinje cell. And how could that be done? This was his suggestion:

[It] is necessary to demonstrate that the mossy fibre–granule cell–Purkinje cell arrangement could operate as a pattern recognition device. The notion fundamental to this is that the mossy fibre–granule cell articulation is essentially a pattern separator. That is, it amplifies discrepancies between patterns that are rather similar . . .

[A] mossy fibre input has been learnt by a given Purkinje cell if, and only if, the input is transformed into impulses in a bundle of parallel fibres all of whose synapses with that Purkinje cell have been facilitated. (p. 440)

But there were two crucial complications:

First, the number of parallel fibres into which a mossy fibre input is translated increases very sharply with the number of active mossy fibres unless the threshold of the granule cells also increases . . . [Economy] arguments suggest that the granule cell threshold should be controlled in a suitable way. An inhibitory interneurone could achieve this, and the Golgi cells are interpreted as fulfilling this role.

The second point is that [variation in parallel fibre activity] will still exist. Whether or not a Purkinje cell should respond to a given mossy fibre input cannot therefore be decided by a fixed threshold mechanism. . . . The natural way to implement [the necessary threshold-setting mechanism] is to allow the parallel fibres to drive an interneurone which inhibits the Purkinje cell: and it will be shown that the various stellate inhibitory cells can be associated with this function, *although their dendritic and axonal distributions are at first sight unsuitable*. (p. 440; italics added)

The theory offered many surprises. These included functional interpretations for previously mysterious structures, and various counter-intuitive claims too (cf. "although their dendritic and axonal distributions are at first sight unsuitable").

One such claim was that a single pair of olivary and Purkinje cells might perform in two different ways at different times. In other words, its input–output relations depend on the type of learning involved—for maintaining posture or for active movement (pp. 466–9).

Another was that the *only* modifiable synapses in the cerebellum are those between the parallel fibres and the Purkinje cells. This was pure theory: these synapses weren't experimentally shown to be modifiable until very much later (Ito 1982, 1984: 115–30). Similarly, Marr's claims that the synapses between the Golgi cells and the mossy and parallel fibres are *not* modifiable were argued on purely computational grounds (p. 454).

At the core of the theory was a new Hebbian learning rule, not included in Brindley's classification:

[If] a parallel fibre is active at about the same time as the climbing fibre to a Purkinje cell with which that parallel fibre makes synaptic contact, then the efficacy of that synapse is increased towards some fixed maximum value. ("At about the same time" is an intentionally inexact phrase: the period of sensitivity needs to be something like 50–100 msec.) (pp. 455–6)

This rule was "Hebbian" in the broad sense. Whereas Hebb had posited learning when a pre-synaptic and a post-synaptic cell are co-active, Marr demanded two co-active pre-synaptic cells, irrespective of activity in the post-synaptic cell. Moreover, only one of the two synapses involved was modifiable.

The function of the climbing fibres, on this theory, was to carry feedback (about the intended movement) from the olivary nucleus to the Purkinje cells. In other words, the cerebellum was being described as a system for 'supervised' learning (see Chapter 12.ii.d).

Perhaps even more surprising, Marr's interpretations often involved precise numbers: of cells, synapses, and excitation levels. Right at the beginning of the paper, he pointed out that "Behind this general structure lie some relatively fixed numerical relations." For example:

Each Purkinje cell has about 200,000 (spine) synapses with the parallel fibres crossing its dendritic tree, and almost every such parallel fibre makes a synaptic contact. The length of each parallel fibre is 2–3 mm . . . Each basket cell axon runs for about 1 mm transversely, which is about the distance of 10 Purkinje cells . . . There is one Golgi cell per 9 or 10 Purkinje cells, and its axon synapses (in glomeruli) with all the granule cells in that region, i.e. around 4500. (pp. 442–3)

The list continued, on and on . . . and on. But why? Who cares?

Marr cared, because he assumed that these numbers, being "relatively fixed", must have some functional (computational) significance. Eccles and his fellow neuro-anatomists, who had helped discover these numbers in the first place, had doubtless made the same assumption. But Marr tried to show, in detail, just what the functional significance is.

He even *predicted* many as yet unknown numbers. Some of these quantitative predictions suggested minimum–maximum limits between which the thresholds of the modifiable synapses must vary, or time periods during which 'simultaneous' events must happen (see above). Others concerned activity levels within cell populations; or the number of synapses on a single cell; or overall cell numbers; or numbers of simultaneously active cells. For instance:

*   "The number of granule cells active at any one time (say in any 50 msec period) is a small fraction (less than $\frac{1}{20}$) of all granule cells" (p. 469).
*   "[Given that there are 200,000 parallel fibres for each Purkinje cell], the maximum desirable number of facilitated synapses on any one Purkinje cell [is] 140,000, and
*   the minimum number of parallel fibres active in any learned event [is] 500" (p. 457).
*   "Calculations based on slightly tenuous assumptions suggest that each Purkinje cell receives connections from about 7000 mossy fibres" (p. 443).

Similarly, in defining his crucial concept of a *codon* (a subset of a collection of active mossy fibres, feeding into a granular *codon cell*), Marr committed himself to specific quantities. He gave mathematical equations for the exact number of codons of a given size that are associated with a given number of active mossy fibres, and for the number of mossy fibres that may influence a single Purkinje cell (pp. 444–5). These numbers were needed, he said, in order to understand "the codon sampling statistics" that enabled the cerebellum to distinguish similar patterns.

These hypotheses were presented as being based solely on neurophysiology and mathematics. But AI had apparently been an inspiration too. For towards the end of the paper, he pointed out that a codon is "closely related to the feature analysis ideas current in the machine intelligence literature" (p. 469)—and he cited Uhr and Vossler (see Section iv.d).

In making his numerical predictions, Marr relied on the general principle of computational "economy", or "efficiency". For instance:

It is [evident] that the maximum codon size used depends critically on the number of claws to each [granule] cell. Given this, the factor that will determine the number of claws to each cell will be *economy of structure*; and the relevant question is what is *the least number* of claws such that [the system has the properties required]. (p. 447; italics added)

The number of patterns a Purkinje cell can learn decreases sharply as the number of active parallel fibres involved in each increases. It is therefore essential to the *efficient* functioning of the system that the codon size should depend on the amount of mossy fibre activity ... The codon size must be maximal, subject to conditions [previously specified]. This ensures that the number of modifiable synapses used for each learned event is minimal, *and hence that the capacity is maximal.* (pp. 449–50; italics added)

Throughout the paper, Marr assumed that evolution must have found the mathematically optimal way of doing things.

Bearing in mind Crick's contrast (Section ii.a) between biology's "slick tricks" and the mathematician's "powerful general principles", it wasn't obvious that Marr's optimality assumption was warranted (cf. Boden 1988: 60–3). Indeed, it's still not obvious, despite an intriguing attempt by a computer scientist in the late 1980s to defend an information-maximizing role for the brain (see Section ix.a, below). But that's just to say that Marr's quantitative predictions might have been wrong. What's more important is that he made them at all.

The paper bristled with testable hypotheses, as well as novel explanations of known facts. Some were so central that "If this is not true, the theory collapses", while the disproof of others would be "embarrassing but not catastrophic" (p. 468). In the two companion papers, Marr elevated such distinctions to a fine art, marking each prediction by up to three asterisks. Unasterisked predictions were those which lay "strictly outside the range of the theory, but about which the theory provides a strong hint" (D. C. Marr 1970: 231; cf. 1971: 77).

Evidently, Brindley's "pep talks" had worked. Marr offered many hostages to experimental fortune. But did the experimenters respond, as Brindley had hoped?

d. Audience reaction

A few did. Rashevsky's intellectual successors—computationally inclined and mathematically competent neuroscientists—found Marr's ideas hugely exciting. Cowan remembers that "the 1969 cerebellum paper created a sensation", and that all three papers generated "a tremendous amount of interest among neurobiologists" (J. A. Anderson and Rosenfeld 1998: 114).

These people, of course, were already familiar with mathematical theories of associative memory. Indeed, Cowan saw Marr's core cerebellar learning rule as a development of Wilfred Taylor's pioneering research in the 1950s (Chapter 12.ii.b), and essentially similar to his colleague Greene's (1965) theory of retrieval, published in Rashevsky's journal four years earlier (see Chapter 12.v.c). According to Cowan: "It requires only inhibitory neurons to control thresholds, and intrinsic climbing fibers to train synapses (both suggested but not implemented by Greene), to complete the picture" (Vaina 1991: 204).

It's not clear whether Marr knew of Greene's paper. In general, he cited others less often than he should have done. I've been persuaded of this by several private conversations over the years, and it's now been remarked in print more than once—by Cowan and Arbib, for instance (J. A. Anderson and Rosenfeld 1998: 110, 231–2). But Cowan also allows that "David had a very original mind, so it's easily credible that David just invented the whole thing himself" (J. A. Anderson and Rosenfeld 1998: 111). However that may be, Marr had added ample neuroscientific flesh onto Greene's mathematical core. (Greene hadn't even mentioned the cerebellum.)

One might have expected neurobiologists *in general* to appreciate that fact. The Rashevsky school did. But many others didn't. As remarked above, no purely theoretical paper had appeared in the *Journal of Physiology* before—and not all of the readers welcomed the novelty. Cowan, again: "a lot of the theoretical biologists felt that [Marr's first paper] was so speculative that it couldn't possibly be right and weren't impressed with it" (J. A. Anderson and Rosenfeld 1998: 115).

Marr's theory of the cerebellum was indeed speculative, for it couldn't (yet) be directly confirmed by single-unit recording or neurochemistry, nor tested as a functioning computer model. And, as Brindley had shown (see Section iii.c), *behavioural* tests were inadequate in principle.

Even some people already working on the abstract mathematics of memory saw it as premature. For instance, Longuet-Higgins described Marr's work as "excellent"—not a term he ever used lightly—but said (in his reply to the Lighthill Report) that "for some time to come" it would be less valuable for the cognitive sciences than GOFAI theories of cognition (Longuet-Higgins 1973: 36–7). (By the end of Marr's short life, Longuet-Higgins believed that that time *had* now come: see subsection f, below.)

Furthermore, Marr's work was difficult to read. Even highly sympathetic critics have remarked on the numbing profusion of neuro-anatomical detail, the "pedantry" of the style, and the "over-elaborate" mathematics (Willshaw and Buckingham 1990).

The mathematics was indeed taxing. In the neocortex paper, Marr defensively said, "The results [of these technical preliminaries] are mainly of an abstract or statistical nature, and despite the length of the formulae, are essentially simple," and "The reader who is not familiar with this notation should not be put off by it. All the important arguments of the paper have been written out in full" (Marr 1970: 193, 167). Many readers were put off, nonetheless. And those who weren't, didn't always master what they were reading. So Willshaw has recently complained that more people have cited Marr than have understood him (Willshaw and Buckingham 1990).

Eventually, however, Marr's early work set a theoretical and experimental agenda that's still driving some neuroscientific research today. Masao Ito, in his now classic book on the cerebellum, recalls that Marr's theory assumed a special type of synaptic plasticity in cerebellar cortex, and says: "[Being] impressed by the close conformity of this assumption to the neuronal circuitry [worked out in our laboratory in the early 1970s], I was prompted to verify [it] experimentally" (Ito 1984, p. ix). (Ito also praises Marr for his three-level account of explanation, outlined in Chapter 7.iii.b—Ito 1984: 2.)

Bruce McNaughten (of the University of Arizona) sees Marr's papers as "guiding lights" for himself and various other neuroscientists. In his judgement, they showed "astounding prescience", given the sparsity of experimental evidence available at the time. Their insights, he says, have been largely substantiated, and "have brought order to a number of otherwise disconnected data on the anatomy, biophysics and information transmission of [the brain]" (Vaina 1991: 121, 126).

Marr's theory of supervised learning in the cerebellum was especially influential, and—although some doubts remain (Vaina 1991: 47–8)—is still widely regarded as essentially sound. That's not to say that every current model of the cerebellum is like his. We'll see in Section ix.b–c, for instance, that Pellionisz's account relies on a very different formal analysis. And Grossberg's model is different again (see Section vi). But Marr's central insights are now accepted—with two qualifications.

One of these concerns the precise mathematical form of his learning rule. It was later proved that the algorithm as he defined it would eventually lead to a state in which all the weights are at a maximum value (Sejnowski 1977). (Nevertheless, it was an ancestor of some now current connectionist learning rules.) The other concerns his neurological interpretation of it.

Marr proposed that the parallel-to-Purkinje synapse is strengthened by co-activity in the parallel and climbing fibres (see above). James Albus had been developing a similar idea—which he likened to a perceptron (Albus 1971). Soon afterwards, he implemented it as the "CMAC" mechanism (Cerebellar Model Arithmetic Computer); this was initially developed to control a robot arm, but is today used by others for various purposes (Albus 1975; van der Smagt 1998). But if Albus shared Marr's commitment to computational explanations, he differed on the neurophysiology.

In Albus's account of the cerebellum, joint activation of the climbing and parallel fibres would *decrease* the efficiency of the parallel-to-Purkinje synapse. Moreover, he saw the learning as driven (via the feedback in the climbing fibres) by *error*, where Marr had seen it as driven by *intention*—that is, by the "context" alerted by the cerebrum. At that time, it seemed that Marr was right; now it's known that they both were. So people speak today of the "Marr–Albus" model.

As with back propagation (Chapter 12.vi.d), there are several priority disputes here. Quite apart from Brindley's and Eccles's vague suggestions about cerebellar learning, formal equivalents of Marr's learning rule had already been defined.

One example was Greene's (1965) work, mentioned above, which Cowan describes as "virtually identical with David Marr's cerebellum model, except it's not applied to the cerebellum". What's more, Grossberg had formulated the same rule in 1964—and he *had* applied it to the cerebellum (Grossberg 1969). He'd even made the same prediction, that learning would occur at the synapses of the parallel fibres and Purkinje cells. However, because of the difficulties in getting his work accepted by journals of psychology or neurophysiology, his cerebellum paper was published in *Studies in Applied Mathematics*—not something which neuroscientists were likely to see.

e. Beyond the cerebellum

Because of the cerebellum's relatively simple histological structure (one climbing fibre to each Purkinje cell, for example), Marr had been able to represent all the neuronal pathways involved. But that wasn't possible for the cerebrum or hippocampus. His theories about these were even more "speculative" as a result.

The paper on neocortex, Marr later told Terrence Sejnowski, was the one he was "most proud of" (Vaina 1991: 297). It focused on the unsupervised learning of high-level concepts: sophisticated examples of Pitts and McCulloch's "universals" and Konorski's "gnostic units" (see Sections iii–iv).

Marr described neocortex as "classifying cortex", in contrast with the "memorizing cortex" of the hippocampus. He claimed that whereas the hippocampus stores patterns in the "language" in which they were input to it, the cerebrum can also generate "a new language" in which to describe (classify) them (1971: 73–4).

The vocabulary of both these languages consisted of codons. Whereas his cerebellar theory had assumed the existence of preformed codons, Marr suggested that the neocortex can learn codons as a result of environmental input (compare Taylor's machine, described in Chapter 12.ii.b). He defined a method whereby certain synapses could calculate conditional probabilities. Uttley had discussed such synaptic calculations too (see Section iii.b), but whereas he'd focused on the *frequency* of events, Marr focused on their *similarities*. He described this learning method as a form of "numerical taxonomy", in which the differences between objects are computed, and then clusters are formed which minimize these differences (1970: 175).

The neocortex paper was full of intriguing ideas—not least, that sleep is necessary for the formation of new codons (Marr 1970: 215). But most of its intellectual cheques still can't be cashed—nor even presented to the cashier. As Sejnowski has pointed out, increased computer power will be needed to evaluate Marr's theory of neocortex properly (Vaina 1991: 298). (Cowan believes that the central learning method will work if the outside world is reasonably coherent, and likens it to classification techniques recently developed by Grossberg and by Teuvo Kohonen—Vaina 1991: 208.)

As for the hippocampus, a huge amount of computational work has been done in the last quarter-century, including much based on 1970s research on cognitive maps (Gluck 1996; Burgess *et al.* 1998). Nevertheless, a very recent review states that "Models of the hippocampal system and episodic memory still have many elements in common with Marr's basic view of hippocampal function, [even though] recent efforts have added considerable elaboration and refinement" (Gluck *et al.* 2003: 274). And Willshaw's judgement (around 1990) was that "Even 20 years after publication, Marr's theory remains the most complete computational model of the hippocampus" (Vaina 1991: 120).

"Complete" doesn't mean "correct", however, and Willshaw himself rejects a number of Marr's claims (Willshaw and Buckingham 1990). The fundamental problem, he remarks, is that the hippocampus is very difficult to analyse, because—unlike the cerebellum—it's not completely connected. In general, Marr's computational arguments didn't sufficiently constrain his theory of hippocampus: other models are possible too.

Moreover, neurones may have computational properties that Marr didn't fully consider, which undermine some of his key arguments. For example, he suggested (giving no algorithm) that some neurones may have a *dual* threshold, in which activation requires inputs from both a minimal number and a minimal proportion of the cell's many synapses. He even discussed what types of local circuitry could do the relevant arithmetic. But whereas he said his model would be "impaired" if only one threshold-type were available, Willshaw sees dual threshold-setting strategies as *essential* for incompletely connected nets. Such units will behave differently from 'classical' neurones, and a particular cell's activity will depend on the specific numbers involved.

With this in mind, Willshaw argues—and has confirmed, by extensive simulation— that Marr's reasons for positing a third neurone layer in the hippocampus were mistaken. These reasons concerned storage capacity in associative memories (a long-standing interest of Willshaw's: Chapter 12.v.c).

Willshaw showed that, in principle, if the dual thresholds and connectivities are set appropriately then two layers will suffice. Not only that: in most conditions, his own two-layer simulation (based on Grossberg's "competitive" nets: see Section vi) performed as well—stored as many memories, with similar recall—as the three-layer version, even though it contained many fewer synapses. Willshaw has also compared the behaviour of various types of dual-threshold mechanism. But he has the advantage of considerable computer power. Marr, when he wrote and 'published' his Fellowship dissertation in the late 1960s, didn't.

Both Marr and Willshaw, of course, were committed to computational theorizing about the brain. But (as remarked already) by no means all neuroscientists are. In the 1960s, that was only to be expected. Brenner, for instance, recalls:

[In] those days biologists were in general very antagonistic to computers. They thought that anybody interested in computing was choosing an easy way out of a responsible job. In other words, if you didn't work at the bench you weren't worth anything! (Brenner 2001: 139–40)

However, many people who specialize in hands-on brain research still, in the twenty-first century, have scant regard for the computational approach. Gross, for example, who is more knowledgeable than most about the history of his field, says this:

I'm a bit dubious about the role of theory *or at least of theoreticians* in biology. So far, I'd suggest, no pure theoretician ever made any contribution to biology at all, and indeed they disappeared so quickly that we have all but lost their names. . . . [The] great theoreticians in biology were above all empiricists who steeped their waking days and nights in observing, collecting, experimenting—such as Aristotle, Darwin, Bernard, Mendel, Freud, Sherrington and in our day Sperry and Hubel and Wiesel. So far, theory without empirical slogging in the same head has yielded nothing. Maybe biology is changing. Maybe one can be a pure computational biologist and never touch a plant and animals but only a computer . . . we'll see. Crick and Watson are the only exceptions I can think of. Maybe they herald a new way in biology. But Crick in neuroscience makes the rule. He is very smart, very well read in neuroscience, gets all the latest data fed to him and for decades has produced zilch but an absurd idea about dreams and a silly localization of consciousness. (C. G. Gross, personal communication; italics added; quoted by permission)

On this view, Marr's work on cerebellum was influential largely because it was so dependent on Hebb and Brindley—both of whom were empiricists first and theoreticians second. (Their books cite many empirical papers with themselves as sole/first author, plus others written with others.) Marr's later work on the brain is much less widely known by neuroscientists (he's not mentioned, for example, in the widely used textbook by Kandel *et al.* 1995). In Gross's opinion, that's as it should be.

This is a special case of a general type of cross-disciplinary complaint. Similar criticisms have been made by biologists of the "pure computational biologist" Stuart Kauffman, as we'll see in Chapter 15.ix.c. And professional psychologists commonly rebuke—and even more commonly ignore—AI modellers for their lack of attention to the experimental data. AI researchers, in turn, often lament others' blithe ignorance of computational concepts and constraints. As for philosophers, they're more than ready to point out philosophical naivety on the part of scientists of all stripes. (Maybe this is because the scientists so often express contempt for philosophy: see ix.d below, and 16, preamble.)

In general, one needs to decide to what extent such attacks are mere defensive territoriality, and to what extent they have real substance. Such decisions are especially relevant if one claims that interdisciplinarity is crucial—which is a large part of the message of this book.

f. A change of tack

The lack of computer power was partly responsible for Marr's unexpected change of direction in the early 1970s. Instead of following up his seminal papers on the brain, he turned to consider the retina and primary visual cortex instead. He studied these until his premature death (aged 35), from leukaemia, in 1980.

His computational theory of vision was outlined in Chapter 7.v.b–d. Here, what's of interest is how he related it to neuroscience.

His first paper on vision, in 1974, aimed to explain how the primate retina computes subjective lightness—which is almost independent of the objective illumination. His Abstract stated that "The operation of the midget bipolar–midget ganglion channel is analysed in detail, and a functional interpretation of various retinal structures is given" (D. C. Marr 1974a: 1377). And his paper ascribed distinct computational functions to (for instance) the small stratified amacrine cells (unistratified and bistratified); the diffuse amacrine cells (narrow-field and wide-field); the amacrine/bipolar synapses . . . and so on.

So far, so familiar: this is just what one would expect from the author of the papers discussed above. And here too, Brindley had part-inspired him (D. C. Marr 1974a: 1379). But Marr's intellectual strategy was already changing.

Instead of starting with a description of the retina *as such* (compare: cerebellum), he started with a lengthy mathematical discussion of the abstract problem of computing lightness. Next, he asked, "Is this method relevant to retinal function?" Only then did he offer a section on 'The Anatomy of the Retina', before making various predictions about the functions of the retinal circuitry. Possibly, this 'mathematical' strategy put the experimenters off: not one contemporary reference to Marr's lightness paper is recorded in the Science Citation Index (Vaina 1991: 226).

From then on, the abstract definition of visual computation increasingly took precedence in Marr's work. The neuro-anatomy, by contrast, almost faded into the background.

Marr explicitly defended his new approach. He did this first (at a vision meeting in December 1973) in terms of two 'Levels of Understanding', and later (in a 1976 grant proposal and in his book of 1982) in terms of his three-level theory of psychological explanation *in general*: see Chapter 7.iii.b. (Perhaps it wasn't a totally new approach: Willshaw points out that the hippocampus paper had havered between *Any such theory must be like this* and *It must be like this, because the hippocampus is like this*—Willshaw and Buckingham 1990.)

Marr's final verdict on the lightness paper was that the neuroscience was mistaken (it turns out that lightness is computed in visual cortex: Zeki 1993), but that the methodology was sound. As he put it: "[It] showed the possible style of a correct analysis. Present is *a clear understanding of what is to be computed, how it is to be done*" (D. C. Marr 1982; italics added)—where the "how" was computational, not neurological.

The turning point—away from neuro-anatomy, towards abstract computational analysis of the task—dated from May 1972. That was when Marr first met Minsky and Papert, an engineer and mathematician then doing pioneering work in GOFAI in MIT's AI Laboratory (see Chapter 10.i.g, ii.a, and v.f).

The occasion was a small interdisciplinary workshop on Marr's theory of cortex, organized at Boston University's School of Medicine by the physiologist Benjamin Kaminer. The participants included many biological luminaries, such as Barlow, Hubel, Wiesel, Kuffler, Crick, and Brenner (and Blomfield, too). But Kaminer remembers that Marr spent much of his time with Minsky and Papert (Vaina 1991: 312).

Their research was very different from his, and they had scant respect for connectionism in general (see Chapters 10.iv.b and 12.iii). The latter point was perhaps a 'plus' for Marr (see below). But, by his standards, they knew next to nothing about the brain. Nor could they outdo him in mathematics, despite their mathematical expertise. So what could they possibly teach him?

The pivotal moment, according to Tomaso Poggio (personal communication), occurred when Marr took a few hours off to give a seminar on the cerebellum at MIT. Minsky remarked from the floor that we have to think about "the [abstract] problem of motor control" before we can ask the right questions about the cerebellar hardware. Marr later described this to Poggio as a crucial insight. For it would underlie his definition of the "computational" level of explanation as psychologically basic (Chapter 7.iii.b).

In addition, Minsky and Papert together persuaded him that computer simulation could help develop and test computational theories. This persuasion was a classic case of business mixed with pleasure. Crick recalls that "The boat trip after [Kaminer's] meeting was David's Road to Damascus . . . He became converted to AI and before long moved to MIT" (Vaina 1991: 314).

Just as Saul already knew of Jesus before his Damascene conversion, so Marr already knew of AI—indeed, Uhr and Vossler's work had been one of only ten citations in his cerebellum paper (see above). Moreover, machine translation and other aspects of NLP had long been established in Cambridge, England (Chapter 9.x.a–d). But AI simulation in general was much more advanced in Cambridge, Massachusetts.

So, in 1973, Marr left the one Cambridge for the other. Initially, he'd intended no more than a three-month visit. In the event, he never left: in 1976 he was officially appointed to the Department of Brain and Cognitive Sciences. (While at MIT, he soon added a passion for flying to his long-standing passion for the clarinet.)

Minsky's remark about "the problem of motor control" had prompted Marr to reassess his own earlier work. In December 1973, he wrote to Brindley:

I do not expect to write any more papers in theoretical neurophysiology—at least not for a long time; but I do not regard the achievements of your 1969, or my papers as negligible. At the very least, they contain techniques that anyone concerned with biological computer architecture should be aware of, and I shall be very surprised if my 1969 or 1971 papers turn out to be very wrong. (quoted in Vaina 1991: 2)

Why had he given up on "papers in theoretical neurophysiology"? Writing to a potential translator at much the same time, he said:

It would be fun to have some of [my early papers] translated into Russian. My present opinion of my earlier work is, however, that even if it is correct it does not take one much further in the study of how the brain works than, for example, the study of more obviously physical phenomena like ... the conduction of nervous impulses. The reason why I believe this is that this part of my work has to do more with computer architecture than with biological computer *programs*! I have studied how some basic "machine-code" instructions can be implemented in nervous tissue; but these studies tell you rather little about how the brain uses these facilities—e.g., what is the overall structure of a particular motor program for picking an object up, or for throwing a ball. *It is the second kind of question that I am now interested in.* (quoted in Vaina 1991: 3)

Having moved to MIT's AI Laboratory, Marr soon abandoned the problem of motor control for the problem of *vision* instead. Another letter to Brindley (in October 1973) declared:

I turned to vision when I arrived here, hoping that insight into the functions you had to perform to recognize something, together with the detailed neurophysiological knowledge and an unexcitable disposition, would be capable of illuminating many questions that are surely not yet vulnerable to the microelectrode. (quoted in Vaina 1991: 2)

The switch to vision was influenced by his MIT colleague Berthold Horn, who was then working on the (mathematical) computation of lightness. Marr's transitional paper (1974*a*) opened with a discussion of Horn's algorithm, and the claim that if it were implemented in the retina that would "make sense" of many of the known biological facts.

That paper was "transitional" not only in (still) saying a good deal about the neuroanatomy, but also in preceding Marr's second switch: from lightness to stereopsis. This interest was grounded in recent neurophysiological work by Barlow, and in Bela Julesz's studies of random-dot stereograms (Barlow *et al.* 1967; Julesz 1971). It owes something also to Parvati Dev, although her model of stereopsis was—dismissively—cited only in a footnote at the end of the paper (J. A. Anderson and Rosenfeld 1998: 231).

Even at MIT, however, Marr was initially frustrated by lack of computer power. In an early lab report on stereopsis, he said:

[complex parallel algorithms] are very expensive to simulate, and it is extremely difficult to derive analytically, from a system with complex non-linear components, quantities that could be measured experimentally. (D. C. Marr 1974*b*: 232)

He discussed various ways in which different 'families' of theories based on stereo disparity might be implemented. But he couldn't actually implement them. (Computer models of scene analysis, by contrast, had been up and running in the AI Lab for some time: Chapter 10.iv.b.)

A few years later, that changed. Now, he was able—with Poggio, then based in Tübingen—to implement two powerful theories of stereopsis (D. C. Marr and Poggio 1976, 1979). And these papers, unlike the one on lightness, received enormous attention from experimentalists. They formed the core of an influential research programme on vision that continued up to—and after—his death, and which spread around the world (see Chapter 7.v.f).

Although Marr did relate his final theory of stereopsis to visual neuroscience, it hadn't been *driven by* the biology, as his cerebellar theory had been. On the contrary:

We feel that an important feature of this theory is that *it grew from an analysis of the computational problems* that underlie stereopsis, and is devoted to a characterization of the processes capable of solving it *without* specific reference to the [neural] machinery in which they run. (D. C. Marr and Poggio 1979: 324; italics added)

Previous theories of depth vision, said Marr, for all their ingenuity in dreaming up algorithms, were fundamentally wrong-headed: "not one of them computed *the right thing*" (D. C. Marr 1982: 122; italics added). As for connectionism in general, much of it was useless for psychological purposes:

Again, the primary unresolved issue is *what* functions you want to implement, and *why*. In the absence of this knowledge, a neural net theory, unless it is closely tied to the known anatomy and physiology of some part of the brain and makes some unexpected predictions [like his own early work], is of no value. (D. C. Marr 1975c: 876)

Evidently, Minsky's lessons—in the boat and the seminar room—had been well learnt. Neuroscience needed cognitive science, not just the other way around.

And people listened. It was Marr above all who made this point clear—both in his work on vision and in his models of the brain. In a glowing obituary of his younger friend, Longuet-Higgins put it like this:

If neurophysiology was *a theoretical vacuum* when he entered it, it is now seething with lively controversy about the validity of his ideas on the visual system...Even if no single one of Marr's detailed hypotheses ultimately survives, which is unlikely, the questions he raises can no longer be ignored and the methodology he proposes seems to be *the only one that has any hope of illuminating the bewildering circuitry of the central nervous system.* (Longuet-Higgins 1982: 992)

14.vi. Realism Rampant

While Marr was formulating his theory of associative memory in the 1960s, an even more data-driven approach was already being developed by Grossberg. He combined neuroscience with an extraordinarily wide range of findings from experimental psychology. Indeed, he started from the behavioural data and brought in the neuroscience to explain it.

(Grossberg's technical work isn't easy to read. His neologisms don't help: for instance, he speaks of "syncytia" and "subsyncytia" where others would speak of networks, neurone pools, or cell assemblies. For an especially accessible summary, see his *American Scientist* article: 1995.)

a. A voice in the wilderness

Grossberg had first written about his project in 1957, as a freshman at Dartmouth—only one year after the famous Dartmouth summer gathering (see Chapter 6.iv.b). As a graduate student at Stanford in the late 1950s, he'd circulated manuscripts containing many detailed hypotheses about integrated brain processes. These even contained an equivalent of the Marr–Albus rule—and also, in what Grossberg termed his Generalized Additive Model, a near-equivalent of what were later named "Hopfield" nets (12.v.f).

Unlike Marr, however, Grossberg had to wait a long time for recognition. By the late 1980s, he was being described as "one of the most visible scientists working in neural networks for nearly twenty years"—that is, from about 1970 (J. A. Anderson and Rosenfeld 1988: 243). But he'd been doing important work since the late 1950s. Between then and the early 1970s he was treated as an eccentric, though brilliant, outsider. Hence Paul Werbos's remark (quoted in Chapter 12.v.g): "We were not people they trusted, neither Steve nor I."

He had great difficulty in getting his early papers published. Although a 500-page monograph appeared from the Rockefeller Institute for Medical Research, where he was a newly arrived graduate student (Grossberg 1964), journal editors proved less amenable. For instance, it wasn't until 1980 that a paper appeared in *Psychological Review*. Before then, they'd been sent back to him without review, as not being the sort of thing the journal published.

One problem was the relentless interdisciplinarity. The bristling mix of psychology, mathematics, neurophysiology, and neuro-anatomy—though not, yet, computing: see below—would have daunted all but the most eclectic cybernetician (see 1.iii.h). Another was the very richness of Grossberg's fare. He provided so many novel ideas in a single paper that any one risked being obscured by the others.

Another hurdle, Grossberg recalls, was the low value that psychologists in the 1950s–1960s gave to abstract mathematical argument:

One had to function primarily as an experimentalist. Even Bill Estes, I was told, had a lot of trouble getting his modelling papers published at first, even though he was already a distinguished experimentalist. (J. A. Anderson and Rosenfeld 1998: 174)

Much as Marr's early brain models seemed merely "speculative" to empirically minded biologists, so Grossberg's early theories of mental processing—despite being grounded in behaviour—failed to impress the psychologists.

It didn't help that the mathematics was taxing. ("The problem is that, although I would often have an idea first...I would develop it too mathematically for most readers"—J. A. Anderson and Rosenfeld 1998: 179.) In any one paper, Grossberg would offer many interacting non-linear differential equations—with no way, at that time, of leaving a computer to do the sums.

He provided proofs, to be sure. But some of these were even more difficult than Rosenblatt's proof of the convergence theorem (see Chapter 12.ii.f). Many readers simply didn't believe them. And his qualitative predictions of the system dynamics and observable behaviour which, he said, would result from his equations were far from obvious to most people. Often, they were counter-intuitive.

Even Stanislaw Ulam and his fellow physicists at Los Alamos, who implemented the equations on their huge computer in 1966, had been sceptical initially. Grossberg, again: "At first they didn't believe the theorems either, but then they ran the networks on the computer, and the simulations did exactly what the theorems said they should" (J. A. Anderson and Rosenfeld 1998: 179).

(Two decades earlier, machine demonstration wouldn't have convinced Los Alamos. In 1944 the newfangled IBM punched-card machines weren't trusted by the scientists designing the implosion atom bomb, and their wives were asked to do the relevant sums using mechanical hand calculators: MacKenzie and Spinardi 1995. Even the EDSAC, a few years later, made numerous errors because of overheated valves, so calculations had to be repeated several times.)

It wasn't only the mathematically inept who wilted. As we saw in Chapter 12.v.g, Grossberg's undergraduate dissertation—which he'd sent, unsolicited, to numerous renowned researchers in 1965–6—"perplexed" the leading mathematical psychologists at Stanford. They realized that there was a good mind there, and invited him to join them as a graduate student accordingly. But they recognized the work's true quality only much later.

The key difficulty was that *the general nature* of Grossberg's mathematics was unfamiliar, quite unlike what was typically used by psychologists or neurophysiologists at the time. Specifically, it was "nonlinear, nonlocal, and nonstationary" (1980: 351).

Put another way, he was discussing complex dynamical systems, adapting from moment to moment—not chains of atomistically conceptualized events (Markov processes, for instance). Dynamical approaches in cognitive science came to prominence from the late 1980s on (see viii–ix below; 15.viii.c, ix, and xi.b; and 16.vii.c). In the late 1950s, when Grossberg circulated his first work, dynamical systems theory was distinctly outré in psychological circles. So the researchers who'd invited him to Stanford didn't actually take up his ideas about additive networks. Instead, they continued working on stimulus sampling and Markov chains, the preferred intellectual approach at that time (Chapter 6.i.a).

Even as an undergraduate, Grossberg addressed virtually every problem area in psychology. The breadth of his ambition would be evident in the subtitle of his 1982 book: *Neural Principles of Learning, Perception, Development, Cognition, and Motor Control*. In his eyes, these are not utterly different phenomena, grounded in independent modules somehow bolted together in the brain. Rather, they are emergent characteristics of a shared body of fundamental processing principles, which generate and stabilize a hierarchically structured connectionist network. This didn't make things easy for academics trained in narrowly defined specialist areas.

In effect, then, he was trying to keep Rosenblatt's promises (see Chapter 12.ii.e–f). But this was controversial, as well as ambitious. For Rosenblatt had made many of those who *did* value mathematical analysis distinctly wary of any sort of connectionism (12.iii).

Even broadly sympathetic brain scientists such as Cowan were deeply doubtful about Rosenblatt's ambitious claims. And in the computational psychology of the 1960s–1970s GOFAI was the preferred approach. Grossberg attracted scepticism accordingly. Today, he's lauded as one of "a few hardy researchers" who persevered with neural-network models (O'Reilly and Munakata 2000: 9). Then, he was seen as a maverick.

Even so, MIT appointed him Assistant Professor of Applied Mathematics in 1967 (when he still hadn't completed his Ph.D.), and Associate Professor in 1969. He was on the conventional tenure track... until the unexpected recession of the early 1970s put MIT into financial difficulty. Because of OPEC's sudden hike in the oil prices in 1973, plus political influence from the counter-culture, there was less R & D money available in the USA than before (see 11.i.b).

Since MIT was hugely dependent on R & D funding, it was hit harder than most universities were. Berthold Horn recalls it as a time when "There was almost no funding for a while" (quoted in Crevier 1993: 117). One result was that almost all of the untenured junior faculty were unexpectedly let go, as the euphemism has it.

In this context, a knight on a white charger—or anyway, a knight—championed Grossberg in 1973. Sir James Lighthill, whose recent critique of GOFAI had set back AI research in the UK and cast a shadow also in the USA (11.iv.a–b), was asked by MIT for his opinion on Grossberg's work. As the Lucasian Professor of Applied Mathematics at Cambridge, he would understand the material if anyone could—and it was clear that many couldn't.

MIT had approached Lighthill because they were bemused by the extraordinarily mixed response to their first call for references. As Grossberg remembers it:

[When asked whom to approach for recommendations] I naively gave them a list of about fifty names of distinguished people across the fields of psychology, neuroscience, and mathematics. I got a very wide range of letters. A number of letters said I deserved a Nobel prize and I am a genius. I also had other letters that said, in effect: "Who the hell does he think he is trying to model the mind?" (J. A. Anderson and Rosenfeld 1998: 182)

Lighthill's answer to the query, says Grossberg, was "a glowing three- or four-page letter which basically said that I was doing exactly what AI should have done".

But to champion isn't necessarily to rescue. Despite this encomium from MIT's chosen court of appeal, they still didn't offer him tenure. His work was felt to be too controversial, given the unhappy financial climate, and he was warned that he'd soon have to leave. However, he was rescued—not by another knight but by Boston University, who offered him a full professorship. He left MIT for Boston in 1975.

Soon after Grossberg's arrival there, he founded the interdisciplinary Center for Adaptive Systems (CAS)—where he remains today (2006). The Boston group covered a wide spectrum, for he insisted that every CAS member be trained in at least three of four disciplines: computer science, mathematics, psychology, and neurobiology.

Much of his later work was done in collaboration with his wife, Gail Carpenter, a leading researcher in her own right. Her first work had involved mathematical analyses of single cells, using the Hodgkin–Huxley equations (J. A. Anderson and Rosenfeld 1998: 200 ff.), but her interests soon broadened. Like him, she was invited as one of only fourteen opening speakers at DARPA's crucial workshop on connectionism in 1987 (12.vii.b).

By that time, clearly, he was accepted as a leader in the field. Indeed, he'd just become the first President, and co-founder, of the International Neural Network Society (12.vii.a). The wilderness had been left far behind.

b. Adaptation—and feature-detectors

What Grossberg was doing from the start, and what Lighthill thought AI should have been doing, was explaining mind and behaviour by bringing in the brain. Starting from behavioural data, he looked for neurophysiological explanations. And he tried to do this as *realistically* as possible. Toy problems were avoided: the focus was on "problems requiring *real-time adaptive* responses of individuals to *unexpected* changes in *complex* environments" (Grossberg 1988, p. viii; italics added). Blocks world, this was not.

He also tried to do it as *parsimoniously* as possible. He spoke of "minimal anatomies", meaning the simplest neurological model that could account for the observed behavioural constraints. A few general principles—i.e. mathematical theorems, some of which he saw as "particularly simple and lucid" (1976*a*: 253)—were used to explain a profusion of empirical facts. Many superficially diverse (and seemingly unrelated) aspects of behaviour were unified, as unexpected emergent properties of an underlying dynamical system.

The data were drawn from psychology, Grossberg's first love as an undergraduate at Dartmouth, and from various levels of neuroscience. Besides explaining various *psychological functions* in terms of common processing principles, he aimed to show how a variety of *neural structures* can emerge from the same underlying neurological source. That is, he wasn't merely looking to current neuroscience for ideas that would explain behaviour. He was also looking to behaviour—in particular, to the way in which the organism adapts moment by moment to a changing world—to suggest hypotheses about as yet unknown neural mechanisms.

With hindsight, one can list many experimental confirmations of biological predictions made by his ART computer model (see below). For example, since 1988 there have been at least a dozen confirmations of his claim that attention and concept matching involve a top-down modulatory on-centre/off-surround network. Some of these, though not all, were experiments done by his Boston team. So he can surely escape Gross's charge (above) about "theoreticians" having contributed nothing to biology.

Consider, for instance, what he said about feature-detectors. By the mid-1970s, neurophysiologists—including Barlow (Barlow and Pettigrew 1971)—had discovered that even if cortical orientation detectors are in some sense innate, they aren't genetically *determined*. (People didn't yet realize that the fact that newborn kittens already possess vertical/horizontal detectors *does not* prove that these must be coded in the genes: see Section ix.a.)

Recent experiments had shown not only that normal experience fine-tunes the detectors already present at birth, but also that unusual visual input after birth can produce corresponding types of detector or, through disuse, destroy them. The relevant differences in *behaviour* had been noted in the 1960s, but now people were studying the neurology. If a newborn kitten wears goggles that present it with only vertical lines in the right eye and only horizontal lines in the left, the orientation cells in its brain will develop/atrophy accordingly. In a word, even low-level feature-detectors can be *learnt*—and unlearnt.

Grossberg asked how this is possible, and why the visual cortex has the columnar structure that it does. Tellingly, his answers were published, soon after his arrival at Boston, in a journal of cybernetics, not neuroscience (Grossberg 1976*a,b*). He described a form of Hebbian learning, or synaptic modification, justified by proofs of mathematical theorems. This two-part paper was built on the account of brain processing he'd been developing for almost twenty years—and it graduated from "additive" to "adaptive resonance" theory (see below).

The first part of the paper concentrated on feature-detectors, and offered a functional classification of the different types. Grossberg defined equations that would enable single cells in the cortex to learn different patterns, having compensated for various types of real-world noise. They might do this by enhancing contrast in the input, for instance, or by computing relative rather than absolute light intensities.

He added that the equations implied that cortical cells learning similar patterns would become grouped together. He indicated how the range, or degree of resolution, of coded similarities would depend on the numbers of neurones involved (Chapter 12.v.c). And he argued that pattern-discriminators can be hierarchically organized—as in Hubel and Wiesel's hypercomplex cells (which he mentioned specifically—1976*a*: 250), Gross's monkey's-hand-detectors, and perceptual concepts in general. In addition, he offered some testable hypotheses about the role of the retinal amacrine cells.

Characteristically, the paper used powerful mathematics inspired by specific findings in neuroscience. For example, it relied on S-shaped (sigmoid) learning rules, which modelled the cells' responses at high/low extremes of stimulation. In thus avoiding over(under)shoots, this mimicked some properties of biological neurones—and other cells, too.

Grossberg used sigmoids for a reason. He'd published a theorem in 1968 which proved that sigmoid learning enables cells both to suppress noise and to enhance contrast sensitivity, thus avoiding what he called the "noise-saturation dilemma". The idea was later used by the PDP connectionists in two famous papers (McClelland and Rumelhart 1981; Rumelhart *et al.* 1986*a,b*).

The proof concerning sigmoid cells wasn't a stand-alone result, but part of an analysis of *all* the ways in which cells could signal one another. That analysis also showed that *on-centre/off-surround interactions* (between cells that obey the membrane equations describing neurones) are *essential* to avoid the noise-saturation dilemma. In other words, it's no accident that the nervous system contains so many on-centre/off-surround units, and so much lateral inhibition (so Hartline had discovered cells whose existence was largely explained by the mathematics of system dynamics: see iii.c, above). Accordingly, Grossberg's model of pattern learning simulated centre–surround cells and lateral inhibition to increase the contrast in the input.

Third, the information that modified the simulated synapses wasn't carried by single responses, as in the networks of Chapter 12. Instead, it was coded by changes in a cell's rate of firing, averaged over small amounts of time. This reflected the neuroscientists' long-held conviction that the *frequency* of neuronal firing is crucial for coding (see Sections iii.b and ix.g). Again, Grossberg's learning rules—which defined recurrent networks (12.viii.b)—distinguished between short-term and long-term memory.

Fifth, some of his equations presupposed facts about local connection structures in the retina and the brain. And last, but by no means least, the model showed how a pattern can be *distributed* over a large number of individual neurones.

All very intriguing—at least, if one could understand it. But as we've seen, at first many couldn't. So why should they believe him? ("Who the hell does he think he is...?") How could they be persuaded that his theory really did imply what he said it did? Perhaps it was just hot air?

Well, no. For another computer model, besides the Los Alamos experiment, had already suggested there was gold in the Grossberg hills. This model, which Grossberg commented on in his 1976 paper, was focused on feature-detectors. It was inspired by additive theory, and in particular by some equations that Grossberg had published, and criticized, early in the decade (Grossberg 1972). Its designer was Christoph von der Malsburg (1942–), who'd implemented it a few years earlier (von der Malsburg 1973).

Von der Malsburg adapted Grossberg's equations in building a two-layer connectionist network that modelled the development of striate cortex from a partly random starting point. Only "partly" random, because the system started off with more local structure than the networks discussed in Chapter 12: as remarked above, Grossberg's theory presupposed certain facts about brain connectivities. The results were startling.

When the 162-unit "retinal" layer was repeatedly presented with lines in various orientations, the 338-unit "cortical" layer gradually organized itself into a two-dimensional structure that matched Hubel and Wiesel's descriptions of visual cortex to a remarkable degree. Units that responded to the same orientation were grouped together in clusters. In addition, neighbouring units responded to similar orientations, so that the clusters were systematically arranged across the "cortex": see the seventh line down in Figure 14.2.

These results ensued even though the input had contained significant amounts, and interacting types, of noise. (This model *illustrated* the emergence of ocular columns rather than *explaining* it: a few years later, von der Malsburg (1979) gave an explicit analysis.)

F<small>IG</small>. 14.2. View onto the [simulated] cortex after 100 steps of learning. [Each small line shows the orientation to which the relevant unit responds most strongly.] Reprinted with permission from von der Malsburg (1973, fig. 13)

Moreover, von der Malsburg found that non-standard stimulation had 'neuro-anatomical' effects comparable to those which occur in real animals. It's evident from Figure 14.1 that the input for the standard learning history depicted there had ranged across many different orientations. When the network's input was restricted to only vertical or only horizontal stimuli, the resulting orientation detectors were skewed and/or absent accordingly. And this skewing became irreversible after a certain "critical" period. Both these results had recently been observed in newborn kittens (C. Blakemore and Cooper 1970).

(As we'll see in Section viii.a, Von der Malsburg was also working on a model of neural self-organization *in the embryo*, prior to any perceptual experience: Willshaw and von der Malsburg 1976.)

This simulation proved that Grossberg's account, at least in its modified form, wasn't merely hot air. However, it didn't model every aspect of additive theory—never mind its later substitute (Grossberg 1976b). For instance, von der Malsburg used only very simple stimuli, namely motionless straight lines, whereas Grossberg had also discussed more complex cases.

Indeed, Grossberg was arguing—as always—in highly general (i.e. theoretically parsimonious) terms. His equations, he said, fitted many different parts of the brain, including olfactory, auditory, and cerebellar cortex, and also the hippocampus. They were part of a general psychophysiological theory he applied also to attention, arousal (including consciousness), analgesia, motor action, communication, reasoning, and psychopathology.

Even an outline simulation of that would require a much more ambitious computer model.

c. ARTful simulations

Together with his wife, Grossberg eventually produced one. They called it ART-2 (Carpenter and Grossberg 1987).

This system learnt to recognize input patterns, both binary and analogue, in real time and despite noise. The categories involved could be familiar, near-familiar, or novel, and they could self-organize on several hierarchical levels. Where GOFAI systems at that time were testing pre-set perceptual hypotheses (see Chapter 10.iv.b), ART-2 formed, learned, and stabilized its own. Moreover, it did this in real time, even after having learnt many different categories, because learnt categories could be accessed *directly*.

ART-2 was based on adaptive resonance theory, introduced in Grossberg's 1976b paper. The core idea was this:

The functional unit of cognitive coding [is] an adaptive resonance, or amplification and prolongation of neural activity, that occurs when afferent data and efferent expectancies reach consensus through a matching process. The resonant state embodies the perceptual event, or attentional focus . . . (Grossberg 1980: 1)

The "expectancies" concerned included not only what we'd normally call expectations, but also predictions, questions, goals (intentions), and even learnt or 'instinctive' motor patterns such as those modelled by Kilmer (see Section iv.a).

On this theory, activity patterns are passed between two networks until they match. At that point, they stabilize. Inputs are held in STM, and compared with previously learnt categories ("codes") in LTM. Near-matches lead to refinement of the pre-existing codes. If there's no match, previously uncommitted LTM (if any) is used to encode the pattern currently held in STM.

This way of putting it may suggest that the patterns are passed *to and fro* between the networks, which could take a long time. But ART-2's activity patterns oscillated at most once, and self-stabilization was mathematically guaranteed. Unlike the networks described in Chapter 12, it didn't require repeated presentations of the input pattern, but "quickly learned to group fifty inputs into thirty-four stable recognition categories after a single presentation of each input" (Carpenter and Grossberg 1987: 151). It did this by using fast parallel search, the nature of which changed adaptively during learning.

ART-2's predecessor, ART-1 (now there's a surprise!), could handle only binary input patterns. Ideas from ART-1 had been borrowed by a number of fellow connectionists for use in their own work. ART-2 was less borrowable, because its many-levelled analogue-friendly representations were much more complex.

Nevertheless, patents were taken out on both systems, and on their successors ART-3 and ARTMAP. By the late 1990s "a lot of people" were using them for industrial R & D (J. A. Anderson and Rosenfeld 1998: 190). Whereas ART-1 and ART-2 dealt with perceptual categories, the later systems went beyond these to include aspects of attention, goal seeking, and motor action.

One might describe ART theory as an updated, and enormously complexified, version of TOTE units (Chapter 6.iv.c), in the sense that the core idea was *testing*:

[Within] an ART system, adaptive pattern recognition is a special case of the more general cognitive process of discovering, testing, searching, learning, and recognizing hypotheses. Applications of ART systems to problems concerning the adaptive processing of large abstract knowledge bases are thus a key goal for future research. (Carpenter and Grossberg 1987: 152)

In other words, Barlow's "incautious" suggestion about intelligence was being deliberately taken up (see Section ii.b).

Strictly speaking, ART-2 wasn't "one" model, but a general class of models. So were ART-1, ART-3, etc. For the neuropsychological theory was expressed as a set of differential equations, whose variables could be instantiated in different ways and where one or more could be experimentally dropped, or "ablated" (pp. 154–5).

Grossberg's theory had initially been developed without any reference to computers, as we've seen. When he arrived at Stanford in 1961, the leading simulator there tried to model his theory, but failed (personal communication). Later, when suitable technology became available, Grossberg turned to simulation himself. Indeed, he not only used that methodology, but proselytized about it.

Besides helping to clarify one's theory, he said, it can help advance it. For it prompts thinking about how the data could arise as "emergent properties of a real-time process engaged moment-by-moment by the external environment" (1988, p. viii). One-off simulations, however, are irrelevant. (Compare Drew McDermott's critique of one-off programs in GOFAI: 11.iii.a.) Copious, and systematic, modelling is required:

Only through the sustained analysis of many hundreds or even thousands of such experiments can one accumulate enough data constraints to discard superficial modelling ideas and to discern a small number of fundamental design principles and circuits. (Grossberg 1988, p. viii)

Through this approach, "a series of design paradoxes, or trade-offs, come into view which balance many data and computational requirements against one another".

Neuroscience in particular, he insisted, needs computer simulation:

The dynamics of large ensembles of neurons are as yet difficult to observe directly. Even if direct observation were possible, it would not explain *how* the interactions among neurons generate the emergent properties that subserve intelligent behavior. Additional methodologies are needed to investigate *how* the collective properties of a neural network are related to its components. Computer simulations of neural networks are crucial tools in the current explosion of work in brain science.

... Once a mathematical model of neural functioning is formulated, [it may be possible to prove] theorems concerning [its] stability or convergence behavior ... [However, neural processes] involving large and hierarchically organized systems of nonlinear ordinary differential equations are characteristically difficult to analyse through purely formal procedures. For these systems there may be no way to determine the output of the model when given a certain input short of "running" the model in a numerical computer simulation. Thus "experiments" can be run on a model, in ways that are similar in some respects to experiments run on human or animal subjects. (Grossberg and Mingolla 1986: 195; italics added)

This method enables neuroscientists to do things they simply couldn't do before:

Simulations can be cheaper and faster [than experiments on organisms], but more importantly they permit a much more precise level of control of many variables than could ever be realized in physiological or behavioral experimental paradigms. [They can even lead to] general design insights which can at times be formalized into mathematical proofs that would otherwise have been difficult to discover. (Grossberg and Mingolla 1986: 196)

In short, computer modelling can promise significant results that wetware experiments can't achieve.

Why did he spell this out at such length, as late as 1986? After all, cognitive scientists had been saying this sort of thing for years. The essential point had even been made, with respect to *mechanical* simulation, by some early 1930s behaviourists (see Chapter 5.iii.c).

The reason was that, even in the late 1980s, many *neuroscientists* were still suspicious of computer-based methods.—Computational (information-processing) questions, yes; computer models of the brain, *No!*—James Anderson recalls that at Rockefeller University in the early 1970s, "psychologists were far more open and responsive to new ideas [i.e. connectionism] than were neuroscientists" (J. A. Anderson and Rosenfeld 1998: 254). Similarly, Arbib found in the mid-1980s—when Grossberg was writing the words just quoted—that "the experimentalists in neuroscience, instead of being excited at having a distinguished program of brain modelling on campus [UMass, Amherst: see Section vi.b–c], saw it as either irrelevant or threatening" (J. A. Anderson and Rosenfeld 1998: 229).

Most brain scientists today are less suspicious (although Gross's comment on Marr smacks of the old school: v.e, above). They're increasingly using computer models in their work. Indeed, virtually all neuroscience—excepting nitty-gritty neurochemistry and theoretically unmotivated brain scanning (see Section x.b)—now involves computational questions about brain function, whether or not simulation is involved. This applies to work on feature detection, cognitive maps, motor control, action selection, language, schizophrenia . . . and consciousness.

As Arbib has remarked, his computationally oriented *Brain Theory Newsletter*, which attracted only a few hundred subscribers in 1980, would be selling in the tens of thousands were it still running today (J. A. Anderson and Rosenfeld 1988: 231). (In fact, it was incorporated into the journal *Cognition and Brain Theory* in 1982: see Chapter 6.v.c.)

d. Avoiding the black box

Inevitably, neuroscientists' computer models will become increasingly realistic. But there's a danger here. Lighthill commended the young Grossberg to MIT for doing what he thought AI should have done: taking account of the brain. The message of his notorious Report (Lighthill 1973) was that biological realism is a Good Thing—and, by implication, the more of it the better. But that implication was questionable.

Even in the 1940s, Ashby (1948: 383) foresaw that advanced versions of the Homeostat might behave in ways "too complex and subtle for the designer's understanding" (see Chapter 4.vii.d). Similarly, von Neumann predicted at the Hixon Symposium that complex automata would be almost as mysterious to their builders as animals are to biologists (Jeffress 1951: 109–10). And a widespread worry in the late 1980s, as we saw in Chapter 12.vi.f, was that large connectionist systems, once they became available, would be unintelligible.

Given enough computer power, one can stuff extra data parameters into one's simulations indefinitely. But if the result is in effect a black box, the exercise has scant scientific value. In the mid-1980s, one British computer scientist put the worry like this:

Connectionism and its second coming is the next unavoidable issue [for AI] . . . [If] subsymbolic networks are necessary to support intelligent systems, then the conceptual transparency of the resulting AI systems is likely to be *on a par with that of the brain—i.e. somewhere close to zero.* Homogeneous networks of subsymbolic elements will severely test our abilities to understand our models. So if subsymbolic network architectures are in some sense necessary then that might be bad news for AI. (Partridge 1986–7: 16; italics added)

Similarly, Willshaw, in defending early Marr against charges of oversimplification, has pointed out "the danger that any completely specified model will become just as difficult to analyse as the brain itself" (Willshaw and Buckingham 1990: 116).

How does Grossberg's research stand up to the charge of black-boxery? It must be said at once that his work appears forbiddingly complex to non-specialists. Consider this statement of his commitment to realistic modelling, for instance:

We [at CAS] typically begin by analysing a huge interdisciplinary data-base within a prescribed problem area. In our work on preattentive vision, for example, we have studied data from many parts of the vision literature—data about boundary competition, texture segmentation, surface perception, depth perception, motion perception, illusory figures, stabilized images, hyperacuity, brightness and color paradoxes, multiple scale filtering, and neurophysiology and anatomy from retina to prestriate cortex. (Grossberg 1988, p. viii)

All these data from visual psychology feature in a single paper on the perception of colour and 3D form (Grossberg 1987*b*). The same paper also draws on hypotheses about the cortical and retinal cells involved—and the neurotransmitters, too.

But the paper—like other CAS publications—isn't a mere ragbag of empirical facts. On the one hand, it draws many theoretical morals. For example, it argues that "the popular hypothesis of independent modules in visual perception is both wrong and misleading. Specialization exists, to be sure, but its functional significance is not captured by the concept of independent modules" (Grossberg 1987*b*: 4). It allows for learned (largely top-down) as well as automatic (bottom-up) processing in object recognition. It offers a new theory of stereopsis. And it makes many testable predictions.

On the other hand—and even more to the point, with respect to black-boxery—the paper rests on a set of mathematically proven theoretical principles. These unify the huge variety of detailed behavioural constraints simulated in the model.

A more recent publication, which explains depth vision and various visual illusions in terms of highly specific parts of the brain, makes the simplicity beneath the complexity even more evident (Grossberg and Howe 2003). To appreciate that simplicity is to understand the model. And the general principles underlying Grossberg's work are clearly intelligible to some people outside his own group at Boston. For—as we've seen—ideas from ART-1 have been used in other connectionists' models, and ART systems are even being used commercially.

One way of putting all this is to point out that it would be misleading to say that the behavioural constraints (about visual illusions, for instance) are "built into" the ART models—still less, that "extra data parameters" are repeatedly "stuffed into" them. Rather, the behavioural details *emerge* from, are *generated* by, the underlying system dynamics. This didn't happen by magic, of course. The design of the system dynamics—the "minimal anatomies", and combinations thereof—was guided by Grossberg's intuitions about what sort of system could generate those (seemingly unrelated) behavioural data. (Intuitions which, as we've seen, even the Los Alamos physicists didn't share.)

The complexity is due to his commitment to biological realism. That is, to his continual addition of further behavioural and neurological constraints (as in the colour/form and stereopsis examples mentioned above) and, when necessary, further theoretical principles. The intention is to *test and expand* the existing theory, not to *complicate and obscure* it by adding innumerable ad hoc hypotheses. Thus far, Grossberg's team, and a fair number of outsiders too, are still able to see the theoretical wood as well as the observable-data trees.

Even so, this approach is not for the faint-hearted. And it remains to be seen whether, in the new century, the CAS group will ever reach a point where even they can't really understand what's going on. The same applies, of course, to 'realistic' brain modelling in general. Anyone who aimed for a *completely* specified model would risk sacrificing understanding to observation. In Marr's terminology, Type-I explanation would threaten to give way to Type-II (see Chapter 7.iii.b).

In short, biological realism could conceivably go too far—especially if the modeller doesn't build the system up from a theoretically parsimonious base. This is one way of expressing the familiar claim that the brain may be too complex for the human mind to understand it. Whether that's true is still an open question, though we shouldn't stop trying yet.

Future "unintelligible" simulations whose performance successfully matched the neuropsychological data would be existence proofs that the brain isn't essentially *mysterious*. But lack of mystery isn't enough for scientific understanding. (The "mystery" of consciousness involves additional issues: see Sections x–xi.)

14.vii. Whole Animals

Kenneth Craik, in the early 1940s, pointed out the importance of "selection pressures" in evolving internal "models" of the environment (see Chapter 4.vi.b). The interests of the animal as a whole were crucial, he said, in shaping such models. But he could give few specific examples.

A fortiori, he couldn't give neuroscientific chapter and verse in describing the *whole* animal. No one at that time could—and it's not even clear that anyone tried. Thinking about the whole animal, as opposed to specific sensory or motor mechanisms, wasn't a practical aim. Moreover, the environmentalist influence of behaviourism kept ethology out of the mainstream (5.ii.c).

Eventually, that changed. The last four decades of the twentieth century saw the gradual rise of what's now called computational neuro-ethology (CNE), a term coined in the late 1980s. CNE's strongest disciplinary links are with neuroscience, ecological psychology, and A-Life. But it has implications for cognitive science in general (Keeley 2000*b*).

One pioneering example, in which the *general* implications were given especial prominence, is discussed in this section. Further examples—focused on crickets, hoverflies, cockroaches, and lampreys—will be described in Chapter 15.vii. All these CNE programmes have generated novel hypotheses about the nervous system, whether in humans or in other animals.

a. CNE—what is it?

CNE is Computational. It's Neural. And it's Ethology. Ethology considers the whole animal's behaviour in its natural habitat, focusing on species-specific adaptations to a particular environmental niche. Neuro-ethology (named in the early 1980s) studies the underlying neural mechanisms. And CNE interprets these in computational terms.

The earliest CNE work was done in the 1960s. The scientist concerned—namely, Arbib—continued his research and expanded his theory over the next forty years, as we'll see (subsections b–c, below). But there was *no* steady growth in interest within the cognitive science community as a whole.

On the contrary, we'll see in Chapter 15.vii that most of the influential CNE research—Arbib excepted—wasn't done until the late 1980s. Randall Beer's "manifesto" (his word) for the field was published in 1990, soon after Christopher Langton's definition of "Artificial Life" (15.xi.b). Partly because Beer allied himself with the newly emerging A-Life community, CNE was now seen in the context of A-Life as much as of neuroscience.

One might wonder why CNE didn't blossom long before then. After all, ethology as such had been growing since the First World War (Chapter 5.ii.c).

One reason was that ethology was overshadowed by behaviourism for many years. Indeed, hardly any of the early work was translated into English before the late 1950s (C. H. Schiller 1957). The behaviourists focused on general mechanisms, studied in laboratory conditions (5.i.a). They saw Konrad Lorenz's fieldwork, for instance, as birdwatching—not science. This wasn't fair: Lorenz had provided systematic observation and even some experiments, not mere anecdotes. But, the power of fashion being what it is, the ethologists were sidelined.

Another reason was that AI hadn't progressed sufficiently. Robotics, in particular, was hugely difficult. Beer's six-legged cockroach robots, for instance, were a triumph of mechanical engineering no less than CNE. By the mid-1990s, however, the *programmable* Khepera robot—designed by Francesco Mondada in 1992–3—could be bought off the shelf. In addition, VLSI techniques could be used to provide (for instance) speedy sensory processing based on neuroscientific data.

A third reason was that neurophysiology itself wasn't far enough advanced until after mid-century (Chapter 2.viii.e). Moreover, it was typically conducted in a reductionist spirit that lost sight of the animal as a whole.

The 'Frog's Eye' work of the late 1950s is a good illustration. It could be regarded as the first exercise in CNE—but if so, this was more by accident than design. Lettvin and his colleagues were driven to consider the frog *as a frog* only by the surprising feature-detectors they found (see Section iii.a). They hadn't set out with the intention of analysing the visual system in terms of predator and prey.

Similarly, in recalling his own "unheard of" results, Hubel has said: "It is hard, now, to think back [to the late 1950s] and realize just how free we were from *any idea of what cortical cells might be doing in an animal's daily life*" (1982: 516; italics added).

If they'd started from an ethological viewpoint, their findings would have been less of a surprise. Indeed, Lettvin's friend Selfridge had already argued, on computational grounds, that ecologically relevant feature-detectors must exist. But he didn't know which, or where, they were: it was Lettvin's team who put the N into the CNE. And it was a research student in their laboratory who would make that "N" *whole*.

b. A wizard from Oz

One of the many people enthused by the 'Frog's Eye' research was Arbib (1940–), an English-born Australian mathematician who arrived at MIT as a graduate student in 1961. He started his Ph.D. with Wiener (later continuing with the probability theorist Henry McKean). Meanwhile, he worked as a research assistant in McCulloch's laboratory until he left in 1963 to go to Stanford (via Europe and Australia).

This seemingly innocuous association with McCulloch had to be kept from Wiener, who became "apoplectic with rage" when, on Arbib's last day, he discovered it (Arbib 2000: 203–4). His vitriolic hatred of McCulloch has been 'explained' in various ways. McCulloch himself told Arbib a story about having involved Wiener's daughter with his Mafia contacts, for her own protection. This was fascinating, says Arbib, but—as with many of McCulloch's stories—probably largely fiction.

The most likely explanation, Arbib now believes, is that Wiener (in 1952) had asked McCulloch to tell him about the brain, and McCulloch's mini-lecture had (characteristically) contained as much conjecture as fact. When, after three years' work, Wiener triumphantly presented a meeting of neurophysiologists with an elegant mathematical theory accounting for everything McCulloch had told him, he was laughed out of court. He hadn't realized that McCulloch was one of those people whose talk is peppered with speculation dressed up as fact—albeit hugely stimulating, insightful, intelligent speculation. (Turing had realized this, but concluded that McCulloch was "a charlatan": Chapter 4.iii.a.)

Nor did Wiener realize it now. Instead, he assumed that McCulloch had maliciously set him up for this fall. With better personal antennae, he would have avoided the humiliation—and the dreadful decline of the young Pitts, whose life was shattered by the feuding (see 4.iii.d).

Lettvin (1999) mentions "a viciously phrased letter" sent by Wiener breaking all connections with McCulloch's group. And he says:

It had nothing to do with any substantive cause but was the result of a deliberate and cynical manipulation designed to sever Wiener's connection with McCulloch and his group. The details are not edifying. Wiener was victimized as much as the group. (Lettvin 1999)

With no further information on the "details", or on who was doing the "manipulation", it's not clear whether this angry letter was part-prompted by the events described by Arbib. But that there was a rift, and that the effect on Pitts was (as Lettvin puts it) "devastating", isn't open to doubt.

The newly arrived Arbib was already more than familiar with computational ideas. He'd found superior proofs for McCulloch and Pitts' 1943 paper while still an undergraduate (Chapter 4.iii.f).

He very soon co-authored a paper on neural modelling with McCulloch and Cowan, then a 'mature' graduate student alongside Arbib, for the seminal conference on self-organizing systems in Chicago: see Chapter 12.iii.e (McCulloch *et al.* 1962). And shortly after that, at one of Oliver Waddington's maverick meetings on theoretical biology, he would define a self-reproducing automaton based on von Neumann's ideas (see Chapter 15.v.b). So his maths was in excellent shape: he went to McCulloch to learn neurophysiology.

The prime lesson Arbib took from the lab's work on feature-detectors wasn't a conversion to single-cell recording, but the realization that perception serves the interests of the organism concerned. Presumably, then, it's integrated with the creature's motor mechanisms in ecologically appropriate ways. He was also inspired by the Mars robot RETIC (see Section iv.a), because of its concern with the action of *the whole animal*. Kilmer, in fact, was the junior colleague at MIT who had the most influence on him (and later joined Arbib's brain research group at the University of Massachusetts, Amherst).

In short, Arbib's aim was to progress from 'What the Frog's Eye Tells the Frog's Brain' to "What the Frog's Eye Tells the Frog" (Arbib 1981a: 23). But whereas the 'Frog's Eye' author Maturana would soon interpret the whole-animal insight in an anti-computational—and highly unorthodox—way, Arbib stayed within the computational camp.

Indeed, he soon became one of its most prominent champions. In 1964 he published an introductory text on the mathematics of automata theory, including McCulloch–Pitts networks (Arbib 1964). To his amazement—he was only 24, and still a Ph.D. student when he wrote it—it got the lead review in *Scientific American*, and a mention in an article on Wiener in the *New York Review of Books*. He now says this was "not so much because of the merit of the book, but the fact that Wiener had just died, and so cybernetics was a hot topic at that time" (J. A. Anderson and Rosenfeld 1998: 223). However that may be, it sold very well.

Eight years later, he added neuroscientific flesh to the computational bones in a widely read book about the brain, tellingly subtitled *An Introduction to Cybernetics as Artificial Intelligence and Brain Theory* (1972). In effect, this was a declaration of intent for his interdisciplinary brain research group at Amherst, where he went after six years at Stanford.

The project was set back by the drying-up of funds for connectionism in the mid-1970s—and Kilmer even deserted the field as a result. After DARPA's crisis meeting in 1987, however, the funding pendulum swung back again (Chapter 12.vii.b). Arbib soon went to the University of Southern California to direct their Brain Project.

The 1980s were hugely busy years for Arbib, but not hugely influential. As remarked above (Section vi.c), most 'pure' neuroscientists didn't take connectionist modelling seriously—especially before PDP became widely visible at the end of the decade. Arbib's pioneering *Brain Theory Newsletter* was discontinued before the wider research community caught up with him.

Eventually, however, his influence grew. His interdisciplinary research was suddenly 'discovered', as other neuroscientists began to express their sensori-motor theories as computer models—and as whole-animal approaches in general gained ground.

In the 1990s, he was Editor of the *Handbook of Brain Theory and Neural Networks* (with a second edition in 2003). He wrote a paper for the AI community recommending biological design principles in robotics (Arbib and Liaw 1995). And he co-authored a book uniting neuroscience and (animal and human) psychology with the computational theory he'd been developing since the early 1970s (Arbib *et al.* 1997). By the turn of the millennium, he was using the recent discovery of 'mirror neurones' to part-explain the evolution of language in the context of his long-established approach (for his latest statement, see Arbib 2005).

In sum, he's been a key player across the whole of cognitive science for over forty years.

c. *Rana computatrix* and its scheming cousins

To find out just what the frog's eye tells the frog, Arbib and his students (starting with Richard Didday) did some behavioural and neurophysiological experiments on the real animals. Their main activity, however, was developing a computer model—more accurately, a still-continuing *series* of models—of sensori-motor integration in the frog (Didday 1976; Arbib 1972, 1982, 1987, 2002*b*; Arbib and Cobas 1991; Cobas and Arbib 1992; Arbib and Lee 1993; Liaw and Arbib 1993; Arbib *et al.* 1997).

Arbib named it *Rana computatrix*. One might naturally assume (many people did, myself included) that this was in memory of Grey Walter's *machina speculatrix*

(Chapter 4.viii.a–b). And indeed, Arbib had already read *The Living Brain* when he first started work on *Rana*. But he wasn't aware of the connection:

I had read [Grey Walter's book]. But what was interesting was that I had not consciously realized the inspiration for the name *Rana computatrix* until some years later, when Dan Dennett asked if *machina speculatrix* were the inspiration—and I suddenly realized it surely was! (Arbib, personal communication)

So, like Paul McCartney's unrecognized memory of 'Yesterday' (Chapter 1.iii.g), Arbib was recalling someone else's work—in a quite specific way—without knowing that he was doing so. If he hadn't been asked about it, he might still be convinced of his originality today.

Putting the name of his computerized frogs on one side, what about the ideas implemented in them? To some extent, Arbib's early inspiration came from Valentino Braitenberg (see 15.vii.a). He says now (personal communication) that he was "excited" by Braitenberg's ideas on the cerebellum, and "enjoyed" a paper foreshadowing the book on *Vehicles* (Braitenberg and Onesto 1960; Braitenberg 1965). But the main intellectual prompting, believe it or not, came from SHAKEY (see 10.ii.a and iii.c): "My real interest in robots was developed by the work of Nils Nilsson and colleagues at SRI on SHAKEY, from around the time I was at Stanford" (personal communication).

One of the lessons learnt from SHAKEY, and from his own later experience too, was the enormous difficulty involved in not merely ("merely"?) designing robots, but actually building them. After all, SHAKEY itself had got its name not from any of the influential ideas implemented in it, but from the fact that—considered as a mechanical object moving on the floor—it wobbled (see 10.ii.a).

Accordingly, Arbib's frog model was only ever implemented as computer programs. He did consider building hardware robot frogs, and published a paper on robot design (Arbib and Liaw 1995). But his robots were all simulated, not real. (One of his students, Tony Lewis, did eventually build a robot salamander, with a flexible spine and a GA-generated neural controller: M. A. Lewis *et al.* 1994.)

Amphibia weren't Arbib's only interest. From the early 1970s, he also worked on mammals. For instance, his team modelled stereo vision, and motor control by the cerebellum (Arbib *et al.* 1974). Much later, as we'll see, he widened his scope still further, applying the ideas initially developed with respect to frogs to the highest reaches of human thought.

Arbib's artificial frog carried out highly distributed information processing, so developing McCulloch's ideas about "redundancy of command". Initially, Arbib used competing, winner-takes-all, networks (see Chapter 12.ix.a). These modelled the dynamic perception–action cycle, wherein perception leads to action which leads to a new perception . . . and so on. Perceptual recognition depended on a "slide-box" defining a fixed set of categories. And (like RETIC) the system specified a finite set of possible actions, one of which would be activated after competition between the relevant networks.

By 1980, however, fixed networks and slide-boxes had given way to multi-layered perceptual and motor "schemas" (Arbib 1981*a,b*). Arbib later said that schema theory "is rooted in the *Critique of Pure Reason* of Immanuel Kant" (Arbib 1995: 11 and sect. 3). But it was mediated by McCulloch's use of Kantian ideas, and the 'Frog's Eye' paper's identification of "genetically built in" neural contexts as "the physiological synthetic *a priori*".

These Kantian ideas, in those behaviourist days, were unorthodox, not to say shocking. However, neuroscientists had Henry Head's example to encourage them (4.vi.a). Accordingly, Arbib—like Bartlett before him (5.ii.b)—borrowed Head's word: what Craik had termed models, he termed schemas. He hadn't used this term initially, but started doing so after friends pointed out the similarity to Bartlett and Jean Piaget—both of whom had spoken of "schemas" (Arbib 1985: 3; S. Gallagher 2004: 53). Indeed, Piaget's constructivism, with its emphasis on continual change in accommodating to the environment, became another important influence.

Schemas were dynamical structures, continually adapted as a result of experience. Each perceptual schema tested for, or (as Grossberg would say) expected, a certain type of object. The simplest ones were feature-detectors under another name. The more complex ones, schema "assemblages", defined general classes. Subtle variations between class members could be identified by "fine-tuning" of the schema's parameters. Similarly, each motor schema specified a general type of action. The simplest of all caused some utterly invariable movement. Usually, however—and unlike the Mars robot—they allowed for fine-tuning to generate movements adapted to circumstances: same action, different movements.

Schemas could be closely associated and/or hierarchically nested—and could be thought of in both neurological and AI terms:

For [brain theory], the analysis of interacting computing agents called *schema instances* is intermediate between the overall specification of some behavior and the neural networks that subserve it. For [distributed AI], schemas provide a form of knowledge representation which differs from frames and scripts by being of a finer granularity. A schema is more like a molecule than an atom in that schemas may well be linked to others to provide yet more comprehensive schemas. (Arbib and Cobas 1991: 143)

Arbib saw schema theory as a way of coping with the problem posed by Pitts and McCulloch in 1947: how to move between a universal and its instances. It was also a way of answering Marr's early question, how the cerebellum can make the 'right' detailed movements when a general type of action is intended. Indeed, Arbib's student Curtis Boylls had started to model the cerebellum and brainstem motor nuclei in the early 1970s. His main aim was to discover how they could convert a general motor schema (such as grasping something) into specific behaviour (such as grasping *this* thing).

Even as a tadpole, *Rana computatrix* wasn't a toy frog living in a toy world. Arbib's early model of vision was relatively realistic when compared with GOFAI's blocks world, or even Marr (1982), for it computed depth using the optic flow caused by bodily movement. That is, it modelled 'animate' vision.

Moreover, the possible actions were based on ethologists' studies of just what real frogs/toads do in subtly different circumstances. A frog will snap at prey that's perceived as being nearby, for instance, but jump towards prey that's further away; and, in certain circumstances, it will make a detour around a barrier placed between itself and the prey (Arbib 1982; Cobas and Arbib 1992; Arbib and Lee 1993; Collett 1979, 1982).

Neuroscientific realism was attempted too. As early as the late 1960s, Arbib's first student, Didday—like Kilmer, originally an engineer—had started to program a model of the frog's neuro-anatomy (Didday 1976). By the 1990s, *Rana computatrix* included many more neurological details (Arbib *et al.* 1997).

Arbib soon generalized his research beyond frogs. For instance, he provided a theory of the "cognitive maps" first ascribed by Edward Tolman to rats (Chapter 5.iii.b) and later located in the hippocampus by John O'Keefe and Lynn Nadel (1978). This model of approach and avoidance in *Rattus computator* (Lieblich and Arbib 1982) eventually grew to include (visual) Gibsonian affordances—such as *go straight ahead, hide, eat, drink,* and *turn (left/right)*—which often led the rat into currently invisible parts of space, sometimes by a complex route (Guazzelli *et al.* 1998).

After rats, monkeys: *Macaca computatrix* simulated the neural schemas controlling visual saccades in macaque monkeys. It included attentional saccades to new light stimuli, and memory saccades to points of previous interest (Dominey and Arbib 1992).

As for humans, they weren't ignored either. Arbib (1981*b*) published an influential diagram of the functioning of the human hand that would have delighted the heart of Charles Bell (2.viii.f). This integrated motor schemas for moving the arm, rotating the wrist, and pre-shaping the hand to 'fit' the object to be grasped:

[When] we reach for an object, the brain has to simultaneously figure out the control of the arm and the shaping of the hand. It's not that you get there and then figure out what to do with the hand. The brain is preshaping the hand at the same time that it's coming up with the trajectory . . . [In 1981 I gave] a fairly simple diagram of the interaction between perceptual schemas—figuring out where the object is, what size it is, and what orientation it is in—and the [motor] schemas for control of the arm, control of the hand, and so on, and how they all interacted. (J. A. Anderson and Rosenfeld 1998: 233)

This *Handbook of Physiology* diagram, he now says, is "probably the most successful thing I've done, in terms of the number of reproductions that have occurred in textbooks and papers" (J. A. Anderson and Rosenfeld 1998: 233).

A few years later, he moved from diagram to program. That is, he developed a schema-based program for controlling a (virtual) three-fingered hand (Arbib *et al.* 1985). The hand was designed to perform the central task of the intellectual life: picking up a coffee cup. And, characteristically, the paper discussed the possible neural implementation of the interacting schemas involved.

When Arbib later applied schema theory to high-level cognition and language, however, the neuroscience was largely notional (Arbib *et al.* 1987). The same was true, in spades, when he trod in the footsteps of Hans Driesch (Chapter 2.vii.b), and of John Z. Young and Longuet-Higgins too, and gave the Gifford Lectures on natural theology (Arbib and Hesse 1986: see Chapter 7.i.g).

He co-authored these with Mary Hesse, a distinguished English philosopher of science—and a regular visitor to the Epiphany Philosophers group in the Cambridge apple orchard (Preface, ii). The vagueness was hardly surprising. A schema (assemblage) for a belief system such as Christianity or Islam, or even for component ideas such as salvation or justice, is difficult enough to define in conceptual terms—as philosophers know to their cost. Relating those concepts to detailed neuroscience simply wasn't possible.

Indeed, it may never be possible, *in principle*. Schema theory is expressed at the functional level, between observed behaviour and neural implementation—and functionalists have good reasons for denying any systematic equivalence between conceptual thought and its neural implementation (see Chapter 16.iii–v). One doesn't have to accept the thesis of multiple realizability undiluted—and Arbib certainly wouldn't do so—to grant that many concepts (schemas) can't be mapped onto, or identified with, specific neural mechanisms.

In other words, Arbib used functionally defined 'architectural' theories to interpret *and to guide* his neurology. (In this, he was comparable to Aaron Sloman, who'd been developing a wide-ranging architectural theory, though without the neurology, for many years: see Chapters 5.i.f and 16.ix.c.)

As the century drew to its close Arbib, with Peter Erdi and Eccles's co-author Janos Szentagothai, published a major book on neural organization. (For diverse peer commentaries, see *Behavioral and Brain Sciences* 2000.) Among many, many, other things, this described their computational models of the olfactory system and hippocampus, including schema-based cognition (Arbib *et al.* 1997, ch. 6). They used a dynamical systems approach to model spatio-temporal processes at various different levels, from the subcellular level up (see Section ix.b). And they drew on:

*   anatomy (brain structures defined at many levels of analysis);
*   physiology (of the cell, the synapse, and EEG);
*   psychology (addressing a wide variety of behaviour and cognition, in animals and humans);
*   mathematics (abstract analysis of the functions and dynamics involved);
*   and various areas of AI.

This was interdisciplinarity in action.

The millennial arrow in Arbib's quiver was a development of Giacomo Rizzolatti's idea that mirror neurones, which he'd discovered in 1995, may have enabled the evolution of language (Rizzolatti and Arbib 1998; Arbib 2002*a,b*). A mirror neurone fires not only when a monkey *performs* a certain hand movement, but also when it *observes* that same hand movement performed by another monkey, or by a human.

Arbib listed three possible uses of mirror neurones, including enabling the animal to anticipate, and to imitate, what another animal is doing. Monkeys aren't good at imitation, but people are. In 1994 PET imaging (see Section ix.b) implied that a part of Broca's area 'lights up' when the person makes hand movements; and in 1995, mirror neurones for grasping were indicated there too. Moreover, comparative anatomy suggests that this sub-area may share its evolutionary origin with the 'grasping' area in the monkey's motor cortex. In short, the language area in our brains seems to involve cells capable of aiding gestural communication. Possibly, it may also contain, or be linked to, mirror neurones for facial expressions and speech movements.

(As Arbib pointed out, this hypothesis *does not* address the origins of syntax, negation, or reference to past events. So Chomsky's scepticism about evolutionary explanations isn't allayed: see Chapter 9.iv.e.)

In sum, when Arbib could relate schemas to actual neurology, he did. When he couldn't, he tried to provide computer simulations and/or formal theories that were at least compatible with neuroscience. When he couldn't do *that*, he tolerated vagueness.

But even his Gifford Lectures were delivered with countless glances towards the brain. He was trying to show that a general type of neural processing could underlie a huge range of behaviour—from feeding in frogs to manual dexterity, and even prayer, in *Homo sapiens*.

Arbib's perceptual and motor schemas were analogous to the "T" and "O" of the TOTE units posited in the visionary *Plans and the Structure of Behaviour* (see Chapter 6.iv.c). The creators of both sides of this analogy indulged in optimistic hand-waving. But *Rana computatrix* and its computational cousins incorporated vastly more neuroscientific detail, and much subtler computational insights, than the *Plans* authors could have dreamt of.

Besides being an entire research industry in himself, Arbib has enabled others to en-gage in a fast-growing neuroscientific/neurocomputational industry. In part, he's done that by his editing of comprehensive interdisciplinary tomes. Very recently, however, he and two colleagues have made available their own object-oriented programming language, NSL (Neural Simulation Language), together with an interesting variety of examples of its use (Weitzenfeld *et al.* 2002).

As we saw in Chapter 10.v.d, an object-oriented language enables the programmer to think at a 'natural' conceptual level. Even more importantly, it can embody a good deal of unseen knowledge within the types of "object" provided. That being so, the objects will—up to a point—behave (and interact) in the naturally expected, appropriate, ways without the programmer having to tell them precisely how to do so. In the case of NSL, the hidden knowledge is drawn from neuroscience. For the language was specially designed to enable people to model *macroscopic* brain structures, as well as lower-level neural networks. Indeed, anatomical structures and schemas can be represented on a number of hierarchical levels. Thanks to NSL, *Rana*'s cousins will surely multiply.

14.viii. Representations Galore

By the end of the 1950s, the notion of mental/cerebral models was a near-universal presupposition in the nascent cognitive science. (Though not in Gibsonian psy-chology: see Chapter 7.v.e.) In neuroscientific circles, analogue examples were hale and hearty—though still few and far between. In psychology, linguistics, and AI, formal–symbolic representations had the higher profile.

Today, the idea is still prominent in neuroscience. A recent review of research on hand movements, for instance, declared that "*internal models are fundamental for understanding a range of processes* [involved in motor planning, control, and learning]", and cited many others working on that assumption (D. M. Wolpert and Ghahramani 2000: 1217; italics added). In short, neuroscientists differ over what the internal models are like, but *that there are such models* has long been widely agreed. (Widely, but not universally: see subsection d, below.)

I'm taking for granted, here, that "representation"—considered as a theoretical term (what the positivists called an intervening variable)—is interpreted realistically by cognitive scientists, even by those who deny that they exist. That is, it's not merely a helpful shorthand, or *façon de parler*. Philosophers of science who take an instrumentalist view of scientific concepts and theories *in general* would disagree. For them, to say that representations (or electrons, or molecules, or genes . . .) exist is simply to say that it's explanatorily/predictively helpful to talk about them, *not* to say that there are real, independently existing, entities out there which these terms are denoting. But such radical instrumentalism is unusual among working scientists, and I shall ignore it here.

For the neuroscientist (or other cognitive scientist), then, there are two questions: Do cerebral representations really exist, yes or no? And if so, what form/s do they take?

These questions aren't straightforward. For *just what could count* as a "cerebral model" or "representation" wasn't clear when Craik introduced the term—and it still isn't. Craik and Minsky both tried to clarify the concept (Chapter 4.vi.b), and—later—so did John R. Anderson and Sloman (7.v.a). But it's extremely slippery. For instance, the late 1980s AI scientists famed for *attacking* representations relied on them nonetheless—albeit of a (situation-bound) type different from those used in GOFAI (see 13.iii.b).

This section indicates how, in the fifty years after Craik, cognitive scientists tried to pin the term down. In addition, it looks at the evidence for (various kinds of) representation in the brain.

a. What's the problem?

Disputes about the nature of representation, or internal models, were—and still are—both empirical and philosophical. Craik himself realized this, and it became increasingly evident in the fifty years after his death.

On the one hand, Craik's (and Mayhew's) computational question of *how* this or that can be represented in the brain remained. Cognitive scientists offered various answers, including some very different from those described in Sections iii–vii above. The representational mechanisms they suggested included the following, each of which has been mentioned in one or more of the preceding chapters:

* GOFAI symbols, seen as stable, manipulable, and copyable entities;
* 'iconic' representations of various kinds;
* deictic (situationist) representations;
* activity patterns in PDP networks;
* "fluid" concepts;
* feature-detectors; and
* anticipatory schemas.

They also included examples introduced in this chapter (three of them in subsections b–c below):

* neurophysiological models reflecting real-world dynamics;
* systematic mappings of one functional/behavioural space onto another;
* emulator systems that 'march ahead' of the processes being modelled; and
* evanescent representations used 'online' and then discarded.

On the other hand, philosophers argued about what *counts* as a representation, in principle. Those sympathetic to cognitive science didn't always accept every item on the list above as a genuine representation.

And many who weren't sympathetic rejected all of them. They weren't denying that such neural-computational mechanisms might actually exist. Rather, they were claiming that *no* scientific theory could ever explain representation (or intentionality, or meaning) as such. Those disputes are outlined in subsection d.

b. From probabilities to geometries

A theory of neural representation fundamentally different from those described so far was initiated in the late 1960s by Pellionisz.

The notion of internal models was central: "[the theory] hinges on the assumption that the relation between the brain and the external world is determined by *the ability of the CNS to construct an internal model of the external world* using an interactive relationship between sensory and motor *expressions*" (Pellionisz and Llinas 1985: 246; italics added). But there were several hierarchical levels of "expressions", with specified ways of translating between them.

As a young man in Budapest, Pellionisz developed Farley's pioneering work on two-dimensional neural nets (see Chapter 12.ii.b). Still in his early twenties, he defined a network of over 64,000 neurones, consisting mostly of two neuronal fields of 31,510 units each (Pellionisz 1968, 1970). Besides this huge jump in scale (Farley had managed only 1,296 units), his system differed from Farley's in being modelled on specific aspects of the brain. It was, in fact, the first computer simulation of the cerebellum. (Marr's work had been formal analysis, not simulation.)

Even as a student (an engineer at Budapest's University of Technology), Pellionisz saw *geometry* as the key to the brain. But as we'll see, he didn't mean Euclid. In an early publication (his first in English), he declared: "[I] analyse the transfer function of the neuronal information preprocessing system of the cerebellar granular layer [in terms of a] geometrical model of the regular neuronal arrangement . . . illustrated by a pattern transformation method" (Pellionisz 1968). In other words, the key idea was that one pattern could be transformed into another, by way of a systematic geometrical method.

While still a teenager, he'd been inspired by reading the Hungarian translation of von Neumann's posthumous book (1958; see 12.i.c) in about 1960. But whereas von Neumann had tentatively suggested that thermodynamics is the language of the brain, for Pellionisz it was geometry:

On literally the last page of his book, the man who knew the mathematical language of computers so well . . . confessed in public on his death bed that we have no idea of the mathematical language of the brain (but he was sure it is different from Boolean algebra, the mathematics of man-made computers).

I always excelled in creating new stuff—thus I was flabbergasted in the era of "Cybernetics", ready to create "thinking machines"—that the "top expert" has no idea of its basic mathematics. And he dies, though he feels that he is coming close.

My life was cut out, right there. We have to discover the mathematical language of brain structure and function (I never thought of the two as separate entities). Looking at the geometry of the brain, I was damn sure the mathematical language was geometrical. (personal communication)

Pellionisz's mathematical "teachers", he now says, were von Neumann, Einstein, and Roger Penrose (1931–). The last two described forms of geometry more general than Euclid's, which Pellionisz eventually used to describe the brain. (So Penrose has contributed constructively, if indirectly, to cognitive science after all: see Chapter 16.v.a, and Section x.d below.)

But he needed neurological teachers, too. Like Marr at much the same time, he relied on the recent surge in histological knowledge of the cerebellum. Indeed, he took this from the horse's mouth. His adviser for his Master's thesis on 'Neuronal Modeling' was Szentagothai, of the Semmelweiss Medical University—co-author of the definitive book on cerebellar anatomy (Eccles *et al.* 1967).

Pellionisz's first model of associative memory was described in some detail in 1970 (the 1968 publication was merely an abstract, based on a fuller talk he'd given in Hungary). It simulated the connections between four distinct cell-types: mossy fibres, granule cells, Purkinje cells, and basket cells. And it even included the relative distances, and the precise numbers of connections.

The different cell-types were represented as four two-dimensional matrices, with formulae expressing the excitatory and inhibitory connections between them. Having computed the activity resulting from mossy-fibre inputs, Pellionisz noted "the emergence of concentrated excitatory spots as a result of the local averaging effected by the granule cells" (1970: 78). And he pointed out that certain unexpected features of the simulation depended directly on the histological data incorporated in it. In other words, Farley's (neurally inspired) connectionism had endured a *rite de passage* to become computational neuroscience.

A further *rite de passage* ensued when Pellionisz, after working for a few years with Szentagothai in Budapest, began his collaboration (at NYU Medical School's Department of Physiology and Biophysics) with the experimentalist Rodolfo Llinas, an expert on the cerebellum. With Llinas's help, he developed his early geometrical approach and applied it to a wide range of neurological data. Following some exploratory simulations (Pellionisz and Szentagothai 1973; Pellionisz *et al.* 1977), his final theory of representation was defined in abstract terms by the late 1970s (Pellionisz 1979).

When they first applied it to the cerebellum, to explain the control of bodily skills, Pellionisz and Llinas took the brain's representational system as given (1979, 1980, 1982; Pellionisz 1983*a,b*). Later, they asked how it could be learnt (Pellionisz and Llinas 1985). They also applied the theory to vision, and to the balance-sensitive semicircular canals (Pellionisz and Llinas 1981, 1985; Pellionisz 1985). In short, they were talking about representation *in general*.

Their theory differed radically from other models of associative memory. Where those relied on Hebbian rules defined over isolated—and identical—neurone pairs, they described representation and learning in terms of systematic mappings between large *sets* of neurones. And "large" meant large: already by the mid-1970s, their computer network had jumped from 64,000 components to over 1.7 million. These included over 8,000 Purkinje units, and others modelling cerebellar and brainstem nuclei. Moreover, each Purkinje unit had its own distinctive role, determined by its precise position in the neural "space" concerned (see below).

The reason was that they were addressing the *coordination* of movements. They accepted Brindley's idea that the cerebral cortex generates broadly defined motor "intentions", which the cerebellum converts into more detailed instructions (the output of the Purkinje cells). But they weren't interested in single body movements, still less in individual muscles. Rather, they considered movements generated within integrated behavioural suites, nicely adapted to the sensory information available at each moment.

They wanted to explain, for example, George Stratton's (1896, 1897*a,b*) and Ivo Kohler's (1962) intriguing findings that people can gradually compensate for distorting spectacles of various kinds. (See also the 'homeostatic' model discussed in Chapter 15.viii.d.)

Even if the visual image is turned upside down (i.e. if the retinal image is turned right side up), the person's movements and perception, hugely dislocated at first, become quasi-normal within a week or so. Only *quasi*-normal: both Stratton and Kohler had noted bizarre visual aberrations. And only a week "or so": Stratton (1897*b*) actually wore the inverting mechanism for only eighty-one hours out of the 200, although for the rest of the time he was blindfolded. (His detailed accounts of his changing experiences—including their relation to various movements, to bizarre dislocations in his body image, and at first to near-overwhelming feelings of depression—are fascinating reports of the phenomenology involved.)

Cats can compensate for distorted input, too (Melvill-Jones and Davies 1976). But some species never do. Snap little goggles, shifting the light 7 degrees to the right, over a chick's eyes as it hatches from the eggshell, and it will starve to death if it's given only sparsely scattered grains of corn to peck at (Hess 1956).

Why is it that people, unlike chicks, can adapt to such sensory distortions? Stratton, at the end of the nineteenth century, had sketched the answer as follows:

Vision as a whole and by itself is indeed neither inverted nor upright. . . . [Upright] vision must mean a vision which gives us objects upright with reference to some non-visual experiences which are taken, for the time being, as the standard of direction. Upright vision, in the final analysis, is *vision in harmony with touch and motor experience*; and the only problem of upright vision is one concerning *the necessary conditions for a reciprocal harmony in our visual and tactual or motor perceptions*. (1897*a*: 184–5; italics added)

The different sense perceptions, Stratton said, "are organized into one harmonious spatial system", where the harmony consists in having our experiences meet our expectations. All that's needed is "a reliable cross-reference" between the senses, not any absolute marker of locations in reality.

Pellionisz and Llinas, too, were concerned with a complex representational *system*, not just a clutch of isolated representations. And they, too, considered the match/mismatch between actual and expected perceptions in guiding bodily action.

To integrate perception and action, they developed an arcane corner of mathematics: a form of non-Euclidean geometry known as tensor analysis. They defined systematic coordinate transformations, or "tensors". These provided a way of passing automatically between highly complex sensory and motor representations, looping through the environment to check on the sensory predictions in play.

The transformations had to be automatic because, as Karl Lashley (1951*a*) had pointed out, nervous conduction is too slow to explain bodily skills as chains of sensory-motor reflexes (Chapter 5.iv.a). They had to be complex, because any 'one' action involves a large number of cooperative and antagonistic muscles, and any 'one' perception involves many sensory receptors—in the retina, inner ear, tendons, and muscles (including the muscle spindles mentioned in Chapter 2.viii.d). There are countless ways of using one's muscles to pick up a coffee cup, and the skilled coffee-drinker finds the most efficient: *these* specific muscle movements are executed, given *those* sensory data about one's current bodily attitude and the position of the cup.

That was already well known. But where Arbib had used schema theory to compute how this could be done (see Section vii.c), Pellionisz and Llinas used tensor geometry.

Tensor network theory wasn't for the faint-hearted. The pure mathematics involved had been developing since Einstein's use of it early in the century, and had "a legendary reputation for difficulty and complexity" (J. A. Anderson *et al.* 1990*a*: 352). (When I mentioned this assessment to my young astrophysicist son-in-law, he agreed: "It bloody is!") Accordingly, their papers bristled with references to "non-Riemannian overcomplete CNS hyperspaces" and "spectral representation of the covariant metric tensor and its proper inverse (or Moore–Penrose generalized inverse) as expressed by their eigendyads".—Light reading, this was not.

Fortunately (for me, at least), the mathematical details needn't concern us. What's important is that tensor theory maps one geometry onto another. The projective geometry used in blocks world (Chapter 10.iv.b) maps Euclidean 2D space onto 3D space, in a very simple (non-perspectival) way. But tensor theory can coordinate multidimensional hyperspaces.

For an intuitive example of a hyperspace, consider the set of people eligible to apply for an imaginary scholarship. Applicants must be male, and must also satisfy at least two of these criteria: between 18 and 32 years old, but the younger the better; born in Derbyshire or Yorkshire; father (preferably) or grandfather a clergyman or a soldier; educated at one of twelve named schools; intending to study science at university; orphaned before the age of 15, and the earlier the better.

(This hyperspace isn't quite so "imaginary" as you may think. In 1961 I spent some hours in a large reference library searching for scholarships to study in the USA: of the forty or so that I found, many listed criteria just as bizarre as these—and only six of the forty were open to women.)

The 'geometry' of this scholarship space has many dimensions, namely the criteria just listed. Some are binary, some continuous; some are mandatory, some optional; and some are disjunctions, with two or twelve disjuncts (e.g. the two counties and twelve schools). An applicant at the 'centre' of the space would satisfy every criterion. Someone on the 'periphery' (besides being male) would satisfy only two, and those only weakly. Try thinking about just how you would locate fifty different applicants within this multidimensional space, deciding on their 'positions', 'neighbours', and 'distances'—and your mind may well begin to boggle. Imagine mapping them onto fifty points in some even more complex space, such as the criteria defining an ideal husband (GSOH etc.), and you might as well give up.

In tensor network theory, such decisions—for highly complex, and diverse, neuronal spaces—were computed precisely, coherently, and testably. And algorithms (sets of neural connections) were defined for transforming one set of hyperspatial coordinates into another, so that the *appropriate* movements were made at each moment, given the sensory information available. The GOFAI robot SHAKEY (Chapter 10.iii.c) had computed only simple 2D-to-3D mappings, and chose from a highly restricted set of movements. But a tensor network robot could generate a much wider range of behaviour, informed by a more complex array of sensory inputs (Pellionisz 1983a).

On this view, a cerebral 'intention' is sculpted into a more precise representation of action by the cerebellum, starting from the current bodily attitude. This attitude is known (represented) by mapping the proprioceptive input from the many tendons and muscle spindles onto 'muscle space'. The series of many-muscled movements that ensues is computed by finding a path through muscle space so as to arrive at the end point (e.g. holding the coffee cup).

Each movement 'predicts'—is associatively mapped onto—a particular set of proprioceptive (and visual and/or tactile) inputs. If the *actual* sensory input is different, perhaps because the arm is blocked by some obstacle, a different muscular 'starting point' is found by automatic mapping, and a new muscular 'pathway' computed accordingly. When such failures of prediction occur systematically (as in wearing distorting spectacles), an adaptive process of *meta-organization* gradually refigures and recoordinates the sensori-motor spaces concerned. This process was mathematically defined, and was described by the authors as the first precise definition of the "largely intuitive" concept of self-organization (Pellionisz and Llinas 1985, sect. 4.1).

An important general point, here, is that this theory posited *hierarchies* of dual geometries in the nervous system, with different tensors linking each pair of contiguous levels. This idea was used (for example) to explain mammalian gaze control—which isn't a simple matter, since both eye movements and neck movements are involved. (Hoverflies lack both, so don't need a complex mechanism of gaze control: Chapter 15.vii.b.) Pitts and McCulloch (1947) had discussed some relevant visual computations. But Rafael Lorente de No (1933b) had already suggested that vision isn't enough, that gaze control also involves feedback from the inner ear (see Chapter 4.v.c). Pellionisz (1985) took this added complexity on board—later adding the neck muscles (Pellionisz and Peterson 1988). He even sketched, though couldn't simulate, the frog's *entire* nervous system in (multilevel) tensorial terms (1983b), depicting the animal as what he might have called—but didn't—*Rana geometrica*.

The theory was applied to self-organization in the embryo, too. People had long observed spontaneous oscillations, or coordinated twitches, in the legs of chick embryos. Now, Pellionisz and Llinas (1985) suggested that these apparently useless movements enable the chick to develop its (initially vague) proprioceptive space in coordination with its (fixed) muscular space. That is, tensor coordinations enable established neural representations to be used, on successive levels, to organize less well-developed ones. (A related idea, of "representational trajectories", was being developed informally in connectionist psychology: Chapter 12.viii.d.)

Tensor theory was tested by biological experiments in Llinas's lab, by computer simulation, and (later) by robotics. By the late 1980s, it was beginning to receive independent experimental support (e.g. Gielen and van Zuylen 1986). James Anderson included the meta-organization paper in his influential *Neurocomputing 2*, and invited Pellionisz to co-edit the volume (J. A. Anderson et al. 1990a). And Pellionisz received the first international award for neurocomputing (Germany's Alexander von Humboldt Prize) in 1990.

One might think this would have 'licensed' Pellionisz as a professional researcher, but it didn't. (Llinas didn't need a licence, being already established when their collaboration began.) Like Grossberg and 'cerebellar Marr' (see Sections v–vi), Pellionisz suffered from neuroscientists' suspicion of fancy mathematics. Even mathematically competent researchers such as Arbib and Shun-Ichi Amari were sceptical, although Amari later adopted somewhat similar ideas (Arbib and Amari 1985; Amari and Wu 1999; cf. Pellionisz and Llinas 1985, endnote).

Pellionisz wasn't offered a tenured academic job, and left NYU in 1989—soon to be employed by NASA, to design software controls for F-15 jet fighters (Pellionisz *et al.* 1992). Later, he worked in the IT industry, and founded a company using tensor network theory to do personalized pattern matching—applying his four-level PureMatch software in finance and medicine, for example (personal communication). Whether he'd have had more to offer to neuroscience, who can tell?

c. Emulation and subjectivity

Tensor network theory—once it was applied to the body's geometry, in the late 1970s—was an early example of what's now called an emulator system (Grush 2004). It's partly because of neural emulators that our walking can be continually adjusted (unlike the path of a ballistic missile, or the hoverfly's flight towards its mate: see Chapter 15.vii.b).

An emulator is an internal model of a feedback system, which works in a similar way but faster. Using feed-forward processes, it continually anticipates the system's responses *and the resultant sensory information*. Given a mismatch between expected and actual input, the emulator can adjust the output controls immediately, instead of just a little late. Without the emulator, and especially if the stimulus is continually changing, the output would be jerky rather than smooth.

Emulators have been used in control engineering from about 1960. Craik couldn't cite them, because they hadn't yet been built. However, they're a paradigm case of the type of representation he had in mind. Since their appearance on the engineering scene, they've been found in the nervous system.

Tensor network theory, as remarked above, defines a hierarchy of emulator systems. Another example was described by Mayhew and colleagues, in a study of the cerebellar control of visual saccades—in which the eyes move to focus on, and track, a particular object (Dean *et al.* 1994). Yet another concerned the control of hand movements made in the dark (Wolpert *et al.* 1995). In view of the controversy over the very existence of cerebral representations, it's worth remarking that those authors saw their results as "direct support for the existence of an internal model" (Wolpert *et al.* 1995: 1880).

The same authors soon argued that emulators and other representations are involved in sensori-motor integration *in general*, where the organism combines multiple sources of information to estimate its own state and that of the environment (Ghahramani *et al.* 1997). And after further research in this area by themselves and many others, they concluded:

The computational study of motor control is fundamentally concerned with the relationship between sensory signals and motor commands. . . .

Computational approaches have started to provide unifying principles for motor control. Several common themes have already emerged. First, internal models are fundamental for understanding a range of processes such as state estimation, prediction, context estimation, control and learning. Second, optimality [i.e. optimal control theory] underlies many theories of movement planning, control and estimation and can account for a wide range of experimental findings. Third, the motor system has to cope with uncertainty about the world and noise in its sensory inputs and motor commands, and the Bayesian approach provides a powerful framework for optimal estimation in the face of such uncertainty. We believe [thanks to a wide range of experimental evidence] that these and other unifying principles will be found to underlie the control of *motor systems as diverse as the eye, arm, speech, posture, balance and locomotion.* (Wolpert and Ghahramani 2000: 1212, 1217; italics added)

Such neurological theories support Stratton's *psychological* insight that it was the difference between the actual and expected sensory inputs which enabled him to adapt to an inverted visual world. A hundred years after Stratton's experiments, Jennifer Freyd (with Geoffrey Miller) offered further psychological evidence for neural processes emulating the world.

She posited dynamic mental models having "representational momentum", to explain her finding that people who perceive a moving object tend to misremember it as having moved further than it actually did (Freyd 1987; G. F. Miller and Freyd 1993). Such models have evolved, she said, to help us track real-world motion by anticipating it rather than merely following it.

Like Roger Shepard and Stephen Kosslyn before her (see Chapter 7.v.a), she reported a linear relation between representation and reality. The size of the mistake in the memory of final position depended directly on (1) the speed of movement and (2) the time elapsed between perception and memory probe. That's just what one would expect, if there were some physically realistic internal model that "keeps moving in the mind"—or, as Craik would have put it, keeps running in the brain—after the actual movement has ceased.

Freyd's evidence may be persuasive to neuroscientists working in the Craikian tradition, and to psychologists—such as Shepard himself—who argue that "second-order isomorphism" has evolved in our perceptual systems (see 7.v.a). But it wouldn't convince committed formalists such as Jerry Fodor and Zenon Pylyshyn. They'd interpret it in GOFAI-friendly terms, as they did Shepard's and Kosslyn's data on mental imagery (Pylyshyn 1973). On that view, her results show that the perception and/or memory of movement is "cognitively penetrable", owing to top-down symbolic inference (see Chapter 7.v.a). Her representational hypothesis would be vindicated only if a Craikian emulator of the type she implied were actually identified by neuroscience.

Lashley (1951a) had argued that there must be some central anticipatory mechanism controlling skilled movement, but could say very little about just how it might work. Today, neural emulation is the key to our understanding of bodily skills. Ito, who in the 1960s co-authored (with Eccles and Szentagothai) *The Cerebellum as a Neuronal Machine*, now says:

With respect to cerebral control functions, cerebellar chips appear to [form] *an internal model of a controller or a control object.* If, while a chip and the system to be modeled are supplied with common input signals, differences in their output signals are returned to the chip as error signals, the chip will gradually assume *dynamic characteristics equivalent to those of the system to be modeled.*

Cerebellar chips [in one area] are connected to the cerebral motor cortex in such a way that these chips constitute *a model that mimics the dynamics of the skeleto-muscular system* (Ito 1984). The motor cortex thus becomes capable of performing a learned movement with precision *by referring to the model* in the cerebellum and not to the skeleto-muscular system. [The] failure to perform a precise reaching movement without visual feedback could be due to the loss of such internal models. (Ito 1999: 111; italics added)

Indeed, the emulator control of skilled (automatic) movement is seemingly only one function of the cerebellum. Brain-imaging techniques have recently suggested that it might also be crucial in skilled thinking (Schmahmann 1997). So Ito adds this:

During thought repetition, a cerebellar chip may form a model of the parietolateral cortex or the prefrontal cortex. A repeatedly learned thought may thus be performed quickly yet accurately even without reference to the consequences of the thought or without conscious attention. (p. 111)

Craik, had he lived, might have seen this suggestion as confirming his claim that thinking involves analogue symbols, too.

Neural emulators—like their control-engineered cousins—are online representations, used to improve the efficiency of behaviour while it's happening. In human beings they can also be run 'offline', so that people can practise physical activities by imagining them. This explains why (as a Norwegian diving instructor told me many years ago, to my amazement at the time) top-level divers living beyond easy reach of an Olympic swimming pool can improve their performance by diving 'in their heads'.

The emulators themselves are enduring memories. The connections that implement them are retained, once they've been learnt. But their adaptive function is to represent momentary 'present-tense' events: where the muscles are *now*, and what the input data are *now*. In other words, they are subject-dependent representations, not objective ones.

Some representations posited by cognitive science are even less objective than this. For they are evanescent structures, coding intermediate results in some multi-stage computational process. Neuroscientific evidence is difficult to come by, for obvious reasons—although high-resolution fMRI might help, especially if combined with electrophysiology (Logothetis 2002). But both computational arguments and psychological evidence are relevant:

* For instance, Aaron Sloman argued in the late 1970s that high-level visual interpretation requires temporary data structures, and implemented them in his POPEYE project (see Chapter 10.iv.b).
* Marr's theory of vision distinguished the viewer-dependent $2\frac{1}{2}$D sketch from representations at the level of object models (Chapter 7.v.c).
* And Perrett's group explained the monkey's face recognition in terms of a limited set of "viewer-centred descriptions" (see Section iv.d, above).
* As for psychological evidence, Ann Treisman (1988) explained her experimental data on perception and problem solving by positing temporary "object files", generated 'on the fly' (i.e. during processing) in working memory.

The crucial point here is that there are relatively 'subjective' and relatively 'objective' representations. In the last decades of the twentieth century, neuroscientists—and some philosophers—increasingly accepted this. They tried to specify the differences, and to explain how objectivity could develop from subjectivity, whether in evolution or in ontogeny (see Chapter 12.x.e–f and Cussins 1990).

In short, by the new millennium cognitive scientists had posited representations galore. Neuroscience now used the term liberally, offering a wide range of hypothetical examples (Parker *et al.* 2002).

d. The philosophers worry

That's not to say, however, that the *philosophers* regarded all of these 'representations' as the genuine article. Their quarrels weren't with the biological evidence, but with its intentional interpretation—and (predictably) with each other.

Craik's philosophy of mind had defined "symbols" and "models" in analogue terms. GOFAI philosophers, by contrast, tried to restrict such concepts to formalist examples (hence their top-down interpretation of imagery, mentioned above). For them, connectionism was concerned with "implementation details", not representation. PDP philosophers favoured a third position (see Chapters 12.x.a–c and 16.iv.e). They argued that activation patterns distributed over many computational units, none of which can be assigned a constant meaning or regarded as a symbol for a nameable concept, are genuine representations too. And some allowed talk of *subjective* representations, as precursors to full-blooded objective concepts (see 12.x.f).

Some philosophers preferred a very wide definition: *any* internal mechanism that was reliably correlated with an external event and causally implicated in ecologically appropriate behaviour (Dretske 1984, 1995). This would include single-cell feature-detectors (which aren't "models" in the Craikian sense, although the circuitry activating them may be), and also the relatively subjective, evanescent, examples noted above. Both of those would be excluded, however, by the familiar definition—familiar to philosophers—of a representation as a (non-semantically identifiable) internal state that causes behaviour to be adaptively coordinated with some environmental feature *even when that feature is absent.*

Moreover, the reference to "an identifiable internal state" excludes cases (highlighted by the dynamical approach) where there's a complex interaction of body, brain, and environment. Of the philosophers who recognized this interactive causation, some allowed that certain aspects of the causal nexus play a special role in adaptive behaviour and should be counted as representations accordingly (Clark and Grush 1999; Clark and Wheeler 1999). Others—including 'Frog's Eye' Maturana (wearing his autopoietic philosopher's hat)—denied the existence of *any* representations in the brain. Far from there being representations galore, there are none at all. (The reasoning behind this unfashionable, even counter-intuitive, view is sketched in Chapters 15.vii and 16.vii and x.c.)

Those philosophers of mind who had no special interest in cognitive science focused on the centuries-old concept of representation, or intentionality, *as such.* They ignored the specifics of neuroscience and AI, although some did ground intentionality in evolutionary biology (see Chapter 16.x.d).

Finally, neo-Kantians didn't simply ignore science but explicitly dismissed it. Or rather, they wished to keep it strictly in its place—which was to answer *empirical* questions. According to them, there can be no naturalistic explanation of intentionality—

not even in neuroscience, never mind AI. They weren't merely saying that there's a distinction between empirical and philosophical questions. Rather, they were saying that science is irrelevant, so much so that it can't even give us helpful clues on philosophical matters. It would follow, for instance, that *even if* scientists discovered innate 'syntactic' mechanisms, Chomsky's *philosophical* claims wouldn't be supported, still less proved.

Clearly, that wasn't the view of the many philosophers of cognitive science I've cited throughout this book, who *did* think that cognitive science could help us to a more adequate philosophy of mind. But even they had to admit that they had no knock-down arguments against neo-Kantianism (see 2.vi and 16.vi–viii).

On one thing, however, all the philosophers agreed. Namely, that empirical cognitive scientists used the words "representation" and "model" in largely intuitive senses—obscuring important distinctions and begging controversial questions.

Craik and Minsky each tried to define these concepts in *philosophical* terms, as we've seen (Chapter 4.vi.b). So did Newell and Simon (16.ix.b). But most of their scientific colleagues didn't. They simply picked them up from the Zeitgeist and ran with them. If they offered 'definitions' at all, these concerned the mechanism rather than the concept.

From the late 1970s on, gallons of philosophical ink were spilled on trying to define representation more strictly. Indeed, it's still spilling, unabated. I've no intention of tracking each drop. Several rivulets are traced in other chapters, however. And two major streams of disagreement—concerning *reportability* and *intentionality*—will be followed in Chapter 16.

(*How to find the rivulets*: David Kirsh's critique of non-representational robotics is outlined in Chapter 13.iii.c. Andy Clark's defence of PDP, and Adrian Cussins's of increasing objectivity, appear in Chapter 12.x. Fodor's 'classical' representationalism is outlined in Chapter 16.iv.c–d; Timothy van Gelder's dynamical critique in Chapter 16.vii.c; and Maturana's autopoietic theory in Chapters 15.vii.b and 16.x.c. Finally, Michael Morris's argument that *no* brain mechanism is a representation is given in Chapter 16.viii.a.)

Here, let's just note that neo-Kantian philosophers insisted on reportability as a criterion of representation. For them, language is essential for mind, intelligence, and thought (16.vi–viii). They would never write books called *Animal Thinking* or *Animal Minds* (Griffin 1984, 1992), and regarded "cognitive ethology" (Ristau 1991) as a contradiction in terms. However, experimental psychology, ethology, AI, and neuroscience developed within the opposing philosophical tradition (see Chapter 2). They drew no sharp line between 'mindful' *Homo sapiens* and 'mindless' dumb animals.

The slipperiness of the concept of representation affects how people think about cognitive science itself. Since the concept was often used in defining the field, disagreements about *what representation is* were reflected in judgements about its scope, and its success.

If one accepts Fodor's sense of representation, for instance, connectionism is either a *refutation* of cognitive science (as Dreyfus claimed) or a mere implementational *adjunct* to it (as Fodor himself believed). Accepting a more catholic definition of representation, connectionism is an interesting *example* of cognitive science (and further putative examples have been given in this section). Similarly, dynamical theories and situated robotics (for instance) must be excluded if, by definition, cognitive science posits representations in explaining behaviour.

This is why, as I said right at the start, the boredom barometer would shoot through the roof if one compared every definition of cognitive science. And it's why I deliberately avoided the elusive term "representation" in my own definition of the field (Chapter 1.ii.a).

It's not even clear that trying to legislate on "the best" definition of representation is sensible. The same applies to life (see 16.x), and to theoretical terms in general (Quine 1951, 1960; Putnam 1962*a*). Comparing putative definitions carefully is one thing, and can be highly illuminating. But announcing *just one* to be the only defensible option is quite another. That's especially so in scientific contexts. As Craik put it, when discussing perception:

> [Scientists and philosophers alike] fail to see that their remedy of exact definition may be *impossible and unattainable* by the very nature of the physical world and of human perception, and that their definition should be corrected in the way of *greater extensiveness and denotative power, rather than greater analytical, intensive, or connotative exactitude.* (p. 4; italics added)

The catch-all sense of 'representation' (which I deliberately haven't tried to *define*) is therefore useful. It acknowledges the intriguing variety of intentional and quasi-intentional mechanisms that generate adaptive behaviour. Philosophers can help clarify the differences, for they're already familiar with a wide range of relevant distinctions and implications. But that variety becomes apparent through empirical research, not philosophical diktat.

14.ix. Computation Challenged

In the final decades of the century, seven research themes in neuroscience challenged certain aspects of the computational approach. Those challenges were largely met—sometimes, by modelling in radically novel ways.

All seven were older ideas revivified by new data. The first four—self-organization, dynamical systems, epigenesis, and neural selection—were closely interrelated, as we'll see. The fifth concerned grandmother cells; the sixth, neurochemicals; and the last, time.

a. Structure without description

At the mid-century Hixon symposium, McCulloch had argued that the information stored in our genes could determine the connections between at most 10,000 neurones, "[even] if that was all it had to do" (W. S. McCulloch 1951: 85). "As we have 10^{10} neurons", he continued, "we can inherit only the general scheme of the structure of our brains. The rest must be left to chance." And by "chance", he said, he really meant learning from experience.

Only a few years afterwards, he and his 'Frog's Eye' colleagues discovered a detailed 'mapping' between the frog's retina and tectum (see Section iii.a). Soon, others found organized columns of orientation detectors in the visual cortex of cats—and in newborn kittens too (Section iv.d). This was a puzzle, for the kittens couldn't have "learnt from experience" in the womb. Their detailed brain structure, it was therefore assumed, *must* have been cleverly programmed by the genes—perhaps by means of genetically specified chemical 'labels' to guide the developing neurones. (McCulloch's worry about the amount of information required for such a program was seemingly forgotten.) After all, brains can't get organized by magic.

In the final quarter-century, however, a number of people showed that (as Ashby had suggested in the 1940s: Chapter 4.vii.c) there may be *general* mechanisms in the developing brain which structure it spontaneously, without either 'external' learning or specific genetic instructions. Specifically, the claim was that the various feature-detector cells, *and* their systematic arrangement in neighbouring clumps and columns in the cortex, were self-generated. Highly influential versions of this claim were due to the biologists Willshaw and von der Malsburg, the engineer Kohonen, and the computer scientist Ralph Linsker.

Willshaw and von der Malsburg (1976) asked how patterned structures might arise within an unorganized mass of cells in the *embryo's* brain. Von der Malsburg had already shown that specific visual inputs could lead to columns of feature-detectors (Section vi.b, above). But the womb doesn't provide specific visual inputs. So now, he and Willshaw generalized the point.

They showed that for any two interconnected sensory cell layers, simple Hebbian rules will make the second develop a 'map' of the first. The one constraint was the presence of short-range excitation and long-range inhibition (as in the 'Mexican hat' detectors described in Chapter 7.v.d). Both of these, of course, had long been observed in the brain.

They proved—and demonstrated by computer modelling—that those mappings will arise in an orderly way, owing to properties of the system as a whole:

[We] have shown that the mappings are set up in a system-to-system rather than a cell-to-cell fashion. The pattern of connections develops in a step-by-step and orderly fashion, the orientation of the mappings being laid down in the earliest stages of development. (Willshaw and von der Malsburg 1976: 431)

They described two independent factors, operating at different scales. One was an optimizing process ensuring a topographical mapping, since "neighbouring presynaptic cells come to connect to neighbouring postsynaptic cells" (p. 433). The other used boundary and initial conditions to affect the position, size, and orientation of the final mapping. They varied the boundary conditions in their computer models, showing how this affected the network's self-structuring.

As if in response to McCulloch's worry, they pointed out that theories of self-organization "have the advantage of requiring only an extremely small amount of information to be specified genetically". Indeed, they noted that the amount of information that would be required for pre-wiring is even greater than inspection of adult brains suggests.

A frog's retina and tectum, for instance, grow at different rates and in different ways, so "the only way the systems could remain matched throughout development [is] for the synaptic relations between retina and tectum to be constantly changing" (p. 432). A rigid pre-wiring plan wouldn't provide the flexibility required. To 'settle' this plasticity into one form or another, the *intra-cerebral* environment was crucial. Their models showed how "the same genetical program for the topographic part can in different situations lead to quite different retinal points being connected to a given tectal location, without the need for [chemical] relabelling".

In saying this, Willshaw and von der Malsburg weren't denying that the brain may use chemical labelling to guide *this* neurone to connect with *that* one. Their point, rather, was that Hebbian self-organization is functioning too (perhaps to 'tune' the results of initial chemical guidance), and that it's potentially powerful enough to work alone. Moreover, a 'chemical' method could in principle specify *any* mapping, however bizarre. Their theory generated only 'natural' (topographical) mappings. (Compare Allen Newell and Herbert Simon's decision to constrain their production systems, converting a universal programming language into a 'weaker' but more biologically realistic framework: see Chapter 7.iv.b.)

A few years later, Kohonen (1982)—who'd been modelling associative memories since the 1960s (see Chapter 12.v.e)—generalized these ideas still further. His analysis applied in principle not only to mappings of spatial features but also to "completely abstract or conceptual items"—*provided that* "their signal representations or feature values are expressible in a metric or topological space that allows their ordering".

The good news was that Barlow's hunch (Section ii.b), that logical reasoning may be grounded in the mechanisms that make efficient perception possible, was reinvigorated. (It had survived also in Grossberg's and Arbib's work, of course.) The bad news was that establishing a suitable "topological space" for concepts (*plough, betrayal, cousin, justice*...) is easier said than done. Margaret Masterman's computerized thesaurus (Chapter 9.x.a) and Robert Abelson's matrix of "themes" (Chapter 7.i.c) might be seen as early attempts, and theories of semantic primitives too (Chapter 9.viii.c). But much greater subtlety and richness would be required to define the "topology" of a plausible self-organizing concept-mapper.

Kohonen (1988) soon used his analysis to build a device for speech recognition. It took speech (single words, or strings of words separated by pauses) as input, analysed it into similar but varying phonemes, and output the words on a typewriter. (In Finnish, unlike English, sound and spelling match closely.) The phoneme classes weren't built in; and they weren't identical with the linguist's phonemes (Kohonen called them "quasi-phonemes", accordingly). Nevertheless, the system learnt to recognize the 'same' phoneme in differing phonetic contexts, and spoken by individuals with different voices and accents. It developed a large vocabulary, and was the best speech-recognizer at the time.

This 'science-fictional' automatic secretary was described in the widely read pages of *Computer Magazine*. The same issue also featured an elegant pebble thrown into the theoretical waters by Linsker. And this made even more of a splash.

Based at IBM's T. J. Watson Research Center, Linsker (1949–) had defined—and implemented—highly abstract models of multi-layer feed-forward networks (1986, 1988, 1990). These showed that simple Hebbian rules, given *random* activity, could lead to structured orientation detectors. Again, the newborn kitten's visual skills were less surprising than they'd seemed.

So far, so familiar. What was original here was Linsker's formulation, using Shannon's information theory, of the "infomax" principle. He applied this both to individual cells and to the network as a whole:

The organizing principle I propose is that the network connections develop in such a way as to maximize the amount of information that is preserved when signals are transformed at each processing stage, subject to certain constraints . . . The constraints or costs may reflect, for example, biochemical and anatomical limitations on the formation of connections, or on the character of the allowed transformations. (Linsker 1988: 529, 536)

In other words, the nervous system has found *the best possible* way of analysing complex sensory input, given that flesh and blood is doing this at all. The broad idea wasn't new (Marr had relied on it in his theory of edge detection: see Chapter 7.v.b–d). But Linsker had apparently found a powerful way to generalize it.

Admittedly, he said, much work would be required to apply the infomax principle to real biology. For example, it wasn't clear whether it could be generalized to cover 'circular' networks involving feedback—as in top-down action by schemas, or recurrent networks. Moreover, only empirical neuroscience could show which "constraints" had been favoured by evolution, or perhaps "by other principles not yet identified". (His reference to "other principles" *besides* evolution was a sign of the times: A-Life had recently revived interest in such principles—see Chapter 15.iii–iv and viii–ix.)

Two additional bonuses were remarked at the close of his paper. "Infomax" made it much more likely that an isolated mutation, in the evolution of a complex neural system, would be adaptive. The apparent 'necessity' for several *simultaneous* mutations evaporates if each neural level can spontaneously adapt to a small alteration in another. The same may apply to other bodily organs too. For the paper's final squib was the suggestion that infomax might apply to dynamical systems of many different types.

In other words: neural development, goal seeking, immune response, and biological evolution might all share similar self-organizing properties. This exciting idea had recently hit the news-stands (see Chapter 15.ix.a), and Linsker's mathematics was taken as strong support.

The "splash" caused by Linsker's ideas is evident in Cowan's rueful comment a few years later:

In the late 1970s, Christoph von der Malsburg and I worked together on the formation of orientation detectors. I had the idea that maybe this idea of a natural tendency to form stripes and blobs [he'd already cited Turing: see Chapter 15.iv.b] was the key to understanding it, and all you needed was a two-layer network stimulated by noise, and it would automatically make the correct feature detectors.

Christoph didn't believe me; he said, "That's magic."

I said, "No, it's just spontaneous symmetry breaking."

It turns out I was right. We never did it, and I have been kicking myself ever since because Ralph Linsker did it. That's exactly what Linsker discovered: that stripes and blobs will spontaneously form in a map. That's the origin of center–surround orientation detectors in the visual cortex. (J. A. Anderson and Rosenfeld 1998: 120–1)

Neuroscientists, besides further analysing Linsker's general results (D. J. C. Mackay and Miller 1990), built more 'realistic' models. For example, a different Hebbian rule was used; and ocular dominance columns were generated partly by neural competition between cells connected to the two eyes (Linsker had considered only one eye): K. D. Miller *et al.* (1989). This work also showed *why* certain sorts of abnormal visual experience result in distinct neural pathologies.

Harry Barrow—a pioneer of AI vision (see Chapter 10.iv.b)—also built on von der Malsburg's 1973 paper, and later on Linsker's work. He and Alistair Bray modelled the origin of the simple and complex cells of visual cortex, to find out which types of pattern were 'preferred' by the brain. They used natural images, whereas von der Malsburg had used artificial line patterns.

Their networks developed systematically structured groups of orientation-sensitive edge- and bar-detectors, or of detectors for colour or position-invariant orientation (Barrow 1987; Barrow and Bray 1992*a,b*, 1993, 1996; Bray and Barrow 1996). (They suspected that a 'composite' version of their model would develop *all* these feature-detectors, but didn't have enough computational power to test this.) They asked why the networks developed as they did, and showed how the results would differ if various parameters were altered.

In short, they explored the mathematical core of several *superficially distinct* self-organizing systems. Barrow and Bray, Linsker, and Willshaw and von der Malsburg were all asking important questions about the inevitability of crucial types of brain mechanism. These mechanisms didn't have to be sketched by some all-powerful Designer in the sky. They didn't even have to be nicely prefigured in the genes. They arose naturally, as a result of self-organizing principles dependent on relatively low-level properties of neural tissue. (If D'Arcy Thompson had still been alive, he'd have been fascinated: see Chapter 15.iii.a.)

b. Dynamics in the brain

The studies described above weren't the only examples of late twentieth-century neuro-science taking self-organization seriously. A number of people (including Arbib, as we've seen) were now drawing not only on Hebb but also on dynamical systems theory. (In general, a dynamical system cycles between a finite number of holistic 'attract-ors'—Ashby's "equilibrium" states—in a more or less regular fashion.) They weren't alone: people in other areas of science were also exploring these ideas (Chapter 15.viii.c, ix, and xi; and 16.vii.c).

Some philosophers were doing so, too. They described the mind/brain in terms of spontaneous self-organization, instead of the input–output systems favoured by classical cognitivism. Besides those to be mentioned in Chapter 16.vii and x, these writers included Susan Oyama (1985). Her unorthodox approach to the developmental sciences (including evolutionary biology) resembled Romantic views that had 'died the death' in scientific circles with the publication of *The Origin of Species* more than a century before (Chapter 2.vi). Nevertheless, it now influenced a number of hard-headed developmental scientists and A-Life researchers.

Oyama criticized GOFAI cognitivism—and most neuroscience—for treating the mind/brain as a given, static, structure. Instead, she argued, it's an ever-changing system, developed by a lengthy process of self-organization. Unless we take this seriously, we can't understand *adult* cognition. It followed that developmental neur-oscience/psychology isn't an optional extra, to be studied by those who like that sort of thing and safely ignored by everyone else. On the contrary, it's central to neuroscience/psychology *in general*.

She wasn't making an empirical claim, although she did cite empirical evidence in support. Rather, she was making a philosophical point. To ask, like McCulloch for example, whether the structuring information is in the genes or in the stimulus, or even in some epigenetic interaction between the two (see subsection c), is to imply that the system—genome, embryo, brain, cortex, feature-detector ... —accepts its structure from something else. This, she argued, is unsatisfying *in principle*, because it leaves the ultimate origin of structure unexplained. We must show how the system can structure itself, owing to its inherent nature. If not, then the mystery remains.

The psychologist Esther Thelen (1941–2004) agreed (1989: 556–7). Putting empirical flesh onto Oyama's philosophical bones, she studied integrated motor actions, including those described by Sherrington (Chapter 2.viii.d). She focused on the nature and development of rhythmic patterns such as walking, kicking, rocking, and communicative waving, and the sudden shift between (for example) crawling/toddling, suckling/ingesting, and sleep/waking (1981, 1985, 1989; Thelen *et al.* 1987; A. Fogel and Thelen 1987; Thelen and Smith 1994).

Where Marr had ascribed bodily skills to specific neurone circuits, and Pellionisz and Llinas to tensor transformations, Thelen ascribed them to stable attractors in dynamical systems (see Chapter 7.vi.g).

Everyone allows, Thelen said, that behaviour arises from a 'system'. But they usually do this only in the Discussion section, after admitting failures in their main-effect theories (1989: 557). She had scant respect for these empty admissions, which "can dilute systems concepts to the point of vacuousness" (1989: 557). What we need, she said, is a mathematically precise account showing how complex systems can produce emergent order without a prior description of it, and how one type of order is 'chosen' over another.

For Thelen, what are normally regarded as mentally guided, pre-prepared, actions are best thought of as forms of self-organizing bodily skill. She ascribed such skills to stable attractors in dynamical systems (see Chapter 15.viii.c–d). Even a small change may tip the system into the basin of a different attractor, so we should expect sudden behavioural shifts. But not *any* change will do: we must discover the 'switches' involved.

On this view, sensori-motor development involves the construction and stabilization of new attractors, and the 'setting' of control parameters that shift the system from one to another. The system complexity is hierarchical, since new attractors are generated by "a cascade of successive bifurcations or phase-shifts" (1989: 570). (Hence the impossibility of going *directly* from a trot to a gallop.)

For example, consider Piaget's "A-not-B error", in which infants search for a toy in its original place even though they've seen it being moved (Chapter 5.ii.c). Piaget himself had explained this in cognitive terms, speaking of the developing "object concept". But Thelen didn't. Instead, she analysed perseverative reaching by differential equations representing the interacting constraints involved (Thelen *et al.* 2001).

These constraints included alternative movement-suites such as reaching and looking (*seeing*, for Thelen, is a form of motor activity); the distance between object and child; the perceptual salience of the hiding place; the time passed before the child is allowed to reach for it; the attractiveness (for the child) of the object; and the past history of the child's actions in respect of it. Each of these had previously been shown to affect whether the child succeeded (which is a puzzle, if some object concept 'inside the head' is supposed to be responsible). But Thelen united them in a coherent way.

Her approach to action was very different from GOFAI planning. This had already been criticized as over-intellectualized by people within the AI stable. Now, Thelen provided psychological meat, and measurement. (How far dynamical theorizing can be applied to paradigm cases of 'rational' planning remained controversial: see 12.vii.c.)

Although her theory was couched at the psychological level, Thelen related her findings to neural mechanisms whenever she could. Among the neuroscientists she cited with approval was Berkeley's Walter Freeman. He'd been the first to apply chaos theory (a variant of dynamical systems theory) to the brain.

'Chaos' in the mathematical sense is very different from 'chaos' in the everyday sense: it sometimes looks random, but it isn't. Chaotic phenomena (which include the weather and the heartbeat) are deterministic, but in practice unpredictable. The reason is that they're hugely influenced by tiny variations in initial conditions. Nevertheless, their overall behaviour shows a number of holistic patterns, or dynamical attractors.

Freeman had been drawn to dynamical theorizing by Turing's 'Morphogenesis' paper and by Ilya Prigogine's work on dissipative structures in physics (J. A. Anderson and Rosenfeld 1998: 31–2). From the 1960s on, he'd experimented—unfashionably—with EEG recordings. Most of his peers paid little attention: as he put it later, there was "a virtual obsession with unit recording" (J. A. Anderson and Rosenfeld 1998: 27).

By the late 1980s, however, some were ready to listen. (Both biological-Turing and Prigogine had by then become names to conjure with: Chapter 15.iv and viii.d.) A widely read paper on the neurobiology of smell, written with Christine Skarda, appeared as a target article for interdisciplinary peer review in *Behavioral and Brain Sciences* (Skarda and Freeman 1987). It claimed that a particular scent is represented (after Hebbian learning) not by single-cell feature-detectors but by a spatial pattern of EEG-recorded activity distributed over the brain's olfactory bulb, and involving "every neuron of the bulb".

And smell was just the beginning: "We hypothesize that chaotic behavior serves as the essential ground state for the neural perceptual apparatus [in general]," and "[We see] implications of our neural model for behavioral theories [too]" (Skarda and Freeman 1987: 161). This claim received an even wider audience four years later, when it appeared in *Scientific American* (Freeman 1991).

As a young man, Freeman had avoided computer models:

[This] was, of course, when the perceptron was in vogue and a dozen other network type of devices . . . which I looked upon as, well, interesting gadgets, but they had nothing to do with how nervous systems work. (J. A. Anderson and Rosenfeld 1998: 27–8)

Later, he did do computer modelling alongside his extensive animal experiments (Freeman 1979, 1987). But further simulation (and experiments) would be needed: "[Our claims can't be advanced] until engineers have built hardware models based on our equations and determined whether they behave the way parts of brains do" (Skarda and Freeman 1987: 170).

Many neuroscientists weren't persuaded—some, because they were still in the grip of the "obsession" with unit recording. In 1993 Freeman said:

We've found the same dynamical patterns as in the olfactory system in the visual and auditory and somatosensory cortices. But they [other neuroscientists, such as Koch and Francis Crick] are still interpreting the process in terms of coupled single cells. They characterize it as "the binding problem"—how to get together a bunch of feature detectors. They're not thinking of the processes in terms of a dynamical systems approach. (J. A. Anderson and Rosenfeld 1998: 38)

Others were sceptical for different reasons. Chaos theory, Donald Perkel (1987) suggested, was a "fad", one in a long line of technological–mathematical metaphors for the brain (he cited under-sea cables, telegraphs and telephones, holograms, tensors, maps, non-linear networks, and computer programs) that were superficial or even misleading. Chaos theory had recently become popular. Indeed, a widely read account by a science journalist had fed the "news-stands" excitement mentioned above with respect to Linsker (Gleick 1987).

Despite his use of the word "fad", Perkel couldn't accuse Freeman of jumping on the bandwagon. For Freeman had been using chaos theory for years. But he did accuse him of being seduced by the mathematics:

[The] beauty, versatility, and power of the mathematical approach may have led its aficionado to find areas of application in the spirit of the proverbial small boy with a hammer, who discovers an entire world in need of pounding. (Perkel 1987)

He recalled Sherrington's (1940) description of the brain as an "enchanted loom", wherein "millions of flashing shuttles weave a dissolving pattern though never an abiding one; a shifting harmony of subpatterns". This newly fashionable approach, said Perkel, *might* be the right way to describe the loom's activity. But only "more fine-grained experimental evidence and correspondingly realistic simulation studies" could show this to be so.

The target authors had already made the last point themselves (1987: 170–1). Now, in their *BBS* Reply, they countered that their theory was "no more or less metaphorical than any other use of descriptive equations properly selected" (p. 187). If it seemed promising, they said, it would continue to be used; if not, some other theoretical tool would be found. That's how science works.

For our purposes, what's interesting is that Skarda and Freeman had considered self-organization at all. But they soon had company: eventually, Freeman co-edited two special numbers of the *Journal of Consciousness Studies*, in which several authors defended comparable views (Núñez and Freeman 1999). Thanks to both dynamical systems theory and connectionism, the last quarter-century saw this topic being taken seriously in neuroscience, much as it was in theoretical biology and A-Life. Even Ashby was now respectable.

'Brain-as-program' had suffered a stunning blow (though not a knockout: see 12.ix.b). But computational neuroscience *in general* hadn't. In fact, this change couldn't have happened without it.

c. Epigenesis

The computational work described in subsections a–b rode a coach and horses through the familiar notion of 'innateness'. For it destroyed the cosy assumption that behaviour already present at birth *must* be genetically specified. (Sometimes, such behaviour is partly due to 'external-world' learning in the womb. It was known by the 1980s that newborn babies can recognize their mother's voice—and also the intonation patterns of her language: DeCasper and Fifer 1980; Mehler *et al.* 1988; Kisilevsky *et al.* 1989.)

Most people had previously thought that behaviour must be due either to genetic specification or to post-natal learning. The long-standing debate on 'innate ideas', for example, reflected this assumption (see Chapters 2.vi.a and 9.ii.c and vii.c–d). But there'd been an important exception: namely, Jean Piaget (Chapter 5.ii.c and 7.vi.g).

Piaget had argued since the 1920s that behaviour is neither purely innate nor purely environmental, but a subtle "epigenetic" interaction of the two. (He borrowed this term from the biologist Conrad Waddington: see 15.iii.b.) He'd offered psychological evidence, based on observations of babies and children. But convincing neurobiological evidence wasn't then available. Cognitive neuroscience didn't get off the ground until Piaget's old age (see Section x.b, below), and the *developmental* version blossomed only after his death.

By end-of-century, however, developmental cognitive neuroscience was an established discipline, with its own textbooks (e.g. M. H. Johnson 1993*a*, 1996). By that time, many examples of epigenesis, in brain as well as behaviour, had been discovered.

As we saw in Chapter 7.vi.g, six co-authors—including neuroscientist Mark Johnson, connectionists Jeffrey Elman and Kim Plunkett (12.viii.b–e), and Piaget's ex-colleague Annette Karmiloff-Smith—wrote a book called *Rethinking Innateness*, accordingly (Elman *et al.* 1996). This book challenged the widely held belief that cognition rests on inherited, unchangeable, 'modules'.

That view had been championed by two towering figures in the early years of cognitive science, Jerry Fodor and Chomsky (see 7.vi, 9.vii.c–d, and 16.iv.d). Meanwhile, evolutionary psychologists such as Steven Pinker and Leda Cosmides had applied Chomsky's and Fodor's ideas to many different types of behaviour, which they supposed to be innate in a strong sense (Chapters 7.vi.d–e and 8.ii.d–e). The modular, "Swiss army knife", picture of the mind/brain was rife.

Nevertheless, the evidence against 'simple' modularity was now strong—not modules, but modularization, was the key (Karmiloff-Smith 1992). Granted, something like Fodorian modules exist in the adult brain. But they weren't there at the beginning, and their specific content depends crucially on what happens during development.

Let's consider the example of face recognition, a skill possessed not only by humans but also by birds and mammals. Building on research on imprinting in chicks by Gabriel Horn (one of my anatomy supervisors for human dissection at Cambridge!), Johnson constructed a biologically plausible model of the forebrain region involved (O'Reilly and Johnson 1994). This generated a range of 'behaviours' like those observed in imprinting experiments, and provided various predictions for further experimental study.

He and Horn also did experiments on chicks (Johnson and Horn 1986, 1988). These showed that imprinting is grounded in two independent neural mechanisms, one subcortical and one cortical, with *no* neural connections growing between them. The first, present at birth, makes the chick turn towards stimuli—natural or artificial—patterned like certain elements of a hen's head and neck. (This mechanism is *not* strictly species-specific: it also works if the stimulus is a duck's face, or even a polecat's.) The second develops later, enabling the chick to distinguish its mother from other hens.

Soon afterwards, with the psychologist John Morton (Director of the MRC's Cognitive Development Unit in London), Johnson turned to study face recognition in young babies (M. H. Johnson and Morton 1991; Morton and Johnson 1991; Johnson 1993*b*). And "young" means young: some of the baby subjects were only half an hour old (and other researchers had even got down to 10 minutes). The method used was to record the baby's head and eye movements in tracking—or ignoring—various stimuli: faces, jumbled 'faces', and non-faces. Here also, his experiments suggested two independent neural mechanisms, with different behavioural effects and time courses.

The first is subcortical, as it is in the chick. It develops in the womb, and its function is to make the newborn baby orient to, and fixate on, facelike stimuli: two dark blobs above another blob, in the general spatial relations of eyes and mouth. 'Scrambled' or malpositioned stimuli are ignored (as they are by the chick). Moreover, any face will do: mother, puppy, teddy bear . . . Given three blobs in the relevant positions, the newborn will orient towards them.

The other mechanism is cortical. It develops two or three months after birth, 'bootstrapping' on the first. But this is a behavioural bootstrapping, not a matter of neural connectivities. The blob system does its job simply by predisposing the neonate to look towards faces. What happens after that doesn't affect it.

The newborn baby shows no preference for pictures of faces over crude facelike blob designs, and doesn't seem to notice any difference. Gradually, however, a preference for faces—and for individual faces—emerges (sic). Since the baby's visual attention is drawn to faces, which in ecologically normal conditions will belong to other humans (especially the mother), he/she will receive many stimuli that can be provided only by these particular experiences. As Johnson and Morton put it, the baby is born into a "species-typical environment" which leads to the construction of more complex mechanisms from the seed of the "primal components"—e.g. the blob–attention circuit.

The environmental face stimuli configure the developing cortical circuitry before it gains control over behaviour in the third month of life—after which *that* circuit disposes the infant to look at faces. (The subcortical circuit appears to 'fade' around the fifth week of life, when preferential face tracking declines sharply—M. H. Johnson and Gilmore 1996: 343.) Gradually, the baby learns to prefer human faces to teddy-bear faces, and eventually to individuate them—normally starting with recognition of the mother.

Presumably, detailed facial-feature-detectors, such as those identified in monkeys by Perrett (Section iv.d), develop as a result of this brain–environment interaction. These may even include cells specialized to recognize expressions of disgust (Richard M. Young *et al.* 1997). But if so, these work only on human, or human-like, faces.

Facial-feature-detectors are typically species-relevant, because the infant is normally surrounded by members of its own species. (This adds a new dimension to Ludwig Wittgenstein's famous remark, "If a lion could speak, we would not understand him"—see Chapter 16.ii.b.) But the developing brain can be fooled by abnormal experience. For instance, if a horned sheep is reared with sheep without horns, it doesn't develop horn-detector cells—as it does if reared naturally, with other horned sheep (Kendrick and Baldwin 1987).

If one must use the term "innate", then what's innate here is the *predisposition* to orient towards crudely facelike stimuli. "Knowledge" of faces-as-such, or of their individual features, certainly isn't innate. By contrast, there's no innate predisposition to look at telephones. To be sure, some of the feature-detectors that will later be involved in learning to recognize telephones are present at birth. They aren't "learnt by experience" (although they are later tuned by experience). But as we saw in subsection a, they aren't specified by "heredity" either.

Granted, certain aspects of brain circuitry do appear to be genetically controlled. These include the shapes of the neurones' dendritic trees; how much these trees branch; where, and to which cell classes, a neurone's axon projects to; how strong the initial connections are; and what neurotransmitters are effective (J. A. Anderson *et al.* 1990*a*: 296). So there's no question of the brain's starting out as a randomly wired system, as in Linsker's work, or von der Malsburg's. But that's a far cry from saying that observable behaviours are innate.

In short (and as we've already seen in Chapter 7.vi.g), nativism and environmentalism are equally misleading. The adult mind/brain is formed by an epigenetic interplay, on many successive (bootstrapping) levels, of brain, behaviour, and environment.

The "environment", of course, includes the *cultural* environment. This provides cognitive technologies (6.ii.c) ranging from language and writing to shipboard instruments and navy rituals (8.iii). These don't merely *aid* the mind but, to a large extent, come to *constitute* it (see A. J. Clark 2003*a*). They become internalized within the brain, so that the brains of two adults from different cultures will be more different from each other than are the brains of any two newborn babies.

Various examples of behavioural "scaffolding" have been reviewed by Horst Hendriks-Jansen (1996), within a wide-ranging critique of GOFAI cognitivism. Although neural epigenesis can't be captured by GOFAI, or even by most connectionism, it has been simulated in more 'realistic' computer models. Some of these focus on neuroscientific details such as orientation detectors, but others deal with socially significant human behaviour (A. W. Young and Burton 1999). (For a very different, non-epigenetic, model of face recognition, see Duvdevani-Bar *et al.* 1998.)

One caveat remains. Oyama's critique of the developmental sciences applied to most epigenetic theories too (1985, ch. 3). The "heart" of the theories she was rejecting is the assumption that

form, or its modern agent, information, exists before the interactions in which it appears and must be transmitted to the organism either through the genes or by the environment. (1985: 25)

"Compromises", she said, "don't help because they don't alter this basic assumption." We should abandon the nature/nurture duality and concentrate on *how the system creates form (order, structure, information) by its own spontaneous activities*. To be sure, it's sometimes convenient to regard genes, or feature-detectors, or interneurones . . . as simply *there*. But they weren't put there by some pre-existing force, nor even by two such forces interacting. They emerged within the living system, as a result of general principles and specific biological/ecological constraints.

d. Neural selection

In the 1950s, the UCL-based anatomist J. Z. Young questioned the widespread assumption—shared from Hartley to Hebb—that learning happens by facilitating the neural pathway that's just been excited. Instead, it might happen "by reducing the effectiveness of the other" (J. Z. Young 1964: 282).

If so, then "learning occurs by elimination of the unused pathways", involving "a reduction of the initial redundancy of connexions". Like Darwinian evolution, learning produces "adaptation by random small changes and shuffling demand selection among large numbers" (p. 288). But it's less random, being guided by "rapid and precise feedbacks [providing] precise information about the results of each action".

The "mechanism" (he continued) was an enigma, calling for "inquiry into the fundamental nature of the information-gathering circuits and the types of models [in Craik's sense] that they produce" (p. 298). The functional aspects, he said, had been illuminated by the learning machine built by his UCL colleague Wilfred Taylor (Chapter 12.i.d). But the molecules were a mystery.

Over the next thirty years, this idea was taken up in a number of fields. By the late 1980s, neurochemists had clarified the mechanism, in terms of various neuro-transmitters and other molecules. Anatomists had shown that synapses continue to proliferate immediately after birth, but soon become fewer—sometimes, they're reduced to under 50 per cent.

Psychological experiments in the 1960s had implied that this pruning results from non-use (e.g. C. Blakemore and Cooper 1970). It was later confirmed that if brain circuitry is deprived of its normal input it will degenerate and/or be taken over for some other function. So, for instance, auditory cortex may develop systematically structured *visual*-feature-detectors (Sur *et al.* 1988, 1990). And (as remarked in Chapter 7.vi.i) the auditory cortex of congenitally deaf children becomes organized so as to compute visual input for linguistic purposes. (Much the same thing happens in A-Life evolutionary robots, where unused neural 'whisker drivers' can get taken over for visual purposes: Chapter 16.vi.c.)

Several neuroscientists integrated these results by explaining the brain's plasticity in terms of "neural selection". Foremost among them were Jean-Pierre Changeux (1936–), of the Institut Pasteur and Collège de France, and Gerald Edelman (1929–), initially at Rockefeller University, then UC San Diego.

Changeux did pioneering work on the molecular mechanism of neural selec-tion (Changeux *et al.* 1973; Changeux and Danchin 1976), and later discussed the development of various neuropsychological "levels", including language (Changeux 1980, 1985; Changeux and Dehaene 1989). Edelman was interested in the synaptic chemistry too, but his main goal was to explain cognition—and consciousness (G. M. Edelman 1978, 1987, 1989; Edelman and Tononi 2000). Already a Nobel prizewinner (in 1972) for his work in immunology, his neuroscience attracted great interest—and withering criticism too, as we'll see.

Neither man shared McCulloch's 'informational' worry about genetic coding: Changeux, for instance, said it was only an "*apparent* paradox" (Changeux and Dehaene 1989). His reason was the huge number of *combinations* of sets of genes: "On strictly theoretical grounds, no opposition thus exists to a full genetic coding of brain organization" (Changeux *et al.* 1984: 125). Nevertheless, neither believed in a pre-set genetic coding of specific connectivities. *Selection*, not *instruction*, was the name of the game. A profusion of connectivities is initially available, only some of which will be stabilized during development or learning.

Young hadn't been able to outline a plausible mechanism. Now, advances in immunology offered a powerful analogy. Edelman himself had shown that all possible antibodies are potentially present at birth, but that only those which match—as he put it, "recognize"—the antigens happening to invade the body are "selected" for clonal reproduction (G. M. Edelman 1992: 74–8).

Applying these ideas to the brain: the hugely diverse "pre-representations" (Changeux's term), or items in the "primary repertoire" (Edelman's), are selectively stabilized or destroyed/damped down by the specific input. And as in Grossberg's ART, this happens by some sort of resonance, or reverberation, between percept and pre-representation. Feedback (for Edelman: "re-entrant connections", influenced by "values" which modulate the Hebbian rule being used) leads to the loss of unused connections and the strengthening of the others. Moreover, Hebb would be happy: besides single connections, *groups* of co-acting neurones (particularly stressed by Edelman) are formed. These are organized in systematic ways—often, as topographical maps.

What's of special interest here is what these neuroscientists said about computational views of the mind/brain. Both rejected functionalism, classical cognitive science, and mainstream connectionism. Yet Changeux looked forward to more realistic connectionist models, and Edelman produced some of the most complex neurocognitive models so far.

Changeux located his work in a long philosophical tradition (epigenetic theory: Chapter 7.vi.g–i), and related it to recent philosophy of mind and cognitive science (Changeux *et al.* 1973; Changeux 1985, ch. 7). Like Piaget, he believed that one can't understand adult cognition unless one understands how it developed. So he had no time for GOFAI, nor for most connectionism, because—in both cases—the instructions/structure are provided 'externally' by the modeller.

Moreover, he rejected the dogma of orthodox cognitive science: multiple realizability (Chapter 16.iii and iv.c–d). For Changeux, the molecular mechanisms at the synapse put functional constraints on development at higher levels, so even cognitive psychologists shouldn't ignore them (Changeux and Dehaene 1989; cf. M. H. Johnson and Karmiloff-Smith 1992).

He explicitly dismissed Fodor's vision of psychology as a theoretically autonomous "special science" (Chapter 16.iv.c), and rebutted Philip Johnson-Laird's assertion that "the physical nature [of the brain] places no constraints on the pattern of thought . . . any future themes of the mind [being] completely expressible within computational terms" (Changeux and Dehaene 1989: 63). His aim, by contrast, was "to *reconstruct* (rather than reduce) a function from (rather than to) its neural components"—and to do this by recognizing the brain's inherent powers of organization.

If most of the then current cognitive science was on the wrong track, Changeux said, suitably realistic computer systems wouldn't be. Connectionist models of highly evolved functions, such as NetTalk or the past-tense learner (Chapter 12.vi.e–f), were "far too simple and even naive compared to what the human brain actually uses for . . . multimodal performances with deep cultural impregnation" (Changeux and Dehaene 1989: 99). Instead, we needed "brain-style computers based on the actual architectural principles of the human brain and possessing some of its authentic [self-organizing] competences rather than simply mimicking some of its surface performances". Hopfield's cooperative networks (12.v.f) were interesting, he allowed, but didn't include selection; von der Malsburg's work was more biologically oriented, so more promising.

Changeux himself went some way towards producing "brain-style" simulations. Besides modelling his wetware ideas (Gouze *et al.* 1983), he implemented a selective network for learning temporal patterns, such as birdsong (Dehaene *et al.* 1987). But

Edelman tried to go much further, modelling "multimodal performances" if not "deep cultural impregnation".

From 1980 on, Edelman's team implemented a still-ongoing series of computer models called Darwin-I, Darwin-II, and so on, plus various 'spin-off' systems (Reeke *et al.* 1990; Krichmar and Edelman 2002). These models focused on perceptual categorization, and on the integration of perception and action. Some included neuronal group selection and/or re-entrant circuitry, but others didn't. Similarly, some had "values" modulating the Hebbian rules, while others didn't. And whereas some were 'pure' simulations, others were robots (the NOMADs: Neurally Organized Multiply/Mobile Adaptive Devices).

Darwin-I, in 1981, was a pure simulation, and focused on pattern categorization. Darwin-III, nine years later, simulated the development of sensori-motor coordination in reaching behaviour (Edelman and Reeke 1990: 47–61). The first incarnation of NOMAD—alias Darwin-IV—appeared in 1992, and added visual tracking, conditioning, and pattern categorization (G. M. Edelman 1992: 91 ff., 192–3). Darwin-VII was completed by 2002, and by mid-2004 Darwin-IX was well under way, adding artificial whiskers (simulating rat neurology) which can both learn and discriminate different textures. (As this book goes to press, the most recent published account describes Darwin-X; this uses a simulated hippocampus to aid navigation and spatial exploration: Krichmar *et al.* 2005.)

Even the relatively early models drew heavily on neuroscientific data. For instance, one simulated the innervation of the back and palm of a monkey's hand, and the way in which their different tactile 'experiences' lead to the formation of differently organized groups of receptive cells (cortical maps) in the monkey's brain (Pearson *et al.* 1987).

The most recent systems are comparable to Grossberg's in size and complexity—and run similar risks of opacity. As some fellow neuroscientists have said:

One reason that many feel uncomfortable with learning by selection is that it makes a proclaimed virtue of complexity and imprecision. There is no sense at all of elegance, proportion, and abstract beauties of simple systems, well understood. Therefore the whole modeling approach sometimes seems slightly mystical. It is difficult to see what is actually going on in the simulations, that is, to see what the mechanisms assumed are doing to the behavior of the network. However, there is no requirement that nature is necessarily easy to understand. (J. A. Anderson *et al.* 1990a: 299)

To put the point in another way, it is as though we can get only so far by using what Marr called Type-I theories (Chapter 7.iii.b). The post-selection model itself is, to a large extent, a Type-II theory.

The current NOMAD (as I write), alias Darwin-VII, which isn't 'evolutionary', has about 20,000 units and 450,000 synapses, and models eighteen cortical and subcortical brain regions (Krichmar and Edelman 2002: 819). It attempts the "multimodal processing" requested by Changeux, for it integrates visual and auditory signals in categorizing the objects in its environment. The team is planning to add tactile whiskers, and probably a 'hippocampus' too (A. Seth, personal communication, June 2004; see Krichmar *et al.* 2005). Other plans are afoot to add re-entrant connections *between* neural regions, as well as within them; to explore the spiking dynamics of *individual* neurones, instead of merely averaging over small groups; and—for practical, not theoretical, purposes—to construct *hybrid* systems in which a NOMAD is linked to a conventional digital computer (Krichmar and Edelman 2002: 827).

The NOMAD of 2002 has a gripper that picks up visually distinguishable cubes and "tastes" their electrical conductivity. It "values" high conductivity, so can be conditioned to pick up striped cubes and avoid spotted ones (only the striped cubes have high conductivity); and it shows higher-order learning, by chaining conditioned stimuli. This NOMAD-ic creature learns to integrate auditory and visual signals in locating the cubes, since the striped cubes emit high tones and the spotted cubes low ones. It has horizontal- and vertical-feature-detectors, which influence categorization differently according to the temporal order of its 'experience'.

I just said that the NOMADs were robots. But Edelman didn't say that. For him, they were "noetic" (perceptual) devices, quite "unlike" the more familiar cybernetic and robotic artefacts (1992: 192). His caption of a diagram showing what is clearly a wheeled robot was this:

While NOMAD looks like a robot, it does not operate like a robot under the strict control of a program. It operates like a noetic device, one that is neurally organized and works according to selectionist principles. (1992: 193)

The implication was that anything properly termed a robot is a GOFAI system of the sort described in Chapter 10.iii.c. He simply ignored the fact that some robots were already being based on very different principles, from Arbib's *Rana computatrix* to evolved robot cockroaches and fish (see Chapter 15.vi.c, vii, and viii.a).

Similarly, he repeatedly denied that his computer models were "computational", or even "connectionist". Darwin-VII, for instance, was said to promise "the development of intelligent machines that follow neurobiological rather than computational principles in their construction" (Krichmar and Edelman 2002: 818).

His models were presented as being based on new principles, originated by him. For instance, when he reported the model of cortical maps of the hand (Pearson *et al.* 1987), his contemporaries—his potential rivals—were ignored. Von der Malsburg wasn't mentioned, and nor were Grossberg, Arbib, Longuet-Higgins, Kohonen, or any other brain-modeller (although one survey was cited: Meinhardt 1982). The great Turing was acknowledged, for his work on morphology (Chapter 15.iv); and so was the luminary Hebb. All the other references were to experimental data from neuroscience (plus one to a computer model of crystallography by one of Edelman's colleagues).

Yet, in the judgement of three neuroscientific experts, "It is likely that the assumptions made in [Edelman's] complex models contain the essential core analyzed by Amari and by Willshaw and von der Malsburg" (J. A. Anderson *et al.* 1990a: 125). In short, whereas Changeux made a point of locating his own ideas in their intellectual context (both scientific and philosophical), Edelman didn't. When he mentioned that context at all it was usually to contrast, or even to scorn, not to give credit. As Daniel Dennett put it, "he seems to have a low opinion of everyone else in cognitive science" (1993a: 285).

This insistence on his own heroic creativity (Chapter 1.iii.e) is part of the reason why Edelman was criticized with such acerbity. For instance, Barlow's review of *Neural Darwinism* opened by noting that Darwinian theories were "much in the air these days", and had been used to good effect in neuroscience (he cited Changeux's work on birdsong). After this coded snub, the review went into high gear. Edelman's version of these by now familiar ideas was described as obscure, incomplete, one-sided, and mistaken on crucial facts—and his suggested selection process "would probably wreck the function of the whole brain". The piece ended with a gentlemanly caveat carrying a vicious sting in its tail:

But epoch-making books are often panned unfairly by bigoted reviewers, so readers are strongly urged to study "this magisterial work" (as the publisher's blurb calls it) and decide for themselves whether it ushers in a new era in neuroscience, or whether it's just a hopeless muddle. (Barlow 1988)

Other stings were in store. Crick (1989b) soon complained that the theory should be called "Neural Edelmanism", since it bore no significant resemblance to Darwin's theory—and nor could he see any real benefit in applying evolutionary theory to self-organization in the brain. These were clashes of Titans: two great names infuriated by another. (Edelman cited both reviews in a popular book, crying "Vive le sport!": 1992: 94–7, 257.) But other people, too, were offended by Edelman's self-serving style. For instance, Dennett—having already complained of Edelman's "low opinion of everyone else" (see above)—tartly remarked that his work "would be a nicer try if he didn't present it as if it were such a saltation in the wilderness" (Dennett 1995a: 397 n.).

Edelman's self-congratulatory rhetoric reached its zenith in his trade books (1989, 1992), whose core messages were that he'd pretty well solved the problem of consciousness, and that "The brain is not a computer." Only members of his own trade unions, biochemists and neuroscientists, were allowed any authority or credited with any common sense. He recycled familiar criticisms of classical cognitive science and early connectionism—so ungenerously, that several readers of my acquaintance were led to defend positions they'd rejected long before.

In particular, he belittled the philosophers at length and en masse (1992, ch. 15). And he quoted an anonymous university president thus:

Why is it that you physicists always require so much expensive equipment? Now, the Department of Mathematics requires nothing but money for paper, pencils, and waste paper baskets, and the Department of Philosophy is better still. It doesn't even ask for waste paper baskets. (G. M. Edelman 1992: 159)

Just good knockabout stuff? We all like a good fight. (And the story about the university president is genuinely funny.) Remember "Vive le sport!", and forget that Wittgenstein, for instance, consigned his entire previous œuvre to the waste-paper basket. If Chomsky could famously get away with mocking his opponents (9.vii.b), why reproach Edelman?

The short answer is that whereas Chomsky's mockery was an amusing frill tacked onto a careful argument, Edelman's was riddled with unexamined assumptions, misattributions, and crude misjudgements of the views he was ridiculing. For instance, when triumphantly noting (1992: 263; see also Edelman and Reeke 1990: 68) that Hilary Putnam had abandoned his own baby, functionalism, he failed to mention—even failed to realize?—that *in the very same book* Putnam had also given up on realism (see Chapter 16.vi).

The long answer is given in the discussion of consciousness below (Sections x–xi) and in Chapter 16 throughout. We'll see that Edelman's brisk adoption of three basic assumptions—realism, evolution of mind, and naturalism—was too easy (1992, chs. 12, 15). Certainly, these are accepted by virtually all neuroscientists (though not Maturana), and by most cognitive scientists too—myself included. But they're problematic nonetheless. Adopting them without argument, or dismissing opposing positions as mere "cotton candy" (as Edward Feigenbaum does: Chapter 11.ii.f), isn't good enough.

e. Grandmother cells

"It is with a sense of proud shame that I accept fathership of the 'grandmother' cell"—so wrote Lettvin (1991), recounting a bizarre story he'd made up in 1969 for an undergraduate lecture. Too complicated to summarize here, it featured an imaginary Russian neurosurgeon and the mother-obsessed Mr Portnoy of Philip Roth's just-published novel *Portnoy's Complaint*. It also featured imaginary cells (not just one, but "18,000") in Portnoy's brain which fired whenever he thought of his mother, and others which fired in respect only of his grandmother.

His terminology had caught on so quickly that Barlow used it a few years later without even bothering to define it (Barlow 1972). Later, however, he did define it: talk of grandmother cells describes

the notion that there are cells in the brain that become active when and only when a grandmother, or some other arbitrary but specific feature, is present to the senses. (Barlow 1995: 421)

The idea was already familiar when Lettvin gave his lecture, if the memorable terminology wasn't. (Konorski, as we've seen, had called them "gnostic" units—although Lettvin hadn't heard of Konorski's work at the time—C. G. Gross 2002: 84.) And its popularity has waxed and waned over the years.

In the 1920s, Lashley argued that "It is very doubtful that the same neurons or synapses are involved even in two similar reactions for the same stimulus", and that his experimental results (on maze learning in rats) were "incompatible with . . . any theories which assume that particular neural integrations are dependent upon definite anatomical paths specialized for them" (Lashley 1929a: 3). Twenty years later, Hebb (1949) muddied the water by positing specific "cell assemblies" *as well as* broadly ranging "phase sequences" (Chapter 5.iv.c). But Lashley reiterated his earlier opinion:

All of the cells of the brain are constantly active and are participating, by a sort of algebraic summation, in every activity. There are no special cells reserved for special memories. (Lashley 1950: 477)

Pitts and McCulloch (1947) didn't go quite that far. Indeed, even Lashley had admitted some specialization in the visual system: "[Equipotentiality] probably holds only for the association areas and for functions more complex than simple sensory or motor coordination" (1929a: 25). But they did contrast their mosaic theory with the "specialized neuron" earlier posited by "the school of Hughlings Jackson" (see Section iii.a).

Yet only ten years later, they helped describe the frog's bug-detectors. And a plethora of increasingly exotic feature-detectors soon followed, as we've seen (iv.b). By 1970, the grandmothers were flourishing: it was widely believed that there may be cells which respond to faces, perhaps even cells which respond to one's grandmother. (Not just one, but many: remember Lettvin's "18,000".)

They were flourishing in the computer labs, too. Many modellers in the 1950s–1970s thought in terms of computerized grandmother cells, and Uhr and Vossler compared their version of Pandemonium to the cells newly described by Hubel and Wiesel (Section iv). Those GOFAI programs were connectionist in spirit, as we've seen (12.ii.d). But they were localist in nature (one unit, one concept), as most early neural network models were.

However, the numbers didn't add up. Lettvin's original story had envisaged 18,000 grandmother cells per grandmother. Yet Mayhew's remark at the opening of this chapter implies there may be only one. Of course, he was speaking informally—but the point stands. The 'Frog's Eye' findings (about the numbers of retinal bug-detectors per bug, and the four correlated sheets of collicular cells) clearly didn't imply *just one* brain cell per bug. However, most neuroscientists who believed in grandmother cells didn't think there were thousands of them, even if there were more than one. But it was soon pointed out—in 1968 (see Page 2000: 445)—that there wouldn't be room in the brain for even *one* neurone devoted to *every* arbitrary combination of features, such as a yellow Volkswagen (C. S. Harris 1980) or a purple Rolls-Royce.

This point was trumpeted to neuroscientists in 1972 (see below), but didn't spread to AI modellers until later. James Anderson remembers "one of the high points" of the first connectionist conference (in 1979) as

Jerry Feldman [an AI pioneer of localist networks] standing up in front of the group and computing how many cells you'd need to have enough grandmother cells to do vision. The numbers became astronomical extremely quickly. He realized he might have to rethink the problem. (J. A. Anderson and Rosenfeld 1998: 371)

Feldman wasn't the only one. The impossible numbers helped buttress the notion of distributed representation, in which each unit contributes to *many* representations—and, in the ideal case, *every* unit contributes to each representation. That too was an old idea, which went to ground in the late 1960s but later hit the headlines (see Chapter 12). Its sudden popularity seemed to ring the grandmother cells' death-knell: drowned by the 'D' in PDP.

A less extreme version (describing 'sparse' distributed representation) had appeared in 1972, in a paper Barlow wrote for the first volume of *Perception*, newly founded by his long-time colleague Gregory. Although hugely influential, its effect was less to promote interest in distributed representation than to feed the "virtual obsession with unit recording" about which Freeman would complain later. It drew heavily on recent experimental data (feature-detectors, for example), but was primarily devoted to abstract arguments about computational efficiency—Barlow's main concern ever since his Ratio Club days.

His revolutionary paper laid down five "dogmas" for perceptual neuroscience. The first was that theories should be couched at the level of individual cells, not "globalist" EEG potentials or holographic memories (Barlow 1972, sect. 6.1). Despite Lashley's (1950) evidence for mass action (as he'd put it: "every instance of recall requires the activity of literally millions of neurons"), experimenters should concentrate on single cells, rather than averaging over many.

The "unreliability" inferred from recent work on stochastic activity in the nervous system (Chapter 2.viii.f) had been overestimated. Although there is some intrinsic noise, its level is "extraordinarily low":

Individual nerve cells were formerly thought to be unreliable, idiosyncratic, and incapable of performing complex tasks without acting in concert and thus overcoming their individual errors. This was quite wrong, and we now realise their apparently erratic behaviour was caused by our ignorance, not the neuron's incompetence. (Barlow 1972, sect. 3.5)

A single neurone, he argued, could code a complex reality (compute a complex decision). But there was a hierarchy here: "at the higher levels, fewer and fewer cells are active, but each represents a more and more specific happening in the sensory environment" (sect. 8). The entire visual scene may be represented by as few as 1,000 neurones. Each of these codes, or corresponds to, "a pattern of external events of the order of complexity of the events symbolized by a word" (he recalled the adage that "A picture is worth a thousand words"—p. 371 and sect. 8.3).

Nevertheless, Barlow spurned grandmother cells. He called them "pontifical cells", a term borrowed from Sherrington (1940), who in turn had taken it from William James. James had spoken of "one central or pontifical [cell] to which our consciousness is attached," in which the physical activities of all other brain cells are "combined" (1890: 179). In his Gifford Lectures,

[Sherrington] introduced the notion of "one ultimate pontifical nerve-cell . . . the climax of the whole system of integration" and immediately rejected the idea in favour of the concept of mind as a "million-fold democracy whose each unit is a cell". (Barlow 1972, sect. 12.4)

(The imagery is telling: it's no accident that a book published in England in 1940 should show an unargued preference for "democracy".) Barlow's rejection of pontifical cells was more carefully reasoned.

He allowed that people who see perception as "a cooperative or emergent activity of many cells" would question his first dogma, saying that "carried to its logical conclusion" it implies there must be a single brain cell corresponding to "each and every recognizable object or scene". And he agreed that this is implausible. That was partly because (as noted above) there aren't enough cells, and partly because a single cell's activity couldn't convey the "richness" of any given perception:

The "grandmother cell" might respond to all views of grandmother's face, but how would that indicate that it shares features in common with other human faces, and that, on a particular occasion, it occurs in a specific position surrounded by other recognizable objects? (Barlow 1972, sect. 12.4)

Instead, we must posit many "cardinal cells", of whom only a few speak at once: "each makes a complicated statement, but not, of course, as complicated as that of the pontiff if he were to express the whole of perception in one utterance". (Remember the intermediate-level demons in Pandemonium.)

But the neuronal hierarchy is very different from the Catholic Church, since the cardinals outnumber the worshippers in the pews. In other words, there are more cortical neurones capable of being influenced by vision than there are receptors in the retina. Far from there being a *single* cell that recognizes one's grandmother, "the college of these cardinals . . . must include a substantial fraction of the 10^{10} cells of the human brain".

In short, Barlow was recommending 'sparse' distributed representations, in which *only some* of the system's units are active. This was consistent with his second dogma: "The sensory system is organized to achieve as complete a representation of the sensory stimulus as possible with the *minimum number* of active neurons" (italics added).

The paper closed with a frank admission, a telling comparison, and a tacit obeisance to Craik:

> But clever neurons are not enough. The simplest computer program with its recursive routines and branch points has more subtlety than the simple hierarchy of clever neurons that I have here proposed as the substrate of perception.
>
> I think one can actually point to the main element that is lacking. [What is needed is] a corresponding model . . . for our own motor actions and their consequences. Such motor and sensory models could then interact and play exploratory games with each other, providing an internal model for the attempts of our ever-inquisitive perceptions to grasp the world around us. A higher-level language than that of neuronal firing might be required to describe and conceptualize such games, but its elements would have to be reducible to, or constructible from, the interactions of neurons.

Even tensor geometry and dynamical approaches, then, wouldn't undermine Barlow's first dogma—for they study systems composed, at base, of individual units.

Although experimental neuroscientists read Barlow's paper avidly, computer modellers—especially those from an engineering background—didn't. As we saw in Chapter 12.v–vii, a variety of people in the 1970s worked on 'whole-network' distributed representations—including PDP, which grabbed the spotlight in the 1980s. Barlow's sparse theory was sidelined, and grandmother cells apparently forgotten.

But could that really be right? After all, those feature-detectors *had* been found in visual cortex. Perhaps the most basic sensory features had cells all to themselves, while more complex and/or ecologically arbitrary patterns didn't? (Lettvin's story had actually started by imagining the discovery of *mother* cells, which are biologically plausible in a way that grandmother cells—and "yellow Volkswagen cells"—are not.)

The PDP group themselves, when talking about *the brain*, had steered a mid-course between Lashley and Lettvin:

> Clearly, there is specificity. Cells do not respond to all conceivable stimuli or even a small subset of them but are quite specific in their responses. At the same time they are not, in our opinion, so specific as to be what Barlow's dogmas would lead one to expect . . .
>
> We suggest that truth lies somewhere in the middle of the two extreme views and that there is a moderate amount of distribution in the cortex, so that any single cell responds to many things but nowhere near all things. [Both experiments and informational considerations point to] a considerable range of specificities from quite specific to quite broad. (Hinton and Anderson 1981: 43–4)

Geoffrey Hinton (1981) had even proved that features are most efficiently coded by a few very coarsely tuned units, not a multitude of finely tuned ones (12.iv.h). In short, the brain isn't a fully distributed system (as their computer models were)—but it doesn't contain 'pure' grandmother cells either. (Anderson still holds that "single-unit responses report only the dimmest shadows of the real mechanisms of neural computation"—J. A. Anderson and Rosenfeld 1998: 262.)

At the millennium, grandmother made her comeback. *Behavioral and Brain Sciences* published a spirited "localist manifesto", plus thirty-eight pages of peer commentary and response (Page 2000). Mike Page (at Cambridge's Cognition and Brain Sciences Unit) complained that localist representations "have acquired a bad reputation in some quarters" largely because "PDP/distributed" and "localist" had come to be seen as mutually exclusive. As he used the terms, however, "Localist models are characterized by the presence of localist representations rather than the absence of distributed representations" (2000: 443). (Compare Minsky's K-line theory, which combined single-unit significance and distributed processing: 12.iii.d.)

Localism in that sense, Page argued, is more biologically plausible than pure localism—and pure PDP, too. Various tasks—including symbolic modelling and memory for serial order—are tractable only for a localist system, so that "in the domain of psychological modelling, localist modelling is to be preferred". And he offered mathematical arguments showing why, given our psychological abilities, neuroscientists should *expect* to find both grandmother cells and distributed cell assemblies.

One of the many peer reviewers was Barlow himself, who declared:

Almost all representations have both distributed and localist aspects, depending upon what properties of the data are being considered. With noisy data, features represented in a localist way can be detected very efficiently, and in binary representations they can be counted more efficiently than those represented in a distributed way. Brains operate in noisy environments, so the localist representation of behaviourally important events is advantageous, and fits what has been found experimentally. Distributed representations require more neurons to perform as efficiently, but they do have greater versatility. (Barlow and Gardner-Medwin 2000: 467)

Barlow was now having second thoughts about just how sparse a sparse representation should be (Barlow 2001*a*,*b*; Gardner-Medwin and Barlow 2001). He'd originally emphasized *economy* in neural coding, having "eagerly stepped on the boat" of Shannon's and Fred Attneave's (1954, 1959) studies of redundancy. This approach was popular for a while, but then "information theory dropped out of the limelight in neuroscience" until a reawakening in the 1980s. His new, or newly revived, "take-home message for the neuroscientist" was "Think probabilities" (Barlow 2001*b*)—see Section iii.b, above.

He now argued (against Shannon and Attneave) that coding should convert the hidden redundancy into an explicit, immediately recognizable, form rather than reducing or eliminating it. Fast behavioural response needs cortical representations with minimum overlap (i.e. very few elements activated by both inputs) when two inputs need to be distinguished, but not when they don't—in learning to respond similarly to both, for instance. So the second dogma bit the dust:

Economy in the number of neurons used for the representation of sensory information is a bad idea, and the reverse is what actually seems to happen. On the other hand economy in the number of *active* neurons would make redundancy explicit rather than hidden, and would make each impulse represent an important, informative, event. (2001*b*, sect. 4)

Thinking probabilities, he said, can help us to understand *why* representations in the cortex "appear to use extravagant numbers of cells" (Gardner-Medwin and Barlow 2001: 477). In other words, we should drop the assumption that the more patterns a memory can hold, the better (see 12.v.c).

The truth about grandmother cells, whatever it turns out to be, will help decide which 'brainlike' models we favour. But it can't undermine computationalism as such. Much of the argument has concerned the computational power of different information-processing systems, and both 'unit-based' and distributed models have been implemented. Computational neuroscience can love its grandmothers or leave them.

f. Modelling modulation

For the first half-century of brain modelling, the general assumption was that a neurone can affect only those others which are connected to it, either directly or indirectly. To be sure, chemicals were involved: neurotransmitters at the synapse enabled one neurone to excite or inhibit the next. But these were thought of as mechanistic nuts and bolts, theoretically invisible at the functional (psychological) level. At that level, it was thought, linkage is all.

Or almost all: neural excitability *in general* might be affected—in changes of mood, for instance—by hormones diffusing throughout the brain. (As Norbert Wiener had put it, hormonal influences were messages "to whom it may concern"—Dupuy 2000: 116; Wiener 1948: 129–30.) Indeed, this fact led to complaints that AI can't model moods or emotions, because it can't model diffuse hormonal influences (Haugeland 1978: see Chapter 7.i.d). But *specific* neuronal functions—feature-detectors, for instance—were thought to depend on connections, not chemicals.

The first nail in this assumptive coffin was driven by the discovery of neuromodulators in the 1970s–1980s. These substances can temporarily alter a neurone's properties: in effect, they substitute one broadly Hebbian rule for another. They inspired various computer models (Fellous and Linster 1998). But, like neurotransmitters, they were thought to function only *locally*. All the then known neuromodulators were large organic molecules. Because they can diffuse only slowly inside the cell, and can't pass through cell membranes, they're stored and released near the synapse and can act only at the synapse itself. And "synapse", after all, is just Sherrington's word for "connection".

Around 1990, however, it was found that at least one neuromodulator is a very small inorganic molecule: nitric oxide, or NO (Garthwaite *et al.* 1988; Holscher 1997; O'Shea *et al.* 1998). This is generated, given the right enzymatic trigger, by (the whole cell body of) specialized neurones. But because it's so readily diffusible, even across cell membranes, NO can't be localized (or stored: it has to be synthesized 'on the spot', when needed).

It diffuses in all directions, and its effects—which depend on the concentration at the point concerned—endure until it decays. (The rate of decay can be varied by enzymes.) Consequently, NO works on all the cells within a given volume of cortex, *whether they're synaptically connected or not*. The resulting functional dynamics of the system (network) are highly complex, and very different from those of 'pure' connectionist systems, for volume signalling replaces point-to-point signalling.

This finding (later repeated for carbon monoxide and hydrogen sulphide) inspired researchers at Sussex's interdisciplinary Centre for Computational Neuroscience and Robotics to design artificial networks of a radically new type, where linkage is *not* all. For instance, the size of the diffusion volume matters, and so does the shape of the source (simulated as a hollow sphere, not a point source). Some of their models were kept as close to the biological data as possible, thanks to Michael O'Shea, a leading researcher on NO in invertebrate nervous systems (Philippides *et al.* 1998, 2005*a*). Others were more abstract, defining a general class of connectionist system called a GasNet.

In GasNets, some nodes scattered across the network can release diffusible 'gases', which modulate the intrinsic properties of other nodes and connections in various ways, depending on concentration. So one and the same node behaves differently at different times. Given certain gaseous conditions, a particular node will affect another despite there being no direct synaptic link. In other words, it's the *interaction* between the gas and the electrical connectivities in the system which is crucial. And, since the gas is emitted only on certain occasions, and diffuses and decays at varying rates, this interaction is dynamically complex.

General properties of neuromodulation could be explored using these systems. Indeed, the 'Gas' could be switched off, leaving the 'Net' unmodulated. When that was done, the resulting ("NoGas") dynamics were less subtle, the sensory morphology less efficient, and evolution (using the techniques described in Chapter 15.vi.c) much slower.

A specific behaviour might involve two *unconnected* sub-nets, which 'worked together' because of the modulatory effects. For instance, a variety of oscillator circuits, with various spiking frequencies and behavioural functions, were generated when GasNets were evolved to function as controllers for autonomous robots (Husbands 1998; Husbands *et al.* 1998; T. Smith *et al.* 2002; for the GasNet group's most recent paper, see Philippides *et al.* 2005*b*).

Feature-detectors (implemented as partially *unconnected* 'networks') arose which could recognize a triangle and use it as a navigation aid. As we'll see in Chapter 15.vi.c, the Sussex A-Life group had already evolved a wholly connective network to do this. Now, they were using neuromodulation to evolve such mechanisms faster, and to make them function more efficiently. *Post hoc* dynamic analysis showed that the modulated feature-detector involved a motor-node with a 2-cycle oscillation, which stimulated gas emission from two other nodes (T. Smith *et al.* 2002, sect. 7). Without the gas, the connectivities in that particular network couldn't do the job.

Space and time, as well as connectivities, affected the behaviour of GasNets. In that sense, even the highly abstract versions were more 'realistic' than conventional connectionist systems—for the brain, too, isn't just circuitry. But they were also highly unrealistic. In current GasNets, the modelled diffusion is instantaneous, not gradual (still less, carefully timed). And there's no interaction between different 'gases', as considered by Turing in his work on embryology (Chapter 15.iv.a). In principle, however, the physics and pharmacology could be modelled much more faithfully (cf. Wood and Garthwaite 1994).

By the time you read this, there will surely be many more simulations of neuromodulation. For, having long avoided neurochemistry, 'psychological' modellers had finally woken up to it.

Arbib was an example. After years of considering only connections (see Section vii, above), he confessed in 1993:

[Until] recently, there's been a burgeoning world of neurochemistry which I've tried to stay away from, quite distinct from the world of relatively large-scale neural networks. Now, suddenly, we're seeing that we need to tailor the learning rules. Now, if you tailor the learning rules, you have to get into the biochemistry and understand how the playing off of the different mechanisms can change the time parameters . . . I now for the first time really have an integrated view in my work which shows the need to unify the neurochemistry with the systems modelling. (J. A. Anderson and Rosenfeld 1998: 235)

Similarly, David Rumelhart abandoned the 'pure' PDP that he'd helped to popularize in the 1980s (see Chapter 12.vi). Interviewed in the same year as Arbib, he said:

[One] of the things I've been working on lately is trying to factor neurochemistry into my models, neuromodulators and things like that, which I take to be much more important than we've realized. I have a paper on the role of neuromodulators. It's a generalization of our conventional [PDP] networks, but which have neuromodulators included. (J. A. Anderson and Rosenfeld 1998: 289)

In short, there'd been a sea change in computational neuroscience. Connections had been married to chemicals, understood not as invisible nuts and bolts but as dynamic functions within neural networks.

At the turn of the millennium, an even more exciting change was in sight, as these ideas moved from pure simulation into the 'wet' world. For instance, in January 2005 a UK team started trying to evolve non-linear computers combining both wetware (diffusing chemicals and/or cultured neurones) and silicon (initially, up to sixty-four miniature electrodes arranged in grids/circuits). If successful, these systems would be capable of reliably achieving a user-defined computation.

This highly interdisciplinary project drew on several sources. One was the Sussex work on the evolution of GasNets, another Sussex's experience in culturing neurones *in vitro*. A third source was research by the physical chemist Annette Taylor, of the University of Leeds, on the dynamical properties of diffusing chemical waves (Chapter 15.iv). And fourth and fifth, previous work by AI scientists and toxicologists at the University of the West of England in Bristol—focused respectively on non-linear computing and the biochemistry of cell culture.

The general aim was to use GA techniques to evolve networks of chemicals and/or neurones which would compute a particular logical function. For instance, when electrodes X and Y were both stimulated artificially, the resulting activity in the final-generation 'wet' network should reliably result in the firing of some other electrode, Z—thus implementing an and-gate.

During the GA evolutionary process, several factors are randomly varied. They include the nature and concentration of various chemicals (e.g. NO, dopamine, and serotonin); the location, at different nodes in the neural network, of the release of neuromodulators; and the amount/timing of growth factor provided while the cells were being cultured. (Both vertebrate and invertebrate neurones are being studied.)

It's too early (as of June 2005) to report any firm results. But you can check up on what's happened by accessing the team's web site (<http://www.informatics.sussex.ac.uk/ccnr/research.html>). Meanwhile, the hope is that this research will throw light on non-linear computation in both engineered and organic machines, as well as teaching us more about neurones, neuromodulators, reinforcement learning, brain development, and medical prosthetics. ("Watch this space!", as they say.)

g. Time blindness—and glimmers of light

Rumelhart continued the conversation just quoted by mentioning the importance of *time*:

> If you are interested in dynamics, it turns out that there's a whole host of timescales involved. Different neuromodulators have different temporal properties; the neurons have different temporal properties, and learning has still others. In the nervous system, these things are all overlapping timescales, not independent ones like in physics. (J. A. Anderson and Rosenfeld 1998: 289)

The GasNet modellers, too, stressed the importance of different timescales in their systems. Indeed, they were among those 'A-Lifers' who had claimed for some years that conventional connectionism was hugely unrealistic—perhaps even biologically irrelevant—because it ignored temporal constraints on neural coding (I. Harvey 1992*b*). (As we've seen, this time blindness was found only in *conventional* connectionism; it didn't apply, for instance, to Grossberg's work.)

Timing matters not merely at the level of neural spikes, but at grosser levels too. For instance, if someone tries to learn two different movements in quick succession, they interfere; if separated by a few hours, they don't (Wolpert and Ghahramani 2000). From about 1970 on, experiments on conditioning showed repeatedly that timing (the onset and offset of the conditioned and unconditioned stimulus) was crucial.

Accordingly, some neuropsychologists even claimed that *associationism itself* was unrealistic, largely because of its 'timelessness'. Charles (Randy) Gallistel, a psychophysicist at UCLA, went so far as to call the associative bond "the phlogiston of psychology" (1997: 85). He not only claimed that *some* learning (e.g. gauging one's position by dead-reckoning, which many insects can do, or learning the centre of rotation of the night sky) isn't associative. He even insisted that "the process of association formation—as traditionally understood—is not involved in *any* form of learning that has been experimentally investigated, including classical and instrumental conditioning" (1997: 81; italics added).

Gallistel linked this claim with a commitment to Chomskyan modules (see Chapters 7.vi.d−e and 16.c−d). In his words: "A bird's gotta learn what a bird's gotta learn, and ditto [as regards language] for humans" (1997: 88). So, bearing in mind the findings of CNE (Chapter 15.vii), he said:

> [There] is no unitary learning process at the computational level of analysis. There are many different learning mechanisms or modules . . . different problem-specific learning mechanisms [which we may call] "instincts to learn". Each such mechanism has a structure that enables it to compute certain facts about the world. This specialization of computational structure renders the mechanism hopelessly ill-suited for computing other facts about the world. For that, one needs other mechanisms, with a different computational structure. (1997: 82)

So, much as GOFAI had moved from general-purpose mechanisms to expert knowledge (Chapter 10.iv), neuroscience should do so too:

Despite long-standing and deeply entrenched views to the contrary, the brain can no longer be viewed as an amorphous plastic tissue that acquires its distinctive competencies from the environment acting on general-purpose cellular-level learning mechanisms. Cognitive neuroscientists, as they trace out the functional circuitry of the brain, should be prepared to identify adaptive specializations as the most likely functional units they will find. *At the circuit level*, special-purpose circuitry is to be expected everywhere in the brain, just as it is now routinely expected in the analysis of sensory and motor function. (Gallistel 1995: 1266; italics added)

He did allow that generality at the *cellular* level is "possible". All specialist learning mechanisms may employ the same set of elementary computational operations. But they may not: "At a time when we cannot specify the instruction set underlying any computation of any substantial complexity, we are in no position to say whether different complex computations use the same elementary computational operations" (1997: 83).

Even classical conditioning, which links *arbitrary* stimulus–response pairs, can't be explained by associationism. For Hebbian rules can't handle the "fascinating" results found in the last quarter-century, the "golden age" of experimental conditioning:

[It appears that *all*] learning involves computing and storing the values of variables. In the case of conditioning, the crucial remembered variables are *the temporal intervals between events* [CS and US onsets and offsets]. But this conception of conditioning is as different from the traditional associative conception as the oxygen theory is different from the phlogiston theory. If this conception prevails, then the associative bond will indeed prove to have been the phlogiston of psychology. (Gallistel 1997: 85; italics added)

In a word, associative theories should be replaced by cognitive theories (Gallistel and Gibbon 2002). These assume that "the crucial remembered variables" are represented in the animal's brain, and that representations and decision rules together determine whether, and when, a response will occur.

No golden age had been needed to persuade neuroscientists that timing was often crucial. They'd known this since before mid-century (and one of Barlow's five dogmas concerned "high impulse frequency"). However, McCulloch in the early 1940s had deliberately ignored temporal properties he knew to be important: his "temporal propositional expressions" concerned temporal sequence, not time as such. His example was followed by virtually all connectionist modellers thereafter. (Early exceptions included Joseph Licklider 1951, 1959, and Braitenberg 1961.)

A few people had remarked that time could be incorporated in principle. Marr, for example, said in the 1960s:

There is no reason why a context dependent system should not be run at different speeds, and if the extra postulate were made of some general intensity control acting uniformly over the effector circuits, a movement learnt at one speed could be performed at another. (D. C. Marr 1969: 466)

Moreover, Marr had inferred specific hypotheses about time intervals from his formal model of the cerebellum (see Section v.c). But most computer modellers ignored time entirely. Even 'temporal' connectionism, aka recurrent networks, considered temporal order rather than measurable time.

This wasn't inevitable: there's no principled reason why time can't be taken seriously in computational work. Although von Neumann computers operate on a 'clock' of uniform time-steps, it's not difficult to simulate real time in most practical circumstances. The block that prevented this was more psychological than computational.

In the early 1990s, Inman Harvey (1992*b*) urged network modellers to stop ignoring the real-value time lags on signals passing from one node to another. And in the late 1990s, he pointed out that models of *asynchronous* networks behave very differently from their synchronous cousins (Harvey and Bossomaier 1997). Some connectionists com- bined recurrent networks with simple models of synchronous neural oscillations (e.g. Lane and Henderson 1998). Others took time lags into account by experimenting with 'pulsed' networks. Here, the significance of a signal is coded by its timing; for instance, two pulses separated by a particular time interval may excite a certain unit, whereas pulses separated by any other time interval may inhibit it (Maass and Bishop 1999). But these weren't intended as serious models of real brains.

There were some exceptions (all in the last quarter-century), mostly coming from the neuroscientific stable. These included Grossberg, von der Malsburg, Arbib, Freeman, Changeux, and Edelman.

For instance, Grossberg modelled real-time perceptual processes, and used codes based on the frequency of unit firing. Arbib did so too, and simulated a number of interacting rhythms and arhythmias defined at the single-cell, correlated-cell, network, and (various) brain-regional levels (Arbib *et al.* 1997, ch. 6). His model hippocampus was one of the most time-informed, as well as one of the most complex, 'brainlike' systems of the period (see Section vii.c). By the new century, enough general interest had been raised to justify a special (double) number of *Adaptive Behavior* devoted to timing issues (Di Paolo 2002*a*).

There are technical obstacles if one wants to *emulate* time. Whereas one can if necessary take months to *simulate* a millisecond, emulation—where the computer generates real-time events at real moments in time—requires "a millisecond" to mean a millisecond. Some computer operating systems don't allow the real-time precision that would be needed to emulate highly time-sensitive processes. But others can already implement robot controllers driven by spike-timing measured to a few tens of milliseconds (Di Paolo 2002*b*: 143). So even emulation isn't always impossible.

As a critique of past work, then, this arrow meets its mark. Most modellers have ignored time, even including the biologists among them. This wasn't because they knew no better. Connections are clearly important, and people were focusing on those: one can't do everything at once.

By the turn of the century, however, it was clear to everyone—not just to the neuroscientists—that if one wants to model the brain, this couldn't continue. As Sejnowski had put it in 1993, "the real problem in the next five, ten years is coming to grips with real time signals [and imperfections too]" (J. A. Anderson and Rosenberg 1998: 331). Seven years later, one of the striking features of the models described by O'Reilly and Munakata (2000) was that many displayed close attention to time. As one would expect, given the authors' commitment to biological realism, the temporal features involved were based on the spiking times of real neurones (see ii.d, above). It was time, at last, for time.

14.x. Cartesian Correlations

This section outlines some theories of consciousness developed within cognitive science. Only *some*: as we'll see, there was an explosion of work on consciousness in the last two decades of the century. This happened quite suddenly. As late as 1979, the person who first gave cognitive psychology its name (Chapter 6.v.b) declared that psychology wasn't "ready" for consciousness (Neisser 1979). And Ernest Hilgard, who'd pioneered work on attention and hypnosis (7.i.h), said in the mid-1980s that most experimental psychologists were still "hostile" to the topic (Hilgard 1977/1986: 294).

At this point, the philosophical health warning given in Chapter 2.iii.a bears repeating. References to conscious experiences, or sensations, trip easily off the tongue—even the tongue of the man on the Clapham omnibus. "And why not? What could be more familiar? We've all had more experiences than we've had hot dinners."

Well, perhaps... In Section xi, we'll ask whether this common-sense notion is for scientific purposes better avoided. Why? Not because questions about the brain's role in consciousness can't be definitively answered (subjectivity, and all that). But because, given the usual interpretation of *experience*, certain 'obvious' questions about mind and brain *can't even be sensibly asked*.

Meanwhile, I'll continue to use our everyday vocabulary as though it were unproblematic. This is appropriate, for nearly all neuroscientists (like the passengers on the omnibus) do so too.

a. Consciousness comes in from the cold

Ever since the seventeenth century (specifically, ever since Descartes), philosophers discussed conscious experience and its relation to events in the brain: causal? parallel? epiphenomenal? identical...? And in the years surrounding the rise of cognitive science, they devoted countless pages to this topic (see Chapter 16.i–vi). But neuroscientists didn't.

Admittedly, neuroscientists asked just what types of consciousness occur when the brain is affected in various ways. In the late nineteenth and early twentieth centuries, they had only clinical evidence to go on. Later (starting with the invention of the EEG: Chapter 4.vi–vii), they could do systematic experiments of various kinds. With the rise of computational ideas, they looked at consciousness in terms of the information processing going on—and the ways in which this might break down. And when brain scanning became possible, various abnormalities of consciousness—including those seen in schizophrenia, multiple personality, and hypnosis—were studied by them accordingly (see references in subsection c, below).

Moreover, they realized that there's no such thing as *the* problem of consciousness. On the contrary—and as the philosophers were pointing out at length (16.iv.a–b and ix.c)—there are many such problems. For the words 'conscious' and 'consciousness' are used to refer to various phenomena, ranging from minimal alertness, through selective attention (6.i.a–c), to self-awareness via higher-order thoughts, or HOTs (7.i.g–i) and sensory experience.

So neuroscientists in the first three-quarters of the century experimented extensively on sleep and waking, various levels of anaesthesia, attention and orientation, conscious memory-recall, and perception—normal, illusory, and subliminal. They studied the brain's role in pain, anxiety, hallucination, and depression. And they described clinical syndromes involving various types of breakdown in conscious unity, such as 'multiple personality' (Chapters 5.ii.a and 7.i.h) and hemi-neglect—in which brain-damaged patients ignore one half of their body, even attributing it to someone else.

The Montreal neurosurgeon Wilder Penfield (1891–1976), an ex-student of Sherrington, studied brain–consciousness correlations in a relatively direct way. In around 1950 he discovered, unexpectedly, that he could elicit memory-like experiences by stimulating his patients' brains during surgery for temporal lobe epilepsy. (Only memory-*like*: the normal associative penumbra, they said, was missing.) One person, for instance, reported: "I heard my mother talking on the phone, telling my aunt to come over that night," and another remembered seeing her nephew and niece in her dining-room: "they were getting ready to go home, putting on their coats and hats ... [and] my mother was talking to them" (Penfield and Perot 1963; see also Penfield 1952; Penfield and Rasmussen 1957; Penfield 1958). But he couldn't do this systematically: the stimulations served surgery, not curiosity.

With the advent of single-cell recording, researchers occasionally came up with results that tempted interpretation in terms of conscious experience. As early as the mid-1950s, for example, it was found that cells in the cat's auditory system (the cochlear nucleus) that are responding to 'meaningless' clicks will immediately stop responding if the cat is shown a mouse, allowed to smell a fish, or shocked on its paw (Galambos *et al.* 1956). The explicit inference was that the cat's *attention* had been captured by the new stimulus. But the tacit implication was that the cat stopped *consciously hearing* the clicks when it had more important fish to fry—or anyway, to *smell*. (That's why this experiment was sometimes cited by New Look psychologists as a support for "perceptual defense": e.g. Bruner and Klein 1960—see Chapter 6.ii.a.)

However, this experiment had been designed to study conditioning, not consciousness. It was found that the auditory cells would eventually stop responding (habituate) if the clicks continued—*unless* the clicks were 'meaningful', having been used as conditioned stimuli warning the animal of an impending electric shock.

The emphasis, in the click experiment, on matters other than consciousness was typical of its time. Conditioning, and even attention, could be studied, for both could be defined in objective terms. What neuroscientists didn't do, until relatively recently, was to consider the origin of conscious experience, or sensations, as such: what philosophers called 'raw feels' or 'sense data' around mid-century, and '*qualia*' today. (Remember that health warning!) They might ask *when* conscious experience occurs, relative to events happening in the brain. After mid-century, they started to ask *what computations* were involved in its construction. But they fought shy of asking *how it's generated by the stuff inside our skulls.*

Occasionally, an acclaimed neuroscientific guru—such as Sherrington (1940) or Eccles (1953, ch. 8)—would dare to speculate on this issue. Their discussion would be self-confessedly bemused and/or embarrassingly naive from the philosophical point of view (see Chapter 16.i.a). But their colossal reputations couldn't be harmed by sniggers behind the reader's hand. Neuroscientists in a more vulnerable professional position—and psychologists, too—tended to avoid the topic entirely.

As late as 1981, John Haugeland remarked that "the term itself is almost a dirty word in the technical literature" of cognitive science (Haugeland 1981a: 32). Even the maverick McCulloch, when he'd declared contra Sherrington that "Mind no longer goes 'more ghostly than a ghost'" (see Chapter 4.iii.e), hadn't grasped this particular nettle firmly.

A few courageous souls tried to fling the nettle onto the bonfire. So Barlow, in his 'five dogmas' paper, declared robustly that

Thinking is brought about by neurons, and we should not use phrases like "unit activity reflects, reveals, or monitors thought processes", because the activities of neurons, quite simply, *are* thought processes. (1972, sect. 5)

But the nettle was slippery—maybe even fireproof. A few pages later, Barlow referred to "the activity of nerve cells accompanying [the personal, subjective, aspect of my] experience", which last is "something that one must be content to leave on one side for the moment" (1972, sect. 10.1). And in the next paragraph, having granted that not all aspects of visual perception are conscious, he insisted—in defence of his fourth "dogma"—that "an element of [conscious] perception can possess a simple neural cause without it necessarily being the case that all simple neural events cause perception". He didn't stop to ask how, if X is *identical* with Y, it can also be said to *accompany* Y, or to *cause* (or *bring about*) Y. We'll pick up this fire-resistant nettle in Chapter 16.i.d.

Much as Barlow was happy to talk about "cause" in this context, so neuroscientists in general were happy to talk about "correlations". However, there were a few intriguing experiments suggesting that mind–brain correlations are much less straightforward than Descartes—and our own contemporaries—imagined.

Most of these were done by Benjamin Libet, who started work on the neuronal basis of specific experiences in humans in the early 1960s (Libet *et al.* 1964, 1967; Libet 1965). For instance, he placed electrodes on the surface of people's sensory cortex during brain surgery, and asked them what—if anything—they felt when he delivered a stimulus; and he measured the brain waves elicited when they experienced (or thought of) something. He discovered that as much as half a second of neuronal activity is required before an experience can occur, although perceptual information is registered subliminally in half that time. Apparently, then, *qualia* are *constructed* by the brain.

This was already known by experimental psychologists. For instance, Paul Kolers—at Harvard's Center for Cognitive Studies—had been working since the late 1950s on "metacontrast", in which presentation of one visual stimulus very soon after another can block or even transform conscious awareness of the first (Kolers and Rosner 1960; Kolers 1972; Kolers and Grnau 1976; Kahneman 1968). But psychologists' experiments on visual masking had nothing to say about the neural mechanisms involved (see subsection b, below).

Libet's results (and Kolers's) didn't arouse widespread interest at the time—still less, heated discussion of their implications for the philosophy of mind (but see P. S. Churchland 1981 and Libet 1981). By the end of the 1980s, however, that had changed.

Suddenly, the neural basis of experience was a hot topic for scientists, as well as for philosophers. Indeed, the two communities started interacting much more than they'd done before, and contributed jointly to many collections of papers on the various aspects of consciousness (e.g. Pope and Singer 1978; Marcel and Bisiach 1988; B. Milner and Rugg 1992; Bock and Marsh 1993; Posner and Raichle 1994).

Discussion was further boosted, for the general reader as well as the professionals, by the publication in 1991 of two best-selling books: Tor Nørretranders's *The User Illusion: Cutting Consciousness Down to Size* and—especially—Dennett's provocatively titled *Consciousness Explained*. As we'll see in Section xi.b, this solved, or rather dissolved, the problem of mind–body correlations by giving a computational/behavioural analysis of consciousness, *including qualia*. But, like Nørretranders's book too, it also leant heavily on neuroscience—including recent work by Libet and by Marcel Kinsbourne (a co-author with Dennett in 1992) that cast further doubt on the *timing* of consciousness.

A biennial interdisciplinary conference entitled 'Toward a Science of Consciousness' was instituted by the newly formed Center for Consciousness Studies in Tucson (Arizona) in 1994. It attracted huge attention, and soon prompted the formation of the Association for the Scientific Study of Consciousness. Journals such as *Cognition and Consciousness* and the *Journal of Consciousness Studies* (tellingly subtitled *Controversies in Science and the Humanities*), and various 'consciousness' email discussion groups, were founded—and flourished. Quantity didn't guarantee quality: many of the papers were weak, to put it politely. By century's end, however, there was a wide variety of more 'respectable' work, both philosophical and neuroscientific (for a taster, see Dehaene 2001).

In the new century, robotics and AI got into the act in a big way (Haikonen 2003; Torrance 2003; Chrisley *et al.* 2005*a*,*b*). Of course, people had been asking philosophical questions about robot consciousness for years—not least, as a result of Turing's 1950 paper. But now things seemingly became more serious.

Between 2001 and 2005, there was a flurry of international meetings on "machine consciousness": in Cold Spring Harbor, Skövde (Sweden), Memphis, Birmingham, Turin, Antwerp, and London. An increasing number of papers on this topic graced the pages of the *Journal of Consciousness Studies*, which then decided to devote a double-number special issue to it (O. Holland 2003). Conferences named for other topics, such as 'Adaptation in Artificial and Biological Systems', ran symposia on machine consciousness. And a dedicated web site was established (<http://www.machineconsciousness.org>).

Some of this interest was purely philosophical. But much of it was linked to practically oriented work on "robot companions" (see Chapter 13.vi.d) and/or the evolution of linguistic communication in robots (Steels 2003). AI/A-Life researchers were now receiving research grants to develop "conscious robots" (Owen Holland, personal communication). And a commercial company, Nokia, was providing funds for this goal too.

Or at least, Nokia were happy to have the project described in that way. Just what they—or anyone else—really understood by the term "machine consciousness" was obscure, and philosophically up for grabs. Often, people tried to avoid the philosophical problems by defining "machine consciousness" in a non-committal way. For instance, the Call for Papers for AISB's symposium at Bristol in 2006 defined "MC" as "the study and creation of artefacts which have mental characteristics *typically associated with* consciousness such as (self-)awareness, emotion, affect, phenomenal states, imagination, etc." (italics added).

Some researchers, such as Igor Aleksander, were even describing their laptops as conscious. However, when challenged he'd retreat into objectivity, saying that he was studying the mechanisms underlying attention, choice, discrimination, and reflexive monitoring—i.e. a large part of what we mean by consciousness, but *not* "consciousness as it is in us" (personal communication). Aleksander's (2000) five criteria of consciousness were: having a sense of place; imagination; focusing of attention; planning future actions; and being guided by emotions. Given the discussion of Chapter 7.d–f, it's clear that the last criterion didn't necessarily include "experiencing feelings of emotion". In short, whether or not people were hoping/intending to attribute *qualia* to their future robots and/or laptops was ambiguous. (As for whether this would have been a nonsensical hope, see Section xi below, and Chapter 16.v.b.)

But that sudden surge of work on machine consciousness was a very late development. Meanwhile, from the late 1980s, the public interest in *human and animal* consciousness grew. By the early 1990s it had become so widespread that the Royal Society in London organized a brief press conference—*not* an academic meeting—on 'Consciousness—Its Place in Contemporary Science' in February 1995. Various recent findings and theories were described to the visiting journalists, with some pride: these were fascinating matters. But the Royal Society's scientists rang loud warning bells:

[The] remarkable consensus of the speakers [was] that, despite all the recent books, TV programmes and other hype, science really did not understand *anything* about consciousness—what it is, how it evolved, how it is generated by the brain, or even what it is for! (K. Sutherland 1994: 285)

We'll return to this point in subsection c. First, let's look briefly at some of the neuroscientific findings that had contributed to this intellectual feeding frenzy.

b. Cognitive neuroscience

The key contributions had come from the discipline dubbed "cognitive neuroscience" by George Miller in the late 1970s (Gazzaniga 1988: 230). He'd realized the need for this after seeing the bewildering display of cognitive deficits and dissociations suffered by Gazzaniga's clinical patients at Cornell Medical College.

To be sure, he was already aware that there were relevant questions to be asked. For instance, he—and many other cognitive psychologists in the West—knew of the pioneering work on cerebral control and aphasias done by the Russian neuropsychologist Alexander Luria. The first English translation had appeared in the 1930s, and—after a long gap—by 1970 there were several more, including Luria's first best-selling trade book *The Mind of a Mnemonist* (Luria 1932, 1960, 1966a,b, 1968, 1970; Luria and Yudovich 1959). Miller had actually cited Luria's work on language development in *Plans and the Structure of Behavior* (p. 140 n.), and his MGP co-author Pribram had written a preface to *Higher Cortical Functions in Man* (Luria 1966a) a few years later. But he'd never actually seen for himself the phenomena that Luria was describing.

On visiting Gazzaniga's clinic, Miller was surprised, as well as bewildered. One reason was that his previous knowledge of clinical deficits was very general. Even Luria's professionally oriented books, like the papers in the clinical journals, were descriptive rather than analytic. And in those days, clinicians and theoreticians rarely mixed (i.a above).

Psychological studies of memory, for instance, typically ignored clinical amnesias. As late as 1980, the second edition of Roberta Klatzky's (1980) widely used textbook on *human* memory said nothing about the topic (although it did give brief discussions of "mnemonists" and of the effects of alcohol and drugs: pp. 307–10, 321–5). The experimental psychologist Brenda Milner (1959) was an exception: she'd described one such case (but in psychiatric/medical journals), and her research was followed up in the 1960s by a handful of her peers. In general, however, clinical work—as William Hirst recalls—was "treated lightly" by them:

Cognitive psychologists clearly felt more comfortable writing about laboratory experiments than about experiments with brain-damaged patients and did not feel compelled to account for neuropsychological results if they contradicted their theoretical position. (Hirst 1988*b*: 250)

Miller now tried to make amends. With Gazzaniga's help, he soon founded the Cognitive Neuroscience Institute (at Rockefeller University) to focus on the theoretical issues raised by clinical syndromes, and by neuroscience in general.

He made no apologies for asking psychologists to consider the brain—which Fodor and Putnam, and most early cognitive psychologists, had declared to be unnecessary (16.iii–iv). Nor did he worry about being called a reductionist. As he wrote in a letter to Gazzaniga in 1978, psychology needed neuroscience:

Since I have always thought of scientific psychology as a branch of biology, this objection [reductionism] carries little weight with me. It would carry greater weight, however, with such distinguished scientists as [the behaviourist] B. F. Skinner or [the functionalist] H. A. Simon. (Gazzaniga 1988: 239)

Gazzaniga was equally unapologetic, but in the opposite direction:

[The] idea that *neuroscience needed cognitive science* has prevailed. I put it that way because that, after all, is what is at stake. The molecular approach in the absence of the cognitive context limited the fashionable neuroscientist to pursuing answers to biologic questions in a manner not unlike that of the kidney physiologist. Although such approaches represent an admirable enterprise, they do, when put in that light, make it impossible for the neuroscientist to attack the central integrative questions of mind–brain research. (Gazzaniga 1988: 241; italics added)

Miller was still a cognitivist, of course. But, twenty years after his manifesto and mission station (6.iv.c–d), his enthusiasm for computers and AI had waned. Writing to Fritz Machlup in January 1983, he said:

I find it difficult to communicate clearly my enormous debt to people like Simon, Newell, and Minsky for helping cognitive psychology take a great leap forward, yet at the same time to communicate with equal clarity that the human brain/mind system is enormously more complicated than, and different from, any contemporary computational system. (G. A. Miller 1983*a*: 57)

Contemporary computers manipulate symbols, and symbol manipulation is certainly one kind of intelligence. But whether that kind of intelligence is the only kind remains an open question. (p. 57)

[As] the difficulties become more and more apparent, the tendency for the computer exploration to expunge psychologism (just as logic did a century earlier [see Chapter 2.ix.b]) and move off on its own seems stronger and more attractive. That might be a good development, but it would spell the doom of cognitive science as a new discipline encompassing both artificial intelligence and psychology. (p. 58)

The newly named discipline grew quickly. Initially based in clinical neurology (e.g. Shallice 1982, 1988*a,b*; Weiskrantz 1986, 1988; Vallar and Shallice 1990), it soon embraced non-patients too. At first, the focus was on vision, motor control, and language. By century's end, motivation and emotion—and their intimate links with cognition—were included also (Rolls 1990; Damasio 1994, 1999, 2001; LeDoux 1996, 2002; Gazzaniga 1995, pt. ix). (So "cognitive" neuroscience, like "cognitive" science itself, is *not* limited to cognition: 1.ii.a.)

The field soon spawned its own communication channels, the *Journal of Cognitive Neuroscience* and *Neuropsychologia*. Besides, hefty review volumes started appearing in the early 1990s (Marcel 1993; Bisiach 1993; Gazzaniga 1995, 1999, 2000; M. H. Johnson 1993*a*; Squire and Kosslyn 1998; O'Reilly and Munakata 2000; Parker *et al.* 2002). In the late 1990s, the interdisciplinary study of development in normal and brain-damaged children was flying (see Chapter 7.vi.i and Section ix.c above). By the turn of the century, then, cognitive neuroscience had made its mark.

According to its name-bestower, the central goal was to understand consciousness:

> I consider consciousness to be the constitutive problem for psychology, just as life is the constitutive problem for biology . . . Psychologists want to discover the molecular logic of the conscious state. [That is to say]: the set of principles that, in addition to the principles of physics, chemistry, and biology, operate to govern the behavior of inanimate matter in conscious systems. (G. A. Miller 1978: 233)

Whether this "set of principles" includes anything over and above abstract (functional, computational) principles of brain and behaviour will be discussed in Section xi. But it's worth noting that Miller's examples here did *not* include experience (*qualia*), and that—in the same letter—he swiftly substituted first "knowledge" and then "epistemic system" for the troublesome "consciousness" (1978: 235–6).

Cognitive neuroscience didn't start with Miller, for the name was newer than the activity. Arbib's schema theory (see Section vii.c, above), Eric Lenneberg's (1967) survey of neural-anatomical specializations for language, and Libet's 1960s research had all been early exercises in the genre. Indeed, so had the late 1960s neurophysiological studies of visual masking in cats (P. H. Schiller 1968). In this and later single-cell work on visual masking (Felsten and Wasserman 1980; Keysers and Perrett 2002), the assumption was that the underlying mechanisms in cats, monkeys, and people might be broadly similar—and that the consciousness was similar too.

Miller, citing Minsky and Marr, stressed the need to pass from purely correlational (brain–behaviour) studies to theories of cerebral information processing and representation (Gazzaniga 1988: 240). Accordingly, many brain researchers now adopted the computational approach he'd pioneered twenty years earlier (Chapter 6.iv.c), analysing behaviour and thinking into distinct functional components.

Clinicians had long noticed that brain damage often leads to some very specific loss: trouble with producing (though not with understanding) abstract nouns, for instance, or 'blindness' to and/or neglect of one side of one's own body. Charles Darwin had remarked the puzzling case of a man who, after suffering a stroke, could tell the time by looking at a clock-face but not by reading the words. Over 100 years later, in the

1980s, some curious cases of mental dissociation—and of abnormal memory—were familiar even to the general reader, thanks to fascinating trade books written by clinical neurologists or psychologists (Luria 1968, 1972; Sacks 1985, 1989).

As for the 1980s professionals, they no longer spoke merely of aphasia, amnesia, ataxia, or apraxia (i.e. disturbances of speech, memory, movement, or action). Instead, they distinguished many different *kinds* of disability previously lumped together under these terms. For instance, they looked out for the various types of action error distinguished by Timothy Shallice and Donald Norman (Chapter 12.ix.b), or for the wide variety of memory disorders (e.g. Cermak 1982). And besides trying to match these with particular brain lesions, they tried to explain and systematize them in terms of (formal or informal) theories of the information processing going on.

Some people took these functional dissociations as proof of the existence of 'modules', inborn mechanisms specialized for the relevant types of information processing. They were wrong: modules-in-the-adult don't necessarily imply modules-in-the-developing-brain (Section ix.c–d above, and Chapter 7.vi.i). However, this wasn't widely recognized at the time—except by *developmental* neuroscientists.

By the 1980s, too, Libet's research programme had delivered some even more intriguing results. In 1979 he reported (1) that the subjective time-of-occurrence of a sensory experience was shifted backwards, so that someone might experience something *as* happening at time-of-stimulation, even though the experience itself occurred half a second later (Libet *et al.* 1979). Yet more startling, his group did work on voluntary action that showed (2) not only that 'spontaneous', 'instantaneous', conscious decisions (like sensory *qualia*) take time to be constructed, but that the relevant motor areas of the brain are *already* being activated before the decision is experienced (Libet *et al.* 1983).

The first finding cast yet more doubt on the common-sense notion of temporal 'cor-relations'. Prompted by a critique from the British philosopher Ted Honderich (1984), Libet (1985*a*) wrote a paper for the *Journal of Theoretical Biology* specifically relating it to "mind–brain theories" (for Honderich's response, see his 1986, 2005). Eccles—with Karl Popper—argued that the result was inexplicable by any neurophysiological pro-cess, so provided further evidence for his long-held dualism (K. R. Popper and Eccles 1977: 364).

The second finding threatened popular notions of free will. Libet was immedi-ately invited to write a paper for *Brain and Behavior Sciences*. This attracted huge attention—and an unusually large number of peer reviews accompanying it in the journal (Libet 1985*b*). By the end of the century, the *Journal of Consciousness Stud-ies* had published a special issue on "the neuroscience of free will" (Libet 1999). Because of the general interest in the topic, this was made available also as a stand-alone book.

Most late twentieth-century neuroscientists still concerned themselves with only one of the many problems of consciousness indicated above. But a few developed comprehensive theories interrelating them all (e.g. Baars 1988; Cotterill 1998; Velmans 1996, 2000; Weiskrantz 1997; G. M. Edelman 1989; Edelman and Tononi 2000; Damasio 1999).

These writers usually took pains to distance "the brain" from "the computer". Nevertheless, most of them used computational ideas—even if they protested (like Edelman) that they weren't doing so.

They all agreed that a huge amount of cognitive processing occurs without consciousness. This had long been known, of course. But new technologies, including brain scanning, now provided further evidence. For instance, manual responses resulting from highly complex operations (integrating perceptual, semantic, and motor processes, plus the interpretation of verbal instructions) show up in motor cortex before the person is aware of them (Dehaene *et al.* 1998*b*). Similarly, they all agreed that attention is necessary for consciousness (although 'blindsight' showed that it's not sufficient: see subsection c), and that certain—ill-specified—types of task apparently can't be achieved unconsciously.

Their theories relating the different levels or aspects of consciousness, and explaining the 'need' for consciousness in certain cases, varied. However, certain key ideas cropped up repeatedly. One of these was the concept of a general workspace, explored at length by Bernard Baars (1988, esp. ch. 2; 2001) and adopted by others too (e.g. Dennett 1991, ch. 9; Dehaene and Naccache 2001).

The core notion here was that 'higher' (including deliberative) consciousness requires a *global* mechanism—broadly comparable to a GOFAI blackboard: Chapter 10.v.e—making information from many different areas simultaneously available. Fodor (1983) had implied as much in respect of the "higher mental processes", but had despaired of explaining such non-modular effects (see Chapters 7.iii.d and 16.iv.d). However, he was interested in why *this* person arrives at *this* belief on *this* occasion. The neuroscientists, by contrast, wanted to know how conscious thought in general is possible. Stanislas Dehaene and Lionel Naccache put it like this:

[Various recent theories suggest that] besides specialized processors, the architecture of the human brain also comprises *a distributed neural system or "workspace" with long-distance connectivity that can potentially interconnect multiple specialized brain areas in a coordinated, though variable manner* . . . Through the workspace, modular systems that do not directly exchange information in an automatic mode can nevertheless gain access to each other's content . . .

If the workspace hypothesis is correct, it becomes an empirical issue to determine which modular systems make their contents globally available to others through the workspace. Computations performed by modules that are not interconnected through the workspace would never be able to participate in a conscious content, regardless of the amount of introspective effort. (Dehaene and Naccache 2001: 13)

Significantly, their term "architecture" was ambiguous: it was primarily computational, and only secondarily neural.

In short, computational ideas were guiding neuroscientific experiments on the many aspects of consciousness, and informing the interpretation of the results.

c. The $64,000 question

The $64,000 question concerned the brain's role in generating conscious experience. Among the topics most discussed by *fin de siècle* thinkers interested in this problem were experiments on alternating perceptions, perceptual binding, and blindsight.

Alternating perceptual experiences seemed to be correlated with specific processes in the brain. It was already known that if a pattern of vertical bars is continuously presented to one eye, and a pattern of horizontal bars to the other, what a human subject actually sees isn't a vertical/horizontal grid, but a regular alternation between the two (simultaneously presented) patterns. When Nikos Logothetis and Jeff Schall (1989) put a monkey in the same experimental situation, its discriminatory behaviour—raising its left or right paw—suggested that it, too, saw first the one pattern and then the other.

Moreover, a group of cells in the monkey's visual cortex was active only when its behaviour implied that it was *aware* of the vertical pattern. A different group was continuously active while the vertical pattern *was being presented as a stimulus* to the eye. The same applied in respect of the horizontal patterns. This implied that a distinct group of cortical neurones is active *only when a particular type of visual phenomenology is being experienced.*

'Binding' is an aspect of conscious unity—remarked on, for instance, by Descartes—whereby a multifaceted object is experienced as 'one' thing. How do auditory, tactile, and visual features get to be attributed to one cat, for instance? And how are the many different visual features integrated, given that each is computed in a different part of visual cortex (Zeki 1993)?

Crick (with Christof Koch) suggested that distinct perceptual features will be attributed to one and the same thing (one and the same Gestalt) if the relevant neurone groups are not only simultaneously activated but are also resonating at the same frequency—perhaps 40 Hertz (Crick and Koch 1990). Von der Malsburg had already suggested something similar as the basis of the cocktail-party effect: hearing one's own name stand out within a confusing buzz of conversation (von der Malsburg and Schneider 1986). But now there seemed to be experimental confirmation. A recent issue of *Nature* had reported that the neurones in different visual-feature-detector columns in cats oscillate synchronously when responding to the same object (Gray *et al.* 1989).

These results had led others to ask whether the researchers had "uncovered the cellular basis of consciousness" (Baringa 1990: 856)—a question which Crick and Koch answered with a firm "Yes!" (The firmness soon became a wobble, as they weren't able to replicate these findings in monkeys: Nørretranders 1991: 432 n. 46.)

As for blindsight, this was first discovered in monkeys (Weiskrantz 1972). Animals with certain lesions in the visual cortex lost much of their discriminatory capacity, but not all of it. Soon, the condition was seen in humans too (Poppel *et al.* 1973; Weiskrantz *et al.* 1974; Weiskrantz 1986, 1997; Moore *et al.* 1998). But here, there was a further surprise. Patients with similarly located brain lesions claimed to be blind—*to have no conscious experience*—with respect to a particular part of their visual field. Asked to point to an object, or to move their eyes in its direction, they'd protest that this was ridiculous: they couldn't see it, so how could they point to it? Nevertheless, if they could be persuaded to 'guess', they usually indicated the right place—sometimes, even positioning their fingers to match the object's shape.

Apparently, much of the sensori-motor processing that Arbib tried to describe (Section vii.c) was going on. But the sensory information wasn't getting through to the 'higher' brain centres so as to become introspectively accessible. (It became clear later that blindsight isn't normal vision near the threshold of consciousness, but involves non-normal processing: Azzopardi and Cowey 1997.)

In the 1990s, an extraordinary clinical case of blindsight was reported (Goodale and Milner 1992; A. D. Milner and Goodale 1993, 1995), which reached the general public early in the new century (Goodale and Milner 2003). The unfortunate woman concerned, who'd suffered carbon monoxide poisoning, appeared to be blind. She couldn't recognize familiar faces, nor identify simple geometrical shapes. Nevertheless, she could put her fingers into the correct attitude and position when reaching out to pick something up. The key demonstration of her visual agnosia involved a plastic card and a narrow rotating slit in a box. She couldn't match the position of the slit perceptually; that is, she couldn't hold the card in the same orientation as the slit. However, if asked to "post" the card through the slit (in many different orientations), she had no problems at all.

In short, she was apparently able to see (see-for-action) things unconsciously which she couldn't see (see-for-perception) consciously. (Her clinicians used this as just one piece of evidence in arguing that there are two anatomically and functionally distinct visual pathways in the brain: one for perception, one for action—see 10.iv.b.)

When the new millennium dawned, then, 'consciousness' was no longer a dirty word in neuroscience. On the contrary, it was the acme of trendiness. In 2000 the British neuroscientist Jeffrey Gray (1934–2004) was asked to review no fewer than three recent books on consciousness at the same time—with several others being reviewed elsewhere in the same publication.

But Gray knew the difference between trendiness and intellectual progress. Despite having been a guest editor for the *Journal of Consciousness Studies* a few years earlier (J. A. Gray 1995), he didn't welcome these three new volumes unreservedly:

There is a great deal of fuss these days about consciousness. Yet, 10 years ago or so, the fussometer reading would have hovered around the zero mark, where it had been stuck for most of the past century. Just a few philosophers worried about a problem that was totally ignored by the scientists. (J. A. Gray 2000: 36)

Why the vaulting fussometer reading? In Gray's opinion, it wasn't due to any scientific or philosophical advance. (Given that he'd introduced the cautionary Royal Society press conference mentioned above, this judgement was only to be expected—cf. J. A. Gray 1995.) Rather, he said, it was due to the recent technologies of brain imaging.

If the technologies were recent, the idea wasn't. James (1890: i. 97) mentioned the Italian physiologist Angelo Mosso, who'd noticed regional pulsations of the cortex, in patients with bits of their skulls missing after operations, which seemed to depend on active thinking. He'd suggested, and James agreed, that this was caused by changes in blood circulation. But studying such changes in subjects with intact skulls wasn't possible before the 1980s.

From 1982, it became possible to measure activity in specific parts of the brain—as opposed to EEGs taken across the whole brain (Frith 1992; Frackowiak *et al.* 1997; Raichle 1998; Frith *et al.* 1999). PET scanning (positron emission tomography) and fMRI (functional magnetic resonance imaging) identified the areas—perhaps only a few millimetres across—with the highest level of glucose metabolism or, what comes to much the same thing, the greatest increase in blood flow. These methods depicted the brain *in action*, for they showed (in some cases, several times a second) how the brain's energy consumption varies while someone thinks about or perceives something, or performs a motor act. What action the energy was being consumed *for*, however, wasn't as clear as it was usually assumed to be (see below).

(For the record, it was taken for granted at that time that most of the brain's energy consumption is for information processing at the synapses. We now know that this takes only 40 per cent of the energy, while another 40 per cent is consumed simply in passing messages along the axons. The rest is used for housekeeping/maintenance: Attwell and Laughlin 2001; Laughlin 2004.)

Cognitive psychologists and clinicians—and the USA's science administrators who named the 1990s "The Decade of the Brain"—were enthused. Questions previously way beyond the reach of neuroscience now seemed to be answerable. For instance:

* Brain scans done when people were asked to think about beliefs and desires supported the early 1980s claim that autism and Asperger's syndrome involve failings in "theory of mind" (Chapter 7.vi.f).

* Studies of schizophrenics implied that their hallucinations of 'inner voices' are due to a failure or delay of communication between different parts of the brain (Cahill *et al.* 1996; Silbersweig *et al.* 1996).

* Schizophrenic, and experimentally induced, delusions about alien control of one's own body were explained in a similar way (Frith *et al.* 2000; S.-J. Blakemore *et al.* 2003; cf. Chapter 7.i.i).

* Hypnotic states and multiple personality were correlated with changes in brain activation (Aikens and Ray 2001; Reinders *et al.* 2003; cf. 7.i.h−i).

* 'Oceanic feelings' and intimations of the numinous, described long before by James (1902), were associated with a certain area in cerebral cortex (Alper 1996; Saver and Rabin 1997; Cook and Persinger 1997; cf. 7.i.i).

* Brain-imagers even managed to explain why one can't tickle oneself (S.-J. Blakemore *et al.* 1998).

A host of other examples had been described, all linking conscious experiences (and voluntary movements) of various kinds with cerebral activity. Similar effects were observed in non-human mammals—but using implanted electrodes rather than brain-scanners. In one widely reported study on rats, signals from motor cortex which normally cause the rat to press a lever attached to a water-dispensing robot were transmitted directly to the robot after the lever had been disconnected (Chapin *et al.* 1999). Because the robot arm responded more quickly than the paw did, the animal soon learned that it needn't actually press the lever any more. In short, the rats had learnt to move robot arms 'by thought alone'.

This caused huge excitement, because of the potential for helping paralysed people. A few months later, the same experimenters (at Duke University, North Carolina) sent brain signals over the Internet to enable an owl monkey to control a 3D robot arm 600 miles away, at MIT (Nicolelis 2001). They planned to include visual and tactile feedback, to see whether closed-loop 'circular causation' could integrate the monkey's real and virtual worlds. And very soon after that, another group used similar techniques to enable a rhesus monkey to learn to play a video game controlled by a joystick—without actually touching the joystick (Serruya *et al.* 2002). (These experiments raise interesting questions regarding 'cognitive technology' and the concept of self: see 16.vii.d.)

Linking rats or monkeys to robots is a major exercise, not undertaken lightly. But brain-scanning studies on humans are less tricky, and are multiplying merrily. Emails reporting some new example reach me several times a week. One that arrived when I was first drafting this chapter announced an fMRI study that used John R. Anderson's computational model of goal-directed problem solving (Chapter 7.iv.iii) to predict blood-oxygen levels "in brain areas involved in the manipulation and planning of goals" (Fincham *et al.* 2002). Since then, many others have followed it (although very few are related to specific computational models, as that one was). The journalists have learnt to keep an eye out for them: brain-scan studies of 'religious' experience (see 8.vi.b, list-item 2), for instance, were widely featured in the media. As for the professional journals, by the time you read this book they will have carried thousands of PET/fMRI reports.

Their theoretical significance, however, is debatable. Brain imaging has even been dubbed "a neo-phrenological fad", because of the difficulty of interpreting it in terms of psychological functions (Uttal 2001). There are four main problems.

One is that such studies are only correlational: *causal* hypotheses have to be inferred from them, and this is a risky business. For instance, a neurone may be using energy to *excite* another cell, or to *inhibit* it. How can we decide? Recently, Logothetis's team at Tübingen's Max Planck Institute for Biological Cybernetics have combined fMRI with single-neurone recording and/or local injections of neurotransmitters (Logothetis 2002). Such studies can tell us whether what's going on is excitatory or inhibitory. But they're technically very difficult, and can't be done on human beings. They're therefore irrelevant to the 'sexy' discoveries reported in the journals—and in the newspapers.

Another difficulty recalls Gregory's complaint about "howling centres" (see Section iii.a, above). If a part of the brain increases its activity when the person thinks about X, that doesn't mean it's a "centre"—still less, *the* "centre"—for thinking about X: many other things will be going on elsewhere. In short, Gregory's critique applies to PET and MRI no less than to classic ablation experiments.

Third, brain-imaging results (unlike the results of brain lesions) aren't stable. They're highly sensitive to context, such as the order in which stimuli are presented and perhaps what the person was thinking about beforehand.

And fourth, one can't sensibly suggest *just what's being done* by the high activity (even if one knew it was excitatory activity) without a theory at the cognitive/psychological level, specifying just what computations might be involved when the thought in question occurs. Usually, no such theory is available. (The research programme described in Chapter 7.vi.i is an exception: there, a significant effort is made to relate neurological details to a specific theory of cognitive processing.) In short, most brain imaging is an a-theoretical fishing expedition: more natural history than science. As Gazzaniga had put it, before this new 'industry' burgeoned, neuroscience *needs* cognitive science.

Responsible researchers know all this, of course, and are careful. But the irresponsible ones—and a fortiori the journalists—seemingly don't, and aren't.

Gray, however, was less worried by such theoretical difficulties than by the naive optimism that brain scanning had banished the mind–body problem. These new gadgets, he declared, had gone to the neuroscientists' heads:

The power of this method has convinced scientists that now they can watch the brain in action, they can forget the hoary old issues raised by philosophers and get on with the *real* job by the simple expedient of describing what the brain does. (J. A. Gray 2000)

But the hoary old issues hadn't gone away. What he'd described some years earlier as "the most basic issue—the theoretical link between the occurrence of conscious experience and the neural substrate of the brain" (J. A. Gray 1993: 263) was still unresolved.

As Gray remarked, the philosophers had been way ahead of the scientists on this one. Or anyway, they'd spent incalculably more hours discussing it. Whether they'd got much nearer to solving it was another matter.

They'd long ago considered what it would mean for our understanding of consciousness if we knew in detail every correlation between events in the brain and conscious events. And they'd (mostly) concluded that this alone would tell us nothing fundamental about the nature of consciousness, or about *how it's possible* for the brain to generate it (see 16.i). In a word, or three: correlation isn't causation.

Moreover, functionalism—or so it was widely believed (Chapter 16.v.b)—hadn't helped. Most people felt that an experiential quality, such as redness or sourness, can't be the same as an abstract computational or causal function (important exceptions are discussed in Section xi, below). Even the arch-functionalist Fodor had washed his hands of consciousness:

Nobody has the slightest idea how anything material could be conscious. Nobody even knows what it would be like to have the slightest idea how anything material could be conscious. So much for the philosophy of consciousness. (Fodor 1992)

Gray agreed. He concluded that modern neuroscience has no idea of how neuronal firing *can* cause conscious thoughts or percepts. Nor can it explain why these evolved at all: given the closure of physics, it's hard to see how conscious states could possibly affect behaviour. In short, this was Princess Elizabeth all over again (see 2.iii.b).

If Gray was so sceptical, why were other neuroscientists so hopeful? What experimental evidence, *over and above* the discovery of many previously unknown mind–brain correlations, had persuaded them that they were about to solve the problem?

The most persuasive examples came not from human brain imaging but from single-unit recording in animals. For instance, experiments on rhesus monkeys (Newsome and Salzman 1993) showed a remarkably systematic correlation between activity in certain orientation-selective cells in the animal's visual cortex and its perceptual discriminations of movement—*and*, common sense would have us say, its experiences of seeing movement.

Before the experiment proper, the monkey was trained to move its eyes in the direction of the perceived movement. Then, stimulating a certain cell circuit caused the monkey to behave as though it saw motion in a certain direction. Moreover, these results were predictable. In other words, the relevant cells are spatially organized in a way that corresponds to the geometry of the monkey's discriminatory behaviour. Assuming that the monkey has conscious states, this evidence suggested not only a causal relationship between individual cells and experiences, but also a systematic mapping from spatial structures in the brain to perceived phenomenal structure. It even enabled the neuroscientists to predict that if such-and-such a cell group were to be stimulated, then such-and-such an experience *must* ensue.

This hypothetico-deductive reasoning is typical of scientific explanation. So, surely, this constituted a scientific explanation of (certain aspects of) the monkey's visual experiences?

Well, yes and no. This experiment couldn't have assuaged Gray's worry. What it showed was that, *given* that certain neurone groups give rise to (cause, generate, are correlated with . . .) visual experiences, *these* neurones lead to *those* experiences, whereas others lead to others. This, of course, was a lot better than nothing. But it didn't answer Gray's question about how it's possible for *any* neuronal activity to lead to *any* conscious experience. It confirmed the previous belief (the common-sense hunch, inherited from Descartes) that there are systematic mind–brain correlations there to be found. But it didn't explain how this can be so.

Moreover, it was (and is) still impossible—or anyway, deeply problematic—to understand causal relations going in the opposite direction. To be sure, some cognitive scientists claimed to explain how conscious experiences can have neuromuscular effects (Velmans 2002). These accounts were intelligible if interpreted functionally in informational, computational, terms. But if some extra (correlated) realm of conscious mental events was posited, Princess Elizabeth's problem remained.

In brief, work on consciousness was riddled with "paradox and cross purposes" (N. Block 2001).

d. Philosophical contortions

Gray wasn't the only one to be worried by the mind–brain problem. Several neuro-scientists, and philosophers sympathetic to neuroscience, suggested more or less radical 'solutions' to it.

Crick (1994), for instance, proffered the "astonishing" hypothesis that conscious states are processes in the brain—or, as he put it, that " 'You', your joys and your sorrows . . . , your sense of personal identity and free will, are in fact no more than the behaviour of a vast assembly of nerve cells and their associated molecules." He claimed, for example, that "freewill is located in or near the anterior cingulate sulcus" (p. 268).

However, whether we take "freewill" to connote a conscious act of choice or a control mechanism for scheduling potentially conflicting motives (see Chapter 7.i.g), it's a category mistake to describe it as being "located" anywhere. What was truly astonishing, at least to the philosophers, was that Crick thought this hypothesis was either new or scientifically straightforward. In fact, it was deeply problematic—like the 'identity theory' of the 1950s (see Chapter 16.i.d), or the more recent 'eliminative materialism' (16.iv.e).

Eliminative materialism had been presaged by Willard Quine in the 1950s (Quine 1953*b*; 1960: 264), but was developed further by Paul Churchland (1979). In the 1970s, he'd formulated his theory without any knowledge of neuroscience or connectionism (see Chapter 12.x.a). Ten years later, he was a colleague of Crick's in San Diego, a devotee of PDP modelling, and—even more to the point—an admirer of Pellionisz and Llinas.

He adapted their theory of coordinate transformations in the cerebellum to define his concept of the "state-space sandwich", which he speculatively applied to the discrimination of tastes—and, he suggested, also to smells, sounds, and colours

(P. M. Churchland 1986*a,b*). He admitted that the last three seemed to be much more complex than tastes, but pointed out that they'd each been described in terms broadly comparable to those of Pellionisz and Llinas. That is, they'd been theorized by appeal to coordinate transformations, "variously called 'multivariate analysis', 'multidimensional scaling', or 'across-fibre pattern coding', and so forth" (1986*b*: 303).

Churchland wasn't assuming a mere collection of unrelated correlations between individual neural events and experiences. For his abstract state-space (like tensor geometry) depicted systematic metrical and similarity relations between the individual points within it. He posited a four-dimensional "taste-space", based on the four types of taste-receptor found in the tongue. The points in the space represented specific distributions of four neuronal spiking frequencies, in a system of four distinct sets of nerve fibres. Crucially, they also represented specific tastes *as experienced*: they could be interpreted *either* as specific neural events *or* as specific tastes.

The *empirical* hypothesis was that people's introspective reports on the similarities and dissimilarities between different tastes would map onto the similarity metric defined by the neural taste-space. (Or anyway, onto something broadly like it: for simplicity's sake, Churchland was ignoring the multidimensional contribution of smell to what we normally regard as the subjective experience of "taste".) In that case, we'd be able to predict that a newly discovered (never-tasted) substance, having such-and-such observable effects on the relevant neural mechanisms, will taste very like x, something like y, and not at all like z—indeed, we could say that it *must* taste like this. Other discriminatory spaces, perhaps with many more than four dimensions, could (he suggested) be similarly defined, and the relevant phenomenology mapped accordingly.

The *philosophical* claim—a highly provocative one—was that to have a sensory experience *simply is* to have one's brain visit a particular point in the relevant sensory hyperspace. We should stop thinking of 'experiences' as separately existing events, in some subjective mental space—what Dennett (1991) later called "the Cartesian theatre". Experiences (as commonly understood) were *eliminated*, the only truly existing things/events being brain states. If neuroscience ever progressed as far as he'd imagined, said Churchland, we'd interpret our taste-language as referring not to mysterious 'inner' events but to the material, neuroscientific, reality (this claim is spelt out further in Chapter 16.iv.e).

Some philosophers were highly sympathetic, and developed their own naturalistic theories broadly on Churchland's model (e.g. Flanagan 1992). But others regarded it as—literally—absurd. Churchland had anticipated their arguments in his first book (see 16.iv.e). Now, he offered the taste-space sandwich as a neurologically plausible example illustrating how his 'absurdities' could be understood.

In other words, Churchland—unlike Crick, who also identified conscious experiences with brain states—realized that eliminative materialism wasn't a straightforward scientific theory. It required careful philosophical argument and scene setting *even to make sense*.

Crick wasn't the only neuroscientist to treat eliminative materialism as a straightfor-ward scientific hypothesis. Edelman did so too. His account of *qualia* was similar in spirit to Churchland's (G. M. Edelman and Tononi 2000, ch. 13). But, true to rhetorical form (see ix.d above), he scorned discussions of *qualia* as "ridiculous", generating "esoteric exertions" from philosophers arguing "unnecessarily" about zombies. He failed to see that his own position was comparable with that of Dennett or Sloman, for whom zombies are—necessarily—*non-sense* (Chapter 16.iv.b). His subtle neuro-physiology may have answered the question "Why *this quale* rather than *that* one?" But, despite his assertion to the contrary, it didn't explain why *any* experience should be either "associated with", "represented by", "produced/generated by", or "defined as" (expressions he used interchangeably) the neural mechanisms concerned.

Another quasi-scientific account of experience was sketched by the mathematical physicist Penrose. He suggested that the brain's generation of consciousness could be explained by CQG: a new theory of "complete quantum gravity". This theory, he said, was needed to explain certain fundamental puzzles in quantum physics. In addition, however, it would show how *non-random* quantum events could happen, and how these could implement information processing that's *not* describable as Turing-computation (see Chapter 16.v.a and ix). This, in turn, he believed to be involved in human creativity (Penrose 1989) and in consciousness (1994*a,b*).

Penrose admitted that he had no idea what this imaginary theory might be. Under-standably, then, he was widely accused of trying to solve one utter mystery by twinning it with another. (For an excellent rebuttal based on other grounds, see Sloman 1992.) Moreover, CQG—if we ever achieved it—would apply to all matter indiscriminately: to mud, just as to neuroprotein. So what's special about the brain? His answer, in a word, was *microtubules*. These are very tiny tubes in the neurone's protoplasm, described by the neurologist Stuart Hameroff (1994).

At first, Penrose said merely that neurones have microtubules, and that CQG quantum effects isolated within them might underlie consciousness. (MacKay (1957) had argued long before that quantum effects would be swamped by a combination of neurone size and redundancy. So perhaps the microtubules would prevent the swamping?) When it was pointed out that virtually all nucleated cells have microtubules—so are oak trees conscious, too?—Penrose replied that those within neurones are different (Grush and Churchland 1995; Penrose 1994*a,b*; Penrose and Hameroff 1995).

Specifically, they're arranged in parallel, not radially from centrioles. They're relatively stable. There are more of them, arranged in more complex networks, than in other cells. They show greater genetic variability. They transport chemical vesicles along the axon and dendritic processes. And there are neuron-specific proteins associated with them. There's even some evidence, he said, that they transmit "complicated signals" and "act rather like a cellular automaton", and that they're responsible for learning in paramaecium (Penrose 1994*b*: 248).

Even supposing all that to be true, however, the question remained how these facts relate to consciousness as such. Certainly, the "stability" of neuronal microtubules might make one think of enduring memories. But memory storage isn't consciousness. Neither is learning, whether in paramaecium or people.

Another argument put forward by Penrose was that the "curious relationship between conscious events and time" (in meta-contrast, for instance: Chapter 6.ii.c) might be explained if the relevant quantum events took about half a second—in which case "about 1,000 to 10,000" neurones might be involved in one conscious event (which "more or less agrees with other estimates"). But this begged the question. Taking for granted that conscious experiences happen *at some time*, it focused on specifying *which* time. But why they should happen *at all* remained unclear, not to say unintelligible.

Penrose had said that mysterious quantum effects might be relevant—which they might: who knows? And he'd said that microtubules might isolate and guide these effects—which, again, they might (though he gave no good reasons to believe this). But, as two Alice fans put it, his theory was "no better supported than any one of a gazillion caterpillar-with-hookah hypotheses" (Grush and Churchland 1995).

In short, Gray's "hoary old issues" were left untouched. Penrose was trying to blind people with (dodgy) science much as Eccles had done fifty years earlier, when he spoke of the will acting through "critically poised neurones" (see Chapters 2.viii.f and 16.i.a), and as Descartes—in his talk about the pineal gland—had done earlier still (2.iii.b).

The same could be said, perhaps even more strongly, with respect to Evan Walker's (2000) account of "the physics of consciousness"—another of the many quantum-based 'explanations' published around that time. Starting from the unpredictability of single neurone-firings, Walker suggested that quantum tunnelling at the synapses could integrate many such firings into "a single quantum mechanical conscious existence" (p. 228). This happens, he said, by means of the electrons travelling from one synapse to another via RNA molecules—and guided by free will. I can't comment on his theory as *physics*—although the Cavendish Laboratory's Matthew Donald (2001) was far from impressed. But Walker was begging crucial philosophical questions, just as Eccles and Penrose had done before him.

A very different scientific (and philosophical) revolution was envisaged by David Chalmers (1966–). Chalmers allowed that near-millennium neuroscience was well on the way to solving all the problems of consciousness—except one (Chalmers 1995, 1996*a*). "The hard problem" (how conscious experience can arise from the brain), he said, hadn't been touched.

The crux of Chalmers's "prototheory" of consciousness was Shannon's concept of information. As we saw in Chapter 4.v.d, this was defined not semantically, as a meaning or intentional content, but as one of a set of possibilities. For the telephone engineer Shannon, information was a transmittable state. For Chalmers, too, "physically realized information is only information in so far as it can be *processed*" (1996*a*: 181). But for him, information had a dual aspect:

Physical realization is the most common way to think about information embedded in the world, but it is not the only way information can be found. We can also find information realized in our *phenomenology*. States of experience fall directly into information spaces in a natural way. There are natural patterns of similarity and difference between phenomenal states, and these patterns yield the difference structure of an information space. Thus we can see phenomenal states as realizing information states within those spaces. (1996*a*: 283–4)

[So we can state "the double-aspect principle":] Whenever we find an information space realized phenomenally, we find the same information space realized physically. And when an experience realizes an information state, the same information state is realized in the experience's physical substrate. (p. 284)

This was reminiscent of Churchland's emphasis on structured phenomenal spaces, related to neural mechanisms by state-space sandwiches. But whereas Churchland thought that an adequate scientific theory would eliminate experiences from the furniture of the world, Chalmers didn't. He insisted that phenomenal qualities are "irreducible".

As for whether information is a metaphysically basic concept, underlying both physics and phenomenology, or simply "a useful construct in characterizing the psychophysical laws", this question was left open (Chalmers 1996a: 286). So too was the question whether it's "a *primitive* feature of the physical world in the way that mass and charge are primitive". Possibly, those physicists who'd already suggested that physics ultimately deals in pure information were correct: "It could be that in some way the physical is derivative on the informational, and the ontology of this view could be worked out very neatly" (p. 287). It could even be that experience is "ubiquitous": that the very simplest systems—not just slugs or thermostats, but atoms too—have a phenomenal aspect (p. 293).

This speculative panpsychism was trumped by a gesture towards a thoroughgoing idealism: perhaps the phenomenal isn't just one of two equal-status "aspects", but the ultimate *basis* of reality. If so, then

Every time a feature such as mass or charge is realized, there is an intrinsic property behind it: a phenomenal or protophenomenal property, or a *microphenomenal* property for short. We will have a set of basic microphenomenal spaces, one for each fundamental physical property, and it is these spaces that will ground the information spaces that physics requires. The *ultimate* differences are these microphenomenal differences. (p. 305; second italics added)

Chalmers discussed all these seemingly "outrageous, or even crazy" ideas, arguing that it's not *obvious* that they're misguided (pp. 293–310). But he didn't commit himself to any one of them.

Chalmers's book drew much attention, not least because he'd refused to sweep the hard problem under the carpet. (Health warning, again: perhaps there's no hard problem to be swept? See Section xi.) But his theory was hardly better received than Penrose's. His speculations about panpsychism and the metaphysics of information didn't help. They were even weirder than Brian Smith's views on computation and intentionality (Chapter 16.ix.e). In most people's opinion, Chalmers's quasi-psychic account of information was mere mystificatory hand-waving, analogous to the preceding *fin de siècle* myth: *élan vital* (see Chapter 2.vii.b).

Searle—unlike Penrose, Walker, or Chalmers—didn't try to sketch a new scientific theory of consciousness. However, he repeatedly stated, "to the point of tedium" (Searle 2001: 513), that there must be some scientific explanation to be found.

He'd already said this about intentionality (Searle 1980). But his comments about the brain's "causal role" in generating meaning had been scientifically empty (Chapter 16.v.d), and so were his comments about how brains cause consciousness—"the central mental notion", without which concepts like meaning, subjectivity, and intelligence can't be understood (Searle 1992, ch. 4).

One might agree with him that "Consciousness ... is caused by neurobiological processes and is as much a part of the natural biological order as any other biological features such as photosynthesis, digestion, or mitosis" (1992: 90). But where does this get us? We're still "very far" from having an adequate biological theory of consciousness (p. 91). Probably, he said, its neurobiology is at least as restricted as that of digestion—but we can't be sure. Perhaps, on other planets, non-carbon-based molecules (though not silicon) underlie conscious experience.

Ten years later, he was still bemused: perhaps even more so, since he was now stressing the paradoxes involved in "the neurobiology of free will" (Searle 2001). But he argued that even if consciousness is based in quantum indeterminacy, conscious thought and decisions needn't be random too—and they clearly aren't. With respect to all three puzzles (free will, consciousness, and intentionality), he said:

[We] *know* that the causal features of the system level phenomena are entirely explainable by the behaviour of the micro phenomena [in the brain] . . . [The] causal relations have the same *formal* structure as the causal relations between molecular movements and solidity. (Searle 2001: 513; first italics added)

This "knowledge", however, amounted to no more than the fact that we now have even more evidence than Descartes did to believe that the brain causes consciousness. What we want to know is *How?* Searle offered no new scientific ideas, dodgy or not: no microtubules, no dual-aspect information, no special types of computation . . . nothing. He simply said that the problem of consciousness is a scientific problem, and encouraged the scientists to get on with it.

Some pessimists agreed with the diagnosis but withheld the encouragement. The philosopher Colin McGinn (1950–) had said in the early 1980s that "We should persist in the hope that some day philosophy (or perhaps science) will find the answer [to the mind–body problem]" (1981: 36). Now, in the early 1990s, he saw any such hope as doomed (McGinn 1989, 1991).

He argued that a scientific theory of mind–brain would be *so* radically different from all known science as to be inconceivable by human beings. Neither our (perceptual) concepts of physical matter nor our (introspective) concepts of consciousness can describe any intelligible *union* of them. Moreover, these are the only conceptual realms open to us. Any third-type concepts that did explain the mind–body relation would *necessarily* be so different that we couldn't entertain them. We simply don't have the cognitive capacity to understand these matters—much as dogs can't understand arithmetic or physics:

We know that brains are the *de facto* causal basis of consciousness . . . [But] we are cut off by our very cognitive constitution from achieving a conception of that natural property of the brain (or of consciousness) that accounts for the psychophysical link. (McGinn 1989: 350)

In a God's eye view, said McGinn, there's no ultimate scientific or metaphysical mystery here. For us, however, there is. We can glimpse the link between brain and consciousness only superficially, as temporal correlations. Interpreting those correlations causally is forever beyond us.

Most readers admitted that McGinn might, conceivably, be right. To deny that would be hubris. But cognitive scientists felt that we've no convincing reason to think he was. Our knowledge of brain mechanisms is so scanty, and the history of neuroscience so short, that it would be defeatist to give up now. McGinn's intellectual pessimism was scornfully termed "the new mysterianism"—as opposed to the "old" (neo-Kantian) mysterianism, which *denied* that consciousness is naturalistic (Flanagan 1992: 8–9).

These valiant—or hopeless—attempts to explain mind–body correlations all came up against a long-familiar problem, memorably posed by Thomas Nagel (1937–) in his paper 'What Is It Like To Be a Bat?' (Nagel 1974). In a nutshell, science is objective whereas consciousness is subjective.

We all know, said Nagel, "what it is like" to be conscious, and this—subjective, inner—experience is inaccessible to science *in principle*. Neurones, neurotransmitters, microtubules, quantum events, computation, information . . . you can forget them all. These objectively describable matters may be necessary conditions for consciousness, but they simply *cannot* provide adequate explanations of it. At best, he said, science might offer an "objective phenomenology" describing the structural aspects of subjective experiences, including those (of bats, for instance) incomprehensible to human beings. (Nagel offered no examples: possibly, one might include AI work showing that identity can be computed without computing shape, which implies that a creature might be able to experience object-identity without being able to recognize shape: Ullman 1979.)

In light of these difficulties, some philosophers—and the occasional neuroscientist, such as Maturana—tried to renegotiate the central concept at issue here. We'll see that some of these negotiators still hoped for a scientific theory of consciousness, whereas others declared this to be impossible in principle—even (*pace* McGinn) for God.

14.xi. Descartes to the Tumbrils?

Now's the time, at last, to consider the anti-Cartesian heresy that *the very idea of mind–body correlations* is so deeply confused that neuroscientists can't be required to explain it. They're off the hook because, as this idea is commonly understood, *there's nothing there to be explained*.

Dennett, having discussed conscious experience ('*qualia*') with various neuroscientists for years, recently exclaimed: "To put it bluntly, nobody outside of philosophy should take a stand on the reality of *qualia* under the assumption that they know what they're saying" (in Gazzaniga 1997: 178). I agree, for reasons outlined below.

But I'd add (and he'd concur) that people *inside* philosophy usually don't know what they're saying either. The difference is that they're well aware that the very notion of conscious experiences—without which, there can be no talk of Cartesian correlations—is radically problematic. They may use the notion, of course. But they cross their fingers behind their backs while doing so.

Cognitive scientists also cross their fingers while speaking of introspection—the way in which (so it's normally said) we gain access to our conscious experiences. For "introspection" is a weasel word. That's been evident since the late nineteenth century, when experimental psychologists following Wilhelm Wundt, James Titchener, or the Würzburg school inspected their own consciousness in very different ways (2.x). Indeed, disagreements about *what counts* as introspection largely account for the phenomenally rapid rise of behaviourism after 1913 (Chapter 5, preamble). Now, it's even clearer that there are various kinds of introspection, involving different underlying mechanisms (Prinz 2004c).

In short, the common assumptions that we all understand what's meant by "experience", and how to gain access to it, are much too quick.

a. Describing the mind, or inventing it?

You'll have noticed that the neuroscientists we've been considering spoke of conscious experiences in animals as well as humans. Even their paper titles sometimes made this explicit (Logothetis 1998). They may have crossed their fingers sometimes, when ascribing *specific* experiences to cats or monkeys. But they saw no problem of principle in saying that some animals have conscious sensations. (You probably agree—although we'll see in Chapter 16.viii.b that some philosophers writing today do not.) A fortiori, they saw no problem of principle in speaking about experiences occurring in adult human minds.

Admittedly, cognitive psychologists had long ago discovered that sincere introspective reports—of visual imagery, for example—sometimes have to be taken with a pinch of salt (see Chapter 7.v.a). And Paul Churchland (1979) had argued that introspection is just as "theory-laden", just as prone to top-down conceptual influence, as ordinary perception is. This was troubling to people who took seriously Descartes's view that we have infallible direct access to our own consciousness. However, *that we do have* introspectible conscious experiences was taken for granted.

We open our eyes and have visual experiences, run our hands over our bodies and have sensations of touch . . . and so on. The experiences may sometimes be surprising, as they are in cases of illusion or meta-contrast. But they occur. Who could deny *that*? It's just common sense.

"Common sense", however, may consist largely of old wives' tales—or old philosophers' tales. If so, we need to be clear about just which tales the old philosophers were telling, and whether they were fact or fiction.

Descartes's *originality* in discussing mind–brain correlations (Chapter 2.iii.a) can be glossed in two very different ways:

 * On the one hand, he might have been the first to say that conscious experiences (understood as inner mental states accessible only to the subject concerned), *already* a familiar topic of conversation, are correlated with specific states of the brain—as opposed to the heart, or the liver, or even some supernatural realm.

 * On the other hand, he might have been the first to say that *there are conscious experiences (in the sense just defined)*—adding, for good measure, that they're closely related to the brain. In that case, the newly minted notion of experiences wouldn't have been part of common sense at the time, even though it became part of common sense very soon afterwards.

The question arises, then, as to which of these two readings is correct. In other words, did Descartes *describe* the conscious mind? Or did he rather *invent* it—and if the latter, would neuroscience profit by abandoning his invention?

b. A computational analysis

Virtually all twentieth-century neuroscientists—though not Maturana (see Chapter 15.vii.b)—tacitly opted for the first interpretation of Descartes's position. To be sure, they didn't posit mental *substance*. Indeed, most of them probably hadn't even read Descartes. But they didn't need to: their common sense (deeply imbued with Cartesian ideas) told them that there are conscious experiences. Moreover, the recent experimental advances sketched in Section x had persuaded them that these experiences could at last be studied by neuroscience. That is, they could be correlated with specific brain events, considered either as their causal basis or as identical with them.

One philosopher who agreed with them wholeheartedly was Chalmers:

[We must] *take consciousness seriously* . . . [I assume] that consciousness exists, and that *to redefine the problem as that of explaining how certain cognitive or behavioural functions are performed* is unacceptable. This is what I mean by taking consciousness seriously.

Some say that consciousness is an "illusion", but I have little idea what this could even mean. It seems to me that we are surer of the existence of conscious experience than we are of anything else in the world . . . I find myself absorbed in an orange sensation, and *something is going on*. There is something that needs explaining, even after we have explained the processes of discrimination and action: there is the *experience*. (Chalmers 1996*a*, p. xii; second italics added)

Among the views Chalmers was rejecting here were those of the computationalist philosophers Dennett (1988; 1991, ch. 12; 1993*b*) and, by implication, the less well-known Sloman (1999; Sloman and Chrisley 2003). Although Dennett wished to outlaw talk of *qualia* and Sloman didn't, their analyses were similar in spirit.

Each of them had argued that the notion of 'experience' can be cashed out *completely* in terms of discriminatory computations and behaviour, both actual and potential. (Having been influenced in their youth by Gilbert Ryle, they were well aware of the relevance of *dispositional* concepts: Chapter 16.i.c.) They both allowed that these matters are grounded in neurophysiology. But they analysed the concept of experience at the level of the computational architecture involved. Sloman, for instance, saw *qualia* as states in the virtual machine that constitutes the mind.

'Zombies'—creatures looking and behaving exactly like us, but having no conscious experiences—were therefore said by them to be logically impossible (see Chapter 16.v.b). More to the point, the traditional question of mind–brain correlations *simply doesn't arise*, for there are no ontologically distinct experiences over and above bodily/cerebral events (Chapter 16.iv.b).

Both Dennett and Sloman glossed *qualia* as self-reports generated by an information-processing system with a complex, and reflexive, computational architecture. 'Subjectivity' was interpreted as the fact that these reports are directly accessible to the highest level of the system itself. They may sometimes be betrayed by involuntary movements, such as facial expressions; and they may be deliberately communicated, by words or body language, to observers. In essence, however, they are reflexive reports: the system talking *to* itself, *about* itself. The directness, or lack of 'evidence', of first-person experiential statements is due to the particular kind of (reflexive) computation involved. It's *not* due to any 'privileged access' to a mysterious inner realm of mental being (or Cartesian theatre).

The "Cartesian theatre", with its lonely audience of one, was repeatedly criticized by Dennett, whose 1991 book was hugely influential. That's not to say that it was hugely endorsed: as we'll see below, many—probably most—of his readers couldn't swallow his denial of *qualia*. That is, they couldn't accept his *interpretation* of *qualia*-talk.

They might be willing to accept his theory of "multiple drafts", in which conscious experiences are continuously (re)constructed from many computational sources (hence their paradoxical timing properties: Dennett and Kinsbourne 1992). They might even be willing to accept his account of the self as a "center of narrative gravity", an intentionalist-stance projection *of* the system *to* the system—which, once constructed, can be used to further voluntary action and moral choice (cf. Chapters 7.i.g and 16.iv.b). But that was a far cry from accepting his account of *qualia*.

Like Nagel (1974), they thought it obvious that there's *something that it's like to be a bat*, and that this experience—grounded in the bat's sensory and behavioural skills and informed by its (language-free) concepts—can't possibly be understood by us. Or like Frank Jackson (1982, 1986), they felt that a scientist (Mary) encaged from birth in a windowless black-and-white room might know everything there is to know about the physical aspects of vision, but she'd *learn something quite new* if she opened the door and stepped out into the garden.

Dennett's response (1991: 398–401) was to pour scorn on the easy assumption that we know what we mean when we say that Mary could know *everything* about the physical aspects of vision. (If she really did, she wouldn't be at all surprised that she was now able to make discriminations, and utter sentences, that she'd never made/uttered before.) He compared this easy assumption to the careless notion that imagining that someone is "filthy rich" enables one to work out the implications of asserting that *she owns everything*. Like the infamous Chinese Room (Chapter 16.v.c–d)—and, for that matter, Nagel's bat (Dennett 1991: 441–8; Akins 1993)—Jackson's Mary is "a classic provoker of Philosophers' Syndrome: mistaking a failure of imagination for an insight into necessity".

Dennett expected no easy victories: "I know that this will not satisfy many of Mary's philosophical fans, and that there is a lot more to be said" (p. 401). Indeed, he foresaw that many (most?) of his readers would feel that consciousness hadn't so much been explained, as explained away. (Several reviewers would remark that his book title invited prosecution under the Trade Descriptions Act.)

So he constructed a fictional fall guy called Otto, who made all the obvious, common-sense objections—only to be answered, not to say mocked, in computational–behavioural terms. For example:

[OTTO:] It seems to me that you've denied the existence of the most indubitably real phenomena there are: the real seemings that even Descartes in his *Meditations* couldn't doubt.

[DENNETT:] In a sense, you're right: that's what I'm denying exist. Let's [consider] the neon color-spreading phenomenon. There seems to be a pinkish glowing ring on the dust jacket. [This refers to a visual illusion, the diagram printed on the shiny white dust jacket of Dennett's book. There's a grid of vertical/horizontal black lines. Red part-lines (in place of black) form a circle around the middle. The viewer 'sees' a glowing pink ring, half an inch wide, bounded by the inner and outer ends of the red part-lines.]

[OTTO:] There sure does.

[D.C.D.:] But there isn't any pinkish glowing ring. Not really.

[OTTO:] Right. But there sure seems to be!

[D.C.D.:] Right.

[OTTO:] So where is it, then?

[D.C.D.:] Where's what?

[OTTO:] The pinkish glowing ring.

[D.C.D.:] There isn't any; I thought you'd just acknowledged that.

[OTTO:] Well yes, there isn't any pinkish ring out there on the page, but there sure seems to be.

[D.C.D.:] Right. There seems to be a pinkish glowing ring.

[OTTO:] So let's talk about *that* ring.

[D.C.D.:] Which one?

[OTTO:] The one that *seems to be.*

[D.C.D.:] There is no such thing as a pink ring that merely seems to be.

[OTTO:] Look, I don't just *say* that there seems to be a pinkish glowing ring; *there really does seem* to be a pinkish glowing ring!

[D.C.D.:] I hasten to agree. I never would accuse you of speaking disingenuously! You really mean it when you say there seems to be a pinkish glowing ring.

[OTTO:] Look. I don't just mean it. I don't just *think* there seems to be a pinkish glowing ring; *there really* seems to be a pinkish glowing ring!

[D.C.D.:] Now you've done it. You've fallen in a trap, along with a lot of others. You seem to think there's a difference between thinking (judging, deciding, being of the firm opinion that) something seems pink to you and something *really seeming* pink to you. But there is no difference. There is no such phenomenon as really seeming—over and above the phenomenon of judging in one way or another that something is the case. (Dennett 1991: 363–4)

As this book goes to press, Otto still lives. But so does Dennett, and he's unabashed. His briefer account of consciousness (Dennett 2005) takes note of his earlier critics, but gives essentially the same answer: *qualia* don't need to be explained, since there are no *qualia*. (For Sloman, there are: they exist as states in virtual machines of a complex kind—see 16.ix.c.)

c. The other side of the river

Computationalists such as Dennett (and Churchland, and Sloman) weren't the only people to criticize the very notion of 'qualia'. The critics also included thinkers from a very different philosophical tradition—so different, indeed, that many neuroscientists were unaware of them.

These radical voices were mostly neo-Kantian, or 'Continental', philosophers (see Chapters 2.vi and 16.vi–viii). However, they also included 'ordinary language' philosophers such as John Austin (1962*b*) and the later Wittgenstein (1953). And by the end of the century they even included Putnam, who'd provided the philosophical base for functionalist cognitive science in the 1960s but who'd changed his mind fundamentally thereafter (see Chapter 16.iii and vi).

Like Dennett, but on very different (non-computationalist) grounds, such people argued that the concept of conscious experiences, *considered as private, inner, events utterly distinct from physical events*, is unintelligible. They saw it as so fundamentally confused as to be nonsense: that is, non-sense.

They blamed Descartes, whom (as remarked in Chapter 2.iii.a) they saw as having *invented* the conscious mind. Aristotle and the medieval philosophers had never mentioned it—and neither had the Renaissance humanist Michel de Montaigne. To be sure, Montaigne used French words (*l'âme, l'esprit*) which Descartes would use too: but he didn't use them in the same—post-*cogito*—sense. In some languages, moreover, no equivalents exist. There's no vocabulary in Ancient Greek—or in Chinese or Croatian either—that maps onto Cartesian concepts of mind and consciousness (K. V. Wilkes 1988). The familiar notion (familiar, that is, to you and to me) that anyone with any common sense, anywhere, must *know* that conscious minds exist, and that conscious experiences happen, is therefore suspect.

In short, it's no accident that Descartes is famed for tussling with the mind–body problem, for it was he who was responsible for it—*not* "he who noticed it"—in the first place. Our common-sense notions of the mind and conscious experience date back to Descartes. They weren't part of common sense before him, and (on this view) they shouldn't be part of common sense today—nor of science, either. Moreover, racking one's brains over mind–body correlations is a waste of time: *there are no mind–body correlations*, as this notion is normally understood.

Nor, for the neo-Kantian, is there any 'identity' between brain states and experiences. On this view, experience (and mind) isn't a product of brain processes but a feature of certain interactions between an agent and its surroundings. These interactions—such as speaking to one's daughter, or sorting apples and oranges into two heaps—are largely due to the internal physical structure of the agent's whole body, and especially of the brain. (So it's no surprise to the non-Cartesian that 'correlations' are found, if people are so philosophically misguided as to look for them.) But brain processes *enable* these interactions, they aren't *constitutive* of them.

Let's take Putnam as our example here, because (unlike some others in this philosophical camp) he's relatively tough-minded. Quite apart from his pivotal role in orthodox functionalism (Chapter 16.iii), he can't be accused of ignorance of cognitive science, nor of lack of sympathy with science in general. The fact that he, of all people, eventually reached this unorthodox position shows (at least) that it deserves to be considered seriously *even by the Ottos of this world*—which, I'd wager, includes most readers of this book.

By the turn of the century, Putnam was saying that "neither the standard problems in the philosophy of mind, nor the 'philosophical positions' they give rise to are really intelligible" (Putnam 1999: 112). This applied also to concepts like "internal phenomenal state" and "sense datum" (and, from his own earlier writings, "functional state").

He wasn't asking, he said,

whether the phrase "internal psychological condition" or "internal psychological state" *ever* has an intelligible use—of course it has!—*but whether we understand what is being claimed when it is said that, e.g.* believing that there are churches in Vienna *is an internal psychological state with the same causal-explanatory role [with respect to behaviour] as the noninternal state* of knowing that there are churches in Vienna. (Putnam 1999: 113)

Similarly, he wasn't denying that we have (as we say) experiences of *yellow*. (Much as Dennett wasn't denying that we have, as we say, experiences of a pinkish glowing ring.) But, whether we're looking at a yellow door or merely hallucinating one, there's no yellow *experience*, *sense datum*, or *quale*:

Yellow isn't a property something we experience *has* (or a property the experience *is*) as talk of qualia suggests; it is a property the experience *ascribes* to the door. The experience *portrays the environment as pertaining yellow* (it [intentionally] "refers to" yellow, as it were); it isn't a yellow (or "subjectively yellow") particular or universal. As William James put it, the quality is "in" the experience "intentionally", but the experience doesn't have it "as an attribute". Confusing having redness or hotness "intentionally" with being red or hot "adjectively" (i.e. as attributes) is [a common philosophical fallacy]. (Putnam 1999: 154)

Of course, sometimes we seem to see a yellow door when there isn't one there. In general, we're subject to perceptual illusions, and even to hallucinations. Positing (illusory) "experiences" appears, prima facie, to explain this. But—according to Putnam—this assumption, too, is mistaken. We don't see an experience, we see real things. That is, our perception is *direct*, not mediated by inner experiences. When we see mistakenly, the explanation lies in the causal transactions between the real thing and the brain. Illusory aspects of the world are, for Putnam, in a sense "veridical". But that's just to say that they are the natural effects on our perceptual system of real things, having certain physical properties and perceived in certain physical/biological conditions.

In the Müller-Lyer illusion, for example (see Figure 6.3), the real-lines-drawn-on-real-paper cause processes in visual cortex that resemble those which occur when we're presented with really unequal lines. And in schizophrenic hallucinations, the misperceptions are caused by chemical imbalances and/or learnt patterns in the brain. Those patterns, in turn, can be traced back to earlier causal transactions with real things—including pictures and printed text in real books, and real words spoken by real storytellers (remember Arbib's hierarchy of schemas).

Certainly, then, we may say that the brain constructs the illusory interpretation. What—on this view—we must *not* say is that it constructs, or causes, or correlates with...some mysterious mental something-or-other, some inner 'experience'. In short, none of the experimental or neurophysiological facts is denied by Putnam. What's denied is that they are to be explained by positing *qualia*, somehow interposed between the brain and the thing being (mis)perceived.

This may strike you (as it would Otto) as crazy. You may want to echo Chalmers's remark, quoted above: "It seems to me that we are surer of the existence of conscious experience than we are of anything else in the world." Is Putnam mad? And is Dennett mad, too? *What on earth* are they up to?

d. Lions and lines

As an analogy, imagine that a great scientist–philosopher on his deathbed was heard to say, just before his last gasp, that navigators would benefit from considering the lions running around the earth.

His mourners found, to their delight, that he'd already drawn their pathways on the globe in his study. Evidently, the lions were quite convivial, for they met regularly (the best-attended gatherings took place at the poles). Sure enough, over the next few centuries the navigators—and eventually also the tourists—found it very helpful to consider the lions. They were even recognized as individuals: four were given special names, and the rest were named by numbers.

Field expeditions aimed at capturing one of these animals, however, always failed. Perhaps the lions were very shy, or very small, or running unseen in microtunnels...or perhaps (because of some unsuspected glitch of quantum physics?) forever invisible. The arguments raged on and on. But no one doubted that the lions were there. After all, their pathways had been measured precisely, and they were mentioned every day—in conversations that undeniably helped people to reach their intended destinations. *Of course* the lions existed!

But, of course, they didn't. The dying man had been misheard: he'd said "lines", not "lions". Moreover, he hadn't meant *real* lines, such as might be made by paw marks. No, these were imaginary lines, of latitude and longitude (what Dennett would call *abstracta*, or "exquisitely useful fictions"—1991: 367; see Chapter 16.iv.b).

There was no harm in his followers' speaking about "lions", if this language was used purely *geographically*. But if it was interpreted as referring to actual lions (perhaps very small ones!), with paws, whiskers, and purrs, it was highly misleading—not to say false. In Ryle's (1949) terms, belief in such lions was a category mistake (Chapter 16.i.c).

The mistake was by now so entrenched, however, that the few people who questioned the lions' existence (as *lions*) were accused of denying the facts underlying map-making and navigation. In truth, they were doing no such thing. But they were undercutting every single one of the ingenious theories (shy, small, quantum-invisible . . .) constructed to answer questions about the elusive lions' real nature. More: they were showing why *no such questions need arise*, and why *no such theory could ever be successful.*—And all without denying *any* of the facts that made lion talk so useful, and which had led the long-dead philosopher to posit the lines/lions in the first place.

One of the points at which this analogy (like all analogies) breaks down is that Descartes himself had *not* been misheard. He did believe in 'lions'—that is, in mental states considered as aspects of a special type of existence, just as real (indeed, just as substantial) as physical matter.

By the twentieth century, his followers had abandoned his belief in mental substance. (The lions could leave no paw marks.) But they still assumed the metaphysically distinct existence of conscious mental states, as events within a distinct level of reality. (The lions' purrs could be heard.) They called them experiences, sensations, raw feels, sense data, *qualia* . . . and woe betide anyone who questioned their existence *as so conceptualized*. The brave people who did question them were wrongly supposed to be denying the facts that had led to such talk in the first place—such as the fact that we sometimes think we're seeing a yellow door, or a pinkish glowing ring, when there isn't one.

Putnam, and Dennett too, is on the side of the *lines*. For both of them, although for somewhat different reasons, the claims of Crick and Churchland that so-called conscious experiences are in truth *the very same thing as brain states* are literally nonsensical. (Churchland, then, had been quite right to say that this wasn't a straightforward scientific theory, inviting tests rather than philosophical analysis.)

What's more, for Putnam (and Dennett) the question of mind–brain correlations, *as it's normally understood*, simply doesn't arise. There are no such correlations. There couldn't be any such correlations, because there are no *qualia* or inner conscious 'experiences' (again, *as these are normally understood*) that might be correlated, or not, with specific brain events. There's no need to puzzle over how to explain mind–brain correlations, or how to 'strengthen' them—pressed, perhaps, by Princess Elizabeth?—into physics-defying *causes*. For there's no correlation there to be explained.

Certainly, Putnam admitted, we need to explain the huge variety of behavioural discriminations displayed by people and animals. And he allowed that neuroscience—indeed, *computational* neuroscience—will be needed to do so (Putnam

1999: 48). Unlike Dennett, however, he didn't try to say what the neurocomputational architecture must be like. His interest was simply to show that the neo-Cartesian assumption shared by virtually all neuroscientists, *that there are mental (conscious) events and physical (brain) states, and that we should try to explain the correlations between them*, must be rejected.

If Putnam (or Dennett) is right, it's no wonder that neuroscientists face huge difficulties in explaining how the brain causes conscious experiences—and that increasingly bizarre 'scientific' hypotheses are dreamt up accordingly. This is just what one would expect, if their basic Cartesian assumption (italicized above) is philosophically confused.

If we decide to reject this assumption, none of the intriguing discoveries of neuroscience (nor of cognitive psychology) need be denied. But they must be interpreted in a very different way.

We can still say that rhesus monkeys—and, perhaps, humans—discriminate motion perception by cortical cells arranged around the compass, and that these discriminations are nicely predictable in neuroscientific terms. What we can't say is that the monkeys, or the humans, *have inner visual experiences* of motion-in-a-certain-direction.

Again, we can still say that the blindsight patient can make certain visual discriminations without being able to report them, and that this is because their information processing doesn't include reflexive computations of the kind necessary for self-report. We can even say that their common-sense belief system leads them to protest at the 'absurdity' of being asked to *guess*. What we can't say is that they lack what the normal person possesses: *inner experiences (qualia)* of objects seen in 3D space.

Last (to take an example discussed in Chapter 7.v.a), we can still say that when people imagine seeing landscapes or elephants, they're able to answer certain questions about them. We can also say that this capacity may involve such-and-such parts of, and/or such-and-such computational processes in, the brain. But we can't say that this behaviour involves *inner mental pictures, or conscious experiences (images).*

The fundamental problem, then, is whether the neuroscience of consciousness should concern itself with *lions*, or with *lines*.

Despite neuroscience's huge intellectual debt to Descartes, who first defined it as a respectable activity (see Chapter 2.iii and viii), perhaps it should dismiss him as the storyteller of consciousness? Admittedly, lifting Descartes onto the philosophical tumbrils would be just as revolutionary as what his compatriots did in 1789—perhaps even more so. For our common sense (and Otto) cries out: "Rubbish! *Of course* neuroscience studies lions!" But the stubborn persistence of Gray's hoary old problems, not to mention the respect due to serious thinkers (albeit some of a very different philosophical persuasion), should at least give us pause.

e. Hung jury

My own view is that *if* we are ever to understand mind–brain correlations, some radical change in both neuroscience and the philosophy of mind will be required. So it's no accident that the end-of-century presses hummed with strange theories such as those of Penrose, Walker, and Chalmers—and others I've not had space to discuss, including various forms of panpsychism.

Some years ago, I argued that the conceptual change concerned would need to be comparable in depth to Maxwell's field theory (Boden 1998*b*). This theory reinterpreted physicists' seemingly contradictory talk of "waves" and "particles". These utterly different, even mutually exclusive, phenomena had been needed to account for different sets of experiments. Maxwell explained both in terms of an underlying theory mentioning neither, and also showed why they'd been used in distinct types of experimental situation. An analogous advance, I said, would be required to explain how the brain can generate *any* conscious experience. (Perhaps mind-as-virtual-machine is the core of the solution?)

However, I also said that this is a big "if". And I asked whether a fundamentally non-Cartesian neo-Kantian (as opposed to a Dennettian) approach might be better.

Then, I answered that it wouldn't. Not only would it mean that the 'mind–body correlations' we all talk about so confidently would have to be systematically redescribed (lines, not lions), but the nature of science in general would be redefined—not to say undermined. Instead of being a *realistic* enterprise, it would be a construction of human minds (and societies), no more 'truthful' than any other. As I said in the brief remarks on postmodernism in Chapter 1.iii.b, I regard this neo-Kantian view as fundamentally irrational. Even in the hands of a skilled analytic philosopher like John McDowell, or the middle Putnam, it's unconvincing (see Chapter 16.viii.b and vi, respectively).

But the late Putnam is different, for he now insists on the "natural realism" of science. He even defends the "direct realism" of perception—where, as we saw above, direct doesn't mean infallible. So—for him—*that* troubling philosophical bird is scotched.

Significant problems remain, however, about whether intentionality (meaning)—which is a crucial aspect of consciousness—can ever be naturalized. One can't responsibly offer an answer to our question about *qualia* without considering this question also.

For my part, if I were confident that intentionality can be scientifically explained—perhaps by Ruth Millikan's (1984) evolutionary approach: see Chapter 16.x.d—I'd gladly abandon the lions, or locate them in virtual reality. We've seen, after all, what trouble they cause. But intentionality is a huge, and far-reaching, philosophical problem—to be considered at length in Chapter 16.

Here, let me just say that my own preference—I have no knock-down argument—is for an evolutionary–computational account of these issues, not a neo-Kantian one (see Boden 1972). Any such account takes science to be a realist enterprise, an assumption I cannot bring myself to drop. One well-developed example, here, is Dennett's. He sketched an early version of Millikan's approach in his first book (1969), and holds that intentionality is something we *ascribe* to certain sorts of complex computational system. We do so because it helps us to predict, manage, and understand their behaviour (16.iv.a–b).

For Putnam, both middle and late, and for other neo-Kantians too, intentionality is a real property. Moreover, for them it is *prior* to science—not something that can be explained by it (Chapter 16.vi–viii). It follows, on that view, that neuroscience can never explain meaning. At most, it can describe what goes on in our brains when we have meaningful thoughts.

In sum, the existence of *qualia*—and therefore of mind–body correlations, *as normally understood*—isn't as "obvious" as it's typically assumed to be. And if their existence is in doubt, their explanation is necessarily problematic. The jury's still out. The discussion is often bedevilled by disagreements rooted in a clash of intuitions wholly impervious to argument (a fact noted by Dennett: see Chapter 16.iv.b). But one thing's sure: mind–brain correlations aren't a purely *scientific* problem. The philosophy has to be sorted out too.

15

A-LIFE IN EMBRYO

Artificial life is widely assumed to be excitingly new. Exciting, it is. But new, it isn't.

It was born in the 1940s, fathered by 'the two Williams': Grey Walter and Ross Ashby (Chapter 4.viii). It developed healthily in the 1950s, nurtured (as we'll see) by Alan Turing and John von Neumann. By 1960, the core ideas—situated behaviour, self-organization, adaptation, and evolution—had been sketched, and all but the last had even been simulated.

Then, A-Life (like connectionism) lapsed into a Sleeping Beauty phase. Nothing much happened for a while. Or rather, it did—but few people noticed. Even Herbert Simon's ant walked largely unseen (Chapter 7.iv.a). In the early 1980s, the sleeper stirred. The highly public awakening happened in 1987, at an interdisciplinary naming party that introduced "A-Life" not only to science but to the outside world. Today, the outside world accepts A-Life in a spirit both more mystified (S. R. L. Clark 1995) and more welcoming (Kember 2003; Whitelaw 2004) than the public perception of classical AI (see 13.vii.b). Even the postmodernists had learnt to love A-Life (see 1.iii.d).

The cyberneticists in general had assimilated life to mind. They believed that the life–mind gap could be bridged by a single type of explanation. Ashby, for example, held that his dynamical theories—and his artefacts—could advance both biology and psychology. Self-organization and adaptation, discrimination and teleology, were seen as equally central concerns.

But their explanatory monism went underground with the breakaway of GO-FAI (see 4.ix.b). Most early computational psychologists focused on logical-symbolic modelling. Propositional contents ruled the day, while metabolism, adaptation, and self-organization were forgotten—or, in the latter case, assimilated to learning. Psychology, it seemed, had nothing to do with biology. Indeed, functionalist philosophy made this explicit as "multiple realizability" (16.iii.b). Cybernetics itself, meanwhile, became more closely associated with control engineering than with psychology or biology.

Later, things changed. By the end of the century, computer modelling was focused increasingly on life, and (in some cases) on the unification of life with mind. Ideas from many different sources were used in explaining the structural changes in biological (embryonic and evolutionary) and psychological development.

The biochemical origins of life were discussed too, sometimes in highly abstract, functional, terms (e.g. Drexler 1989). And people tried to model, and/or to build, chemical analogues of the origins of life at the molecular level (e.g. Luisi and Varela

1989; Szostak *et al.* 2001; Luisi *et al.* 2004). Apart from a few brief remarks at the close of Section x.b, I'll ignore these chemical discussions.

The theoretical focus of A-Life is the central feature of living things, namely self-organization. So to define self-organization is to identify the basic themes of this chapter:

* Self-organization is the spontaneous emergence (and maintenance) of order, out of an origin that's ordered to a lesser degree.
* It concerns not mere superficial change, but fundamental structural development.
* The development is spontaneous, or autonomous, in that it results from the intrinsic character of the system (often in interaction with the environment), rather than being imposed on it by some external force or designer.
* In self-organizing systems, higher-level properties result from interactions between simpler ones—and there may be many different levels of organization involved.

The phrase "spontaneous emergence" in that definition may sound like magic. It may even *feel* like magic:

When I wrote the program [i.e. Biomorph: see vi.b, below], I never thought that it would evolve anything more than a variety of tree-like shapes . . . Nothing in my biologist's intuition, nothing in my 20 years' experience of programming computers, and nothing in my wildest dreams, prepared me for what actually emerged on the screen. . . . I distinctly heard the triumphal chords of *Also Sprach Zarathustra* (the "2001 theme") in my mind. I couldn't eat, and that night "my" insects swarmed behind my eyelids as I tried to sleep. (Dawkins 1986: 60)

If even the hard-headed Richard Dawkins reacted in this way to something unexpected on his utterly unmagical computer, how much more have people wondered at the examples of self-organization in the biological realm.

Emergence isn't imaginary: surprising phenomena, even radically *new* phenomena, do happen. In particular, unexpected high-level patterns can result from interactions between very simple lower-level processes (A. J. Clark 2001: 112–17). But it isn't magic, either—even though "emergent" is sometimes used as a buzzword with near-mystical overtones. (For a wide range of definitions, see Stephan 1998, 2003.) The aim of A-Life, with respect to emergence, is to provide the explanation without losing the wonder.

This chapter outlines the diverse origins of A-Life, and recounts how they were eventually woven together. It starts with a brief sketch of the historical links between the concepts of life, mind, and self-organization. Then, Section ii distinguishes A-Life from biomimetics. Section iii relates it to early mathematical biology, and to neo-Kantian biology too.

The story of A-Life as such begins with Section iv. (More accurately, it began in Chapter 4.viii, as remarked above.) Sections iv and v outline the relevance of reaction–diffusion equations and self-replicating automata, respectively. Evolutionary networks are discussed in Section vi.

The sensori-motor integration of the whole animal is the topic of Section vii, which outlines research in computational neuro-ethology (CNE), much of which appears in A-Life journals. Sections viii and ix describe a wide range of work on complex systems, including some done by physicists and computer scientists, and some done in a neo-Kantian spirit. The way in which A-Life was eventually recognized as a single field is described in Section x. Finally, Section xi discusses two state-of-the-art examples that are potentially relevant to cognitive science in general.

15.i. Life, Mind, Self-Organization

The historical roots of cognitive science are highly diverse, as we've seen. In particular, questions about mental processes and living things have long been intertwined. Recently, that intertwining has become even more closely tangled. For it's now clear that self-organization—the core feature of life—is an important concept in neuroscience and psychology too (see Chapters 12 and 14—especially 12.ii and v, and 14.vi and viii.c–d).

a. Life and mind versus life-and-mind

It's widely assumed—though rarely explicitly argued—that life is *necessary* for mind. This assumption, and the closely related dispute about the possibility of "strong A-Life", will be explored at length in Chapter 16.x. Meanwhile, let's merely acknowledge that all the minds we know of are found in living things. That's why "life" and "mind" are so commonly associated.

René Descartes didn't make the 'necessity' assumption (Chapter 2.ii–iii). He didn't believe in life-and-mind. Rather, he thought that the association between minds and living (human) bodies is fundamentally unintelligible, arranged by the whim of God. On his view, life and mind are radically different: living organisms, including human bodies, can be described in scientific (mechanistic) terms whereas minds can't. But as we saw in Chapter 2.v–vii, his successors didn't all agree with him.

Some disagreed by doubting the possibility of a physicalist biology. The chemist Justus von Liebig, for instance, held that chemical processes are "altered" by a vital force so as to produce compounds "altogether different" from those in inanimate matter. Others disagreed by claiming that life and mind can be explained in essentially similar ways—perhaps mechanistic (e.g. Julien de La Mettrie), perhaps not (e.g. Johann von Goethe and the Naturphilosophen).

The late nineteenth-century advances in general physiology and neurophysiology didn't settle the dispute. Non-physicalist accounts of both life and mind were still proposed after the turn of the century. It was said, for instance, that embryonic development is directed by vitalist entelechies, that adaptive evolution results from a creative life force, and that purposive behaviour involves a specifically psychic energy (see Chapter 2.vii.b and 2.x.b).

In short, in the early twentieth century life and mind were still thought by many people to lie beyond the bounds of empirical science. And life-and-mind was still a highly controversial idea.

b. Self-organization, in and out of focus

The four concepts just mentioned—embryogenesis, adaptation, evolution, and purpose—are fundamentally akin. Indeed, the word "evolution" used to mean embryogenesis—which is why Charles Darwin wrote of "descent with modification" in *The*

Origin (Bowler 1975). All four are examples of self-organization. So another way of putting the point made above is to say that, well into the twentieth century, self-organization—whether biological or psychological—was felt to lie outside science. A fortiori, it was out of reach of machine-based theories.

By the mid-twentieth century, the cybernetic movement had made more people willing to believe that these phenomena could be scientifically explained. Cybernetic explanations weren't mechanistic in the Cartesian sense outlined in Chapter 2.ii.a. They focused not on matter or energy, but on abstractions such as information, computation, adaptation, or equifinality (see Chapter 4.v–viii). These were assumed to be instantiated in physical systems, notably in organisms.

The cyberneticians, then, saw no essential mystery in self-organization. Moreover, they discussed its *origin*, as well as its *maintenance*. Ross Ashby's "ultrastable" Homeostat, for instance, showed how randomness could give way to order. And Norbert Wiener, in discussing quantum physics at the fifth Macy seminar, spoke of order arising from chaos. So whereas Claude Bernard and Walter Cannon had focused on the body's self-maintenance (see Chapter 2.vii.a), their followers in the 1940s–1950s were interested also in the generation of new order, whether in adaptation or (more rarely) in development.

Nevertheless, they weren't all happy with the term "self-organization". Ashby sometimes used it, and "self-coordinating" too (Ashby 1960: 10). But he also complained that such phrases were "fundamentally confused and inconsistent" and "probably better allowed to die out" (Ashby 1962: 269). He saw them as potentially mystifying because they imply that there's an organizer when in fact there isn't. Some modern researchers agree, carefully avoiding 'self-organization' and referring instead to *organization*: the ability to act as a unified whole (see Chapter 14.viii.c).

Yet the concept of self-organization hasn't died out. It's often mentioned in relation to neural networks—which Marvin Minsky and Seymour Papert (1969) saw as potentially related to "how the genetic program computes organisms" (see 12.iii–v). Neuroscientists use it in explaining the development of topographical maps in the brain (14.vi.b and ix.a). And it's employed by many scientists in A-Life—not to imply some mysterious 'inner organizer', but to focus on *the spontaneous origin and development* of organization at least as much as on its maintenance. As I use it in this chapter, the term carries that bias.

Self-organization is the central feature of life (see Section ii.b, below). It's no accident, then, that it's especially prominent in A-Life—as opposed to AI, psychology, or neuroscience. Those three disciplines focus on particular examples of behaviour/cognition, which they occasionally explain in terms of self-organization. By contrast, A-Life tends to focus on self-organization *as such*, using particular behaviours (such as flocking, ant trails, walking, or bodily control via perceptuo-motor mechanisms) as illustrations of it. (Similarly, the cyberneticists treated reflexes as illustrations of circular causation, rather than behavioural problems to be studied in their own right.)

The initial excitement about cybernetics in biology was soon swamped. It lost out to another way of mapping abstractions onto physical reality: the genetic code. This was hypothesized by Erwin Schrödinger in the 1940s, and attributed to DNA's double helix by Francis Crick and James Watson in 1953. It was finally decrypted, as amino-acid triplets coding for proteins, in the 1960s. As a result, the holism of cybernetics was swiftly overshadowed by a strongly reductionist molecular biology (Judson 1979).

This approach soon detailed many specific mechanisms involved in self-organization. These included gene regulation, immune reactions, and the 'programming' of neurones to make connections of specific types. But self-organization *as such* wasn't a concept of molecular biology.

It was to be born again, in a new guise, towards the end of the century. (Ironically, some of these revivalist ideas drew on the molecular biology of the immune system: Chapter 14.ix.d.) At that time, too, the cyberneticists' faith that similar types of explanation apply to both life and mind was renewed. Whether that faith was *justified* will be discussed in Chapter 16.x.

15.ii. Biomimetics and Artificial Life

Not every physical artefact that mimics living things is an example of A-Life. To show why, let's consider two examples of artificial fish—one ancient, one modern.

a. Artificial fish

In 1776 Henry Cavendish (1731–1810) nominated Captain James Cook for election to the Royal Society. Having just completed his second great voyage of discovery, Cook had exciting tales to tell of exotic fish and alien seas. But so did Cavendish. For, in the very same year, he had built an artificial electric fish and lain it in an artificial sea (Wu 1984; Hackman 1989).

Its body was made of wood and sheepskin, and its electric organ was two pewter discs, connected by a brass chain to a large Leyden battery; its habitat was a trough of salt water. Cavendish's aim was to prove that "animal electricity" is the same as the physicist's electricity, not an essentially different (vital) phenomenon. His immobile 'fish' wouldn't have fooled anyone into thinking it was a real fish, despite its fish-shaped leather 'body'. But, and this was the point, it did deliver a real electric shock, indistinguishable from that sent out by a real torpedo fish.

A hundred years later, a latter-day Cavendish might have interpreted his fish's electric organs (sensors and effectors) as Darwinian adaptations, with specific functions in the organism's biology. Today's neo-Darwinians are happy to do the same.

(However, some of them see no possibility of a *determinate* answer to the question "What is it for?" Daniel Dennett argues that attributions of biological function can *never* be wholly closed, wholly cut and dried: "It is not just that I can't tell, and they can't tell; *there is nothing to tell*"—1987: 312; italics added. His view on this matter is a biological version of his non-realist stance on psychological intentionality, outlined in Chapter 16.iv.b. It certainly doesn't match the practice of working biologists—neither in general (Kitcher and Kitcher 1988; Amundson 1988), nor in the particular case of electric fish (Keeley 1999). I'll take it for granted in this chapter, then, that biological/adaptive function can often be assigned with confidence, even if there are some difficult cases.)

Just over 200 years later, the fourth international conference on A-Life opened with a paper—and an uncannily lifelike demonstration—on the computer simulation of fish (Terzopoulos *et al.* 1994).

Whereas Cavendish's 'fish' had been a solitary object lying inert in a dish of water, these denizens of the VDU were constantly in motion, sometimes forming hunter–hunted pairs or co-moving schools. Each one, swimming around constantly, was an autonomous (independently acting) creature, with simple perceptual abilities that enabled it to respond to the world and to its fellows. Unlike the crickets and cockroaches discussed in Section vii below, or Jacques de Vaucanson's anatomically detailed duck (2.iv.a), they weren't robots: they were software creatures existing in a computer-generated virtual world.

To achieve these lifelike simulations, the modern authors started with digitized still photographs of real fish representing shape, colour, and texture. They converted these into movies not by hand-crafted image-by-image animation (as in Walt Disney's *Snow White*) but by defining rules enabling a computer to change the images automatically. These rules were based on the anatomy, mechanics, and ethology of various species of real fish, and also on real-world hydrodynamics.

The major bodily movements of the animations, with their associated changes in body shape, resulted from twelve internal muscles (conceptualized as springs). The computerized fish learned to control these in order to ride the (simulated) hydrodynamics of the surrounding sea water. A host of minor movements arose from the definitions of seventy-nine other springs and twenty-three nodal point masses, whose (virtual) physics resulted in subtly lifelike locomotion.

But realistic bodily movement doesn't suffice for realistic behaviour: where should the (animated) fish move to, how fast, how urgently, and when? These questions were answered by ethologically credible rules defining the movement trajectories and environmental conditions appropriate for various behaviours. These included feeding, hunting, fleeing, nuzzling, and reproducing. Interactive aspects, such as predator–prey behaviour, resulted from rules defining the mutual responses of two or more autonomous creatures.

Cavendish had intended his artificial fish to deliver an intellectual shock, as well as a real one. His aim was to demystify a vital phenomenon, to show the continuity between the physical and the organic—and, of course, to display the physical principle underlying the living behaviour.

He thought this shocking hypothesis to be so important that he invited some colleagues into his laboratory to observe the experiment—so far as we know, the only occasion on which he did so (Wu 1984: 602). Certainly, such an invitation from the taciturn Cavendish was a remarkable event: an acquaintance said that he "probably uttered fewer words in the course of his life than any man who ever lived to fourscore years, not at all excepting the monks of la Trappe" (Lord Brougham, quoted in *Encyclopaedia Britannica* 1990: 975). (It's been suggested that Cavendish's unsociability was due to Asperger's syndrome; the same posthumous 'diagnosis' has been made of Albert Einstein and Isaac Newton: Sacks 2001; Baron-Cohen and James 2003.)

Our contemporary fish-makers had a similar intellectual aim. Their digitized fish weren't intended as mere toys, or tools for Hollywood animators—although similar techniques are now being used for both home entertainment and commercial cinema.

Rather, they were intended to clarify "the interplay of physics, locomotion, perception, behavior, and learning in higher animals" (Terzopoulos *et al.* 1994: 17). That is, they were examples of research in A-Life.

Should we say, then, that A-Life originated with Cavendish, two centuries ago?—No. Cavendish's work was an early foray into biomimetics, not A-Life.

b. What is A-Life?

Biomimetics studies physical artefacts that mimic the material stuff of life, putting these studies to scientific or engineering use. (The term "biomimesis" also has another meaning: as defined by Warren McCulloch in 1961, it is "the imitation of one form of life by another"—for instance, a stick insect that looks just like a leaf: Bensaude-Vincent forthcoming. That meaning isn't relevant here.)

For instance, biomimetics explores the irritability and excitability of (non-biological) membranes. Or it builds iron-wire-in-nitric-acid models of nervous transmission (Lillie 1936). Or it analyses living structures for purposes of architectural design: water lily leaves for Joseph Paxton's magnificent Crystal Palace, perhaps. Or it models artificial sensors on sensors in plants and, especially, animals (Barth *et al.* 2003). Specific material properties are taken seriously, so that if some chemical element were missing, or some physical measure different, the biomimetic model would ideally be correspondingly altered.

A-Life is more abstract than this, even though the virtual-fish world does include data about actual fish and does attempt precise simulation of hydrodynamics. It draws its distinctive ideas from computer science and complex dynamics (see Sections v–viii). It uses computer modelling—including robotics—as its methodology. And it aims to explain not only particular vital phenomena but also the properties of life in general. As Christopher Langton (1949–) said when he first defined it in 1986, it studies "life as it could be" rather than merely "life as we know it" (Langton 1989*b*: 2).

The core of A-Life is self-organization, which (as remarked in the preamble) can involve interactions occurring on several levels. In living organisms, the relevant interactions include chemical diffusion, perception and communication, adaptation, and evolution.

The emergent properties are also of many different kinds. They include universal characteristics of life, such as autonomy, reproduction, and metabolism. They also cover distinct lifestyles, such as parasitism, and particular behaviours, such as flocking, hunting, or evasion. Bodily morphology is included: for instance, the anatomy of sense organs, motor organs (such as fins), and control mechanisms. So, too, are widespread developmental processes, such as cell differentiation.

The formation of single cells, the simplest living things, is a fundamental example of biotic self-organization. Many would say it's *the* fundamental example (see Section vii.b). But they'd allow that the spontaneous formation of new chemical compounds in prebiotic 'soups' is also relevant to biology.

Self-organization *tout court* doesn't suffice for life. It occurs in some purely chemical systems—such as the Belousov–Zhabotinsky diffusion reaction, wherein organized ring patterns of different colours arise within an initially homogeneous fluid. Life requires some or all of the types of self-organization just mentioned. ("Some or all", because there's no universally agreed definition of the concept of life (Boden 1996, pt. iv): see also 16.x.b–c.) Accordingly, work in A-Life has ranged across all these aspects of living organisms.

The aim of A-Life is to understand life, not to construct new life forms. Admittedly, a tiny handful of A-Life researchers hope to synthesize new life, even *virtual* life (see Section ix.b). In general, however, A-Life doesn't try to go one better than biomimetics. Its late twentieth-century artefacts, like Cavendish's leather fish, are meant to exemplify *theories about life*—not life itself.

15.iii. Mathematical Biology Begins

Someone might seek to distance A-Life still further from Cavendish's research, by suggesting another sense in which it is "more abstract".

Cavendish's experiment couldn't have been done without the artificial fish in its bath of conducting fluid, because his aim was to reproduce the same physical phenomenon (electrical conductivity) that occurs in some living things. Biomimetics requires physical mimesis. But—so someone might argue—A-Life's artefacts, namely computers, are dispensable. If artefacts are needed at all, then just three are enough: pencil, paper, and armchair. For A-Life is a recent variety of mathematical biology, and mathematics is an abstract science.

Our imaginary interlocutor is right in saying that A-Life is a form of mathematical biology. That is, it uses (broadly) mathematical concepts and techniques to analyse and explain properties of living things. But it doesn't follow that A-Life's artefacts are dispensable, as we'll see.

a. Of growth and form

Isolated examples of mathematically expressed biological research were scattered in the pre-twentieth-century literature. But mathematical biology as an all-encompassing and systematic approach was attempted only after the turn of the century. Its instigator was the zoologist Sir D'Arcy Thompson (1860–1948) of the University of St Andrews, whose visionary work *On Growth and Form* was first published in 1917. Richly packed with fascinating examples, it ran to nearly 800 pages. A much enlarged (1,116 pages) second edition appeared in 1942, six years before the author's death.

D'Arcy Thompson—he's hardly ever referred to merely as "Thompson"—was born only a year after the publication of *On the Origin of Species*, and was already middle-aged when Queen Victoria died in 1901. He survived through both world wars, dying at almost 90 years old in 1948. That was the year in which the Manchester MADM computer, for which Turing was the first programmer, became operational.

If he had an exceptional span in life years, he also had an extraordinary span in intellectual skills. He was a highly honoured classical scholar, who translated *Historia Animalium* for the authoritative edition of Aristotle (D. W. Thompson 1910). In addition, he was a biologist and mathematician, and was offered chairs in Classics and Mathematics as well as in Zoology.

While still a teenager (if 'teenagers' existed in Victorian Scotland), he edited a brief book of essays from the Museum of Zoology in Dundee (D. W. Thompson 1880). But he soon graduated to larger tomes. In his early twenties, he edited and translated a German biologist's scattered writings on how flowers of different types are pollinated by insects—in a 670-page volume for which Darwin himself wrote the Preface (D. W. Thompson 1883).

Forty years later, he was commenting on ancient Egyptian mathematics in *Nature* (D. W. Thompson 1925), and analysing the catches made by fishermen trawling off Aberdeen (D. W. Thompson 1931). His essays ran from classical biology and astronomy, through poetry and medicine, to 'Games and Playthings' from Greece and Rome, and included popular pieces written for *Country Life*, *Strand Magazine*, and *Blackwood's Magazine* (D. W. Thompson 1940). His last book, out a few months before he died, was *A Glossary of Greek Fishes*: a "sequel" to his volume on all the birds mentioned in ancient Greek texts (D. W. Thompson 1895, 1947). Clearly, then, D'Arcy Thompson was a man of parts.

He was no mere list-maker, as some of the titles above might suggest. On the contrary, he was a great intellect and a superb wordsmith. His major book has been described by the biologist Peter Medawar as "beyond comparison the finest work of literature in all the annals of science that have been recorded in the English tongue" (Medawar 1958: 232). And his intoxicating literary prose was matched by his imaginative scientific vision.

Although Darwin had written the Preface for his first 'real' book, D'Arcy Thompson became increasingly critical of Darwinian theory. An early intimation of this was in his paper 'Some Difficulties of Darwinism', given in 1894 to an Oxford meeting of the British Association for the Advancement of Science (one of Babbage's many brainchildren: Chapter 3.i.a).

His book, over twenty years later, explained at length why he felt Darwinism to be inadequate as an explanation of the living creatures we see around us. Like some maverick modern biologists (see Section viii), he regarded natural selection as strictly secondary to the origin of biological form—which must be explained in a different way.

He integrated a host of individual biological facts within a systematic vision of the order implicit in living organisms. That is, he used various ideas from mathematics not only to describe, but also to explain, fundamental features of biological form. He wasn't content, for example, to note that patterns of leaf sprouting on plants may often be described by a Fibonacci number series (such as 0, 1, 1, 2, 3, 5, 8, 13, 21, . . .). He converted this finding from a mathematical curiosity into a biologically intelligible fact, by pointing out that this is the most efficient way of using the space available.

Significantly, he often combined 'pure' mathematical analysis with the equations of theoretical physics. In this way, he tried to explain not only specific anatomical facts (such as the width and branching patterns of arteries, relative to the amount of blood to be transported), but also why certain forms appear repeatedly in the living world.

D'Arcy Thompson referred to countless examples of actual organisms, but he had in mind also *all possible* life forms. As he put it:

[I] have tried in comparatively simple cases to use mathematical methods and mathematical terminology to describe and define the forms of organisms . . . [My] study of organic form, which [I] call by Goethe's name of Morphology, is but a portion of that wider Science of Form which deals with the forms assumed by matter under all aspects and conditions, and, in a still wider sense, with forms which are theoretically imaginable. (1917/1942: 1026)

For D'Arcy Thompson, then, the shapes of animals and plants aren't purely random: we can't say "Anything goes". To the contrary, developmental and evolutionary changes in morphology are constrained by underlying general principles of physical and mathematical order.

As his own acknowledgement made clear, D'Arcy Thompson's work was closely related to Goethe's rational morphology (see Chapter 2.vi.e). Like Goethe, whom he quoted with approval several times in his book, he sought an abstract description of the anatomical structures and transformations found in living things—indeed, in all possible things. So he discussed the reasons for the spherical shape of soap bubbles, for instance.

His reference to "forms which are theoretically imaginable" recalls Goethe's remark (quoted in 2.vi.e) that

With such a model [of the archetypal plant and its transformations] . . . one will be able to contrive an infinite variety of plants. They will be strictly logical plants—in other words, even though they may not actually exist, they could exist. [Compare: "life as it could be".] They will not be mere picturesque and imaginative projects. They will be imbued with inner truth and necessity. And the same will be applicable to all that lives. (quoted in Nisbet 1972: 45)

And like Goethe, he believed that certain forms were more natural, more likely, than others. In some sense, he thought, there are "primal phenomena".

Also like Goethe—though here, the analogy becomes more strained—he asked questions about the physical mechanisms involved in bodily growth. But his philosophical motivation for those questions was importantly different. Although D'Arcy Thompson was sympathetic to some of the claims of the Naturphilosophen, he wasn't a fully paid-up member of their club. Indeed, he opened his book by criticizing Immanuel Kant and Goethe, complaining that they had ruled mathematics out of natural history (Thompson 1917/1942: 2).

In part, he was here expressing his conviction that "the harmony of the world is made manifest in Form and Number, and the heart and soul and all the poetry of Natural Philosophy are embodied in the concept of mathematical beauty" (pp. 1096–7). This conviction wasn't shared by his professional colleagues: "Even now, the zoologist has scarce begun to dream of defining in mathematical language even the simplest organic forms" (p. 2). But in part, he was saying that physics—real physics—is crucially relevant for understanding "Form".

The idealist Goethe had seen different kinds of sap as effecting the growth of sepal or petal, but for him those abstract possibilities had been generated by the divine intelligence self-creatively immanent in Nature. D'Arcy Thompson, by contrast, argued that it is real physical processes, instantiating strictly physical laws, which generate the range of morphological possibilities. Certainly, those laws conform to abstract mathematical relationships—to projective geometry, for example. But biological forms are made possible by underlying material–energetic relations.

Accordingly, D'Arcy Thompson tried to relate morphology to physics, and to the dynamical processes involved in bodily growth. He suggested that very general physical (as opposed to specific chemical or genetic) constraints could interact to make some biological forms possible, or even necessary, while others are impossible.

Had he lived today, D'Arcy Thompson would doubtless have relished the work on the self-organization of feature-detectors described in Chapter 14.vi.b and ix.a. For this explains why we should expect to find systematic neuro-anatomical structure in the brain, as opposed to a random ragbag of individually effective detector cells. Moreover, the "why" isn't a matter of selection pressures, but of spontaneous self-organization. But that recent research required computational concepts and computing power (not to mention anatomical data) that he simply didn't have. He could use only the mathematics and physics available in the early years of the century.

Although D'Arcy Thompson wasn't the first biologist to study bodies, he might be described as the first biologist who took *embodiment* seriously. The physical phenomena he discussed included diffusion, surface forces, elasticity, hydrodynamics, gravity, and many others. And he related these to specific aspects of bodily form.

His chapter 'On Magnitude', for example, argued both that size can be limited by physical forces and that the size of the organism determines which forces will be the most important. Gravity is crucial for mice, men, and mammoths, but the form and behaviour of a water beetle may be conditioned more by surface tension than by gravity. A bacillus can in effect ignore both, being subject rather to Brownian motion and fluid viscosity. Again, his discussion of 'The Forms of Cells' suggested, among many other things, that the shape and function of cilia follow naturally from the physics of their molecular constitution.

(A very recent discovery about axon sizes would have delighted his heart: Faisal *et al.* 2005. The thinner the axons are, the more neurones can be packed into a given volume. But no axons thinner than about 0.1 of a micron have ever been found, even in species with tiny bodies and tinier brains. Computer simulations show that the molecular 'noise' involved in the opening and shutting of sodium/potassium ion channels becomes overwhelming below that point, causing random spontaneous action potentials. In brief, the *biophysics* constrains the minimal diameter of an axon. A copper wire measuring 0.001 of a micron could transmit currents efficiently. But equally tiny axons, if they are to transmit information reliably, are *impossible*.)

Perhaps the best-known chapter of *On Growth and Form*, and the one which had the clearest direct influence, was entitled 'On the Theory of Transformations, or the Comparison of Related Forms'. This employed a set of two-dimensional Cartesian grids to show how differently shaped skulls, limb bones, leaves, and body forms are mathematically related. One form could generate many others, by enlargement, skewing, and rotation.

So, instead of a host of detailed comparisons of individual body parts bearing no theoretical relation with each other, anatomists were now being offered descriptions having some analytical unity.

To be sure, these purely topological transformations couldn't answer questions about more radical alterations in form. The gastrulation (self-invagination) of an embryo, for example, couldn't be explained in this way (see Section iv.a, below). And only very few zoologists—of whom Medawar was one—tried to use D'Arcy Thompson's specific method of analysis. But his discussion inspired modern-day allometrics: the study of the ratios of growth rates of different structures, in embryology and taxonomy.

b. More admiration than influence

One didn't need to be doing allometrics to admire D'Arcy Thompson. By mid-century, he was widely revered as a scientist of exceptional vision (Hutchinson 1948; Le Gros Clark and Medawar 1945). The second edition of *On Growth and Form* was received with excitement in 1942, the first edition (of only 500 copies) having sold out twenty years before. Reprints had been forbidden by D'Arcy Thompson himself, while he worked on the revisions, and second-hand copies had been fetching ten times their original price.

However, only a decade after the long-awaited second edition, the advent of molecular biology turned him overnight into a minority taste. Ironically, much the same had happened to his muse Goethe, whose still-unanswered biological questions simply stopped being asked when Charles Darwin's theory of evolution came off the press in 1859 (see Chapter 2.vi.f). By the end of the 1960s, only a few biologists regarded D'Arcy Thompson as more than a historical curiosity.

One of these was Conrad Waddington (1905–75), a developmental biologist at the University of Edinburgh (whose theory of "epigenesis" influenced Jean Piaget: see Chapter 7.vi.g). Waddington continually questioned the reductionist assumption that molecular biology can—or rather, will—explain the many-levelled self-organization of living creatures. It's hardly surprising, then, that D'Arcy Thompson was often mentioned in his 'invitation only' seminars on theoretical biology, held in the late 1960s at the Rockefeller Foundation's Villa Serbelloni on Lake Como (Waddington 1966–72). (The seminars gathered together a fascinating group of people, including several mentioned in this book: Michael Arbib, Ted Bastin, David Bohm, Brian Goodwin, Christopher Longuet-Higgins, John Maynard Smith, and Howard Pattee.)

But Waddington, too, was a maverick, more admired than believed. His theory of epigenesis couldn't be backed up by convincing empirical evidence, whether in the developing brain or in the embryo as a whole. Only after his death did his ideas gain ground. Significantly, the proceedings of the first A-Life conference were dedicated to him (Langton 1989a, p. xiii).

D'Arcy Thompson's most devoted admirers, however, had to concede that it was difficult to turn his vision into robust theoretical reality. Despite his seeding of allometrics, his direct influence on biology was less strong than one might expect, given the excitement (still) experienced on reading his book.

Even subsequent attempts to outline a mathematical biology eschewed his methods. Joseph Woodger's (1929, 1937) axiomatic biology, for instance, owed more to mathematical logic and the positivists' goal of unifying science (see Chapter 9.v.a) than to D'Arcy Thompson. And Turing's mathematical morphology employed numerically precise differential equations, not geometrical transformations—even though *Of Growth and Form* was cited at the end of Turing's paper (see Section iv.a). In short, D'Arcy Thompson figured more as inspirational muse than as purveyor of specific biological theory or fact.

The reason why his influence on other biologists, although "very great", was only "intangible and indirect" (Medawar 1958: 232) is implied by his own summary comment.

At the close of his final chapter, he recalled the intriguing work of a naval engineer who, in 1888, had described the contours and proportions of fish "from the shipbuilder's point of view". He suggested that hydrodynamics must limit the form and structure of swimming creatures. But he admitted that he could give no more than a hint of what this means, in practice. In general, he said:

Our simple, or simplified, illustrations carry us but a little way, and only half prepare us for much harder things . . . *If the difficulties of description and representation could be overcome*, it is by means of such co-ordinates in space that we should at last obtain an adequate and satisfying picture of the processes of deformation and the directions of growth. (1917/1942: 1090; italics added)

c. Difficulties of description

The "difficulties of description and representation" remained insuperable for more than half a century after the publication of those first 500 copies of D'Arcy Thompson's book.

Glimpses of how they might be overcome arose (in the early 1950s) a few years after his death. Actually overcoming them took even longer. Or perhaps one should rather say it *is taking* even longer. We're now in a position to appreciate the potential of his ideas better than his contemporaries could, and to put more computational flesh onto the bones he sketched so many years ago (Chaplain *et al.* 1999). For this early exercise in mathematical biology resembled current work in A-Life in various ways.

Were D'Arcy Thompson alive today, he'd be very interested in A-Life:

* He'd be fascinated by the exhibition of the virtual swimming fish, with its detailed interplay of hydrodynamics and bodily form.
* He'd be intrigued by the A-Life evolution of decidedly unlifelike behaviour, as a result of a specific mistake in the simulated physics (Sims 1994).
* He'd recognize the potential relevance for morphology of computational work on diffusion gradients (see Section iv),
* and he'd sympathize with those who see evolution as grounded in general principles of order (Section viii).
* He'd appreciate the artificial evolution of a range of networks controlling lamprey swimming (Section vii.d).
* As remarked above, he'd appreciate the recent work on self-organization in the brain (14.vi.b and ix.a).
* He'd agree with A-Lifers who stress the dynamical dialectic between environmental forces and bodily form and behaviour.
* He might have embarked on a *virtual biomimetics*: a systematic exploration of the effects of (simulated) physical principles on (simulated) anatomies.
* And he'd certainly share A-Life's concern with life as it could be—his "theoretically imaginable forms"—rather than life as we know it.

Nevertheless, there are three important, and closely related, differences between D'Arcy Thompson's work and research in A-Life. Each of these reflects his historical situation—specifically, the fact that his work was done before the invention of computers. One difference concerns the practical usefulness of computer technology, and shows why (contrary to the suggestion noted above) A-Life's artefacts are not, in fact, dispensable. The other two concern limitations on the mathematical concepts available when D'Arcy Thompson was writing: in his words, the difficulties of description and representation that needed to be overcome.

First, D'Arcy Thompson was able to consider only broad outlines, largely because he had to calculate the implications of his theories using hand and brain alone. Today, theories with richly detailed implications can be stated and tested with the help of superhuman computational power. The relevant theories concern (for instance) the hydrodynamics of fish; the interactions between various combinations of diffusion gradients; and processes of evolution and co-evolution, occurring over many thousands of generations.

In addition, we can now study chaotic phenomena (which include many aspects of living organisms), where tiny alterations to the initial conditions of a fully deterministic system may have results utterly different from those in the non-altered case. These results can't be predicted by approximation, or by mathematical analysis. The only way to find out what they are is to watch the system—or some computer specification of it—run, and see what happens. In all these cases, the "help" A-Life gets from computers isn't an optional extra, but a practical necessity.

Second, D'Arcy Thompson's theory, though relatively wide in scope, didn't encompass the most general feature of life: self-organization as such. Instead, it considered many specific examples of self-organization. This isn't surprising. Prior to computer science and information theory, no precise language was available in which this could be discussed.

And third, although he did consider deformations produced by physical forces, D'Arcy Thompson focused more on structure than on process. This is characteristic of pre-computational theories in general. Prior to computer science, with its emphasis on the exact results of precisely specified procedures, scientists lacked ways of expressing—still less, of accurately modelling (and tracking)—the details of change.

Uniform physical changes could be described by linear differential equations, to be sure. And Charles Babbage could lay down rules, or programs, determining indefinitely many "miraculous" discontinuities (see Chapter 3.i.b). But much as Babbage (as he admitted) couldn't program the transformation of caterpillar into butterfly, so D'Arcy Thompson's mathematics couldn't describe the morphological changes and dynamical bifurcations that occur in biological development.

One might have expected that cybernetics would provide some of the necessary advances in descriptive ability. Not only was it a form of mathematical biology, but it used computer modelling as a research technique (see Chapter 4.v–vii). As the study of "circular causal systems", it drew on mainstream ideas about metabolism and reflexology, not on the morphological questions that interested D'Arcy Thompson. But it considered some central biological concerns now at the core of A-Life: adaptive self-organization, the close coupling of action and perception, and the autonomy of embodied agents.

It even made some progress. For instance, Ashby's (1952) "design for a brain", and his Homeostat, depicted brain and body as dynamical physical systems (see section xi, below). And Grey Walter's (1950*b*) tortoises, explicitly intended as "an imitation of life", showed that lifelike behavioural control can be generated by a very simple system (see Chapter 4.vii).

However, the cybernetics of the 1950s was hampered by lack of computational power, and by the diversionary rise of symbolic AI. Only much later—and partly because of lessons learned by symbolic AI—could cybernetic ideas be implemented more convincingly. Even so, some recent (dynamical) approaches suffer the limitation remarked in Chapter 4.vi.a with respect to cybernetics: they can't easily represent hierarchical structure, or detailed structural change.

As it turned out, it was physics and computer science—not cybernetics—which, very soon after D'Arcy Thompson's death in 1948, produced mathematical concepts describing the generation of biological form. Indeed, two of the founding fathers of computer science and AI, Turing and von Neumann, were also the two founding fathers of A-Life. (Von Neumann's intellectual range was even greater than Turing's, including chemical engineering for example: Ulam 1958.)

Around mid-century, they each developed accounts of self-organization, showing how simple processes could generate complex systems involving emergent order. They might have done this during D'Arcy Thompson's lifetime, had they not been preoccupied with defence research. While Turing was code breaking at Bletchley Park, von Neumann was in Los Alamos, cooperating in the Manhattan Project to design the atom bomb.

The end of the war freed some of their time for more speculative activities. Both turned to abstract studies of self-organization. Their new theoretical ideas eventually led to a wide-ranging mathematical biology, which could benefit from the increasingly powerful technology that their earlier work had made possible.

15.iv. Turing's Biological Turn

Turing's towering presence in the history of cognitive science is without parallel. He pioneered the theory of computer science and the design of digital computers, and outlined the research programme of AI—including connectionism (Chapters 4.i–ii, 10.i.f, and 12.i.b). In addition, his tongue-in-cheek challenge that came to be known as the Turing Test provoked a huge philosophical response, and still prompts attention today (Chapter 16.ii.c).

What's been less widely recognized, until recently, is his important contribution to mathematical biology—and, indirectly, to neuroscience.

a. A mathematical theory of embryology

The origin of biological form was still hugely mysterious at mid-century. Babbage's suggestion (Chapter 3.i.b) that metamorphosis might be due to some predetermined change in the laws of physics wasn't satisfactory. Scientists wanted to be able to explain the emergence of new forms *without* positing any change in the underlying physics. Better still, they wanted to do this by reference to specific aspects of the known laws of physics. But how was that to be done? Despite D'Arcy Thompson's provocative writings, no one had any clear idea. It was Turing who sketched a preliminary answer.

For the last few years of his life, Turing's energy went primarily into what he called "my mathematical theory of embryology". Indeed, after writing the first Manchester programming manual in 1950, he neglected his duties in the computing laboratory there as a result of his new interest.

His attitude perplexed and irritated his university colleagues, and the engineers and government officials charged with improving and applying the novel technology. For them, this was an exciting time. The world's first commercially available electronic computer (the Ferranti Mark I), which had been largely designed by Turing himself, was delivered to the Manchester laboratory early in 1951 (see Chapter 3.v.b). But he took little interest in it. His concentration was directed instead on his biological ideas. In 1952, when his Manchester colleagues were enjoying their new toy, he published a mathematical paper on morphogenesis (A. M. Turing 1952).

That was only two years before his untimely death: he committed suicide in 1954 (see 4.i.a). He was still working on these ideas when he died, and Max Newman's brief summary written for the Royal Society's memoir appeared in his mother's book (S. S. Turing 1959: 139–44). Drafts of, and extensive notes for, three further mathematical papers on 'The Morphogen Theory of Phyllotaxis' were found in his rooms. (They were partly corrected/prepared by N. E. Hoskin and B. Richards, with notes by Robin Gandy; details are available online in the Turing archive, <http://www.turingarchive.org>, AMT/C/8–10 and 24–7.) Generally, however, when people today refer to "Turing on morphogenesis" they mean his 1952 paper.

In his published account of morphogenesis, Turing followed the path pioneered by D'Arcy Thompson. Indeed, Thompson's book, which he had read before the war, was one of only six references he cited. This paper, in which he discussed the role of fundamental physical processes in the generation of biological form, is now recognized as an early essay in A-Life.

It is also, in principle, an early essay in neural dynamics—and is seen as such by Jack Cowan and Walter Freeman, for instance (see 12.ii.b and 14.x.a–b). The reason is that the crux of the 1952 paper was the problem of how self-organization (pattern development) *in general* is possible. Turing (1947*b*) had already asked how organization could arise in unorganized neural networks, but there he'd assumed some outside interference, or training procedure (12.i.b). Now, he considered the origin of organization without outside interference.

In particular, he asked how homogeneous cells can develop into differentiated tissues, and how these tissues can arrange themselves in regular patterns on a large scale, such as stripes or segments.

Embryologists of the 1950s knew well enough that ordered complexity arises from the undifferentiated egg. But how this is possible was a mystery. They appealed to vaguely conceptualized "morphogenetic fields", controlled by hypothetical, presumably biochemical, forces called "organizers". But what an organizer was, and how it could produce novel structure, they couldn't say.

Turing didn't claim to know the chemical details either. His achievement was to prove, in mathematical terms, that relatively simple chemical processes could, in principle, generate order from homogeneous tissue. The processes concerned involve chemicals whose mutual interactions as they diffuse throughout the system can (sometimes) destroy or build each other.

Since no one knew just which chemicals these might be, Turing referred to them simply as morphogens (from the Greek: form originators). He showed that the interactions between two or more morphogens, each initially distributed uniformly across the system, could eventually produce waves of differing concentrations. This can happen in non-living systems (chemicals diffusing in a bowl, for instance) as well as living creatures. But Turing suggested that in an embryo or developing organism, a succession of these processes might prompt the appearance of ordered structures such as spots, stripes, tentacles, or segments.

This may seem like magic: how can difference arise from homogeneity? Turing allowed that a perfectly homogeneous system in stable equilibrium would never differentiate. But if the equilibrium is unstable, even very slight disturbances could trigger differentiation.

Some disturbances, he pointed out, are inevitable, given that—as D'Arcy Thompson had insisted—living matter is subject to the laws of physics. For example, random Brownian motion within the cell fluids must vary the pairwise interactions between the molecules. And molecules will be slightly deformed as they pass through the cell wall. Even minute disturbances like these could upset the initial (unstable) equilibrium.

Turing gave simple differential equations defining possible interactions between two morphogens. Various terms specified the initial concentrations of the two substances, their rates of diffusion, and the speed at which one could destroy (or build up) the other. He showed that certain numerical values of these terms would result in ordered structures with biological plausibility.

For instance, irregular dappling or spot patterns could result from two morphogens diffusing on a plane surface (see Figure 15.1). Diffusion waves within a twenty-cell ring could give rise to regularly spaced structure, reminiscent of the embryonic beginnings of circular patterns of cilia, tentacles, leaf buds, or petals—or segments, if the ring were broken. (Such a ring is, in effect, a cellular automaton: see Section v.a.)

Turing suggested that order could be generated in three dimensions also. For example, diffusion waves could cause embryonic gastrulation, in which a sphere of homogeneous cells develops a hollow (which eventually becomes a tube). And interactions between

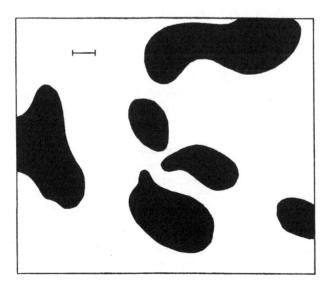

FIG. 15.1. Dappling produced by diffusion equations. Reprinted with permission from A. M. Turing (1952: 60)

more than two morphogens could produce travelling waves, such as might underlie the movements of a spermatozoon's tail.

b. History's verdict

Turing's imaginative paper was read with excitement by embryologists. It proved beyond doubt what D'Arcy Thompson had suggested: that self-organization of a biologically plausible kind could, in principle, result from relatively simple chemical processes. But it didn't, and couldn't, prove that real biological structures actually do emerge in this way. Only experimental developmental biology can show that. (Compare the point made in Chapter 7.iii.c, that only psychological experiments can confirm the psychological reality of a computer model of mind.)

This question couldn't be experimentally addressed at mid-century, because the necessary biochemical techniques weren't yet available. Today, it's still being debated.

For instance, the biologist Maynard Smith (1920–2004), who'd hoped to find that Turing waves underlie the segmentation of the fruit fly *Drosophila*, later admitted that this seems not to be the case: apparently, a different gene controls each segment (personal communication). However, he allowed that Turing waves may be responsible for segmentation in those animals, such as earthworms, in which most of the segments are identical, and/or that they were crucial at an earlier stage of evolution, before specialized segment genes appeared. Moreover, he said, the *numbers* of petals in flowers, for example, may be specified by wavelength ratios of Turing waves (though as yet there's no evidence for this).

Again, the geneticist Gabriel Dover (personal communication) argues that even in earthworms the generation of iterated segments is driven by strictly *local* processes. On his view, it's like adding yet another square at the end of a toilet roll, as opposed to taking a sheet of flimsy paper and spreading waves across it to generate stripes of perforation. And finally, the embryologist Lewis Wolpert (personal communication) denies that Turing waves can explain the development of fingers (an example of repeated structure).

In developmental neuroscience, however, recent work (described in Chapter 14.x.a) suggests that self-organization in the brain—of feature-detectors, for instance—does occur by means of mechanisms like those which Turing described. Cowan, for example, sees "a very close relationship" between Turing's ideas and the brain (J. A. Anderson and Rosenfeld 1998: 118). And this, he says, is no accident:

The same epigenetic mechanism for pattern formation, the tendency to make stripes and blobs, is ubiquitous in nature. Cloud patterns, animal coat markings, hallucination patterns [in migraine, for example], maps—all that is sitting there. The brain is no different in many respects from any other physical organization. There's a tendency for pattern formation to occur because it's got all the same kinds of machinery in it. (J. A. Anderson and Rosenfeld 1998: 121)

D'Arcy Thompson would surely have agreed.

If Turing's paper didn't immediately affect embryology, nor did it spawn A-Life as a computational discipline. For he faced a methodological obstacle comparable to one facing D'Arcy Thompson before him.

Turing had access to the Manchester University computer, and used it (or, occasionally, a desk calculator) to calculate the successive steps of interaction between the morphogens. But this was a time-consuming and tedious matter. Moreover, in the absence of computer graphics, the computer's numerical results had to be laboriously converted to a (more easily intelligible) visual representation by hand. Turing himself remarked that much better machines would be needed to follow up his ideas.

Now, half a century later, such machines exist. Powerful computers have recently been used to vary the numerical parameters in Turing's own equations, to calculate the results, (often) to apply two or more equations successively, and to convert all these numerical results into graphical form (Turk 1991). In this way, Turing's equations have generated spot patterns, stripes, and reticulations resembling those seen in various living creatures.

Some of the structures produced in these computational experiments are shown in Figures 15.2–15.4. The large and small spots in the upper half of Figure 15.2 result from changing the size parameter in Turing's reaction–diffusion equation. If the large-spot pattern is frozen, and the small-spot equations then run over it, we get the 'cheetah spots' at bottom left. The 'leopard spots' at bottom right are generated by a similar two-step process, except that the numbers representing the concentrations of chemicals in the large spots are altered before the small-spot equation is run.

These cascades of reaction–diffusion systems kicking in at different times (reminiscent of the switching on and off of genes: see Section viii.b) result in more naturalistic patterns than does a simultaneous superposition of the two equations. Similar cascades of a 5-morphogen system generate the lifelike patterns of Figure 15.3. And the addition of rules linking stripe equations to 3D contours produces the moulded zebra stripes of Figure 15.4. This work was done for the purposes of computer graphics, not theoretical biology. But it shows that simple rules can produce lifelike complexity.

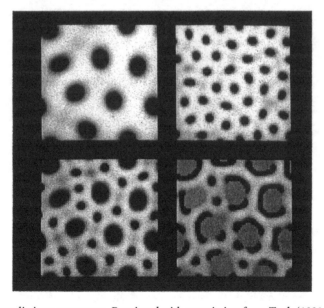

Fɪɢ. 15.2. Naturalistic spot patterns. Reprinted with permission from Turk (1991: 292)

FIG. 15.3. Complex naturalistic patterns. Reprinted with permission from Turk (1991: 292)

FIG. 15.4. Zebra with 3D-algorithmic stripes. Reprinted with permission from Turk (1991: 293)

Other A-Life work on the physical principles underlying morphogenesis combines computer simulation with biomimetics. For example, chemical experiments (first done by Stéphane Leduc, soon after 1900) show that osmosis can produce lifelike forms, even in inorganic solutions. If pieces of calcium chloride are dropped into a saturated solution of sodium phosphate, the surface film of calcium chloride molecules functions as a 'membrane' separating a growing form (made of calcium chloride plus water) from the surrounding phosphate solution. This happens because osmotic pressure (more water molecules outside than inside) causes water to pass through the surface film, which increases in area as a result.

Computer simulations of osmosis show that, if the rates of production and disintegration of molecules are in balance, it will in general produce persisting (dynamically self-maintaining) topological unities (Zeleny *et al.* 1989). Because these are regarded as possible precursors of autonomous cells bounded by cell walls, this type of research is sometimes called "synthetic biology" (see Section x.b, below).

Some A-Life research has explored the morphological implications of differential cell growth, but without specifying actual physical mechanisms such as diffusion. For instance, the Dutch biologist Aristid Lindenmayer (1925–89) and others have followed up D'Arcy Thompson's ideas by defining various algorithms (known as L-systems) for describing cell growth in plants (Lindenmayer 1968, 1977; Jean 1984).

A Fibonacci number series, for example, may be employed without asking how it's instantiated in chemistry. When L-systems are expressed as computer graphics, a wide range of lifelike plant forms results. These can aid biologists by showing which forms can or can't be generated by certain types of rule, applied in one order or another. (L-systems are also widely used in computer animation.)

The focus on morphogenesis as a result of bottom-up, localized, physical processes is very much in the spirit of Turing, and D'Arcy Thompson too. Now, it's respectable. Early in the century, it wasn't. Indeed, in 1907 the French Académie des sciences refused to publish Leduc's account of osmotic precipitation, because it "touched too closely on the then-discredited notion of spontaneous generation" (Zeleny *et al.* 1989: 126). Self-organization, apparently, was seen by the Académie as either myth or magic.

Finally, exciting work (outlined in Chapter 14.ix.f) has recently begun on using A-Life techniques to evolve computers partly composed of diffusing waves of chemicals. The hope is that logical functions, such as *and-gates*, can be implemented in this way—and that these non-linear machines may be able to do other things which today's digital computers cannot.

So history's verdict on Turing's biological paper is one of approval. The editor of a collection of Turing's work has even stated that this paper has been cited more often "than the rest of his works taken together" (P. T. Saunders 1992).

I suspect that that isn't true, since the Turing Test paper has been cited countless times by philosophers and computer scientists—not to mention members of the general public. But there's no way of knowing for sure, because citation indices don't normally search the bibliographies given in books, often omit non-scientific journals, and never search the 'public' media.

What is true, however, is that by the time that statement was made, many more people were aware of the biological paper than was the case even ten years earlier. I'd noticed this myself. Having greatly admired it when I came across it as a medical student in the mid-1950s, I was surprised to find over the next twenty-five years that knowledge of it was pretty limited even among my scientific colleagues. (As for philosophers, they'd very rarely heard of it, never mind read it.) Now, references to this mid-century paper are ten a penny.

15.v. Self-Replicating Automata

Self-organization, and its biological relevance, interested von Neumann (1903–57) no less than Turing.

His work on this topic was even wider in scope than Turing's. For it included the problems of reproduction and evolution—including the puzzles of how complex a system must be in order to be able to replicate itself, and how a self-replicator could generate descendants more complex than itself. (Turing (1950) had mentioned the possibility of a child machine inheriting features from its parents, but hadn't detailed how this might be possible.)

a. Self-organization as computation

Von Neumann's approach differed from Turing's in treating self-organization in terms of information, not physics. His aim was to define the necessary conditions for self-replication in general: reproduction as it could be, not merely reproduction as we know it.

That "merely", however, may be misleading. The biological mechanism of heredity was still a mystery when von Neumann first described it in informational terms. When the double helix was discovered five years later (in 1953), and the genetic code decrypted some years after that, it turned out that biological reproduction did indeed fit von Neumann's abstract description.

He thought of replication as a computational process, requiring a minimal level of computational complexity. Specifically, he realized that part of the self-replicating system must function both as instructions and as data—namely, as a self-description. Turing had noted in 1936 that one and the same tape state could be interpreted sometimes as data and sometimes as instructions, but he didn't apply this insight to self-replication (see Chapter 4.i.d).

Von Neumann even defined a universal replicator, a computational system capable of reproducing any system, itself included. (Biological organisms aren't universal replicators: cats give birth only to kittens.) He also remarked that errors in copying the self-description could lead to evolution, which might thus be studied computationally.

In other words, "evolution" was being seen as an abstract principle. It could be applied not only to genes and/or species but to computational structures too—and, as eventually became clear, to the immune system and neural development (Chapter 14.ix.d). Formulated by Donald Campbell as BVSR, or Blind Variation and Selective Retention, it was even applied to concepts and cultural practices too (8.v.b). However, all that was in the future. In von Neumann's day, evolution was mostly thought of as strictly biological.

In the 1930s, von Neumann was already sharing his fascination with these topics with the mathematician Stanislaw Ulam (1909–84), who had raised them in coffee-house conversations as early as 1929. And they continued to discuss them after the war, when they were both working at Los Alamos.

Ulam has been described by a colleague as "probably the only close friend von Neumann ever had", and also as "the more original mathematician of the two—but notorious for not doing the detailed work" (Rota 1989). He published almost nothing throughout his life, but had many seminal ideas for which others (who went off and did the detailed work) got the credit.

One of these was a pioneering chess program (for a 6×6 board: no bishops) that took an average of twelve minutes per move, but which spent that time on a comprehensive two-move lookahead (McCorduck 1979: 157). Another earned him the dubious distinction of being joint patent-holder with Edward Teller for a crucial process in the production of the H-bomb—and the same colleague, Gian-Carlo Rota (1989), judges that Ulam's ideas were the more important contribution.

It's entirely possible, then, that Ulam seeded von Neumann's mind with the problem of self-reproduction and/or its outline solution. Certainly, they discussed these matters repeatedly:

Our usual conversations [in about 1954] were either about mathematics or about his new interest in a theory of automata. These conversations had started in a sporadic and superficial way before the war at a time when such subjects hardly existed. After the war and before his illness [in 1955] we held many discussions on these problems. I proposed to him some of my own ideas about automata consisting of cells in a crystal-like arrangement. (Ulam 1989: 19; cf. Ulam 1962)

However, Ulam went on to say that von Neumann's "more concrete ideas developed [only] after his involvement with electronic machines".

In the 1930s, von Neumann couldn't formulate the problems clearly—still less, solve them. In 1943, however, he read McCulloch and Walter Pitts' "neural logic" paper and was introduced to computing on a visit to England (see Chapters 4.iv.a and 3.v.e, respectively). Thus inspired by the likely biological relevance of Turing machines, he claimed in 1946 to have formulated the problem of self-replication rigorously.

Two years later, he had the outlines of a solution—and of a general theory of automata, which he took to include both natural and artificial systems. That theory underlies much of what is now called A-Life.

He presented his outline in a talk at the Hixon symposium in 1948. This was the same interdisciplinary meeting at which Karl Lashley described his ideas on serial order in behaviour (Chapter 5.iv.a), and it was largely thanks to Lashley's urging that von Neumann later wrote up his talk for publication (von Neumann 1951). The Hixon paper sketched the general form of a self-reproductive system, and pointed out the possibility of heritable copy errors, or "mutations". It also described an (imaginary) mechanical system capable of self-assembly from physical parts.

This "kinematic" model caught some people's imagination, inspiring a simple mechanical model of self-replication that used interlocking tiles (L. S. Penrose 1959) and speculations by NASA scientists about robotic self-assembly in space (Levy 1992: 32–42). Von Neumann himself, however, was dissatisfied with it. It was limited to a specific type of physical system, and it left crucial processes unexplained (for example, recognizing, picking up, and welding the parts): "By axiomatising automata in this manner, one has thrown half of the problem out of the window, and it may be the more important half" (quoted in Burks 1970, p. xv).

He therefore turned (at Ulam's suggestion) to study systems defined not by physics, but by logic. Because Ulam had compared them to crystal growth in two dimensions, von Neumann called them crystalline regularities, or tessellations (tilings). Now, they are known as cellular automata, or CAs.

A CA is a computational "space" made up of many discrete cells, each of which can be in one of several states, and each of which changes (or retains) its state according to specific rules. More strictly: a CA is a space plus a specification of an initial space state. The same space may give very different results, given different initial states. CAs are artificial constructs, products of the logical imagination. But (as von Neumann foresaw) they can be used to simulate natural systems, if these involve equivalent principles of change.

Before the initial state can be specified, randomly or otherwise, the CA space itself must be defined. This requires decisions on many different options. The person defining the space must specify the number of dimensions; the size (sometimes infinite) of the space; the set of possible states per cell; the 'shape' (boundaries) and 'neighbours' of each cell; the timing (synchronous or asynchronous) of the cells' state changes; and the rules governing those state changes.

These rules are expressed in logical terms. For instance, a cell in state 6 (out of 12 possible states) must change to state 8 if and only if at least two of its neighbours are in state 10, and/or at least one is in state 3. The rules may apply universally across the space, or they may (in effect) vary according to region. And they can be of various general types. For instance, they may be fixed or mutable (changing systematically with time and/or subject to random mutation). And they may be deterministic, indeterministic, or probabilistic. On top of all this, a CA may be 'pure', or it may be subject to a certain amount of noise, or error.

Clearly, the potential variety of CAs is vast. Among the myriad possibilities are some systems with immense computational power. Indeed, as von Neumann suggested (but couldn't prove), his universal constructor is equivalent to a universal Turing machine, which can in principle perform any possible computation (Chapter 4.i.c).

Correlatively, CAs can in principle be used to simulate any natural system governed by local laws (as opposed to action at a distance), and having no essential discontinuities that can't be approximated discretely (as in the differential calculus). In short, CAs can be used to model dynamical systems describable by the laws of physics.

As usual, practicalities fall short of possibilities. For one thing, a natural system that could in principle be simulated by a CA may be so complex that such a simulation can never be found. Even if the CA rules can be specified, the initial state cannot. Putting this point in David Marr's terminology (Chapter 7.iii.b), many natural phenomena may be explicable only by Type 2 theories, which are extremely difficult to find. Von Neumann himself, for example, suggested that:

[It] is futile to look for a precise logical concept, that is, for a precise verbal description, of "visual analogy". It is possible that the connection pattern of the visual brain itself is the simplest logical expression of this principle. (von Neumann 1951)

For another thing, most CAs that have actually been studied are relatively simple—although "simple", here, is a relative term. In practice, CAs normally have only a few dimensions: a 1D line, a 2D grid (as in von Neumann's case), or a 3D volume. Even CAs modelling hyper-dimensional fitness landscapes (see Section viii.b), or the networks called CTRNs (Section xi.b), don't approach the number of possible dimensions. (Whether they're large enough, nevertheless, to represent all the relevant evolutionary factors is an empirical question.)

Many CAs have an initial state defined by only a few cells. Von Neumann's universal replicator was an exception. It had a "body" (the central construction unit) of 32,000 cells, plus a "tail" (the instruction tape) 150,000 cells long.

Similarly, many spaces assume a strictly limited number of neighbours per cell, each located immediately adjacent to the cell. For von Neumann, each cell had four neighbours: those sharing square boundaries with it. For John Conway (1937–), whose three-rule game of Life was the first A-Life entertainment to hit the popular scientific press (M. Gardner 1970), cells have eight neighbours: the four boundary-sharers, plus the cells touching the four diagonals. (Conway's Life is not to be sneezed at: in principle, it can generate a Universal Turing Machine.) However, some CAs allow *any* cell in the space to count as a neighbour, thus allowing for large-scale parallel processing (see Section vi.a).

In practice, too, the state-change rules also are relatively simple. Most CA rules are both deterministic and unvarying. The rules are usually applied across all regions of the space, each cell deciding its next state according to the same rule set. (Again, some recent work discussed below is an exception.) As for the number of possible states per cell, von Neumann specified twenty-nine when defining his universal replicator (these were based on the logical functions used in his design for the EDVAC computer). But many CAs, Life included, use only two.

Finally, most CAs are defined with respect to distinct time steps. That is, all the cells change (or not) simultaneously. Asynchronous CAs, in which the rules may be identical but the cells have independent "clocks", can produce very different behaviour (Ingerson and Buvel 1984; Bersini and Detours 1994).

Not all CAs can self-replicate, for many lack the necessary complexity. But self-replication isn't the only lifelike property that makes CAs interesting. Even highly restricted systems can generate unpredictable order, which sometimes emerges only after hundreds of iterations. In general, the bottom-up (cell-by-cell) parallel processing characteristic of CAs can give rise to higher-level order which the GOFAI approach outlined in Chapter 10 would have tried to impose top-down.

For example, the fascination of Life is its ability to generate unexpected, and sometimes systematically repeated, 2D forms such as "gliders". A simple 1D system (a single line of cells, generating another line below it . . . and so on) may produce lifelike patterns reminiscent of the markings on cone shells (Wolfram 1984*a*). A set of three simple rules can enable computer-animated creatures (bats, dinosaurs, or milk bottles) to "flock" realistically (Reynolds 1987). And some recent A-Life models show complexities of co-evolutionary behaviour that are still more intriguing (see Section vi.c).

b. Why the delay?

Given the intellectual power of von Neumann's pioneering research, why wasn't it followed up immediately? D'Arcy Thompson's vision of a systematic mathematical biology had been reinforced. But where was the rush to turn vision into reality?

A few unorthodox biologists saw the point—McCulloch, of course, included. Automata theory was discussed at Waddington's ground-breaking meetings on theoretical biology (where D'Arcy Thompson was explicitly remembered). At one of these, the young Arbib presented a pencil-and-paper definition of a self-replicating program

(Arbib 1966, 1969). At another, Longuet-Higgins presented his ideas about the mathematics of associative memory (see Chapter 12.iv.c). And Gordon Pask, ahead of the pack as usual, experimented with simple evolutionary machines—not programmed as abstract cellular automata, but implemented as solutions of chemical salts (Pask 1961: 105–8).

But such experiments couldn't then be taken much further, and most computer scientists remained unimpressed. Almost all mid-century computer modellers focused their attention on other forms of cybernetics, or on GOFAI, or on simple neural networks. A-Life wasn't identified as a scientific endeavour until many years later (see Section x.a). It was later still that "chemical" computing, broadly anticipated by Pask, became a hot topic (Adamatzky *et al.* 2003; Sienko *et al.* 2003).

There are three reasons for this. One is the lack of computer power at the time. Von Neumann's A-Life research, like Turing's (and D'Arcy Thompson's), was a pencil-and-paper exercise, an abstract study in computational logic. He couldn't explore its implications by computer modelling, because the necessary technological advances were still far in the future. Accordingly, the first people to take up his ideas were logicians or mathematicians.

Even they were—for the second reason—very few in number. Von Neumann's CA research didn't immediately excite widespread attention because most of it remained unpublished until well after his death (he died of cancer early in 1957). A few brief hints were included in his (unfinished) lectures on *The Computer and the Brain*, published soon after he died (von Neumann 1958). And an account of the main ideas (written by the originator of BASIC) appeared in *Scientific American* (Kemeny 1955). However, this paper on 'Man Viewed as a Machine' contained very little detail. And Ulam, who had suggested CAs to von Neumann in the first place, published a relevant paper in the early 1960s (Ulam 1962)—but this, too, was very sketchy as compared with von Neumann's still-unpublished writings.

Von Neumann's major manuscript on automata, begun in 1952, was never finished. In 1949, a year after his Hixon talk, he lectured on 'The Theory and Organization of Complicated Automata' to an audience of mathematicians and computer scientists at the University of Illinois. There, he discussed various general types of automata. These included not only CAs, but also continuous systems—such as might be used to model the physical diffusion processes discussed by Turing. Others were quasi-neurological automata inspired by McCulloch and Pitts, involving thresholds, excitation/inhibition, and refractory periods, or fatigue. (He also revised his mid-1940s estimate of the number of individual units needed for self-replication, from the high 10,000s to about a million.) But these lectures, too, remained unpublished at the time.

The person most responsible for disseminating—and, initially, for developing—von Neumann's work on self-organization after he died was the philosopher and logician Arthur Burks (1915–   ). He did many years of editorial work to bring von Neumann's papers to public view. He circulated mimeographed notes of the 1949 Illinois lectures to those who asked for them, and in 1961 tried to interest behavioural scientists in general (Burks 1961). But von Neumann's own manuscripts weren't made widely available until a decade after his death (Burks 1966, 1970).

Burks was an ex-colleague of von Neumann's, having worked with him since the mid-1940s. They had cooperated on the design of the EDVAC, the ENIAC, and the Princeton computer (and jointly originated flow diagrams as a tool for program design: Aspray 1990). And, although he never discussed CAs personally with von Neumann (Levy 1992: 60), Burks had been thinking along similar lines for a long time. Indeed, it was he who coined the term "cellular automata".

His 'Logic of Computers' group at the University of Michigan was the first place where CAs were actually run on computers. Some of these runs were simulations of physical or biological processes (such as heart fibrillation), while others aimed to advance the theory of CAs in general. He developed, and even corrected, some of von Neumann's ideas while editing the manuscripts. For example, he proved his suggestion that a universal constructor, equivalent to a universal Turing machine, is possible; and he improved on his ("double-path") design for constructing instruction loops of arbitrary length within the CA space. In short, his intellectual contribution to the field was considerably more than the term "editor" normally implies.

Burks tried to interest the AI community in von Neumann's work, but he came up against the third reason why it was neglected. The fundamental difference in approach—bottom-up rather than top-down, and largely abstract to boot—was a formidable barrier.

Add the fact that MIT and Stanford (and later Yale) were recognized, not least by their own inhabitants, as the main centres of AI, and Michigan stood very little chance. It probably didn't help, in meetings dominated by ex-physicists and AI hackers, that Burks described himself as a philosopher. I remember attending several AI meetings in the early 1970s where he made an effort to talk about CAs and von Neumann to others, including people he hadn't met before. But it was obvious, and quite distressing in human terms, that he wasn't being taken seriously.

Inside Michigan, it was a different matter. Burks's students were introduced to von Neumann's ideas some years before they were published. They, too, helped to advance the field (Burks 1970).

In the early 1960s, one (James Thatcher) completed an alternative design for a universal constructor. Another, Edgar Codd (who would pioneer relational databases in the late 1970s), showed that a universal replicator could be axiomatized with only eight cell states instead of von Neumann's twenty-nine (Codd 1968, ch. 4). He did this by modelling candidate CAs on computers, iteratively redefining them when the computer tests failed—a methodology foreseen by von Neumann, but not available to him.

Later, another Michigan student, Richard Laing (1977), showed that a self-replicating machine needn't be 'magically' provided with a self-description, but could build one by inspecting itself. (It was Laing who'd been so critical of Mortimer Taube's anti-AI *Computers and Common Sense* in the early 1960s, and who'd been simulating large neural networks at that time too: see Chapter 11.ii.b and Richard Laing 1961*a*.) And later still, Langton (1984)—an ex-student of Burks's—would produce the first functioning implementation of a self-replicating automaton (see Section ix.a).

15.vi. Evolution Enters the Field

The most important work done in Burks's department was John Holland's (1929–). In his early twenties, Holland (1929–) had co-authored a model of Donald Hebb's theory of self-organizing cell assemblies (Rochester *et al.* 1956), which discovered unexpected difficulties in the theory (see Chapters 5.iv.b and 12.ii.d). But his prime inspiration came from von Neumann, not Hebb.

He turned towards automata theory as a very young man. Partly because von Neumann himself wasn't publishing on the topic, Holland's 'Survey' (1959) and two other general papers were cited in Minsky's bibliography of AI (Feigenbaum and Feldman 1963: 494). But Holland was much more than a competent surveyor, for he pushed the field forward in a very important way. Only a few years after von Neumann's death, he defined a new type of CA that answered the master's question of how an automaton could give rise to something more complex than itself (J. H. Holland 1962).

He did this by following up von Neumann's hint that a CA could support evolution, if it had an imperfect copying mechanism and some principle of selection. In other words, Holland was using biological ideas to extend CAs, rather than CAs to study biology.

His subsequent research elaborated these ideas. Eventually, they became highly influential. But this outcome was long delayed. The importance of Holland's approach wasn't widely recognized for over twenty years.

a. Holland, and mini-trips elsewhere

Holland's new systems, which he called iterative circuit computers, were more general than von Neumann's CAs. Not only did they allow neighbourhood relations and transition functions to change over time, but they weren't strictly localistic. Holland allowed for direct communication between *any* two cells (all cells could count as "neighbours"), so that his systems performed essentially parallel processing within a huge, and very noisy, space.

This meant that

* programs (of any length) stored in contiguous cells could be moved (or copied) to another set of cells in one step.
* In addition, they could be combined, by being located in adjacent cell sets.
* Or they could be split into two parts—
* which in turn could be combined with other programs or program parts.
* Specific types of noise, or copy error, were allowed to produce novel combinations of program parts.
* Moreover, given a measure of program efficiency, the best of the new combinations could be selected so as to generate programs better adapted to their task.

The analogy to biological genetics and evolution was intentional. Holland had first experienced self-improving computer programs in the late 1940s, when he worked with Arthur Samuel and John McCarthy on a neural-net simulation of IBM's first commercial calculator. Samuel, who "devoted his spare time to the subject of machine learning" (Samuel 1959), would soon (in 1952) complete a GOFAI checkers-player that learnt to play better than he did himself (see Chapter 10.i.e). The neural network he developed with Holland was much less impressive, but it did show a crude form of learning.

Holland's subsequent research benefited also from his Michigan-based familiarity with von Neumann's ideas. In particular, he was inspired by von Neumann's hints about evolution in automata.

By the early 1960s, Holland was already thinking in terms of concepts borrowed from biological genetics: crossover, linkage, and dominance (J. H. Holland 1962). By the late 1960s, he had developed a comprehensive logical theory of adaptive systems, and had formally defined the first genetic algorithm (GA) (J. H. Holland 1975). GAs enable a program to improve not by top-down hand coding, but by bottom-up evolution. (The program may or may not be capable also of individual learning.) They involve parallel search over a large number of candidates at each generation.

Even before Holland showed how these effects could be produced, several researchers (independently) had already raised the possibility of computerized—as opposed to mechanical (see L. S. Penrose 1959)—evolution (D. B. Fogel 1998). Most of them were inspired by Ashby and Grey Walter rather than von Neumann, whose work on CAs wasn't yet widely known (see Section v.b).

Perhaps the first was George Friedman at UCLA, who in the mid-1950s outlined how a "selective feedback computer" might evolve electrical circuits to maintain the internal temperature of a mobile robot (Friedman 1956). This goal was more taxing than it sounds. The robot's environment included not only hotter and colder areas, but areas where the temperature changed suddenly or slowly, and areas of very high temperature signalled by a flash of light (one of which changed position unpredictably). The parameters of the robot's circuits could be altered by a switchbox, and the most successful circuits at a given stage were selected and then randomly varied by instructions from the "Control".

At much the same time, Hans Bremermann (then at the University of Washington, Seattle, but later at Berkeley) began a series of theoretical studies of the evolution of "goal-seeking, self-organizing" systems (Bremermann 1962; Bremermann *et al.* 1966). He tried to analyse a range of evolutionary systems, showing how changing the variables would lead to different types of result. And in the early 1960s, Ingo Rechenberg in Germany actually evolved engineering designs based on wind-tunnel experiments (Rechenberg 1964).

These efforts had come out of the cybernetic stable, being focused on Shannon-type predictive systems. But others were influenced also by the nascent research on AI planning (Chapter 10.i.d).

For instance, the engineer Lawrence Fogel proposed the development of "goal-seeking" AI, in which artificial evolution looked after the tactics while artificial intelligence set the purposive strategy (L. J. Fogel *et al.* 1966). In the late 1960s, he joined with some central figures of the cybernetics–AI community—including Minsky, Allen Newell, McCarthy, and the ubiquitous McCulloch—in exploring the potential for evolving program-controlled sensory/motor prostheses for human patients and air pilots (Fogel and McCulloch 1970: 271–95). These instruments, he suggested, would autonomously adapt to their "experience" yet also be controllable by the human user. The user could thus concentrate on the higher-level goals, leaving the detailed execution—and eventually the setting of low-level sub-goals—to the machine.

This work, whether implemented or merely speculative, was less sophisti-cated—though more visible—than Holland's. Large-scale parallelism wasn't avail-able; indeed, the current generation was sometimes restricted to only two programs or genotypes. And it relied on mutation rather than crossover (but see below). In other words, although it was termed "evolutionary programming", it didn't follow the biological analogy so closely as Holland's "genetic" method did. (The term "GAs" is sometimes used to cover all types of evolutionary computing, but is sometimes restricted to methods similar in spirit to Holland's.)

The early evolutionary programming didn't sweep the field. Besides its lack of power, it was sometimes associated with extravagant claims that invited hostility from other AI modellers. The proposal for evolutionary prosthetics, for instance, aroused scepticism from some discussants.

Thus Minsky commented: "The question is not, 'Is tree search better?,' because this is a tree search. The question is, 'Is the language of mutation and offspring as good as the language of subgoal, and so forth?'" (L. J. Fogel and McCulloch 1970: 284). His own answer (pp. 287, 288) appeared to be "No", since he doubted whether Fogel's evolutionary technique was an advance on Albert Uttley's conditional probability machine (Chapter 12.ii.c). Doubts were expressed also by McCarthy and Papert, both of whom saw random variation as a much less promising way of improving programs than the more 'rational' AI approach (pp. 292–4).

Nevertheless, AI work on evolutionary computing continued in a few places outside Michigan, from the early 1960s onwards. And some of this exemplified what would now be termed A-Life, since it was directed to biological concerns.

The Stanford biophysicist–philosopher Pattee (1926–), another regular Wadding-ton invitee, had already applied automata theory to the origin of life (Pattee 1966). Now, he simulated evolution in an entire ecosystem (Conrad and Pattee 1970). But his model produced no significant results. Years later, he commented:

[The] populational behavior of the biota appeared to be chaotic, although we did not know the significance of chaos at that time. However, in spite of biotic behavior that was unendingly novel in the chaotic sense, it was also clear that the environment was too simple to produce interesting emergent behavior. (Pattee 1989: 71)

Holland's (then unrecognized) achievement had been to show how "interesting emer-gent behaviour" could indeed be generated.

A GA program makes random changes (copy errors) in a specific batch of code (equivalent to the genotype), evaluates the results, and selects suitable candidates for breeding the next generation. As Holland pointed out, the "suitable" programs aren't always the "best": occasionally, his GA would select a low-performance program for breeding. This probabilistic approach enabled potentially useful program parts hidden within even highly inefficient programs to be (sometimes) preserved in the gene pool.

Von Neumann hadn't specified just which sorts of random copy error would be most evolutionarily fruitful. He appeared to assume that point mutations, wherein a genetic unit is substituted by another, suffice in an asexual population. By contrast, Holland stressed crossover or recombination between a pair of parents in a sexual population, wherein a set of adjacent units on one string—compare: adjacent genes on a chromosome—is swapped for a set of adjacent units on another string. Indeed, he allowed for multiple crossovers, wherein two or three gene sets are exchanged between the two strings.

Holland pointed out that crossover (as defined above) enables useful building blocks to be preserved during self-replication. A point mutation can affect any gene, anywhere. But this type of crossover is unlikely to make changes within small sets of adjacent units, because there are fewer potential breakage points within a short string than within a long one. Moreover, it can build up high-level blocks consisting of two or more lower-level blocks, and these hierarchical structures also can be preserved throughout evolution.

(It turned out later that Holland's restrictive definition of crossover was unnecessary. 'Uniform crossover'—wherein a coin is tossed at each locus in the paired gene strings, and either mother or father is chosen at random—makes widespread changes, yet can drive successful evolution: I. Harvey, personal communication.)

As Holland realized, however, strict adjacency is both improbable and unnecessary. It's unlikely that potentially cooperative lower-level blocks will always lie together. Even if they start out that way, some future crossover may separate them. For that matter, even a single block may contain functionally irrelevant units lying between the relevant ones. And the statistics are borne out by the biology: adaptively co-functional (sets of) genes needn't lie next to one another on the chromosome.

Accordingly, he specified building blocks (sub-strings within a program) as flexible "schemata" rather than fully determinate strings. And he defined a new type of GA to deal with them.

He represented schemata as fixed-length strings of "0s", "1s", and asterisks—where an asterisk means "it doesn't matter whether this unit is a 0 or a 1". For example, the schema "0010****" can be instantiated by any 8-unit string starting with 0010 (00101111, 00100000, 00101010, and so on); the schema "**0010**" is instantiated by 00001000, 11001010, 10001001, and so on. Holland's program could identify any string in which the relevant building block appears, and could manipulate sequences or hierarchies of building blocks while ignoring insertions of 'irrelevant' units. That was just as well, since such insertions would often be produced by the GA's own crossover activities. In short, his GA was efficient and highly robust.

Moreover, Holland proved the "Schema Theorem": that schemata, which enabled cooperation between different building blocks, would increase their representation within the population exponentially with time. Much later, it became clear that he should have stressed that they would do this *ceteris paribus*. This get-out clause is important, because as soon as schemata do start increasing, their relative fitness will change.

In other words, the conditions under which the Schema Theorem holds break down almost immediately. Partly for this reason, many now consider that the Schema Theorem didn't give a good picture of what was going on in Holland's models. Nevertheless, it was influential for a time, since it appeared to offer a mathematical proof of their evolutionary efficiency.

By the mid-1980s, Holland had found a way of making GAs even more powerful—and he'd applied them to classification, inductive reasoning, and philosophy of science (J. H. Holland *et al.* 1986). In artificial evolution, unlike the biological case, it's in principle possible to protect the most useful building blocks from being dropped, or damaged, during replication. But how is this to be done?

Given that many different strings (rules) are being searched in parallel, and that several contributory rules are acting together in performing the task, it's no trivial matter to discover which are especially useful. Holland devised an ingenious method for automatically advantaging the most adaptive rules within a complex evolved program. This was the bucket-brigade algorithm, whereby appropriate fractions of the overall "credit" are passed back to individual rules (J. H. Holland *et al.* 1986: 70–3).

b. Awaiting the computers

One might ask the same question about Holland's research as about von Neumann's. Why didn't it receive immediate recognition? Why have GAs been widely used in AI and A-Life only since the 1980s, given that they were first defined in the 1960s?—The answers are very similar.

Holland's interests in cellular automata, like von Neumann's, were more theoretical than technological. In his early work he did discuss the advantages and disadvantages of actually building machines as iterative circuit computers (Burks 1970). But he didn't implement his GA after he'd defined it. That was done by his student David Goldberg in 1983, to solve problems about controlling the flow of gas in pipelines (Goldberg 1987). (Other Michigan students had already used the ideas in theoretical studies—of adaptive game playing, for instance: Bagley 1967.) In any case, the computational power available before the 1980s hardly sufficed for interesting GA research.

Again, Holland—unlike some of his intellectual descendants—didn't initially seek to publicize his work, still less to turn it into the latest AI 'cult'. Although he published papers in the 1960s, his book on the formal theory of adaptive systems didn't appear until 1975.

Even that wasn't widely read. I recall being hugely impressed by the paper he gave at a twenty-person weekend meeting held on the edge of Dartmoor in 1981 (cf. Selfridge *et al.* 1984). I'd never heard of him, and when I got home from the Devon countryside I asked my AI colleagues in Sussex why they weren't shouting his name to the rooftops. Some replied that his work wasn't usable—and some had never heard of him either.

Finally, Holland's approach, like von Neumann's, was fundamentally at odds with the then dominant culture in AI. Burks had failed to persuade GOFAI researchers of the potential of CAs and bottom-up processing, but at least he'd focused on deterministic automata. Holland's programs, by contrast, relied crucially on rampant (and richly parallel) random change. It's perhaps not surprising that highly skilled GOFAI programmers, well aware that to omit a single LISP bracket was to crash their system, failed to see the intellectual promise of his work. (It probably didn't help that he was based in Michigan: theoretical linguistics wasn't the only discipline to exclude those outside the magic circle—see Chapter 9.viii.a.)

But that's not to say that mainstream AI was ignoring the possibility of evolving programs. As remarked above, such efforts had started in the early 1960s. Twenty years later, a number of people were using some sort of evolutionary method—variation plus selection—to evolve GOFAI programs.

John Koza (1943–) was a leading example (Koza 1989, 1992*a*). His technique enabled programs of increasing length and complexity to be generated, because larger program parts could be substituted for (and nested within) smaller ones. Besides applying this approach for optimizing AI problem solving, Koza used it to produce programs acting as sensori-motor controllers for simulated robots, which evolved the capacity to follow irregularly shaped "walls" (Koza 1992*b*).

However, Koza's method was less richly parallel than Holland's, and also less 'free'. Instead of allowing changes at any point in the code, it was constrained to swapping syntactically sensible LISP expressions. Koza therefore avoided the difficulty, which Holland had faced, of deciding how to identify and preserve useful mini-sequences—including 'interrupted' sequences containing blocks of non-relevant material.

Similar remarks apply to Karl Sims's (1962–) *interactively* evolved image-generating programs (developed on the Connection Machine at MIT's Media Lab: see 13.v.a), in which the selection is done by a human (Sims 1991). Although some variations can affect the very heart of the LISP code, the set of genetic operators ensures that only 'sensible' programs result.

Gradually, interest in evolutionary computation increased. By the mid-1980s, it had grown enough to support the first international conference on genetic algorithms (Grefenstette 1985). And this interest grew apace. A few years later, simple evolutionary programs such as Dawkins's Biomorph had even entered the home, influencing children's views on the concept of life (Dawkins 1986: 51–74; Turkle 1995).

By the late 1980s, AI work on evolutionary programming had burgeoned. Holland's ideas were better known, and huge computational power was becoming available. A number of interesting implementations of GAs—some directly inspired by Holland's work, some not—were appearing.

One was produced by Daniel Hillis (1956–) at MIT (Hillis 1992). To do so, he took advantage of the enormously powerful Connection Machine that he'd designed in the early 1980s, having been greatly influenced by Minsky's ideas on parallelism (12.iii.d). The first, quarter-sized, version had been bought by MIT's Media Lab in 1986, but the full-sized version was available soon afterwards. This had no fewer than 65,536 processors, with a maximum of sixteen 'steps' for any one to connect with any other (Hillis 1985).

Using that machine, Hillis was able to study populations of up to a million individuals, with up to 256 "chromosomes" per individual. For illustrative purposes, Hillis used his GAs to optimize sorting networks. Sorting algorithms put some set of elements into some rational order. For instance, they arrange a random set of numbers into an increasing series. In general, they are easy to test: their "fitness" can be readily assessed. Accordingly, Hillis was able to show a clear improvement as a result of applying GAs.

Other AI workers had used GAs to optimize computer programs. Hillis deliberately made the task—one might think—more difficult, by introducing "parasites", or "predators". These were sets of up to twenty test cases, which evolved to become more and more difficult as the sorting algorithms improved. Hillis found, as he expected (and as Darwinian theory predicted), that the parasites would make the improvement more rapid than it had been without them.

The reason is that competitors can dislodge a population from a local peak in its fitness landscape (see Section viii.b), so that it has the chance of finding a higher peak elsewhere. As Hillis remarked, this is similar in principle to the technique of simulated annealing used in connectionism to move a system away from a local minimum (Chapter 12.vi.b).

Whereas Hillis wasn't primarily interested in the biological relevance of co-evolution, others were. In the mid-1980s, people began to use GAs as experimental test beds for theoretical biology. The phenomena they studied would include parasitism, symbiosis, predator–prey behaviour, the evolution of sensori-motor anatomy, evolutionary change in fitness criteria, and the speeding-up of evolution by the introduction of competitors.

In some cases, mutation rates and population size were systematically varied, to investigate their effect on the pattern of evolution. Simulated species were tracked for thousands of generations, to see what types of order emerged, and why. In general, such GA work showed that very simple rules can give rise to richly ordered, and often unexpected, behaviour.

It was found, for instance, that the "punctuated" evolution posited by some biologists (Eldredge and Gould 1972; Gould and Eldredge 1993) can emerge in computer models—and can be explained in precise, and purely Darwinian, terms. Background mutations (ones not causing a change in the phenotype) would sometimes accumulate over many hundreds, even thousands, of generations—only to cause a sudden change in the phenotype when a few 'last-minute' mutations occurred and interacted with them. In short, hidden gradualism was producing observable saltations.

Early research of this type was done in the mid-1980s by Thomas Ray (1954–) at the University of Delaware. Ray was a botanist, specializing in the ecology of tropical rainforests (Ray 1992, 1994). Accordingly, his Tierra system was a virtual world of co-evolving "species", which competed and cooperated in ways broadly similar to biological species.

Ray's individual "creatures" were strings of self-replicating computer code, competing for space in computer memory. Some strings evolved the ability to borrow the self-replication facilities of other strings; and some host strings became resistant to these parasites, by evolving parts that prevented such attachments. After many generations, a variety of species, including parasites, hyper-parasites (outwitting the host's evolved resistance), and symbionts, usually emerged. Ray's work would eventually spawn a still-continuing worldwide experiment in computational ecology, in which virtual creatures are free to evolve in a "Digital Reserve" formed by pooling the idle time of computers linked by the Internet (Ray 1996; Ray and Hart 1998). (Ray himself is convinced that virtual creatures such as these can be genuinely alive: see Section ix.b, and Chapter 16.x.)

c. The saga of SAGA

GAs were developed in a new direction in the early 1990s, by Inman Harvey (1946–) at the University of Sussex (I. Harvey 1992a). Harvey extended GAs to cover open-ended, potentially indefinite evolution of increasingly complex artefacts, with genotypes of potentially unlimited length. ("Artefacts", because he didn't claim that his GA models were detailed theories of real life.)

Harvey's main aim (personal communication) was to discover how to set optimal mutation rates, given genotype spaces of different kinds. For example, some might involve "ridges" or "neutral" (*sic*) networks. In addition, he wanted to work with genomes that could vary in length.

From his point of view, Holland's approach had two drawbacks. First, Holland had considered only fixed-length genotypes, and his Schema Theorem guaranteed increasing fitness only for these. Second, he'd implicitly assumed that there was typically a large amount of genetic diversity within the population being evolved. Harvey argued that both these assumptions are inaccurate for natural evolution, and needn't apply to artificial evolution.

The use of fixed-length genotypes wasn't compatible with open-ended evolution (see subsection d, below). The search space of a fixed-length genome is finite—perhaps astronomical, but finite. Radically novel forms simply can't arise. Moreover, biological organisms that are more complex tend to have more genes. The reason is that a longer genome allows for more complex gene–gene interactions. It follows, said Harvey, that open-ended artificial evolution, too, requires variable-length genomes.

In relying on fixed-length genomes, Holland had supposed that the finite search space included some (relatively) optimal solution, which would eventually be found. For specific problems, this assumption is justified—in principle, if not always in practice.

But, as Harvey pointed out, biological evolution isn't like that. Species don't evolve in order to solve pre-given problems—although we sometimes speak as if they do, saying for example that blinking evolved "to protect the eye". Indeed, new fitness criteria arise as a result of evolution: no eyes, no 'need' for blinking. In other words, new 'problems' and new types of 'relevance' are spontaneously created by evolution.

This point had been implicit in Pask's early models of the emergence of new types of sensor: crystalline threads sensitive to auditory pitch or magnetic fields, for instance (see Chapter 4.v.e). But Pask hadn't specifically related it to evolutionary models, whereas Harvey did.

So, too, did Peter Cariani (1956–), a philosopher of A-Life in Pattee's group at SUNY Binghamton (and also an auditory neurophysiologist, now at the Massachusetts Eye and Ear Infirmary), and Peter Kugler and M. T. Turvey of the University of Connecticut's Ecological Psychology group (Cariani 1992; Kugler and Turvey 1987; Kugler *et al.* 1990). All of them insisted that biological evolution is open-ended—and, in general, leads to increasingly complex structures.

The second problematic assumption noted by Harvey concerned genetic diversity. Holland's Schema Theorem had implicitly presupposed wide genetic variation in the artificial population, but this doesn't apply to natural populations. The individuals within a biological species have very similar genotypes. In other words, the population is relatively *converged* within genotype space.

Further, Harvey argued, even if one were to begin with a highly diverse population, it would rapidly converge so that individual genotypes became more similar. The population genetics would then have settled down into a dynamic equilibrium between the forces of selection and variation. Thereafter, evolution must take place in the context of low genetic diversity. (Harvey's original argument focused on populations with variable-length genomes, but he later showed that it applies also to most fixed-length GA problems.)

The dynamic balance between selection and variation—and therefore the potential for further evolution—will be affected by altering the mutation rate. This in turn will depend on various factors, including the form of selection used, the shape of the fitness landscape, and—he realized later—whether crossover as well as mutation is allowed.

With these points in mind, Harvey developed the SAGA algorithm (Species Adaptation GA), to enable "hill-crawling" rather than hill climbing (I. Harvey 1992*a*, 1994). The technical arguments driving SAGA were tested out on Stuart Kauffman's models of fitness landscapes (see Section viii.b).

On rugged landscapes, sudden large-scale genetic change is likely to decrease fitness drastically. Fruitful interbreeding over many generations therefore requires not a ragbag of hugely different individuals but a genetically converged population: in other words, a species. SAGA's technical parameters were defined, and set, accordingly. Taking account of the three factors mentioned above, Harvey tried to determine the optimal mutation rates and set his models so as to approximate to them. For instance, it followed from Kauffman's point remarked above that, given a highly converged population, only small variations in length would tend to be viable. Similarly, crossover should result only in small variations in the genome.

In terms analogous to those used in Chapter 1.ii.b to categorize types of AI, Harvey's work counts as "technological"—not "biological"—A-Life. His work was done for engineering purposes, to make artificial evolution more powerful. Nevertheless, his ideas were potentially relevant to biology, for example to the question whether biological evolution has (sometimes? usually?) settled on near-optimal mutation rates.

In either case, the importance of SAGA was that it wasn't restricted to optimization on pre-defined problems. It could simulate the long-term incremental evolution of entire populations, and the generation of increasingly complex forms.

SAGA was used in the early 1990s by Harvey and his colleagues, and a few years later by others also, to evolve both virtual and physical 'species'. These were used to explore (for example) the rate of evolutionary improvement given varying mutation rates, the emergence of sensori-motor coupling between system and environment, and the effects of different levels of environmental noise. (Later, the Sussex group would evolve robot brains from GasNets, as well as from purely connectionist systems: Chapter 14.ix.f.)

For instance, a simulation of co-evolving predator and prey was used to measure evolutionary progress in strategies for hunting and escape (Cliff and Miller 1995). Evolutionary 'progress' is a problematic concept, not least because species evolve in complementary ways. As the Red Queen said to Alice, they have to run as fast as they can to stay in the same place. This causes difficulty: although an animal's speed can be directly measured, its efficiency cannot, because predator and prey are always in equilibrium. In a simulation, however, improvement can be measured by 'reviving' ancestor generations and pitting them against later ones.

As for physical artefacts, populations of robots embedded in specific environments evolved control mechanisms coupling them to those particular environments (see below).

Computer *hardware*, too, was developed by incremental (fixed-length) evolution using SAGA. Adrian Thompson (1970–) achieved the first evolved hardware robot controller in the mid-1990s. He worked with a reconfigurable home-made circuit on board a real robot, where the sonar echoes were simulated (during evolution) according to how the robot moved in a simulation of the corridor. After evolution the simulation was switched off, and the real sonar echoes in the real environment were used, to confirm that the robot did indeed work (A. Thompson 1995). Later, when suitable reconfigurable chips became available, he evolved circuits in simulation and then downloaded them into actual silicon chips (A. Thompson *et al.* 1996; Thompson 1998).

We noted in Section iii.c, when discussing D'Arcy Thompson, that a mistake in the simulated physics of a virtual environment may be picked up by the software creatures evolving in it (cf. Sims 1994). Similarly, contingent features of a laboratory task environment can become locked into evolving material artefacts—sometimes, bearing a startling resemblance to actual biological mechanisms. The Sussex work on evolutionary robotics showed that this can happen in negative or positive ways, decreasing or increasing complexity, respectively.

For instance, one of a robot's two 'eyes', and all of its 'whiskers', sometimes lost their connections to the controlling neural network (the 'brain'), if—in the environment provided—neither stereopsis nor touch was essential to perform the task (Cliff *et al.* 1993). (Compare the fact that auditory cortex in the congenitally deaf, or in laboratory animals deprived of auditory input, comes to be used for visual computation: Chapter 14.ix.d.)

More positively, the 'brains' of robots encountering a triangle of white cardboard sometimes evolved a 'feature-detector' analogous to those discovered in monkey brains in the 1960s by David Hubel and Torstein Wiesel (14.iv.a). This was a mini-network sensitive to a light–dark gradient at a particular orientation. It evolved as part of a visuo-motor mechanism, its connections to motor units enabling the robots to use the white triangle as a navigation aid (I. Harvey *et al.* 1994; P. Husbands *et al.* 1995).

In another experiment, the group evolved a neural circuit capable of controlling sixteen robot motors—two for each of the artificial octopod's legs—in coordination. As a result, the robot could move coherently and 'appropriately'. What's more, the evolved neural network took account of the signals from the robot's four infra-red sensors, ten light sensors, and various sensory bumpers and whiskers (Jacobi 1998).

Likewise, in the evolution of computer hardware at Sussex, circuits evolved to compensate for—and sometimes even to exploit—physical properties of the chip that were irrelevant to the chip fabricator's model of what was going on (A. Thompson 1997). Being irrelevant (in that sense), they were unknown, and not part of the chip specification.

Up to a point, such opportunistic noise tolerance is an advantage. But the exploitation of accidental details can be dangerous. In biological terms, a species that exploits a highly specialized niche is especially vulnerable to environmental change. Likewise, if the evolved logic circuit were to be copied into a different silicon chip, it might not work. And the robots' feature-detector would be useless if the triangle were black, or had sides sloping at a different angle.

Environmental 'noise' was therefore deliberately provided by the Sussex group, to prevent such overspecialization—and to follow up their intuition that evolution is easier in noisy environments. Biological evolution has noise provided anyway, in the sense that organisms live in variable environments. Even animals that eat only one type of food don't insist on its absolute purity. And many vertebrates have evolved comprehensive sets of feature-detectors, sensitive to both light–dark and dark–light gradients, in any orientation (Chapter 14.iv.b).

Evolution was modelled also by exploiting genetic drift across neutral (*sic*) networks. Some units in a genome may be irrelevant, in the sense that mutations have no effect on fitness. In the current genetic context, they have no causal influence on the phenotype. But, given mutations elsewhere in the genome, they might be useful in the future (see the discussion of "punctuated" evolution in Tierra, above).

Many organisms contain large amounts of 'junk DNA'. But it's hard to believe that evolution favours useless burdens—so perhaps junk DNA does convey some advantage? The molecular biologist Manfred Eigen suggested in the late 1980s that junk DNA could enable wider exploration of the genetic space, by providing fitness-neutral pathways that sometimes lead to high-fitness structures that would otherwise be inaccessible. In other words, neutral networks could prevent the species from getting stuck on a local optimum in the fitness landscape. Instead, it could move 'sideways' through the space, altering the genome without altering the fitness.

Harvey and Thompson (1996) combined Eigen's idea with (a fixed-length version of) SAGA, and with the recognition that the way of encoding the genome did indeed allow potentially useful "junk". Using GA parameters optimized to take advantage of this phenomenon, they evolved hardware circuits and also studied various abstract properties of evolution in general (see also Barnett 1998).

At much the same time as Harvey was working on SAGA, Steve Grand was independently developing a new type of computer world called Creatures, inhabited by evolving virtual organisms of various types (Grand *et al.* 1997; Grand and Cliff 1998; Cliff and Grand 1999). Although this was initially intended for home entertainment—and swept the world on its release in 1996—it could also be used as a test bed for theories of mental architecture, and for various commercially useful applications. For instance. it was used to simulate customers' behaviour in a bank, to enable the management to plan where to position the counters, advice tables, and automatic tellers (R. Saunders 2000).

The story of Creatures is told in Chapters 13.vi.d and 16.x.b. Here, its interest is that it was initiated outside academia, by someone who wasn't a trained computer scientist—and that it enthused a number of artists and cultural commentators. Artificial evolution had entered the real world.

d. Open-ended evolution

By the late 1990s, evolutionary computing was an established methodology in AI and A-Life:

* It merited a handbook of over 600 pages to describe the various techniques (Baeck *et al.* 2000).
* It could be used for systematic experiments concerning both life and mind.
* At the level of evolutionary biology, it could investigate the effects of different types and rates of mutation, on different types of fitness landscapes, given different amounts of environmental noise, and allowing for fitness-neutral genetic drift. (These were the questions driving SAGA, as we've seen.)
* And at the psychological level, simple relations between environment, neural mechanisms, and movement and behavioural strategies could be explored in evolutionary terms.

All very well . . . However, a significant problem remained, one which had been highlighted by Harvey and by the Pattee group mentioned above—namely, the open-endedness of biological evolution.

They'd pointed out, for example, that phylogenetic evolution has seen the appearance of novel types of sensory organ. As well as producing *improved* eyes, ears, or noses, it has produced *first-time* eyes, ears, or noses. How can that be?

The new sensory organs don't spring out of nothing, of course. They are adaptations of pre-existing structures, which may or may not have had some other function (e.g. the bones of the inner ear evolved from jaw parts, which in turn evolved from gill slits). Because novel forms of perception afford novel problems/challenges and novel solutions, they can't be generated by a GA with a fixed set of parameters—not even by a length-variable SAGA system. If the parameters foreseen by the programmer as potentially relevant don't happen to include (e.g.) light, then no eye can possibly emerge in the simulation.

In the new century, two of Harvey's younger colleagues managed to circumvent this problem—not by overturning the logic of the argument, but by bypassing it. They did this by working with physical hardware as well as computer programs. In brief, Jon Bird (1968–) and Paul Layzell (1964–) used Thompson's SAGA-based technique to evolve oscillator circuits—and unexpectedly ended up with a novel sensor: a primitive radio receiver (Bird and Layzell 2002).

We encountered this research in Chapter 4.v.e, in the context of Pask's self-organizing artificial "ear". As explained there, the evolution of the radio wave sensor depended on unforeseen parameters, such as proximity to a PC monitor and the aerial-like properties of printed circuit boards. Bird and Layzell concluded:

We have argued that there are three key properties that devices must embody in order for selection pressure to form them into novel sensors:

* they are situated in the physical world;
* they consist of primitives with no fixed functional roles;
* and the primitives are sensitive to a wide range of environmental stimuli. (Bird and Layzell 2002: 1841)

The good news was that if those conditions are satisfied, it needn't matter if the A-Life programmer doesn't—and couldn't—foresee every physical factor required for a novel sensor to arise. (Which isn't to say, of course, that a novel sensor actually will arise.) The bad news was that simulations, i.e. programs running on computers, aren't enough: truly open-ended evolution can happen only in physical devices, situated in the physical world. (D'Arcy Thompson would have approved!)

As I write this (in early 2005), comparable experiments are virtually non-existent. The small proportion of A-Life workers who do deal with "hardware" issues are almost all concerned with biochemistry (see Section x.b, below). Although they plan/interpret their experiments in the context of biological evolution, they don't use evolution in the experiments themselves. One exciting exception is a just-initiated project on evolving organic–silicon computers (described in 14.ix.f). Besides relying on GAs, this work is exploiting—and investigating—the physical dynamics of diffusion reactions. In such (still rare) cases, Turing and von Neumann and Holland *and D'Arcy Thompson too* are all marching in step. Most current evolutionary A-Life, however, recalls only the first three.

Not surprisingly, then, open-ended evolution was recently identified by the A-Life community as *the* major challenge facing them in the new century. That judgement appeared in spring 2003, when the journal *Artificial Life* published a paper bravely titled 'Collective Intelligence of the Artificial Life Community on its Own Successes, Failures, and Future' (Rasmussen *et al.* 2003*b*).

In a sense, they'd done this already. For the paper was a report drawing on the 'A-Life VII' conference in Portland, Oregon, and an earlier report on that had appeared in the same journal three years before (Bedau *et al.* 2000). Indeed, that one had nine authors, whereas the new version listed only four (two appeared on both lists). So why bother? And why call the report of the *smaller* group a "Collective" opinion of the A-Life "Community"? The reason was that this 2003 paper was based on an analysis of an extensive Internet consultation of A-Life researchers. This was intended for conversion into a continuing presence on the Web. In Francis Bacon's terminology (Chapter 2.ii.b), it was an attempt to establish an electronic Solomon's House, to serve the A-Life community in a more disciplined way than an email list or bulletin board.

The 250 delegates to 'A-Life VII' had been asked to consider the "grand challenges" for the field. After discussion through questionnaires, and through meetings held in Portland, the final verdict listed fourteen "open problems" (Bedau *et al.* 2000). The more detailed email survey added an order of priority (i.e. number of votes). The most important issues were said to be: open-ended evolution; better theory, to underpin/unify ad hoc experimentation; the definition of life; creating life (see Section x.b); and understanding dynamical hierarchies. You'll have noticed which one was top of the list.

15.vii. From Vehicles to Lampreys

Evolutionary robotics, as we've just seen, might throw light on how neural mechanisms can evolve to generate ecologically appropriate behaviour. But how neural is "neural"?

It's one thing for a robot's control mechanism to be broadly analogous to real biology, as the artificial triangle-detector was analogous to the orientation cells in visual cortex. It's quite another for a robot (or a computer simulation), whether evolutionary or not, to model actual neurology in any detail. If that could be done, however, it should provide *both* more lifelike robots/simulations *and* a way of testing, and generating, specific biological hypotheses. In other words, robots or simulations produced with real animals firmly in mind could give us computational neuro-ethology, or CNE (Chapter 14.vii)—a term coined by one of the scientists responsible for the triangle-detecting robot (see subsection b, below).

One of the first people to realize this was Arbib, who took the whole animal—and its detailed neurology—into account when modelling integrated sensori-motor mechanisms. In the 1960s, he was doing this virtually alone: his *Rana computatrix* had no implemented CNE cousins. But as the years passed, the four obstacles remarked in Chapter 14.vii.a lessened. From the early 1980s, practical work in CNE flowered.

Frogs were joined, as subjects for computer modelling, by many other animals. And 'computer modelling' increasingly came to mean robots as well as simulations—with predictable disagreements about which methodology was superior. Some of these artificial animals, whether implemented as robots or as programs, weren't designed but evolved—with interesting implications for neuroscience.

Claims to be modelling the "whole" animal, of course, had to be taken with a pinch of salt. Even apparently 'simple' animals often turn out to be much more complex than they seem.

For example, every one of the hundreds of hairs found on a spider's leg (eight legs, remember) bears a number of sophisticated mechanoreceptors and chemoreceptors (Barth 2002; Humphrey *et al.* 2003). (I say "a spider", but that's shorthand: there are eight times as many species of spider as of mammals.) And whereas the nematode worm has 302 neurones (whose action and development have been individually mapped by Sydney Brenner's team at Cambridge: 14.v.b), ants have approximately half a million.

a. Valentino's vehicles

While Arbib was concentrating on increasingly realistic frogs and largely speculative humans, the MIT neuro-anatomist Valentino Braitenberg (1926–) was defining a whole family of imaginary animals (Braitenberg 1984). These were toys, in the sense that they were descriptions of possible mechanisms *stripped down to the bare essentials*. (They also part-inspired the design of actual toys, such as the robot dogs manufactured in end-of-century Japan: R. A. Brooks, personal communication.) However, the "essentials" were all drawn from neuroscience—including the work on feature-detectors discussed in Chapter 14.iv.

Braitenberg wasn't the first to imagine models of whole animals. Friedman, for example, had speculated on a homeostatic robot (see Section vi.a). Edward Tolman and Clark Hull had sketched the "schematic sowbug" and other notional robots (Chapter 5.iii.c). Grey Walter had not only imagined his tortoises, but built them (Chapter 4.viii.a–b). And McCulloch had helped program a controller for a Mars robot (Chapter 14.v.a). But Braitenberg's discussion was much more systematic, concerning not a single model animal but an increasingly complex series of them. He showed persuasively how a wide range of behaviours could be generated by a relatively simple set of basic (and neurologically plausible) mechanisms.

He'd been interested in physical simulations since the late 1950s, when he described systems of coupled oscillators as functional models of cell assemblies and correlation-based models of the cerebellum (1961). In the mid-1960s, he wrote a lively paper giving designs for simple creatures that would move around in response to smell and vision. The three key concepts were *taxis* (automatic movement whose direction and speed is determined by the position and intensity of the stimulus), *kinesis* (unoriented movement, whose speed depends on stimulus intensity), and *decussation* (wherein one side of the brain is connected to the opposite side of the body). It was this paper, together with Braitenberg's earlier theory of movement timing by the cerebellum, which had inspired Arbib to work on *Rana computatrix* in the first place (Arbib 2002*b*: 13).

By the early 1980s, neuroscientific work on learning, and on the integration of perception and action, had blossomed. In an intriguing and highly accessible book—which introduced many people to cognitive science—Braitenberg (1984) now defined a series of increasingly complex creatures, or "Vehicles".

Vehicles were described by Braitenberg himself as a "fantasy" (1984: 95). They weren't implemented: not as robots, and not as programs either. They count as a form of CNE nevertheless, because his book was an abstract exercise in comparative neuropsychology.

In effect, his Vehicles were paper-and-pencil designs for what are now called situated robots (see Section viii.a, below). They provided a phylogenetic scale, going from simple to complex and ending with robots supposedly capable of "running after a dream" (p. 83). Each chapter started from, and added to, the design achieved in the chapter before. So, like biological evolution, the sequence had no viability gaps.

The first Vehicle had (i.e. was imagined as having) one motor, and one sensor—for temperature, light, chemicals, or sound waves. Braitenberg pointed out that even this creature, which simply moved in the direction in which it happened to be pointing, with a speed proportional to the stimulus intensity, would in practice do things that weren't built into the rules. For it would encounter friction, which would affect both its speed and its direction. In general, the behaviour of an *implemented* (embodied) Vehicle would depend not only on its design but also on its dynamic interaction with the environment.

Vehicle 2, which had *two* motors and *two* sensors, was a combination of two instances of Vehicle 1. Although its motors, too, moved faster with increased stimulation of the sensors, its behaviour was more complex. Moreover, the behaviour differed according to just how the two component Vehicles were combined. Each sensor might be connected to the motor on its own side, or to the one on the opposite side (the simplest case of decussation). In the first case (Vehicle 2a), the creature would tend to avoid the source of stimulation. In the second (Vehicle 2b), it would turn towards the source and move forward until hitting it.

The other twelve designs, each with distinct varieties as in Vehicles 2a and 2b, were synthesized by adding a relatively simple computational feature to the one before. These features included inhibition, thresholds, lateral inhibition, association, and one-way facilitation. As Braitenberg remarked, the last of these—which an engineer would regard as "the main technical problem to be solved"—had been modelled by Uttley in the 1950s, and was now being used in many models of conditioning (p. 114).

Even Vehicle 1, Braitenberg had claimed, would be described by an observer as *alive*. If someone were to watch Vehicle 2 for a while, their descriptions would be psychological as well as biological:

Let Vehicles 2a and 2b move around in their world for a while and watch them. Their characters are quite opposite. Both DISLIKE sources. But 2a [the same-sided version] becomes restless in their vicinity and tends to avoid them, escaping until it safely reaches a place where the influence of the source is scarcely felt. Vehicle 2a is a COWARD, you would say. Not so Vehicle 2b [the decussated version]. It, too, is excited by the presence of sources, but resolutely turns towards them and hits them with high velocity, as if it wanted to destroy them. Vehicle 2b is AGGRESSIVE, obviously. (Braitenberg 1984: 9)

As this passage shows, the Vehicles were mischievously described in psychological terms—such as LOVE (as early as Vehicle 3!), VALUES, LOGIC, CONCEPTS, IDEAS, FORESIGHT, EGOTISM, and OPTIMISM. These anthropomorphic properties were said to be just as "obvious" as AGGRESSION.

Clearly, Braitenberg's tongue was wedged firmly in his cheek. But there was method in his mischief. He was aiming not only to show how highly complex behaviour could be generated by fundamentally simple computational processes, but also to ground this fact in known neurological mechanisms.

Like Pitts and McCulloch (1947), but with the benefit of nearly forty more years of brain research, he related his abstract designs to the specifics of neuro-anatomy. He referred, for instance, to the pyramidal and stellate cells of the cerebellum, and to the orientation columns in visual cortex. He also referred to non-vertebrate examples, such as the compound eyes of insects (pp. 100 ff., 105 ff.).

He even offered computational explanations of some puzzling biological facts. For example, he argued that decussation hasn't evolved by chance as a selective advantage. Rather, it's a natural consequence of a mathematical fact: if each computational unit is connected to all the others then, given bilateral symmetry, there will be more connections crossing the median plane than staying on one side of it (pp. 95–9). Similarly, he suggested that the feature-detectors for orientation and direction that had so amazed their discoverers, Hubel and Wiesel, "may be just a special case of a general principle of the cortex" (p. 140).

Not surprisingly, Braitenberg's work later became highly influential in CNE, and in other areas of A-Life (see Section viii.a). But whereas he'd discussed different types of animal in highly general terms, most of this late-century research modelled the neurology and behaviour of particular species.

The animals chosen were many and various. Besides frogs, which Arbib had been simulating since the late 1960s, several types of fish were considered. Multi-finned fish were modelled by computer graphics programs that simulated how they swam, taking into account both muscle movements and detailed hydrodynamics (see Section ii.a, above). The "robotuna" was a swimming robot fish, whose tail propulsion depended on the production of (and interactions with) specific vortices in the water (Triantafyllou and Triantafyllou 1995). Lampreys were modelled also (see subsection d, below). So were rats: for example, how their navigation abilities depended on the spatial maps in the hippocampus (Mataric 1991).

b. Of hoverflies

Lest you think that only vertebrates were so honoured, insects were studied too. Indeed, the relative simplicity of insects' behaviour and neurology made them a prime focus for CNE. The many insect species involved included hoverflies, crickets, stick insects, and cockroaches—and the CNE methods used to study them differed significantly, as we'll now see.

A male hoverfly will track a female as she flies through the air. Once she settles on a flower, he shoots forward at a very high rate of acceleration, lands on the female, and mates with her. His change of direction, on seeing her settle, depends on the particular approach angle subtended by the target female at the time. If the flower suddenly moves in the breeze, too bad: the male's movement is ballistic. Like a bullet, its trajectory can't be adjusted in mid-flight. The reason is that, unlike the guided missiles developed by the early cyberneticians (see Chapter 4.vii), the fly has no internal mechanism enabling its mating path to be influenced by feedback from the target.

The male's tracking of the female while she's in flight does vary as her position changes. But it doesn't show the flexible selection and variation of pathways often seen in humans and other mammals. Instead, the male fly's tracking flight path is determined by a small set of simple and inflexible rules, hardwired into the insect's nervous system. These rules transform a specific visual signal into a specific muscular response.

One doesn't normally think of anatomical sex differences as including the eyes. But in hoverflies (more specifically, in *Syritta pipiens*), they do. The male has a specialized frontally directed high-resolution set of ommatidia (eyelets) in its compound eye that is functionally a fovea, like that in human eyes. The female doesn't. As she flies around, her eyes aren't good enough to tell her that, just beyond her visual range, lurks the male who is tracking her. In other words, whereas he's capable of a primitive form of visual *attention*, she isn't.

For the male to do the tracking, he needs to keep her foveated—i.e. in the centre of his visual field. Just like humans, the male *Syritta* uses a ballistic eye movement (a saccade) to foveate a target that appears in its peripheral vision. Once the target is in the fovea, it's kept there by means of smooth tracking movements (more precisely, by a succession of tiny ballistic movements, which are damped by the resistance of the air). The male regulates the distance to the female by keeping the vertical height of her image on his eye within a set range. He has feature-detectors in his retina that respond in an appropriate fashion: the closer the target image is to the right height, the more actively they fire.

The "right" height, of course, refers implicitly to the size of a female hoverfly. Just as a frog will snap at a moving object only if its radius of curvature lies within certain limits (14.iv.a), so a male hoverfly will accept as a target only something that happens to be much the same size as a female. But this is all that's needed to induce tracking: if an appropriately sized speck of ash happens to be floating in the air, it may be tracked instead.

These facts were discovered in the mid-1970s by biologists Michael Land and Thomas Collett at the University of Sussex (Land 1975; Collett and Land 1975; Collett 1980). They were later turned from NE into CNE by David Cliff (1966–), one of the Sussex 'A-Lifers' responsible for the triangle-detecting robot (Cliff 1991*a*).

Given the question raised in Chapter 1.i.a, whether the study of insects can teach us anything about human minds, it's worth noting that both Land (1975) and Cliff saw the hoverfly as a simple analogue of animate vision in humans. In "animate" vision, important perceptual information is provided by the creature's own movements (Ballard 1989, 1991). It's worth noting, too (with respect to the attacks on GOFAI described in Chapter 13.iii.b), that it was Cliff who suggested reading "AI" as "artificial insects".

Much as a human's gaze, or visual attention, is captured by certain visual stimuli (a phenomenon discussed in Pitts and McCulloch's 'Universals' paper: see Chapter 12.i.c), so is the hoverfly's. In both cases, behaviour—including eye movements—is influenced accordingly. But once the hoverfly's eyes have centred, its action selection is more rigidly constrained.

Because its eyes are fixed in its head and it has no neck movements, there's no distinction between direction of gaze and body orientation. Moreover, although it can fly in any direction—including sideways, backwards, and stationary hovering—its flight direction is generated by only three components. On top of all this its eyes are less complex, so more easily simulated, than human eyes. Each has about 30,000 photoreceptors, as against 127 million in the human case. In short, 'realistic' simulation of the hoverfly's visuo-motor system was a reasonable goal.

Cliff designed a neural network named *Syritta computatrix*, or SYCO, in homage to Grey Walter and Arbib—and Alfred Hitchcock. This modelled how the hoverfly uses image sampling to lock the most important stimulus onto the area of greatest visual acuity in its eyes; how it tracks that stimulus in motion; and how the visual information is used to determine its flight.

Things weren't made unnaturally easy for SYCO: varying air turbulence and cross-winds were simulated too. As Cliff pointed out, this form of computer vision was very different from that of GOFAI, which typically used toy worlds, static receptors with uniform resolution, and static images (see 10.iv.b).

To simulate the rigid flight control described above, Cliff relied not on exhaustive hand design but on training. SYCO used backprop (see 12.vi.c) to learn to match specific retinal images with specific flight directions. The intention was to compare the 'neurology' of the successful network (once it emerged) with the neurology of the real fly, and to choose between alternative explanations—or generate new ones—accordingly. Cliff constantly evaluated and improved his CNE model by comparing it with the NE data, and managed to answer some previously posed theoretical questions.

He pointed out that "Experiments impossible in living tissue or impracticable in robotics can routinely be performed on SYCO" (1991*a*: 95). But SYCO was just one system. So Cliff broadened his argument in a "provisional manifesto" for CNE in general (Cliff 1991*b*). He explicitly endorsed Humberto Maturana's whole-animal approach (see Section viii.b), including his argument that "the nervous system should not be treated as an input–output device" (1991*b*: 32). Unlike Maturana, however, Cliff was happy to describe his own project as a computational one:

Computational neuroethology replaces the *in vacuo* approach of connectionism with a (simulated) *in vivo* approach; and in doing so, the semantics of the model are automatically grounded.

So, computational neuroethology can be provisionally defined as the study of neuroethology using the techniques of computational neuroscience. (1991*b*: 34)

(He was happy to use the term *computational* because he thought of it then as fairly "neutral"—as it is in this book. Today, because of the later developments in A-Life, he'd prefer *synthetic*: personal communication.)

Although Cliff here described this definition as "provisional", he'd been using the term CNE in unpublished work since 1988 (1991*b*: 34). It first left the press in 1990, when Randall Beer used it in the subtitle of his book *Intelligence as Adaptive Behavior: An Experiment in Computational Neuroethology*. By the mid-1990s, the new coinage had been widely adopted.

Besides vaunting CNE as a way of using A-Life to do neuroscience, Cliff drew a moral for other areas of cognitive science. He criticized the biologically unrealistic assumptions of both GOFAI and connectionism. And he ended with this anti-anthropocentric flourish:

All current connectionist models of language are wildly premature. Language will be best understood as a very high layer in a subsumption architecture: how it interacts with lower layers could be of vital importance, and we should study these lower layers first.

If we are still yet to discover the subsymbolic or neural basis underlying the dance-language of bees, how then are we supposed to study such aspects of human language at anything but the most gross level of neuroanatomy (i.e. studies of lesioned patients)? We simply do not know enough. (1991*b*: 37–8)

It didn't follow that the studies of language described in Chapter 7.ii and 9 were a waste of time, for they hadn't addressed the "neural basis" of language. But connectionist work on past-tense learning, for instance, was seen by Cliff as fundamentally wrong-headed (Chapter 12.vi.e).

As for Arbib's talk of linguistic schemas (Chapter 14.vii.c), this (by implication) was possibly on the right track—but highly untimely nevertheless. The message was that to understand language, we must first understand hoverflies.

c. Playing cricket

Cliff had defended *Syritta computatrix* partly by reference to the "impracticalities" of robotics (see above). Nevertheless, some influential CNE research relied on this methodology.

One example was Barbara Webb's (1965–) work on crickets, begun in Edinburgh's AI Department and continued under Psychology's flag at Stirling (and now back in Edinburgh again, in the School of Informatics). Like Cliff, she was offering not only a contribution to neuroscience but also a critique of symbolic and connectionist AI. Her theoretical focus was on "behaviour and physics" rather than "representation" or "general abilities" (1994: 52–3; 1996; 2001*a,b*).

Francis Crick pointed out that evolution employs "slick tricks" when it can (see Chapter 14.ii.a). Turning from Crick to crickets, it turns out that these creatures use an especially slick trick to manage their courting behaviour. The female finds the male by locating the source of his song (produced by drawing one wing across the other). *Good! So she can hear!* Well, yes and no.

A human Juliet hearing her Romeo can first recognize his song, distinguishing it from all the other sound patterns she can identify, and then decide in which direction she should go to meet him—or to avoid him, if they've had a tiff. She doesn't have to go straight there: if her ardour isn't overwhelming, she can run a few errands on the way. (The female cricket may not respond if she's already mated, or if there are predators around—but if she goes, she goes.) Moreover, if Romeo sings half an octave lower than he did yesterday, or half as fast, Juliet can still find him. She can even learn to appreciate a new song, should he decide to sing one.

Not so, the cricket. As remarked in Chapter 14.iv.d, insects can't learn new patterns. Even more to the point, the female cricket can recognize only one song, sung at only one speed and frequency—give or take . . . (cf. Webb and Scutt 2000: 247 ff.). What she takes as the "speed" of the song is the rate of repetition of the syllabic patterns within it—which varies for different species of cricket. When she hears her conspecific's song, she can do only one thing: move towards it. Unlike the hoverfly, however, she can walk *or* fly to meet her mate—and if she walks, she can do so more or less directly.

Her ears respond to many different sounds, but she doesn't rely on that fact to recognize the male's song. She doesn't need, and doesn't possess, a brain stuffed full of auditory-feature-detectors for coding a wide range of sounds. Instead, she has a custom-built mechanism—a single 'feature-detector'—sensitive to only one frequency. (She also has a simple mechanism for hearing and fleeing bat ultrasound: Webb, personal communication.)

Her automatic mate-finder isn't a *neural* feature-detector, like those described in Section iv. Instead, it's implemented as a fixed-length tube in her thorax (the trachea), which is connected to the ears on her front legs and to her spiracles.

When the male cricket sings, the sound waves travelling through the air in the tube interact with those in the air outside, resulting in phase shifts that produce vibrations of different intensity at each eardrum. But the length of the tubes in the female's body is an exact proportion of the physical wavelength of the sound emitted by the male. So the *physics* ensures that phase cancellation happens only for sounds having the frequency of the male cricket's chirp. Similarly, the *physics* ensures that the intensity difference depends wholly on the direction of the sound source. The female doesn't have to work that out, nor decide what to do about it: she's neurally hardwired to move in the direction concerned. (Notice that there's no need to assume she's 'conscious' of any sound.)

So much was suspected by the early 1980s. The cricket's sound-directed movement (phonotaxis) and its neural underpinnings were already "one of the insect systems most thoroughly studied in neuroethology" (Webb 1994: 45). Researchers knew, for instance, that there are about fifty auditory receptors (though they weren't yet individually identified), and several pairs of auditory interneurones in the pro-thoracic ganglion (Webb and Scutt 2000: 248). In 1985 the behaviour and neuro-anatomy were described for the public, in the *Scientific American* (Huber and Thorson 1985).

All very intriguing . . . But despite the congratulatory tone of the *Scientific American* article, the underlying mechanism still wasn't understood. Various sound cues had been suggested, but how they might be discriminated by the cricket was unclear. Equally, whether—and how—they interacted in the nervous system wasn't known. Another mystery was how the cricket's hearing could be integrated with any other modality, such as vision (this isn't too surprising, since most of the phonotaxis experiments were done in the dark).

It was generally assumed that the *recognition* of the song happened in the brain, and that the neural system representing the song was distinct from the *localization* system. Nevertheless, this wasn't certain. So a neurological paper published in the *Journal of Comparative Physiology* of 1988 admitted:

It is still entirely unclear whether or how [the direction and the characteristic sound] of the calling song are processed independently of one another in the brain, or how the brain triggers and controls phonotactic walking. (quoted in Webb 1994: 45)

In the mid-1990s, the same journal published a paper (modelling the role of the pro-thoracic ganglion) which simply took recognition for granted, and which used spiking rates to predict the direction of movement without specifying how this correlation might be implemented (Webb and Scutt 2000: 248; Webb 2001c). In short, a great deal of neuro-ethological hand-waving was going on.

In the early 1990s, Webb decided to address these questions by devising "the simplest possible robotic mechanism that could support the observed behaviour of the cricket" (Webb 1994: 46). That is, she turned the NE into CNE—and she expressed the "C" as physical robots, not (like Cliff) as computer simulations.

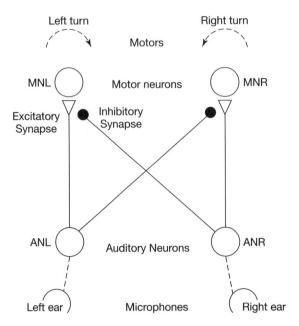

FIG. 15.5. The four-neurone system for phonotaxis. Auditory neurones (AN), modelling the AN1 pair in the cricket pro-thoracic ganglion, receive auditory input. They have a high recovery level, i.e. continue producing spikes over a raised membrane potential while the stimulus continues, and a relatively slow time constant. Each connects to a motor neurone (MN), which will be excited at the onset of firing, and to the opposite AN–MN synapse, which will be inhibited at the onset of firing. Thus, repeated onsets of one AN before the other are required to drive the respective MN above threshold, which causes a turn. Redrawn with permission from Webb and Scutt (2000: 251)

As for seeking the simplest possible mechanism, she was closely guided by the neuro-anatomy but didn't try to model every neurone that had been observed. Eventually, she got down to as few as four (see Figure 15.5).

Unlike the artificial cockroaches described below, Webb's early 1990s robots looked nothing like insects. They had wheels, not legs, and were built on a Lego base or a commercial Khepera robot. After all, the theoretical questions didn't concern the movements of the cricket's legs, but the sensori-motor circuitry making her move towards the song.

Like Arbib's *Rana computatrix* (see 14.vii), Webb's machines grew closer to the biological data as the series progressed. By the turn of the millennium, her robots modelled the neural processes of summation, firing latency, thresholding, recovery, decay, and comparison. Changes in membrane potential were simulated by a small set of state rules, defined in terms of above/below resting, above/below threshold, and above/below recovery (Webb and Scutt 2000: 250).

The neurones were connected to the motors on either side in such a way that recognition of the ideal signal would cause the robot to turn towards it (compare Braitenberg's Vehicles). And Webb tracked the activity levels of the individual neurones, as well as observing the whole robot's behaviour.

Most of these robots didn't move in a straight line—and nor do female crickets. The machines, like the insects, zigzagged within 60 degrees of the optimal direction, and weren't put off by obstacles unless the sound source was very close to the starting point. Later, Webb's group showed (with the help of specially designed VLSI chips from Caltech) that combining hearing with vision enabled the creature to move to the target more directly (Webb and Harrison 2000).

The internal parameters were systematically varied, to test the neural hypotheses involved. Significantly, it turned out that the robot's success was generated by a decision mechanism much simpler than had previously been thought necessary. That is, CNE had provided—and tested—a more elegant theory than straight NE. A-Life had advanced our knowledge of real life.

Webb's work showed conclusively (for example):

* that the various neurological theories which had assumed that there *must be* distinct systems for recognition and localization were mistaken;
* that this was so *even though* the female is able to choose the more attractive of two simultaneous songs (a fact that had previously been thought to support the distinct-systems hypothesis);
* that the neural firing rate was *not* (as had often been assumed) the crucial feature of the signal;
* and that one and the same mechanism could both compare latencies and identify the temporal patterning of the song.

In short, she had clarified many neuroscientific questions, answered some, and raised some more. Answering the unanswered questions, she said, would require work both in robotics and in 'wet' neuro-ethology—and, ideally, close cooperation between them.

To critics who sneered at the use of wheeled robots to model graceful crickets, Webb had an effective reply—a modern version of Vaucanson's remarks about what he'd learnt by building his flute-player (Chapter 2.iv.b). In general, she said, the practical weaknesses of robotic CNE could be seen as theoretical strengths:

The precision of sensing and motor control is generally much worse in the robot than would be expected for the cricket, so the success of the mechanism is unlikely to be due to abilities that the cricket could not really have. The robot constitutes a *subset* of the capabilities of the cricket rather than an *abstraction* of them. (Webb 1994: 51)

By contrast, purely programmed simulations could be too powerful to be biologically convincing:

Computer modelling of *realistic* physics of phonotaxis in a moving animal or robot is quite difficult. "Ideal" sound propagation can be described by fairly simple equations, but in any real situation, with a directed speaker, a floor surface, reflecting walls and so on, it becomes extremely difficult to calculate with any accuracy . . .

Using a computer model rather than a robot would, in this case, substantially weaken the justification for extending the results of testing the hypothesis to the cricket. For example, it would be hard to claim that, insofar as sensors and motors differ, those in the model are substantially *worse* in accuracy than in the cricket: it would be more likely that the mechanism only works because of the idealised conditions. (p. 53)

In some studies, Webb's group worked backwards from robot to insect, instead of from insect to robot (Lund *et al.* 1997). In other words, they asked whether, and how, sensori-motor robots could be made to respond to the songs of *real* crickets.

In the earlier experiments, the "ideal" sound signal had been partly determined by parameters such as robot size, processing time in the circuitry and software, movement speed, and turning inertia. As a result, the robot had sought a signal very different from that preferred by biological crickets. For instance, the songs it responded to were ten to twenty times slower than those of the real animals, and the area that could be experimentally explored was only a fifth of the size (relative to the size of the creature) typically used in NE experiments on crickets.

By the mid-1990s, technological progress in robotics offered Webb speedier processing, and a more fast-moving creature. As well as confirming that her earlier neural hypotheses could indeed explain response to real cricket songs, these new experiments enabled her to test ideas about *why* crickets, given their sensory anatomy, prefer certain frequencies rather than others.

In effect, then, Webb was now in a position to say not only "The cricket behaves as it does because its neurology is what it is", but also "If the cricket's neurology were like *this*, she'd behave *thus and so*." Indeed, she could even have said "No cricket could have a neurology like *that*, because if she did she'd never find a mate." It's no surprise, then, that Webb's work was sometimes published in A-Life journals. For A-Life studies not only "life as we know it" but also "life as it could be".

Early in the new century, Webb's team reported a model of phonotaxis that included a much larger number of the known neural properties, and a wider behavioural repertoire (Reeve and Webb 2002). Drawing on biophysical theories of computation developed by Christof Koch (1999), they included (for instance) a variable exponential decay of membrane potential; a biologically plausible 'weight' determining the standard change of conductance in particular ion channels; and short-term mechanisms for facilitating or depressing this change. Whereas the earlier models had focused on one pair of auditory neurones (marked AN in Figure 15.5), they now included several other types. The speed of the cricket's movement was varied, as well as its direction, and these were integrated with both visual and auditory inputs. For example, in brighter light the robot was less likely to stop when it was moving towards the sound.

Webb's new robots included not only wheeled systems, but also six-legged creatures similar to those discussed below. Significantly, these insect-like robots can move across natural terrain, such as summertime and frost-covered grass, not just the unnaturally flat laboratory floor (Horchler *et al.* 2004; Reeve *et al.* 2005). As well as responding to real cricket songs, they rely on biologically based neural simulators (and specially crafted VLSI chips) which are even closer to the actual neurophysiology than in Webb's earlier experiments. (Her team hope to make them closer still, by feeding the spiking patterns from real cricket neurones into the model.) This example illustrates how far "neuromorphic engineering" had come since Carver Mead's pioneering VLSI retina (see 12.v.f).

We shouldn't look down our noses too hard at the cricket. For humans, too, sometimes rely on inbuilt anatomical tricks, and on 'quick-and-dirty' responses to limited aspects of the sensory information.

Indeed, one set of human responses probably involves another fixed-length physical mechanism for sound location. In cats, the time difference between the moments at which the signal arrives in the two ears is coded by the connection lengths of specific neural circuits in the brain, and this tiny (microsecond) delay is used as a cue in locating the source (Smith *et al.* 1993). The same is very likely true of people. However, the mammalian system—because it depends on a range of time differences rather than a specific phase interaction—isn't limited to a particular sound frequency, as the cricket is; and mammals have a series of delay lines of different lengths rather than just one fixed delay.

A closer analogy, perhaps, is the strong tendency people show to turn to face a sound source (Webb, personal communication). They don't need to recognize the sound, or to locate it accurately. They simply keep turning until the two ears have the same input. Whether or not this is done by dedicated neural circuitry, the point is that the information in the sensory signals doesn't have to be 'explicitly' decoded to enable the response.

So, much as William Harvey's studies of slugs, newts, and eels helped him understand the human blood system (Chapter 2.ii.a), CNE studies of crickets can help us think about our own psychology. But whereas blood circulation in newts is very similar to ours, although slower, perception and action selection in insects is profoundly different. Humans (Juliet, for instance) can do much, much more.

As for stick insects and cockroaches, their neuro-anatomy was simulated in six-legged robots modelled on NE findings about real beasts. The theoretical focus, here, was on walking *as such*.

These hexapod robots, using their multi-segmented legs, could clamber over obstacles rather than having to avoid them, and could even climb stairs. Some could recoil from an obstacle. They'd walk backwards steadily for a while before turning, instead of repeatedly colliding with the obstacle while edging along its surface as Grey Walter's *Machina speculatrix*, and the simplest Vehicles too, had done. (Beer sometimes referred to his cockroaches as *Periplaneta computatrix*, in homage to Grey Walter.)

Such results weren't trivial. The choice of the next movement—*which* leg should be moved, and *how*?—had to be just right. What was the correct placement, force, timing, and sequence? And how should the legs interact? They had to be largely independent, because there might well be a pebble near the 'foot' of the second leg and none near the sixth. But there had to be some degree of integration. If the second leg were lifted higher than usual, the balance of the insect's body would be affected—so the other legs would have to compensate. Moreover, each leg was jointed: how were the several segments to be coordinated?

Some of the most influential work of this type was done in the 1980s by Beer (1961–) and colleagues, including mechanical engineers (directed by Roger Quinn), at Case Western Reserve University, Cleveland. Initially, they were inspired by Braitenberg's Vehicles, by Rodney Brooks's situated robotics (viii.a, below), and by Phil Agre and David Chapman's Pengi (13.iii.b). (Later, they'd be enthused also by ideas about dynamical systems: see xi.b below.) Beer's group designed movement controllers for artificial cockroaches (R. D. Beer 1990; Beer and Chiel 1991). In the late 1980s, they began to experiment with GAs to *evolve* the controllers, rather than designing them by hand (Beer, personal communication).

Most of their robots were physical machines. Others, especially when they got into evolutionary robotics, were abstract simulations, functioning within a virtual environment. In both cases, their "neuro-anatomy" initially reflected NE data from several insect species, not just cockroaches. But they later studied the behaviour of the death-head cockroach in particular detail, and used their robot results to suggest further NE hypotheses (Quinn and Ritzmann 1998).

Beer's robots modelled the control of walking, on uneven terrain and with occasional obstacles. They had six pacemaker neurones, all of which automatically produced rhythmic bursts of activity, and each of which controlled the swing and stance of a single leg. Kinaesthetic feedback from leg-position sensors and local leg reflexes was included, and some models also had two sensory "antennae". In addition, the leg circuits were coupled together. The result was a family of robots with a range of physically stable gaits very like those seen in real cockroaches.

In the simulated creatures, walking was done with a purpose (Beer and Chiel 1991). That is, it was triggered and guided by several internal drives, or motives: eating, seeking food, edge following, and wandering. Some of these were mutually incompatible. After all, if a human nursemaid has only two hands (see Chapter 7.i.f), a cockroach has only six legs. Since it can't be everywhere at once, it must choose what to try to do next (as well as how to do it). In short, Beer's creatures faced the same fundamental problem as the Mars robot described in Chapter 14.v.a, but had a more advanced neural architecture with which to do so.

d. Evolving lampreys

The nervous systems of Beer's early cockroaches were hand-designed—as was the artificial stick insect, some of whose features Beer incorporated in his own work (Cruse *et al.* 1995). But some of his later creations were evolved, using GAs (R. D. Beer and Gallagher 1992). Starting with largely random networks—and therefore with paralysed cockroaches—he ended with happily perambulating creatures.

Cliff and his colleagues, as we've seen (Section vi.c), also did pioneering work in evolutionary robotics. Myriad 'neural' circuits were brought into existence at each generation. But only those which—like the artificial orientation detector—contributed to the robot's "fitness" (in this case, its navigational ability) had a high probability of surviving into the next.

Intriguing results such as these inspired some professional neuroscientists to start using GAs to study real animals. By the late 1990s, David Willshaw (some of whose earlier work was described in Chapters 12.v.c and 14.ix.a) was doing evolutionary CNE—though not evolutionary robotics.

With two AI researchers from Edinburgh, he compared the efficiency of artificially evolved neural networks with real ones, namely those used for controlling swimming in lampreys (Ijspeert *et al.* 1997). (These eel-like fish have a much earlier claim to fame: King Henry I of England is said to have died from "a surfeit of lampreys" in 1135.)

A lamprey doesn't have paired fins, but swims by rhythmic undulations of its long body. By 1990, biological NE had discovered a great deal about the neural circuitry, transmitters, and membrane properties involved. What it couldn't discover was what alternatives, if any, are biologically possible. What non-existent, but viable, lampreys might biological evolution have generated, had the genetic mutations been different? Even straightforward CNE simulation of the lamprey's neuromuscular system, which had been achieved in the early 1990s, couldn't answer that question.

Willshaw and his AI colleagues now used GAs to show that there are many network architectures capable of controlling this type of swimming. Moreover, they found that some of their artificial networks, *although* composed of 'neurones' closely based on the biological data, were much more efficient than those found in real lampreys. Some had a frequency range five times larger. Even when the connections (and their type: excitatory or inhibitory) were fixed to be identical with those in real lampreys, some networks evolved—within only 100 generations—with frequency ranges three times larger.

This was interesting, in view of Marr's earlier claim that biological evolution can be trusted to find the *optimal* solution for a given computational task (see Chapter 7.iv.g). It was also interesting as yet another description of "life as it could be". Evidently, CNE could provide both science and biologically plausible science fiction.

15.viii. Parallel Developments

Computer-based research on self-organization accumulated slowly. Most of it, including Holland's and the CNE examples described above, could be properly developed only when vast computational power became available in the 1980s.

The germinal ideas were growing, nevertheless. Some focused on the *results* of self-organization, some on the self-organizing *process*. In general, however, this research led to new ways of thinking about life and mind. Or rather, it used new theories and data to go back to pre-GOFAI days, building on Turing's and von Neumann's 'biological' writings and even revivifying ideas from the late eighteenth century.

These 'new' ideas were arising both within and outside cognitive science. Various people were engaged in fundamentally similar enterprises, often without realizing it. The potentially unifying themes were self-organization (emergent order), complex dynamics, distributed control, the close coupling of perception and action (and 'whole-animal' approaches in general), and autonomous systems embedded in their environment. But the unification still hadn't happened.

By 1986–7, when A-Life was first identified as a unitary field (see Section ix.a), relevant research lay scattered across mathematics, physics, and biology—and almost every discipline of cognitive science.

* In AI, for example, evolutionary modelling was studying the emergence of novel systems of increasing order (see Section vi, above), and work on agents was exploring the nature of distributed cognition (see Chapter 13.d).
* Researchers in AI and computational psychology, partly inspired by PDP connectionism (12.vi–viii), were focusing on self-organization in distributed systems (Rumelhart, McClelland, *et al.* 1986; Forrest 1991).
* Much of this work was drawing on the growing body of research in dynamical systems (see below, and Section xi).
* Neuroscientists, too, were looking at the origin and the functioning of integrated self-organized systems (Chapter 14.vi–ix, and Section vii above).

* And a few philosophers and/or biologists-turned-philosophers were exploring theories of life and mind very different in spirit from orthodox empiricism (Chapters 14.ix.b and 16.x.c, and subsection b below).
* Only theoretical linguistics, preoccupied by the "Linguistic Wars" (Chapter 9.ix.a), was immune.

This section, and the next, illustrates both the diversity and the commonality of this widespread work.

a. Artificial ants

Emergent order was being studied by a new approach in robotics, in which Herbert Simon's ant (Chapter 7.iv.i) was at last coming into its own.

As described by Simon twenty years earlier, the ant's apparently complex behaviour resulted from local interactions with its environment. Similarly, the newly built "situated" robots possessed a small number of simple reflexes, movements triggered by specific environmental cues (R. A. Brooks 1986; Beer and Gallagher 1992). These weren't movements of the entire creature, but of a single part, such as a leg. Distinct body parts would each have their own inbuilt reflex mechanism, triggered independently of the others. But these separate reflexes functioned in parallel, so that the creature as a whole 'walked' successfully.

The integrated behaviour of these biologically inspired robots, like Beer's artificial cockroaches (see Section vii.c, above), was organized bottom-up. At the lowest level of reflexes, there were no interconnections. (In the language of NK networks, N was small and K was zero: see Section ix.b, below.) But several levels of reflex mechanism might be provided, as in Brooks's "subsumption" architecture. The extra levels came from addition rather than reconstruction: the robot remained effective ('viable') at every stage.

Subsumption architectures were very different from GOFAI, and Brooks's insect-like robots not at all like SHAKEY (Chapters 10.iii.c and 13.iii.b). Earl Sacerdoti's (1974, 1975a,b) ingenious contingency planning and last-minute adjustments were given no place here. There was some flexibility, to be sure. Higher-level reflexes could inhibit lower-level ones in certain circumstances, given certain environmental cues. Apparently purposive behaviour could even emerge as a result (Mataric 1991; Maes 1991a,b). But there was no top-down behavioural control, still less any detailed modification of the execution of lower-level modules.

In short, these systems were very different from GOFAI robots. Arguably, they supported explanations of adaptive/purposive behaviour, even in animals (including humans), very different from what orthodox functionalism had recommended (Hendriks-Jansen 1996: 135–53).

Situated robots, and similarly conceived software "agents" (Riecken 1994), were proudly described by their builders as "autonomous" (e.g. Maes 1991c). What this actually meant was that their behaviour resulted from their own inner structure, not from any externally imposed program or plan. But there was a risk (dare one say "a hope"?) that people would read much more into it. This sort of autonomy is a far cry from human autonomy, or freedom—which, counter-intuitive though this may seem, is better modelled by GOFAI than by A-Life (Boden 1995a; see also Chapter 7.i.g).

These A-Life efforts, and their overly anthropomorphic interpretation, had been anticipated by Braitenberg. As we saw in Section vii.a, he'd discussed designs for "Vehicles" that were, in effect, unimplemented designs for situated robots. They'd also been anticipated by Grey Walter, though in a less neurologically explicit, and less systematic, fashion (see Chapter 4.viii.a–b).

As befitted the director of an institute of biological cybernetics, Braitenberg had acknowledged the influence of Wiener (Braitenberg 1984: 2). But his brief book didn't mention Grey Walter, whose tortoises were the first situated robots, and which generated approach–avoidance behaviour comparable to Braitenberg's LOVE and COWARDICE. Even Brooks, who did salute Grey Walter's work, criticized him for "the lack of mechanisms for abstractly describing behaviour at a level below the complete behaviour, so that an implementation could reflect those simpler components" (Brooks 1991*b*, sect. 3.3).

This criticism wasn't entirely merited: in a manuscript of 1961, Grey Walter had analysed his tortoises' behaviour in precisely this way (see Chapter 4.viii.b). But the piece had remained unpublished. It wasn't until the end of the century that Grey Walter was recognized in the A-Life conference Proceedings as "the pioneer of real artificial life" (O. Holland 1997). And it wasn't until the millennium had passed that an international A-Life meeting, on biologically based robotics, was explicitly dedicated to his memory (Damper and Cliff 2002).

Situated robots were compared by Brooks himself to Jacob von Uexküll's animal species, each with its own life-world (Chapter 5.ii.c). And—despite Brooks's disclaimer (1991*a*, sect. 7.3)—they might have been likened to Simon's Production Systems of the early 1970s (see 7.iv.b). For these, too, had relied on automatic responses to predefined cues. But Simon had focused on human reasoning, and had posited a rich internal environment of symbolic representations.

Simon had also stressed the role of the external environment, such as pencil marks on paper. But this aspect of his work was ignored by the situated roboticists of the 1980s. It was taken up (and much extended) later by philosophers and psychologists interested in situated cognition and embodiment, who recognized the importance of cultural artefacts for prompting human behaviour (see Chapters 8.iii.a and 16.vii.d).

The A-Life roboticists were inspired not by human reasoning or language, but by the world-driven behaviour of relatively simple animals. Many of their robots looked something like insects, having long bodies and six or eight jointed legs. A few, including those developed by Webb and Beer, even modelled insect neurology in some detail. Accordingly, some wags repeated Cliff's remark that "AI" now stood for "artificial insects".

The point of this remark, besides its power to raise a laugh, was tacitly to classify A-Life as a form of AI. In my view, that's correct: just as connectionism is a form of AI, so is A-Life. However, the jibe wasn't welcome in the A-Life community, most of whom went to great lengths to distance their work from "AI".

In part, this new AI methodology was driven by failure. It was trying to avoid the notorious failure of classical AI to solve the frame problem (see Chapters 10.iii.e and 13.i). But in part, it—and evolutionary robotics, too—was grounded in a different conception of intelligence, one which took seriously the fact that human and animal minds evolved in real-world environments.

This approach (outlined in 13.iii.b) insisted that advanced behavioural abilities emerge from simple ones, and that the path from simple to complex has no viability gaps (though it needn't appear continuous: see the comment on punctuated evolution in Section vi.b). The simple abilities are closely coupled with environmental conditions, so the frame problem can't arise.

Moreover, the view was that even advanced mental abilities function in similar sorts of ways, so that the concepts of computation and representation are irrelevant for explaining cognition. The leading evolutionary roboticists described their work in terms of dynamical systems (see subsections c–d, below, and Section xi), explicitly rejecting the types of explanation familiar in classical AI and (most) connectionism (Beer and Gallagher 1992; Husbands *et al.* 1995). Similarly, Brooks saw situated robotics as offering "intelligence without representation" (Brooks 1987/1991*a*) and also "without reason" (1991*b*).

Brooks wasn't speaking, here, only of insect intelligence. Despite some (ambiguous) backtracking with respect to linguistic and logical thinking, he was soon to initiate a long-term robotics project modelling the early development of human cognition—and, with Daniel Dennett's cooperation, even consciousness (Brooks and Stein 1994; Dennett 1994*a*). Dennett (1978*c*) had suggested many years before that AI researchers should try to build a *whole* cognitive system: now, here was his chance.

The Cog robot, which has already given rise to a commercially marketed baby doll, combined inbuilt reflexes with connectionist learning. It was an early exercise in what's now called "epigenetic robotics", in which a robot's control system and behaviour develop differently according to the particularities of its environment.

So Cog incorporated some of the findings in developmental psychology, linguistics, and cognitive neuroscience described in previous chapters. For example, psychologists had shown in the mid-1970s that joint attention develops by mutual gaze, gaze following, and pointing (6.ii.c). Brooks's team used these features to model the development of joint attention—not mother–baby this time, but human–robot—in Cog (Scassellati 1996, 2001).

In this still ongoing project, now directed by Brooks's students Cynthia Breazeal (Ferrell) and Brian Scassellati, the artificial denizens of "The Cog Shop" became ever more "human" (for the most recent reports, see the web site at <http://www.ai.mit.edu/projects/humanoid-robotics-group/cog/cog.html>). This was true not only of their appearance, but also of their behaviour—especially, their "social" interactions. The robot was not only given "needs" (drives, motivations) to induce caretaking responses on the part of the human interacting with it, but also babylike responses that signalled to the carer whether or not those needs had yet been met.

(More recently, the MIT group have worked on a puppy-like furry robot called Leonardo, which—up to a point—can imitate the facial expressions of the human being. Babies can do this too, probably by using the "mirror neurones" discovered in the 1990s: 14.vii.c. And Scassellati, now at Yale, is building Nico, a creature that will recognize and respond to gestures as well as speech. This is no mere trivial toy. Like his medical funders, Scassellati hopes to use Nico to identify, and perhaps even to cure, early childhood autism: see 7.vi.f. Indeed, he hopes eventually to be able to use humanoid robots to *test* theories in developmental psychology, much as computer simulations are so used today.)

Given that description, this work may sound reminiscent of the "social" AI described in Chapter 7.i.c. But its methodology was very different. For Brooks's team avoided top-down control, symbolic computation, planning, and objective world modelling. In short, the GOFAI approach was being turned on its head: *situated action*, including social interaction, was the buzz phrase (see 13.iii.b).

Predictably, there was resistance. That is, there was theoretical resistance (13.iii.c). Behavioural resistance was less predictable. On encountering Cog, and despite its highly *non-human* appearance, people tended to engage with it as though it were human—or anyway, animate. The key feature seemed to be the robot's ability to catch the human's eye and maintain mutual gaze and/or attention.

This animistic response was even stronger with respect to Kismet, a Cog-like creature equipped with floppy pink ears, large blue eyes, and an amazing set of eyelashes (Breazeal and Scassellati 2002). The robot's ears, eyes, and eyelashes could be put into different physical positions, spontaneously interpreted by people as sadness, anger, tiredness, interest, excitement, happiness, or even disgust. With the addition of "needs" and associated feedback signals to its carers, Kismet was able to trigger still more of the emotional responses normally elicited by (and evolved for) vulnerable fellow humans. The carers even lapsed into 'motherese', the syntactically and intonationally distinct dialect that mothers use when speaking to babies and infants (see 9.vii.c).

Even Sherry Turkle—not only a trained psychoanalyst, but a long-time colleague of the MIT AI community (and ex-wife of Papert)—reacted in that way. She didn't need Kismet or Leonardo to make her do so, for the more primitive Cog sufficed:

Cog's mobile torso, neck, and head stand on a pedestal. Trained to track the largest moving object in its field (because this will usually be a human being), Cog "noticed" me soon after I entered its room. Its head turned to follow me and I was embarrassed to note that this made me happy. I found myself competing with another visitor for its attention. At one point, I felt sure that Cog's eyes had "caught" my own. My visit left me shaken—not by anything that Cog was able to accomplish but by my own reaction to "him". For years, whenever I had heard Rodney Brooks speak about his robotic "creatures", I had always been careful to mentally put quotation marks around the word. But now, with Cog, I had found that the quotation marks disappeared. Despite myself, and despite my continuing skepticism about this research project, I had behaved as though in the presence of another being. (Turkle 1995: 266)

Turkle's scepticism was based largely on postmodernist philosophical grounds. The same was true of her MIT colleague, the feminist philosopher of science Evelyn Fox Keller, who queried the "human" responses to Kismet (Fox Keller forthcoming). Others looked askance at Cog for different reasons.

Quite apart from scepticism over the "consciousness" of the humanoid robot envisaged by Brooks, many felt that he'd failed to learn the lesson that both Lashley (1951*a*) and Noam Chomsky (1959*b*) had tried to teach Burrhus Skinner: that S–R reflexology isn't enough. In particular, Brooks's claim that representations are unnecessary for intelligence—and especially for human intelligence—was soon challenged (13.iii.c). The disagreement eventually led to a better understanding of the diversity of mechanisms that can be called "representations" (14.viii).

The proof of the pudding...? Some ten years after its start, Cog is now virtually on ice (for a recent account, see Brooks *et al.* 2000). It was always a 'spare-time' project, and most of the original team—then, graduate students—have left MIT to take jobs elsewhere. (Brooks himself has been swamped by administration since becoming director of the AI Lab.) A comparable project, well funded by Sony, has begun under the direction of Luc Steels (1952–) at the Free University of Brussels. Their pudding may turn out to have a more satisfying taste.

So, too, may the concoction due to Alan Mackworth's group, in their study of robot soccer (Sahota and Mackworth 1994). Unlike some workers in the area (Chapter 11.iii.b), they're guided by a generally important theoretical aim. Namely, they want to explore the need for, and the possibility of, an *integration* of classical and situated robotics. Describing the first as the Hegelian "thesis", and the second as its "antithesis", they offer an architecture for *reactive deliberation*—seemingly, an oxymoron. Their robots integrate sensory perception, real-time decision making, planning, plan recognition, learning, and coordination (11.iii.b).

Less 'sexy', but interesting nonetheless, is the recent robotics work based on insect navigation. To mention just one example (similar research is being done at Sussex), a team that includes Brooks's ex-student Maja Mataric (1965–) is modelling the way-finding strategies of desert ants, so that their robots will be able to return to 'base camp' after going out to explore unfamiliar territory (Roumeliotis *et al.* 2000). This project was inspired by Collett's discovery that desert ants, unlike their more familiar cousins, don't use pheromones to find their way home (Collett *et al.* 1998). Rather, they rely on path integration, backed up by visual landmarks. The landmarks are needed because if integration were used alone, tiny errors would accumulate and eventually lead the creatures astray.

In sum, the Brooksian dogma—though fruitful at its inception—is looking less and less persuasive. Nevertheless, his advice that roboticists should study insects is still being followed, if often interpreted in a broader sense.

b. New philosophies of biology

The 'biological' view of intelligence wasn't restricted to AI, for it had already developed within other areas of cognitive science.

In psychology, James Gibson's (1966) ecological theory was the outstanding example (see Chapters 5.ii.c and 7.v.e). Although it had been regarded, not least by Gibson himself, as radically opposed to AI, it was now beginning to influence AI research. Recent computational work on animate vision, for instance, had assumed that the creature's own movements through its environment provided crucial perceptual information (Ballard 1989, 1991).

Neuroscientists had discovered sensori-motor circuits illustrating the close coupling between animal and environment. And computer models of such systems had been developed since the late 1960s by CNE researchers, as we've seen. Moreover, sophisticated models of self-organization in mammalian brains had been developed by Stephen Grossberg, Christoph von der Malsburg, and Andras Pellionisz and Rodolfo Llinas (see Chapter 14.vi and viii.b–c).

Developmental psychology and cognitive neuroscience had indicated the close epigenetic interplay between inbuilt dispositions and experience (Chapters 12.viii.c–e and 14.viii.c). And some linguists had attributed fundamental aspects of both syntax and semantics to their grounding in our material embodiment (Boden 1981*a*).

Philosophy, too, was involved. Non-Cartesian accounts of mind, stressing embodiment and rejecting the representational theory of perception, had long been available within the neo-Kantian tradition (see Chapter 2.vi). And they'd featured in the philosophy of cognitive science since the 1960s, when Hubert Dreyfus first put forward his critique of GOFAI (11.ii.a and 16.vii.a). He'd drawn heavily on the ideas of the continental phenomenologists Maurice Merleau-Ponty and Martin Heidegger, and of the later Wittgenstein.

Dreyfus's critique had initially been dismissed by most cognitive scientists, whose work lay within the alternative, empiricist, tradition (see 11.ii.a–b). When Terry Winograd was converted by Dreyfus (and Fernando Flores) in the early 1980s, this was widely seen as a betrayal by one of the masters of GOFAI (11.ii.g). In the 1990s, these ideas were stirring again. Heidegger, for instance, was becoming prominent in certain philosophers' thoughts about cognitive science—including some highly sympathetic to situated and evolutionary robotics (van Gelder 1992/1995; M. W. Wheeler 1996, 2005).

For many years, Heidegger had been regarded by analytical philosophers with utter contempt. Hardly any of them bothered to say anything about him—although Anthony Quinton complained about his "ponderous and rubbishy woolgathering" (a judgement he still stands by: personal communication). At the millennium, however, the heat was turned up. Heidegger was called "a dismal windbag, whose influence has been completely disastrous" (Blackburn 2000), and "the greatest catastrophe in the history of philosophy" (Edwards 2004: 9).

Both those remarks were made by highly eminent analytic philosophers, disturbed (especially in Edwards's case) by the fact that by the end of the century, Heideggerian ideas had spread apace. Just why he was now thought by some respectable Anglo-Saxon philosophers, and by some cognitive scientists, to be relevant is discussed in Chapter 16.vii.

Besides downplaying the mind's embeddedness in the physical world, Cartesianism had encouraged analytic methods in biology (see Chapter 2). Not until after the mid-twentieth century was there a wide range of scientific studies providing ideas about self-organization. By the late 1980s, a number of people had developed holistic philosophies of life informed by these scientific advances.

Pattee, a participant at the first A-Life meeting (see Section x.a), was an influential case in point. He'd always drawn on both cybernetics and biology. In the 1960s he'd edited a collection of early papers on computer and mathematical models of life (Pattee *et al.* 1966), and soon simulated an ecosystem himself (see Section vi.b). But his main project was to articulate a philosophy of life.

He described organisms as self-organizing dynamical systems realized in a physical medium (Pattee 1985, 1989). And he was especially concerned to show how stable "symbols"—in particular, genes—could arise within dynamically changing physical systems (2001). He saw this symbolism as the necessary ground for the genotype–phenotype distinction, and indeed for the phenomenon of life itself.

Others with similar views included two of Pattee's students: Robert Rosen (1985, 1991) and Cariani (1992). (Rosen was initially a student of Nicholas Rashevsky, whose Chicago biophysics group gave him a lasting interest in metabolism as the basis of life.) All three rejected the Cartesian view of organisms as "mechanisms"—that is, as analytically decomposable systems. They allowed that computer simulations of the self-organizing dynamics involved can be useful. But they insisted that these must be very different from symbolic AI models. In particular, the emergence of new dynamical wholes, and of new types of function (such as new powers of perception), are problematic for traditional AI.

The emergence of new dynamical wholes was the key concept in an unorthodox philosophy of biology (and cognition) that had been maturing for thirty years. It was developed by the Chilean neuro-anatomist Maturana (1928–) and his student, a Chilean psychologist, Francisco Varela (1946–2001). Although originated by Maturana, the theory—and the A-Life work inspired by it—was eventually associated just as strongly with Varela (Di Paolo 2004; McMullin 2004).

Maturana had been a member of the cybernetics group in the 1950s, and co-authored the hugely influential 1959 paper 'What the Frog's Eye Tells the Frog's Brain'. In the 1960s, he defined both life and cognition in terms of "autopoiesis": the continuous self-production of an autonomous entity (Maturana 1969; see 16.x.c).

The word was coined from the Greek: *autos*, or self, and *poiein*, or creation. But the idea, as Varela acknowledged later (Weber and Varela 2002), had been largely inspired by Kant. Specifically, it was grounded in Kant's vision of organisms as systems in which the parts are formed for and from the others, giving a dynamic whole without which the existence of the parts would be meaningless (see Chapter 2.vi.b).

This 1960s biological approach advanced steadily, though always in the shadow of the mainstream. The 1970s saw an account of autopoiesis appear in the *International Journal of Man–Machine Studies*—hardly a neo-Kantian enterprise (Maturana 1975). And the same decade produced the first computer simulation of *the* fundamental autopoietic event: the spontaneous formation of the cell membrane (Zeleny 1977, 1981; McMullin and Varela 1997). Now, in the mid-1980s, Maturana's ideas had been included in a highly respected series on the philosophy of science (Maturana and Varela 1972/1980). (Soon, they would be applied to cognitive science as such: Varela *et al.* 1991.)

The theory of autopoiesis was fundamentally opposed to non-holistic, logic-al–computational accounts of life and mind. This may seem strange, for the 'Frog's Eye' paper had been prompted by computational ideas, and had led to their wider use in neurophysiology (Chapter 14.iv). But in his mature philosophy, Maturana regarded artefacts of any sort, and especially of the logical–computational variety, as misleading analogies of living things.

Even admittedly convenient terms such as *input, output*, and (ironically) *feature-detector* were discouraged. Talk of *representations*, too, was to be avoided. Instead, one should conceptualize the brain as a dynamical living system, wherein stimuli aren't "inputs" but "perturbations". As his colleague Varela put it:

The LGN [lateral geniculate nucleus] is usually described as a "relay" station to the cortex. However . . . most of what the neurons in the LGN receive comes not from the retina (less than 20%), but from other centers inside the brain. . . . What reaches the brain from the retina is only a gentle perturbation on an ongoing internal activity, which can be modulated . . . but not instructed. This is the key. To understand the neural processes from a nonrepresentationist point of view, it is enough just to notice that whatever perturbation reaches from the medium will be informed according to the internal coherences of the system. (Varela 1987: 59–60)

Maturana wasn't the only neuroscientist to object to informational/computational terminology. Jean-Pierre Changeux, for instance, saw it as a mistake to describe brain processes as being determined by the sensory input:

[Variations] in a physical parameter in the environment are translated into variations in nerve impulses . . . A chain of successive reactions, explicable in strictly physico-chemical terms, *regulates the spontaneous activity of the oscillator, which pre-exists all interaction with the outside world.* The impulses produced are therefore *independent from the physical stimulus* to which the organ is sensitive. The sense organs behave like *regulators* of molecular clocks. The stimulation that they receive from the outside world sets them forward or backward and corrects their timing. *There is no direct analogy, however, between the physical stimulus received from the environment and the nervous signal produced.* (1985: 82–3; italics added)

In other words, Changeux (like Maturana) favoured a dynamical, even 'non-mechanistic', account of the brain, while allowing that brain processes could be described at the physico-chemical level if one wished.

But Maturana went beyond neuroscience, to develop a philosophy of biology in general. He objected to descriptions of the nervous system as "mediating between perception and action". Instead, the living autopoietic unity—cell or organism—was conceptualized as a self-organizing dynamical system, closely coupled with its environment.

So closely coupled, indeed, that they could even be regarded as a single system. For Maturana, the "environment" isn't objectively present independently of the organism. Rather, it's constituted in opposition to and cooperation with the activity of the living thing. What the observer chooses to count as 'one' system is a matter of scientific convenience. In the 1990s, some philosophers of cognitive science would draw on such ideas to smear the boundaries of the 'mind'/'self' in a similar way (Chapter 16.vii.d).

As for the computer modelling of autopoiesis, this too became more prominent in the 1990s, though still a minority taste (McMullin 2004: 283–92). Increasingly, it was pursued in the context of biochemical theories.

For example, Naoki Ono and Takashi Ikegami in Japan developed an approach called LAC, or Lattice Artificial Chemistry. They employed abstract chemical schemas to generate "proto-cells" showing boundedness, self-maintenance, reproduction, and evolution. Initially, they could deal with only one or two dimensions (Ono and Ikegami 1999, 2000, 2001); later, they modelled a 3D lattice (Madina *et al.* 2003).

Their experiments would start from an assortment of randomly moving (abstract) molecules: some hydrophilic, some hydrophobic, and some autocatalytic. Given parameter values within a certain range, structures such as vesicles would emerge, and spontaneously split into two. The authors see this as a study of the origins of life, which must have involved molecules in the primeval soup coming together into "compartments" somehow separated from what was now the first "environment". This sort of self-organization is the crux of autopoietic theory.

c. Dynamical systems

It's not surprising that Maturana's work received a more sympathetic hearing in the late 1980s than it had done in the preceding years. For by that time, more people were avoiding talk of representations and/or studying self-organizing dynamical systems.

For instance, PDP-based connectionism ascribed meaning to holistic patterns of activation, not to individual units (Chapter 12.vi). A few neuroscientists and psychologists were starting to explain brain and behaviour in terms of dynamical attractors (14.ix.b). Some 'neurological' work in CNE (R. D. Beer 1995*a,b*) was interpreted using dynamical systems theory. The evolution of the robot orientation detector (Section vi.c, above) was described by the research team in terms of state space attractors, not input–output functions (P. Husbands *et al.* 1995). And some philosophers were arguing that dynamical, not computational, concepts are needed to explain cognition (van Gelder 1992/1995; Port and van Gelder 1995).

A dynamical system is one whose states change according to definite rules. The rules may be deterministic or probabilistic; linear or non-linear; fixed or alterable; uniform or region-/unit-specific; and time-sensitive or synchronously "clocked". A self-organizing dynamical system is one in which new forms of order spontaneously emerge.

A von Neumann computer is a dynamical system, although it's rarely described as such (Giunti 1997). So is a pendulum, whose system dynamics are relatively simple. In addition, the term is commonly applied to massively parallel systems, where it's easier to identify overall patterns of activity than to keep track of individual units. Often, these patterns are stable or recurrent, in which case they are described in terms of attractors. An attractor is a point, region, or state cycle within the phase space (the set of all possible states) of a dynamical system, such that the system tends to approach it from any randomly chosen starting point and to return to it when perturbed.

Dynamical systems go under various names. They're called circular causal systems, reaction–diffusion systems, neural networks, cellular automata, NK networks, and CTRNs (Continuous-Time Recurrent Networks). And they even include gases.

At the most fundamental level, all these are similar (although the defining rules are different):

* That's why von Neumann's CAs could be used to model the laws of physics (see Section v.a),
* why CTRNs can model any dynamical system (see xi.b, below),
* and why thermodynamics can be used to specify connectionist networks (see 12.v.f and vi.b).
* It's why Ashby could include plants growing, plants ageing, and machines moving—as well as brains living—as examples of dynamical systems (Chapter 4.viii.c).
* And it's why D'Arcy Thompson's and Turing's claims about the formative powers of physics and chemistry count as early studies of the dynamics of development.

By the late 1980s, dynamical systems were being studied by many different disciplines. Some of the authors were biological scientists trying to explain specific phenomena. Such phenomena included the brain's control of (and switching between) different bodily gaits (Thelen 1985; Thelen and Smith 1994), the coupling of organism and environment by way of movement (Kugler and Turvey 1987, 1988), and the evolution of new systems of perception, or "measurement" (Kugler *et al.* 1990; Turvey and Shaw 1995). Others were physicists, mathematicians, or computer scientists seeking abstract descriptions of dynamical systems in general. (Some interesting recent examples are discussed in the following subsection, and in Sections ix.a–b and xi.)

Whether the growing sympathy for dynamical systems was entirely merited is controversial. It's not clear, for example, that theories couched in terms of dynamical systems can explain all the specifics of human thinking. It's one thing to say that the mind/brain is a dynamical system (compare: made of atoms). It's quite another to say that dynamical concepts (compare: atomic theory) suffice to explain everything that it does.

This isn't a question of accepting magic, but of rejecting reductionism. Even chemistry requires concepts not needed in physics; and genetics, neurophysiology, and psychology require yet more (see Chapter 16.iv.c). Nor is it a question of the *philosophical* problem of meaning, or intentionality. Even if one takes for granted that a dynamical system could express meanings and/or behave meaningfully, it doesn't follow that the mathematical concepts of dynamical systems theory suffice to represent structured linguistic propositions or inferences.

In other words, this is a current version of the problem that mid-century cybernetics had with propositional content (see Chapters 4.v.d and ix.b, and 16.vii.c).

15.ix. Order and Complexity

The emergence of order in cellular automata had been investigated for many years before the naming of A-Life, both at Michigan (see Sections v–vi) and elsewhere. A group at MIT's computing laboratory, directed by Ed Fredkin, had been working on CAs since the 1970s, and an ex-Michigan member had recently outlined a dedicated CA machine (Toffoli 1984). But especially interesting work of this type had been done by Stephen Wolfram (1959–) and, in particular, Stuart Kauffman (1939–).

Both Wolfram and Kauffman used CAs to explore complexity—including life—*as such*. If it had any interest at all, then, their work was relevant to a huge range of topics, both within biology and outside it.

a. The four classes of CA

By the late 1980s, the computer scientist Wolfram was extending the study of cellular automata in various ways. He'd noted that even very simple CAs could generate lifelike visual patterns, such as shell markings (see Section v.a), so might be relevant in explaining specific biological phenomena. But he also used CAs as grounds for claiming that a certain level of complexity, involving both order and novelty, is needed for life in general—and for computation, too (Wolfram 1983, 1984*a,b*, 1986).

Taking advantage of the newly available computational power, Wolfram modelled many different CAs, and observed them in action. On the basis of these experiments—experiments, not theorems—he suggested a classification into four classes (a classification anticipated in outline by Turing in 1950: see 16.ii.a).

Wolfram's four classes of CA were: I—those which eventually reach stasis; II—those which settle into rigidly periodic behaviour; III—those which remain forever chaotic; and IV—those which achieve order that's stable without being rigid, so that the structural relations between consecutive states are varied yet intelligible. The final category, he said, included life.

Later, in the early 1990s, Langton would add numbers to this qualitative classification. He did extensive, and hugely tedious, CA modelling to explore the conditions in which each of Wolfram's four classes would in fact emerge. And he came up with an intriguing result.

His experiments led him to claim that life is possible only within a very narrowly restricted range of numerical values on a simple statistical measure of complexity, which he called the "lambda parameter". This type of complexity was located "at the edge of chaos": near the phase transition between Wolfram's Class IV and his Class III (Langton 1990, 1992). As Langton pointed out, this means that there's not only a lower limit to the complexity of living things—as von Neumann had argued in the 1940s—but an upper limit also (Langton 1990: 36–7).

Another person inspired by Wolfram's classification of CAs was Andrew Wuensche (1943–). Independently of Langton, Wuensche defined a different measure of complexity, the Z-parameter (Wuensche and Lesser 1992; Wuensche 1999). This was a natural extension of his remarkable "reverse algorithm", developed in the late 1980s for computing the *predecessors* of CA states—which in effect provided CA systems with "memory" (Wuensche 1997). (You may remember that McCulloch and Pitts had argued that it's in principle impossible to describe neural nets retrospectively: see 4.iii.f.) When the two complexity measures were compared, it turned out that there was a strong relationship between them. But the Z-parameter was the more general, being able to avoid certain exceptions to lambda.

Largely because Wolfram's qualitative classification of CAs had whetted people's appetite for quantitative measures of complexity, both Langton's work and Wuensche's attracted significant attention.

In fact, however, the cyberneticist Ashby had beaten them to it. He and a colleague, Crayton Walker, had defined a broadly comparable measure over twenty years earlier (Walker and Ashby 1966). Their concept of "internal homogeneity" was equivalent to Langton's lambda, for CAs having only two possible states. In the absence of a flourishing CA modelling community, however, this earlier theoretical approach hadn't been much noticed.

b. K for Kauffman

Even before Wolfram and Langton, Kauffman had investigated the nature and limits of complexity with various biological questions in mind. His first papers had appeared in the late 1960s, when he was a young member of the mathematical biology group in Chicago.

Twenty years later, his work was a provocative mix of mathematics, biochemistry, and biology—integrated by a highly unfashionable, though ancient, philosophical framework. For some readers, it was intoxicating. For others, infuriating. Mainstream biologists still rejected the general approach, especially the downplaying of natural selection (see below). They also questioned the fit between Kauffman's mathematics and biological fact.

Kauffman was originally trained in philosophy, physiology, and psychology (his teachers included Stuart Sutherland and McCulloch). He then entered medical school.

His interests at that time (the early 1960s) were in brain function and cell differentiation in the embryo.

The then current research on neural functioning was largely inspired by the 1940s models of McCulloch and Pitts, which represented neurones as Boolean on–off units (see Chapters 4.iii.e and 12.i–ii). As for cell differentiation, this still wasn't understood. Turing's ideas on diffusion reactions (see Section iv.a) remained speculative and non-specific. However, the recent discovery of regulatory genes (Jacob and Monod 1961) suggested the possibility that embryonic development might be directed by specific genes becoming active at successive stages. Since the role of regulatory genes is to 'switch' other genes on or off, Kauffman decided to use Boolean networks to model both kinds of activity.

He contacted McCulloch to discuss these ideas, and was introduced by him to the developmental biologist Goodwin (1931–), with whom he would later co-author several papers. Goodwin was a student of Waddington at the time, and through him Kauffman was invited to speak at Waddington's seminar (Kauffman 1969). There could have been no more suitable a context for using mathematical analysis to illuminate epigenesis.

In these studies of the metabolic base of cell differentiation, Kauffman was investigating the action of what Turing had called morphogens. But his questions, like Turing's, were very general ones. He didn't ask which protein causes a liver cell to develop, and which produces a neurone. Rather, he wanted to know (for example) why the morphogens, whatever they happen to be, generate 264 distinct human cell types—as opposed to 100, 1,000, or even millions. (Just how one can count "cell types" is a thorny question, to which we'll return.)

His (primitive) computer models explored the behaviour of randomly connected Boolean NK networks. Such a network has N units, each with K inputs. Each unit follows some (randomly assigned) rule, which determines its on–off activity according to the (binary) values of the K inputs.

Kauffman found that NK networks often settle into, or cycle between, a restricted number of stable patterns. That is, of the many—perhaps astronomically many—possible states of the system, only a small number can be reliably maintained. Moreover, when $K = 2$ there's a simple ratio between the value of N and the number of attractors: the latter is the square root of the former.

It was then thought that the number of genes in the human genome is about 100,000 (now, it's believed there are only 30,000–60,000). Kauffman pointed out that with $N = 100,000$, there would be 317 types of stability—which, he said, approximates the number of human cell-types. (With $N = 60,000$, the number of attractors would be about 245; with $N = 30,000$, it would be 173.) In other words, it was perhaps no accident (on this view) that Santiago Ramón y Cajal had been able to find more than one type of brain cell, but fewer than 100 (see 2.viii.c).

There were three intriguing implications. First, that there may be something special about networks where $K = 2$—or (more generally) that the numerical values of N and K may be significant in determining network behaviour. Second, that these abstract properties may be instantiated in a wide variety of contexts, wherever order emerges in systems of interconnected units. And third, that the power of natural selection may therefore be limited: the number of cell-types (for instance) may depend as much on the self-organizing properties of the embryo as on the *post hoc* effects of selection. Over the 1970s and 1980s, Kauffman explored all three implications.

As for the significance of K = 2, Kauffman soon proved that very richly interconnected networks, where K approaches N − 1, produce chaos (Wolfram's Class III), not order. More surprisingly, he showed that, in an NK network of this simple type, chaos will result if K is larger than 5. In general, the precise numerical values of N and K are crucial determinants of the overall dynamics of the system.

In one sense, he admitted, both N and K are irrelevant. *All* NK networks will achieve order eventually, because they have a finite number of states. So if any state reoccurs, the preceding trajectory through the phase space must be traversed again . . . and again. But "finite" can mean astronomical, and "eventually" can mean billions of years. So there's a real distinction between NK networks that will achieve order within some reasonable time and those which won't.

What counts as a "reasonable" time depends on the context, and in living organisms will involve many different constraints. But whatever the context, the numerical values of N and K affect the system's properties as a whole. Moreover, numerically tiny differences in N or K may result in huge system differences: either a phase transition into a different type of order (analogous to a new embryonic stage, for example), or a collapse into chaos.

In his later work, Kauffman supported Langton's suggestion that life exists "at the edge of chaos". And he proved, what had long been regarded as intuitively obvious, that with increasing complexity an organized system becomes more resistant to change. As he put it, selection becomes less able to alter the system's fundamental properties (see below).

As for generalizing this approach, Kauffman was nothing if not ambitious, seeing a wide range of phenomena as essentially comparable. He first applied his ideas to autocatalytic networks (in which various chemicals cooperate in generating each other), which he thought of as primitive metabolisms (Kauffman 1969, 1971). Later, he extended them to the functioning, and the evolution, of systems defined at many different levels.

These included enzymes and antibodies; genomes; embryos; neural networks; species; ant colonies; ecosystems—and even market economies and human cultures (Kauffman 1983, 1986a,b, 1989, 1993). For instance, he used NK networks to model fitness landscapes and co-evolution, and to define the circumstances in which genetic variation and natural selection would be most likely to lead to an increase in fitness.

Kauffman's ideas influenced various groups working in A-Life, including some who used NK networks for evolving artificial species (see Section vi.c). But his seminal claims weren't all well founded.

For instance, near the end of the century Harvey showed that Kauffmann's analysis of NK networks is false in the general case. It applies only to synchronously updated random Boolean networks, not to asynchronously updated ones. For these, the picture is so different that "there is no sense in which the behaviour of synchronous [NK networks] gives any clue towards [their] behaviour" (Harvey and Bossomaier 1997: 74).

In other words, Kauffmann's work provides yet another example of a mathematician ignoring real time. Von Neumann's CAs, too, were defined synchronously (see Section v.a). The Turing machine assumed a sequence of atemporal steps, defined in practice by the digital computer's—essentially arbitrary—clock (see Chapter 4.iv). And neural network theorists have—until very recently—typically ignored real-time properties, despite their importance in actual brains. Temporal *order* might be taken into account (by Uttley or Jeffrey Elman, for instance: see Chapter 12.ii.c and viii.b). But precise temporal intervals usually weren't (14.ix.g). Only at the turn of the millennium, for instance, did a connectionist textbook pay attention to them (O'Reilly and Munakata 2000, e.g. sects. 2.5 and 6.7).

One might argue that all these mathematical idealizations are necessary in the initial stages of research, and that their results will at least approximate those of the relevant real systems. Harvey allowed the first of these defences, but not—or not always—the second. He also showed that, contrary to what was widely assumed, asynchronous networks can be both analysed and simulated (pp. 74–5). In short, while his critique cast serious doubt on Kauffman's models of genetic regulation, for instance, it also showed how his ideas could be built on in unexpected ways.

c. Morphology revived

Kauffman's work on evolution had been initially prompted by Maynard Smith, a leading evolutionary biologist. In the early 1980s, Maynard Smith persuaded Kauffman that he should no longer ignore evolution, and suggested that he combine the mathematics of fitness landscapes with his own NK approach (Maynard Smith, personal communication). But the models of evolution that ensued merely confirmed Kauffman in his commitment to a highly unorthodox conclusion—one already held by Waddington and Goodwin.

Maynard Smith and Goodwin were colleagues for many years at the University of Sussex. Their relations were warm, friendly, and highly respectful on both sides. (We were members of a small dining club, so I witnessed them interacting many times.) Despite the mutual respect, however, their philosophical commitments were, and remained, fundamentally different.

Discussing the improbability of a species' settling on a high-but-narrow peak in the fitness landscape, Kauffman declared:

Selection typically cannot reach and hold an adapting population in arbitrarily located or overprecise volumes of parameter space. Thus it is not foolish to suppose that the forms we see are largely those which are easily generated by the underlying developmental mechanisms. (Kauffman 1993: 642)

As for what is "easily generated" in development, he said:

[My work, and Goodwin's, suggests that], contrary to intuition, morphogenesis may be deeply robust. Organisms, rather than being tinkered-together contraptions, may exhibit a nearly inevitable and stable order. Real morphogenesis is due not to the unfolding of any single developmental mechanism but to the beautifully ordered unfolding in time and space of some richly integrated combination of simpler mechanisms such as cell-sorting, sheet-folding, positional discontinuities, and reaction–diffusion mechanisms. . . . [These] *constrain* the morphologies which emerge to a small subset, each of which occupies a large volume of state space and parameter. Rather than causing complexity, integration of developmental mechanisms may generically yield simplicity and order. (Kauffman 1993: 637)

In short, natural selection is important, but not fundamental. If we want to know why certain imaginable morphologies have not appeared, even though they are represented at some point in the fitness landscape (the co-evolutionary phase space), the answer will sometimes be that fundamental developmental constraints prevent it. Like snowballs in hell, they are theoretically conceivable but dynamically impracticable.

This view was stated even more strongly in the concluding passages of Kauffman's *magnum opus*, *The Origins of Order* (1993):

This book is an effort to continue in [D'Arcy] Thompson's tradition with the spirit now animating parts of physics. It seeks origins of order in the generic properties of complex systems. Those properties [I have] discussed have ranged [over many levels of biological self-regulation, including] the origins of spatial integration in multicellular systems when products of individual cells can reach their neighbours . . .

I have tried [to characterize] the interaction of selection and self-organization. To some great extent, evolution is a complex combinatorial optimization process in each of the co-evolving species in a linked ecosystem, where the [fitness] landscape of each actor deforms as the other actors move . . . In sufficiently complex systems, much of the order seen in organisms is precisely the spontaneous order in the systems of which we are composed. Such order has beauty and elegance, casting an image of permanence and underlying law over biology. Evolution is not just "chance caught on the wing". It is not just a tinkering of the ad hoc, of bricolage, of contraption. It is emergent order honored and honed by selection. (Kauffman 1993: 644)

The reference to tinkering was a direct challenge to François Jacob, the co-discoverer of regulatory genes. He had famously described evolution as *bricolage*, or tinkering (Jacob 1977).

As if such downplaying of natural selection weren't maverick enough, Kauffman "continued in D'Arcy Thompson's tradition" even to the extent of paying homage to Goethe as well as to Kant. Many biologists had been entranced by D'Arcy Thompson's writing, as we've seen, but most readers had ignored his praise of Goethe and Kant. Even in the late eighteenth and nineteenth centuries, their holistic approach to biology had been scorned by most empiricists—though not, or not entirely, by Hermann von Helmholtz (see 2.vi.e). Now, Kauffman was resurrecting these long-dead philosophers.

His unfashionable remarks about them weren't mere fleeting asides. To the contrary, they were rhetorically prominent declarations of biological faith. The opening pages of his most important book expressed his kinship with the pre-Darwinian tradition of rational morphology; and the Epilogue began by praising Kant's philosophy of organisms (Kauffman 1993: 3–5, 643). As we saw in Chapter 2.vi.e–f, the rational morphologists had seen organisms as holistic self-organizing systems, shaped according to (unspecified) universal laws of form—an a-historical approach that was to be suddenly eclipsed by Darwinism.

Kauffman's appreciation of this outmoded theoretical tradition was largely due to his interactions with Waddington and Goodwin. Waddington had tried to combine D'Arcy Thompson's developmental approach with evolution and genetics. But his ideas about epigenetic landscapes were uncomfortably vague, and his concern with whole organisms was overshadowed by molecular biology. Eventually, his student Goodwin offered a theory (of developmental transformations within morphogenetic fields) that was better grounded in experimental data and more mathematically precise than Waddington's (Webster and Goodwin 1996, pt. II).

Goodwin identified even more strongly with the rationalist tradition in biology than Waddington had done. His major book took its motto from Goethe, and devoted many pages (written by his colleague Gerry Webster) to a philosophical discussion of these ideas (Webster and Goodwin 1996: vii, pt. I).

In his early research, Goodwin had used cognitive and informational concepts to try to capture some of the high-level, holistic features of organisms (Goodwin 1974, 1976; Boden 1981*b*). Later, and especially in his research with Kauffman, he relied more on dynamical mathematics. As an experimental biologist, he was closer to the actual data than Kauffman.

Kauffman's mathematical work on fitness landscapes, for example, showed that fundamental changes in body plan will become less probable as development or evolution proceeds, and that certain transformations will be possible (certain attractors available) only on certain trajectories. Consequently, certain types of organ will tend to be found together, in recognizable morphological groups (Kauffman 1993, ch. 6). But Goodwin asked just which transformations actually occur, and how they come about.

In an early discussion of slime moulds, for instance, he pointed out that a single chemical may have very different morphogenetic "meanings" in different developmental contexts (Goodwin 1976). And his later work hypothesized specific transformations and morphogenetic fields in a wide range of examples, from slime moulds to the pentadactyl limb (Webster and Goodwin 1996, pt. II).

Just as computational psychologists need to test their models against the empirical data, so mathematical biologists need to do this too. Accordingly, Kauffman sought evidence from various areas of biology to confirm his mathematical analyses. When discussing the significance of very low values of K, for example, he remarked that each gene or active molecule in living organisms (from bacteria to primates) appears to be directly regulated by only a few other molecules—often, as few as two. In general, his book referred to a large number of biological studies in support of his approach.

However, the evidence wasn't always available, nor even consistent. His remark just cited, for instance, is questionable: many genes can interact directly with at least ten others. If chaos theoretically ensues when K is greater than 5, this fact is an embarrassment. So is the fact that individual neurones may receive thousands of inputs from others, yet the brain generates highly ordered cognition. Again, his claim that *Drosophila* segmentation can be explained by reaction–diffusion models inspired by Turing (Kauffman 1993: 566–600) is disputed (see Section iv.b). And as for counting cell-types, it's not obvious that the best way to classify cells, as natural kinds, is by their morphological properties, as is done in histology textbooks (and by Ramón y Cajal): hidden properties, programmed in by the genes, may be just as important (Maynard Smith, personal communication).

Maynard Smith, for instance, allowed that Kauffman's work on autocatalytic protein networks perhaps throws light on the origin of life (Maynard Smith and Szathmáry 1995: 69–72). But he also argued that many questions about evolution call for more biological data, not more theoretical models. On his view, computer models of actual genetic systems are likely to be more useful than simulations of evolution in general (Maynard Smith 1996).

In short, sceptics often complained that Kauffman's models (and other A-Life abstractions, such as Ray's), though admittedly intriguing, were too far from the biological coalface to be really useful. This opposition was stoked by Kauffman's provocative views on natural selection, and his 'old-fashioned' philosophical holism.

d. Discussions in the desert

Not everyone was situated at the biological coalface, however. Many people re-membered that, although a real neurone is very different from an abstractly defined McCulloch–Pitts unit (see Chapter 14.ii), McCulloch's simplification had inspired important empirical research. So, too, might Kauffman's. And for some readers, his dynamical approach seemed to offer a way of thinking about self-organization in many different disciplines besides biology.

That's why, in 1985, Kauffman was invited to join the Santa Fe Institute (SFI), in the research group on Complex Systems. SFI had been set up one year before, by a group of physicists and mathematicians at nearby Los Alamos. Their research focus was on non-linear dynamical systems.

Following in the footsteps of the physicists Schrödinger (1887–1961) and, especially, Ilya Prigogine (1917–2003), they asked how fundamental physical principles, concern-ing entropy and equilibrium, can allow—or even give rise to—self-sustaining order, especially biological order. Schrödinger (1944) had argued that living things survive not thanks to food, but thanks to negative entropy. And Prigogine himself had described his Nobel-winning work on "dissipative structures"—irreversible far-from-equilibrium processes—as a study of the "self-organization" of matter (Prigogine *et al.* 1969; Prigogine and Stengers 1984).

For instance, he'd shown that a heated liquid may form rotating "convection cells", patterns of physical activity that hugely outstrip the range of the intermolecular forces that cause them. It may be that he called this "self-organization" because he spent a good deal of time with the embryologists in Brussels (Dupuy 2000: 130). If so, biological morphology may have been close to his mind even when he was doing 'pure' thermodynamics.

The group at Los Alamos also worked on the theories of chaos and complexity. As Kauffman's work on NK networks had illustrated, these dynamical phenomena are intimately related. They'd been studied within mathematical physics since the mid-1960s, but had been recognized by the wider scientific community only recently—partly as a result of the activities of the Los Alamos group (Waldrop 1992; Kellert 1993). It was in light of this historical context that Kauffman had described his own work as "an effort to continue in [D'Arcy] Thompson's tradition *with the spirit now animating parts of physics*" (1993: 644; italics added).

The Los Alamos physicists encouraged a widely interdisciplinary approach. Their Center for Nonlinear Studies had initiated a series of conferences on various aspects of self-organization. One of these, focusing on emergent order and distributed computa-tion, included non-physicists such as Holland, Kauffman, Hillis, Douglas Hofstadter, Stephanie Forrest, Stevan Harnad, and Paul Churchland (Forrest 1991). The new institute, SFI, was intended as a focus for interdisciplinary research concerned with complex systems *in general*.

One early SFI activity was a joint physics–economics workshop, held there in 1987. Two of the economists involved (Kenneth Arrow and Brian Arthur) had long stressed the non-linearity of economic systems—something ignored by classical economics. But that's not to say that all was plain sailing when these matters were discussed. The initial physics–economics culture clash has been tellingly described by Mitchell Waldrop (1992, ch. 4). Although this meeting ended fruitfully, even pleasantly, some later SFI workshops "degenerated into shouting matches and sulking" (p. 143). Interdisciplinarity isn't always easy—or good-tempered.

In short, the personnel at SFI were nothing if not open-minded. Over the next decade, they welcomed colleagues in physics, mathematics, computer science, AI, neuroscience, biology, psychology, economics, and philosophy. All were invited to discuss the emergence of order in complex systems. Many chose to focus on aspects of life and cognition.

As so often happens, the open-mindedness of the early days eventually became muted. By the end of the 1990s, both Kauffman and Langton had been sidelined by SFI, apparently for speculating too wildly (C. J. Langton, personal communication). The puzzle of life had been replaced in SFI's affections by puzzles about predicting the stock market.

15.x. Naming and Synthesis

A-Life was born healthy at mid-century, as we've seen. But the pattering of its tiny feet caused little disturbance. The infant field wasn't even mentioned for over thirty years. Some of its core *topics* were widely discussed, to be sure—for example, in Hofstadter's hugely popular *Gödel, Escher, Bach* (see Chapter 12.x.a). And the Media Lab's Vivarium (a simulated aquarium), alongside its work on animation, was attracting funding from Hollywood and computer manufacturers, and drawing huge numbers of visitors (13.v.a). But artificial life as such was invisible. Then, suddenly, it seemed to be everywhere.

The suddenness was due to a public act of naming in 1986, and a christening party in 1987. Before then, the term "A-Life" didn't exist. The seeming ubiquity was due to the wide range of disciplines represented at the party, and to media interest prompted by the infant's name—even more outrageous, evidently, than Fifi Trixibelle.

a. The party

Studies of self-organization prior to the 1990s had many underlying similarities, as we've seen. But most people at the time were aware of the links only dimly, if at all. Indeed, they were largely ignorant of each other's activities.

One wouldn't expect ethologists, for instance, to have come across Craig Reynolds's paper on flocking, published in the journal of *Computer Graphics* (Reynolds 1987). His model showed that the group behaviour typical of flocks of birds or schools of fish can emerge from three simple rules—so simple, that they might conceivably be 'hardwired' into animals' brains. Computer animation experts soon picked up on this work. It was used, for example, to generate the lifelike 'schools' of virtual fish described at the outset of this chapter, and to choreograph the flocks of bats in the film *Batman*. But people interested in real fish, and how they manage to do what they do, were hardly likely to have *Computer Graphics* on their reading lists.

This isn't an isolated example. Much of the relevant research was inaccessible to potentially interested readers, because of disciplinary boundaries.

The unification (such as it is) occurred in the late 1980s, when the term A-Life was coined by Langton. He first used it in print in 1986 (Langton 1986). But he'd thought of it in 1985, during a meeting organized at Los Alamos by the physicists Doyne Farmer and Norman Packard, and by the mathematicians Buton Wendroff and Alan Lapedes (Rasmussen *et al.* 2003*b*: 210). The official title of that meeting had been a mouthful: 'Evolution, Games, and Learning: Models for Adaptation in Machines and Nature'. The term *artificial life* was more pithy, more unifying—and, of course, much more provocative.

Langton was a Los Alamos physicist, especially interested in the relation between complexity and life. He had a wide intellectual background, and an iconoclastic attitude to disciplinary boundaries (Levy 1992: 93–103; Kelly 1994: 343–6). This would have counted against him in most academic institutions, but was no disadvantage in Los Alamos. The physicists there had highly interdisciplinary sympathies, as we've seen.

By 1986, Langton was embarking on his painstaking research involving the lambda parameter (see Section ix.a). But he'd already completed a number of important studies in what he now decided to call A-Life.

After working with Burks at Michigan, he'd recently achieved the first implementation of a self-replicator (Langton 1984). (Arbib's 1966 design for a self-replicating CA, mentioned in Section v.b, wasn't implemented.)

He'd also produced various models of ant trails (Langton 1986: 135–6). He showed that a disorganized group of ants, each one following a very simple rule, could give rise to many different trails (e.g. circles, spirals, straight lines . . .). Tiny rule changes could result in observably different larger-scale phenomena. And these phenomena were reminiscent of real insects. For instance, ants dropping pheromones as they walked could eventually converge on a small number of (relatively straight) paths leading to the best food sources.

This was a purely abstract study: Langton was no biologist. Later, however, others would use real-life data in modelling the effects on ant trails of different chemical concentrations, walking speeds, and antennae spans—another example of CNE (Sharpe and Webb 1998). Indeed, Langton's paper led A-Lifers to a general interest in "stigmergy", wherein social behaviour is mediated by individuals responding to environmental changes caused by other individuals (Bonabeau 1999). This phenomenon was named by a French biologist in the 1950s, and many empirical data had been collected. Now, biologists had the possibility of analysing it in rigorous terms.

Mathematicians were interested, too. The notion of "Langton's ant" was generalized in the CA/A-Life community, and led to many follow-up analyses(e.g. Gale and Propp 1994; Gale *et al.* 1995; Stewart 1994; Gajardo *et al.* 2002). In addition, AI work on agents and distributed cognition sometimes drew on Langton's ideas (e.g. O. Holland and Melhuish 1999: see 13.iii.d).

Already connected with the Santa Fe Institute, Langton helped found their A-Life research group. And it was he who organized "the first workshop on A-Life". This event took place in 1987, in Los Alamos. It's remembered by many who attended as a party, even a celebration, as much as a workshop.

The invitations had been issued by Langton, who sent them not only to a wide variety of selected individuals but also to Internet newsgroups in many different disciplines. He recalls this as an anxious period: he had no idea how many people would be intrigued by his invitation, nor how many would see the point if they actually came. After the event, he described it as producing "a growing sense of excitement and camaraderie—even profound relief—as previously isolated research efforts were opened up to one another for the first time" (Langton 1989*a*, p. xvi).

Langton's memory of the meeting is doubtless coloured by his own role in it. But the workshop did catalyse an increasing level of interest, and even cooperation, between scientists from different disciplines. Beer (personal communication), for instance, now remembers it as hugely important for consolidating his nascent interest in dynamical systems—first triggered by Hofstadter and Winograd, and recently strengthened by a best-seller on chaos theory (see xi.b, below). Many others today would tell similar stories. The general sense of excitement fostered by this meeting is clear from the many informal remarks of participants quoted by science journalists reporting on the new field (Levy 1992; Kelly 1994).

It also led to a great deal of 'hype'. Some of this was generated by the participating scientists themselves. Langton later admitted that there was "more good science and less 'gee-whiz' " in the second volume of A-Life Proceedings than in the first (Langton *et al.* 1992, p. xiv).

But much of the hype was generated by 'outsiders'. The people attending included a generous sprinkling of famous names, associated with a wide range of sciences. That was likely to be noticed. And Langton's vision of the future of the field was highly provocative: "The ultimate goal [of A-Life is] to create 'life' in some other medium, ideally a *virtual* medium" (see below). Such a claim was bound to—indeed, was intended to—attract attention. However, it also drew scepticism and scorn.

As for the name bestowed on the field, this too had its drawbacks. Irrespective of specific remarks made by Langton or anyone else, the term "artificial life", like "artificial intelligence" before it, invites literal interpretation from the media. I've encountered journalists from good newspapers who evidently think that the question "Does system X really use A-Life technology?" means the same as "Does system X contain creatures that are really alive?" In short, the meeting was a journalist's dream.

Langton was far from unhappy about that. And within a few years, A-Life had gained the attention of the counter-culture. For it exemplified a new image of science, more 'romantic' than the traditional image was.

The philosopher Stephen Clark (1995) contrasted "Faustean" and "Magean" science, the first being intelligible and largely predictable (though often bringing unwanted side effects) and the second opaque, almost like magic. Where GOFAI—and science in general—had been seen as dealing in preconceived plans and pre-defined goals, with "little use for spontaneity, trial-and-error, unplanned discovery, vaguely-defined ends, or informality" (Edwards 1996: 168), A-Life showed unpredictable evolution, emergence, and responsiveness to environmental details.

Stephen Clark cited it, accordingly, as an example of Magean science. Many professional artists apparently agreed with him. They were inspired by what they saw (often, wrongly) as the aims, and philosophical claims, of work in A-Life. By the turn of the century, hardly more than a decade after Langton's naming party, there was a still-growing body of art based broadly on these ideas (Whitelaw 2004). And students of contemporary culture, such as Sarah Kember, were making grand claims for A-Life as a cultural discourse describing "posthuman" ("cyborgian") life (2003, p. vii and *passim*).

But Clark saw Magean science as threatening, too. Just because of its opacity and unpredictability, it was unintelligible and beyond control—which is to say, it risked going out of control (cf. Kelly 1994). Soon, science-fiction novels/films such as Michael Crichton's (2002) *Prey* would give vivid expression to those cultural fears.

Worrying or not, however, A-Life became the talk of the town—much as Langton had hoped. The tone of "the talk of the town" wasn't always one of apocalyptic anxiety. Often, the *gee whiz* factor won out. Consider (for example) this University of Toronto press release of 2000, which describes the then latest version of the artificial fish mentioned in Section ii.a, above:

Surviving jaws: virtual merman 'thinks' his way to safety

Using a mythical merman and hungry sharks, a University of Toronto computer science professor and two former graduate students have pushed the notion of artificial intelligence and virtual life to a new level.

In his creation of a virtual underwater world, Professor Demetri Terzopoulos has fashioned more than just a cool screen saver—he has given his animated characters the ability to think. A hungry shark circles ominously, looking for a nice meal, while a nervous merman searches for a place to hide. When the shark swims away, the merman dashes from behind large rocks to open water with the shark in hot pursuit. Will his cleverly devised plan allow him to reach safety or not?

"This is more than artificial intelligence", says Terzopoulos. "It's artificial life. Computer graphics, animation and virtual reality have advanced dramatically over the past decade. We are now able to create characters that are self-animating with functional bodies and brains that have behaviour, perception, learning and cognition centres."

Terzopoulos and his former students have developed the cognitive modelling language that enables animated characters to reason. For example, it enabled the virtual merman to formulate a plan of action by reasoning about his situation given certain knowledge, such as the fact that he cannot outrun sharks but can use underwater rocks to hide. "With cognitively empowered graphical characters, the animator need only specify a behaviour outline and, through reasoning, the character will automatically work out a detailed sequence of actions."

The potential for future applications are immense, Terzopoulos says. Cognitive modelling and the cognitive modelling language can become powerful tools for scientists, animators and game developers. His paper, co-authored with John Funge and Xiaoyuan Tu, was published at the 1999 ACM SIGGRAPH conference, the premier forum for research in computer graphics. (University of Toronto web site, <http://www.utoronto.ca/>, 29 February 2000)

If this was the official, relatively restrained, account you can perhaps imagine the enthusiastically garbled versions that would appear in the less scrupulous newspapers.

(Excessive hype, from within and outside the field, is still a problem—and an embarrassing one, for those seeking professional legitimacy. In the A-Life community's turn-of-century self-assessment, the key failure was said to be "No rigor", and the greatest advantage of founding an International Society for A-Life—set up in 2003—was "reducing hype" by improving intellectual standards: Rasmussen *et al.* 2003*b*.)

Despite the hyperbole and misunderstandings, however, there was—and is—a serious intellectual core. The common scientific interest aroused at the workshop was afterwards described by Langton as

an emerging consensus ... of the "essence" of Artificial Life [based on] themes such as *bottom-up* rather than *top-down* modelling, *local* rather than *global* control, *simple* rather than *complex* specifications, *emergent* rather than *prespecified* behavior, *population* rather than *individual* simulation, and so forth. (Langton 1989*a*, p. xvi)

These themes are evident in the seminal work of Turing, von Neumann, Burks, and Holland (and in the wide range of later research mentioned in Sections vii–viii). And, if interpreted without reference to computer models, most of them can be seen in D'Arcy Thompson's writings, too.

Such people's pioneering work also fits the definition given in Langton's Call for Papers in 1986:

Artificial life is the study of artificial systems that exhibit behavior characteristic of natural living systems. It is the quest to explain life in any of its possible manifestations, without restriction to the particular examples that have evolved on earth. This includes biological and chemical experiments, computer simulations, and purely theoretical endeavors. Processes occurring on molecular, social, and evolutionary scales are subject to investigation. The ultimate goal is to extract the logical form of living systems. (quoted in Levy 1992: 113)

D'Arcy Thompson and von Neumann, for instance, would have been quick to allow that the logical form of living systems isn't immediately evident. Discovering it might add something new to our accepted concept of life. In Langton's terms, we could come to see that "life as it could be" is a wider category than "life as it is".

For example, the crucial values of Langton's lambda parameter or Wuensche's Z (see Section ix.a) might come to be included within the definition of life.

Again, if reproduction is regarded as essential to life, then life might be defined as requiring von Neumann's dual (data/instructions) functionality. Von Neumann showed also that a universal replicator is possible: so perhaps some extra-terrestrial living organism could give birth to puppies as well as kittens—and to tadpoles and grass seeds, too? One might object that such a heredity-cheating creature could never have evolved. But that's relevant here only if one regards evolution itself as an essential criterion of life. Some biologists and philosophers do (Bedau 1996; Moreno and Ruiz-Mirazo 2004: 251–5; Luisi *et al.* forthcoming, 3). However, this decision isn't without drawbacks (see Chapter 16.x.d).

As for the logical form of living *bodies*, most biologists would probably say that there's no such thing. But not all: we've seen that recent work in D'Arcy Thompson's (and Goethe's) footsteps suggests that—given the laws of physics—only certain types of morphology are actually possible.

b. Simulation or realization?

For Langton, the logical form of living systems is such as to allow the possibility of *virtual* life. Contrary to D'Arcy Thompson, the constraints of physical embodiment are a distraction best avoided:

The ultimate goal of the study of artificial life would be to create "life" in some other medium, ideally [*sic*] a *virtual* medium where the essence of life has been abstracted from the details of its implementation in any particular model. (Langton 1986)

Ray agrees. He argues that digitized creatures existing in cyberspace (very like his, described in Section vi) could be genuinely alive (Ray 1992, 1994).

This philosophical claim (discussed in Chapter 16.x) is called strong A-Life, in analogy to strong AI. Most A-Lifers reject it, believing that life in principle requires some sort of physical body and metabolism. Pattee, for instance, always insisted on a clear distinction between simulating life and realizing it, because he saw the laws of physics as crucial in explaining biological organization (Pattee 1972, 1976). So the journalists who excitedly reported that the production of cyberlife is the goal of A-Life *in general* were mistaken. So, too, were those who said that A-Life *in general* was aiming to produce "new forms of life", whether virtual or physical.

But their mistakes were understandable. For Langton, the instigator and organizer of the inaugural event, explicitly held both views. Moreover, as the official subtitle of the meeting he chose 'An Interdisciplinary Workshop on the Synthesis [*sic*] and Simulation of Living Systems' (Langton 1989*a*). And, in a paper published shortly before the first A-Life meeting, he'd said:

We would like to build models that are so life-like that they cease to become *models* of life and become *examples* of life themselves. (Langton 1986)

Another point in mitigation was mentioned above: the likelihood that "artificial life" would be interpreted by outsiders to mean "life made artificially".

In fact, only a few insiders understood the term in this way. Indeed, some soon adopted less philosophically provocative (and somewhat less inclusive) language. They spoke of "adaptive systems", for example, or of "animats"—a conflation of "animal" and "robot" (Meyer and Wilson 1991). That enabled them to work on computer models of life without implying that simulation was intended as realization.

Because of the high profile of the 1987 meeting and its immediate successors, however, the term *artificial life*, like the term *artificial intelligence*, is here to stay. One must remember, therefore, that using artefacts to learn about life—even "life as it could be"—isn't the same thing as aiming to make life artificially.

Whether life could be made artificially remains an interesting question nevertheless. So does the question whether virtual life is in principle possible. So, too, do disputed claims about whether only life can support mind, whether adaptive autonomy (without representation) can suffice to do so, whether evolution is necessary for life and/or intelligence, and whether minds are distinct from living brains.

These are all partly philosophical questions (see Chapter 16.x), to which scientific work is dialectically relevant. That's why reports of experiments in this area often include discussions of the *definition* of life, and phrases like "the definition of life one wishes to use" (Luisi and Oberholzer 2001: 353).

If they were purely scientific questions then one, at least, would have been answered at the dawn of the new millennium. London's *Sunday Times* of 30 January 2000, under the headline 'First Artificial DNA Can Create New Forms of Life', reported that researchers in Texas "have made the world's first synthetic DNA [and] mapped out the exact way it will be configured to create synthetic organism one (SO1), the microbe destined to be the world's first man-made creature".

Three years later, the *New Scientist* excitedly reported that Craig Venter (director of the Human Genome project) is trying to build "the minimal genome" (by stripping down the genome of the simplest existing organism), and that the Harvard nanotechnologist George Church is aiming to do much the same thing (Ainsworth 2003). But the *New Scientist* had the philosophical edge over the *Sunday Times*, for its journalist acknowledged the controversial nature of the *concept of 'life'*. These matters aren't so straightforward as his *Sunday Times* rival had assumed.

That's evident from the conceptual arguments (i.e. not just the empirical disagreements) that arise within so-called "wet" A-Life. This form of A-Life *does* aim to create new life. Specifically, it aims to create living/lifelike systems out of chemical/biochemical substances. Sometimes, it even starts from what Kant would have termed "crude inorganic matter" (see 2.vi.b), although sometimes it starts from an already living thing.

The notion that this is in principle possible dates from the mid-1920s, in a paper by the Russian biochemist Alexander Oparin. Remaining untranslated, this had little impact at the time. But Oparin's late 1930s book caused huge excitement, not to say scandal (Oparin 1936/1938). The first Oparin-inspired experiments to succeed in synthesizing molecules characteristic of life, namely amino acids, from simple substances present on the primitive earth were done half a century ago (S. L. Miller 1953; Miller and Urey 1959), and sugars had followed by the late 1960s (Reid and Orgel 1967).

Today, such experiments continue. But most of them don't count as wet A-Life. For that, the aim must be not merely to synthesize unorganized biomolecules, but to create *entire, albeit simple, living things*. Several groups around the world are attempting this. And they're doing so in interestingly different ways, depending on their philosophical views about the nature of life as such.

Venter's team in Maryland—who drew the attention of the *New Scientist*—are working top-down. That is, they're starting from a living, independently replicating, cell (*Mycoplasma genitalium*) that has the smallest known genome: only 517 genes (Fraser *et al.* 1995; Hutchison *et al.* 1999). They aim to identify any non-essential genes (the so-called junk DNA), and to construct a new, stripped-down, genome that will generate a recognizably *living* organism. By the eve of the new century, they'd identified 480 protein-forming genes, of which between 265 and 350 seemed to be essential (including 100 of unknown function). If and when they achieve their aim, it's unlikely that there will be much disagreement over whether the result really is a living organism. After all, it will be a cleaned-up copy of a naturally evolved life form.

Things may be—indeed, they already are—rather different in cases where researchers aim (*à la* Oparin) to concoct life from the bottom up, as opposed to working top-down by removing redundant DNA. The problematic philosophical issues are more evident when people start from a non-biological base and try to build living systems anew. Such people often restrict themselves to molecules we already recognize as "biochemical".

But sometimes they don't, hoping to find very general biochemical principles which cover substances other than the ones found in terrestrial organisms. (Life as it could be.)

The biochemists who are seeking to model and/or produce physical analogues of the molecular origins of life don't always agree about just what molecules count as "biological" and what as "pre-biological", or about what is a "basically autonomous system" but *not* a living thing, or about how "biomolecules" can become sufficiently organized to merit the label of "full-fledged living beings" (Moreno and Ruiz-Mirazo 2004: 245–55).

Similarly, they disagree on just how the criterion of a *boundary* is to be interpreted, and what its philosophical significance is, when chemical compartments (micelles, vesicles . . .) arise within which specific chemical reactions can take place. For example, liposomes can form vesicles; if these contain an enzyme for replication, an RNA template, and ingredients for RNA synthesis, then the system may self-replicate (Biebricher *et al.* 1982). Some biochemists—notably Pier Luigi Luisi's group in Rome—interpret such chemical boundaries in terms of the philosophy of autopoiesis (Bachman *et al.* 1990, 1992; Walde *et al.* 1994; Luisi *et al.* 2004).

Luisi, a long-time friend and admirer of Varela, relates his experiments to an autopoietic interpretation of "the minimal cell" (Luisi and Varela 1989; Luisi and Oberholzer 2001: 349–50; Luisi 2002). This concept has been discussed for almost fifty years. For example, Harold Morowitz (1967) argued that the minimal cell need be no more than nine-tenths the size of *M. genitalium*. (For more recent estimates, see Islas *et al.* 2004; Szathmáry 2005.) But new biochemical techniques have led to a new spurt of interest, including two international conferences in 2004.

The minimal cell is the smallest possible living system, having "the minimal and sufficient number of components to be called alive" (Luisi *et al.* forthcoming, 3). It isn't thought to be a particular system, which *had to be* generated if life was to appear at all. Rather, the assumption is that there's a broad family of potential minimal cells. Which of these was/were actually realized is a question more for biological history than for abstract biological theory.

As for the term "living", this is usually understood in wet A-Life to mean: capable of self-maintenance (metabolism), self-reproduction, and evolution. Pre-cellular systems, or proto-life, may fulfil only one or two of these three criteria. And/or they may be reproducible for only a few generations, or they may be able to reproduce only part of themselves. Luisi calls such systems "limping life" (Luisi *et al.* forthcoming, 22). Workers in wet A-Life typically argue that they were essential stepping stones on the way to the emergence of fully fledged life.

Researchers differ in their commitment to what I just called biological history, as opposed to biological theory. So there's no consensus over whether we should be trying (like the Texas researchers who impressed the *Sunday Times*) to replicate the biomolecules already familiar to us. Some argue that wet A-Life may, or even should, attempt to build "chemical automata" made up of *alternative* chemical components (Drexler 1989; Bro 1997; Moreno and Ruiz-Mirazo 2004: 255).

In addition, of course, there are many empirical disagreements. These include one instigated at the University of Copenhagen by Peter Nielsen, the "inventor" of PNAs (polyamide-linked nucleic acids). PNAs are man-made "chimaera" molecules, having some of the structural properties of DNA and some of the chemical properties of proteins (Nielsen *et al.* 1991). Significantly, they follow the Watson–Crick base-pairing rules (Egholm *et al.* 1993). So the question arises whether PNA chemistry can replace RNA chemistry for the purposes of wet A-Life (Rasmussen *et al.* 2003a).

Here, as elsewhere, science and philosophy can interact. Work in wet A-Life often includes explicit discussion of how to define "life" (e.g. Moreno and Ruiz-Mirazo 2004: 254–7). And for those who believe that life and cognition are *essentially* linked, it may affect their philosophy of mind as well.

15.xi. After the Party

After the party was over, life went on. And A-Life went on, too. Only seventeen years have passed since Langton's epochal gathering. Nevertheless, a large body of work has been done.

A few examples drawn from the 1990s, or later, have already been mentioned. These include the evolutionary robots outlined in Section vi.c, the primitive open-ended evolution sensors of Section vi.d, the various models of neuro-ethology discussed in Section vii.b–d, and the biochemical A-Life briefly mentioned above. By and large, however, this chapter has followed the time-line, from the early work to the official founding/naming of the field. It may be appropriate, then, to end it by describing two more instances of *very* recent research.

These will bring things full circle, all the way back to mid-century cybernetics. But the circle is an ascending spiral, not a plane figure: besides lifting cybernetic understanding to a new level, these two examples open up exciting pathways for further study. Moreover, they raise fundamental issues that are relevant to cognitive science *in general*.

a. Resurrection of the Homeostat

Over a century ago, George Stratton (1896, 1897a,b) earned immortality by an experiment in which he wore inverting lenses for eighty-one hours, spread over eight days (see 14.viii.b). At first, his movements were utterly inappropriate to the physical environment around him. But his visuo-motor system gradually adapted until he was able to negotiate his world almost as successfully as before. Moreover, he no longer had to make allowances for the fact that the world looked upside down because, after the adaptation, the world *didn't* look upside down, but broadly "normal".

This phenomenon has intrigued psychologists and neurophysiologists ever since, and various similar experiments—some with left–right inverting spectacles—have been done. Now, further examples are cropping up in a wide range of VR applications (see 13.vi.b), including novel forms of motor and/or sensory prostheses (A. J. Clark 2003a).

Ashby's interest in it was mentioned in passing in Chapter 4.viii.c, and Andras Pellionisz's neurologically based explanation was described in Chapter 14.viii.b. Recently, another—more abstract—model of recovery from inversion has exploited Ashby's notion of a homeostatic dynamical system.

Ezequiel Di Paolo (2000b) evolved simulated light-seeking robots that could gradually recover phototaxis after their visual field was inverted. Putting it like that, however, may be misleading. For—like Stratton himself—they weren't evolved *to recover from visual inversion.* In other words, this adaptation didn't feature in the fitness function. Rather, they were evolved to perform phototaxis. But the 8-unit control network that enabled them to do this was of a type which *also* enabled them to recover when later subjected to the radical challenge of visual inversion.

Phototaxis was achieved by evolving the robots' control network, or "brain"—broadly, as described in Section vi.c, above. Two of the eight units ("neurones") in the network were attached to the left and right motors, and two to the sensors ("eyes"). In addition, every unit was connected to every other. But the excitatory/inhibitory nature of the connections could be varied by mutation, as could the numbers assigned to units and connections (see below). At each generation, six lights, of various intensities, were shown at different times and at different places. The robots that, on average, approached them most closely and most quickly, and that stayed by the light when they got there, were selected as parents for the next generation. (Noise was added to the system, to make the evolving circuits more robust.)

So far, so conventional (if such a still-recent approach can be called conventional). The difference from previous work was the nature of the units in the network. The key point is that each one of these was locally homeostatic. In other words, each artificial neurone had a permissible range of activation rates (not evolved, but assigned to it by Di Paolo). If it went above or below its limits, the weight on the incoming (presynaptic) connections would be changed so as to lower or increase its activity, respectively.

Unlike blood temperature then, where homeostasis restores a specific value, a neurone's acceptable activation rate could have various values—all lying within the relevant range. Given the numerical limits of each of the eight individual neurones, the numbers in the rule of presynaptic weight change were evolved. That is, the fitness function—besides measuring success in the behaviour of light seeking, as described above—measured success in homeostasis of all eight units.

In perfect homeostasis, no neurone would ever move outside its limits, so weight changes would never need to be made. But it turned out that neurones with very powerful homeostasis were bad news: they tended to do nothing, so that the robot hardly moved. For phototaxis to evolve, the unit homeostasis had to be less than perfect.

The result of 1,000–2,000 generations of evolution would be a light-seeking robot with a neural controller wherein each unit would recover its permissible activation rate if perturbed by some unusual input. An "unusual input" would normally be something like an especially strong (or weak, or distant) light source. But the robots' sensory anatomy enabled the malign roboticist to provide yet more unusual inputs.

Each robot had a left eye and a right eye. That is, its two sensors were placed relatively far apart on its body, so that the difference between the two visual inputs would depend on (in functionalist terms, would "encode") the specific location of the light source. A near-equivalent to Stratton's upside-down inversion, then, was to switch the anatomical positions of the two eyes, while leaving their neuronal connections intact.

As you might expect, each immediately post-operative robot was unable to approach the light. Indeed, it would move in the *opposite* direction. Di Paolo found, however, that about half of his robots eventually managed to adapt to their new situation, ending up with perfect phototaxis as before.

Although the behavioural outcome (reaching the light) was identical, the behavioural strategy wasn't. In one case, for instance, the pre-operative robot would begin by moving slightly to one side and then to the other, in order to centre the light in its visual field, and would then move directly towards it. After adaptation to inversion, the same robot moved in an arc of a circle to locate the source. (Di Paolo also found that the longer the period before inversion, the longer it took for adaptation to occur; he compared this to the "critical periods" for neural adaptation that have been observed in animals—2000*b*: 445.)

Senator Proxmire wouldn't have been impressed (see Chapter 6.iv.f). Granted, phototaxic robots could conceivably be useful. But if a researcher has managed to produce light-seeking robots, why waste the taxpayers' money in messing around with their sense organs so that they fall into confusion?

Put like that, the Proxmire complaint sounds persuasive. But, as was so often the case in his lifetime, the Senator would have been wrong. For the light-seeking robots were just an example. Famous though Stratton's experiment is, it wasn't Di Paolo's primary concern. What he was really engaged in was something even more interesting. Explicitly paying homage to Ashby, he was resurrecting the Homeostat (Chapter 4.viii.c). Specifically, he was demonstrating the sort of self-organization that Ashby had called ultrastability. Homeostasis of the system as a whole resulted from changes triggered by local homeostasis of its component units.

Moreover, his work was physiologically promising, in the sense that his general methodology allowed for the modelling of many other sorts of cerebral plasticity besides weight change (Di Paolo 2000*b*: 447–8). For instance, it could be used to model neuronal depression or potentiation, or modulatory synapses. It could even be used to simulate diffusive chemical neuromodulation—a phenomenon recently studied by his Sussex colleagues (cf. 14.ix.f).

Above all, for our purposes here, his chosen example was *psychologically* interesting, and highly relevant to cognitive science. Indeed, he sees his phototactic robots, like his earlier work on the evolution of communication (13.iii.e), as evidence supporting *a radical critique* of orthodox (functionalist) cognitive science (2000*b*: 448).

The reason is that his robots' behavioural homeostasis—that is, their successful adaptation to environmental perturbation—wasn't guided by any specific behavioural goals. Rather, it resulted from the underlying dynamical principles of the system. (This point was confirmed by the fact that adaptation also occurred when he made various "lesions" in one of the sensor or motor "organs"—2000*b*: 446.) In short, their behaviour was in effect *purposeful*, but wasn't directed by *purposes*.

b. Analysing dynamics

Di Paolo's work was experimental, not analytic. Each of his successfully adapting robots provided an existence proof of a homeostatic dynamical system with intriguing psychological overtones. (Only overtones: unlike the neurologist Pellionisz, he wasn't trying to say just how, in fact, Stratton's brain might have recovered from visual inversion.) But he didn't know just how many 8-unit networks were in principle capable of this adaptation. Nor did he know just what type, or types, they could be. He didn't even know whether as many as eight neurones were strictly necessary.

That's not unusual. In general, cognitive scientists who manage to get interesting behaviour out of a dynamical system typically don't know just why it occurred. (Up to a point, the same is true of GOFAI; but GOFAI researchers learnt long ago that they should try to analyse their programs, not just glory in them: see Chapter 11.iii.) Dynamicists rarely know just what it was about *this* dynamical system which made it capable of generating *that* behaviour.

Very recently, however, some elegant mathematical work has enabled researchers to say something along these lines, at least for very small networks of a certain general type. The systems concerned are called CTRNs, or CTRNNs: continuous-time recurrent (neural) networks.

This is a very broad class. (The "brains" controlling Di Paolo's robots were an extended version, as we'll see.) The first person to pay significant attention to them was the cockroach-building Beer (R. D. Beer 1995c). Beer first defined them in 1991 (he doesn't know whether anyone else had done so already: personal communication). He used the double-N label: CTRNNs.

When they're called CTRNs (by Di Paolo, for instance), the N for "neural" is omitted. Arguably, the single-N label is better (and I'll use it in what follows). For, as Beer himself pointed out when he defined the term in the first place, the units *need not* represent neurones, nor even neuronal cell assemblies. They can stand for *any* essential variable, *any* convenient dynamical building block, in the dynamical system being modelled.

In general physiology, these would be key aspects of metabolism. In neuroscience and psychology, they'd be features of neural function and behaviour, respectively. And in sociology or management studies, they'd be properties of the social communications concerned. Even Pask's ideas about the cybernetics of the criminal underworld (4.v.e) could be represented by a CTRN. (It's significant that his ambitious "conversation theory" was intended as a *general* account of feedback systems, in which the conversationalists identified by the rebarbative notation could be cells, people, or organizations.)

Today, CTRNs are increasingly used by people doing dynamical modelling in A-Life and neuroscience. One reason for this is a mathematical proof achieved in the early 1990s (Funahashi and Nakamura 1993). Much as a universal Turing machine can model any discrete computation, so these networks can in principle represent any dynamical phenomenon.

A CTRN is made up of units computing continuously in time. In effect, it's a cellular automaton in which the time-steps are infinitesimal in duration. (If it's implemented on a von Neumann computer, a CTRN will in fact involve *discrete* state transitions.) But each unit has its own numerical time-constant, which determines just how quickly/slowly it will give its output in response to a significant input. The units aren't binary, but sigmoidal. (Di Paolo's units are unusual in having a *range* of "permissible" activation rates that's smaller than the range of possible activation rates, and a homeostatic mechanism for bringing them back within the permissible range as appropriate.)

What counts as a significant input is determined by the unit's numerical "bias" parameter. This is broadly equivalent to the threshold of a McCulloch–Pitts neurone. But the bias (and the other numerical parameters of a CTRN) doesn't have to be an integer. Unlike McCulloch–Pitts thresholds, it can be defined to whatever degree of precision is required.

Each connection is either excitatory or inhibitory, and carries some numerical weight. The weights on the connections between any two units can be asymmetrical, in which case unit A will influence unit B more strongly than B influences A. (So a CTRN could model the asymmetrical communications between the various crew members aboard ship: see Chapter 8.iii.b.) Often, all the weights stay the same once they've been assigned. But sometimes, for instance in Di Paolo's phototaxic robots, the individual weights can change during the system's "lifetime".

Finally, the networks are completely connected. That is, each unit is connected to every other unit—and even to itself, by a recursive feedback loop. Because the network is completely connected, any change in one unit *directly* affects every other unit. (A CTRN is thus similar to a continuous Hopfield net, described in Chapter 12.v.f; but Hopfield units don't have self-loops, and Hopfield connections are all symmetrical.) Being completely connected makes a CTRN a dynamical system par excellence:

Much of the unique flavor of dynamical systems is captured by the idea of *coupling*. [Two] variables are coupled when the way each *changes* at any given time depends directly on the way the other *is* at that time. In other words, coupled variables simultaneously, interdependently co-evolve, just like arm angle and engine speed in the centrifugal [Watt] governor. *Genuinely dynamical systems exhibit high degrees of coupling: every variable is changing all the time, and all pairs of variables are, either directly or indirectly, mutually determining the shapes of each other's changes.* (van Gelder 1997: 437; final italics added)

Complete connection is mathematically interesting, and it enables holistic study of dynamical systems in which every individual part influences every other. But it's rarely biologically plausible. For instance, the number of connections found in mammalian nervous systems scales roughly linearly as the number of neurones increases, instead of exploding as it does in fully connected networks (C. F. Stevens 1989).

Nevertheless, CTRNs can be used to model real biological circuits, for if a weight is set to zero then the connection concerned is in effect deleted. (Similarly, a huge bias parameter would deactivate the relevant unit.) So every conceivable network, whether fully connected or not, can be represented as a CTRN.

Beer's first artificial cockroaches, described in Section vii.c above, were situated robots inspired by Brooks and by Pengi. One might almost say that they were engineered rather than controlled. But Beer soon began to use dynamical system controllers to improve their walking.

His dynamical turn had been inspired by two near-simultaneous publications. One was James Gleick's (1987) popular book on chaos theory (personal communication), and the other was the Winograd–Flores book (see 11.ii.g)—which introduced Beer to the ideas of Maturana and Varela. From that time on, they were the major intellectual influences on his work.

Initially, Beer was most excited by how specific dynamical systems could help him build better robots. For example, he evolved a CTRN leg controller that repeatedly adapted its parameters to compensate for the fact that the (simulated) leg was *growing*, so changing the values of the physical constraints involved (J. G. Gallagher and Beer 1993, sect. 6). In 1991–2 he wrote his dynamicist's "manifesto" (personal communication) for the journal *Artificial Intelligence* (R. D. Beer 1995a). But he became increasingly interested in the abstract properties of his robot mechanisms *considered as dynamical systems.*

By the start of the new century, these general principles were his main focus. He now wanted to discover just which dynamical attractors were possible for different classes of CTRNs. This was a mathematical exercise (described below), but one with practical implications.

For instance, suppose he found that some class of CTRNs involved cyclic (oscillating) attractors. Intuitively, it seems that such networks would be specially apt for controlling rhythmic activities, like walking. If so, then one particular CTRN oscillator would need to be built into any actual robot. Of course, Beer could simply have picked out one example from the class of oscillators to use as the gait controller for an artificial cockroach leg. But his choice would have been largely arbitrary. So, instead, he picked a "parent" CTRN from this general class and then used GAs to evolve even more suitable descendants.

In other words, he was able to search the wide range of possibilities generated from the chosen seed by random mutation. (There was another advantage, too: biological evolution didn't have the luxury of picking something out of an explicitly analysed class, so Beer wanted to convince himself that these stable oscillators *could actually be evolved*.)

Beer's mathematics wasn't wholly a priori. As Gleick's book had pointed out at length, one can't predict (i.e. analytically prove) just how a complex dynamical system will behave. So Beer had to rely on experimental induction. That is, he defined/evolved a range of CTRNs, and studied their behaviour *post hoc* to see if he could find any regularities. Langton had done much the same thing when defining his "lambda" parameter in the mid-1980s, though using CAs rather than CTRNs (see ix.a, above). But whereas he'd looked at scores of different systems, Beer—with his students Sean Psujek and Jeffrey Ames—looked at several millions (Psujek *et al.* 2004).

Why so many? Well, the rich connectivity of CTRNs means that the range of possible architectures is huge. (An architecture is defined as a set of directed connections.) For a 3-unit network there are 64 architectures, for a 4-unit net there are 4,096, and 5 units surpass the half-million mark (to 528,384). Then, the numbers rapidly become unmanageable. Moreover, any one architecture has multifarious examples, because each parameter (the three connection weights, the bias, and the time constant) can have various numerical values.

Consequently, the Case Western team studied only CTRNs of up to five units. (For the 5-unit networks, they couldn't inspect every possibility so looked at a 1 per cent sample instead.)

In evolving CTRNs (with a population size of 100, and 250 generations), they soon discovered that convergence was common. In other words, CTRNs (like NK networks) are *apt* to fall into stable limit cycles—so are potentially promising as models of living/nervous systems. (Compare Wolfram's intuitions about which type of CA is capable of life and/or computation.) But they wanted to discover which architectures are most likely to converge to this or that stable oscillation.

For instance:

* What types of CTRN make oscillations possible?
* Or closed feedback loops?

* And when do these oscillators and/or loops involve connections to the "motor" neurones? (In their experiments, three units were "motor" neurones for the two opposing swing muscles and the foot, and the others—if any—were unassigned.)
* What classes will oscillate between three (or four . . .) states, rather than two?
* What's the minimum number of connections necessary to achieve successful walking?
* And what types of oscillator are best for achieving speed and stability in walking?
* Can any CTRNs produce several different oscillators?
* Or a combination of oscillator/s and point attractor/s?
* If so, what's needed for these oscillators to be smoothly coordinated?
* Last, but not least: for any given architecture, are there particular clusters of parameter-values that are especially promising?

They discovered, for example, which class of 3-unit CTRNs is best suited for controlling walking. There were three behaviourally distinct groups: 29 out of the 64 3-unit possibilities generated either no movement at all or only a single step; 8 produced slow, jerky, walking; and 27 resulted in fast, rhythmically smooth, stepping. On inspecting the connectivity patterns of the three groups, they found key differences in the existence and connectivity of feedback loops between the motor neurones for the foot and the two swing muscles. They then asked whether the same graph-theoretic properties could be used to *predict* similar classes of behavioural output in 4- and 5-unit nets.—Yes, they could. (There were some exceptions, but these could be explained: see 2004, sect. 4, paras. 7–8.)

In addition, some types of network were found to be *more easily evolvable* than others (2004, sect. 5). And qualitative descriptions like those mentioned above were supplemented by quantitative measures of fitness, distinguishing each individual motor neurone. These measures enabled the team to study an architecture's average results as well as its best ones. Some types were found to be more generally useful than others, even though their best performances were lower. (Di Paolo, you'll remember, had selected the robots which approached the lights most closely *on average*. But he couldn't say what type of controller was most likely to give a good performance-average.)

Even more recently, Beer's group have observed how the phase portrait of a dynamical system changes when the numerical value of one of the parameters is changed (R. D. Beer 2005). A "movie" of the effects of varying the bias in a 2- or 3-unit network, for example, shows how the system changes smoothly from one attractor to another, or diverges from any equilibrium. Plotted as a diagram of a two- or three-dimensional space, a particular region (or regions) can be identified where most of the "interesting" things happen, and others can be seen to be sterile. (Compare the pioneering fourfold classification of CAs mentioned in ix.a, above.)

This abstract analysis has intriguing biological and psychological implications. For example, given a specific CTRN, one can sometimes predict how its behaviour will change as the parameters are varied in certain ways. And Beer has recently used these ideas to explain the behaviour of a "minimally cognitive agent" with as many as fourteen artificial neurones (personal communication). Conceivably, such analysis might even help biologists to show why certain behaviours have evolved (i.e. are easily evolvable) from one origin rather than another.

However, "intriguing" and "conceivably" is as much as can be said, at present. And maybe you shouldn't hold your breath. The very recent (2005) analysis just described is certainly exciting, for it can be applied to CTRNs of *any* size. But readily intelligible 'Euclidean' diagrams can be drawn only for two or three dimensions. The larger CTRNs are like the multidimensional hyperspaces (including the imaginary scholarship rules) discussed in Chapter 14.viii.b: not beyond mathematical understanding, but fiendishly difficult.

As for Beer's *complete* descriptions of the space of architectures, these have been applied only to very tiny CTRNs. Even artificial CTRNs may have more units than can be dealt with in this way, and real organisms are huge by comparison: the minimalist. *C. elegans* has 302 neurones (14.v.b), and mammalian brains have many millions.

These biological facts may not be quite so devastating as they seem, for the Case Western researchers discovered that *more* need not be *better*: once there are five or more connections (connections, not units), the best fitness for the robot-walking task hardly alters (Psujek *et al.* 2004, sect. 3). That may be why, as remarked above, biological networks—even including the cerebral cortex—are very sparsely connected in comparison with full-blooded CTRNs. Nevertheless, the numbers remain awesome. (It's not clear that increased computer power will help much; but possibly quantum computers?—see 16.ix, preamble.)

In other words, the combinatorial explosion threatens work in dynamical systems just as it threatens GOFAI. So one shouldn't expect Beer to provide further exhaustive analyses of CTRNs: five units today, eight tomorrow, perhaps the elegant 302 next month . . . Rather, one should see his work—including the varying-parameter phase portraits—as a promising exploration of *the kinds of effects, properties, or concepts* that may be needed in thinking about realistic (i.e. larger) systems.

For example, rhythmic behaviour in general—not just walking, and not just cock-roaches—depends on oscillator attractors, which in turn depend on there being closed loops in the architecture. With hindsight, that may be obvious. But it wasn't obvious, even to committed dynamical theorists (such as Di Paolo: personal communication), before Beer's analysis.

Much the same might be said about psychological AI. Consider, for instance, research on mental architecture (Chapter 7.i.e–f), on anxiety-ridden speech (7.ii.c), on the complex webs of relevance involved in everyday thinking (7.iii.d), or even on the development of the past tense (12.vi.e and x.d). The hope of the cognitive scientists concerned is not to map the computations involved in precise detail, but to understand *what sort of system* a personality (or a neurosis, or an associative memory, or an infant learning to speak) is, and *in what sorts of ways* it can be manifested, in *what sorts of circumstances*. Most computational psychologists believe that concepts drawn from classical and connectionist AI can answer those questions. By contrast, dynamicists believe that dynamical concepts and insights can best help us understand what sort of system an organism, or a mind, is. Similar bets, different horses.

In sum, dynamical systems are still more theoretically opaque than either GOFAI or orthodox connectionism. It's not clear how far Beer's analysis can be applied to the dynamics evolved by Nature. And, for all Di Paolo's 8-unit ingenuity, Ashby may have been right to fear that future Homeostats would be "too complex and subtle for the designer's understanding" (1948: 383). But some provocative beginnings have been made. If many open questions still remain, and others can't yet even be asked, that's par for the course in any creative science.

PHILOSOPHIES OF MIND AS MACHINE

Philosophies of mind and machine have been prominent since the 1630s. But for René Descartes and virtually all of his successors, that "and" really meant "not". Mind was contrasted with machines, not likened to them (see Chapter 2).

Philosophies of mind *as* machine didn't appear until over 200 years later (Chapter 4.ii and vi). Today, they still have many philosophical adversaries. And even those philosophers who are sympathetic to mind-as-machine disagree about *just which* machines these might be. Nor would they accept Herbert Simon's claim that, in writing the Logic Theorist (6.iii.c and 10.i.b), he and his colleagues had "solved the venerable problem of how a system composed of matter can have the properties of mind" (Simon 1991: 190).

Evidently, then, philosophical problems don't get solved in a hurry. One might even say that they don't get solved at all—for if they do, they're relabelled as "scientific" problems. AI often suffers a similar fate (see 13.vii.a). Having bravely located firm ground in unmapped and marshy territory, its successes are renamed "computer science" and what was once exploration becomes tourism. Moreover, some philosophers insist that their aim is not to solve problems but to *dissolve* them, "to show the fly the way out of the fly bottle" (Wittgenstein 1953, para. 309; cf. A. J. T. D. Wisdom 1952: 259).

Soluble or not, philosophical problems have already featured in our discussion. Chapter 2 outlined two broad philosophical movements with competing influences on the life sciences. And other chapters have included arguments on specific topics, such as:

* free will (Chapters 5.ii.a and 7.i.g);
* innate ideas and nativism (7.vi, 8.v−vi, 9.vii, 12.viii.c−e, 12.x.e, and 14.x.c);
* word meaning (4.iii.c, 7.ii.d, 9.iv and x.d);
* concepts (5.iv.c, 8.i.b, 9.x.d, and 12.x);
* non-conceptual content (12.x.f);
* representations (7.i.h, 7.v.a, 12.ix.e, and 14.viii−ix);
* scientific explanation (7.iii and 12.x.f);
* consciousness (7.i.g−h and 14.x−xi);
* relativism and scientific objectivity (1.iii.b);
* tacit knowledge (13.ii.b); and
* life (15.vii.b).

Some issues in the philosophy of religion were implicit in Chapter 8.vi, and even metaphysics has been mentioned (13.i.b–c and 7.iv.e—and see the discussion of computation in Section ix.e, below).

A fortiori, cognitive science is related to the philosophy of mind. That's the general topic of this chapter. Section i outlines the competing philosophies of mind dominant at the inception of cognitive science, and the various philosophical impasses involved. To understand those is to appreciate why, in the late 1950s and early 1960s, cognitive science seemed to be such a liberation.

Section ii considers the influence of the Turing Test on philosophy and AI. Various forms of functionalism are discussed in Sections iii and iv. Some objections to functionalism feature in Section v, and the founder's recantation in Section vi. Further neo-Kantian critiques are discussed in Sections vii and viii. Differing concepts of computation, and their relation to intentionality, are explored in Section ix. Finally, Section x deals with the relevance of life to mind.

Aperitif. Before starting, however, we need to ask why non-philosophers should bother with philosophy at all.

Many think they should not. For instance, one reviewer of an early outline of this book suggested that the proposed chapter on philosophy be entirely omitted. Cognitive science, he/she evidently believed, can get on very well without it. (And this *despite* the fact that most leaders in the field find it exciting largely because they believe that it solves the notorious mind–body problem.)

Even philosophers admit that "First-rate philosophy can be profoundly irritating, especially to non-philosophers" (Bringsjord and Zenzen 2002: 241). The slow-ness/insolubility mentioned above is one common source of irritation. But the authors just quoted were thinking of another:

[Following Socrates' example, the philosopher] starts with innocent enquiry, pushes and probes, makes some seemingly innocuous inference . . . and boom!—suddenly she has shown that what scientists and engineers take for granted *shouldn't* be taken for granted. (ibid.)

Occasionally, to be sure, what was taken for granted is later confirmed by the philosopher concerned, with stronger arguments provided to buttress it. Even then, the common-sense belief may have been transformed in the process (see the discussion of free will in Chapter 7.i.g). That doesn't make for comfortable reading. But sometimes, what was previously assumed is rejected altogether. In that case, the question is left worryingly open. And there's no hope for a neat resolution in the next number of *Nature*.

Hard-headed scientists, as a result, are likely to agree that "There is nothing so absurd that it has not been said by some philosopher"—a remark of Cicero's, quoted by Descartes and by many others after him. (For Cicero and Descartes, of course, "philosophy" included what we would call science. It's ironic, then, that the biologist Lewis Wolpert, while repeatedly dismissing philosophy in the most contemptuous terms, celebrates the seeming absurdity, or "unnaturalness", of science: L. Wolpert 1992.)

Friedrich Nietzsche, some 2,000 years after Cicero, even remarked that a philosopher is "a terrible explosive, endangering everything" (Kauffmann 1969: 281). That wasn't a criticism, for he wasn't hoping for (epistemological) safety. Most practising scientists, by contrast, are.

Myriad beliefs normally taken for granted are questioned (*questioned*, not necessarily rejected) in the philosophy of cognitive science. For instance: that mind requires life; that zombies are logically possible; that human freedom can't be scientifically explained; and that dogs and monkeys have conscious experiences—and intentions, too. The list could go on, and on . . .

That's irritating enough, perhaps. But one challenge in particular annoys the "scientists and engineers": the denial that a real world exists independently of human minds, and that science can find out about it.

This counter-intuitive claim isn't new. It was lying in wait at the heart of Descartes's philosophy (he toyed with it himself, in suggesting that his whole life might be a dream: 2.iii.b), and was later forefronted by the neo-Kantians (2.vi and vii.c). Nor is it tucked away safely inside dusty philosophy books. To the contrary, it's often wheeled up to the scientists' front door. The "strong programme" in the sociology of knowledge defends this claim, and has led to vituperative "science wars" in which gallons of metaphorical blood have been angrily spilt (see 1.iii.b).

Scientists, however, have little patience for the anti-realist challenge. (That may not be true of quantum physicists, but in my experience it applies to most life scientists.) Even when they're discussing and/or dismissing philosophy as such, they often don't mention this aspect of it. Or if they do, they don't give it more than a glancing thought. Gerald Edelman is a case in point, as we've seen (14.ix.d).

Philosophers of cognitive science usually are willing to argue about realism (although sometimes they deliberately sidestep the question: see vii.d, below). But they almost always end by accepting it, *even while admitting that they can't decisively prove it*. That's what I did, provisionally, in Chapter 1.i.c and iii.b; and it's what Jerry (Jerrold) Fodor does too (see viii.b, below). Indeed, seven of the ten sections below ignore the realism/anti-realism dispute. It features only in Sections vi–viii.

So why ask non-philosopher readers to consider it at all? One reason is that some experimental cognitive scientists, the followers of Humberto Maturana and Francisco Varela (x.c, below), support views which—whether they realize it or not—are fundamentally anti-realist. (Occasionally they do realize it, and explicitly accept them: I. Harvey 2005.)

Another is that the counter-cultural opposition mentioned at various points in this book (e.g. 1.iii.b–d, 6.1.d, and 11.ii.f) is grounded not only in political suspicion of science's applications, but (often) also on neo-Kantian scepticism about its realist foundations. The counter-culture can be properly understood, and persuasively (if not definitively) rebutted, only if that is appreciated.

A third is that an interesting new twist on the realism/anti-realism debate is related to the recent AI technology of virtual reality, or VR (13.vi). Some philosophers' reactions to this draw on work in cognitive science (see viii.c, below).

The main reason, however, is that anti-realism is grounded in an alternative, and in non-scientific circles highly influential, philosophy of mind. If that philosophy is correct, then cognitive science is fundamentally wrong-headed—indeed, impossible. For to drop realism is to drop naturalism: the belief that there could, in principle, be a science of mind. "Cognitive science", on this alternative view, should be a very different kind of enterprise: hermeneutic, not naturalistic (Harré 2002). Intentionality is to be presupposed, not explained.

In sum, no serious consideration of cognitive science can ignore philosophy. Quite apart from the specific philosophical issues that arise *within* the field (concepts, representations, zombies, free will, self . . . and, of course, the mind–body problem), its very possibility, as a coherent intellectual enterprise, is in doubt. Those doubts must be understood, even if one ultimately rejects them—as I do myself. Moreover, the huge attraction of the field lies largely in its promise to help solve one of the greatest philosophical puzzles of all: the mind–body problem.

This chapter, then, explains why that promise seemed to many people to be convincing—and why many others are *still* not convinced.

16.i. Mid-Century Blues

In the 1940s and 1950s, four philosophies of mind competed for attention from English-language philosophers. ('Continental' philosophers favoured fundamentally different accounts, not taken seriously in analytic circles at the time: see Section vii.) Mind-as-machine wasn't one of them.

GOFAI, connectionist, and cybernetic approaches were already hovering in the wings, to be sure (see Chapters 4–15). But professional philosophers were looking straight ahead, on-stage.

There, the forerunners in the talent competition—in order of appearance—were epiphenomenalism, logical positivism, logical behaviourism, and identity theory. Alan Turing was cast as a support dancer for behaviourism's solo: not until the 1960s would he move onto centre-stage.

Each key contestant attracted many fans. But few onlookers sat comfortably in their seats. As we'll see, they had to admit to—or systematically repress (Feigl 1958/1967: 3)—worrying flaws in every performance.

a. Interactionist squibs

Besides these four, there was an also-ran: interactionism. This was dismissed out of hand by most mid-century thinkers, who shared Princess Elizabeth's scepticism about the mind's ability to move the brain (see Chapter 2.iii.b). An occasional philosopher, however, insisted that interactionism—and even substance dualism—hadn't actually been refuted (Ewing 1954: 114).

Some scientist–philosophers went even further. The chemist Michael Polanyi, for instance, suggested that "some enlarged laws of nature may make possible the realization of operational principles acting by consciousness", and that the mind might "exercise power over the body merely by sorting out the random impulses of the ambient thermal agitation" (Polanyi 1958: 397, 403 n.).

Similarly, the leading neuroscientist John Eccles (1903–97) argued that conscious choice is a matter of essentially mental influences causing changes in the brain (see Chapter 2.viii.f). This applied only to the cerebrum: as we saw in Chapter 14.v.c, Eccles was happy to describe the cerebellum as "a Neuronal Machine".

Eccles avoided speculations about the pineal gland, preferring to tell fairy stories in modern scientific language ("fields", "quantum effects", and "critically poised neurones"). But he encountered much the same logical difficulties that had bedevilled Descartes long before.

He said, for instance: "[My] hypothesis assumes that the 'will' or 'mind influence' has itself a spatio-temporal patterned character in order to allow it this operative effectiveness [on specific cells]" (Eccles 1953: 277–8). This was given almost as an aside: *how it is possible* for an essentially mental influence to be spatio-temporal wasn't discussed. Eccles's move, here, invites Peter Medawar's classic comment on Teilhard de Chardin's even woollier remarks about consciousness (especially how individual minds can be absorbed in the one Omega consciousness): "And so our hero escapes from his appalling predicament: with one bound, Jack was free" (Medawar 1961: 104).

Some twenty years later, a distinguished philosopher of science—one of the tiny handful of people elected to both the British Academy and the Royal Society (others are the cognitive scientists Uta Frith and Philip Johnson-Laird)—would join Eccles in promoting interactionism (Popper and Eccles 1977). Karl Popper's obituarist for the Academy, a close colleague for many years, said of him, "At heart he was a Cartesian interactionist" (Watkins 1997: 679), and mentioned his saying that he believed in what was then being dubbed "the ghost in the machine". This was at a talk given in Oxford, where Gilbert Ryle's ghost-denying approach was then dominant (see subsection c, below).

By the late 1970s, there were many other ghost-deniers besides Ryle. All of them upstaged Popper. Despite his high standing in the profession, the audience for his interactionist cameo was small, and the applause muted.

b. Puffs of smoke and nomological danglers

The first 'spot' in the contest was epiphenomenalism, singing a hymn to body–mind causation but trashing mind–body influence. This competitor had been ushered on stage by Thomas Huxley long before, in the 1870s (Chapter 2.viii.b). He'd costumed mental events as puffs of smoke from a steam-engine: unmissable, but ineffectual.

Now, eighty years later, many scientifically minded people—including most neuro-physiologists—were still applauding it. Many expressed it in much the same (Cartesian) terms as Huxley had done. But the denial of mind-to-body—and even mind-to-mind—causation remained counter-intuitive, and the posited body-to-mind causation remained a metaphysical mystery.

Even when put less provocatively, in twentieth-century (phenomenalist) terms of lawful regularities rather than causal interaction, the doctrine was still problematic. Mental events are regularly connected with the physical story, but—given that the physical story is apparently seamless—they don't lead back into it. They don't even lead without interruption to each other, for the mental story isn't seamless. Nor is it regular: thought x is not predictably followed by thought y.

Herbert Feigl (1902–88) noted these difficulties:

Causality [i.e. regular succession] in the mental series is by far too spotty to constitute a "chain" of events sufficiently regular to be deterministic by itself. Epiphenomenalism in a value-neutral scientific sense may be understood as the hypothesis of a one-to-one correlation of [mental states] to (some, not all) [cerebral states], with determinism (or as much of it as is allowed by modern physics) holding for the [cerebral] series, and of course the "dangling" nomological relations connecting [the cerebral and mental events]. (Feigl 1958/1967: 15)

Feigl's nomological danglers were perhaps less puzzling than Huxley's body-to-mind causation, but they were a philosophical embarrassment nonetheless. They were especially problematic for those people—including many leading linguists and psychologists, as well as Feigl himself—who shared the logical positivists' faith in the unity of science (Chapter 9.v.a).

The logical positivists had bounded onto the boards in the 1930s. They saw their role not as cutting the mind–body Gordian knot, but as preventing it from being tied in the first place. They argued that the mind–body problem, like all metaphysical puzzles, is a "pseudo-problem" generated by misunderstanding of language. Although they held that all knowledge is based in sense experience, they insisted that their approach wasn't just another form of post-Cartesian idealism. And they briskly denied any fundamental difficulty in relating mind and body:

The answer to the question whether sense-contents are mental or physical is that they are neither; or rather, that the distinction between what is mental and what is physical does not apply to sense-contents. It applies only to objects which are logical constructions out of them. But what differentiates one such logical construction from another is the fact that it is constituted by different sense-contents or by sense-contents differently related . . . [There] is no philosophical problem concerning the relationship of mind and matter, other than the linguistic problems of defining certain symbols which denote logical constructions in terms of symbols which denote sense-contents . . . Being freed from metaphysics, we see that there can be no *a priori* objections to the existence of either causal [i.e. Humean] or of epistemological connections between minds and material things. (Ayer 1936: 123–4)

So far, so dismissive. Too dismissive, for many people. Even Feigl, who'd been involved with logical positivism since its inception in 1920s Vienna, refused to accept that mind–body is a mere pseudo-problem. (His own solution was a version of the identity theory: see below.)

c. Dispositions and category mistakes

A third philosophical contestant appeared in the late 1940s, eliciting both vociferous applause from the gallery and angry hisses from the stalls: logical behaviourism. This too, though for different reasons, saw the mind–body problem as an illusion based in misunderstandings of language.

The most well-known script used by this contender was Gilbert Ryle's book *The Concept of Mind* (1949), whose advance flyer had been posted a few years earlier (Ryle 1946). The book was immediately hailed in *Mind* as "probably one of the two or three most important and original works of general philosophy which have been published in English in the last twenty years"—*despite* (said this reviewer) its having fundamental flaws (Hampshire 1950: 237). Ryle's crisp sentences and humorous polemic were exceptionally widely read. There were five reprints within five years, and many more thereafter.

Ryle (1900–82) was an Oxford philosopher who often traversed the English countryside in the 1930s to hear Ludwig Wittgenstein (1889–1951) lecture in Cambridge.

Ryle was deeply influenced by him, and they became close friends. But the amity didn't last. Wittgenstein's 'later' work had circulated privately for some years (Chapter 9.x.d). To his annoyance, however, Ryle's book was published first and received enormous attention.

Wittgenstein's chagrin wasn't merely priority pique, nor simply plagiarism paranoia. (He was quick to accuse others of plagiarizing, and misrepresenting, his work: the young Richard Braithwaite, for example, was forced to write a letter to *Mind* 'apologizing' to him on this count: Wittgenstein 1933; Braithwaite 1933.) He felt that Ryle had not only benefited from his ideas, but also misrepresented—or anyway, altered—them. For instance, Ryle relied heavily on "dispositions", a term that Wittgenstein disliked (see below).

Their earlier friendship dissolved into thin air. Wittgenstein dismissed Ryle as not "serious". For his part, Ryle deplored Wittgenstein's "unhealthy" and "pedagogically disastrous" influence on his students (Monk 1990: 275, 495).

Ryle dismissed Descartes's dualistic "myth" of "the ghost in the machine" as radically incoherent and/or a source of countless examples of infinite regress. For example, it explained "intelligent" action in terms of "intelligent" thought—hardly a helpful move.

Descartes was cast as the villain in Ryle's script. However, the *Mind* reviewer pointed out that Descartes wasn't the only one to be blamed. As he put it, "the myth of the mind as a ghost within the body is one of the most primitive and natural" of all beliefs (Hampshire 1950: 239). Moreover, the words for mind, soul, or spirit in most European languages had the same etymology as the words for ghost "long before Descartes or modern mechanics". It was true, however, that Descartes's account had been especially clear, and especially influential.

But if Descartes had been talking nonsense, said Ryle, he hadn't been talking utter nonsense:

A myth is of course not a fairy story. It is the presentation of facts belonging to one category in the idioms appropriate to another. To explode a myth is accordingly not to deny the facts but to re-allocate them. And this is what I am trying to do. (Ryle 1949: 8)

For Ryle, everyday psychological terms don't denote occurrences within some non-material world. (Compare Wittgenstein: "When I think in language, there aren't 'meanings' going through my mind in addition to the verbal expressions: the language is itself the vehicle of thought" and "Speech with and without thought is to be compared with the playing of a piece of music with and without thought"—1953, paras. 329, 341.)

Rather, such terms should be analysed as shorthand descriptions of actual *and possible* behaviour. That is, they concern behavioural "dispositions"—comparable to the brittleness of glass.

To say that glass is brittle is to say that it's disposed to break, in certain circumstances. A given piece of glass may never actually break, but it's brittle nonetheless. *If* it were dropped onto a hard surface, *then* it would break. Analogously, Ryle argued, to say that someone is jealous, or happy, or thinking about a problem . . . is to say that they are disposed to behave in certain ways, given certain circumstances. Even to say that someone is hot, or itchy, or seeing red (in either sense!) is to remark on their actual and/or potential behaviour.

He allowed that people can keep their thoughts and feelings to themselves. But he analysed this as a counterfactual claim. What it meant, he argued, was simply that *if* the circumstances had been different (if their interlocutor were less indiscreet and/or more trustworthy, or if they were on the point of death, or . . .), *then* they would have expressed them.

Words denoting "propositional attitudes" (Bertrand Russell's term: B. Russell 1918–19)—such as *know, believe, desire, prefer, fear,* and *hope*—were treated in the same way. To believe that *p* is to be disposed to say that *p*, and to behave in ways that would be appropriate (given the person's other beliefs and desires) if *p* were true. As in this example, psychological concepts are generally defined in terms of others. So this approach put a new gloss on what would previously have been thought of as mind–mind (or even mind–body) causation. Pity, for instance, is—among other things—the disposition to help and/or comfort ('behaviour') someone seen to be in distress ('circumstances'). So to say that Mary helped Jane "because" she pitied her is to mark a conceptual link, not a causal (contingent) one.

As for intelligence (said Ryle), this isn't some ghostly power causally responsible for behaviour, but rather a certain manner of behaving. Intelligent behaviour is carried out carefully, thoughtfully, skilfully, appropriately, and flexibly. Largely, the intelligence manifests itself only as tacit *knowledge how*. Sometimes, a person can comment on aspects of their behaviour, and answer questions or give advice, by verbal statements of *knowledge that*. Intelligence in general, however, isn't a matter of explicit reasoning—whether kept to oneself or shouted to the rooftops. In a nutshell:

Overt intelligent performances are not clues to the workings of minds; they are those workings. Boswell described Johnson's mind when he described how he wrote, talked, ate, fidgeted and fumed. (Ryle 1949: 58)

In general, then, Ryle didn't try to forbid talk of mental states, but he gave an unfamiliar analysis of their nature. (Similarly, computational analyses of human freedom admit that people have a psychological characteristic which animals don't, but gloss this in an unorthodox way: see Chapter 7.i.g.)

All very plausible, possibly, for third-person remarks about mental states. But what about first-person statements? Surely these are reports of inner happenings?—Ryle was obdurate. He said (for instance) that first-person statements about one's feelings of emotion are "avowals" rather than reports:

Avowing "I feel depressed" is doing one of the things, namely one of the conversational things, that depression is the mood to do. It is not a piece of scientific premiss providing, but a piece of conversational moping. That is why, if we are suspicious, we do not ask "Fact or fiction?", "True or false?", "Reliable or unreliable?", but "Sincere or shammed?" The conversational avowal of moods requires not acumen but openness. It comes from the heart, not the head. It is not discovery [by Cartesian direct access], but voluntary non-concealment. (Ryle 1949: 102)

(Compare Wittgenstein: "'So you are saying that the word "pain" really means crying?'—On the contrary, the verbal expression of pain replaces crying and does not describe it"—1953, para. 244.)

For all his hard-headedness in rejecting the Ghost, Ryle had no interest in saving the Machine:

[The] influence of the bogy of mechanism has for a century been dwindling because, among other reasons, during this period the biological sciences have established their title of "sciences". The Newtonian system is no longer the sole paradigm of natural science. Man need not be degraded to a machine by being denied to be a ghost in a machine. He might, after all, be a sort of animal, namely, a higher mammal. There has yet to be ventured the hazardous leap to the hypothesis that perhaps he is a man. (Ryle 1949: 328)

Ryle's reference, here, to a non-Newtonian paradigm in biology is puzzling. He may have been thinking of Claude Bernard's work on self-organization, or Walter Cannon's on homeostasis (Chapter 2.vii.a). Presumably, he wasn't referring to their 1940s equivalent, cybernetics. For his cybernetic contemporaries were arguing that all organisms are fundamentally like *some* machines, albeit machines unlike anything imagined by Descartes (see Chapter 4.v–vii).

Logical behaviourism was very different from the psychological variety discussed in Chapter 5.i and iii. Psychological behaviourists, in essence, were experimental epiphenomenalists. They claimed that mental (conscious) states are ontologically distinct from the body; that proper knowledge of mental states can be had only by the person whose states they are; and that, for scientific purposes, third-party concepts of mental states can be defined in the non-intentional terms of *stimulus* and *response*. As noted above, Ryle rejected every one of these claims.

So did Wittgenstein. He wrote:

"Are you not really a behaviourist in disguise? Aren't you at bottom really saying that everything except human behaviour is a fiction?"—If I do speak of a fiction, then it is of a *grammatical* fiction.

How does the philosophical problem about mental processes and states and about behaviourism arise?—The first step is the one that altogether escapes notice. We talk of processes and states and leave their nature undecided. Sometimes perhaps we shall know more about them—we think. But that is just what commits us to a particular way of looking at the matter. For we have a definite concept of what it means to know a process better. (The decisive movement in the conjuring trick has been made, and it was the very one we thought quite innocent.)—And now the analogy which was to make us understand our thoughts falls to pieces. So we have to deny the yet uncomprehended process in the yet unexplored medium. And now it looks as if we had denied mental processes. And naturally we don't want to deny them. (Wittgenstein 1953, paras. 307–8)

And yet you again and again reach the conclusion that the sensation itself is a *nothing*.—Not at all. It is not a *something*, but not a *nothing* either! The conclusion was only that a nothing would serve just as well as a something about which nothing could be said. (para. 304)

Some cognitive scientists today, including Daniel Dennett and Aaron Sloman (both of whom encountered Ryle while studying at Oxford), see themselves as following in Ryle's footsteps (Dennett 1969, pp. xi, 2000; Sloman 1996c, Acknowledgements). This may seem strange. For his neo-Wittgensteinian assumption that talk of non-material processes can be understood only dualistically was very different from cognitive science—which is mentalistic, although not dualistic. However, there are two good reasons for arguing that Ryle should be seen as a precursor of cognitive science, not just a predecessor (see 9.ii).

First, Ryle's subtle analyses of psychological concepts, and of the rich interconnections between them, were highly illuminating as accounts of mental architecture. His account of emotions (1949: 83–115), for instance, was an inspiration for Sloman's work on the topic (Chapter 7.i.f). And in exhibiting the conceptual (rational) links between the many varieties of belief and desire, he prefigured Dennett's theory of intentional systems (see Section iv.b).

Second, Ryle's term "disposition" can be interpreted in a way that fits very well with cognitive science. Whether Ryle himself understood the term in this way is another matter.

Most talk of "dispositions" is systematically ambiguous, denoting behaviour and/or its underlying causes. This is why Wittgenstein preferred to avoid the term (1953, para. 149 and pp. 191–2). The psychologist William McDougall had explicitly defined the word in both these senses, and had kept them clearly distinct (McDougall 1923; see 5.ii.a). Ryle didn't.

His term is usually read *descriptively*, as denoting *an observable tendency to behave in a certain manner* (O'Shaughnessy 1970). This sense was dominant in the 1950s, partly because analytic philosophers of mind didn't see it as their role to offer explanations. That, Wittgenstein had taught them, was the business of brain scientists. However, Ryle's term was later read also as *explanatory*, denoting *the mechanism responsible for the relevant behaviour* (Squires 1970).

Whether Ryle himself understood it in this way is doubtful. But cognitive scientists in general favour the dual reading. And, while any "mechanism" which Ryle might have intended would have been neurophysiological, they allow computational explanations too. (Whether they all expect these to map neatly onto the propositional attitudes is another question: see Section iv.)

If Ryle's bravura performance was hailed by many of his contemporaries, it was fiercely criticized by others. Even the adulatory *Mind* reviewer was ambivalent. Quite apart from his analytic philosophical method in general (which he later confessed was the *real* topic of the book, mind–body being a "notorious and large-size Gordian knot" that he could use as an illustrative example—Ryle 1971: 12), there were two major problems.

First, it wasn't clear that dispositional analyses could be fully spelt out in practice, not least because so many psychological concepts are defined in terms of others. Moreover, many people felt that consciousness was still missing—or, to put it in linguistic terms, that philosophical behaviourism didn't give a satisfactory account of first-person psychological statements. Ryle's analyses of sensations, and Wittgenstein's dismissal of private languages (1953, paras. 243–76) and "the beetle in the box" (1953, para. 293), failed to persuade them that Cartesian mental processes are dispensable. On the contrary, they seemed—to many critics—to be recalcitrantly real. (A dialogue between Ryle/Wittgenstein and the obstinate "Otto" might have been very similar to that given in Chapter 14.xi.b.)

d. Questions of identity

In 1956 a fourth challenger entered the limelight—and, especially for scientific audiences, threatened to win the contest. This was mind–brain identity theory.

The new theory accepted dispositional analyses for most psychological predicates, adopting the dual interpretation that posits explanatory mechanisms in the brain. But sensations, or experiences, were not analysed in this way. Nor were they understood as epiphenomena. Denying any special realm of conscious events, identity theory held that sensations are *identical* with brain states. The brain state doesn't cause the sensation, nor is it correlated with it. Rather, it constitutes it. Ontologically there is only one thing. But it can be described, and known, in two ways.

Identity theory was first published in a scientific journal, the *British Journal of Psychology*, by Ullin Place (1924–2000). (As usual, there were precursors: Feigl, for instance, had been thinking along similar lines for some time.) Place insisted that the theory was "a reasonable scientific hypothesis, not to be dismissed on logical grounds alone" (Place 1956: 44). Only future work in neurophysiology could confirm its truth, he said, although suggestive empirical evidence was already available.

"Logical" objections were sure to be made, of course, and Place anticipated two. The first was that our introspective access to our conscious states proves that they are events of a different order. Place called this "the phenomenological fallacy". He argued that it rested on "the mistaken idea that descriptions of the appearances of things are descriptions of the actual state of affairs in a mysterious internal environment" (1956: 44).

The second was that mental and physical states are radically different, so can't possibly be identical. Or, expressed less 'metaphysically': the meanings of mental and physical terms are very different, which is why people can talk about their experiences without knowing that the brain is in any way involved.

Distinguishing "the 'is' of definition" from "the 'is' of composition", Place replied that things falling under very different concepts may in fact be compositionally identical. Some such identities are superficial and contingent (Mary's table *is* an old packing case). But some are deep and systematic, and discoverable only by science. People had talked about lightning for thousands of years, but scientists eventually discovered that lightning *is* electric charges in motion. It follows, Place argued, that the fact that our concepts of mental states are logically very different from our concepts of brain states doesn't prove that they pick out non-identical things.

Stated thus, Place was right. Feigl made the same point by appealing to Gottlob Frege's distinction between *sense* and *reference* (Chapter 2.ix.b). But even two "very different" concepts may share an important feature. All Place's examples relied on shared spatio-temporal dimensions. (Think of how one discovers that a table is a packing case, or that lightning is electric discharge.) And, *pace* Eccles, conscious states—notoriously—are not in space.

Accordingly, the identity theory was widely regarded as paradoxical. Even David Armstrong, who published the best-developed version of it—known as central-state materialism—ten years later, admitted: "Certainly I myself found the theory paradoxical when I first heard it expounded" (Armstrong 1968: 73).

The next move—taken by Jack Smart (1920–), in whose Adelaide department Place had written his paper—was intended to resolve the paradox. Smart suggested that we wield Occam's razor and *decide* to treat experiences and brain states as identical. In that case, an experience could be said to have the same location as the brain process which constitutes it. He even said:

[We] may *easily* adopt a convention (which is not a change in our present rules for the use of experience words but an addition to them) whereby it would make sense to talk of an experience in terms appropriate to physical processes. (Smart 1959; italics added)

Many critics could hear Jack (the giant killer) bounding again. Norman Malcolm (1911–90), for instance, insisted that Smart *was* suggesting a change in usage, and one that would be fundamentally incoherent (Malcolm 1964). But at least Smart, unlike his successor Francis Crick (14.x.d), recognized that any such change would be a scientifically grounded philosophical decision, not a scientific discovery.

Whether the scientific grounding required could ever be delivered was unclear. The identity theorists held that each type of mental state is, as a matter of fact, identifiable with a certain type of brain state. Yet there were many reasons to doubt this—some scientific, some philosophical.

One was Karl Lashley's (1950) recent discovery that memories seem not to be located in any specific part of the cortex (Chapter 5.iv.a). Nowadays, the term "brain state" could include some highly distributed pattern of activity (see Chapters 12 and 14), making Lashley's findings less problematic. At mid-century, however, "brain state" usually intended the activity of a single neurone or localized cell assembly. In those terms, hardly any positive evidence was available. Single-cell feature-detectors were yet to be discovered (see Chapter 14.iv). It had recently been found that complex conscious states can be elicited by stimulating points in the temporal cortex. But their content was predictable only, if at all, in the individual concerned (Penfield 1952, 1958).

Natural language, too, presented a problem. Except on an extreme Humboldtian view (Chapter 9.iv.b), two people can sometimes express the same thought. But if they speak in different tongues (different phonemes, etc.), their brain states must differ in many ways. Even dogs gave pause for thought. Presumably dogs, and possibly Martians, can feel (non-conceptualized) pain, much as a baby can. But a dog's or Martian's brain is very different from a baby's—which would imply (according to the identity thesis) that it *cannot* feel pain (David Lewis 1980).

Besides 'scientific' hesitations such as these, more fundamental attacks were mounted by Wittgensteinian philosophers. Malcolm (1964), for instance, argued that only someone whose culture includes milkmen can have the sudden thought that they haven't put out the milk bottles. This thought therefore can't be identified with a brain state, for that could conceivably occur (perhaps due to microsurgery) in anybody's head. Any brain state could even happen, as Smart himself had pointed out (1963), in a brain *in vitro*. But for Malcolm, the notion that a brain—in a head or in a vat—can *think* is literally absurd. Thoughts can be sensibly ascribed only to embodied and enculturated persons.

At the end of the 1950s, then, none of the four main rivals had monopolized the applause, and all were receiving some highly critical reviews. Despite the near-universal rejection of mental substance, Descartes's mid-twentieth-century heirs faced much the same mind–body problems as he had done (see Chapter 2.iii.b).

Some voices off-stage—across the Channel—were arguing that "mind" and "body" are Cartesian illusions anyway (see Sections vii–viii below, and Chapter 14.xi). But most of the analytic audience wasn't listening. Those few—notably Ryle and Wittgenstein—who had heard them were expressing this view in different terms.

(About ten years later, at a meeting held in France, Ryle admitted that his book "could be described as a sustained essay in phenomenology, if you are at home with that label"—1962: 188. In the discussion, which also included A. J. Ayer and Willard Quine, one of the leading Continentals replied: "[What Mr Ryle] was saying was not so strange to us, and the distance, if there is a distance, is one that he puts between us rather than one I find there"—Merleau-Ponty 1960: 65. And the 'analytic' reviewer of the meeting's Proceedings apparently agreed: C. M. Taylor 1964b. All this isn't too surprising, for as a young man Ryle had been an exponent of Edmund Husserl, and had reviewed Martin Heidegger's *Being and Time* with respect. The respect was coloured with scepticism, however. Ryle's review opened with "This is a very difficult and important book . . . though . . . I suspect that this advance is an advance towards disaster," and closed by saying, "Phenomenology is at present heading for bankruptcy and disaster and will end either in self-ruinous Subjectivism or in a windy mysticism"—Ryle 1929: 355, 370.)

As for mind-as-machine, the topic had barely arisen. On the rare occasions when mind—or rather, adaptive behaviour—was likened to machine in the philosophical journals (Ashby 1947; J. O. Wisdom 1951), it was cybernetic machines which were in question (see Chapter 4.v–vii). Digital computers were hardly mentioned until 1950, when a provocative squib in *Mind* written by Alan Turing prompted a flurry of discussion about the possibility of machine intelligence.

16.ii. Turing Throws Down the Gauntlet

Turing himself was more interested in the machine than in the intelligence. In other words, his *Mind* paper was primarily intended as a sketch of a research programme for AI (see 10.i.f), not as a conceptual analysis of "intelligence". However, virtually all the philosophical authors who responded to it addressed what they saw as Turing's behaviourism, rather than his imagined machines.

Moreover, those who did mention his machines didn't do so sympathetically, or in detail. Not until the 1960s would philosophers of mind start to use computational ideas in a constructive manner (Sections iii–iv, below).

a. Sketch of a future AI

Well before mid-century, Turing had been considering mind in a very different way from the writers discussed above (see Chapters 3.v.b and 4.ii.b). From the late 1930s, he'd believed that many, perhaps all, thoughts were formal–computational in nature. And soon after the war, he wrote a technical report on how to make "thinking machinery"—and there outlined what was later called the Turing Test. But the Official Secrets Act prevented publication.

Some of Turing's philosopher contemporaries were aware of his general position. In October 1949 he took part in a Manchester seminar on 'The Mind and the Computing Machine', with the philosophers Dorothy Emmet and Wolfe Mays (1912–2005), and the chemist–philosopher Polanyi. But none of them was persuaded (Manchester Philosophy Seminar 1949; Polanyi 1958). Other discussants on this occasion included the mathematician Max Newman, the zoologist John Z. Young, and the psychologist Frederic Bartlett. They were more ready to consider his ideas seriously, and to try to relate them to what was known about the brain (he'd mentioned both logical and 'neurone-based' models: Chapter 10.i.a).

Many philosophers, however, still knew of Turing only as a mathematical logician and/or a slave of Manchester's newfangled calculating machine, vulgarly dubbed a "giant brain" by the newspapers. This was the world's first modern computer—but why should they be interested in *that*? Many more hadn't heard of him at all.

In 1950 Turing's philosophical profile suddenly became more prominent, when he published a paper in *Mind* on 'Computing Machinery and Intelligence'. It ignored all the mind–body 'isms' outlined in Section i—although it did rebut several potential objections, including the argument from Gödel's theorem. (For the record, that argument had originated with Emil Post some thirty years earlier, and had featured in a paper of 1941 rejected by the *American Journal of Mathematics* as too "historical": Post 1965.) Nevertheless, Turing's 1950 essay was rich fare. It served up two intellectually tasty dishes—one of which was left untouched by most of its philosophical readers.

Primarily, the *Mind* paper comprised an outline programme for AI research, based on Turing's still-secret report. It provided a tutorial on digital computers, describing programs in terms of "tables of instructions" (rules for linking Input plus Internal State to Output). And it suggested how these machines might be useful in thinking about thinking.

It said, for example, that if we want to find analogies between computers and brains, we should focus on "mathematical analyses of function", not superficial similarities such as using electricity. Charles Babbage, after all, had "all the essential ideas" but hadn't used electricity. And it suggested that intelligence could be modelled by these machines (even though Turing knew there are *some* questions they can't answer: Chapter 4.i.c).

The paper provided "recitations tending to produce belief", as opposed to "convincing arguments of a positive nature", about what practical advances might be made—and *how*. So in the august pages of *Mind*, Turing discussed possible AI work on game playing, perception, language, and learning, giving tantalizing hints about what had already been achieved (for details, see Chapter 10.i.a and f). He even envisaged cloning—which he said would *not* be regarded as "constructing a thinking machine".

Most philosophers at the time paid no serious attention to Turing's substantive argument about the possibility of AI. They focused, rather, on his tongue-in-cheek claim that it would be natural to attribute thought to a computer that succeeded in the "imitation game" (a chess version of which he'd already carried out on a "paper machine": A. M. Turing 1947a: 23).

The game (soon dubbed the Turing Test: see subsection c, below) had been proposed in order to avoid dismissive arguments based on the meanings of words. These were wheeled out, nonetheless, some citing the authority of the *OED* (Mays 1952: 149). Turing himself resolutely refused to define either "machine" or "thinking":

I propose to consider the question, "Can machines think?" This should begin with definitions [framed to reflect] the normal use of the words, but this attitude is dangerous. [It implies that the meaning and the answer are] to be sought in a statistical survey such as a Gallup poll. But this is absurd. Instead of attempting such a definition I shall replace it by another [namely, the

Turing Test], which is closely related to it and is expressed in relatively unambiguous words. (A. M. Turing 1950: 433)

He added two provocative predictions, based on his expectation of practical advance in AI. The first was that "in about fifty years' time" machines would be able to play the imitation game with at least a 30 per cent probability of fooling "an average interrogator" for five minutes (p. 442). (What he actually wrote was: "[the] interrogator will not have more than 70 per cent chance of making the right identification". This is very often misread as predicting that the computer would *fool* the interrogator in 70 per cent of cases; but, as Blay Whitby pointed out to me, it means that the interrogator would be *correct* up to 70 per cent of the time.)

This prediction gave grounds for the second:

The original question, "Can machines think?" I believe to be too meaningless to deserve discussion. Nevertheless, I believe that at the end of the century the use of words and general educated opinion will have altered so much that one will be able to speak of machines thinking without expecting to be contradicted. (p. 455)

This was slippery. It could be read as a philosophical claim (an "educated opinion"), or as an empirical prediction about "the use of words". Moreover, the term "contradicted" was ambiguous. To allow someone—without reproof—to speak of computers thinking, or even to do so oneself, isn't necessarily to believe that computers actually think (see Sections v–vii, below).

By 1952, Turing's ideas had been featured in the newspapers, and he was debating them on BBC radio (for the BBC transcripts, see Copeland 1999). And the first philosophical replies had been published (Pinsky 1951; Mays 1951, 1952). One of these, written by Mays, would have appeared even earlier, had it not been censored by Ryle:

When Turing published his paper in *Mind* in 1950, Gilbert Ryle sent me the proofs and asked me to comment on them. I did, but Ryle rejected it on the ground that it was too polemical. (I used the word consciousness too many times, which in those days was heresy. It's now well-trodden ground by Searle, Dreyfus, and others.) Instead he asked Minsky to reply [who, in the end, didn't]. My reply got published in *Philosophy* some time later. (Mays, personal communication)

(Mays wasn't anti-machines: he'd co-designed an electrical logic device in the late 1940s—see Preface, ii. He'd also arranged for two digital computers to be shown at a 1950 philosophy conference, which could transform strings of up to 2,048 binary digits into logical symbolism: Mays *et al.* 1951. However, he *was* anti-formalist and anti-behaviourist with respect to the philosophy of mind. Indeed, he later founded the British Society for Phenomenology.)

Initially, these papers were a trickle, not a flood. In the 1950s, the attention of scientifically minded philosophers was largely captured by the identity theory (see Section i.d). One Cambridge apple orchard housed a group of philosophers, and others, deeply influenced by Turing. But they were more concerned to put his scientific ideas into effect than to publish philosophical papers about them (see Preface, ii, and Chapter 9.x.d).

It wasn't until the 1960s—thanks partly to functionalism and partly to results in early AI—that Turing began to be cited regularly by philosophers of mind. Usually, they ignored the main thrust of the piece and focused only on the Turing Test. But a few picked up Turing's gauntlet, by asking whether certain aspects of mind could, in fact, be simulated in computers.

Keith Gunderson, then at Princeton, was one of the first to pay Turing the compliment of disagreeing with him about his *main* point: that AI could (at least) simulate intelligence.

Gunderson had been alerted to AI/functionalism in the 1950s, even before either of those names existed. The first wake-up call was Turing's classic essay. The next was Michael Scriven's (1953) paper denying consciousness to robots (because they aren't alive: see Section x, below). The third was Paul Ziff's (1959) discussion of feelings in robots. And the last was a seminar given at Princeton by Hilary Putnam in 1959 (personal communication). Now, in the early 1960s, Gunderson published a paper on the imitation game (1964*b*), another touching on language in computers (1964*a*, sect. III), and another—in reply to Ziff—on emotions in robots (1963).

With respect to Turing's core claim, Gunderson was sceptical. He distinguished various forms of "computerophily" (1964*a*: 211), in which philosophers claimed that this or that human capacity could be simulated, perhaps even instantiated, in computers. On the one hand, he shared Descartes's doubts about the provision of language and *general* intelligence to machines (see 2.iii.c), although he did consent to leave open the question of whether computers could ever use language successfully (1964*a*: 222). On the other, in thinking about feelings in robots he'd begun to develop the distinction between "program receptive" and "program resistant" properties—later spelt out at length (Gunderson 1971; cf. 1985: 166–247). Some psychological phenomena, he said, could never be understood in AI/functionalist terms, whereas others could: emotions and problem solving, respectively (see 7.i.d).

Computer scientists, by contrast, had appreciated the main point of Turing's piece in *Mind* immediately. By 1963 it was already "one of the best-known papers" on AI, and was included in the influential collection on *Computers and Thought* (Feigenbaum and Feldman 1963: 9). Interest in it continued to grow. By the century's end, it had been anthologized scores of times, and cited on countless occasions. It still features regularly, both in professional journals (of philosophy and cognitive science) and in the media.

b. The gauntlet spurned

For ten years or more, most philosophers (as opposed to computer scientists) considered Turing's paper as a version of behaviourism, not as a substantive claim about thought. (Honourable exceptions included Mays, Gunderson, Scriven, Ziff, and John Lucas: see Section v.a–c.) In that sense, the philosophical community declined to pick up the gauntlet he'd thrown down. Indeed, many of these early replies said nothing specific about computers—which were still fairly inaccessible (see 3.v).

Even after computers became familiar, anti-Turing arguments—for instance, that robots must be non-conscious "zombies"—usually focused on his approach in the abstract (Section v.b, below). Its detailed implementation was ignored.

Ned Block's (1942–) influential argument about the giant look-up table was seemingly an exception (N. Block 1982). "Seemingly", because it ignored the intractable problem of the combinatorial explosion—which no computer scientist would have done (Dennett 1988). Interpreted literally, the Turing Test allowed the possibility that the computer's individual answers had been pre-stored, each one directly triggered by a particular question. But that wouldn't count as intelligence, which as Descartes had pointed out (Chapter 2.iii.c) is able to generate new answers to new questions.

However, Block's objection was well aimed only at the letter of Turing's paper, not at its spirit. In his wartime work—trying to determine what coding machine could be responsible for the observed 'behaviour'—Turing had indeed been seeking look-up tables: *this* letter of the message alphabet corresponds to *that* letter of the code alphabet. (It had turned out that there were two successive look-up tables, not one.) But by 1950, when his paper appeared in *Mind*, computing had already moved on. Programs now defined internal computations and structured spaces of generative possibilities, not mere lists of input–output pairs. In other words, Turing had excluded look-up tables implicitly, if not explicitly.

Despite Turing's coyness about defining thought, his paper implied that some conceivable digital computers could actually think. It was one of his philosophical colleagues at Manchester who first published substantive reasons for denying this.

Mays argued—like John Searle, thirty years later—that computer programs are all syntax and no semantics (Mays 1951, 1952). Or, as he put it:

But if we grant that logical machines are complex pieces of symbolism, a development of the visual aids to thinking which we have known for centuries, in order that the signs may acquire a significance they need to be given a specific logical or mathematical interpretation. As Whitehead [in his *Universal Algebra* of 1898, pp. 3–5] tells us, though we can study the art of practical manipulation of these signs without needing to assign any meaning to them, abstract calculi only possess a serious scientific value when they can be given an important interpretation.

Neglect of the pragmatic or instrumental aspects of such machines, leads to the tendency to attribute to them a capacity for thinking which they have only by proxy. The transformation of formulae according to a fixed set of logical rules, is not, however, a sufficient criterion of thinking. Unless the resultant formulae or patterns of symbols are retranslated in terms of their referents, the transformation remains a meaningless array of marks. [We need a human] intelligence to programme the machine and interpret the end-result . . . (Mays 1952: 159)

He failed to point out that Turing himself had made the same point in his *Mind* contribution—not once, but twice. We can speak of computers making "mistakes", he said, only "when some meaning is attached to the output signals from the machine". Similarly: "[it needs to] be shown that the machine has *some* thought with *some* subject matter. Nevertheless, 'the subject matter of a machine's operations' does seem to mean something, at least to the people who deal with it" (1950: 445).

Mays added a scientific remark, citing Gestalt psychology and Lashley's psycho-physiology (Chapter 5.i.b and iv.a). Thinking is very unlike logic, he said, since "the nervous system operates as an organic whole". And he blamed Wittgenstein for putting AI on the wrong track:

The basic assumption in applying the calculating machine analogy to the mind is that thinking operates in the form of an atomic system. It accepts Wittgenstein's view of the world as a structure of atomic facts . . . The progenitor is, of course, Wittgenstein [in the *Tractatus*] . . . Indeed one might say that modern digital computors [*sic*] are electrified pieces of Wittgensteinian logic. (Mays 1952: 161)

Here, Mays was right. It had been Warren McCulloch's youthful infatuation with logical atomism which eventually led him and Walter Pitts to propose AI, and influenced the design of the von Neumann computer (see Chapter 4.iii–iv).

But that was the early Wittgenstein. The later Wittgenstein saw those ideas as fundamentally misguided: thought was *not* a matter of atomistic logic (Chapter 9.x.d). Nor was language: on the contrary, it's part of our biology, one of our "forms of life"—so "If a lion could speak, we would not understand him" (1953: 223). As for machines, he said:

Could a machine think?—Could it be in pain?—Well, is the human body to be called such a machine? It surely comes as close as possible to being such a machine.

But a machine surely cannot think!—Is that an empirical statement? No. We only say of a human being and what is like one that it thinks. We also say it of dolls and no doubt of ghosts too. Look at the word "to think" as a tool. (Wittgenstein 1953, paras. 359, 360)

One might say that Turing agreed. For he was arguing that if a machine behaved like us, we would say it was thinking. But here the analogy ended. Wittgenstein was presupposing the presence of a human body and human "forms of life" as criteria for thinking, whereas Turing explicitly discounted these (see Section vii.a).

Mays was one of the very few commentators to pick up Turing's *main* philosophical challenge. Most philosophers were diverted by the Turing Test.

c. The Turing Test: Then and now

The phrase "Turing Test" wasn't used by Turing himself, and it suggests rather more philosophical weight than he'd intended. His friend Robin Gandy (1919–95) recalled the paper being written as lighthearted "propaganda", inviting giggles as much as serious philosophical critique (Gandy 1996: 125).

We saw, above, that Turing's teasing discussion of the imitation game had been aimed at avoiding endless definitions of intelligence. In fact, it spawned a minor—and still continuing—philosophical industry doing just that (e.g. Millican and Clark 1996). The lengthy entry in the online *Stanford Encyclopedia of Philosophy*, for example, lists over forty essential publications—and six Internet sources too.

Some saw the Test as indicating (if not *proving*) not intelligence *as such* but embodied, enculturated, human intelligence—which no computer could achieve, even though it might be genuinely intelligent in some other way (R. French 1990; Boden 1995b). And some elaborated it as the Total Turing Test (Harnad 1989), and even as the Total Total Turing Test (Harnad 1991, 1994; Bringsjord 1995). The TTT included full sensori-motor behaviour, or robotics. And the TTTT added "Turing indistinguishability right down to the neurons and molecules"—cloning, again? Most commentators, however, were content to discuss the original pen-pal version.

Despite this huge body of literature, one can't say briefly what the Turing Test *is*. The all-human version of the imitation game was clear enough: an interrogator exchanges written messages with a man and a woman, trying to distinguish them (only the man is allowed to lie). However, Turing's paper suggested three different 'machine' versions.

The first two involved a computer, an interrogator, and another human be-ing—whose output text is compared with the computer's to decide which is which.

The computer plays the part of (1) a man pretending to be a woman, or (2) a man; the other human being is (1) a woman, or (2) a man. The third version involved only a computer and a human interrogator. The human knows that he/she is conversing with a computer, and must decide whether it is thinking. In discussions of the paper, the first definition was almost always forgotten. The second was often ignored. Usually, philosophers—like Turing himself—focused on the third.

Further confusion arose after AI started to achieve practical results. Then, people often considered a fourth case, where the interrogator does *not* know that a computer might be involved.

Sometimes, this situation arose accidentally. One famous example involved a proto-type of the ELIZA program described in Chapter 10.iii.a, interacting with the NewFAI researcher Danny Bobrow. Slightly different reports of this dialogue have appeared in print. The version given in (Kaneff 1970: 32–3), and reproduced below, was personally confirmed to me by Bobrow. (However, as the editor of the *SIGART Newsletter* sagely remarked, "it's a good enough story that it probably deserves repetition even if there is not a word of truth in it!": *SIGART Newsletter* 1968.)

A computer salesman had a teletype interchange with Bobrow—or so he thought:

SALESMAN: DO YOU THINK THAT I CAN USE THE COMPUTER THIS MORNING?
TELETYPE RESPONSE: WHY DO YOU ASK?
SALESMAN: I MIGHT BE ABLE TO MAKE SOME ADDITIONAL TELCOMP SALES.
TELETYPE RESPONSE: WHY AREN'T YOU SURE?
SALESMAN: MY PROSPECTIVE CUSTOMERS HAVE NEVER SEEN A DEMONSTRATION OF THE SYSTEM.
TELETYPE RESPONSE: DOES THIS MEAN VERY MUCH TO YOU?
SALESMAN: OF COURSE IT DOES.
TELETYPE RESPONSE: YOU SEEM VERY POSITIVE ABOUT THAT.
SALESMAN: THIS IS RIDICULOUS.
TELETYPE RESPONSE: WHAT DO YOU MEAN THIS IS RIDICULOUS?
SALESMAN: IS THERE ANY REASON WHY I CAN'T USE THE COMPUTER NOW?
TELETYPE RESPONSE: TELL ME MORE ABOUT YOUR PROSPECTIVE CUSTOMERS.
SALESMAN: PLEASE DIAL ME UP ON 491–1850

At that point, no teletype response occurred. (The salesman hadn't put a full stop at the end of his curt request, and ELIZA always needed a full stop to trigger a response.) And, of course, no phone call was forthcoming either. Eventually, the furious salesman phoned Bobrow direct: "Why are you being so snotty to me?" Bobrow, still half-asleep, naturally asked: "What do you mean, why am I being so snotty to you?" The reply was unprintable.

Journalists and others, some of whom should have known better, celebrated this episode as ELIZA's having "passed the Turing Test". However, it didn't fit any of the definitions of the imitation game given in Turing's paper.

More to the point, the fourth-case situation sometimes arose by design. A few years after Bobrow was so rudely awakened, an experiment was carried out to "validate" Kenneth Colby's computer model of paranoia (Chapter 7.i.a). A random group of psychiatrists were asked to give diagnostic interviews by teletype, and some were connected not to a mental patient but to PARRY (Colby *et al.* 1972). Another group were asked to rate the transcripts for paranoia. And a third group were sent the transcripts, told that some had involved a computer, and asked to say which.

Sure enough, no one in the first group realized they were interviewing a program. The "weak" and "strong" versions of Colby's model were diagnosed, by the first two groups, as slightly and highly paranoid, respectively. (Occasionally, PARRY's linguistic limitations led to a diagnosis of physical brain damage.) And the *man or machine?* guesses of the third group achieved only chance level.

Colby himself had the wit to see that this wasn't a Turing Test as the master had defined it. He pointed out that, had the interviewers been warned, they would have asked different questions, designed to discover whether the interviewee was a program. In a follow-up experiment, he did warn the judges beforehand—and they were mistaken almost as often (Heiser *et al.* 1980). Again, however, this wasn't the unrestricted test imagined by Turing: the psychiatrists' freedom to question was limited by medical ethics (*one* communicant was a human patient) and professional focusing (on paranoia, and its simulation). And again, Colby was explicit about this, referring to *"Turing-Like* Tests" and stressing the *limitations* of such experiments in assessing simulations. Similarly, John Clippinger (1977, ch. 9) pointed out that the plausibility tests done on ERMA weren't reliable as validations of his psychological theory (Chapter 7.ii.c). Other commentators were often less careful, describing the programs in these (and similar) examples as having passed the Turing Test.

The high media profile given to this type of case not only brought the Turing Test into the layman's vocabulary, but also reinforced its presence in the vocabulary of AI. The early AI researchers had been inspired by the scientific vision of Turing's 1950 paper, and its many challenges to their ingenuity. In addition, most had read it as providing a criterion of their success. Especially in the 1960s and early 1970s, AI work was often judged in terms of some version of the Turing Test. Colby's experimental "validation" was a case in point. And most philosophers agreed that the Test was an appropriate criterion. Hubert Dreyfus, for instance, declared it to be "just what was needed" for evaluating AI (H. L. Dreyfus 1972, p. xxi).

Today, outsiders still use TT comparisons to evaluate AI's achievements—usually, to criticize them. By 1980 however, serious AI professionals were much less likely to judge their work in these terms. Bernard Meltzer, first Editor of the journal *Artificial Intelligence*, had already argued that the Turing Test was not only irrelevant but constraining, since it *limited* AI to considering human intelligence (Meltzer 1971). From 1980 onwards, AI researchers in general came to agree with him.

John McCarthy, in a presidential pep talk to his fellow AAAI members, specifically said that improving standards in AI research *did not* require passing, or even considering, the Turing Test: it was a "challenge", not a "scientific criterion" (1984: 8; see 11.iii.b). A few years later, when AI was no longer in its infancy and so-called "failures" had multiplied, several writers described evaluation by Turing Test as positively damaging to AI (Hayes and Ford 1995; Whitby 1996a; Sloman 2002). The main exception concerned people writing games and entertainments in the 1990s, who talked about "believable" agents and valued virtual-reality systems that could 'fool' the user to some extent (Chapter 13.vi).

To be sure, in 1991 the Boston Computer Society sponsored a public run of the Turing Test, the American Association for AI reported it in their magazine (R. Epstein 1992), and the event was subsidized by the National Science and Sloan Foundations. This was the first in an annual competition (visible live on the Web since 1999) funded by Hugh

Loebner. The yearly prize is $2,000, for the "best" entry. But the rules state that $25,000 awaits "the first computer whose responses [are] indistinguishable from a human's" (see the official web site: <http://www.loebner.net/prizef/loebner-prize.html>). And entrants at that level will also be able to compete for the Grand Prize of $100,000, which requires "audio-visual capabilities" too.

(One might regard this as somewhat niggardly. For Ed Fredkin had set the same sum aside in 1980 for the first program to beat a reigning world champion in chess—no audio-visuals required. The Fredkin prize was won in 1997, by IBM's Deep Blue: Hamilton and Hedberg 1997; Michie 1997. Then Deep Blue was disassembled, never to play again: Hsu 2002.)

By the end of the century, the Loebner competition had moved around the world, to be sponsored by Flinders University in South Australia and then by the London Science Museum. However, the event always was—and still is—more publicity than science.

The original planning committee involved philosophers as well as computer scientists (Shieber 1994). Early members included Dennett, Quine, and (initially) Bernard Cohen; and the philosopher Block was one of the first referees. (Quine might almost be an honorary computer scientist, for two of his papers on logic spawned a widely used method for minimizing the number of gates or interconnections in computer circuits and chips: Quine 1952, 1955.)

A genuine Turing Test being out of the question, the competition used a highly restricted version: only one topic per conversation, and a non-probing conversational style on the part of the human judge. The programs (and hidden humans) were rated on how "human-like" they were, as well as on whether or not they were human. It turned out, to the judges' surprise, that some programs were mistaken for people—and vice versa!

Dennett eventually resigned as chairman of the prize committee, because they couldn't (then) be persuaded to toughen the test—to include fancier syntax and pronominal reference, for instance (Dennett 1998: 27–9). A few years later, the restrictions on topic hopping and probing were lifted. And the judges now include experts in NLP and other areas of AI.

But the competition still attracts little interest from AI professionals (although the 1997 prize was won by a team led by Yorick Wilks, a leading NLP researcher whose "preference semantics" was mentioned in Chapter 9.x.d). It won't make anyone's fortune. Even the intermediate prize will probably remain forever unclaimed, and achieving apparently human performance *in the general case* is impossible in practice (and perhaps also in principle).

More important, it's now recognized, at least by insiders, that the Turing Test isn't an appropriate way of judging AI's progress. This was made crystal clear in Turing's suggested landmark year. The Loebner prizewinner in 2000 was a program much nearer to ELIZA than to SHRDLU or LUNAR. And NLP research had already progressed hugely since all of those (Chapter 9.x–xi). In any event, AI's aim isn't *épater les bourgeois*. Highly human-like performance is only rarely required, even in technological AI.

The exceptions include certain applications of speech processing. Someone needing an artificial voice will normally want one that sounds as human-like as possible—not to deceive their interlocutors, but to put them at their ease. (I say "normally" because there is at least one such person who's said he doesn't want a human-sounding speech synthesizer: see 9.xi.g.) These voices may be not only humanlike, but plausibly idiosyncratic: early-millennial speech-synthesizers can pronounce written text in many different local accents (9.xi.g). Similarly, someone wanting a computerized "companion" may well prefer to hear a pleasing human-like voice, and familiar regional pronunciation (see 13.vi.d). In the general case, however, the Turing Test is irrelevant for AI.

Or rather, the Turing Test *as philosophers normally understand it* is irrelevant. In a recent issue of the *AI Magazine* devoted to a twenty-five-year retrospective on AI, one of the invited articles did indeed discuss the Turing Test. It even declared that "the test still stands as a grand challenge for artificial intelligence, it is part of how we define ourselves as a field, it won't go away" (P. R. Cohen 2005: 61). But its title was a give-away: 'If Not [*sic*] Turing's Test, Then What?' For the writer was arguing that "Turing's test is not irrelevant, *though its role has changed over the years* . . . [Among] AI researchers, the question is no longer, 'What should we do to pass the test?' but, 'Why can't we pass it?'" (italics added). These shifting attitudes to the TT in AI's professional community, despite its unchanged role in the minds of the general public (and most philosophers), have been noted also by Whitby (1996*a*) and by Robert French (2000). In short: although the test "won't go away", passing it isn't regarded as AI's core goal, or as the key criterion of its success.

After AI programs had entered the public domain, the AI community asked whether passing the Turing Test—even in the attenuated sense of a program's evoking trust in the user—is actually desirable (cf. Boden 1977, ch. 15). Professional codes of practice were suggested, to prevent the production of deceptively 'human' systems and to limit marketing hype (Council for Science and Society 1989; Whitby 1988). Marketing hype continued, nonetheless.

But the potential for deception depends on the sophistication of the user. An early medical-assistant program (dealing with issues in medical ethics) explicitly reminded its users that they might have relevant knowledge which it didn't, and that only they could make the decision required (Sieghart and Dawson 1987). Whether such warnings are appropriate at any given point in AI's history depends on what the users (in this case, medics and paramedics in Great Britain) can be assumed to know about AI.

This is why Turing's prediction about the 30 per cent fooling of "the average interrogator" failed. It might, perhaps, have been borne out in the underdeveloped world. But, by the year 2000, people in technological societies already knew too much about what computers can and can't do. A computer salesman accidentally connected to ELIZA today wouldn't ever reach the stage of raging at Bobrow.

Admittedly, language generation—needed for the imitation game—is still the least impressive aspect of NLP research (see Chapter 9.xi.c). Deception in other modes is sometimes possible. A real-time jazz improvisation system can fool many people for much longer than five minutes (Paul Hodgson, personal communication). A program can compose music indistinguishable (by all but experts) from that of Bach, and other famous composers (see Chapter 13.iv.b). And I've found that a program that interprets marks of expression in musical scores (a more complex problem than one might think: Longuet-Higgins 1994) can play Chopin's romantic Fantaisie-Impromptu in C sharp minor—which lasts about four minutes—so convincingly that even musically experienced AI researchers often can't tell that a computer is 'at the keyboard'. To be sure, the Turing Test (*sic*) allows *any* comparison, other than bodily form. But Turing's imitation game (*sic*) allowed only comparisons made via teletyped language.

By contrast, Turing's prediction about changing word usage came true long before he expected. By the turn of the millennium, the new usages had spread far beyond science. People now speak about computers in psychological terms as a routine, everyday, matter.

For instance, as Scriven (1953: 235) foresaw, they speak about "consulting" the computer, not just "using" it. Indeed, changes have occurred in the opposite direction too, computing terminology being used to describe human thought. My own favourite example (you probably have yours) came from the mouth of a British Museum curator of Pacific artefacts—hardly a paradigm member of the computer culture. While talking about how to tell indigenous peoples about the ways in which Aids spreads, she said of one suggested approach that "People just can't compute it."

But was psychological language being understood literally, when applied to computers? Had spontaneous belief and/or educated opinion changed, along with the use of words?

The sociologist Sherry Turkle tried to find out. In the late 1970s, when some children had already had regular access to computers for some years, she found that these youngsters were happy to describe computers as (really) "thinking" and "intelligent", but not as having "feelings" or "emotions", nor as being "alive" (Turkle 1984).

Some years later, when A-Life technology was widely available, she found that both children and adults still insisted that computers aren't "really alive, as we are" (Turkle 1995). But they vacillated (as well as disagreeing) over whether robots, and/or VDU-screen creatures, are alive merely as ants are, sort of alive, or not alive (though admittedly in control of their own movement). Sometimes, these distinctions were brushed aside with the remark "It's just a machine".

That is, the people interviewed by Turkle tried to draw a principled distinction between computers and human beings—though not necessarily in terms of *thinking*. But they often did so by giving mutually inconsistent definitions in different cases. The rest of this chapter outlines how philosophers of cognitive science have tried to address the man–machine distinction more rigorously—whether to insist on it or to deny it.

16.iii. Functionalist Freedoms

Functionalism was first defined as such by Putnam (1926–), in 1960. (He didn't actually use that word, but he did define mental events in terms of causal–computational functions, as we'll see.) But it had been more or less explicit in other thinkers too—who weren't professional philosophers, and whose work was read hardly at all by the philosophical community.

One shouldn't assume, then, that there were no functionalists before Putnam. However, all self-styled philosophical "functionalists" were inspired by Putnam's early work.

That word "early" is important. Many years later, Putnam would roundly reject his earlier approach, as we'll see in Section vi. Here, and in Sections iv–v also, "Putnam" means early Putnam.

a. Just below the surface

Functionalist approaches to mind were sketched as early as 1943 by McCulloch and Pitts, and also by Craik (see Chapters 4.iii–iv and vi, and 14.viii). I don't mean merely that their psychological research was functionalist in spirit. In addition, all of them engaged in explicitly *philosophical* argument, relating their work to the views of Kant, for instance, and to the nature of universals, cause, and number.

Craik, for example, repeatedly described his book as offering a new "philosophy" (see 4.vi.c). His position was as much brain-as-machine as mind-as-machine, for he identified "the nature of thought" as the use of cerebral models whose *physical* properties paralleled those of external reality. Anticipating the inevitable anti-materialist objections, he declared:

[People who scorn materialism] deserve criticism for blindly refusing to consider whether things are always what they seem—whether their own powers of introspection are able to tell them the secrets of their own mental processes and whether many of their own acts are not very similar to those of the machines, natural and man-made, which lie all round them and which they will not take the trouble to understand. (Craik 1943: 98)

[There] is something wonderful in the idea that man's brain is the greatest machine of all, imitating within its tiny network events happening in the most distant stars, predicting their appearances with accuracy, and finding in this power of successful prediction and communication the ultimate feature of consciousness. (p. 99)

And he drew a robust epistemological moral:

Further, I see no difficulty in understanding how anything so "different" from physical objects and concepts and reasoning can tell us something more about those physical objects; for I see no reason to suppose that the processes of reasoning *are* fundamentally different from the mechanism of physical nature. On our model theory neural or other mechanisms can imitate or parallel the behaviour and interaction of physical objects and so supply us with information on physical processes which are not directly observable to us [and which may even be imaginary]. Our thought, then, has objective validity because it is not fundamentally different from objective reality but is specially suited for imitating it . . . (p. 99)

It followed that the classic questions of philosophy were to be interpreted in a new way. One example (out of many):

Our question . . . is not to ask what kind of thing a number is, but to think what kind of mechanism could represent so many physically possible or impossible, and yet self-consistent, processes as number does. (p. 55)

Despite claiming to have glimpsed "the secrets of [our] own mental processes", however, Craik didn't explicitly address the hot topics outlined above in Section i. Or rather, he made many remarks about the *topics*, but very few references to the contemporary philosophical *literature*. (By 1943, of course, only two of the four key contestants had appeared on stage: see Section i.b.)

It's not clear that many philosophers of the 1940s read his work, although the Cambridge-based and scientifically minded Braithwaite and Margaret Masterman occasionally mentioned him. If they had, they might have found it highly suggestive. But they might not have afforded it the status of "yet another philosophy" (Craik

1943, p. vii) because, by philosophers' standards, Craik's argument wasn't sufficiently rigorous.

A functionalist approach was implicit also in Turing's *Mind* paper of 1950. And, in his anticipation of various objections, he said a little about some questions in the philosophy of mind. Even so, he didn't engage with the philosophical literature, and he deliberately avoided discussion of the currently competing 'isms'. Philosophers did read his paper, of course—indeed, they flocked to it in droves. But as we've seen, they focused on the semi-joking behaviourism, not on the wholly serious functionalism.

As a result, these early functionalist ideas were initially picked up by psychologists and computer scientists, not philosophers. By 1960, the Logic Theorist had been proving the theorems of *Principia Mathematica* for a full four years (see Chapters 6.iii.c and 10.i.b). Moreover, its authors explicitly denied that they were making any claims about the structure of the brain:

Discovering what neural mechanisms realize these information processing functions in the human brain is a task for another level of theory construction. *Our theory is a theory of the information processes involved in problem solving and not a theory of neural or electronic mechanisms for information processing.* (Newell *et al.* 1958a; italics added)

In the terminology soon to be introduced by another functionalist philosopher, this was a statement of *multiple realizability*. That is, a psychological (computational) theory can be expressed, and may be true, without making any commitment whatever as to the physical mechanisms that implement it. Put baldly: psychologists can ignore the brain. Put more precisely:

whether the physical descriptions of the events subsumed by [psychological] generalizations have anything in common is, in an obvious sense, *entirely irrelevant* to the truth of the generalizations, or to their interestingness, or to their degree of confirmation, or, indeed, to any of their epistemologically important properties. (Fodor 1974: 14–15; italics added)

Machine-inspired "structural explanation" had already entered both experimental and Freudian psychology (J. A. Deutsch 1953, 1960; K. M. Colby 1955). Internal models, enabling one to "go beyond the information given", were posited by the 'New Look' cognitive psychologists (Chapter 6.ii). And the provocative *Plans and the Structure of Behavior* was coming off the press (Chapter 6.iv). These writings teemed with philosophically relevant remarks. Even so, hardly any philosophers at the time had their eye on the computational ball.

Masterman did (and Braithwaite, too). But she was thinking about conceptual structure and language, not mental processes as such (Chapter 9.x.d). Across the Atlantic, Putnam did too—and he *was* concerned with the metaphysics of mind.

Many years later, he said, "I may have been the first philosopher to advance the thesis that the computer is the right model for the mind" (Putnam 1988, p. xi). Given the views of McCulloch, Craik, and Turing, perhaps he should rather have said he was the first *card-carrying* philosopher to do so. However, as a card-carrier he could offer the rigorous argument, and the attention to philosophical 'isms', which Craik couldn't. It's not surprising, then, that Putnam's work, unlike Craik's, *was* seen as "yet another philosophy". Functionalism, at last, had appeared above the surface of the water.

b. The shackles loosened

Putnam's statement of functionalism in 1960 was hugely influential. His paper on 'Minds and Machines' didn't catch philosophers' attention through informing them of technical results. For AI—including Putnam's own recent work on automatic theorem proving (10.iii.b)—wasn't even mentioned, though Noam Chomsky was. Rather, it excited them because it promised an escape from the mid-century philosophical impasse described in Section i.

Indeed, it dismissed the previous philosophies as "empty"—and irrelevant to scientific psychology. This was Putnam's verdict even on Smart's identity theory, which he saw as the best of a bad bunch. And, as though to put icing on the cake, his paper picked up Turing's gauntlet at last—treating the challenge not as a threat but as an inspiration.

Putnam studied at Princeton, taught philosophy of science at MIT in the early 1960s, and crossed the Charles River to Harvard in 1965. He was especially interested in the philosophy of logic and mathematics, and in computational logic too. For instance, it was he who proved that Chomsky's transformational grammar was equivalent to a Turing machine (see Chapter 9.vi.d). Moreover, his (LT-inspired) paper on computational theorem proving was cited in Marvin Minsky's 'Steps Toward Artificial Intelligence' (10.i.g), and later helped lead to resolution theorem proving in AI (10.iii.b).

Situated in Princeton and MIT, the leading centre for automata theory and AI, Putnam was well aware that these disciplines might have implications for psychology. Initially, he shared the positivists' belief in the "unity of science" (see Chapter 9.v.a, and Oppenheim and Putnam 1958). But he soon recanted, for his functionalist writings depicted psychology as a science in its own right.

Putnam argued that there's nothing metaphysically special about mind. Every mind–body puzzle, including subjectivity, would also arise in an automaton that could inspect some of its own states and reason inductively about them. Even introspective infallibility would be paralleled, if the machine's Turing table led it to print *I am in state A* whenever it entered state A: since this would be automatic, there would be no room for "mistake". As for souls, he tartly remarked that if dualism proves that people have souls, it proves that Turing machines do too—and he was as loath to ascribe souls to computers as Descartes had been in respect of "worms, gnats, [and] caterpillars" (see Chapter 2.iii.e).

So far, so negative. But to establish these conclusions, Putnam made five substantive claims:

* First, that human beings and other organisms are finite automata, describable by Turing machine tables—including input–output rules specifying the action of sensory/motor organs.
* Second, that these descriptions focus on the machine's (abstract) "logical" states, not its (physical) "structural" states.
* Third, that a given logical state "may be physically realized in an almost infinite number of different ways".
* Fourth, that the logical and structural descriptions of an automaton are analogous to psychological and physiological descriptions of a person.
* And last, that "further exploration of this analogy may make it possible to further clarify the notion of a 'mental state' ".

In sum:

The *functional organization* (problem solving, thinking) of the human being or machine can be described in terms of the sequences of mental or logical states respectively (and the accompanying verbalizations), without reference to the "physical realization" of these states. (Putnam 1960, sect. 3; italics added)

The "further exploration" was provided a few years later (1967a,b). Abstract Turing machines, Putnam then declared, could in principle be implemented in Descartes's mental substance, if it existed. Actual mental states, however, are in fact implemented in brain states. These may vary. The physiology of pain (for instance) in a dog, or an octopus, differs from ours. But pain as such is defined psychologically, or functionally, not physiologically.

It follows, he said, that psychological predicates can't be understood as Rylean dispositions (see Section i.c). On the contrary, functionalist analyses explain such dispositions. As for the Turing Test, this is unreliable because two very different programs might generate exactly the same behaviour.

Putnam admitted that his appeal to Turing machines was vague—not least because he held that *everything* is a Turing machine, under *some* description. (Both he and Searle would later use this point to attack functionalism: see Sections v–vi.) Even so, Turing machines—or programs—are much easier to study than brains are:

This hypothesis, in spite of its admitted vagueness, is far *less* vague than the 'physical–chemical state' hypothesis [i.e. identity theory] is today, and far more susceptible to investigation of both a mathematical and an empirical kind. Indeed, to investigate this hypothesis is just to attempt to produce "mechanical" models of organisms—and isn't this, in a sense, just what psychology is about? (Putnam 1967b: 435)

Moreover, computer models appeared to carry scant metaphysical baggage. The causal relations between hardware and software seemed to be clear, unlike those between "mind" and "body". (This happy assumption would eventually be questioned: see Section ix.c–f.)

The relevant computational models, said Putnam, should simulate not only perception and motor action but also reasoning, preferences, values, and beliefs. The implication was that a scientific psychology would broadly conserve the psychological relations implicit in everyday language.

These include specific conceptual links, such as those between *pity, help,* and *comfort* (see Section i.c), and the general architecture of rationality that's presupposed by the language of intention, accident, and excuse (cf. Austin 1957). But whereas the ordinary-language philosophers saw these relations as conceptual, Putnam saw them as explanatory. He even hoped to find some "normal form" for psychological theories, so that animals' hunger, aggression, and pain (though not their physiology) could be explained in much the same way as ours.

Functionalism, then, was not only a philosophy of mind but also an empirical theory—or rather, a 'schema' for generating empirical theories—about mental states and processes. Such a combination was highly suspect in certain circles, being seen as an example of scientism: the view that scientific facts are metaphysically fundamental, and that all important questions have some scientific answer (Stenmark 2001; cf. N. Maxwell 1984).

Putnam, following his teacher Quine (1908–2000), who favoured a "naturalized epistemology" (Quine 1952), saw philosophy as continuous with science. But Wittgensteinians didn't. He'd already crossed swords with them on the subject of dreaming, arguing—contra Malcolm—that novel scientific discoveries (in this case, REM sleep) may affect the meanings of everyday concepts (1962*a*).

Moreover, he'd already answered Turing's question (see Section ii) in much the same way. Although it's "false" or "improper" to say *baldly* that machines think, or use language, the strong analogy between digital computers and human behaviour makes it reasonable to extend the meanings of those words to cover both (Putnam 1960). Now, in his discussion of 'Psychological Predicates' (later retitled 'The Nature of Mental States'), Putnam was imbuing *all* psychological vocabulary with ideas drawn from computer science.

For philosophers not dismayed by charges of scientism, functionalism was a heady liberation. It promised to loosen the shackles imposed by the problems mentioned in Section i, and some remarked in Section ii.

Not all post-1960 versions of it were solely, or even partially, inspired by Putnam. Fodor's functionalism, for instance, owed much to Chomsky and Jerome Bruner as well as to Putnam (his tutor). Kenneth Sayre's maverick version leant on cybernetics more than on logic and AI (Sayre 1965, 1969, 1976). And my own followed Masterman and *Plans and the Structure of Behavior* (Boden 1965; 1970; 1972, esp. 52–9; and see Preface, ii).

But Putnam's functionalism, which focused on the metaphysical questions without referring to specific empirical data, was hugely important for many philosophers. It licensed talk about mind, while avoiding metaphysically mysterious events, processes, or substance. And it offered many other freedoms, too.

For instance:

* It underwrote the explanatory links between (and within) mind and body that we normally take for granted.
* Thanks to multiple realizability, it saved the main claim of identity theory (that each mental state is identical with some brain state), without stating—falsely—that all mental states of type *x* type are identical with brain states of type *y*.
* It was consistent with materialism but left space for an autonomous, non-reductionist, psychology—intentionality being grounded in causal–computational mechanisms.
* It ruled out psychological behaviourism,
* while saving the subtle conceptual analyses of logical behaviourism.
* But it reinterpreted these analyses as explanatory, identifying mental states in terms of their mutual causal relations.
* It valued what Dennett (1981*a*) would later call "folk" psychology, seeing our everyday talk as a proto-theory, a starting point for scientific study of both human and animal minds.

* Above all, it promised a novel research programme (for both philosophy and psychology) that was free to exploit the rigour, richness, precision, and testability of computational theories and models.

It's no wonder, then, that—for audiences leaning towards science—this new player swept its predecessors from the philosophical stage. Turing was posthumously promoted from spear-carrier to matinée idol, and functionalism soon became the orthodox position in the philosophy of cognitive science.

It wasn't universally applauded, to be sure. Quine, for example, held that psychology is *not* autonomous, because belief–desire vocabulary is in principle reducible to non-intentional terms (Quine 1953*b*, 1960). Others complained that functionalism can't allow for consciousness. Yet others were radically opposed to the entire 'scientistic' project. Some of these objections will be discussed in Sections v–viii.

First, let's look at some of the ways in which functionalism was expressed by its supporters.

16.iv. Three Variations on a Theme

Even orthodoxy allows reinterpretation. Functionalism is "the thesis that the essence of our psychological states resides in the abstract causal roles they play in a complex economy of internal states mediating environmental inputs and behavioral outputs" (P. M. Churchland and Churchland 1981: 121). But this general thesis developed into many mutually disputatious positions.

The most influential philosophers of functionalism, *post* Putnam, were Fodor, Dennett, Paul and Patricia Churchland, and Andy Clark (whose PDP-based philosophy was discussed in Chapter 12.x.e).

Other important accounts, functionalist in spirit if not in philosophical jargon, were offered by Chomsky (Chapter 9.ii–vii), Allen Newell and Herbert Simon (Section ix.b, below), and—later—Minsky (Chapters 7.i.g, 10.i.f, and 12.iii.d). Their scientific research influenced all the philosophers described in this section. (Sloman developed an interesting version in the 1970s too, but this was less influential than it deserved: see Section ix.c, below.)

The late 1960s saw extended statements of functionalism being provided by Fodor (1935–) and Dennett (1942–). Both men eventually became famous—even notorious. But their near-simultaneous books were very different. In part, the differences reflected the fact that Fodor was a cognitive scientist with a strong interest in philosophy, whereas Dennett was a philosopher with a nascent interest in cognitive science.

Fodor was an interdisciplinary animal. He had a joint appointment in psychology and philosophy at MIT, had attended both Chomsky's and Bruner's seminars in the late 1950s, and gave detailed tutorial seminars on Chomsky's *Syntactic Structures* at that time (Chapters 6.i.e and ii.b–c, and 9.viii.c). His first book, *Psychological Explanation* (1968), defended this new type of research. It supported Putnam's claims about multiple realizability and the autonomy of psychology, and spelt out the notion of "functional equivalence" in theoretical psychology. With respect to the philosophy of mind in general, it accused Ryle and Wittgenstein of conflating mentalism and dualism (see Section i.c).

His arguments against behaviourism were well received by sympathetic souls. But for those who'd already rejected it, there were no eye-openers here. Later, Fodor would offer surprises aplenty, with the publication of *The Language of Thought* (1975). Before then, the surprises came rather from Dennett.

a. Content and consciousness

When Dennett's book hit the scene in 1969, he himself was a surprise. Teaching in California since leaving Oxford in 1965, he was still unknown. His first three papers had appeared only a few months earlier, and were critiques of others—including Dreyfus (Dennett 1968)—rather than free-standing statements of his own position. Now, *Content and Consciousness* addressed the topics that would preoccupy him throughout his career: intentionality, consciousness, free will, evolution, and the comparison between human and animal minds.

Integrating challenging philosophical claims with wide-ranging scientific argument, many readers found it an exhilarating book. I remember writing to this total stranger across the seas, to say so—one of only four times I've ever done that. Anti-functionalists, it must be said, demurred. The *Times Literary Supplement* ran a damning review by a Wittgensteinian, who described it as "not a happy piece of work" and complained about the "many references to scientific findings and, it must be confessed, scientific speculations" ([Hamlyn] 1970). (This anonymous review was written by the philosopher of psychology David Hamlyn: Dennett, personal communication.) But many others were excited.

The central-state materialists welcomed this new version of centralism. So Smart, asked by Ryle to review it for *Mind*, decided to write a Critical Notice instead, and a fellow Australian did likewise (Smart 1970; R. L. Franklin 1970). In the USA, Thomas Nagel (1970) wrote a sympathetic, though unconvinced, review—and Gilbert Harman immediately ran a seminar on the book at Princeton. (Harman had already published on the psychological implications of Chomsky's work, and he later co-founded Princeton's Program/Laboratory for Cognitive Science with George Miller: Harman 1988.)

Dennett was soon invited to speak to Harman's group (his first public presentation) and elsewhere, so ending his "five anonymous years of apprenticeship" at Irvine. (He moved to Tufts University in the early 1970s, where he's remained ever since.) His 1969 book, and the paper he read at Princeton in December 1970, set much of the agenda in the philosophy of cognitive science for years ahead.

Dennett was nothing if not heretical. Despite being deeply influenced by Ryle (his ex-tutor) and Wittgenstein, he disagreed with them in important ways. He posited a "sub-personal" (functionalist) level of explanation as the key to understanding mental phenomena, and saw meaning as rooted not in public language and society but in mechanisms evolved in human and animal brains. He disagreed with Putnam, too. He leant heavily on Putnam's logicist account of minds as Turing machines, and often used AI examples in discussing sub-personal mechanisms (e.g. Dennett 1969: 101–13). However, the "functions" he was most interested in weren't purely logical, or even causal, but at base evolutionary (see below).

Moreover, he refused to identify thoughts with either functional states or brain states. To identify *x* with *y* is to say that both terms refer to the same thing—but Dennett argued that mentalistic terms don't refer to anything at all. *Thought, belief, intention, fear*, even *pain*: all these are non-referential (like the word *sake*, in "She did it for my sake"). Specifically, they are items in a mode of discourse we use to ascribe meaning, not to discover it. To be sure, "people *do* have beliefs, intentions, and so forth" (and machines don't: "a computer is no more *really* an information processor than a river *really* [desires to reach the sea]") (Dennett 1969: 89–90). But meaning, or content, isn't an extra item over and above bodily behaviour or brain states.

Here, Dennett was leaning on Quine (1908–2000) and on Wilfrid Sellars (1912–89). Sellars had distinguished the "manifest" and "scientific" images of man (1956). The former sees people as having beliefs, desires, and intentions, while the latter describes them in terms of physics and neurophysiology. These two images were seen by Sellars as distinct logical "spaces"—of reasons and of causes, or norms and facts. They provided different ways of talking about people—with no guarantee of a neat mapping between the two. And no guarantee, either, of reliability: Sellars specifically attacked "the myth of the Given", according to which experiential "data" are an unassailable foundation for knowledge (see Section viii.b).

As for Quine, he'd already argued that the language of belief and desire is an "essentially dramatic idiom" (Quine 1960: 219). Although Quine was Dennett's other philosophical mentor (besides Ryle) (Dennett 1994*b*), Dennett rejected his view that intentional language is in principle reducible to physics. If we stopped using this language, said Dennett, no existing thing need be ignored. But we could no longer interpret anything as meaningful.

Such interpretations assume rationality. When we say that someone carrying an umbrella wants to keep dry, we presuppose a rationally interrelated set of meanings which result in "appropriate" behaviour. This insight wasn't new: we saw in Chapter 5.iii.a that Thomas Dewey and Ralph Perry had said it long before. And (following Wittgenstein) Elizabeth Anscombe (1957) and Charles Taylor (1964*a*) had said it more recently.

But whereas Anscombe and Taylor had taken it for granted that meaning is intrinsic to people and/or language, Dennett asked what it is about human beings that makes it 'natural' to ascribe meaning to them. His highly unfashionable answer, in a nutshell, was: brain states and evolution.

The "crux" of any centralist theory of mind, he said, is "the ascription of content or meaning to particular central states of the brain" (Dennett 1969: 71). For Dennett, this couldn't be done (as both Putnam and Smart had tried to do) by assigning the content of mental states to the relevant brain states: the semantic traffic goes in the opposite direction.

The semantic content of neural states, for Dennett, is based in evolutionary function (pp. 48 ff.). And since evolution works on whole organisms, "one can only ascribe content to a neural event, state or structure when it is a link in a demonstrably appropriate chain between the afferent and the efferent" (p. 78). In calling a retinal cell a bug-detector, for example (see Chapter 14.iv.a), it's crucial that it's linked to a motor circuit that causes the frog's sticky tongue to shoot out appropriately (*sic*) or its leg muscles to make it jump to the right (*sic*) place.

Moreover, neural content is the basis of linguistic meaning:

[Philosophers have asked] whether events and states of the nervous system could be assigned meanings or ascribed contents, and assigning meanings was seen as associating events or states with verbal expressions. Verbal expressions, however, are not the ultimate vehicles of meaning, for they have meaning only in so far as they are the ploys of ultimately non-linguistic systems. (Dennett 1969: 88)

To posit meanings in the brain wasn't to posit languages in the brain. Dennett criticized scientists' talk of "brain-writing", and of neural "codes" and "languages", because it usually relied on unexplained mechanisms for understanding such languages (pp. 86–8). This, he said, is to postulate "little men" in the brain. (Bruner and Ulric Neisser had already argued that computational theories could avoid homunculism: see Chapters 6.ii.c and 7.i.b. But they hadn't focused on the problem of meaning, or intentionality, as such.)

As for comparing the brain to "a community of correspondents", this—said Dennett—was "the most far-fetched and least useful" of all the common analogies, since it appears "merely to replace the little man in the brain with a committee". (Ironically, he would later favour Minsky's "society of mind" in his mature theory of consciousness (Dennett 1991): see 12.iii.d.)

This Rylean disdain for homunculi extended also to philosophers' talk of *acts of will*, *volitions*, and the *self* (and to ego, id, and super-ego), when presented as "explanations" of mental life. In Dennett's words: "The solitary audience in the theatre of consciousness, the internal decision-maker and source of volitions or directives, the reasoner, if taken as *parts* of a person, serve only to postpone analysis" (p. 190).

However, to assign a neural basis to linguistic meaning wasn't to say that mental states could be neatly mapped onto neurophysiology: "[we cannot] find precisely worded *messages* for neural vehicles to carry ... [but this] is merely an inability to map the fundamental onto the derived, and as such should not upset us" (p. 88). It will upset us, however, if we hope that neuroscience will eventually identify the neural vehicles of particular propositional attitudes.

Besides multiple realizability and the complexity of the brain, said Dennett, there's the difficulty that intentional vocabulary "has no real precision of its own". We talk about beliefs, for instance—but just what is it to believe a proposition? How many of its logical implications are involved? And just what does a child believe who announces "Daddy is a doctor"?

These questions, for Dennett, can have no clear answers. His account of belief resembled Wilhelm von Humboldt's of language (Chapter 9.iv.b). That is, what we call the "same" belief is probably different in every individual—and, even more important, can't be precisely pinned down in any particular case.

Despite insisting that ordinary-language terms like *belief* and *desire* are unavoidably imprecise, Dennett didn't deny their usefulness. His view on *conscious* and *aware* was very different:

[The philosopher who seeks to understand consciousness is] forced to do psychology rather than "pure" philosophy.
... Can a machine be conscious? This question cannot be answered until we arrive at a conclusion about what it is to be conscious, and ordinary language does not tell us ... [We need] more than a solution to a conceptual problem via analysis of language, for language is deficient in this area. The way out is an analysis of phenomena at the sub-personal level, and although this leads one into areas many philosophers would prefer to avoid, the alternative is the perpetuation of traditional confusions. (p. 130)

The question *What is consciousness?* has no single answer, because this term is irredeemably confused, "an unhappy conglomerate of a number of separable concepts" (p. 114). Philosophers have assumed ever since Descartes that it's the most accessible concept of all. In fact, said Dennett, it's useless for philosophical or scientific purposes and should be replaced by new concepts that make the relevant distinctions clear.

Among the cases marked by the everyday term are those where someone can give an introspective report of their current thought, and those where they're evidently affected by some stimulus but can't report it (like the car-driver who negotiates a corner while engrossed in conversation). Dennett offered two new concepts, defined by sub-personal criteria, to distinguish such cases.

Following Putnam's example of the introspective automaton (see Section iii.b), he suggested that some functional systems—such as human brains—include a "speech centre". If a mental state is input to that centre, the immediate (Turing-tabled) output is a verbal expression whose content is the same as the content of the input state. This expression, when input to other subsystems, will cause further changes. Often, it leads to speech movements; but these may be inhibited, as in reading silently or simply keeping one's own counsel. For such a system, two senses of awareness can be defined:

(1) A is aware-1 that *p* at time *t* if and only if *p* is the content of the input state of A's "speech centre" at time *t*.
(2) A is aware-2 that *p* at time *t* if and only if *p* is the content of an internal event in A at time *t* that is effective in directing current behaviour. (pp. 118–19)

The sub-personal processes involved in awareness-2, said Dennett, may be very like those in AI models of perception, language, and problem solving (p. 151).

As for whether dogs and other dumb (*sic*) animals are conscious, we must distinguish the two senses. Understood as aware-2, animals are indeed conscious—sometimes, in highly discriminating ways. But understood as aware-1, they are not: in this sense of the term, only language-using creatures can be conscious.

So far, so good. Dennett's distinction was useful when applied to human beings, and even seemed to make retrospective sense of Descartes's views on animal consciousness (Chapter 2.ii.d–e). But there was a sting in the tail—or rather, two.

First, it followed that computers could be conscious. A computer can clearly be aware-2 of things. And it can also be aware-1, if it is set up so as to give verbal reports of some of its internal states. Dennett was unabashed:

[This] may seem to be an intolerable situation, but only if one clings to the folklore that has accrued to the ordinary word "aware" ... [There] is no important residue in the ordinary concept of awareness that is not subsumed under either awareness-1 or awareness-2. There is no room ... for a concept of awareness-3, which would apply only to people and rule out all imaginable machines. (p. 121)

Second, it followed that there are no conscious epiphenomena, no experiential qualities, pure sensations, raw feels, or *qualia*. These denizens of the mind, about which there had been so much argument (see Section i), were mere metaphysical phantoms. This conclusion had been implicit in Putnam's functionalism, but Dennett (like Ryle before him) was resolutely explicit.

Even pain, he insisted, is a wholly functionalist phenomenon:

Where discriminating is an analysable personal activity, like discriminating good apples from bad by checking for colour and crispness, we can distinguish the qualities from the discriminating of them. But in the case of distinguishing sensations as painful, the act of discrimination itself is the only clue to the localization (in space and time) of the presumed quality. Insisting that, above and beyond our ability to distinguish sensations as painful, there is the quality of painfulness, is thus insisting on an unintelligible extra something. (p. 92)

His comment here was reminiscent of Wittgenstein's, that "a nothing would serve just as well as a something about which nothing could be said" (see Section i.c). For Dennett, if there is no further function there is indeed nothing more to be said. But at the personal level, a verbal expression of pain is "a *bona fide report*", not—*pace* Wittgenstein—"an outcry of sorts" (p. 112 n.). (Some years later, Dennett would argue that the concept of pain is actually incoherent—and that *this* is the reason why one can't build a computer to feel pain: 1978*a*. See also Wall 1974, 1979.)

b. From heresy to scandal

Dennett's multiply heretical volume was strong meat—and a strong sauce would soon be added. Most early discussions of his position considered not just his book but also (sometimes, only) the paper first given at Princeton (Dennett 1971). The book had named two explanatory styles, or "stances" (e.g. p. 101): personal (intentional) and sub-personal (functional/physical). The paper named three: intentional, design, and physical.

The physical stance seeks explanations couched in physical, or purely neuro-physiological, vocabulary. This is often infeasible in practice, even where artefacts are concerned, and is functionally uninformative in principle. Indeed, neurophysiologists were already using non-physical concepts, such as bug-detectors and orientation detectors (Chapter 14.iv).

The design stance, by contrast, "relies on the notion of *function*, which is purpose-relative or teleological" (Dennett 1971: 88). It's a form a reverse engineering, which assumes that the system is well, even optimally, designed to perform certain tasks. Artefacts, computers included, are typically understood in this way. The design stance can operate at various levels of abstraction, such as electronic circuitry, transistors and switches, multipliers and dividers, strategic units (like planners and line-finders), or programmed instructions. In other words, there may be many different levels of virtual machine (see Section ix.c).

But Dennett held that natural systems can be understood in this way too. Indeed, the design stance is "the proper direction for theory builders [in psychology and neuroscience] to take whenever possible" (1971: 96). Not only does it avoid unanalysed ascriptions of intelligence, but it can compensate for the *lack* of rationality that bedevils actual intentional systems.

The intentional stance was defined as before. But there were four reinforcements of points already made in *Content and Consciousness*.

First, Dennett declared that "Intentional theory is vacuous as psychology because it presupposes and does not explain rationality or intelligence", and wherever a theory uses intentional language, "there a little man is concealed" (Dennett 1971: 99, 96).

Second, he enlarged upon the fact that people's "rationality" is limited, which causes both principled and practical difficulties in assigning beliefs and desires to individuals. Third, he stressed the instrumental status of intentional discourse, as interpretation not discovery: "a particular thing is an intentional system *only in relation to the strategies of someone who is trying to explain and predict its behaviour*" (Dennett 1971: 87; italics added). And last, he pointed out that it was already impossible, even for the system's designer, to explain or predict the behaviour of the best chess-playing computers except by using intentional vocabulary.

To say of the computer that *It's trying to get its queen out early* is thus not an avoidable metaphor, but the only practicable way of understanding its performance. This description is (pragmatically) justified, even though there may be no specific programmed instruction, or program goal, to *get the queen out early*. The instructions and goals, whatever they are, were designed (*sic*) to have this overall effect. Nevertheless, even the programmer can't remember and/or run through the relevant rules so as to understand what's going on or predict what's likely to happen next.

Previously, Dennett would have said that no computer can *really* try to get its queen out early (see above). Now, he insisted:

Lingering doubts about whether the chess-playing computer *really* has beliefs and desires are misplaced; for the definition of intentional systems I have given does not say that intentional systems *really* have beliefs and desires, but that one can explain and predict their behavior by *ascribing* beliefs and desires to them, and whether one calls what one ascribes to the computer beliefs or belief-analogues or information complexes or Intentional whatnots makes no difference to the nature of the calculation one makes on the basis of the ascription. (Dennett 1971: 91)

All very moderate, perhaps . . . until one realized that human beings were being classified as intentional systems *in just the same sense*. His ready admission that even the best computers were crude and narrow-minded, by human standards, didn't alter that.

Understandably, Dennett's provocative views sparked lively debate, both inside and outside the functionalist camp. His scandalous dismissal of *qualia*, in particular, aroused huge dissent (see Chapter 14.xi.b). Nagel, for instance, believed that no objective science could possibly explain the subjectivity of consciousness. Hence his unconvinced review of Dennett's first book (1970), and his memorable paper 'What Is It Like To Be a Bat?' (1974)—later answered by one of Dennett's students with 'What Is It Like To Be Boring and Myopic?' (Akins 1993).

Today, with respect to *qualia*, Dennett remains in the minority. But he's not alone. Sloman, for many years, has analysed consciousness in terms of a complex virtual machine—and *qualia* as computational phenomena within it (Sloman 1978, 1999; Sloman and Chrisley 2003).

At the end of the century, with most of his professional colleagues and lecture audiences still unpersuaded on *qualia*, Dennett would tell the (true) story of a card trick called 'The Tuned Deck'. This trick bemused even groups of expert magicians for many years. As an old man, its inventor revealed the secret: instead of creating *one exceptionally ingenious* card trick, he had craftily hopped about between *many familiar* ones. Once his expert audience had discounted trick A, because he'd fooled them into expecting it just before using trick B, he could use trick A at any time without their noticing it—and similarly for tricks B to Z.

'Pure' experience, for Dennett, is eternally elusive for much the same reason. Once we have detailed *all* the many functions—verbal naming and description, discriminatory behaviour, emotional responses, conceptual associations . . .—involved in seeing red, or smelling lilac, *nothing more remains* to be explained. By the same token, he now says, if you took a pill to get rid of only your toothache's *qualia* but not its causal effects—neural stimuli, wincing, rubbing your jaw, lack of concentration, bad temper . . .—then *you wouldn't even know.*

Dennett has become convinced, however, that disagreements on this issue are often rooted in a clash of intuitions quite impervious to argument (lecture at University of Sussex, May 2000). He's penned various imaginary conversations with a common-sense sceptic called Otto (one was given in Chapter 14.xi.b). But Otto is never persuaded, despite Dennett's responding to every one of his outraged objections.

Dennett put other issues on the agenda, too. His comments on the imprecision of belief led some functionalists to recommend that cognitive science abandon folk-psychological language entirely (P. M. Churchland 1979; Stich 1983). And his instrumentalism with respect to mental states—along with his "flagship" renunciation of it (in terms of "real patterns", and "*abstracta*" such as centres of gravity) (1981*b*; cf. 1987: 3)—caused gallons of ink to be spilled.

(Most of that ink was philosophical. But psychologists' pens were eventually wielded too. The intentional stance was applied in empirical studies of the "Theory of Mind" in young children, non-human primates, and other species: see Chapter 7.vi.f.)

Several of Dennett's later works received enormous attention (e.g. 1978*b*, 1987, 1988, 1991, 1996). This was especially true of his "multiple drafts" theory of consciousness, which pictured the mind as a virtual machine vastly more subtle (and more scientifically grounded) than his aware-1/aware-2 distinction had done (see 14.xi). A notorious aspect of his theory of consciousness was his attack on *qualia*; this had been a long time a-brewing, for it was first delivered late in 1978 and was widely circulated in draft before being published ten years later. Only his analysis of free will (1984*a*), in my opinion, received less regard than it deserved (see Chapter 7.i.g).

Today, the debate still rages: for an indication, see *Behavioral and Brain Sciences* (1988); Dahlbom (1993). But despite the elaboration of his views on consciousness, there's been no fundamental change in Dennett's position. On rereading *Content and Consciousness* after twenty years, he commented: "I am struck more by my doctrinal consistency than by my developments. Most of the changes seem to me to be extensions, extrapolations, and further arguments, not shifts" (1987, p. xi) (see also Dennett 1994*b*).

c. Must angels learn Latin?

In the 1970s, while Dennett's early writings were setting the philosophy of mind alight, Fodor was continuing his own attack on the Ryle–Wittgenstein approach (Fodor 1975: 2). His first book had appeared shortly before Dennett's, as we've seen. But his visibility soared in 1975, when his volume on *The Language of Thought* offered a form of functionalism very different from Dennett's.

It differed partly in its philosophical claims and concerns, for it was unambiguously realist about mental states and it ignored consciousness and freedom. But it differed also in its attitude to the practice of cognitive science.

Dennett had used chess programs and memory research as examples, but he wasn't especially concerned to help cognitive scientists do their job. Or rather, his help consisted in trying to dispel the conceptual confusions he felt they shared with many philosophers. Only much later would he collaborate closely with psychologists, ethologists, neuro-scientists, and roboticists (see e.g. Dennett 1994a, 1996, and Chapter 15.vii.a). Fodor, by contrast, was committed to computational psychology and to Chomskyan linguistics, as such.

His early work referred to highly detailed research on perception, concept learning, decision making, problem solving, and—above all—psycholinguistics. He was a leading champion of Chomsky, and had already co-authored a textbook on the psychology of language (Fodor *et al.* 1974). He was also inspired by GOFAI, though more for its reliance on formal symbolism than for its modelling of specific mechanisms. And he adopted—and never relinquished—the implication of Putnam's 1960s papers, that "we have no reason to doubt that it is possible to have a scientific psychology that vindicates commonsense belief/desire explanation" (Fodor 1987b: 16).

In short, Fodor took the dual role of functionalism—as philosophy and theoretical psychology—especially to heart. That had been evident in his first book (1968), but it was even clearer in *The Language of Thought*. This aimed to provide clear theoretical underpinnings for the cognitive revolution already under way. It was a paradigm of 'classical', formalist, cognitive science. For it described intentional processes as syntactic operations defined on mental representations (see also Fodor 1978b).

This view was boldly presented as the only "even remotely plausible" candidate for a scientific psychology. There was no middle way, either then or later. At century's end, Fodor had developed serious doubts about whether all mental processes are computations. Nevertheless, he still held the same opinion about what a scientific psychology must be like: "[computational psychology] is far the best theory of cognition that we've got; indeed, the only one we've got that's worth the bother of a serious discussion" (Fodor 2000b: 1).

The book quickly became influential, even infamous. Like Newell and Simon before him (1972), Fodor insisted that the mind (or brain) is "literally" a computational system. Functionalist talk shouldn't be, though it often was, interpreted as a mere *façon de parler* (1975: 51, 76). And he drew two major implications, summarized as "no computation without representation" and "there's no point in angels learning Latin" (pp. 34, 86).

The first of these, if not the second, might prompt one to ask "Why the fuss?" After all, cognitive psychologists, Chomskyan linguists, and functionalist philosophers had been positing computation and internal representations for over fifteen years.—Yes, said Fodor, but they hadn't been clear about what these are. They hadn't realized that "Computation presupposes a medium of computation: a representational system," and that "The pressing question is what properties does the system of internal representations have." This wasn't a purely philosophical question: the answer must fit the facts, as any scientific theory must do (pp. 27, 33, 156).

Even this may seem fairly anodyne. But Fodor also said:

[What] I am proposing to do is resurrect the traditional notion that there is a "language of thought" [later often referred to as LOT] and that characterizing that language is a good part of what a theory of the mind needs to do. (1975: 33)

That was guaranteed to raise many hackles. As we saw in Chapter 9.ii–iv, philosophers in past centuries had posited an internal language of thought to explain adult reasoning and/or to explain how children manage to learn their mother tongue. To adopt the first goal would upset the Wittgensteinians, who saw no room for any level of explanation between natural language and neurophysiology (which they assumed wouldn't employ any computational notions): see Section v.f, below. To adopt the second would outrage not only Wittgensteinians, with their commitment to public language, but also empiricists—who had already attacked Chomsky's nativism con brio (Chapter 9.vii.c). Never one to fear controversy, Fodor adopted both.

His central claim about mental representations was uncompromising, and very different from Dennett's:

[Modern cognitive psychology assumes that] the computational states ascribable to organisms can be directly explicated as relations between the organism and *formulae*; i.e. formulae in the internal code . . . [And, even more importantly, it assumes that] for any propositional attitude of the organism (e.g. fearing, believing, wanting, intending, learning, perceiving, etc., that *P*) there will be a corresponding computational relation between the organism and some formula(e) of the internal code such that (*the organism has the propositional attitude if the organism is in that relation*) is nomologically necessary. (p. 75)

These formulae, said Fodor, are required not primarily for (public) communication, but for representing concepts, goals, and the like. Even animals, if they can represent such things, must have a language of thought. Internal representations are real structures, apt for discovery rather than mere interpretation. They are composed of atomic primitives, whose computational (semantic) properties determine the properties of the whole—as in the formulae of symbolic logic, or the sentences of natural language.

As for just what those primitives might be, they certainly included the basic concepts of the *innate* language of thought. And if his early 1960s work in generative semantics (Chapter 9.viii.c) had been correct, these would suffice (see below). However, worries about computational efficiency had now convinced him that natural-language words can't, in practice, be analysed into semantic atoms every time they're used. Rather, they're still present (as *functional* primitives) in the adult's internal code:

[There] is no process for [analytic] definition *at all*; i.e., both the defined expression and its definition appear as items in the primitive vocabulary of the representational system. (1975: 133)

Learning a definition principally involves learning a meaning postulate. It thus adds to the constraints (not on computing memory but) on long-term memory; it adds a rule of inference to the list that is stored there. That is why . . . abbreviatory definition and other recoding schemes make formulae easier to understand: Computing memory is expensive, but long-term memory is cheap. (p. 150)

In other words, an adult's language of thought includes all the words they know in their mother tongue. Hence Fodor's wry salute to his philosophical opponents: "There may, then, really be some point to the late Wittgensteinian insistence upon the surface richness of natural languages" (p. 156).

He also conceded that recent psychological research on the nature of concepts seemed to support the later, rather than the earlier, Wittgenstein (see Chapters 8.i.b and 9.x.d). That is, concepts appeared to be stored as stereotypes, exemplars, or even images rather than definitions. But he saw this as "terribly difficult", asking "How, for example, does one *access* an exemplar? If your concept of a dog is . . . a representation of a stereotypic dog, how do you go about determining what *falls under* the concept?" (p. 153). Ten years later, PDP connectionism seemed to many people to have answered his question (see 12.x). But Fodor was unconvinced, and later defended a strictly atomistic theory of concepts (Fodor 1998*a*).

Just what representations are actually involved in thinking was still unclear. Fodor allowed that they might include images (1975: 174–94). But although visual imagery could be a "vehicle of reference" it couldn't be a "vehicle of truth"—in other words, a thought. Thought, he said, is the ascription of properties to things, and is essentially propositional and rational: "The sequence of events that causally determines the mental state of an organism will be describable as a sequence of steps in a derivation if it is describable in the vocabulary of psychology at all" (p. 198).

This claim carried another surprise. Putnam had implied that all mental states, and a fortiori all propositional attitudes, are grist to the functionalist mill. But Fodor denied this. For him, psychology required computational causes. It included perception and concept learning, since these (or so it was thought at the time: see below) involved the generation and testing of hypotheses. And it might include some Freudian examples (p. 200), such as neurotic beliefs semantically generated by defence mechanisms (Chapter 7.i.a). But much of our most systematic and interesting mental life, he argued, was probably excluded.

Sensation, he said, isn't caused by computation. Creative thoughts might not be. Many associative processes probably aren't. And nor are emotional influences on perception and belief. Such matters might be explained by biology (neurophysiology), but not by psychology. In sum, functionalism's scope had been significantly reduced: "It may be that we are laboring in quite a small vineyard" (p. 202).

What about the notoriety? This arose because Fodor claimed that the innate language of thought was "as powerful as any language that one can ever learn" (p. 82).

He confessed that this claim was "scandalous", but insisted that no serious alternative had been proposed: "The only coherent [account of language learning] is one which presupposes a very extreme nativism" (p. 96). Human babies don't have adult concepts—such as *air-plane*—ready-formed. But they are born with the representational capacity (concepts and combinatorial operations) to express the meaning of any concept that can be learned later (p. 152). More accurately, they're either born with these concepts and operations or they mature, biologically, so as to attain them. What they do *not* do, is learn them.

Fodor's argument for this highly counter-intuitive position leant crucially on Bruner's research on learning and Chomsky's views on innate grammar (see Chapters 6.ii.b and 9.vii). Following Bruner, he saw concept learning as "essentially inductive extrapolation, [which] presupposes a format for representing the experiential data [and] a source of hypotheses for predicting future data" (p. 42). Since language learning involves learning not only what counts as a sentence but also the semantics of the vocabulary, it follows (he said) that "one cannot learn a first language unless one already has a system capable of representing the predicates in that language" (p. 64). Every word must be translatable into the innate language of thought, even though (as remarked above) it isn't normally translated into it.

The very possibility of an innate language, whatever its representational power, was highly controversial. Ryleans would scent an infinite regress of understanding (Dennett's "little man"), and Wittgensteinians would reject any pre-cultural private language. Fodor countered both objections by a brain–computer analogy:

though the machine must have a compiler if it is to use the input/output language, it doesn't *also* need a compiler for the machine language. What avoids an infinite regression of compilers is the fact that the machine is *built* to use the machine language. Roughly, [the formulae of the machine language] correspond directly to computationally relevant physical states and operations of the machine . . . (p. 66)

On this view, what happens when a person understands a sentence [in a language he has learned] must be a translation process basically analogous to what happens when a machine "understands" (viz., compiles) a sentence in its programming language. (p. 67)

He used this computer analogy also to explain why babies have to learn their mother tongue—even though, according to him, they already have the capacity to represent the meaning of every dictionary word. People need natural languages for much the same reasons that they need programming languages (see 10.v). We don't have world enough, or time, to express everything in terms of primitives—or to remember it, if we did.

And this brings us, finally, to the angels:

If an angel is a device with infinite memory and omnipresent attention—a device for which the performance/competence distinction [see 7.iii.a] is vacuous—then, on my view, there's no point in angels learning Latin; the conceptual system available to them by virtue of having done so can be no more powerful than the one they started out with. (1975: 86)

Only a philosopher would have sailed so close to the wind as to risk this seeming absurdity. Like Dennett's denial of *qualia*, it exemplified the fourth cause of (non-philosophers') irritation identified in the preamble to this chapter. Fodor himself admitted blandly, even happily, that it looked like a *reductio ad absurdum* (p. 82).

Many critics noted that his argument was based in ignorance, for he assumed that no non-inductive theory of concept learning was conceivable. Philosophers also sought other flaws in his argument (e.g. Dennett 1977: 273; P. S. Churchland 1978; Sloman 1987b). As for the psychologists, most simply ignored this self-confessedly "scandalous" claim.

But both communities—and eventually, neuroscientists too—hotly debated many of his other claims (see, for example, Chapters 7.v.i, 12.x, 14.ix.e, and 15.vi–viii). That is, they questioned the existence, and the nature, of each of the following:

* an innate language of thought;
* real, individuated, representations;
* formal–computational (and compositional) internal states;
* mental states whose semantic relations depend on their syntactic structure;
* scientifically describable carriers of specific propositional attitudes;
* a psychological science approximated by common-sense belief–desire psychology;
* non-computational associative thought;
* and non-conceptual semantic content.

Angels aside, then, Fodor's influence was enormous. Even Dennett, who disagreed with him on many points, said: "Fodor challenges us to find a better theory, and I fully expect that challenge to be met, but when better theories emerge they will owe a good deal to Fodor's reconnaissance" (Dennett 1977: 280).

d. Fodorian frills

If Fodor's reconnaissance had led to surprises, so did his next scouting party. In 1980 he undertook "a meditation upon the consequences of assuming that mental processes are formal processes" (Fodor 1980a: 64).

He concluded that a "rational" psychology (Chomsky's term: see Chapter 9.ii–iv) must be "methodologically solipsist" (Putnam's term, but first used by Russell). And this implied that psychology will *never* do what it's normally intended to do—namely, explain how we gain knowledge of, or even beliefs about, the world.

Putnam's part in this philosophical drama was to have set the stage by distinguishing two general approaches to intentionality (Putnam 1975a). These differ over whether the meaning (and therefore the identity) of every mental state depends on the existence, and nature, of anything outside the subject whose state it is.

For proponents of "wide" meaning, such as the recently converted Putnam himself, it does. (As he put it, " 'meanings' just ain't in the *head*!"—Putnam 1975a: 227.) For champions of "narrow" meaning, it doesn't. They are *methodologically* solipsist, said Putnam (p. 220): for them, the content of our mental states could be just the same even if nothing else existed. (Theories of wide and narrow meaning are often termed externalist and internalist, respectively.)

Fodor, *qua* computational psychologist, danced on the narrow side of the stage. Formal computation, being purely syntactic, excludes semantic concepts such as truth and reference, so can say nothing about the world. Terry Winograd's SHRDLU exemplified this, he said, being a simulated robot within a purely virtual reality (see Chapters 9.xi.b and 10.iv.a). The dire implication was that psychology can't ever tell us how we see tigers, smell lilacs, understand speech, or learn that politicians aren't to be trusted.

There seemed to be a chink of light, for Fodor granted that a "natural" (neuro-physiological) psychology could discover the physical causes of sensation. However, to know the causes of *perception* is impossible in the general case. Science might tell us what a cat is, but not what a mat is—for the concept *mat*, like most concepts, doesn't name a natural kind. Someone's perception (belief, fear, desire . . .) that the cat is on the mat could be *naturally* explained only by explaining every causal detail involved. To do this for all objects of thought would require "the theory of *everything*" (Fodor 1980a: 70). As Fodor remarked, Leonard Bloomfield had used similar reasons in arguing that a naturalistic semantics is impossible (see Chapter 9.v.b). In short, despite its principled failure to engage with the real world, "computational psychology is the only one that we are going to get" (1980a: 66).

All was not lost, however: *philosophy* could come to the rescue. If we complemented a functionalist psychology by a theory of semantics then we could, after all, bring the lilacs and politicians into the picture. For we could then say how perception and belief are possible.

The rest of Fodor's career, with one important exception, was—and still is—devoted to defending his previous views (on intentional realism, "Rationalist" nativism, and the adult's rich language of thought) and to developing a philosophical semantics to fit (e.g. Fodor 1981*a,b*, 1987*b*, 1998*a*, 2000*a*).

The exception concerned the notion of mental modularity. When he wrote *The Language of Thought*, Fodor still followed the 'New Look' psychologists (6.ii) in seeing perception and learning as the formation and testing of hypotheses. "For all we know", he believed, "cognition may be saturated with rationality through and through" (1975: 173)—a view shared by Zenon Pylyshyn (1973), for whom even mental imagery reflected beliefs (7.v.a). By the early 1980s, however, the New Look was under fire from several directions.

For example:

* Chomsky had posited "mental organs", of which the language-acquisition device—underlying the language "faculty"—was only one (see Chapter 9.viii.b).
* GOFAI was increasingly relying on domain-specific knowledge (see 10.iv).
* David Marr, speaking of "the sensorium of sight" (Marr and Nishihara 1978), had detailed a variety of fast-acting, automatic, bottom-up processes in low-level vision (7.v.b–d).
* Developmental psychologists were discovering to their surprise that newborn babies have "knowledge" of—or, better, predispositions to attend to—various aspects of the world (including faces: see 14.ix.c). (To their surprise, because Jean Piaget, no less than the behaviourists, had denied this: Boden 1994*a*.)
* And Pylyshyn (1980) had defined the mind's "fixed functional architecture", whatever this turned out to be, as inescapable. It was "cognitively impenetrable", or uninfluenced by beliefs (see 7.v.a).

All these cognitive scientists agreed that the inborn propensities could generate actual ideas, or mental contents, *only* given the relevant environmental triggering. In sum, Immanuel Kant's views on the inherent structuring of the mind had seemingly been broadly vindicated (see Chapter 9.ii.c).

Fodor's response was *The Modularity of Mind* (1983). Like *The Language of Thought*, this book interested psychologists as well as philosophers. (It later received a detailed rebuttal from the developmental psychologist Annette Karmiloff-Smith: 1992.)

With his usual flair for scandal, Fodor picked up Chomsky's term and recommended "faculty psychology". He wasn't resurrecting the outmoded nineteenth-century faculty psychology, with its notions of Will, Perseverance, and Morality. Rather, by faculty psychology he meant "the view that many fundamentally different kinds of psychological mechanisms must be postulated in order to explain the facts of mental life" (1983: 1).

Following Pylyshyn, he defined mental modules as innate input systems whose computations were automatic, and couldn't be influenced top-down by concepts or beliefs. (One doesn't become immune to the Müller-Lyer illusion by learning that it is an illusion.) If rationality implies belief, then *perception*—despite being computationally complex at all levels—wasn't "saturated with rationality *through and through*" after all.

Cognition, however, was—and Fodor now saw this as a major problem (Chapter 7.iii.d). He argued that since any belief (hope, desire . . .) can be influenced by indefinitely many others, there can be no principled theory of problem solving or belief fixation. (Or, one might add, of 'New Look' perception, or of neurotic repression: see Chapters 6.ii.a and 7.i.a.) The best we can hope for is an informed natural history: a relatively unsystematic collection of problem-solving procedures and suggestive anecdotes. This would be an advance on common sense primarily in recognizing that propositional attitudes are (literally) GOFAI-computational.

But are they?—As we saw in Chapter 12, GOFAI would soon be challenged by PDP connectionism. (And Fodor would defend his position accordingly: Fodor and Pylyshyn 1988.)

e. Eliminative materialism

The earliest philosophers of PDP-based functionalism were the Churchlands. Their views on concepts, mental states, and consciousness were indicated in Chapters 12.x.b–c and 14.x.d.

As we saw there, they differed from Putnam, Dennett, and Fodor not only in rejecting GOFAI (abandoning logical deductions and intentional inferences for vector-to-vector transformations), but also in trying to avoid mentalistic language entirely. "Folk psychology" was to be discarded, not reduced. But their eliminative materialism was buttressed by PDP and computational neuroscience, not originated by them. To the contrary, it had long-standing philosophical roots.

Paul Churchland's earliest work pre-dated the resurgence of connectionism by several years. It had argued that advances in neurophysiology could conceivably lead us to think about minds in such a radically different way that beliefs, desires, fears, hopes . . . would be "eliminated", or "disappear" (P. M. Churchland 1979, 1981).

Even then, this idea wasn't new. The possibility had been implicit in Sellars's 1950s notion of the manifest image (see above), and had been made explicit in the early 1960s by Paul Feyerabend (1963a,b) and Richard Rorty (1965). Moreover, philosophers of science such as Russ Hanson (1958) and Thomas Kuhn (1962), and New Look psychologists such as Bruner and Richard Gregory (Chapter 6.ii), had already suggested that our empirical theories affect our perceptions.

The novelty in Churchland's approach was that he made a serious attempt to imagine how scientific knowledge could alter perception—including introspection—in a fundamental, all-encompassing, way.

Rorty too, in a book published in the very same year as Churchland's, considered the possibility of radically different types of perception. He imagined alien beings who never reported having *perceptions, ideas*, or other *mental representations*. Instead, they spoke in terms of neurophysiology. So, for instance, "When their infants veered toward hot stoves, mothers cried out, 'He'll stimulate his C-fibers' "; and a visual illusion might make them say "How odd! It makes neuronic bundle G-14 quiver, but when I look at it from the side I can see that it's not a red rectangle at all" (Rorty 1979: 71).

But Rorty wasn't arguing for scientific realism: far from it. Besides trying to banish the Cartesian concept of mind (partly by showing that intelligent creatures could do without it), he was denying that any form of knowledge, science included, is more fundamental than any other. According to Rorty, philosophy isn't an investigation but a "conversation", more like literary criticism than logic or science.

To Churchland, that was anathema. He insisted on the realist nature of science—he called his book *Scientific Realism and the Plasticity of Mind*—and used science (i.e. New Look psychology: Chapter 6.ii) to show that our perception actually is theory-laden.

Moreover, he tried hard to imagine *what it might be like* to have a radically different type of experience.

So, for instance, he described how he had trained himself to *see* the night sky so that certain astronomical facts, concerning the geometry of the solar system, were directly perceived, not inferred. These facts are usually regarded as counter-intuitive, precisely because they conflict with our normal perceptions. He explored other cases too, suggesting how children in an imaginary culture—whose 'ordinary' conception of reality was the one embodied in modern physics—might be taught to perceive the world in a very different, and more veridical, manner:

It is important for us to try to appreciate, if only dimly, the extent of the perceptual transformation here envisaged. These people do not sit on the beach and listen to the steady roar of the pounding surf. They sit on the beach and listen to the aperiodic atmospheric compression waves produced as the coherent energy of the ocean waves is audibly redistributed in the chaotic turbulence of the shallows ... They do not warm themselves next the fire and gaze at the flickering flames. They absorb some EM energy in the 10^{-5} range emitted by the highly exothermic oxidation reaction, and observe the turbulences in the thermally incandescent river of molecules forced upwards by the denser atmosphere surrounding.

These observational descriptions, so arcane to us, are in no way arcane to the people under discussion. This is the only idiom they know. (P. M. Churchland 1979: 29–30)

As this passage suggests, a perceptual upheaval might involve not just seeing things differently, but seeing different things. In other words, our perception of physical reality might be informed by a different physical ontology. (And, *pace* Rorty, this alternative ontology might be *nearer the truth about reality*.) Instead of merely making us see/think of various "material objects" in a new way, science might lead us to discard such categories entirely:

The "facts", as currently conceived and observed by us, form the starting place for theoretical inquiry, but its successful pursuit may well reveal that we should vacate that starting place as hastily as possible. Large-scale intellectual progress will involve the wholesale rejection of old *explananda* as frequently as it involves the wholesale introduction of new *explanantia*. (P. M. Churchland 1979: 44)

And if the ontology of physical objects wasn't sacrosanct, neither was the ontology of mind.

It was even more difficult, however, to imagine how introspection might be systematically transformed, for the relevant neuroscientific theory wasn't yet available. (His "phase-space sandwiches" would come later: see Chapter 14.x.d.) Nevertheless, said Churchland, introspection is as theory-laden as any other form of perception. And the common-sense theory of mind—a "thoroughly thumb-worn theory whose cultural assimilation is complete" (p. 2)—didn't appear to map neatly onto what was already known about the brain.

Moreover, he predicted, this failure of mapping would probably deepen as neuroscience advanced:

[The] prospect we face is that a detailed neurophysiological conception of ourselves might simply displace our mentalistic [belief–desire] self-conception in much the same way that oxidation theory (and modern chemistry generally) simply displaced the older phlogiston theory of matter transformation. That we are long in the habit of making non-inferential introspective judgments in the terms of the theory to be displaced affects the matter not at all. (p. 5)

A halfway house, empty of physics and neurochemistry, would be a functionalist account. But this might be very different from what orthodox functionalists assumed:

Even as a functional characterization of ourselves, [folk psychology] may turn out to be taxonomically cockeyed, radically incomplete, and altogether too confused to merit continued use, when compared to the much superior functional characterizations that an adequate theory of the central nervous system can be expected to provide. (p. 113)

In short, folk psychology may be a *false* theory, rather than an approximate and/or incomplete one. (Incomplete, because—as Churchland pointed out—it works only for normal human agents, not for brain-damaged patients or animals.) To protest that we know it to be true through introspection (Cartesian "direct access") is futile. If our perception of physical things is theory-laden, and perhaps mistaken, so also is introspection.

These ideas hadn't yet quite gelled in Churchland's mind. Eliminative materialism was presented in his book merely as a coherent philosophical option, although his sympathies were clear. Two years later, however, he argued that it should definitely be adopted (P. M. Churchland 1981).

If there was no place for belief and desire in the philosophy of mind, there was no place for them in a scientific psychology either. But what was to replace them? What was the psychological equivalent of oxygen? Or, if we define ourselves (functionally) as "epistemic engines", how should we conceptualize the "epistemic states" through which we pass?

Churchland denied that these states can be helpfully seen as sentences, processed by some "ideal sentential automaton" (ISA). Orthodox functionalism, and logical positivism too, had adopted a "vision of rational intellectual activity as consisting essentially in a dance of propositional states, a dance whose form preserves certain propositional relations" (1979: 126). But he argued that the *normative* aspects of thought and knowledge ("epistemic virtue") can't be grounded in such an account. Since very young children don't use sentences, we can't explain the development of rationality in sentential terms—despite Fodor's "recent and noteworthy attempt to make a go of a linguistic interpretation of the infant's cognitive activities" (p. 131).

The fundamental constraints on what constitutes rational activity must therefore lie elsewhere: "we must try to penetrate to that deeper intellectual kinematics of which our manipulation of sentences is just the occasional and superficial reflection" (p. 141). Beyond assuming that they must contribute to evolutionary survival, he didn't pretend to know what these constraints are. But he used thermodynamic analogies in speculating about them.

He described animals as "informational sponges", absorbing information from the physical environment *and somehow using that information to get more information.* As for how that added value might be achieved, he imagined an intertidal creature, whose survival depends on its predicting the superficially chaotic variations in water level. This could be done, he said, by means of three internal oscillatory parameters, whose interactions "model" the relevant dimensions of the outside world—and he sketched the mathematics involved. (Compare Kenneth Craik's approach, described in Chapters 4.b–c and 14.viii.)

Churchland's closing words showed how far he already was from GOFAI functionalism. They also showed why he would later be so sympathetic to PDP connectionism, and to dynamical (vector transformation) approaches derived from neuroscience:

And finally, it appears likely that the thermodynamics of "irreversible" processes will provide the underlying framework . . . for whatever genuine progress gets made here. For it is this theory that renders physically intelligible such things as the process of synthetic evolution in general, and the Sun-urged growth of a rose in particular. And what is human knowledge but a cortically embodied flower, fanned likewise into existence by the ambient flux of energy and information? (p. 151)

From the late 1980s on, accordingly, he and his wife treated those areas of cognitive science as sources of ideas about what cognition is, and how it works—two questions they saw as inseparable.

16.v. Counter-moves

So the functionalists reigned supreme, bickering among themselves about the rules of the club but encountering no serious opposition?—Not at all.

Outsiders protested that the functionalist enterprise was flawed in various ways. Indeed, one influential outsider was an ex-insider: Putnam resigned from the club some thirty years after founding it. As a result, the heady relief of the 1960s, on apparently escaping from the metaphysical impasse outlined in Section i, was gradually tempered.

'Friendly' objections, raised by people who spoke the same language as the functionalists, are discussed in this section. One influential critique was Block's (1978) 'Troubles With Functionalism'—but functionalism's troubles were many, and only a few can be mentioned here.

'Hostile' complaints grounded in a radically different perspective, a development of the neo-Kantianism sketched in Chapter 2.vi, are outlined in Sections vii–viii. Wittgensteinians were represented in both camps, as we'll see. So, too, was Putnam (Section vi).

a. Gödel to the rescue?

Disputes based in Gödel's theorem were voiced from the start. The general form of such objections is that computers are inherently limited in a way in which human minds are not.

Kurt Gödel had proved in 1931 that for any consistent logical system rich enough to contain elementary arithmetic, there is at least one meaningful sentence that cannot be proved in the system, but which humans can see to be true. And Turing himself had shown, independently of Gödel, that there are some well-defined questions that a given Turing machine can't answer (see 4.i.c).

For Turing, however, there was a crucial difference between a *given* Turing machine and *any* Turing machine. Accordingly, he rebutted "the mathematical objection" in *Mind*. He pointed out that "questions which cannot be answered by one machine may be satisfactorily answered by another". Moreover, no one had *proved* that we aren't subject to the same limitations. Certainly, we can sometimes see that a machine is giving the wrong answer (or failing to answer at all), which makes us feel superior. But we make mistakes too. And we can't triumph simultaneously over all machines: "There might be men cleverer than any given machine, but then again there might be other machines cleverer again, and so on" (A. M. Turing 1950).

Lucas (1929–) disagreed. A few years later, at a meeting of the Oxford Philosophical Society in 1959, he argued that "The Gödelian formula is the Achilles' heel of the cybernetical machine" (J. R. Lucas 1961: 116). This claim was the core of his seminal paper on 'Minds, Machines, and Gödel', published in *Philosophy* in 1961.

Gödel's proof, said Lucas, showed not that minds are "superior" to machines, but that they're different. However, describing the difference in positive terms was tricky:

We are trying to produce a model of the mind which is mechanical—which is essentially "dead"—but the mind, being in fact "alive," can always go one better than any formal, ossified, dead system can. Thanks to Gödel's theorem, the mind always has the last word. (J. R. Lucas 1961: 116)

Despite the canny scare quotes, the terms *dead* and *alive* didn't really help.

Towards the end of his paper, however, Lucas expressed the difference by reference to the "unity" of consciousness. The crux of Gödel's theorem is self-reference, and it can be escaped only by moving up a level—at which point, an equivalent logical difficulty arises. But, said Lucas, in recognizing that when a conscious being knows something he also knows that he knows it, and knows that he knows that he knows . . . we aren't positing "an infinite sequence of selves and super-selves and super-super-selves". Rather, we're recognizing that a conscious mind, since it has no true parts, "can both consider itself and its performance and yet not be other than that which did the performance . . . [It] is already complete, and has no Achilles' heel" (p. 125). This argument took the place of the *proof* that Turing had asked for to show that humans aren't subject to the same limitations as machines.

Most readers saw Lucas's argument as a weapon in the battle against functionalism. But he hadn't originally intended it in that way: "[I was] innocent of functionalism rather than anti-functionalist. Like many other isms, it just passed me by" (personal communication). He'd been led to Gödel's theorem not by functionalism but by the problem of free will (see also J. R. Lucas 1970: 114–23).

That's why his paper ended in the claim that mechanism is false. He allowed that a super-computer might come to behave unpredictably, as if it had "a mind of its own". But in that case, "it would cease to be a machine, within the meaning of the act" (1961: 127). It followed, he said, that "no scientific inquiry can ever exhaust the infinite variety of the human mind" and that our concepts of freedom and morality are safe from the challenge of mechanism.

(Functionalists argued, to the contrary, that a computational approach can not only allow for what we call freedom but also explain how it's possible: Boden 1972, ch. 7 and pp. 330–4; 1978; Sloman 1974; Dennett 1984*a*.)

For every person who was convinced by Lucas's paper, someone else wasn't. It became hugely influential, and is still attracting readers in the new century (it's on his web site, and is viewed by around fifty people a week—see J. R. Lucas 2000: 5). At a meeting held at the University of Sussex in 1990 to mark the fortieth anniversary of Turing's

Mind paper, Lucas looked back on how the argument had progressed (J. R. Lucas 1996). Wryly recalling the many replies that had been "lacking in either courtesy or caution", he pointed out that a large number of different objections had been raised by his critics (including cognitive scientists Dennett, Clark Glymour, and Douglas Hofstadter). He didn't regard any as decisive, and ended as he'd done before: roundly proclaiming the demise of mechanism.

Lucas had written the paper at Princeton, where he went (in 1957–8) specifically to work on Gödel. One of the people he encountered there was Putnam, but he didn't manage to convince him: in his 1960 paper, Putnam dismissed the Gödelian objection as a fallacy. Thirty years later, however, he saw it as a problem—not a knock-down formal argument, but an epistemological reminder that our reason enables us to go beyond whatever it can formalize (Putnam 1988: 118; cf. Putnam 1997: 39).

Roger Penrose, too, used Gödel's theorem to attack strong AI and computational psychology (R. Penrose 1989: 102–8). As a Fields medallist (the mathematical equivalent of a Nobel prizewinner), his voice received a very wide hearing—not least, in the media worldwide. (It doubtless helped that his conclusion, the superiority of human minds over computers, was what the vast majority of people wanted to hear.) Nevertheless, like Lucas before him he was challenged repeatedly, and at length (e.g. Sloman 1992).—This show, as they say, will run and run.

b. Consciousness and zombies

A second objection to functionalism that was anticipated (teasingly) by Turing concerned consciousness. As various people would later point out, there's no such thing as *the* problem of consciousness (see Chapter 14.xi.a, and Section iv.a below). The most hotly disputed of the *several* problems of consciousness concerns qualitative experience, which was discussed in Chapter 14.x–xi. Here, we need add only two points.

The first is that the mid-century arguments about consciousness (Section i, above) acquired new versions in the context of functionalism. Often, they were couched in terms of "zombies": android robots behaving *exactly like us*, but lacking all sensation or feeling. The objection typically ran as follows: since such creatures are obviously logically possible, functionalism must be false.

Dennett (given his views outlined in Section iv.a–b, and in Chapter 14.xi.b too) didn't see this as obvious at all. On the contrary, he said, the concept of zombies is incoherent, and belief in their possibility "ridiculous" (Dennett 1991, ch. 10.4; 1995*b*). Sloman (1999) argued, similarly, that nothing could have the same computational architecture as us (necessary for it to behave exactly like us), yet lack sensation.

An early variant of the zombie argument was due to Ziff (1920–2003). He claimed that "no robot could sensibly be said to feel anything", because we could program it to act any way we liked (Ziff 1959). So, for instance, we could make it act "tired" when lifting a feather but not a ton, or blue things but not green things. Again, the functionalist would ask whether these apparent "possibilities" assume a virtual machine that is at base incoherent. (A robot might tap into reserves of energy when lifting green things, because of some general motivational attitude to green; the ton-weight case, ignoring 'emergencies', is more problematic.)

Sometimes, it was claimed not that androids *needn't* be conscious, but that they *couldn't* be conscious. Scriven (1953), for instance, argued that no robot could be conscious, because it wouldn't be alive (see Section x.a). A few years later (Scriven 1960), however, he recanted: "I now believe that it is possible so to construct a supercomputer as to make it wholly unreasonable to deny that it had feelings."

The second point to be noted here is that the problem of *qualia* caused even some functionalists to admit defeat. If Dennett and the Churchlands—and Sloman (see Section ix.c)—thought they'd solved it, Chalmers and Fodor didn't.

Chalmers (1995, 1996*a*) called it the "hard" problem, which neither functionalism nor neuroscience could solve (see 14.x.d). But at least, he suggested, there was a glimmer of light. If we were to admit a "dual-aspect" notion of *information* as a fundamental feature of the natural world, then conscious experience could be accommodated in science. Sceptics complained that this wasn't explaining consciousness, but slipping it in through the cellar door—a millennial version of *élan vital* (see Chapter 2.vii.b).

As for Fodor, he had no hope of an answer. Saying that consciousness couldn't be a functional ("relational") phenomenon, he declared:

Nor do we know, even to a first glimmer, how a brain (or anything else that is physical) could manage to be a locus of conscious experience. This last is, surely, among the ultimate metaphysical mysteries; don't bet on anybody ever solving it. (Fodor 1995*a*: 83)

This was about as pessimistic as Colin McGinn's (1989, 1991) claim that consciousness is as far beyond our cognitive capacity to understand as algebra is for dogs. Indeed, for "pessimistic" one might read "defeatist". (Whether neuroscience might one day enable us to understand *qualia* was discussed in Chapter 14.x–xi.)

c. That room in China

A third cluster of counter-moves concerned intentionality. These also were made from the inception of functionalism, and were mentioned (in passing) by Turing himself (1950).

One of these, emanating from Searle (1932–) at the University of California at Berkeley in 1980, soon became notorious. This was a sparkling intellectual hatchet job, featuring the Chinese room.

People unfamiliar with Searle's other writings often assume that he had no sympathy for cognitive science in general and AI in particular. That's not so. "Since the beginnings of the discipline", he has said, "I have been a practicing 'cognitive scientist' " (1992: 197). And he's described AI—and computer functionalism—as "one of the most exciting developments in the entire two-thousand-year history of materialism" (1992: 43).

Moreover, he welcomed the fact that NLP researchers studying conversation had used his own work on speech acts (see 9.xi.f), not least because it enabled him to improve the theory. As he put it:

Now the beauty of AI, and this I really do admire, is that it forces you to pose those questions [about rules for understanding language] precisely and forces you to state your theory precisely. In fact the things I've written about—metaphors and indirect speech acts and so on—a great deal of it has been programmed by people working in various AI labs. So I think, in fact, that AI is an immensely useful tool in the study of language and the study of the mind . . . (in Pagels *et al.* 1984: 356)

However, there's cognitive science and then there's cognitive science. In Searle's view, "most mainstream cognitive scientists simply repeated the worst mistakes of the behaviorists [by ignoring] the essential features of the mind" (1992, p. xii). Those features, he said, are intentionality (the focus of his 1980 paper) and consciousness (the focus of his 1992 book).

The central point of Searle's provocative paper—the semantic emptiness of AI programs, considered as uninterpreted symbols—wasn't new. As we've seen, Fodor had embraced this in his "meditation" on formalism published a few months before, and Mays had noted it as early as 1949 (Manchester Philosophy Seminar 1949; Mays 1951, 1952). What Searle originated was an elegant slogan ("all syntax and no semantics"), a terminological distinction ("weak" versus "strong" AI), and an extraordinarily seductive—and slippery—thought experiment.

The terminological distinction quickly entered the discourse of cognitive science. However, it wasn't usually understood in just the way that Searle defined it. Both "weak" and "strong", in this context, have been interpreted in different ways. Searle put it like this:

According to weak AI, the principal value of the computer in the study of the mind is that it gives us a very powerful tool. For example, it enables us to formulate and test hypotheses in a more rigorous and precise fashion. But according to strong AI, the computer is not merely a tool in the study of the mind; rather, the appropriately programmed computer really *is* a mind, in the sense that computers given the right programs can be literally said to *understand* and have other cognitive states. (Searle 1980: 417)

So defined, weak AI—as Searle was quick to point out—is on a par with using computers to test theories about rainstorms. It claims no more theoretical affinity with minds than with anything else.

"Weak AI" is usually understood, however, as the claim that some of the computational features—processes, structures, virtual machines, architectures—involved are substantive theoretical terms in psychological explanations. As explained in Chapter 1.ii.a, this is the core thesis of cognitive science. Searle implied that to go this far and no further (that is, to claim explanatory power while drawing the line at strong AI) was disreputable, an example of muddled—even pusillanimous—thinking based on some fuzzy "computer metaphor". For him, *programs are tools* and *programs are minds* were the only intellectually honourable positions—and he praised Newell and Simon for their "straightforwardness", accordingly (see Section ix.b).

As for "strong" AI, this too is ambiguous. There are at least eight readings of it: some true, some false, some requiring empirical investigation (Sloman 1992). So even if the thought experiment 'worked', anyone who said baldly that strong AI had been dismissed would be too quick.

Searle's thought experiment entered the discourse of cognitive science—and everyday discourse, too. It was featured on TV and radio around the globe, including the BBC's annual Reith Lectures (Searle 1984), and in many newspapers and popular science magazines.

This, it seems, was a surprise to Stevan Harnad, the editor of *BBS*—where (on the say-so of the associate editor, Pylyshyn) Searle's paper was first published, alongside twenty-seven peer commentaries. He hadn't been enthusiastic:

I cannot say that I was especially impressed . . . [It] seemed to be yet another tedious "Granny Objection" about why/how we are not computers. . . . Across the ensuing years, further commentaries and responses continued to flow as, much to my surprise, Searle's paper became *BBS*'s most influential target article (and still is, to the present day) as well as something of a classic in cognitive science. (At [a Rochester Conference in 1982] Pat Hayes went so far as *to define cognitive science* as "the ongoing research program of showing Searle's Chinese Room Argument to be false"—"and silly", I believe he added at the time. (Harnad 2002: 294–5; italics added)

That last addendum was a dig at Searle, not Hayes. But one could be forgiven for thinking otherwise, having read Harnad's own writings on the topic:

As the arguments and counter-arguments kept surging across the years I chafed at being the only one on the planet not entitled (*ex officio*, being the umpire) to have a go, even though I felt that I could settle Searle's wagon if I had a chance . . . [When I eventually joined the online discussion on "comp.ai"] I found comp.ai choked with such a litany of unspeakably bad anti-Searle arguments that I found I had to spend all my air-time defending Searle against these non-starters instead of burying him, as I had intended to do. (p. 296)

To Harnad's annoyance, even Searle believed him to be on his side, and urged him "to keep fighting the good fight". Now, however, Harnad himself admits that on the one "essential point" they'd actually agreed all along (p. 296). This historical vignette is just one illustration of the slipperiness of Searle's thought experiment.

Aspiring writers may take comfort in knowing that a paper disparaged by one editor was published nonetheless—and sped around the world immediately. But to match this phenomenon they'll need to think up an example that's equally seductive.

Just in case you can't already recite it in your sleep, here it is:

Situation: Searle in a windowless room; a slot through which paper slips with "squiggles" and "squoggles" on them are occasionally passed in; a box of slips carrying similar doodles, of various shapes; and a rule book, saying that if a squiggle is passed in then Searle should find a blingle-blungle and pass it out, or perhaps go through a long *sequence* of doodle pairings before passing some particular shape out.

Denouement: Unknown to Searle-in-the-room, the doodles are Chinese writing; the rule book is a Chinese NLP program, comparable to Wendy Lehnert's question-answerer (see Chapter 9.xi.d); and the Chinese people outside the room are happily using Searle to answer their questions about some topic or other.

Punchline: Searle entered the room unable to understand Chinese and, no matter how long he stays there, he still won't understand a word of it when he comes out.

Conclusion: Formal computation alone (which is what Searle-in-the-room is doing) can't generate intentionality. Therefore strong AI is impossible.

Arguably, Searle was here attacking a straw man. Newell and Simon, and Minsky and John McCarthy too, certainly believed that AI systems of a certain complexity would be *really* intelligent, and *really* have beliefs. But, as we'll see in Section ix.b, they didn't have abstract Turing-computation in mind.

Searle, however, did (1980: 417). He argued that programs are mere shufflers of meaningless shapes. Indeed, he said, any AI program could in principle be interpreted as (mapped onto) many different activities: tax laws, dance routines . . . whatever. As regards the program itself, the choice between "meanings" is arbitrary. Any meaning it appears to have is derived entirely from us. So although we can't help speaking of computers in intentional terms (as Dennett had said), they don't "intrinsically" merit such ascriptions—whereas people do. So strong AI is an illusion.

As for weak AI, Searle doubted that brains generally implement formal computations. If they don't, then (GOFAI-based) computational psychology—far from being, as Fodor had claimed, the only psychology on offer—can't even *begin* to explain our mental life. But even if they do, he said, the Chinese room argument showed that *something more* is needed for intentionality.

The Chinese room, like the Turing Test, spawned a minor philosophical industry. A goodly number of readers agreed that the Chinese room argument proved just what he said it proved, while just as many saw it as fundamentally wrong-headed. Critique and reply alternated in many different places, and in varied forms. For example, the Chinese *room* was transformed into the Chinese *gym*, to deal with connectionism (Searle 1982, 1984, 1990*a,b*, 1992). And, twenty years after the original publication, an entire volume of new critiques was specifically commissioned by a major university press (Preston and Bishop 2002).

The barrage of responses to Searle's paper mostly ignored the mysterious "something more"—with good reason, as we'll see below. Instead, they focused on his novel way of expressing the old empty-symbolism argument.

Many objections were versions of a position he'd anticipated—and rejected—as "the Robot reply". This said that Searle-in-the-room would acquire understanding of Chinese if sensori-motor mechanisms were added. In other words, symbols (or concepts) must be grounded in world-engaging activities in order to be meaningful, and to refer to individually discriminable things (e.g. Fodor 1980*b*; Dretske 1984, 1995; Harnad 1989, 1990).

Some philosophers of language in the 1980s held that causal processes aren't enough, that they have to be grounded in evolutionary history—in which case only *evolved* robots, at most, could possess intentionality (see Section x.d). And a number of critics argued that causation is intimately linked with *computation* as it is understood in practice, so that Searle's—and Turing's—abstract concept of computation is largely irrelevant (see Section ix).

d. Neuroprotein and intentionality

Searle, too, had stressed the importance of causation—but in a most unilluminating way. This concerned the "something more" needed for intentionality.

According to him, neuroprotein has "causal powers" capable of generating intentionality, whereas metal and silicon don't. As he put it some years later: "I think it is empirically absurd to suppose that we could duplicate the causal powers of neurons entirely in silicon" (1992: 66). The brain, he granted, is a digital computer—because *everything*, including your bedroom wallpaper, can be seen as a digital computer.

(We needn't discuss that point here: for rebuttals, see Chrisley 1995; Chalmers 1996*b*; Copeland 1996.) But its formal structure isn't the point:

Whatever else intentionality is, it is a biological phenomenon, and it is as likely to be as causally dependent on the specific biochemistry of its origin as lactation, photosynthesis, or any other biological phenomena. No one would suppose that we could produce milk and sugar by running a computer simulation of the formal sequences in lactation and photosynthesis, but where the mind is concerned many people are willing to believe in such a miracle . . . (1980: 424)

We could even simulate brain functions (he said), by having the Chinese room contain a system of water pipes and adjustable valves corresponding to neurones and synapses. But still there'd be no understanding, because "[the brain's] causal properties, its ability to produce intentional states, wouldn't even have been simulated, never mind reproduced" (p. 421).

One might be tempted to describe this as a fairy story about the brain, except that fairy stories have some positive content. Even Penrose's shaky speculations about microtubules would offer more than this (Chapter 14.x.d).

Searle's move here, besides being empty, ignored the nature of explanation in neuroscience. Certainly, we have strong empirical evidence—as Descartes did too—that the brain is causally implicated in mental life. But at the level of *material stuff* (compare: lactose, chlorophyll), this is intuitively unintelligible.

In other words, we're in much the same philosophical position as the alien in the science-fiction story, if not *quite* so surprised by the brute facts:

"They're made out of meat."

"Meat?"

"Meat. They're made out of meat."

"Meat?"

"There's no doubt about it . . . They're completely meat."

"That's impossible. What about the radio signals? The messages to the stars."

"They use the radio waves to talk, but the signals don't come from them. The signals come from machines."

"So who made the machines? That's who we want to contact."

"They made the machines. That's what I'm trying to tell you. Meat made the machines."

"That's ridiculous. How can meat make a machine? You're asking me to believe in sentient meat."

"I'm not asking you, I'm telling you . . . We probed them. They're meat all the way through."

"No brain?"

"Oh there is a brain all right. It's just that the brain is made out of meat."

"So . . . what does the thinking?"

"You're not understanding, are you? The brain does the thinking. The meat."

"Thinking meat! You're asking me to believe in thinking meat!"

"Yes, thinking meat! Conscious meat! Loving meat. Dreaming meat. The meat is the whole deal! Are you getting the picture?"

"Omigod. You're serious then. They're made out of meat." (Bisson 1991)

Even neuroscientists can empathize with the alien's amazement. For in so far as we understand how neuroprotein (aka meat) grounds intentionality, we do so in terms of functional concepts. We speak of *messages, excitation, thresholds, codes, bug-detectors, orientation detectors*, and other types of *computation* (see Chapter 14). The neurotransmitters (meat juices) aren't interesting primarily for their biochemistry, but for the fact that they function to alter thresholds, for instance. Even the crucial sodium pump, whose material chemistry is thoroughly understood, is significant for its functional properties, in allowing an electrical impulse (message) to propagate along the axon. Similarly, we know a great deal about the chemistry of the synapse—but what's psychologically relevant is how this affects the neurone's message-passing functions (Chapter 14.ix.d and f).

Many of us may share Searle's unargued hunch about the inadequacy of metal and silicon. Neuroprotein may indeed be the only substance on earth that can support intentionality. But if so, why? Perhaps only neuroprotein allows the combination of chemical complexity, stability, and flexibility that's required for implementing the virtual machines that generate behaviour and thought? To say this, however, is—again—to make a functional point: what matters is what a particular metabolite enables the brain to *do* (see 14.ix.f).

Although this aspect of Searle's paper didn't engender much debate, it did encourage the fourth counter-move against functionalism: scepticism about multiple realizability.

e. How multiple is multiple?

Putnam's initial claim had been that all the functional properties of human and animal minds are in principle realizable in indefinitely many ways. But whether they were all multiply realizable in practice was another matter.

If philosophers were apparently licensed to ignore neuroscience, other cognitive scientists weren't. Indeed, even some philosophers felt that they would have to take notice of the brain after all, if it turned out that computations typical of brains could do things which other types of computation (in practice) could not.

This is why PDP connectionism, despite its many 'unnatural' features (such as back propagation), was often seen as relevant to *philosophical* theories of concepts and thinking (Chapter 12.x). It's why even connectionism was soon criticized for being insufficiently close to neurobiology, and urged to take facts about the brain on board (14.ii.d). It's why the Churchlands used ideas drawn from neuroscience, such as "phase-space sandwiches", to help justify their version of eliminative materialism (see 14.x.d). And, in the context of computer science, it's also why some AI scientists see Searle's argument as fundamentally irrelevant to their research (see Section ix).

Encouragement, however, isn't initiation: it didn't need Searle to make cognitive scientists consider neurophysiology. John von Neumann himself had surmised that the "logic" of the brain was very different from that of the von Neumann computer, and in 1947 McCulloch and Pitts had said much the same—offering hypotheses about what the alternative might be (see Chapter 12.i.c). Newell and Simon had long designed AI models that reflected specific properties of the brain, and Marr (among others) had asked what types of computation certain neural mechanisms could perform

(Chapters 7.iv.b and 14.v). In short, the functionalist dogma of multiple realizability was already being taken with a large pinch of salt.

Even so, the explanatory focus was more on *the types of computation* involved than on their biochemical embodiment. As the years passed, it became increasingly clear that neuroscience can provide crucial information about the functional organization of a cognitive system (the virtual machine), not just about how it happens to be materially implemented (Bechtel and Mundale 1999; Keeley 2000*a*). And at the turn of the twenty-first century, a connectionist textbook appeared which paid greater attention to neurobiology than the fledgling PDP had done (O'Reilly and Munakata 2000: see 14.ii.d).

f. Subconsciousness attacked

The fifth counter-move was more radical. The functionalists described brains as implementing computations, and even Searle was willing to entertain this as a hypothesis. But the more deeply committed Wittgensteinians saw it as absurd.

Their problem wasn't specifically with computation, although they thought that "supposing the brain to be a computer is mere fashion" (Anscombe 1974: 235). They were suspicious of *any* attempt to explain mental life in terms of sub-personal mechanisms that weren't consciously acknowledged by the subject. Even Freudian theory was suspect for this reason. Brain mechanisms there must be, of course—but those were the neurophysiologist's business.

This attitude was evident in the uncompromising rejection of New Look psychology by two of Wittgenstein's leading disciples, Anscombe (1919–2001) and Malcolm. At a meeting held in 1971 at the University of Kent at Canterbury, Gregory (1974*b*; cf. 1975) summarized his 'New Look' view of perceptions as hypotheses (see 6.ii.e). Anscombe would have none of it:

Now an hypothesis is something answerable to evidence, to data. To what data could the perceptual hypotheses that Gregory speaks of be answerable, but to perceptions? *Are* these perceptions then in turn hypotheses, and so on *ad infinitum*? (Anscombe 1974: 213)

What is framing the hypothesis? Is [it] the conception which Gregory suggests, that the framer is some mechanism that produces hypotheses that are answerable to input? For *we* who perceive don't know what the input is.

Hypotheses are *that* things are the case ... [It] looks as if Gregory's theory involved our perceiving apparatus itself as entertaining judgments. Hypotheses are predictive by way of inference. They also involve the logical constants, can be negative, universal, particular, conjunctive, and disjunctive. Perception can be all of these things except for being disjunctive ... By perception's being alternative I would mean two perceptions presenting themselves as alternatives at the same time. (p. 218)

When there being a tree here ... is the description of what is plainly the case, plainly and visibly to the perceiver—it is absurd to call the perceptual belief an hypothesis. (p. 243)

Rather than speaking of hypotheses, she suggested, he should have spoken of "models", "patterns", or "schematic sketches of possibilities".

Gregory responded with some asperity:

[How] much surprise are we allowed before a word becomes philosophically suspect? ... I do not want a normal word, because I do not want to express a normal thought ... The "framer" I regard as brain mechanisms . . . Surely we do not *have* to say that total human beings alone have the prerogative to frame hypotheses? (Animals might do so—and are *they* conscious? Computers might do so—and surely *they* are not conscious?) Again, why should we be captured by the inertia of language? (Gregory 1974*b*: 232, 234)

Anscombe fought back: "I don't mind a bit if he uses the word 'hypothesis' in a novel way . . . I merely want to understand what he says and assess its value" (pp. 236–7). Gregory had found out some interesting things, she conceded, "but how are his enquiries logically facilitated by having the scientific hypothesis paradigm rather than by thinking of patterns and models?"

This defence would have rung more true if her initial paper had been more constructive in spirit. As Gregory put it, "[Do] these philosophers have rival theories of perception up their sleeves? If so, they must surely justify their linguistic preferences not from common usage but by reference to their theories" (Gregory 1974*b*: 234).

Malcolm (1971) went even further than Anscombe, rejecting the cognitive approach wholesale. Much as he'd earlier denied that scientific evidence could justify our saying that dreams occur at certain times during sleep (see Section iii.b), so he now dismissed cognitive structures and processes as "myth". He wasn't saying that *these* structures and processes (hypotheses, perhaps) are implausible whereas *those* (models, perhaps) are better founded. Rather, he was outlawing all theorizing about subconscious psychological mechanisms.

A powerful counter-attack, analogous to Putnam's (1962*a*) critique of Malcolm's views on dreaming, appeared immediately (M. Martin 1973). It argued that scientists typically extend linguistic usage, and that cognitive psychologists had good grounds for positing various types of unconscious process and structure. But Malcolm wasn't persuaded.—It's small wonder, then, that Fodor introduced his own approach in the mid-1970s by observing that "many philosophers [still believe] that Ryle and Wittgenstein killed this sort of psychology some time about 1945, and there is no point to speculating on the prospects of the deceased" (Fodor 1975: 2).

16.vi. Betrayal

Already under fire from these many directions, functionalism then had to face the unkindest cut of all. Putnam himself started to have doubts in the early 1970s, and explicitly disowned his brainchild some fifteen years later.

Ironically, GOFAI suffered a similar trauma at much the same time, being betrayed in 1986 by the erstwhile wunderkind Terry Winograd (Chapter 11.ii.g). But Putnam's disaffection with functionalism was the more shocking. For whereas Winograd had picked up someone else's ball before running with it, Putnam had sewn the ball together before throwing it onto the pitch.

In 1973, at a meeting on 'Computers and the Mind' at Berkeley, Putnam still defended the autonomy of psychology (Putnam 1975*b*). But he rejected his previous claims that the whole human being is a Turing machine, and that a mental state is a state of a Turing machine, saying he had been "too much in the grip of the reductionist outlook".

That was more a complaint about oversimplification than a rejection of the spirit of functionalism. By the late 1980s, however, rejection had won out:

The desire that grips Fodor, then, as it once gripped me, is the desire to make belief–desire psychology "scientific" by simply identifying it outright with computational psychology . . . Mentalism is just the latest form taken by a more general tendency in the history of thought, the tendency to think of concepts as scientifically describable ("psychologically real") entities in the mind or brain. And it is this entire tendency that, I shall argue, is misguided. (Putnam 1988: 7)

Functionalism, construed as the thesis that propositional attitudes are just computational states of the brain, cannot be correct. (p. 73)

One might compare this change of viewpoint with Bruner's contemporaneous flight from experimentation and computation to interpretation and narrative (Chapters 6.ii.c and 8.ii.a). Or one might see it as a Damascene conversion typical of the "counter-cultural" trend so prominent in the 1970s–1980s (see 1.iii.c–d). But it was much more carefully argued than most such examples (including Bruner's), and much longer drawn out than the episode on the road to Damascus.

The three pillars supporting Putnam's recantation were universal realizability, externalism, and meaning holism. Each one contradicted a central aspect of his 1960s position. And each one had arisen from *friendly* territory, in the sense that they developed while Putnam still had faith in functionalism.

In thinking further about the nature of language, he didn't improve functionalism but rejected it. Indeed, he ended up in what was traditionally seen as the enemy camp.

a. Friendly fire

Universal realizability was multiple realizability gone wild. The doctrine of mind–brain multiple realizability, said Putnam, "still seems to me to be as true and important as it ever did" (1988, p. xii). But now, he added a mind-computation version, which forbade the identification of mental states with specific computations.

Using an argument so highly technical that it was relegated to an Appendix, he claimed that *any* program is implemented in *any* object. So mental states supervene on computations much as they supervene on brain states: they can't be identified with each other.

This—if correct—put paid to computational psychology: if every pebble implements Newell and Simon's GPS, and every leaf Chomsky's grammar, why bother?

Searle later added this argument to his own arsenal (1992: 208–9). But others saw it as invalid, on the grounds that it ignored the causal aspects of computation. That, of course, was a heretical view—indeed, for most philosophers it qualified as *nonsense*. For their normal assumption was—and still is—that computation is a purely formal notion, not a causal one. This disagreement will be discussed at length in Section ix.

The second pillar of Putnam's self-critique was externalism. This is the type of philosophical semantics—whose popularity grew in the late 1970s and 1980s—which favours "wide" content (see Section iv.d).

Sometimes, wide content is understood in naturalistic terms, such as agent–environment relations (G. Evans 1982), natural kinds (Kripke 1980), or evolutionary history (Millikan 1984). And Putnam himself argued that the meaning of nouns like "water" is fixed by the physical constitution of the natural kind con-cerned—whether we know about it or not. (This was the import of his much-disputed "Twin Earth" example—1975*a*: 223–7.)

But he also saw interpretative norms as crucial. Indeed, he even argued that causal accounts are inherently non-naturalistic, because what we count as "the" cause (like "the" explanation) depends on our interests (Putnam 1982).

Wittgensteinians, of course, had long stressed the relevance of cultural practices and norms in assigning meaning to language. Now, Malcolm's milk bottles had come home to roost (see Section i.d). If we can't individuate concepts and beliefs without reference to the physical and cultural environment, then neither brain states nor computational states *as such* can carry the content that identity theorists and functionalists, respectively, had ascribed to them. As Putnam put it: " 'meanings' just ain't in the *head*!" (1975*a*: 227). (Compare Wittgenstein: "If God had looked into our minds he would not have been able to see there whom we were speaking of"—1953: 217.)

As for the third pillar, meaning holism, this was a neo-Quinean position which held that the meaning of an individual term (concept) isn't independent of the overall framework of beliefs in which it's involved.

It's not just that meanings are fuzzy, incapable of definition by necessary and sufficient conditions. Worse, they're essentially interconnected. No one can think *that there are churches in Vienna*, for example, unless they possess concepts like *city, building, religion* . . . and so on. The reason is that the meaning of the concept *church* involves the meanings of many other concepts.

Quine had developed this theory in his attempt to find an adequate philosophy of science—indeed, to ground philosophy *in* science. But it could also be seen as a modern version of the humanist Humboldt's view that language is inseparable from culture (Chapter 9.iv.b). You and I may use words we're happy to regard as 'equivalent', translating the Cook Island Maori *tamariki* as the English *children*, for example. But Western and Polynesian beliefs about children are subtly different, so (on this view) our concepts are different too. This scientific/humanist ambiguity explains why Putnam eventually crossed over to 'the other side' (see subsection b).

Dennett, you'll remember (I hope!), had argued that *beliefs* can't be pinned down, for this very reason. But it hadn't turned him away from a neo-functionalist view of cognitive science. However, it worried Putnam more than Dennett. It was puzzling over this fact, Putnam said later, which was "the beginning of the end of my attachment to functionalism" (1999: 118).

Eventually, it led Putnam to the conclusion that the very notion of a "functional state" is "not fully intelligible". (It also led him to conclude that the notion of an "internal phenomenal state" isn't fully intelligible either, another point on which he agreed with Dennett: Chapter 14.xi.c–d.)

Mental states (propositional attitudes), Putnam now argued, are similarly intercon-nected, and similarly impossible to pin down. No two people thinking "The cat sat on the mat" are thinking exactly the same thought, because their beliefs about cats and mats aren't identical. When we say that they *are*, we're relying on cultural norms that decide *what we are willing to count as* the "same" thought. (*Tamariki/children*, again.) It follows that no single computational account, howsoever complex, can fit every instance of "one and the same thought".

Whether Putnam, Quine, or anyone else (even Humboldt) ever took this doctrine literally is arguable. Putnam specifically allowed that two terms may have the same *denotation* (though not the same *sense*, or *meaning*): "rational animal" and "featherless biped", for example. But what's important here is that Fodor's functionalism, and GOFAI modelling in general, took a very different approach to meaning.

Even PDP models, which allowed for fuzziness and family resemblances in individual concepts (12.x.a–b), didn't reflect mutualities of meaning between beliefs. Indeed, Putnam's holism implied that *no* scientific theory could use meanings (concepts, mental states) as scientific objects, playing an explanatory role. (For Fodor's rebuttal, see Fodor 1987*b*, ch. 3.) That is, psychology—because it deals with meanings, or *content*—can't be naturalized:

> There is nothing in [my mature position on perception and conception] that is "antiscientific" in the sense of *standing in the way* of serious attempts to provide better models, *both neurological and computational*, of the brain processes upon which our perceptual and conceptual powers depend, processes concerning which we still know very little. Moreover, it is *a profound mistake* to equate serious science with Cartesianism cum materialism that has for three centuries tried to wrap itself in the mantle of science. (Putnam 1999: 48; first italics in the original, others added)

At the end of the century, Putnam remained unrepentant. Cognitive science, he said, had turned out to be more science fiction than science:

> There is no harm in speculating about scientific possibilities that we are not presently able to realize; but is the possibility of an "ideal psychological theory" of this sort [i.e. expressed in a "normal form": see Section iii.b] anything more than a "we know not what"? . . . Do we have any conception of what such a theory might look like? Even if we had a candidate . . . how would we go about verifying that it does implicitly define the unreduced psychological properties? One hears a lot of talk about "cognitive science" nowadays, but one needs to distinguish the putting forward of a scientific theory, or the flourishing of a scientific discipline with well-defined questions, from the proffering of promissory notes for possible theories that one does not know even in principle how to redeem. (Putnam 1997: 36)

One must take care, however, to note just what he meant by cognitive science. He *wasn't* saying that the research described in Chapters 6–15 is worthless. Nor was he complaining (like N. Stuart Sutherland: see Chapter 1.ii.c) that some self-styled cognitive scientists focus not on empirical work but on the philosophical disagreements within the field.

Rather, he was saying that those (like Fodor) who sought a computational theory that would map neatly onto individual propositional attitudes, and that would identify their content independently of our interpretative practices and the context of use, were doomed to failure. The "narrow" side of the semantic stage, for Putnam, can support no real action. And even wide content—*pace* Dretske, Dennett, or Ruth Millikan—can't be analysed in naturalistic, non-normative, terms.

b. Crossing the river

Putnam's 'friendly' three-pronged attack on functionalism, described above, was only part of his reason for abandoning it. By the late 1980s, he'd become committed also to a fundamentally hostile (neo-Kantian) critique. As we'll see in Section vii, various versions of this had been lurking in the background—specifically: across the English Channel—for well over a century. They reached Putnam via the later Wittgenstein, and eventually turned his ideas upside down.

Putnam described his apostasy as "a matter of being torn between opposing views of the nature of philosophy" (1988, p. xii). This opposition is evident, for instance, in the difference between the "early" and "late" Wittgenstein. One way of describing it is in terms of attitudes towards metaphysical realism.

The typical empiricist–scientific position is that a real world exists independently of us, with its own intrinsic properties and a determinate built-in structure. Our thoughts represent it, and are *true* in so far as they correspond to it. The business of science is to describe the world and our relation to it, and philosophy—including the philosophy of mind—must be constrained by the findings of science. As Bertrand Russell put it, in criticizing the later Wittgenstein:

[Philosophers should resist] the desire to separate [philosophy] sharply from empirical science. I do not think such a separation can usefully be made. A philosophy which is to have any value should be built upon a wide and firm foundation of knowledge that is not specifically philosophical. (B. Russell 1956*b*: 329)

This form of realism, however, appears to put a metaphysical gulf between thought (i.e. representation) and world (reality). The typical neo-Kantian position, outlined in Chapter 2.vi, aims to avoid that difficulty.

It holds that things in the empirical world aren't mind-independent, but constituted by the concepts informing human minds. For Kant himself, there was also the noumenal world. Some of his followers dropped this unknowable thing-in-itself, and moved on to idealism: the view that *all* reality is constituted by human thought. Some implications for the philosophy of language and biology were noted in Chapters 9.iv and 15.viii.b, respectively. As for the philosophy of mind, the implication highlighted by the Continental phenomenologists in the first half of the twentieth century was that there is no metaphysically fundamental Subject–Object or mind–body (or even body–environment) distinction. The notion that there is, they argued, is a Cartesian illusion.

Illusion or not, this Cartesian view underlay most early cognitive science—some cyberneticians perhaps excepted (see Chapter 4.v–vii). In general, the mind was conceptualized without reference to the body, and Subject and Object were taken as given. There was no suggestion that the nature of mental states depends on the fact that we have bodies (as opposed to brains). Certainly, sensory transducers and motor effectors entered this functionalist story. But the prime theoretical focus lay inside the head. Mental life was seen as a sensori-motor sandwich (2.iii.a, 10.iv.b), a repeated cycle of *perception, thought, action*. And *thought* was where the interest lay.

This is why most early AI (as we saw in Chapters 9.x–xi and 10) confined itself to text-based examples, and discarded much of the real-world information even when sensori-motor aspects were included. In effect, seeing ('scene analysis') was treated as a species of reasoning. Even robotics, in GOFAI, concentrated on abstract representation and planning. The messy details of the real world, and of bodily movement in it, were excluded whenever possible. By contrast, many workers in situated robotics, such as

Barbara Webb for instance, would welcome these messy details as a way of testing their theories (see Chapter 15.vii.c).

Putnam had started out as an unabashed realist. Before formulating functionalism, he'd even believed that all sciences reduce to physics (see Section iii.b). But, like Wittgenstein before him, he changed his mind. By the early 1980s, he was convinced that there are no metaphysically intrinsic properties, no mind-independent facts, or "ready-made world" (Putnam 1982).

Properties (he now argued) are "essential" only *relative to a description*. If one crushes a statue, it is no longer a statue. But think of it as a piece of clay, and the crushing doesn't matter: shape isn't essential. In the final chapter of *Representation and Reality* (1988), Putnam developed this neo-Kantian theme. He avoided out-and-out idealism, allowing that minds are in the world, not the fundamental creators of it. But he avoided straightforward realism, too. He called his theory *"internal* realism":

> [My idea] is not the view that it's all *just* language. We can and should insist that some facts are there to be discovered and not legislated by us. But this is something to be said when one has adopted a way of speaking, a language, a "conceptual scheme". To talk of "facts" without specifying the language to be used is to talk of nothing; the word "fact" no more has its use fixed by the world itself than does the word "exist" or the word "object". (1988: 114)

He conceded that *within* any given conceptual scheme we need to distinguish between the natural world and our judgements about it. But it is we who determine the nature of our concepts, including *mind, body, self,* and *material object*. And it is we who construct normative and/or interest-relative notions such as *good, rational, justified,* and even *cause*. Since a fundamentally mind-independent (naturalistic) account of the world is impossible, mind and language are philosophically prior to science and can't be explained by it.

In short, Putnam had committed the ultimate betrayal: he'd crossed the metaphysical divide to join the neo-phenomenological camp. To be sure, he soon moved to the outskirts of the encampment: by the turn of the millennium, he had modified his "internal realism" of the 1980s to a less extreme (less idealistic) "common-sense realism", according to which we are *directly* aware of real, independently existing, things (Putnam 1999). But still, he had no truck with the notion of internal phenomenal states (*qualia*, sense data), nor with talk of "correlations" between mental and physical states (see 14.x.c–d).

Putnam wasn't alone in pitching his tent in the neo-Kantian compound. One philosopher of cognitive science was already there waiting, and others would soon arrive—as we'll now see.

16.vii. Neo-Phenomenology—From Critique to Construction

Some philosophers used neo-phenomenological ideas not to deny the possibility of a scientific psychology, but to suggest new ways of doing it. Susan Oyama (1985) was a case in point. As we saw in Chapter 14.ix.b, she conceptualized the mind not as a mysteriously pre-structured input–output machine, but as the outcome of a series of developmental shifts in a self-organizing system. However, her key interest was *the origin of form*, of which 'knowledge' is one example. Here, we'll focus on philosophers coming to neo-phenomenology from a concern with *the nature of representation*.

a. Where Dreyfus was coming from

If Putnam had done a philosophical volte-face, one important critic of cognitive science had been consistent in his neo-Kantianism since the early 1960s. Indeed, Dreyfus was already launching his polemic against AI and computational psychology while Putnam was still formulating functionalism (Dreyfus 1965, 1967), and his highly influential *What Computers Can't Do* appeared soon afterwards (1972).

Dreyfus's fourfold diagnosis of AI's failings was discussed in Chapter 11.ii.a–c. Here, our interest is in the philosophy underlying it. This was drawn primarily from the phenomenologists Heidegger (1889–1976) and Maurice Merleau-Ponty (1908–61), from the (closely related) Gestalt psychologists (5.ii.b), and from the later Wittgenstein. Another influence was Polanyi, whose philosophy of science had stressed the scientist's "tacit" knowledge and practical skills (Polanyi 1958; cf. H. L. Dreyfus and Dreyfus 1986; and see 13.ii.b)—and who'd countered Turing face to face in the 1940s by positing unformalizable knowledge (see Preface, ii).

All these thinkers inspired Dreyfus's rejection of AI's ontological assumption "that what there is, is a set of facts each logically independent of all the others" (H. L. Dreyfus 1972: 68). The early Wittgenstein, who had strongly influenced McCulloch (4.iii.c), had put it like this: "The world is the totality of facts . . . The world divides into facts . . . What is the case, the fact, is the existence of atomic facts" (1922: 31). But the later Wittgenstein had roundly rejected it, commenting for instance that "It makes no sense at all to speak absolutely of the 'simple parts of a chair' " (1953, para. 47). Dreyfus echoed this remark:

Even a chair is not understandable in terms of any set of facts or "elements of knowledge". To recognize an object as a chair . . . involves a whole context of human activity of which the shape of our body, the institution of furniture, the inevitability of fatigue, constitute only a small part. And these factors are no more isolable than the chair. (Dreyfus 1972: 122)

The phenomenologists also inspired Dreyfus's claim that the objects we experience—and to which we have direct access: no representations required—are inherently meaningful, or significant. They exist within a field of human interests (what the Gestalt psychologists called a "life-world") that gives them *relevance*, and which informs and guides our thinking.

Heidegger, in particular, had highlighted the immediate "situation"—in which the phenomenological subject has its Being, and *creates* the "beings" in the world and its coping response to them. The person's situation isn't a part of some objective external environment, but a lived context for practical action. Similarly, Merleau-Ponty had said "Motility [is not] a handmaid of consciousness, transporting the body to that point in space of which we have formed a representation beforehand . . . [We must] avoid saying that our body is *in* space, or *in* time. It *inhabits* space and time" (1945/ 1962: 139).

It follows, Dreyfus insisted, that we don't need to posit a third level of discourse, between phenomenological experience and neurophysiology. He followed Merleau-Ponty's example in saying that "what the learner acquires in experience [of the world] is not *represented* in the mind at all but is *presented* to the learner as a more and more finely discriminated situation" (1998: 3). Indeed, he argued—like Malcolm: see Section v.f—that it's not even clear what expressions like "mental processing" or "mental operations" could possibly mean (1972: 91). He was especially critical of GOFAI, but he rejected *any* theory of internal representations. In short: "To avoid inventing problems and mysteries we must leave the physical world to the physicists and neurophysiologists, and return to our description of the human world which we immediately perceive" (1972: 183).

As for where the meanings come from, he said, they're grounded in the body. Here, he was following Merleau-Ponty, who had written extensively on the difference between one's own body and mere physical objects. The body, in the phenomenologist's sense, cannot be described in purely naturalistic terms since it is the *origin* of meaning, or value:

My body is the fabric into which all objects are woven, and it is, at least in relation to the perceived world, the general instrument of my "comprehension."
   It is my body which gives significance not only to the natural object, but also to cultural objects like words. (Merleau-Ponty 1945/1962: 235)

Hence, said Dreyfus (1967), computers must have bodies in order to be intelligent. But this requires more than being a robot: "no one understands how to begin to program primitive robots, even those which move around, to *have a world*" (H. L. Dreyfus 1972: 211; italics added).

On this view, we cope with the world by exercising our "bodily skills", which range from pattern recognition, through language, to tool use. (Tool use was a favourite example not only of Heidegger, Merleau-Ponty, and Polanyi but also of George Miller in his cognitive science manifesto: Miller *et al.* 1960. It's instructive to compare what Heidegger and Miller said about what's involved in using a hammer: Chapter 6.iv.c.)

Skilled behaviour, for Dreyfus, isn't the result of computation or information processing, and nor is it the execution of a plan. We may (consciously) follow rules or make plans when we're learning the skill, but "there is a moment when we finally transfer control to the body" (H. L. Dreyfus 1972: 160). Then, we just do it.

For cognitive scientists reading Dreyfus's critique, there was no "just" about it. They wanted to know *how* behaviour, skilled or unskilled, happens. (Later, they would also want to know how bodily skills *develop*: 12.viii.c–e and 14.ix.a–d.) The intellectual chasm between them and Dreyfus was huge:

The alternative [i.e. phenomenological] view has many hurdles to overcome. The greatest of these is that it cannot be presented as an alternative scientific explanation . . . [We] shall have to propose a different *sort* of explanation, a different sort of answer to the question "How does man produce intelligent behavior" or even a different sort of question, for the notion of "producing" behavior instead of simply exhibiting it is already colored by the [Cartesian/scientific] tradition. For a product must be produced in some way; and if it isn't produced in some definite way, the only alternative seems to be that it is produced magically. (H. L. Dreyfus 1972: 144)

[The alternative answer] takes the form of a phenomenological description of the behavior involved. It, too, can give us understanding if it is able to find the general characteristics of such behavior . . . Such an account can even be called an explanation if it goes further and tries to find the fundamental features of human activity which serve as the necessary and sufficient conditions for all forms of human behavior. (p. 145)

Compare Merleau-Ponty:

[Phenomenology] is a matter of describing, not of explaining or analysing . . . The whole universe of science is built upon the world as directly experienced, and if we want to subject science itself to rigorous scrutiny and arrive at a precise assessment of its meaning and scope, we must begin by reawakening the basic experience of the world of which science is the second-order expression. (1945/1962, p. viii)

Cognitive scientists could happily allow that phenomenologists often provide subtle descriptions of experience. One excellent example, published a few years later (with the book motto drawn from Heidegger), concerned the changes in consciousness and bodily skills that happen when one learns to improvise jazz on the piano (Sudnow 1978/2001). But they wanted to know, in scientific terms, *how* the body enables this to happen.

Dreyfus's insistence that it's simply a matter of physical energies affecting the brain, while true at one level of discourse, didn't help. As Sutherland (1974) soon pointed out, "leaving it to the neurophysiologists" might itself involve the use of what Dreyfus had called third-level concepts. Indeed, Marr had already developed a computational theory of the cerebellum—specifically aimed to show how bodily skills can be learnt, smoothed, and controlled (Chapter 14.iv.c). To be fair, this was a connectionist account, not a GOFAI one. Nevertheless, it was a computational theory (see Section ix, below).

Given this intellectual chasm, Dreyfus-versus-X—where X is virtually any pre-1990s cognitive scientist—was less a meeting of minds than the simultaneous snarling of different species of beast. Dreyfus's book caused some bewilderment even among philosophers, for the relation between Anglo-American and Continental philosophers in the early 1970s was more like ships passing in a foggy night than opponents engaging in the arena.

In other words, most Anglo-Americans (with notable exceptions, such as Mays and Taylor) paid scant attention to anything said by the Continentals. When they did, they weren't impressed.

The realist Russell had expressed this intellectual chasm in uncompromising terms. Dismissing the later Wittgenstein as "completely unintelligible", he complained, "I have not found in Wittgenstein's *Philosophical Investigations* anything that seemed to me interesting and I do not understand why a whole school finds important wisdom in its pages" (B. Russell 1956*b*: 319). In his obituary of Wittgenstein a few years before, he'd been more tactful. The later philosophy was mentioned only indirectly: "Getting to know Wittgenstein was one of the most exciting intellectual adventures of my life. In later years there was a lack of sympathy between us . . . " and—as the closing words—"Of the development of his opinions after 1919 I cannot speak" (B. Russell 1951: 297, 298). Despite this conventional observance of *de mortuis nil nisi bonum*, the message was clear. In Russell's eyes, Wittgenstein had betrayed everything he (with Russell) had previously stood for.

On the other side, Heidegger and Merleau-Ponty were no less hostile to realism. And Heidegger was hostile to clarity too. He notoriously declared that "Making itself intelligible is suicide for philosophy" (Blackburn 2000).

Russell wasn't the only one to bridle at such a sentiment. By the 1940s, Heidegger was "a figure of fun, too absurd to be taken seriously as a threat" by analytic philosophers (Dummett 1975: 437), and accused by Anthony Quinton (personal communication) of "ponderous and rubbishy woolgathering". The one-time editor of *Mind* Simon Blackburn (2000) has provided a hilarious critique of Heidegger, and speaks of "the purgatory of trying to read [him]" (2005: 78). Similarly, Paul Edwards delivers some savage sarcasm in his attempt "to stem this [current] tide of unreason" (2004: 9). Noting that Heidegger's work has appeared in about twenty languages, he remarks that there's no Hebrew translation: "This is perhaps just as well—the Jews have surely suffered enough already" (p. 11).

Given such entrenched hostility within the Anglo-American camp, it's doubly strange that this German philosopher should now be so fashionable in some areas of cognitive science. For besides eschewing "intelligibility", he scorned the cyberneticists' view of man-as-machine as the ultimate metaphysical illusion. (He was also a savage critic of technology in general.) One might say that he foresaw what Donna Haraway (1986/1991) would call the "cyborg" culture, and fought against it with all the contempt he could muster.

Ryle had been unusual among analytic philosophers, having read fairly widely in phenomenology and being at least partly sympathetic to it (Ryle 1929, 1962). And Taylor (1971) had recently followed his critique of behaviourism (see Chapter 5.iii.b) by drawing on Heidegger, and on hermeneutics, to argue that science—because it's constructed by human subjectivity—can't hope to explain meaning.

But most Anglo-American philosophers at that time—and as noted above, many even now—judged Heidegger to be "a prolix charlatan and poisoner of good sense", whose prose style was "an abomination", mere "bombastic, indecipherable jargon" (Steiner 1978: 13, 16). When analytic philosophers such as Putnam strayed into the enemy territory of anti-realism, they were more likely to start from paths laid down by Wittgenstein (or even Quine) than Heidegger. So, for instance, they would follow Wittgenstein (and Ryle) in seeing "folk psychology" not as a psychological theory, adequate or otherwise, but as a network of conceptual necessities, constitutive of the very concept of mind.

The prime exception, and the person who was most responsible for bringing Heidegger's work back into Anglophone philosophy, was Rorty (although Dreyfus would come a close second). Rorty's first attempt at resurrecting Heidegger was in a conference paper read in 1974, and published soon afterwards (Rorty 1976). But the most influential was his book *Philosophy and the Mirror of Nature* (1979). His accusation there that Descartes had "invented" the mind (cf. 2.iii.a) was largely due to Heidegger. So, too, was his focus on *practice* as opposed to *cognition*.

b. Hands-on Heideggerians

Rorty's efforts were by no means universally welcomed. But nor were they in vain. By the end of the century, more people were willing to take anti-Cartesian, even Heideggerian, ideas seriously—and to use them *constructively* in generating new approaches in cognitive science. (Whether they were faithful to the originals in so doing is another matter: see below.) Even Dreyfus was now recommending a "reconciliation" between phenomenology and AI (H. L. Dreyfus and Dreyfus 1990).

Some of these anti-Cartesians were scientists doing hands-on research in computer science, AI, A-Life, psychology, or neuroscience. They included a number of people mentioned elsewhere in this book (including in Section x, below): Winograd, Michael Arbib, Brian Smith, Philip Agre, Rodney Brooks, Inman Harvey, Randall Beer, Howard Pattee, Esther Thelen—and Maturana and Varela, whose assault on Cartesian realism dated back to the 1960s.

None of these people was originally inspired by Dreyfus, though some cited him in support after developing their own position. Indeed, with the part-exception of Pattee they weren't initially inspired by the philosophical literature at all. Their unorthodox views arose, rather, from reflecting on their own scientific research.

Winograd, for instance, had been frustrated by the limitations on his GOFAI wizardry in NLP (Chapter 9.xi.b). Later, he was influenced partly by Dreyfus (a Stanford colleague from the mid-1970s) and Heidegger, and also by the Frankfurt School, Maturana and Varela, and Fernando Flores (11.ii.g). Indeed, he and Flores co-authored a book in which they explicitly rejected classical realism (Winograd and Flores 1986: 30–1). So did Varela, who had recently applied autopoietic theory to cognitive psychology (Varela *et al.* 1991).

The psychologist Thelen saw the mind/brain as a dynamical system closely coupled with the environment, making this seemingly obvious brain–environment distinction problematic (Thelen 1985; Thelen and Smith 1994) (see Chapter 14.ix.b). Harvey, too, had questioned Cartesian realism, in connection with his work in evolutionary robotics (Husbands *et al.* 1995; Harvey 2000, 2005). And Smith's research—a heady mix of hard-headed computer science and highly speculative metaphysics—had surpassed all but Maturana and Varela in its commitment to criticizing realism.

Others such as Brooks (1987) had rejected representation (and reason too: 1991*b*), if not realism. (That's not to say that Brooks *argued for* realism: he simply assumed it, not being concerned with metaphysics.)

The situated robots described by Brooks, Harvey, Beer, and Arbib were all featured in previous chapters (13.iii.b, 14.vi.c, and 15.vi.c and vii.c). Here, we need add only that our discussion (above) suggests that situated robots are ill-named. To be sure, they respond relatively directly to specific environmental cues, in virtue of their physical constitution. That's why they were sometimes described in terms of phenomenological ideas about the body and the lived situation (e.g. Haugeland 1995). But they aren't situated in the full sense, because they aren't genuinely *embodied*.

As Merleau-Ponty would have been quick to point out, being an autonomous mobile robot isn't the same as having a body. This follows, too, from the philosophical biology of Maturana and Varela (Chapter 15.vii.b, and Section x.c, below). As for Agre and Smith, they're considered in Sections ix.c and ix.e, below.

c. Flights from the computer

Other late-century anti-Cartesians were philosophers of cognitive science—such as John Haugeland, Timothy van Gelder, and Clark. In general, they were sympathetic to the 'insider' attacks just mentioned. Like these scientists, they identified with cognitive science as an explanatory enterprise. If it was faulty, they wanted to fix it—not abandon it.

Today, Haugeland (at the University of Pittsburgh) is a leader of the neo-phenomenological movement in the philosophy of cognitive science. He took Dreyfus's work seriously from the start, and wrote a paper with him for the Kent meeting where Malcolm and Anscombe had also spoken (see Section v.f) (H. L. Dreyfus and Haugeland 1974). But his phenomenological sympathies were then largely implicit. They became fully explicit only in the 1990s.

Whereas Dreyfus's attacks on AI were unremittingly hostile, in both substance and tone, Haugeland's critique was more measured, and more constructive. In his first monograph, he complained:

Debunkers want to shoot AI down before it takes off, regardless of any actual research, as if the very idea were somehow crazy or incoherent. But the prospects for cheap victories strike me as slender. (1985: 249)

So, for example, he doubted whether computational psychology, or "cognitivism" (1978), could conceptualize moods, and ego involvement (see Chapter 7.i.d). But instead of dismissing cognitivism outright, he said that such phenomena "deserve a lot more study, even in AI" (1985: 238). Similarly, while suggesting that GOFAI couldn't model imagery, he allowed that some non-GOFAI computers could (p. 228).

In the 1980s, Haugeland saw cognitive science as "a great intellectual revolution", based on "a profound and distinctive empirical hypothesis" (pp. 2, 249). Its "paradigm", namely GOFAI—defined so as to include the claim that thought *is* symbol manipulation (pp. 112–13)—was "the most powerful and successful approach to psychology ever known" (p. 250). It had already made some important discoveries, and might—or, as the sceptics claimed, might not—make many more:

By disciplining theory, the computational model also liberates it: issues can be formulated and accounts developed that were hitherto essentially inconceivable. Certainly these early explorations have encountered unanticipated complexities and difficulties, but even they constitute a wealth of unprecedented empirical knowledge. (p. 211)

We hate to withhold judgment: scepticism is intellectual anemia. How much more fun, more vigorous, more apparently forthright to take sides! Yet, sometimes, the results are just not in. I am not really convinced that GOFAI is impossible; on the other hand, I'm certainly far from persuaded that it's inevitable. I am dubious: near as I can see, after thirty years, the hard questions remain open. (p. 254)

In short: "no 'disproofs' of AI are proposed, but at most some issues to ponder" (p. 213).

One of the issues to ponder was intentionality. In questioning whether one can "make sense" of AI programs, Haugeland wasn't merely saying (like Searle in 1980) that AI meanings are derivative, but (like Dreyfus) that meaning originates in the lived context and is expressed by embodied, situated, action.

In the 1980s, he put this in Dennettian—not phenomenological—terms. To ascribe meaning, he said, we must "ascribe beliefs, goals, and faculties [to mice as well as to men] so as to maximize a system's overall manifest competence . . . [namely] the ability to achieve goals by making rational decisions in the light of knowledge and perception" (1985: 214–15, 264). Even then, however, phenomenology was in his mind (and his endnotes). He wasn't sorry, he said, to have neglected phenomenology in his text—but he regretted not having more space to discuss "the 'background' of everyday practice" (1985, p. ix). This concept was prominent in Continental philosophy. It had also been stressed by Searle, who defined it as "a set of nonrepresentational mental capacities ['know-how'] that enable all representing to take place" (1983*b*: 143).

Ten years later, Haugeland's phenomenological sympathies were more explicit—and more committed. Previously, he'd felt no qualms in ascribing intelligence to mice (1985: 214). Now, in a glowing review of Dreyfus's unrepentant update, *What Computers Still Can't Do* (H. L. Dreyfus 1992), he declared: "There's only *one* world, *this* one—and it's ours . . . In my own view (and I suspect also in Dreyfus'), there is no such thing as animal or divine intelligence" (Haugeland 1996: 127). Even if there were, he said, that would only extend the scope of who "we" are—for "the *world* just is the realm of the meaningful; in other words, it is where intelligence abides" (p. 127). And he explicitly rebutted the realist's challenge:

But what about the physical universe: countless stars and galaxies, vast mindless forces, fifteen billion years of atoms and the void? Isn't that the *real* world, a fleeting speck of which we happen to throw an interpretation over, and regard as meaningful? No, that's backwards. The physical universe is *a part of* our world . . . it is *not* primary. Accordingly, cognitive science would be trying to build from the roof down if it began its investigation of intelligence with our understanding of physics. The foundations of intelligence lie not in the abstruse but in the mundane. (p. 127)

Card-carrying functionalists couldn't be expected to agree. They preferred his earlier dismissal of "zany idealism" (1985: 39).

As for Haugeland's opinion of functionalists, he now accused them of presupposing "tendentious and substantive metaphysics" (1996: 120). In other words: Descartes's metaphysics, minus his substance dualism.

Haugeland's metaphysics was phenomenological. It started out from the subject's lived presence in the immediate situation, and was unremittingly externalist with regard to meaning. It followed, he argued, that the scope and boundaries of intelligent systems are not what is usually assumed—by functionalists, as by common sense. Study of "the brain alone" is insufficient for understanding intelligence, and even "an entire individual organism may not be, by itself, encompassing enough" (1996: 121).

Compare Dreyfus:

But what about *my* experience, one may ask; my private set of facts, surely that is in my mind? This seems plausible only because one is still confusing this human world with some sort of physical universe. My personal plans and my memories are inscribed in the things around me just as are the public goals of men in general. My memories are stored in the familiar look of a chair or the threatening air of a street corner where I was once hurt . . . After all, personal threats and attractions are no more subjective than general human purposes. (H. L. Dreyfus 1972: 178)

Through the 1990s, Haugeland's philosophy grew increasingly Heideggerian (dare one say "zany"?). It eventually included claims such as "The general form of free human commitment—of care or faith—is love" (Haugeland 1998: 2). Smith might have sympathized, as we'll see (cf. Section ix.e). But most cognitive scientists wouldn't.

Van Gelder (1962–), then based at the Australian National University and Indiana University (now, in Melbourne), exhorted the philosophers of the 1990s to consider not the computer, but the Watt governor (1992/1995, 1998; van Gelder and Port 1995). The old opposition between AI and cybernetics had recently been reawakened by wide-ranging work in dynamical systems (see 4.v–vii, 14.viii–ix, and 15.viii.c, ix, and xi). Van Gelder took this work as his model.

In arguing (as Dreyfus had done) that the brain is an analogue system of constantly varying energy levels, he tried to be constructive. Unlike most dynamical theorists, he allowed for internal representations. But he described them in a novel way. Challenging Fodor's claim that formalist psychology was "the only one that we are going to get" (see Section iv.d), he asked, 'What Might Cognition Be, If Not Computation?'—and answered: state transitions in dynamical systems (1992/1995). He described this hypothesis as a vindication of David Hume's dream of a scientific psychology based on mathematical laws like those of physics (see Chapter 2.x.a).

Unfortunately, to try to be constructive isn't necessarily to succeed. Several philosophers protested that van Gelder was saying little of any substance.

For instance, he insisted that he wasn't "vainly attempting to do *without* complex internal structures [but was] dramatically reconceiving how they might be instantiated" (1998: 626). As remarked in Chapter 15.vii.c however, it's doubtful whether the dynamical approach can distinguish specific propositional content. Even Clark, who credited the dynamical approach with significant explanatory power, complained that the system's internal functional organization (and the adaptive roles of its components) is obscured by the highly abstract vocabulary of dynamical theory (A. J. Clark 1997: 118–28).

Likewise, Sloman argued that dynamical theory can't represent any interestingly complex mental state (1993, sects. 7–8). If one construes a mind/brain (or a computer) as having a single "atomic" state—a point in a high-dimensional, numerically defined, vector space—with a trajectory in the phase space of possible global states, one loses all the important structure in the system. One can't identify the many coexisting, more or less independent, subsystems (and sub-subsystems . . .), or their various, and largely asynchronous, interactions.

Of course, said Sloman, one can try to characterize these in detail—perhaps by constantly varying the number of dimensions, adjusting the numerical parameters, and/or defining attractors on different levels. But once one starts doing this, the distinction between a dynamical systems view and a more conventional AI view begins to disappear. It's no accident, then, that computer scientists speak of computational substructures and sub-processes, not trajectories of global patterns of bits or electric pulses. We need the concept of the virtual machine (see Section ix.c). In short, whereas Hume can be forgiven for taking physics as the model for psychology, today's philosophers cannot.

Yet others objected that the notion of a "dynamical system" is so wide that it includes whirlpools and windmills—and digital computers, too (Giunti 1997). Unless one can say just which types of dynamical system are specifically cognitive, one isn't saying much (Chrisley 1998).

Moreover, current computer models of dynamical systems aren't actually continuous, even if—like CTRNs (15.ix.b)—they're called continuous. Rather, they involve discrete state transitions (they're implemented in von Neumann machines). So, contra van Gelder, they *cannot* do things which classical computation can't do.

Finally, van Gelder defined "computation" in terms of Turing machines, arguing (correctly) that this abstract notion can't support various aspects of real information-processing systems, such as temporality. But, as both Sloman (1993) and Ronald Chrisley (1999) pointed out, there are good reasons for adopting a very different definition of computation, immune to van Gelder's criticisms (see Section ix.c and f).

To dismiss highly general philosophical claims for a certain theoretical approach isn't to say that it's never useful (especially if enriched by others: cf. Sloman's remarks, above). A meeting on 'Dynamic Representations in Cognition', co-organized by van Gelder at Indiana University in 1991, considered situated robotics, neuroscience, and psychology (Port and van Gelder 1995). The approach seemed helpful, for instance, in explaining the sudden shifts, or "phase transitions", between different gaits (walk, trot, canter, gallop), or the development of skills such as reaching, grasping, and coordinated kicking (Thelen 1995; see 14.ix.b).

One paper even applied dynamics to Simon's old stamping ground: decision making (Townsend and Busemeyer 1995; cf. Busemeyer and Townsend 1993). It defined a "decision field theory", to reflect the fact that decision making is temporal (unlike logic or probability theory). Continually shifting considerations—preferences, values, expectations, inferences, potential conclusions—"push" and "pull" against each other as the mind moves through the decisional state space; and the trajectory varies with the time available. (Similarly physics-based language was used by William James, in a startlingly apposite passage comparing the phenomenology of choice with the elasticity of a material object—quoted in Townsend and Busemeyer 1995: 102.)

Prima facie, this example seems to contradict the criticisms (above) about dynamics being unable to deal with propositional content and cognitive structure. But to show that the appearance and interaction of "atomic" states X, Y, and Z can be dynamically described isn't to use those states' specific internal structure to explain *why* they arise or *why* they influence each other in the ways they do. The decisional (semantic) trajectories defined by Townsend and Busemeyer were suggested by their prior understanding of rationality, not by the 'physics' of their field theory. That is, the various possible options, and the reasons for and against, weren't actively generated but were taken as given. What was being studied was how the mind then vacillated between them.

Analogously, Haugeland (1978, sect. 6) suggested long ago that a holographic system could act as an associator of visual patterns, and thus underlie (human) chess playing. The first member of a pair of chess positions would be "an important common substructure" in chess, and the other a move (or two alternative moves) that's generally powerful in response to such structures. That's plausible, perhaps, as a description of what goes on in the practised player's mind. But it's not clear that it can explain why response X or response Y was tried out in the first place. That, too, would have to be explained as a series of pattern associations (see 12.x.g).

d. A computational philosophy of embodiment

Clark, like van Gelder, was impressed by the late-century research in dynamical systems—and also by work in animate vision and A-Life (14.vii and 15.v). By the early 1990s, he was already well known as a philosopher of connectionism (Chapter 12.x.e). Then, he 'intermitted' from Sussex for seven years to work at the University of Washington, St Louis—a world-leading base for neuroscience. (Now, he's back in the UK, at the University of Edinburgh.) The intellectual environment in Washington helped him to shine a biological searchlight onto connectionism, and to relate the philosophical question of what counts as a "representation" to various types of neural mechanism (see Chapter 14.viii.d).

The diversity of biological representations, and their difference from the static entities posited by Fodor and GOFAI, was one theme of his book *Being There* (1997). This punning title linked Heidegger with both Woody Allen and Peter Sellers (not easy!). But it was the provocative subtitle, *Putting Brain, Body, and World Together Again*, which announced the main theme.

Clark aimed to counter the disembodied, individualistic picture inherited from Descartes. He did more than insist that "'meanings' just ain't in the head". He described his position as "active externalism", because it focused on the active role of the environment in driving cognitive processes (what he later called "cognitive technology": A. J. Clark 2001, ch. 8)—see Chapter 12.x.e. Indeed, he went further: he saw the (natural and cultural) environment as *constituting* minds and mental capacities.

The empirical roots of this position were deep and long-standing (Chapter 5.ii.b–c). Gibsonian "ecological" psychologists had long stressed the role of the agent's actions (and interests) in exploring its environment, and the specific bodily adaptations involved. Jean Piaget had applied similar insights in the context of "epigenetic" development. Even GOFAI workers had often remarked, though also often forgotten, that the external world provides a rich source of information, lessening the need for detailed on-board memory.

All these ideas had recently been revivified by an upsurge of research on enactive cognition, cognitive and developmental neuroscience, situated and evolutionary robotics, and dynamical systems (see Chapters 8.iii, 14.vi–vii and ix.d, and 15.vi–viii). The picture of the mind/brain as a *self-organizing* system had been reinforced by connectionism on the one hand and developmental neuroscience on the other. Moreover, dynamical approaches had stressed the difficulty, even arbitrariness, of locating the boundary between closely coupled systems.

Clark was the leading philosopher of cognitive science to draw on these later studies in detail. Others included Horst Hendriks-Jansen (1996) and Michael Wheeler (1996, 2005), both also from the Sussex stable. Clark distinguished a range of explanations found in this literature, and defined criteria for deciding when one is appropriate rather than another. But as well as offering such methodological morals, he painted a picture of "active, embodied, cognition"—and of "the *extended* mind".

The latter idea had already been bruited by neo-Kantians such as Dreyfus and Haugeland (see above), and by phenomenologists in general (Rowlands 2003). It had been argued also by postmodernist and/or feminist philosophers such as Donna Haraway (1986/1991). It had even been supported by 'pure' functionalists. Newell and Simon, for instance, had spoken of the environment as providing an "external memory" (see 7.iv.a). Indeed, the notion that individual minds, or selves, are constituted—constituted, not merely influenced—by their social–cultural relations with other selves was a commonplace of much social psychology and social–political philosophy (Hollis 1977—and see Chapter 13.iii.e).

But this view of the mind hadn't been prominent in the functionalist camp. In his first book, Clark had quoted the PDP group's comment that "the external environment [is] a key extension to our mind", and described it as "highly revealing" (A. J. Clark 1989: 133). There, he had related it to the mind's creation of a multiplicity of *virtual machines*. Later, he developed it at length, drawing insightfully on the new empirical evidence mentioned above. And he gave a *computational* justification for it.

So he described language and culture as "natural" artefacts. Like ants' pheromone trails and termites' nests, they are part of the natural history (cf. "forms of life") of the creatures who made them. Their function, said Clark, is to restructure the computational tasks we face into a form better suited to the human brain—a basically connectionist, pattern-recognizing, system.

He'd already argued this (in terms of "representational trajectories") in relation to PDP research in general, as we saw in Chapter 12.viii.d. Now, he focused on the specifically human examples. We naturally construct a system, or "scaffold", of cultural situations. That system makes our situated (i.e. largely automatic) actions more efficient, given the computational resources we possess. (This is why chairs are imbued with human, and even personal, meanings: see above.)

These social and physical artefacts, Clark argued, are so closely coupled with our intelligence that they are in effect integral to it. They inform, and are informed by, our thoughts, memories, decisions, emotions, and desires. In a very real sense, then, a human mind, or self, isn't encapsulated within one individual's skull, or even (skin-delimited) body, but is dynamically extended across the human environment:

[The] boundary between the intelligent systems and the world [is] more plastic than [has] previously been supposed—in many cases, selected extra-bodily resources constitute important parts of extended computational and cognitive processes. Taken to extremes, this seepage of the mind into the world threatens to reconfigure our fundamental self-image by broadening our view of persons to include, at times, aspects of the local environment. (A. J. Clark 1997: 213–14)

Clark's "externalism" was very similar to Clifford Geertz's 'non-psychological' anthropology (Chapter 8.ii.a). But it was even more shocking, for whereas Geertz had spoken of the "mind" being located outside the head, Clark said this also about the "self"—even closer to the bone, as one might say. Minus the scientific trimmings, his view soon reached a much wider philosophical audience, in a prizewinning paper co-authored with his Washington colleague Chalmers (Clark and Chalmers 1998). The prize was awarded by *The Philosopher's Annual* for one of "the year's ten best papers in philosophy"—*any* area of philosophy. Not everyone was convinced, however: for a critique, see (Butler 1998, ch. 6).

The implication for cognitive science was that we must think of, and therefore study, mental capacities or skills in a new way. The skill of *ship navigation*, for instance, emerges from (is situated within) a complex coupling of individual personalities, social roles and conventions, maps and instruments, mariners' knowledge, problem solving (often spread across several crew members), and a variety of bodily skills (for details, see Chapter 8.iii).

Similarly, said Clark, "much of what we commonly identify as our mental capacities may . . . turn out to be properties of the wider, environmentally extended systems of which human brains are just one (important) part" (1997: 214). Those systems, of course, include computers. Clark's friend van Gelder (2005) has explicitly linked current PC functionality to changes in the "self-constitution" of the rationality of the people who use them—changes which he regards as largely welcome (see Chapter 10.i.h).

A few years later, Clark would have mentioned the intriguing performance artist Stelarc, who came as Visiting Artist to Sussex's Centre for Computational Neuroscience and Robotics. Stelarc was the first person to explore the possibilities and implications of a wide range of man–machine interconnections (M. Smith 2005). One project initiated with CCNR, for instance, was to use electrical signals from his muscles to control a specially built robot many miles away. Indeed, he'd already done this in simple ways (and we saw in Chapter 14.x.b that comparable results have been achieved by linking monkeys' brains to far-distant robots).

The distinction/link between 'real' and 'virtual' worlds (Chapter 13.vi) is a prime interest not only of Stelarc (1986, 1994, 2002a, b), but also of Clark. He recently returned to it in his popular book describing human beings as "natural-born cyborgs"—in which he did, indeed, mention Stelarc (A. J. Clark 2003a: 115–38; cf. Haraway 1986/1991). Mainly, however, he showed how various novel types of computing technology—wearable computers, environmental computers, extrasensory prosthetics, virtual reality, augmented reality . . . —were destined to change people's thinking, experience, and self-image as much as language, writing, and printing already have done.

For all his sympathy with the new methodologies, however, Clark refused to share their all-encompassing scorn for the old orthodoxy:

The true lesson . . . is not that we somehow succeed *without* representing (or, worse, without computing). Rather, it is that the *kinds* of internal representation and computation we employ are selected so as to complement the complex social and ecological settings in which we must act. (A. J. Clark 1997: 221)

In one important respect—and despite his occasional references to Heidegger, Merleau-Ponty, and Dreyfus—Clark doesn't 'fit' into this section. Putnam, Dreyfus, and Haugeland had all rejected realism. Even Putnam's "internal realism" held that our concepts of mind, body, and self are prior to science, not grounded in it.

By contrast, Clark's theory of "the extended mind" rested as much on scientific research as on philosophical argument. (Russell would have approved.) Like the vast majority of cognitive scientists, he presupposed realism. Indeed, he specifically declined to engage in the realism/anti-realism debate:

Heidegger was opposed to the idea that knowledge involves a relation between minds and an independent world . . . —a somewhat metaphysical question on which I take no stand. (A. J. Clark 1997: 171)

Varela *et al.* use their reflections as evidence against realist and objectivist views of the world. I deliberately avoid this extension, which runs the risk of obscuring the scientific value of an embodied, embedded approach by linking it to the problematic idea that objects are not independent of the mind. My claim, in contrast, is simply that the aspects of real-world structure which biological brains represent will often be tightly geared to specific needs and sensori-motor capacities. The target of [my] critique is thus *not* the idea that brains represent aspects of a real independent world, but rather the idea of those representations as *action-neutral* and hence as requiring extensive additional computational effort to drive intelligent responses. (p. 173)

Other *fin de siècle* analytic philosophers were less wary of these metaphysical issues. By the end of the century, anti-realist voices were increasingly heard in the general philosophy of mind—as we'll now see.

16.viii. Mind and "Nature"

By the dawn of the twenty-first century, the "opposing views of the nature of philosophy" still caused radical dissent—even mutual contempt. And the contempt wasn't new.

Realists and anti-realists exemplify what William James (1907) called "tough" and "tender-minded" philosophers, respectively—and little love is lost between them. As James put it, "The tough think of the tender as sentimentalists and soft-heads. The tender feel the tough to be unrefined, callous, or brutal" (p. 13). James's diagnosis was confirmed by Russell's tone of exasperation, when criticizing the later Wittgenstein (see Section vii.a, above).

Later still, it was illustrated by the "science wars" (see 1.iii.b), and by arguments about postmodernism in general (Gross and Levitt 1994/1998; Gross *et al.* 1996; Sokal and Bricmont 1998). Some examples in social psychology and anthropology were mentioned in Chapters 6.i.d and 8.ii.b–c, respectively.

Not least, this mutual mistrust has imbued discussions of mind-as-machine. For example, Guy Robinson (1972) ridiculed functionalism and strong AI by borrowing Thomas Hobbes's contemptuous rhetoric: "When men write whole volumes of such stuff, are they not mad, or intend to make others so?" Since Hobbes himself was a committed mechanist (see 2.iii.b), Robinson was being more than a little mischievous. But he was also being sincere.

By the millennium, many philosophers—in private, if not in print—either shared Robinson's reaction or rejected it with equal passion. Nevertheless, the two approaches had recently become more closely intertwined in some people's work than anyone would have imagined, thirty years earlier.

The new recruits to neo-Kantianism included some hard-headed—but not tough-minded—philosophers raised (like Putnam himself) in the analytic tradition. They respected science but not scientism, and regarded functionalism, and most of cognitive science, as passing illicitly from the former to the latter. Indeed, they shared Rorty's hope for "the disappearance of psychology as a discipline distinct from neurology" (1979: 121). They saw mind (the realm of rationality) as distinct from nature (the realm of the physical and life sciences), and scientific data as philosophically irrelevant. Russell didn't live to see this: presumably, he was turning in his grave.

a. No representations in the brain

One example of this new approach appeared as a volume in a popular series of student-oriented textbooks on 'Problems of Philosophy'. Gregory McCulloch (1952–2002) described the mind as inseparable both from the body and from its surroundings (Mc-Culloch 1995).

Gregory (unlike Warren!) McCulloch leaned heavily on Wittgenstein and Jean-Paul Sartre, and to a lesser extent on Heidegger and Dreyfus. He rejected "mentalism"—by which he meant functionalism, and in particular the language-of-thought hypothesis. Fodor's position, he said, was probably not even the only available scientific psychology—and, for sure, it gave "no account at all of the folk psychological [everyday] mind" (McCulloch 1995: 155). (Clark's 2001 textbook on the philosophy of cognitive science was similar in some ways, but lacked the anti-realism.)

Michael Morris (1955–), a colleague of Clark's (and mine) at Sussex, sympathized with this position. He, too, insisted on the inseparability of the world and our conceptions of it. He avoided idealism, he said, because his theory was *externalist* about meaning. (He later changed his mind about that: see below.) And he had no time for functionalist defences of folk psychology. The notion of folk psychology, on his view, is a scientistic "myth"—one that is "wrong in every particular" (Morris 1992: 111). To say that the death penalty acts as a deterrent is to make a psychological claim (one that's often made by the folk). To regard people as having beliefs and desires, he insisted, is not.

To treat beliefs and desires as theoretical entities, said Morris, is to adopt a fundamentally mistaken (Cartesian) conception of the mind. When we say "She did it because she thought it would help", we are not—*pace* Fodor and his functionalist friends—identifying a causal relationship, but a rational one (pp. 108–9).

As for brain processes, these can cause other bodily happenings, but not behaviour. Or at least, brain processes can't cause the kind of behaviour for which *persons* are *responsible* (namely, *actions*):

But what *does* cause the behaviour? Nothing. The relevant kind of behaviour [i.e. behaviour which someone is responsible for] is not caused; it is just done. This does not violate such principles as that every event has a cause. People's deeds are not events . . . Events happen; they occur; they are caused. Deeds are just done . . . [This] does not cut deeds adrift from the order of event causation. There is no problem, for example, with thinking that that bit of neuron activity's causing *my hand to rise* is partially constitutive of my action when *I raise my hand*. (p. 136; italics added)

In short, propositional attitudes aren't causes, as they were said to be by functionalist-Putnam. Rather, they are aspects of a system (a human mind) imbued with values. An orthodox Wittgensteinian might put this by saying that they're semantic pieces in the game of rationality, whose rules aren't causal laws but epistemic norms of coherence, evidence, and warrant. Morris prefers to say that there are only two fundamental norms, or "values": namely, moral goodness and truth—intelligible only in combination and contrast with each other. (Hence the title of his book: *The Good and the True*.) The precondition for both of them, he argues, is our personal responsibility.

It follows from this that beliefs and desires can't be explained by *any* naturalistic account. Neuroscience, computational psychology, evolutionary semantics (see Section x.d) . . . none of these can account for rationality. Our evaluation of truth (and our respect for whatever is properly termed *knowledge*) is a philosophical precondition for science, not something that can be explained by it.

This anti-naturalistic position was rooted in the mid-century (pre-functionalist) Wittgensteinian arguments that *intentions*, since they are characterized in terms of the action intended, can't be causes (Melden 1961; cf. Boden 1972). But, writing some thirty years later (and with Clark and myself as colleagues), Morris felt that he had to say something about cognitive science.

This made him unusual, since most neo-Kantian philosophers—including the much more well-known John McDowell (1942–), discussed below—simply ignored its detailed claims. By the same token, it made him interesting for our purposes here. For surely, there should be rather more detailed intellectual contact between philosophy of mind and neuroscience than all-encompassing references to "the brain". Granted that science can't answer philosophical questions, philosophers (of mind) should attempt to situate those scientific theories (of the brain) which at least *appear* to be relevant.

Morris argued that mental representations *as understood in functionalist cognitive science* don't exist—not in Fodor's realist sense, nor in Dennett's instrumentalist sense either (Morris 1991). In other words, there are no *non-semantically individuated objects or states which have meaning, or content*, because no naturalistically identified event can be inherently meaningful. Whereas Dreyfus had implied that cognitive science is a waste of time, Morris didn't. But he argued that cognitive science—*qua* science, not metaphysics of mind—can manage perfectly well without positing representations in the sense defined above (1991: 28–30).

For example, he said, it can investigate certain abstract properties of cognitive capacities, such as the differences between connectionism and other AI architectures. This needn't have any direct bearing on the human mind, but it may—perhaps in respect of the nature of concepts, for instance. (Here, Morris was tacitly agreeing with Churchland's claim that neuroscience may be relevant even to a normative epistemology: Chapter 12.x.c.)

Again, it can study the capacities of actual cognitive systems, using computational concepts of various kinds to discover *functional* relations between *semantically* individuated states.

And third, it can offer partial explanations of psychological capacities, if it can find non-semantically identified types of event that correspond with semantically identified ones—whose meanings are then *stipulatively* assigned to the relevant scientifically identified events. That is: given that we already understand what it is for the cat to sit on the mat, we can decide (*sic*) to assign this meaning to certain neural or computational states. Accordingly, Morris would have no quarrel, for instance, with Clark's claim that neuroscience has discovered various types of representation. (He did suggest, however, that representations would actually be found only for low-level cognitive capacities.)

(For the record, Morris now thinks that he *was* idealist then, even though he believed that he wasn't: personal communication. His view today, in a nutshell that I shan't attempt to crack open, is this: "(1) That the world as it is in itself is entirely independent of our conceptions of it. (2) That there is no correspondence between our conceptions of the world and the world as it is in itself. (3) That most so-called realists—i.e. the ones who believe in correspondence—are in fact committed to idealism (because only an idealist can think that the world as it is in itself has to correspond to our conceptions of it). And (4) that we can perfectly well describe the world as it is in itself—though not by means of any correspondence, of course." What's interesting here, given that Morris is broadly of the anti-Russellian camp, is his continued effort to repudiate idealism. That, clearly, he sees as a step too far.)

b. Mind as second nature

Similar evaluations of science and scientism, and so of the "proper" scope of cognitive science (though without the attention to theoretical details) informed McDowell's highly influential John Locke Lectures.

These were given in Oxford in 1991, and published a few years later (McDowell 1994*a*). McDowell, then at the University of Pittsburgh but originally from the Oxford stable, shared Rorty's view that Descartes had "invented" the mind, not described it (cf. Chapter 2.iii.a).

He was strongly influenced by Sellars's dualism of logical spaces, as Dennett had been too (see Section iv.a). But he rejected Dennett's functionalist analysis of consciousness (see also McDowell 1994*b*). For McDowell, conscious experience is an irreducibly *rational* category—like belief, intentional action, and all other mental phenomena. Accordingly, he rejected any suggestion that the laws of natural science could explain human psychology.

He also refused to follow Nagel—or Everyman—in ascribing subjective consciousness to bats. Subjectivity (he argued) requires meaning, or content, which requires concepts. Without concepts, there can be no particular view of the world, nor any understanding of its general properties. So, despite what most cognitive scientists assume, there can be no such thing as non-conceptual content (12.x.f). And, for McDowell, there can be no such thing as animal experience. He allowed—contra Descartes (Chapter 2.ii.d–e)—that it would be "outrageous" to deny that animals have feelings (such as pain) or "perceptual sensitivity to their environment". But these aren't rational phenomena, so aren't experiences. (Whether they can be anything over and above the patterns of animal behaviour to which Descartes himself drew attention is unclear.)

McDowell wasn't alone in drawing a sharp distinction between humans and animals: most neo-phenomenologists did so. Haugeland, for instance, changed his mind about intelligence in mice as his position became more Heideggerian. And Wittgensteinians in general argued that animals have no concepts, or thoughts. At the Kent meeting, for instance, Peter Geach (1916–) announced to a crowded lecture hall that "You can't really say that dogs are intelligent." This elicited fierce growls from a blind philosopher's normally silent guide dog—and hoots of laughter from the rest of the audience (and Peter, too).

(Wheeler was an exception, here. Although his view of cognitive science was strongly influenced by Heidegger, who had specifically denied *Dasein* to animals, Wheeler argued that they, too, construct their own "worlds": 1996, 2005. Clark, similarly, rejected Heidegger's "thoroughly social", language-based, view of embodied action—A. J. Clark 1997: 171.)

To claim, as McDowell did, that there is no non-conceptual content is to say there are no mind-independent data (no "Given") to act as the foundation of our knowledge of the world. In other words, no purely causal processes could justify, or even provide rational grounds for, our empirical beliefs. Although causal mechanisms (of perception, for example) are necessary for our having beliefs, no scientific psychology—on this view—can explain how these beliefs emerge.

When cognitive scientists study low-level vision, for instance, they typically describe causal (neurophysiological) processes in computational–intentional terms (see Chapter 7.v.b–d). They say that DOG detectors and the like provide information (cf. 'content') about light-intensity gradients, edges, and depth. For McDowell, such talk may be scientifically defensible: "It would be dangerous to deny, from a philosophical armchair, that cognitive psychology is an intellectually respectable discipline, *at least so long as it stays within its proper bounds*" (1994a: 55; italics added). But, he insisted, it's also "a recipe for trouble", since it implicitly encourages the myth of the Given. Perceptual mechanisms as such can provide no meaning, no semantic content, whatsoever.

Moreover, the brain can't generate—still less, justify—beliefs, hunches, inferences, or knowledge. All these are matters of normative rationality (the "demands of reason" imply that we "ought" to believe the truth, and to reflect on our own ideas). They therefore lie forever outside science. In that sense, McDowell declared, cognitive science is impossible.

Like Putnam and Morris, McDowell wanted to avoid idealism—including Rorty's version of it (see Section iv.e). He tried to do this not by resurrecting Kant's noumenal world, but by positing a direct (non-representational), and *natural*, relation between mind and the empirical world.

What he meant by "natural", however, was highly unusual. Certainly, his form of naturalism—like the usual forms, which he termed "bald" naturalism—implied *not supernatural*. (On this point, he aligned himself with Aristotle, who had seen rationality as part of human nature: see Chapter 2.ii.a.) But he distinguished between "nature" and "disenchanted nature".

The latter (modernist) notion is the world as described by the laws of science, conceptualized as existing independently of us. The former (as he defined it) includes also "second nature": the habits of rational thought and action into which human infants are initiated by adults as they mature, and which involve the emergence of "resonance" or "responsiveness" to meaning.

Babies, he said, are the only animals (*sic*) to possess the natural, inborn, capacity to acquire concepts, epistemic norms, and other cultural practices—including the interpretation of behaviour in terms of rational relations between beliefs and desires. This "natural capacity" can't be explained by science, because the logical space of causal laws (as opposed to rational norms) has no room for meaning. If neuroscientists discovered innate brain mechanisms underlying some Chomskyan language acquisition device, they wouldn't have explained our capacity for language. And evolutionary biology couldn't do this either:

It would be one thing to give an evolutionary account of the fact that normal human maturation includes the acquisition of a second nature, which involves responsiveness to meaning; it would be quite another thing to give a constitutive account of what responsiveness to meaning is. (pp. 123–4)

So far, so familiar: neo-Kantians in general drive a wedge between reason and cause. But McDowell, using the concept of second nature, claimed to be able to reconcile them. Philosophy, he said, must be sharply distinguished from science. Nevertheless, philosophers inclined—mistakenly, but understandably—to a *bald* (scientistic) naturalism could rest easy:

This should defuse the fear of supernaturalism [including various forms of Platonism]. Second nature could not float free of potentialities that belong to a normal human organism. This gives human reason enough of a foothold in the realm of law to satisfy any proper respect for modern natural science. (p. 84)

Whether Russell, for example, would have seen this as satisfying a "proper respect for science" is highly doubtful. For sure, Fodor didn't. He praised McDowell's book for raising "a number of our deepest perplexities"—but defiantly added, "Which, however, is not to say that I believe a word of it" (Fodor 1995*b*: 3).

Nor was it to say that he could actually disprove it, on pain of contradiction. (This was a paradigm case of the "profoundly irritating" nature of first-rate philosophy: see preamble, above.) Claiming that the two logical spaces hadn't been reconciled, Fodor identified several aspects of McDowell's position that he felt were mistaken, unjustified, or merely metaphorical. What is it, for instance, to "resonate" to meaning? And he stated his own, alternative (but not strictly provable), claims.

McDowell, he said, was "as good a contemporary representative of this [anti-naturalistic] philosophical sensibility as you could hope to find". But, he insisted, "it's all wrong-headed. Science isn't an enemy, it's just us" (Fodor 1995*b*: 8).

c. Mind and VR-as-nature

In the preamble to this chapter, I said that a new twist on the realism/anti-realism debate is related to developments in VR (13.vi). This "twist" combines debate about the existence/non-existence of a real world outside us with debate about just what Fodor's word "us" (in the quotation above) really means. The latter question isn't asking what a particular individual's self is "really" like (a problem raised with respect to the different types of VR avatar in Chapter 13.vi.e). Rather, it's asking what it is to be a human being, and to have (or if you prefer, to be the origin of) genuinely human experience.

These deep philosophical questions have been triggered by the fictional exploration of VR in the hugely popular film of 1999, *The Matrix*. (And I really do mean *hugely* popular: a Google search on 10 January 2006 elicited 104 million items.) The human beings in this story are supposed to be experiencing a purely virtual world, while their living but inert bodies are farmed for energy by the machines in charge. The machines are clever enough to fool the people (the brains?) into thinking that they are walking, talking, and eating in a world much like our own.

Among the leading philosophers of mind who've written about the implications of *The Matrix* are Clark, Chalmers, McGinn, and Dreyfus—with his son Stephen (H. L. Dreyfus and Dreyfus 2002; cf. Dreyfus 2000, 2003). Led by Dreyfus (who remarked that, for the first time, the students were now assigning significant homework to their professors), they all contributed online papers to the Warner Brothers *Matrix* web site. Searle and Dennett gave interviews on the web site, too.

Besides these big guns, the cinematic version of Descartes's worry about the *malin génie* (see 2.iii.b) has prompted many other philosophical discussions, some specifically placed in the context of "popular culture" (Irwin 2002). A clear-headed account that's especially helpful for non-philosophers has been written by Matt Lawrence (2005). But the most relevant in the context of this book are the initiating paper by Dreyfus and son and the later essay by Clark.

Dreyfus uses the distinction between telepresence and VR in arguing that the experience of the *Matrix* inhabitants isn't authentic, and—even more "worrying"—isn't fully human either. In telepresence, such as the distance surgery outlined in Chapter 13.vi.b, radio signals from the real world reach the human participant. The person may have the experience of looking at a TV screen, and of hearing sounds through earphones. So far, so VR. But this virtual reality is closely based on, indeed causally linked with, a genuine reality. Moreover, as Dreyfus points out, that reality can fight back—as reality does, all too often. For instance, the surgeon may receive signals that are not only uncontrollable but highly unwelcome, if the tele-operated robotic tool (located many miles away) encounters unexpected flesh or bone in the real patient's body.

In purely fictional VR, however, as depicted in *The Matrix*, any apparent fighting back on the part of the virtual reality is pre-programmed by the designer. It follows, says Dreyfus, that certain sorts of creativity, which are characteristic of real humans in real environments, are blocked off. Radical changes in the person's mode of interaction with the VR world simply aren't possible, because the possibilities have been strictly limited by the VR program. In that sense, the experience of the *Matrix* inhabitants isn't fully human, even though it may be indistinguishable from genuine experience almost all of the time.

Clark's (2003*b*) paper is more closely linked to cognitive science as such. He sets out the filmgoer's three interpretative options: *The Matrix* as "dream, simulation, or hybrid". Citing various experimental findings (some of which were mentioned in earlier chapters), he defends the cognitive science view of human experience as shifting, constructed, and embodied—and dependent on specific brain mechanisms, including the neurophysiology of wakefulness and dreaming. And he draws specific comparisons, including some important contrasts, between our experiences when awake or dreaming and our experiences when involved with VR. The cinematic humans do indeed have experiences "much like our own", as remarked above. But they aren't entirely like our own, and nor are they compatible with what's known about the neurophysiology involved. (The film is science fiction, to be sure: but human physiology was supposed to be unchanged.)

On the issue that's central to this section, Clark finally opts for a version of realism. That is, he argues that *if* it were the case (which it isn't, not quite) that the humans depicted in *The Matrix* had *exactly* the same experiences of perceptual and bodily engagement with a real, and often resistant, world as we do when we're fully awake, *then* they'd be "embodied intelligences through and through", each having/being "a real body, realized in the non-standard medium of bits of information". In other words, having the experience of engaging with a resistant world, by way of one's senses and motor actions, is (so Clark claims) *what it is* for the mind to be "embodied".

However, it's by no means evident that Clark's arguments (or those of another would-be-realist interpreter of *The Matrix*: Chalmers 2003) would satisfy a more down-to-earth realist. In short, the realism/anti-realism dispute hasn't been settled yet. Maybe it never will be—a possibility that makes the "scientists and engineers" mentioned in the preamble distinctly uncomfortable.

16.ix. Computation as a Moving Target

For all his passionate dissent from McDowell's anti-realist position, Fodor didn't pretend to have a knock-down argument. This isn't surprising: as we've seen, we're dealing here with "opposing views of the nature of philosophy".

It's little wonder, then, that shock and excitement—not to mention scepticism—greeted the claim that one can dissolve this long-standing metaphysical impasse by reflecting on the nature of *computation* (see subsection e, below).

It's almost as wonder-worthy, you might think, that anyone bothered to reflect on the nature of computation in the first place. Hadn't Turing settled that question, once and for all? Wasn't he the one (or he and Alonzo Church the two) who'd claimed that the intuitive notion of "computation" could be formally defined (see Chapter 4.i.c)? Unless someone was going to find *extra* meanings hidden in mathematicians' intuitive use of the term (as, in a sense, Penrose tried to do: 1989; cf. Sloman 1992), what more was there to say?

Well, quite a lot—as we'll now see. And much of that has relevance for what we're prepared to count as "cognitive science", because (by definition) one of the two intellectual footpaths followed by cognitive science is computation (see Chapter 1.ii.a).

a. Three senses of computation

The seminal definition, in this context, was Turing's (see Chapter 4.i.c). He saw computation as the step-by-step symbol manipulation carried out by a Turing machine, conceptualized in abstract mathematical terms. Although he held that these operations could in principle be implemented in physical mechanisms, his concept of computation was a purely formal one.

His definition remains fundamental to theoretical computer science, and no cognitive scientist would dispute its importance. Indeed, it's still the only *rigorous* account of what computation is.

However, as the practices of AI and computer science became increasingly varied, two additional senses arose. (A recent paper distinguishes about a dozen different senses, but for our purposes here that's over-egging the pudding: B. C. Smith 2002.) Neither of these is well defined, as we'll see. But to ignore them is to miss much of the intellectual excitement that's driven the field over the past thirty years or more.

Turing (1936) defined computation in formal terms, as an exercise in pure mathematics. His definition was a formative influence on cognitive science, in the sense that McCulloch and Pitts (1943) used it in proving the logical equivalence of their abstractly defined neural networks and Turing machines (Chapter 4.iii.e). But when (a few years later) digital computers appeared on the scene, enabling logical computations to be implemented in electronic machinery, the notion of computation became less clear.

The notion of proof itself—and GOFAI's stress on theorem-provers (Chapter 10.iii.b)—was questioned (even by lawyers) when mechanized 'proofs' of more than a million inferences came on the scene (MacKenzie 1991*b*, 1993). These programs, it was argued, could be informally checked but not formally verified. Yet (so one eminent software expert declared) their conclusions should be accepted as *proofs* nonetheless:

We must beware of having the term "proof" restricted to one, extremely formal, approach to verification. If proof can only mean axiomatic verification with theorem provers, most of mathematics is unproven and unprovable. The "social" processes of proof are good enough for engineers in other disciplines, good enough for mathematicians, and good enough for me. (cited in MacKenzie 1993: 177)

For our purposes, what's important is that the notion of computation absorbed some of the features of the machines in which it was implemented. Certainly, a program could still be thought of as a series of well-formed formulae of some logical calculus, or programming language—in other words, as a set of uninterpreted mathematical expressions. But it could also be considered in a very different way, as essentially connected with causal processes in computers (see subsections b–c, below).

With the rise of functionalism in the 1960s, Turing's mathematical definition, already featured in the pages of *Mind* in 1950, became more familiar to philosophers. Indeed, it was for many years the only concept of computation they considered. Several philosophers discussed it at length, and used it as a defining criterion of cognitive science and/or AI (e.g. Haugeland 1985; Fodor 1980a; Pylyshyn 1980; Copeland 1993). We've already seen that it's the notion relied on by the authors of various key papers:

* by Mays (1952) in rejecting the Turing Test;
* by Putnam (1960) in defining functionalism;
* by Fodor (1975, 1980a) in defining cognitive psychology, and in arguing for methodological solipsism;
* by Searle (1980) in saying that AI programs are "all syntax and no semantics";
* and by van Gelder (1992/1995) in criticizing computationalism and recommending dynamical models instead.

The relevance of Turing's definition wasn't argued by these philosophers, so much as assumed by them. Putnam, for instance, took it for granted in his discussion of multiple realizability. And Searle treated it as an unassailable premiss in his attack on strong AI. He even took it to be a premiss shared by his philosophical arch enemies Newell and Simon, the high priests of strong AI. He had good reason for this. Nevertheless, we'll see (in subsection b) that one might read Newell and Simon's work in a different way.

By the 1990s, however, some philosophers and logicians were puzzling over the definition of "computation", and asking *just which* actual and notional machines were equivalent to Turing machines, and which were not (Bringsjord 1994; Calude *et al.* 1998; Copeland and Sylvan 1999).

AI practitioners were moving beyond Turing-computation too. For instance, certain types of neural network (in which the weights could be represented as irrational numbers) were said to be capable of analogue computation (Siegelmann 1995, 1999). Indeed, even *without* such maverick methods, AI scientists weren't interpreting computation as rigidly as outsiders assumed. Sloman—a skilled AI programmer, as well as a philosopher—has said:

No programmer or computer engineer has, to my knowledge, ever thought of programs in [Searle's] way, and as a programmer myself I have never thought of programs that way. (This criticism of Searle was made by many computer scientists and AI people around 1980.) (Sloman, personal communication)

Sloman goes even further, for he sees Turing machines as largely "irrelevant" for the history of computers and AI (Sloman 2002). Both of these, he argues, are natural developments of two types of artefact: machines for controlling machines (including mechanical looms, musical boxes, and Hollerith machines), and machines for operating on abstract entities (numbers, logical terms, or other symbols). These existed long before Turing was even a gleam in his mother's eye (see Chapters 2.iv and ix, 3, and 4). They blossomed—in power, sophistication, and range of application—in the 1940s–1950s because of electronic technology, not because of Turing's mathematics. Much as Babbage didn't need Turing machines to design a close equivalent of a von Neumann computer (Chapter 3.iii.b), so the modern pioneers of computing and AI didn't strictly *need* them either. The people for whom Turing machines were crucial were those doing *theoretical* computer science.

That's all true. (And Turing himself, before learning about electronics, thought that physical computers were infeasible because they'd have to be as big as the Albert Hall: Chapter 4.i.d.)

One must add, however, that Turing used his theoretical ideas when designing the ACE, and that the designers of MADM—the first stored-program general-purpose electronic digital computer—did so too (3.v.b–c). One must add, also, that AI and computational psychology—and the von Neumann computer itself—were recognized as exciting possibilities in the mid-1940s partly because of a marriage between Turing machines and Russell's logic (4.iii–iv). Furthermore, Turing's paper in *Mind*, which was deeply influenced by his earlier work, was intended as a manifesto for AI and was hugely influential in the young AI community (see Section ii, above). In short, although computers and AI would have happened even without Turing machines, the way in which they did happen was deeply influenced by them.

What's especially relevant in this context is that *philosophers* focused on Turing's formal definition of computation. This said nothing about implementation, whether the physical or the virtual machine. But he'd made provocative claims in the philosophers' trade journal about (future) AI's ability to model the mind, and was (wrongly) believed by them to have offered a behaviouristic Test for successful AI. Putnam, when defining functionalism, used the abstract concept of Turing machines and didn't even mention AI.

Moreover, philosophers in general could cope with mathematical logic, but knew little about computers and even less about the experience of using them. For these reasons, they assumed (and most still do) that Turing-computation was *essential* to the inception and practice of symbolic AI. Having this (first) sense of computation in mind, then, they usually see "computation" as typified by GOFAI and often regard "cognitive science" as essentially symbolic.

The second sense of computation is less restrictive, for it amounts to *Whatever methods are actually used in computer modelling*. Because the list of practicable methods has lengthened with the advance of AI/A-Life, just what this second sense picks out has changed over the years—and will continue to do so.

GOFAI still qualifies, of course. But so do the several kinds of connectionism (Chapter 12); the coordinate transformations of tensor network theory (14.viii.b–c); the various styles of evolutionary programming (15.vi); and the many types of cellular automata (15.v and viii). Situated robotics and dynamical models inspired by physics are covered too (15.vii–viii and xi).

Indeed, new and notional forms of computation are now being discussed. Some of these involve "quantum computers" (D. Deutsch 1985), which—in case they are practically feasible—could do Turing-computation much faster than conventional computers do. (In November 2001, scientists at MIT reported having used a "quantum computer" to find the factors of 15; some of the large banks, dependent as they are on prime-number cryptography, were sufficiently worried to employ consultants to look into these issues immediately: C. T. Ross, personal communication.)

Others involve what Jack Copeland and Diane Proudfoot have called hypercomputers (Copeland 1998b; Copeland and Proudfoot 1999; Copeland 2000). These, like Turing's imaginary O-machines mentioned in Chapter 4.i.c, would compute the uncomputable (the key terms being interpreted in the intuitive and Church–Turing senses, respectively). The philosopher Selmer Bringsjord, for instance, has argued that hypercomputation is already being performed by human minds, and that it might one day be performed by computers too (Bringsjord 1992; Bringsjord and van Heuveln 2003: 64–7; Bringsjord and Zenzen 2003). Although he criticizes Penrose's statement of the argument from Gödel, he provides a "modal" version of it which also rejects current AI as the core of a science of mind (Bringsjord and Arkoudas 2004).

At present, ideas about hypercomputers are more speculative than practical (Bringsjord 2001; Copeland 2002a; M. Scheutz 2002). But if they're implemented in some future technology, they too will be included under this second, catch-all, sense of "computation".

The more inclusive notion of "computational" concepts has become increasingly prominent since the mid-1980s, in discussions of mind as machine. But it was there right from the start. When cognitive science first emerged, most of the people involved were active both in logical-computational theorizing and in dynamical cybernetics—which they saw as part of the same general enterprise. It wasn't until some years later that these two types of computer modelling were restricted to different research communities (see 4.v–viii).

McCulloch and Pitts, for instance, saw their seminal connectionist paper of 1947 as a new way of developing the ideas they'd put forward in 1943 (Chapter 12.i.c). So did their contemporaries. The sociological separation of GOFAI and connectionist AI happened later. Defining their collective project in a way that excluded the non-GOFAI activities (as Fodor, for instance, did) would have struck them as bizarre.

But perhaps it's better to be bizarre than to be boring? This undiscriminating definition of computation may seem to have no bite, for it doesn't imply any particular philosophical position. Indeed, some philosophers prefer to define cognitive science in terms of formal representations and Turing-computation largely *because* they then know just what it is that they're talking about. If all non-symbolic enterprises fall by the referential wayside, too bad.

What's more, the second definition may even seem perverse. For the proponents of some of the methodologies just listed made a point of distancing themselves from classical AI, as we've seen. They insisted that their models involved a "new form of computation, one clearly based upon principles that have heretofore not had any counterpart in computers" (Norman 1986*b*: 534); and they spoke of "*sub-symbolic processing*" (Smolensky 1988), of "intelligence *without* representation" (R. A. Brooks 1991*a*), and of "cognition *not* computation" (van Gelder 1992/1995). Why, then, did a growing number of people come to think of computation—and therefore cognitive science—in such an all-inclusive way?

The answer is that these modelling approaches—like different programming languages (see 10.v)—involve significantly different virtual machines, with different capabilities.

* A PDP system, for example, has strengths and weaknesses largely complementary to those of a GOFAI program.
* Localist connectionist networks are different again.
* Cellular automata differ from classic sequential programs, and their performance varies according to the types of rule they embody.
* One evolutionary model may be significantly different from another, and all can do things which non-evolutionary programs in practice can't.
* Dynamical systems, such as CTRNs, are unlike classical AI, and can traverse their state space in diverse ways.

Moreover, one virtual machine can be implemented in a different virtual machine . . . and so on. We saw in Chapter 12.viii–ix, for example, that a *sequential* virtual machine can be emulated by a *parallel* virtual machine implemented in a *serial* computer (which in turn may involve several 'layers' of programming languages, ultimately implemented in the machine code).

A host of interesting questions have already arisen about the types of virtual machine that could model human minds—or, more provocatively, what type(s) of virtual machine(s) the mind might actually *be*. Others haven't yet been asked, because they (will) concern types of computer model still to be developed. As remarked above, these may involve highly unconventional machines, including hypercomputers (Calude *et al.* 1998).

To restrict "computation" or "cognitive science" only to Turing computation and GOFAI is as unreasonable as restricting "physics" to the mathematics available to Galileo. When Galileo said (in 1623) that the Book of Nature is "written in the language of mathematics", he didn't mean that the concepts used in *his* mathematics could solve all the problems of physics, but that physicists need to travel along mathematical pathways. And indeed, modern physics isn't definable in terms of only one type of theory (van Fraassen 1980; Cartwright 1983).

Analogously, cognitive science employs a variety of concepts and models, which are "computational" in a general sense. (Even dynamicists speak of the "computations" performed by their models.)

The third sense of computation is philosophically the most controversial. It's also the most unclear. Indeed, it's not so much one sense as a group of rather different senses, all informed by the same general aim.

This aim is to present computation as intentional, and meaning as computational—in part, by focusing on the causal relations involved. The third-style approach conflates *what computers do* with *what minds do*, by way of an account of intentionality that supposedly applies to both. In the rest of this section, we'll consider four—or, more accurately, three and a bit—examples of this third sense of the term.

b. Physical symbol systems

The earliest, and most influential, third-style example was due to Newell and Simon (Newell and Simon 1961, 1976; Newell 1980).

Their AI programs specified formal computational systems of the type defined by Turing. As psychologists, however, they were interested in computations that are grounded in the world and capable of directing behaviour in it. Accordingly, they outlined a semantic theory designed to avoid what Putnam and Fodor called methodological solipsism, and to support what Searle would term the Robot reply.

Their central claim was that intentionality is achieved, in both minds and computers, by implementing certain types of formal computation. Such implementations, or Physical Symbol Systems (PSSs), were seen as "the necessary and sufficient means for general intelligent action". In other words, the mind *is* a PSS. Moreover, said Simon (towards the end of the century):

[Some programs can] demonstrably understand the meanings (at least some of the meanings) of their symbols, and [they] have goals . . . In contrast [to a simulation of digestion], a computer simulation of thinking thinks . . . We need not talk about computers thinking in the future tense; they have been thinking (in smaller or bigger ways) for forty years. They have been thinking "logically" and they have been thinking "intuitively"—even "creatively". (Simon 1995a)

In arguing their case, Newell and Simon defined computation in causal terms. A *symbol*, they said, is a physical pattern with causal effects. The meaning of a symbol is the set of changes it enables the information-processing system to effect, either *to* or *in response to* some object or process (outside or inside the system itself). Analogously, concepts such as *representation, interpretation, designation, reference, naming, standing for*, and *aboutness* were causally defined. For example:

Two notions are central to this structure of expressions, symbols, and objects: designation and interpretation.

> *Designation.* An expression designates an object if, given the expression, the system can either affect the object itself or behave in ways depending on the object.

In either case, access to the object via the expression has been obtained, which is the essence of designation.

> *Interpretation.* The system can interpret an expression if the expression designates a process and if, given the expression, the system can carry out the process.

Interpretation implies a special form of dependent action: given an expression, the system can perform the implicated process, which is to say, it can evoke and execute its own processes from expressions that designate them. (Newell and Simon 1976: 116)

Whereas the meaning of an atomic symbol depends on its causal history and effects, that of a complex symbolic expression depends on the meaning of its components *and the relations between them*. It was the complex expressions which interested them, for they saw these as the inner core of mentality. But how were these relations depicted? Within any implemented symbol structure, they said, the component symbols are "related in some physical way (such as one token being next to another)".

However, implementation details can vary (multiple realizability, again). So computational psychologists may choose to concentrate on symbol types, rather than symbol tokens. In practice, then, Newell and Simon usually focused on the formal properties of their psychological theories, not on their physical implementation. ("Usually", because they did sometimes take account of neuroscientific evidence: see Chapter 7.iv.b.)

This is why they're typically seen—by Searle (1980), for instance—as proponents of the first sense of computation, not the third. On that view, PSS theory holds that causation grounds the interpretation of computations once implemented, not that it's involved in defining computation *as such*.

c. From computation to architecture

In emphasizing the causal basis of meaning in computational systems, Newell and Simon weren't alone: Sloman did so too.

He dismissed causal theories of meaning (like theirs) that assume some physical relation between a symbol and its referent, because we can refer to non-existing things (e.g. 1986*a*). But he did stress the *virtual* causal processes required for understanding. Moreover, whereas Newell and Simon had taken the concept of cause for granted, Sloman argued that we don't yet understand computation largely because we don't yet understand causation.

Sloman initiated his third-style approach in the 1970s, and is still developing it (Sloman 1978, 1992, 1996*a,b,c*, 1999, 2002). Throughout this period, he was seeking a new way of thinking about computation, even a new concept of computation—but not a new *definition* of it. In his view, we shouldn't expect to find any precise definition of computation that could function as *the* core notion of cognitive science.

Nor should we expect to find a precise definition of mind (or any other category of folk psychology—such as intelligence, for instance). Philosophers who propose candidates for this illusory honour prompt endless disputes about whether this or that system *is* a mind, *really*. Sloman, by contrast, always stressed the complexity, variety, and many degrees of overlap between different actual or conceivable systems.

He now recommends that we "abandon the notion that the concept of computation is the only or even the central foundation for the study of mind" (1996*a*: 215). He expects the concept to be eventually "replaced by a new taxonomy of designs, covering a more general class of architectures and mechanisms" (1996*a*: 216). This taxonomy will include both minds and mindlike systems, with no clear boundary between them. Computational mechanisms—including many types of virtual machine—are of course involved. But so are physical mechanisms, including neurochemical influences (14.ix.f).

In short, Sloman is a philosopher who takes the design stance seriously. He offers "not a philosophical argument about 'correct' concepts to use, but an engineering argument about the appropriateness of different (animal or artificial) designs for different tasks" (1996*a*: 216).

As remarked in Chapter 2.i.a, the word *engineering* can be a powerful turn-off. Sloman is well aware of this. But he was already arguing in the 1970s, in his book *The Computer Revolution in Philosophy*, that philosophers need to understand computation if they are to solve a raft of philosophical problems (1978, esp. ch. 6). These concern every area of philosophy, but especially the philosophy of mind, science, and language (Sloman 1978, chs. 2–3 and 10; 1986*a*).

He defiantly predicted: "[Within] a few years philosophers, psychologists, educationalists, psychiatrists, and others will be professionally incompetent if they are not well-informed about these [AI's] developments" (1978, p. xiii). At that time, the content of his claim was considered as provocative as its tone. Now, even with "engineering" added, it would be taken rather more seriously.

Many philosophers, however, would still discount it. For they think they already know what computation is—because Turing told us. Sloman allows that Turing's definition is the only precise one, and the only one associated with a clear and solid body of theory. It suffices for dealing with abstract mathematical matters (such as decidability, incompleteness, and complexity), where neither time nor causality is involved.

He allows, too, that Turing's definition has acted as a powerful "catalyst" to cognitive science. But—as Newell and Simon implied in defining PSSs—that notion of computation isn't really what we're interested in, when we ask whether minds are computational machines.

Partly in response to Searle's attempt to divorce computation from intentionality, Sloman argued in the 1980s that computers of appropriate complexity could understand. As for what "appropriate" meant, or "understanding" either, this required further research. But the computers of that time already embodied simple examples of some of the processes needed, such as building and comparing structured representations. The important question, he said, isn't "When does a machine understand something?"—which misleadingly implies some clear cut-off point at which understanding ceases. Rather, it is "What [many] things does a machine—whether biological or not—need to be able to do, in order to be able to understand?" (Sloman 1986b).

Although Sloman shared Newell and Simon's impulse to think about computation in terms of causal relations, he differed from them on two crucial points. First, he criticized them for ignoring causal processes within the virtual, as opposed to the physical, machine. For the philosophy of mind, as for psychology, it's the virtual machine which counts (see Section iv). Sloman's later writings explored the increasingly diverse computational processes and ontologies that are instantiated in AI and other forms of computer science. Even hands-on professionals, he said, don't fully understand what their machines are doing.

Second, he didn't take the concept of cause as a philosophical given. On his view, no current analysis captures all our intuitions about causality. It's closely linked with notions of possibility and conditionals, which is why Humean analyses fail. But "possible worlds" accounts, he believed, aren't satisfactory either (Sloman 1996c). Nor are the (closely related) attempts in recent theoretical computer science to understand concurrent interacting processes, such as are found on the Internet or in control systems for complex machinery (Sloman, personal communication). But such interactions need to be understood, if we are to achieve a satisfactory notion of causation *as such*.

Besides considering everyday examples, Sloman discussed the causal relations within a wide variety of virtual machines, and between the virtual and the physical machine (Sloman 1992, 1996b,c). One of his conclusions was that a virtual process can properly be said to cause a physical one, so that *qualia*—which he analyses in computational terms (Chapter 14.xi)—aren't epiphenomenal, but really do cause changes in the brain. But while he has suggested various desiderata for a comprehensive theory of causation, he's under no illusion that he has found one. Nor has he achieved the comprehensive taxonomy of designs necessary, he believes, for understanding both minds and computers. His third-style work is a promissory note rather than a definitive theory.

For many years, Sloman's publications were few and far between, and rarely in mainstream journals. His ideas were highly valued, nevertheless, by a small but discriminating international audience. As we saw in Chapter 10.iv.b, his work on computer vision involved several ideas that were ahead of their time, being rediscovered later by others. He was personally acquainted with most of the AI pioneers, and played an important role in the institutional development of cognitive science in England (where he worked at the University of Sussex for almost thirty years).

In the 1990s, various aspects of his research suddenly became more visible. His long-standing work on motivation and emotion (7.i.f) attracted increasing international attention, as did his approach to consciousness—which, like Dennett's, explicated it in terms of a complex virtual machine. And his ideas on computation, though not yet gathered into the long-awaited book, became more readily accessible through increased publication in print, and on the Web. In short, his influence grew. Many philosophers, however, were still wary of his (principled) refusal to offer crisp definitions, and bemused by his semi-technical discussions of computer architectures and virtual machines.

Sloman, then, is one of those—few, but growing in number—who regard computation (even within GOFAI) as a deep philosophical puzzle. And, as the other side of the coin, he's convinced that even computer scientists don't properly understand what computers are already doing—never mind what they might do in future.

d. The bit in "three and a bit"

This doubly problematic position—that computation is a puzzle, and that even hands-on users don't understand it—is held also by the AI scientist Agre (1997). His 1980s critique of GOFAI planning (see 13.iii.b) was the seed of his later ideas on computation. These are far from what one might expect from a hands-on AI professional.

Like Sloman (and, or so I've argued above, Newell and Simon too), Agre sees computation and intentionality as essentially connected. But he reaches this conclusion in a very different way, being a member of the neo-phenomenological camp discussed in Section vii.

Drawing on postmodernist authors such as Jacques Derrida as well as on Heidegger and Wittgenstein, Agre developed an "interactionist" approach to computation that is radically opposed to classical functionalism. He applied his interactionist vocabulary—or metaphors: his account was far from clear—not only to the computer as the human observer experiences it, but even to the electronic circuitry involved (e.g. Agre 1997: 92–102). And he insisted that computation can't be understood without considering the technical specifics of real examples:

Computational principles, on my view, relate the analysis of a form of interaction to the design of machinery. These principles are irreducibly intuitive, taking precise form only in the context of particular technical proposals. (Agre 1997: 20)

The tokens of mechanized logic are not exactly meaningless, since they are useless without the meanings that they are routinely treated as carrying. (p. 96)

Must combinational logic be understood within the mentalistic framework of abstraction and implementation? The happy answer of no can be sustained only by cultivating an awareness of the choices that lie implicit within day-to-day technical practice . . . In order to formulate an alternative, let us start by speaking about digital circuitry in a different way, rescuing it from the discourse of abstraction and implementation and reappropriating it within the alternative discourse of intentionality and embodiment. (pp. 103–4)

[One] should not confuse the articulated conceptions that inform technical practice with the reality of that practice. As [Brian Cantwell] Smith (1987) has pointed out, any device that engages in any sort of interaction with its environment will exhibit some kind of indexicality . . . AI-research needs an account of intentionality that affords clear thinking about the ways in which artifacts can be involved in concrete activities in the world. (pp. 241–2)

Despite his potentially off-putting references to Derrida (hardly an iconic hero of cognitive science: 8.ii.b), Agre—like Sloman—was later invited to contribute to a volume seeking to go beyond Turing-computation (Agre 2002). Nevertheless, I've chosen to describe him, unflatteringly, as "the bit in 'three and a bit' ". Why? The reason is not just his almost perverse unclarity, but the fact that his writing is heavily dependent on another AI researcher, who surpasses him in thoroughness—and, it must be said, in perversity too.

e. A philosophy of presence

As the last quotation indicates, one of the 'technical' authors who strongly influenced Agre was the Canadian computer scientist and philosopher Brian Cantwell Smith (1950–).

Originally trained at MIT, Smith's work was mostly done at Xerox PARC and Stanford (in the Philosophy Department), where he helped found Stanford's CSLI (Center for the Study of Language and Information). Later, he moved to the cognitive science programme at Bloomington, Indiana, and thence to Duke University, North Carolina. Now, he's Dean of Information Science at the University of Toronto.

His interests were always wide, combining intellectual interdisciplinarity with so-ciopolitical commitments. For example, he was co-founder—with Winograd—of Computer Professionals for Social Responsibility, and CPSR's first President (see Chapter 11.i.c). But no one could have foreseen that this MIT graduate would at-tempt to turn the notion of computation upside down—or, if you prefer, to enrich it immeasurably. I offer you that choice, because Smith's work is still hugely controversial.

Smith's research, developed over the past quarter-century, is the most far-reaching attempt to address philosophical puzzlement about the nature of computation. One might say that there's nothing it doesn't *attempt* to reach. (Some would add that there's nothing it *does* reach.) It is rich fare, though still a minority taste. Among the gourmets attracted to it, however, are some respected philosophers of cognitive science—and the computer scientist Matthias Scheutz, editor of the *Computationalism*: New Directions volume mentioned above (see B. C. Smith 2002).

This puzzlement grew out of Smith's hands-on experience as a computer professional, where knowledge of the material implementation often didn't enable him to understand what was going on. As he put it:

[With] respect to computer systems *we already know the answers to all the physiological questions* (we have the source code and wiring diagram), without that necessarily leading to any serious understanding, at the right explanatory level, of "what the program is doing". (B. C. Smith 1996: 148)

He drew the obvious moral: "we should be cautious about any hope that by under-standing the brain we will thereby automatically understand the mind" (p. 148).

As for the notion that *Of course we know what it is that computers are doing, because Turing told us*, Smith dismisses that as far too quick. For him, what Silicon Valley treats as computation *is* computation. And the denizens of Silicon Valley, he says, construe "computation" in at least six distinct ways—each of which allows for further distinctions within it. These different meanings are: formal (uninterpreted) symbol manipulation; effective computability; execution of algorithms; digital state machines; information processing; and physically embodied symbol systems. Certainly, one may be tempted to chide the Valley residents for their unclarity—and he does. But for Smith it is what their computers actually do which is important. Turing-computability (the first item on his sixfold list) doesn't capture that.

Smith's mature position—which he dubs "a philosophy of presence"—is another example of interactionist, neo-phenomenological AI. But it's more philosophically sophisticated than most, and much more provocative. Indeed, many professional philosophers (though by no means all, as we'll see) would regard "provocative" as an overly polite description.

Deeply influenced by his technical experience of Silicon Valley's various artefacts, Smith stresses the causal relations between program states (as Sloman had done too), the controllability of various devices, their basis in physics, and their being inescapably situated in the real (and the virtual) world. He sees computation as "inherently participatory", and computers as having "intentional capacities ultimately grounded in *practice*"—analogous to the human practices stressed by Heidegger and Dreyfus (B. C. Smith 1996: 305, 149).

His first publications, in the early 1980s, were on the semantics of program-ming languages. There, he distinguished Turing-computability from (implemented) computation as such, which *makes things happen* in computers. In other words, com-putation is a causal concept. He was thinking primarily not of the program's electronic implementation, but of causation in the *virtual* machine.

On that view, the expressions of any given programming language aren't mere empty syntax. They refer (*sic*) to virtual objects and causal processes such as variables (whose values may or may not differ at different times), numbers, procedures, strings, and larger structures containing these. Computer scientists, said Smith, should define their basic terms (such as "variable") accordingly (B. C. Smith 1982). He even designed a novel programming language, and a new semantics for LISP, on this principle.

This early work was seen by some readers as pure computer science—and it was included in an influential collection on knowledge representation in AI (B. C. Smith 1985). But Smith already had two philosophical aims: to put intentional flesh onto dry logicist bones, and to show how self-reflective knowledge is possible. His unorthodox approach whetted some philosophers' appetites, and his promised book was eagerly awaited.

Meanwhile, his ideas were developing in discussion with colleagues at Xerox PARC and Stanford who were working on situational semantics for AI and linguistics (for instance, Stanley Rosenschein, Jon Barwise, and John Perry), and with philosophers such as Haugeland, Adrian Cussins, and Bruno Latour.

His work began to influence the philosophy of cognitive science. It was cited, for instance, in an attempt to escape from the Chinese room (Boden 1988: 247–51); in an argument that non-conceptual content registers the world without articulating it as objects and properties (Cussins 1990); and in a rebuttal of Putnam's claim that computational states are universally realizable (Chrisley 1995). In 1991 Smith indicated his current position in a radical critique of Douglas Lenat's encyclopedia project (see 13.i.c and ii.a):

[Lenat views] representation as explicit—as a matter of just writing things down. I take it as an inexorably tacit, contextual, embodied faculty, that enables a participatory system to stand in relation to what is distal, in a way that it must constantly coordinate with its underlying physical actions. (B. C. Smith 1991: 285)

But this was a mere taster. Five more years were to pass before the promised book appeared (a brief "introduction" appeared six years after that: B. C. Smith 2002).

Significantly, the title had changed—twice. Originally heralded in 1982 as *Is Computation Formal?*, it was flagged in 1991 as *A View from Somewhere: An Essay on the Foundations of Computation and Intentionality*. It was eventually published (in 1996), however, as *On the Origin of Objects*. In short, the enquiry into computation had become an enquiry into metaphysics.

Smith described the book as an overview, a philosophical "story stripped of its computational heritage" (1996, p. x). What he meant by that was that (like Chomsky in the 1950s: Chapter 9.vi.a) he was holding back most of the technical research that led to his ideas. However, four foundational volumes on *The Origins of Computation and Intentionality* were announced as "in preparation", and (again, like Chomsky) drafts were already informally available.

Now convinced that one must understand metaphysics to understand computation (and vice versa), Smith had developed a novel approach to the nature of objects, individuation, particularity, subjectivity, and meaning. All these matters, he argued, are inescapable if one wants to say *what computers actually do*.

On his view, neither physical objects nor intentional subjects are metaphysically given (as they were for Descartes). They all arise from the "participatory engagement" of distinguishable regions of the metaphysically basic dynamic flux. This flux, described as "riotous fomenting fields" (1996: 281), is the subject matter of field-theoretic physics, and—having no objects—involves neither individuality nor particularity. Objects emerge, or are constructed, as a result of dynamic participatory relations. Smith used the analogy of an English garden, whose order-in-wildness has to be both constructed and continuously maintained (p. ix).

So where others were broadening *Dasein* from humans to animals (see Section viii.b), Smith broadened constructive dynamical interaction from animals to rocks, and even to atoms. All these, he argued, are regions of the flux which achieve particularity and identity through participatory activity. But only some involve *minds*.

Intentionality is at base a form of registration (alias representation: cf. Bennett 1976) that requires a relatively high degree of disconnectedness, or autonomy, as well as connectedness. (The neural emulators and other types of representation discussed in Chapter 14.viii exemplify various combinations of connectedness/disconnectedness.) This type of engagement gives rise to subjectivity and objectivity alike. And its participatory aspect can't be escaped:

[One] of the most important facts about the inherently participatory picture of registration being painted here is that the form of objectivity available to it cannot be achieved by mistakenly trying more and more to completely disconnect from the world, in a vain attempt to achieve the infamous view from nowhere [Nagel 1986]. On the contrary, the ability to register—*the ability to make the world present, and to be present in the world*, which is after all what this is a theory of—requires that one inhabit one's particular place in the deictic flux, and participate appropriately in the enmeshing web of practices, so as to sustain the kinds of coordination that make the world come into focus with at least a degree of stability and clarity. (pp. 305–6)

In other words, 100 per cent disconnection—pure objective contemplation—is neither possible nor useful: Smith spoke of "the utter futility [that results] when one pulls word and world too far apart" (p. 306).

Smith's third-style position on computation was even more evident now than it had been in the early 1980s. He explicitly denied any fundamental metaphysical distinction between intentionality in human beings and computers:

The constitutive patterns of partial connection and partial disconnection, of interwoven separation and engagement, while in detail so infinitely various as to defy description, nevertheless reveal regularities across a wide range of cases—from high-level political struggles for autonomy to simple error-correction regimens in low-level computer circuitry. (pp. 347–8)

One's mind can't fail to boggle on reading Smith's book. Quite apart from specific claims such as those quoted above, his overall philosophical agenda—to reconcile the "opposing views of the nature of philosophy" outlined in Section vi.b, above—couldn't be more ambitious. In a recent paper intended as an "introduction" to his book, Smith says as much himself:

For sheer ambition, physics does not hold a candle to computer or cognitive . . . science. Hawking and Weinberg are wrong. It is we, not the physicists, who must develop a theory of everything. (B. C. Smith 2002: 53)

And "everything", here, really does mean everything. In his book, Smith claims to have retained the major insights of both Continental and empirical–analytic traditions, without any of their problematic ontological assumptions. He also claims that his metaphysics gives us norms as well as facts, and that his account of objectivity can be "ultimately construed as a way of *living right*, rather than merely as a way of speaking truthfully" (1996: 108). (Compare Morris on the good and the true: Section viii.a, above.)

As one might expect, Smith's readers were mostly bemused. And even those who appreciated his effort were often highly ambivalent. For instance, the Washington University computer scientist Ronald Loui, who reviewed the book in *Artificial Intelligence*, described it thus:

[It is] an important book, even a beautiful book. It reasserts its author as one of the deepest and [most] erudite thinkers of computing. It is also, to this reviewer, an intellectually uninteresting book and thoroughly frustrating to read. (Loui 1998: 353)

After listing some of the "frustrations", which included Smith's decision to spend so much time on metaphysics ("I would have demanded an apology rather than an admission": Loui 1998: 353), he continued:

Along the way, gems are dropped that reflect an incredibly rich understanding of computing. I consider these gems to be more valuable than the details of the main line of Smith's investigation.... *If Brian Smith were not paralysed by metaphysical problems*, he could deliver the seminal work on the philosophy of computing just by collecting and displaying these gems of his insight. (Loui 1998: 354; italics added)

Nevertheless, *On the Origin of Objects* was published carrying high praise from philosophers and computer scientists alike. Haugeland, for instance, saw not paralysis but fundamental, all-engrossing, metaphysical insight. He was quoted on the dust jacket as saying: "Smith recreates our understanding of objects essentially from scratch—and changes, I think, everything."

As remarked at the outset of this chapter, however, philosophical problems don't get solved in a hurry. One shouldn't expect that Smith's highly unorthodox metaphysics would—or will—convince everyone. Some may disdain its 'engineering' origins in his practical experience of computing (see Chapter 2.i.a). Even setting such snobbery aside, however, agreement won't come easily.

Phenomenologists won't readily endorse his use of field-theoretic physics to arrive at conclusions largely similar to theirs. For as we've seen, they believe that a naturalistic philosophy of mind is impossible. Similarly, Searle attacks Smith's ascriptions of intentionality to computation, arguing that syntax (pattern) isn't really "in the physics" but is "observer-relative"—so that Smith's work provides no escape from the Chinese room (Searle 1990a; 1992: 209).

Even those who—like Smith himself—reject formalism while favouring a mind-as-machine credo may disagree with him in various ways. My own view, for instance, is that he helped himself to the "dynamic flux"—his version of Kant's noumenal world—without proper licence. He claims that he's pulled this concept up by its own bootstraps (in the final sixty pages), to form a "constructivist" metaphysics of objects and intentionality (cf. also pp. 188–9). But it seems to me that he begs this fundamental philosophical question instead of answering it.

In addition, many readers will be bemused—perhaps even repelled—by the extraordinary mix of dry argumentation and intoxicating (or embarrassing) rhetoric. The rhetoric is evident in his promise to do the metaphysical bootstrapping just mentioned:

I have been assuming an underlying space of physical feature fields—fields that were never paid for, and which anyway are about to be discarded ... The fields will go. But the patterns will remain: restless patterns of stabilization and coordination, of invention and description and activity and design, of struggle and submission and conquest and peace, sometimes collaborative, sometimes singular. All these participatory activities arise out of ineffable connection and subside back into it, at a different place or different time or under different circumstances, often benefitting from the perspective of abstraction and registration, but never escaping from the located, the directed, and the exquisitely particular. (1996: 312–13)

The crucial point, however, is that it's not only Smith's *rhetorical style* that's surprising. (Surprising, that is, to analytic philosophers and computer scientists. Devotees of Heideggerian metaphysics would be less bemused, for he too constantly appealed to freedom, commitment, poetry, and love.) In short, Smith's *account of computation* is very different indeed from what most philosophers expect.

Different or not, Smith is adamant that understanding computation is crucial for cognitive science:

It is sobering, in retrospect, to realize that *our preoccupation with the fact that computers are computational* has been the major theoretical block in the way of understanding how important computers are. They are computational, of course; that much is tautological. But only when we *let go of the conceit that that fact is theoretically important* . . . will we finally be able to see, without distraction, and thereby, perhaps, at least partially to understand, how a structured lump of clay can sit up and think. (B. C. Smith 2002: 52–3)

f. The moral of the story

The general moral, here, is that the rise of third-style discussions (such as those of Sloman, Agre, and Smith) strongly suggests that we don't yet know what computation—understood intuitively as what computers do—*is*. In particular, we can't assume without question that it's utterly distinct from intentionality. It follows—if the mind is a computational machine—that we don't yet know what the mind is, either.

The core thesis of cognitive science—that mental processes are computational—should therefore be interpreted transparently, not opaquely (Chrisley 1999). In other words, the claim isn't that mind can be explained by our current ideas about computation, but that it's explicable by *whatever theory turns out to be the best account of what computers do*.

"Computers", as suggested above, may turn out to include hypercomputers, computing functions that aren't Turing-computable. So Copeland (2000, 2002*b*) has argued for much the same position as Chrisley, in opposing "wide" to "narrow" mechanism. Narrow mechanism sees the mind as some sort of Turing machine, whereas wide mechanism allows for types of computation that aren't Turing-computational—but which are executed in *machines* nonetheless. If wide mechanism is true, then classical functionalism bites the dust. But mind-as-machine doesn't.

There's a clear analogy, here, with physicalism in metaphysics. Physicalists don't normally tie themselves down to today's physics, but have in mind *whatever turns out to be the best account of the physical world*. The best physics may turn out to be surprising (yet again). And, as Chrisley points out, the best theory of computation may turn out to be surprising too.

It's clear from the rise of the second (inclusive) and third (causal/intentional) senses discussed above that the concept of computation has been a moving target since its introduction in the 1930s. Turing's definition is still the clearest, to be sure, and is understandably still cited by philosophers. But even Searle now allows (largely because of Smith) that the mathematics and electronics haven't been clearly connected, and that "there is little theoretical agreement among the practitioners on such absolutely fundamental questions as . . . What exactly is a computational process?" (1992: 205).

In sum: the still widespread notion that philosophers can give—or just assume—a quick definition of computation, and then get on to the *really* interesting philosophical issues, takes far too much for granted.

16.x. What's Life Got To Do With It?

All the minds we know about are found in living things. But why?

* Couldn't there be mind and meaning without life?
* And what is life?
* Given that it involves self-organization, just what sort of self-organization is it?
* Must it involve evolution, for instance?
* Is embodiment essential? And what's that? Is mere physicality enough for embodiment?
* Or is metabolism needed too—and, again, what's that?
* What's the link, if any, between metabolism and mind?
* Could life be generated artificially, and if not why not?
* Is strong (i.e. virtual) A-Life impossible in principle?
* If so, does it follow that strong AI is an illusion too?

These questions took a long time to surface in cognitive science. Or perhaps one should rather say to *resurface*. For in the very early days, life and mind were both discussed—and were treated largely on a par.

The cyberneticians of the 1940s applied their theories of self-controlling machines to both living organization and purpose (see 4.v–vii). Life wasn't regarded as necessary for teleology, for self-guided missiles were said to exemplify goal seeking (Rosenblueth *et al.* 1943). However, since life and mind were supposed to consist in fundamentally similar principles of control, it didn't seem surprising that all the minds we know about are grounded in living things.

McCulloch had been a member of this life-and-mind movement since its inception in the 1930s. But he'd been deeply interested in logical analyses of language even before then (see 4.iii.c). In the event, his paper of 1943 turned attention away from life in favour of mind.

Control by feedback gave way to logic-based computation. The NewFAI computer-modelling community saw mind and meaning as matters for propositional logic, their origin in adaptive behaviour and living embodiment being downplayed. Even unreconstructed cyberneticians now sometimes spoke of mind without mentioning life: when William Ross Ashby wrote a paper on cybernetics for *Mind*, he concentrated on defending materialism against mind–body dualism, not on exploring the philosophy of self-organization (Ashby 1947).

In short, by mid-century the link between life and mind—though still widely accepted, even taken for granted—wasn't explicitly considered. The people on the NewFAI side of the emerging cybernetics/symbolic schism (4.ix) ignored life and spoke only of mind. The people on the other side said very little about mental phenomena beyond perception and goal seeking: self-reflection and reasoning were mostly ignored.

That life/mind split within cognitive science lasted for several decades. Much the same was true in philosophy, especially the analytic variety. It's still the case, in 2006, that the link between life and mind is ignored (beyond mere lip-service) by the mainstream. However, these issues are now arousing interest—largely as a result of the rise of A-Life.

a. Life in the background

Even in the early 1960s, there were a few exceptions to what I've just said. Most important, with respect to their historical role in cognitive science, were the people impressed by cybernetics who were developing holistic philosophies—and computer models—of life and/or mind (15.vi–vii).

The most influential of these (all discussed later in this section) were to be the Chilean neuroscientist Maturana (with Varela and Milan Zelený), whose similarity to Dreyfus was mentioned in Section vii.b above, and the philosopher Howard Pattee, with his students Robert Rosen and Peter Cariani. Both Maturana and Pattee were working on the concept of life from the 1960s. Although they had some early disciples, their ideas didn't become widely known until the 1990s. By the new millennium, however, Maturana and Varela's ideas had been published in semi-popular form, and an entire issue of the journal *BioSystems* was devoted to Pattee's work and influence (Rocha 2001).

All of the above were either scientists (like Maturana) or philosophers very close to science, so close that they sometimes got involved in scientific work (Pattee, for instance). None were "pure" philosophers.

In general, the mid-century philosophers who were interested in the puzzle of life-and-mind came from the Continental, not the analytic, side of the fence. As a result, they were largely ignored by the scientific community. Moreover, they were exceptions even within their own, neo-Kantian, tradition. To be sure, phenomenologists in general took the human being's embodied living-in-the-world as philosophically basic. And Wittgenstein saw language as part of our "natural history", declaring: "What has to be accepted, the given, is—so one could say—*forms of life*" (1953: 226). Most of them had scant interest, however, in what biologists mean by life—which includes oak trees and barnacles, as well as human beings.

One exception to this was the existentialist theologian Hans Jonas (1903–93), who developed a new philosophy of biology in the 1950s. Unlike Maturana and Varela, he wasn't interested in biology for its own sake, but as an aspect of what he saw as the disastrous cultural denouement of Descartes's materialism (Jonas 1966: 58–63).

An ex-pupil of Heidegger, Jonas fled Germany for England when the German Association for the Blind expelled its Jewish members (Jonas 1966, p. xii). From the mid-1950s to the mid-1970s he worked at the New School of Social Research, in New York. Despite still regarding Heidegger as "the most profound and . . . important [proponent] of existential philosophy" (p. 229), he rejected his philosophical dichotomy between humans and other living things—and his pro-Nazi sympathies, too. He explained the latter in terms of "the absolute formalism of [Heidegger's] philosophy of decision", in which "not *for* what or *against* what one resolves oneself, but *that* one resolves oneself becomes the signature of authentic Dasein" (Jonas 1990: 200). And that, in turn, he saw as a result of the stripping-away of value from nature, its "spiritual denudation" by Descartes and modern science (Jonas 1966: 232).

It was in response to this disenchantment of nature that Jonas published various essays on life in the post-war years, and collected them as *The Phenomenon of Life* in 1966. They outlined a framework for a biology that would admit value as an intrinsic feature of life in general. ("Outlined" and "framework" are important here: he discussed almost no specific examples.)

Embodiment, and in particular metabolism, was seen by Jonas as philosophically crucial (1966: 64–91). Not only was life essential for the emergence of mind (pp. 99–107), but *all* self-organized matter was, in a sense, ensouled—though where Maturana and Varela spoke of life as involving *cognition*, Jonas spoke of life as involving *self-concern*. (He lauded Heidegger for having "shattered the entire quasi-optical model of a primarily *cognitive* consciousness, focusing instead on the wilful, striving, feeble, and mortal ego"—1996: 44.) As he put it:

One way of interpreting [the ascending scale of life] is in terms of scope and distinctness of experience, of rising degrees of world perception.... Another way, concurrent with the grades of perception, is in terms of progressive freedom of action.... [One] aspect of the ascending scale is that in its stages the "mirroring" of the world becomes ever more distinct and self-rewarding, beginning with the most obscure sensation somewhere on the lowest rungs of animality, even with the most elementary stimulation of organic irritability as such, in which somehow already otherness, world, and object are germinally "experienced," that is, made subjective, and responded to.

[We spoke, above, of freedom.] One expects to encounter the term in the area of mind and will, and not before: but if mind is prefigured in the organic from the beginning, then freedom is. And indeed our contention is that *even metabolism, the basic level of all organic existence, exhibits it: that it is itself the first form of freedom.* (Jonas 1966: 2–3; italics added)

Even in "the blind automatism of the chemistry carried on in the depths of our bodies", there is "a principle of freedom . . . foreign to suns, planets, and atoms". For living organisms have a special type of identity and continuity: a stable dynamic form made of an ever-changing material substrate. In short, "mind is prefigured in organic existence as such" (p. 5). Plants, too, have "metabolic needs", although they stand in an "immediate" relationship to their environment. And metabolism is the necessary base of all forms of mediation: perception, motility (action), emotion, and—ultimately—conscious imagination and self-reflection. (These phenomena emerge as a result of evolution: Darwin, *despite* his materialist assumptions, had enabled us to understand this: pp. 38–58.) Life and mind are ontologically inseparable: "the organic even in its lowest forms prefigures mind, and . . . mind even on its highest reaches remains part of the organic" (p. 1).

In other words, Jonas was offering an answer to the question of *why* all the minds we know about are found in living things. At the same time, he was offering an answer to the question of *what life is*.

He explicitly refused to speculate about the origins of life (p. 4), even though this was already being discussed by biochemists (Chapter 15.x.b). His interests were ontological, not scientific: metabolism was "the break-through of being" from mere physicality to "the indefinite range of possibilities which hence stretches to the farthest reach of subjective life" (p. 3). (Accordingly, he retained a Heideggerian hostility to technological theories/analogies of life or mind: pp. 108–26.)

The book was reissued (by several different publishers) in 1979, 1982, and 2001, and translated into German in 1994. So one can't say that it was wholly ignored. Indeed, because of his stress on the intrinsic value of life and humankind's responsibility towards it, Jonas's work—especially his volume on ethics (Jonas 1984)—has recently become better known thanks to the environmentalist movement.

In the 1950s and 1960s, however, his philosophy of biology was ignored by analytical philosophers and mainstream biologists alike. (And by cyberneticists too, whose analysis of living purpose and rocket teleology *in the very same terms* he'd rejected as "spurious and mainly verbal"—1966: 111.) The same was true of Maturana and Varela's early work, but they have now earned a clear, if still marginal, place in the history of cognitive science. Jonas has not (but see Di Paolo forthcoming). He's relevant here not as a protagonist in that historical drama, but as a mid-century philosopher who tried to argue the case that mind requires life, rather than taking it for granted.

Another philosopher who'd done this was Henri Bergson (Chapter 2.vii.c). By the end of the twentieth century, Bergson's views on "creative evolution" were being revived in some philosophical circles—especially in "process" philosophy/theology (Sibley and Gunter 1978; Papanicolaou and Gunter 1987). This emulated Bergson alongside the even greater hero Alfred North Whitehead (4.iii.b). But some unorthodox scientists were taking an interest too. The physicist–philosopher Henri Bortoft (1996) put Bergson second only to Goethe as a precursor of current dynamical theories in science and philosophy (see 2.vii.c). And a few neo-Bergsonian philosophers even tried to relate his ideas to cognitive science and/or A-Life.

For example, Gilles Deleuze (1925–) revived certain aspects of Bergson's philosophy by stating them in terms of ideas about dynamical systems (Deleuze 1966/1988). I'm saying that at second hand, I must confess, for Deleuze himself is nigh unreadable by anyone more accustomed to analytical philosophy. Much as Richard Montague's work couldn't spread among linguists until a clear account of it had been provided by Barbara Partee (see 9.ix.c), so Deleuze's has been made accessible to cognitive scientists by his expositor Manuel DeLanda (2002).

Although he rejected Bergson's dualist interpretation of *élan vital*, Deleuze offered a "re-enchantment" of matter, nevertheless. He even (confusingly) used the term "spirituality" in talking about matter and life. But this wasn't intended as transcendent spirituality: rather, it referred to the abstract principles of self-organization, and the structured spaces of possibilities, that are inherent in matter/energy.

He saw matter not as inert stuff subject to external influences, but as the source of formative material *processes*. A soap-bubble, for instance, actively minimizes the surface tension at every point (it dynamically "computes" its own shape). The dynamical structures generated by matter were said to be constrained, in part, by abstract topological principles describing connectivities and attractors of various kinds (compare Stuart Kauffman's work on NK networks, and Randall Beer's on CTRNs: 15.ix.b and xi.b).

On this view, life was a special case of matter, and mind a special case of life. It followed that there's no *special* difficulty about giving a naturalistic, even a materialistic, account of mind or intentionality, even though spelling one out in detail may be highly challenging.

However, these intriguing analogies weren't helpful in furthering scientific understanding. (Or anyway, they haven't been helpful yet: DeLanda's relatively accessible version of Deleuze appeared only two years ago, and it remains to be seen whether many scientists will take it up.) Admittedly, the ever-maverick neuroscientist Karl Pribram (1987)—accused in the early 1960s of actually *believing* the MGP manifesto (6.iv.c)—described the cerebral basis of some cognitive processes in Bergsonian terms. But that's not to say that he *used* Bergson's ideas to make discoveries which otherwise would not have been made. Rather, he pointed out an analogy between Bergson's views on memory and his own (long-standing) holographic/holonomic theory (cf. 12.v.c).

Cognitive scientists who weren't already sympathetic to dynamical systems and/or Kauffman's approach to A-Life weren't likely to be interested in Bergson's work at all, even if they encountered it. And that was unlikely: as remarked in Chapter 2.vii.c, it had been more or less forgotten since mid-century—especially by philosophers of an analytic cast of mind.

For over thirty years, then, the concept of life was usually ignored in discussions of mind as machine. To be sure, the psychologist Miller raised the topic—but he immediately dropped it like a hot potato. He was, he said, "unclear" whether epistemic (cognitive) systems should be defined as animate or inanimate. The advantage of defining them as animate was that "we cut artificial intelligence free to develop in its own way, independent of the solutions that organic evolution happens to have produced" (G. A. Miller 1978: 9). (This remark pre-dated the concept of strong A-Life by a decade: clearly, Miller thought it obvious that computers and life are incompatible.) But whether it made "any real difference" in conceptualizing the study of *mental* processes was "unclear".

Most analytic philosophers tacitly assumed some life–mind linkage—which would imply that if computers aren't alive then they aren't psychological systems. They evidently thought this point so obvious that, even when they bothered to state it explicitly, they didn't offer any arguments for it.

Scriven, for instance, confidently declared—without giving reasons—that "Life is itself a necessary condition of consciousness" and that "Robots . . . are composed only of mechanical and electrical parts, and cannot be alive" (Scriven 1953: 233). Lucas hinted at a similar position in his own reply to Turing's 1950 paper (see Section v.a). Geach insisted that AI systems can't have beliefs and intentions because they're "certainly not alive" (Geach 1980: 81). And some, such as Searle (1980, 1992) and Ruth Millikan (1984), explicitly linked intentionality with biology (neurochemistry and evolution, respectively). But even they didn't discuss the nature of life as such.

Two exceptions that proved (i.e. tested) the rule were Putnam's (1964) paper on 'Robots: Machines or Artificially Created Life?' and Geoffrey Simons's (1983) book *Are Computers Alive?*

Despite its title, Putnam's paper focused mainly not on life, but on consciousness. At one point, Putnam endorsed Ziff's claim that it's an "undoubted fact" that if a robot isn't alive then it can't be conscious. But he was relying on "the semantic rules of our language", not on any quasi-explanatory relationship between life and mind.

He also said (this time, disagreeing with Ziff) that something which is clearly a mechanism might be alive. Again, however, this was linguistic philosophy in action. Sometimes, Putnam heretically recommended *changes in meaning* due to new scientific data, as he did when countering Malcolm's account of dreaming (Putnam 1962a). But in the paper on robots and life, he was talking only about what *current usage* allowed one to say (or imagine) without contradiction. The nearest he got to discussing a substantive claim about life was to scorn the suggestion that the primary difference between a robot and a living organism is the "softness" or "hardness" of the body parts (1964: 691).

Much later, Putnam's paper was discussed at length, and accused of incoherently combining Aristotelian and Cartesian views (Matthews 1977). At the time, however, it didn't prompt philosophical interest in the concept of life.

Simons, writing twenty years after Putnam, used concepts drawn from GOFAI and cybernetics to claim that computers can be *really* alive, and *really* intelligent. He specifically denied that the genesis of the system is relevant to whether it's alive: "A mechanically *assembled* [i.e. not evolved or self-constructed: see below] system may reasonably be regarded as living . . . " (1983: 23). However, his argument was neither deep nor convincing, and (deservedly) attracted little attention.

b. Functionalist approaches to life

With the rise of A-Life in the late 1980s, the nature of life became an inevitable topic for computational research. Inevitable, but in practice not central: most A-Life workers focused on other questions, maintaining a diplomatic silence on this one. Some of their colleagues, however, were more bold.

The relevant discussions were guided by two radically opposed philosophies. (Sounds familiar?—see Section vi.b.) These were functionalism and metabolic holism, a special case of dynamical systems theory.

Functionalism, in this context, is the view that the characteristics of life (see Chapter 15.ii.b) can be described by informational concepts. So self-organization involves the appearance of new levels of order, abstractly defined. Autonomy, emergence, development, adaptation, responsiveness, and evolution concern various types of structure, process, and control. Even reproduction (on this view) can be defined informationally, as self-copying.

The one exception is the concept of metabolism, which concerns not information but energy. Thoroughgoing A-Life functionalists weren't worried by this, as we'll see. But their opponents argued that they should be.

It's often assumed (wrongly) that all A-Life workers are thoroughgoing functionalists. This is largely because Christopher Langton, following John von Neumann's lead, wrote this position into his definition of the field in 1986 (Chapter 15.ii.b and ix).

Moreover, he drew the obvious implication: a licence for strong A-Life. If living self-organization is definable in logical terms, then a virtual "creature" implemented in computer memory that satisfied these abstract criteria—whatever they are—would be genuinely alive. ("Whatever they are", because definitions differed. For instance, Langton suggested including the lambda parameter, Andrew Wuensche the Z-parameter: Chapter 15.viii.a.)

Some A-Life colleagues were quick to join Langton in this claim. Thomas Ray, for instance, declared:

The intent of [my] work is to synthesize rather than simulate life . . . To state such a goal leads to semantic problems, because life must be defined in a way that does not restrict it to carbon-based forms. It is unlikely that there could be general agreement on such a definition . . . Therefore, I shall simply state my conception of life in its most general sense. I would consider a system to be living if it is self-replicating, and capable of open-ended evolution [generating] structures and processes that were not designed-in or preconceived by the creator. (Ray 1992: 372)

As we saw in Chapter 15.vi.b, Ray's Tierra system did indeed generate phenomena not designed-in by Ray. These included co-evolving parasites, hyper-parasites, cheaters, and symbionts. Ray's response was a curious combination of modesty and hubris:

[The] results presented here are based on evolution of the first creature that I designed, written in the first instruction set that I designed. Comparison with the [virtual] creatures that have evolved shows that the one I designed is not a particularly clever one . . . It would appear then that it is rather easy to create life. (p. 393)

As for the problematic concept of metabolism, Ray said two things. On the one hand, the computer consumes physical energy too. On the other, the equivalent of metabolism can be functionally defined:

In studying the natural history of synthetic organisms, it is important to recognize that they have a distinct biology due to their non-organic nature. In order to fully appreciate their biology, one must understand the stuff of which they are made. To study the biology of creatures of the RNA world would require an understanding of organic chemistry and the properties of macro-molecules. To understand the biology of digital organisms requires a knowledge of the properties of machine instructions and machine language algorithms. (p. 397)

I will discuss the inoculation of evolution by natural selection into the medium of the digital computer. This is not a physical/chemical medium; it is a logical/informational medium . . . Evolution is then allowed to find the natural forms of living organisms in the artificial medium. These are not models of life, but independent instances of life. (Ray 1994: 179)

For some broadly functionalist A-Life scientists, this was a step too far—and *much* too far for most philosophers (e.g. Harnad 1994; Olson 1997). Those A-Life colleagues were content to interpret most of the characteristics of life in informational terms—but not metabolism, which is irredeemably physical. However, since they defined metabolism as mere energy dependency, their rejection of Ray's position was intuitive rather than strongly argued (see below).

Many A-Life colleagues simply avoided the question, by way of the "diplomatic silence" mentioned above. They were interested in studying specific aspects of life, such as evolution or flocking, not in discussing its general nature. They were even less interested in considering the "strong A-Life" scenarios sketched by Langton and Ray—and later by Steve Grand (1958–).

Grand's first claim to fame was that he designed the hugely popular computer game Creatures. This swept the world in the early 1990s (see Chapter 13.vi.d), and was still being widely celebrated—for broadly counter-cultural reasons—in the new century (Kember 2003: 91–105).

Creatures enabled the user to evolve unusually sophisticated computer creatures (Grand and Cliff 1998). Their neural-network brains supported simple learning, and included 'neuromodulators' as well as several types of 'neurone'. The creatures also had a simulated biochemistry, with the potential to model a large number of metabolic and hormonal functions, from digestion to ovulation. As a piece of lifelike software engineering, it was way beyond the general state of the art when it appeared, and is still impressive. Indeed, it could conceivably be used as a powerful test bed for AI models of motivation and emotion such as those discussed in Chapter 7.i.e–f (Boden 2000*b*).

Grand's current technical aim is to build an "imaginative" robot called Lucy, whose intelligence will emerge "naturally"—and holistically—from its 100,000-neurone hardware (Whitby and Grand 2001). As he points out, this attempt to build Dennett's (1978*c*) "whole iguana" is very different from MIT's Cog project, with which Dennett himself was involved (see 15.vii.a).

The Cog robot was carefully designed module by module, bits of its "intelligence" being successively bolted on. Grand, by contrast, wants an already integrated intelligence to emerge from a relatively unorganized base. Rather than providing Lucy's brain with spatial maps or orientation columns, for instance, he hopes that these would emerge spontaneously (much as ocular dominance columns arose in the work of Christoph von der Malsburg and Ralph Linsker: see Chapter 14.vi.b and ix.a). And the robot would learn to perform "voluntary" actions by associating the image (representation, model) of the desired action with the muscle movements required to achieve it (compare Marr's theory of the cerebellum: 14.v.c).

The Lucy project is startlingly ambitious—I'm tempted to say, utterly impracticable. But the A-Life expert David Cliff (personal communication) believed Creatures to be utterly impracticable too, when first consulted by the games company to whom Grand had offered it. Given what Grand had told them it could do, it must—so Cliff thought—be either hype and/or a superficial con trick, carefully tailored to present a convincing 'demonstration'. (Even the impressive SHRDLU, you'll remember, could handle only the one conversation without tripping over its toes: 9.xi.b.) And the fact that it had been two-finger-typed on Grand's bedroom computer wasn't promising. Not until he got down into the machine code was Cliff convinced—at which point he suggested how it could be improved still further, using some of the ideas discussed in Chapters 7.i.f and 15.vi–ix.

Grand didn't know about those ideas already, because he's an autodidact. As such, he's undeterred by received academic opinion. And he's a highly creative computer engineer, who's already designed one apparently impossible system that does just what it was intended to do. He's thus in an entirely different class from the self-publicizing roboticists Kevin Warwick and Hugo de Garis, on whose 'research'—technical no less than philosophical—I forbear to comment, for fear of scorching the page.

I wouldn't bet a large sum of money on Lucy. And I don't agree with those cultural commentators who claim that Grand is "one of the 18 scientists most likely to revolutionise our lives in the coming century" (ICA 2000). Nevertheless, as Richard Dawkins has remarked, "If anybody can pull off a spectacular breakthrough, it'll probably be him" (Whitby and Grand 2001: 13). (For the most recent status report on Grand's progress, see his web site at <http://cyberlife-research.com>.)

At the turn of the century, Grand (2000, 2003) made a number of highly provocative claims about the philosophical significance of his own past and future work. He sees his virtual creatures as more than merely *lifelike*: they are "sort of alive", or even "a sort of life". When challenged on this point, he insists (personal communication). Grand is an autodidact in philosophy too, but here there's no good reason to give him the benefit of the doubt. Whereas Creatures (considered as technology) clearly does what he said it would do, his philosophical arguments are challengeable—and, in my view, as mistaken as Ray's. Strong A-Life is no more plausible in Creatures than in Tierra—and even Grand's predicted robot Lucy wouldn't count as genuinely *alive* (see the discussion of metabolism, below).

Where the general public were concerned, Lucy made something of a splash. Although it must be said—and often is said, by other roboticists—that if Grand hadn't fitted a furry gorilla-face onto the head, and if he'd called it Robot 37 instead of Lucy, people wouldn't have been quite so interested. (Similarly, the young Minsky's robot arm aroused no attention until he put a shirtsleeve on it: see 1.iii.h.) Quite apart from the overexcitement of the journalists (the same old story!), a number of commentators have picked up on it as an expression of wider cultural concerns.

The anthropologist Lucy (*sic*!) Suchman, for example, who cast doubt on GOFAI planning some twenty years ago and focused on *human–machine communication* soon after that (13.iii.b), described her robotic namesake as one among the disturbing category of the "almost human" (Castañeda and Suchman forthcoming; cf. Suchman 2004).

Besides the familiar anthropologists' fare of totems and other things "doing duty as persons", these include children, non-human primates, and AI/A-Life machines. The cultural status of children has been a focus of commentary at least since Jean-Jacques Rousseau (1712–78), and twentieth-century developmental psychology has helped fuel this fire. As for primates, advances in field ethology have led to the culturally problematic Great Ape programme (7.vi.f). The eighteenth-century automata (2.i.b) challenged contemporary notions of the person (Riskin 2003). Now, as Suchman pointed out, various actual and imaginary AI projects are exciting comment not only in the philosophy of mind but in our wider culture too.

Lucy (which Suchman discusses at length) is only one example of the "almost human" produced by AI/A-Life. Cog, and especially its successor Kismet (see 13.vi.d), are others. The feminist philosopher Evelyn Fox Keller (forthcoming), for instance, sees some "serious anxieties" with respect to providing Kismet and the like with facial expressions that reliably elicit emotional reactions in human viewers. (She's particularly worried by the plan to use robots like Brian Scasselatti's Nico to *test* theories in human developmental psychology.) Still other almost-humans—all media darlings in their day—include ELIZA, expert systems, AI agents ("softbots"), VR avatars, Turing's computer conversationalist, Stanley Kubrick's HAL, and Steven Spielberg's David.

The behaviour—and man–machine interactions—of many of these systems is far more human-like than Lucy's is. But because Grand, besides providing the superficial furry face, speaks of *life* as well as *mind*, his work aroused more outside comment than most. In addition, his A-Life system is not virtual/intellectual (as softbots are) but *embodied*—or at least, *material*. It's therefore of interest to those commentators, including phenomenologists and many feminist philosophers, for whom the downplaying of embodiment in the analytic–scientific tradition has been a fundamental mistake (Haraway 1997: 186, 302–3, and *passim*; Kember 2003: 105–15, 198 ff., and *passim*).

Suchman (like Haraway, and also Clark: vii.d, above) takes personhood, in whatever culture, to be constituted not by an individual person-in-the-mind but by the nexus of social relations and interactions available. On that view, the cultural status of robots, and other AI/A-Life systems, is determined less by their seeming intelligence than by the pattern of interactions we choose to engage in with them. But the influence is

1438 16.x: WHAT'S LIFE GOT TO DO WITH IT?

reciprocal: in so far as we do engage with them, we modify our own self-image in various subtle ways (cf. 13.vi.d).

c. The philosophy of autopoiesis

Some A-Life researchers dismissed all these science-fictional scenarios *because they were fundamentally opposed to functionalism in the first place*. Among these were the proponents of Maturana's theory of "autopoiesis".

This was perhaps the best-developed philosophy of metabolic holism. (The competing candidate is the work of Pattee's group: see below.) It even inspired several computer models of biochemical autopoiesis (Zelený 1977; Zelený *et al.* 1989), and a wide range of work in A-Life (McMullin 2004). This included work in "wet" A-Life, in which biochemical autopoiesis *as such* was studied too (see Chapter 15.x.b; Bachman *et al.* 1990; Walde *et al.* 1994).

Originated in the 1960s, Maturana's theory was strongly influenced by Heinz von Foerster's cybernetics. Despite the fact that an English translation was published over a quarter-century ago in the highly respected Boston Studies in the Philosophy of Science (Maturana and Varela 1972/1980), it has remained a minority taste. It's clear from my personal acquaintance that many philosophers have never even heard of it.

One reason, no doubt, is its rebarbative vocabulary and unrelenting abstraction. Also, it has some highly counter-intuitive implications, as we'll see. Nevertheless, it offers a principled way of grounding mind in life. Rather than arguing (like Searle) that neuroprotein happens to cause intentionality, as chlorophyll happens to cause photosynthesis, this view grounds intentional categories in an essentially autopoietic biology.

For Maturana and Varela (1972/1980), life is "autopoiesis in the physical space". Autopoiesis in general is defined as the continuous self-production of an autonomous entity. As they put it (you're advised to take a deep breath here):

An autopoietic machine is a machine organized (defined as a unity) as a network of processes of production (transformation and destruction) of components that produces the components which: (i) through their interactions and transformations continuously regenerate the network of processes (relations) that produced them; and (ii) constitute it (the machine) as a concrete unity in the space in which they (the components) exist by specifying the topological domain of its realization as such a network. (Maturana and Varela 1972/1980: 79)

Or more colloquially, an autopoietic system "pulls itself up by its own bootstraps and becomes distinct from its environment through its own dynamics, in such a way that both things are inseparable". This type of self-organization can occur in the world of human communication, in which case we have some kind of social institution (cf. Teubner 1987, 1993). But when it happens in the physical world, we have a living organism.

The autopoiesis concerned here is a special case of homeostasis (see Chapter 4.v.c), where what's preserved isn't one feature, such as blood temperature, but the organization of the system as a unitary whole. This requires the self-creation of a unitary physical system, by the spontaneous formation of a boundary—at base, the cell membrane—and the continuous generation and maintenance of the body's own components.

For Maturana and Varela, *body* and *embodiment* are autopoietic categories. So too are *cognition, communication, meaning,* and *language,* all of which they defined in terms of the interactions of living things. In the more accessible version of their theory that appeared around 1990 (Maturana and Varela 1987, 1992), and in Varela's book co-authored with cognitive psychologists (Varela *et al.* 1991), they focused on human language, understanding, society, and consciousness—all described as necessarily rooted in our biology.

In fact, they were overly liberal with their ascriptions of intentionality (Boden 2001), for they declared that "Living systems are cognitive systems, and living as a process is a process of cognition" (1972/1980: 13). Taken seriously, this extends knowledge even to algae and oak trees. One can—and should—express the idea that algae and acorns are pre-adapted to their environment without using the concept of *knowledge.* Such over-liberality was an occupational hazard for cyberneticians: as we saw in Chapter 4.v.e, Gregory Bateson had similarly attributed *knowledge* to redwood forests, and *mind* to whirlpools and oscillating electrical circuits.

From the autopoietic viewpoint, both strong A-Life and strong AI are absurdities. For computers aren't autopoietic systems. Even self-assembling robots, if assembled from manufactured parts as opposed to being self-organized by some alien biochemistry, wouldn't be alive. (Nor would they have bodies.) Consequently, robotic intelligence is impossible too.

Autopoietic theory is a special case of the general (anti-functionalist) position that metabolism is essential for life. Believers in strong A-Life (such as Ray, quoted above), when confronted with this view, typically pointed out that computers consume energy too. They sometimes added that the "physics and chemistry" of their virtual creatures is constituted by the computer's memory and operating system.

A number of philosophers, some of whom weren't committed to autopoietic theory, replied that metabolism is more than mere energy dependency. Rather, it's the self-production and self-maintenance of the physical body by energy budgeting, involving self-equilibrating energy exchanges of some *necessary* complexity (Pattee 1989; Cariani 1992; Sober 1992; Boden 1999). They argued that strong A-Life is possible only if virtual systems can metabolize *in the sense just given,* or if metabolism is inessential for life. But neither alternative is tenable.

Living 'tin-can' robots are also excluded by this approach. Only robots powered by complex biochemical cycles of synthesis and breakdown would be truly alive, and truly embodied. This is the basis of the intuition scorned by Putnam, that "softness" and "hardness" matter (see above).

Elliott Sober (1948–) cited other biological properties, besides metabolism, in arguing against strong A-Life (Sober 1992). Digestion and predation (for example) each relate an organism to something outside itself, where that "something" is essentially physical. Both can be realized in multiple ways (defined by biochemistry and behaviour), but in every instance some physical organism has to interact with—hunt, eat, transform—another. Like metabolism itself, these features can be usefully simulated by A-life models. But they can't be replicated, so strong A-Life is impossible.

Sober's argument would be endorsed by autopoietic theorists. But to see metabolism as essential for life isn't necessarily to accept autopoietic philosophy. For this has some surprising implications, which many people reject. One was noted above, namely, the conflation of life and cognition. Another was remarked in Chapter 15.viii.b: the embargo on terms such as *input, output, function, feature-detector*, and *representation*. Two more concern features often listed in definitions of life: reproduction and evolution.

Maturana and Varela's claim that the formation of the cell membrane is *the* fundamental phenomenon of biology, and that life involves the "total subordination of [all the processes of change within] the system to the maintenance of its unity" (1972/1980: 97), implied that reproduction isn't essential for life. For them, this process is not (as functionalists claim) informational self-copying, but the formation of new autopoietic unities from previous ones. It follows that life is prior to reproduction (pp. 105–7). This wasn't a merely conceptual point, but a substantive biological hypothesis: that the earliest living organisms needn't have been able to reproduce (Boden 2000*b*).

Evolution, also, was seen by them as inessential, because it requires reproduction. (Inessential for life, but not for what's normally regarded as knowledge: they admitted that only evolution can generate the complex organisms typically credited with cognition.)

This conclusion, though unusual, is less controversial. For, *pace* Ray, and many theoretical biologists too (e.g. Maynard-Smith 1996), there are three independent arguments against defining life in terms of evolution. First, populations, not individual organisms, would be paradigm cases of life. Second, creationism would be conceptually incoherent, not just false. And third, a population in evolutionary equilibrium wouldn't count as alive.

d. Evolution, life, and mind

Some philosophers of A-Life, nevertheless, took evolution (together with metabolism) to be the sort of self-organization which characterizes life. Pattee was an early example, followed by his students Rosen (1985, 1991) and Cariani (1992, 1997). He'd modelled co-evolution in the 1960s (see Chapter 15.vi.a). Subsequently, he focused on the emergence of new phenotypic structures and functions.

A crucial example, for Pattee (1985), was novel types of "measurement", or classification. These were understood as ranging from enzyme activity to sensory perception—as in the evolution of new sensory organs (see Chapters 4.v.e and 15.vi.d). Pattee's concept of measurement was intriguingly similar to Smith's "participatory registration"—but, unlike Smith, he retained the first definition of computation distinguished in Section ix.a. So he specifically dismissed strong A-Life, arguing that measurement requires physical interaction, which can't be realized by formal computational systems. A fortiori, no novel biological functions can emerge in formal evolutionary systems (15.vi.d). He did allow, however, that "weak" A-Life modelling (simulation) could help clarify central biological and psychological concepts.

In the 1990s, another philosopher of A-Life argued that evolution is an essential criterion. Mark Bedau (1954–) explicitly accepted the three counter-intuitive implications mentioned above, because of the explanatory power gained by defining life in evolutionary terms (Bedau 1996). And this explanatory potential, he said, was augmented by A-Life. In presenting his account of "supple adaptation" (alias evolution), he argued that A-Life modelling can deepen our understanding of life as such, because it helps us to study evolution in dynamic and quantitative terms. Moreover, he extended his evolutionary argument from life to mind (Bedau 1999, in preparation).

A-Life philosophers weren't the only ones to link life and mind. Others, too, had grounded knowledge and meaning in biological evolution. Dennett had sketched an evolutionary account of meaning in *Content and Consciousness*, although philosophers then were more interested in other aspects of his work (see Section iv.a). By the mid-1980s, however, two influential examples of teleological or evolutionary semantics had appeared.

The philosopher of science David Papineau (1947–) argued that the content of beliefs depends on how they guide actions to satisfy desires, whose content is basically determined by natural selection (Papineau 1984, 1987). Similarly, Millikan (1933–) grounded intentionality in evolutionary history (Millikan 1984). Her book title was deliberately provocative: *Language, Thought, and Other Biological [sic] Categories*. This was guaranteed to raise philosophical hackles in devotees of the later Wittgenstein, and neo-Kantians in general (see Sections vi–viii, above).

Millikan upset many science-inclined naturalists too, by giving more philosophical weight to evolution than to neuroscience. Thus she argued that a perfect simulacrum of a human being, magically constituted in the middle of a swamp by a sudden combination of the relevant molecules, would have *no* beliefs, desires, or other intentional properties (1984: 93, 337–8; 1996; cf. Boorse 1976). It would, of course, utter the very same words as a human being would, if engaged in 'conversation'. For all the language-relevant events in the swamp-man's brain (and ears, and lips . . .) are, by hypothesis, identical with those of a person. But it wouldn't be a genuine conversation—for, on the swamp-man's side, no meanings or intentions would be being expressed. (In her defence, one could point out that we accept thermodynamics *even though* it allows the theoretical possibility of a snowball in hell: is swamp-man any more implausible?)

This imaginary example highlighted her central—and controversial—claim, that current meanings depend in part on events that happened millions of years ago. Millikan was saying, in effect, that Searle had been wrong about the "something more" that's needed *in principle* for intentionality. According to her, it's not neurochemistry as such that grounds meaning—nor even neurochemistry in interaction with the body and environment. Only evolutionary history can fix the system's semantics.

If Millikan's (or Papineau's) version of biological semantics is correct, then no 'ready-made' AI system, nor even a self-organizing—but non-evolutionary—A-Life system, could enjoy mind, intelligence, or meaning.

However, evolutionary semantics was later related to research in evolutionary robotics (Boden 2001). We saw in Chapter 15.vi.c that a robot's neural-network 'brain' may evolve 'feature-detectors' analogous to those found in mammalian visual cortex. So a mini-network may evolve that's sensitive to a light–dark gradient at an orientation matching one side of a white cardboard triangle, and that's used by the robots as a navigation aid (Harvey *et al.* 1994; Husbands *et al.* 1995). Such examples challenge Searle's (1980) view that the "meaning" of a computer model must always be derivative, and arbitrary to boot (see Section v.c).

One could debate whether the feature-detector means "light–dark gradient sloping up and to the right" as opposed to "left side of the white triangle". But similar difficulties attend the ascription of non-conceptual content to animals. (Are bug-detectors really *bug*-detectors, whether for the frog or for the frog's brain?—see Chapter 12.x.f and Cussins 1990: 416–17.)

The important point is that the various meanings one might want to ascribe to the robot aren't arbitrary. Nor are they derivative, based only in the human purposes involved in their design. They aren't based purely on causal regularities, either. They spring to mind as candidate meanings because the mini-networks concerned have evolved, within that task environment, to discriminate certain visual features and guide the robot's movements accordingly. That is, they're environmentally, enactively, and evolutionarily grounded.

However, to say these A-Life "meanings" aren't arbitrary isn't to say they're genuine. There's no consensus among A-Life researchers on whether evolutionary robotics could produce real intentionality. For the pure A-Life functionalist, it could: the triangle-detector is a primitive case, and more advanced (animal-like) examples would embody richer meanings. For Maturana, it couldn't: evolution and intentionality can occur only in biological organisms—so quasi-evolved robots can quasi-embody only quasi-meanings.

Nor is there a consensus among philosophers unconnected with A-Life, for the nature of life, mind, and the life–mind relation remain controversial.

Not everyone accepts an evolutionary semantics, for example. A causal semantics can't support the common-sense intuition that mind can arise *only* from life, unless the relevant causal relations can be shown to arise *only* in living things. And a model-theoretic semantics can't support it at all.

The competing A-Life methodologies of the early 1990s were systematically compared, and related to earlier philosophies of life, by Peter Godfrey-Smith (1994). He distinguished three dimensions of variation: internalism and externalism; asymmetrical and symmetrical externalism; and weak and strong versions of the continuity of life and mind.

Internalist approaches see life as autonomous self-organization, wherein internal constraints govern the history and interactions of the constituent units of the system. Examples include autopoietic theory and Stuart Kauffman's autocatalytic networks (15.viii.b). Externalist approaches explain the system's internal structure primarily as a result of its adaptive interactions with the environment. Work on evolutionary robotics is one example.

The asymmetric externalist emphasizes the organism's adaptive responses to its environment. By contrast, the symmetric externalist pays attention also to the active role of the adaptive organism in shaping that environment. Examples are situated robotics, and Ray's or Pattee's models of co-evolution, respectively.

Finally, the weak continuity theorist sees mind as emerging only from life, but as significantly different from it, whereas the strong continuity theorist regards mind and life as ontologically similar, sharing basic organizational principles. Descartes wasn't a continuity theorist at all, for he saw mind and living bodies as utterly distinct (see Chapter 2.iii and Matthews 1977). Examples of strong continuity theorists include the Naturphilosophen (Chapter 2.vi), the cybernetics movement (Chapter 4.v–vii), and autopoietic theorists. Arguably, they also include philosophers of non-conceptual content (Chapter 12.x.f) and participatory computation (see Section ix.e, above). And someone who argues that not all living things are cognitive systems (see above) is supporting weak continuity in that respect.

However, "mind" covers a number of abilities, and some of these may be strongly continuous with life whereas others aren't. Language has often been seen as a cut-off point. For instance, we saw in Chapter 2.ii.a, g, that Aristotle was a strong continuity theorist for perception and autonomous movement, but perhaps not for human reason (cf. Matthews 1992). Heideggerians who confine *Dasein* to human beings, or Wittgensteinians who ascribe intentionality only to linguistic concepts, count thus far as weak continuity theorists. But some neo-phenomenologists (such as Clark and Wheeler) ascribe intentionality to non-human animals, too.

Analogously, many AI connectionists allow that GOFAI insights will be needed to model the 'logical' aspects of human thinking (see Chapter 12.viii–ix), whereas some dynamical theorists deny this (Section vii.c, above). And nouvelle AI (a label recalling the minimalism of nouvelle cuisine) insists that AI must be grounded in 'lower' abilities, like those of our evolutionary precursors, whether or not it has to add GOFAI methods on top.

In sum, the relation between life and mind is still highly problematic. That applies to work in AI/A-Life, and to philosophy too. The common-sense view is that the one (*life*) is a precondition of the other (*mind*). But there's no generally accepted way of proving that to be so.

17

WHAT NEXT?

This chapter might have joined chapter XI of *Through the Looking Glass* as one of the shortest in the English language. For in response to "What next?", what is there to say but "Who knows!"? Fundamental advances, in particular, are unforeseeable. As Captain Cook's biographer put it, "Genius, of whatever sort, takes us unawares: is not, even in retrospect, deducible" (Beaglehole 1974: 3).

One doesn't have to adopt a Romanticist view of creativity, nor a literal interpretation of "genius", to agree with that. Creative ideas are unpredictable for a number of very different reasons, not all of which will be mentioned here (but see Boden 1990*a*, ch. 9). And some are more unpredictable than others. Even a carefully designed technological artefact will have some unpredictable features, as we saw in Chapter 8.v.b (Ziman 2000*b*). A relatively speculative creative idea can be more surprising still. (And its social impact is even less predictable: Tim Berners-Lee himself couldn't have foreseen that in a mere three years, from 1993 to 1996, the Web would grow from 130 sites to over 600,000 — Battelle 2005: 40.)

More precisely, *historically* creative ideas, never generated by anyone before, are unforeseeable. Creative ideas that are new only for the person concerned can sometimes be foreseen, and even deliberately brought about, by other people — think of Socratic dialogue, for example. In what follows, I'll use "creative" to mean historically creative.

17.i. What's Unpredictable?

One source of unpredictability is serendipity: the finding of something valuable without its being specifically sought. Since this is unexpected by definition, prediction simply isn't on the cards when it's involved. The classic case in science is Alexander Fleming's noticing the dirty dish of agar-jelly, which led eventually to the discovery of penicillin. In cognitive science, the part-accidental discovery of several visual-feature-detectors is another illustration (Chapter 14.iv.a−b). In both cases, of course, a good deal of careful and systematic work had to follow the initial observation, before anything worth calling a "discovery" could be achieved.

Another source of unpredictability is change in the wider cultural context. The generation — and still more, the acceptance — of new scientific ideas can be discouraged, encouraged, and even part-guided by social–political factors (see 1.iii.b−d and 2.ii.b−c).

The positive reception of heterarchy (10.iv.a) and distributed cognition (13.iii.d–e), for instance, was influenced by political ideology, and of expert systems by nationalism and economics (11.v). Even if cultural changes could be predicted (which they can't), their effects on contemporary scientific thinking could not.

This applies also to shifts in the power relations between the various groups/disciplines within cognitive science, and so to *what will be seen as cognitive science* in the future:

Some 25 years after its various beginnings there still is no such thing as a core cognitive science. Depending on where one looks, which departments one queries, who one's friends are, the core of cognitive science will be asserted to be neurophysiology, psychology, artificial intelligence, linguistics, or some more vague concept like human/machine interaction or symbolic or connectionist modeling. The result may not have been cognitive science, but it has been exciting and scientifically fruitful. It has created a community of interests and increased interdisciplinary communication. But as of now there are still viable independent cognitive sciences such as neurophysiology, linguistics, and psychology that flourish with or without the cognitive science label or affiliation. *It is difficult to say at this point where this will lead.* (G. Mandler 1996: 23; italics added)

A specially important type of cultural change concerns new technology. Sometimes, the new instruments are needed in order to do things that obviously needed doing—for instance, developing micro-electrodes to record from the cell body of a single neurone—or anyway, a very small number thereof (2.viii.e). Such cases are relatively predictable: it was clear that people would try, and probable that someone would eventually succeed. Other technological tools may come as more of a surprise, at least to people in other areas of science: a few biophysicists may have been able to predict brain-scanning techniques, but psychologists couldn't.

Technology can be used to prompt new concepts, as well as to find new data. Indeed, technical ideas have been transmuted into psychological theories on many occasions (Gigerenzer 1991*b*, 1994). Moreover, machines have been used as analogies for the brain for hundreds of years (Fryer 1978)—jukeboxes included (2.viii.f). The latest, of course, is the computer itself. So sceptics often say that it's just the latest in a long line of such analogies, to be displaced eventually by some unforeseen invention coming who knows when.

Well, yes and no. In cognitive science, the computer isn't merely a superficial analogy, a metaphor fished out of the memory—perhaps for purposes of popularization—after the real scientific work has been done. On the contrary, it provides substantive concepts in psychological and neuroscientific theories.

The computational concepts concerned were diverse even at the outset (4.ix), and have multiplied over the years (16.ix). Besides symbolic, connectionist, and evolutionary AI, they include dynamical systems described by differential equations (14.vi, 15.viii–ix and xi). The future may well hold unpredictable new machines, even less imaginable today than quantum computers are. But cognitive scientists believe that only *some sort of computational machine* will be relevant. For their key claim is that mind can be explained (not by today's ideas about computation, but) by *whatever theory turns out to be the best account of what computers do* (Chrisley 1999; see 16.ix.f). In that sense, they would endorse Philip Johnson-Laird's remark that "The computer is the last metaphor; it need never be supplanted" (1983: 10).

Other psychological reasons for unpredictability apply to all instances of creativity. They include the rich idiosyncrasy of human minds and the relative—though only *relative*—freedom of creative thinking (13.iv; cf. 7.i.g). Introspectively, it may seem as though almost anything can happen—at least, according to the molecular biologist François Jacob:

Day science employs reasoning that meshes like gears . . . One admires its majestic arrangement as that of a da Vinci painting or a Bach fugue. One walks about it as in a French formal garden . . . Night science, on the other hand, wanders blindly. It hesitates, stumbles, falls back, sweats, wakes with a start. Doubting everything . . . It is a workshop of the possible . . . where thought proceeds along sensuous paths, tortuous streets, most often blind alleys. (Jacob 1988: 296)

If the creative scientist himself "wanders blindly" much of the time, so much less can his thoughts be foreseen by other individuals—who don't even know his *present* thinking in much detail.

Moreover, some new ideas strike us as paradoxical, not to say crazy, even *after* they've occurred. That often happens in instances of "transformational" creativity, in which one or more dimensions of the previously accepted style of thinking is/are radically altered or dropped (Boden 1990*a*, chs. 3–4). The more basic the dimension, the more fundamental the conceptual change will be. In such cases, it's hard for the new idea to be understood, and even harder than usual for it to gain acceptance. A fortiori it's harder to predict.

In particle physics, that's par for the course. Freeman Dyson reported an encounter between Niels Bohr and Wolfgang Pauli, who'd given a lecture on his new theory:

Bohr rose to speak. "We are all agreed", he said to Pauli, "that your theory is crazy. The question which divides us is whether it is crazy enough. My own feeling is that it is not crazy enough." (Dyson 1958: 74)

And Dyson commented:

The objection that they are not crazy enough applies to all the attempts which have so far been launched at a radically new theory of elementary particles. It applies equally to crackpots. Most of the crackpot papers which are submitted to *The Physical Review* are rejected, not because it is impossible to understand them, but because it is possible. Those which are impossible to understand are usually published. When the great innovation appears, it will almost certainly be in a muddled, incomplete and confusing form. To the discoverer himself it will only be half-understood; for everybody else it will be a mystery. *For any speculation that does not at first glance look crazy, there is no hope.* (italics added)

Cognitive science as a whole is less rococo, less conceptually bizarre, than particle physics. But several seemingly crazy ideas have found a respected place within it, after being fiercely resisted as "absurd"—perceptual defence, for example (Chapters 6.ii and 16.v.f), and object-oriented programming (10.v.d and 13.v.d). And remember the punchline of the quip about *Plans and the Structure of Behavior*: ". . . and Pribram believed it!" (6.iv.c). When people said that, they weren't dismissing those new ideas as worthless. They were allowing that they were weird-but-interesting, so worth thinking about. (Karl Pribram was made the fall guy because he'd recently defended a holographic theory of memory—hardly the usual bread-and-butter fare: 12.v.c. Nor is Bergsonian philosophy, but Pribram later dallied with that as well: 16.x.a. His reputation as a maverick was deserved. However, even those who called him "crazy Karl Pribram" later admitted that his strange ideas had "caught on", and that "his neurophysiological speculations are decades beyond other physiological work"—Walter Weimer, interview in Baars 1986: 309–10.)

Many future contributions, too, will seem weird initially—though just how weird they'll need to be remains to be seen. The "particle physics" of the field is conscious experience. This has already prompted many highly counter-intuitive theories, including some crackpot publications. I argued in Chapter 14.xi.e that a currently undreamt-of (i.e. crazy) approach will be needed to explain it.

Close runners-up in order of difficulty, and so in licensed craziness, are intentionality and computation. We saw in Chapter 16.ix.e how an extraordinary (crazy?) theory of those-two-together has come from an AI scientist–philosopher who thinks that "For sheer ambition, physics does not hold a candle to computer or cognitive . . . science" (B. C. Smith 2002: 53).

Sometimes, experts declare future progress to be not so much unpredictable as impossible. This view was implicit in Thomas Watson's notorious remark in 1943, as IBM chairman, that "I think there is a world market for maybe five computers." (He died in 1956, so never knew just how wrong he was. But he wasn't alone: Howard Aiken, of all people, said, "there will never be enough problems, enough work, for more than one or two of these computers"—Edwards 1996: 66.) And it was explicit in the advice given to Konrad Zuse in 1937 by Kurt Pannke, a manufacturer of specialized calculators:

"Someone informed me", Dr. Pannke began, "that you have invented a computing machine. Now, I don't want to discourage you from continuing to work as an inventor and from developing new ideas, but I must go ahead and tell you one thing: in the field of computing machines, practically everything has been researched and perfected to the last detail. There's hardly anything left to invent . . ." (Zuse 1993: 42)

(To be fair to Pannke, he later changed his mind. He provided money to fund Zuse's home-based research, and recommended his machine to the German military—fortunately, with no effect: see 11.i.a.)

17.ii. What's Predictable?

I can't imagine anyone suggesting that there's "hardly anything left" to be discovered in cognitive science. But I've just allowed that creative ideas can't be predicted, only awaited. So perhaps I should now present you with an empty page, and leave it at that? After all, that's a respected rhetorical device. Laurence Sterne did it 250 years ago, when he declined to describe a beautiful woman in *The Life and Opinions of Tristram Shandy*, leaving it to the reader's imagination instead.

I don't have the courage to follow Sterne's example. But it wouldn't be appropriate in any case. For there is something that can be said.

All scientific research, in whatever domain, is located within some identifiable conceptual space where further creative exploration (and transformation) is clearly possible, and where some dimensions seem especially rich in potential with respect to current unsolved problems (Boden 1990a, 2004). Peer review, especially of proposals for future research, depends on that fact. We can't predict the detailed outcome of such explorations and transformations, much as David Livingstone couldn't foresee his discovery of the Victoria Falls. But we can reasonably expect that if we follow *these* dimensions of the space (compare: the Zambesi River, the mountains glimpsed ahead . . .), we'll find something of interest. That is, we can have intellectually defensible, if not infallible, hunches about where future discoveries are likely to occur.

In so far as such predictions are possible, I've indicated mine already. The previous chapters have told "the story so far"—but always with an eye to possible future episodes. So the relatively small volume of *recent* work that I've mentioned was chosen not just because it's recent, nor even because it's intriguing. It was selected because I think it's promising, capable of development in ways that seem likely to be fruitful.

One way of justifying our hunches about where interesting new ideas are likely to arise is to rely on sub-hunches about *how* those ideas might be generated. In other words, some specific exploratory pathways are recognizable as familiar ones, because they've often been found to be fruitful.

* For instance, once a simple deterministic space has been defined, it's very likely that people will eventually try to complexify it in certain ways. So when John von Neumann defined the basic cellular automaton, he knew very well that probabilistic and even evolutionary CAs would be explored later (Chapter 15.v–vi).

* Again, once problem solving had been seen as a simple hierarchy (6.iii), it was inevitable that more complex and/or 'open-execution' plan hierarchies would be explored (10.iii.c). It was even a good bet that theories of problem solving would eventually be transformed by hierarchy's being made less pure (10.iv.a), or perhaps deliberately dropped (13.iii.b, 15.viii.a).

* Third, when Alan Turing wrote his morphology paper, he knew that increasingly complex systems of reaction–diffusion equations would be explored, once computer power allowed (15.iv).

(Similar remarks apply to creativity in artistic contexts. So, for example, it was nigh inevitable that post-Renaissance composers would progressively complexify tonal harmony. And it was always on the cards that someone—it happened to be Arnold Schoenberg—would eventually transform the space of tonal music by dropping the home-key constraint altogether: C. Rosen 1976; Boden 1990a, ch. 4.)

In short, the common notion that creative thought is unpredictable because it's chaotic (in the everyday sense) is mistaken. There's significant method in creative madness. It's our tacit recognition of this fact which enables us to identify certain work as promising, even though we can't spell out the promises.

17.iii. What's Promising?

The recent empirical research that I see as promising in these terms includes the following (listed here in no particular order):

* hybrid systems I: symbolic/connectionist (Chapters 12.iii.d and ix.b, 15.viii.a)
* hybrid systems II: situated/deliberative (7.iv.b and 13.iii.c)
* hierarchical networks (12.viii.b and ix.b)
* connectionist work on the role of imagery of words (12.ix.e)
* statistical approaches to NLP (9.x, preamble, and 9.x.f)
* integration of connectionist learning with detailed neurophysiological data (14.ii.d)
* modular and/or time-based neural networks (12.ix.a, 14.ix.g)
* programmed/evolutionary neuromodulation (14.ix.f)
* AI-evolved organic–silicon computing networks (14.ix.f)
* computational neuro-ethology (14.vii, 15.vii)
* insect navigation strategies (15.viii.a)
* types of cerebral representation, especially emulators (14.viii)
* the epigenesis of thought and language in normal and brain-damaged children (7.vi.g–i)
* models of clinical apraxia and aphasia (12.ix.b, 14.x.b)
* theories of control in hypnosis (7.i.h)
* brain-scanning, *provided that* it's related to specific psychological theories (14.x.b)
* developmental trajectories (12.viii.c–e and x.e)
* fast/simple heuristics (7.iv.f–g)
* the origin of specific bounds on human rationality (7.iv.h)
* cognitive technology, including virtual reality (10.i.h, 13.vi, 16.vii.d)
* computational theories of creativity (9.iv.f, 13.iv)
* evolutionary modelling (14.ix.d and f, 15.vi)
* achieving open-ended evolution and/or creativity (13.iv.c, 15.vi.d)
* mathematical analyses of dynamical systems (14.vi and ix.b, 15.ix.b and xi.b)
* homeostasis in CTRNs (15.xi.a)
* distributed cognition and agents (8.iii, 12.ii–vi and x, 13.iii.c, 15.viii–ix)
* computational architectures integrating knowledge, motivation, and emotion (7.i.e–g and 7.iv.b–c).

If forced to choose only one of these items, I'd pick the last: work on integrated mental architectures. Indeed, I did that on the fiftieth anniversary of the 1953 discovery of the double helix, when the British Association invited several people to write 200 words for their magazine *Science and Public Affairs* on "what discovery/advance/development in their field they think we'll be celebrating in 50 years' time". This choice reflected my own long-standing interests in personality and psychopathology (Preface, ii). But it wasn't idiosyncratic: two years later, the UK's computing community voted for "The Architecture of Brain and Mind" as one of the seven "Grand Challenges" for the future (<http://www.ukcrc.org.uk/grand_challenges/index.cfm>). One member of the five-man committee carrying this project forward is Aaron Sloman, who's been thinking about architectural issues since the 1970s (7.i.f, 10.iv.b, and 16.ix.c). If progress is to be made on this front, my hunch is that his team will be in a good position to make it.

The Grand Challenges grew out of the UK government's 'Foresight' Programme (instituted in 2003 for a ten-to-twenty-year planning horizon), and in particular out of its Cognitive Systems Project. Naturally, government ministers aren't falling over themselves to help solve the problems of cognitive science for their own sake. For them, applications are all—whether in health, education, business, transport, arts and entertainment, or (of course) the military. But as the Project's official Report (DTI 2004) makes clear, scientific and technological motives are often very closely related (and can be satisfied only by interdisciplinary thinking). It should be no surprise, then, that architectures to support "emotional" robots and "social" human–computer interactions are now being investigated at the behest of Whitehall—and, naturally, of the Pentagon too.

The strength and range of the list of "promises" given above show that cognitive science is still a fruitful "scientific research programme" (Lakatos 1970). *Mind-as-machine*, in both its incarnations (1.ii.a), has generated many suggestive theories. These have been amended—and sometimes dropped—on the basis of further advances in our understanding, but in many cases the central insights remain. Marr's work on vision is one obvious example (7.v.b–f), but others have been described in previous chapters.

The field's potential won't be unlocked without new psychological–computational *theories*. Greater computer power may well be necessary, but it won't be sufficient. Even quantum computers and hypercomputers won't suffice to fill the bill (16.ix.a). Fundamental scientific advance will need more Ideas, not just more Bytes. Likewise, more and/or fancier PET/fMRI brain scanning won't suffice either, even though it will often be useful (14.ii.d and ix.c).

On a higher plane of abstraction, I've discussed some recent philosophical research concerning

* the nature of computation (16.ix)
* the variety of virtual machines (16.ix.a)
* conscious experience (14.x–xi)
* the nature of intentionality (16.x.d)
* the origin of conceptual content (12.ix.e and x.f)
* the nature/existence of non-conceptual content (12.x.f and 16.viii.b)
* mind and/as embodiment (16.vii)
* the boundary between self and world (16.vii.d)
* the nature of life, and its relation to mind (15.i and 16.x)
* the resolution/reintegration of neo-Kantian and analytic philosophical viewpoints (16.vii.b–d, ix.d–f, and x.a).

All of these matters will be key foci of effort and controversy in the foreseeable future. Indeed, they're so difficult, and so deep, that I expect them to remain key foci well over 100 years from now. For as remarked at the outset of Chapter 16, philosophical problems don't get solved in a hurry.

Nor, in these cases, will they get solved in disciplinary isolation. They'll require fundamental *and reciprocal* advances in up to five fields: philosophy, psychology, anthropology, neuroscience, and AI/A-Life. (Theoretical linguistics, as opposed to the philosophy of language, is less relevant here—unless we include *cognitive* linguistics: see 7.ii, preamble, 9.ix.g.)

17.iv. What About Those Manifesto Promises?

In Chapter 6.iv.c I said that a good way of judging how far cognitive science has succeeded is to compare it with the hopes/promises expressed in *Plans and the Structure of Behavior* (Miller *et al.* 1960). By the turn of the millennium, virtually all of MGP's promises had been at least partially met. The "satellite images", and the Newell Test, outlined in Chapter 7.vii surveyed many different examples.

To mention just two:

* hypnosis has been demystified (along with multiple personality and religious experience): (7.i.h–i, 8.vi.b, and 14.x.c), and
* MGP's distinction between Plans as animal instincts and as human purposes is now far better understood (7.i and iv, 14,vi.c, and 15.vii).

Although discussions of these matters have been hugely complicated since they wrote their manifesto, today's answers are broadly consistent with theirs. For TOTE units were—deliberately—defined so abstractly that they covered *both* the inbuilt sensori-motor skills of crickets and hoverflies *and* the deliberative (and hypnotic) planning of human beings.

Neither "demystified" nor "far better understood" implies that all the relevant questions have been answered. Far from it. But we're much clearer now about just how MGP's questions can be profitably put.

Consider, for example, their nature–nurture distinction mentioned above. This simplistic duality has given way—within cognitive science, if not yet in the minds of the general public—to an epigenetic view of development. This view was already waiting in the wings before cognitive science got started (5.ii.c). Now, it's prominent in disciplines as varied as psychology, neuroscience, philosophy, A-Life, and robotics (7.vi, 14.vii and ix.c, and 15.viii.a).

There's no reason why this process should cease now. And it doesn't require every psychological question to be answerable by a simulation. For MGP's futuristic remarks concerned a general approach to the mind: computational theorizing, not necessarily computer simulation as such. We'll surely see many new computer models (some of which will reflect new findings in neuroscience). We'll probably see radically new *types* of model (16.ix). Functioning computer models can test a theory's implications and coherence more rigorously than any other method (7.iii.c). But the novel theoretical concepts are what's important, in understanding what sort of system, or virtual/physical machine, the mind/brain is.

One thing is beyond dispute: that the rich subtlety of human minds is even more awe-inspiring than the arch-humanist Wilhelm von Humboldt (9.iv) believed it to be. Indeed, I've already identified this realization as the major result of computational psychology as a whole (7.vii.a). It follows that it will never be possible to capture every psychological detail, whether in a theory or a simulation. Predicting, explaining, or interpreting the specific thoughts/actions of individual people will always be largely "idiographic" (7.iii, preamble), a matter for the unargued intuitions of psychologists *qua* human beings, not for their deliberate conclusions *qua* scientists.

However, that doesn't spell disappointment for MGP. For on the one hand, idiographic insights can often be enriched, and sharpened, by considering general mechanisms. Remember, for instance, the varied ways of expressing different types of anxiety in speech (7.ii.c). On the other hand, the prediction/explanation of highly particular personal matters wasn't what MGP were aiming for. (Nor is this the aim of scientific psychology in general: 7.iii.d.) Rather, they hoped to understand how such phenomena are *possible*.

It's not only MGP's questions which can now be posed more fruitfully. The familiar puzzles that opened this story (1.i.a), many of them centuries old, have all been illuminated—and some even solved—by the successors of the visionary manifesto.

More answers will doubtless be found: the future of cognitive science will be as exciting as its past. But to say *what they'll be* would be like an eighteenth-century Admiralty Board foreseeing James Cook's extraordinary achievements in navigation and map-making: impossible.

REFERENCES

The page references attached to the quotations in my text almost all cite the original sources of journal papers; where a reprint is mentioned in the list of References, however, they refer to that. Similarly, page references to books cite the first edition unless otherwise specified.

ANON. (1973), 'The Darkvale Report on Applied Mathematics: A Cardboard Conference', *AISB European Newsletter*, 14 (July), 33–5. Pub. under a pseudonym, but written by M. B. Clowes: see 11.iv.b.

AARONSON, D., GRUPSMITH, E., and AARONSON, M. (1976), 'The Impact of Computers on Cognitive Psychology', *Behavioral Research Methods & Instrumentation*, 8: 129–38.

AARSLEFF, H. (1970), 'The History of Linguistics and Professor Chomsky', *Language*, 46: 570–85.

ABELSON, H., and DISESSA, A. (1980), *Turtle Geometry: The Computer as a Medium for Exploring Mathematics* (Cambridge, Mass.: MIT Press).

ABELSON, R. P. (1963), 'Computer Simulation of "Hot" Cognition', in Tomkins and Messick (1963: 277–98).

—— (1973), 'The Structure of Belief Systems', in Schank and Colby (1973: 287–339).

—— (1976), 'Script Processing in Attitude Formation and Decision-Making', in J. S. Carroll and J. W. Payne (eds.), *Cognition and Social Behavior* (Hillsdale, NJ: Lawrence Erlbaum), 33–46.

—— (1981*a*), 'The Psychological Status of the Script Concept', *American Psychologist*, 36: 715–29.

—— (1981*b*), 'Constraint, Construal, and Cognitive Science', *Proceedings of the Third Annual Conference of the Cognitive Science Society*, Berkeley, 1–9.

—— and CARROLL, J. D. (1965), 'Computer Simulation of Individual Belief Systems', *American Behavioral Scientist*, 8: 24–30.

—— and REICH, C. M. (1969), 'Implicational Molecules: A Method for Extracting Meaning from Input Sentences', *Proceedings of the International Joint Conference on Artificial Intelligence*, Washington, 641–7.

—— and ROSENBERG, M. J. (1958), 'Symbolic Psycho-Logic: A Model of Attitudinal Cognition', *Behavioral Science*, 3: 1–13.

ACKLEY, D. H., and LITTMAN, M. (1992), 'Interactions Between Learning and Evolution', in Langton et al. (1992: 1–23).

—— HINTON, G. E., and SEJNOWSKI, T. J. (1985), 'A Learning Algorithm for Boltzmann Machines', *Cognitive Science*, 9: 147–69; repr. in Anderson and Rosenfeld (1988: 638–49).

ADAMATZKY, A., DE LACY COSTELLO, B., MELHUISH, C., and RATCLIFFE, N. (2003), 'Liquid Brains for Robots', *AISB Quarterly*, no. 112 (Spring 2003), 5.

ADAMS, F. (2003), 'Semantic Paralysis', *Behavioral and Brain Sciences*, 26: 666–7.

ADAMS, H. (1900), 'The Dynamo and the Virgin', in H. Adams, *The Education of Henry Adams* (Boston: Houghton Mifflin, 1918), 379–90.

ADDISON, J., and STEELE, R. (1712), 'The Pleasures of the Imagination', *The Spectator*, vol. iii, nos. 411–42: 535–82.

ADORNO, T. W., FRENKEL-BRUNSWIK, E., LEVINSON, D. J., and SANFORD, R. N. (1950), *The Authoritarian Personality* (New York: Harper).

ADRIAN, E. D. (1914), 'The All-or-None Principle in Nerve', *Journal of Physiology*, 47: 460–74.

—— (1926), 'The Impulses Produced by Sensory Nerve Endings', *Journal of Physiology*, 61: 49–72.

—— (1934), 'Electrical Activity of the Nervous System', *Archives of Neurology and Psychiatry*, 32: 1125–36.

—— (1936), 'The Electrical Activity of the Cortex', *Proceedings of the Royal Society of Medicine*, 29: 197–200.

—— (1947), *The Physical Background of Perception* (Oxford: Clarendon Press).

—— and BRONK, D. W. (1928), 'The Discharge of Impulses in Motor Nerve Fibres. Part I. Impulses in Single Fibres of the Phrenic Nerve', *Journal of Physiology*, 66: 81–101.

—— and ZOTTERMAN, Y. (1926), 'The Impulses Produced by Sensory Nerve Endings. Part 2: The Response of a Single End-Organ', *Journal of Physiology*, 61: 151–71.

AGIN, G. J., and BINFORD, T. O. (1973), 'Computer Description of Curved Objects', *Proceedings of the Third International Joint Conference on Artificial Intelligence*, Stanford, Calif., 629–40.

Agre, P. E. (1988), *The Dynamic Structure of Everyday Life*, MIT AI Lab Technical Report 1085 (Cambridge, Mass.: MIT).

——— (1990), Review of Lucy Suchman, *Plans and Situated Actions: The Problem of Human–Machine Communication*, *Artificial Intelligence*, 43: 369–84.

——— (1993), 'The Symbolic Worldview: Reply to Vera and Simon', *Cognitive Science*, 17: 61–9.

——— (1995), 'Computational Research on Interaction and Agency', *Artificial Intelligence*, 72: 1–52; also in Agre and Rosenschein (1995: 1–52).

——— (1997), *Computation and Human Experience* (Cambridge: Cambridge University Press).

——— (1998–2003), 'Networking on the Network: A Guide on Professional Skills for PhD Students', available at <http://polaris.gseis.ucla.edu/pagre/>.

——— (2002), 'The Practical Logic of Computer Work', in Scheutz (2002: 129–42).

——— and Chapman, D. (1987), 'Pengi: An Implementation of a Theory of Activity', *AAAI-87*, Seattle, 268–72.

——— ——— (1991), 'What Are Plans For?', in Maes (1991c: 17–34).

——— and Rosenschein, S. J. (eds.) (1995), *Computational Theories of Interaction and Agency* (Cambridge, Mass.: MIT Press, 1996); first pub. as *Artificial Intelligence*, 72 and 73 (Jan. and Feb. 1995).

——— and Rotenberg, M. (1997), *Technology and Privacy: The New Landscape* (Cambridge, Mass.: MIT Press).

——— and Schuler, D. (1997), *Reinventing Technology, Rediscovering Community: Critical Explorations of Computing as a Social Practice* (London: Ablex).

Aida, H., Tanaka, H., and Moto-oko, T. (1983), 'A PROLOG Extension for Handling Negative Knowledge', *New Generation Computing*, 1: 87–91.

Aikens, D., and Ray, W. J. (2001), 'Frontal Lobe Contributions to Hypnotic Susceptibility', *International Journal of Clinical and Experimental Hypnosis*, 49: 320–9.

Ainsworth, C. (2003), 'The Facts of Life', *New Scientist*, 31 May, 28–31.

AISB Quarterly (2003), 'Father Hacker's Guide for the Young AI-Researcher', no. 112 (Spring 2003), 12.

Akins, K. A. (1993), 'What Is it Like to Be Boring and Myopic?', in Dahlbom (1993: 124–60).

Albert, B., Hames, R., Hill, K., Martins, L. L., Peters, J., Turner, T., and Borofsky, R. (2001), 'Roundtable Forum: Ethical Issues Raised by Patrick Tierney's *Darkness in El Dorado*. An Open Letter to our Anthropological Colleagues from the Roundtable's Participants (August 30, 2001)—Plus 19 Background Reports', *Public Anthropology: Engaging Issues, 2001*, available at <http://www.publicanthropology.org>, webmaster R. Borofsky.

Albus, J. S. (1971), 'A Theory of Cerebellar Function', *Mathematical Bioscience*, 10: 25–61.

——— (1975), 'A New Approach to Manipulator Control: The Cerebellar Model Articulation Controller (CMAC)', *Journal of Dynamic Systems, Measurement, and Control*, 97: 270–7.

Aleksander, I. (2000), *How to Build a Man: Dreams and Diaries* (London: Weidenfeld & Nicolson).

——— and Taylor, J. (eds.) (1992), *Artificial Neural Networks II: Proceedings of the International Conference on Artificial Neural Networks* (Amsterdam: Elsevier North-Holland).

Allen, C., and Bekoff, M. (1997), *Species of Mind: The Philosophy and Biology of Cognitive Ethology* (Cambridge, Mass.: MIT Press).

Allen, J. F., and Frisch, A. M. (1982), 'What's in a Semantic Network?', *Proceedings of the Twentieth Meeting of the Association for Computational Linguistics*, Toronto, 19–27.

Allport, D. A. (1980), 'Patterns and Actions', in G. Claxton (ed.), *Cognitive Psychology: New Directions* (London: Routledge & Kegan Paul), 26–64.

Allport, F. H. (1955), *Theories of Perception and the Concept of Structure* (New York: Wiley).

Allport, G. W. (1942), *The Use of Personal Documents in Psychological Science*, Social Science Research Council Bulletin 49 (New York).

——— (1946), 'Personalistic Psychology as Science: A Reply', *Psychological Review*, 53: 132–5.

——— (1951), *The Individual and his Religion: A Psychological Interpretation* (London: Constable). Based on the Lowell Lectures, Boston, 1947.

——— (1954), *The Nature of Prejudice* (Cambridge, Mass.: Addison-Wesley).

——— (1955), *Becoming: Basic Considerations for a Psychology of Personality* (New Haven: Yale University Press).

——— (1960), 'The Open System in Personality Theory', *Journal of Abnormal and Social Psychology*, 61: 301–10.

Almond, P. (2005), 'Get Fell In You 'Orrible Little Robot', *Sunday Times*, 20 Feb., Focus, 1.21.

ALONSO, E. (2002), 'AI and Agents: State of the Art', *AI Magazine*, 23: 3 (Fall), 25–9.

ALPER, M. (1996), *The 'God' Part of the Brain: A Scientific Interpretation of Human Spirituality and God* (New York: Rogue Press).

ALPERIN, J., BROWN, A., HUANG, J., and SANDY, S. (2001), *Bolt, Beranek, and Newman Inc.: A Case History of Transition*, MIT Structure of Engineering Revolutions Final Project Papers, available at <http://www.mit.edu/6.933/www/Fall2001/BBN.pdf>.

ALTICK, R. D. (1957), *The English Common Reader: A Social History of the Mass Reading Public 1800–1900* (Chicago: Chicago University Press).

Alvey Committee (1982), *A Programme for Advanced Information Technology: The Report of the Alvey Committee* (London: HMSO).

AMAREL, S. (1962), 'An Approach to Automatic Theory Formation', in M. C. Yovits and S. Cameron (eds.), *Self-Organizing Systems: Proceedings of an Interdisciplinary Conference*, Illinois, 5–6 May (Oxford: Pergamon Press), 107–75.

—— (1968), 'On Representations of Problems of Reasoning about Actions', in Michie (1968a: 131–72).

AMARI, S.-I. (1967), 'A Theory of Adaptive Pattern Classifiers', *IEEE Transactions on Electronic Computers*, 16: 299–307.

—— (1977), 'Neural Theory of Association and Concept-Formation', *Biological Cybernetics*, 26: 175–85.

—— and WU, S. (1999), 'Improving Support Vector Machine Classifiers by Modifying Kernel Functions', *Neural Networks*, 12: 783–9.

AMUNDSON, R. (1988), 'Logical Adaptationism', *Behavioral and Brain Sciences*, 11: 505–6.

ANDERSON, A. R. (ed.) (1964), *Minds and Machines* (Englewood Cliffs, NJ: Prentice-Hall).

ANDERSON, J. A. (1972), 'A Simple Neural Network Generating an Interactive Memory', *Mathematical Biosciences*, 14: 197–220; repr. in Anderson and Rosenfeld (1988: 181–92).

—— and ROSENFELD, E. (eds.) (1988), *Neurocomputing: Foundations of Research* (Cambridge, Mass.: MIT Press).

—— —— (eds.) (1998), *Talking Nets: An Oral History of Neural Networks* (Cambridge, Mass.: MIT Press).

—— PELLIONISZ, A., and ROSENFELD, E. (eds.) (1990a), *Neurocomputing 2: Directions for Research* (Cambridge, Mass.: MIT Press).

—— ROSSEN, M. L., VISCUSO, S. R., and SERENO, M. E. (1990b), 'Experiments with Representation in Neural Networks: Object Motion, Speech, and Arithmetic', in H. Haken and M. Stadler (eds.), *Synergetics of Cognition* (Berlin: Springer), 54–69; repr. in Anderson et al. (1990a: 705–16).

ANDERSON, JAMES R., KUWAHATA, H., KUROSHIMA, K., LEIGHTY, K. A., and FUJITA, K. (2005), 'Are Monkeys Aesthetists? Rensch (1957) Revisited', *Journal of Experimental Psychology: Animal Behavior Processes*, 31: 71–8. (Rensch 1957 is 'The Intelligence of Elephants', *Scientific American*, 196: 44–9; this discussed elephants' discrimination of, and memory for, geometrical patterns.)

ANDERSON, JOHN R. (1976), *Language, Memory, and Thought* (Hillsdale, NJ: Erlbaum).

—— (1978), 'Arguments Concerning Representations for Visual Imagery', *Psychological Review*, 85: 249–77.

—— (1979), 'Further Arguments Concerning Representations for Mental Imagery: A Response to Hayes-Roth and Pylyshyn', *Psychological Review*, 86: 395–406.

—— (1980), *Cognitive Psychology and its Implications*, 5th edn. (San Francisco: Freeman, 2000).

—— (1982), 'Acquisition of Cognitive Skill', *Psychological Review*, 89: 369–406.

—— (1983), *The Architecture of Cognition* (Cambridge, Mass.: Harvard University Press).

—— (ed.)(1993), *The Rules of the Mind* (Hillsdale, NJ: Lawrence Erlbaum).

—— (1995), *Learning and Memory: An Integrated Approach* (Chichester: Wiley).

—— and BOWER, G. H. (1972), 'Recognition and Retrieval Processes in Free Recall', *Psychological Review*, 79: 97–123.

—— —— (1973), *Human Associative Memory* (Washington: Winston).

—— and LEBIERE, C. (2003), 'The Newell Test for a Theory of Cognition', *Behavioral and Brain Sciences*, 26: 587–640.

—— KLINE, P., and BEASLEY, C. M. (1977/1980), *A Theory of the Acquisition of Cognitive Skills*, ONR Technical Report 77-1 (Ithaca, NY: Yale University Press, 1977); pub., updated, as 'Complex Learning Processes', in R. E. Snow, P. A. Federico, and W. E. Montague (eds.), *Aptitude, Learning, and Instruction*, ii (Hillsdale, NJ: Erlbaum Associates, 1980), 199–232.

ANDERSON, JOHN R., BOYLE, C. F., FARRELL, R., and RESIER, B. J. (1987), 'Cognitive Principles in the Design of Computer Tutors', in P. Morris (ed.), *Modelling Cognition* (Chichester: Wiley), 93–133.

_____ BOYLE, C. F., CORBETT, A., and LEWIS, M. W. (1990), 'Cognitive Modeling and Intelligent Tutoring', *Artificial Intelligence*, 42: 7–49.

_____ CORBETT, A., FINCHAM, J., HOFFMAN, D., and PELLETIER, R. (1992), 'General Principles for an Intelligent Tutoring Architecture', in J. W. Regian and V. J. Shute (eds.), *Cognitive Approaches to Automated Transfer* (Hillsdale, NJ: Lawrence Erlbaum), 81–106.

ANDRÉ, E. (ed.) (1999), *Animated Interface Agents: Making them Intelligent, Applied Artificial Intelligence*, special issue, 13/4–5.

ANDREAE, J. H. (1963), 'STeLLA: A Scheme for a Learning Machine', *Proceedings of the 2nd IFAC [International Federation for Automatic Control] Congress*, Basle, 497–502; officially pub. in V. Broida, D. H. Barlow, and O. Schafer (eds.), *Automatic and Remote Control*, ii: *Theory* (London: Butterworths, 1964), 497–502.

_____ (1969a), 'A Learning Machine with Monologue', *International Journal of Man–Machine Studies*, 1: 1–20.

_____ (1969b), 'Learning Machines—A Unified View', in A. R. Meetham and R. A. Hudson (eds.), *Encyclopedia of Information, Linguistics, and Control* (Oxford: Pergamon Press).

_____ (1977), *Thinking with the Teachable Machine* (London: Academic Press).

_____ (1987), 'Design of a Conscious Robot', *Metascience*, 5: 41–54.

_____ (1998), *Associative Learning for a Robot Intelligence* (London: Imperial College Press).

_____ and CLEARY, J. G. (1976), 'A New Mechanism for a Brain', *International Journal of Man–Machine Studies*, 8: 89–119.

ANGYAN, A. J. (1959), 'Machina Reproducatrix', in Blake and Uttley (1959: ii. 933–44).

ANSCOMBE, G. E. M. (1957), *Intention* (Oxford: Blackwell).

_____ (1974), 'Comment on Professor R. L. Gregory's Paper', in S. C. Brown (ed.), *Philosophy of Psychology* (London: Macmillan), 211–20, 235 ff., 243–4.

ANTONIOU, G., and WILLIAMS, M.-A. (1997), *Nonmonotonic Reasoning* (Cambridge, Mass.: MIT Press).

APTER, J. T. (1946), 'Eye Movements Following Strychninization of the Superior Colliculus of Cats', *Journal of Neurophysiology*, 9: 73–85.

APTER, M. J. (1970), *The Computer Simulation of Behaviour* (London: Hutchinson University Library).

ARBIB, M. A. (1961), 'Turing Machines, Finite Automata and Neural Nets', *Journal of the Association for Computing Machinery*, 8: 467–75.

_____ (1964), *Brains, Machines, and Mathematics* (New York: McGraw-Hill; 2nd edn., Berlin: Springer-Verlag, 1987).

_____ (1966), 'A Simple Self-Reproducing Universal Automaton', *Information and Control*, 9: 177–89.

_____ (1969), 'Self-Reproducing Automata: Some Implications for Theoretical Biology', in Waddington (1966–72: ii. 204–26).

_____ (1972), *The Metaphorical Brain: An Introduction to Cybernetics as Artificial Intelligence and Brain Theory* (New York: Interscience).

_____ (1981a), 'Visuomotor Coordination: From Neural Nets to Schema Theory', *Cognition and Brain Theory*, 4: 23–39.

_____ (1981b), 'Perceptual Structures and Distributed Motor Control', in V. B. Brooks (ed.), *Handbook of Physiology: The Nervous System II* (Bethesda, Miss.: American Physiological Society), 1449–80.

_____ (1982), 'Modelling Neural Mechanisms of Visuomotor Coordination in Frog and Toad', in S. Amari and M. A. Arbib (eds.), *Competition and Cooperation in Neural Nets*, Lecture Notes in Biomathematics 45 (Berlin: Springer-Verlag), 342–70.

_____ (1985), *In Search of the Person: Philosophical Explorations in Cognitive Science* (Amherst: University of Massachusetts Press).

_____ (1987), 'Levels of Modelling of Visually Guided Behavior', *Behavioral and Brain Sciences*, 10: 407–65.

_____ (1992), 'Schema Theory', in S. Shapiro (ed.), *The Encyclopedia of Artificial Intelligence*, ii (Chichester: Wiley), 1427–43.

_____ (1994), Review of A. Newell, *Unified Theories of Cognition*, in W. Clancey, S. Smoliar, and M. Stefik (eds.), *Contemplating Minds* (Cambridge, Mass.: MIT Press), 21–39.

_____ (1995), 'Schema Theory: From Kant to McCulloch and Beyond', in R. Moreno-Diaz and J. Mira-Mira (eds.), *Brain Processes, Theories and Models: An International Conference in Honor of W. S. McCulloch 25 Years After his Death* (Cambridge, Mass.: MIT Press), 11–23.

_____ (2000), 'Warren McCulloch's Search for the Logic of the Nervous System', *Perspectives in Biology and Medicine*, 43: 193–216.

_____ (2002*a*), 'The Mirror System, Imitation, and the Evolution of Language', in C. Nehaniv and K. Dautenhahn (eds.), *Imitation in Animals and Artefacts* (Cambridge, Mass.: MIT Press), 229–80.

_____ (2002*b*), 'From *Rana Computatrix* to *Homo Loquens*', in R. Damper and D. Cliff (eds.), *Biologically-Inspired Robotics: The Legacy of W. Grey Walter*, Proceedings of the EPSRC/BBSRC International Workshop WGW-02, Bristol, 12–31; rev. version: '*Rana Computatrix* to Human Language: Towards a Computational Neuroethology of Language', *Philosophical Transactions of the Royal Society*, ser. A, 361, special issue: *Biologically Inspired Robotics* (2003), 2345–79.

_____ (2003), 'Brain, Meaning, Grammar, Evolution [*sic*]', *Behavioral and Brain Sciences*, 26: 668–9.

ARBIB, M. A. (2005), 'From Monkey-Like Action Recognition to Human Language: An Evolutionary Framework for Neurolinguistics', incl. peer commentary and reply, *Behavioral and Brain Sciences*, 28: 105–7.

_____ and AMARI, S.-I. (1985), 'Sensorimotor Transformations in the Brain (with a Critique of the Tensor Theory of Cerebellum)', *Journal of Theoretical Biology*, 112: 123–55.

_____ and COBAS, A. (1991), 'Schemas for Prey-Catching in Frog and Toad', in Meyer and Wilson (1991: 142–51).

_____ and HESSE, M. B. (1986), *The Construction of Reality*, The Gifford Lectures, 1983 (Cambridge: Cambridge University Press).

_____ and LEE, H. B. (1993), 'Anuran Visuomotor Coordination for Detour Behavior: From Retina to Motor Schemas', in J.-A. Meyer, H. Roitblat, and S. W. Wilson (eds.), *From Animals to Animats 2*, Proceedings of the Second International Conference on Simulation of Adaptive Behavior (Cambridge, Mass.: MIT Press), 42–51.

_____ and LIAW, J.-S. (1995), 'Sensorimotor Transformations in the Worlds of Frogs and Robots', *Artificial Intelligence*, 72: 53–79.

_____ BOYLLS, C. C., and DEV, P. (1974), 'Neural Models of Spatial Perception and the Control of Movement', in W. D. Keidel, W. Handler, and M. Spreng (eds.), *Cybernetics and Bionics* (Munich: Oldenbourg), 216–31.

_____ IBERALL, T., and LYONS, D. (1985), 'Coordinated Control Programs for Movements of the Hand', in A. W. Goodwin and I. Darian-Smith (eds.), *Experimental Brain Research*, suppl., 10: 111–29.

_____ CONKLIN, E. J., and HILL, J. C. (1987), *From Schema Theory to Language* (Oxford: Oxford University Press).

_____ ERDI, P., and SZENTAGOTHAI, J. (1997), *Neural Organization: Structure, Function, and Dynamics* (Cambridge, Mass.: MIT Press).

ARKIN, R. C. (1998), *Behavior-Based Robotics* (Cambridge, Mass.: MIT Press).

ARMER, P. (1960/1963), 'Attitudes Toward Intelligent Machines', *Symposium on Bionics, 1960*, WADD [Wright Air Development Division, US Air Force] Technical Report 60/600, 13–19; repr. in Feigenbaum and Feldman (1963: 389–405); page references are to this version.

ARMSTRONG, D. M. (1968), *A Materialist Theory of the Mind* (London: Routledge & Kegan Paul).

ARNAULD, A. (1662), *The Art of Thinking: Port-Royal Logic*, trans. J. Dickoff and P. James (Indianapolis: Bobbs-Merrill, 1964).

ASADA, M., KITANO, H., NODA, I., and VELOSO, M. (1999), 'RoboCup: Today and Tomorrow—What We Have Learned', *Artificial Intelligence*, 110: 193–214.

_____ VELOSO, M., TAMBE, M., NODA, I., KITANO, H., and KRAETSCHMAR, G. K. (2000), 'Overview of RoboCup-98', *AI Magazine*, 21: 9–19.

ASCH, S. E. (1956), 'Studies of Independence and Conformity: A Minority of One Against a Unanimous Majority', *Psychological Monographs*, 70, no. 9.

ASCOTT, R. (1964), 'The Construction of Change', *Cambridge Opinion*, 1, special issue: *Modern Art in Britain*, 37–42; repr. as ch. 1 of Ascott (2003: 97–107).

_____ (1966/1967), 'Behaviourist Art and the Cybernetic Vision', *Cybernetica*, 9 (1966), 247–64; 10 (1967), 25–56; repr. as ch. 3 of Ascott (2003: 109–56).

_____ (1990), 'Is there Love in the Telematic Embrace?', *Art Journal* (College Arts Association of America), 49: 241–7; repr. as ch. 15 of Ascott (2003: 232–46).

_____ (1998/2003), 'Technoetic Aesthetics: 100 Terms and Definitions for the Post-Biological Era', in Ascott (2003: 375–82); first pub. in Japanese, trans. E. Fujihara, in R. Ascott, *Art and Telematics: Toward the Construction of a New Aesthetics* (Tokyo: NTT Publishing), 1998.

Ascott, R. (2003), *Telematic Embrace: Visionary Theories of Art, Technology, and Consciousness* (London: University of California Press).

Ashby, W. R. (1940), 'Adaptiveness and Equilibrium', *Journal of Mental Science*, 86: 478–83.

—— (1947), 'The Nervous System as a Physical Machine; with Special Reference to the Origin of Adaptive Behaviour', *Mind*, 56: 44–59.

—— (1948), 'Design for a Brain', *Electronic Engineering*, 20: 379–83.

—— (1952), *Design for a Brain: The Origin of Adaptive Behaviour* (London: Wiley; 2nd edn., rev., London: Chapman, 1960).

—— (1956), *An Introduction to Cybernetics* (London: Chapman & Hall).

—— (1960), *Design for a Brain: The Origin of Adaptive Behaviour*, 2nd edn., rev. (London: Chapman & Hall).

—— (1962), 'Principles of the Self-Organizing System', in H. von Foerster and G. W. Zopf (eds.), *Principles of Self-Organization* (New York: Pergamon Press), 255–78.

—— (*c.*1968), unpub. MS (marked as being "already with the translators"), archives of the Burden Institute, Bristol.

Ashley, K. D. (1990), *Modeling Legal Argument* (Cambridge, Mass.: MIT Press).

—— and Keefer, M. (1996), 'Ethical Reasoning Strategies and their Relation to Case-Based Instruction: Some Preliminary Results', *Proceedings of the Eighteenth Annual Conference of the Cognitive Science Society* (Mahwah, NJ: Lawrence Erlbaum), 483–8.

Aspray, W. (1990), *John von Neumann and the Origins of Modern Computing* (Cambridge, Mass.: MIT Press).

Astington, J. W., Harris, P. L., and Olson, D. R. (eds.) (1988), *Developing Theories of Mind* (Cambridge: Cambridge University Press).

Atanasoff, J. V. (1940), 'Computing Machine for the Solution of Large Systems of Linear Algebraic Equations', in Randell (1982: 315–35).

Atkinson, A. P., and Wheeler, M. W. (2004), 'The Grain of Domains: The Evolutionary-Psychological Case Against Domain-General Cognition', *Mind and Language*, 19/2: 147–76.

Atkinson, R. C. (1999), 'The Golden Fleece, Science Education, and U.S. Science Policy', *Proceedings of the American Philosophical Society*, 143: 407–17.

Atran, S. (1990), *Cognitive Foundations of Natural History: Towards an Anthropology of Science* (Cambridge: Cambridge University Press).

—— (1998), 'Folkbiology and the Anthropology of Science: Cognitive Universals and Cultural Particulars', *Behavioral and Brain Sciences*, 21: 547–609.

—— (2002), *In Gods We Trust: The Evolutionary Landscape of Religion* (Oxford: Oxford University Press).

Attneave, F. (1954), 'Some Informational Aspects of Visual Perception', *Psychological Review*, 61: 183–93.

—— (1959), *Applications of Information Theory to Psychology: A Summary of Basic Concepts, Methods, and Results* (New York: Holt, Rinehart and Winston).

Attwell, D., and Laughlin, S. B. (2001), 'An Energy Budget for Signalling in the Grey Matter of the Brain', *Journal of Cerebral Blood Flow and Metabolism*, 21: 1133–45.

Augier, M., and March, J. G. (eds.) (2004), *Models of a Man: Essays in Memory of Herbert A. Simon* (Cambridge, Mass.: MIT Press).

Aunger, R. (ed.) (2000), *Darwinizing Culture: The Status of Memetics as a Science* (Oxford: Oxford University Press).

Austin, J. L. (1957), 'A Plea for Excuses', *Proceedings of the Aristotelian Society*, 57: 1–30.

—— (1962*a*), *How to Do Things with Words*, The William James Lectures, 1955 (Oxford: Clarendon Press).

—— (1962*b*), *Sense and Sensibilia*, reconstructed from the manuscript notes by G. J. Warnock (Oxford: Oxford University Press).

Avis, J., and Harris, P. L. (1991), 'Belief–Desire Reasoning Among Baka Children: Evidence for a Universal Conception of Mind', *Child Development*, 62: 460–7.

Ayer, A. J. (1936), *Language, Truth, and Logic* (London: Victor Gollancz).

Azzopardi, P., and Cowey, A. (1997), 'Is Blindsight Like Normal, Near-Threshold Vision?', *Proceedings of the National Academy of Sciences USA*, 94: 14190–4.

Baars, B. J. (1986), *The Cognitive Revolution in Psychology* (London: Guilford Press).

—— (1988), *A Cognitive Theory of Consciousness* (Cambridge: Cambridge University Press).

___ (2001), *In the Theatre of Consciousness: The Workspace of the Mind* (Oxford: Oxford University Press).

BABBAGE, C. B. (1826), 'On a Method of Expressing by Signs the Action of Machinery', *Philosophical Transactions of the Royal Society*, 116: 250–65.

___ (1830), *Reflections on the Decline of Science in England, and on Some of its Causes* (London: printed for B. Fellowes).

___ (1832/1835), *On the Economy of Machinery and Manufactures*, 4th edn. (London: Charles Knight, 1835; 1st edn., 1832); first pub. as several chapters in the *Encyclopaedia Metropolitana* (1829).

___ (1838), *The Ninth Bridgwater Treatise: A Fragment*, 2nd edn. (New York: New York University Press; 1st edn., London: Murray, 1837); repr. as M. Campbell-Kelly (ed.), *The Works of Charles Babbage*, ix (London: Pickering & Chatto, 1991).

___ (1851), *The Exposition of 1851; or, Views of the Industry, the Science and the Government, of England* (London: John Murray).

___ (1864), *Passages from the Life of a Philosopher*, ed. M. Campbell-Kelly (London: William Pickering, 1994; first pub. London: Longman, Green).

BACHANT, J., and McDERMOTT, J. P. (1984), 'R1 Revisited: Four Years in the Trenches', *AI Magazine*, 5/3 (Fall), 21–32.

BACHMAN, P. A., WALDE, P., LUISI, P. L., and LANG, J. (1990), 'Self-Replicating Reverse Micelles and Chemical Autopoiesis', *Journal of the American Chemical Society*, 112: 8200–1.

___ LUISI, P. L., and LANG, J. (1992), 'Autocatalytic Self-Replicating Micelles as Models for Prebiotic Structures', *Nature*, 357: 57–9.

BACH-Y-RITA, P. (1984), 'The Relationship Between Motor Processes and Cognition in Tactile Visual Substitution', in W. Prinz and A. F. Sanders (eds.), *Cognition and Motor Processes* (Berlin: Springer-Verlag), 149–60.

___ (2002), 'Sensory Substitution and Qualia', in A. Noe and E. Thompson (eds.), *Vision and Mind: Selected Readings in the Philosophy of Perception* (Cambridge, Mass.: MIT Press), 497–514.

BACON, F. (1605), *The Advancement of Learning*, in J. Spedding, R. L. Ellis, and D. D. Heath (eds.), *The Works of Francis Bacon* (Clair Shores, Mich.: Scholarly Press, 1969).

___ (1620), *The New Organon*, ed. and trans. L. Jardine and M. Silverthorne (Cambridge: Cambridge University Press, 2000). Bacon's Latin subtitle (*sive indicia vera de interpretatione naturae*) is variously translated as *True Suggestions for the Interpretation of Nature* (W. Wood, William Pickering, 1844) and *Or a True Guide to the Interpretation of Nature* (G. W. Kitchin, Oxford University Press, 1855) etc.

___ (1624), *New Atlantis*, in J. Spedding, R. L. Ellis, and D. D. Heath (eds.), *The Works of Francis Bacon* (Clair Shores, Mich.: Scholarly Press, 1969).

BADDELEY, A. (2001), 'Memories of Memory Research', in G. C. Bunn, A. D. Lovie, and G. D. Richards (eds.), *Psychology in Britain: Historical Essays and Personal Reflections* (London: BPS Books/Science Museum), 344–52.

___ and HITCH, A. D. (1974), 'Working Memory', in G. H. Bower (ed.), *The Psychology of Learning and Motivation*, viii (New York: Academic Press), 47–89.

BAECK, T., FOGEL, D. B., and MICHALEWICZ, Z. (eds.) (2000), *Evolutionary Computation*, 2 vols. (Oxford: Oxford University Press/Institute of Physics Publishing).

BAERNSTEIN, H. D., and HULL, C. L. (1931), 'A Mechanical Model of the Conditioned Reflex', *Journal of General Psychology*, 5: 99–106.

BAGLEY, J. D. (1967), 'The Behavior of Adaptive Systems which Employ Genetic and Correlation Agorithms', Ph.D. diss., University of Michigan.

BAKER, G., and MORRIS, K. J. (1996), *Descartes' Dualism* (London: Routledge).

BAKER, J. K. (1975), 'The DRAGON System: An Overview', *IEEE Transactions on Acoustics, Speech, and Signal Processing*, ASSP-23/1: 24–9.

BALDWIN, J. M. (1915), *Genetic Theory of Reality: Being the Outcome of Genetic Logic as Issuing in the Aesthetic Theory of Reality Called Pancalism, with an Extended Glossary of Terms* (London: G. P. Putnam's Sons).

BALLARD, D. H. (1989), 'Reference Frames for Animate Vision', *Proceedings of the Eleventh International Joint Conference on Artificial Intelligence*, 1635–41.

___ (1991), 'Animate Vision', *Artificial Intelligence*, 48: 57–86.

___ HINTON, G. E., and SEJNOWSKI, T. J. (1983), 'Parallel Visual Computation', *Nature*, 306: 21026.

BALLING, J. D., and FALK, J. A. (1982), 'Development of Visual Preference for Natural Environments', *Environment and Behavior*, 14: 5–28.

BALLONOFF, P. A. (ed.) (1974), *Mathematical Models of Social and Cognitive Structures: Contributions to the Mathematical Development of Anthropology*, papers presented to the 1972 meeting of the American Anthropological Association, Toronto (Urbana: University of Illinois Press).

BALLS, M. (1983), 'Alternatives to Animal Experimentation', *Alternatives to Laboratory Animals*, 11: 56–62.

BALZ, A. G. A. (1951), *Cartesian Studies* (New York: Columbia University Press).

BAR-HILLEL, Y. (1951), 'The State of Machine Translation in 1951', *American Documentation*, 2: 229–37; repr. in Bar-Hillel (1964: 153–79).

—— (1953), 'Machine Translation', *Computers and Automation*, 2: 1–6.

—— (1959), 'Discussion on the Paper by Mr. R. H. Richens', in Blake and Uttley (1959: i. 303–7).

—— (1960), 'The Present Status of Automatic Translation of Languages', *Advances in Computers*, 1: 91–163; appendix repr. in Bar-Hillel (1964: 174–9).

—— (1963), *Measures of Syntactic Complexity* (Jerusalem: Hebrew University); rev. in Booth (1967: 29–50).

—— (1964), *Language and Information* (Reading, Mass.: Addison-Wesley).

—— (1970), *Aspects of Language: Essays and Lectures on Philosophy of Language, Linguistic Philosophy, and Methodology of Linguistics* (Jerusalem: Magnes Press).

BARINGA, M. (1990), 'The Mind Revealed?', *Science*, 249: 856–8.

BARKER, E. (1989), *New Religious Movements: A Practical Introduction* (London: HMSO).

—— and WARBURG, M. (eds.) (1998), *New Religions and New Religiosity* (Aarhus: Aarhus University Press).

BARKER, R. G. (1963), 'On the Nature of the Environment', *Journal of Social Issues*, 19: 17–38.

—— (1968), *Ecological Psychology: Concepts and Methods for Studying the Environment of Human Behavior* (Stanford, Calif.: Stanford University Press).

BARKOW, J. H. (1973), 'Darwinian Psychological Anthropology: A Biosocial Approach', *Current Anthropology*, 14: 373–88.

—— (1975), 'Prestige and Culture: A Biosocial Interpretation', *Current Anthropology*, 16: 553–72.

—— (1976), 'Attention Structure and Internal Representations', in M. R. Chance and R. R. Larsen (eds.), *The Social Structure of Attention* (London: Wiley), 203–19.

—— (1983), 'Begged Questions in Behavior and Evolution', in G. C. L. Davey (ed.), *Animal Models of Human Behavior: Conceptual, Evolutionary, and Neurobiological Perspectives* (Chichester: Wiley), 205–22.

—— (1989), *Darwin, Sex, and Status: Biological Approaches to Mind and Culture* (Toronto: University of Toronto Press).

—— (1992), 'Beneath New Culture is Old Psychology: Gossip and Social Stratification', in Barkow *et al.* (1992: 627–37).

—— COSMIDES, L., and TOOBY, J. (1992), *The Adapted Mind: Evolutionary Psychology and the Generation of Culture* (Oxford: Oxford University Press).

BARLOW, H. B. (1953), 'Summation and Inhibition in the Frog's Retina', *Journal of Physiology*, 119: 69–88.

—— (1959), 'Sensory Mechanisms, the Reduction of Redundancy, and Intelligence', in Blake and Uttley (1959: ii. 535–9).

—— (1961), 'The Coding of Sensory Messages', in W. H. Thorpe and O. L. Zangwill (eds.), *Current Problems in Animal Behaviour* (Cambridge: Cambridge University Press), 331–60.

—— (1972), 'Single Units and Sensation: A Neuron Doctrine for Perceptual Psychology?', *Perception*, 1: 371–94.

—— (1988), 'Neuroscience: A New Era?', review of Gerald Edelman, *Neural Darwinism*, *Nature*, 331: 571.

—— (1995), 'The Neuron Doctrine in Perception', in Gazzaniga (1995: 415–35).

—— (2001*a*), 'The Exploitation of Regularities in the Environment by the Brain', *Behavioral and Brain Sciences*, 24, special issue: *The Work of Roger Shepard: The Case for Cognitive Universals*, 602–7.

—— (2001*b*), 'Redundancy Reduction Revisited', *Network: Computation in Neural Systems*, 12, special issue: *Statistics of Natural Stimuli*, 241–53.

—— and GARDNER-MEDWIN, A. (2000), 'Localist Representation Can Improve Efficiency for Detection and Counting', *Behavioral and Brain Sciences*, 23: 467–8.

—— and LEVICK, W. R. (1965), 'The Mechanism of Directionally Selective Units in Rabbit's Retina', *Journal of Physiology*, 178: 477–504.

—— and PETTIGREW, J. D. (1971), 'Lack of Specificity of Neurones in the Visual Cortex of Young Kittens', *Journal of Physiology*, 218: 98–100.

—— HILL, R. M., and LEVICK, W. R. (1964), 'Retinal Ganglion Cells Responding Selectively to Direction and Speed of Image Motion in the Rabbit', *Journal of Physiology*, 173: 377–407.

—— BLAKEMORE, C., and PETTIGREW, J. D. (1967), 'The Neural Mechanism of Binocular Depth Discrimination', *Journal of Physiology*, 193: 327–42.

BARNETT, L. (1998), 'Ruggedness and Neutrality—The NKp Family of Fitness Landscapes', *Proceedings of Sixth International Conference on Artificial Life*, 18–27.

BARON, R. J. (1981), 'Mechanisms of Human Facial Recognition', *International Journal of Man–Machine Studies*, 15: 137–78.

BARON-COHEN, S. (1995), *Mindblindness: An Essay on Autism and Theory of Mind* (Cambridge, Mass.: MIT Press).

—— (2000*a*), 'Theory of Mind and Autism: A Fifteen-Year Review', in Baron-Cohen *et al.* (2000: 3–20).

—— (2000*b*), 'Autism: Deficits in Folk Psychology Exist Alongside Superiority in Folk Physics', in Baron-Cohen *et al.* (2000: 73–82).

—— (2002), 'The Extreme Male Brain Theory of Autism', *Trends in Cognitive Science*, 6: 248–54.

—— (2003), *The Essential Difference: Men, Women and the Extreme Male Brain* (London: Allen Lane).

—— and JAMES, I. (2003), 'Einstein and Newton Showed Signs of Asperger's Syndrome', *New Scientist*, 3 May, 10. This is based on an interview with the 'authors', but was written by a journalist.

—— LESLIE, A. M., and FRITH, U. (1985), 'Does the Autistic Child Have a "Theory of Mind"?', *Cognition*, 21: 37–46.

—— RING, H., MORIARTY, J., SCHMITZ, B., COSTA, D., and ELL, P. (1994), 'The Brain Basis Theory of Mind: The Role of the Orbitofrontal Region', *British Journal of Psychiatry*, 165: 640–9.

—— TAGER-FLUSBERG, H., and COHEN, D. J. (eds.) (2000), *Understanding Other Minds: Perspectives from Developmental Cognitive Neuroscience*, 2nd edn. (Oxford: Oxford University Press).

BARR, A., and FEIGENBAUM, E. A. (eds.) (1981), *The Handbook of Artificial Intelligence*, i (London: Pitman).

—— —— (eds.) (1982), *The Handbook of Artificial Intelligence*, ii (London: Pitman).

BARROW, H. G. (1987), 'Learning Receptive Fields', *Institute of Electronics and Electrical Engineers First Conference on Neural Nets*, San Diego, iv. 115–21.

—— (1989), 'AI, Neural Networks and Early Vision', *AISB Quarterly*, no. 69: 6–25.

—— and BRAY, A. J. (1992*a*), 'A Model of Adaptive Development of Complex Cortical Cells', in Aleksander and Taylor (1992: 881–4).

—— —— (1992*b*), 'Activity-Induced "Colour Blob" Formation', in Aleksander and Taylor (1992: 884–7).

—— —— (1993), 'An Adaptive Neural Model of Early Visual Processing', in F. H. Eeckman and J. M. Bower (eds.), *Computation and Neural Systems* (Amsterdam: Kluwer Academic), 171–5.

—— —— (1996), *Models of Synaptic Development in Early Visual Cortex*, Cognitive Science Research Paper 332 (Brighton: University of Sussex).

—— and CRAWFORD, G. F. (1972), 'The Mark 1.5 Edinburgh Robot Facility', in Meltzer and Michie (1972: 465–80).

—— and POPPLESTONE, R. J. (1971), 'Relational Descriptions in Picture Processing', in Meltzer and Michie (1971: 377–96).

—— and SALTER, S. H. (1969), 'Design of Low-Cost Equipment for Cognitive Robot Research', in Meltzer and Michie (1969*b*: 555–66).

BARSALOU, L. W. (1999*a*), 'Perceptual Symbol Systems', *Behavioral and Brain Sciences*, 22: 577–609.

—— (1999*b*), 'Language Comprehension: Archival Memory or Preparation for Situated Action?', *Discourse Processes*, 28: 61–80.

BARSKY, R. F. (1997), *Noam Chomsky: A Life of Dissent* (Cambridge, Mass.: MIT Press).

BARTH, F. G. (2002), *A Spider's World: Senses and Behavior*, trans. A. Biederman-Thorson (London: Springer).

BARTH, F. G., HUMPHREY, J. A. C., and SECOMB, T. W. (eds.) (2003), *Sensors and Sensing in Biology and Engineering* (Vienna: Springer-Verlag).

BARTHES, R. (1968/1977), 'The Death of the Author', in R. Barthes, *Image, Music, Text*, ed. and trans. S. Heath (New York: Hill & Wang), 142–8; this essay was originally written in 1968.

BARTLETT, F. C. (1932), *Remembering: A Study in Experimental and Social Psychology* (Cambridge: Cambridge University Press).

—— (1946), Obituary notice: K. J. W. Craik, *British Journal of Psychology*, 36: 109–16.

—— (1958), *Thinking: An Experimental and Social Study* (London: Unwin University Books).

BARWISE, J., and PERRY, J. (1984), *Situations and Attitudes* (Cambridge, Mass.: MIT Press).

BASALLA, G. (1988), *The Evolution of Technology* (Cambridge: Cambridge University Press).

BASTIN, E. W. (1960), 'Abstract Concerning a Self-Organizing Machine (CASPAR) Having a Hierarchical Structure of Algebraic Levels', in M. C. Yovits and S. Cameron (eds.), *Self-Organizing Systems* (Oxford: Pergamon Press), 316–18.

—— (1969), 'A General Property of Hierarchies', in C. H. Waddington (ed.), *Toward a Theoretical Biology*, ii: *Sketches* (Edinburgh: Edinburgh University Press), 252–65.

BATES, A. M., BOWDEN, B. V., STRACHEY, C. S., and TURING, A. M. (1953), 'Digital Computers Applied to Games', in Bowden (1953: 286–310).

BATES, E., BENIGNI, L., BRETHERTON, I., CAMAIONI, L., and VOLTERRA, V. (1979), *The Emergence of Symbols: Cognition and Communication in Infancy* (London: Academic Press).

BATESON, G. (1936), *Naven: A Survey of the Problems Suggested by a Composite Picture of the Culture of a New Guinea Tribe drawn from Three Points of View* (Cambridge: Cambridge University Press).

BATESON, G. (1971), 'The Cybernetics of "Self": A Theory of Alcoholism', *Psychiatry*, 34: 1–18.

—— (1972), *Steps to an Ecology of Mind* (New York: Ballantine Books).

—— (1979), *Mind and Nature: A Necessary Unity* (London: Wildwood House).

—— JACKSON, D. D., HALEY, J., and WEAKLAND, J. H. (1956), 'Toward a Theory of Schizophrenia', *Behavioral Science*, 1: 201–27.

BATTELLE, J. (2005), *The Search: How Google and its Rivals Rewrote the Rules of Business and Transformed our Culture* (London: Nicholas Breasley).

BATTERSBY, C. (1978), 'Hume's Easy Philosophy: Ease and Inertia in Hume's Newtonian Science of Man', D.Phil. thesis, University of Sussex.

—— (1979), 'Hume, Newton and "The Hill Called Difficulty" ', in S. C. Brown (ed.), *Philosophers of the Enlightenment*, Royal Institute of Philosophy Lectures, 12 (Hassocks: Harvester Press), 31–55.

BAUER, F. L., and WOSSNER, H. (1972), 'The "Plankalkul" of Konrad Zuse: A Forerunner of Today's Programming Languages', *Communications of the Association for Computing Machinery*, 15: 678–85.

BAWDEN, A., GREENBLATT, R. D., HOLLOWAY, J., KNIGHT, T., MOON, D., and WEINREB, D. (1977), *LISP Machine Progress Report*, AIM-444 (Cambridge, Mass.: MIT AI Lab); officially pub. as 'The LISP Machine', in P. H. Winston and R. H. Brown (eds.), *Artificial Intelligence: An MIT Perspective*, ii (Cambridge, Mass.: MIT Press), 345–76.

BAYLOR, G. W., and GASCON, J. (1974), 'An Information Processing Theory of the Development of Weight Seriation in Children', *Cognitive Psychology*, 6: 1–40.

—— and LEMOYNE, G. (1975), 'Experiments in Seriation with Children: Towards an Information Processing Explanation of the Horizontal Decalage', *Canadian Journal of Behavioral Science*, 7: 4–29.

—— GASCON, J., LEMOYNE, G., and POTHIER, N. (1973), 'An Information Processing Model of Some Seriation Tasks', *Canadian Psychologist*, 14: 167–96.

BEACH, F. A., HEBB, D. O., MORGAN, C. T., and NISSEN, H. W. (eds.) (1960), *The Neuropsychology of Lashley* (New York: McGraw-Hill).

BEACH, K. (1988), 'The Role of External Mnemonic Symbols in Acquiring an Occupation', in M. M. Gruneberg and R. N. Sykes (eds.), *Practical Aspects of Memory*, i (New York: Wiley), 342–6.

BEAGLEHOLE, J. C. (1974), *The Life of Captain James Cook* (London: A & C Black).

BEANEY, M. (ed.) (1997), *The Frege Reader* (Oxford: Blackwell).

BEAUDOIN, L. P. (1994), 'Goal Processing in Autonomous Agents', Ph.D. thesis, School of Computer Science, University of Birmingham, available at <http://www.cs.bham.ac.uk/research/cogaff/>.

BECHTEL, W., and MUNDALE, J. (1999), 'Multiple Realizability Revisited: Linking Cognitive and Neural States', *Philosophy of Science*, 66: 175–207.

—— ABRAHAMSEN, A., and GRAHAM, G. (1998), 'The Life of Cognitive Science', in W. Bechtel and G. Graham (eds.), *A Companion to Cognitive Science* (Oxford: Blackwell), 1–104.

BECK, A. T. (1976), *Cognitive Therapy and the Emotional Disorders* (New York: Meridian Books).

BECKER, J. D. (1973), 'A Model for the Encoding of Experiential Information', in Schank and Colby (1973: 396–446).

—— (1975), 'The Phrasal Lexicon', in R. C. Schank and B. L. Nash-Webber (eds.), *Theoretical Issues in Natural Language Processing: An Interdisciplinary Workshop in Computational Linguistics, Psychology, Linguistics, and Artificial Intelligence*, 10–13 June, Cambridge, Mass. (Arlington, Va.: Association for Computational Linguistics), 70–3.

BECKETT, A. (forthcoming), 'Stafford Beer and Chile's Internet' (provisional title), in M. W. Wheeler *et al.* (forthcoming).

BEDAU, M. A. (1996), 'The Nature of Life', in Boden (1996: 332–57).

—— (1999), 'Supple Laws in Biology and Psychology', in V. Hardcastle (ed.), *Where Biology Meets Psychology: Philosophical Essays* (Cambridge, Mass.: MIT Press), 287–302.

—— (in preparation), *The Suppleness of Evolution, Life and Mind* (Cambridge, Mass.: MIT Press).

—— MCCASKILL, J. S., PACKARD, N. H., RASMUSSEN, S., ADAMI, C., GREEN, D. G., IKEGAMI, T., KANEKO, K., and RAY, T. S. (2000), 'Open Problems in Artificial Life', *Artificial Life*, 6: 363–76.

BEDINI, S. A. (1964), 'Automata in the History of Technology', *Technology and Culture*, 5: 24–42.

BEER, R. D. (1990), *Intelligence as Adaptive Behavior: An Experiment in Computational Neuroethology* (Boston: Academic Press).

—— (1995*a*), 'A Dynamical Systems Perspective on Agent–Environment Interaction', *Artificial Intelligence*, 72: 173–215.

—— (1995*b*), 'Computational and Dynamical Languages for Autonomous Agents', in Port and van Gelder (1995: 121–48).

—— (1995*c*), 'On the Dynamics of Small Continuous-Time Recurrent Neural Networks', *Adaptive Behavior*, 3: 469–509.

—— (2004), 'Autopoiesis and Cognition in the Game of Life', *Artificial Life*, 10: 309–26.

—— (2005), talk given at the Workshop on Dynamical Systems and Adaptive Behaviour, University of Sussex, 8 Mar.

—— and CHIEL, H. J. (1991), 'The Neural Basis of Behavioral Choice in an Artificial Insect', in J.-A. Meyer and S. W. Wilson (eds.), *From Animals to Animats*, Proceedings of the First International Conference on Simulation of Adaptive Behavior (Cambridge, Mass.: MIT Press), 247–54.

—— and GALLAGHER, J. C. (1992), 'Evolving Dynamical Neural Networks for Adaptive Behavior', *Adaptive Behavior*, 1: 91–122.

BEER, S. (1959), *Cybernetics and Management* (London: English Universities Press).

—— (1999), 'On the Nature of Models: Let Us Now Praise Famous Men and Women, from Warren McCulloch to Candace Pert', *Informing Science*, 2/3: 69–82.

BEHAR, R. (1996), 'Anthropology that Breaks your Heart', in R. Behar, *The Vulnerable Observer: Anthropology that Breaks your Heart* (Boston: Beacon Press), 161–77.

Behavioral and Brain Sciences (1988), 'Open Peer Commentary [on Dennett's *Intentional Stance*]', 11: 505–34.

—— (2000), 'Open Peer Commentary [on Arbib *et al.* (1997)]', 23: 533–71.

BELL, C. (1811), 'Idea of a New Anatomy of the Brain' (privately published); repr. in W. Dennis (ed.), *Readings in the History of Psychology* (New York: Appleton-Century-Crofts, 1948), 113–24; first generally pub. in *Journal of Anatomy and Physiology*, 3 (1869), 153–66.

—— (1834), *The Hand: Its Mechanism and Vital Endowments as Evincing Design*, The Fourth Bridgwater Treatise on the Power, Wisdom, and Goodness of God, as Manifested in the Creation, 3rd edn. (London: William Pickering; 1st edn., 1833).

BELLIN, D., and CHAPMAN, G. (eds.) (1987), *Computers in Battle: Will they Work?* (Boston: Harcourt Brace Jovanovich).

BELLUGI, U., and STUDDERT-KENNEDY, M. (eds.) (1980), *Signed and Spoken Language: Biological Constraints on Linguistic Form*, report of the Dahlen Workshop on Sign Language and Spoken Language, Berlin 1980 (Weinheim: Verlag Chemie).

—— MARKS, S., BIHRLE, A. M., and SABO, H. (1988), 'Dissociation Between Language and Cognitive Functions in Williams Syndrome', in D. Bishop and K. Mogford (eds.), *Language Development in Exceptional Circumstances* (Edinburgh: Churchill Livingstone), 132–49.

BELLUGI, U., BIHRLE, A., JERNIGAN, T., TRAUNER, T., and DOHERTY, S. (1991), 'Neuropsychological, Neurological, and Neuroanatomical Profile of Williams Syndrome', *American Journal of Medical Genetics Supplement*, 6: 115–25.

BELLWOOD, P. (1987), *The Polynesians: Prehistory of an Island People*, rev. edn. (London: Thames & Hudson).

BENNETT, J. A. (1976), *Linguistic Behavior* (Cambridge: Cambridge University Press).

BENSAUDE-VINCENT, B. (forthcoming), 'Machinery and Biomimetism: Two Alternative Paradigms of Nanotechnologies', in J. Riskin (ed.), *The Sistine Gap: Essays in the History and Philosophy of Artificial Life* (Stanford, Calif.: Stanford University Press).

BENT RUSSELL, S. (1913), 'A Practical Device to Simulate the Working of Nervous Discharges', *Journal of Animal Behavior*, 3: 15–35.

——(1917a), 'Compound Substitution in Behavior', *Psychological Review*, 24: 62–73.

——(1917b), 'Advanced Adaptation in Behavior', *Psychological Review*, 24: 413–25.

BERANEK, L. (2005), 'BBN's Earliest Days: Founding a Culture of Engineering Creativity', *IEEE Annals of the History of Computing*, 27/2: 52–64.

BERGSON, H. (1911a), *Creative Evolution*, trans. from French A. Mitchell (London: Macmillan; first pub. 1907).

——(1911b), *Matter and Memory*, trans. from French N. M. Paul and W. S. Palmer (London: Allen & Unwin); first pub., subtitled *Essai sur la relation du corps á l'esprit* (1906).

——(1935), *The Two Sources of Morality and Religion*, trans. from French R. A. Audra and C. Brereton (London: Macmillan; first pub. 1932).

——(1946), *The Creative Mind: Essays*, trans. from French M. L. Andison (New York: Philosophical Library; first pub. 1934).

BERKELEY, E. C. (1949), *Giant Brains or Machines that Think* (London: Chapman & Hall).

——(1956), *Small Robots—Report* (Newtonville, Mass.: Berkeley Enterprises), available at <http://www.blinkenlights.com/classiccmp/berkeley/report.html>.

BERLIN, I. (1999), *The Roots of Romanticism* (London: Chatto & Windus).

BERLIN, O. B. (1972), 'Speculations on the Growth of Ethnobotanical Nomenclature', *Journal of Language and Society*, 1: 63–98.

——(1974), 'Further Notes on Covert Categories and Folk Taxonomies: A Reply to Brown', *American Anthropologist*, 76: 327–9.

——(1978), 'Ethnobiological Classification', in E. Rosch and B. B. Lloyd (eds.), *Cognition and Categorization* (Hillsdale, NJ: Lawrence Erlbaum), 9–26.

——and KAY, P. (1969), *Basic Color Terms: Their Universality and Evolution* (Berkeley: University of California Press).

——BREEDLOVE, D., and RAVEN, P. H. (1973), 'General Principles of Classification and Nomenclature in Folk Biology', *American Anthropologist*, 75: 214–42.

BERLINER, H. J., and BEAL, D. F. (eds.) (1990), *Computer Chess, Artificial Intelligence*, special issue, 43: 1–123.

BERNARD, C. (1865), *An Introduction to the Study of Experimental Medicine*, trans. H. C. Green (1927; New York: Dover, 1957).

BERNE, E. (1964), *Games People Play: The Psychology of Human Relationships* (New York: Castle).

——(1972), *What do you Say After you Say Hello? The Psychology of Human Destiny* (New York: Grove Press).

BERNERS-LEE, T. (1989), 'Information Management: A Proposal', internal report, CERN, Switzerland; repr. in Packer and Jordan (2001: 189–205).

——(2000), *Weaving the Web: The Original Design and Ultimate Destiny of the World Wide Web by its Inventor*, written with Mark Fischetti (New York: HarperBusiness).

BERNSTEIN, A., and ROBERTS, M. DE V. (1958), 'Computer vs. Chess-Player', *Scientific American*, 198: 96–105.

——ROBERTS, M. DE V., ARBUCKLE, T., and BELSKY, M. S. (1958), 'A Chess Playing Program for the IBM 704', *Proceedings of the 1958 Western Joint Computer Conference*, Los Angeles, 157–9.

BERNSTEIN, J. (1981a), 'Profile of Marvin Minsky', *New Yorker*, 14 Dec., 48–126.

——(1981b), *The Analytical Engine: Computers—Past, Present and Future*, newly rev. (New York: William Morrow; 1st edn. 1963).

BERSINI, H., and DETOURS, V. (1994), 'Asynchronicity Induces Stability in Cellular Automata Based Models', *Proceedings of Fourth International Conference on Artificial Life*, 382–7.

BERTENTHAL, B. I., CAMPOS, J. J., and BARRETT, K. (1984), 'Self-Produced Locomotion: An Organizer of Emotional, Cognitive, and Social Developments in Infancy', in R. N. Emde and R. Harmon (eds.), *Continuities and Discontinuities in Development* (London: Plenum Press), 175–210.

BESNARD, P. (1989), *An Introduction to Default Logic* (London: Springer-Verlag).

BEURLE, R. L. (1956), 'Properties of a Mass of Cells Capable of Regenerating Pulses', *Philosophical Transactions of the Royal Society*, ser. B, 240: 55–94.

——— (1959), 'Storage and Manipulation of Information in the Brain', *Journal of the Institution of Electrical Engineers*, 5: 75–82.

BEVER, T. G., LACKNER, J. R., and KIRK, R. (1969), 'The Underlying Structures of Sentences Are the Primary Units of Immediate Speech Processing', *Perception and Psychophysics*, 5: 225–31.

BIEBRICHER, C. K., DICKMANN, S., and LUCE, R. (1982), 'Structural Analysis of Self-Replicating RNA Synthesized by QBeta Replicase', *Journal of Molecular Biology*, 154: 629–48.

BINFORD, T. O. (1971), 'Visual Perception by Computer', paper presented to the Institute of Electronics and Electrical Engineers Conference on Systems and Control, Miami, Dec.

BINSTED, K. (1996), 'Machine Humour: An Implemented Model of Puns', Ph.D. thesis, University of Edinburgh.

——— and RITCHIE, G. D. (1997), 'Computational Rules for Punning Riddles', *Humor: International Journal of Humor Research*, 10: 25–76.

——— PAIN, H., and RITCHIE, G. D. (1997), 'Children's Evaluation of Computer-Generated Punning Riddles', *Pragmatics and Cognition*, 5: 305–54.

BIRD, J., and LAYZELL, P. (2002), 'The Evolved Radio and its Implications for Modelling the Evolution of Novel Sensors', *Proceedings of Congress on Evolutionary Computation*, CEC-2002, 1836–41.

BISHOP, G. H. (1946), 'Nerve and Synaptic Conduction', *Annual Review of Physiology*, 8: 355–74.

BISIACH, E. (1993), 'Mental Representation in Unilateral Neglect and Related Disorders', *Quarterly Journal of Experimental Psychology*, 46A: 435–61.

BISSON, T. (1991), 'Alien/Nation', *Omni*, Apr.

BLACK, J. B., and BOWER, G. H. (1980), 'Story Understanding as Problem Solving', *Poetics*, 9: 223–50.

——— and WILENSKY, R. (1979), 'An Evaluation of Story Grammars', *Cognitive Science*, 3: 213–30.

BLACK, M. (1959), 'Linguistic Relativity: The Views of Benjamin Lee Whorf', *Philosophical Review*, 68: 228–38.

——— (1970), 'Comment on Chomsky's "Problems of Explanation in Linguistics" ', in R. Borger and F. Cioffi (eds.), *Explanation in the Behavioural Sciences* (Cambridge: Cambridge University Press), 452–61.

BLACKBURN, S. W. (2000), Review of M. Heidegger, *Contributions to Philosophy, New Republic*, 30 Oct.

——— (2005), *Truth: A Guide for the Perplexed*, The Gifford Lectures 2004 (London: Allen Lane).

BLACKMORE, S. J. (1999), *The Meme Machine* (Oxford: Oxford University Press).

BLACKWELL, A., and RODDEN, K. (eds.) (2003), *Sketchpad: A Man–Machine Graphical Communication System—Ivan Edward Sutherland*, Technical Report no. 574 (UCAM-CL-TR-574), University of Cambridge Computer Laboratory, available at <http://www.cl.cam.ac.uk/TechReports/UCAM-CL-TR-574>; electronic version of Ph.D. thesis, MIT, Jan. 1963.

BLAKE, D. V., and UTTLEY, A. M. (eds.) (1959), *The Mechanization of Thought Processes*, 2 vols., Proceedings of a Symposium held at NPL on 24–7 Nov. 1958, National Physical Laboratory Symposium no. 10 (London: HMSO).

BLAKEMORE, C., and COOPER, G. F. (1970), 'Development of the Brain Depends on the Visual Environment', *Nature*, 228: 477–8.

BLAKEMORE, S.-J., WOLPERT, D. M., and FRITH, C. D. (1998), 'Central Cancellation of Self-Produced Tickle Sensation', *Nature Neuroscience*, 1: 635–40.

——— OAKLEY, D. A., and FRITH, C. D. (2003), 'Delusions of Alien Control in the Normal Brain', *Neuropsychologia*, 41: 1058–67.

BLOCH, B. (1949), 'Obituary of Leonard Bloomfield', *Language*, 25: 87–94.

BLOCH, M. E. F. (1991), 'Language, Anthropology and Cognitive Science', *Man*, 26: 183–98.

——— (1998), *How We Think They Think: Anthropological Approaches to Cognition, Memory and Literacy* (Oxford: Westview Press).

BLOCH, M. E. F. (2000), 'A Well-Disposed Social Anthropologist's Problems with Memes', in Aunger (2001: 189–203).

BLOCK, H. D. (1962), 'The Perceptron: A Model for Brain Functioning. I', *Reviews of Modern Physics*, 34: 123–35.

——(1970), 'A Review of "Perceptrons"', *Information and Control*, 17: 501–22.

——KNIGHT, B. W., and ROSENBLATT, F. (1962), 'Analysis of a Four-Layer Series Coupled Perceptron. II', *Reviews of Modern Physics*, 34: 135–42.

BLOCK, N. (1978), 'Troubles with Functionalism', in C. W. Savage (ed.), *Perception and Cognition: Issues in the Foundations of Psychology*, Minnesota Studies in the Philosophy of Science, 9 (Minneapolis: University of Minnesota Press), 261–325.

——(ed.) (1981), *Imagery* (Cambridge, Mass.: MIT Press).

——(1982), 'Psychologism and Behaviorism', *Philosophical Review*, 90: 5–43.

——(1983), 'Mental Pictures and Cognitive Science', *Philosophical Review*, 4: 499–541.

——(2001), 'Paradox and Cross Purposes in Recent Work on Consciousness', *Cognition*, 79: 197–219.

BLOMFIELD, S. J. W., and MARR, D. C. (1970), 'How the Cerebellum May Be Used', *Nature*, 227: 1224–48.

BLOOMFIELD, L. (1913), Review of Wilhelm Wundt, *Elemente der Völkerpsychologie*, *American Journal of Psychology*, 24: 449–53.

——(1914), *An Introduction to the Study of Language* (New York: Henry Holt).

——(1926), 'A Set of Postulates for the Science of Language', *Language*, 2: 153–64.

——(1931), Obituary of Albert Paul Weiss, *Language*, 7: 219–21.

——(1933), *Language* (London: Allen & Unwin).

——(1939), *Linguistic Aspects of Science*, vol. i/4 of O. Neurath (ed.), *International Encyclopedia of Unified Science* (Chicago: University of Chicago Press).

BLOOR, D. (1973), 'Wittgenstein and Mannheim and the Sociology of Mathematics', *Studies in the History and Philosophy of Science*, 4: 173–91.

BLUM, H. (1973), 'Biological Shape and Visual Science, Part I', *Journal of Theoretical Biology*, 38: 205–87.

BOAL, I., CLARK, T. J., MATTHEWS, J., and WATTS, M. (2005), 'Blood for Oil?', *London Review of Books*, 27/8 (21 Apr.), 12–16.

BOAS, F. (ed.) (1911), *Handbook of American Indian Languages*, i (Washington: Smithsonian Institution).

——(1927), *Primitive Art* (Oslo: H. Aschehoug).

——(1949), 'Hero's *Pneumatica*: A Study of its Transmission and Influence', *Isis*, 40: 38–48.

BOBROW, D. G. (1968), 'Natural Language Input for a Computer Problem-Solving System', in Minsky (1968: 146–226).

——(1972), 'Requirements for Advanced Programming Systems for List Processing', *Communications of the Association for Computing Machinery*, 15: 618–27.

——(ed.) (1984), *Qualitative Reasoning About Physical Systems* (Amsterdam: North-Holland).

——and BRADY, J. M. (1998a), Editorial (to 100th volume), *Artificial Intelligence*, 100: 1–3.

————(1998b), *Artificial Intelligence 40 Years Later*, *Artificial Intelligence*, special issue, 103: 1–362.

——and COLLINS, A. M. (eds.) (1975), *Representation and Understanding: Studies in Cognitive Science* (New York: Academic Press).

——and HAYES, P. J. (1985), 'Artificial Intelligence—Where Are We?', *Artificial Intelligence*, 25: 375–415.

——and RAPHAEL, B. (1964), 'A Comparison of List-Processing Computer Languages: Including a Detailed Comparison of COMIT, IPL-V, LISP 1.5, and SLIP', *Communications of the Association for Computing Machinery*, 7: 231–40.

——and WINOGRAD, T. (1977), 'An Overview of KRL, a Knowledge Representation Language', *Cognitive Science*, 1: 3–46.

BOCK, G., and MARSH, J. (eds.) (1993), *Experimental and Theoretical Studies of Consciousness* (Chichester: Wiley).

BODEN, M. A. (1959), 'In Reply to Hart and Hampshire', *Mind*, ns 68: 256–60.

——(1962), 'The Paradox of Explanation', *Proceedings of the Aristotelian Society*, ns 62: 159–78.

——(1965), 'McDougall Revisited', *Journal of Personality*, 33: 1–19.

——(1966), 'Optimism', *Philosophy*, 41: 291–303.

——(1970), 'Intentionality and Physical Systems', *Philosophy of Science*, 37: 200–14.

——(1972), *Purposive Explanation in Psychology* (Cambridge, Mass.: Harvard University Press).

——(1977), *Artificial Intelligence and Natural Man* (New York: Basic Books; 2nd edn., expanded, London: MIT Press; New York: Basic Books, 1987).

____ (1978), 'Human Values in a Mechanistic Universe', in G. Vesey (ed.), *Human Values: Royal Institute of Philosophy Lectures 1976–77* (Brighton: Harvester Press), 135–71.

____ (1979), *Piaget*, Fontana Modern Masters (London: Fontana; 2nd edn., enlarged, 1994).

____ (1980), 'Real-World Reasoning', in L. J. Cohen and M. B. Hesse (eds.), *Applications of Inductive Logic* (Oxford: Oxford University Press), 359–75.

____ (1981*a*), 'Implications of Language Studies for Human Nature', in M. A. Boden, *Minds and Mechanisms: Philosophical Psychology and Computational Models* (Ithaca, NY: Cornell University Press), 174–90.

____ (1981*b*), 'The Case for a Cognitive Biology', in M. A. Boden, *Minds and Mechanisms: Philosophical Psychology and Computational Models* (Ithaca, NY: Cornell University Press), 88–112.

____ (1981*c*), 'The Structure of Intentions', *Journal for the Theory of Social Behaviour*, 3: 23–46.

____ (1983), 'Artificial Intelligence and Animal Psychology', *New Ideas in Psychology*, 1: 11–33; first pub. as Cognitive Science Research Paper no. 16, University of Sussex, 1981.

____ (1984*a*), 'On Modelling Institutions (Comment on Fararo and Skvoretz, "Institutions as Production Systems")', *Journal of Mathematical Sociology*, 10 (1984), 205–10.

____ (1984*b*), 'Artificial Intelligence and Social Forecasting', *Journal of Mathematical Sociology*, 9: 341–56.

____ (1985), 'Artificial Intelligence and Legal Responsibility', part of a panel discussion on 'Legal Implications of AI', *Proceedings of the Ninth International Joint Conference on Artificial Intelligence*, Los Angeles, 1267–80.

____ (1987), *Artificial Intelligence and Natural Man*, 2nd edn., enlarged (London: MIT Press).

____ (1988), *Computer Models of Mind: Computational Approaches in Theoretical Psychology* (Cambridge: Cambridge University Press).

____ (1990*a*), *The Creative Mind: Myths and Mechanisms* (London: Weidenfeld & Nicolson; 2nd edn., expanded, rev., London: Routledge, 2004).

____ (ed.) (1990*b*), *The Philosophy of Artificial Intelligence* (Oxford: Oxford University Press).

____ (1991), 'Horses of a Different Color?', in W. Ramsey, D. E. Rumelhart, and S. P. Stich (eds.), *Philosophy and Connectionist Theory* (Hillsdale, NJ: Lawrence Erlbaum), 3–19.

____ (1994*a*), *Piaget*, 2nd edn., enlarged (London: HarperCollins).

____ (1994*b*), 'What Is Creativity?', in Boden (1994*c*: 75–118).

____ (1994*c*), *Dimensions of Creativity* (Cambridge, Mass.: MIT Press).

____ (1994*d*), 'Multiple Personality and Computational Models', in A. Phillips-Griffiths (ed.), *Philosophy, Psychology, and Psychiatry* (Cambridge: Cambridge University Press), 103–14.

____ (1995*a*), 'Artificial Intelligence and Human Dignity', in J. Cornwell (ed.), *Nature's Imagination: The Frontiers of Scientific Vision* (Oxford: Oxford University Press), 148–60; repr., with amendments, as 'Autonomy and Artificiality', in Boden (1996: 95–108).

____ (1995*b*), 'Could a Robot Be Creative—And Would We Know?', in K. M. Ford, C. Glymour, and P. J. Hayes (eds.), *Android Epistemology* (Cambridge, Mass.: MIT Press), 51–72.

____ (ed.) (1996), *The Philosophy of Artificial Life* (Oxford: Oxford University Press).

____ (1997), 'Commentary on Simon's paper on "Machine Discovery"', in Zytkow (1997: 201–3).

____ (1998*a*), 'Creativity and Artificial Intelligence', *Artificial Intelligence*, 103: 347–56.

____ (1998*b*), 'Consciousness and Human Identity: An Interdisciplinary Perspective', in J. Cornwell (ed.), *Consciousness and Human Identity* (Oxford: Oxford University Press), 1–20.

____ (1999), 'Is Metabolism Necessary?', *British Journal for the Philosophy of Science*, 50: 231–48.

____ (2000*a*), 'Crafts, Perception, and the Possibilities of the Body', *British Journal of Aesthetics*, 40: 289–301.

____ (2000*b*), 'Autopoiesis and Life', *Cognitive Science Quarterly*, 1: 1–29.

____ (2001), 'Life and Cognition', in J. Branquinho (ed.), *The Foundations of Cognitive Science* (Oxford: Oxford University Press), 11–22.

____ (2002), 'Artificial Intelligence and the Far Future', in G. F. R. Ellis (ed.), *The Far-Future Universe: Eschatology from a Cosmic Perspective* (London: Templeton Press), 207–24.

____ (2004), *The Creative Mind: Myths and Mechanisms*, 2nd edn., expanded, rev. (London: Routledge; 1st edn., London: Weidenfeld & Nicolson, 1990).

____ (forthcoming), 'Aesthetics and Interactive Art', in R. L. Chrisley, C. Makris, R. W. Clowes, and M. A. Boden (eds.), *Art, Body, Embodiment* (Cambridge: Cambridge Scholars Press).

____ KOWALSKI, R. A., SERGOT, M. J., WILLICK, M. S., BLOOMBECKER, J., and WILKS, Y. A. (1985), 'Artificial Intelligence and Legal Responsibility', part of a panel discussion on 'Legal Implications of AI', *Proceedings of the Ninth International Joint Conference on Artificial Intelligence*, Los Angeles, 1267–80.

BODEN, M. A., BUNDY, A., and NEEDHAM, R. M. (eds.) (1994), *Artificial Intelligence and the Mind: New Breakthroughs or Dead Ends?*, *Philosophical Transactions of the Royal Society*, ser. A, special issue, 349/1689.

BODENHEIMER, F. S. (1958), *The History of Biology: An Introduction* (London: William Dawson).

BOILEN, S., FREDKIN, E., LICKLIDER, J. C. R., and McCARTHY, J. (1963), 'A Time-Sharing Debugging System for a Small Computer', *Proceedings of the Spring Joint Computer Conference*, Detroit, 23: 51–7.

BONABEAU, E. (ed.) (1999), *Stigmergy*, *Artificial Life*, special issue, 5/2: 95–202.

BONASSO, R. P. (1984), *ANALYST: An Expert System for Processing Sensor Returns*, MTP-83 W 00002 (McLean, Va.: MITRE Corporation).

BOND, A. H., and GASSER, L. (eds.) (1988), *Readings in Distributed Artificial Intelligence* (San Francisco: Morgan Kaufmann).

BONDARENKO, A., DUNG, P. M., KOWALSKI, R., and TONI, F. (1997), 'An Abstract, Argumentation-Theoretic Approach to Default Reasoning', *Artificial Intelligence*, 93: 63–101.

BOOLE, G. (1847), *The Mathematical Analysis of Logic: Being an Essay Towards a Calculus of Deductive Reasoning* (repr. Oxford: Blackwell, 1948).

——(1854), *An Investigation of the Laws of Thought: On Which Are Founded the Mathematical Theories of Logic and Probabilities* (Cambridge: Macmillan).

BOORSE, C. (1976), 'Wright on Functions', *Philosophical Review*, 85: 70–86.

BOOTH, A. D. (ed.) (1967), *Machine Translation* (Amsterdam: North-Holland).

BORGES, J. (1966), *Other Inquisitions 1937–1952* (New York: Washington Square Press).

BORING, E. G. (1942), *Sensation and Perception in the History of Experimental Psychology* (New York: Appleton-Century-Crofts).

——(1946), 'Mind and Mechanism', *American Journal of Psychology*, 59: 173–92.

——(1957), *A History of Experimental Psychology*, 2nd edn., rev. (New York: Appleton-Century-Crofts).

BORKO, H. (ed.) (1962), *Computer Applications in the Behavioral Sciences* (Englewood Cliffs, NJ: Prentice-Hall); based on a course entitled 'The Use of Electronic Computers in Psychological Research' given at the University of Southern California.

BORTOFT, H. (1996), *The Wholeness of Nature: Goethe's Way of Science* (Edinburgh: Floris Press).

BOUCHER, D. (ed.) (1997), *The British Idealists* (Cambridge: Cambridge University Press).

BOUCHON-MEUNIER, B., YAGER, R. R., and ZADEH, L. A. (eds.) (1995), *Fuzzy Logic and Soft Computing* (Singapore: World Scientific).

BOURDIEU, P. (1977), *Outline of a Theory of Practice*, trans. R. Nice (Cambridge: Cambridge University Press).

——(1990), *The Logic of Practice*, trans. R. Nice (Cambridge: Polity Press).

BOURGUIGNON, E. (1976), *Possession* (San Francisco: Chandler & Sharp).

BOUTILIER, C., SHOHAM, Y., and WELLMAN, M. P. (eds.) (1997), *Economic Principles of Multi-Agent Systems*, *Artificial Intelligence*, special issue, 94/1–2: 1–268.

BOWDEN, B. V. (ed.) (1953), *Faster than Thought: A Symposium on Digital Computing Machines* (London: Pitman).

——(1972), 'Preface', in Meltzer and Michie (1972, pp. v–ix).

BOWER, G. H. (1970), 'Organizational Factors in Memory', *Cognitive Psychology*, 1: 18–46.

——(1978), 'Experiments on Story Comprehension and Recall', *Discourse Processes*, 1: 211–31.

BOWER, J. M., and BEEMAN, D. (1994), *The Book of GENESIS: Exploring Realistic Neural Models with the GEneral NEural SImulation System* (New York: Springer-Verlag).

BOWLER, P. J. (1975), 'The Changing Meaning of "Evolution"', *Journal of the History of Ideas*, 36: 95–114.

BOWLES, M. D. (1996), 'US Technological Enthusiasm and British Technological Skepticism in the Age of the Analog Brain', *Annals of the History of Computing*, 18/4: 5–15.

BOYD, R., and RICHERSON, P. J. (1985), *Culture and the Evolutionary Process* (Chicago: University of Chicago Press).

——— (forthcoming), *Nature of Cultures* (Chicago: University of Chicago Press).

BOYER, P. (1987), 'The Stuff "Traditions" Are Made Of: On the Implicit Ontology of an Ethnographic Category', *Philosophy of the Social Sciences*, 17: 49–65.

——(1990), *Tradition as Truth and Communication: A Cognitive Description of Traditional Discourse* (Cambridge: Cambridge University Press).

_____ (1991), 'Causal Thinking and its Anthropological Misrepresentation', *Philosophy of the Social Sciences*, 22: 187–213.

_____ (1992), 'Explaining Religious Ideas: Elements of a Cognitive Approach', *Numen*, 39: 27–57.

_____ (ed.) (1993), *Cognitive Aspects of Religious Symbolism* (Cambridge: Cambridge University Press).

_____ (1994), *The Naturalness of Religious Ideas: A Cognitive Theory of Religion* (London: University of California Press).

_____ (1996), 'Cognitive Limits to Conceptual Relativity: The Limiting Case of Religious Ontologies', in J. J. Gumperz and S. C. Levinson (eds.), *Rethinking Linguistic Relativity* (Cambridge: Cambridge University Press), 203–31.

_____ (2003), 'Religious Thought and Behavior as By-Products of Brain Function', *Trends in Cognitive Sciences*, 7: 119–24.

BOYLE, R. (1744), 'A Free Inquiry into the Vulgarly Received Notion of Nature Made in an Essay Addressed to a Friend, To Which Is Pre-Fixed the Life of the Author by Thomas Birch', in T. Birch (ed.), *The Works of the Honourable Robert Boyle*, iv (London: A. Millar).

BRACHMAN, R. J. (1977), 'What's in a Concept: Structural Foundations for Semantic Networks', *International Journal of Man–Machine Studies*, 9: 127–52.

_____ (1979), 'On the Epistemological Status of Semantic Networks', in Findler (1979: 3–50).

_____ (1983), 'What IS-A Is and Isn't: An Analysis of Taxonomic Links in Semantic Networks', *IEEE Computer*, 16/10: 30–6.

_____ and LEVESQUE, H. J. (eds.) (1995), *Readings in Knowledge Representation* (Los Altos, Calif.: Morgan Kauffman).

_____ _____ (eds.) (2004*), Knowledge Representation and Reasoning* (London: Morgan Kaufmann).

_____ _____ and REITER, R. (eds.) (1991), *Knowledge Representation, Artificial Intelligence*, special vol., 49 (Amsterdam: Elsevier; reissued Cambridge, Mass.: MIT Press, 1992).

BRADBURY, J. (2003), 'Reach for the Virtual Skies', *The Guardian*, 13 Nov., available at <http://www.guardian.co.uk/computergames/story/0.11500.1083538.00.html>.

BRADY, J. M. (1985), 'Artificial Intelligence and Robotics', *Artificial Intelligence*, 26: 79–122.

BRAITENBERG, V. (1961), 'Functional Interpretation of Cerebellar Histology', *Nature*, 190: 539–40.

_____ (1965), 'Taxis, Kinesis, and Decussation', in N. Wiener and J. P. Schade (eds.), *Cybernetics of the Nervous System*, Progress in Brain Research 17 (Amsterdam: Elsevier), 210–22.

_____ (1984), *Vehicles: Essays in Synthetic Psychology* (Cambridge, Mass.: MIT Press).

_____ and ONESTO, N. (1960), 'The Cerebellar Cortex as a Timing Organ', *Congresso Instituto Medicina Cibernetica* (Naples), 239–55.

BRAITHWAITE, R. B. (1933), Letter to the Editor, *Mind*, 42: 416.

_____ (1955), *Theory of Games as a Tool for the Moral Philosopher*, Inaugural Lecture, Cambridge, 2 Dec. 1954 (Cambridge: Cambridge University Press).

_____ (1971), 'An Empiricist's View of the Nature of Religious Belief', in B. Mitchell (ed.), *The Philosophy of Religion* (Oxford: Oxford University Press), 72–91.

BRAND, S. (ed.) (1968), *The Whole Earth Catalog*, 1st edn. no longer available; facs., with updates and additional essays: 'Thirtieth Anniversary Celebration: The Whole Earth Catalog', *Whole Earth*, no. 95 (Winter 1998).

_____ (1972), 'SPACEWAR: Fanatic Life and Symbolic Death Among the Computer Bums', *Rolling Stone*, 7 Dec.

_____ (1988), *The Media Lab: Inventing the Future at MIT* (London: Penguin).

BRANNIGAN, A. (1981), *The Social Basis of Scientific Discoveries* (Cambridge: Cambridge University Press).

BRANSFORD, J. D., and JOHNSON, M. K. (1972), 'Contextual Prerequisites for Understanding: Some Investigations of Comprehension and Recall', *Journal of Verbal Learning and Verbal Behaviour*, 11: 717–26.

_____ _____ (1973), 'Considerations of Some Problems of Comprehension', in W. G. Chase (ed.), *Visual Information Processing* (London: Academic Press), 383–438.

BRATKO, I., and MICHIE, D. M. (1980), 'An Advice Program for a Complex Chess Programming Task', *Computer Journal*, 23: 353–9.

_____ and MULEC, P. (1981), 'An Experiment in Automatic Learning of Diagnostic Rules', *Informatika*, 4: 18–25.

_____ KOPEC, D., and MICHIE, D. M. (1978), 'Pattern-Based Representation of Chess End-Game Knowledge', *Computer Journal*, 21: 149–53.

BRATKO, I., MOZETIC, I., and LAVRAC, N. (1988), 'Automatic Synthesis and Compression of Cardiological Knowledge', in J. Hayes, D. M. Michie, and J. Richards (eds.), *Machine Intelligence 11* (Oxford: Oxford University Press), 435–54.

――――――― (1989), *KARDIO: A Study in Deep and Qualitative Knowledge for Expert Systems* (Cambridge, Mass.: MIT Press).

BRAY, A. J., and BARROW, H. G. (1996), *Simple Cell Adaptation in Visual Cortex: A Computational Model of Processing in the Early Visual Pathway*, Cognitive Sciences Research Paper 331 (Brighton: University of Sussex).

BREAZEAL, C., and SCASSELLATI, B. (1999), 'A Context-Dependent Attention System for a Social Robot', *Proceedings of the Sixteenth International Joint Conference on Artificial Intelligence*, Stockholm, 1146–51.

―――― (2000), 'Infant-Like Social Interactions Between a Robot and a Human Caretaker', *Adaptive Behavior*, 8, special issue: *Simulation Models of Social Agents*, ed. C. Dautenhahn, 49–74.

―――― (2002a), 'Robots that Imitate Humans', *Trends in Cognitive Science*, 6: 481–7.

―――― (2002b), 'Challenges in Building Robots that Imitate People', in C. Nehaniv and K. Dautenhahn (eds.), *Imitation in Animals and Artefacts* (Cambridge, Mass.: MIT Press), 363–90.

BREHM, J. W., and COHEN, A. R. (1962), *Explorations in Cognitive Dissonance* (New York: Wiley).

BREMERMANN, H. J. (1962), 'Optimization Through Evolution and Recombination', in M. C. Yovits, G. T. Jacobi, and G. D. Goldstein (eds.), *Self-Organizing Systems* (Washington: Spartan Books), 93–106.

―――― ROGSON, M., and SALAFF, S. (1966), 'Global Properties of Evolutionary Processes', in H. H. Pattee, E. A. Edelsack, L. Fein, and A. B. Callahan (eds.), *Natural Automata and Useful Simulations: Proceedings of a Symposium on Fundamental Biological Models* (Washington: Spartan Books), 3–42.

BRENNER, S. (2001), *A Life in Science* (London: BioMed Central).

BRESNAN, J. W. (1978), 'A Realistic Transformational Grammar', in M. Halle, J. W. Bresnan, and G. A. Miller (eds.), *Linguistic Theory and Psychological Reality* (Cambridge, Mass.: MIT Press), 1–59.

―――― and KAPLAN, R. M. (1982), 'Lexical-Functional Grammar: A Formal System for Grammatical Representation', in J. W. Bresnan (ed.), *The Mental Representation of Grammatical Relations* (Cambridge, Mass.: MIT Press), 173–281.

BRETT, G. S. (1962), *Brett's History of Psychology*, abridged 1-vol. edn., ed. R. S. Peters (London: Allen & Unwin).

BRICE, C. R., and FENNEMA, C. L. (1970), 'Scene Analysis Using Regions', *Artificial Intelligence*, 1: 205–26.

BRINDLEY, G. S. (1967), 'The Classification of Modifiable Synapses and their Use in Models for Conditioning', *Proceedings of the Royal Society*, ser. B, 168: 361–76.

―――― (1969), 'Nerve Net Models of Plausible Size that Perform Many Simple Learning Tasks', *Proceedings of the Royal Society*, ser. B, 174: 173–91.

BRINGSJORD, S. (1992), *What Robots Can and Can't Be* (Dordrecht: Kluwer).

―――― (1994), 'Computation, Among Other Things, Is Beneath Us', *Minds and Machines*, 4: 469–88.

―――― (1995), 'Could, How Could We Tell If, and Why Should— Androids Have Inner Lives?', in K. M. Ford, C. Glymour, and P. J. Hayes (eds.), *Android Epistemology* (Cambridge, Mass.: MIT Press), 93–121.

―――― (2001), 'In Computation, Parallel Is Nothing, Physical Everything', *Minds and Machines*, 11: 95–9.

―――― and ARKOUDAS, K. (2004), 'The Modal Argument for Hypercomputing Minds', *Theoretical Computer Science*, 317: 167.

―――― and FERRUCCI, D. A. (2000), *Artificial Intelligence and Literary Creativity: Inside the Mind of BRUTUS, a Storytelling Machine* (Mahwah, NJ: Lawrence Erlbaum).

―――― and VAN HEUVELN, B. (2003), 'The "Mental Eye" Defence of an Infinitized Version of Yablo's Paradox', *Analysis*, 63: 61–70.

―――― and ZENZEN, M. (2002), 'Toward a Formal Philosophy of Hypercomputation', *Minds and Machines*, 12: 241–58.

―――――― (2003), *Superminds: People Harness Hypercomputation, and More* (Dordrecht: Kluwer).

BRO, P. (1997), 'Chemical Reaction Automata', *Complexity*, 2/3: 38–44.

BROADBENT, D. E. (1952a), 'Listening to One of Two Synchronous Messages', *Journal of Experimental Psychology*, 44: 51–5.

―――― (1952b), 'Failures of Attention in Selective Listening', *Journal of Experimental Psychology*, 44: 428–33.

―――― (1958), *Perception and Communication* (Oxford: Pergamon Press).

―――― (1970a), 'Frederic Charles Bartlett, 1886–1969', *Biographical Memoirs of Fellows of the Royal Society*, 16: 1–13.

_____(1970b), 'Sir Frederic Bartlett: An Appreciation', *Bulletin of the British Psychological Society*, 23 (Jan.), 1–3.

_____(1970c), 'In Defence of Empirical Psychology', *Bulletin of the British Psychological Society*, 23: 87–96.

_____(1971), *Decision and Stress*, The Paul M. Fitts Lectures, University of Michigan, 1969 (London: Academic Press).

_____(1973), *In Defence of Empirical Psychology* (London: Methuen).

_____(1979), Foreword, in new edn. of Miller *et al.* (1960) (New York: Adams, Bannister, Cox), pp. xxvi–xxvii.

_____(1980), 'The Minimization of Models', in A. J. Chapman and Jones (1980: 113–27).

_____(1993), 'Comparison with Human Experiments', in D. E. Broadbent (ed.), *The Simulation of Human Intelligence*, Wolfson College Lectures 1991 (Oxford: Blackwell), 198–217.

BROMLEY, A. G. (1982), 'Charles Babbage's Analytical Engine, 1838', *Annals of the History of Computing*, 4: 196–217.

_____(1990), 'Difference Engines and Analytical Engines', in W. Aspray (ed.), *Computing Before Computers* (Ames: Iowa State University Press), 59–98.

_____(1991), *The Babbage Papers in the Science Museum: A Cross-Referenced List* (London: Science Museum).

BROOK, A. (1994), *Kant and the Mind* (Cambridge: Cambridge University Press).

BROOKS, F. P. (1999), 'What's Real About Virtual Reality?', *IEEE Computer Graphics and Applications*, 19/6: 16–27.

BROOKS, R. A. (1986), 'A Robust Layered Control System for a Mobile Robot', *Institute of Electronics and Electrical Engineers Journal of Robotics and Automation*, 2: 14–23.

_____(1987), *Intelligence Without Representation*, AI Lab Memo (Cambridge, Mass.: MIT); officially pub. as Brooks (1991a).

_____(1991a), 'Intelligence Without Representation', *Artificial Intelligence*, 47: 139–59.

_____(1991b), 'Intelligence Without Reason', *Proceedings of the Twelfth International Joint Conference on Artificial Intelligence*, Sydney, 1–27; first pub. as MIT AI Lab Memo 1293 Cambridge, Mass.: MIT, (Apr. 1991).

_____(1992), 'Artificial Life and Real Robots', in F. J. Varela and P. Bourgine (eds.), *Toward a Practice of Autonomous Systems: Proceedings of the First European Conference on Artificial Life* (Cambridge, Mass.: MIT Press), 3–10.

_____(1999), *Cambrian Intelligence: The Early History of the New AI* (Cambridge, Mass.: MIT Press).

_____and MAES, P. (eds.) (1994), *Artificial Life IV: 4th International Workshop on the Synthesis and Simulation of Living Systems: Selected Papers* (Cambridge, Mass.: MIT Press).

_____and STEIN, L. A. (1994), 'Building Brains for Bodies', *Autonomous Robots*, 1: 7–25.

_____BREAZEAL, C., MARJANOVIC, M., SCASSELLATI, B., and WILLIAMSON, M. (2000), 'The Cog Project: Building a Humanoid Robot', in G. Nehaniv (ed.), *Computation for Metaphors, Analogy, and Agents* (Berlin: Springer-Verlag), 52–87.

BROWN, J. S., and BURTON, R. R. (1978), 'Diagnostic Models for Procedural Bugs in Basic Mathematical Skills', *Cognitive Science*, 2: 155–92.

_____and VANLEHN, K. (1980), 'Repair Theory: A Generative Theory of Bugs in Procedural Skills', *Cognitive Science*, 4: 379–426.

BROWN, P., and LAMBERT, N. (in preparation), *White Heat and Cold Logic: A History of Computer Art* (Cambridge, Mass.: MIT Press).

BROWN, R. (1958), *Words and Things* (Glencoe, Ill.: Free Press).

_____(1965), *Social Psychology* (New York: Free Press).

_____(1970), 'The First Sentences of Child and Chimpanzee', in R. Brown (1970: 208–31).

_____(1973), *A First Language: The Early Stages* (Cambridge, Mass.: Harvard University Press).

_____and HERRNSTEIN, R. J. (1981), 'Icons and Images', in N. Block (1981: 19–50).

BROWN, T. (1974), 'From Mechanism to Vitalism in Eighteenth-Century English Physiology', *Journal of the History of Biology*, 7: 179–216.

BROWNE, J. (2002), *Charles Darwin*, ii: *The Power of Place* (London: Jonathan Cape).

_____and TAYLOR, A. (1989), 'The Inherent Dangers of "Naive" Legal Knowledge Bases', *AISB Quarterly*, no. 68.

BRUCE, C., DESIMONE, R., and GROSS, C. G. (1981), 'Visual Properties of Neurons in a Polysensory Area in Superior Temporal Sulcus of the Macaque', *Journal of Neurophysiology*, 46: 369–84.

BRUCE, D. (1994), 'Lashley and the Problem of Serial Order', *American Psychologist*, 49: 93–103.

BRUNER, J. S. (1944), 'Public Opinion and America's Foreign Policy', *American Sociological Review*, 9: 50–6.

_____ (1951), 'Personality Dynamics and the Process of Perceiving', in R. R. Blake and G. V. Ramsey (eds.), *Perception: An Approach to Personality* (New York: Ronald Press), 121–47; repr. in Bruner (1974: 89–113).

_____ (1957a), 'Going Beyond the Information Given', in H. Gruber, K. R. Hammond, and R. Jessor (eds.), *Contemporary Approaches to Cognition* (Cambridge, Mass.: Harvard University Press), 41–69; repr. in Bruner (1974: 218–38).

_____ (1957b), 'On Perceptual Readiness', *Psychological Review*, 64: 123–52.

_____ (1959), 'Inhelder and Piaget's *The Growth of Logical Thinking*', *British Journal of Psychology*, 50: 363–70.

_____ (1960), *The Process of Education* (Cambridge, Mass.: Harvard University Press).

_____ (1961), Preface, in W. McDougall, *Body and Mind: A History and a Defence of Animism* (Boston: Beacon Press; first edn., London: Methuen, 1911), pp. vii–xvii.

_____ (1962), *On Knowing: Essays for the Left Hand* (Cambridge, Mass.: Harvard University Press).

_____ (1964), 'The Course of Cognitive Growth', *American Psychologist*, 19: 1–15.

_____ (1966a), *Studies in Cognitive Growth* (New York: Wiley).

_____ (1966b), *Towards a Theory of Instruction* (New York: W. W. Norton).

_____ (1970), 'The Nature and Uses of Immaturity', *American Psychologist*, 27/8: 1–60.

_____ (1974), *Beyond the Information Given: Studies in the Psychology of Knowing*, selected, ed. and introd. J. M. Anglin (London: Allen & Unwin).

_____ (1978), 'Learning How to Do Things with Words', in J. S. Bruner and A. Garton (eds.), *Human Growth and Development*, Wolfson College Lectures 1976 (Oxford: Clarendon Press), 62–84.

_____ (1980), 'Jerome S. Bruner', in G. Lindzey (ed.), *A History of Psychology in Autobiography*, vii (San Francisco: Freeman), 74–151.

_____ (1983), *In Search of Mind: Essays in Autobiography* (New York: Harper & Row).

_____ (1988), 'Founding the Center for Cognitive Studies', in Hirst (1988a: 90–9).

_____ (1990), *Acts of Meaning* (Cambridge, Mass.: Harvard University Press).

_____ (1992), 'Another Look at New Look 1', *American Psychologist*, 47: 780–3.

_____ (1996), Foreword, in Shore (1996, pp. xiii–xvii).

_____ (1997), 'Will Cognitive Revolutions Ever Stop?', in Johnson and Erneling (1997: 279–92).

_____ (2002), *Making Stories: Law, Literature, Life* (New York: Farrar, Straus & Giroux).

_____ and GOODMAN, C. C. (1947), 'Value and Need as Organizing Factors in Perception', *Journal of Abnormal and Social Psychology*, 42: 33–44.

_____ and KLEIN, G. S. (1960), 'The Functions of Perceiving: New Look Retrospect', in B. Kaplan and S. Wapner (eds.), *Perspectives in Psychological Theory* (New York: International Universities Press), 61–77; repr. in Bruner (1974: 114–24).

_____ and POSTMAN, L. (1947a), 'Tension and Tension-Release as Organizing Factors in Perception', *Journal of Personality*, 15: 300–8.

_____ _____ (1947b), 'Emotional Selectivity in Perception and Reaction', *Journal of Personality*, 16: 69–77.

_____ _____ (1949a), 'On the Perception of Incongruity: A Paradigm', *Journal of Personality*, 18: 206–23.

_____ _____ (1949b), 'Perception, Cognition, and Behavior', *Journal of Personality*, 18: 14–31.

_____ and POTTER, M. C. (1964), 'Interference in Visual Recognition', *Science*, 144: 424–5.

_____ POSTMAN, L., and RODRIGUES, J. (1951), 'Expectation and the Perception of Color', *American Journal of Psychology*, 64: 216–27.

_____ GOODNOW, J., and AUSTIN, G. (1956), *A Study of Thinking* (New York: Wiley).

BRYSON, A. E., and HO, Y.-C. (1969), *Applied Optimal Control: Optimization, Estimation, and Control* (London: Ginn).

BUCHANAN, B. G. (1972/1973), Review of Hubert Dreyfus, *What Computers Can't Do*, *Computer Reviews*, 18–21; first pub. as Stanford AI Project Memo AIM-181 (Computer Science Department, 1972).

_____ (2003), Interview in 'Bruce Buchanan Retires', *LINKS: Newsletter of the Department of Computer Science*, University of Pittsburgh, 6/1 (Spring), 2–4.

_____ and HEADRICK, T. E. (1970), 'Some Speculation About Artificial Intelligence and Legal Reasoning', *Stanford Law Review*, 23 (Nov.), 40–62.

____ and MITCHELL, T. M. (1978), 'Model-Directed Learning of Production Rules', in D. A. Waterman and F. Hayes (eds.), *Pattern-Directed Inference Systems* (New York: Academic Press), 297–312.

____ and SHORTLIFFE, E. H. (1984), *Rule-Based Expert Programs: The MYCIN Experiments of the Stanford Heuristic Programming Project* (Reading, Mass.: Addison-Wesley).

____ and SRIDHARAN, N. S. (1973), 'Analysis of Behavior of Chemical Molecules: Rule Formation on Non-Homogeneous Classes of Objects', *Proceedings of the Third International Joint Conference on Artificial Intelligence*, Los Angeles, 67–76.

____ SUTHERLAND, G. L., and FEIGENBAUM, E. A. (1969a), 'Heuristic DENDRAL: A Program for Generating Explanatory Hypotheses in Organic Chemistry', in Meltzer and Michie (1969a: 209–54).

____ ____ ____ (1969b), 'Rediscovering Some Problems of Artificial Intelligence in the Context of Organic Chemistry', in Meltzer and Michie (1969b: 253–80).

____ FEIGENBAUM, E. A., and SRIDHARAN, N. S. (1972), 'Heuristic Theory Formation: Data Interpretation and Rule Formation', in Meltzer and Michie (1972: 267–90).

BULLOCK, S., and CLIFF, D. (1997), 'The Role of "Hidden Preferences" in the Artificial Co-evolution of Symmetrical Signals', *Proceedings of the Royal Society*, ser. B, 264, 505–11.

BUNDY, A. M. (1981), 'What Is the Well-Dressed AI Educator Wearing Now?', *AI Magazine*, 3/1: 13–14.

____ (1984), 'Superficial Principles: An Analysis of a Behavioural Law', *AISB Quarterly*, 49: 20–2.

____ (1988), 'The Use of Explicit Plans to Guide Inductive Proofs', in E. Lusk and R. Overbeek (eds.), *Proceedings of the Ninth Conference on Automated Deduction*, Lecture Notes in Computer Science no. 310 (London: Springer), 111–20.

____ and CLUTTERBUCK, R. (1985), 'Raising the Standards of AI Products', *Proceedings of the Ninth International Joint Conference on Artificial Intelligence*, Los Angeles, 1289–94.

____ SILVER, B., and PLUMMER, D. (1985), 'An Analytical Comparison of Some Rule-Learning Programs', *Artificial Intelligence*, 27: 137–82.

____ SMAILL, A. D., and HESKETH, J. (1990), 'Turning Eureka Steps into Calculations in Automatic Program Synthesis', in S. L. H. Clarke (ed.), *Proceedings of UK IT 90*, Southampton, Mar. (London: IEEE), 221–6.

BURGESS, N., DONNETT, J. A., and O'KEEFE, J. (1998), 'Using a Mobile Robot to Test a Model of the Rat Hippocampus', *Connection Science*, 10: 291–300.

BURKS, A. W. (1961), 'Computation, Behavior, and Structure in Fixed and Growing Automata', *Behavioral Science*, 6: 5–22.

____ (1966), *Theory of Self-Reproducing Automata* (Urbana: University of Illinois Press).

____ (1970), *Essays on Cellular Automata* (Urbana: University of Illinois Press).

BURSTALL, R. M., and POPPLESTONE, R. J. (1966), 'POP-2 Reference Manual', in Dale and Michie (1968: 205–46).

BURSTEIN, M. H., and McDERMOTT, D. V. (2005), 'Ontology Translation for Interopability Among Semantic Web Services', *AI Magazine*, 26/1: 71–82.

BURT, D. M., and PERRETT, D. I. (1995), 'Perception of Age in Adult Caucasian Male Faces: Computer Graphic Manipulation of Shape and Colour Information', *Proceedings of the Royal Society*, ser. B, 259: 137–43.

BURTON, R. R. (1982), 'Diagnosing Bugs in a Simple Procedural Skill', in Sleeman and Brown (1982: 157–84).

BUSEMEYER, J. R., and TOWNSEND, J. T. (1993), 'Decision Field Theory: A Dynamic-Cognitive Approach to Decision Making in an Uncertain Environment', *Psychological Review*, 100: 432–59.

BUSH, V. (1931), 'The Differential Analyser: A New Machine for Solving Differential Equations', *Journal of the Franklin Institute*, 212: 447–88.

____ (1945), 'As We May Think', *Atlantic Monthly*, 176 (July), 101–8; repr. in Packer and Jordan (2001: 135–53).

BUSS, D. M. (ed.) (forthcoming), *Handbook of Evolutionary Psychology* (New York: Wiley).

BUTLER, K. (1998), *Internal Affairs* (Dordrecht: Kluwer).

BUTTERFIELD, H. (1931), *The Whig Interpretation of History* (London: Bell).

BUTTERWORTH, G. (1991), 'The Ontogeny and Phylogeny of Joint Visual Attention', in Whiten (1991: 223–32).

BUXTON, H. W. (1880), *Memoir of the Life and Labours of the Late Charles Babbage Esq. F.R.S.*; first pub., ed. A. Hyman, as vol. 13 in the Charles Babbage Institute Reprint Series for the History of Computing (Cambridge, Mass.: MIT Press, 1988).

Byrne, A. W., and Whiten, A. (1988), *Machiavellian Intelligence: Social Expertise and the Evolution of Intellect in Monkeys, Apes, and Humans* (Oxford: Oxford University Press).

Byrne, M. D. (1995), 'Integrating, Not Debating, Situated Action and Computational Models: Taking the Environment Seriously', in A. Ram and K. Eiselt (eds.), *Proceedings of the Sixteenth Annual Conference of the Cognitive Science Society*, Aug. 1994 (Hove: Lawrence Erlbaum), 118–23.

Cahill, C., Silbersweig, D. A., and Frith, C. D. (1996), 'Psychotic Experiences Induced in Deluded Patients Using Distorted Auditory Feedback', *Cognitive Neuropsychiatry*, 1: 201–11.

Calder, A. J., Rowland, D., Young, A. W., Nimmo-Smith, I., Keane, J., and Perrett, D. I. (2000), 'Caricaturing Facial Expressions', *Cognition*, 76: 105–46.

Calude, C., Casti, J. L., and Dinneen, M. (eds.) (1998), *Unconventional Models of Computation* (London: Springer).

Campbell, D. T. (1956), 'Perception as Substitute Trial and Error', *Psychological Review*, 63: 331–42.

—— (1960), 'Blind Variation and Selective Retention in Creative Thought as in Other Knowledge Processes', *Psychological Review*, 67: 380–400.

—— (1965), 'Variation and Selective Retention in Sociocultural Evolution', in H. R. Barringer, G. I. Blanksten, and R. W. Mack (eds.), *Social Change in Developing Areas: A Reinterpretation of Evolutionary Theory* (Cambridge, Mass.: Schenkman), 19–49.

—— (1974a), 'Unjustified Variation and Selective Retention in Scientific Discovery', in F. J. Ayala and T. Dobzhansky (eds.), *Studies in the Philosophy of Biology: Reduction and Related Problems* (London: Macmillan).

—— (1974b), 'Evolutionary Epistemology', in P. A. Schilpp (ed.), *The Philosophy of Karl Popper* (La Salle, Ill.: Open Court), 412–63.

—— (1975), 'On the Conflicts Between Biological and Social Evolution and Between Psychology and Moral Tradition', presidential address to the American Psychological Association, Aug. 1975, *American Psychologist*, 30: 1103–26. (For peer commentaries, see *American Psychologist*, 31: 341–80.)

—— (1976), 'Reprise', *American Psychologist*, 31: 381–4. Afterword to Campbell (1975).

—— and Cziko, G. A. (1990), 'Comprehensive Evolutionary Epistemology Bibliography', *Journal of Social and Biological Sciences*, 13: 41–81.

Campbell-Kelly, M. (1994), Introduction, in Babbage (1864: 7–36).

—— (2002), 'The EDSAC Simulator', in R. Rojas and U. Hashagen (eds.), *The First Computers: History and Architectures* (Cambridge, Mass.: MIT Press), 397–416.

Campos, J. J., Langer, A., and Krowitz, A. (1970), 'Cardiac Responses on the Visual Cliff in Pro-Locomotor Infants', *Science*, 170: 196–7.

—— Hiatt, S., Ramsay, D., Henderson, C., and Svejda, M. (1978), 'The Emergence of Fear on the Visual Cliff', in M. Lewis and L. A. Rosenbloom (eds.), *The Development of Affect* (London: Plenum), 149–82.

—— Bertenthal, B. I., and Kermoian, R. (1992), 'Early Experience and Emotional Development: The Emergence of Wariness of Heights', *Psychological Science*, 3: 61–4.

—— Thein, S., and Owen, D. (2003), 'A Darwinian Legacy to Understanding Human Infancy: Emotional Expressions as Behavior Regulators', *Annals of the New York Academy of Sciences*, 1000: 110–34.

Candy, L. (ed.) (2005), *Creativity and Cognition 2005*, Proceedings of the Fifth Conference on Creativity and Cognition, 12–15 Apr., Goldsmiths College (New York: ACM Press).

—— and Edmonds, E. (eds.) (2002), *Explorations in Art and Technology* (London: Springer).

Canfield, K. (1993), 'The Microstructure of Logocentrism: Sign Models in Derrida and Smolensky', *Postmodern Culture*, 3: 1–39.

Cannon, W. (1926), 'Some General Features of Endocrine Influence on Metabolism', *American Journal of the Medical Sciences*, 171: 1–20.

—— (1932), *The Wisdom of the Body* (New York: W. W. Norton).

Cantor, N. (2000), *Official Statement from the University of Michigan on the Book 'Darkness in El Dorado'*, University of Michigan, 13 Nov., available at <http://www.umich.edu/~urel/darkness.html>.

Capper, P., and Susskind, R. E. (1988), *Latent Damage Law—The Expert System* (London: Butterworths).

Card, S. K., Moran, T. P., and Newell, A. (1983), *The Psychology of Human–Computer Interaction* (Hillsdale, NJ: Lawrence Erlbaum).

Carey, S. (1978), 'The Child as Word Learner', in M. Halle, J. W. Bresnan, and G. A. Miller (eds.), *Linguistic Theory and Psychological Reality* (Cambridge, Mass.: MIT Press), 264–93.

—— (1985), *Conceptual Change in Childhood* (Cambridge, Mass.: MIT Press).

_____ and GELMAN, R. (eds.) (1991), *Epigenesis of the Mind: Essays in Biology and Knowledge* (Hillsdale, NJ: Lawrence Erlbaum).

CARIANI, P. (1992), 'Emergence and Artificial Life', in Langton *et al.* (1992: 775–97).

_____ (1993), 'To Evolve an Ear: Epistemological Implications of Gordon Pask's Electrochemical Devices', *Systems Research*, 10/3: 19–33.

_____ (1997), 'Emergence of New Signal-Primitives in Neural Systems', *Intellectica*, 2: 95–143.

CARNAP, R. (1934), *The Logical Syntax of Language*, trans. A. Smeaton (London: Routledge, 1937).

CARPENTER, B. E., and DORAN, R. W. (1977), 'The Other Turing Machine', *Computer Journal*, 20: 269–79.

_____ _____ (eds.) (1986), Introduction to *A. M. Turing's ACE Report of 1946 and Other Papers*, ed. B. E. Carpenter and R. W. Doran (Cambridge, Mass.: MIT Press), 1–19.

_____ and GROSSBERG, S. (1987), 'ART 2: Self-Organization of Stable Category Recognition Codes for Analog Input Patterns', *Applied Optics*, 26: 4919–30; repr. in J. A. Anderson *et al.* (1990a: 151–62).

_____ COHEN, M. A., and GROSSBERG, S. (1987), 'Computing with Neural Networks: The Role of Symmetry', *Science*, 235: 1226–7.

CARROLL, J. B. (1956), *Language, Thought, and Reality: Selected Writings of Benjamin Lee Whorf* (Cambridge, Mass.: MIT Press).

CARROLL, L. (1977), *Lewis Carroll's Symbolic Logic. Part I: Elementary, 1896. Fifth Edition; Part II: Advanced*, ed. W. W. Bartley (Hassocks: Harvester Press); never previously published.

_____ and CASAGRANDE, J. B. (1966), 'The Function of Language Classifications in Behavior', in E. E. Maccoby, T. M. Newcomb, and E. L. Hartley (eds.), *Readings in Social Psychology*, 3rd edn. (London: Methuen), 18–31.

CARRUTHERS, P. (2003), 'Moderately Massive Modularity', in A. O'Hear (ed.), *Mind and Persons: Royal Institute of Philosophy Lectures* (Cambridge: Cambridge University Press), 69–91.

_____ and CHAMBERLAIN, A. (eds.) (2000), *Evolution and the Human Mind: Modularity, Language, and Meta-Cognition*, papers from five workshops and a conference at the University of Sheffield, 1996–8 (Cambridge: Cambridge University Press).

_____ and SMITH, P. K. (eds.) (1996), *Theories of Theories of Mind* (Cambridge: Cambridge University Press).

CARSON, R. (1960), *The Silent Spring* (Boston: Houghton Mifflin).

CARTER, R. (1998), *Mapping the Mind* (London: Weidenfeld & Nicolson).

CARTWRIGHT, N. (1983), *How the Laws of Physics Lie* (Oxford: Oxford University Press).

CASSELL, J., PELACHAUD, C., BADLER, N., STEEDMAN, M., ACHORD, B., BECKET, T., DOUVILLE, B., PREVOST, S., and STONE, M. (1994), 'Animated Conversation: Rule-Based Generation of Facial Expression, Gesture and Spoken Intonation for Multiple Conversational Agents', *Proceedings of SIGGRAPH '94*, Computer Graphics Annual Conference, Orlando, 413–20, available at <http://www.iut.univ-paris8.fr/~pelachaud>.

_____ SULLIVAN, J., PREVOST, S., and CHURCHILL, E. (eds.) (2000), *Embodied Conversational Agents* (Cambridge, Mass.: MIT Press).

CASSIRER, E. (1944), *An Essay on Man* (New Haven: Yale University Press).

CASTAÑEDA, C., and SUCHMAN, L. (forthcoming), 'Robot Visions', in S. Ghamari-Tabrizi (ed.), *Companions with Haraway: Thinking Together* (provisional title).

CASTELFRANCHI, C. (1998), 'Modelling Social Action for AI Agents', *Artificial Intelligence*, 103: 157–82.

CASTI, J. L. (1998), *The Cambridge Quintet: A Work of Scientific Speculation* (London: Little, Brown).

CAUTE, D. (2003), *The Dancer Defects: The Struggle for Cultural Supremacy During the Cold War* (Oxford: Oxford University Press).

CAVALIERI, P., and SINGER, P. (eds.) (1993), *The Great Ape Project: Equality Beyond Humanity* (London: Fourth Estate).

CAVALLI-SFORZA, L. L. (2000), *Genes, Peoples, and Languages*, trans. M. Seielstad (London: University of California Press).

_____ and FELDMAN, M. W. (1973), 'Cultural Versus Biological Inheritance: Phenotypic Transmission from Parents to Children', *American Journal of Human Genetics*, 25: 618–37.

_____ _____ (1981), *Cultural Transmission and Evolution: A Quantitative Approach* (Princeton: Princeton University Press).

CENDROWSKA, J., and BRAMER, M. (1984), 'Inside an Expert System: A Rational Reconstruction of the MYCIN Consultation System', in T. O'Shea and M. Eisenstadt (eds.), *Artificial Intelligence: Tools, Techniques, Applications* (New York: Harper & Row), 453–97.

CERMAK, L. S. (ed.) (1982), *Human Memory and Amnesia* (Hillsdale, NJ: Lawrence Erlbaum).

CERUZZI, P. E. (1991), 'When Computers Were Human', *Annals of the History of Computing*, 13: 237–44.

CHAGNON, N. A., and IRONS, W. (eds.) (1979), *Evolutionary Biology and Human Social Behavior: An Anthropological Perspective* (North Scituate, Mass.: Duxbury Press). Based on papers presented to two symposia of the American Anthropological Association's 1976 annual meeting.

CHALMERS, D. J. (1990), 'Syntactic Transformations on Distributed Representations', *Connection Science*, 2: 53–62.

——— (1995), 'Facing Up to the Problem of Consciousness', *Journal of Consciousness Studies*, 2: 200–19.

——— (1996a), *The Conscious Mind: In Search of a Fundamental Theory* (Oxford: Oxford University Press).

——— (1996b), 'Does a Rock Implement Every Finite-State Automaton?', *Synthese*, 108: 309–33.

——— (2003), 'The Matrix as Metaphysics', 20 Mar., available at <http://whatisthematrix.warnerbros .com>.

CHAMBERS, R. (1844), *Vestiges of the Natural History of Creation* (London: J. Churchill); many modern edns.

CHAN, J. (1990), 'The Generation of Affect in Synthesized Speech', *Journal of the American Voice I/O Society*, 8: 1–19.

CHANG, M. (1993), interviewed in M. Rosenfeld, 'The Computer Couch: Psychiatry's New Technological Tools', *Washington Post*, 27 Feb., p. A.1.

CHANGEUX, J.-P. (1980), 'Is There a Compromise between Chomsky and Piaget?', in Piatelli-Palmarini (1980: 184–202).

——— (1985), *Neuronal Man: The Biology of Mind*, trans. L. Garey (New York: Pantheon); first pub. as *L'Homme neuronal* (Paris: Fayard, 1983).

——— and DANCHIN, A. (1976), 'Selective Stabilization of Developing Synapses as a Mechanism for the Specification of Neuronal Networks', *Nature*, 264: 705–12.

——— and DEHAENE, S. I. (1989), 'Neuronal Models of Cognitive Functions', *Cognition*, 33: 63–109.

——— COURRÈGE, P., and DANCHIN, A. (1973), 'A Theory of the Epigenesis of Neural Networks by Selective Stabilization of Synapses', *Proceedings of the National Academy of Sciences, USA*, 70: 2974–8.

——— HEIDMANN, T., and PATTE, P. (1984), 'Learning by Selection', in P. Marler and H. S. Terrace (eds.), *The Biology of Learning* (Berlin: Springer-Verlag), 115–33.

CHAPIN, J. K., MOXON, K. A., MARKOWITZ, R. S., and NICOLELIS, M. A. (1999), 'Real-Time Control of a Robot Arm Using Simultaneously Recorded Neurons in the Motor Cortex', *Nature Neuroscience*, 2: 664–70.

CHAPLAIN, M. A. J., SINGH, G. D., and MCLACHLAN, J. C. (eds.) (1999), *On Growth and Form: Spatiotemporal Pattern Formation in Biology* (Chichester: Wiley).

CHAPMAN, A. J., and JONES, D. M. (eds.) (1980), *Models of Man* (Leicester: British Psychological Society).

CHAPMAN, R. (2001), 'SPARK and Abstract Interpretation: A White Paper', available at <http://www.sparkada.com/downloads/92> (Sept.).

——— BURNS, A., and WELLINGS, A. (1994a), 'Integrated Program Proof and Worst-Case Timing Analysis of SPARK Ada', *Proceedings of the Workshop on Language, Compiler, and Tool-Support for Real-Time Systems*, previously available from the FTP Archive of the University of York's 'Real-Time Systems Research Group', Computer Science Department (no longer online).

——— ——— ——— (1994b), 'Static Worst-Case Timing Analysis of Ada', *Ada Letters*, 14/5: 88–91.

CHARNIAK, E. (1972), *Toward a Model of Children's Story Comprehension*, AI-TR-266 (Cambridge, Mass.: MIT AI Lab).

——— (1973), 'Jack and Janet in Search of a Theory of Knowledge', *Proceedings of the Third International Joint Conference on Artificial Intelligence*, Los Angeles, 337–43.

——— (1974), '*He Will Make You Take It Back*': A Study in the Pragmatics of Language (Castagnola: Istituto per gli Studi Semantici e Cognitivi).

——— and MCDERMOTT, D. V. (1985), *Introduction to Artificial Intelligence* (Reading, Mass.: Addison-Wesley).

CHASE, W. G., and SIMON, H. A. (1973), 'Perception in Chess', *Cognitive Psychology*, 4: 55–81.

CHATER, N., REDINGTON, M., NAKISA, R., and OAKSFORD, M. (1997), 'Rationality the Fast and Frugal Way', *Proceedings of the Nineteenth Annual Conference of the Cognitive Science Society*, Los Angeles, 96–101.

CHAUVOIS, L. (1957), *William Harvey: His Life and Times; His Discoveries; His Methods* (London: Hutchinson).

CHENEY, D. L., and SEYFARTH, R. L. (1990), *How Monkeys See the World: Inside the Mind of Another Species* (Chicago: University of Chicago Press).

CHENG, P., and SIMON, H. A. (1995), 'Scientific Discovery and Creative Reasoning with Diagrams', in S. M. Smith, T. B. Ward, and R. A. Finke (eds.), *The Creative Cognition Approach* (Cambridge, Mass.: MIT Press), 205–28.

CHERRY, E. C. (1953), 'Some Experiments on the Recognition of Speech, with One and with Two Ears', *Journal of the Acoustical Society of America*, 25: 975–9.

CHISHOLM, R. M. (1967), 'Intentionality', in P. Edwards (ed.), *The Encyclopedia of Philosophy*, iv (New York: Macmillan), 201–4.

CHOMSKY, A. N. (1956), 'Three Models for the Description of Language', *I.R.E. Transactions on Information Theory*, IT-2: 113–24; repr. in R. D. Luce, R. Bush, and E. Galanter (eds.), *Readings in Mathematical Psychology*, ii (New York: Wiley, 1965), 105–24.

—— (1957), *Syntactic Structures* ('s-Gravenhage: Mouton).

—— (1959a), 'On Certain Formal Properties of Grammars', *Information and Control*, 2: 137–67; repr., amended, in Luce *et al.* (1963/1965: ii: 323–418).

—— (1959b), Review of B. F. Skinner, *Verbal Behavior*, *Language*, 35: 26–58.

—— (1961), 'On the Notion "Rule of Grammar"', in R. Jakobsen (ed.), *Structure of Language and its Mathematical Aspect* (Providence, RI: American Mathematical Society), 6–24.

—— (1962), 'A Transformational Approach to Syntax', in A. A. Hill (ed.), *Proceedings of the Third Texas Conference on Problems of Linguistic Analysis in English, 1958* (Austin: University of Texas), 124–58.

—— (1964), *Current Issues in Linguistic Theory* (The Hague: Mouton).

—— (1965), *Aspects of the Theory of Syntax* (Cambridge, Mass.: MIT Press).

—— (1966), *Cartesian Linguistics: A Chapter in the History of Rationalist Thought* (New York: Harper & Row).

—— (1967), 'Recent Contributions to the Theory of Innate Ideas', *Synthese*, 17: 2–11.

—— (1968), *Language and Mind* (New York: Harcourt, Brace & World).

—— (1969), 'Comments on Harman's Reply', in Hook (1969: 152–9).

—— (1970), 'Problems of Explanation in Linguistics', in R. Borger and F. Cioffi (eds.), *Explanation in the Behavioural Sciences* (Cambridge: Cambridge University Press), 425–51, 462–70.

—— (1972a), Preface to the Enlarged Edition, in A. N. Chomsky, *Language and Mind*, enlarged edn. (New York: Harcourt Brace Jovanovich), pp. vii–xii.

—— (1972b), *Problems of Knowledge and Freedom: The Russell Lectures* (London: Fontana/Collins).

—— (1973), 'Psychology and Ideology', in A. N. Chomsky, *For Reasons of State* (London: Collins), 104–50.

—— (1975), *The Logical Structure of Linguistic Theory* (New York: Plenum Press).

—— (1980a), *Rules and Representations* (Oxford: Blackwell).

—— (1980b), 'Rules and Representations', *Behavioral and Brain Sciences*, 3: 1–61. A précis of Chomsky (1980a), with peer commentary and response.

—— (1982), *Some Concepts and Consequences of the Theory of Government and Binding* (Cambridge, Mass.: MIT Press).

—— (1986), *Knowledge of Language: Its Nature, Origin, and Use* (London: Praeger).

—— (1993), 'A Minimalist Program for Linguistic Theory', in K. L. Hale and S. J. Keyser (eds.), *The View from Building 20: Essays in Linguistics in Honor of Sylvain Bromberger* (Cambridge, Mass.: MIT Press), 1–52.

—— (1995), *The Minimalist Program* (Cambridge, Mass.: MIT Press).

—— (1996a), 'Language and Thought: Some Reflections on Venerable Themes', in A. N. Chomsky, *Powers and Prospects: Reflections on Human Nature and the Social Order* (London: Pluto), 1–30.

—— (1996b), 'Language and Nature', in A. N. Chomsky, *Powers and Prospects: Reflections on Human Nature and the Social Order* (London: Pluto), 31–54.

—— (1996c), 'Writers and Intellectual Responsibility', in A. N. Chomsky, *Powers and Prospects: Reflections on Human Nature and the Social Order* (London: Pluto), 55–69.

—— (1999), 'Language Design', in S. Griffiths (ed.), *Predictions: Thirty Great Minds on the Future* (Oxford: Oxford University Press), 30–2.

—— (2001), *September 11* (Crows Nest, NSW: Allen & Unwin).

CHOMSKY, A. N. (2005), 'Simple Truths, Hard Choices: Some Thoughts on Terror, Justice and Self-Defence', Royal Institute of Philosophy Annual Lecture 2004 (19 May), *Philosophy*, 80: 5–28.

—— and LASNIK, H. (1977), 'Filters and Controls', *Linguistic Inquiry*, 8: 425–504.

CHOMSKY, A. N. and MILLER, G. A. (1958), 'Finite State Languages', *Information and Control*, 1: 91–112.

CHRISLEY, R. L. (1990), 'Cognitive Map Construction and Use: A Parallel Distributed Processing Approach', in D. S. Touretsky, G. E. Hinton, and T. J. Sejnowski (eds.), *Proceedings of the 1990 Connectionist Models Summer School* (San Mateo, Calif.: Morgan Kaufmann), 287–302.

——(1993), 'Connectionism, Cognitive Maps, and the Development of Objectivity', *Artificial Intelligence Review*, 7: 329–54.

——(1995), 'Why Everything Doesn't Realize Every Computation', *Minds and Machines*, 4: 403–20.

——(1998), 'What Might Dynamical Intentionality Be, If Not Computation? Peer Commentary on T. van Gelder', *Behavioral and Brain Sciences*, 21/4: 634–5.

——(1999), 'Transparent Computationalism', in M. Scheutz (ed.), *Proceedings of the Workshop 'New Trends in Cognitive Science 1999: Computationalism—The Next Generation'* (Vienna: Conceptus-Studien). Officially published in M. Scheutz (ed.), *New computationalism* (Berlin: Academia Verlag, 2000), 105–20.

——(ed.) (2000), *Artificial Intelligence: Critical Concepts*, 4 vols. (London: Routledge).

——and HOLLAND, A. (1995), 'Connectionist Synthetic Epistemology: Requirements for the Development of Objectivity', in L. F. Niklasson and M. B. Bodén (eds.), *Current Trends in Connectionism* (Hillsdale, NJ: Lawrence Erlbaum).

——CLOWES, R. W., and TORRANCE, S. (2005*a*), *Next-Generation Approaches to Machine Consciousness*, Cognitive Sciences Research Paper CSRP-574 (Brighton: University of Sussex).

——————(eds.) (2005*b*), *Next-Generation Approaches to Machine Consciousness*, Proceedings of the AISB-05 Symposium on Next-Generation Approaches to Machine Consciousness: Imagination, Development, Intersubjectivity, and Embodiment (Hatfield: University of Hertfordshire).

CHRISTIE, R., and JAHODA, M. (eds.) (1954), *Studies in the Scope and Method of 'The Authoritarian Personality'* (Glencoe, Ill.: Free Press).

CHURCH, A. (1937), 'A. M. Turing: On Computable Numbers, with an Application to the *Entscheidungsproblem*', *Journal of Symbolic Logic*, 2: 42–3.

CHURCHLAND, P. M. (1979), *Scientific Realism and the Plasticity of Mind* (Cambridge: Cambridge University Press).

——(1981), 'Eliminative Materialism and the Propositional Attitudes', *Journal of Philosophy*, 78: 67–90.

——(1986*a*), 'Cognitive Neurobiology: A Computational Hypothesis for Laminar Cortex', *Biology and Philosophy*, 1: 25–51.

——(1986*b*), 'Some Reductive Strategies in Cognitive Neurobiology', *Mind*, 95: 279–309.

——(1989), *A Neurocomputational Perspective: The Nature of Mind and the Structure of Science* (Cambridge, Mass.: MIT Press).

——(1995), *The Engine of Reason, the Seat of the Soul: A Philosophical Journey into the Brain* (Cambridge, Mass.: MIT Press).

——and CHURCHLAND, P. S. (1981) 'Functionalism, Qualia, and Intentionality', *Philosophical Topics*, 12: 121–45.

CHURCHLAND, P. S. (1978), 'Fodor on Language Learning', *Synthese*, 38: 149–59.

——(1981), 'On the Alleged Backwards-Referral of Experiences and its Relevance to the Mind–Body Problem', *Philosophy of Science*, 48: 165–81.

——(1986), *Neurophilosophy: Towards a Unified Theory of the Mind–Brain* (Cambridge, Mass.: MIT Press).

——and SEJNOWSKI, T. J. (1992), *The Computational Brain* (Cambridge, Mass.: MIT Press).

CIOFFI, F. (1970), 'Freud and the Idea of a Pseudo-Science', in R. Borger and F. Cioffi (eds.), *Explanation in the Behavioural Sciences* (London: Cambridge University Press), 471–99.

CLARK, A. J. (1989), *Microcognition: Philosophy, Cognitive Science, and Parallel Distributed Processing* (Cambridge, Mass.: MIT Press).

——(1990), 'Connectionist Minds', *Proceedings of the Aristotelian Society*, 90: 83–102.

——(1991*a*), 'In Defence of Explicit Rules', in W. Ramsey, S. Stich, and D. Rumelhart (eds.), *Philosophy and Connectionist Theory* (Hillsdale, NJ: Lawrence Erlbaum), 115–28.

——(1991*b*), 'Systematicity, Structured Representations and Cognitive Architecture: A Reply to Fodor and Pylyshyn', in T. Horgan and J. Tienson (eds.), *Connectionism and the Philosophy of Mind* (Dordrecht: Kluwer), 198–218.

——(1993), *Associative Engines: Connectionism, Concepts, and Representational Change* (Cambridge, Mass.: MIT Press).

____ (1997), *Being There: Putting Brain, Body, and World Together Again* (Cambridge, Mass.: MIT Press).

____ (2001), *Mindware: An Introduction to the Philosophy of Cognitive Science* (Oxford: Oxford University Press).

____ (2003a), *Natural-Born Cyborgs: Why Minds and Technologies Are Made to Merge* (Oxford: Oxford University Press).

____ (2003b), 'The Twisted Matrix: Dream, Simulation or Hybrid?', 19 Dec., available at <http://whatisthematrix.warnerbros.com>.

____ (2004), 'Is Language Special? Some Remarks on Control, Coding, and Coordination', *Language Sciences*, 26, special issue: *Distributed Cognition and Integrational Linguistics*, 717–26.

____ (2005), 'Material Symbols: From Translation to Coordination in the Constitution of Thought and Reason', talk given to the COGS Seminar at the University of Sussex, 15 Feb.

____ (in preparation), 'Word, Niche, and Super-Niche: How Language Makes Minds Matter More', draft, forthcoming in J. Acero and F. Rodriguez (eds.), *Theoria*, special issue: *Language and Thought: Empirical and Conceptual Viewpoints*.

____ and CHALMERS, D. J. (1998), 'The Extended Mind', *Analysis*, 58: 7–19.

____ and GRUSH, R. (1999), 'Towards a Cognitive Robotics', *Adaptive Behavior*, 7: 5–16.

____ and KARMILOFF-SMITH, A. (1993), 'The Cognizer's Innards: A Psychological and Philosophical Perspective on the Development of Thought', *Mind and Language*, 8: 487–519.

____ and THORNTON, C. (1997), 'Trading Spaces: Computation, Representation, and the Limits of Uninformed Learning', *Behavioral and Brain Sciences*, 20: 57–90.

____ and WHEELER, M. W. (1999), 'Genic Representation: Reconciling Content and Causal Complexity', *British Journal for the Philosophy of Science*, 50: 103–35.

CLARK, D. M. (1996), 'Panic Disorder and Social Phobia', in D. M. Clark and Fairburn (1996: 121–53).

____ and FAIRBURN, C. G. (eds.) (1996), *The Science and Practice of Cognitive Behaviour Therapy* (Oxford: Oxford University Press).

CLARK, K. L. (1978), 'Negation as Failure', in H. Gallaire and J. Minker (eds.), *Logic and Databases* (New York: Plenum Press), 293–322.

CLARK, S. R. L. (1995), 'Tools, Machines, and Marvels', in R. Fellowes (ed.), *Philosophy and Technology* (Cambridge: Cambridge University Press).

CLARK, W. A., and FARLEY, B. G. (1955), 'Generalizations of Pattern Recognition on a Self-Organizing System', *Proceedings of the 1955 Western Joint Computer Conference*, Mar., 7: 86–91.

CLARKE, A. C. (1968), *2001: A Space Odyssey*. (New York: New American Library). Based on a screenplay by Stanley Kubrick and Arthur C. Clarke.

CLARKE, J. A. (1970), *Gabriel Naude, 1600–1653* (Hamden, Conn.: Anchor Books).

CLARKSON, G., and SIMON, H. A. (1960), 'Simulation of Individual and Group Behavior', *American Economic Review*, 50: 920–32.

CLIFF, D. (1991a), 'The Computational Hoverfly: A Study in Computational Neuroethology', in Meyer and Wilson (1991: 87–96).

____ (1991b), 'Computational Neuroethology: A Provisional Manifesto', in Meyer and Wilson (1991: 29–39).

____ and GRAND, S. (1999), 'The Creatures Global Digital Ecosystem', *Artificial Life*, 5: 77–93.

____ and MILLER, G. F. (1995), 'Tracking the Red Queen: Measurements of Adaptive Progress in Co-evolutionary Simulations', in F. Moran, A. Moreno, J. J. Merelo, and P. Chacon (eds.), *Advances in Artificial Life*, Proceedings of the Third European Conference on Artificial Life (ECAL95) (Berlin: Springer-Verlag), 200–18.

____ HARVEY, I., and HUSBANDS, P. (1993), 'Explorations in Evolutionary Robotics', *Adaptive Behavior*, 2: 73–110.

CLIFFORD, J., and MARCUS, G. E. (1986), *Writing Culture: The Poetics and Politics of Ethnography* (Berkeley: University of California Press).

CLIFFORD, J. L. (ed.) (1947), *Dr Campbell's Diary of a Visit to England in 1775* (Cambridge: Cambridge University Press).

CLIPPINGER, J. H. (1975), 'Speaking with Many Tongues: Some Problems in Modelling Speakers of Actual Discourse', *Proceedings on Theoretical Issues in Natural Language Processing-1 (TINLAP)*, Cambridge, Mass., 68–73.

CLIPPINGER, J. H. (1977), *Meaning and Discourse: A Computer Model of Psychoanalytic Discourse and Cognition* (Baltimore: Johns Hopkins University Press).

CLOAK, F. T. (1966), 'Cultural Microevolution', *Research Previews*, 13/2: 7–10 (Chapel Hill: University of North Carolina Press).

——— (1968), 'Is a Cultural Ethology Possible?', *Research Previews*, 15/1: 37–47; a further-developed paper, though with the same title, appeared in 1975: see Cloak (1975a).

——— (1973), 'Elementary Self-Replicating Instructions and their Works: Toward a Radical Reconstruction of General Anthropology through a General Theory of Natural Selection', paper presented to the Ninth International Congress of Anthropological and Ethnographical Sciences, 1973; circulated as Congress pre-prints, and available from the author through the 1970s; published in 2000 at <http://www.thoughtcontagion.com/cloak1973.htm>.

——— (1975a), 'Is a Cultural Ethology Possible?', *Human Ecology*, 3: 161–81; an earlier version, with the same title, had appeared in 1968: see Cloak (1968).

——— (1975b), 'That a Culture and a Social Organization Mutually Shape Each Other Through a Process of Continuing Evolution', *Man–Environment Systems*, 5: 3–6; first presented in Mar. 1974 to the Symposium on 'Generative Models in Social Organization' at the Annual Meeting of the Central States Anthropological Society, Chicago; and in May 1974 to the Annual Meeting of the Animal Behavior Society in Urbana, Ill.

——— (1986), 'The Causal Logic of Natural Selection: A General Theory of Biology', in R. Dawkins and M. Ridley (eds.), *Oxford Surveys in Evolutionary Biology*, ii (Oxford: Oxford University Press), 132–86.

CLOCKSIN, W. F. (1984), 'An Introduction to PROLOG', in T. O'Shea and M. Eisenstadt (eds.), *Artificial Intelligence: Tools, Techniques, and Applications* (London: Harper & Row), 1–22.

——— and MELLISH, C. S. (1981), *Programming in PROLOG* (Berlin: Springer-Verlag).

CLOWES, M. B. (1967), 'Perception, Picture Processing and Computers', in N. L. Collins and D. M. Michie (eds.), *Machine Intelligence 1* (Edinburgh: Edinburgh University Press), 181–97.

——— (1969), 'Pictorial Relationships—A Syntactic Approach', in Meltzer and Michie (1969a: 361–83).

——— (1971), 'On Seeing Things', *Artificial Intelligence*, 2: 79–116.

——— and PARKS, J. R. (1961), 'A New Technique in Character Recognition', *Computer Journal*, 4: 121–6.

COBAS, A., and ARBIB, M. A. (1992), 'Prey-Catching and Predator-Avoidance in Frog and Toad: Defining the Schemas', *Journal of Theoretical Biology*, 157: 271–304.

COCKCROFT, E. (1974), 'Abstract Expressionism: Weapon of the Cold War', *Artforum*, 15 (June), 39–41.

CODD, E. F. (1968), *Cellular Automata* (New York: Academic Press).

COHEN, H. (1979), *Harold Cohen: Drawing*, exhibition catalogue (San Francisco: San Francisco Museum of Modern Art).

——— (1981), *On the Modelling of Creative Behavior*, RAND Paper P-6681 (Santa Monica, Calif.: RAND Corporation).

——— (1995), 'The Further Exploits of AARON Painter', in Franchi and Guzeldere (1995: 141–60).

——— (2002), 'A Million Millennial Medicis', in Candy and Edmonds (2002: 91–104).

COHEN, I. B. (1988), 'Babbage and Aiken; with Notes on Henry Babbage's Gift to Harvard and Other Institutions of a Portion of his Father's Difference Engine', *Annals of the History of Computing*, 10/3: 171–219.

——— and WELCH, G. W. (eds.) (1999), *Makin' Numbers: Howard Aiken and the Computer* (Cambridge, Mass.: MIT Press).

COHEN, L. J. (1981), 'Can Human Irrationality Be Experimentally Demonstrated?', *Behavioral and Brain Sciences*, 4: 317–45.

COHEN, P. R. (2005), 'If Not Turing's Test, Then What?', *AI Magazine*, 26/4 (25th anniversary issue), 61–7.

——— and FEIGENBAUM, E. A. (eds.) (1982), *The Handbook of Artificial Intelligence*, ii (London: Pitman).

——— and PERRAULT, C. R. (1979), 'Elements of a Plan-Based Theory of Speech Acts', *Cognitive Science*, 3/3: 177–212.

——— MORGAN, J., and POLLACK, M. E. (eds.). (1990), *Intentions in Communication* (Cambridge, Mass.: MIT Press).

COLBY, B. N. (1966), 'Ethnographic Semantics', *Current Anthropology*, 7: 3–32.

——— (1975), 'Culture Grammars', *Science*, 187: 913–19.

——— (1996), 'Cognitive Anthropology', in D. Levinson and M. Ember (eds.), *Encyclopedia of Cultural Anthropology* (New York: Henry Holt), 213–15.

——— FERNANDEZ, J., and KRONENFELD, D. (1980), 'Toward a Convergence of Cognitive and Symbolic Anthropology', *American Ethnologist*, 8: 422–50.

COLBY, K. M. (1955), *Energy and Structure in Psychoanalysis* (New York: Ronald Press).

——— (1963), 'Computer Simulation of a Neurotic Process', in Tomkins and Messick (1963: 165–80).

_____ (1964), 'Experimental Treatment of Neurotic Computer Programs', *Archives of General Psychiatry*, 10: 220–7.

_____ (1967), 'Computer Simulation of Change in Personal Belief Systems', *Behavioral Science*, 12: 248–53.

_____ (1975), *Artificial Paranoia: A Computer Simulation of Paranoid Processes* (New York: Pergamon).

_____ (1976), 'Clinical Implications of a Simulation Model of Paranoid Processes', *Archives of General Psychiatry*, 33: 854–7.

_____ (1977), 'An Appraisal of Four Psychological Theories of Paranoid Phenomena', *Journal of Abnormal Psychology*, 86: 54–9.

_____ (1981), 'Modeling a Paranoid Mind', *Behavioral and Brain Sciences*, 4: 515–34.

_____ and COLBY, P. M. (1990), *Overcoming Depression: Manual (with Program-Disks)* (Malibu, Calif.: Malibu Artifactual Intelligence Works (25307 Malibu Rd, Malibu, Calif. 90265)).

_____ and GILBERT, J. P. (1964), 'Programming a Computer Model of Neurosis', *Journal of Mathematical Psychology*, 1: 405–17.

_____ and SMITH, D. C. (1969), 'Dialogues Between Humans and an Artificial Belief System', *Proceedings of the International Joint Conference on Artificial Intelligence*, Washington, 319–24.

_____ and STOLLER, R. J. (1988), *Cognitive Science and Psychoanalysis* (Hillsdale, NJ: Lawrence Erlbaum).

_____ WATT, J. B., and GILBERT, J. P. (1966), 'A Computer Method of Psychotherapy: Preliminary Communication', *Journal of Nervous and Mental Diseases*, 142: 148–52.

_____ WEBER, S., and HILF, F. D. (1971), 'Artificial Paranoia', *Artificial Intelligence*, 2: 1–36.

_____ HILF, F. D., WEBER, S., and KRAMER, H. C. (1972), 'Turing-Like Indistinguishability Tests for the Validation of a Computer Simulation of Paranoid Processes', *Artificial Intelligence*, 3: 199–222.

_____ GOULD, R. L., and ARONSON, G. (1989), 'Some Pros and Cons of Computer-Assisted Psychotherapy', *Journal of Nervous and Mental Disease*, 177: 105–8.

COLE, M., and BRUNER, J. S. (1971), 'Cultural Differences and Inferences About Psychological Processes', *American Psychologist*, 26: 867–76.

COLELLA, V. S., KLOPFER, E., and RESNICK, M. (2001), *Adventures in Modeling: Exploring Complex, Dynamic Systems with StarLogo* (New York: Teachers College Press).

COLLETT, T. S. (1979), 'A Toad's Devious Approach to its Prey: A Study of Some Complex Uses of Depth Vision', *Journal of Comparative Physiology*, ser. A, 131: 179–89.

_____ (1980), 'Angular Tracking and the Optometer Response: An Analysis of Visual Reflex Interaction in a Hoverfly', *Journal of Comparative Physiology*, ser. A, 140: 145–58.

_____ (1982), 'Do Toads Plan Routes? A Study of Detour Behavior of *B. Viridis*', *Journal of Comparative Physiology*, ser. A, 146: 261–71.

_____ and LAND, M. F. (1975), 'Visual Control of Flight Behaviour in the Hoverfly, *Syritta pipiens L.*', *Journal of Comparative Physiology*, 99: 1–66.

_____ COLLETT, T., BISCH, S., and WEHNER, R. (1998), 'Local and Global Vectors in Desert Ant Navigation', *Nature*, 394: 269–72.

COLLIER, B. (1970), 'The Little Engines that Could've: The Calculating Machines of Charles Babbage', Ph.D. thesis, Harvard University; facs. with new introd. and new preface, Harvard Dissertations in the History of Science (New York: Garland, 1990).

_____ (1990), Preface to the Garland Edition, in the 1990 facs. of Collier (1970), no pagination.

COLLINI, S. (2004), 'Stainless Splendour', *London Review of Books*, 26/14 (22 July), 6–10.

COLLINS, A., and MICHALSKI, R. S. (1989), 'The Logic of Plausible Reasoning: A Core Theory', *Cognitive Science*, 13: 1–49.

COLLINS, A. M., and QUILLIAN, M. R. (1972), 'Experiments on Semantic Memory and Language Comprehension', in L. W. Gregg (ed.), *Cognition in Learning and Memory* (New York: Wiley), 117–38.

_____ WARNOCK, E., AIELLO, N., and MILLER, M. (1975), 'Reasoning from Incomplete Knowledge', in Bobrow and Collins (1975: 383–416).

COLLINS, H. M. (1989), 'Computers and the Sociology of Scientific Knowledge', *Social Studies of Science*, 19: 613–24.

_____ (1990), *Artificial Experts: Social Knowledge and Intelligent Machines* (Cambridge, Mass.: MIT Press).

_____ and KUSCH, M. (1998), *The Shape of Actions* (Cambridge, Mass.: MIT Press).

COLLINS, P., and HOGG, J. M. (2004), 'The Ultimate Distributed Workforce: The Use of ICT for Seafarers', *AI and Society*, 18: 209–41.

COLMAN, A. M. (2003), 'Depth of Strategic Reasoning in Games', *Trends in Cognitive Sciences*, 7/1: 2–4.

COLMERAUER, A., and ROUSSEL, P. (1993), 'The Birth of Prolog', *2nd ACM SIGPLAN Conference on History of Programming Languages*, Cambridge, Mass., 37–52.

COLTHEART, M., RASTLE, K., PERRY, C., LANGDON, R., and ZIEGLER, J. (2001), 'DRC: A Dual Route Cascaded Model of Visual Word Recognition and Reading Aloud', *Psychological Review*, 108: 204–56.

COLTON, S., BUNDY, A. M., and WALSH, T. (2000), 'On the Notion of Interestingness in Automated Mathematical Discovery', *International Journal of Human–Computer Studies*, 53: 351–75.

COMEAU, C., and BRYAN, J. (1961), 'Headsight Television System Provides Remote Surveillance', *Electronics*, 10 Nov., 86–90.

COMRIE, L. J. (1946), 'Babbage's Dream Come True', *Nature*, 158: 567–8.

CONLISK, J. (2004), 'Herbert Simon as Friend to Economists Out of Fashion', in Augier and March (2004: 191–5).

CONRAD, M., and PATTEE, H. H. (1970), 'Evolution Experiments with an Artificial Ecosystem', *Journal of Theoretical Biology*, 28: 393–409.

COOK, C. M., and PERSINGER, M. (1997), 'Experimental Induction of the "Sensed Presence" in Normal Subjects and an Exceptional Subject', *Perceptual and Motor Skills*, 85: 683–93.

COOPER, R., FOX, J., FARRINGDON, J., and SHALLICE, T. (1996), 'Towards a Systematic Methodology for Cognitive Modelling', *Artificial Intelligence*, 85: 3–44.

—— SHALLICE, T., and FARRINGDON, J. (1995), 'Symbolic and Continuous Processes in the Automatic Selection of Actions', in Hallam (1995: 27–37).

COPE, D. (1991), *Computers and Musical Style* (Oxford: Oxford University Press).

—— (2000), *The Algorithmic Composer* (Madison, Wis.: A-R Editions).

—— (2001), *Virtual Music: Computer Synthesis of Musical Style* (Cambridge, Mass.: MIT Press).

—— (2006), *Computer Models of Musical Creativity* (Cambridge, Mass.: MIT Press).

COPELAND, B. J. (1993), *Artificial Intelligence: A Philosophical Introduction* (Oxford: Blackwell).

—— (1996), 'What Is Computation?', *Synthese*, 108: 335–59.

—— (1997), 'The Broad Conception of Computation', *American Behavioral Scientist*, 40: 690–716.

—— (1998a), 'Turing's O-Machines, Searle, Penrose, and the Brain', *Analysis*, 58: 128–38.

—— (1998b), 'Super Turing-Machines', *Complexity*, 4: 30–2.

—— (1999), 'A Lecture and Two Radio Broadcasts by Alan Turing on Machine Intelligence', in K. Furukawa, D. M. Michie, and S. Muggleton (eds.), *Machine Intelligence 15* (Oxford: Oxford University Press), 445–76.

—— (2000), 'Narrow Versus Wide Mechanism', *Journal of Philosophy*, 97: 1–32.

—— (2001), 'Colossus and the Dawning of the Computer Age', in R. Erskine and M. Smith (eds.), *Action This Day* (London: Bantam Press), 342–69.

—— (2002a), 'Effective Computation by Humans and Machines', *Minds and Machines*, special issue: *Hypercomputing*, 13: 281–300.

—— (2002b), 'Narrow Versus Wide Mechanism', in Scheutz (2002: 59–86).

—— (ed.) (2004), *The Essential Turing: Seminal Writings in Computing, Logic, Philosophy, Artificial Intelligence, and Artificial Life—plus the Secrets of Enigma* (Oxford: Clarendon Press).

—— (ed.) (2005), *Alan Turing's Automatic Computing Engine: The Master Codebreaker's Struggle to Build the Modern Computer* (Oxford: Oxford University Press).

——and ASTON, G. (n.d.), *AlanTuring.net* (reference articles), available at <http://www.alanturing.net/turing_archive>.

—— and PROUDFOOT, D. (1998), 'Enigma Variations', *Times Literary Supplement*, 3 July, 6.

—— —— (1999), 'Alan Turing's Forgotten Ideas in Computer Science', *Scientific American*, 280 (Apr.), 99–103.

—— —— (2000), 'What Turing Did After he Invented the Universal Turing Machine', *Journal of Logic, Language, and Information*, 9: 491–509.

—— —— (2005), 'Turing and the Computer', in Copeland (2005: 107–48).

—— and SYLVAN, R. (1999), 'Beyond the Universal Turing Machine', *Australasian Journal of Philosophy*, 77: 46–67.

COPLEY, P., and GARTLAND-JONES, A. (2005), 'Musical Form and Algorithmic Solutions', in Candy (2005: 226–31).

CORDEMOY, G. DE (1668), *Un Discours physique de la parole, Œuvres philosophiques, avec une étude bio-bibliographique*, ed. P. Clair and F. Girbal (Paris: Presses Universitaires de France, 1968), 196–256.

CORDESCHI, R. (1991), 'The Discovery of the Artificial: Some Protocybernetic Developments 1930–1940', *AI & Society*, 5: 218–38.

—— (2000), 'Early-Connectionism Machines', *AI & Society*, 14: 314–30.

—— (2002), *The Discovery of the Artificial: Behavior, Mind and Machines Before and Beyond Cybernetics* (Dordrecht: Kluwer).

CORNELIUS, P. (1965), *Languages in Seventeenth- and Early Eighteenth-Century Imaginary Voyages* (Geneva: Librairie Droz).

CORNOCK, S., and EDMONDS, E. A. (1970/1973), 'The Creative Process Where the Artist Is Amplified or Superseded by the Computer', *Leonardo*, 6 (1973), 11–16; based on their paper in *Proceedings of International Symposium on Computer Graphics*, Brunel University, Apr. 1970.

CORNWALL-JONES, K. (1990), 'The Commercialisation of Artificial Intelligence', D.Phil. thesis, Science Policy Research Unit, University of Sussex.

COSMIDES, L. (1989), 'The Logic of Social Exchange: Has Natural Selection Shaped How Humans Reason?', *Cognition*, 31: 187–276.

—— and TOOBY, J. (1992), 'Cognitive Adaptations for Social Exchange', in Barkow *et al.* (1992: 163–228).

—— and TOOBY, J. (1994), 'Beyond Intuition and Instinct Blindness: The Case for an Evolutionarily Rigorous Cognitive Science', *Cognition*, 50: 41–77.

—— —— and BARKOW, J. H. (1992), 'Evolutionary Psychology and Conceptual Integration', in Barkow *et al.* (1992: 3–15).

COSS, R. G. (1965), *Mood Provoking Visual Stimuli: Their Origins and Applications* (Los Angeles: University of California Press).

—— (1977), 'Constraints on Innovation: The Role of Pattern Recognition in the Graphic Arts', in S. L. Carmean and B. L. Grover (eds.), *Creative Thinking*, Proceedings of the Eighth Western Symposium on Learning (Bellingham: Western Washington University), 24–44.

—— (1981), 'Reflections on the Evil Eye', in A. Dundes (ed.), *The Evil Eye: A Folklore Casebook* (New York: Garland), 181–91.

—— (2003), 'The Role of Evolved Perceptual Biases in Art and Design', in E. Voland and K. Grammer (eds.), *Evolutionary Aesthetics* (London: Springer), 69–130.

—— and GOLDTHWAITE, R. O. (1995), 'The Persistence of Old Designs for Perception', in N. S. Thompson (ed.), *Perspectives in Ethology*, xi: *Behavioral Design* (New York: Plenum Press), 83–148.

—— and MOORE, M. (1990), 'All that Glistens: Water Connotations in Surface Finishes', *Ecological Psychology*, 2: 367–80.

—— —— (1994), 'Preschool Children Recognize the Utility of Differently Shaped Trees: A Cross-Cultural Evaluation of Aesthetics and Risk Perception', in M. Francis, P. Lindsey, and J. S. Rice (eds.), *The Healing Dimensions of People–Plant Relations* (Davis, Calif.: Center for Design Research, Department of Environmental Design, University of California), 407–23.

COTTERILL, R. (1998), *Enchanted Looms: Conscious Networks in Brains and Computers* (Cambridge: Cambridge University Press).

COTTINGHAM, J. (1978), ' "A Brute to the Brutes?" Descartes' Treatment of Animals', *Philosophy*, 53: 551–9.

—— (1986), *Descartes* (Oxford: Blackwell).

COUGHLAN, J. M., and FERREIRA, S. J. (2002), 'Finding Deformable Shapes Using Loopy Belief Propagation', *Proceedings of the European Conference on Computer Vision*, 3: 453–68.

Council for Science and Society (1989), *Benefits and Risks of Knowledge-Based Systems*, report of a working party for the Council for Science and Society (Oxford: Oxford University Press).

COWAN, N. (2001), 'The Magical Number Four in Short-Term Memory: A Reconsideration of Mental Storage Capacity', *Behavioral and Brain Sciences*, 24: 87–114.

COX, S., LINCOLN, M., TRYGGVASON, J., NAKISA, M., WELLS, M., TUTT, M., and ABBOTT, S. (2002), 'Tessa, a System to Aid Communication with Deaf People', *Proceedings of the Fifth International ACM Conference on Assistive Technologies (Edinburgh)* (New York: ACM Press), 205–12.

CRAIK, K. J. W. (1943), *The Nature of Explanation* (Cambridge: Cambridge University Press).

—— (1947/1948), 'Theory of the Human Operator in Control Systems: I. The Operation of the Human Operator in Control Systems: II. Man as an Element in a Control System', *British Journal of Psychology*, 38: 56–61, 142–8.

CRE (1988), *Report of a Formal Investigation into St. George's Hospital Medical School* (London: Commission for Racial Equality, Elliot House, 10–12 Allington Street, London SW1E 5EH).

CREVIER, D. (1993), *AI: The Tumultuous History of the Search for Artificial Intelligence* (New York: Basic Books).

CRICHTON, M. (2002), *Prey* (London: HarperCollins).

CRICK, F. H. C. (1989*a*), 'The Recent Excitement About Neural Networks', *Nature*, 337: 129–32.

—— (1989*b*), 'Neural Edelmanism', *Trends in Neurosciences*, 12: 240–8.

—— (1994), *The Astonishing Hypothesis: The Scientific Search for the Soul* (London: Simon & Schuster).

—— and ASANUMA, C. (1986), 'Certain Aspects of the Anatomy and Physiology of the Cerebral Cortex', in McClelland and Rumelhart *et al.* (1986: 333–71).

—— and KOCH, C. (1990), 'Towards a Neurobiological Theory of Consciousness', *Seminars in Neuroscience*, 2: 263–75.

CRONBACH, L. J. (1957), 'The Two Disciplines of Scientific Psychology', *American Psychologist*, 12: 671–84.

CRONIN, H. (1991), *The Ant and the Peacock: Altruism and Sexual Selection from Darwin to Today* (Cambridge: Cambridge University Press).

CRONLY-DILLON, J. R. (1991), 'The Origin of Invertebrate and Vertebrate Eyes', in J. R. Cronly-Dillon and R. L. Gregory (eds.), *Evolution of the Eye and Visual System* (Basingstoke: Macmillan), 15–51.

CRUSE, H., BRUNN, D. E., BARTLING, C., DEAN, J., DREIFERT, M., KINDERMANN, T., and SCHMITZ, F. (1995), 'Walking: A Complex Behavior Controlled by Simple Networks', *Adaptive Behavior*, 3: 385–418.

CUNNINGHAM, A., and JARDINE, N. (eds.) (1990), *Romanticism and the Sciences* (Cambridge: Cambridge University Press).

CURTIS, P. (1992), *Mudding: Social Phenomena in Text-Based Virtual Realities*, Xerox PARC Technical Report CSL-92-4 (Apr.); repr., abridged, in Packer and Jordan (2001: 317–34).

CURTISS, S. (1977), *Genie: A Psycholinguistic Study of a Modern-Day 'Wild Child'* (New York: Academic Press).

CUSSINS, A. (1990), 'The Connectionist Construction of Concepts', in Boden (1990*b*: 368–440).

—— (1992), 'Content, Embodiment, and Objectivity: The Theory of Cognitive Trails', *Mind*, 101: 651–88.

CUTLER, A. (ed.) (1982), *Slips of the Tongue and Language Production* (Berlin: Mouton).

DAHLBOM, B. (ed.) (1993), *Dennett and his Critics: Demystifying Mind* (Oxford: Blackwell).

DALE, E., and MICHIE, D. M. (eds.) (1968), *Machine Intelligence 2* (Edinburgh: Oliver & Boyd).

DALY, M. (1989), 'Vivisection in Eighteenth-Century Britain', *British Journal for Eighteenth-Century Studies*, 12: 57–68.

DAMASIO, A. R. (1994), *Descartes' Error: Emotion, Reason and the Human Brain* (New York: Putnam).

—— (1999), *The Feeling of What Happens: Body and Emotion in the Making of Consciousness* (New York: Harcourt Brace).

—— (2001), 'Reflections on the Neurobiology of Emotion and Feeling', in J. Branquinho (ed.), *The Foundations of Cognitive Science* (Oxford: Clarendon Press), 99–108.

DAMER, B. (1998), *Avatars! Exploring and Building Virtual Worlds on the Internet* (Berkeley: Peachpit Press).

DAMPER, R. I., and CLIFF, D. (eds.) (2002), *Biologically-Inspired Robotics: The Legacy of W. Grey Walter*, Proceedings of the EPSRC/BBSRC International Workshop WGW-02 (Bristol: Hewlett-Packard Laboratories).

D'ANDRADE, R. (1995*a*), *The Development of Cognitive Anthropology* (Cambridge: Cambridge University Press).

—— (1995*b*), 'Moral Models in Anthropology', *Current Anthropology*, 36: 399–408.

—— (2000), 'The Sad Story of Anthropology 1950–1999', *Cross-Cultural Research*, 34: 219–32.

—— and STRAUSS, C. (1992), *Human Motives and Cultural Models* (Cambridge: Cambridge University Press).

DARPA (1988), *DARPA Neural Network Study: October 1987–February 1988* (Fairfax, Va.: AFCEA International Press).

DARTNALL, T. (ed.) (2002), *Creativity, Cognition, and Knowledge: An Interaction* (London: Praeger).

DARWIN, C. R. (1859), *On the Origin of Species by Means of Natural Selection; or, The Preservation of Favoured Races in the Struggle for Life* (London: John Murray).

—— (1871), *The Descent of Man, and Selection in Relation to Sex*, 2 vols. (London: John Murray). The quotations in Ch. 8 are from the 2nd edn. (1874/1901).

DARWIN, F. (ed.) (1887), *The Life and Letters of Charles Darwin; With an Autobiographical Chapter*, 3 vols. (London: John Murray).

DA SOLA POOL, I., ABELSON, R. P., and POPKIN, S. L. (1965), *Candidates, Issues, and Strategies: A Computer Simulation of the 1960 and 1964 Presidential Elections* (Cambridge, Mass.: MIT Press).

DAUTENHAHN, K. (ed.) (1999), *Human Cognition and Social Agent Technology* (Amsterdam: John Benjamins).

—— (ed.) (2002), *Socially Intelligent Agents: Creating Relationships with Computers and Robots* (Boston: Kluwer Academic).

DAVEY, A. C. (1978), *Discourse Production: A Computer Model of Some Aspects of a Speaker* (Edinburgh: Edinburgh University Press).

DAVID, E. E., and SELFRIDGE, O. G. (1962), 'Eyes and Ears for Computers', *Proceedings of the Institute of Radio Engineers*, 50: 1093–1101.

DAVIDSON, D. (1980), *Essays on Actions and Events* (Oxford: Oxford University Press).

DAVIES, D. (2004), *Art as Performance* (Oxford: Blackwell).

DAVIES, M., and STONE, T. (eds.) (1995a), *Folk Psychology: The Theory of Mind Debate* (Oxford: Blackwell).

—— —— (eds.) (1995b), *Mental Simulation: Evaluations and Applications* (Oxford: Blackwell).

DAVIS, D. B. (1985), 'Assessing the Strategic Computing Initiative', *High Technology*, 5: 41–9.

DAVIS, M., and PUTNAM, H. (1960), 'A Computing Procedure for Quantification Theory', *Journal of the Association for Computing Machinery*, 7: 201–15; this is an amended version of M. Davis and H. Putnam, *A Computational Proof Procedure*, AFOSR [Air Force Office of Scientific Research] TR 59-124 (Troy, NY: Rensselaer Polytechnic Institute, 1959).

DAVIS, R. (1976/1982), *The Application of Meta-Level Knowledge to the Construction, Maintenance, and Use of Large Knowledge Bases*, STAN-CS-76-552, Department of Computer Science (Stanford, Calif.: Stanford University); repr. in Davis and Lenat (1982: 227–484).

—— and LENAT, D. B. (eds.) (1982), *Knowledge-Based Systems in Artificial Intelligence* (London: McGraw-Hill).

DAWKINS, R. (1976), *The Selfish Gene* (Oxford: Oxford University Press).

—— (1982), *The Extended Phenotype: The Gene as the Unit of Selection* (Oxford: Freeman).

—— (1986), *The Blind Watchmaker* (Harlow: Longman).

—— (1993), 'Viruses of the Mind', in Dahlbom (1993: 13–27).

DEAN, P., MAYHEW, J., and LANGDON, P. (1994), 'Learning and Maintaining Saccadic Accuracy: A Model of Brainstem–Cerebellum Interactions', *Journal of Cognitive Neuroscience*, 6: 117–38.

DEAN FODOR, J. (1977), *Semantics: Theories of Meaning in Generative Grammar* (Hassocks: Harvester Press).

DECASPER, A. J., and FIFER, W. P. (1980), 'Of Human Bonding: Newborns Prefer their Mothers' Voices', *Science*, 208: 1174–6.

DEESE, J. (1962), 'On the Structure of Associative Meaning', *Psychological Review*, 69: 161–75.

DEHAENE, S. I. (1997), *The Number Sense: How the Mind Creates Mathematics* (Oxford: Oxford University Press).

—— (ed.) (2001), *The Cognitive Neuroscience of Consciousness*, Cognition, 79, special issue, 1–2 (Apr.).

—— and NACCACHE, L. (2001), 'Towards a Cognitive Neuroscience of Consciousness: Basic Evidence and a Workspace Framework', *Cognition*, 79: 1–37.

—— CHANGEUX, J.-P., and NADAL, J. P. (1987), 'Neural Networks that Learn Temporal Sequences by Selection', *Proceedings of the National Academy of Sciences, USA*, 84: 2727–31.

—— DEHAENE-LAMBERTZ, G., and COHEN, L. (1998a), 'Abstract Representations of Numbers in the Animal and Human Brain', *Trends in Neuroscience*, 21: 355–61.

—— NACCACHE, L., LE CLEC, H. G., KOECHLIN, E., MUELLER, M., DEHAENE-LAMBERTZ, G., VAN DE MOORTELE, P. F., and LE BIHAN, D. (1998b), 'Imaging Unconscious Semantic Priming', *Nature*, 395: 597–600.

—— SPELKE, E., PINEL, P., STANESCU, R., and TSIVKIN, S. (1999), 'Sources of Mathematical Thinking: Behavioral and Brain-Imaging Evidence', *Science*, 284: 970–4.

DELANDA, M. (2002), *Intensive Science and Virtual Philosophy* (London: Continuum).

DE LATIL, P. (1953), *La Pensée artificielle* (Paris: Gallimard); trans. Y. M. Golla as *Thinking by Machine: A Study of Cybernetics* (London: Sidgwick & Jackson, 1956).

DELEUZE, G. (1966/1988), *Bergsonism*, trans. from French H. Tomlinson and B. Habberjam (New York: Zone Books; first pub. 1966).

DELISLE BURNS, B. (1958), *The Mammalian Cerebral Cortex* (London: Edward Arnold).

—— (1968), *The Uncertain Nervous System* (London: Edward Arnold).

DEMAZEAU, Y., and MÜLLER, J.-P. (eds.) (1990), *Decentralized A.I.: First European Workshop on Modelling Autonomous Agents in a Multiagent World*, Cambridge, Aug. 1989 (Amsterdam: North-Holland).

DE MEY, M. (1982), *The Cognitive Paradigm: Cognitive Science, a Newly Explored Approach to the Study of Cognition Applied in an Analysis of Science and Scientific Knowledge* (Dordrecht: Reidel).

DENNETT, D. C. (1968), 'Machine Traces and Protocol Statements', *Behavioral Science*, 13: 155–62.

——— (1969), *Content and Consciousness: An Analysis of Mental Phenomena* (London: Routledge & Kegan Paul).

——— (1971), 'Intentional Systems', *Journal of Philosophy*, 68: 87–106.

——— (1977), 'Critical Notice: *The Language of Thought* by Jerry Fodor', *Mind*, 86: 265–80; later retitled 'A Cure for the Common Code?', repr. in Dennett (1978*b*: 90–108).

——— (1978*a*), 'Why You Can't Make a Computer that Feels Pain', *Synthese*, 38: 415–56.

——— (1978*b*), *Brainstorms: Philosophical Essays on Mind and Psychology* (Cambridge, Mass.: MIT Press).

——— (1978*c*), 'Why Not the Whole Iguana?', *Behavioral and Brain Sciences*, 1: 103–4.

——— (1978*d*), 'Beliefs About Beliefs', *Behavioral and Brain Sciences*, 1: 568–70.

——— (1978*e*), 'Where Am I?', in Dennett (1978*b*: 310–23).

——— (1981*a*), 'Three Kinds of Intentional Psychology', in R. Healey (ed.), *Reduction, Time, and Reality: Studies in the Philosophy of the Natural Sciences* (Cambridge: Cambridge University Press), 37–61.

——— (1981*b*), 'True Believers: The Intentional Strategy and Why it Works', in A. F. Heath (ed.), *Scientific Explanation: Papers Based on Herbert Spencer Lectures Given in the University of Oxford* (Oxford: Oxford University Press), 53–75.

——— (1984*a*), *Elbow Room: The Varieties of Free Will Worth Wanting* (Cambridge, Mass.: MIT Press).

——— (1984*b*), 'Cognitive Wheels: The Frame Problem of AI', in C. Hookway (ed.), *Minds, Machines and Evolution: Philosophical Studies* (Cambridge: Cambridge University Press), 129–52.

——— (1986), 'Information, Technology and the Virtues of Ignorance', *Daedalus*, 115/3: 135–53.

——— (1987), *The Intentional Stance* (Cambridge, Mass.: MIT Press).

——— (1988), 'Quining Qualia', in Marcel and Bisiach (1988: 42–77).

——— (1990), 'Memes and the Exploitation of Imagination', *Journal of Aesthetics and Art Criticism*, 48: 127–35.

——— (1991), *Consciousness Explained* (London: Allen Lane).

——— (1993*a*), Review of Allen Newell, *Unified Theories of Cognition*, *Artificial Intelligence*, 59: 285–94.

——— (1993*b*), Review of John Searle, *The Rediscovery of the Mind*, *Journal of Philosophy*, 90: 193–205.

——— (1994*a*), 'The Practical Requirements for Making a Conscious Robot', in Boden *et al.* (1994: 133–46).

——— (1994*b*), 'Dennett, Daniel C.', in S. Guttenplan (ed.), *A Companion to the Philosophy of Mind* (Oxford: Blackwell), 236–44.

——— (1995*a*), *Darwin's Dangerous Idea: Evolution and the Meanings of Life* (London: Allen Unwin).

——— (1995*b*), 'The Unimagined Preposterousness of Zombies: Commentary on Moody, Flanagan, and Polger', *Journal of Consciousness Studies*, 2: 322–6.

——— (1996), *Kinds of Minds: Towards an Understanding of Consciousness* (London: Weidenfeld & Nicolson).

——— (1998), 'Can Machines Think?', in D. C. Dennett, *Brainchildren: Essays in Designing Minds* (Cambridge, Mass.: MIT Press), 3–29. Includes two Postscripts.

——— (2000), 'Re-introducing "The Concept of Mind"', in G. Ryle, *The Concept of Mind* (repr. London: Penguin), pp. ix–xix.

——— (2003*a*), *Freedom Evolves* (New York: Viking).

——— (2003*b*), 'Beyond Beanbag Semantics', *Behavioral and Brain Sciences*, 26: 673–4.

——— (2005), *Sweet Dreams: Philosophical Obstacles to a Science of Consciousness* (Cambridge, Mass.: MIT Press).

——— and KINSBOURNE, M. (1992), 'Time and the Observer: The Where and Why of Consciousness in the Brain', *Behavioral and Brain Sciences*, 15: 183–247.

DERRIDA, J. (1977*a*), 'Signature, Event, Context', trans. S. Weber and J. Mehlman, *Glyph: Johns Hopkins Textual Studies*, 1: 172–97; repr. in J. Derrida, *Limited Inc.* (Evanston, Ill.: Northwestern University Press, 1988), 1–24.

——— (1977*b*), 'Limited Inc. a b c', trans. S. Weber, *Glyph: Johns Hopkins Textual Studies*, 2: 162–254; repr. in J. Derrida, *Limited Inc.* (Evanston, Ill.: Northwestern University Press, 1988), 29–110.

DESCARTES, R. (1637), *Discourse on the Method of Rightly Conducting the Reason and Seeking for Truth in the Sciences*, trans. E. S. Haldane and G. R. T. Ross (Cambridge University Press, 1911), i; and trans. G. E. M. Anscombe and P. T. Geach (London: Nelson, 1954).

—— (1641/1642), *Meditations on First Philosophy*, 2nd edn. (1642), trans. E. S. Haldane and G. R. T. Ross (Cambridge: Cambridge University Press, 1911), i.

—— (1643), 'Correspondence with Princess Elizabeth of Bohemia', in G. E. M. Anscombe and P. T. Geach (eds., trans.) *Descartes* (London: Nelson, 1954), 274–8.

—— (1646), 'Letter to the Marquess of Newcastle (23rd November)', in A. Kenny (ed., trans.), *Descartes: Philosophical Letters* (Oxford: Clarendon Press, 1970), 205–8.

—— (1649), *The Passions of the Soul*.

Des Chene, D. (forthcoming), 'Abstracting from the Soul: The Mechanics of Locomotion', in J. Riskin (ed.), *The Sistine Gap: Essays in the History and Philosophy of Artificial Life* (Stanford, Calif.: Stanford University Press).

de Solla Price, D. J. (1964), 'Automata and the Origins of Mechanism and Mechanistic Philosophy', *Technology and Culture*, 5: 9–23.

—— (1975), *Gears from the Greeks: The Antikythera Mechanism, a Calendar Computer from ca.80 B.C.* (New York: Science History Publications).

—— and Beaver, D. B. (1966), 'Collaboration in an Invisible College', *American Psychologist*, 21: 1011–18.

Deutsch, B. G. (1974), 'The Structure of Task-Oriented Dialogs', *Proceedings of the IEEE Symposium on Speech Recognition*, Pittsburgh, 250–3.

Deutsch, D. (1985), 'Quantum Theory, the Church–Turing Principle, and the Universal Quantum Computer', *Proceedings of the Royal Society*, ser. A, 400: 97–117.

Deutsch, J. A. (1953), 'A New Type of Behaviour Theory', *British Journal of Psychology*, 44: 304–17.

—— (1954), 'A Machine with Insight', *Quarterly Journal of Experimental Psychology*, 6: 6–11.

—— (1960), *The Structural Basis of Behavior* (Chicago: University of Chicago Press).

Dewey, T. (1896), 'The Reflex Arc Concept in Psychology', *Psychological Review*, 3: 357–70.

Dick, S. J. (1993), 'The Search for Extraterrestrial Intelligence and the NASA High Resolution Microwave Survey (HRMS): Historical Perspectives', *Space Science Reviews*, 64: 93–139.

Dickson, D. (1984), *The New Politics of Science* (New York: Pantheon Books).

—— (1986), 'Research Fares Well in New French Budget', *Science*, 233: 718.

Didday, R. L. (1976), 'A Model of Visuomotor Mechanisms in the Frog Optic Tectum', *Mathematical Biosciences*, 30: 169–80.

Diderot, D. (1751), 'Lettre sur les sourds et muets, à l'usage de ceux qui entendent et qui voient (avec additions pour servir d'éclaircissement)', in *Œuvres complètes de Diderot*, i (Paris: Garnier Frères, 1875; facs., Liechtenstein: Kraus, 1966), 343–428. The latest of the modern edns. is M. Hobson and S. Harvey (eds.), *Lettre sur les aveugles* (Paris: Flammarion, 2000).

Dienes, Z., and Perner, J. (1999), 'A Theory of Implicit and Explicit Knowledge', *Behavioral and Brain Sciences*, 22: 735–808.

—— —— (2004), 'Executive Control Without Conscious Awareness: The Cold Control Theory of Hypnosis', working paper, Department of Psychology, University of Sussex; abridged version, entitled 'The Cold Control Theory of Hypnosis', in G. Jamieson (ed.), *Hypnosis and Conscious States: The Cognitive Neuroscience Perspective* (Oxford: Oxford University Press, forthcoming). Page references are to the draft MS (2004).

Dijksterhuis, E. J. (1956), *The Mechanization of the World Picture*, trans. C. Dikshoorn (Oxford: Clarendon Press, 1961).

Dilthey, W. (1883), *Introduction to the Human Sciences: An Attempt to Lay a Foundation for the Study of Society and History*, trans. from German by R. J. Betanzos (Detroit: Wayne State University Press, 1988).

Dinneen, G. P. (1955), 'Programming Pattern Recognition', *Proceedings of the Western Joint Computer Conference*, Mar., 7: 94–100.

d'Inverno, M., Luck, M., and Wooldridge, M. (1997), 'Cooperation Structures', *Proceedings of the Fifteenth International Joint Conference on Artificial Intelligence*, Nagoya, Japan, 600–5.

Di Paolo, E. A. (1998), 'An Investigation into the Evolution of Communication', *Adaptive Behavior*, 6: 285–324.

—— (1999), 'On the Evolutionary and Behavioral Dynamics of Social Coordination: Models and Theoretical Aspects', D.Phil. thesis, School of Cognitive and Computing Sciences, University of Sussex.

—— (2000a), 'Behavioral Coordination, Structural Congruence and Entrainment in a Simulation of Acoustically Coupled Agents', *Adaptive Behavior*, 8: 25–46.

—— (2000b), 'Homeostatic Adaptation to Inversion of the Visual Field and Other Sensorimotor Disruptions', in J.-A. Meyer, A. Berthoz, D. Floreano, H. Roitblat, and S. W. Wilson (eds.), *From*

Animals to Animats 6, Proceedings of the Sixth International Conference on Simulation of Adaptive Behavior (Cambridge, Mass.: MIT Press), 440–9.

Di Paolo, E. A. (ed.) (2002*a*), *Plastic Mechanisms, Multiple Time Scales, and Lifetime Adaptation, Adaptive Behavior*, special issue, 10/3–4: 141–265.

—— (2002*b*), 'Evolving Spike-Timing Dependent Synaptic Plasticity for Robot Control', in R. Damper and D. Cliff (eds.), *Biologically-Inspired Robotics: The Legacy of W. Grey Walter*, Proceedings of the EPSRC/BBSRC International Workshop WGW-02, Bristol, 142–9; rev. version: 'Evolving Spike-Timing-Dependent Plasticity for Single-Trial Learning in Robots', *Philosophical Transactions of the Royal Society of London*, ser. A, 361, special issue: *Biologically Inspired Robotics* (2003), 2299–3319.

—— (2004), 'Unbinding Biological Autonomy: Francisco Varela's Contributions to Artificial Life', *Artificial Life*, 10/3: 231–4. Editorial introd. to a special issue of *Artificial Life*, a memorial for Francisco Varela.

—— (forthcoming), 'Autopoiesis, Adaptivity, Teleology, Agency', *Phenomenology and the Cognitive Sciences*, 4/4, special issue: S. Torrance (ed.), *Enactive Experience*.

DiSessa, A. (1988), 'Knowledge in Pieces', in G. Forman and P. B. Pufall (eds.), *Constructivism in the Computer Age* (Hillsdale, NJ: Lawrence Erlbaum), 49–70.

Doan, A., and Halevy, A. Y. (2005), 'Semantic-Integration Research in the Database Community: A Brief Survey', *AI Magazine*, 26/1: 83–94.

Dominey, P. F., and Arbib, M. A. (1992), 'A Cortico-Subcortical Model for Generation of Spatially Accurate Sequential Saccades', *Cerebral Cortex*, 2: 153–75.

Donald, M. (1991), *Origins of the Modern Mind: Three Stages in the Evolution of Culture and Cognition* (Cambridge, Mass.: Harvard University Press).

Donald, M. J. (2001), Review of Evan Harris Walker, *The Physics of Consciousness*, *Psyche: An Interdisciplinary Journal of Research on Consciousness*, 7/15 (Oct.); available at <http://psyche.cs.monash.edu.au/v7/psyche-7-15-donald.html>.

Dong, A. (2003), 'A Cybernetics-Based Design Process for Intelligent Rooms', *Proceedings of the 37th Conference of the Australia and New Zealand Architectural Science Association*, Sydney, i. 267–74.

Doran, J. E. (1968*a*), 'Experiments with a Pleasure-Seeking Automaton', in Michie (1968*a*: 195–216).

—— (1968*b*), 'New Developments of the Graph Traverser', in Dale and Michie (1968: 119–36).

—— (1969), 'Planning and Robots', in Meltzer and Michie (1969*b*: 519–32).

—— (1970), 'Systems Theory, Computer Simulations and Archaeology', *World Archaeology*, 1: 289–8.

—— (1977), 'Knowledge Representation for Archaeological Inference', in Elcock and Michie (1977: 433–54).

—— and Hodson, F. R. (eds.) (1975), *Mathematics and Computers in Archaeology* (Cambridge, Mass.: Harvard University Press).

—— and Michie, D. M. (1966), 'Experiments with the Graph Traverser Program', *Proceedings of the Royal Society*, ser. A, 294: 235–59.

—— and Palmer, M. (1995), 'The EOS Project: Integrating Two Models of Paleolithic Social Change', in N. Gilbert and R. Conte (eds.), *Artificial Societies* (London: UCL Press), 103–289.

Dorril, S. (2000), *MI6: Fifty Years of Special Operations* (London: Fourth Estate).

Dougherty, J. W. D. (ed.) (1985), *Directions in Cognitive Anthropology* (Urbana: University of Illinois Press).

Dowty, D., Wall, R., and Peters, S. (1981), *An Introduction to Montague's Semantic Theory* (Dordrecht: Reidel).

Doya, K. (2000), 'Complementary Roles of Basal Ganglia and Cerebellum in Learning and Motor Control', *Current Opinion in Neurobiology*, 10: 732–9.

Doyle, J. (1979), 'A Truth Maintenance System', *Artificial Intelligence*, 12: 231–72.

—— and Dean, T. (1997), 'Strategic Directions in Artificial Intelligence', *AI Magazine*, 18/1: 87–101.

Doyon, A., and Liagre, L. (1956), 'Méthodologie comparée du biomécanisme et de la mécanique comparée', *Dialectica*, 10: 292–335.

Drachmann, A. G. (1963), *The Mechanical Technology of Greek and Roman Antiquity: A Study of the Literary Sources* (Copenhagen: Munksgaard).

Drescher, G. L. (1989), 'A Mechanism for Early Piagetian Learning', *Proceedings of the Tenth International Joint Conference on Artificial Intelligence*, Milan, 290–4.

—— (1991), *Made-Up Minds: A Constructivist Approach to Artificial Intelligence* (Cambridge, Mass.: MIT Press).

DRESHER, E., and HORNSTEIN, N. (1976), 'On Some Supposed Contributions of Artificial Intelligence to the Scientific Study of Language', *Cognition*, 4: 321–78.

DRETSKE, F. I. (1984), *Knowledge and the Flow of Information* (Oxford: Blackwell).

—— (1995), *Naturalizing the Mind* (Cambridge, Mass.: MIT Press).

DREXLER, K. E. (1989), 'Biological and Nanomechanical Systems: Contrasts in Evolutionary Complexity', in Langton (1989a: 501–19).

DREYFUS, H. L. (1965), *Alchemy and Artificial Intelligence*, Research Report P-3244, Dec. (Santa Monica, Calif.: Rand Corporation).

—— (1967), 'Why Computers Must Have Bodies in Order to be Intelligent', *Review of Metaphysics*, 21: 13–32.

—— (1972), *What Computers Can't Do: A Critique of Artificial Reason* (New York: Harper & Row).

—— (1979), *What Computers Can't Do: A Critique of Artificial Reason*, 2nd edn., rev. (New York: Harper & Row).

—— (1981), 'From Micro-Worlds to Knowledge Representation: AI at an Impasse', in Haugeland (1981: 161–204); excerpted, with minor revisions, from the Introduction to Dreyfus (1979).

—— (1992), *What Computers Still Can't Do: A Critique of Artificial Reason* (Cambridge, Mass.: MIT Press).

—— (1998), 'Intelligence Without Representation', paper on the University of Houston web site: <http://www.hfac.uh.edu/cogsci/dreyfus.html>, accessed 3 Apr. 2001.

—— (2000), 'Telepistemology: Descartes' Last Stand', in K. Goldberg (ed.), *The Robot in the Garden: Telerobotics and Telepistemology in the Age of the Internet* (Cambridge, Mass.: MIT Press), 48–63.

—— (2001), *On the Internet* (London: Routledge).

—— (2003), 'Existentialist Phenomenology and the Brave New World of *The Matrix*', *Harvard Review of Philosophy*, 11 (Fall), 18–31; amended version of Dreyfus and Dreyfus (2002).

—— and DREYFUS, S. E. (1962), 'Comments', in M. Greenberger (ed.), *Management and the Computer of the Future* (Cambridge, Mass.: MIT Press; New York: Wiley), 321–2. These comments do not appear as a separate item, but are included in the 'General Discussion' on pp. 315–24.

—— —— (1986), *Mind Over Machine: The Power of Human Intuition and Expertise in the Era of the Computer* (Oxford: Blackwell).

—— —— (1988), 'Making a Mind Versus Modelling the Brain: Artificial Intelligence Back at a Branch Point', in Graubard (1988: 15–43).

—— —— (1990), 'Towards a Reconciliation of Phenomenology and AI', in D. Partridge and Y. A. Wilks (eds.), *The Foundations of Artificial Intelligence* (Cambridge: Cambridge University Press), 396–410.

—— —— (2002), 'The Brave New World of the Matrix', 20 Nov., available on <http://whatisthematrix.warnerbros.com>; for an amended version, see Dreyfus (2003).

—— and HAUGELAND, J. (1974), 'The Computer as a Mistaken Model of the Mind', in S. C. Brown (ed.), *Philosophy of Psychology* (London: Macmillan), 247–58.

DREYFUS, S. E. (1962), 'Artificial Intelligence: Deus ex Machina', review of M. Taube, *Computers and Common Sense*, *The Second Coming Magazine*, 1/4: 52–5.

DRIESCH, H. A. E. (1908), *The Science and Philosophy of the Organism*, The Gifford Lectures, 2 vols. (London: Black).

DTI (2004), *FORESIGHT: Cognitive Systems Project, Applications and Impact* (London: Department of Trade and Industry, Office of Science and Technology).

DUMMETT, M. (1975), 'Can Analytical Philosophy Be Systematic, and Ought it to Be?', paper presented to the Congress on the Philosophy of Hegel, Stuttgart, 1975; repr. in M. Dummett, *Truth and Other Enigmas* (London: Duckworth), 437–58.

DUNCKER, K. (1945), *On Problem Solving*, trans. from German by L. S. Lees (Washington: American Psychological Association); first pub. in German in Psychological Monographs, 58/5 (Baltimore: Johns Hopkins Press, 1935).

DUPUY, J.-P. (2000), *The Mechanization of the Mind: On the Origins of Cognitive Science*, trans. M. B. DeBevoise (Princeton: Princeton University Press; first pub. 1994).

DURHAM, W. H. (1976), 'The Adaptive Significance of Cultural Behavior', *Human Ecology*, 4: 89–121.

—— (1979), 'Toward a Coevolutionary Theory of Human Biology and Culture', in N. A. Chagnon and W. G. Irons (eds.), *Evolutionary Biology and Human Social Behavior: An Anthropological Perspective* (North Scituate, Mass.: Duxbury Press), 39–58.

—— (1991), *Coevolution: Genes, Culture, and Human Diversity* (Stanford, Calif.: Stanford University Press).

DURKHEIM, É. (1912/1915), *The Elementary Forms of the Religious Life: A Study in the Sociology of Religion*, trans. from French J. W. Swade (London: Allen & Unwin, 1915; first pub. 1912).

DUSSER DE BARENNE, J. G., and McCULLOCH, W. S. (1938), 'Functional Organization of the Sensory Cortex of the Monkey', *Journal of Neurophysiology*, 1: 69–85.

DUVDEVANI-BAR, S., EDELMAN, S., HOWELL, J., and BUXTON, H. (1998), *A Similarity-Based Method for the Generalization of Face Recognition Over Pose and Expression, International Conference on Automatic Face and Gesture Recognition (Nara)*, 118–23.

DWORETZKY, J. P. (1981), *Introduction to Child Development* (St Paul, Minn.: West Publishing Co.).

DYER, M. G. (1983), *In-Depth Understanding: A Computer Model of Integrated Processing for Narrative Comprehension* (Cambridge, Mass.: MIT Press).

DYSON, F. J. (1958), 'Innovation in Physics', *Scientific American*, 199 (Sept.), 74–82.

ECCLES, J. C. (1953), *The Neurophysiological Basis of Mind: The Principles of Neurophysiology* (Oxford: Clarendon Press).

—— (1964), *The Neurophysiology of Synapses* (New York: Springer-Verlag).

—— (1986), 'Do Mental Events Cause Neural Events Analogously to the Probability Fields of Quantum Mechanics?', *Proceedings of the Royal Society*, ser. B, 227: 411–28.

—— ITO, M., and SZENTAGOTHAI, J. (1967), *The Cerebellum as a Neuronal Machine* (New York: Springer-Verlag).

ECKART, C., and YOUNG, G. (1936), 'The Approximation of One Matrix by Another of Lower Rank', *Psychometrika*, 1: 211–18.

EDELMAN, G. M. (1978), 'Group Selection and Phasic Reentrant Signaling: A Theory of Higher Brain Function', in G. M. Edelman and V. B. Mountcastle (eds.), *The Mindful Brain: Cortical Organization and the Group-Selective Theory of Higher Brain Function* (Cambridge, Mass.: MIT Press), 51–100.

—— (1987), *Neural Darwinism: The Theory of Neuronal Group Selection* (New York: Basic Books).

—— (1989), *The Remembered Present: A Biological Theory of Consciousness* (New York: Basic Books).

—— (1992), *Bright Air, Brilliant Fire: On the Matter of the Mind* (New York: Basic Books).

—— (1998), 'Representation Is Representation of Similarities', *Behavioral and Brain Sciences*, 21: 449–67.

—— and REEKE, G. N. (1990), 'Is it Possible to Construct a Perception Machine?', *Proceedings of the American Philosophical Society*, 134: 36–73.

—— and TONONI, G. (2000), *Consciousness: How Matter Becomes Imagination* (London: Allen Lane).

EDELMAN, S. (2003), 'Generative Grammar with a Human Face?', *Behavioral and Brain Sciences*, 26: 675–6.

EDGLEY, R. (1970), 'Innate Ideas', in G. Vesey (ed.), *Knowledge and Necessity* (London: Macmillan), 1–33.

EDMONDS, E. A. (1981), 'Adaptive Man–Computer Interfaces', in M. J. Coombs and J. L. Alty (eds.), *Computing Skills and the User Interface* (London: Academic Press), 389–426.

—— (1986), 'Negative Knowledge: Toward a Strategy for Asking in Logic Programming', *International Journal of Man–Machine Studies*, 24: 597–600.

—— (2002), 'Structure in Art Practice: Technology as an Agent for Concept Development', *Leonardo*, 35: 65–71.

—— (2003), *On New Constructs in Art* (Sydney: Creativity and Cognition Studios, University of Technology).

—— and GUEST, S. (1977a), *A Guide to SYNICS*, Research Note, Leicester Polytechnic, School of Mathematics, Computing and Statistics.

—— —— (1977b), 'An Interactive Tutorial System for Teaching Programming', *Proceedings of the IERE Conference on Computer Systems and Technology*, Sussex, 263–70.

—— and LEE, J. (1974), 'An Appraisal of Some Problems of Achieving Fluid Man/Machine Interaction', *Proceedings of the European Computing Congress on Online Computing Systems*, Uxbridge, Middx., 635–45.

EDWARDS, P. (2004), *Heidegger's Confusions* (New York: Prometheus).

EDWARDS, P. N. (1996), *The Closed World: Computers and the Politics of Discourse in Cold War America* (Cambridge, Mass.: MIT Press).

EGHOLM, M., *et al.* (10 authors, incl. P. E. Nielsen) (1993), 'PNA Hybridizes to Complementary Oligonucleotides Obeying the Watson–Crick Hydrogen-Bonding Rules', *Nature*, 365: 566–8.

EISENSTADT, M. (1976), 'Processing Newspaper Stories: Some Thoughts on Fighting and Stylistics', *Proceedings of the AISB Summer Conference*, Edinburgh, July, 104–17.

EKMAN, P. (1979), 'About Brows: Emotional and Conversational Signals', in M. von Cranach, K. Foppa, W. Lepenies, and D. Ploog (eds.), *Human Ethology: Claims and Limits of a New Discipline* (Cambridge: Cambridge University Press), 169–248.

—— (1985/2001), *Telling Lies: Clues to Deceit in the Marketplace, Politics and Marriage* (New York: W. W. Norton; 2nd edn., expanded, 2001).

—— (1992), 'Facial Expressions of Emotions: New Findings, New Questions', *Psychological Science*, 3: 34–8.

—— (1998), Introduction, Afterword, and Commentary, in C. Darwin, *The Expression of the Emotions in Man and Animals* (HarperCollins: London).

—— and DAVIDSON, R. J. (eds.) (1994), *The Nature of Emotion: Fundamental Questions* (Oxford: Oxford University Press).

ELCOCK, E. W. (1968), 'Descriptions', in Michie (1968a: 173–80).

—— (1988), *ABSYS: The First Logic Programming Language—An Obituary and a Retrospective*, Technical Report 210 (Toronto: University of Western Ontario).

—— and GRAY, P. M. D. (1988), *ABSYS, Equation Solving, and Logic Programming*, Technical Report 213, July (Toronto: University of Western Ontario).

—— and MICHIE, D. M. (eds.) (1977), *Machine Intelligence 8* (Chichester: Ellis Horwood).

—— FOSTER, J. M., GRAY, P. M. D., McGREGOR, J. J., and MURRAY, A. M. (1971), 'ABSET, a Programming Language Based on Sets: Motivation and Examples', in Meltzer and Michie (1971: 467–92).

ELDREDGE, N., and GOULD, S. G. (1972), 'Punctuated Equilibria: An Alternative to Phyletic Gradualism', in T. J. M. Schopf (ed.), *Models in Paleobiology* (San Francisco: Freeman Cooper), 82–115.

ELLEN, R. F. (2003), 'Biological Knowledge Across Cultures', in H. Selin (ed.), *Nature Across Cultures: Views of Nature and Environment in Non-Western Cultures* (London: Kluwer Academic), 47–74.

ELLENBERGER, H. F. (1970), *The Discovery of the Unconscious: The History and Evolution of Dynamic Psychiatry* (London: Allen Lane).

ELLIS, W. D. (1938), *A Source Book of Gestalt Psychology* (London: Kegan Paul).

ELMAN, J. L. (1990), 'Finding Structure in Time', *Cognitive Science*, 14: 179–212.

—— (1993), 'Learning and Development in Neural Networks: The Importance of Starting Small', *Cognition*, 48: 71–99.

—— (2004), 'An Alternative View of the Mental Lexicon', *Trends in Cognitive Sciences*, 8: 301–6.

—— BATES, E. A., JOHNSON, M. H., KARMILOFF-SMITH, A., PARISI, D., and PLUNKETT, K. (1996), *Rethinking Innateness: A Connectionist Perspective on Development* (Cambridge, Mass.: MIT Press).

Encyclopaedia Britannica (1990), 'Entry on Henry Cavendish', *The New Encyclopaedia Britannica*, ii: *Micropaedia*, 15th edn. (London: Encyclopaedia Britannica), 974–5.

ENEVER, T. (1994), *Britain's Best Kept Secret: Ultra's Base at Bletchley Park* (Stroud: Alan Sutton).

ENGELBART, D. C. (n.d.), Quoted on Mouse Site, at <http://sloan.stanford.edu/MouseSite/1968Demo.html>, accessed 25 June 2004.

—— (1962), *Augmenting Human Intellect: A Conceptual Framework*, Report no. AFOSR 3233, prepared for the Airforce Office of Scientific Research (Menlo Park, Calif.: Stanford Research Institute); repr., abridged, in Packer and Jordan (2001: 64–90).

—— (1963), 'A Conceptual Framework for the Augmentation of Man's Intellect', in P. W. Howerton and D. C. Weeks (eds.), *Vistas in Information Handling*, i (Washington: Spartan Books), 1–29.

ENQUIST, M., and ARAK, A. (1994), 'Symmetry, Beauty, and Evolution', *Nature*, 372: 169–72.

EPSTEIN, J. M., and AXTELL, R. (1996), *Growing Artificial Societies: Social Science from the Bottom Up* (Cambridge, Mass.: MIT Press).

EPSTEIN, R. (1992), 'The Quest for the Thinking Computer', *AI Magazine*, 13/2: 80–95.

ERDELYI, M. H. (1974), 'A New Look at the New Look: Perceptual Defense and Vigilance', *Psychological Review*, 81: 1–25.

ERMAN, L. D., and LESSER, V. R. (1980), 'The HEARSAY-II Speech Understanding System: A Tutorial', in W. Lea (ed.), *Trends in Speech Recognition* (Englewood Cliffs, NJ: Prentice-Hall), 361–81.

ERNST, G. W., and NEWELL, A. (1967), 'Some Issues of Representation in a General Problem Solver', *Proceedings of the Spring Joint Computer Conference*, 30: 583–600.

ESTES, W. K. (1950), 'Toward a Statistical Theory of Learning', *Psychological Review*, 57: 94–107.

—— (1959), 'The Statistical Approach to Learning Theory', in S. Koch (ed.), *Psychology: A Study of a Science*, ii: *General Systematic Formulations, Learning, and Special Processes* (New York: McGraw-Hill), 380–491.

ETCOFF, N. (1999), *Survival of the Prettiest: The Science of Beauty* (London: Little, Brown).

Euromap Report (1998), *The Euromap Report: Challenge and Opportunity for Europe's Information Society*, sponsored by the Telematics Applications Programme of DG XIII (Luxembourg: LINGLINK, Anite Systems).

EVANS, D. (2001), *Emotion: The Science of Sentiment* (Oxford: Oxford University Press).

——— and CRUSE, P. (eds.) (2004), *Emotion, Evolution, and Rationality* (Oxford: Oxford University Press).

EVANS, G. (1982), *The Varieties of Reference* (Oxford: Clarendon Press).

EVANS, T. G. (1968), 'A Program for the Solution of Geometric-Analogy Intelligence Test Questions', in Minsky (1968: 271–353). Based on his Ph.D. diss., MIT, Department of Electrical Engineering, 1963.

EWING, A. C. (1934), *Idealism: A Critical Survey* (London: Methuen; 3rd edn., 1961).

——— (1954), 'The Relation Between Mind and Body as a Problem for the Philosopher', *Philosophy*, 29: 112–21.

EYSENCK, H. J., ARNOLD, W. J., and MEILI, R. (1975), *Encyclopedia of Psychology* (London: Fontana/Collins; first pub. London: Search Press, 1972).

FAHLMAN, S. E. (1974), 'A Planning System for Robot Construction Tasks', *Artificial Intelligence*, 5: 1–50; first pub. as MIT AI Lab Memo AI-TR-283 (Cambridge, Mass.: MIT, May 1973).

——— and LEBIERE, C. (1990), 'The Cascade-Correlation Learning Architecture', in D. S. Touretsky (ed.), *Advances in Neural Information-Processing Systems II* (San Mateo, Calif.: Morgan Kauffman), 524–32.

FAISAL, A. A., WHITE, J. A., and LAUGHLIN, S. B. (2005), 'Ion-Channel Noise Places Limits on the Miniaturization of the Brain's Wiring', *Current Biology*, 15: 1143–9.

FARARO, T. J., and SKVORETZ, J. (1984), 'Institutions as Production Systems', *Journal of Mathematical Sociology*, 10: 117–82.

FARLEY, B. G., and CLARK, W. A. (1954), 'Proposed Machine Simulation of Self-Organizing Systems Implemented as Large McCulloch–Pitts Networks', *Transactions of the Institute of Radio Engineers*, PG-IT 4: 76–84.

——— ——— (1961), 'Activity in Networks of Neuron-Like Elements', in C. Cherry (ed.), *Information Theory* (London: Butterworths), 242–51.

FARRELL, B. A. (1981), *The Standing of Psychoanalysis* (Oxford: Oxford University Press).

FATIMA, S. S., WOOLDRIDGE, M., and JENNINGS, N. R. (2004), 'An Agenda Based Framework for Multi-Issue Negotiation', *Artificial Intelligence*, 152: 1–45.

FAUCONNIER, G. R., and TURNER, M. (2002), *The Way We Think: Conceptual Blending and the Mind's Hidden Complexities* (New York: Basic Books).

FAUVEL, J., and WILSON, R. J. (1994), 'The Lull Before the Storm: Combinatorics and Religion in the Renaissance', *Bulletin of the Institute of Combinatorics and its Applications*, 11: 49–58.

FEIGENBAUM, E. A. (1961), 'The Simulation of Verbal Learning Behavior', *Proceedings of the Western Joint Computer Conference*, 19: 121–32.

——— (1969), 'Artificial Intelligence: Themes in the Second Decade', in A. J. H. Morrell (ed.), *Information Processing 68*, ii (Amsterdam: North-Holland), 1008–22.

——— (1977), 'The Art of Artificial Intelligence: Themes and Case Studies of Knowledge Engineering', *Proceedings of the Fifth International Joint Conference on Artificial Intelligence*, Cambridge, Mass., 1014–29.

——— (2005), 'Stories of AAAI—Before the Beginning and After: A Love Letter', *AI Magazine*, 26/4 (25th anniversary issue), 30–5.

——— BUCHANAN, B. G., and LEDERBERG, J. (1971), 'On Generality and Problem-Solving: A Case Study Using the DENDRAL Program', in Meltzer and Michie (1971: 165–90).

——— and FELDMAN, J. A. (eds.) (1963), *Computers and Thought* (New York: McGraw-Hill).

——— and MCCORDUCK, P. (1983), *The Fifth Generation: Artificial Intelligence and Japan's Computer Challenge to the World* (London: Addison-Wesley).

——— and SIMON, H. A. (1962), 'A Theory of the Serial Position Effect', *British Journal of Psychology*, 53: 307–20.

——— ——— (1984), 'EPAM-like Models of Recognition and Learning', *Cognitive Science*, 8: 305–36.

——— and WATSON, R. W. (1965), *An Initial Problem Statement for a Machine Induction Research Project*, Stanford AI Memo no. 30 (5 Apr.); repr. as the appendix to Lederberg (1987).

——— MCCORDUCK, P., and NII, H. P. (1988), *The Rise of the Expert Company: How Visionary Companies Are Using Artificial Intelligence to Achieve Higher Productivity and Profits* (London: Macmillan).

FEIGL, H. (1958/1967), *The 'Mental' and the 'Physical': The Essay and a Postscript* (Minneapolis: University of Minnesota Press; *Essay* first pub. 1958).

FELDMAN, J. A., and BALLARD, D. H. (1982), 'Connectionist Models and their Properties', *Cognitive Science*, 6: 205–54.

—— and NARAYANAN, S. (2004), 'Embodiment in a Neural Theory of Language', *Brain and Language*, 89: 385–92.

FELLOUS, J.-M., and LINSTER, C. (1998), 'Computational Models of Neuromodulation', *Neural Computation*, 10: 771–805.

FELSTEN, G., and WASSERMAN, G. S. (1980), 'Visual Masking: Mechanisms and Theories', *Psychological Bulletin*, 88: 329–53.

FENG, J. (ed.) (2004), *Computational Neuroscience: A Comprehensive Approach* (London: Chapman & Hall).

FERNALD, M. R. (1912), 'The Diagnosis of Mental Imagery', *Psychological Monographs*, 14/58: 1–169.

FERRARI, M. (ed.) (2002), *The Varieties of Religious Experience: Centenary Essays* (Thorverton: Imprint Academic).

FESTINGER, L. (1957), *A Theory of Cognitive Dissonance* (Evanston, Ill.: Row, Peterson).

—— RIECKEN, H. W., and SCHACHTER, S. (1956), *When Prophecy Fails: A Social and Psychological Study of a Modern Group that Predicted the Destruction of the World* (Minneapolis: University of Minnesota Press).

FEYERABEND, P. (1963a), 'Materialism and the Mind–Body Problem', *Review of Metaphysics*, 17: 49–67.

—— (1963b), 'Mental Events and the Brain', *Journal of Philosophy*, 60: 295–6.

FIESLER, E. (1994), 'Comparative Bibliography of Ontogenetic Neural Networks', *Proceedings of the International Conference on Artificial Neural Networks*, Sorrento, 793–6.

FIKES, R. E., and NILSSON, N. J. (1971), 'STRIPS: A New Approach to the Application of Theorem Proving to Problem Solving', *Artificial Intelligence*, 2: 189–208.

—— HART, P. E., and NILSSON, N. J. (1972a), 'Learning and Executing Generalized Robot Plans', *Artificial Intelligence*, 3: 251–88.

—— —— —— (1972b), 'Some New Directions in Robot Problem Solving', in Meltzer and Michie (1972: 405–30).

FINCHAM, J. M., CARTER, C. S., VAN VEEN, V., STENGER, V. A., and ANDERSON, J. R. (2002), 'Neural Mechanisms of Planning: A Computational Analysis Using Event-Related fMRI', *Proceedings of the National Academy of Sciences, USA*, 99: 3346–51.

FINDLER, N. V. (ed.) (1979), *Associative Networks: Representation and Use of Knowledge by Computers* (London: Academic Press).

FINGER, S. (2000), 'Edgar D. Adrian: Coding in the Nervous System', in S. Finger, *Minds Behind the Brain: A History of the Pioneers and their Discoveries* (Oxford: Oxford University Press), 239–58.

FINKE, R. A., and SCHMIDT, M. J. (1977), 'Orientation-Specific Color After-Effects Following Imagination', *Journal of Experimental Psychology: Human Perception and Performance*, 3: 599–606.

FISCHER, M. D. (1994), *Applications in Computing for Social Anthropologists* (London: Routledge).

FISHER, S. (1990), 'Virtual Interface Environments', in Laurel (1990: 423–38); repr. in Packer and Jordan (2001: 239–46).

FITT, S. (2003), *Documentation and User Guide to UNISYN Lexicon and Post-Lexical Rules*, Technical Report, Centre for Speech Technology Research (Edinburgh: University of Edinburgh), available, together with lexicon and software, at <http://www.cstr.ed.ac.uk/projects/unisyn/unisyn_release.html>.

—— and ISARD, S. D. (1999), 'Synthesis of Regional English Using a Keyword Lexicon', *Proceedings of Eurospeech 99*, Budapest, 823–6.

FLAMM, K. (1987), *Targeting the Computer* (Washington: Brookings Institution).

FLANAGAN, O. (1992), *Consciousness Reconsidered* (Cambridge, Mass.: MIT Press).

—— (1994), 'Multiple Identity, Character Transformation, and Self-Reclamation', in G. Graham and G. L. Stephens (eds.), *Philosophical Psychopathology* (Cambridge, Mass.: MIT Press), 135–62.

FLAVELL, J. H. (1962), *The Developmental Psychology of Jean Piaget* (New York: Van Nostrand Reinhold).

FLECK, J. (1982), 'Development and Establishment in Artificial Intelligence', in N. Elias, H. Martins, and R. Whitley (eds.), *Scientific Establishments and Hierarchies*, special issue of *Sociology of the Sciences*, 6: 169–217.

FLEW, A. G. N. (1950/1955), 'Theology and Falsification', *University*, 1950 (this was an ephemeral student publication); first repr. in A. Flew and A. MacIntyre (eds.), *New Essays in Philosophical Theology* (London: SCM Press, 1955), 96–9.

FLOR, N., and HUTCHINS, E. L. (1991), 'Analyzing Distributed Cognition in Software Teams: A Case Study of Team Programming During Perfective Software Maintenance', in J. Koehnemann-Belliveau,

T. Moher, and S. Robertson (eds.), *Empirical Studies of Programmers: Fourth Workshop* (Norwood, NJ: Ablex), 36–64.

FLUGEL, J. C. (1951), *A Hundred Years of Psychology, 1833–1933; with Additional Part on Developments 1933–1947* (London: Duckworth).

FODOR, J. A. (1968), *Psychological Explanation: An Introduction to the Philosophy of Psychology* (New York: Random House).

—— (1970), 'Three Reasons for Not Deriving "Kill" from "Cause to Die" ', *Linguistic Inquiry*, 1: 429–38.

—— (1974), 'Special Sciences, or the Disunity of Science as a Working Hypothesis', *Synthese*, 28: 77–115; repr. in (Fodor 1975: 9–25).

—— (1975), *The Language of Thought: A Philosophical Study of Cognitive Psychology* (Hassocks: Harvester Press).

—— (1978a), 'Tom Swift and his Procedural Grandmother', *Cognition*, 6: 229–47.

—— (1978b), 'Propositional Attitudes', *The Monist*, 61: 501–23.

—— (1979), 'Reply to Johnson-Laird', *Cognition*, 7: 201–95.

—— (1980a), 'Methodological Solipsism Considered as a Research Strategy in Cognitive Psychology', *Behavioral and Brain Sciences*, 3: 63–72.

—— (1980b), 'Searle on What Only Brains Can Do', *Behavioral and Brain Sciences*, 3: 431–2.

—— (1981a), 'Three Cheers for Propositional Attitudes', in J. A. Fodor, *Representations: Philosophical Essays on the Foundations of Cognitive Science* (Brighton: Harvester Press), 100–203.

—— (1981b), 'The Present Status of the Innateness Controversy', in J. A. Fodor, *Representations: Philosophical Essays on the Foundations of Cognitive Science* (Brighton: Harvester Press), 257–316.

—— (1983), *The Modularity of Mind: An Essay in Faculty Psychology* (Cambridge, Mass.: MIT Press).

—— (1984), 'Observation Reconsidered', *Philosophy of Science*, 51: 23–43.

—— (1987a), 'Modules, Frames, Fridgeons, Sleeping Dogs, and the Music of the Spheres', in Garfield (1987: 26–36).

—— (1987b), *Psychosemantics: The Problem of Meaning in the Philosophy of Mind* (Cambridge, Mass.: MIT Press).

—— (1992), 'The Big Idea: Can there Be a Science of Mind?', *Times Literary Supplement*, 3 July.

—— (1995a), Review of Paul Churchland, *The Engine of Reason, the Seat of the Soul*, *Times Literary Supplement*, 25 Aug.; repr. in Fodor (1998b: 83–9).

—— (1995b), Review of John McDowell, *Mind and World*, *London Review of Books*, 17/8 (20 Apr.), 10–11; repr. in Fodor (1998b: 3–8).

—— (1997), Review of Jeff Elman *et al.*, *Rethinking Innateness*, *London Review of Books*, 20/12; repr. in Fodor (1998b: 143–51).

—— (1998a), *Concepts: Where Cognitive Science Went Wrong* (Oxford: Clarendon Press).

—— (1998b), *In Critical Condition: Polemical Essays on Cognitive Science and the Philosophy of Mind* (Cambridge, Mass.: MIT Press).

—— (1998c), 'The Trouble with Psychological Darwinism (a Review of Books by S. Pinker and H. Plotkin)', *London Review of Books*, 20/2 (15 Jan.), 11–13; repr. in Fodor (1998b: 203–14).

—— (1998d), 'There and Back Again: A Review of Annette Karmiloff-Smith's *Beyond Modularity*', in Fodor (1998b: 127–42).

—— (2000a), *A Theory of Content—and Other Essays* (Cambridge, Mass.: MIT Press).

—— (2000b), *The Mind Doesn't Work That Way: The Scope and Limits of Computational Psychology* (Cambridge, Mass.: MIT Press).

—— and BEVER, T. G. (1965), 'The Psychological Reality of Linguistic Segments', *Verbal Learning and Verbal Behavior*, 4: 414–20.

—— and GARRETT, M. F. (1966), 'Some Reflections on Competence and Performance (with Peer-Commentary)', in J. Lyons and R. J. Wales (eds.), *Psycholinguistic Papers* (Edinburgh: Edinburgh University Press).

—— and McLAUGHLIN, B. P. (1990), 'Connectionism and the Problem of Systematicity: Why Smolensky's Solution Doesn't Work', *Cognition*, 35: 183–204.

—— and PYLYSHYN, Z. W. (1981), 'How Direct Is Visual Perception? Some Reflections on Gibson's "Ecological Approach" ', *Cognition*, 9: 139–96.

—— —— (1988), 'Connectionism and Cognitive Architecture: A Critical Analysis', *Cognition*, 28: 3–71.

_____ JENKINS, J. J., and SAPORTA, S. (1967), 'Psycholinguistics and Communication Theory', in F. E. X. Dance (ed.), *Human Communication Theory: Original Essays* (New York: Holt, Rinehart & Winston), 160–201.

_____ BEVER, T. G., and GARRETT, M. F. (1974), *The Psychology of Language: An Introduction to Psycholinguistics and Generative Grammar* (New York: McGraw-Hill).

FOGEL, A., and THELEN, E. (1987), 'The Development of Expressive and Communicative Action in the First Year: Reinterpreting the Evidence from a Dynamic Systems Perspective', *Developmental Psychology*, 23: 747–61.

FOGEL, D. B. (ed.) (1998), *Evolutionary Computation: The Fossil Record* (New York: IEEE Press).

FOGEL, L. J., and McCULLOCH, W. S. (1970), 'Natural Automata and Prosthetic Devices', in D. M. Ramsay-Klee (ed.), *Aids to Biological Communication: Prosthesis and Synthesis*, ii (New York: Gordon & Breach), 221–62; repr. in D. B. Fogel (1998: 255–95).

_____ OWENS, A. J., and WALSH, M. J. (1966), *Artificial Intelligence Through Simulated Evolution* (New York: Wiley).

FORBUS, K. D., GENTNER, D., MARKMAN, A. B., and FERGUSON, R. W. (1998), 'Analogy Just Looks Like High Level Perception: Why a Domain-General Approach to Analogical Mapping Is Right', *Journal of Experimental and Theoretical AI*, 10: 231–57.

FORD, K. M., and HAYES, P. J. (eds.) (1991), *Reasoning Agents in a Dynamic World: The Frame Problem*, Proceedings of the First International Workshop on Human and Machine Cognition (Greenwich, Conn.: JAI Press).

_____ and PYLYSHYN, Z. W. (eds.) (1996), *The Robot's Dilemma Revisited: The Frame Problem in Artificial Intelligence* (Norwood, NJ: Ablex).

FORGY, C. L. (1981), *OPS5 User's Manual*, Technical Report, Computer Science Department (Pittsburgh: Carnegie Mellon University).

_____ (1984), *The OPS83 Report*, Technical Report, Computer Science Department (Pittsburgh: Carnegie Mellon University).

_____ and McDERMOTT, J. (1977), 'OPS, a Domain-Independent Production-System Language', *Proceedings of the Fifth International Joint Conference on Artificial Intelligence*, Cambridge, Mass., 933–9.

FORREST, S. (ed.) (1991), *Emergent Computation: Self-Organizing, Collective, and Cooperative Phenomena in Natural and Artificial Computing Networks* (Cambridge, Mass.: MIT Press); first pub. as a special issue of *Physica*, ser. D, 42 (1990).

FORSTER, E. M. (1909), 'The Machine Stops', *Oxford and Cambridge Review*, Nov., 8: 83–122.

FOSSEY, D. (1983), *Gorillas in the Mist* (London: Hodder & Stoughton).

FOSTER, J. M. (1967), *List Processing* (London: Macdonald).

_____ (1968), 'Assertions: Programs Written Without Specifying Unnecessary Order', in Michie (1968a: 387–91).

_____ and ELCOCK, E. W. (1969), 'ABSYS 1: An Incremental Compiler for Assertions: An Introduction', in Meltzer and Michie (1969a: 423–9).

FOUTS, R. S., HIRSCH, A. D., and FOUTS, D. H. (1982), 'Cultural Transmission of a Human Language in a Chimpanzee Mother–Infant Relationship', in H. E. Fitzgerald, J. A. Mullins, and P. Page (eds.), *Psychobiological Perspectives*, iii (New York: Plenum Press), 159–93.

FOX KELLER, E. (forthcoming), 'Booting-Up Baby', in J. Riskin (ed.), *The Sistine Gap: Essays in the History and Philosophy of Artificial Life* (Stanford, Calif.: Stanford University Press).

FRACKOWIAK, R. S. J., FRISTON, C. J., FRITH, C. D., DOLAN, R., and MAZZIOTTA, J. C. (eds.) (1997), *Human Brain Function* (San Diego: Academic Press).

FRANCHI, S., and GUZELDERE, G. (eds.) (1995), *Constructions of the Mind: Artificial Intelligence and the Humanities*, Stanford Humanities Review, special issue, 4/2: 1–345.

FRANKLIN, J., DAVIS, L., SHUMAKER, R., and MORAWSKI, P. (1987), 'Military Applications', in S. C. Shapiro and D. Eckroth (eds.), *Encyclopedia of Artificial Intelligence*, i (Chichester: Wiley), 604–14.

FRANKLIN, R. L. (1970), 'Critical Notice of D. C. Dennett's *Content and Consciousness*', *Australasian Journal of Philosophy*, 48: 264–73.

FRANKLIN, S., and GRAESSER, A. (1997), 'Is it an Agent, or Just a Program? A Taxonomy for Intelligent Agents', in J.-P. Müller, M. Wooldridge, and N. R. Jennings (eds.), *Intelligent Agents III: Agent Theories, Architectures, and Languages* (Berlin: Springer-Verlag), 21–35.

FRANKS, B., and COOPER, R. (1995), 'Why Some Hybrid Solutions Aren't Really Solutions (and Why Others Aren't Really Hybrid)', in Hallam (1995: 61–71).

FRASCINA, F. (ed.) (2000), *Pollock and After: The Critical Debate* (London: Routledge).

FRASER, C. M., *et al.* (25 authors, incl. J. C. Venter) (1955), 'The Minimal Gene Complement of *Mycoplasma Genitalium*', *Science*, 270: 397–403.

FRAZER, J. G. (1890), *The Golden Bough: A Study of Comparative Religion*, 2 vols. (London: Macmillan).

FREEMAN, W. J. (1979), 'EEG Analysis Gives Model of Neuronal Template-Matching Mechanism for Sensory Search with Olfactory Bulb', *Biological Cybernetics*, 35: 221–34.

——(1987), 'Simulation of Chaotic EEG Patterns with a Dynamic Model of the Olfactory System', *Biological Cybernetics*, 56: 139–50.

——(1991), 'The Physiology of Perception', *Scientific American*, 264: 78–85.

FREGE, G. (1884), *The Foundations of Arithmetic*, trans. J. L. Austin (Oxford: Oxford University Press, 1950).

FRENCH, R. (1990), 'Subcognition and the Limits of the Turing Test', *Mind*, 99: 53–65.

——(2000), 'The Turing Test: The First Fifty Years', *Trends in Cognitive Sciences*, 4: 115–21.

FRENCH, R. D. (1975), *Antivivisection and Medical Science in Victorian Society* (Princeton: Princeton University Press).

FRENKEL-BRUNSWIK, E. (1948), 'Dynamic and Cognitive Categorization of Qualitative Material. II. Interviews of the Ethnically Prejudiced', *Journal of Psychology*, 25: 261–77.

——(1949), 'Intolerance of Ambiguity as an Emotional and Perceptual Personality Variable', *Journal of Personality*, 18: 108–43.

FREUD, A. (1937), *The Ego and the Mechanisms of Defence* (London: Hogarth Press).

FREUD, S. (1895), *Project for a Scientific Psychology*, trans. J. Strachey, Standard edn., i (London: Hogarth Press, 1966), 295–397.

——(1900), *The Interpretation of Dreams*, Standard edn., iv and v (London: Hogarth Press, 1953).

——(1901), *The Psychopathology of Everyday Life*, trans. A Tyson, Standard edn., vi (London: Hogarth Press, 1960).

——(1909), *Analysis of a Phobia in a Five-Year-Old Boy*, trans. J. Strachey, Standard edn., x (London: Hogarth Press, 1955), 3–152.

——(1913), *Totem and Taboo*, trans. J. Strachey, Standard edn., xiii (London: Hogarth Press, 1955), 1–161.

——(1914), *On the History of the Psychoanalytic Movement*, Standard edn., xiv (London: Hogarth Press, 1957), 1–66.

——(1925), 'A Note Upon the "Mystic Writing Pad" ', Standard edn., xix (London: Hogarth Press, 1976), 227–32.

——(1926), *The Question of Lay Analysis*, trans. J. Strachey, Standard edn., xx (London: Hogarth Press, 1959), 183–258.

——(1927), *The Future of an Illusion*, trans. W. D. Robson-Scott and J. Strachey, Standard edn., xxi (London: Hogarth Press, 1961), 5–58.

——(1930), *Civilisation and its Discontents*, trans. J. Riviere and J. Strachey, Standard edn., xxi (London: Hogarth Press, 1961), 64–145.

——(1932), *Moses and Monotheism*, trans. J. Strachey, Standard edn., xxiii (London: Hogarth Press, 1964), 7–140.

FREYD, J. J. (1987), 'Dynamic Mental Representations', *Psychological Review*, 94: 427–38.

FRICK, F. C., and MILLER, G. A. (1951), 'A Statistical Description of Operant Conditioning', *American Journal of Psychology*, 64: 20–36.

FRIEDMAN, G. (1956), 'Selective Feedback Computers for Engineering Synthesis and Nervous System Analogy' (Extracts from M.Sc. thesis, UCLA), in D. B. Fogel (ed.), *Evolutionary Computation: The Fossil Record* (New York: IEEE Press, 1998), 30–84.

FRIENDLY, M. (1988), *Advanced LOGO: A Language for Learning* (Hillsdale, NJ: Lawrence Erlbaum).

FRITH, C. D. (1992), *The Cognitive Neuropsychology of Schizophrenia* (Hove: Erlbaum Press).

——and FRITH, U. (2000), 'The Physiological Basis of Theory of Mind: Functional Neuroimaging Studies', in Baron-Cohen *et al.* (2000: 334–56).

——PERRY, R., and LUMER, E. (1999), 'The Neural Correlates of Conscious Experience: An Experimental Framework', *Trends in Cognitive Science*, 3: 105–14.

——BLAKEMORE, S.-J., and WOLPERT, D. M. (2000), 'Abnormalities in the Awareness and Control of Action', *Philosophical Transactions of the Royal Society*, ser. B, 355: 1771–8.

FRITH, U. (1989), *Autism: Explaining the Enigma* (Oxford: Blackwell; 2nd edn., rev., 2003).

_____ (ed.) (1991), *Autism and Asperger Syndrome* (Cambridge: Cambridge University Press).

FRUDE, N. (1983), *The Intimate Machine: Close Encounters with the New Computers* (London: Century).

FRY, D. P., and DENES, P. (1953), 'Experiments in Mechanical Speech Recognition', in W. Jackson (ed.), *Communication Theory* (New York: Academic Press), 206–12.

FRYER, D. M. (1978), 'Wheels Within Wheels: An Examination of the Nature of Psychological Explanation via a Theoretically Oriented History of Some Mechanical Models', Ph.D. thesis, University of Edinburgh.

_____ and MARSHALL, J. C. (1979), 'The Motives of Jacques de Vaucanson', *Technology and Culture*, 20: 257–69.

FULLER, S. (1995), 'Commentary on Harnad's "The Post-Gutenberg Galaxy"', *Times Higher Educational Supplement*, 12 May.

FUNAHASHI, K.-I., and NAKAMURA, Y. (1993), 'Approximation of Dynamical Systems by Continuous Time Recurrent Neural Networks', *Neural Networks*, 6: 801–6.

FUNT, B. V. (1980), 'Problem-Solving with Diagrammatic Representations', *Artificial Intelligence*, 13: 201–30.

_____ (1983), 'A Parallel-Process Model of Mental Rotation', *Cognitive Science*, 7: 67–93.

FURTH, M., CHANG, C. C., and CHURCH, A. (1974), 'Richard Montague, Philosopher, Los Angeles', *University of California: In Memoriam* (University of California).

FURUKAWA, K., MICHIE, D. M., and MUGGLETON, S. (eds.) (1994), *Machine Intelligence 13: Machine Intelligence and Inductive Learning* (Oxford: Clarendon Press).

GABRIEL, R. (1987), 'LISP', in S. C. Shapiro and D. Eckroth (eds.), *Encyclopedia of Artificial Intelligence*, i (Chichester: Wiley), 508–28.

GAINES, B. R., and ANDREAE, J. H. (1966), 'A Learning Machine in the Context of the General Control Problem', *Proceedings of the 3rd IFAC [International Federation for Automatic Control] Congress, London* (London: Butterworths), 342–9.

GAJARDO, A., MOREIRA, A., and GOLES, E. (2002), 'Complexity of Langton's Ant', *Discrete Mathematics*, 117: 41–50.

GALAMBOS, R., SHEATZ, G., and VERNIER, V. G. (1956), 'Electrophysiological Correlates of a Conditioned Response in Cats', *Science*, 123: 376–7.

GALANTER, E. (1988), 'Writing *Plans*', in Hirst (1988*a*: 36–44).

_____ and GERSTENHABER, M. (1956), 'On Thought: The Extrinsic Theory', *Psychological Review*, 63: 218–27.

GALE, D., and PROPP, J. (1994), 'Further Ant-ics', *Mathematical Intelligencer*, 16/1: 37–42.

_____ _____ SUTHERLAND, S., and TROUBETZKOY, S. (1995), 'Further Travels with my Ant', *Mathematical Intelligencer*, 17/3: 48–56.

GALLAGHER, J. G., and BEER, R. D. (1993), 'A Qualitative Dynamical Analysis of Evolved Locomotion Controllers', in J.-A. Meyer, H. Roitblat, and S. Wilson (eds.), *From Animals to Animats 2*, Proceedings of the Second International Conference on Simulation of Adaptive Behavior (Cambridge, Mass.: MIT Press), 71–80.

GALLAGHER, S. (2004), 'The Minds, Machines, and Brains of a Passionate Scientist: An Interview with Michael Arbib', *Journal of Consciousness Studies*, 11/12: 50–67.

GALLISTEL, C. R. (1990), *The Organization of Learning* (Cambridge, Mass.: MIT Press).

_____ (1995), 'The Replacement of General-Purpose Theories with Adaptive Specializations', in Gazzaniga (1995: 1255–67).

_____ (1997), 'Neurons and Memory', in Gazzaniga (1997: 72–89).

_____ and GIBBON, J. (2002), *The Symbolic Foundations of Conditioned Behaviour* (Hillsdale, NJ: Lawrence Erlbaum).

GALLWEY, W. T. (1974), *The Inner Game of Tennis* (New York: Random).

GANDY, R. (1996), 'Human Versus Mechanical Intelligence', in Millican and Clark (1996: 125–36).

GÄRDENFORS, P. (ed.) (1992), *Belief Revision* (Cambridge: Cambridge University Press).

GARDNER, B. T., and GARDNER, R. A. (1971), 'Two-Way Communication with an Infant Chimpanzee', in A. Schrier and F. Stollnitz (eds.), *Behavior of Nonhuman Primates*, iv (New York: Academic Press), 117–84.

_____ _____ (1974), 'Comparing the Early Utterances of Child and Chimpanzee', in A. Pick (ed.), *Minnesota Symposium on Child Psychology*, viii (Minneapolis: University of Minnesota Press), 3–23.

GARDNER, H. (1985), *The Mind's New Science: A History of the Cognitive Revolution* (New York: HarperCollins).

GARDNER, M. (1958), *Logic Machines and Diagrams* (New York: McGraw-Hill). (The chapter 'The *Ars Magna* of Ramón Lull' repr. as ch. 3 of M. Gardner, *Science: Good, Bad, and Bogus* (Buffalo, NY: Prometheus, 1989).)

——(1970), 'The Fantastic Combinations of John Conway's New Solitaire Game "Life"', *Scientific American*, 223/4: 120–3.

GARDNER, R. A., and GARDNER, B. T. (1969), 'Teaching Sign Language to a Chimpanzee', *Science*, 165: 664–72.

GARDNER-MEDWIN, A. R., and BARLOW, H. B. (2001), 'The Limits of Counting Accuracy in Distributed Neural Representations', *Neural Computation*, 13: 477–504.

GARFIELD, E. (1975), 'Highly Cited Articles: 19. Human Psychology and Behavior', in E. Garfield, *Essays of an Information Scientist* (Philadelphia: Institute for Scientific Information), 262–8.

GARFIELD, J. L. (ed.) (1987), *Modularity in Knowledge Representation and Natural Language Understanding* (Cambridge, Mass.: MIT Press).

GARFINKEL, H. (1956), 'Conditions of Successful Status Degradation Ceremonies', *American Journal of Sociology*, 61: 420–4.

——(1967), *Studies in Ethnomethodology* (Englewood Cliffs, NJ: Prentice-Hall).

GARNER, W. R. (1988), 'The Contribution of Information Theory to Psychology', in Hirst (1988a: 19–35).

GARTHWAITE, J., CHARLES, S. L., and CHESS-WILLIAMS, R. (1988), 'Endothelium-Derived Relaxing Factor Release on Activation of NMDA Receptors Suggests Role as Intracellular Messenger in the Brain', *Nature*, 336: 385–8.

GARTLAND-JONES, A., and COPLEY, P. (2003), 'The Suitability of Genetic Algorithms for Musical Composition', *Contemporary Music Review*, 22/3: 43–55.

GAULD, A., and SHOTTER, J. (1977), *Human Action and its Psychological Investigation* (London: Routledge & Kegan Paul).

GAZDAR, G. J. M. (1979a), *English as a Context-Free Language*, University of Sussex (Linguistics Group) Research Memo (Brighton).

——(1979b), 'Review of R. P. Stockwell's "Foundations of Syntactic Theory"', *Journal of Linguistics*, 15: 197–8.

——(1981a), 'On Syntactic Categories', in H. C. Longuet-Higgins, J. Lyons, and D. E. Broadbent (eds.), *The Psychological Mechanisms of Language* (London: Royal Society and British Academy), 53–69.

——(1981b), 'Speech Act Assignment', in A. K. Joshi, B. L. Webber, and I. A. Sag (eds.), *Elements of Discourse Understanding* (Cambridge: Cambridge University Press), 64–83.

——(1982), 'Phrase Structure Grammars', in P. I. Jacobson and G. K. Pullum (eds.), *The Nature of Syntactic Representation* (Dordrecht: Reidel), 131–86.

——(1987), 'Generative Grammar', in J. Lyons (ed.), *New Horizons in Linguistics: 2* (London: Penguin), 122–51.

——(1998), 'Applicability of Indexed Grammars to Natural Languages', in U. Reyle and C. Rohrer (eds.), *Natural Language Parsing and Linguistic Theories* (Dordrecht: Kluwer), 69–94.

——KLEIN, E., PULLUM, G., and SAG, I. A. (1985), *Generalized Phrase Structure Grammar* (Oxford: Blackwell).

——SPARCK JONES, K., and NEEDHAM, R. M. (eds.) (2000), *Computers, Language and Speech: Integrating Formal Theories and Statistical Data* (London: Royal Society).

GAZZANIGA, M. S. (1988), 'Life with George: The Birth of the Cognitive Neuroscience Institute', in Hirst (1988a: 230–41).

——(ed.) (1995), *The Cognitive Neurosciences* (Cambridge, Mass.: MIT Press).

——(ed.) (1997), *Conversations in the Cognitive Neurosciences* (Cambridge, Mass.: MIT Press).

——(ed.) (1999), *The New Cognitive Neurosciences* (Cambridge, Mass.: MIT Press); rev. edn. of Gazzaniga (1995).

——(ed.) (2000), *Cognitive Neuroscience: A Reader* (Oxford: Blackwell).

——and LEDOUX, J. E. (1978), *The Integrated Mind* (New York: Plenum Press).

GEACH, P. T. (1980), 'Some Remarks on Representations', *Behavioral and Brain Sciences*, 3: 80–1.

GEERTZ, C. (1963/1966), 'Religion as a Cultural System', in M. P. Banton (ed.), *Anthropological Approaches to the Study of Religion*, Proceedings of the Association of Social Anthropologists Conference, Jesus College, Cambridge, 1963 (London: Tavistock, 1966), 1–46.

——(1966), 'The Impact of the Concept of Culture on the Concept of Man', in J. R. Platt (ed.), *New Views of the Nature of Man* (Chicago: University of Chicago Press), 93–118; repr. in Geertz (1973a: 33–54).

_____ (1973a), *The Interpretation of Cultures: Selected Essays* (New York: Basic Books).

_____ (1973b), 'Thick Description: Toward an Interpretive Theory of Culture', in Geertz (1973a: 3–32).

GEHRING, W. J., and IKEO, K. (1999), 'Pax 6: Mastering Eye Morphogenesis and Eye Evolution', *Trends in Genetics*, 15: 371–7.

GELERNTER, H. L. (1959), 'Realization of a Geometry-Theorem Proving Machine', *Proceedings of an International Conference on Information Processing* (Paris: UNESCO House), 273–82.

_____ HANSEN, J. R., and GERBERICH, C. L. (1960a), 'A FORTRAN Compiled List Processing Language', *Journal of the Association for Computing Machinery*, 7: 87–101.

_____ _____ and LOVELAND, D. W. (1960b), 'Empirical Explorations of the Geometry-Theorem Proving Machine', *Proceedings of the Western Joint Computer Conference*, New York, 17: 143–7.

GELLNER, E. (1985), *The Psychoanalytic Movement: Or the Coming of Unreason* (London: Paladin).

_____ (1988), 'The Stakes in Anthropology', *American Scholar*, 57: 17–30.

_____ (1992), *Postmodernism, Reason, and Religion* (London: Routledge).

GENTNER, D. (1983), 'Structure-Mapping: A Theoretical Framework for Analogy', *Cognitive Science*, 7: 155–70.

_____ BREM, S., FERGUSON, R. W., MARKMAN, A. B., LEVIDOW, B. B., WOLFF, P., and FORBUS, K. D. (1997), 'Conceptual Change via Analogical Reasoning: A Case Study of Johannes Kepler', *Journal of the Learning Sciences*, 6: 3–40.

GEORGE, F. H. (1959), 'Cybernetic Models and their Applications/Cybernetics and Inductive Programming I & II', *Process Control and Automation*, 6: 92–7, 422–7, 478–82.

_____ (1961), *The Brain as a Computer* (Oxford: Pergamon Press).

GERBER, E. R. (1985), 'Rage and Obligation: Samoan Emotion in Conflict', in G. White and J. Kirkpatrick (eds.), *Person, Self, and Experience: Exploring Pacific Ethnopsychologies* (Berkeley: University of California Press), 121–67.

GERGEN, K. J. (1973), 'Social Psychology as History', *Journal of Personality and Social Psychology*, 26: 309–20.

_____ (1991), *The Saturated Self: Dilemmas of Identity in Contemporary Life* (New York: Basic Books).

_____ (1994), 'Exploring the Postmodern: Perils or Potentials?', *American Psychologist*, 49: 412–16.

_____ (2001), 'Psychological Science in a Postmodern Context', *American Psychologist*, 56: 803–13.

GETTING, P. A. (1989), 'Reconstruction of Small Neural Networks', in Koch and Segev (1989: 171–94).

GHAHRAMANI, Z., WOLPERT, D. M., and JORDAN, M. I. (1997), 'Computational Models of Sensorimotor Integration', in P. Morasso and V. Sanguineti (eds.), *Self-Organization, Computational Maps, and Motor Control* (Amsterdam: Elsevier), 117–47.

GIBSON, E. J., and WALK, R. D. (1960), 'The Visual Cliff', *Scientific American*, 202 (Apr.), 64–71.

GIBSON, J. J. (1950), *The Perception of the Visual World* (Cambridge, Mass.: Riverside Press).

_____ (1966), *The Senses Considered as Perceptual Systems* (Westport, Conn.: Greenwood Press).

_____ (1977), 'The Theory of Affordances', in R. Shaw and Bransford (1977: 67–82).

_____ (1979), *The Ecological Approach to Visual Perception* (London: Houghton Mifflin).

_____ (1982), 'Autobiography', in E. S. Reed and Jones (1982: 7–22).

GIBSON, W. (1984), *Neuromancer* (London: Gollancz).

_____ (2001), FOREWORD, in Packer and Jordan (2001, pp. ix–xii).

GIELEN, C. C. A. M., and VAN ZUYLEN, E. J. (1986), 'Coordination of Arm Muscles During Flexion and Supination: Application of the Tensor Analysis Approach', *Neuroscience*, 17: 527–39.

GIGERENZER, G. (1991a), 'How to Make Cognitive Illusions Disappear: Beyond "Heuristics and Biases"', *European Review of Social Psychology*, 2: 83–115.

_____ (1991b), 'From Tools to Theories: A Heuristic of Discovery in Cognitive Psychology', *Psychological Review*, 98: 254–7.

_____ (1994), 'Where Do New Ideas Come From?', in Boden (1994c: 53–74).

_____ (1996), 'On Narrow Norms and Vague Heuristics: A Reply to Kahneman and Tversky', *Psychological Review*, 103: 592–6.

_____ (2000), *Adaptive Thinking: Rationality in the Real World* (Oxford: Oxford University Press).

_____ (2002a), *Reckoning with Risk: Learning to Live with Uncertainty* (London: Penguin).

_____ (2002b), 'In the Year 2054: Innumeracy Defeated', in P. Sedlmeier and T. Betsch (eds.), *Etc.: Frequency Processing and Cognition* (Oxford: Oxford University Press), 55–66.

_____ (2004a), 'Striking a Blow for Sanity in Theories of Rationality', in Augier and March (2004: 389–410).

GIGERENZER, G. (2004*b*), 'Fast and Frugal Heuristics: The Tools of Bounded Rationality', in D. J. Koehler and N. Harvey (eds.), *Blackwell Handbook of Judgment and Decision Making* (Oxford: Blackwell), 62–88.

——(2004*c*), 'Mindless Statistics', *Journal of Socio-Economics*, 33: 587–606.

——and GOLDSTEIN, D. G. (1996*a*), 'Reasoning the Fast and Frugal Way: Models of Bounded Rationality', *Psychological Review*, 103: 650–69.

————(1996*b*), Mind as Computer: The Birth of a Metaphor', *Creative Research Journal*, 9: 131–44.

————(1999), 'Betting on One Good Reason: The Take the Best Heuristic', in Gigerenzer and Todd (1999: 75–95).

——and MURRAY, D. J. (1987), *Cognition as Intuitive Statistics* (Hillsdale, NJ: Lawrence Erlbaum).

——and SELTEN, R. (eds.) (2001), *Bounded Rationality: The Adaptive Toolbox* (Cambridge, Mass.: MIT Press).

——and STURM, T. (2004), 'Tools = Theories = Data? On Some Circular Dynamics in Cognitive Science', to appear in M. G. Ash and T. Sturm (eds.), *Psychology's Territories: Historical and Contemporary Perspectives from Different Disciplines* (Mahwah, N. J.: Erlbaum, forthcoming). Page-references are to MS version.

——and TODD, P. M. (eds.) (1999), *Simple Heuristics that Make Us Smart* (Oxford: Oxford University Press).

GILLISPIE, C. C. (1951), *Genesis and Geology: A Study in the Relations of Scientific Thought, Natural Theology, and Social Opinion in Great Britain, 1790–1850* (Cambridge, Mass.: Harvard University Press).

GILOVICH, T., VALLONE, R., and TVERSKY, A. (1985), 'The Hot Hand in Basketball: On the Misconception of Random Sequences', *Cognitive Psychology*, 17: 295–314.

GIORGI, A. (1970), *Psychology as a Human Science: A Phenomenologically Based Approach* (London: Harper & Row).

——(2000), 'Psychology as Human Science Revisited', *Journal of Humanistic Psychology*, 40/3: 56–73.

——(2005), 'Remaining Challenges for Humanistic Psychology', *Journal of Humanistic Psychology*, 45/2: 204–16.

GIUNTI, M. (1997), *Computation, Dynamics, and Cognition* (Oxford: Oxford University Press).

GLANVILLE, J. (1661–76), *The Vanity of Dogmatizing: The Three 'Versions'*, ed. S. Medcalf (Hove: Harvester Press, 1970).

GLEICK, J. (1987), *Chaos: Making a New Science* (New York: Viking Press).

GLOBUS, G. (1992), 'Derrida and Connectionism: Différance in Neural Nets', *Philosophical Psychology*, 5: 183–97.

GLUCK, M. A. (ed.) (1996), *Computational Models of Hippocampal Function in Memory, Hippocampus*, special issue, 6/6.

——MEETER, M., and MYERS, C. E. (2003), 'Computational Models of the Hippocampus Region: Linking Incremental Learning and Episodic Memory', *Trends in Cognitive Sciences*, 7: 269–76.

GOBET, F. (1998), 'Memory for Meaningless: How Chunks Help', *Proceedings of the 20th Meeting of the Cognitive Science Society* (Mahwah, NJ: Lawrence Erlbaum), 398–403.

GÖDEL, K. (1931), 'On Formally Undecidable Propositions of *Principia Mathematica* and Related Systems', trans. E. Mendelson for M. Davis (ed.), *The Undecidable: Basic Papers on Undecidable Propositions, Unsolvable Problems, and Computable Functions* (Hewlett, NY: Raven Press, 1965), 5–38; first pub. in German in *Monatshefte für Mathematik und Physik*, 38 (1931), 173–98.

GODFREY-SMITH, P. (1994), 'Spencer and Dewey on Life and Mind', in Brooks and Maes (1994: 80–9).

GOETHE, J. VON (1790), *An Attempt to Interpret the Metamorphosis of Plants (1790), and Tobler's Ode to Nature (1782)*, trans. A. Arber, Chronica Botanica, 10/2 (Waltham, Mass.: Chronica Botanica, 1946), 63–126.

GOFFMAN, E. (1959), *The Presentation of Self in Everyday Life* (Garden City, NY: Doubleday).

——(1967), *Interaction Ritual: Essays on Face-to-Face Behavior* (Chicago: Aldine).

GOLD, E. M. (1967), 'Language Identification in the Limit', *Information and Control*, 16: 447–74.

GOLDBERG, D. (1987), 'Computer-Aided Pipeline Operation Using Genetic Algorithms and Rule Learning. Part I: Genetic Algorithms in Pipeline Optimization', *Engineering with Computers*, 3: 35–45.

GOLDMAN, A. I. (1993), 'The Psychology of Folk Psychology', *Behavioral and Brain Sciences*, 16: 15–28.

——(forthcoming), *Simulating Minds: The Philosophy, Psychology, and Neuroscience of Mindreading* (New York: Oxford University Press).

GOLDSTEIN, I. P. (1975), 'Summary of MYCROFT: A System for Understanding Simple Picture Programs', *Artificial Intelligence*, 6: 249–88.

GOLEMAN, D. (1996), *Emotional Intelligence: Why it Can Matter More than IQ* (London: Bloomsbury).

GOLLER, C. (1999), 'Learning Search-Control Heuristics for Automated Deduction Systems with Folding Architecture Networks', *Proceedings of the European Symposium on Artificial Neural Networks*, Bruges, 45–50.

——and KÜCHLER, A. (1996), 'Learning Task-Dependent Distributed Representations by Back-Propagation Through Structure', *Proceedings of the International Conference on Neural Networks: ICNN-96*, 347–52.

GONZALES, F. M., PRESCOTT, T. J., GURNEY, K., HUMPHRIES, M., and REDGRAVE, P. (2000), 'An Embodied Model of Action Selection Mechanisms in the Vertebrate Brain', in J.-A. Meyer, A. Berthoz, D. Floreano, H. Roitblat, and S. W. Wilson (eds.), *From Animals to Animats 6*, Proceedings of the Sixth International Conference on Simulation of Adaptive Behavior (Cambridge, Mass.: MIT Press), 157–66.

GOOD, I. J. (1968), 'A Five-Year Plan for Automatic Chess', in Dale and Michie (1968: 89–118).

GOODALE, M. A., and MILNER, A. D. (1992), 'Separate Visual Pathways for Perception and Action', *Trends in Neuroscience*, 13: 20–3.

————(2003), *Sight Unseen: An Exploration of Conscious and Unconscious Vision* (Oxford: Oxford University Press).

GOODALL, J. (1986), *The Chimpanzees of Gombe: Patterns of Behavior* (Cambridge, Mass.: Belknap Press).

GOODENOUGH, W. H. (1956), 'Componential Analysis and the Study of Meaning', *Language*, 32: 195–216.

——(1965), 'Yankee Kinship Terminology: A Problem in Componential Analysis', *American Anthropologist*, 67: 259–87.

——(1990), 'Evolution of the Human Capacity for Beliefs', *American Anthropologist*, 92: 597–612.

GOODFIELD, G. J. (1960), *The Growth of Scientific Physiology: Physiological Method and the Mechanist–Vitalist Controversy, Illustrated by the Problems of Respiration and Animal Heat* (London: Hutchinson).

GOODMAN, N. (1951), *The Structure of Appearance* (Cambridge, Mass.: Harvard University Press).

——(1968), *Languages of Art: An Approach to a Theory of Symbols*, John Locke Lectures, Oxford 1962 (Indianapolis: Bobbs-Merrill).

——(1969), 'The Emperor's New Ideas', in Hook (1969: 138–42).

GOODWIN, B. C. (1974), 'Embryogenesis and Cognition', in W. D. Keidel, W. Handler, and M. Spreng (eds.), *Cybernetics and Bionics* (Munich: Oldenbourg), 47–54.

——(1976), *Analytical Physiology of Cells and Developing Organisms* (London: Academic Press).

GOPNIK, M., and CRAGO, M. B. (1991), 'Familial Aggregation of a Developmental Language Disorder', *Cognition*, 39: 1–50.

GORDON, R. M. (1986), 'Folk Psychology as Simulation', *Mind and Language*, 1: 158–71.

GOULD, S. J., and ELDREDGE, N. (1993), 'Punctuated Equilibrium Comes of Age', *Nature*, 366: 223–7.

GOUZE, J. L., LASRY, J. M., and CHANGEUX, J.-P. (1983), 'Selective Stabilization of Muscle Innervation During Development: A Mathematical Model', *Biological Cybernetics*, 46: 207–15.

GRAHAM, G. (1999), *The Internet: A Philosophical Enquiry* (London: Routledge).

GRAND, S. (2000), *Creation: Life and How to Make It* (London: Weidenfeld & Nicolson).

——(2003), *Growing Up with Lucy: How to Build an Android in Twenty Easy Steps* (London: Weidenfeld & Nicolson).

——and CLIFF, D. (1998), 'Creatures: Entertainment Software Agents with Artificial Life', *Autonomous Agents and Multi-Agent Systems*, 1: 39–57.

————and MALHOTRA, M. (1997), 'Creatures: Artificial Life Autonomous Software Agents for Home Entertainment', in W. L. Johnson and B. Hayes-Roth (eds.), *Proceedings of the First International Conference on Autonomous Agents* (New York: ACM Press), 22–9.

GRANIT, R. (1966), *Charles Scott Sherrington: An Appraisal* (London: Nelson).

GRAUBARD, S. (ed.) (1988), *The Artificial Intelligence Debate: False Starts, Real Foundations* (Cambridge, Mass.: MIT Press); first pub. as 'Artificial Intelligence', *Daedalus*, 117/1 (Winter 1988), from the Proceedings of the American Academy of Arts and Sciences.

GRAY, C. H. (1995), *A Cyborg Handbook* (London: Routledge).

GRAY, C. M., KONIG, P., ENGEL, A. E., and SINGER, W. (1989), 'Oscillatory Responses in Cat Visual Cortex Exhibit Inter-Columnar Synchronization Which Reflects Global Stimulus Properties', *Nature*, 338: 334–7.

GRAY, J. A. (1993), 'Consciousness, Schizophrenia, and Scientific Theory', in Bock and Marsh (1993: 263–81).

GRAY, J. A. (1995), 'Consciousness—What Is the Problem and How Should it Be Addressed?', *Journal of Consciousness Studies*, 2: 5–9.

———— (2000), 'Review on "Consciousness" ', *Times Higher Educational Supplement*, 13 Oct., 36.

GRAZIANO, M. S., and GROSS, C. G. (1993), 'A Bimodal Map of Space: Tactile Receptive Fields in the Macaque Putamen with Corresponding Visual Receptive Fields', *Experimental Brain Research*, 97: 96–109.

GREEN, B. F., WOLF, A. K., CHOMSKY, C., and LAUGHERY, K. R. (1961), 'BASEBALL: An Automatic Question-Answerer', *Western Joint Computer Conference*, 19: 219–24; repr. in Feigenbaum and Feldman (1963: 207–16).

GREEN, C. C. (1969), 'Theorem-Proving by Resolution as a Basis for Question-Answering', in Meltzer and Michie (1969*a*: 183–205).

GREEN, D. M., and SWETS, J. A. (1966), *Signal Detection Theory and Psychophysics* (New York: Wiley).

GREENBLATT, R. D. (1974), *The LISP Machine*, AI-WP-79 (Cambridge, Mass.: MIT AI Lab).

———— KNIGHT, T., HOLLOWAY, J., MOON, D., and WEINREB, D. (1984), *The Lisp Machine*, in D. R. Barstow, H. E. Shrobe, and E. Sandewall (eds.), *Interactive Programming Environments* (London: McGraw-Hill), 326–52.

GREENE, P. H. (1962), 'On Looking for Neural Networks and "Cell Assemblies" that Underlie Behavior—I: Mathematical Model; II: Neural Realization of the Mathematical Model', *Bulletin of Mathematical Biophysics*, 24: 247–75, 395–411.

———— (1965), 'Superimposed Random Coding of Stimulus–Response Connections', *Bulletin of Mathematical Biophysics*, 7: 191–202.

GREENFIELD, P. M., and BRUNER, J. S. (1969), 'Culture and Cognitive Growth', in D. A. Goslin (ed.), *Handbook of Socialization Theory and Research* (Chicago: Rand McNally), 633–54.

GREENWALD, A. G. (1992), 'New Look 3: Unconscious Cognition Reclaimed', *American Psychologist*, 47: 766–79.

GREFENSTETTE, J. J. (ed.) (1985), *Proceedings of the First International Conference on Genetic Algorithms and their Applications* (Hillsdale, NJ: Lawrence Erlbaum).

GREGORY, R. L. (1959), 'Models and the Localization of Function in the Central Nervous System', in Blake and Uttley (1959: ii. 669–80).

———— (1961), 'The Brain as an Engineering Problem', in W. H. Thorpe and O. L. Zangwill (eds.), *Current Problems in Animal Behaviour* (Cambridge: Cambridge University Press), 307–30.

———— (1964/1974), 'A Technique for Minimizing the Effects of Atmospheric Disturbance on Photographic Telescopes', *Nature*, 203: 274–6; repr., with prefatory comment, in Gregory (1974*a*: 501–18).

———— (1966), *Eye and Brain: The Psychology of Seeing* (London: Weidenfeld & Nicolson).

———— (1967), 'Will Seeing Machines Have Illusions?', in N. L. Collins and D. M. Michie (eds.), *Machine Intelligence 1* (Edinburgh: Edinburgh University Press), 169–80.

———— (1970), *The Intelligent Eye* (London: Weidenfeld & Nicolson).

———— (1974*a*), *Concepts and Mechanisms of Perception* (London: Duckworth).

———— (1974*b*), 'Perceptions as Hypotheses', in S. C. Brown (ed.), *Philosophy of Psychology* (London: Macmillan), 195–210, 231–5.

———— (1975), 'Do We Need Cognitive Concepts?', in M. S. Gazzaniga and C. Blakemore (eds.), *Handbook of Psychobiology* (London: Academic Press), 607–28.

———— (1977), *Eye and Brain: The Psychology of Seeing*, 3rd edn., with added chapter on 'Eye and Brain Machines', 226–40, 250 (London: Weidenfeld & Nicolson).

———— (2001), 'Adventures of a Maverick', in G. C. Bunn, A. D. Lovie, and G. D. Richards (eds.), *Psychology in Britain: Historical Essays and Personal Reflections* (Leicester: British Psychological Society), 381–92.

———— and WALLACE, J. G. (1963), *Recovery from Early Blindness: A Case Study*, Experimental Psychology Society Monograph no. 2 (Cambridge: Cambridge University Press).

———— ———— and CAMPBELL, F. W. (1959/1974), 'Changes in the Size and Shape of Visual After-Images Observed in Complete Darkness During Changes of Position in Space', *Quarterly Journal of Experimental Psychology*, 11: 54–5; repr., with prefatory comment, in Gregory (1974*a*: 295–8).

GRENE, M. G., and DEPEW, D. J. (2004), *The Philosophy of Biology: An Episodic History* (Cambridge: Cambridge University Press).

GREY WALTER, W. (1950*a*), 'An Electro-Mechanical Animal', *Discovery*, Mar., 90–5.

———— (1950*b*), 'An Imitation of Life', *Scientific American*, 182/5: 42–5.

———— (1953), *The Living Brain* (London: Duckworth).

_____ (1956), 'The Imitation of Mentality', *Proceedings of the Royal Institution*, 36: 365–9.

_____ (*c*.1968), 'The Future of Machina Speculatrix', MS, Burden Neurological Institute archives, Bristol.

GRICE, H. P. (1957), 'Meaning', *Philosophical Review*, 66: 377–88.

_____ (1967/1975), 'Logic and Conversation', in P. Cole and J. L. Morgan (eds.), *Syntax and Semantics*, iii: *Speech Acts* (New York: Academic Press), 41–58. This is part of the unpublished, but widely circulated, MS of the William James Lectures, Harvard University, 1967.

_____ (1969), 'Utterer's Meaning and Intentions', *Philosophical Review*, 68: 147–77.

GRIFFIN, D. R. (1976), *The Question of Animal Awareness: Evolutionary Continuity of Mental Experience* (New York: Rockefeller University Press).

_____ (1978), 'Prospects for a Cognitive Ethology', *Behavioral and Brain Sciences*, 4: 527–38.

_____ (1984), *Animal Thinking* (Cambridge, Mass.: Harvard University Press).

_____ (1992), *Animal Minds: Beyond Cognition to Consciousness* (Chicago: University of Chicago Press).

GROSS, C. G. (1968), Review of J. Konorski, *Integrative Activity of the Brain, Science*, 160: 652–3.

_____ (1998), *Brain, Vision, Memory: Tales in the History of Neuroscience* (Cambridge, Mass.: MIT Press).

_____ (2002), 'Genealogy of the "Grandmother Cell"', *Neuroscientist*, 8: 84–90.

_____ BENDER, D. B., and ROCHA-MIRANDA, C. E. (1969), 'Visual Receptive Fields of Neurons in Inferotemporal Cortex of the Monkey', *Science*, 166: 1303–6.

_____ ROCHA-MIRANDA, C. E., and BENDER, D. B. (1972), 'Visual Properties of Neurons in Inferotemporal Cortex of the Macaque', *Journal of Neurophysiology*, 35: 96–111.

GROSS, P. R. (1998), 'Bashful Eggs, Macho Sperm, and Tonypandy', in N. Koertge (ed.), *A House Built on Sand: Exposing Postmodernist Myths About Science* (Oxford: Oxford University Press), 59–65.

_____ and LEVITT, N. (eds.) (1994/1998), *Higher Superstition: The Academic Left and its Quarrels with Science* (Baltimore: Johns Hopkins University Press; new edn., 1998).

_____ _____ and LEWIS, M. W. (eds.) (1996), *The Flight from Science and Reason* (New York: New York Academy of Sciences).

GROSSBERG, S. (1964), *The Theory of Embedding Fields with Applications to Psychology and Neurophysiology* (New York: Rockefeller Institute for Medical Research).

_____ (1969), 'On Learning of Spatiotemporal Patterns by Networks with Ordered Sensory and Motor Components. I: Excitatory Components of the Cerebellum', *Studies in Applied Mathematics*, 48: 105–32.

_____ (1971), 'Pavlovian Pattern Learning by Non-Linear Neural Networks', *Proceedings of the National Academy of Sciences*, 68: 828–31.

_____ (1972), 'Neural Expectation: Cerebellar and Retinal Analogs of Cells Fired by Learnable or Unlearned Pattern Classes', *Kybernetik*, 10: 49–57.

_____ (1974), 'Classical and Instrumental Conditioning by Neural Networks', in R. Rosen and F. Snell (eds.), *Progress in Theoretical Biology*, iii (New York: Academic Press), 51–141.

_____ (1976*a*), 'Adaptive Pattern Classification and Universal Recoding: I. Parallel Development and Coding of Neural Feature Detectors', *Biological Cybernetics*, 23: 121–34; repr. in Anderson and Rosenfeld (1988: 245–58).

_____ (1976*b*), 'Adaptive Pattern Classification and Universal Recoding: II. Feedback, Expectation, Olfaction, and Illusions', *Biological Cybernetics*, 23: 187–202.

_____ (1980), 'How Does a Brain Build a Cognitive Code?', *Psychological Review*, 87: 1–5.

_____ (1982), *Studies of Mind and Brain: Neural Principles of Learning, Perception, Development, Cognition, and Motor Control* (Boston: Reidel).

_____ (1987*a*), 'Competitive Learning: From Interactive Activation to Adaptive Resonance', *Cognitive Science*, 11: 23–63; repr. in Grossberg (1988: 213–50).

_____ (1987*b*), 'Cortical Dynamics of Three-Dimensional Form, Color, and Brightness Perception. I: Monocular Theory. II: Binocular Theory', *Perception and Psychophysics*, 41: 87–158; repr. in Grossberg (1988: 1–126).

_____ (ed.) (1988), *Neural Networks and Natural Intelligence* (Cambridge, Mass.: MIT Press).

_____ (1995), 'The Attentive Brain', *American Scientist*, 83: 438–49.

_____ and HOWE, P. D. L. (2003), 'A Laminar Cortical Model of Stereopsis and Three-Dimensional Surface Perception', *Vision Research*, 43: 801–29.

_____ and MINGOLLA, E. (1986), 'Computer Simulations of Neural Networks for Perceptual Psychology', *Behavior Research Methods, Instruments, and Computers*, 18: 601–7; repr. in Grossberg (1988: 195–211).

GROSZ, B. (1977), 'The Representation and Use of Focus in a System for Understanding Dialogs', *Proceedings of the Fifth International Joint Conference on Artificial Intelligence*, Cambridge, Mass., 67–76.

GROSZ, B. (1996), 'Collaborative Systems', *AI Magazine*, 17/2: 67–85.

——— and KRAUS, S. (1996), 'Collaborative Plans for Complex Group Action', *Artificial Intelligence*, 86: 269–357.

——— ——— (1999), 'The Evolution of Shared Plans', in M. J. Wooldridge and A. Rao (eds.), *Foundations of Rational Agency* (Norwell, Mass.: Kluwer Academic), 227–62.

——— and SIDNER, C. L. (1979), 'Attention, Intentions, and the Structure of Discourse', *Computational Linguistics*, 12: 175–204.

GRÜNBAUM, A. (1984), *The Foundations of Psychoanalysis: A Philosophical Critique* (Berkeley: University of California Press).

——— (1986), 'Précis of *The Foundations of Psychoanalysis: A Philosophical Critique*', with peer commentary, *Behavioral and Brain Sciences*, 9: 217–84.

GRUSH, R. (2004), 'The Emulation Theory of Representation: Motor Control, Imagery, and Perception', *Behavioral and Brain Sciences*, 27: 377–442.

——— and CHURCHLAND, P. S. (1995), 'Gaps in Penrose's Toilings', *Journal of Consciousness Studies*, 2: 10–29.

GUAZZELLI, A., CORBACHO, F. J., BOTA, M., and ARBIB, M. A. (1998), 'Affordances, Motivation, and the World Graph Theory', *Adaptive Behavior*, 6: 435–71.

GUETZKOW, H. (ed.) (1962), *Simulation in Social Science: Readings* (Englewood Cliffs, NJ: Prentice-Hall).

GUHA, R. V., and LENAT, D. B. (1989), *The World According to CYC—Part 3*, Microelectronics and Computer Technology Corporation Technical Report ACT-AI-455-89 (Austin, Tex., Dec.).

GUILBAUT, S. (1983), *How New York Stole the Idea of Modern Art: Abstract Expressionism, Freedom, and the Cold War*, trans. A. Goldhammer (Chicago: University of Chicago Press). This wasn't published in French until 1988.

GULLAHORN, J. T., and GULLAHORN, J. E. (1963), 'A Computer Model of Elementary Social Behavior', in Feigenbaum and Feldman (1963: 375–86).

——— ——— (1964), 'Computer Simulation of Human Interaction in Small Groups', *American Federation of Information Processing Societies, Spring 1964 Joint Computer Conference*, 25: 103–13 (Washington: Spartan Books).

——— ——— (1965), 'The Computer as a Tool for Theory Development', in Hymes (1965: 401–26).

GUNDERSON, K. (1963), 'Interview with a Robot', *Analysis*, 23: 136–42.

——— (1964*a*), 'Descartes, La Mettrie, Language and Machines', *Philosophy*, 39: 193–222.

——— (1964*b*), 'The Imitation Game', *Mind*, 73: 234–45.

——— (1968), 'Robots, Consciousness, and Programmed Behaviour', *British Journal for the Philosophy of Science*, 19: 109–22.

——— (1971), *Mentality and Machines* (New York: Doubleday).

——— (1981), 'Paranoia Concerning Program-Resistant Aspects of the Mind—and Let's Drop Rocks on Turing's Toes Again', *Behavioral and Brain Sciences*, 4: 537–9.

——— (1985), *Mentality and Machines*, 2nd edn., expanded (Minneapolis: University of Minnesota Press).

GURNEY, K. N., PRESCOTT, T. J., and REDGRAVE, P. (1998), 'The Basal Ganglia Viewed as an Action Selection Device', *Proceedings of the Eighth International Conference on Artificial Neural Networks*, Sweden, 1033–8.

GUSTON, D. H. (2000), *Between Politics and Science: Assuring the Integrity and Productivity of Research* (Cambridge: Cambridge University Press).

GUY, K., GEORGHIOU, L., QUINTAS, P., HOBDAY, M., CAMERON, H., and RAY, T. (1991), *Evaluation of the Alvey Programme for Advanced Information Technology*, Report commissioned by the Department of Trade and Industry and the Science and Engineering Research Council (London: HMSO).

GUZELDERE, G., and FRANCHI, S. (eds.) (1994), *Bridging the Gap: Where Cognitive Science Meets Literary Criticism (Herbert Simon and Respondents)*, Stanford Humanities Review, special suppl., 4/1: 1–164.

GUZMAN, A. (1967), *Some Aspects of Pattern Recognition by Computer*, AI-TR-224 (Cambridge, Mass.: MIT AI Lab).

——— (1968), *Computer Recognition of Three-Dimensional Objects in a Visual Scene*, AI-TR-228 (Cambridge, Mass.: MIT AI Lab).

——— (1969), 'Decomposition of a Visual Field into Three-Dimensional Bodies', in A. Grasselli (ed.), *Automatic Interpretation and Classification of Images* (New York: Academic Press), 243–76.

HAACK, S. (1996), 'Towards a Sober Sociology of Science', in P. R. Gross, N. Levitt, and M. W. Lewis (eds.), *The Flight from Science and Reason* (New York: New York Academy of Sciences), 259–65.

HACKMAN, W. D. (1989), 'Scientific Instruments: Models of Brass and Aids to Discovery', in D. Gooding, T. Pinch, and S. Schaffer (eds.), *The Uses of Experiment* (Cambridge: Cambridge University Press).

HAIKONEN, P. O. (2003), *The Cognitive Approach to Conscious Machines* (Thorverton: Imprint Academic).

HAIRE, M., and GRUNES, W. F. (1950), 'Perceptual Defenses: Processes Protecting an Organized Perception of Personality', *Human Relations*, 3: 403–12.

HALDANE, E. S., and ROSS, G. R. T. (trans. and eds.) (1911), *Philosophical Works of Descartes*, 2 vols. (Cambridge: Cambridge University Press).

HALL, M. (1833), 'On the Reflex Function of the *Medulla Oblongata* and *Medulla Spinalis*', *Philosophical Transactions of the Royal Society*, 123: 635–65.

HALLAM, J. (ed.) (1995), *Hybrid Problems, Hybrid Solutions* (Oxford: IOS Press).

HALLOWELL, A. I. (1955), 'The Self and its Behavioral Environment', in A. I. Hallowell, *Culture and Experience*, Publications of the Philadelphia Anthropological Society, 4 (Philadelphia: University of Pennsylvania Press), 75–110.

HALPERN, J. Y., and PEARL, J. (2005*a*), 'Causes and Explanations: A Structural-Model Approach. Part I: Causes', *British Journal for the Philosophy of Science*, 56: 863–87.

—— —— (2005*b*), 'Causes and Explanations: A Structural-Model Approach. Part II: Explanations', *British Journal for the Philosophy of Science*, 56: 889–911.

HALVORSEN, P.-K., and LADUSAW, W. A. (1979), 'Montague's "Universal Grammar": An Introduction for the Linguist', *Linguistics and Philosophy*, 3: 185–223.

HAMEROFF, S. R. (1994), 'Quantum Coherence in Microtubules: A Neural Basis for Emergent Consciousness?', *Journal of Consciousness Studies*, 1: 91–118.

HAMILTON, C. M., and HEDBERG, S. (1997), 'Modern Masters of an Ancient Game', *AI Magazine*, 18/4: 11–12.

[HAMLYN, D. M.] (1970), 'Status of the Mental', *Times Literary Supplement*, 2 Dec., 152.

HAMPSHIRE, S. (1950), 'Critical Notice on Gilbert Ryle's *The Concept of Mind*', *Mind*, NS 59: 237–55.

HANKS, S., and MCDERMOTT, D. V. (1986), 'Default Reasoning, Nonmonotonic Logics, and the Frame Problem', *Proceedings of the American Association for Artificial Intelligence 1986*, Philadelphia, 328–33.

HANSEN-LOVE, O. (1972), *La Révolution Copernicienne du langage dans l'œuvre de Wilhelm von Humboldt* (Paris: Vrin).

HANSON, N. R. (1958), *Patterns of Discovery* (Cambridge: Cambridge University Press).

HARAWAY, D. J. (1986/1991), 'A Cyborg Manifesto: Science, Technology, and Socialist-Feminism in the Late Twentieth Century', in D. J. Haraway, *Simians, Cyborgs, and Women: The Reinvention of Nature* (London: Routledge, 1991), 149–81; rev. version of an essay written in 1986.

—— (1997),   *Modest-Witness@Second-Millennium.FemaleMan-Meets-OncoMouse: Feminism and Technoscience* (London: Routledge).

HARDING, S. (1986), *The Science Question in Feminism* (Milton Keynes: Open University Press).

HEDBERG, S. R. (2005), 'Celebrating Twenty-Five Years of AAAI: Notes from the AAA-05 and IAAI-05 Conferences', *AI Magazine*, 26/3: 12–16.

HARDY, S. (1984), 'A New Software Environment for List-Processing and Logic Programming', in T. O'Shea and M. Eisenstadt (eds.), *Artificial Intelligence: Tools, Techniques, and Applications* (London: Harper & Row), 110–36.

HAREL, I., and PAPERT, S. (eds.) (1991), *Constructionism: Research Reports and Essays 1985–1990, by the Epistemology and Learning Research Group* (Norwood, NJ: Ablex).

HARMAN, G. H. (1963), 'Generative Grammars Without Transformation Rules: A Defense of Phrase Structure', *Language*, 39: 597–616.

—— (1965), 'The Inference to the Best Explanation', *Philosophical Review*, 74: 88–95.

—— (1967), 'Psychological Aspects of the Theory of Syntax', *Journal of Philosophy*, 64: 75–87.

—— (1969), 'Linguistic Competence and Empiricism', in Hook (1969: 143–51).

—— (1988), 'Cognitive Science?', in Hirst (1988*a*: 258–68).

HARMON, P., and KING, D. (eds.) (1985), *Expert Systems: Artificial Intelligence in Business* (Chichester: Wiley).

HARNAD, S. (1978), Editorial, *Behavioral and Brain Sciences*, 1: 1–2.

—— (1985), 'Hebb, D. O.—Father of Cognitive Psychobiology 1904–1985', *Behavioral and Brain Sciences*, 8: 765.

—— (1989), 'Minds, Machines, and Searle', *Journal of Theoretical and Experimental Artificial Intelligence*, 1: 5–25.

HARNAD, S. (1990), 'The Symbol Grounding Problem', *Physica D*, 42: 335–46.

_____ (1991), 'Other Bodies, Other Minds: A Machine Incarnation of an Old Philosophical Problem', *Minds and Machines*, 1: 43–54.

_____ (1994), 'Artificial Life: Synthetic vs. Virtual', in C. G. Langton (ed.), *Artificial Life III* (Reading, Mass.: Addison-Wesley), 539–52.

_____ (1995*a*), 'The Post-Gutenberg Galaxy: How to Get There from Here', *Times Higher Educational Supplement*, 12 May.

_____ (1995*b*), 'There's Plenty of Room in Cyberspace: Sorting the Esoterica from the Exoterica (Response to Fuller)', *Times Higher Educational Supplement*, June.

_____ (2002), 'Minds, Machines, and Searle 2: What's Right and Wrong About the Chinese Room Argument', in Preston and Bishop (2002: 294–307).

HARRÉ, R. M. (1986), 'An Outline of the Social Constructionist Viewpoint', in R. M. Harré (ed.), *The Social Construction of Emotions* (Oxford: Blackwell), 2–14.

_____ (1994), 'The Second Cognitive Revolution: From Computation to Discourse', in A. Phillips-Griffiths (ed.), *Philosophy, Psychology, and Psychiatry* (Cambridge: Cambridge University Press), 25–40.

_____ (2002), *Cognitive Science: A Philosophical Introduction* (London: Sage).

_____ and SECORD, P. F. (1972), *The Explanation of Social Behaviour* (Oxford: Blackwell).

HARRIES-JONES, P. (1995), *A Recursive Vision: Ecological Understanding and Gregory Bateson* (Toronto: Toronto University Press).

HARRIS, C. S. (1980), 'Insight or Out of Sight? Two Examples of Perceptual Plasticity in the Human Adult', in C. S. Harris (ed.), *Visual Coding and Adaptability* (Hillsdale, NJ: Lawrence Erlbaum), 95–149.

HARRIS, P. L. (1991), 'The Work of the Imagination', in Whiten (1991: 141–172).

_____ and KOENIG, M. (forthcoming), 'Imagination and Testimony in Cognitive Development: The Cautious Disciple?', in I. Roth (ed.), *Imaginative Minds* (London: British Academy).

HARRIS, R. A. (1993), *The Linguistic Wars* (New York: Oxford University Press).

HARRIS, Z. S. (1951), *Methods in Structural Linguistics* (Chicago: University of Chicago Press).

_____ (1952), 'Discourse Analysis', *Language*, 28: 1–30.

_____ (1957), 'Co-occurrence and Transformation in Linguistic Structure', *Language*, 33: 283–340.

_____ (1968), *Mathematical Structures of Language* (New York: Interscience).

HARRISON, A. (1978), *Making and Thinking: A Study of Intelligent Activities* (Hassocks: Harvester Press).

HARRISON, H., and MINSKY, M. L. (1992), *The Turing Option* (London: Viking Press).

HART, P. E. (1975), 'Progress on a Computer-Based Consultant', *Proceedings of the Fourth International Joint Conference on Artificial Intelligence*, Tbilisi, 831–41.

HARTLEY, D. (1749), *Observations on Man: His Frame, his Duty, and his Expectations* (London); facs. ed. T. L. Huguelet (repr. Gainesville, Fla.: Scholars' Facsimiles and Reprints, 1966).

HARTLINE, H. K. (1938), 'The Response of Single Optic Nerve Fibres of the Vertebrate Eye to Illumination of the Retina', *American Journal of Physiology*, 121: 400–15.

HARTREE, D. R. (1949), *Calculating Instruments and Machines* (Urbana: University of Illinois Press).

HARVEY, B. (1997), *Computer Science Logo Style*, 3 vols., 2nd edn. (Cambridge, Mass.: MIT Press).

HARVEY, I. (1992*a*), 'Species Adaptation Genetic Algorithms: A Basis for a Continuing SAGA', in F. J. Varela and P. Bourgine (eds.), *Toward a Practice of Autonomous Systems: Proceedings of the First European Conference on Artificial Life* (Cambridge, Mass.: MIT Press), 346–54.

_____ (1992*b*), *Untimed and Misrepresented: Connectionism and the Computer Metaphor*, COGS Research Paper no. 245 (Brighton: University of Sussex); repr. *AISB Quarterly*, no. 96 (1996), 20–7.

_____ (1994), 'Evolutionary Robotics and SAGA: The Case for Hill Crawling and Tournament Selection', in C. G. Langton (ed.), *Artificial Life III* (Reading, Mass.: Addison-Wesley), 299–326.

_____ (2000), 'Robotics: Philosophy of Mind Using a Screwdriver', in T. Gomi (ed.), *Evolutionary Robotics: From Intelligent Robots to Artificial Life*, iii (Ottawa: AAI Books), 207–30.

_____ (2005), 'Evolution and the Origins of the Rational', in Zilhao (2005: 113–31).

_____ and BOSSOMAIER, T. (1997), 'Time Out of Joint: Attractors in Asynchronous Random Boolean Networks', in P. Husbands and I. Harvey (eds.), *Fourth European Conference on Artificial Life* (Cambridge, Mass.: MIT Press), 67–75.

_____ and THOMPSON, A. (1996), 'Through the Labyrinth Evolution Finds a Way: A Silicon Ridge', in T. Higuchi and M. Iwata (eds.), *Proceedings of the First International Conference on Evolvable Systems: From Biology to Hardware, 1996* (ICES96) (Berlin: Springer-Verlag), 406–22.

—— Husbands, P., and Cliff, D. (1994), 'Seeing the Light: Artificial Evolution, Real Vision', in D. Cliff, P. Husbands, J.-A. Meyer, and S. W. Wilson (eds.), *From Animals to Animats 3*, Proceedings of the Third International Conference on Simulation of Adaptive Behavior (Cambridge, Mass.: MIT Press), 392–401.

Harvey, W. (1628), *Movement of the Heart and Blood in Animals: An Anatomical Essay*, trans. K. J. Franklin, in *William Harvey: The Circulation of the Blood, and Other Writings* (London: Dent, 1963), 1–112.

—— (1649), *The Circulation of the Blood*, trans. K. J. Franklin, in *William Harvey: The Circulation of the Blood, and Other Writings* (London: Dent, 1963), 113–210.

Haugeland, J. (1978), 'The Nature and Plausibility of Cognitivism', *Behavioral and Brain Sciences*, 1: 215–26.

—— (ed.) (1981a), *Mind Design: Philosophy, Psychology, Artificial Intelligence* (Cambridge, Mass.: MIT Press).

—— (1981b), 'Analog and Analog', in J. I. Biro and R. W. Shahan (eds.), *Mind, Brain, and Function: Essays in the Philosophy of Mind* (Norman: University of Oklahoma Press), 213–26.

—— (1985), *Artificial Intelligence: The Very Idea* (Cambridge, Mass.: MIT Press).

—— (1995), 'Mind Embodied and Embedded', in Y.-H. Houng and J.-C. Ho (eds.), *Mind and Cognition* (Taipei: Academia Sinica), 3–38; repr. in Haugeland (1998: 207–37).

—— (1996), 'Body and World: A Review of *What Computers Still Can't Do* (Hubert L. Dreyfus)', *Artificial Intelligence*, 80: 119–28.

—— (ed.) (1997), *Mind Design II: Philosophy, Psychology, Artificial Intelligence*, rev., enlarged (Cambridge, Mass.: MIT Press).

—— (1998), *Having Thought: Essays in the Metaphysics of Mind* (Cambridge, Mass.: Harvard University Press).

Haugen, E. (1957), 'The Semantics of Icelandic Orientation', *Word*, 13: 447–59.

Hayes, J. E., and Levy, D. N. L. (1976), *The World Computer Chess Championship, Stockholm 1974* (Edinburgh: Edinburgh University Press).

—— and Michie, D. M. (eds.) (1983), *Intelligent Systems: The Unprecedented Opportunity* (Chichester: Ellis Horwood).

Hayes, K. J., and Hayes, C. (1951), 'Intellectual Development of a Home-Raised Chimpanzee', *Proceedings of the American Philosophical Society*, 95: 105–9.

Hayes, P. J. (1973), 'Some Comments on Sir James Lighthill's Report on Artificial Intelligence', *AISB European Newsletter*, 14 (July), 36–54.

—— (1979), 'The Naive Physics Manifesto', in D. M. Michie (ed.), *Expert Systems in the Micro-Electronic Age* (Edinburgh: Edinburgh University Press), 242–70.

—— (1985a), 'The Second Naive Physics Manifesto', in J. Hobbs and R. C. Moore (eds.), *Formal Theories of the Commonsense World* (Norwood, NJ: Ablex), 1–20.

—— (1985b), 'The Ontology of Liquids', in J. Hobbs and R. C. Moore (eds.), *Formal Theories of the Commonsense World* (Norwood, NJ: Ablex), 71–107.

—— (1992), 'Summary of: K. Ford & P. Hayes (1992): *Reasoning Agents in a Dynamic World: The Frame Problem*', *Psycoloquy*, 3/59, available at <http://psycprints.ecs.soton.ac.uk>.

—— and Ford, K. M. (1995), 'Turing Test Considered Harmful', *Proceedings of the Fourteenth International Joint Conference on Artificial Intelligence*, Montreal, 972–7.

Hayes-Roth, B. (1980), *Human Planning Processes*, Report no. R-2670-ONR (Santa Monica, Calif.: RAND Corporation).

—— and Hayes-Roth, F. (1978), *Cognitive Processes in Planning*, Report no. R-2366-ONR (Santa Monica, Calif.: RAND Corporation).

Hayes-Roth, F., Waterman, D. A., and Lenat, D. B. (eds.) (1983), *Building Expert Systems* (Reading, Mass.: Addison-Wesley).

Hayward, R. (2001a), 'The Tortoise and the Love-Machine: Grey Walter and the Politics of Electro-Encephalography', in C. Borck and M. Hagner (eds.), *Mindful Practices: On the Neurosciences in the Twentieth Century, Science in Context*, special issue, 14: 615–41.

—— (2001b), ' "Our Friends Electric": Mechanical Models of Mind in Postwar Britain', in G. C. Bunn, A. D. Lovie, and G. D. Richards (eds.), *Psychology in Britain: Historical Essays and Personal Reflections* (London: BPS Books/Science Museum, 2001), 290–308.

Head, H., and Holmes, G. (1911), 'Sensory Disturbances from Cerebral Lesions', *Brain*, 34: 102–254.

Heal, J. (1986), 'Replication and Functionalism', in J. Butterfield (ed.), *Language, Mind and Logic* (Cambridge: Cambridge University Press), 135–50.

HEAL, J. (1994), 'Simulation Versus Theory Theory: What Is at Issue?', in Peacocke (1994: 129–44).

—— (1996), 'Simulation, Theory and Content', in Carruthers and Smith (1996: 75–89).

—— (1998), 'Co-cognition and Off-Line Simulation: Two Ways of Understanding the Simulation Approach', *Mind and Language*, 13: 477–98.

HEARNSHAW, L. S. (1964), *A Short History of British Psychology 1840–1940* (London: Methuen).

HEARST, M., and HIRSH, H. (2000), 'AI's Greatest Trends and Controversies', *IEEE Intelligent Systems and their Applications*, 15: 8–17.

HEBB, D. O. (1949), *The Organization of Behavior: A Neuropsychological Theory* (New York: Wiley).

—— (1955), 'Drives and the CNS (Conceptual Nervous System)', *Psychological Review*, 62: 243–54.

—— (1958), *A Textbook of Psychology* (Philadelphia: Saunders).

—— (1960), 'The American Revolution', *American Psychologist*, 15: 735–45.

—— (1980), *Essay on Mind* (Hillsdale, NJ: Lawrence Erlbaum).

HEERWAGEN, J. H., and ORIANS, G. H. (1986), 'Adaptations to Windowlessness: A Study of the Use of Visual Decor in Windowed and Windowless Offices', *Environment and Behavior*, 18: 623–39.

—— —— (1993), 'Humans, Habitats, and Aesthetics', in S. R. Kellert and E. O. Wilson (eds.), *The Biophilia Hypothesis* (Washington: Island Press), 138–72.

HEIDER, F. (1958), *The Psychology of Interpersonal Relations* (New York: Wiley).

—— and SIMMEL, M. L. (1944), 'An Experimental Study of Apparent Behavior', *American Journal of Psychology*, 57: 243–59.

HEILIG, M. L. (1955), 'The Cinema of the Future', trans. U. Feldman, *Presence: Teleoperators and Virtual Environments*, 1 (1992), 279–94; first pub. in Mexico as 'El Cine del Futuro', *Espacios*, nos. 23–4 (Jan.–June).

HEIMS, S. J. (1991), *The Cybernetics Group* (Cambridge, Mass.: MIT Press).

HEISER, J. F., COLBY, K. M., FAUGHT, W. S., and PARKINSON, R. C. (1980), 'Can Psychiatrists Distinguish a Computer Simulation of Paranoia from the Real Thing? The Limitations of Turing-Like Tests as Measures of the Adequacy of Simulations', *Journal of Psychiatric Research*, 15: 149–62.

HELD, R., and HEIN, A. (1963), 'Movement-Produced Stimulation in the Development of Visually-Guided Behavior', *Journal of Comparative Physiological Psychology*, 56: 872–6.

HELMHOLTZ, H. VON (1850), 'On the Rate of Transmission of the Nerve Impulse', trans. Mrs A. G. Dietze, in W. Dennis (ed.), *Readings in the History of Psychology* (New York: Appleton-Century-Crofts), 197–8.

—— (1853), 'On Goethe's Scientific Researches', trans. H. W. Eve, in H. Helmholtz, *Popular Lectures on Scientific Subjects*, new edn. (London: Longman, Green, 1884), 29–52.

—— (1854), 'On the Interaction of the Natural Forces', trans. J. Tyndall, in H. von Helmholtz, *Popular Lectures on Scientific Subjects*, new edn. (London: Longman, Green, 1884), 137–74.

—— (1860), 'Theory of Colour Vision', trans. B. Rand, in W. Dennis (ed.), *Readings in the History of Psychology* (New York: Appleton-Century-Crofts), 199–205.

HEMPEL, C. G. (1952), 'Fundamentals of Concept Formation in Empirical Science', in O. Neurath (ed.), *International Encyclopedia of Unified Science*, ii/7 (Chicago: University of Chicago Press).

HENDRIKS-JANSEN, H. (1996), *Catching Ourselves in the Act: Situated Activity, Interactive Emergence, Evolution, and Human Thought* (Cambridge, Mass.: MIT Press).

HERDER, J. G. (1772), *Treatise Upon the Origin of Language*, trans. anon. (London: 1827).

HERLE, A., and ROUSE, S. (eds.) (1998), *Cambridge and the Torres Strait: Centenary Essays on the 1898 Anthropological Expedition* (Cambridge: Cambridge University Press).

HERMELIN, B., and O'CONNOR, N. (1989), 'Intelligence and Musical Improvisation', *Psychological Medicine*, 19: 447–57.

HERON, W. (1957), 'The Pathology of Boredom', *Scientific American*, 196: 52–6.

HERSEY, G. L. (1996), *The Evolution of Allure: Sexual Attractiveness from the Medici Venus to the Incredible Hulk* (Cambridge, Mass.: MIT Press).

HERTWIG, R., and GIGERENZER, G. (1999), 'The "Conjunction Fallacy" Revisited: How Intelligent Inferences Look Like Reasoning Errors', *Journal of Behavioral Decision-Making*, 12: 275–305.

—— and TODD, P. M. (2003), 'More Is Not Always Better: The Benefits of Cognitive Limits', in D. Hardman and L. Macchi (eds.), *Thinking: Psychological Perspectives on Reasoning, Judgment and Decision Making* (Chichester: Wiley), 213–31.

HESS, E. H. (1956), 'Space Perception in the Chick', *Scientific American*, 195/1: 71–82.

HEWITT, C. (1969), 'PLANNER: A Language for Proving Theorems in Robots', *Proceedings of the International Joint Conference on Artificial Intelligence, Washington, D.C.*, 295–301.

—— (1972), *Description and Theoretical Analysis (Using Schemata) of PLANNER: A Language for Proving Theorems and Manipulating Models in a Robot*, AI-TR-258 (Cambridge, Mass.: MIT AI Lab).

HEYES, C. M. (1998), 'Theory of Mind in Nonhuman Primates', with peer commentary, *Behavioral and Brain Sciences*, 21: 101–48.

—— and DICKINSON, A. (1990), 'The Intentionality of Animal Action', *Mind and Language*, 5: 87–104.

—— —— (1995), 'Folk Psychology Won't Go Away: Response to Allen and Bekoff', *Mind and Language*, 10: 329–32.

—— and HULL, D. L. (eds.) (2001), *Selection Theory and Social Construction: The Evolutionary Epistemology of Donald T. Campbell* (Albany: State University of New York Press).

HIEB, M. R., and MICHALSKI, R. S. (1993), 'Multiple Inference in Multi-Strategy Task-Adaptive Learning: Dynamic Interlaced Hierarchies', *Informatica: An International Journal of Computing and Informatics*, 17: 399–412.

HILGARD, E. R. (1965), *Hypnotic Susceptibility* (New York: Harcourt, Brace & World).

—— (1973), 'A Neo-Dissociation Interpretation of Pain Reduction in Hypnosis', *Psychological Review*, 80: 396–411.

—— (1974), 'Autobiography', in G. Lindzey (ed.), *A History of Psychology in Autobiography*, vi (Englewood Cliffs, NJ: Prentice-Hall), 129–60.

—— (1977/1986), *Divided Consciousness: Multiple Controls in Human Thought and Action* (New York: Wiley-Interscience; 2nd edn., 1986).

—— (1980), 'The Trilogy of Mind: Cognition, Affection, and Conation', *Journal of the History of the Behavioral Sciences*, 16: 107–17.

—— (1986), 'Interview with Ernest R. Hilgard', in Baars (1986: 285–98).

—— (1992), 'Dissociation and Theories of Hypnosis', in E. Fromm and M. R. Nash (eds.), *Contemporary Hypnosis Research* (New York: Guilford Press), 69–101.

—— and MARQUIS, D. G. (1940), *Conditioning and Learning* (New York: Appleton-Century-Crofts).

HILL, DEBORAH (1993), *Finding Your Way in Longgu: Geographical Reference in a Solomon Islands Language*, Cognitive Anthropology Research Group, Working Paper no. 21 (Nijmegen: Max Planck Institute for Psycholinguistics).

HILL, D. R. (ed. and trans.) (1974), *The Book of Knowledge of Ingenious Mechanical Devices, by Ibn al-Razzazz al-Jazari*, trans. from Arabic (Dordrecht: Reidel).

HILLER, L. A., and ISAACSON, L. M. (1959), *Experimental Music: Composition with an Electronic Computer* (New York: McGraw-Hill).

HILLIS, W. D. (1985), *The Connection Machine* (Cambridge, Mass.: MIT Press).

—— (1992), 'Co-evolving Parasites Improve Simulated Evolution as an Optimization Procedure', in Langton *et al.* (1992: 313–24). Amended version of a paper presented to the Emergent Systems Conference in Santa Fe, 1989.

HINTON, G. E. (1976), 'Using Relaxation to Find a Puppet', *Proceedings of the AISB Summer Conference*, University of Edinburgh, 148–57.

—— (1977), 'Relaxation and its Role in Vision', Ph.D. thesis, University of Edinburgh.

—— (1980), 'Inferring the Meaning of Direct Perception', *Behavioral and Brain Sciences*, 3: 387–8.

—— (1981), 'Shape Representations in Parallel Systems', *Proceedings of the Seventh International Joint Conference on Artificial Intelligence*, Vancouver, 1088–96.

—— (1988/1990), 'Representing Part–Whole Hierarchies in Connectionist Networks', in Hinton (1990: 47–75); first pub. in *Proceedings of the 10th Annual Conference of the Cognitive Science Society*, Montreal (Hillsdale, NJ: Lawrence Erlbaum).

—— (1989), 'Connectionist Learning Procedures', *Artificial Intelligence*, 40: 185–234.

—— (ed.) (1990), *Connectionist Symbol Processing, Artificial Intelligence*, special issue, 46: 1–2.

—— and ANDERSON, J. A. (1981), *Parallel Models of Associative Memory* (Hillsdale, NJ: Lawrence Erlbaum).

—— and SEJNOWSKI, T. J. (1986), 'Learning and Relearning in Boltzmann Machines', in Rumelhart, McClelland, *et al.* (1986: 282–317).

—— and SHALLICE, T. (1989/1991), *Lesioning a Connectionist Network: Investigations of Acquired Dyslexia*, Technical Report CRG-TR-89-3, Connectionist Research Group, University of Toronto; officially pub. as 'Lesioning an Attractor Network: Investigations of Acquired Dyslexia', *Psychological Review*, 98 (1991), 74–95.

HINTON, G. E., MCCLELLAND, J. L., and RUMELHART, D. E. (1986), 'Distributed Representations', in Rumelhart, McClelland, *et al.* (1986: 77–109).

HIRSCH, H. V. (1972), 'Visual Perception in Cats After Environmental Surgery', *Experimental Brain Research*, 15: 409–23.

HIRSCHFELD, L. A., and GELMAN, S. A. (eds.) (1994), *Mapping the Mind: Domain-Specificity in Culture and Cognition* (Cambridge: Cambridge University Press).

HIRST, W. (ed.) (1988*a*), *The Making of Cognitive Science: Essays in Honor of George A. Miller* (Cambridge: Cambridge University Press).

——— (1988*b*), 'Cognitive Psychologists Become Interested in Neuroscience', in Hirst (1988*a*: 242–56).

HOBART, R. E. (1934), 'Free-Will as Involving Determination and Inconceivable Without It', *Mind*, 43: 1–27.

HOBBES, T. (1651), *Leviathan; or, The Matter, Forme and Power of a Commonwealth Ecclesiasticall and Civil*, repr. ed. A. D. Lindsay (London: J. M. Dent, 1957).

HODGES, A. (1983), *Alan Turing: The Enigma* (London: Burnett Books).

——— (1997), *Turing: A Natural Philosopher* (London: Phoenix).

HODGKIN, A. L., and HUXLEY, A. F. (1952), 'A Quantitative Description of Membrane Current and its Application to Conduction and Excitation in Nerve', *Journal of Physiology*, 117: 500–44.

HODGSON, D. (1991), *The Mind Matters* (Oxford: Oxford University Press).

——— (2005), 'A Plain Person's Free Will', *Journal of Consciousness Studies*, 12: 3–19, 76–95; peer commentaries, 20–75.

HODGSON, P. W. (2002), 'Artificial Evolution, Music and Methodology', *Proceedings of the Seventh International Conference on Music Perception and Cognition*, Sydney (Adelaide: Causal Productions), 244–7.

——— (2005), 'Modeling Cognition in Creative Musical Improvisation', D.Phil. thesis, Department of Informatics, University of Sussex.

HOFSTADTER, D. R. (1979), *Gödel, Escher, Bach: An Eternal Golden Braid* (New York: Basic Books).

——— (1983/1985), 'Waking Up from the Boolean Dream, or Subcognition as Computation', and 'Post Scriptum', in Hofstadter (1985*a*: 631–54, 654–65); the first item was first pub. as 'Artificial Intelligence: Subcognition as Computation', in F. Machlup and U. Mansfield (eds.), *The Study of Information: Interdisciplinary Messages* (New York: Wiley, 1983), 263–85.

——— (1985*a*), *Metamagical Themas: Questing for the Essence of Mind and Pattern* (New York: Viking Press).

——— (1985*b*), 'Metafont, Metamathematics, and Metaphysics', in Hofstadter (1985*a*: 260–96). This item was *not* previously pub. in his *Scientific American* column.

——— (2001*a*), 'Staring Emmy Straight in the Eye—And Doing My Best Not to Flinch', in Cope (2001: 33–82).

——— (2001*b*), 'A Few Standard Questions and Answers', in Cope (2001: 293–305).

——— (2002), 'How Could a COPYCAT Ever Be Creative?', in Dartnall (2002: 405–24).

——— and DENNETT, D. C. (eds.) (1981), *The Mind's I: Fantasies and Reflections on Self and Soul* (Brighton: Harvester Press).

——— and FARG (The Fluid Analogies Research Group) (1995), *Fluid Concepts and Creative Analogies: Computer Models of the Fundamental Mechanisms of Thought* (New York: Basic Books).

——— and McGRAW, G. (1995), 'Letter Spirit: Esthetic Perception and Creative Play in the Rich Microcosm of the Roman Alphabet', in Hofstadter and FARG (1995: 407–66).

——— and MITCHELL, M. (1993/1995), 'The Copycat Project: A Model of Mental Fluidity and Analogy-Making', in K. Holyoak and J. Barnden (eds.), *Advances in Connectionist and Neural Computation Theory*, ii: *Analogical Connections* (Norwood, NJ: Ablex, 1993), 31–112; also pub. in Hofstadter and FARG (1995: 205–300), two chs.

——— MITCHELL, M., and FRENCH, R. M. (1987), *Fluid Concepts and Creative Analogies: A Theory and its Computer Implementation*, CRCC Technical Report 18, Center for Research on Concepts and Cognition, Indiana University, Bloomington.

HOIJER, H. (1954), 'The Sapir–Whorf Hypothesis', in H. Hoijer (ed.), *Language in Culture* (Chicago: University of Chicago Press).

HOLENDER, D., and DUSCHERER, K. (2004), 'Unconscious Perception: The Need for a Paradigm Shift', *Perception and Psychophysics*, 66: 872–81.

HOLLAN, J. D., HUTCHINS, E. L., and KIRSH, D. (2000), 'Distributed Cognition: Toward a New Foundation for Human–Computer Interaction Research', *ACM Transactions on Computer–Human Interaction*, 7: 174–96.

HOLLAND, J. H. (1959), *Survey of Automata Theory*, Report 2900-52-R, Willow Run Laboratories (Ypsilanti: University of Michigan).

—— (1962), 'Outline for a Logical Theory of Adaptive Systems', *Journal of the Association for Computing Machinery*, 9: 297–314.

—— (1975), *Adaptation in Natural and Artificial Systems* (Ann Arbor: University of Michigan Press).

—— HOLYOAK, K. J., NISBETT, R. E., and THAGARD, P. R. (1986), *Induction: Processes of Inference, Learning, and Discovery* (Cambridge, Mass.: MIT Press).

HOLLAND, O. (1997), 'Grey Walter: The Pioneer of Real Artificial Life', in C. G. Langton and K. Shimohara (eds.), *Artificial Life V: Proceedings of the Fifth International Workshop on the Synthesis and Simulation of Living Systems* (Cambridge, Mass.: MIT Press), 34–41.

—— (2002), 'Grey Walter: The Imitator of Life', in R. Damper and D. Cliff (eds.), *Biologically-Inspired Robotics: The Legacy of W. Grey Walter*, Proceedings of the EPSRC/BBSRC International Workshop WGW-02, Bristol, 32–48; rev. version: 'Exploration and High Adventure: The Legacy of Grey Walter', *Philosophical Transactions of the Royal Society of London*, ser. A, 361, special issue: *Biologically Inspired Robotics* (2003), 2085–121.

—— (ed.) (2003), *Machine Consciousness* (Exeter: Imprint Academic), *Journal of Consciousness Studies*, special issue, 10/4–5.

—— and MELHUISH, C. (1999), 'Stigmergy, Self-Organization and Sorting in Collective Robotics', *Artificial Life*, 5: 173–202.

HOLLIS, M. (1977), *Models of Man: Philosophical Thoughts on Social Action* (Cambridge: Cambridge University Press).

HOLSCHER, C. (1997), 'Nitric Oxide, the Enigmatic Neuronal Messenger: Its Role in Synaptic Plasticity', *Trends in Neuroscience*, 20: 298–303.

HOLTON, G. J. (1992), 'How to Think About the "Anti-Science" Phenomenon', *Public Understanding of Science: An International Journal of Research in the Public Dimensions of Science and Technology*, 1: 103–28.

—— (1993), *Science and Anti-Science* (Cambridge, Mass.: Harvard University Press).

HOLYOAK, K. J., and THAGARD, P. (1989), 'Analogical Mapping by Constraint Satisfaction', *Cognitive Science*, 13: 295–356.

HOMANS, G. C. (1961), *Social Behavior: Its Elementary Forms* (New York: Harcourt).

HONAVAR, V., and UHR, L. (eds.) (1994), *Artificial Intelligence and Neural Networks: Steps Toward Principled Integration* (Boston: Academic Press).

HONDERICH, T. (1984), 'Is the Mind Ahead of the Brain? Benjamin Libet's Evidence Examined', *Journal of Theoretical Biology*, 110: 115–29.

—— (1986), 'Mind, Brain and Time: Rejoinder to Libet', *Journal of Theoretical Biology*, 118: 367–75.

—— (2005), 'On Benjamin Libet: Is the Mind Ahead of the Brain? Behind It?', in T. Honderich, *On Determinism and Freedom* (Edinburgh: Edinburgh University Press, 2005), 71–95.

HOOK, S. (ed.) (1969), *Language and Philosophy* (New York: New York University Press).

HOPCROFT, J. E., and ULLMAN, J. D. (1979), *Introduction to Automata Theory, Languages, and Computation* (Reading, Mass.: Addison-Wesley).

HOPFIELD, J. J. (1982), 'Neural Networks and Physical Systems with Emergent Collective Computational Abilities', *Proceedings of the National Academy of Sciences, USA*, 79: 2554–8; repr. in Anderson and Rosenfeld (1988: 460–4).

—— (1984), 'Neurons with Graded Response have Collective Computational Properties Like Those of Two-State Neurons', *Proceedings of the National Academy of Sciences, USA*, 81: 3088–92; repr. in J. A. Anderson and Rosenfeld (1988: 579–84).

—— and TANK, D. W. (1986), 'Computing with Neural Circuits: A Model', *Science*, 233: 625–33.

HORCHLER, A., REEVE, R., WEBB, B., and QUINN, R. (2004), 'Robot Phonotaxis in the Wild: A Biologically Inspired Approach to Outdoor Sound Localization', *Advanced Robotics*, 18/8: 801–16.

HORNIK, K., STINCHCOMBE, M., and WHITE, H. (1989), 'Multilayered Feed-Forward Networks Are Universal Approximators', *Neural Networks*, 2: 359–66.

HORVITZ, E., and BARRY, M. (1995), 'Display of Information for Time-Critical Decisions', *Proceedings of the Eleventh Conference on Uncertainty in Artificial Intelligence*, Montreal, Aug. 1995 (San Francisco: Morgan Kaufmann), 296–305.

HORWOOD, W. (1987), *Skallagrigg* (London: Viking Press).

HOVLAND, C. I. (1952), 'A "Communication Analysis" of Concept Learning', *Psychological Review*, 59: 461–72.

HOVLAND, C. I. (ed.) (1957), *The Order of Presentation in Persuasion* (New Haven: Yale University Press).

———and HUNT, E. B. (1960), 'The Computer Simulation of Concept Attainment', *Behavioral Science*, 5: 265–7.

HOWARD, I. P. (1953), 'A Note on the Design of an Electro-Mechanical Maze Runner', *Durham Research Review*, no. 4 (Sept.), 54–61.

HOWE, J. A. M. (1994), 'Artificial Intelligence at Edinburgh University: A Perspective', written Nov. 1994; placed on Edinburgh's Division of Informatics web site, Aug. 1998, <http://www.dai.ed.ac.uk>.

———O'SHEA, T., and PLANE, F. (1980), 'Teaching Mathematics Through LOGO Programming: An Evaluation Study', in R. Lewis and E. D. Tagg (eds.), *Computer-Assisted Learning: Scope, Progress, and Limits* (Amsterdam: North-Holland), 85–102.

HSU, F. (2002), *Behind Deep Blue: Building the Computing that Defeated the World Chess Champion* (Princeton: Princeton University Press).

HUBEL, D. H. (1982), 'Exploration of the Primary Visual Cortex, 1955–78', *Nature*, 299: 515–24.

———(1988), *Eye, Brain, and Vision* (Oxford: W. H. Freeman).

———and WIESEL, T. N. (1959), 'Receptive Fields of Single Neurones in the Cat's Striate Cortex', *Journal of Physiology*, 148: 579–91.

——— ———(1962), 'Receptive Fields, Binocular Interaction and Functional Architecture in the Cat's Striate Cortex', *Journal of Physiology*, 160: 106–54.

——— ———(1968), 'Receptive Fields and Functional Architecture of Monkey Striate Cortex', *Journal of Physiology*, 195: 215–43.

——— ———(1974), 'Sequence Regularity and Geometry of Orientation Columns in the Monkey Striate Cortex', *Journal of Comparative Neurology*, 158: 267–94.

HUBER, F., and THORSON, J. (1985), 'Cricket Auditory Communication', *Scientific American*, 253/6: 47–54.

HUDSON, L. (1972), *The Cult of the Fact: A Psychologist's Autobiographical Critique of his Discipline* (London: Jonathan Cape).

———(1975), *Human Beings: An Introduction to the Psychology of Human Experience* (London: Jonathan Cape).

HUFFMAN, D. A. (1971), 'Impossible Objects as Nonsense Sentences', in Meltzer and Michie (1971: 295–325).

———(1977a), 'A Duality Concept for the Analysis of Polyhedral Scenes', in Elcock and Michie (1977: 475–92).

———(1977b), 'Realizable Configurations of Lines in Pictures of Polyhedra', in Elcock and Michie (1977: 493–509).

HUGHES, R. (1991), 'The Decline of the City of Mahagonny', in R. Hughes, *Nothing If Not Critical: Selected Essays on Art and Artists* (London: Harvill), 3–28.

HUHNS, M. N., and SINGH, M. P. (eds.) (1997), *Readings in Agents* (San Francisco: Kaufmann).

———and STEPHENS, L. M. (1999), 'Multiagent Systems and Societies of Agents', in G. Weiss (1999: 79–120).

HULL, C. L. (1920), *Quantitative Aspects of the Evolution of Concepts, Psychological Monographs*, 28/1.

———(1930), 'Knowledge and Purpose as Habit Mechanisms', *Psychological Review*, 37: 511–25.

———(1933), *Hypnosis and Suggestibility: An Experimental Approach* (New York: Appleton-Century-Crofts).

———(1934), 'The Concept of the Habit-Family-Hierarchy and Maze-Learning', *Psychological Review*, 41: 33–54, 134–52.

———(1937), 'Mind, Mechanism, and Adaptive Behavior', *Psychological Review*, 44: 1–32.

———(1943), *Principles of Behavior* (New York: Appleton-Century-Crofts).

———HOVLAND, C. I., ROSS, R. T., HALL, M., PERKINS, D. T., and FITCH, M. (1940), *Mathematico-Deductive Theory of Rote Learning: A Study in Scientific Methodology* (New Haven: Yale University Press).

HULL, D. L. (1982), 'The Naked Meme', in H. C. Plotkin (1982: 273–327).

———(1988a), *Science as a Process: An Evolutionary Account of the Social and Conceptual Development of Science* (Chicago: University of Chicago Press).

———(1988b), 'Interactors Versus Vehicles', in H. C. Plotkin (ed.), *The Role of Behavior in Evolution* (Cambridge, Mass.: MIT Press), 19–50.

HUMBOLDT, W. VON (1836), *On Language: The Diversity of Human Language-Structure and its Influence on the Mental Development of Mankind*, trans. P. Heath (Cambridge: Cambridge University Press, 1988).

HUME, D. (1739), *A Treatise of Human Nature*, ed. A. D. Lindsay (London: J. M. Dent, 1911).

―― (1748), *An Enquiry Concerning Human Understanding and Concerning the Principles of Morals*, ed. L. A. Selby-Bigge and P. H. Nidditch (Oxford: Clarendon Press, 1975).

HUMPHREY, J. A. C., BARTH, F. G., REED, M. A., and SPAK, A. (2003), 'The Physics of Arthropod Medium-Flow Sensitive Hairs: Biological Models for Artificial Sensors', in Barth *et al.* (2003: 129–44).

HUMPHREY, N., and DENNETT, D. C. (1989), 'Speaking for Our Selves: An Assessment of Multiple Personality Disorder', *Raritan*, 9: 68–98; repr. in D. C. Dennett, *Brainchildren: Essays in Designing Minds* (Cambridge, Mass.: MIT Press, 1998), 31–55.

HUMPHRIES, M. D., GURNEY, K., and PRESCOTT, T. J. (forthcoming), 'Is there an Integrative Center in the Vertebrate Brain-Stem? A Robotic Evaluation of a Model of the Reticular Formation Viewed as an Action Selection Device', *Adaptive Behavior*.

HUNT, E. B. (1962), *Concept Learning: An Information Processing Problem* (New York: Wiley).

―― and HOVLAND, C. I. (1961), 'Programming a Model of Human Concept Formulation', *Proceedings of the Western Joint Computer Conference*, 19: 145–55.

―― and QUINLAN, J. R. (1968), 'A Formal Deductive Problem-Solving System', *Journal of the Association for Computing Machinery*, 15: 625–46.

―― MARIN, J., and STONE, P. J. (1966), *Experiments in Induction* (London: Academic Press).

HUNTER, M. (1981), *Science and Society in Restoration England* (Cambridge: Cambridge University Press).

―― (1994), *The Royal Society and its Fellows 1660–1700: The Morphology of an Early Scientific Institution* (Oxford: Alden Press).

HURFORD, J. R. (2003), 'The Neural Basis of Predicate-Argument Structure (with peer-commentary)', *Behavioral and Brain Sciences*, 26: 261–316.

HUSBANDS, P. N. (1998), 'Evolving Robot Behaviours with Diffusing Gas Networks', in P. N. Husbands and J.-A. Meyer (eds.), *Evolutionary Robotics*, Proceedings of EvoRob-98, Paris, Apr. (Berlin: Springer-Verlag), 71–86.

―― (forthcoming), 'The Ratio Club', in M. W. Wheeler *et al.* (forthcoming).

―― HARVEY, I. and CLIFF, D. (1995), 'Circle in the Round: State Space Attractors for Evolved Sighted Robots', *Journal of Robotics and Autonomous Systems*, 15: 83–106.

―― SMITH, T., JAKOBI, N., and O'SHEA, M. (1998), 'Better Living Through Chemistry: Evolving Gas Nets for Robot Control', *Connection Science*, 10: 185–210.

HUTCHINS, E. L. (1980), *Culture and Inference: A Trobriand Case-Study* (Cambridge, Mass.: Harvard University Press).

―― (1995), *Cognition in the Wild* (Cambridge, Mass.: MIT Press).

―― and HINTON, G. E. (1984), 'Why the Islands Move', *Perception*, 13: 629–32.

―― HOLLAN, J. D., and NORMAN, D. A. (1986), 'Direct Manipulation Interfaces', in Norman and Draper (1986: 87–124).

HUTCHINS, W. J. (1986), *Machine Translation: Past, Present, Future* (Chichester: Ellis Horwood).

―― (1994), 'Research Methods and System Designs in Machine Translation: A Ten-Year Review, 1984–1994', in *Machine Translation, Ten Years On, 12–14 November 1994* (Cranfield: Cranfield University).

―― (1995), 'Machine Translation: A Brief History', in E. F. K. Koerner and R. E. Asher (eds.) (1995), *Concise History of the Language Sciences: From the Sumerians to the Cognitivists* (Oxford: Pergamon), 431–45.

HUTCHINSON, G. E. (1948), 'In Memoriam, D'Arcy Wentworth Thompson', *American Scientist*, 36: 577–606.

HUTCHISON, C. A., PETERSON, S. N., GILL, S. R., CLINE, R. T., WHITE, O., FRASER, C. M., SMITH, H. O., and VENTER, J. C. (1999), 'Global Transposon Mutagenesis and a Minimal *Mycoplasma Genome*', *Science*, 286: 2165–9.

HUXLEY, A. (1999), 'Obituary: Professor Sir Alan Hodgkin', *The Independent*, 4 Jan., *Monday Review*, 6.

HUXLEY, T. H. (1874), 'On the Hypothesis that Animals Are Automata, and its History', in T. H. Huxley, *Method and Results: Essays* (London: Macmillan, 1898), 199–250.

HYMAN, R. A. (1982), *Charles Babbage: Pioneer of the Computer* (Oxford: Oxford University Press).

―― (ed.) (1989), *Science and Reform: Selected Works of Charles Babbage* (Cambridge: Cambridge University Press).

HYMES, D. (ed.) (1964), *Language in Culture and Society* (New York: Harper & Row).

―― (ed.) (1965), *The Use of Computers in Anthropology* (The Hague: Mouton).

―― (1972), Review of Noam Chomsky, *Language*, 48: 416–27.

―― and FAUGHT, J. (1981), *American Structuralism*, new edn. (The Hague: Mouton).

ICA (2000), *What Is Life? How Can We Build A Soul? Steve Grand in Conversation*, flyer for an event on 14 Nov. (London: Institute of Contemporary Arts).

IDE, N., and VERONIS, J. (1998), 'Word–Sense Disambiguation: The State of the Art', *Computational Linguistics*, 24: 1–40, introd. to special issue: *Word–Sense Disambiguation*.

IJSPEERT, A. J., HALLAM, J., and WILLSHAW, D. (1997), 'Artificial Lampreys: Comparing Naturally and Artificially Evolved Swimming Controllers', in P. Husbands and I. Harvey (eds.), *Fourth European Conference on Artificial Life* (Cambridge, Mass.: MIT Press), 256–65.

ILLICH, I. D. (1971), *Deschooling Society* (London: Calder & Boyars).

INGERSON, T. E., and BUVEL, R. L. (1984), 'Structure in Asynchronous Cellular Automata', *Physica D*, 10: 59–68.

INGLIS, F. (2000), *Clifford Geertz: Culture, Custom, and Ethics* (Cambridge: Polity Press & Blackwell).

INHELDER, B., and PIAGET, J. (1958), *The Growth of Logical Thinking from Childhood to Adolescence: An Essay on the Construction of Formal Operational Structures*, trans. A. Parsons and S. Milgram (London: Routledge & Kegan Paul; first French edn., 1955).

_____ _____ (1964), *The Early Growth of Logic in the Child: Classification and Seriation*, trans. E. A. Lunzer and D. Papert (London: Routledge & Kegan Paul; first French edn., 1959).

IRWIN, W. (ed.) (2002), *The Matrix and Philosophy: Welcome to the Desert of the Real*, Popular Culture and Philosophy (Chicago: Open Court).

ISARD, M., and MacCORMICK, J. (in preparation), 'Multi-Scale Loopy Belief Propagation for Stereo Vision and Optic Flow'. This will probably be submitted to *Advances in Neural Information Processing Systems*.

ISLAS, S., BECERRA, A., LUISI, P.-L., and LAZCANO, A. (2004), 'Comparative Genomics and the Gene Complement of a Minimal Cell', *Origins of Life in the Evolutionary Biosphere*, 34: 243–56.

ITARD, J.-M.-G. (1801), *The Wild Boy of Aveyron*, trans. G. and M. Humphrey (New York: Appleton-Century-Crofts, 1962).

ITKONEN, E. (1996), 'Concerning the Generative Paradigm', *Journal of Pragmatics*, 25: 471–501.

ITO, M. (1982), 'Experimental Verification of Marr-Albus' Plasticity Assumption for the Cerebellum', *Acta Biologica Academiae Scientiarum Hungaricae*, 33: 189–99.

_____ (1984), *The Cerebellum and Neural Control* (New York: Raven Press).

_____ (1999), 'Cerebellum', in R. A. Wilson and Keil (1999: 110–11).

IVEY, W. (2000), *The National Endowment for the Arts, 1965–2000: Chronology* (New York: National Endowment for the Arts).

JABLONKA, E. (2000), 'Lamarckian Inheritance Systems in Biology: A Source of Metaphors and Models in Technological Evolution', in Ziman (2000c: 27–40).

JACKENDOFF, R. (2002), *Foundations of Language: Brain, Meaning, Grammar, Evolution* (Oxford: Oxford University Press).

_____ (2003), 'Précis of *Foundations of Language: Brain, Meaning, Grammar, Evolution*', with peer commentaries, *Behavioral and Brain Sciences*, 26: 651–707.

JACKSON, B. S. (1995), *Making Sense in Law: Linguistic, Psychological and Semiotic Perspectives* (Liverpool: Deborah Charles).

JACKSON, F. C. (1982), 'Epiphenomenal Qualia', *Philosophical Quarterly*, 32: 127–36.

_____ (1986), 'What Mary Didn't Know', *Journal of Philosophy*, 83: 291–5.

JACOB, F. (1977), 'Evolution and Tinkering', *Science*, 196: 1161–6.

_____ (1988), *The Statue Within: An Autobiography* (New York: Basic Books).

_____ and MONOD, J. (1961), 'Genetic Regulatory Mechanisms in the Synthesis of Proteins', *Journal of Molecular Biology*, 3: 318–56.

JACOBI, N. (1998), 'Running Across the Reality Gap: Octopod Locomotion Evolved in a Minimal Simulation', in P. Husbands and J.-A. Meyer (eds.), *Evolutionary Robotics* (Berlin: Springer-Verlag), 39–58.

JACOBS, R. A., JORDAN, M. I., NOWLAN, S. J., and HINTON, G. E. (1991), 'Adaptive Mixtures of Local Experts', *Neural Computation*, 3: 79–87.

JAKOBSON, R. O. (1942–3), *Six Lectures on Sound and Meaning*, trans. J. Mepham (Hassocks: Harvester Press, 1978).

_____ FANT, C. G. M., and HALLE, M. (1952), *Preliminaries to Speech Analysis* (Cambridge, Mass.: MIT Press).

JAMES, W. (1880), 'Great Men and their Environment', *Atlantic Monthly*, 46: 441–59.

_____ (1890), *The Principles of Psychology*, 2 vols. (New York: Henry Holt).

_____ (1902), *The Varieties of Religious Experience* (New York: Collier).

_____ (1907), *Pragmatism: A New Name for Some Old Ways of Thinking* (London: Longman).

_____ (1911), 'Percept and Concept: The Import of Concepts', in W. James, *Some Problems of Philosophy* (New York: Longman, Green), 47–74.

JAMNIK, M. (2001), *Mathematical Reasoning with Diagrams: From Intuition to Automation* (Stanford, Calif.: CSLI Press).

_____ BUNDY, A., and GREEN, I. (1998), *Verification of Diagrammatic Proofs*, Research Paper 924, University of Edinburgh, Division of Informatics.

JARDINE, L. (2003), *The Curious Life of Robert Hooke: The Man who Measured London* (London: HarperCollins).

JARDINE, N. (1991), *The Scenes of Inquiry: On the Reality of Questions in the Sciences* (Oxford: Clarendon Press).

JAYNES, J. (1970), 'The Problem of Animate Motion in the Seventeenth Century', *Journal of the History of Ideas*, 31: 219–34.

JEAN, R. V. (1984), *Mathematical Approaches to Pattern and Form in Plant Growth* (New York: Wiley).

JEFFRESS, L. A. (ed.) (1951), *Cerebral Mechanisms in Behavior: The Hixon Symposium*, Proceedings of a meeting held in Pasadena, 1948 (New York: Wiley).

JENKINS, J. J. (1986), 'Interview with James J. Jenkins', in Baars (1986: 239–52).

JESPERSON, O. (1922), *Language: Its Nature, Development, and Origin* (London: Allen & Unwin).

_____ (1924), *The Philosophy of Grammar* (London: Allen & Unwin).

_____ (1937), *Analytic Syntax* (London: Allen & Unwin).

JEVONS, W. S. (1870), 'On the Mechanical Performance of Logical Inference', *Philosophical Transactions of the Royal Society*, 160: 497–518.

JOHN of SALISBURY (1159), *The Metalogicon: A Twelfth-Century Defense of the Verbal and Logical Arts of the Trivium*, ed. and trans. D. D. McGarry (Berkeley: University of California Press, 1955).

JOHNSON, D. M., and ERNELING, C. E. (eds.) (1997), *The Future of the Cognitive Revolution* (Oxford: Oxford University Press).

JOHNSON, M. D. (1987), *The Body in the Mind: The Bodily Basis of Meaning, Imagination, and Reason* (Chicago: University of Chicago Press).

JOHNSON, M. H. (ed.) (1993*a*), *Brain Development and Cognition: A Reader* (Oxford: Blackwell).

_____ (1993*b*), 'Constraints on Cortical Plasticity', in M. H. Johnson (1993*a*: 703–21).

_____ (1996), *Developmental Cognitive Neuroscience: An Introduction* (Oxford: Blackwell).

_____ and GILMORE, R. O. (1996), 'Developmental Cognitive Neuroscience: A Biological Perspective on Cognitive Change', in R. Gelman and T. K.-F. Au (eds.), *Perceptual and Cognitive Development* (San Diego: Academic Press), 333–72.

_____ and HORN, G. (1986), 'Dissociation of Recognition Memory and Associative Learning by a Restricted Lesion of the Chick Forebrain', *Neuropsychologia*, 24: 329–40.

_____ _____ (1988), 'The Development of Filial Preferences in the Dark-Reared Chick', *Animal Behaviour*, 36: 675–83.

_____ and KARMILOFF-SMITH, A. (1992), 'Can Neural Selectionism be Applied to Cognitive Development and its Disorders?', *New Ideas in Psychology*, 10: 35–46.

_____ and MORTON, J. (1991), *Biology and Cognitive Development: The Case of Face Recognition* (Oxford: Blackwell).

JOHNSON, T. (1985), *Natural Language Computing: The Commercial Applications* (London: Ovum).

JOHNSON-LAIRD, P. N. (1975), 'Models of Deduction', in R. J. Falmagne (ed.), *Reasoning: Representation and Process in Children and Adults* (Hillsdale, NJ: Lawrence Erlbaum), 7–54.

_____ (1978), 'What's Wrong with Grandma's Guide to Procedural Semantics: A Reply to Jerry Fodor', *Cognition*, 6: 262–71.

_____ (1981), 'Computers, Cognition, and Mental Models', *Cognition*, 10: 139–43.

_____ (1983), *Mental Models: Towards a Cognitive Science of Language, Inference, and Consciousness* (Cambridge: Cambridge University Press).

_____ (1989/1993), 'Jazz Improvisation: A Theory at the Computational Level'. Working Paper, MRC Applied Psychology Unit, Cambridge, 1989; final version pub. in P. Howell, R. West, and I. J. Cross (eds.), *Representing Musical Structure* (London: Academic Press, 1993), 291–326.

_____ and SHAFIR, E. (1994), 'The Interaction Between Reasoning and Decision-Making: An Introduction', in P. N. Johnson-Laird and E. Shafir (eds.), *Reasoning and Decision-Making* (Oxford: Blackwell), 1–10.

JOHNSON-LAIRD, P. N. and WASON, P. C. (1970), 'Insight into a Logical Relation', *Quarterly Journal of Experimental Psychology*, 22: 49–61.

JONAS, H. (1966), *The Phenomenon of Life: Toward a Philosophical Biology* (New York: HarperCollins). Page references are to the 2001 edn. (Evanston, Ill.: Northwestern University Press).

——(1984), *The Imperative of Responsibility: In Search of an Ethics for the Technological Age* (Chicago: Chicago University Press).

——(1990), 'Heidegger's Resoluteness and Resolve: An Interview', in G. Neske and E. Kettering (eds.), *Martin Heidegger and National Socialism: Questions and Answers* (New York: Paragon House), 197–203.

——(1996), *Of Mortality and Morality: A Search for the Good after Auschwitz*, ed. L. Vogel, Northwestern University Studies in Phenomenology and Existential Philosophy (Evanston, Ill.: Northwestern University Press).

JONES, E. E., KANOUSE, D. E., KELLEY, H. H., NISBETT, R. E., VALINS, S., and WEINER, B. (eds.) (1972), *Attribution: Perceiving the Causes of Behavior* (Morristown, NJ: General Learning Press).

JONSSON, A., MORRIS, P., MUSCETTOLA, N., RAJAN, K., and SMITH, B. (2000), 'Planning in Interplanetary Space: Theory and Practice', *Proceedings of the 5th International Conference on Artificial Intelligence Planning Systems* (Breckenridge, Colo.: AAAI Press), 177–86.

JORDAN, M. I. (1986/1989), 'Serial Order: A Parallel Distributed Processing Approach', in J. L. Elman and D. E. Rumelhart (eds.), *Advances in Connectionist Theory* (Hillsdale, NJ: Lawrence Erlbaum, 1989), 214–49; previously available as Technical Report ICS 8608, Institute of Cognitive Science, UC San Diego, La Jolla, Calif., 1986.

JOYNSON, R. B. (1970), 'The Breakdown of Modern Psychology', *Bulletin of the British Psychological Society*, 23: 261–9.

——(1974), *Psychology and Common Sense* (London: Routledge & Kegan Paul).

JUDSON, H. F. (1979), *The Eighth Day of Creation: The Makers of the Revolution in Biology* (London: Jonathan Cape).

JULESZ, B. (1971), *Foundations of Cyclopean Perception* (Chicago: Chicago University Press).

JUNG, C. G. (1933), *Modern Man in Search of a Soul*, trans. W. S. Dell and C. F. Baynes (London: Kegan Paul, Trench, Trubner).

KAELBLING, L. P. (1988), 'Goals as Parallel Program Specifications', *Proceedings of the Seventh National Conference on Artificial Intelligence, St Paul, Minnesota*, 60–5.

——and ROSENSCHEIN, S. J. (1990), 'Action and Planning in Embedded Agents', *Robotics and Autonomous Systems*, 6: 35–48.

KAHN, D. (1996), *The Codebreakers: The Story of Secret Writing*, 2nd edn., rev. (New York: Scribner).

KAHNEMAN, D. (1968), 'Method, Findings, and Theory in Studies of Visual Masking', *Psychological Bulletin*, 70: 404–25.

——and TVERSKY, A. (1972), 'Subjective Probability: A Judgment of Representativeness', *Cognitive Psychology*, 3: 430–54.

————(1973), 'On the Psychology of Prediction', *Psychological Review*, 80: 237–51.

————(1979), 'Prospect Theory: An Analysis of Decision Under Risk', *Econometrica*, 47: 263–91.

——(1996), 'On the Reality of Cognitive Illusions: A Reply to Gigerenzer's Critique', *Psychological Review*, 103: 582–91.

——SLOVIC, P., and TVERSKY, A. (eds.) (1982), *Judgment Under Uncertainty: Heuristics and Biases* (Cambridge: Cambridge University Press).

KANDEL, E. R., and SQUIRE, L. R. (2000), 'Neuroscience: Breaking Down Scientific Barriers to the Study of Brain and Mind', *Science*, 290: 1113–20.

——SCHWARZ, J. H., and JESSELL, T. M. (1995), *Essentials of Neural Science and Behavior* (London: Prentice-Hall).

KANEFF, S. (ed.) (1970), *Picture Language Machines* (New York: Academic Press).

KANNER, L. (1943), 'Autistic Disturbances of Affective Contact', *Nervous Child*, 2: 217–50.

KANT, I. (1781), *Critique of Pure Reason*, trans. N. Kemp Smith (London: Macmillan, 1934).

——(1790), *Critique of Judgment*, trans. J. C. Meredith (Oxford: Clarendon Press, 1952).

KAPLAN, R. M. (1972), 'Augmented Transition Networks as Psychological Models of Sentence Comprehension', *Artificial Intelligence*, 3: 77–100.

——(1975), 'On Process Models for Sentence Analysis', in D. A. Norman and D. E. Rumelhart (eds.), *Explorations in Cognition* (San Francisco: W. H. Freeman), 117–35.

Kaplan, S. (1992), 'Environmental Preference in a Knowledge-Seeing, Knowledge-Using Organism', in Barkow *et al.* (1992: 581–98).

—— and Kaplan, R. (1982), *Cognition and Environment: Functioning in an Uncertain World* (New York: Praeger).

Kapur, D., and Mundy, J. L. (eds.) (1988), *Geometric Reasoning, Artificial Intelligence*, special issue, 37: 1–412.

Karmiloff, K., and Karmiloff-Smith, A. (1998), *Everything your Baby Would Ask, If Only he or she Could Talk* (London: Ward Lock).

—— —— (2001), *Pathways to Language: From Fetus to Adolescent* (Cambridge, Mass.: Harvard University Press).

Karmiloff-Smith, A. (1979a), *A Functional Approach to Child Language: A Study of Determiners and Reference* (Cambridge: Cambridge University Press).

—— (1979b), 'Micro- and Macro-Developmental Changes in Language Acquisition and Other Representational Systems', *Cognitive Science*, 3: 81–118.

—— (1979c), 'Problem-Solving Procedures in Children's Construction and Representation of Closed Railway Circuits', *Archives de Psychologie*, 47 (1807), 37–59.

—— (1984), 'Children's Problem-Solving', in M. E. Lamb, A. L. Brown, and B. Rogoff (eds.), *Advances in Developmental Psychology*, iii (Hillsdale, NJ: Lawrence Erlbaum), 39–90.

—— (1986), 'From Meta-Processes to Conscious Access: Evidence from Children's Metalinguistic and Repair Data', *Cognition*, 23: 95–147.

—— (1990a), 'Constraints on Representational Change', *Cognition*, 34: 57–83.

—— (1990b), 'Piaget and Chomsky on Language Acquisition: Divorce or Marriage?', *First Language*, 10: 255–70.

—— (1992), *Beyond Modularity: A Developmental Perspective on Cognitive Science* (London: MIT Press).

—— (1994), *Baby it's You* (London: Ebury). Based on the Channel 4 TV series of the same name.

—— (2001), 'Elementary, My Dear Watson, the Clue is in the Genes—Or Is It?', British Psychological Society Centennial Lecture, Nov. 2001. A slightly abridged version of this lecture, which omits the Williams dialogue quoted in Ch. 7.vi.h, is published in *Proceedings of the British Academy*, 117: *2001 Lectures* (Oxford: Oxford University Press and British Academy, 2002), 525–43. However, the dialogue did appear in the summary in *The Guardian*, 6 Nov. 2001, available at <http://education.guardian.co.uk/higher/medicalscience/story/0,9837,588297,00.html>, accessed 26 Oct. 2005.

—— and Inhelder, B. (1975), 'If You Want to Get Ahead, Get a Theory', *Cognition*, 3: 195–212.

—— Brown, J. H., Grice, S., and Paterson, S. J. (2003), 'Dethroning the Myth: Cognitive Dissociations and Innate Modularity in Williams Syndrome', *Developmental Neuropsychology*, 23: 227–42.

—— Johnson, H., Grant, J., Jones, M. C., Karmiloff, Y.-N., Bartrip, J., and Cuckle, C. (1993), 'From Sentential to Discourse Functions: Detection and Explanation of Speech Repairs in Children and Adults', *Discourse Processes*, 16: 565–89.

Kass, M., Witkin, A., and Terzopoulos, D. (1987), 'Snakes: Active Contour Models', *Proceedings of the First International Conference on Computer Vision*, London (Washington: Computer Press of the IEEE), 259–68.

Katz, J. J., and Fodor, J. A. (1963), 'The Structure of a Semantic Theory', *Learning*, 39: 170–210.

Kauffman, S. A. (1969), 'Metabolic Stability and Epigenesis in Randomly Connected Nets', *Journal of Theoretical Biology*, 22: 437–67.

—— (1971), 'Cellular Homeostasis, Epigenesis, and Replication in Randomly Aggregated Macro-Molecular Systems', *Journal of Cybernetics*, 1: 71–96.

—— (1983), 'Developmental Constraints: Internal Factors in Evolution', in B. C. Goodwin, N. Holder, and C. G. Wylie (eds.), *Developmental Evolution* (Cambridge: Cambridge University Press), 195–225.

—— (1986a), 'A Framework to Think About Regulatory Systems', in W. Bechtel (ed.), *Integrating Scientific Disciplines* (The Hague: Nijhoff), 165–84.

—— (1986b), 'Autocatalytic Sets of Proteins', *Journal of Theoretical Biology*, 119: 1–24.

—— (1989), 'Adaptation on Rugged Fitness Landscapes', in D. L. Stein (ed.), *Lectures in the Sciences of Complexity* (Redwood City, Calif.: Addison-Wesley), 527–618.

—— (1993), *The Origins of Order: Self-Organization and Selection in Evolution* (Oxford: Oxford University Press).

Kauffmann, W. (ed.) (1969), *Nietzsche: On the Genealogy of Morals and Ecce Homo* (New York: Vintage).

KAY, A. C. (1972), 'A Personal Computer for Children of All Ages', *Proceedings of the ACM National Conference*, Boston, Aug. (11 pages, unnumbered).

——— (1977), 'Microelectronics and the Personal Computer', *Scientific American*, 237/3: 230–44.

——— (1984), 'Computer Software', *Scientific American*, 251/3 (Mar.), 191–207.

——— (1990), 'User Interface: A Personal View', in B. Laurel (ed.), *The Art of Human–Computer Interface Design* (Reading, Mass.: Addison-Wesley), 191–207; repr. with two ellipses, in Packer and Jordan (2001: 121–31).

——— (1993), 'The Early History of Smalltalk', *2nd ACM SIGPLAN Conference on History of Programming Languages*, Cambridge, Mass., 69–95; repr. in T. Bergin and R. Gibson (eds.), *History of Programming Languages*, ii (New York: ACM Press, 1996), 511–98.

——— and GOLDBERG, A. (1977), 'Personal Dynamic Media', *IEEE Computer*, 10/3: 31–41; repr. in Packer and Jordan (2001: 167–78).

KAY, M. (1980), *Algorithm Schemata and Data Structures in Syntactic Processing*, CSL-80-12 (Palo Alto, Calif.: Xerox Corporation); repr. in B. J. Grosz, K. Sparck Jones, and B. L. Webber (eds.), *Readings in Natural Language Processing* (Los Altos, Calif.: Morgan Kaufmann, 1986), 35–70.

KAY, P., and MAFFI, L. (1999), 'Color Appearance and the Emergence and Evolution of Basic Color Lexicons', *American Anthropologist*, 101: 743–60.

KEELEY, B. L. (1999), 'Fixing Content and Function in Neurobiological Systems: The Neuroethology of Electroreception', *Biology and Philosophy*, 14: 395–430.

——— (2000a), 'Shocking Lessons from Electric Fish: The Theory and Practice of Multiple Realization', *Philosophy of Science*, 67: 444–65.

——— (2000b), 'Neuroethology and the Philosophy of Cognitive Science', *Philosophy of Science (Proceedings)*, 67: S404–17.

KELLERT, S. H. (1993), *In the Wake of Chaos: Unpredictable Order in Dynamical Systems* (Chicago: University of Chicago Press).

KELLOGG, L. A. (1933), *The Ape and the Child* (New York: McGraw-Hill).

KELLY, K. (1994), *Out of Control: The New Biology of Machines* (London: Fourth Estate).

KEMBER, S. (2003), *Cyberfeminism and Artificial Life* (London: Routledge).

KEMENY, J. G. (1955), 'Man Viewed as a Machine', *Scientific American*, 192: 58–67.

——— FANO, R. M., and KING, W. G. (1962), 'A Library for 2000 A.D.', in M. Greenberger (ed.), *Management and the Computer of the Future* (New York: MIT Press/Wiley), 134–78.

KEMPEN, G. (1977), 'On Conceptualizing and Formulating in Sentence Production', in S. Rosenberg (ed.), *Sentence Production* (Hillsdale, NJ: Lawrence Erlbaum), 259–74.

KEMPSON, R. M. (1977), *Semantic Theory* (Cambridge: Cambridge University Press).

KENDRICK, K. M., and BALDWIN, B. A. (1987), 'Cells in Temporal Cortex of Conscious Sheep Can Respond Preferentially to the Sight of Faces', *Science*, 236: 448–50.

KEYSERS, C., and PERRETT, D. I. (2002), 'Visual Masking and RSVP Reveal Neural Competition', *Trends in Cognitive Sciences*, 6: 120–5.

——— ——— (2004), 'Demystifying Social Cognition: A Hebbian Perspective', *Trends in Cognitive Sciences*, 8: 501–7.

KHABAZA, T., and SHEARER, C. (1995), 'Data Mining with Clementine', *IEE Colloquium on Knowledge Discovery in Databases*, IEE Digest no. 1995/021B (London), Feb.

KHATIB, O., YOKOI, K., BROCK, O., CHANG, K., and CASALI, A. (1999), 'Robots in Human Environments: Basic Autonomous Capabilities', *International Journal of Robotics Research*, 18: 684–96.

——— BROCK, O., CHANG, K., CONTI, F., RUSPINI, D., and SENTIS, L. (2002), 'Robotics and Interactive Simulation', *Communications of the Association for Computing Machinery*, 45: 46–51.

KILBURN, T., and WILLIAMS, F. C. (1953), 'The University of Manchester Computing Machine', in Bowden (1953: 117–29).

KILMER, W. L. (1997), 'A Command Computer for Complex Autonomous Systems', *Neurocomputing*, 17: 47–59.

——— McCULLOCH, W. S., and BLUM, J. (1969), 'A Model of the Vertebrate Central Command System', *International Journal of Man–Machine Studies*, 1: 279–309.

KING, M., and PERSCHKE, S. (1987), 'EUROTRA', in M. King (ed.), *Machine Translation Today: The State of the Art* (Edinburgh: Edinburgh University Press), 373–91.

KIPLING, R. (1902), *Just So Stories for Little Children* (London: Macmillan).

KIRKPATRICK, S., GELATT, C. D., and VECCHIS, M. P. (1983), 'Optimization by Simulated Annealing', *Science*, 220: 671–80.

KIRSCH, I., and LYNN, S. J. (1998), 'Dissociation Theories of Hypnosis', *Psychological Bulletin*, 123: 100–15.

KIRSH, D. (ed.) (1991*a*), *Foundations of Artificial Intelligence*. *Artificial Intelligence*, special issue, 47/1–3 (Amsterdam: Elsevier; also pub. Cambridge, Mass.: MIT Press, 1992); identical page numbers.

——— (1991*b*), 'Today the Earwig, Tomorrow Man?', *Artificial Intelligence*, 47: 161–84.

KISILEVSKY, B. S., MUIR, D. W., and LOW, J. A. (1989), 'Human Fetal Responses to Sound as a Function of Stimulus Intensity', *Obstetrics and Gynaecology*, 73: 971–6.

KITANO, H. (1999), 'Preface [to special issue: *RoboCup*]', *Artificial Intelligence*, 110: 189–91.

KITAZAWA, S., KIMURA, T., and YIN, P. B. (1998), 'Cerebellar Complex Spikes Encode Both Destinations and Errors in Arm Movements', *Nature*, 392: 494–7.

KITCHER, P., and KITCHER, P. (1988), 'The Devil, the Details, and Dr. Dennett', *Behavioral and Brain Sciences*, 11: 517–18.

KLAHR, D. (1992), 'Information-Processing Approaches to Cognitive Development', in M. H. Bornstein and M. E. Lamb (eds.), *Developmental Psychology: An Advanced Textbook*, 3rd edn. (Hillsdale, NJ: Lawrence Erlbaum), 273–336.

——— and WALLACE, J. G. (1970), 'An Information Processing Analysis of Some Piagetian Experimental Tasks', *Cognitive Psychology*, 1: 358–87.

——— ——— (1976), *Cognitive Development: An Information Processing View* (Hillsdale, NJ: Lawrence Erlbaum).

——— LANGLEY, P., and NECHES, R. (eds.) (1987), *Production System Models of Learning and Development* (Cambridge, Mass.: MIT Press).

KLATT, D. H. (1977), 'Review of the ARPA Speech-Understanding Project', *Journal of the Acoustical Society of America*, 62: 1345–66.

KLATZKY, R. L. (1980), *Human Memory: Structures and Processes*, 2nd edn. (San Francisco: W. H. Freeman).

KLEENE, S. C. (1956), 'Representations of Events in Nerve Nets and Finite Automata', in Shannon and McCarthy (1956: 3–42).

KLEIN, S., AESCHLIMANN, J. F., BALSIGER, D. F., CONVERSE, S. L., COURT, C., FOSTER, M., LAO, R., OAKLEY, J. D., and SMITH, J. (1973), *Automatic Novel Writing: A Status Report*, Technical Report 186 (Madison: University of Wisconsin Computer Science Department).

KLEINFELD, D., and SOMPOLINSKY, H. (1989), 'Associative Network Models for Central Pattern Generators', in Koch and Segev (1989: 195–246).

KLEMM, F. (1957), *A History of Western Technology*, trans. D. W. Singer (London: Allen & Unwin).

KLIMA, E. S., and BELLUGI, U. (1979), *The Signs of Language* (Cambridge, Mass.: Harvard University Press).

KLINE, P. (1972), *Fact and Fantasy in Freudian Theory* (London: Methuen).

KLUTSCH, C. (2005), 'The Summer 1968 in London and Zagreb: Starting or End Point for Computer Art?', in Candy (2005: 109–17).

KLUVER, B. (1966), 'The Great Northeastern Power Failure', in Packer and Jordan (2002: 33–8); paper presented at the College Art Association, New York, Jan. 1966; first pub. in Swedish, *Dagens Nyheter* (Mar. 1966).

KNEALE, W. C., and KNEALE, M. (1962), *The Development of Logic* (Oxford: Clarendon Press).

KNOWLSON, J. (1975), 'Gesture as a Form of Universal Language', *Journal of the History of Ideas*, 26: 495–508.

KOCH, C. (1990), 'Biophysics of Computation: Toward the Mechanisms Underlying Information Processing in Single Neurons', in Schwartz (1990: 97–113).

——— (1999), *Biophysics of Computation* (Oxford: Oxford University Press).

——— and SEGEV, I. (eds.) (1989), *Methods in Neuronal Modeling: From Synapses to Networks* (Cambridge, Mass.: MIT Press).

KOCH, S. (1961), 'Psychological Science Versus the Science–Humanism Antinomy: Intimations of a Significant Science of Man', *American Psychologist*, 16: 629–39.

——— (1974), 'Psychology as Science', in S. C. Brown (ed.), *Philosophy of Psychology* (London: Macmillan), 3–40.

KOERNER, E. F. K. (1989), 'The Chomskyan "Revolution" and its Historiography: Observations of a Bystander', in E. F. K. Koerner, *Practicing Linguistic Historiography: Selected Essays* (Amsterdam: John Benjamins), 101–46. Extracts, under the title 'The Anatomy of a Revolution in the Social Sciences', are available at <http://www.tlg.uci.edu/~opoudjis/Work/KK.html>.

KOERNER, E. F. K. (1994), 'The Anatomy of a Revolution in the Social Sciences', *Dhumbadji!*, 1/4 (Winter 1994), 3–17; extracts from Koerner (1989); available at <http://www.tlg.uci.edu/~opoudjis/Work/KK.html>.

_____ and ASHER, R. E. (eds.) (1995), *Concise History of the Language Sciences: From the Sumerians to the Cognitivists* (Oxford: Pergamon Press).

KOERTGE, N. (2000), ' "New Age" Philosophies of Science: Constructivism, Feminism and Postmodernism', *British Journal for the Philosophy of Science*, 51, special suppl., 667–84.

KOHLER, I. (1962), 'Experiments with Goggles', *Scientific American*, 206/5: 62–86.

KOHLER, W. (1925), *The Mentality of Apes*, trans. from 2nd rev. edn. E. Winter (London: Harcourt).

_____ (1945), Introduction, in Duncker (1945, pp. iii–iv).

KOHN, M., and MITHEN, S. J. (1999), 'Handaxes: Products of Sexual Selection?', *Antiquity*, 73: 518–26.

KOHONEN, T. (1972), 'Correlation Matrix Memories', *IEEE Transactions on Computers*, C-21: 353–9; repr. in J. A. Anderson and Rosenfeld (1988: 174–80).

_____ (1982), 'Self-Organized Formation of Topologically Correct Feature Maps', *Biological Cybernetics*, 43: 59–69; repr. in J. A. Anderson and Rosenfeld (1988: 511–21).

_____ (1988), 'The "Neural" Phonetic Typewriter', *Computer Magazine*, 21: 11–22.

_____ OJA, E., and LEHTIO, P. (1981), 'Storage and Processing of Information in Distributed Associative Memory Systems', in Hinton and Anderson (1981: 105–43).

KOLERS, P. A. (1968), 'Some Psychological Aspects of Pattern Recognition', in P. A. Kolers and M. Eden (eds.), *Recognizing Patterns: Studies in Living and Automatic Systems* (Cambridge, Mass.: MIT Press), 4–61.

_____ (1972), *Aspects of Motion Perception* (London: Pergamon Press).

_____ and ROSNER, B. S. (1960), 'On Visual Masking (metacontrast); Dichoptic Observation', *American Journal of Psychology*, 73: 2–21.

_____ and VON GRNAU, M. (1976), 'Shape and Color in Apparent Motion', *Vision Research*, 16: 329–35.

KOLODNER, J. L. (1992), 'An Introduction to Case-Based Reasoning', *Artificial Intelligence Review*, 6: 3–34.

_____ (1993), *Case-Based Reasoning* (San Mateo, Calif.: Morgan Kauffmann).

KONORSKI, J. (1967), *Integrative Activity of the Brain: An Interdisciplinary Approach* (Chicago: University of Chicago Press).

KÖRNER, S. (1955), *Kant* (Harmondsworth: Penguin).

KOSSLYN, S. M. (1973), 'Scanning Visual Images: Some Structural Implications', *Perception and Psychophysics*, 14: 90–4.

_____ (1975), 'Information Representation in Visual Images', *Cognitive Psychology*, 7: 341–70.

_____ (1980), *Image and Mind* (Cambridge, Mass.: Harvard University Press).

_____ (1981), 'The Medium and the Message in Mental Imagery: A Theory', *Psychological Review*, 88: 46–66.

_____ (1983), *Ghosts in the Mind's Machine: Creating and Using Images in the Brain* (New York: W. W. Norton).

_____ (1996), *Image and Brain: The Resolution of the Imagery Debate*, new edn. (Cambridge, Mass.: MIT Press).

_____ and POMERANTZ, J. R. (1977), 'Imagery, Propositions, and the Form of Internal Representations', *Cognitive Psychology*, 9: 52–76.

KOWALSKI, R. A. (1979), *Logic for Problem-Solving* (Amsterdam: North-Holland).

_____ (1988), 'The Early Years of Logic Programming', *Communications of the Association for Computing Machinery*, 31: 38–44.

_____ (1992), 'Legislation as Logic Programs', in G. Comyn, N. E. Fuchs, and M. J. Ratcliffe (eds.), *Logic Programming in Action*, Proceedings of the 2nd International Logic Programming Summer School (Berlin: Springer-Verlag), 203–30.

_____ (2001), 'Artificial Intelligence and the Natural World', *Cognitive Processing*, 4: 547–73.

_____ (2002), 'Directions for Logic Programming', in A. C. Kakas and F. Sadri (eds.), *Computational Logic: Logic Programming and Beyond. Essays in Honour of Robert Kowalski* (London: Springer), 26–32.

_____ (2003), 'A Logic-Based Model for Conflict Resolution', MS.

_____ (in preparation), *Computational Logic for Human Affairs* (provisional title). See <http://www.doc.ic.ac.uk/~rak/>.

_____ and KUEHNER, D. (1971), 'Linear Resolution with Selection Function', *Artificial Intelligence*, 2: 227–60.

_____ and Sᴇʀɢᴏᴛ, M. J. (1985), 'Computer Representation of the Law', Part of a Panel Discussion on 'Legal Implications of AI', *Proceedings of the Ninth International Joint Conference on Artificial Intelligence*, Los Angeles, 1267–80.

_____ and Tᴏɴɪ, F. (1996), 'Abstract Argumentation', *Artificial Intelligence and Law*, 4: 275–96.

Kᴏᴢᴀ, J. R. (1989), 'Hierarchical Genetic Algorithms Operating on Populations of Computer Programs', *Proceedings of the 11th International Joint Conference on Artificial Intelligence* (San Mateo, Calif.: Morgan Kauffman), 768–74.

_____ (1992*a*), *Genetic Programming: On the Programming of Computers By Means of Natural Selection* (Cambridge, Mass.: MIT Press).

_____ (1992*b*), 'Evolution of Subsumption Using Genetic Programming', in F. J. Varela and P. Bourgine (eds.), *Toward a Practice of Autonomous Systems: Proceedings of the First European Conference on Artificial Life* (Cambridge, Mass.: MIT Press), 110–19.

Kᴏᴢʟᴏғғ, M. (1973), 'American Painting During the Cold War', *Artforum*, 13 (May), 43–54.

Kᴏᴢᴜʟɪɴ, A. (1990), *Vygotsky's Psychology: A Biography of Ideas* (London: Harvester Wheatsheaf).

Kʀᴀғᴛ, A. (1984), 'XCON: An Expert Configuration System at Digital Equipment Corporation', in Winston and Prendergast (1984: 41–9).

Kʀᴀsɴᴇʀ, L. (1958), 'Studies of the Conditioning of Verbal Behavior', *Psychological Bulletin*, 55: 148–70.

Kʀɪᴄʜᴍᴀʀ, J. L., and Eᴅᴇʟᴍᴀɴ, G. M. (2002), 'Machine Psychology: Autonomous Behavior, Perceptual Categorisation, and Conditioning in a Brain-Based Device', *Cerebral Cortex*, 12: 818–30.

_____ Nɪᴛᴢ, D. A., Gᴀʟʟʏ, J. A., and Eᴅᴇʟᴍᴀɴ, G. M. (2005), 'Characterizing Functional Hippocampal Pathways in a Brain-Based Device as it Solves a Spatial Memory Task', *Neuroscience*, 102: 2111–16.

Kʀɪᴘᴋᴇ, S. (1963), 'Semantical Constructions on Modal Logic', *Acta Philosophica Fennica*, 16: 83–94.

_____ *Naming and Necessity* (Cambridge, Mass.: Harvard University Press).

Kʀᴏᴋᴇʀ, A., and Kʀᴏᴋᴇʀ, M. (eds.) (1997), *Digital Delirium* (New York: St Martin's Press).

Kʀᴏɴᴇɴғᴇʟᴅ, D. B. (2004*a*), 'Cognitive Research Methods', in K. Kempf-Leonard (ed.), *Encyclopedia of Social Measurement* (New York: Academic Press).

_____ (2004*b*), 'Structural Models in Anthropology', in K. Kempf-Leonard (ed.), *Encyclopedia of Social Measurement* (New York: Academic Press).

_____ and Kᴀᴜs, A. (1993), 'Starlings and Other Critters: Simulating Society', *Journal of Quantitative Anthropology*, 4: 143–74.

Kʀᴜᴇɢᴇʀ, J. I., and Fᴜɴᴅᴇʀ, D. C. (2004), 'Towards a Balanced Social Psychology: Causes, Consequences, and Cures for the Problem-Seeking Approach to Social Behaviour and Cognition', *Behavioral and Brain Sciences*, 27: 313–76.

Kʀᴜᴇɢᴇʀ, M. W. (1977), 'Responsive Environments', *American Federation of Information Processing Systems*, 46 (13–16 June), 423–33; repr. in Stiles and Selz (1996: 473–86).

_____ (1991), *Artificial Reality*, 2nd edn. (Reading, Mass.: Addison-Wesley).

Kᴜʙɪᴇ, L. S. (1930), 'A Theoretical Application to Some Neurological Problems of the Properties of Excitation Waves Which Move in Closed Circuits', *Brain*, 53: 166–77.

_____ (1941), 'The Repetitive Core of Neurosis', *Psychoanalytic Quarterly*, 10: 23–43.

_____ (1953), 'The Place of Emotions in the Feedback Concept', in H. von Foerster (ed.), *Cybernetics—Circular, Causal, and Feedback Mechanisms in Biological and Social Systems: Transactions of the Ninth Conference* (New York: Macy Foundation), 48–62; discussion: 62–72.

Kᴜɢʟᴇʀ, P. N., and Tᴜʀᴠᴇʏ, M. T. (1987), *Information, Natural Law, and the Self-Assembly of Rhythmic Movement* (Hillsdale, NJ: Lawrence Erlbaum).

_____ _____ (1988), 'Self-Organization, Flowfields, and Information', *Human Movement Science*, 7: 97–129.

_____ Sʜᴀᴡ, R. E., Vɪɴᴄᴇɴᴛᴇ, K. J., and Kɪɴsᴇʟʟᴀ-Sʜᴀᴡ, J. (1990), 'Inquiry into Intentional Systems I: Issues in Ecological Physics', *Psychological Research*, 52: 98–121.

Kᴜʜɴ, T. S. (1962), *The Structure of Scientific Revolutions* (Chicago: University of Chicago Press).

Kᴜɪᴘᴇʀs, B. J. (1993*a*), 'Reasoning with Qualitative Models', *Artificial Intelligence*, 59: 125–32.

_____ (1993*b*), 'Qualitative Simulation: Then and Now', *Artificial Intelligence*, 59: 133–40.

_____ (2001), 'Qualitative Simulation', in R. A. Meyers (ed.), *Encyclopedia of Physical Science and Technology* (New York: Academic Press).

_____ (2004), 'Why Don't I Take Military Funding?', available at <http://www.cs.utexas.edu/users/kuipers/opinions>.

KUIPERS, B. J., McCARTHY, J., and WEIZENBAUM, J. (1976), 'Computer Power and Human Reason: Comments', *SIGART Newsletter*, no. 58 (June), 4–12.

KULKARNI, D., and SIMON, H. A. (1988), 'The Processes of Scientific Discovery: The Strategy of Experimentation', *Cognitive Science*, 12: 139–75.

KUPER, A. (1999), *Culture: The Anthropologists' Account* (Cambridge, Mass.: Harvard University Press).

KURLAND, D. M., PEA, R. D., CLEMENT, C., and MAWTY, R. (1986), 'A Study of the Development of Programming Ability and Thinking Skills in High School Students', *Journal of Educational Computing Research*, 2: 429–58.

KURZWEIL, R. (1990), *The Age of Intelligent Machines* (Cambridge, Mass.: MIT Press).

KUZNAR, L. A. (1997), *Reclaiming a Scientific Anthropology* (Walnut Creek, Calif.: Altamira).

KYBURG, H. E. (1980), 'Comments and Replies: The Winograd–Boden Session', in L. J. Cohen and M. B. Hesse (eds.), *Applications of Inductive Logic* (Oxford: Clarendon Press), 376–8.

LADD, J. (1987), 'Computers and War: Philosophical Reflection on Ends and Means', in Bellin and Chapman (1987: 297–314).

LAING, R. (1961a), 'Nerve-Net Simulations', *Communications of the ACM*, 4: 405. An ACM 'progress report'.

—— (1961b), 'Automata Theory', *Communications of the ACM*, 4: 406. An ACM 'progress report'.

—— (1962), 'Review of *Computers and Common Sense* by M. Taube', *Behavioral Science*, 7: 238–40.

—— (1975), 'Some Alternative Reproductive Strategies in Artificial Molecular Machines', *Journal of Theoretical Biology*, 54: 63–84.

—— (1977), 'Automaton Models of Reproduction by Self-Inspection', *Journal of Theoretical Biology*, 66: 437–56.

—— (1989), 'Artificial Organisms: History, Problems, and Directions', in Langton (1989a: 49–62).

LAING, R. D. (1960), *The Divided Self: An Existential Study of Sanity and Madness* (London: Tavistock Press).

—— (1967), *The Politics of Experience; and The Bird of Paradise* (Harmondsworth: Penguin).

—— and ESTERSON, A. (1964), *Sanity, Madness and the Family: Families of Schizophrenics* (London: Tavistock Press).

—— PHILLIPSON, H., and LEE, A. R. (1966), *Interpersonal Perception: A Theory and a Method of Research* (London: Tavistock Press).

LAIRD, J. E., and ROSENBLOOM, P. S. (1992), 'In Pursuit of Mind: The Research of Allen Newell', *AI Magazine*, 13/4: 18–45.

—— NEWELL, A., and ROSENBLOOM, P. (1987), 'Soar: An Architecture for General Intelligence', *Artificial Intelligence*, 33: 1–64.

LAKATOS, I. (1970), 'Falsification and the Methodology of Scientific Research Programmes', in I. Lakatos and A. Musgrave (eds.), *Criticism and the Growth of Knowledge* (Cambridge: Cambridge University Press), 91–196.

LAKOFF, G. P., and JOHNSON, M. (1980), *Metaphors We Live By* (Chicago: University of Chicago Press).

LALAND, K. N., and BROWN, G. R. (2002), *Sense and Nonsense: Evolutionary Perspectives on Human Behaviour* (Oxford: Oxford University Press).

LAMBERT, N. (2005), 'The CACHe Project: Its Work and Outcomes', in Candy (2005: 71–5).

LA METTRIE, J. O. DE (1748), *Man a Machine*, trans. G. C. Bussey and M. W. Calkins (Chicago: Open Court, 1912).

LANCELOT, C., and ARNAULD, A. (1660), *A General and Rational Grammar*, Eng. trans. of 1753 (Menston: Scolar Press, 1968).

LAND, M. F. (1975), 'Similarities in the Visual Behaviour of Arthropods and Men', in M. S. Gazzaniga and C. Blakemore (eds.), *Handbook of Psychobiology* (New York: Academic Press), 49–72.

LANDES, J. B. (forthcoming), 'The Anatomy of Artificial Life: An Eighteenth-Century Perspective', in J. Riskin (ed.), *The Sistine Gap: Essays in the History and Philosophy of Artificial Life* (Stanford, Calif.: Stanford University Press).

LANE, P., and HENDERSON, J. (1998), 'Simple Synchrony Networks: Learning to Parse Natural Language with Temporal Synchrony Variable Binding', *Proceedings of the International Conference on Artificial Neural Networks*, Sweden, 615–20.

LANGLEY, P. W. (1978), 'BACON.1: A General Discovery System', *Proceedings of the Second National Conference of the Canadian Society for Computational Studies in Intelligence*, Toronto, 173–80.

—— (1979), 'Descriptive Discovery Processes: Experiments in Baconian Science', Ph.D. diss., Pittsburgh: Carnegie Mellon University, Department of Psychology.

_____ (1981), 'Data-Driven Discovery of Physical Laws', *Cognitive Science*, 5: 31–54.

_____ (1998), 'The Computer-Aided Discovery of Scientific Knowledge', in S. Arikawa and H. Motoda (eds.), *Discovery Science*, Proceedings of the First International Conference on Discovery Science, Fukuoka, Japan (Berlin: Springer-Verlag), 25–39.

_____ BRADSHAW, G. L., and SIMON, H. A. (1981), 'BACON.5: The Discovery of Conservation Laws', *Proceedings of the Seventh International Joint Conference on Artificial Intelligence*, Vancouver, 121–6.

_____ SIMON, H. A., BRADSHAW, G. L., and ZYTKOW, J. M. (1987), *Scientific Discovery: Computational Explorations of the Creative Process* (Cambridge, Mass.: MIT Press).

LANGTON, C. G. (1984), 'Self-Reproduction in Cellular Automata', *Physica D*, 10: 135–44.

_____ (1986), 'Studying Artificial Life with Cellular Automata', *Physica D*, 22: 1120–49.

_____ (ed.) (1989*a*), *Artificial Life*, Proceedings of an Interdisciplinary Workshop on the Synthesis and Simulation of Living Systems, Sept. 1987 (Redwood City, Calif.: Addison-Wesley).

_____ (1989*b*), 'Artificial Life', in Langton (1989*a*: 1–47); rev. in Boden (1996: 39–94).

_____ (1990), 'Computation at the Edge of Chaos: Phase-Transitions and Emergent Computation', *Physica D*, 42: 12–37; repr. in Forrest (1991: 12–37).

_____ (1992), 'Life at the Edge of Chaos', in Langton *et al.* (1992: 41–91).

_____ TAYLOR, C., FARMER, J. D., and RASMUSSEN, S. (eds.) (1992), *Artificial Life II* (Redwood City, Calif.: Addison-Wesley).

LARDNER, D. (1834), 'Babbage's Calculating Engine', in Hyman (1989: 51–109); first pub. in *Edinburgh Review*, 120: 263–327.

LARGE, A. (1985), *The Artificial Language Movement* (Oxford: Blackwell).

LASHLEY, K. S. (1923), 'The Behavioristic Interpretation of Consciousness', *Psychological Review*, 30: 237–72, 329–53.

_____ (1924), 'Studies of Cerebral Functioning in Learning: VI. The Theory that Synaptic Resistance is Reduced by the Passage of the Nerve Impulse', *Psychological Review*, 31: 369–75.

_____ (1929*a*), *Brain Mechanisms and Intelligence: A Quantitative Study of Injuries to the Brain* (Chicago: University of Chicago Press).

_____ (1929*b*), 'Nervous Mechanisms in Learning', in C. Murchison (ed.), *The Foundations of Experimental Psychology* (Worcester, Mass.: Clark University Press), 524–63.

_____ (1931), 'Mass Action in Cerebral Function', *Science*, 73: 245–54.

_____ (1938), 'Experimental Analysis of Instinctive Behaviour', *Psychological Review*, 45: 445–71.

_____ (1942), 'The Problem of Cerebral Organization in Vision', in H. Kluver (ed.), *Visual Mechanisms* (Lancaster, Penn.: Jacques Cattell Press), 301–22.

_____ (1950), 'In Search of the Engram', *Symposium of the Society of Experimental Biology*, no. 4 (Cambridge: Cambridge University Press), 454–82.

_____ (1951*a*), 'The Problem of Serial Order in Behavior', in Jeffress (1951: 112–35). Text of a lecture given at the Hixon Symposium, 1948.

_____ (1951*b*), 'Comments on W. S. McCulloch's "Why the Mind Is in the Head" ', in Jeffress (1951: 70–2).

_____ (1952), [untitled], Vanuxem Lectures delivered at Princeton University, Feb. 1952, in Orbach (1998, pt. II).

_____ (1958), 'Cerebral Organization and Behavior', *Research Publications of the Association for Research in Nervous and Mental Disease*, 36: 1–18; repr. in F. A. Beach *et al.* (1960: 529–43).

_____ CHOW, K. L., and SEMMES, J. (1951), 'An Examination of the Electrical Field Theory of Cerebral Integration', *Psychological Review*, 58: 123–36.

LASKI, J. G., and BUXTON, J. N. (1962), 'Control and Simulation Language', *Computer Journal*, 5: 194–9.

LATOUR, B., and WOOLGAR, S. (1979), *Laboratory Life: The Construction of Scientific Facts*, with introd. Jonas Salk (London: Sage).

LAUDAN, L. (1998), 'Demystifying Underdetermination', in M. Curd and J. A. Cover (eds.), *Philosophy of Science: The Central Issues* (New York: W. W. Norton), 320–53.

LAUGHLIN, S. B. (2004), 'The Implications of Metabolic Energy Requirements in the Representation of Information in Neurons', in M. S. Gazzaniga (ed.), *The Cognitive Neurosciences III* (Cambridge, Mass.: MIT Press).

LAUREL, B. (ed.) (1990), *The Art of Human–Computer Interface Design* (Reading, Mass.: Addison-Wesley).

LAVINGTON, S. H. (1975), *A History of Manchester Computers* (Manchester: NCC Publications).

LAWRENCE, M. (2005), *Like a Splinter in your Mind: The Philosophy Behind the Matrix Trilogy* (Oxford: Blackwell).

LAYNE, J., LAND, M., and ZEIL, J. (1997), 'Fiddler Crabs Use the Visual Horizon to Distinguish Predators from Conspecifics: A Review of the Evidence', *Journal of the Marine Biology Association, UK*, 77: 43–54.

LEAF, M. J. (1972), *Information and Behavior in a Sikh Village: Social Organization Reconsidered* (Berkeley: University of California Press).

LEAKE, D. B. (ed.) (1996), *Case-Based Reasoning: Experiences, Lessons, and Future Directions* (Cambridge, Mass.: MIT Press).

—— (2005), 'AAAI News: Fall News from the American Association for Artificial Intelligence', *AI Magazine*, 26/3: 3–11.

LE CAT, C.-N. (1744), *Traité des sens* (Amsterdam: J. Wetstein).

LEDERBERG, J. (1987), 'How DENDRAL Was Conceived and Born', in *A History of Medical Informatics*, ACM Conference Proceedings (New York: ACM Press), 14–44.

LEDOUX, J. E. (1996), *The Emotional Brain: The Mysterious Underpinnings of Emotional Life* (New York: Simon & Schuster).

—— (2002), *Synaptic Self: How Our Brains Become Who We Are* (New York: Viking Penguin).

LEE, D. D. (1949), 'Being and Value in a Primitive Culture', *Journal of Philosophy*, 8: 401–15.

LEES, R. B. (1957), Review of Chomsky, *Syntactic Structures*, *Language*, 33: 375–407.

LE GROS CLARK, W. E., and MEDAWAR, P. B. (eds.) (1945), *Essays on Growth and Form Presented to D'Arcy Wentworth Thompson* (Oxford: Oxford University Press).

LEHNERT, W. G. (1978), *The Process of Question Answering* (Hillsdale, NJ: Lawrence Erlbaum).

—— and RINGLE, M. H. (eds.) (1982), *Strategies for Natural Language Processing* (Hillsdale, NJ: Lawrence Erlbaum).

LEHRMAN, D. S. (1955), 'The Physiological Basis of Parental Feeding Behaviour in the Ring Dove (*Streptopilia Risoria*)', *Behaviour*, 7: 241–86.

—— (1958*a*), 'Effect of Female Sex Hormones on Incubation Behavior in the Ring Dove (*Streptopilia Risoria*)', *Journal of Comparative and Physiological Psychology*, 51: 142–5.

—— (1958*b*), 'Induction of Broodiness by Participation in Courtship and Nest-Building in the Ring Dove (*Streptopilia Risoria*)', *Journal of Comparative and Physiological Psychology*, 51: 32–6.

LEIBER, J. (1988), ' "Cartesian" Linguistics?', *Philosophia*, 18: 309–46.

LEIBNIZ, G. W. (1685), 'Leibniz: On his Calculating Machine', trans. from Latin by M. Kormes, in D. E. Smith (ed.), *A Source Book in Mathematics* (New York: McGraw-Hill, 1929), 173–81.

—— (1690), *New Essays on Human Understanding*, ed. and trans. P. Remnant and J. Bennett (Cambridge: Cambridge University Press, 1985).

—— (1714), *The Monadology*, in P. P. Wiener (ed.), *Leibniz: Selections* (New York: Charles Scribner's Sons, 1951), 533–51.

—— (1961), *Die Philosophischen Schriften von Gottfried Wilhelm Leibniz*, vii, ed. C. I. Gerhardt (Hildesheim: Georg Olms).

LEITH, P. (1986*a*), 'Fundamental Errors in Legal Logic Programming', *Computer Journal*, 29: 545–52.

—— (1986*b*), 'Legal Expert Systems: Misunderstanding the Legal Process', *Computers and Law*, 49: 26–31.

—— (1988), 'The Application of AI to Law', *AI & Society*, 2: 31–46.

—— (1992), 'Law in Books and Law in Action: The Oral Nature of Law', *Artificial Intelligence Review*, 6: 227–35.

LENAT, D. B. (1977), 'The Ubiquity of Discovery', *Artificial Intelligence*, 9: 257–86.

—— (1983), 'The Role of Heuristics in Learning by Discovery: Three Case Studies', in Michalski *et al.* (1983: 243–306).

—— (1995), 'CYC: A Large-Scale Investment in Knowledge Infrastructure', *Communications of the Association for Computing Machinery*, 38/11: 33–8.

—— and FEIGENBAUM, E. A. (1991), 'On the Thresholds of Knowledge', *Artificial Intelligence*, 47: 185–250.

—— and GUHA, R. V. (1990), *Building Large Knowledge-Based Systems: Representation and Inference in the CYC Project* (Reading, Mass.: Addison-Wesley).

—— and SEELY BROWN, J. (1984), 'Why AM and EURISKO Appear to Work', *Artificial Intelligence*, 23: 269–94.

—— PRAKASH, M., and SHEPHERD, M. (1986), 'CYC: Using Commonsense Knowledge to Overcome Brittleness and Knowledge Acquisition Bottlenecks', *AI Magazine*, 64: 65–85.

LENNEBERG, E. H. (1964), 'A Biological Perspective on Language', in E. H. Lenneberg (ed.), *New Directions in the Study of Language* (Cambridge, Mass.: MIT Press), 65–88.

—— (1967), *Biological Foundations of Language* (New York: Wiley).

LENOIR, T. (1982), *The Strategy of Life: Teleology and Mechanics in Nineteenth Century German Biology* (Dordrecht: Reidel).

LEONHARDT, D. (2000), 'John Tukey, 85, Statistician; Coined the Term "Software" ', obituary, *New York Times*, 29 July, A129.

LEPORE, E., and PYLYSHYN, Z. W. (eds.) (1999), *What Is Cognitive Science?* (Oxford: Blackwell).

LESLIE, A. M. (1987), 'Pretense and Representation: The Origins of "Theory of Mind" ', *Psychological Review*, 94: 412–26.

____ (1991), 'The Theory of Mind Impairment in Autism: Evidence for a Modular Mechanism of Development?', in Whiten (1991: 63–78).

____ (2000), ' "Theory of Mind" as a Mechanism of Selective Attention', in M. S. Gazzaniga (ed.), *The New Cognitive Neurosciences*, 2nd edn., rev. (Cambridge, Mass.: MIT Press), 1235–47.

____ and FRITH, U. (1987), 'Meta-Representation and Autism: How Not to Lose One's Marbles', *Cognition*, 27: 291–4.

LETTVIN, J. Y. (1991), 'On Grandmother Cells (Letter to Horace Barlow)', in Barlow (1995: 434–5).

____ (1999), 'Walter Pitts', in R. A. Wilson and Keil (1999: 651).

____ MATURANA, H. R., McCULLOCH, W. S., and PITTS, W. H. (1959), 'What the Frog's Eye Tells the Frog's Brain', repr. in McCulloch (1965: 230–55); first pub. in *Proceedings of the Institute of Radio Engineers*, 47/11 (1959), 1940–59.

____ ____ PITTS, W. H., and McCULLOCH, W. S. (1961), 'Two Remarks on the Visual System of the Frog', in W. A. Rosenblith (ed.), *Sensory Communication: Contributions to the Symposium on Principles of Sensory Communication*, Endicott House, MIT, 19 July–1 Aug. 1959 (Cambridge, Mass.: MIT Press), 757–76.

LEVINE, J. M. (1992), *The Battle of the Books: History and Literature in the Augustan Age* (Ithaca, NY: Cornell University Press).

LEVINE, R., CHEIN, I., and MURPHY, G. (1942), 'The Relation of the Intensity of a Need to the Amount of Perceptual Distortion: A Preliminary Report', *Journal of Psychology*, 13: 283–93.

LÉVI-STRAUSS, C. (1958/1963), *Structural Anthropology*, trans. from French C. Jacobson and B. Grundfest Schoepf (London: Basic Books, 1963; first pub. 1958).

____ (1964/1969), *The Raw and the Cooked*, trans. from French J. and D. Weightman (New York: Harper & Row, 1969; first pub. 1964).

LEVY, S. (1992), *Artificial Life: The Quest for a New Creation* (New York: Pantheon).

____ (1994), *Hackers: Heroes of the Computer Revolution*, 2nd edn. (New York: Delta).

LEWIS, D. (1972), *We the Navigators: The Ancient Art of Landfinding in the Pacific* (Honolulu: University Press of Hawaii).

LEWIS, D. K. (1969), *Convention: A Philosophical Study* (Cambridge, Mass.: Harvard University Press).

____ (1980), 'Mad Pain and Martian Pain', in N. Block (ed.), *Readings in Philosophy of Psychology*, i (London: Methuen), 216–22; first presented as a talk at Rice University, Apr. 1978.

LEWIS, M. A., FAGG, A. H., and BEKEY, G. (1994), 'Genetic Algorithms for Gait Synthesis in a Hexapod Robot', in Y. F. Zheng (ed.), *Recent Trends in Mobile Robots* (Singapore: World Scientific), 317–31.

LIAW, J.-S., and ARBIB, M. A. (1993), 'Neural Mechanisms Underlying Direction-Selective Avoidance Behavior', *Adaptive Behavior*, 1: 227–61.

LIBET, B. (1965), 'Cortical Activation in Conscious and Unconscious Experience', *Perspectives in Biology and Medicine*, 9: 77–86.

____ (1981), 'The Experimental Evidence for Subjective Referral of a Sensory Experience Backwards in Time: Reply to P. S. Churchland', *Philosophy of Science*, 48: 182–97.

____ (1985a), 'Subjective Antedating of a Sensory Experience and Mind–Brain Theories', *Journal of Theoretical Biology*, 114: 563–70.

____ (1985b), 'Unconscious Cerebral Initiative and the Role of Conscious Will in Voluntary Action', *Behavioral and Brain Sciences*, 8: 529–66.

____ (ed.) (1999), *The Volitional Brain: Towards a Neuroscience of Free Will*, *Journal of Consciousness Studies*, special issue, 6/8–9 (Aug.–Sept.).

____ DELATTRE, L. D., and LEVIN, G. (1964), 'Production of Threshold Levels of Conscious Sensation by Electrical Stimulation of Human Somatosensory Cortex', *Journal of Neurophysiology*, 27: 546–78.

____ ALBERTS, W. W., WRIGHT, E. W., and FEINSTEIN, B. (1967), 'Responses of Human Somatosensory Cortex to Stimuli Below Threshold for Conscious Sensation', *Science*, 158: 1597–1600.

LIBET, B., WRIGHT, E. W., FEINSTEIN, B., and PEARL, D. K. (1979), 'Subjective Referral of the Timing for a Conscious Sensory Experience', *Brain*, 102: 193–224.

——— GLEASON, C. A., WRIGHT, E. W., and PEARL, D. K. (1983), 'Time of Conscious Intention to Act in Relation to Onset of Cerebral Activity', *Brain*, 106: 623–42.

LICHTENSTEIN, E. H., and BREWER, W. F. (1980), 'Memory for Goal-Directed Events', *Cognitive Psychology*, 12: 412–45.

LICKLIDER, J. C. R. (1951), 'A Duplex Theory of Pitch Perfection', *Experientia*, 7: 128–34.

——— (1959), 'Three Auditory Theories', in S. Koch (ed.), *Psychology: The Study of a Science*, i (New York: McGraw-Hill), 94–144.

——— (1960), 'Man–Machine Symbiosis', *Institute of Radio Engineers Transactions on Human Factors in Electronics*, 1: 4–11.

——— (1965), *Libraries of the Future* (Cambridge, Mass.: MIT Press).

——— (1988), 'The Early Years: Founding IPTO', in T. C. Bartee (ed.), *Expert Systems and Artificial Intelligence: Applications and Management* (Indianapolis: Howard W. Sams), 219–27.

——— and TAYLOR, R. W. (1968), 'The Computer as a Communications Device', *Science and Technology: For the Technical Man in Management*, 76: 21–31.

LIEBLICH, I., and ARBIB, M. A. (1982), 'Multiple Representations of Space Underlying Behavior', *Behavioral and Brain Sciences*, 5: 627–59.

LIGHTHILL, J. (1973), 'Artificial Intelligence: A General Survey', in SRC (1973a: 1–21).

LILLIE, R. S. (1936), 'The Passive Iron-Wire Model of Protoplasmic and Nervous Transmission and its Physical Analogies', *Biological Review*, 11: 181–209.

LINDEBOOM, G. A. (1979), *Descartes and Medicine* (Amsterdam: Rodopi).

LINDENMAYER, A. (1968), 'Mathematical Models for Cellular Interaction in Development, Parts I and II', *Journal of Theoretical Biology*, 18: 280–315.

——— (1977), 'Theories and Observations of Developmental Biology', in R. E. Butts and J. Hintikka (eds.), *Foundational Problems in the Special Sciences* (Dordrecht: Reidel), 103–18.

——— (1987), 'Models for Multicellular Development', in A. Kelmenova and J. Kelman (eds.), *Trends, Techniques, and Problems in Theoretical Computer Science* (Berlin: Springer-Verlag), 138–68.

LINDGREN, M. (1990), *Glory and Failure: The Difference Engines of Johann Müller, Charles Babbage, and Georg and Edvard Sheutz*, trans. C. G. McKay (Cambridge, Mass.: MIT Press).

LINDHOLM, C. (1997), 'Logical and Moral Dilemmas of Postmodernism', *Journal of the Royal Anthropological Institute*, NS 3: 747–60.

LINDSAY, P. H., and NORMAN, D. A. (1972), *Human Information Processing: An Introduction to Psychology* (New York: Academic Press).

LINDSAY, R. K., BUCHANAN, B. G., FEIGENBAUM, E. A., and LEDERBURG, J. (1980), *Applications of Artificial Intelligence for Organic Chemistry: The Dendral Project* (New York: McGraw-Hill).

LINSKER, R. (1986), 'From Basic Network Principles to Neural Architecture (Series)', *Proceedings of the National Academy of Sciences, USA*, 83: 7508–12, 8390–4, 8779–83.

——— (1988), 'Self-Organization in a Perceptual Network', *Computer*, 21: 105–17; repr. in J. A. Anderson *et al.* (1990: 528–40).

——— (1990), 'Perceptual Neural Organization: Some Approaches Based on Network Models and Information Theory', *Annual Review of Neuroscience*, 13: 257–81.

LIPTON, P. (1991), *Inference to the Best Explanation* (London: Routledge).

LIVINGSTON LOWES, J. (1930), *The Road to Xanadu: A Study in the Ways of the Imagination* (London: Houghton; rev. 1951).

LLOYD MORGAN, C. (1894), *An Introduction to Comparative Psychology* (London: Scott).

LOCKE, J. (1690), *An Essay Concerning Human Understanding*, 5th edn. (1706); ed. P. Nidditch (Oxford: Oxford University Press, 1984).

LOCKE, W. N., and BOOTH, A. D. (eds.) (1955), *Machine Translation of Languages* (London: Chapman & Hall).

LOGOTHETIS, N. K. (1998), 'Single Units and Conscious Vision', *Philosophical Transactions of the Royal Society*, ser. B, 353: 1801–18.

——— (2002), 'The Neural Basis of the Blood-Oxygen-Level-Dependent Functional Magnetic Resonance Imaging Signal', in Parker *et al.* (2002: 1003–38).

——— and SCHALL, J. (1989), 'Neuronal Correlates of Subjective Visual Perception', *Science*, 245: 761–3.

LONGINO, H. (1990), *Science as Social Knowledge: Values and Objectivity in Scientific Inquiry* (Princeton: Princeton University Press).

LONGUET-HIGGINS, H. C. (1962), 'Two Letters to a Musical Friend', *Music Review*, Aug., 244–8; Nov., 271–80; repr. in Longuet-Higgins (1987: 64–81).

—— (1968a), 'A Holographic Model of Temporal Recall', *Nature*, 217: 104.

—— (1968b), 'The Non-Local Storage of Temporal Information', *Proceedings of the Royal Society*, ser. B, 171: 327–34.

—— (1972), 'The Algorithmic Description of Natural Language', *Philosophical Transactions of the Royal Society*, ser. B, 182: 255–76.

—— (1973), 'Comments on the Lighthill Report and the Sutherland Reply', in SRC (1973: 35–7).

—— (1976), 'The Perception of Melodies', *Nature*, 263: 646–53.

—— (1979), 'The Perception of Music', *Proceedings of the Royal Society*, ser. B, 205: 307–22.

—— (1981), 'Artificial Intelligence—A New Theoretical Psychology?', *Cognition*, 10: 197–200.

—— (1982), 'A Theory of Vision', *Science*, 218: 991–2.

—— (1987), *Mental Processes: Studies in Cognitive Science* (Cambridge, Mass.: MIT Press).

—— (1994), 'Artificial Intelligence and Musical Cognition', in Boden *et al.* (1994: 103–13).

—— and PRAZDNY, K. (1980), 'The Interpretation of a Moving Image', *Philosophical Transactions of the Royal Society*, ser. B, 208: 385–97.

—— and STEEDMAN, M. J. (1971), 'On Interpreting Bach', in Meltzer and Michie (1971: 221–42).

—— LYONS, J., and BROADBENT, D. E. (eds.) (1981), *The Psychological Mechanisms of Language* (London: Royal Society and British Academy).

LORENTE DE NO, R. (1933a), 'Studies on the Structure of the Cerebral Cortex, I: The Area Entrorhinalis', *Journal of Psychology and Neurology* (Leipzig), 45: 381–438.

—— (1933b), 'Vestibulo-Ocular Reflex Arc', *Archives of Neurology and Psychiatry*, 30: 245–91.

—— (1934), 'Studies on the Structure of the Cerebral Cortex, II: Continuation of the Study of the Ammonic System', *Journal of Psychology and Neurology* (Leipzig), 46: 113–77.

—— (1938), 'Analysis of the Activity of the Chains of Internuncial Neurons', *Journal of Neurophysiology*, 1: 207–44.

—— (1947), *A Study of Nerve Physiology* (New York: Rockefeller Institute).

LORENZ, K. (1935–9), 'Companionship in Bird Life: Fellow Members of the Species as Releasers of Social Behavior' (1935); 'The Nature of Instinct: The Conception of Instinctive Behavior' (1937); and 'Comparative Study of Behavior' (1939); all trans. from German in Schiller (1957: 83–128, 129–75, 239–63).

—— (1952), *King Solomon's Ring: New Light on Animal Ways*, trans. M. K. Wilson (London: Methuen).

—— and TINBERGEN, N. (1938), 'Taxis and Instinctive Action in the Egg-Retrieving Behavior of the Greylag Goose', trans. from German in Schiller (1957: 176–208).

LOUI, R. P. (1998), 'B. C. Smith's *On the Origin of Objects* (Book Review)', *Artificial Intelligence*, 106: 353–8.

LOUNSBURY, F. G. (1956), 'A Semantic Analysis of the Pawnee Kinship Usage', *Language*, 32: 158–94.

LOVEJOY, C. O., COHN, M. J., and WHITE, T. D. (1999), 'Morphological Analysis of the Mammalian Postcranium: A Developmental Perspective', *Proceedings of the National Academy of Sciences, USA*, 96 (Nov.), 13247–52.

LOVELACE, A. A. (1843), 'Notes by the Translator', repr. in Hyman (1989: 267–311); first pub. in Menabrea (1843).

LOVELOCK, J. (1988), *The Ages of Gaia: A Biography of our Living Earth* (Oxford: Oxford University Press).

LOWERRE, B., and REDDY, R. (1980), 'The HARPY Speech Understanding System', in W. Lea (ed.), *Trends in Speech Recognition* (Englewood Cliffs, NJ: Prentice-Hall), 340–60.

LUCAS, J. R. (1961), 'Minds, Machines, and Gödel', *Philosophy*, 36: 112–27.

—— (1970), *The Freedom of the Will* (Oxford: Oxford University Press).

—— (1996), 'Minds, Machines, and Gödel: A Retrospect', in P. J. R. Millican and A. J. Clark (eds.), *Machines and Thought: The Legacy of Alan Turing*, i (Oxford: Oxford University Press), 103–24.

—— (2000), 'A Don's Defence', *Oxford Magazine*, no. 174, 5–7.

LUCAS, K. (1905), 'On the Gradation of Activity in a Skeletal Muscle-Fibre', *Journal of Physiology*, 33: 124–37.

—— (1909), 'The "All-or-None" Contraction of the Amphibian Skeletal Muscle-Fibre', *Journal of Physiology*, 38: 113–33.

LUCE, R. D. (1959), *Individual Choice Behavior: A Theoretical Analysis* (New York: Wiley).

LUCE, R. D., BUSH, R. R., and GALANTER, E. (eds.) (1963–5), *Handbook of Mathematical Psychology*, 3 vols. (New York: Wiley).

LUCENA, P. S., GATTASS, M., and VELHO, L. (2002), 'Expressive Talking Heads: A Study on Speech and Facial Expression in Virtual Characters', *Revista SCIENTIA* (São Leopoldo, Brazil: Unisinos); available at <http://www.visgraf.impa.br>.

LUCKHAM, D. (1967), 'The Resolution Principle in Theorem-Proving', in N. L. Collins and D. M. Michie (eds.), *Machine Intelligence 1* (Edinburgh: Edinburgh University Press), 47–61.

LUCY, J. A. (1992), *Language Diversity and Thought: A Reformulation of the Linguistic Diversity Hypothesis* (Cambridge: Cambridge University Press).

LUISI, P. L. (2002), 'Toward the Engineering of Minimal Living Cells', *Anatomical Record*, 268: 208–14.

——— and OBERHOLZER, T. (2001), 'Origins of Life on Earth: Molecular Biology in Liposomes as an Approach to the Minimal Cell', in F. Giovanelli (ed.), *The Bridge Between the Big Bang and Biology* (Rome: Consiglio Nazionale delle Ricerche (CNR) Press), 345–55.

——— and VARELA, F. (1989), 'Self-Replicating Micelles: A Chemical Version of a Minimal Autopoietic System', *Origins of Life and Evolution of the Biosphere*, 19: 633–43.

——— RASI, P. S. S., and MAVELLI, F. (2004), 'A Possible Route to Prebiotic Vesicle Reproduction', *Artificial Life*, 10: 297–308.

——— FERRI, F., and STANO, P. (forthcoming), 'Approaches to Semi-Synthetic Minimal Cells: A Review', *Naturwissenschaften*. Page references are to the 42-page preprint.

LUND, H. H., WEBB, B., and HALLAM, J. (1997), 'A Robot Attracted to the Cricket Species *Gryllus Binmaculatus*', in P. Husbands and I. Harvey (eds.), *Fourth European Conference on Artificial Life* (Cambridge, Mass.: MIT Press), 246–55.

LURIA, A. R. (1932), *The Nature of Human Conflicts, or, Emotion, Conflict and Will: An Objective Study of Disorganization and Control of Human Behaviour*, trans. W. H. Gantt (New York: Liveright).

——— (1960), *The Role of Speech in the Regulation of Normal and Abnormal Behavior* (New York: Irvington).

——— (1966*a*), *Higher Cortical Functions in Man*, trans. B. Haigh, prefaces by Hans-Lukas Teuber and Karl H. Pribram (London: Tavistock).

——— (1966*b*), *Human Brain and Psychological Processes*, trans. B. Haigh (London: Harper & Row).

——— (1968), *The Mind of a Mnemonist: A Little Book About a Vast Memory*, trans. L. Solotaroff (New York: Basic Books).

——— (1970), *Traumatic Aphasia: Its Syndromes, Psychology and Treatment*, trans. M. Critchley (The Hague: Mouton).

——— (1972), *The Man with a Shattered World: The History of a Brain Wound*, trans. L. Solotaroff (New York: Basic Books).

——— and YUDOVICH, F. A. (1959), *Speech and the Development of Mental Processes in the Child: An Experimental Investigation*, trans. O. Kovasc and J. Simon (London: Staples Press).

LYNCH, A. (1996), *Thought Contagion: How Belief Spreads Through Society* (New York: Basic Books).

——— (2002), 'Evolutionary Contagion in Mental Software', in R. J. Sternberg and J. C. Kaufmann (eds.), *The Evolution of Intelligence* (London: Erlbaum), 289–314.

LYONS, J. (1991), *Chomsky*, 3rd edn., expanded (London: Fontana Press).

LYOTARD, J.-F. (1984), *The Postmodern Condition: A Report on Knowledge*, trans. from French by G. Bennington and B. Massumi (Manchester: University of Manchester Press; first pub. 1979).

MAASS, W., and BISHOP, C. M. (eds.) (1999), *Pulsed Neural Networks* (Cambridge, Mass.: MIT Press).

McCALL, E. A. (1965), 'A Generative Grammar of Sign', MA diss., University of Iowa.

McCARTHY, J. (1958), *An Algebraic Language for the Manipulation of Symbolic Expressions*, MIT AI Lab Memo 1 (Cambridge, Mass.: MIT, Sept.).

——— (1959), 'Programs with Common Sense', in Blake and Uttley (1959: 75–91); repr. as sect. 7.1 of McCarthy (1968: 403–10).

——— (1960), 'Recursive Functions of Symbolic Expressions and their Calculation by Machine, Part I', *Communications of the Association for Computing Machinery*, 3: 184–95.

——— (1963), *Situations, Actions and Causal Laws*, Stanford AI Project Memo 2 (Stanford, Calif.: Stanford University); repr. as sect. 7.2 of McCarthy (1968: 410–17).

——— (1968), 'Programs with Common Sense', in Minsky (1968: 403–17). This is a combination of two earlier papers: McCarthy (1959, 1963).

——— (1974), 'Review of "Artificial Intelligence: A General Survey" ', *Artificial Intelligence*, 5: 317–22.

_____ (1977), 'Epistemological Problems of Artificial Intelligence', *Proceedings of the Fifth International Joint Conference on Artificial Intelligence*, Cambridge, Mass., 1038–44.

_____ (1980a), 'Circumscription—A Form of Non-Monotonic Reasoning', *Artificial Intelligence*, 13: 27–40.

_____ (1980b), 'Addendum: Circumscription and Other Non-Monotonic Formalisms', *Artificial Intelligence*, 13: 171–2.

_____ (1983), 'Some Expert Systems Need Common Sense', *Annals of the New York Academy of Science*, 426: 129–37.

_____ (1984), 'We Need Better Standards for Research', *AI Magazine*, 5/3: 7–8.

_____ (1986), 'Applications of Circumscription to Formalizing Common-Sense Knowledge', *Artificial Intelligence*, 28: 86–116.

_____ (1989), Review of B. Bloomfield (ed.), *The Question of Artificial Intelligence*, *Annals of the History of Computing*, available at <http://www-formal.stanford.edu/jmc/index.html>.

_____ (2005), 'The Future of AI–A Manifesto', *AI Magazine*, 26/4 (25th anniversary issue), 39.

_____ and HAYES, P. J. (1969), 'Some Philosophical Problems from the Standpoint of Artificial Intelligence', in Meltzer and Michie (1969a: 463–502).

_____ MINSKY, M. L., ROCHESTER, N., and SHANNON, C. E. (1955), 'A Proposal for the Dartmouth Summer Research Project on Artificial Intelligence', in Chrisley (2000: ii. 44–53).

_____ ABRAHAMS, P. W., EDWARDS, D. J., HART, T. P., and LEVIN, M. I. (1962), *Lisp 1.5 Programmer's Manual* (Cambridge, Mass.: MIT Press).

McCARTY, L. T. (1973), 'Interim Report on the TAXMAN Project: An Experiment in Artificial Intelligence and Legal Reasoning', *Artificial Intelligence Techniques in Legal Problem Solving: Selected Papers Presented at a Stanford Law School Workshop*, 28–9 Apr. 1972, Stanford University mimeo, 37–46. Page references are to the author's hand-prepared draft.

_____ (1977), 'Reflections on TAXMAN: An Experiment in Artificial Intelligence and Legal Reasoning', *Harvard Law Review*, 90: 837–93.

_____ (1980), 'The TAXMAN Project: Towards a Cognitive Theory of Legal Argument', in B. Niblett (ed.), *Computer Science and Law* (Cambridge: Cambridge University Press), 23–43.

McCAWLEY, J. D. (1995), 'Generative Semantics', in Koerner and Asher (1995: 343–8).

McCLELLAND, J. L. (2000), Foreword, in O'Reilly and Munakata (2000, pp. xix–xxiii).

_____ and ROGERS, T. T. (2003), 'The Parallel Distributed Processing Approach to Semantic Cognition', *Nature Reviews Neuroscience*, 4: 310–22.

_____ and RUMELHART, D. E. (1981), 'An Interactive Activation Model of Context Effects in Letter Perception: Part 1. An Account of Basic Findings', *Psychological Review*, 86: 287–330.

_____ _____ and the PDP Research Group (1986), *Parallel Distributed Processing: Explorations in the Microstructure of Cognition*, ii: *Psychological and Biological Models* (Cambridge, Mass.: MIT Press).

McCLOSKEY, M. (1983a), 'Naive Theories of Motion', in D. Gentner and A. L. Stevens (eds.), *Mental Models* (Hillsdale, NJ: Lawrence Erlbaum), 299–324.

_____ (1983b), 'Intuitive Physics', *Scientific American*, 248/4: 122–30.

McCORDUCK, P. (1979), *Machines Who Think: A Personal Inquiry into the History and Prospects of Artificial Intelligence* (San Francisco: W. H. Freeman).

_____ (1991), *AARON's Code: Meta-Art, Artificial Intelligence, and the Work of Harold Cohen* (New York: W. H. Freeman).

McCRAE, R. (1972), 'Innate Ideas', in R. J. Butler (ed.), *Cartesian Studies* (Oxford: Blackwell).

McCULLOCH, G. (1995), *The Mind and its World* (London: Routledge).

McCULLOCH, W. S. (1946), 'Finality and Form in Nervous Activity', The J. A. Thompson Lecture, 2 May; first pub. as 'Finality and Form', American Lecture Series, 11 (Springfield, Ill.: Charles C. Thomas, 1952); repr. in W. S. McCulloch (1965: 256–75).

_____ (1947), 'A Heterarchy of Values Determined by the Topology of Nervous Nets', *Bulletin of Mathematical Biophysics*, 7: 89–93; repr. in W. S. McCulloch (1965: 40–5).

_____ (1948), 'Through the Den of the Metaphysician', in W. S. McCulloch (1965: 142–56); first presented at the Philosophical Club of the University of Virginia, Mar. 1948.

_____ (1949), 'Physiological Processes Underlying Psychoneuroses', in W. S. McCulloch (1965: 373–86); first pub. in *Proceedings of the Royal Society of Medicine*, 42 (1949), 71–84.

_____ (1951), 'Why the Mind Is in the Head', in W. S. McCulloch (1965: 72–141); first pub. in Jeffress (1951: 42–111).

McCulloch, W. S. (1953), 'The Past of a Delusion', in W. S. McCulloch (1965: 276–303); first presented at the Chicago Literary Club, 1953.

_____ (1955), '*Mysterium Iniquitatis* of Sinful Man Aspiring into the Place of God', in McCulloch (1965: 157–64); first pub. in *Scientific Monthly*, 80 (1955), 35–9.

_____ (1961*a*), 'What Is a Number, that a Man May Know it, and a Man, that he May Know a Number?', in W. S. McCulloch (1965: 1–18); first pub. in *General Semantics Bulletin*, nos. 26 and 27 (Lakeville, Conn.: Institute of General Semantics, 1961), 7–18.

_____ (1961*b*), 'Where Is Fancy Bred?', in H. W. Brosin (ed.), *Lectures on Experimental Psychiatry* (Pittsburgh: University of Pittsburgh Press), 311–24; repr. in W. S. McCulloch (1965: 216–29).

_____ (1964), 'What's in the Brain that Ink May Character?', in W. S. McCulloch (1965: 387–97); first presented at the International Congress for Logic, Methodology, and Philosophy of Science, Jerusalem, Aug. 1964.

_____ (1965), *Embodiments of Mind*, ed. S. Papert (Cambridge, Mass.: MIT Press).

_____ and Pitts, W. H. (1943), 'A Logical Calculus of the Ideas Immanent in Nervous Activity', in W. S. McCulloch (1965: 19–39); first pub. in *Bulletin of Mathematical Biophysics*, 5 (1943), 115–33.

_____ Arbib, M. A., and Cowan, J. D. (1962), 'Neurological Models and Integrative Processes', in M. C. Yovits, G. T. Jacobi, and G. D. Goldstein (eds.), *Self-Organizing Systems* (Washington: Spartan Books), 49–59.

McDermott, D. V. (1974), *Assimilation of New Information by a Natural-Language Understanding System*, AI-TR-291 (Cambridge, Mass.: MIT AI Lab).

_____ (1976), 'Artificial Intelligence Meets Natural Stupidity', *SIGART Newsletter of the ACM*, no. 57, Apr.; repr. in Haugeland (1981*a*: 143–60). Page references cite the 1981 edn.

_____ (1978), 'Tarskian Semantics, or, No Notation Without Denotation', *Cognitive Science*, 2: 277–82.

_____ (1982*a*), 'A Temporal Logic for Reasoning About Processes and Plans', *Cognitive Science*, 6: 101–55.

_____ (1982*b*), 'Non-Monotonic Logic II: Non-Monotonic Modal Theories', *Journal of the Association for Computing Machinery*, 29: 33–57.

_____ (1985), 'Reasoning About Plans', in J. Hobbs and R. C. Moore (eds.), *Formal Theories of the Commonsense World* (Norwood, NJ: Ablex), 269–317.

_____ (1987), 'A Critique of Pure Reason', *Computational Intelligence*, 3: 151–60.

_____ (2001*a*), 'An Interview with Drew McDermott (by Kentaro Toyama)', *ACM Crossroads Student Magazine*, as modified 23 Jan., <http://www.acm.org/crossroads/xrds3-1/interview.html>.

_____ (2001*b*), *Mind and Mechanism* (Cambridge, Mass.: MIT Press).

_____ and Doyle, J. (1980), 'Non-Monotonic Logic I', *Artificial Intelligence*, 13: 41–72.

_____ and Hendler, J. (1995), 'Planning: What it Is, What it Could Be: An Introduction to the Special Issue on Planning and Scheduling', *Artificial Intelligence*, 76: 1–15.

McDermott, J. P. (1980), 'R1: An Expert in the Computer Systems Domain', *Proceedings of the 1st Annual National Conference on Artificial Intelligence*, Stanford, Calif., 269–71.

_____ (1982), 'XSEL: A Computer Sales Person's Assistant', in P. J. Hayes, D. M. Michie, and Y.-H. Pao (eds.), *Machine Intelligence 10* (Chichester: Ellis Horwood), 325–37.

McDougall, W. (1897), 'On the Structure of Cross-Striated Muscle, and a Suggestion as to the Nature of its Contraction', *Journal of Anatomy and Physiology*, 31: 410–41, 539–85.

_____ (1901), 'Some New Observations in Support of Thomas Young's Theory of Light and Colour Vision', *Mind*, 10: 52–97, 210–45, 347–82.

_____ (1902–6), 'The Physiological Factors of the Attention-Process', *Mind*, 11 (1902), 316–51; 12 (1903), 289–302; 15 (1906), 329–59.

_____ (1904), 'The Sensations Excited by a Single Momentary Stimulation of the Eye', *British Journal of Psychology*, 1: 78–113.

_____ (1905), *Physiological Psychology*, The Temple Primers (London: Dent).

_____ (1908), *An Introduction to Social Psychology* (London: Methuen; 3rd edn., 1936).

_____ (1911), *Body and Mind: A History and a Defense of Animism* (London: Methuen).

_____ (1923), *An Outline of Psychology* (London: Methuen).

_____ (1926), *An Outline of Abnormal Psychology* (London: Methuen).

_____ (1932), *The Energies of Men: A Study of the Fundamentals of Dynamic Psychology* (London: Methuen).

McDowell, J. (1994*a*), *Mind and World* (Cambridge, Mass.: Harvard University Press).

_____ (1994*b*), 'The Content of Perceptual Experience', *Philosophical Quarterly*, 44: 190–205.

McEliece, R. J., Mackay, D. J. C., and Cheng, J.-F. (1998), 'Turbo Decoding as an Instance of Pearl's "Belief Propagation" Algorithm', *IEEE Journal on Selected Areas in Communications*, 16/2: 140–52.

Macfarlane, A., and Martin, G. (2002), *Glass: A World History* (Chicago: Chicago University Press).

McGill, W. J. (1982), *The Year of the Monkey: Revolt on Campus 1968–69* (London: McGraw-Hill).

—— (1988), 'George A. Miller and the Origins of Mathematical Psychology', in Hirst (1988*a*: 3–18).

McGinn, C. (1981), *The Character of Mind* (Oxford: Oxford University Press).

—— (1989), 'Can we Solve the Mind–Body Problem?', *Mind*, 98: 349–66.

—— (1991), *The Problem of Consciousness* (Oxford: Blackwell).

McGrath, A. (2001), *In the Beginning: The Story of the King James Bible, and How it Changed a Nation, a Language, and a Culture* (London: Hodder & Stoughton).

McGraw, G. (1995), 'Letter Spirit (Part One): Emergent High-Level Perception of Letters Using Fluid Concepts', Ph.D. thesis (Cognitive Science), Indiana University.

Mache, F.-B. (2000), 'The Necessity of and Problems with a Universal Musicology', in Wallin *et al.* (2000: 473–80).

Machlup, F., and Mansfield, U. (eds.) (1983), *The Study of Information: Interdisciplinary Messages* (New York: Wiley-Interscience).

MacKay, D. J. C., and Miller, K. D. (1990), 'Analysis of Linsker's Simulations of Hebbian Rules', *Neural Computation*, 2: 173–87.

MacKay, D. M. (1949/1959), 'On the Combination of Digital and Analogue Computing Techniques in the Design of Analytic Engines', in Blake and Uttley (1959: i. 55–65); first written for private circulation in May 1949.

—— (1951), 'Mindlike Behaviour in Artefacts', *British Journal for the Philosophy of Science*, 2: 105–21.

—— (1956*a*), 'The Epistemological Problem for Automata', in Shannon and McCarthy (1956: 235–50).

—— (1956*b*), 'Towards an Information-Flow Model of Human Behaviour', *British Journal of Psychology*, 47: 30–43.

—— (1957), 'Brain and Will', *The Listener*, 57: 788–9.

—— (1965), 'A Mind's Eye View of the Brain', in N. Wiener and J. P. Schade (eds.), *Cybernetics of the Nervous System*, Progress in Brain Research, 17 (Amsterdam: Elsevier), 321–32.

McKenna, T. M. (1994), 'The Role of Interdisciplinary Research Involving Neuroscience in the Development of Intelligent Systems', in Honavar and Uhr (1994: 75–92).

MacKenzie, D. (1991*a*), 'Nuclear Weapons Laboratories and the Development of Supercomputing', *Annals of the History of Computing*, 13: 179–201; repr. in MacKenzie (1998: 99–130).

—— (1991*b*), 'The Fangs of the VIPER', *Nature*, 352: 467–8; repr. in MacKenzie (1998: 159–64).

—— (1993), 'Negotiating Arithmetic, Constructing Proof', *Social Studies of Science*, 23: 37–65; repr. in MacKenzie (1998: 165–84).

—— (1995), 'The Automation of Proof: A Historical and Sociological Exploration', *Annals of the History of Computing*, 17/3: 7–29.

—— (1998), *Knowing Machines: Essays on Technical Change* (Cambridge, Mass.: MIT Press).

—— and Elzen, B. (1991), 'The Charismatic Engineer', in *Jaarboek voor de Geschiedenis van bedriff en techniek 8*; repr. in Mackenzie (1998: 131–58).

—— and Spinardi, G. (1995), 'Tacit Knowledge and the Uninvention of Nuclear Weapons', *American Journal of Sociology*, 101: 44–99; repr. in MacKenzie (1998: 215–60).

McKeown, K. (1985), 'Discourse Strategies for Generating Natural-Language Text', *Artificial Intelligence*, 27: 1–42.

—— and Pan, S. (2000), 'Prosody Modelling in Concept-to-Speech Generation: Methodological Issues', in Gazdar *et al.* (2000: 1419–30).

McKinney, D. W. (1973), *The Authoritarian Personality Studies: An Inquiry into the Failure of Social Science Research to Produce Demonstrable Knowledge* (The Hague: Mouton).

Mackworth, A. K. (1998), Abstract (circulated by email) of 'Quick and Clean: Designing and Building Vision-Based Robots', talk given at the Computer Science Department, University of York, 29 Apr. 1998. (Cf. Sahota and Mackworth 1994.)

—— (1973), 'Interpreting Pictures of Polyhedral Scenes', *Artificial Intelligence*, 4: 121–38.

—— (2003), 'On Seeing Robots', in A. Basu and X. Li (eds.), *Computer Vision: Systems, Theory, and Applications* (Singapore: World Scientific Press), 1–13.

MacLennan, B. J., and Burghardt, G. M. (1994), 'Synthetic Ecology and the Evolution of Cooperative Communication', *Adaptive Behaviour*, 2: 151–88.

McLUHAN, M. (1964), *Understanding Media: The Extensions of Man* (New York: McGraw-Hill).

McMULLIN, B. (2004), 'Thirty Years of Computational Autopoesis: A Review', *Artificial Life*, 10: 277–95.

_____ and VARELA, F. J. (1997), 'Rediscovering Computational Autopoiesis', in P. Husbands and I. Harvey (eds.), *Fourth European Conference on Artificial Life* (Cambridge, Mass.: MIT Press), 38–47.

McRAE, R. (1972), 'Innate Ideas', in R. J. Butler (ed.), *Cartesian Studies* (Oxford: Blackwell), 32–54.

McREYNOLDS, P. (1980), 'The Clock Metaphor in the History of Psychology', in T. Nickles (ed.), *Scientific Discovery: Case Studies* (Dordrecht: Reidel), 97–112.

MADDOX, B. (2002), *Rosalind Franklin: The Dark Lady of DNA* (London: HarperCollins).

MADINA, D, ONO, N., and IKEGAMI, T. (2003), 'Cellular Evolution in a 3D Lattice Artificial Chemistry', in W. Banzhaf, T. Christaller, P. Dittrich, J. T. Kim, and J. Ziegler (eds.), *Advances in Artificial Life: Proceedings of the Seventh European Conference on Artificial Life*, Dortmund, Sept. 2003 (Berlin: Springer-Verlag), 59–68.

MADISON, P. (1961), *Freud's Concept of Repression and Defense: Its Theoretical and Observational Language* (Minneapolis: University of Minnesota Press).

MAES, P. (1991a), 'A Bottom-Up Mechanism for Behavior Selection in an Artificial Creature', in Meyer and Wilson (1991: 238–46).

_____ (1991b), 'Situated Agents Can Have Goals', in Maes (1991c: 49–70).

_____ (ed.) (1991c), *Designing Autonomous Agents: Theory and Practice from Biology to Engineering and Back* (Cambridge, Mass.: MIT Press).

MAGNUS, R. (1906/1949), *Goethe as a Scientist*, trans. H. Norden from the German edn. of 1906 (New York: H. Schuman, 1949). This includes 'Bibliography by Rudolf Magnus, Supplemented by Gunther Schmid', 249–53.

MAHON, P. (1945/2004), 'History of Hut 8 to December 1941', in Copeland (2004: 267–312). This is the first half of Mahon, *The History of Hut Eight, 1939–1945*; the complete document is available at <http://www.AlanTuring.net/mahon_hut_8>.

MAIER, N. R. F. (1929), *Reasoning in White Rats*, Comparative Psychology Monographs, 6/3 (Baltimore: Johns Hopkins Press).

_____ (1930), 'Reasoning in Humans, I: On Direction', *Journal of Comparative Psychology*, 10: 115–43.

_____ (1931), 'Reasoning in Humans, II: The Solution of a Problem and its Appearance in Consciousness', *Journal of Comparative Psychology*, 12: 181–94.

_____ (1933), 'An Aspect of Human Reasoning', *British Journal of Psychology*, 24: 144–55.

MALCOLM, N. (1964), 'Scientific Materialism and the Identity Theory', *Dialogue*, 3: 115–25.

_____ (1973), 'Thoughtless Brutes', *Proceedings of the American Philosophical Society*, 46: 5–20.

_____ (1971), 'The Myth of Cognitive Processes and Structures', in T. Mischel (ed.), *Cognitive Development and Epistemology* (New York: Academic Press), 385–92.

MALINOWSKI, B. (1915–18), *The Ethnography of Malinowski: The Trobriand Islands 1915–18*, ed. M. W. Young (London: Routledge & Kegan Paul, 1979).

_____ (1922), *Argonauts of the Western Pacific: An Account of Native Enterprise and Adventure in the Archipelagoes of Melanesian New Guinea* (London: Routledge).

MALLEN, G. (2005), 'Reflections on Gordon Pask's Adaptive Teaching Concepts and their Relevance to Modern Knowledge Systems', in Candy (2005: 86–91).

Manchester Philosophy Seminar (1949), Notes on a discussion on 'The Mind and the Computing Machine', held in the Manchester Philosophy Seminar, 27 Oct. Written by a member of the Philosophy Department.

MANDLER, G. (1986), 'Interview with George Mandler [in 1975]', in Baars (1986: 253–69).

_____ (1996), 'The Situation of Psychology: Landmarks and Choicepoints', *American Journal of Psychology*, 109: 1–35.

_____ (2002a), 'Origins of the Cognitive (R)evolution [sic]', *Journal of History of the Behavioral Sciences*, 38: 339–53.

_____ (2002b), 'Psychologists and the National Socialist Access to Power', *History of Psychology*, 5: 190–200.

_____ (2002c), *Interesting Times: An Encounter with the 20th Century 1924–* (Hillsdale, NJ: Lawrence Erlbaum).

MANDLER, J. M. (1984), *Stories, Scripts, and Scenes: Aspects of Schema Theory* (Hillsdale, NJ: Lawrence Erlbaum).

_____ and JOHNSON, N. S. (1976), 'Remembrance of Things Parsed: Story Structure and Recall', *Cognitive Psychology*, 9: 111–51.

____ and MANDLER, G. (1968), 'The Diaspora of Experimental Psychology: The Gestaltists and Others', in D. Fleming and B. Bailyn (eds.), *The Intellectual Migration: Europe and America, 1930–1960* (Cambridge, Mass.: Harvard University Press), 371–419.

____ SCRIBNER, S., COLE, M., and DEFOREST, M. (1980), 'Cross-Cultural Invariance in Story Recall', *Child Development*, 51: 19–26.

MANUEL, F. E. (1968), *A Portrait of Isaac Newton* (Cambridge, Mass.: Harvard University Press).

MARCEL, A. (1993), 'Slippage in the Unity of Consciousness', in Bock and Marsh (1993: 168–80).

____ and BISIACH, E. (eds.) (1988), *Consciousness in Contemporary Science* (Oxford: Oxford University Press).

MARCUS, G. F., VIJAYAN, S., BANDI RAO, S., and VISHTON, P. M. (1999), 'Rule-Learning by Seven-Month-Old Infants', *Science*, 283: 77–80.

MARCUS, M. P. (1979), 'A Theory of Syntactic Recognition for Natural Language', in P. H. Winston and R. H. Brown (eds.), *Artificial Intelligence: An MIT Perspective*, i (Cambridge, Mass.: MIT Press), 193–230.

____ (1980), *A Theory of Syntactic Recognition for Natural Language* (Cambridge, Mass.: MIT Press).

MARCUSE, H. (1964), *One-Dimensional Man: Studies in the Ideology of Advanced Industrial Society* (Boston: Beacon Press).

MARGOSHES, A. (1967), 'Schelling, Friedrich Wilhelm Joseph Von', in P. Edwards (ed.), *The Encyclopedia of Philosophy*, vii (New York: Macmillan), 305–9.

MARKOWITZ, H. (1952), 'Portfolio Selection', *Journal of Finance*, 7: 77–91.

MARR, A. (forthcoming), '"He Thought it the Deuil, Whereas Indeede it was a Meere Mathematicall Inuention": Understanding Automata in the Late Renaissance', in J. Riskin (ed.), *The Sistine Gap: Essays in the History and Philosophy of Artificial Life* (Stanford, Calif.: Stanford University Press).

MARR, D. C. (1969), 'A Theory of Cerebellar Cortex', *Journal of Physiology*, 202: 437–70.

____ (1970), 'A Theory for Cerebral Neocortex', *Proceedings of the Royal Society*, ser. B, 176: 161–234.

____ (1971), 'Simple Memory: A Theory for Archicortex', *Philosophical Transactions of the Royal Society*, ser. B, 262: 23–81.

____ (1974a), 'The Computation of Lightness by the Primate Retina', *Vision*, 14: 1377–88.

____ (1974b), *A Note on the Computation of Binocular Disparity in a Symbolic, Low-Level Visual Processor*, MIT AI Lab Memo no. 327 (Cambridge, Mass.: MIT); repr. in Vaina (1991: 231–8).

____ (1975a), *Analyzing Natural Images: A Computational Theory of Texture Vision*, MIT AI Lab Memo 334 (Cambridge, Mass.: MIT, June 1975).

____ (1975b), *Early Processing of Visual Information*, MIT AI Lab Memo 340 (Cambridge, Mass.: MIT, Dec. 1975); officially pub. in *Philosophical Transactions of the Royal Society*, ser. B, 275 (1976), 483–524.

____ (1975c), 'Approaches to Biological Information Processing', *Science*, 190: 875–6.

____ (1977), 'Artificial Intelligence: A Personal View', *Artificial Intelligence*, 9: 37–48.

____ (1979), 'Representing and Computing Visual Information', in P. H. Winston and R. H. Brown (eds.), *Artificial Intelligence: An MIT Perspective*, ii (Cambridge, Mass.: MIT Press), 17–82.

____ (1982), *Vision: A Computational Investigation into the Human Representation and Processing of Visual Information* (San Francisco: W. H. Freeman).

____ and HILDRETH, E. (1980), 'Theory of Edge-Detection', *Proceedings of the Royal Society*, ser. B, 207: 187–217.

____ and NISHIHARA, H. K. (1978), 'Visual Information Preocessing: Artificial Intelligence and the Sensorium of Sight', *Technology Review*, 81: 2–23.

____ and POGGIO, T. (1976), 'Cooperative Computation of Stereo Disparity', *Science*, 194: 283–7.

____ ____ (1977), 'From Understanding Computation to Understanding Neural Circuitry', *Neuroscience Research Program Bulletin*, 15: 470–88.

____ ____ (1979), 'A Computational Theory of Human Stereo Vision', *Proceedings of the Royal Society*, ser. B, 204: 301–28.

MARSHALL, P., ROGERS, Y., and SCAIFE, M. (2004), 'PUPPET: Playing and Learning in a Virtual World', *International Journal of Continuing Engineering Education and Life-long Learning*, 14: 519–31.

MARTIN, E. (1991), 'The Egg and the Sperm: How Science Has Constructed a Romance Based on Stereotypical Male–Female Roles', *Signs: Journal of Women in Culture and Society*, 16: 485–501.

MARTIN, M. (1973), 'Are Cognitive Processes and Structure a Myth?', *Analysis*, 33: 83–8.

MARX, K. (1890), *Capital: A Critique of Political Economy*, i; trans. B. Fowkes (Harmondsworth: Penguin, 1976). Page references are to the 1976 edn.

MASLOW, A. H. (1962), *Toward a Psychology of Being* (Princeton: Van Nostrand).

Masterman, M. M. (1953), 'Metaphysical and Ideographic Language', in C. A. Mace (ed.), *British Philosophy in Mid-Century: A Cambridge Symposium* (London: Allen & Unwin), 283–357.

____ (1957), 'The Thesaurus in Syntax and Semantics', *Mechanical Translation*, 4: 1–2.

____ (1962), 'Semantic Message Detection for Machine Translation Using an Interlingua', *Proceedings of the 1961 Conference on Machine Translation of Languages and Applied Language Analysis* (London: HMSO), 437–75.

____ (1967), 'Mechanical Pidgin Translation', in Booth (1967: 195–227).

____ (1970), 'The Nature of a Paradigm', in I. Lakatos and A. Musgrave (eds.), *Criticism and the Growth of Knowledge* (Cambridge: Cambridge University Press), 59–89.

____ (1971), 'Computerized Haiku', in J. Reichardt (ed.), *Cybernetics, Art, and Ideas* (London: Studio Vista), 175–83.

____ and McKinnon Wood, R. (1968), 'Computerized Japanese Haiku', in Reichardt (1968: 54–5).

____ Needham, R. M., Sparck Jones, K., and Mayoh, B. (1957/1986), *Agricola Incurvo Terram Dimovit Aratro (Virgil, 'Georgics'): First Stage Translation into English with the Aid of 'Roget's Thesaurus'*, Report (ML84) ML92, Nov. 1957 (Cambridge: Cambridge Language Research Unit); repr. with new introd. by K. Sparck Jones (Cambridge: Computer Laboratory, University of Cambridge, Apr. 1986); the 1986 version is repr. in Wilks (forthcoming).

Mataric, M. J. (1991), 'Navigating with a Rat Brain: A Neurobiologically Inspired Model for Robot Spatial Representation', in Meyer and Wilson (1991: 169–75).

Matthews, G. B. (1977), 'Consciousness and Life', *Philosophy*, 52: 13–26.

____ (1992), '*De Anima* 2.2–4 and the Meaning of Life', in M. C. Nussbaum and A. O. Rorty (eds.), *Essays on Aristotle's 'De Anima'* (Oxford: Clarendon Press), 185–93.

Maturana, H. R. (1969), 'The Neurophysiology of Cognition', in P. Garvin (ed.), *Cognition: A Multiple View* (New York: Spartan Books), 3–24.

____ (1975), 'The Organization of the Living: A Theory of the Living Organization', *International Journal of Man–Machine Studies*, 7: 313–32.

____ and Varela, F. J. (1972/1980), *Autopoiesis and Cognition: The Realization of the Living*, Boston Studies in the Philosophy of Science (Boston: Reidel, 1980); first pub. in Spanish (1972).

____ ____ (1987), *The Tree of Knowledge: The Biological Roots of Human Understanding* (Boston: New Science Library).

____ ____ (1992), *The Tree of Knowledge: The Biological Roots of Human Understanding* (Boston: New Science Library); rev. edn. of Maturana and Varela (1987).

Maude, H. E. (1980), *The Gilbertese Maneaba* (Suva: Institute of Pacific Studies, University of the South Pacific).

Maxwell, J. C. (1868), 'On Governors', *Proceedings of the Royal Society*, 16: 270–83.

Maxwell, N. (1984), *From Knowledge to Wisdom: A Revolution in the Aims and Methods of Science* (Oxford: Blackwell).

____ (2004), *Is Science Neurotic?* (London: Imperial College Press).

May, R. (1961), *Existential Psychology* (New York: Random House).

Mayhew, J. E. W. (1983), 'Stereopsis', in O. J. Bradick and A. C. Sleigh (eds.), *Physical and Biological Processing of Images* (New York: Springer-Verlag), 204–16.

____ and Frisby, J. (1984), 'Computer Vision', in T. O'Shea and M. Eisenstadt (eds.), *Artificial Intelligence: Tools, Techniques, Applications* (New York: Harper & Row), 301–57.

Maynard Smith, J. (1996), 'Evolution—Natural and Artifical', in Boden (1996: 173–8).

____ and Szathmáry, E. (1995), *The Major Transitions in Evolution* (Oxford: W. H. Freeman).

____ ____ (1999), *The Origins of Life: From the Birth of Life to the Origins of Language* (Oxford: Oxford University Press).

Mays, W. (1951), 'The Hypothesis of Cybernetics', *British Journal for the Philosophy of Science*, 2: 249–50.

____ (1952), 'Can Machines Think?', *Philosophy*, 27: 148–62.

____ and Henry, D. P. (1953), 'Jevons and Logic', *Mind*, 62: 484–505.

____ and Prinz, D. G. (1950), 'A Relay Machine for the Demonstration of Symbolic Logic', *Nature*, 165: 197–8.

____ Henry, D. P., and Hansel, C. E. M. (1951), 'Note on the Exhibition of Logical Machines at the Joint Session, July 1950', *Mind*, 60: 262–4.

Mead, C. (1989), *Analog VLSI and Neural Systems* (Reading, Mass.: Addison-Wesley).

____ and Mahowald, M. (1991), 'The Silicon Retina', *Scientific American*, 5 (May), 76–82.

MEDAWAR, P. B. (1958), 'Postscript: D'Arcy Thompson and *Growth and Form*', in R. D. Thompson, *D'Arcy Wentworth Thompson: The Scholar-Naturalist, 1860–1948* (London: Oxford University Press), 219–33.

—— (1961), 'Critical Notice of Pierre Teilhard de Chardin, *The Phenomenon of Man*', *Mind*, 70: 99–106.

MEEHAN, J. (1975), 'Using Planning Structures to Generate Stories', *American Journal of Computational Linguistics*, microfiche 33: 77–93.

—— (1981), ' "TALE-SPIN" and "Micro TALE-SPIN" ', in Schank and Riesbeck (1981: 197–258).

MEEHL, P. E. (1954), *Clinical Versus Statistical Prediction: A Theoretical Analysis and a Review of the Evidence* (Minneapolis: University of Minnesota Press).

MEHL, L. (1959), 'Automation in the Legal World', in Blake and Uttley (1959: ii. 755–87).

MEHLER, J., MORTON, J., and JUSCZYK, P. (1984), 'On Reducing Language to Biology', *Cognitive Neuropsychology*, 1: 83–116.

—— JUSCZYK, P. W., LAMBERTZ, G., HALSTED, N., BERTONCINI, J., and AMIEL-TISON, C. (1988), 'A Precursor of Language Acquisition in Young Infants', *Cognition*, 29: 143–78.

MEINHARDT, H. (1982), *Models of Biological Pattern Formation* (New York: Academic Press).

MELDEN, A. I. (1961), *Free Action* (London: Routledge & Kegan Paul).

MELTZER, B. (1971), 'Personal View: Bury the Old War-Horse!', *SIGART Newsletter of the ACM*, no. 27 (Apr.).

—— and MICHIE, D. M. (eds.) (1969a), *Machine Intelligence 4* (Edinburgh: Edinburgh University Press).

—— —— (1969b), *Machine Intelligence 5* (Edinburgh: Edinburgh University Press).

—— —— (eds.) (1971), *Machine Intelligence 6* (Edinburgh: Edinburgh University Press).

—— —— (eds.) (1972), *Machine Intelligence 7* (Edinburgh: Edinburgh University Press).

MELVILL-JONES, G., and DAVIES, P. R. T. (1976), 'Adaptation of Cat Vestibulo-Ocular Reflex to 200 Days of Optically Reversed Vision', *Brain Research*, 103: 551–4.

MENABREA, L. F. (1843), *Sketch of the Analytical Engine Invented by Charles Babbage, Esq.*, trans. A. Lovelace (London: R. & J. E. Taylor); repr. in Hyman (1989: 243–66); first pub. in *Scientific Memoirs*, 3: 666–731.

MERLEAU-PONTY, M. (1945/1962), trans. C. Smith, *Phenomenology of Perception* (London: Routledge & Kegan Paul).

—— (1960), 'Phenomenology and Analytic Philosophy', trans. M. B. Smith, in M. Merleau-Ponty, *Texts and Dialogues: On Philosophy, Politics, and Culture*, ed. H. J. Silverman and J. Barry (London: Humanities Press, 1992), 59–72; French original pub. later, in Ryle (1962: 93–6).

MERRY, M. (ed.) (1985), *Expert Systems 85* (Cambridge: Cambridge University Press).

MERZ, J. T. (1904–12), *A History of European Thought in the Nineteenth Century*, 4 vols. (London: Blackwood).

METROPOLIS, N., and WORLTON, J. (1980), 'A Trilogy of Errors in the History of Computing', *Annals of the History of Computing*, 2: 49–59.

MEYER, J.-A., and WILSON, S. W. (eds.) (1991), *From Animals to Animats*, Proceedings of the First International Conference on Simulation of Adaptive Behavior (Cambridge, Mass.: MIT Press).

MIALET, H. (2003), 'Is the End in Sight for the Lucasian Chair? Stephen Hawking as Millennium Professor', in K. C. Knox and R. Noakes (eds.), *From Newton to Hawking: A History of Cambridge University's Lucasian Professors of Mathematics* (Cambridge: Cambridge University Press), 425–59.

MICHALSKI, R. S. (1973), 'Discovering Classification Rules Using Variable-Valued Logic System VL1', *Proceedings of the Third International Joint Conference on Artificial Intelligence*, Los Angeles, 162–72.

—— (ed.) (1983), *Proceedings of the International Machine Learning Workshop*, Monticello, Ill., June (Urbana: University of Illinois Department of Computer Science).

—— and CHILAUSKI, R. L. (1980), 'Learning by Being Told and Learning from Examples: An Experimental Comparison of the Two Methods of Knowledge Acquisition in the Context of Developing an Expert System for Soybean Disease Diagnosis', *International Journal of Policy Analysis and Information Systems*, 4, special issue: *Knowledge Acquisition and Induction*, 125–61.

—— and LARSON, J. B. (1978), *Selection of Most Representative Training Examples and Incremental Generation of VL1 Hypotheses: The Underlying Methodology and Description of Programs ESEL and AQ11*, Report 867 (Urbana: University of Illinois Department of Computer Science).

—— CARBONELL, J. G., and MITCHELL, T. M. (eds.) (1983), *Machine Learning: An Artificial Intelligence Approach* (Palo Alto, Calif.: Tioga).

—— BRATKO, I., and KUBAT, M. (1998), *Machine Learning and Data Mining: Methods and Applications* (Chichester: Wiley).

MICHIE, D. M. (1966), 'Game-Playing and Game-Learning Automata', in L. Fox (ed.), *Advances in Programming and Non-Numerical Computation* (Oxford: Pergamon Press), 183–96; repr. in Michie (1974: 31–49).

——(ed.) (1968*a*), *Machine Intelligence 3* (Edinburgh: Edinburgh University Press).

——(1968*b*), 'Computer—Servant or Master', *Theoria to Theory*, 2: 242–9; repr. in Michie (1974: 66–74).

——(1970), 'Future for Integrated Cognitive Systems', *Nature*, 228: 717–22; repr. in Michie (1974: 75–90).

——(1972), *A Six Year Project to Develop an Intelligent Problem-Solving System* (Edinburgh University: Department of Machine Intelligence and Perception). Modified version of a seven-year project proposed to the SRC.

——(1973*a*), 'Understanding the Machine', *Computer Weekly*, 5 Apr.; repr. in Michie (1982: 248–51).

——(1973*b*), 'Machine Intelligence at Edinburgh', *Management Informatics*, 2: 7–12.

——(1973*c*), 'Machines and the Theory of Intelligence', *Nature*, 241: 507–12.

——(1973*d*), 'Machine Intelligence in the Cycle Shed', *New Scientist*, 57 (22 Feb.), 422–3.

——(1974), *On Machine Intelligence* (Edinburgh: Edinburgh University Press).

——(ed.) (1979), *Expert Systems in the Micro-Electronic Age* (Edinburgh: Edinburgh University Press).

——(1982), *Machine Intelligence and Related Topics: An Information Scientist's Weekend Book* (London: Gordon & Breach).

——(1997), 'Slaughter on Seventh Avenue', *New Scientist*, 7 June, 26–9.

——(2002), 'The Very Early Days', a talk to the seminar 'Artificial Intelligence: Recollections of the Pioneers', Science Museum, London, 11 Oct., available at <http://www.aiai.ed.ac.uk/events/ccs2002>.

——and CHAMBERS, R. A. (1968), 'BOXES: An Experiment in Adaptive Control', in Dale and Michie (1968: 137–52).

——and JOHNSTON, R. (1984), *The Creative Computer: Machine Intelligence and Human Knowledge* (London: Viking Penguin).

——and ROSS, R. (1969), 'Experiments with the Adaptive Graph Traverser', in Meltzer and Michie (1969*b*: 301–18).

——SPIEGHALTER, D. J., and TAYLOR, C. C. (eds.) (1994), *Machine Learning, Neural and Statistical Classification* (Chichester: Ellis Horwood).

MICHOTTE, A. E. (1963), *The Perception of Causality*, trans. T. R. Miles and E. Miles (London: Methuen).

MILGRAM, S. (1963), 'Behavioral Study of Obedience', *Journal of Abnormal and Social Psychology*, 67: 371–8.

——(1974), *Obedience to Authority: An Experimental View* (New York: Harper & Row).

MILLER, G. A. (1947*a*), 'The Masking of Speech', *Psychological Bulletin*, 44: 105–29.

——(1947*b*), 'Sensitivity to Changes in the Intensity of White Noise and its Relation to Masking and Loudness', *Journal of the Acoustical Society of America*, 19: 609–19.

——(1951), *Language and Communication* (New York: McGraw-Hill).

——(1952), 'Finite Markov Processes in Psychology', *Psychometrika*, 17: 149–67.

——(1953), 'What Is Information Measurement?', *American Psychologist*, 8: 3–11.

——(1956*a*), 'The Magical Number Seven, Plus or Minus Two: Some Limits on our Capacity for Processing Information', *Psychological Review*, 63: 81–97.

——(1956*b*), 'Information and Memory', *Scientific American*, 195/2 (Aug.), 42–46.

——(1962), *Psychology: The Science of Mental Life* (New York: Harper & Row).

——(1967), 'Project Grammarama', in G. A. Miller, *The Psychology of Communication: Seven Essays* (New York: Basic Books), 125–87.

——(1978), 'Letter to M. S. Gazzaniga, Late May 1978', in Gazzaniga (1988: 6–9).

——(1983*a*), 'Addendum to the Prologue: Letter from George Miller (January 18, 1983)', in F. Machlup and U. Mansfield (eds.), *The Study of Information: Interdisciplinary Messages* (New York: Wiley-Interscience), 57–9.

——(1983*b*), 'A Model Science', *London Review of Books*, 3 (16 Nov.), 18–19.

——(1986), 'Interview with George A. Miller', in Baars (1986: 200–23).

——and CHOMSKY, A. N. (1963), 'Finitary Models of Language Users', in Luce *et al.* (1963–5: ii. 419–91).

——and FRICK, F. C. (1949), 'Statistical Behavioristics and Sequences of Responses', *Psychological Review*, 56: 311–24.

——GALANTER, E., and PRIBRAM, K. H. (1960), *Plans and the Structure of Behavior* (New York: Henry Holt).

——and Isard, S. (1964), 'Free Recall of Self-Embedded English Sentences', *Information and Control*, 7: 293–303.

——and Johnson-Laird, P. N. (1976), *Language and Perception* (Cambridge: Cambridge University Press).

——and Selfridge, J. A. (1950), 'Verbal Context and the Recall of Meaningful Material', *American Journal of Psychology*, 63: 176–85.

——Bruner, J. S., and Postman, L. (1954), 'Familiarity of Letter Sequences and Tachistoscopic Identification', *Journal of General Psychology*, 50: 129–39.

Miller, G. F. (2000), *The Mating Mind: How Sexual Choice Shaped the Evolution of Human Nature* (London: William Heinemann).

——and Freyd, J. J. (1993), *Dynamic Mental Representations of Animate Motion: The Interplay Among Evolutionary, Cognitive, and Behavioral Dynamics*, Cognitive Science Research Paper CSRP-290 (Brighton: University of Sussex).

Miller, K. D., Keller, J. B., and Stryker, M. P. (1989), 'Ocular Dominance Column Development: Analysis and Simulation', *Science*, 245: 605–15.

Miller, S. L. (1953), 'A Production of Amino Acids Under Possible Primitive Earth Conditions', *Science*, 117: 528–9.

——and Urey, H. C. (1959), 'Organic Compound Synthesis on the Primitive Earth', *Science*, 130: 245–51.

Millican, P. J. R., and Clark, A. J. (eds.) (1996), *Machines and Thought: The Legacy of Alan Turing*, i (Oxford: Oxford University Press).

Millikan, R. G. (1984), *Language, Thought, and Other Biological Categories: New Foundations for Realism* (Cambridge, Mass.: MIT Press).

——(1996), 'On Swampkinds', *Mind and Language*, 11: 105–17.

Milner, A. D., and Goodale, M. A. (1993), 'Visual Pathways to Perception and Action', in T. P. Hicks, S. Molotchnikoff, and T. Ono (eds.), *Progress in Brain Research*, 95 (Amsterdam: Elsevier), 317–37.

—— ——(1995), *The Visual Brain in Action* (Oxford: Oxford University Press).

Milner, B. (1959), 'The Memory Defect in Bilateral Hippocampal Lesions', *Psychiatric Research Reports*, 11: 43–52.

——and Rugg, M. (eds.) (1992), *The Neuropsychology of Consciousness* (New York: Academic Press).

Milner, P. M. (1999), 'Cell Assemblies: Whose Idea?', *Psycoloquy*, 10/53, <http://www.cogsci.ecs.soton.ac.uk/psycoloquy>.

Miner, R. C. (1998), 'Verum-Factum and Practical Wisdom in the Early Writings of Giambattista Vico', *Journal of the History of Ideas*, 59: 53–73.

Minsky, M. L. (1956*a*), 'Some Universal Elements for Finite Automata', in Shannon and McCarthy (1956*a*: 117–28).

——(1956*b*), *Notes on the Geometry Problem, I and II*, Artificial Intelligence Project, Dartmouth College, Hanover, Vt. (Aug.).

——(1956*c*), *Heuristic Aspects of the Artificial Intelligence Problem*, Group Report 34-55, ASTIA Document AD 236885 (Lexington, Mass.: MIT Lincoln Laboratories, Dec.).

——(1959), 'Some Methods of Artificial Intelligence and Heuristic Programming', in Blake and Uttley (1959: i. 5–27).

——(1961*a*), 'A Selected Descriptor-Indexed Bibliography to the Literature on Artificial Intelligence', *Institute of Radio Engineers Transactions on Human Factors in Electronics*, vol. HFE-2, 39–55; repr. in Feigenbaum and Feldman (1963: 453–523).

——(1961*b*), 'Steps Toward Artificial Intelligence', *Proceedings of the Institute of Radio Engineers*, 49: 8–30; repr. in Feigenbaum and Feldman (1963: 406–50); first pub. as an MIT Technical Report (1956)—see Minsky (1956*c*).

——(1961*c*), 'Descriptive Languages and Problem Solving', *Proceedings of the Western Joint Computer Conference*, Los Angeles, 215–18.

——(1965), 'Matter, Mind, and Models', *Proceedings of the International Federation of Information Processing Congress*, 1 (Washington: Spartan Books), 45–9.

——(ed.) (1968), *Semantic Information Processing* (Cambridge, Mass.: MIT Press).

——(1969), 'Form and Content in Computer Science', 1969 Turing Award Lecture, in R. L. Ashenhurst and S. Graham (eds.), *ACM Turing Award Lectures: The First Twenty Years 1966–1985* (London: Addison-Wesley, 1987), 219–42.

MINSKY, M. L. (1975), 'A Framework for Representing Knowledge', in Winston (1975: 211–77); first pub. as MIT AI Lab Memo 306 (Cambridge, Mass.: MIT, 1974).

——— (1977), 'Plain Talk About Neurodevelopmental Epistemology', *Proceedings of the Fifth International Conference on Artificial Intelligence* (Cambridge, Mass.: MIT), 1083–92.

——— (1979), 'The Society Theory of Thinking', in P. H. Winston and R. H. Brown (eds.), *Artificial Intelligence: An MIT Perspective*, i. (Cambridge, Mass.: MIT Press), 421–52.

——— (1980), 'K-Lines: A Theory of Memory', *Cognitive Science*, 4: 117–33.

——— (1984*a*), 'The Problems and the Promise', in Winston and Prendergast (1984: 243–51).

——— (1984*b*), 'From Feedback to Debugging', in O. G. Selfridge, E. L. Rissland, and M. A. Arbib (eds.), *Adaptive Control of Ill-Defined Systems* (New York: Plenum Press), 115–26.

——— (1985), *The Society of Mind* (New York: Simon & Schuster).

——— (1994), 'Negative Expertise', *International Journal of Expert Systems*, 7: 13–19.

——— (in preparation), *The Emotion Machine*, incomplete draft, available at <http://web.media.mit.edu/ -minsky/E1/eb1.html>. Quotations are taken from the version dated 29 Apr. 2002.

——— and PAPERT, S. A. (1969), *Perceptrons: An Introduction to Computational Geometry* (Cambridge, Mass.: MIT Press).

——— ——— (1988), 'Prologue: A View from 1988' and 'Epilogue: The New Connectionism', in M. L. Minsky and S. A. Papert, *Perceptrons: An Introduction to Computational Geometry*, 2nd edn. (Cambridge, Mass.: MIT Press), pp. viii–xv, 247–80.

——— and RIECKEN, D. (1994), 'A Conversation with Marvin Minsky About Agents', in Riecken (1994: 23–9).

——— and SELFRIDGE, O. G. (1961), 'Learning in Random Nets', in C. Cherry (ed.), *Information Theory: Fourth Symposium (Royal Institution)* (London: Butterworths), 335–47.

MIROWSKI, P. (2002), *Machine Dreams: Economics Becomes Cyborg Science* (Cambridge, Mass.: MIT Press).

MITCHELL, D. E., GIFFIN, F., MUIR, D., BLAKEMORE, C., and VAN SLUYTERS, R. C. (1976), 'Behavioral Compensation of Cats After Early Rotation of One Eye', *Experimental Brain Research*, 25: 109–13.

MITCHELL, M. (1990/1993), 'COPYCAT: A Computer Model of High-Level Perception and Conceptual Slippage in Analogy-Making,' Ph.D. thesis, University of Michigan, 1990; official version pub. as *Analogy-Making as Perception* (Cambridge, Mass.: MIT Press, 1993).

MITCHELL, T. M. (1979), 'An Analysis of Generalization as a Search Problem', *Proceedings of the Sixth International Joint Conference on Artificial Intelligence*, Tokyo, 577–82.

——— CARUANA, R., FREITAG, D., McDERMOTT, J., and ZABOWSKI, D. (1994), 'Experience with a Learning Personal Assistant', in Riecken (1994: 81–91).

——— UTGOFF, P. E., and BANERJI, R. (1983), 'Learning by Experimentation: Acquiring and Refining Problem-Solving Heuristics', in Michalski *et al.* (1983: 163–90).

MITHEN, S. J. (1996*a*), 'Domain-Specific Intelligence and the Neanderthal Mind', in P. Mellars and K. Gibson (eds.), *Modelling the Early Human Mind* (Cambridge: McDonald Institute for Archaeology).

——— (1996*b*), *The Prehistory of the Mind: A Search for the Origins of Art, Religion, and Science* (London: Thames & Hudson).

——— (2003), 'Handaxes: The First Aesthetic Artefacts', in E. Voland and K. Grammer (eds.), *Evolutionary Aesthetics* (London: Springer), 261–75.

MOFFATT, J. (1916), 'Aristotle and Tertullian', *Journal of Theological Studies*, 17: 170–1.

MOLINA, A. H. (1989), *The Social Basis of the Microelectronics Revolution* (Edinburgh: Edinburgh University Press).

MOLLON, J. D. (n.d.), 'History of the EPS: Beginnings', available at <http://www.eps.ac.uk/ society/begin.html>, accessed 8 Nov. 2004.

MONK, R. (1990), *Ludwig Wittgenstein: The Duty of Genius* (London: Jonathan Cape).

——— (1996), *Bertrand Russell 1872–1921: The Spirit of Solitude* (London: Jonathan Cape).

——— (2000), *Bertrand Russell 1921–70: The Ghost of Madness* (London: Jonathan Cape).

MONTAGUE, R. (1970), 'Universal Grammar', *Theoria*, 36: 373–98; repr. in R. Montague, *Formal Philosophy*, ed. R. H. Thomason (New Haven: Yale University Press, 1974), 222–46.

——— (1973), 'The Proper Treatment of Quantification in Ordinary English', in J. Hintikka, J. Moravcsik, and P. Suppes (eds.), *Approaches to Natural Language: Proceedings of the 1970 Stanford Workshop on Grammar and Semantics* (Dordrecht: Reidel), 221–42; repr. in R. Montague, *Formal Philosophy*, ed. R. H. Thomason (New Haven: Yale University Press, 1974), 247–70.

MONTAIGNE, M. DE (1580), 'Apologie de Raimond Sebond', in M. Rat (ed.), *Essais de Montaigne: Livre second* (Paris: Classiques Garnier, 1958), 115–317.

MONTEMERLO, M. D. (1992), 'The AI Program at the National Aeronautics and Space Administration: Lessons Learned During the First Seven Years', *AI Magazine*, 134: 49–61.

MOORE, D. L. (1977), *Ada, Countess of Lovelace: Byron's Legitimate Daughter* (London: John Murray).

MOORE, G. E. (1919), 'External and Internal Relations', *Proceedings of the Aristotelian Society*, NS 20: 40–62.

——— (1936), 'Is Existence a Predicate?,' *Proceedings of the Aristotelian Society: Supplementary Volume*, 15: 154–88.

MOORE, O. K., and ANDERSON, S. B. (1954*a*), 'Modern Logic and Tasks for Experiments on Problem Solving Behavior', *Journal of Psychology*, 38: 151–60.

——— ——— (1954*b*), 'Search Behavior in Individual and Group Problem Solving', *American Sociological Review*, 19: 702–14.

MOORE, T. H., RODMAN, R., and GROSS, C. G. (1998), 'Man, Monkey, and Blindsight', *The Neuroscientist*, 4: 227–30.

MORENO, A., and RUIZ-MIRAZO, K. (2004), 'Basic Autonomy as a Fundamental Step in the Synthesis of Life', *Artificial Life*, 10: 235–60.

MORGAN, M. J. (1977), *Molyneux's Question: Vision, Touch and the Philosophy of Perception* (Cambridge: Cambridge University Press).

——— (1984), 'Computational Theories of Vision (A Critical Notice of Marr's *Vision*)', *Quarterly Journal of Experimental Psychology*, 36A: 157–65.

MOROWITZ, H. J. (1967), 'Biological Self-Replicating Systems', *Progress in Theoretical Biology*, 1: 35–58.

MORTON, J. (1969), 'The Interaction of Information in Word Recognition', *Psychological Review*, 76: 165–78.

——— (1970), 'A Functional Model of Memory', in D. A. Norman (ed.), *Models of Human Memory* (New York: Academic Press), 203–54.

——— (1981), 'Will Cognition Survive?', *Cognition*, 10: 227–34.

——— and JOHNSON, M. H. (1991), 'Conspec and Conlern: A Two-Process Theory of Infant Face Recognition', *Psychological Review*, 98: 164–81.

MORRIS, M. R. (1991), 'Why There Are No Mental Representations', *Minds and Machines*, 1: 1–30.

——— (1992), *The Good and the True* (Oxford: Clarendon Press).

MORRISON, A. P. (ed.) (2002), *A Casebook of Cognitive Therapy for Psychosis* (Hove: Brunner-Routledge).

MOSER, U., SCHNEIDER, W., and VON ZEPPELIN, I. (1968), 'Computer Simulation of a Model of Neurotic Defence Mechanism (Clinical Paper and Technical Paper)', *Bulletin of the Psychological Institute of the University of Zurich*, 2: 1–77.

——— VON ZEPPELIN, I., and SCHNEIDER, W. (1969), 'Computer Simulation of a Model of Neurotic Defence Processes', *International Journal of Psychoanalysis*, 50: 53–64.

MOTO-OKA, T. (ed.) (1982), *Proceedings of the International Conference on Fifth Generation Computer Systems*, Tokyo (Amsterdam: North-Holland).

——— and KITSUREGAWA, M. (1985), *The Fifth Generation Computer: The Japanese Challenge*, trans. from Japanese F. D. R. Apps (Chichester: Wiley).

MOYNIHAN, T. (2003), 'The Sweet Sounds of MIDI', written 21 May 2003, available at <http://www.g4tv.com/screensavers/features/44197/The_Sweet_Sounds_of_MIDI.html>.

MOZER, M. C., and SMOLENSKY, P. (1989), 'Using Relevance to Reduce Network Size Automatically', *Connection Science*, 1: 3–16.

MUELLER, A. (2004), 'We Can Be Heroes', *Independent on Sunday*, 3 Oct., 12–13.

MUNZ, P. (1993), *Philosophical Darwinism: On the Origin of Knowledge by Means of Natural Selection* (London: Routledge).

MURDOCK, G. P. (1956), 'How Culture Changes', in H. L. Shapiro (ed.), *Man, Culture, and Society* (New York: Oxford University Press), 247–60.

MURPHY, K. P., WEISS, Y., and JORDAN, M. I. (1999), 'Loopy Belief Propagation for Approximate Inference: An Empirical Study', *Proceedings of the Fifteenth Conference on Uncertainty in Artificial Intelligence*, Stockholm, 467–75.

MUSSA-IVALDI, F. A. (1999), 'Modular Features of Motor Control and Learning', *Current Opinion in Neurobiology*, 9: 713–17.

NAGEL, T. (1970), Review of D. C. Dennett, *Content and Consciousness*, *Journal of Philosophy*, 69: 220–4.

NAGEL, T. (1974), 'What Is It Like To Be a Bat?', *Philosophical Review*, 83: 435–50.

—— (1986), *The View from Nowhere* (Oxford: Oxford University Press).

NAGHSHINEH-POUR, R., WILLIAMS, N., and RAM, B. (1999). 'Logistics Issues in Autonomous Food Production Systems for Extended Duration Space Exploration', in P. A. Farrington, H. B. Nembhard, D. T. Sturrock, and G. W. Evans (eds.), *Winter Simulation Conference 1999*, Phoenix, Ariz. (New York: IEEE), 1253–7.

NAKE, F. (2005), 'Computer Art: A Personal Recollection', in Candy (2005: 54–62).

NARASIMHAN, R. (1964), 'Labelling Schemata and Syntactic Descriptions of Pictures', *Information and Control*, 7: 151–79.

—— (1966), 'Syntax-Directed Interpretation of Classes of Pictures', *Communications of the Association for Computing Machinery*, 9: 166–73.

—— (2004), *Artificial Intelligence and the Study of Agentive Behaviour* (New Delhi: Tata McGraw-Hill).

NARMOUR, E. (1989), *The Analysis and Cognition of Basic Melodic Structures: The Implication-Realization Model* (Chicago: University of Chicago Press).

—— (1992), *The Analysis and Cognition of Melodic Complexity: The Implication-Realization Model* (Chicago: University of Chicago Press).

Nature (2005*a*), 'Three Cheers', 437 (1 Sept.), 2.

—— (2005*b*), 'Still Not Deterred', 437 (1 Sept.), 2–3.

NAUDÉ, G. (1625/1657), *The History of Magick by Way of Apology, For All the Wise Men Who Have Unjustly Been Reputed Magicians, from the Creation to the Present Age*, trans. J. Davies (London: John Streater, 1657); facs. of the 1625 edn.: *Apologie pour tous les grands personnages qui ont este faussement soupconnez de magie* (Farnborough: Gregg, 1972).

—— (1627/1644), *Advice on Establishing a Library*, trans. A. Taylor (Berkeley: University of California Press).

NAZZI, T. (2003), 'Early Language Development in Williams Syndrome', talk given at 'Cognitive Development, Genes and the Brain: From Piaget to Cognitive Neuroscience', workshop in honour of Prof. Annette Karmiloff-Smith, Institute of Child Health, London, 20 Sept. 2003.

NEGROPONTE, N. (1970), *The Architecture Machine: Toward a More Human Environment* (Cambridge, Mass.: MIT Press).

—— (1975), *Soft Architecture Machines* (Cambridge, Mass.: MIT Press).

NEISSER, U. (1959), *Hierarchies in Pattern Recognition*, Group Report 54–9 (Oct.) (Cambridge, Mass.: Lincoln Laboratory, MIT).

—— (1963), 'The Imitation of Man by Machine', *Science*, 139: 193–7.

—— (1967), *Cognitive Psychology* (New York: Appleton-Century-Crofts).

—— (1976), *Cognition and Reality: Principles and Implications of Cognitive Psychology* (San Francisco: W. H. Freeman).

—— (1979), 'Is Psychology Ready for Consciousness? Review of Hilgard's *Divided Consciousness*', *Contemporary Psychology*, 24/2: 99–100.

—— (1986), 'Interview with Ulric Neisser', in Baars (1986: 274–84).

—— (1994), 'Multiple Systems: A New Approach to Cognitive Theory', *European Journal of Cognitive Psychology*, 6/3: 225–41.

—— (1997), 'The Future of Cognitive Science: An Ecological Analysis', in D. M. Johnson and Erneling (1997: 247–60).

NELSON, T. H. (1965), 'Computers, Creativity, and the Nature of the Written Word', a lecture given at Vassar College, Jan., and summarized in L. Wedeles, 'Professor Nelson Talk Analyzes P.R.I.D.E.', *Miscellany News*, 3 Feb. (New York: Vassar College).

—— (1974), *'Computer Lib: You Can and Must Understand Computers Now'* and *'Dream Machines: New Freedoms Through Computer Screens—A Minority Report'*, two papers bound together (Chicago: private publication); excerpts repr. in Packer and Jordan (2001: 156–66).

—— (1977), *The Home Computer Revolution* (South Bend, Ind.: private publication).

NETTL, B. (2000), 'An Ethnomusicologist Contemplates Universals in Musical Sound and Musical Culture', in Wallin *et al.* (2000: 463–72).

NEVATIA, R., and BINFORD, T. O. (1977), 'Description and Recognition of Complex Curved Objects', *Artificial Intelligence*, 8: 77–98.

NEWELL, A. (1955), 'The Chess Machine: An Example of Dealing with a Complex Task by Adaptation', *Proceedings of the Western Joint Computer Conference*, 101–8.

____ (ed.) (1960), *Information Processing Language V Manual*, Rand Memos P-1897 and P-1918 (Santa Monica, Calif.: Rand Corporation); trade publication (Englewood Cliffs, NJ: Prentice-Hall, 1961).

____ (1962), 'Some Problems of Basic Organization in Problem-Solving Programs', in Yovits *et al.* (1962: 393–423).

____ (1969), 'A Step Toward the Understanding of Information Processes', *Science*, 165 (Aug.), 780–2.

____ (1972), 'A Theoretical Exploration of Mechanisms for Coding the Stimulus', in A. Melton and E. Martin (eds.), *Coding Processes in Human Memory* (Washington: Winston), 373–434.

____ (1973*a*), 'Production Systems: Models of Control Structures', in W. G. Chase (ed.), *Visual Information Processing* (London: Academic Press), 463–526.

____ (1973*b*), 'Artificial Intelligence and the Concept of Mind', in Schank and Colby (1973: 1–60).

____ (1973*c*), 'You Can't Play Twenty Questions with Nature and Win', in W. G. Chase (ed.), *Visual Information Processing* (London: Academic Press), 283–308.

____ (1976/1992), 'Fairy Tales', *AI Magazine*, 13/4: 46–8; first given as a speech at Carnegie Mellon University, 1976; printed posthumously.

____ (1980), 'Physical Symbol Systems', *Cognitive Science*, 4: 135–83.

____ (1980/2005), 'AAAI President's Message', *AI Magazine*, 1/1: 1–4; repr. in *AI Magazine*, 26/4 (25th anniversary issue), 24–9.

____ (1982), 'The Knowledge Level', *Artificial Intelligence*, 18: 87–127.

____ (1983), 'Intellectual Issues in the History of Artificial Intelligence' and 'Endnotes to the Papers on Artificial Intelligence', in F. Machlup and U. Mansfield (eds.), *The Study of Information: Interdisciplinary Messages* (New York: Wiley), 187–227, 287–94.

____ (1990), *Unified Theories of Cognition*, The William James Lectures, 1987 (Cambridge, Mass.: Harvard University Press).

____ (1992), Précis of *Unified Theories of Cognition*, with peer commentary, *Behavioral and Brain Sciences*, 15: 425–92.

____ and SHAW, J. C. (1957), 'Programming the Logic Theory Machine', *Proceedings of the 1957 Western Joint Computer Conference*, 230–40.

____ and SIMON, H. A. (1956*a*), *Current Developments in Complex Information Processing*, Technical Report P-850 (Santa Monica, Calif.: Rand Corporation).

____ ____ (1956*b*), 'The Logic Theory Machine: A Complex Information-Processing System', *IRE Transactions on Information Theory*, IT-2/3: 61–79.

____ ____ (1958), 'Chess-Playing Programs and the Problem of Complexity', *IBM Journal of Research and Development*, 2: 320–35; repr. in Feigenbaum and Feldman (1963: 39–70).

____ ____ (1961), 'GPS—A Program that Simulates Human Thought', in H. Billing (ed.), *Lernende Automaten* (Munich: Oldenbourg), 109–24; repr. in Feigenbaum and Feldman (1963: 279–93).

____ ____ (1963), 'Computers in Psychology', in Luce *et al.* (1963–5: i. 361–428).

____ ____ (1965), 'The Logic Theory Machine', *Institute of Radio Engineers Transactions on Information Theory*, IT-2/3: 61–79.

____ ____ (1972), *Human Problem Solving* (Englewood Cliffs, NJ: Prentice-Hall).

____ ____ (1976), 'Computer Science as Empirical Enquiry: Symbols and Search', *Communications of the Association for Computing Machinery*, 19: 113–26; repr. in Boden (1990*b*: 105–32).

____ ____ (1987), 'Postscript: Reflections on the Tenth Turing Award Lecture', in R. L. Ashenhurst and S. Graham (eds.), *ACM Turing Award Lectures: The First Twenty Years* (New York: ACM Press), 314–17.

____ SHAW, J. C., and SIMON, H. A. (1957), 'Empirical Explorations with the Logic Theory Machine', *Proceedings of the Western Joint Computer Conference*, 15: 218–39; repr. with added subtitle, 'A Case Study in Heuristics', Feigenbaum and Feldman (1963: 109–33).

____ SHAW, J. C., and SIMON, H. A. (1958*a*), 'Elements of a Theory of Human Problem-Solving', *Psychological Review*, 65: 151–66.

____ ____ ____ (1958*b*), 'Chess Playing Programs and the Problem of Complexity', *IBM Journal of Research and Development*, Oct., 2: 320–35; repr. in Feigenbaum and Feldman (1963: 39–70).

____ ____ ____ (1958/1962), *The Processes of Creative Thinking*, Technical Report P-1320 (Santa Monica, Calif.: RAND Corporation, 1958); rev. version in H. E. Gruber, G. Terrell, and M. Wertheimer (eds.), *Contemporary Approaches to Creative Thinking* (New York: Atherton Press, 1962), 63–119.

____ ____ ____ (1959), 'A General Problem-Solving Program for a Computer', *Proceedings of the International Conference on Information Processing*, Paris, June, 256–64.

NEWELL, A., SHAW, J. C., and SIMON, H. A. (1962), 'A Variety of Intelligent Learning in a General Problem Solver', in M. C. Yovits and S. Cameron (eds.), *Self-Organizing Systems: Proceedings of an Interdisciplinary Conference*, Chicago, 5 and 6 May 1959 (Oxford: Pergamon Press), 153–89.

_____ TONGE, F. M., FEIGENBAUM, E. A., GREEN, B. F., and MEALY, G. H. (1964), *Information Processing Language V—Manual*, 2nd edn. (Englewood Cliffs, NJ: Prentice-Hall, for the Rand Corporation).

_____ BARNETT, J., FORGIE, J. W., GREEN, C. C., KLATT, D. H., LICKLIDER, J. C. R., MUNSON, J. H., REDDY, D. R., and WOODS, W. A. (1973), *Final Report of a Study Group on Speech Understanding Systems* (Amsterdam: North-Holland).

_____ ROSENBLOOM, P. S., and LAIRD, J. E. (1989), 'Symbolic Architectures for Cognition', in M. I. Posner (ed.), *Foundations of Cognitive Science* (Cambridge, Mass.: MIT Press), 93–131.

_____ YOUNG, R. M., and POLK, T. (1993), 'The Approach Through Symbols', in D. E. Broadbent (ed.), *The Simulation of Human Intelligence*, Wolfson College Lectures 1991 (Oxford: Blackwell), 33–70.

NEWMAN, J. R. (ed.) (1956), *The World of Mathematics*, 4 vols. (New York: Simon & Schuster). Turing's 'Computing Machinery and Intelligence' is in iv. 2099–2123.

NEWQUIST, H. (1994), *The Brain Makers* (Indianapolis: SAMS).

NEWSOME, W. T., and SALZMAN, C. D. (1993), 'The Neuronal Basis of Motion Perception', in Marsh (1993: 217–30).

NICOLELIS, M. A. (2001), 'Actions from Thoughts', *Nature*, 409, suppl. (18 Jan.), 403–7.

NIELSEN, P. E., EGHOLM, M., BERG, R. H., and BUCHARDT, O. (1991), 'Sequence-Selective Recognition of DNA by Strand Displacement with a Thymine-Substituted Polyamide', *Science*, 254: 1497–1500.

NISBET, H. B. (1972), *Goethe and the Scientific Tradition* (London: Institute of Germanic Studies, University of London).

NISBETT, R. E., and ROSS, L. (1980), *Human Inference: Strategies and Shortcomings of Social Judgment* (Englewood Cliffs, NJ: Prentice-Hall).

NOLFI, S., ELMAN, J., and PARISI, D. (1994), 'Learning and Evolution in Neural Networks', *Adaptive Behavior*, 3: 5–28.

NORMAN, D. A. (ed.) (1981), *Perspectives on Cognitive Science*, papers presented to the First Annual Meeting of the Cognitive Science Society, La Jolla, Calif., Aug. 1979 (Norwood, NJ: Ablex).

_____ (1986a), 'Cognitive Engineering', in Norman and Draper (1986: 31–61).

_____ (1986b), 'Reflections on Cognition and Parallel Distributed Processing', in McClelland, Rumelhart, et al. (1986: 531–46).

_____ (1988), *The Psychology of Everyday Things* (New York: Basic Books).

_____ (1994), 'How Might People Interact with Agents', in Riecken (1994: 68–71).

_____ and BOBROW, D. G. (1975), 'On Data-Limited and Resource-Limited Processes', *Cognitive Psychology*, 7: 44–64.

_____ and DRAPER, S. W. (eds.) (1986), *User Centered System Design* (Hillsdale, NJ: Lawrence Erlbaum).

_____ and LEVELT, W. J. M. (1988), 'Life at the Center', in Hirst (1988a: 100–9).

_____ and RUMELHART, D. E. (eds.) (1975), *Explorations in Cognition* (San Francisco: W. H. Freeman).

_____ and SHALLICE, T. (1980/1986), *Attention to Action: Willed and Automatic Control of Behavior*, CHIP Report 99, University of California, San Diego, 1980; officially pub. in R. Davidson, G. Schwartz, and D. Shapiro (eds.), *Consciousness and Self Regulation: Advances in Research and Theory*, iv (New York: Plenum Press, 1986), 1–18.

NORMAN, E. (2002), *Secularisation* (London: Continuum).

NORMAN, J. (2002), 'Two Visual Systems and Two Theories of Perception: An Attempt to Reconcile the Constructivist and Ecological Approaches', *Behavioral and Brain Sciences*, 25: 73–144.

NORMAN, J. M. (2004), *The Origins of Cyberspace: A Library on the History of Computing, Networking and Telecommunications*, catalogue for an auction, 23 Feb. 2005 (New York: Christie's).

_____ (2005), *Auction Result: The Origins of Cyberspace*, 23 Feb., two-page inserted addendum to J. M. Norman (2004) (New York: Christie's).

NØRRETRANDERS, T. (1991), *The User Illusion: Cutting Consciousness Down to Size*, trans. J. Sydenham (New York: Penguin).

NOVAK, R. A. (1972), *Wilhelm von Humboldt as a Literary Critic* (Berne: Herbert Lang).

NÚÑEZ, R., and FREEMAN, W. J. (1999), *Reclaiming Cognition: The Primacy of Action, Intention and Emotion, Journal of Consciousness Studies*, 6, Special issues 11–12, Nov.–Dec. 1999.

NYCE, J. M., and KAHN, P. (1989), 'Innovation, Pragmatism, and Technological Continuity: Vannevar Bush's Memex', *Journal of the American Association for Information Science*, 40: 214–20.

_____ _____ (1991), *From Memex to Hypertext: Vannevar Bush and the Mind's Machine* (San Diego, Calif.: Harcourt Brace Jovanovich).

OAKLEY, B., and OWEN, K. (1989), *Alvey: Britain's Strategic Computing Initiative* (Cambridge, Mass.: MIT Press).

OAKLEY, D. A. (1999), 'Hypnosis and Consciousness: A Structural Model', *Contemporary Hypnosis*, 16: 215–23.

OATLEY, K., and JENKINS, J. M. (1996), *Understanding Emotions* (Oxford: Blackwell).

O'CALLAGHAN, T. A. (2003), 'A Hybrid Legal Expert System', Honours thesis (Feb.), Department of Computer Science, Australian National University, available at <http://cs.anu.edu.au/software/shyster/tom>.

_____ POPPLE, J., and MCCREATH, E. (2003a), 'SHYSTER-MYCIN: A Hybrid Legal Expert System', *Proceedings of the Ninth International Conference on Artificial Intelligence and Law*, Edinburgh, June, 103–4. Abstract only.

_____ _____ _____ (2003b), *Building and Testing the SHYSTER-MYCIN Hybrid Legal Expert System* (Canberra: Australian National University); online technical report, Department of Computer Science, Oct., available at <http://cs.anu.edu.au/techreports/2003/TR-CS-03-01>.

O'CONNELL, D. N., SHOR, R. E., and ORNE, M. T. (1970), *Hypnotic Age Regression: An Empirical and Methodological Analysis, Journal of Abnormal Psychology* Monograph, 76/3, pt. 2 (Washington: American Psychological Association).

O'DONOHUE, W. T. (1993), 'The Spell of Kuhn on Psychology: An Exegetical Elixir', *Philosophical Psychology*, 6: 267–87.

OGDEN, C. K., and RICHARDS, I. A. (1923), *The Meaning of Meaning: A Study of the Influence of Language Upon Thought and of the Science of Symbolism*, International Library of Psychology, Philosophy, and Scientific Method (London: Kegan Paul, Trubner).

O'HEAR, A. (1995), 'Art and Technology: An Old Tension', in R. Fellows (ed.), *Philosophy and Technology* (Cambridge: Cambridge University Press), 143–58.

OHNUKI-TIERNEY, E. (1981), 'Phases in Human Perception/Cognition/Symbolization Processes: Cognitive Anthropology and Symbolic Classification', *American Ethnologist*, 8: 451–67.

OHTA, Y., and TAMURA, H. (1999), *Mixed Reality: Merging Real and Virtual Worlds* (Berlin: Springer-Verlag).

O'KEEFE, J., and NADEL, L. (1978), *The Hippocampus as a Cognitive Map* (Oxford: Clarendon Press).

OLDFIELD, R. C. (1954), 'Memory Mechanisms and the Theory of Schemata', *British Journal of Psychology*, 45: 14–23.

_____ and ZANGWILL, O. L. (1942–3), 'Head's Concept of the Body Schema and its Application in Contemporary British Psychology', *British Journal of Psychology*, 32: 267–86; 33: 58–64, 113–29, 143–9.

OLSON, E. T. (1997), 'The Ontological Basis of Strong Artificial Life', *Artificial Life*, 3: 29–39.

OLVERES, J., BILLINGHURST, M., SAVAGE, J., and HOLDEN, A. (1998), 'Intelligent, Expressive, Avatars', *Proceedings of the First Workshop on Embodied Conversational Characters (WECC-1)*, Lake Tahoe, Calif., Oct., available at <http://www.hit1.washington.edu/publications/r-98-32>.

ONO, N., and IKEGAMI, T. (1999), 'Model of Self-Replicating Cell Capable of Self-Maintenance', in D. Floreano, J. D. Nicaud, and F. Mondada (eds.), *Proceedings of the Fifth European Conference on Artificial Life*, Lausanne, Sept. 1999 (Berlin: Springer-Verlag), 399–406.

_____ _____ (2000), 'Self-Maintenance and Self-Reproduction in an Abstract Cell Model', *Journal of Theoretical Biology*, 206: 243–53.

_____ _____ (2001), 'Artificial Chemistry: Computational Studies on the Emergence of Self-Reproducing Units', in J. Kelemen and P. Sosik (eds.), *Advances in Artificial Life: Proceedings of the Sixth European Conference on Artificial Life*, Prague, Sept. 2001 (Berlin: Springer-Verlag), 186–95.

OPARIN, A. I. (1936/1938), *The Origin of Life*, trans. from the Russian edn. of 1936 S. Morgulis (New York: Macmillan, 1938).

OPPENHEIM, P., and PUTNAM, H. (1958), 'Unity of Science as a Working Hypothesis', in H. Feigl, M. Scriven, and G. Maxwell (eds.), *Concepts, Theories, and the Mind–Body Problem*, Minnesota Studies in the Philosophy of Science, 2 (Minneapolis: University of Minnesota Press), 3–36.

ORBACH, J. (1998), *The Neuropsychological Theories of Lashley and Hebb: Contemporary Perspectives Fifty Years After Hebb's 'The Organization of Behavior'* (Lanham, Md.: University Press of America).

_____ (1999), 'Author's Rationale for Soliciting Commentary', *Psycoloquy*, 18 Sept., call for papers from *Psycoloquy*, an APA-sponsored refereed electronic journal.

O'REILLY, R. C., and JOHNSON, M. H. (1994), 'Object Recognition and Sensitive Periods: A Computational Analysis of Visual Imprinting', *Neural Computation*, 6: 357–89.

_____ and MUNAKATA, Y. (2000), *Computational Explorations in Cognitive Neuroscience: Understanding the Mind by Simulating the Brain* (Cambridge, Mass.: MIT Press).

ORIANS, G. H. (1980), 'Habitat Selection: General Theory and Applications to Human Behavior', in J. S. Lockard (ed.), *The Evolution of Human Social Behavior* (New York: Elsevier), 49–66.

_____ (1986), 'An Ecological and Evolutionary Approach to Landscape Aesthetics', in E. C. Penning-Rowsell and D. Lowenthal (eds.), *Landscape Meaning and Values* (London: Allen & Unwin), 3–25.

_____ and HEERWAGEN, J. H. (1992), 'Evolved Responses to Landscapes', in Barkow *et al.* (1992: 555–79).

ORNE, M. T. (1959), 'The Nature of Hypnosis: Artifact and Essence', *Journal of Abnormal and Social Psychology*, 58: 277–99.

ORTONY, A. (ed.) (1979), *Metaphor and Thought* (Cambridge: Cambridge University Press).

OSGOOD, C. E., SUCI, G. J., and TANNENBAUM, P. H. (1957), *The Measurement of Meaning* (Urbana: University of Illinois Press).

O'SHAUGHNESSY, B. (1970), 'The Powerlessness of Dispositions', *Analysis*, 331: 1–15.

O'SHEA, M., COLBERT, R., WILLIAMS, L., and DUNN, S. (1998), 'Nitric Oxide Compartments in the Mushroom Bodies of the Locust Brain', *NeuroReport*, 9: 333–6.

O'SHEA, T., and SELF, J. (eds.) (1984), *Teaching and Learning with Computers* (Brighton: Harvester Press).

_____ and YOUNG, R. M. (1978), 'A Production Rules Account of Errors in Children's Subtraction', *Proceedings of the AISB Conference*, Hamburg, 229–37.

OYAMA, S. (1985), *The Ontogeny of Information: Developmental Systems and Evolution* (Cambridge: Cambridge University Press).

PACKER, R., and JORDAN, K. (2001), *Multimedia: From Wagner to Virtual Reality* (London: W. W. Norton).

PAFFRATH, J. D. (1984), *Stelarc: Obsolete Body Suspensions 1976–1982* (Davis, Calif.: JP Publishing).

PAGE, M. (2000), 'Connectionist Modelling in Psychology: A Localist Manifesto', with peer commentary, *Behavioral and Brain Sciences*, 23: 443–512.

PAGELLO, E., MENEGATTI, E., BREDENFEL, A., COSTA, P., CHRISTALLER, T., JACOFF, A., POLANI, D., RIEDMILLER, M., SAFFIOTTI, A., SKLAR, E., and TAMOICHI, T. (2004), 'RoboCup 2003: New Scientific and Technical Advances', *AI Magazine*, 25/2 (Summer), 81–98.

PAGELS, H. R., DREYFUS, H. L., McCARTHY, J., MINSKY, M. L., PAPERT, S., and SEARLE, J. R. (1984), 'Panel Discussion: Has Artificial Intelligence Research Illuminated Human Thinking?', in H. R. Pagels (ed.), *Computer Culture: The Scientific, Intellectual and Social Impact of the Computer*, Annals of the New York Academy of Sciences, 426: 138–60; repr. in Chrisley (2000: iii. 334–56).

PAIVIO, A. (1969), 'Mental Imagery in Associative Learning and Memory', *Psychological Review*, 76: 241–63.

_____ (1971), *Imagery and Verbal Processes* (New York: Holt, Rinehart & Winston).

_____ (2005), *Mind and its Evolution: A Dual Coding Theoretical Approach* (Mahwah, NJ: Lawrence Erlbaum).

PAPANICOLAOU, A. C., and GUNTER, P. A. Y. (eds.) (1987), *Bergson and Modern Thought: Towards a Unified Science* (London: Harwood).

PAPERT, S. (1965), Introduction, in W. S. McCulloch (1965, pp. xiii–xx).

_____ (1968), *The Artificial Intelligence of Hubert L. Dreyfus: A Budget of Fallacies*, MIT AI Lab Memo 154 (Cambridge, Mass.: MIT, Jan.).

_____ (1972), 'Teaching Children to Be Mathematicians Versus Teaching About Mathematics', *International Journal for Mathematical Education in Science and Technology*, 3: 249–62.

_____ (1973), *Uses of Technology to Enhance Education*, AI-Memo-298 (Cambridge, Mass.: MIT AI Lab).

_____ (1980), *Mindstorms: Children, Computers, and Powerful Ideas* (Brighton: Harvester Press).

_____ (1988), 'One AI or Many?', in Graubard (1988: 1–14).

_____ (1993), *The Children's Machine: Rethinking School in the Age of the Computer* (New York: Basic Books).

PAPINEAU, D. (1984), 'Representation and Explanation', *Philosophy of Science*, 51: 55–73.

_____ (1987), *Reality and Representation* (Oxford: Blackwell).

PARKER, A., DERRINGTON, A., and BLAKEMORE, C. (eds.) (2002), *The Physiology of Cognitive Processes*, *Philosophical Transactions of the Royal Society*, ser. B, 357, special issue, 957–1146 (London: Royal Society).

PARKER, I., and SHOTTER, J. (eds.) (1990), *Deconstructing Social Psychology* (London: Routledge).

PARKER-RHODES, A. F. (1956), 'An Electronic Computer Programme for Translating Chinese into English', *Mechanical Translation*, 3: 14–19.

——— (1978), *Inferential Semantics* (Hassocks: Harvester Press).

PARNAS, D. L. (1985a), *Software and SDI: Why Communications Systems Are Not Like SDI*, US Senate Testimony, Dec. 1985, available from CPSR, PO Box 717, Palo Alto, CA 94302, USA.

——— (1985b), 'Software Aspects of Strategic Defense Systems', *American Scientist* (Sept.–Oct.), 432–40.

——— (1986), 'Why I Won't Work on SDI: One View of Professional Responsibility', *CPSR [Computer Professionals for Social Responsibility]*, Newsletter 4/2: 1, 8–11.

——— (1987), 'Computers in Weapons: The Limits of Confidence', in Bellin and Chapman (1987: 202–32).

PARSONS, K. (ed.) (2003), *The Science Wars: Debating Scientific Knowledge and Technology* (Amherst, Mass.: Prometheus Books).

PARSONS, T., and VOGT, E. Z. (1962), 'Clyde Kay Maben Kluckhohn 1905–1960', *American Anthropologist*, 64: 140–61.

PARTEE, B. H. (1975), 'Montague Grammar and Transformational Grammar', *Linguistic Inquiry*, 6: 203–300.

PARTRIDGE, D. A. (1986/87), 'Conference Report on a Workshop on the Foundations of AI, Las Cruces, New Mexico', *AISB Quarterly*, 60 (Winter 1986–7), 15–16 (pub. by the Society for the Study of Artificial Intelligence and Simulation of Behaviour).

——— and ROWE, J. (1994), *Computers and Creativity* (Oxford: Intellect Books).

PARUNAK, H. V. D. (1999), 'Industrial and Practical Applications of DAI', in G. Weiss (1999: 337–421).

PASK, G. (1958), 'Electronic Keyboard Teaching Machines', *Journal of the National Association for Education and Commerce*; repr. in R. Glaser and A. Lumsdaine (eds.), *Teaching Machines and Programmed Learning*, i (Washington: National Educational Association, 1960), 336–49.

——— (1959), 'Physical Analogues to the Growth of a Concept', in Blake and Uttley (1959: ii. 877–922); discussion, pp. 923–30.

——— (1961), *An Approach to Cybernetics* (London: Hutchinson).

——— (1963), 'The Use of Analogy and Parable in Cybernetics with Emphasis Upon Analogies for Learning and Creativity', *Dialectica*, 17: 167–202.

——— (1964), 'A Discussion of Artificial Intelligence and Self-Organization', *Advances in Computers*, 5: 109–226.

——— (1969), 'The Architectural Relevance of Cybernetics', *Architectural Design* (Sept.), 7: 494–6.

——— (1971), 'A Comment, a Case-History, and a Plan', in J. Reichardt (ed.), *Cybernetics, Art, and Ideas* (London: Studio Vista), 76–99.

——— (1975a), *The Cybernetics of Human Learning and Performance: A Guide to Theory and Research* (London: Hutchinson Educational).

——— (1975b), *Conversation, Cognition, and Learning: A Cybernetic Theory and Methodology* (Oxford: Elsevier).

——— and SCOTT, B. C. E. (1973), 'CASTE: A System for Exhibiting Learning Strategies and Regulating Uncertainties', *International Journal for Man–Machine Studies*, 5: 17–52.

PATERSON, S. J., BROWN, J. H., GSODL, M. K., JOHNSON, M. H., and KARMILOFF-SMITH, A. (1999), 'Cognitive Modularity and Genetic Disorders', *Science*, 286: 2355–8.

PATTEE, H. H. (1966), 'Physical Theories, Automata, and the Origin of Life', in H. H. Pattee, E. A. Edelsack, L. Fein, and A. B. Callahan (eds.), *Natural Automata and Useful Simulations: Proceedings of a Symposium on Fundamental Biological Models* (Washington: Spartan Books), 73–106.

——— (1972), 'Laws and Constraints, Symbols and Languages', in Waddington (1966–72: iv: 248–58).

——— (1976), 'The Role of Instabilities in the Evolution of Control Hierarchies', in T. R. Burns and W. Buckley (eds.), *Power and Control* (London: Sage).

——— (1985), 'Universal Principles of Measurement and Language Functions in Evolving Systems', in J. Casti and A. Karlqvist (eds.), *Complexity, Language, and Life: Mathematical Approaches* (Berlin: Springer-Verlag), 168–281.

——— (1989), 'Simulations, Realizations, and Theories of Life', in Langton (1989: 63–77).

——— (2001), 'The Physics of Symbols: Bridging the Epistemic Cut', *BioSystems*, 60: 5–21.

——— EDELSACK, E. A., FEIN, L., and CALLAHAN, A. B. (eds.) (1966), *Natural Automata and Useful Simulations: Proceedings of a Symposium on Fundamental Biological Models* (Washington: Spartan Books).

PAVLOV, I. P. (1897/1902), *The Work of the Digestive Glands: Lectures by J. P. Pawlow*, trans. from Russian W. H. Thompson (London: Charles Griffin, 1902; first pub. 1897).

PAVLOV, I. P. (1923/1927), *Conditioned Reflexes: An Investigation of the Physiological Activity of the Cerebral Cortex*, trans. from Russian G. V. Anrep (London: Humphrey Milford, 1927; first pub. 1923).

—— (1925/1928), *Lectures on Conditioned Reflexes*, i: *Twenty Five Years of Objective Study of the Higher Nervous Activity (Behaviour) of Animals*; ii: *Conditioned Reflexes and Psychiatry*, trans. W. H. Gantt, from the 3rd Russian edn. (1925), with five new chapters added for the translation (London: Martin Lawrence).

PEA, R. D., and KURLAND, D. M. (1984), 'On the Cognitive Effects of Learning Computer Programming', *New Ideas in Psychology*, 2: 137–68.

—— and HAWKINS, J. (1987), 'LOGO and the Development of Thinking Skills', in D. N. Perkins, J. Lochhead, and J. Bishop (eds.), *Thinking, Progress in Research and Teaching* (Hillsdale, NJ: Lawrence Erlbaum).

PEACOCKE, C. (1986), 'Explanation in Computational Psychology: Language, Perception, and Level 1.5', *Mind and Language*, 1: 101–23.

—— (ed.) (1994), *Objectivity, Simulation and the Theory of Consciousness: Current Issues in the Philosophy of Mind* (Oxford: Oxford University Press).

—— (1996), 'The Relation Between Philosophical and Psychological Theories of Concepts', in A. Clark and P. Millican (eds.), *Connectionism, Concepts, and Folk Psychology: The Legacy of Alan Turing*, ii, Proceedings of the 1991 Colloquium on Alan Turing, University of Sussex (Oxford: Oxford University Press), 115–38.

PEARL, J. (1988), *Probabilistic Reasoning in Intelligent Systems: Networks of Plausible Inference* (San Francisco: Morgan Kaufmann).

—— (2000), *Causality: Models, Reasoning, and Inference* (Cambridge: Cambridge University Press).

PEARSON, J. C., FINKEL, L. H., and EDELMAN, G. M. (1987), 'Plasticity in the Organization of Adult Cerebral Cortical Maps: A Computer Simulation Based on Neuronal Group Selection', *Journal of Neuroscience*, 7: 4209–23.

PEDERSON, E. (1993), *Geographic and Manipulable Space in Two Tamil Linguistic Systems*, Cognitive Anthropology Research Group, Working Paper no. 20 (Nijmegen: Max Planck Institute for Psycholinguistics).

PELLIONISZ, A. (1968), 'Transfer Function of a Geometrical Neuronal Configuration of the Cerebellar Granular Layer', *Acta Biochemica et Biophysica Academiae Scientiarum Hungaricae*, 3: 462.

—— (1970), 'Computer Simulation of the Pattern Transfer of Large Cerebellar Neuronal Fields', *Acta Biochemica et Biophysica Academiae Scientiarum Hungaricae*, 5: 71–9.

—— (1979), 'Modeling of Neurons and Neuronal Networks', in F. O. Schmitt and F. G. Worden (eds.), *The Neurosciences*, iv: *Study Program* (Cambridge, Mass.: MIT Press), 525–50.

—— (1983*a*), 'Brain Theory: Connecting Neurobiology to Robotics. Tensor Analysis: Utilizing Intrinsic Coordinates to Describe, Understand and Engineer Functional Geometries of Intelligent Organisms', *Journal of Theoretical Neurobiology*, 2: 185–213.

—— (1983*b*), 'Sensorimotor Transformations of Natural Coordinates via Neuronal Networks: Conceptual and Formal Unification of Cerebellar and Tectal Models', in M. A. Arbib and R. Lara (eds.), *Proceedings of Workshop on Visuomotor Coordination in Frog and Toad: Models and Experiments*, Computer and Information Science Technical Report 83-19 (Amherst: University of Massachusetts), 1–20.

—— (1985), 'Tensorial Aspects of the Multidimensional Approach to the Vestibulo-Ocular Reflex and Gaze', in A. Berthoz and G. Melvill-Jones (eds.), *Adaptive Mechanisms in Gaze Control* (Amsterdam: Elsevier), 231–96.

—— and LLINAS, R. (1979), 'Brain Modelling by Tensor Network Theory and Computer Simulation. The Cerebellum: Distributed Processor for Predictive Coordination', *Neuroscience*, 4: 323–48.

—— —— (1980), 'Tensorial Approach to the Geometry of Brain Function: Cerebellar Coordination via Metric Tensor', *Neuroscience*, 5: 1125–36.

—— —— (1981), 'Genesis and Modification of the Geometry of CNS Hyperspace: Cerebellar Space-Time Metric Tensor and "Motor Learning"', *Society for Neuroscience Abstracts*, 7: 641.

—— —— (1982), 'Space-Time Representation in the Brain: The Cerebellum as a Predictive Space-Time Metric Tensor', *Neuroscience*, 7: 2249–970.

—— —— (1985), 'Tensor Network Theory of the Metaorganization of Functional Geometries in the Central Nervous System', in A. Berthoz and G. Melvill Jones (eds.), *Adaptive Mechanisms in Gaze Control* (Amsterdam: Elsevier), 223–32; repr. in J. A. Anderson *et al.* (1990*a*: 356–81).

_____ and PETERSON, B. W. (1988), 'A Tensorial Model of Neck Motor Activation', in B. W. Peterson and F. Richmond (eds.), *Control of Head Movement* (Oxford: Oxford University Press), 1718–86.

_____ and SZENTAGOTHAI, J. (1973), 'Dynamic Single Unit Simulation of a Realistic Cerebellar Network Model', *Brain Research*, 49: 83–99.

_____ LLINAS, R., and PERKEL, D. H. (1977), 'A Computer Model of the Cerebellar Cortex of the Frog', *Neuroscience*, 2: 19–36.

_____ JORGENSEN, C. J., and WERBOS, P. J. (1992), 'Cerebellar Neurocontroller Project, for Aerospace Applications, in a Civilian Neurocomputing Initiative in the "Decade of the Brain"', *International Joint Conference on Neural Networks—92*, Nagoya, Japan, 373–8.

PENFIELD, W. (1952), 'Memory Mechanisms', *Archives of Neurology and Psychiatry*, 67: 178–91.

_____ (1958), *The Excitable Cortex in Conscious Man* (Liverpool: Liverpool University Press).

_____ (1975), *The Mystery of the Mind: A Criticial Study of Consciousness and the Human Brain* (Princeton: Princeton University Press).

_____ and PEROT, P. (1963), 'The Brain's Record of Auditory and Visual Experience', *Brain*, 86: 595–696.

_____ and RASMUSSEN, T. (1957), *The Cerebral Cortex of Man: A Clinical Study of Localization of Function* (New York: Macmillan).

PENROSE, L. S. (1959), 'Automatic Mechanical Self-Reproduction', *New Biology*, 28: 92–117.

PENROSE, R. (1989), *The Emperor's New Mind: Concerning Computers, Minds, and the Laws of Physics* (Oxford: Oxford University Press).

_____ (1994a), *Shadows of the Mind: A Search for the Missing Science of Consciousness* (Oxford: Oxford University Press).

_____ (1994b), 'Mechanisms, Microtubules, and Mind', *Journal of Consciousness Studies: Controversies in Science and the Humanities*, 1: 241–9.

_____ and HAMEROFF, S. (1995), 'What Gaps? Reply to Grush and Churchland', *Journal of Consciousness Studies*, 2: 99–112.

PENTLAND, A. P. (1986), 'Perceptual Organization and the Representation of Natural Form', *Artificial Intelligence*, 28: 293–332.

PENTON-VOAK, I. S., PERRETT, D. I., CASTLES, D. L., KOBAYASHI, T., BURT, D. M., MURRAY, L. K., and MINAWISAMA, R. (1999), 'Menstrual Cycle Alters Face Preference', *Nature*, 399: 741–2.

PEPPERBERG, I. M. (2000), *The Alex Studies: Cognitive and Communicative Abilities of Grey Parrots* (Cambridge, Mass.: Harvard University Press).

PEREIRA, F. (2000), 'Formal Grammar and Information Theory: Together Again?', in Gazdar *et al.* (2000: 1239–52).

PÉREZ-RAMOS, A. (1988), *Francis Bacon's Idea of Science and the Maker's Knowledge Tradition* (Oxford: Clarendon Press).

PERKEL, D. H. (1987), 'Chaos in Brains: Fad or Insight?', *Behavioral and Brain Sciences*, 10: 180–1.

_____ (1988), 'Logical Neurons: Variations on Themes by W. S. McCulloch', *Trends in Neurosciences*, 11: 9–12.

PERKY, C. W. (1910), 'An Experimental Study of Imagination', *American Journal of Psychology*, 21: 422–52.

PERLIS, A. (1982), 'Epigrams on Programming', *SIGPLAN Notices* (Sept.), 7–13.

PERNER, J. (1991), *Understanding the Representational Mind* (Cambridge, Mass.: MIT Press).

_____ (2003), 'Dual Control and the Causal Theory of Action', in N. Eilan and J. Roessler (eds.), *Agency and Self-Awareness* (Oxford: Oxford University Press), 218–43.

PERRETT, D. I., ROLLS, E. T., and CAAN, W. (1982), 'Visual Neurones Responsive to Faces in the Monkey Temporal Cortex', *Experimental Brain Research*, 47: 329–42.

_____ SMITH, P. A. J., POTTER, D. D., MISTLIN, A. J., HEAD, A. S., MILNER, A. D., and JEEVES, M. A. (1985), 'Visual Cells in the Temporal Cortex Sensitive to Face View and Gaze Direction', *Proceedings of the Royal Society*, ser. B, 223: 293–317.

_____ HARRIES, M., MISTLIN, A. J., and CHITTY, A. J. (1990), 'Three Stages in the Classification of Body Movement by Visual Neurons', in H. B. Barlow, C. Blakemore, and E. M. Weston-Smith (eds.), *Images and Understanding* (Cambridge: Cambridge University Press), 94–107.

_____ HEITANEN, J. K., ORAM, M. W., and BENSON, P. J. (1992), 'Organization and Functions of Cells Responsive to Faces in the Temporal Cortex', *Philosophical Transactions of the Royal Society*, ser. B, 335: 31–8.

_____ BURT, D. M., PENTON-VOAK, I. S., LEE, K. J., ROWLAND, D. A., and EDWARDS, R. (1999), 'Symmetry and Human Facial Attractiveness', *Evolution and Human Behavior*, 20: 295–307.

PERRY, C., NADON, R., and BUTTON, J. (1992), 'The Measurement of Hypnotic Ability', in E. Fromm and M. R. Nash (eds.), *Contemporary Hypnosis Research* (New York: Guilford Press), 459–90.

PERRY, R. B. (1918), 'Docility and Purposiveness', *Psychological Review*, 25: 1–21.

―――― (1921*a*), 'A Behavioristic View of Purpose', *Journal of Philosophy*, 18: 85–105.

―――― (1921*b*), 'The Independent Variability of Purpose and Belief', *Journal of Philosophy*, 18: 169–80.

PETERS, P. S., and RITCHIE, R. W. (1973), 'On the Generative Power of Transformational Grammars', *Information Science*, 6: 49–83.

PFEIFER, R., and SCHEIER, C. (1999), *Understanding Intelligence* (Cambridge, Mass.: MIT Press).

PFUNGST, O. (1911/1965), *Clever Hans (The Horse of Mr. von Osten): A Contribution to Experimental Animal and Human Psychology, with an Introduction by Prof. C. Stumpf*, trans. C. L. Rahn, 1st edn. (1911); reissued, ed. R. Rosenthal (New York: Holt, Rinehart & Winston, 1965).

PHILIPPIDES, A., HUSBANDS, P., and O'SHEA, M. (1998), 'Neural Signalling—It's a Gas!', in L. Niklasson, M. Boden, and T. Ziemke (eds.), *ICANN98: Proceedings of the 8th International Conference on Artificial Neural Networks* (London: Springer), 51–63.

―――― ―――― SMITH, T., and O'SHEA, M. (2005*a*), 'Flexible Couplings: Diffusing Neuromodulators and Adaptive Robotics', *Artificial Life*, 11: 139–60.

―――― OTT, S. R., HUSBANDS, P. N., LOVICK, T. A., and O'SHEA, M. (2005*b*), 'Modeling Cooperative Volume Signaling in a Plexus of Nitric Oxide Synthase-Expressing Neurons', *Journal of Neuroscience*, 25/28: 6520–32.

PHILLIPS, W. (1815), *An Outline of Mineralogy and Geology* (London: William Phillips).

PIAGET, J. (1926), *The Language and Thought of the Child*, trans. from French M. Warden (London: Kegan Paul, Trench & Trubner; first edn., 1923).

―――― (1928), *Judgment and Reasoning in the Child*, trans. from French M. Warden (London: Kegan Paul, Trench & Trubner; first edn., 1924).

―――― (1929), *The Child's Conception of the World*, trans. from French J. and A. Tomlinson (London: Kegan Paul; first edn., 1926).

―――― (1932), *The Moral Judgment of the Child*, trans. from French M. Gabain (London: Kegan Paul; first edn., 1932).

―――― (1950), *The Psychology of Intelligence*, trans. from French M. Piercy and D. E. Berlyne (London: Routledge & Kegan Paul; first edn., 1947).

―――― (1952*a*), 'Autobiography', in E. G. Boring (ed.), *A History of Psychology in Autobiography*, iv (New York: Russell & Russell), 237–56.

―――― (1952*b*), *The Origins of Intelligence in Children*, trans. from French M. Cook (New York: International Universities Press; first edn., 1936).

―――― (1960), 'Reply to Bruner', appended to J. S. Bruner, 'Individual and Collective Problems in the Study of Thinking', *Annals of the New York Academy of Science*, 91 (1960), 22–37.

―――― (1967), *Les Modèles et la formalisation du comportement* (Paris: Éditions du Centre National de la Recherche Scientifique).

―――― (1970), *The Child's Conception of Movement and Speed*, trans. from French G. E. T. Holloway and M. J. Mackenzie (London: Routledge & Kegan Paul; first edn., 1946).

PIATELLI-PALMARINI, M. (ed.) (1980), *Language and Learning: The Debate Between Jean Piaget and Noam Chomsky* (Cambridge, Mass.: Harvard University Press).

PICA, P., LEMER, C., IZARD, V., and DEHAENE, S. I. (2004), 'Exact and Approximate Arithmetic in an Amazonian Indigene Group', *Science*, 306: 499–503.

PICARD, R. W. (1997), *Affective Computing* (Cambridge, Mass.: MIT Press).

―――― (1999), 'Response to Sloman's Review of Affective Computing', *AI Magazine*, 20/1 (Mar.), 134–7.

PINKER, S. (1994), *The Language Instinct: The New Science of Language and Mind* (London: Allen Lane/Penguin Press).

―――― (1997), *How the Mind Works* (London: Allen Lane).

―――― (2002), *The Blank Slate: The Modern Denial of Human Nature* (London: Allen Lane).

―――― and PRINCE, A. (1988), 'On Language and Connectionism: Analysis of a Parallel Distributed Model of Language Acquisition', *Cognition*, 28: 73–193.

PINSKY, L. (1951), 'Do Machines Think About Machines Thinking?', *Mind*, ns 60: 397–8.

PITTS, W. H., and McCULLOCH, W. S. (1947), 'How We Know Universals: The Perception of Auditory and Visual Forms', in McCulloch (1965: 46–66); first pub. in *Bulletin of Mathematical Biophysics*, 9 (1947), 127–47.

PLACE, U. T. (1956), 'Is Consciousness a Brain Process?', *British Journal of Psychology*, 47: 44–50.

PLATE, T. A. (1995), 'Holographic Reduced Representations', *IEEE Transactions on Neural Networks*, 6: 623–41.

PLAUT, D. C. (1999), 'Word Reading and Acquired Dyslexia', *Cognitive Science*, 23: 543–68.

——and SHALLICE, T. (1993), 'Deep Dyslexia: A Case Study of Connectionist Neuropsychology', *Cognitive Neuropsychology*, 10: 377–500.

——McCLELLAND, J. L., SEIDENBERG, M. S., and PATTERSON, K. (1996), 'Understanding Normal and Impaired Word Reading: Computational Principles in Quasi-Regular Domains', *Psychological Review*, 103: 56–115.

PLOTKIN, G. D. (1969), 'A Note on Inductive Generalization', in Meltzer and Michie (1969*b*: 153–63).

——(1971), 'A Further Note on Inductive Generalization', in Meltzer and Michie (1971: 101–24).

PLOTKIN, H. C. (ed.) (1982), *Learning, Development, and Culture: Essays in Evolutionary Epistemology* (Chichester: Wiley).

——(1997), *Evolution in Mind: An Introduction to Evolutionary Psychology* (London: Allen Lane).

——(2002), *The Imagined World Made Real: Towards a Natural Science of Culture* (London: Allen Lane).

PLUNKETT, K., and MARCHMAN, V. (1991), 'U-Shaped Learning and Frequency Effects in a Multi-Layered Perceptron: Implications for Child Language Acquisition', *Cognition*, 38: 1–60.

————(1993), 'From Rote Learning to System Building: Acquiring Verb-Morphology in Children and Connectionist Nets', *Cognition*, 48: 21–69.

——and SINHA, C. (1992), 'Connectionism and Developmental Theory', *British Journal of Developmental Psychology*, 10: 209–54.

POLANYI, M. (1958), *Personal Knowledge: Towards a Post-Critical Philosophy* (London: Routledge, & Kegan Paul).

——(1966), 'The Logic of Tacit Inference', *Philosophy*, 41: 1–18.

——(1967), *The Tacit Dimension* (London: Routledge & Kegan Paul).

POLLACK, J. B. (1990), 'Recursive Distributed Representations', *Artificial Intelligence*, 46/1–2: 77–105.

POLLACK, M. E. (ed.), *Proceedings of the Fifteenth International Joint Conference on Artificial Intelligence*, Nagoya, Aug. 1997 (San Francisco: Morgan Kaufmann).

POLLARD, C., and SAG, I. A. (1994), *Head-Driven Phrase Structure Grammar* (Chicago: University of Chicago Press).

POLYA, G. (1945), *How to Solve It: A New Aspect of Mathematical Method* (New York: Doubleday).

POPE, K. S., and SINGER, J. L. (eds.) (1978), *The Stream of Consciousness: Scientific Investigations into the Flow of Human Experience* (New York: Plenum Press).

POPPEL, E., HELD, R., and FROST, D. (1973), 'Residual Function After Brain Wounds Involving the Central Visual Pathways in Man', *Nature*, 243: 295–6.

POPPER, F. (1993), *Art of the Electronic Age* (London: Thames & Hudson).

POPPER, K. R. (1935), *The Logic of Scientific Discovery*, trans. from German K. R. Popper (London: Hutchinson, 1959).

——(1944/1957), *The Poverty of Historicism* (London: Routledge & Kegan Paul, 1957); rev. versions of essays first pub. in *Economica*, 11 (1944), 86–103, 119–37; 12 (1945), 69–89.

——(1945), *The Open Society and its Enemies* (London: Routledge & Kegan Paul).

——(1963), *Conjectures and Refutations: The Growth of Scientific Knowledge* (London: Routledge & Kegan Paul).

——(1965), *Of Clouds and Clocks: An Approach to the Problem of Rationality and the Freedom of Man*, Second Arthur Holly Compton Memorial Lecture (St Louis, Miss.: Washington University).

——(1968), 'Epistemology Without a Knowing Subject', in B. van Rootselaar and J. F. Staal (eds.), *Proceedings of the Third International Congress for Logic, Methodology and Philosophy of Science*, 25 Aug.–22 Sept. 1965 (Amsterdam: North-Holland), 333–73.

——(1978), 'Natural Selection and the Emergence of Mind', *Dialectica*, 32: 339–55; first Darwin Lecture, delivered at Darwin College, Cambridge, Nov. 1977.

——and ECCLES, J. C. (1977), *The Self and its Brain* (Berlin: Springer-Verlag).

POPPLESTONE, R. J. (1966), 'POP-1: An Outline Language', in Dale and Michie (1966: 185–94).

——(1969), 'An Experiment in Automatic Induction', in Meltzer and Michie (1969*b*: 203–15).

PORT, R. F., and VAN GELDER, T. J. (eds.) (1995), *Mind as Motion: Explorations in the Dynamics of Cognition* (Cambridge, Mass.: MIT Press).

PORTER, R. (2001), *Enlightenment: Britain and the Creation of the Modern World* (London: Penguin).

Posner, M. I., and Raichle, M. E. (1994), *Images of Mind* (San Francisco: W. H. Freeman).

Post, E. L. (1943), 'Formal Reductions of the General Combinatorial Decision Problem', *American Journal of Mathematics*, 65: 197–215.

——— (1965), 'Absolutely Unsolvable Problems and Relatively Undecidable Propositions—Account of an Anticipation', in M. Davis (ed.), *The Undecidable: Basic Papers on Undecidable Propositions, Unsolvable Problems and Computable Functions* (Hewlett, NY: Raven Press), 340–433. Pub. posthumously.

Postal, P. M. (2005), Foreword, in Sampson (2005, pp. vii–xi).

Postman, L., and Bruner, J. S. (1946), 'The Reliability of Constant Errors in Psychophysical Measurement', *Journal of Psychology*, 21: 293–9.

Povinelli, D. J., and Preuss, T. M. (1995), 'Theory of Mind: Evolutionary History of a Cognitive Specialization', *Trends in Neuroscience*, 18: 418–24.

——— and Vonk, J. (2003), 'Chimpanzee Minds: Suspiciously Human?', *Trends in Cognitive Sciences*, 7: 157–60.

Powell, H. (2002), 'Breakfast in the Pacific: Report of the Sixth Natural Language Processing Pacific Rim Symposium, Tokyo', *Trends in Cognitive Sciences*, 6: 111–12.

Power, R. J. D. (1979), 'The Organisation of Purposeful Dialogues', *Linguistics*, 17: 107–52.

Pratt, V. (1987), *Thinking Machines: The Evolution of Artificial Intelligence* (Oxford: Blackwell).

Premack, D. (1971), 'Language in Chimpanzee?', *Science*, 172: 808–22.

——— and Woodruff, G. (1978), 'Does the Chimpanzee Have a "Theory of Mind"?', *Behavioral and Brain Sciences*, 1: 515–26.

Prescott, T. J. (2001), 'The Evolution of Action Selection', MS; amended version appeared as 'Forced Moves or Good Tricks in Design Space? Great Moments in the Evolution of the Neural Substrate for Action Selection', in J. J. Bryson, T. J. Prescott, and A. Seth (eds.), *Proceedings of IJCAI-2005 Workshop on Modeling Natural Action Selection* (Hatfield, Herts.); an extended version will be submitted to *Adaptive Behavior* in 2006.

——— Redgrave, P., and Gurney, K. N. (1999), 'Layered Control Architectures in Robots and Vertebrates', *Adaptive Behavior*, 7: 99–127.

Preston, J., and Bishop, M. (eds.) (2002), *Views into the Chinese Room: New Essays on Searle and Artificial Intelligence* (Oxford: Oxford University Press).

Pribram, K. H. (1969), 'The Neurophysiology of Remembering', *Scientific American*, 220/1: 73–86.

——— (1982), 'Localization and Distribution of Function in the Brain', in J. Orbach (ed.), *Neuropsychology After Lashley* (Hillsdale, NJ: Lawrence Erlbaum), 273–96.

——— (1987), 'Bergson and the Brain: A Bio-logical Analysis of Certain Intuitions', in Papanicolaou and Gunter (1987: 149–74).

——— and Mishkin, M. (1955), 'Simultaneous and Successive Visual Discrimination in Monkeys with Inferotemporal Lesions', *Journal of Comparative Physiological Psychology*, 48: 198–202.

——— Nuwer, M., and Baron, R. J. (1974), 'The Holographic Hypothesis of Memory Structure in Brain Function and Perception', in D. H. Krantz, R. C. Atkinson, R. D. Luce, and P. Suppes (eds.), *Contemporary Developments in Mathematical Psychology*, ii (San Francisco: W. H. Freeman), 416–57.

Price, S., Rogers, Y. A., Scaife, M., Stanton, D., and Neale, H. (2003), 'Using "Tangibles" to Promote Novel Forms of Playful Learning', *Interacting with Computers*, 15: 169–85.

Prigogine, I., Lefever, R., Goldbetter, A., and Herschkowitz-Kauffman, M. (1969), 'Symmetry Breaking Instabilities in Biological Systems', *Nature*, 223: 913–16.

——— and Stengers, I. (1984), *Order Out of Chaos* (New York: Bantam Books).

Prinz, J. J. (2002), *Furnishing the Mind: Concepts and their Perceptual Basis* (Cambridge, Mass.: MIT Press).

——— (2004*a*), 'Which Emotions Are Basic?', in D. Evans and Cruse (2004: 69–88).

——— (2004*b*), *Gut Reactions: A Perceptual Theory of Emotion* (Oxford: Oxford University Press).

——— (2004*c*), 'The Fractionation of Introspection', *Journal of Consciousness Studies*, 11/7–8: 40–57.

Prior, A. N. (1967), *Past, Present, and Future* (Oxford: Oxford University Press).

Pritchard, R. M., Heron, W., and Hebb, D. O. (1960), 'Visual Perception Approached by the Method of Stabilized Images', *Canadian Journal of Psychology*, 14: 67–77.

Psujek, S., Ames, J., and Beer, R. D. (2004), 'Connection and Coordination: The Interplay Between Architecture and Dynamics in Evolved Model Pattern Generators', Research Report, Case Western Reserve University, available at <http://vorlon.cwru.edu/~beer/Papers>. Submitted for publication.

Pullum, G. K. (1996), 'Nostalgic Views from Building 20', *Journal of Linguistics*, 32: 137–47.

Pumfrey, S. (1991), 'Ideas Above his Station: A Social Study of Hooke's Curatorship of Experiments', *History of Science*, 29: 1–44.

Putnam, H. (1960), 'Minds and Machines', in S. Hook (ed.), *Dimensions of Mind: A Symposium* (New York: New York University Press), 148–79.

—— (1961), 'Some Issues in the Theory of Grammar', *Proceedings of the Twelfth Symposium in Applied Mathematics*, American Mathematical Society; repr. in G. Harman (ed.), *On Noam Chomsky: Critical Essays*, 2nd edn. (University of Massachusetts Press, 1982), 80–103.

—— (1962a), 'Dreaming and "Depth Grammar"', in R. J. Butler (ed.), *Analytical Philosophy (First Series)* (Oxford: Blackwell), 211–35.

—— (1962b), 'The Analytic and the Synthetic', in H. Feigl and G. Maxwell (eds.), *Scientific Explanation, Space and Time*, Minnesota Studies in the Philosophy of Science, 3 (Minneapolis: University of Minnesota Press), 33–69.

—— (1964), 'Robots: Machines or Artificially Created Life?', *Journal of Philosophy*, 61: 668–91.

—— (1967a), 'The Mental Life of Some Machines', in H.-N. Castañeda (ed.), *Intentionality, Minds, and Perception* (Detroit: Wayne State University Press), 177–200 (and 206–13); repr. in H. Putnam, *Mind, Language, and Reality: Philosophical Papers*, ii (Cambridge: Cambridge University Press, 1975), 408–28.

—— (1967b), 'The Nature of Mental States', first pub. as 'Psychological Predicates', in W. H. Capitan and D. Merrill (eds.), *Art, Mind, and Religion* (Pittsburgh: University of Pittsburgh Press), 37–48; repr. in H. Putnam, *Mind, Language, and Reality: Philosophical Papers*, ii (Cambridge: Cambridge University Press, 1975), 429–40.

—— (1975a), 'The Meaning of "Meaning"', in H. Putnam, *Mind, Language, and Reality: Philosophical Papers*, ii (Cambridge: Cambridge University Press), 215–71.

—— (1975b), 'Philosophy and our Mental Life', in H. Putnam, *Mind, Language, and Reality: Philosophical Papers*, ii (Cambridge: Cambridge University Press, 1975), 291–303.

—— (1982), 'Why there Isn't a Ready-Made World', *Synthese*, 51: 141–67.

—— (1988), *Representation and Reality* (Cambridge, Mass.: MIT Press).

—— (1997), 'Functionalism: Cognitive Science or Science Fiction?', in D. M. Johnson and Erneling (1997: 32–44).

—— (1999), *The Threefold Cord: Mind, Body, and World* (New York: Columbia University Press).

Pylyshyn, Z. W. (1973), 'What the Mind's Eye Tells the Mind's Brain: A Critique of Mental Imagery', *Psychological Bulletin*, 80: 1–24.

—— (1974), 'Minds, Machines, and Phenomenology: Some Reflections on Dreyfus' "What Computers Can't Do"', *Cognition*, 3: 20–42.

—— (1978), 'Computational Models and Empirical Constraints', *Behavioral and Brain Sciences*, 1: 93–128.

—— (1979), 'Complexity and the Study of Artificial and Human Intelligence', in M. Ringle (ed.), *Philosophical Perspectives in Artificial Intelligence* (Brighton: Harvester Press), 23–56; repr. in Haugeland (1981a: 67–94).

—— (1980), 'Computation and Cognition: Issues in the Foundations of Cognitive Science', *Behavioral and Brain Sciences*, 3: 111–32.

—— (1984), *Computation and Cognition: Toward a Foundation for Cognitive Science* (Cambridge, Mass.: MIT Press).

—— (ed.) (1987), *The Robot's Dilemma: The Frame Problem in Artificial Intelligence* (Norwood, NJ: Ablex).

Quillian, M. R. (1961), 'A Design for an Understanding Machine', talk given at the Colloquium on Semantic Problems in Natural Language, King's College, Cambridge, Sept.; for a published version, see the one-page ACM 'progress report': 'Simulation of Human Understanding of Language', *Communications of the ACM* (1961), 4: 406.

—— (1962), 'A Revised Design for an Understanding Machine', *Mechanical Translation*, 7: 17–29.

—— (1967), 'Word Concepts: A Theory and Simulation of Some Basic Semantic Capabilities', *Behavioral Science*, 12: 410–30.

—— (1968), 'Semantic Memory', in Minsky (1968: 227–70).

—— (1969), 'The Teachable Language Comprehender: A Simulation Program and Theory of Language', *Communications of the Association for Computing Machinery*, 12: 459–76.

Quine, W. v. O. (1951), 'Two Dogmas of Empiricism', *Philosophical Review*, 60: 20–43.

—— (1952), 'The Problem of Simplifying Truth Functions', *American Mathematical Monthly*, 59: 521–31.

—— (1953a), *From a Logical Point of View* (Cambridge, Mass.: Harvard University Press).

QUINE, W. V. O. (1953b), 'On Mental Entities', *Proceedings of the American Academy of Arts and Sciences*, 80: 198–203.

—— (1955), 'A Way to Simplify Truth Functions', *American Mathematical Monthly*, 62: 627–31.

—— (1960), *Word and Object* (Cambridge, Mass.: MIT Press).

—— (1969), 'Linguistics and Philosophy', in Hook (1969: 95–8).

QUINLAN, J. R. (1969), 'A Task-Independent Experience-Gathering Scheme for a Problem-Solver', *Proceedings of the International Joint Conference on Artificial Intelligence*, Washington, 193–7.

—— (1979), 'Discovering Rules by Induction from Large Collections of Examples', in Michie (1979: 168–201).

—— (1983), 'Learning Efficient Classification Procedures and their Application to Chess End Games', in Michalski *et al.* (1983: 463–82).

—— (1986), 'Induction of Decision Trees', *Machine Learning*, 1: 81–106.

—— (1988), 'Decision Trees and Multi-Valued Attributes', in J. E. Hayes, D. M. Michie, and J. Richards (eds.), *Machine Intelligence 11: Towards an Automated Logic of Human Thought* (Oxford: Clarendon Press), 305–18.

—— (1993), *C4.5: Programs for Machine Learning* (San Mateo, Calif.: Morgan Kaufmann).

QUINN, N. (1987), 'Convergent Evidence for a Cultural Model of American Marriage', in D. Holland and N. Quinn (eds.), *Cultural Models in Language and Thought* (Cambridge: Cambridge University Press), 173–92.

—— (1991), 'The Cultural Basis of Metaphor', in J. Fernandez (ed.), *Beyond Metaphor: The Theory of Tropes in Anthropology* (Stanford, Calif.: Stanford University Press), 56–93.

QUINN, R. D., and RITZMANN, R. E. (1998), 'Construction of a Hexapod Robot with Cockroach Kinematics Benefits Both Robotics and Biology', *Connection Science*, 10: 239–54.

RADNITZKY, G., and BARTLEY, W. W. (eds.) (1987), *Evolutionary Epistemology, Rationality and the Sociology of Knowledge* (La Salle, Ill.: Open Court).

RAIBERT, M. H., and SUTHERLAND, I. E. (1983), 'Machines that Walk', *Scientific American* (Jan.), 44–53.

RAICHLE, M. (1998), 'Behind the Scenes of Functional Brain Imaging: A Historical and Physiological Perspective', *Proceedings of the National Academy of Sciences (USA)*, 95: 765–72.

RAMÓN Y CAJAL, S. (1901–17), *Recollections of my Life*, trans. E. Horne Craigie and J. Cano (Cambridge, Mass.: MIT Press, 1989).

RANDELL, B. (1972), 'On Alan Turing and the Origins of Digital Computers', in Meltzer and Michie (1972: 3–20).

—— (ed.) (1973), *The Origins of Digital Computers* (Berlin: Springer-Verlag).

—— (ed.) (1982), *The Origins of Digital Computers*, 3rd edn. (Berlin: Springer-Verlag).

RANSON, S. W., and HINSEY, J. C. (1930), 'Reflexes in Hind Limbs of Cats After Transection of Spinal Cord at Various Levels', *American Journal of Physiology*, 94: 490–1.

RAO, R. P. N., and SEJNOWSKI, T. K. (2001), 'Spike-Timing Dependent Hebbian Plasticity as Temporal Difference Learning', *Neural Computation*, 13: 2221–37.

RAPHAEL, B. (1968), 'SIR: Semantic Information Retrieval', in Minsky (1968: 33–145).

—— (1971), 'The Frame-Problem in Problem-Solving Systems', in N. V. Findler and B. Meltzer (eds.), *Artificial Intelligence and Heuristic Programming* (Edinburgh: Edinburgh University Press), 159–69.

—— (1976), *The Thinking Computer: Mind Inside Matter* (San Francisco: W. H. Freeman).

RASHEVSKY, N. (1938), *Mathematical Biophysics: Physicomathematical Foundations of Biology* (Chicago: University of Chicago Press).

RASKIN, V. (1996), 'Computer Implementation of the General Theory of Verbal Humor', in J. Hulstijn and A. Nijholt (eds.), *Automatic Interpretation and Generation of Verbal Humor (IWCH-96): Twente Workshop on Language Technology* (Enschede: University of Twente), 9–19.

RASMUSSEN, S., CHEN, L., NILSSON, M., and ABE, S. (2003a), 'Bridging Nonliving and Living Matter', *Artificial Life*, 9: 269–316.

——RAVEN, M. J., KEATING, G. N., and BEDAU, M. A. (2003b), 'Collective Intelligence of the Artificial Life Community on its Own Successes, Failures, and Future', *Artificial Life*, 9: 207–35.

RAY, T. S. (1992), 'An Approach to the Synthesis of Life', in Langton *et al.* (1992: 371–408).

—— (1994), 'An Evolutionary Approach to Synthetic Biology: Zen and the Art of Creating Life', *Artificial Life*, 1: 179–210.

—— (1996), 'Evolution of Parallel Processes in Organic and Digital Media', in D. Waltz (ed.), *Natural and Artificial Parallel Computation* (Philadelphia: SIAM Press), 69–91.

___(1996–), *Continuing Report on the Network Tierra Experiment*, continually updated since 1996, available at <http://www.hip.atr.co.jp/~ray/tierra/netreport/netreport.html>.

___ and HART, J. (1998), 'Evolution of Differentiated Multi-Threaded Digital Organisms', in C. Adami, R. K. Belew, H. Kitano, and C. E. Taylor (eds.), *Artificial Life VI* (Cambridge, Mass.: MIT Press), 295–304.

REASON, J. (1990), *Human Error* (Cambridge: Cambridge University Press).

RECHENBERG, I. (1964), 'Cybernetic Solution Path of an Experimental Problem', trans. B. F. Toms, in D. B. Fogel (1998: 301–9).

REDDY, D. R., and NEWELL, A. (1974), 'Knowledge and Representation in a Speech Understanding System', in L. W. Gregg (ed.), *Knowledge and Representation* (Baltimore: Lawrence Erlbaum), 253–86.

___ ERMAN, L. D., FENNELL, R. D., and NEELY, R. B. (1973), 'The HEARSAY Speech Understanding System', *Proceedings of the Third International Joint Conference on Artificial Intelligence*, Los Angeles, 185–93.

REDDY, R. (2005), 'The Origins of the American Association for Artificial Intelligence', *AI Magazine*, 26/4 (25th anniversary issue), 5–12.

REED, C. (1997), 'Representing and Applying Knowledge for Argumentation in Social Context', *AI & Society*, 11: 138–54.

___ and NORMAN, T. J. (eds.) (2004), *Argumentation Machines: New Frontiers in Argument and Computation* (Dordrecht: Kluwer).

___ ___ and GABBAY, D. (forthcoming), *Handbook of Argumentation and Computation*.

REED, D. (1976), *Anna* (London: Secker).

REED, E. S. (1996a), *Encountering the World: Toward an Ecological Psychology* (Oxford: Oxford University Press).

___ (1996b), *The Necessity of Experience* (New Haven: Yale University Press).

___ and JONES, R. K. (1982), *Reasons for Realism: Selected Essays of James J. Gibson* (Hillsdale, NJ: Lawrence Erlbaum).

REED, H. B. (1946), 'Factors Influencing the Learning and Retention of Concepts. I. The Influence of Set', *Journal of Experimental Psychology*, 36: 71–87.

REEKE, G. N., FINKEL, L. H., SPORNS, O., and EDELMAN, G. M. (1990), 'Synthetic Neural Modeling: A Multilevel Approach to the Analysis of Brain Complexity', in G. M. Edelman, W. E. Gall, and W. M. Cowan (eds.), *Signal and Sense: Local and Global Order in Perceptual Maps* (New York: Wiley-Liss), 607–707.

REEVE, R., and WEBB, B. (2002), 'New Neural Circuits for Robot Phonotaxis', in R. Damper and D. Cliff (eds.), *Biologically-Inspired Robotics: The Legacy of W. Grey Walter*, Proceedings of the EPSRC/BBSRC International Workshop WGW-02, Bristol, 225–32; rev. version in *Philosophical Transactions of the Royal Society of London*, ser. A, 361, special issue: *Biologically Inspired Robotics* (2003), 2245–66.

___ WEBB, B., HORCHLER, A., INDIVERI, G., and QUINN, R. (2005), 'New Technologies for Testing a Model of Cricket Phonotaxis on an Outdoor Robot Platform', *Robotics and Autonomous Systems*, 51/1: 41–54.

REHLING, J. A. (2001), 'Letter Spirit (Part Two): Modeling Creativity in a Visual Domain', Ph.D. thesis (Cognitive Science), Indiana University, Bloomington.

___ (2002), 'Results in the Letter Spirit Project', in Dartnall (2002: 273–82).

REICHARDT, J. (ed.) (1968), *Cybernetic Serendipity: The Computer and the Arts*, *Studio International*, special issue, pub. to coincide with an exhibition held at the Institute of Contemporary Arts, 2 Aug.–20 Oct. 1968 (London: Studio International).

REICHARDT, W., and POGGIO, T. (1976), 'Visual Control of Orientation in the Fly. Pt. I. A Quantitative Analysis', *Quarterly Review of Biophysics*, 9: 311–75.

REICHENBACH, H. (1947), *Elements of Symbolic Logic* (London: Macmillan).

REID, C., and ORGEL, L. E. (1967), 'Synthesis of Sugars in Potentially Prebiotic Conditions', *Nature*, 216: 455–6.

REINDERS, A. A. T. S., NIJENHUIS, E. R. S., PAANS, A. M. J., KORF, J., WILLEMSEN, A. T. M., and DEN BOER, J. A. (2003), 'One Brain, Two Selves', *NeuroImage*, 20: 2119–25.

REITER, R. (1980), 'A Logic for Default Reasoning', *Artificial Intelligence*, 13: 81–132.

REITMAN, W. R. (1962), Review of M. Taube, *Computers and Common Sense*, *Science*, 135: 718.

___ (1963), 'Personality as a Problem-Solving Coalition', in Tomkins and Messick (1963: 69–100).

___ (1965), *Cognition and Thought: An Information-Processing Approach* (New York: Wiley).

REITMAN, W. R., GROVE, R. B., and SHOUP, R. G. (1964), 'Argus: An Information-Processing Model of Thinking', *Behavioral Science*, 9: 270–81.

RENFREW, C. (2003), *Figuring it Out: What Are We? Where Do We Come From? The Parallel Visions of Artists and Archaologists* (London: Thames & Hudson).

RESCHER, N. (1964), *Hypothetical Reasoning* (Amsterdam: North-Holland).

RESNICK, M. (1994), *Turtles, Termites, and Traffic Jams: Explorations in Massively Parallel Microworlds* (Cambridge, Mass.: MIT Press).

REX, F. (2005), 'LambdaMOO: An Introduction', available at <http://www.lambdamoo.info>.

REYNA, S. P. (1994), 'Literary Anthropology and the Case Against Science', *Man*, 29: 555–81.

REYNOLDS, C. W. (1987), 'Flocks, Herds, and Schools: A Distributed Behavioral Model', *Computer Graphics*, 21: 25–34.

RHEINGOLD, H. (2000), *Tools for Thought: The History and Future of Mind-Expanding Technology*, expanded edn. (Cambridge, Mass.: MIT Press; first edn., 1985).

——— (2002), *Smart Mobs: The Next Social Revolution* (Cambridge, Mass.: Perseus).

RICH, C., SINDER, C. L., and LESH, N. (2001), 'Collagen: Applying Collaborative Discourse Theory to Human–Computer Interaction', *AI Magazine*, 22/4: 15–25.

RICHARDS, J. E., and RADER, N. (1981), 'Crawling-Onset Age Predicts Visual Cliff Avoidance in Infants', *Journal of Experimental Psychology, Human Perception and Performance*, 7: 382–7.

RICHENS, R. H. (1958), 'Interlingual Machine Translation', *Computer Journal*, 1: 144–7.

——— and BOOTH, A. D. (1955), 'Some Methods of Mechanized Translation', in W. N. Locke and Booth (1955: 24–46).

RICHERSON, P. J., and BOYD, R. (1976), 'A Simple Dual Inheritance Model of the Conflict Between Social and Biological Evolution', *Zygon*, 11: 254–62; later amended as: 'A Dual Inheritance Model of the Human Evolutionary Process: I. Basic Postulates and a Simple Model', *Journal of Social and Biological Sciences*, 1: 127–53.

——— ——— (2000), 'Climate, Culture, and the Evolution of Cognition', in C. M. Hayes and L. Huber (eds.), *The Evolution of Cognition* (Cambridge, Mass.: MIT Press), 329–46.

RICKEL, J., and JOHNSON, W. L. (1998), 'STEVE: A Pedagogical Agent for Virtual Reality', *Proceedings of the Second International Conference on Autonomous Agents*, Minneapolis, St Paul, 30–8.

——— ——— (1999), 'Animated Agents for Procedural Training in Virtual Reality: Perception, Cognition, and Motor Control', *Applied Artificial Intelligence*, 13: 343–82.

——— ——— (2000), 'Task-Oriented Collaboration with Embodied Agents in Virtual Worlds', in Cassell *et al.* (2000: 95–122).

——— LESH, N., RICH, C., SIDNER, C. L., and GERTNER, A. (2002), 'Collaborative Discourse Theory as a Foundation for Tutorial Dialogue', *International Conference on Intelligent Tutoring Systems (ITS)*, Biarritz, 2363: 542–51.

RIDLEY, M. (2003), *Nature via Nurture: Genes, Experience and What Makes Us Human* (London: Fourth Estate).

RIECKEN, D. (ed.) (1994), *Agents, Communications of the Association for Computing Machinery*, special issue, 37/7 (July).

RIEGER, C. J. (1975a), 'Conceptual Overlays: A Mechanism for the Interpretation of Sentence Meaning in Context', *Proceedings of the Fourth International Joint Conference on Artificial Intelligence*, Los Angeles, 143–50.

——— (1975b), 'The Commonsense Algorithm as a Basis for Computer Models of Human Memory, Inference, Belief, and Contextual Language Comprehension', in R. C. Schank and B. Nash-Webber (eds.), *Theoretical Issues in Natural Language Processing (Proceedings of a Workshop of the Association of Computational Linguistics, June)*, Cambridge, Mass. 199–214.

RIESEN, A. H. (1947), 'The Development of Visual Perception in Man and Chimpanzee', *Science*, 106: 107–8.

——— (1958), 'Plasticity of Behavior: Psychological Aspects', in H. F. Harlow and C. N. Woolsey (eds.), *Biological and Biochemical Bases of Behavior* (Madison: University of Wisconsin Press), 425–50.

RIPS, L. J. (1986), 'Mental Muddles', in M. Brand and R. M. Harnish (eds.), *The Representation of Knowledge and Belief* (Tucson: University of Arizona Press).

RISKIN, J. (2003), 'Eighteenth-Century Wetware', *Representations*, 83: 97–125.

_____ (forthcoming), 'The Android's I: A Joint History of Consciousness and Artificial Life', in J. Riskin (ed.), *The Sistine Gap: Essays in the History and Philosophy of Artificial Life* (Stanford: Stanford University Press).

RISSLAND, E. L. (1985), 'AI and Legal Reasoning', *Proceedings of the Ninth International Joint Conference on Artificial Intelligence*, Los Angeles, 1254–60.

_____ and ASHLEY, K. D. (1987), 'A Case-Based System for Trade Secrets Law', *Proceedings of the First International Conference on AI and Law* (New York: ACM Press), 60–6.

_____ and SKALAK, D. B. (1991), 'CABARET: Statutory Interpretation in a Hybrid Architecture', *International Journal of Man–Machine Studies*, 34: 839–87.

_____ ASHLEY, K. D., and LOUI, R. P. (2003), 'AI and Law: A Fruitful Synergy', introd. to special issue on AI and law, *Artificial Intelligence*, 150: 1–16.

RISTAU, C. A. (1991), *Cognitive Ethology: The Minds of Other Animals: Essays in Honor of Donald R. Griffin* (Hillsdale, NJ: Lawrence Erlbaum).

_____ and ROBBINS, D. (1982), 'Language in the Great Apes: A Critical Review', *Advances in the Study of Behavior*, 12: 141–255.

RITCHIE, G. D. (2001), 'Current Directions in Computational Humour', *Artificial Intelligence Review*, 16: 119–35.

_____ (2003a), *The JAPE Riddle Generator: Technical Specification*, Informatics Research Report EDI-INF-RR-0158, Feb. (Edinburgh: University of Edinburgh, School of Informatics).

_____ (2003b), *The Linguistic Analysis of Jokes* (London: Routledge).

_____ and HANNA, F. K. (1984), 'AM: A Case Study in AI Methodology', *Artificial Intelligence*, 23: 249–68.

RIZZOLATTI, G., and ARBIB, M. A. (1998), 'Language Within Our Grasp', *Trends in Neurosciences*, 21/5: 188–94.

ROBERTS, L. G. (1963), 'Machine Perception of Three-Dimensional Solids', Ph.D. thesis (Cambridge, Mass.: MIT Department of Electrical Engineering, 22 May).

_____ (1965), 'Machine Perception of Three-Dimensional Solids', in J. T. Tippett, D. A. Berkowitz, L. C. Clapp, C. J. Koester, and A. Vanderburgh (eds.), *Optical and Electro-Optical Information Processing* (Cambridge, Mass.: MIT Press), 159–98.

ROBINS, B., DAUTENHAHN, K., TE BOEKHORST, R., and BILLARD, A. (2004), 'Effects of Repeated Exposure to a Humanoid Robot on Children with Autism', in S. Keates, J. Clarkson, P. Langdon, and P. Robinson (eds.), *Designing a More Inclusive World* (London: Springer), 225–36.

_____ _____ and DUBOWSKI, J. (2005), 'Robots as Isolators or Mediators for Autistic Children? A Cautionary Tale', *Proceedings of the AISB-05 Symposium on Robot Companions: Hard Problems and Open Challenges in Robot–Human Interaction*, Hatfield, Herts., 82–8.

ROBINSON, G. (1972), 'How to Tell your Friends from Machines', *Mind*, 81: 504–18.

ROBINSON, J. A. (1963), 'Theorem-Proving on the Computer', *Journal of the Association for Computing Machinery*, 10: 163–74.

_____ (1965), 'A Machine-Oriented Logic Based on the Resolution Principle', *Journal of the Association for Computing Machinery*, 12: 23–41.

_____ (1968), 'The Generalized Resolution Principle', in Michie (1968a: 77–93).

ROBSON, J. G. (1983), 'Frequency Domain Visual Processing', in O. J. Braddick and A. C. Sleigh (eds.), *Physical and Biological Processing of Images* (New York: Springer), 73–87.

ROCHA, L. M. (2001), *The Physics and Evolution of Symbols and Codes: Reflections on the Work of Howard Pattee*, *BioSystems*, special issue, 60 (Dublin: Elsevier).

ROCHESTER, N., HOLLAND, J. H., HAIBT, L. H., and DUDA, W. L. (1956), 'Tests on a Cell Assembly Theory of the Action of the BRAIN, Using a Large Digital Computer', *Institute of Radio Engineers Transactions on Information Theory*, 2: 80–93.

ROGERS, T. T., and MCCLELLAND, J. L. (2004), *Semantic Cognition: A Parallel Distributed Processing Approach* (Cambridge, Mass.: MIT Press).

ROGERS, Y. A. (forthcoming), 'Distributed Cognition and Communication', in K. Brown (ed.), *Encyclopaedia of Language and Linguistics*, 2nd edn. (Amsterdam: Elsevier).

_____ and MULLER, H. (forthcoming), 'A Framework for Designing Sensor-Based Interaction to Facilitate Exploration and Reflection', *International Journal of Human–Computer Studies*.

_____ SCAIFE, M., GABRIELLI, S., SMITH, H., and HARRIS, E. (2002a), 'A Conceptual Framework for Mixed Reality Environments: Designing Novel Learning Activities for Young Children', *Presence*, 11: 677–86; available at <http://www.slis.indiana.edu/faculty/yrogers/publications.html>.

ROGERS, Y. A., SCAIFE, M., HARRIS, E., PHELPS, T., PRICE, S., SMITH, H., MULLER, H., RANDALL, C., MOSS, A., TAYLOR, I., STANTON, D., O'MALLEY, C., CORKE, G., and GABRIELLI, S. (2002*b*), 'Things Aren't What they Seem to Be: Innovation Through Technology Inspiration', *DIS2002* (Designing Interactive Systems Conference), London, 373–9.

ROJAS, R. (ed.) (2002), *Encyclopedia of Computers and Computer History* (London: Fitzroy Dearborn).

ROLAND, A., and SHIMAN, P. (2002), *Strategic Computing: DARPA and the Quest for Machine Intelligence, 1983–1993* (Cambridge, Mass.: MIT Press).

ROLLS, E. T. (1990), 'A Theory of Emotion, and its Application to Understanding the Neural Basis of Emotion', *Cognition and Emotion*, 4: 161–90.

—— and TREVES, A. (1998), *Neural Networks and Brain Function* (Oxford: Oxford University Press).

ROMIJN, H. (2002), 'Are Virtual Photons the Elementary Carriers of Consciousness?', *Journal of Consciousness Studies*, 9: 61–81.

ROMNEY, A. K., and D'ANDRADE, R. G. (eds.) (1964), *Transcultural Studies in Cognition, American Anthropologist*, special issue, 6: 3/2, Report of a conference sponsored by the Social Science Research Council Committee on Intellective Processes (Meenasha, Wis.: American Anthropological Association).

RORTY, R. (1965), 'Mind–Body Identity, Privacy, and Categories', *Review of Metaphysics*, 19: 24–54.

—— (1976), 'Overcoming the Tradition: Heidegger and Dewey', *Review of Metaphysics*, 30: 280–305.

—— (1979), *Philosophy and the Mirror of Nature* (Princeton: Princeton University Press).

ROSALDO, R. (1989), 'Grief and a Headhunter's Rage', in R. Rosaldo, *Culture and Truth: The Remaking of Social Analysis* (Boston: Beacon Press), 1–21.

ROSCH/HEIDER, E. (1971), ' "Focal" Color Areas and the Development of Color Names', *Developmental Psychology*, 4: 447–55. (Eleanor Rosch was publishing under the name E. R. Heider at this time.)

—— (1972), 'Universals in Color Naming and Memory', *Journal of Experimental Psychology*, 93: 10–20. (Eleanor Rosch was publishing under the name E. R. Heider at this time.)

—— and OLIVIER, D. C. (1972), 'The Structure of Color Space in Naming and Memory for Two Languages', *Cognitive Psychology*, 3: 337–54. (Eleanor Rosch was publishing under the name E. R. Heider at this time.)

ROSCH, E. (1973), 'On the Internal Structure of Perceptual and Semantic Categories', in T. E. Moore (ed.), *Cognitive Development and the Acquisition of Language* (New York: Academic Press), 111–44.

—— (1975), 'Cognitive Reference Points', *Cognitive Psychology*, 7: 532–47.

—— (1977), 'Human Categorization', in Warren (1977: 95–121).

—— (1978), 'Principles of Categorization', in E. Rosch and B. B. Lloyd (eds.), *Cognition and Categorization* (Hillsdale, NJ: Lawrence Erlbaum), 28–49.

—— and MERVIS, C. B. (1975), 'Family Resemblances: Studies in the Internal Structure of Categories', *Cognitive Psychology*, 7: 573–605.

ROSEN, C. (1976), *Schoenberg* (Glasgow: Collins).

ROSEN, R. (1985), 'Organisms as Causal Systems Which Are Not Mechanisms: An Essay into the Nature of Complexity', in R. Rosen (ed.), *Theoretical Biology and Complexity: Three Essays on the Natural Philosophy of Complex Systems* (Orlando, Fla.: Academic Press), 165–203.

—— (1991), *Life Itself: A Comprehensive Inquiry into the Nature, Origin, and Fabrication of Life* (New York: Columbia University Press).

ROSENBERG, M. J., and ABELSON, R. P. (1960), 'An Analysis of Cognitive Balancing', in Rosenberg *et al.* (1960: 112–63).

—— HOVLAND, C. I., McGUIRE, W. J., ABELSON, R. P., and BREHM, J. W. (1960), *Attitude Organization and Change: An Analysis of Consistency Among Attitude Components* (New Haven: Yale University Press).

ROSENBLATT, F. (1958), 'The Perceptron: A Probabilistic Model for Information Storage and Organization in the Brain', *Psychological Review*, 65: 386–408.

—— (1959), 'Two Theorems of Statistical Separability in the Perceptron', in Blake and Uttley (1959: i. 419–56).

—— (1962), *Principles of Neurodynamics: Perceptrons and the Theory of Brain Mechanisms* (Washington: Spartan Books).

ROSENBLOOM, P. S., LAIRD, J. E., NEWELL, A., and McCARL, R. (1992), 'A Preliminary Analysis of the SOAR Architecture as a Basis for General Intelligence', in D. Kirsh (ed.), *Foundations of Artificial Intelligence* (Cambridge, Mass.: MIT Press), 289–325.

—— —— —— (eds.) (1993), *The SOAR Papers: Research on Integrated Intelligence*, 2 vols. (Cambridge, Mass.: MIT Press).

ROSENBLUETH, A., and WIENER, N. (1950), 'Purposeful and Non-Purposeful Behavior', *Philosophy of Science*, 17: 318–26.

———— and BIGELOW, J. (1943), 'Behavior, Purpose, and Teleology', *Philosophy of Science*, 10: 18–24.

ROSENFELD, A., HUMMEL, R. A., and ZUCKER, S. W. (1976), 'Scene Labelling by Relaxation Operations', *IEEE Transactions on Systems, Man, and Cybernetics*, 6: 420–33.

ROSENHAN, D. L. (1963), 'Discussion: Affect and Attitude Structure in Simulating "Hot" Cognition', in Tomkins and Messick (1963: 299–302).

ROSENSCHEIN, J. S., and GENESERETH, M. R. (1985), 'Deals Among Rational Agents', *Proceedings of the Ninth International Joint Conference on Artificial Intelligence*, Los Angeles, 91–9.

———— and ZLOTKIN, G. (1994), *Rules of Encounter: Designing Conventions for Automated Negotiation Among Computers* (Cambridge, Mass.: MIT Press).

ROSENSCHEIN, S. J. (1981), 'Plan Synthesis: A Logical Perspective', *Proceedings of the Seventh International Joint Conference on Artificial Intelligence*, Vancouver, 331–7.

———— (1985), 'Formal Theories of Knowledge in AI and Robotics', *New Generation Computing*, 4: 345–57.

———— and KAELBLING, L. P. (1986), 'The Synthesis of Digital Machines with Provable Epistemic Properties', in J. Y. Halpern (ed.), *Theoretical Aspects of Reasoning About Knowledge: Proceedings of the 1986 Conference at Monterey, Caifornia* (Los Altos, Calif.: Morgan Kaufmann), 83–98.

———— ———— (1995), 'A Situated View of Representation and Control', *Artificial Intelligence*, 73, special issue: *Computational Research on Interaction and Agency*, 149–73.

ROSENTHAL, D. M. (1986), 'Two Concepts of Consciousness', *Philosophical Studies*, 49: 329–59.

———— (2000), 'Consciousness, Content, and Metacognitive Judgments', *Consciousness and Cognition*, 9: 203–14.

———— (2002), 'Consciousness and Higher-Order Thought', in L. Nadel (ed.), *Encyclopedia of Cognitive Science* (Basingstoke: Palgrave Macmillan), 717–26.

ROSENTHAL, R. (1964), 'The Effect of the Experimenter on the Result of Psychological Research', in B. A. Maher (ed.), *Progress in Experimental Personality Research* (New York: Academic Press), 79–114.

———— and FODE, K. L. (1963), 'The Effect of Experimenter Bias on the Performance of the Albino Rat', *Behavioral Science*, 8: 183–9.

ROSS, J. R. (1968), 'Constraints on Variables in Syntax', Ph.D. thesis, Massachusetts Institute of Technology; officially pub. as *Infinite Syntax!* (Norwood, NJ: Ablex, 1986).

ROSS, T. (1938), 'The Synthesis of Intelligence: Its Implications', *Psychological Review*, 45: 185–9.

ROSZAK, T. (1969), *The Making of a Counter-Culture: Reflections on the Technocratic Society and its Youthful Opposition* (Garden City, NY: Doubleday).

———— (1986), *The Cult of Information: The Folklore of Computers and the True Art of Thinking* (New York: Pantheon).

ROTA, G.-C. (1989), 'The Lost Café', in N. G. Cooper (ed.), *From Cardinals to Chaos: Reflections on the Life and Legacy of Stanislaw Ulam* (Cambridge: Cambridge University Press), 23–32.

ROUMELIOTIS, S. I., PIRJANIAN, P., and MATARIC, M. (2000), 'Ant-Inspired Navigation in Unknown Environments', *Proceedings of the 2000 AAAI International Conference on Autonomous Agents*, Barcelona, 25–6.

ROWDEN, C. (ed.) (1992), *Speech Processing* (London: McGraw-Hill).

ROWLANDS, M. (2003), *Externalism: Putting Mind and World Back Together Again* (Chesham: Acumen).

RUBINSTEIN, M. (2002), 'Markowitz's "Portfolio Selection": A Fifty-Year Retrospective', *Journal of Finance*, 57: 1041–5.

RUMBAUGH, D. (ed.) (1977), *Language Learning by a Chimpanzee: The Lana Project* (New York: Academic Press).

RUMELHART, D. E. (1975), 'Notes on a Schema for Stories', in Bobrow and Collins (1975: 211–36).

———— (1979), 'Some Problems with the Notion that Words Have Literal Meanings', in Ortony (1979: 78–90).

———— and McCLELLAND, J. L. (1982), 'An Interactive Activation Model of Context Effects in Letter Perception: Part 2. The Contextual Enhancement Effect and Some Tests and Extensions of the Model', *Psychological Review*, 89: 60–94.

———— ———— (1986), 'On Learning the Past Tenses of English Verbs', in Rumelhart, McClelland, *et al.* (1986: 216–71).

———— ———— and the PDP Research Group (1986), *Parallel Distributed Processing: Explorations in the Microstructure of Cognition*, i: *Foundations* (Cambridge, Mass.: MIT Press).

RUMELHART, D. E. and NORMAN, D. A. (1982), 'Simulating a Skilled Typist: A Study of Skilled Cognitive-Motor Performance', *Cognitive Science*, 6: 1–36.

——HINTON, G. E., and WILLIAMS, R. J. (1986a), 'Learning Internal Representations by Error Propagation', in Rumelhart, McClelland, *et al.* (1986: 318–62).

—— —— —— (1986b), 'Learning Representations by Back-Propagating Errors', *Nature*, 323: 533–6.

——SMOLENSKY, P., McCLELLAND, J. L., and HINTON, G. E. (1986c), 'Schemata and Sequential Thought Processes in PDP Models', in McClelland, Rumelhart, *et al.* (1986: 7–57).

RUPKE, N. A. (ed.) (1987), *Vivisection in Historical Context* (London: Croom Helm).

RUSO, B., RENNINGER, L., and ALTZWANGER, K. (2003), 'Human Habitat Preferences: A Generative Territory for Evolutionary Aesthetics Research', in E. Voland and K. Grammer (eds.), *Evolutionary Aesthetics* (London: Springer), 279–94.

RUSSELL, B. (1905), 'On Denoting', *Mind*, 14: 479–93.

—— (1918–19), 'The Philosophy of Logical Atomism', *Monist*, 28: 495–527, eight lectures given at the Royal Institute of Philosophy, Gordon Square, 1918; repr. in B. Russell, *Logic and Knowledge* (London: Allen & Unwin, 1956).

—— (1951), 'Ludwig Wittgenstein', *Mind*, 60: 297–8.

—— (1956a), *Portraits from Memory, and Other Essays* (London: Allen & Unwin).

—— (1956b), 'Philosophical Analysis', *The Hibbert Journal: A Quarterly Review of Religion, Theology, and Philosophy*, 54: 319–29.

—— (1957), *Why I Am Not a Christian—And Other Essays on Religion and Related Subjects* (London: Allen & Unwin).

——and WHITEHEAD, A. N. (1910), *Principia Mathematica*, i (Cambridge: Cambridge University Press).

RUSSELL, S., and NORVIG, P. (2003), *Artificial Intelligence: A Modern Approach*, 2nd edn. (London: Pearson Education).

RYAN, B. (1991), 'DYNABOOK Revisited with Alan Kay', *Byte* (Feb.), 203–8.

RYCHENER, M. D. (1976), *Production Systems as a Programming Language for Artificial Intelligence Applications*, Technical Report, Computer Science Department (Pittsburgh: Carnegie Mellon University).

RYCROFT, C. (1968), *A Critical Dictionary of Psychoanalysis* (London: Nelson).

RYLE, G. (1929), 'Critical Notice of Heidegger's *Sein und Zeit*', *Mind*, 38: 355–70.

—— (1946), 'Knowing How and Knowing That', *Proceedings of the Aristotelian Society*, 46: 1–16.

—— (1949), *The Concept of Mind* (London: Hutchinson's University Library).

—— (1962), 'La Phénomenologie contre *The Concept of Mind*', *La Philosophie Analytique: Cahiers de Royaumont*, 4: 65–84 (Paris: Éditions de Minuit); paper given at an Anglo-French conference, Royaumont (the ensuing discussion is on pp. 85–104); Eng. trans.: 'Phenomenology versus *The Concept of Mind*', in G. Ryle, *Collected Papers*, i: *Critical Essays* (London: Hutchinson, 1971), 179–96.

—— (1971), 'Autobiographical Sketch', in O. Wood and G. Pitcher (eds.), *Ryle: A Collection of Critical Essays* (London: Macmillan), 1–15.

SABBAGH, K. (2004), 'The Strange Case of Louis de Branges', *London Review of Books*, 26/14 (July), 13–14.

SACERDOTI, E. D. (1974), 'Planning in a Hierarchy of Abstraction Spaces', *Artificial Intelligence*, 5: 115–36.

—— (1975a), 'The Non-Linear Nature of Plans', *Proceedings of the Fourth International Joint Conference on Artificial Intelligence*, Tbilisi, 206–14.

—— (1975b), *A Structure for Plans and Behavior*, AI Technical Note 109 (Menlo Park, Calif.: SRI).

SACKS, O. (1985), *The Man Who Mistook his Wife for a Hat: And Other Clinical Tales* (New York: Simon & Schuster).

—— (1989), *Seeing Voices: A Journey into the World of the Deaf* (Berkeley: University of California Press).

—— (2001), 'Henry Cavendish: An Early Case of Asperger's Syndrome?', *Neurology*, 57: 1347.

SAFFRAN, J. R., ASLIN, R. N., and NEWPORT, E. L. (1996), 'Statistical Learning by 8-Month-Old Infants', *Science*, 274: 1926–8.

SAHLINS, M. D., and SERVICE, E. R. (eds.) (1960), *Evolution and Culture* (Ann Arbor: University of Michigan Press).

SAHOTA, M., and MACKWORTH, A. K. (1994), 'Can Situated Robots Play Soccer?', *Proceedings of the Canadian Conference on Artificial Intelligence*, Banff, Alberta, 249–54.

SALMON, W. C. (1989), *Four Decades of Scientific Explanation* (Minneapolis: University of Minnesota Press).

SALOMON, G., and PERKINS, D. N. (1987), 'Transfer of Cognitive Skills from Programming: When and How?', *Journal of Educational Computing Research*, 3: 149–69.

SAMPSON, G. R. (1974), 'Is there a Universal Phonetic Alphabet?', *Language*, 50: 236–59.

―――― (1979*a*), *Liberty and Language* (Oxford: Oxford University Press).

―――― (1979*b*), 'The Indivisibility of Words', *Journal of Linguistics*, 15: 39–47.

―――― (1979*c*), 'What Was Transformational Grammar? A Review of Noam Chomsky's *The Logical Structure of Linguistic Theory*', *Lingua*, 48: 355–78.

―――― (1980), *Schools of Linguistics: Competition and Evolution* (London: Hutchinson).

―――― (1983), 'Deterministic Parsing', in M. King (ed.), *Parsing Natural Language* (London: Academic Press), 91–116.

―――― (1997), *Educating Eve: The 'Language Instinct' Debate* (London: Cassell).

―――― (2001), 'Review of *The Spoken Language Translator* (by M. Rayner *et al.*)', *AISB Quarterly*, no. 105 (Spring), 26–7.

―――― (2005), *The 'Language Instinct' Debate: Revised Edition* (London: Continuum). Updated version of Sampson (1997).

―――― and McCARTHY, D. (2004), *Corpus Linguistics: Readings in a Widening Discipline* (London: Continuum).

SAMUEL, A. L. (1959), 'Some Studies in Machine Learning Using the Game of Checkers', *IBM Journal of Research and Development*, 3: 211–29; repr. in Feigenbaum and Feldman (1963: 71–108).

―――― (1967), 'Some Studies in Machine Learning Using the Game of Checkers, II—Recent Progress', *IBM Journal of Research and Development*, 11: 601–17; repr. in F. J. Crosson (ed.), *Human and Artificial Intelligence* (New York: Appleton-Century-Crofts, 1970), 81–116.

SAMUELS, R. (1998), 'Evolutionary Psychology and the Massive Modularity Hypothesis', *British Journal for the Philosophy of Science*, 49: 575–602.

―――― (2000), 'Massively Modular Minds: Evolutionary Psychology and Cognitive Architecture', in Carruthers and Chamberlain (2000: 13–46).

―――― (2002), 'Nativism in Cognitive Science', *Mind and Language*, 17: 233–65.

―――― (forthcoming), *Flexibility and the Architecture of Cognition* (Oxford: Oxford University Press).

―――― STICH, S. P., and TREMOULET, P. D. (1999), 'Rethinking Rationality: From Bleak Implications to Darwinian Modularity', in Lepore and Pylyshyn (1999: 74–120).

SANDEWALL, E. (1972), 'An Approach to the Frame Problem, and its Implementation', in Meltzer and Michie (1972: 167–94).

SANDHOLM, T. W. (1999), 'Distributed Rational Decision Making', in G. Weiss (1999: 201–58).

―――― and VULKAN, N. (1999), 'Bargaining with Deadlines', *Proceedings of the Sixteenth National Conference on Artificial Intelligence*, Orlando, Fla., 44–51.

SAPIR, E. (1921), *Language* (New York: Harcourt, Brace & World).

SARBIN, T. R. (1950), 'Contributions to Role-Taking Theory: I. Hypnotic Behavior', *Psychological Review*, 57: 255–70.

SARLE, W. S. (1995), 'Neural Network Modelling Questions', email of 27 June, later mounted on David Aha's web page of machine learning resources, <http://home.earthlink.net/%7Edwaha/research/machine-learning.html>.

SAUNDERS, F. S. (1999), *Who Paid the Piper? The CIA and the Cultural Cold War* (London: Granta Books).

SAUNDERS, P. T. (ed.) (1992), *Morphogenesis: Collected Works of A. M. Turing* (Amsterdam: Elsevier Science).

SAUNDERS, R. (2000), 'CyberLife: A Possible Architecture for Complex Design Computing Solutions', available at <http://www.arch.usyd.edu.au/~rob/study/CyberLife.html>.

SAUSSURE, F. (1916), *Course in General Linguistics*, trans. W. Baskin, ed. C. Bally and A. Sechenaye (London: Peter Owen, 1960).

SAVAGE-RUMBAUGH, E. S. (1991), 'Language Learning in the Bonobo: How and Why they Learn', in N. A. Krasnegor, D. M. Rumbaugh, R. L. Schiefelbusch, and M. Studdert-Kennedy (eds.), *Biological and Behavioral Determinants of Language Development* (Hillsdale, NJ: Lawrence Erlbaum), 209–33.

―――― and LEWIN, R. (1994), *Kanzi: The Ape at the Brink of the Human Mind* (London: Doubleday).

―――― SHANKER, S. G., and TAYLOR, T. J. (1998), *Apes, Language, and the Human Mind* (Oxford: Oxford University Press).

SAVER, J., and RABIN, J. (1997), 'The Neural Substrates of Religious Experience', *Journal of Neuropsychiatry and Clinical Neurosciences*, 9: 498–510.

SAYRE, K. M. (1965), *Recognition: A Study in the Philosophy of Artificial Intelligence* (Notre Dame, Ind.: University of Notre Dame Press).

Sayre, K. M. (1969), *Consciousness: A Philosophic Study of Minds and Machines* (New York: Random House).

_____ (1976), *Cybernetics and the Philosophy of Mind* (Atlantic Highlands, NJ: Humanities Press).

Scaife, M. (2005), 'External Cognition, Innovative Technologies, and Effective Learning', in P. Gärdenfors and P. Johansson (eds.), *Cognition, Education and Communication Technology* (Hillsdale, NJ: Lawrence Erlbaum), 181–202.

_____ and Bruner, J. S. (1975), 'The Capacity for Joint Visual Attention in the Infant', *Nature*, 253: 265–6.

_____ and Rogers, Y. (1996), 'External Cognition: How Do Graphical Representations Work?', *International Journal of Human–Computer Studies*, 45: 185–213.

_____ _____ (2001), 'Informing the Design of a Virtual Environment to Support Learning in Children', *International Journal of Human–Computer Studies*, 55: 115–43.

Scassellati, B. (1996), 'Mechanisms of Shared Attention for a Humanoid Robot', *Embodied Cognition and Action: Proceedings of the 1996 AAAI Fall Symposium*, Cambridge, Mass. (Menlo Park, Calif.: AAAI Press), 102–6.

_____ (2001), 'Investigating Models of Social Development Using a Humanoid Robot', in B. Webb and T. R. Consi (eds.), *Biorobotics: Methods and Applications* (Cambridge, Mass.: MIT Press), 145–67.

Schaeffer, J., and Lake, R. (1996), 'Solving the Game of Checkers', in R. Nowakowski (ed.), *Games of No Chance: Combinatorial Games at MSRI, 1994* (Cambridge: Cambridge University Press), 119–33.

Schaffer, S. (1983), 'Natural Philosophy and Public Spectacle in the Eighteenth Century', *History of Science*, 21: 1–43.

_____ (1994), 'Making Up Discovery', in Boden (1994c: 14–51).

_____ (1996), 'Babbage's Dancer and the Impresarios of Mechanism', in F. Spufford and J. Uglow (eds.), *Cultural Babbage: Technology, Time and Invention* (London: Faber & Faber), 53–80.

_____ (2003a), 'Paper and Brass: The Lucasian Professorship 1820–39', in K. C. Knox and R. Noakes (eds.), *From Newton to Hawking: A History of Cambridge University's Lucasian Professors of Mathematics* (Cambridge: Cambridge University Press), 241–93.

_____ (2003b), 'OK Computer', available at the Hypermedia Research Centre web site, maintained by the School of Communication and Creative Industries, University of Westminster, <http://www.hrc.wmin.ac.uk/theory-okcomputer.html>.

Schank, R. C. (1973), 'Identification of Conceptualizations Underlying Natural Language', in Schank and Colby (1973: 187–247).

_____ (1982), 'Reminding and Memory Organization: An Introduction to MOPS', in Lehnert and Ringle (1982: 455–94).

_____ and Abelson, R. P. (1977), *Scripts, Plans, Goals, and Understanding* (Hillsdale, NJ: Lawrence Erlbaum).

_____ and Colby, K. M. (eds.) (1973), *Computer Models of Thought and Language* (San Francisco: W. H. Freeman).

_____ and Langer, E. (eds.) (1994), *Beliefs, Reasoning, and Decision-Making: Psycho-Logic in Honor of Bob Abelson. A Festschrift* (Hillsdale, NJ: Lawrence Erlbaum).

_____ and Riesbeck, C. K. (1981), *Inside Computer Understanding: Five Programs Plus Miniatures* (Hillsdale, NJ: Lawrence Erlbaum).

_____ and the Yale AI Project (1975), *SAM—A Story Understander*, Research Report 43 (New Haven: Yale University Department of Computer Science, Aug.).

Scheper-Hughes, N. (1992), *Death Without Weeping: The Violence of Everyday Life in Brazil* (Berkeley: University of California Press).

_____ (1995), 'The Primacy of the Ethical: Propositions for a Militant Anthropology', *Current Anthropology*, 36: 409–40.

Scheutz, G., and Scheutz, E. (1857), *Specimens of Tables, Calculated, Stereo-Moulded, and Printed by Machinery* (London); repr. in M. Campbell-Kelly (ed.), *The Works of Charles Babbage*, ii: *The Difference Engine and Table Making* (London: William Pickering, 1989), 194–221.

Scheutz, M. (ed.) (2002), *Computationalism: New Directions* (Cambridge, Mass.: MIT Press).

Schiller, C. H. (ed.) (1957), *Instinctive Behavior: The Development of a Modern Concept* (New York: International Universities Press).

Schiller, P. H. (1968), 'Single Unit Analysis of Backward Visual Masking and Metacontrast in the Cat Lateral Geniculate Nucleus', *Vision Research*, 8: 855–66.

Schmahmann, J. D. (1997), *The Cerebellum and Cognition* (San Diego: Academic Press).

SCHMIDT, C. F. (1976), 'Understanding Human Action: Recognizing the Plans and Motives of Other Persons', in J. S. Carroll and J. W. Payne (eds.), *Cognition and Social Behavior* (Hillsdale, NJ: Lawrence Erlbaum), 47–68.

SCHMITT, F. O. (1992), 'The Neurosciences Research Program: A Brief History', in F. Samson and G. Adelman (eds.), *The Neurosciences: Paths of Discovery*, ii (Boston: Birkhauser), 1–21.

SCHNEIDER, D. M. (1965), 'American Kinship Terms and Terms for Kinsmen: A Critique of Goodenough's Componential Analysis of Yankee Kinship Terms', *American Anthropologist*, 67: 288–308.

SCHRÖDINGER, E. (1944), *What Is Life? The Physical Aspect of the Living Cell* (Cambridge: Cambridge University Press).

SCHUBERT, L. K. (1978), Review of J. H. Andreae, *Thinking with the Teachable Machine*, *Alberta Journal of Educational Research*, 24: 291–3.

SCHULTZ, W., DAYAN, P., and MONTAGUE, P. R. (1997), 'A Neural Substrate of Prediction and Reward', *Science*, 275: 1593–9.

SCHWARTZ, B., WARD, A., MONTEROSSO, J., LYUBOMIRSKY, S., WHITE, K., and LEHMAN, D. R. (2002), 'Maximizing Versus Satisficing: Happiness is a Matter of Choice', *Journal of Personality and Social Psychology*, 83: 1178–97.

SCHWARTZ, E. L. (ed.) (1990), *Computational Neuroscience* (Cambridge, Mass.: MIT Press).

Scientific American (1915), 'Torres and his Remarkable Automatic Devices', suppl., 80/2079 (6 Nov.), 296–8.

SCRIVEN, M. (1953), 'The Mechanical Concept of Mind', *Mind*, 62: 230–40.

——— (1960), 'The Compleat Robot: A Prolegomena [*sic*] to Androidology', in S. Hook (ed.), *Dimensions of Mind* (New York: New York University Press), 118–42.

SEARLE, J. R. (1969), *Speech Acts: An Essay in the Philosophy of Language* (Cambridge: Cambridge University Press).

——— (1972), 'Chomsky's Revolution in Linguistics', in G. Harman (ed.), *On Noam Chomsky: Critical Essays*, 2nd edn. (Amherst: University of Massachusetts Press, 1972), 2–33; first pub. in *New York Review of Books*, 29 June 1972.

——— (1975), 'Indirect Speech Acts', in P. Cole and J. L. Morgan (eds.), *Syntax and Semantics*, iii: *Speech Acts* (New York: Academic Press), 59–82.

——— (1977), 'Reiterating the Differences: A Reply to Derrida', *Glyph: Johns Hopkins Textual Studies*, 1: 198–208.

——— (1980), 'Minds, Brains, and Programs', *Behavioral and Brain Sciences*, 3: 417–57. Includes peer commentaries, and reply.

——— (1982), 'The Chinese Room Revisited: Response to Further Commentaries on "Minds, Brains, and Programs" ', *Behavioral and Brain Sciences*, 5: 345–8.

——— (1983*a*), 'The Word Turned Upside Down', *New York Review of Books*, 27 Oct., 77 ff.

——— (1983*b*), *Intentionality: An Essay in the Philosophy of Mind* (Cambridge: Cambridge University Press).

——— (1984), *Minds, Brains, and Science: The Reith Lectures* (London: BBC Publications).

——— (1990*a*), 'Is the Brain's Mind a Computer Program?', *Scientific American* (Jan.), 20–5.

——— (1990*b*), 'Is the Brain a Digital Computer?', *Proceedings of the American Philosophical Association*, 64: 21–37.

——— (1990*c*), 'Collective Intentions and Actions', in Cohen *et al.* (1990: 401–15).

——— (1992), *The Rediscovery of the Mind* (Cambridge, Mass.: MIT Press).

——— (2001), 'Free Will as a Problem in Neurobiology', *Philosophy*, 76: 491–514.

——— and VANDERVEKEN, D. (1985), *Foundations of Illocutionary Logic* (Cambridge: Cambridge University Press).

SECORD, J. A. (1985), 'Newton in the Nursery: Tom Telescope and the Philosophy of Tops and Balls', *History of Science*, 23: 127–51.

SEDGWICK, E. K., and FRANK, A. (1995), 'Shame and the Cybernetic Fold', in E. K. Sedgwick and A. Frank (eds.), *Shame and its Sisters: A Silvan Tomkins Reader* (Durham, NC: Duke University Press), 1–28.

SEGAL, G. (1996), 'The Modularity of Theory of Mind', in Carruthers and Smith (1996: 141–57).

SEIDENBERG, M. S. (1997), 'Language Acquisition and Use: Learning and Applying Probabilistic Constraints', *Science*, 275: 1599–1603.

SEJNOWSKI, T. J. (1977), 'Statistical Constraints on Synaptic Plasticity', *Journal of Theoretical Biology*, 69: 385–9.

SEJNOWSKI, T. J. and ROSENBERG, C. R. (1986), *NETtalk: A Parallel Network that Learns to Read Aloud*, Johns Hopkins University Electrical Engineering and Computer Science Technical Report JHU/EECS-86/01; repr. in J. A. Anderson and Rosenfeld (1988: 663–72).

———— (1987), 'Parallel Networks that Learn to Pronounce English Text', *Complex Systems*, 1: 145–68.

———— KOCH, C., and CHURCHLAND, P. S. (1988), 'Computational Neuroscience', *Science*, 241: 1299–1306.

SELFRIDGE, O. G. (1955), 'Pattern Recognition and Modern Computers', *Proceedings of the 1955 Western Joint Computer Conference*, Mar., 7: 91–3.

———— (1956), 'Pattern Recognition and Learning', in C. Cherry (ed.), *Proceedings of the Third London Symposium on Information Theory* (London: Academic Press), 345–53.

———— (1959), 'Pandemonium: A Paradigm for Learning', in Blake and Uttley (1959: i. 511–29).

———— (1981), 'Tracking and Trailing: Adaptation in Movement Strategies', MS.

———— (1984), 'Some Themes and Primitives in Ill-Defined Systems', in Selfridge *et al.* (1984: 21–6).

———— and FEURZEIG, W. (2002), 'Learning in Traffic Control: Adaptive Processes and EAMs', *Neural Networks 2002: Proceedings of the 2002 International Joint Conference on Neural Networks*, Honolulu, iii. 2598–2603.

———— and NEISSER, U. (1960), 'Pattern Recognition by Machine', *Scientific American*, 203/3: 60–8.

———— RISSLAND, E. L., and ARBIB, M. A. (eds.) (1984), *Adaptive Control in Ill-Defined Systems* (New York: Plenum Press).

SELLARS, W. (1956), 'Empiricism and the Philosophy of Mind', in H. Feigl and M. Scriven (eds.), *Foundations of Science and the Concepts of Psychology and Psychoanalysis*, Minnesota Studies in the Philosophy of Science, 1 (Minneapolis: University of Minnesota Press), 253–329.

SENATOR, T. E., GOLDBERG, H. G., WOOTON, J., COTTINI, M. A., KHAN, A. F. U., KLINGER, C. D., LLAMAS, W. M., MARRONE, M. P., and WONG, R. W. H. (1995), 'The Financial Crimes Enforcement Network AI System (FAIS): Identifying Potential Money Laundering from Reports of Large Cash Transactions', *AI Magazine*, 16/4: 21–39.

SENECA (*c.*55/1932), *Letters to Lucilius*, trans. E. P. Barker, 2 vols. (Oxford: Clarendon Press).

SENGUPTA, K. (2000), 'GCHQ Releases the Secret Details of How Bletchley Park Built the First Computer', *The Independent*, 30 Sept., 13.

The Serbelloni File (1972), unpub. letters and minutes/notes concerning the social implications of AI, held at the Department of AI in Edinburgh until destroyed by fire in Dec. 2002.

SERGOT, M. J., SADRI, F., KOWALSKI, R. A., HAMMOND, P., and CORY, H. T. (1986), 'The British Nationality Act as a Logic Program', *Communications of the Association for Computing Machinery*, 29: 370–86.

SERRUYA, M., HATSOPOULOS, N. G., PANINSKI, L., FELLOWS, M. R., and DONOGHUE, J. P. (2002), 'Immediate, Real-Time Use of a Neurally-Based Control Signal for Movement', *Nature*, 416: 141–2.

SEYFARTH, R. M., CHENEY, D. L., and MARLER, P. (1980), 'Vervet Monkey Alarm Calls: Evidence for Predator Classification and Semantic Communication', *Animal Behaviour*, 28: 1070–94.

SHALLICE, T. (1972), 'Dual Functions of Consciousness', *Psychological Review*, 79: 383–93.

———— (1982), 'Specific Impairments of Planning', *Philosophical Transactions of the Royal Society*, ser. B, 298: 199–209.

———— (1988*a*), *From Neuropsychology to Mental Structure* (Cambridge: Cambridge University Press).

———— (1988*b*), 'Information-Processing Models of Consciousness: Possibilities and Problems', in A. Marcel and Bisiach (1988: 305–33).

———— and WARRINGTON, E. K. (1970), 'Independent Functioning of Verbal Memory Stores: A Neuropsychological Study', *Quarterly Journal of Experimental Psychology*, 22: 261–73.

SHANAHAN, M. P. (1997), *Solving the Frame Problem: A Mathematical Investigation of the Common Sense Law of Inertia* (Cambridge, Mass.: MIT Press).

SHANKER, E. A. (2003), 'From Cybernetics to Telematics: The Art, Pedagogy, and Theory of Roy Ascott', in Ascott (2003: 1–95).

SHANKER, S. (1998), *Wittgenstein's Remarks on the Foundations of AI* (London: Routledge).

SHANNON, C. E. (1938), 'A Symbolic Analysis of Relay and Switching Circuits', *Transactions of the American Institute of Electrical Engineers*, 57: 713–23.

———— (1948), 'A Mathematical Theory of Communication', *Bell System Technical Journal*, 27: 379–423, 623–56.

———— (1950*a*), 'Programming a Computer for Playing Chess', *Philosophical Magazine*, 41: 256–75.

———— (1950*b*), 'A Chess-Playing Machine', *Scientific American*, 182/2: 48–51.

____ (1951), 'Presentation of a Maze-Solving Machine', in H. von Foerster (ed.), *Cybernetics: Transactions of the Eighth Conference* (New York: Macy Foundation), 173–80.

____ (1953), 'Computers and Automata', *Proceedings of the Institute of Radio Engineers*, 41: 1234–41.

____ and McCARTHY, J. (eds.) (1956), *Automata Studies* (Princeton: Princeton University Press).

____ and WEAVER, W. (1949), *The Mathematical Theory of Communication* (Urbana: University of Illinois Press).

SHAPIN, S. (1989), 'The Invisible Technician', *American Scientist*, 77: 554–63.

____ (1991), ' "A Scholar and a Gentleman": The Problematic Identity of the Scientific Practitioner in Early Modern England', *History of Science*, 29: 279–327.

____ (1994), *A Social History of Truth: Civility and Science in Seventeenth-Century England* (Chicago: University of Chicago Press).

____ and SCHAFFER, S. (1985), *Leviathan and the Air-Pump: Hobbes, Boyle, and the Experimental Life* (Princeton: Princeton University Press).

SHARPE, T., and WEBB, B. (1998), 'Simulated and Situated Models of Chemical Trail Following in Ants', in R. Pfeifer, B. Blumberg, J.-A. Meyer, and S. W. Wilson (eds.), *From Animals to Animats 5*, Proceedings of the Fifth International Conference on Simulation of Adaptive Behavior (Cambridge, Mass.: MIT Press), 195–204.

SHARPLES, M. (1996), 'An Introduction to Human–Computer Interaction', in M. A. Boden (ed.), *Artificial Intelligence*, part of the *Handbook of Perception and Cognition*, 2nd edn. (London: Academic Press), 293–323.

SHASHA, D., and LAZERE, C. (1995), *Out of their Minds: The Lives and Discoveries of 15 Great Computer Scientists* (New York: Copernicus).

SHAW, J. C., NEWELL, A., SIMON, H. A., and ELLIS, T. O. (1958), 'A Command Structure for Complex Information Processing', *Proceedings of the Western Joint Computer Conference*, Los Angeles, 119–28; repr. in G. Bell and A. Newell (eds.), *Computer Structures: Readings and Examples* (New York: McGraw-Hill), 349–62.

SHAW, R., and BRANSFORD, J. (eds.) (1977), *Perceiving, Acting, and Knowing: Toward an Ecological Psychology*, based on a conference at the Center for Research in Human Learning, University of Minnesota, 23 July–17 Aug. (Hillsdale, NJ: Lawrence Erlbaum).

SHEPARD, R. N. (1984), 'Ecological Constraints on Internal Representation: Resonant Kinematics of Perceiving, Imagining, Thinking, and Dreaming', *Psychological Review*, 91: 417–47.

____ (1994/2001), 'Perceptual Cognitive Universals as Reflections of the World', *Psychonomic Bulletin and Review*, 1 (1994), 2–28; repr., with extensive peer commentary and author's response, in *Behavioral and Brain Sciences*, 24 (2001), special issue: *The Work of Roger Shepard: The Case for Cognitive Universals*, 581–601.

____ (2004), 'How a Cognitive Psychologist Came to Seek Universal Laws', *Psychonomic Bulletin and Review*, 11: 1–23.

____ and CHIPMAN, S. (1970), 'Second-Order Isomorphism of Internal Representations: Shapes of States', *Cognitive Psychology*, 1: 1–17.

____ and COOPER, L. A. (1982), *Mental Images and their Transformations* (Cambridge, Mass.: MIT Press).

____ and METZLER, J. (1971), 'Mental Rotation of Three-Dimensional Objects', *Science*, 171: 701–3.

SHEPHERD, G. M. (1991), *Foundations of the Neuron Doctrine*, History of Neuroscience, 6 (Oxford: Oxford University Press).

SHERIF, M. (1935), *A Study of Some Social Factors in Perception*, Archives of Psychology, no. 187 (New York).

SHERRATT, Y. (2005), *Continental Philosophy of Social and Political Science* (Cambridge: Cambridge University Press).

SHERRINGTON, C. S. (1898), 'Decerebrate Rigidity, and Reflex Co-ordination of Movements', *Journal of Physiology*, 12: 319–32.

____ (1906), *The Integrative Action of the Nervous System*, The Silliman Lectures, Yale, 1904 (London: Constable).

____ (1937), 'Letter to John Fulton', in J. P. Swazey, *Reflexes and Motor Integration: Sherrington's Concept of Integrative Action* (Cambridge, Mass.: Harvard University Press), 76.

____ (1940), *Man On his Nature*, The Gifford Lectures (Cambridge: Cambridge University Press, 1937–8; 2nd edn., Harmondsworth: Penguin, 1955).

____ (1942), *Goethe on Nature and on Science* (Cambridge: Cambridge University Press).

SHERRINGTON, C. S. (1946), *The Endeavour of Jean Fernel: With a List of the Editions of his Writings* (Cambridge: Cambridge University Press).

SHIEBER, S. M. (1994), 'Lessons from a Restricted Turing Test', *Communications of the Association for Computing Machinery*, 37/6: 70–8.

SHIELDS, R. (ed.) (1996), *Cultures of Internet: Virtual Spaces, Real Histories, Living Bodies* (London: Sage).

SHILS, E. A., and FINCH, H. A. (eds.) (1949), *Max Weber on the Methodology of the Social Sciences*, trans. from German by E. A. Shils and H. A. Finch (New York: Free Press; first pub. 1904, 1905, 1917); three essays.

SHIRAI, Y. (1973), 'A Context-Sensitive Line-Finder for Recognition of Polyhedra', *Artificial Intelligence*, 4: 95–120.

SHIVERS, O. (1993), *BodyTalk and the BodyNet: A Personal Information Infrastructure*, Personal Information Architecture Note 1, 12/1/93 (Cambridge, Mass.: MIT Laboratory for Computer Science).

SHOHAM, Y., and TENNENHOLTZ, M. (1995), 'On Social Laws for Artificial Agent Societies: Off-Line Design', *Artificial Intelligence*, 73: 231–52.

SHOLL, D. A., and UTTLEY, A. M. (1953), 'Pattern Discrimination and the Visual Cortex', *Nature*, 171: 387–8.

SHORE, B. (1982), *Sala'ilua: A Samoan Mystery* (New York: Columbia University Press).

—— (1988), 'Interpretation Under Fire', *Anthropological Quarterly*, 61: 161–76.

—— (1996), *Culture in Mind: Cognition, Culture, and the Problem of Meaning* (Oxford: Oxford University Press).

—— (2000), 'Globalization, the Nation States, and the Question of "Culture": The Unnatural History of Nations and Cultures', *The Semiotic Frontline*, public interactive web site for advanced semiotic transactions, <http://www.semioticon.com/frontline.htm>.

—— (2004), 'Minding our Manners: Conversations with Jerome Bruner', MS. The conversations were recorded in the mid-1990s. Page references are to the draft MS of June 2004.

SHORTLIFFE, E. H., AXLINE, S. G., BUCHANAN, B. G., MERIGAN, T. C., and COHEN, N. S. (1973), 'An Artificial Intelligence Program to Advise Physicians Regarding Antimicrobial Therapy', *Computers and Biomedical Research*, 6: 544–60.

—— DAVIS, R., AXLINE, S. G., BUCHANAN, B. G., GREEN, C. C., and COHEN, N. S. (1975), 'Computer-Based Consultations in Clinical Therapeutics: Explanation and Rule Acquisition Capabilities of the MYCIN System', *Computers and Biomedical Research*, 8: 303–20.

SHOTTER, J. (1970), 'The Philosophy of Psychology', *Bulletin of the British Psychological Society*, 80: 207–12.

—— (2001), *Marginal Monsters and Degradation Rituals*, appendix, available at <http://www.pubpages.unh.edu/~jds/Marginalized.htm>, to J. Shotter, 'Towards a Third Revolution in Psychology: From Inner Mental Representations to Dialogically-Structured Social Practices', in D. Bakhurst and S. G. Shaker (eds.), *Jerome Bruner: Language, Culture, Self* (London: Sage, 2001), 167–83.

SHOTTON, M. A. (1989), *Computer Addiction? A Study of Dependency* (London: Taylor & Francis).

SHRAGER, J., and LANGLEY, P. (eds.) (1990), *Computational Models of Discovery and Theory Formation* (San Mateo, Calif.: Morgan Kaufmann).

SHULTZ, T. R. (1991), 'Simulating Stages of Human Cognitive Development with Connectionist Models', in L. Birnbaum and G. Collins (eds.), *Machine Learning: Proceedings of the Eighth International Workshop* (San Mateo, Calif.: Morgan Kaufmann), 105–9.

—— MARESCHAL, D., and SCHMIDT, W. C. (1994), 'Modeling Cognitive Development on Balance Scale Phenomena', *Machine Learning*, 16: 57–92.

—— SCHMIDT, W. C., BUCKINGHAM, D., and MARESCHAL, D. (1995), 'Modeling Cognitive Development with a Generative Connectionist Algorithm', in T. J. Simon and G. S. Halford (eds.), *Developing Cognitive Competence: New Approaches to Process Modeling* (Hillsdale, NJ: Lawrence Erlbaum).

SIBLEY, J. R., and GUNTER, P. A. Y. (eds.) (1978), *Process Philosophy: Basic Writings* (Washington: University Press of America).

SIEGELMANN, H. (1995), 'Computation Beyond the Turing Limit', *Science*, 268: 545–8.

—— (1999), *Neural Networks and Analog Computation: Beyond the Turing Limit* (Boston: Birkhauser).

SIEGHART, P., and DAWSON, J. (1987), 'Computer-Aided Medical Ethics', *Journal of Medical Ethics*, 13: 185–8.

SIEGLER, R. S. (1983), 'Information-Processing Approaches to Development', in W. Kessen (ed.), *Handbook of Child Psychology*, i: *History, Theory, and Methods* (New York: Wiley), 129–211.

—— (1989), 'Mechanisms of Cognitive Development', *Annual Review of Psychology*, 40: 353–79.

SIENKO, T., ADAMATZKY, A., RAMBIDI, N. G., and CONRAD, M. (2003), *Molecular Computing* (Cambridge, Mass.: MIT Press).

SIGAL, L., ISARD, M., SIGELMAN, B. H., and BLACK, M. J. (2003), 'Attractive People: Assembling Loose-Limbed Models Using Non-Parametric Belief Propagation', *Advances in Neural Information Processing Systems*, 16: 1539–46.

SIGART Newsletter (1968), 'ELIZA Passes the Turing Test', no. 10 (June), no pagination.

SILBERSWEIG, D. A., STERN, E., FRITH, C. D., SCHNORR, L., CAHILL, C., JONES, T., and FRACKOWIAK, R. S. J. (1996), 'Disordered Brain Functional Connectivity During Hallucinations in a Schizophrenic Patient', *Biological Psychiatry*, 39: 516–17.

SIMMONS, P. L., and SIMMONS, R. F. (1962), 'The Simulation of Cognitive Processes: An Annotated Bibliography', *Institute of Radio Engineers Transactions on Electronic Computers*, EC-10: 462–83; EC-11: 535–52.

SIMON, H. A. (1947a), *Administrative Behavior: A Study of Decision-Making Processes in Administrative Organization* (New York: Macmillan).

—— (1947b), 'The Axioms of Newtonian Mechanics', *Philosophical Magazine*, 38: 888–905.

—— (1951), Review of N. Rashevsky, *The Mathematical Biology of Social Behavior*, *Econometrica*, 19: 357–8.

—— (1952), 'On the Definition of the Causal Relation', *Journal of Philosophy*, 49: 517–28.

—— (1955), 'A Behavioral Model of Rational Choice', *Quarterly Journal of Economics*, 69: 99–118.

—— (1956), 'Rational Choice and the Structure of the Environment', *Psychological Review*, 63: 129–38.

—— (1957), *Models of Man, Social and Rational: Mathematical Essays on Rational Human Behaviour in a Social Setting* (New York: Wiley).

—— (1962), 'The Architecture of Complexity', *Proceedings of the American Philosophical Society*, 106: 467–82.

—— (1965), *The Shape of Automation for Men and Management* (New York: Harper & Row).

—— (1967), 'Motivational and Emotional Controls of Cognition', *Psychological Review*, 74: 29–39.

—— (1969), *The Sciences of the Artificial*, The Karl Taylor Compton Lectures (Cambridge, Mass.: MIT Press); 2nd and 3rd edns. (1981, 1996): these contain Simon (1962) as the final chapter.

—— (1976), *Administrative Behavior: A Study of Decision-Making Processes in Administrative Organization*, 3rd edn. (New York: Free Press).

—— (1978), 'On How to Decide What to Do', *Bell Journal of Economics*, 9: 494–507.

—— (1991), *Models of my Life* (New York: Basic Books).

—— (1992), 'Herbert Simon Remembers Allen Newell', *AI Magazine*, 13/4: 30. Extract from Simon (1991: 199–201).

—— (1994a), 'Bottleneck of Attention: Connecting Thought with Motivation', in W. D. Spaulding (ed.), *Integrative Views of Motivation, Cognition, and Emotion* (Lincoln: University of Nebraska Press), 1–21.

—— (1994b), 'Literary Criticism: A Cognitive Approach', in Guzeldere and Franchi (1994: 1–27).

—— (1995a), 'Machine as Mind', in K. M. Ford, C. Glymour, and P. J. Hayes (eds.), *Android Epistemology* (Cambridge, Mass.: MIT Press, 1995), 23–40.

—— (1995b), 'Explaining the Ineffable: AI on the Topics of Intuition, Insight, and Inspiration', *Proceedings of the Fourteenth International Joint Conference on Artificial Intelligence*, i. 939–48.

—— (1995c), 'Artificial Intelligence: An Empirical Science', *Artificial Intelligence*, 77: 95–127.

—— (1997a), 'Machine Discovery', in Zytkow (1997: 171–200).

—— (1997b), 'Machine Discovery: Reply to Comments', in Zytkow (1997: 225–32).

—— and FEIGENBAUM, E. A. (1964), 'An Information-Processing Theory of Some Effects of Similarity, Familiarization, and Meaningfulness in Verbal Learning', *Journal of Verbal Learning and Verbal Behavior*, 3: 385–96.

—— and NEWELL, A. (1958), 'Heuristic Problem Solving: The Next Advance in Operations Research', *Operations Research*, 6: 1–10.

—— —— (1986), 'Information Processing Language V on the IBM 650', *Annals of the History of Computing*, 8: 47–9.

SIMONS, G. L. (1983), *Are Computers Alive? Evolution and New Life Forms* (Brighton: Harvester Press).

SIMMONS, R. F. (1965), 'Answering English Questions by Computer: A Survey', *Communications of the Association for Computing Machinery*, 8: 53–70.

—— (1970), 'Natural Language Question-Answering Systems: 1969', *Communications of the Association for Computing Machinery*, 13: 15–30.

Sims, K. (1991), 'Artificial Evolution for Computer Graphics', *Computer Graphics*, 25/4: 319–28.

—— (1994), 'Evolving 3D-Morphology and Behavior by Competition', *Artificial Life*, 1: 353–72.

Singer, C. (1959), *A Short History of Scientific Ideas to 1900* (Oxford: Clarendon Press).

Singh, D. (1993), 'Waist-to-Hip Ratio (WHR): A Defining Morphological Feature of Health and Female Attractiveness', *Journal of Personality and Social Psychology*, 65: 293–307.

Sivilotti, M. A., Mahowald, M. A., and Mead, C. A. (1987), 'Real-Time Visual Computations Using Analog CMOS Processing Arrays', in P. Losleben (ed.), *Advanced Research in VLSI: Proceedings of the 1987 Stanford Conference* (Cambridge, Mass.: MIT Press), 295–312; repr. in J. A. Anderson and Rosenfeld (1988: 703–11).

Skalak, D. B., and Rissland, E. L. (1992), 'Arguments and Cases: An Inevitable Intertwining', *Artificial Intelligence and Law*, 1: 3–44.

Skarda, C. A., and Freeman, W. J. (1987), 'How Brains Make Chaos in Order to Make Sense of the World', with peer commentaries, *Behavioral and Brain Sciences*, 10: 161–95.

Skinner, B. F. (1938), *The Behavior of Organisms: An Experimental Analysis* (New York: Appleton-Century-Crofts).

—— (1948), *Walden Two* (New York: Macmillan).

—— (1950), 'Are Theories of Learning Necessary?', *Psychological Review*, 57: 193–216.

—— (1957), *Verbal Behavior* (New York: Appleton-Century-Crofts).

—— (1959), 'A Case History in Scientific Method', in S. Koch (ed.), *Psychology: A Study of a Science*, ii (New York: McGraw-Hill), 359–79.

—— (1967), 'Autobiography', in E. G. Boring and G. Lindzey (eds.), *A History of Psychology in Autobiography*, v (New York: Appleton-Century-Crofts), 385–414.

—— (1971), *Beyond Freedom and Dignity* (New York: Knopf).

—— (1979), *The Shaping of a Behaviorist: Part Two of an Autobiography* (New York: Knopf).

Slagle, J. R., and Hamburger, H. (1985), 'An Expert System for a Resource Allocation Program', *Communications of the Association for Computing Machinery*, 28: 994–1004.

Slater, A., Mattock, A., and Brown, E. (1990), 'Size Constancy at Birth: Newborn Infants' Responses to Retinal and Real Size', *Journal of Experimental Child Psychology*, 49: 314–22.

Sleeman, D. H., and Brown, J. S. (eds.) (1982), *Intelligent Tutoring Systems* (London: Academic Press).

Sloman, A. (1971), 'Interactions Between Philosophy and Artificial Intelligence: The Role of Intuition and Non-Logical Reasoning in Intelligence', *Artificial Intelligence*, 2: 209–25.

—— (1974), 'Physicalism and the Bogey of Determinism', in S. C. Brown (ed.), *Philosophy of Psychology* (London: Macmillan), 283–304.

—— (1975), 'Afterthoughts on Analogical Representation', in R. C. Schank and B. L. Nash-Webber (eds.), *Theoretical Issues in Natural Language Processing: An Interdisciplinary Workshop in Computational Linguistics, Psychology, Linguistics, and Artificial Intelligence*, Cambridge, Mass., 10–13 June (Arlington, Va.: Association for Computational Linguistics), 164–8.

—— (1978), *The Computer Revolution in Philosophy: Philosophy, Science, and Models of Mind* (Brighton: Harvester Press); out of print, but available—and continually updated—at <http://www.cs.bham.ac.uk/research/cogaff/crp/>.

—— (1982), 'Towards a Grammar of Emotions', *New Universities Quarterly*, 36: 230–8.

—— (1983), 'Image Interpretation: The Way Ahead?', in O. J. Braddick and A. C. Sleigh (eds.), *Physical and Biological Processing of Images* (New York: Springer), 380–402.

—— (1986a), 'Reference Without Causal Links', in J. B. H. du Boulay and L. Steels (eds.), *Seventh European Conference on Artificial Intelligence* (Amsterdam: North-Holland), 369–81.

—— (1986b), 'What Sorts of Machine Can Understand the Symbols They Use?', *Proceedings of the Aristotelian Society*, suppl., 60: 61–80.

—— (1987a), 'Motives, Mechanisms, and Emotions', *Cognition and Emotion*, 1: 217–33; repr. in Boden (1990b: 231–47).

—— (1987b), 'Reference Without Causal Links', in J. B. H. du Boulay, D. Hogg, and L. Steels (eds.), *Advances in Artificial Intelligence—II* (Dordrecht: North-Holland), 369–81.

—— (1989), 'On Designing a Visual System: Towards a Gibsonian Computational Model of Vision', *Journal of Experimental and Theoretical AI*, 1: 289–337.

—— (1992), 'The Emperor's Real Mind: Review of Roger Penrose's *The Emperor's New Mind: Concerning Computers Minds and the Laws of Physics*', *Artificial Intelligence*, 56: 355–96.

_____ (1993), 'The Mind as a Control System', in C. Hookway and D. Peterson (eds.), *Philosophy and the Cognitive Sciences* (Cambridge: Cambridge University Press), 69–110.

_____ (1995), 'Sim_agent help-file', available at <ftp://ftp.cs.bham.ac.uk/pub/dist/poplog/sim/help/sim_agent>. See also 'Sim_agent web-page', available at <http://www.cs.bham.ac.uk/axs/cog_affect/sim_agent.html>.

_____ (1996*a*), 'Beyond Turing Equivalence', in P. J. R. Millican and A. J. Clark (eds.), *Machines and Thought: The Legacy of Alan Turing*, i (Oxford: Oxford University Press), 179–220.

_____ (1996*b*), 'Towards a General Theory of Representations', in D. M. Peterson (ed.), *Forms of Representation: An Interdisciplinary Theme for Cognitive Science* (Exeter: Intellect Books), 118–40.

_____ (1996*c*), 'Actual Possibilities', in L. C. Aiello and S. C. Shapiro (eds.), *Principles of Knowledge Representation and Reasoning: Proceedings of the Fifth International Conference (KR '96)* (San Francisco: Morgan Kaufmann), 627–38.

_____ (1998), 'Ekman, Damasio, Descartes, Alarms and Meta-Management', *Proceedings of the International Conference on Systems, Man, and Cybernetics (SMC98)* (San Diego, Calif.: IEEE Press), 2652–7.

_____ (1999), Review of R. Picard, *Affective Computing*, *AI Magazine*, 20/1 (Mar.), 127–33.

_____ (2000), 'Architectural Requirements for Human-Like Agents Both Natural and Artificial (What Sorts of Machines Can Love?)', in K. Dautenhahn (ed.), *Human Cognition and Social Agent Technology: Advances in Consciousness Research* (Amsterdam: John Benjamins), 163–95.

_____ (2001), 'Beyond Shallow Models of Emotion', *Cognitive Processing: International Quarterly of Cognitive Science*, 2: 177–98.

_____ (2002), 'The Irrelevance of Turing Machines to Artificial Intelligence', in M. Scheutz (2002: 87–127).

_____ (2003*a*), 'How Many Separately Evolved Emotional Beasties Live Within Us?', in R. Trappl, P. Petta, and S. Payr (eds.), *Emotions in Humans and Artifacts* (Cambridge, Mass.: MIT Press), 29–96.

_____ (2003*b*), 'How to Build a Human-Like Mind', *AISB Quarterly*, no. 112 (Spring 2003), 7, 11.

_____ (n.d.), The CogAff group's web site: <http://www.cs.bham.ac.uk/research/cogaff>.

_____ and CHRISLEY, R. L. (2003), 'Virtual Machines and Consciousness', in O. Holland (2003: 133–72).

_____ and CROUCHER, M. (1981), 'Why Robots Will Have Emotions', *Proceedings of the Seventh International Joint Conference on Artificial Intelligence* (Vancouver), 197–202.

_____ and POLI, R. (1995), 'Sim_agent: A Toolkit for Exploring Agent Designs', in M. Wooldridge, J.-P. Müller, and M. Tambe (eds.), *Intelligent Agents*, ii (Berlin: Springer-Verlag), 392–407.

SMART, J. J. C. (1959), 'Sensations and Brain Processes', *Philosophical Review*, 68: 141–56.

_____ (1963), 'Materialism', *Journal of Philosophy*, 60: 651–62.

_____ (1970), 'Critical Notice of D. C. Dennett's *Content and Consciousness*', *Mind*, 79: 616–23.

SMEE, A. (1849), *Elements of Electro-Biology, or the Voltaic Mechanism of Man; of Electro-Pathology, especially of the Nervous System; and of Electro-Therapeutics* (London: Longman, Brown, Green & Longmans).

_____ (1850), *Instinct and Reason Deduced from Electro-Biology* (London: Reeve & Benham).

_____ (1851), *The Process of Thought Adapted to Words and Language; together with a Description of the Relational and Difference Machines* (London: Longman, Brown, Green & Longmans).

SMITH, B. C. (1982), *Reflection and Semantics in a Procedural Language*, Ph.D. diss. and Technical Report LCS/TR-272 (Cambridge, Mass.: MIT).

_____ (1985), 'Prologue to *Reflection and Semantics in a Procedural Language*', in R. J. Brachman and H. J. Levesque (eds.), *Readings in Knowledge Representation* (Los Altos, Calif.: Morgan Kauffman), 31–40.

_____ (1987), *The Correspondence Continuum*, Technical Report No. CSLI-87-71 (Menlo Park, Calif.: Center for the Study of Language and Information, Stanford University); first pub. in *Proceedings of the Sixth Canadian Conference on Artificial Intelligence*, Montreal, May 1986.

_____ (1991), 'The Owl and the Electric Encyclopedia', *Artificial Intelligence*, 47: 251–88.

_____ (1996), *On the Origin of Objects* (Cambridge, Mass.: MIT Press).

_____ (2002), 'The Foundations of Computing', in M. Scheutz (2002: 23–58).

SMITH, D. C. (1977), *Pygmalion: A Computer Program to Model and Stimulate Creative Thought* (Basel: Birkhauser).

SMITH, J. A., and BOYD, K. M. (eds.) (1991), *Lives in the Balance: The Ethics of Using Animals in Biomedical Research. The Report of a Working Party of the Institute of Medical Ethics* (Oxford: Oxford University Press).

SMITH, M. (ed.) (2005), *Stelarc: The Monograph* (Cambridge, Mass.: MIT Press).

SMITH, N. V. (1999), *Noam Chomsky: Ideas and Ideals* (Cambridge: Cambridge University Press).

SMITH, P. H., JORIS, P. X., and YIN, T. C. (1993), 'Projections of Physiologically Characterized Spherical Bushy Cell Axons from the Cochlear Nucleus of the Cat: Evidence for Delay Lines to the Medial Superior Olive', *Journal of Comparative Neurology*, 331: 245–60.

SMITH, R. G. (1980), 'The Contract Net Protocol: High-Level Communication and Control in a Distributed Problem-Solver', *IEEE Transactions on Computers*, C-29/no. 12: 1104–13.

SMITH, S. L., and MILLER, G. A. (1952), 'The Effects of Coding Procedures on Learning and Memory', in *Quarterly Progress Report of Research Laboratory of Electronics, MIT, to Air Force Human Resources Research Laboratories*, Dec. (Cambridge, Mass.: MIT), 7–10.

SMITH, S. M. (1977), 'Coral-Snake Pattern Recognition and Stimulus Generalization by Naive Great Kikadees (*Aves Tyrannidae*)', *Nature*, 265: 535–6.

SMITH, T., HUSBANDS, P., and O'SHEA, M. (2002), 'Neuronal Plasticity and Temporal Adaptivity: GasNet Robot Control Networks', *Adaptive Behavior*, 10: 161–83.

SMOLENSKY, P. (1987a), 'Connectionist AI, Symbolic AI, and the Brain', *AI Review*, 1: 95–110.

—— (1987b), 'The Constituent Structure of Mental States: A Reply to Fodor and Pylyshyn', *Southern Journal of Philosophy*, 26: 137–60.

—— (1988), 'On the Proper Treatment of Connectionism', *Behavioral and Brain Sciences*, 11: 1–74.

SNOW, C. P. (1959), *The Two Cultures: The Rede Lecture* (Cambridge: Cambridge University Press).

SOBER, E. (1992), 'Learning from Functionalism—Prospects for Strong Artificial Life', in Langton *et al.* (1992: 749–66).

—— and WILSON, D. S. (1998), *Unto Others: The Evolution and Psychology of Unselfish Behavior* (Cambridge, Mass.: Harvard University Press).

SOKAL, A., and BRICMONT, J. (1998), *Intellectual Impostures: Postmodern Philosophers' Abuse of Science*, trans. A. Sokal and J. Bricmont from the French edn. of 1997 (London: Profile Books).

SOLAN, Z., HORN, D., RUPPIN, E., and EDELMAN, S. (forthcoming), 'Unsupervised Learning of Natural Languages', *Proceedings of the National Academy of Sciences (USA)*. Page reference is to the 25-page MS.

SOLOMONOFF, R. J. (1958), 'The Mechanization of Linguistic Learning', *Proceedings of the Second International Congress of Cybernetics*, Namur, 180–93.

—— (1964), 'A Formal Theory of Inductive Inference', *Information and Control*, 7: 1–22, 224–54.

SOMMER, R. (1997), 'Further Cross-National Studies of Tree-Form Preference', *Ecological Psychology*, 9: 153–60.

—— and SUMMIT, J. (1995), 'An Exploratory Study of Preferred Tree Form', *Environment and Behavior*, 27: 540–57.

SOPHIAN, C., and ADAMS, N. (1987), 'Infants' Understanding of Numerical Transformations', *British Journal of Developmental Psychology*, 5: 257–64.

SORCE, J. F., EMDE, R. N., CAMPOS, J., and KLINNERT, M. (1985), 'Maternal Emotional Signaling: Its Effect on the Visual Cliff Behavior of 1-year-olds', *Developmental Psychology*, 21: 195–200.

SPARCK JONES, K. (1988), 'A Look Back and a Look Forward', *Proceedings, 11th International Conference on Research and Development in Information Retrieval, ACM-SIGIR, 1988*, ed. Y. Chiaramella (Grenoble: Presses Universitaires de Grenoble), 13–29.

SPELKE, E. S. (1991), 'Physical Knowledge in Infancy: Reflections on Piaget's Theory', in Carey and Gelman (1991: 133–70).

—— (1994), 'Initial Knowledge: Six Suggestions', *Cognition*, 50: 431–5.

SPERBER, D. (1975), *Rethinking Symbolism*, trans. from French A. L. Morton (Cambridge: Cambridge University Press; first pub. 1974).

—— (1985), 'Anthropology and Psychology', *Man*, NS 20: 73–89.

—— (1990), 'The Epidemiology of Beliefs', in C. Fraser and G. Gaskell (eds.), *The Social Psychological Study of Widespread Beliefs* (Oxford: Clarendon Press), 25–44.

—— (1994), 'The Modularity of Thought and the Epidemiology of Representations', in Hirschfeld and Gelman (1994: 39–67).

—— (1997a), *Explaining Culture: A Naturalistic Approach* (Oxford: Blackwell).

—— (1997b), *Mind and Language*, 12: 67–83.

—— (2002), 'In Defense of Massive Modularity', in E. Dupoux (ed.), *Language, Brain, and Cognitive Development: Essays in Honor of Jacques Mehler* (Cambridge, Mass.: MIT Press), 47–57.

—— and HIRSCHFELD, L. (1999), 'Culture, Cognition, and Evolution', in R. A. Wilson and Keil (1999, pp. cxi–cxxxii).

—— and WILSON, D. (1986), *Relevance: Communication and Cognition* (Oxford: Blackwell).

_____ _____ (1996), 'Fodor's Frame Problem and Relevance Theory', *Behavioral and Brain Sciences*, 19: 530–2.

SPERDUTI, A. (1992), *Labeling RAAM*, Technical Report TR-93-029 (Berkeley, Calif: International Computer Science Institute).

_____ and STARITA, A. (1994), 'On the Access by Content Capabilities of the LRAAM', *IEEE International Conference on Neural Networks*, Orlando, Fla., 1143–8.

SPERRY, R. W. (1947), 'Effect of Crossing Nerves to Antagonistic Limb Muscles in the Monkey', *Archives of Neurology and Psychiatry*, 58: 452–73.

SPILLER, N. (ed.) (2002), *Cyber-Reader: Critical Writings for the Digital Era* (London: Phaidon).

SPINOZA, B. (1677), *Ethic*, trans. W. H. White, in J. Wild (ed.), *Spinoza Selections* (New York: Charles Scribner's Sons, 1930), 94–400; also many other edns.

SPRAT, T. (1667), *The History of the Royal-Society of London, for the Improving of Natural Knowledge* (London: printed by T.R. for J. Martyn and J. Allestry); facs. edn.: *History of the Royal Society*, ed. J. I. Cope and H. Whitmore Jones (St Louis: Washington University Press, 1958).

SQUIRE, L. R., and KOSSLYN, S. M. (eds.) (1998), *Findings and Current Opinion in Cognitive Neuroscience* (Cambridge, Mass.: MIT Press).

SQUIRES, R. (1970), 'Are Dispositions Lost Causes?', *Analysis*, 31: 15–18.

SRC (1972), *Policy and Programme Review of Research and Training in Computer Science* (London: Science Research Council/HMSO).

_____ (1973), *Artificial Intelligence: A Paper Symposium*, a Report prepared for the Science Research Council (London: HMSO). With contributions by J. Lighthill, N. S. Sutherland, R. M. Needham, H. C. Longuet-Higgins, and D. M. Michie.

SRIPADA, C. S., and GOLDMAN, A. I. (2005), 'Simulation and the Evolution of Mindreading', in Zilhao (2005: 148–61).

STANFIELD, R. A., and ZWAAN, R. A. (2001), 'The Effect of Implied Orientation Derived from Verbal Context on Picture Recognition', *Psychological Science*, 12: 153–6.

STARK, D. (1999), 'Heterarchy: Distributing Intelligence and Organizing Diversity', in J. Clippinger (ed.), *The Biology of Business: Decoding the Natural Laws of Enterprise* (San Francisco: Jossey-Bass), 153–80.

State of the Art Committee (1978), *Report of the State of the Art Committee to the Advisors of the Alfred P. Sloan Foundation*, 1 Oct. 1978, in F. Machlup and U. Mansfield (eds.), *The Study of Information: Interdisciplinary Messages* (New York: Wiley-Interscience, 1983), 75–80.

STEELS, L. (1998a), 'The Origins of Ontologies and Communication Conventions in Multi-Agent Systems', *Autonomous Agents and Multi-Agent Systems*, 1: 169–94.

_____ (1998b), 'Synthesizing the Origins of Language and Meaning Using Co-evolution, Self-Organization, and Level Formation', in J. R. Hurford, M. Studdert-Kennedy, and C. Knight (eds.), *Approaches to the Evolution of Language: Social and Cognitive Bases* (Cambridge: Cambridge University Press), 384–405.

_____ (1998c), 'The Origin of Syntax in Visually Grounded Robotic Agents', *Artificial Intelligence*, 103: 133–56.

_____ (2003), 'Language Re-entrance and the "Inner Voice"', in O. Holland (2003: 173–85).

_____ and BELPAEME, T. (forthcoming), 'Coordinating Perceptually Grounded Categories Through Language: A Case Study for Colour', *Behavioral and Brain Sciences*. Accepted as target article; to appear with multiple-author commentary.

_____ and BROOKS, R. A. (eds.) (1995), *The Artificial Life Route to Artificial Intelligence: Building Embodied, Situated Agents* (Northvale, NJ: Lawrence Erlbaum).

_____ and KAPLAN, F. (2001), 'AIBO's First Words: The Social Learning of Language and Meaning', *Evolution of Communication*, 4: 3–32.

STEFIK, M. J., and SMOLIAR, S. W. (eds.) (1991), 'Four Reviews of [M. Minsky,] *The Society of Mind* and a Response', *Artificial Intelligence*, 48: 319–96.

_____ _____ (eds.) (1993), 'Eight Reviews of [A. Newell,] *Unified Theories of Cognition* and a Response', *Artificial Intelligence*, 59: 261–413.

_____ _____ (eds.) (1996), '[H. L. Dreyfus,] *What Computers Still Can't Do*: Five Reviews and a Response', *Artificial Intelligence*, 80: 95–192.

_____ AIKINS, J., BALZER, R., BENOIT, J., BIRNBAUM, L., HAYES-ROTH, F., and SACERDOTI, E. D. (1982), 'The Organization of Expert Systems', *Artificial Intelligence*, 18: 135–73.

STEHL, R. (1955), *The Robots Are Amongst Us* (London: Arco).

STEIN, D. (1985), *Ada: A Life and a Legacy* (Cambridge, Mass.: MIT Press).

STEIN, L. A. (1990), 'An Atemporal Frame Problem', *International Journal of Expert Systems*, 3: 371–81.

STEINER, G. (1978), *Heidegger* (London: Fontana).

Stelarc (1984), *Obsolete Body: Suspensions*, ed. with James D. Paffrath (Davis, Calif.: J.P. Publications).

—— (1986), 'Beyond the Body: Amplified Body, Laser Eyes, and Third Hand', *NMA* (*New Music Articles*), 6 (1986–7), 27–30; repr. in Stiles and Selz (1996: 427–30).

—— (1994), 'Prosthetics, Robotics and Remote Existence: Postevolutionary Strategies', *Leonardo*, 24: 591–5; first pub. in *Art Cognition: Pratiques artistiques et sciences cognitives* (Aix-en-Provence: Cypres/École d'Art), 44–56.

—— (2002*a*), *Alternate Interfaces*, with M. Grzinic, B. Massumi, and T. Murray (Melbourne: Faculty of Art & Design, Monash University).

—— (2002*b*), 'From Zombies to Cyborg Bodies: Extra Ear, Exoskeleton and Avatars', in Candy and Edmonds (2002: 115–24).

STENMARK, M. (2001), *Scientism* (Aldershot: Ashgate).

STEPHAN, A. (1998), 'Varieties of Emergence in Artificial and Natural Systems', *Zeitschrift für Naturforschung*, 53: 639–56.

—— (2003), 'Emergence', in L. Nadel (ed.), *Encyclopaedia of Cognitive Science*, i (London: Nature Publications), 1108–15.

STEPHENS, J. M. (1929), 'A Mechanical Explanation of the Law of Effect', *American Journal of Psychology*, 41: 422–31.

STERELNY, K. (2003), *Thought in a Hostile World: The Evolution of Human Cognition* (Oxford: Blackwell).

STERN, R. (1988), Introduction, in F. von Schelling, *Ideas for a Philosophy of Nature*, trans. E. E. Harris and P. Heath (Cambridge: Cambridge University Press, 1988), pp. ix–xxiii.

STEUER, J. (1992), 'Defining Virtual Reality: Dimensions Determining Telepresence', *Journal of Communication*, 42/4: 73–93.

STEVENS, A. L., and RUMELHART, D. E. (1975), 'Errors in Reading: Analysis Using an Augmented Transition Network Model of Grammar', in D. A. Norman and Rumelhart (1975: 136–56).

STEVENS, C. F. (1989), 'How Cortical Interconnectedness Varies with Network Size', *Neural Computation*, 1: 473–9.

STEVENS, S. S. (1946), 'On the Theory of Scales of Measurement', *Science*, 103: 677–80.

—— (ed.) (1951), *Handbook of Experimental Psychology* (New York: Wiley).

—— and GALANTER, E. H. (1957), 'Ratio Scales and Category Scales for a Dozen Perceptual Continua', *Journal of Experimental Psychology*, 54: 377–411.

STEWART, I. (1994), 'The Ultimate in Anty-Particles', *Scientific American*, 271/1 (July), 88–91.

STICH, S. P. (ed.) (1975), *Innate Ideas* (Berkeley: University of California Press).

—— (1983), *From Folk Psychology to Cognitive Science: The Case Against Belief* (Cambridge, Mass.: MIT Press).

—— and NISBETT, R. (1980), 'Justification and the Psychology of Human Reasoning', *Philosophy of Science*, 47: 188–202.

STILES, K., and SELZ, P. (eds.) (1996), *Theories and Documents of Contemporary Art: A Sourcebook of Artists' Writings* (Berkeley: University of California Press).

STOLCKE, A. (1997), 'Linguistic Knowledge and Empirical Methods in Speech Recognition', *AI Magazine*, 18/4: 25–31.

STONE, P. J., BALES, R. F., NAMENWIRTH, J. Z., and OGILVIE, D. M. (1962), 'The General Inquirer: A Computer System for Content Analysis and Retrieval Based on the Sentence as a Unit of Information', *Behavioral Science*, 7: 484–98.

STONE, R. J. (2005), 'Serious Gaming: Virtual Reality's Saviour?', in H. Thwaites (ed.), *Proceedings of the 11th International Conference on Virtual Systems and Multimedia (VSMM): Virtual Reality at Work in the 21st Century: Impact on Society*, Ghent, 3–7 Oct., 773–86; invited keynote speech.

STONE, V. (1978), 'Some Personal Recollections of Sir Frederic Bartlett and his Era', *Bulletin of the British Psychological Society*, 31: 87–90.

STRACEY, M. (1997), 'Rebuilding the Baby: The Reconstruction of the Prototype Manchester Mark I Computer', *AISB Quarterly*, no. 98: 17–18.

STRACHEY, C. (ed.) (1924–32), *The Letters of the Earl of Chesterfield to his Son*, 2 vols. (London: Methuen).

STRACHEY, C. S. (1952), 'Logical or Non-Mathematical Programmes', *Proceedings of the 1952 ACM National Meeting*, Toronto, 46–9.

STRATTON, G. M. (1896), 'Some Preliminary Experiments on Vision', *Psychological Review*, 3: 611–17.

_____ (1897*a*), 'Upright Vision and the Retinal Image', *Psychological Review*, 4: 182–7.

_____ (1897*b*), 'Vision Without Inversion of the Retinal Image', *Psychological Review*, 4: 341–60, 463–81.

STRAUSS, C. (1992), 'What Makes Tony Run? Schemas as Motives Reconsidered', in D'Andrade and Strauss (1992: 197–224).

_____ and QUINN, N. (1997), *A Cognitive Theory of Cultural Meaning*, Publications of the Society for Psychological Anthropology (Cambridge: Cambridge University Press).

STRAWSON, P. F. (1950), 'On Referring', *Mind*, 59: 320–44.

STURM, T., and GIGERENZER, G. (2006), 'How Can We Use the Distinction between Discovery and Justification? On the Weakness of the Strong Programme in the Sociology of Science', in J. Schickore and F. Steinle (eds.), *Revisiting Discovery and Justification* (Dordrecht: Kluwer), 133–58.

STURROCK, J. (1998), 'Le Pauvre Sokal', *London Review of Books*, 20/14 (16 July), 8–9.

SUCHMAN, L. A. (1987), *Plans and Situated Actions: The Problem of Human–Machine Communication* (Cambridge: Cambridge University Press).

_____ (1993), 'Response to Vera and Simon's *Situated Action: A Symbolic Interpretation*', *Cognitive Science*, 17: 71–5.

_____ (2004), 'Figuring Personhood in Sciences of the Artificial', pub. by the Department of Sociology, Lancaster University, at <http://www.comp.lancs.ac.uk/sociology/papers/suchman-figuring-personhood.pdf>.

SUDNOW, D. (1978/2001), *Ways of the Hand: The Organization of Improvised Conduct* (London: Routledge); rev. version, omitting philosophical jargon: *Ways of the Hand: A Rewritten Account* (Cambridge, Mass.: MIT Press, 2001).

SUPPES, P. (1968), 'The Desirability of Formalisation in Science', *Journal of Philosophy*, 65: 651–64.

SUR, M., GARRAGHTY, P. E., and ROE, A. W. (1988), 'Experimentally Induced Visual Projections into Auditory Thalamus and Cortex', *Science*, 242: 1437–41.

_____ PALLAS, S. L., and ROE, A. W. (1990), 'Cross-Modal Plasticity in Cortical Development: Differentiation and Specification of Sensory Neocortex', *Trends in Neuroscience*, 13: 227–33.

SUSSKIND, R. E. (1986/2000), 'The Computer Judge: Early Thoughts', *Modern Law Review*, 49: 125–38; slightly amended version in Susskind (2000: 275–92): page references are to this version.

_____ (1987), *Expert Systems in Law: A Jurisprudential Enquiry* (Oxford: Clarendon Press).

_____ (1993), *Essays on Law and Artificial Intelligence* (Oslo: TANO).

_____ (2000), *Transforming the Law: Essays on Technology, Justice, and the Legal Marketplace* (Oxford: Oxford University Press).

SUSSMAN, G. J. (1973/1975), *A Computer Model of Skill Acquisition* (New York: American Elsevier, 1975), pub. version of a previously circulated Ph.D. thesis, MIT AI Lab Memo AI-TR-297 (Cambridge, Mass: MIT, Aug. 1973).

_____ (1974), 'The Virtuous Nature of Bugs', *Proceedings of the AISB Summer Conference, July 1974*, University of Sussex, 224–37.

_____ and McDERMOTT, D. V. (1972*a*), *Why Conniving Is Better than Planning*, MIT AI Lab Memo 255a (Cambridge, Mass.: MIT).

_____ _____ (1972*b*), *The CONNIVER Reference Manual*, MIT AI Lab Memo 259 (Cambridge, Mass.: MIT).

_____ WINOGRAD, T., and CHARNIAK, E. (1970), *Micro-PLANNER Reference Manual*, MIT AI Lab Memo 203 (Cambridge, Mass.: MIT).

SUTHERLAND, I. E. (1963), 'Sketchpad: A Man–Machine Graphical Communication System', *Proceedings of the Spring Joint Computer Conference*, Detroit, 329–46.

_____ (1965), 'The Ultimate Display', *Proceedings of the International Federation of Information Processing Congress*, New York (Washington: Spartan Books), ii. 506–8, 582–3; repr. in Packer and Jordan (2001: 232–6).

_____ (1968), 'A Head-Mounted Three-Dimensional Display', *Proceedings of the Fall Joint Computer Conference*, San Francisco, 33: 757–64.

SUTHERLAND, J. (2004), *Stephen Spender: The Authorized Biography* (London: Viking Press).

SUTHERLAND, K. (1994), 'Consciousness—Its Place in Contemporary Science', *Journal of Consciousness Studies*, 1: 285–6.

SUTHERLAND, N. S. (1965), Review of C. M. Taylor, *The Explanation of Behaviour*, *Philosophical Quarterly*, 15: 379–81.

_____ (1973), 'Some Comments on the Lighthill Report and on Artificial Intelligence', in SRC (1973: 22–31).

SUTHERLAND, N. S. (1974), 'Computer Simulation of Brain Function', in S. C. Brown (ed.), *Philosophy of Psychology* (London: Macmillan), 259–68.

—— (1976), *Breakdown: A Personal Crisis and a Medical Dilemma* (London: Weidenfeld & Nicolson).

—— (1982), 'The Vision of David Marr', *Nature*, 298: 691–2.

—— (1992), *Irrationality: The Enemy Within* (London: Constable).

—— (1995), *Macmillan Dictionary of Psychology*, 2nd edn., enlarged (Basingstoke: Macmillan).

SUTHERLAND, W. S., MUGGLIN, M. G., and SUTHERLAND, I. E. (1958), 'An Electro-Mechanical Model of Simple Animals', *Computers and Automation*, 7: 6–8, 23–5, 32.

SUTTERLIN, C. (2003), 'From Sign and Schema to Iconic Representation: Evolutionary Aesthetics of Pictorial Art', in E. Voland and K. Grammer (eds.), *Evolutionary Aesthetics* (London: Springer), 131–70.

SUTTON, R. S., and BARTO, A. G. (1981), 'Toward a Modern Theory of Adaptive Networks: Expectation and Prediction', *Psychological Review*, 88: 135–71.

—— —— (1998), *Reinforcement Learning: An Introduction* (Cambridge, Mass.: MIT Press).

SWADE, D. D. (1991), *Charles Babbage and his Calculating Engines* (London: Science Museum).

—— (1993), 'Redeeming Charles Babbage's Mechanical Computer', *Scientific American* (Feb.), 86–91.

—— (1996), ' "It Will Not Slice a Pineapple": Babbage, Miracles, and Machines', in F. Spufford and J. Uglow (eds.), *Cultural Babbage: Technology, Time and Invention* (London: Faber & Faber), 34–51.

—— (2000), *The Cogwheel Brain: Charles Babbage and the Quest to Build the First Computer* (London: Little, Brown).

SWAIN, H. (1999), 'Noam Chomsky', in S. Griffiths (ed.), *Predictions: Thirty Great Minds on the Future* (Oxford: Oxford University Press), 22–9.

SWETS, J. A. (2005), 'The ABC's of BBN: From Acoustics to Behavioral Sciences to Computers', *IEEE Annals of the History of Computing*, 27/2: 15–29.

SWIFT, J. (1726), *Gulliver's Travels*, pt. III: *A Voyage to Laputa, Balnibarbi, Luggnagg, Glubbdubdrib, and Japon*; many edns. Quotations taken from the Penguin edn. of 1967.

Synthese (1967), 'Symposium on Innate Ideas', 17: 1–28. Some follow-ups to that symposium are in Hook (1969).

SZATHMÁRY, E. (2005), 'Life: In Search of the Simplest Cell', *Nature*, 433: 469–70.

SZOSTAK, J. W., BARTELL, D., and LUISI, P. L. (2001), 'Synthesizing Life', *Nature*, 409: 387–90.

TANKARD, J. W. (1979), 'The H. G. Wells Quote on Statistics: A Question of Accuracy', *Historia Mathematica*, 6: 30–3.

TANNER, W. P., and SWETS, J. A. (1954), 'A Decision-Making Theory of Visual Detection', *Psychological Review*, 61: 401–9.

TASSABEHJI, M. (2003), 'Research on the Molecular Biology of Williams Syndrome', talk given at 'Cognitive Development, Genes and the Brain: From Piaget to Cognitive Neuroscience', workshop in honour of Prof. Annette Karmiloff-Smith, Institute of Child Health, London, 20 Sept. 2003.

TATE, A. (1975), 'Interacting Goals and their Use', *Proceedings of the Fourth International Joint Conference on Artificial Intelligence*, Tbilisi, 215–18.

TAUBE, M. (1961), *Computers and Common Sense: The Myth of Thinking Machines* (New York: Columbia University Press).

TAYLOR, C. M. (1964a), *The Explanation of Behaviour* (London: Routledge & Kegan Paul).

—— (1964b), Review of *La Philosophie analytique*, *Philosophical Review*, 1: 132–5.

—— (1971), 'Interpretation and the Sciences of Man', *Review of Metaphysics*, 25: 3–51.

TAYLOR, J. M. (2001), 'The Information Age: Public and Personal', in R. Catlow and S. Greenfield (eds.), *Proceedings of the Royal Institution of Great Britain*, 71 (Oxford: Oxford University Press), 161–88.

TAYLOR, P. A. (2000), 'Concept-to-Speech Synthesis by Phonological Structure Mapping', in Gazdar *et al.* (2000: 1403–16).

TAYLOR, W. K. (1956), 'Electrical Simulation of Some Nervous System Functional Activities', in E. C. Cherry (ed.), *Information Theory, 3* (London: Butterworths), 314–28.

—— (1959), 'Pattern Recognition by Means of Automatic Analogue Apparatus', *Proceedings of the Institution of Electrical Engineers*, 106: 198–209.

—— (1964), 'Cortico-Thalamic Organization and Memory', *Proceedings of the Royal Society*, ser. B, 159: 466–78.

Teledildonics (2003), available at <http://www.betterhumans.com/teledildonics.article>, 2 Sept. 2003, consulted 2 July 2004.

TENENBAUM, J. M., and BARROW, H. G. (1976), *Experiments in Interpretation-Guided Segmentation*, AI Technical Note 123 (Menlo Park, Calif.: SRI).

_____ and WEYL, S. (1975), 'A Region-Analysis Subsystem for Interactive Scene Analysis', *Proceedings of the Fourth International Joint Conference on Artificial Intelligence*, Tbilisi, 682–7.

_____ GARVEY, T. D., WEYL, S., and WOLF, H. C. (1974), *An Interactive Facility for Scene Analysis Research*, AI Technical Note 87 (Menlo Park, Calif.: SRI).

TERRACE, H. S. (1979a), 'How Nim Chimpsky Changed My Mind', *Psychology Today*, 13: 65–76.

_____ (1979b), *Nim: A Chimpanzee Who Learned Sign-Language* (London: Eyre Methuen).

_____ (1981), 'A Report to a Committee, 1980', in T. A. Sebeok and R. Rosenthal (eds.), *The Clever Hans Phenomenon: Communication with Horses, Whales, Apes, and People*, Annals of the New York Academy of Sciences, 364 (New York: New York Academy of Sciences), 94–114.

_____ PETTITO, L. A., SANDERS, R. J., and BEVER, T. G. (1979), 'Can an Ape Create a Sentence?', *Science*, 206: 891–902.

TERZOPOULOS, D., TU, X., and GZESZCZUK, R. (1994), 'Artificial Fishes with Autonomous Locomotion, Perception, Behavior, and Learning in a Simulated Physical World', in Brooks and Maes (1994: 17–27).

TETTAMANTI, M. BUCCINO, G., SACCUMAN, M. C., GALLESE, V., DANNA, M., SCIFO, P., FASIO, F., RIZZOLATTI, G., CAPPA, F. S., and PERANI, D. (2005), 'Listening to Action-Related Sentences Activates Fronto-Parietal Motor Circuits', *Journal of Cognitive Neuroscience*, 17/2: 1–9.

TEUBNER, G. (1993), *Law as an Autopoietic System*, trans. A. Bankowska and R. Adler (Oxford: Blackwell).

_____ (ed.) (1987), *Autopoietic Law: A New Approach to Law and Society* (New York: de Gruyter).

THAGARD, P. (1982), 'From the Descriptive to the Normative in Psychology and Logic', *Philosophy of Science*, 49: 24–42.

_____ (1988), *Computational Philosophy of Science* (Cambridge, Mass.: MIT Press).

_____ (1989), 'Explanatory Coherence', *Behavioral and Brain Sciences*, 12: 435–502.

_____ (1990), 'Concepts and Conceptual Change', *Synthese*, 82: 255–74.

_____ HOLYOAK, K. J., NELSON, G., and GOCHFELD, D. (1988), *Analog Retrieval by Constraint Satisfaction*, Research Paper, Cognitive Science Laboratory, Princeton University (Nov.).

THELEN, E. (1981), 'Kicking, Rocking, and Waving: Contextual Analysis of Rhythmical Stereotypes in Normal Human Infants', *Animal Behavior*, 29: 3–11.

_____ (1985), 'Developmental Origins of Motor Coordination: Leg Movements in Human Infants', *Developmental Psychobiology*, 18: 1–22.

_____ (1989), 'Self-Organization in Developmental Processes: Can Systems Approaches Work?', in M. Gunnar and E. Thelen (eds.), *Systems and Development*, Minnesota Symposium in Child Psychology, 22 (Hillsdale, NJ: Lawrence Erlbaum), 77–117; repr. in M. H. Johnson (1993a: 555–91).

_____ (1995), 'Time-Scale Dynamics and the Development of an Embodied Cognition', in Port and van Gelder (1995: 69–100).

_____ and SMITH, L. B. (1994), *A Dynamic Systems Approach to the Development of Cognition and Action* (Cambridge, Mass.: MIT Press).

_____ KELSO, J. A. S., and FOGEL, A. (1987), 'Self-Organizing Systems and Infant Motor Development', *Developmental Review*, 7: 39–65.

_____ SCHÖNER, G., SCHEIER, C., and SMITH, L. B. (2001), 'The Dynamics of Embodiment: A Field Theory of Infant Perseverative Reaching', *Behavioral and Brain Sciences*, 24: 1–86.

THES (1975), 'A Change in ARPA's Funding-Policy', *Times Higher Education Supplement*, 14 Mar., 9.

THOMAS, I. (1967), 'Interregnum (in the History of Logic)', in P. Edwards (ed.), *The Encyclopedia of Philosophy*, iv (New York: Macmillan), 534–7.

THOMAS, M. S. C., and KARMILOFF-SMITH, A. (2002), 'Are Developmental Disorders Like Cases of Adult Brain Damage? Implications from Connectionist Modelling', *Behavioral and Brain Sciences*, 727–87.

_____ _____ (2003), 'Connectionist Models of Development, Developmental Disorders and Individual Differences', in R. J. Sternberg, J. Lautrey, and T. I. Lubart (eds.), *Models of Intelligence: International Perspectives* (Washington: American Psychological Association), 133–50.

THOMAS, R. K. (2001), 'Lloyd Morgan's Canon: A History of Misrepresentation', *History and Theory of Psychology Eprint Archive*, <http://htpprints.yorku.ca/>.

THOMPSON, A. (1995), 'Evolving Electronic Robot Controllers that Exploit Hardware Resources', in F. Moran, A. Moreno, J. J. Merelo, and P. Chacon (eds.), *Advances in Artificial Life*, Proceedings of the Third European Conference on Artificial Life—ECAL95 (Berlin: Springer-Verlag), 640–56.

THOMPSON, A. (1997), 'Evolving Inherently Fault-Tolerant Systems', *Proceedings of the Institution of Mechanical Engineers*, Part I, 211: 365–71.

——(1998), *Hardware Evolution: Automatic Design of Electronic Circuits in Reconfigurable Hardware by Artificial Evolution* (Berlin: Springer-Verlag).

——HARVEY, I., and HUSBANDS, P. (1996), 'Unconstrained Evolution and Hard Consequences', in E. Sanchez and M. Tomassini (eds.), *Towards Evolvable Hardware: The Evolutionary Engineering Approach* (Berlin: Springer-Verlag), 136–65.

THOMPSON, D. W. (ed.) (1880), *Studies from the Museum of Zoology in University College, Dundee* (Dundee: Museum of Zoology).

——(trans. and ed.) (1883), *The Fertilization of Flowers by Insects*, by Hermann Müller, preface by Charles Darwin (London: Macmillan).

——(1895), *A Glossary of Greek Birds* (Oxford: Clarendon Press).

——(trans.) (1910), *Historia Animalium*, vol. iv of *The Works of Aristotle*, ed. J. A. Smith and W. D. Ross (Oxford: Clarendon Press).

——(1917/1942), *On Growth and Form* (Cambridge: Cambridge University Press; 2nd edn., expanded, 1942; abridged edn., ed. J. T. Bonner, 1961).

——(1925), 'Egyptian Mathematics: The Rhind Mathematical Papyrus—British Museum 10057 and 10058', *Nature*, 115: 935–7.

——(1931), *On Saithe, Ling and Cod, in the Statistics of the Aberdeen Trawl-Fishery, 1901–1929* (Edinburgh: Great Britain Fishery Board for Scotland).

——(1940), *Science and the Classics* (London: Oxford University Press).

——(1947), *A Glossary of Greek Fishes* (London: Oxford University Press).

THOMPSON, H. (1984), 'There Will Always Be Another Moonrise: Computer Reliability and Nuclear Weapons', *The Scotsman*; also in *AISB Quarterly* (Spring–Summer), nos. 53–4: 21–3, available at <http://www.aisb.org.uk/articles/moonrise.html>.

——(1985), 'Empowering Automatic Decision-Making Systems', *Proceedings of the Ninth Joint International Conference on Artificial Intelligence*, Los Angeles, 1281–3.

THOMPSON, R. K. R., ODEN, D. L., and BOYSEN, S. T. (1997), 'Language-Naive Chimpanzees (*Pan troglodytes*) Judge Relations Between Relations in a Conceptual Matching-to-Sample Task', *Journal of Experimental Psychology: Animal Behavior Processes*, 23: 31–43.

THOMSON, J. (1876), 'An Integrating Machine, Being a New Kinematic Principle', *Proceedings of the Royal Society*, 24: 262–5.

THOMSON, W. (1876), 'An Application of Professor James Thomson's Integrator to Harmonic Analysis of Meteorological, Tidal, and Other Phenomena, and to the Integration of Differential Equations', *Royal Society of Edinburgh Proceedings*, 9: 138. This is an announcement of a talk given to the RSE, 6 Mar. 1876; there is no text or abstract.

THORNDIKE, E. L. (1898), *Animal Intelligence: An Experimental Study of the Associative Processes in Animals, Psychological Review Monograph Supplement*, no. 8.

THORNDYKE, P. W. (1977), 'Cognitive Structures in Comprehension and Memory of Narrative Discourse', *Cognitive Psychology*, 9: 77–110.

THORNE, J. P., BRATLEY, P., and DEWAR, H. M. (1968), 'The Syntactic Analysis of English by Machine', in D. M. Michie (ed.), *Machine Intelligence 3* (Edinburgh: Edinburgh University Press), 281–309.

THORNHILL, R. (1998), 'Darwinian Aesthetics', in C. Crawford and D. L. Krebs (eds.), *Handbook of Evolutionary Psychology: Ideas, Issues, and Applications* (Hillsdale, NJ: Lawrence Erlbaum), 543–72.

THORNTON, C. (2000), *Truth from Trash: How Learning Makes Sense* (Cambridge, Mass.: MIT Press).

TIERNEY, P. (2000), *Darkness in El Dorado: How Scientists and Journalists Devastated the Amazon* (New York: W. W. Norton).

TIGER, L. (2003), *The Apes of New York* (Christchurch, NZ: Cybereditions). This publisher provides hard copy by print-on-demand, and electronic versions.

TINBERGEN, N. (1948), 'Social Releasers and the Experimental Method Required for their Study', *Wilson Bulletin*, 60: 6–51.

——(1951), *The Study of Instinct* (Oxford: Clarendon Press).

——and KUENEN, D. J. (1939), 'Releasing and Directing Stimulus Situations in *Turdus m. merula L.* and *T. e. ericetorum Turton*', trans. from German C. H. Schiller, in Schiller (1957: 209–38).

TODD, S. C., and LATHAM, W. (1992), *Evolutionary Art and Computers* (London: Academic Press).

TODES, D. P. (2001), *Pavlov's Physiology Factory: Experiment, Interpretation, Laboratory Enterprise* (Baltimore: Johns Hopkins University Press).

TOFFLER, A. (1980), *The Third Wave* (New York: William Morrow; repr. New York: Bantam Books, 1981). Page references are to Bantam Books edn.

TOFFOLI, T. (1984), 'CAM6: A High-Performance Cellular-Automaton Machine', *Physica D*, 10: 195–204.

TOLMAN, E. C. (1920), 'Instinct and Purpose', *Psychological Review*, 27: 217–33.

_____ (1922), 'A New Formula for Behaviorism', *Psychological Review*, 29: 44–53.

_____ (1923), 'The Nature of Instinct', *Psychological Bulletin*, 20: 200–16.

_____ (1925*a*), 'Behaviorism and Purpose', *Journal of Philosophy*, 22: 36–41.

_____ (1925*b*), 'Purpose and Cognition: The Determiners of Animal Learning', *Psychological Review*, 32: 285–97.

_____ (1932), *Purposive Behavior in Animals and Men* (New York: Appleton-Century-Crofts).

_____ (1935), 'Psychology Versus Immediate Experience', *Philosophy of Science*, 2: 356–80.

_____ (1941), 'Discrimination vs. Learning and the Schematic Sawbug', *Psychological Review*, 48: 367–82.

_____ (1948), 'Cognitive Maps in Rats and Men', *Psychological Review*, 55: 189–208.

_____ (1958), *Behavior and Psychological Man: Essays in Motivation and Learning* (Berkeley: University of California Press).

_____ (1959), 'Principles of Purposive Behavior', in S. Koch (ed.), *Psychology: A Study of a Science*, ii (New York: McGraw-Hill), 92–157.

_____ and BRUNSWIK, E. (1935), 'The Organism and the Causal Texture of the Environment', *Psychological Review*, 42: 43–77.

TOMASELLO, M. (1996), 'Do Apes Ape?', in C. M. Heyes and B. G. Galef (eds.), *Social Learning in Animals: The Roots of Culture* (London: Academic Press), 319–46.

_____ (1999), *The Cultural Origins of Human Cognition* (Cambridge, Mass.: Harvard University Press).

_____ and BATES, E. (2001), *Language Development: The Essential Readings* (Oxford: Blackwell).

_____ KRUGER, A. C., and RATNER, H. H. (1993), 'Cultural Learning', *Behavioral and Brain Sciences*, 16: 495–552.

_____ CALL, J., and HARE, B. (2003*a*), 'Chimpanzees Understand Psychological States: The Question Is Which Ones and to What Extent', *Trends in Cognitive Science*, 7: 153–6.

_____ _____ _____ (2003*b*), 'Chimpanzees Versus Humans: It's Not That Simple', *Trends in Cognitive Sciences*, 7: 239–40.

TOMKINS, S. S. (1963), *Affect, Imagery, Consciousness*, ii: *The Negative Affects* (New York: Springer).

_____ and MESSICK, S. (eds.) (1963), *Computer Simulation of Personality: Frontier of Psychological Research* (New York: Wiley).

TOOBY, J. (2000*a*), *Center for Evolutionary Psychology Report on 'Darkness in El Dorado'* (Santa Barbara: University of California), <http://www.psych.ucsb.edu/research/cep/index.html>.

_____ (2000*b*), 'Jungle Fever', *Slate Magazine*, 25 Oct. 2000, <http://slate.msn.com/id/91946>.

_____ and COSMIDES, L. (1992), 'The Psychological Foundations of Culture', in Barkow *et al.* (1992: 19–136).

TORRANCE, S. (2003), 'Conscious Robots: Could We, Should We, Create Them?', *Journal of Health, Social and Environmental Issues*, 4/2: 41–6.

TORRES Y QUEVEDO, L. (1914), 'Essays on Automatics: Its Definition and Theoretical Extent of its Applications', trans. R. Basu, in Randell (1973: 87–105).

TOULMIN, S. E. (1953), *The Philosophy of Science: An Introduction* (London: Hutchinson).

_____ (1961), *Foresight and Understanding: An Inquiry into the Aims of Science* (Bloomington: Indiana University Press).

_____ (1972), *Human Understanding: The Collective Use and Evolution of Concepts* (Princeton: Princeton University Press).

_____ (1999), *Cosmopolis: The Hidden Agenda of Modernity* (Chicago: Chicago University Press).

TOURETSKY, D. S., and HINTON, G. E. (1985), 'Symbols Among the Neurons: Details of a Connectionist Inference Architecture', *Proceedings of the Fourth International Conference on Artificial Intelligence*, Los Angeles, 238–43.

_____ _____ (1988), 'A Distributed Connectionist Production System', *Cognitive Science*, 12: 423–66.

TOWNSEND, J. T., and BUSEMEYER, J. R. (1995), 'Dynamic Representation of Decision-Making', in Port and van Gelder (1995: 101–20).

TRASK, R. L. (1996), *Historical Linguistics* (London: Edward Arnold).

TREHUB, A. (1991), *The Cognitive Brain* (Cambridge, Mass.: MIT Press).

TREISMAN, A. M. (1960), 'Contextual Cues in Selective Listening', *Quarterly Journal of Experimental Psychology*, 12: 242–8.

—— (1988), 'Features and Objects: The Fourteenth Bartlett Memorial Lecture', *Quarterly Journal of Experimental Psychology*, 40a: 201–37.

TRESCH, M. C., SALTIEL, P., and BIZZI, E. (1999), 'The Construction of Movement by the Spinal Cord', *Nature Neuroscience*, 2: 162–7.

TREVARTHEN, C. (1993), 'The Self Born in Intersubjectivity: The Psychology of an Infant Communicating', in U. Neisser (ed.), *The Perceived Self: Ecological and Interpersonal Sources of Self-Knowledge* (Cambridge: Cambridge University Press), 121–73.

TREVARTHEN, C. B. (1979), 'Communication and Cooperation in Early Infancy: A Description of Primary Intersubjectivity', in M. Bullowa (ed.), *Before Speech: The Beginning of Interpersonal Communication* (Cambridge: Cambridge University Press).

—— (1994), 'Regulation of Human Cognition and its Growth', in Furukawa *et al.* (1994: 247–68).

TRIANTAFYLLOU, M. S., and TRIANTAFYLLOU, G. S. (1995), 'An Efficient Swimming Machine', *Scientific American*, 272: 40–8.

TULVING, E., and THOMPSON, D. M. (1973), 'Encoding Specificity and Retrieval Processes in Episodic Memory', *Psychological Review*, 80: 352–73.

TURING, A. M. (1936), 'On Computable Numbers with an Application to the *Entscheidungsproblem*', *Proceedings of the London Mathematical Society*, ser. 2, 42/3 (30 Nov.), 230–40, 42/4 (23 Dec.), 241–65. This paper is often wrongly dated as 1937, presumably because Turing published a note on corrections, using almost the same title, in 1937: vol. 43, pp. 544–6. Repr. in M. Davis (ed.), *The Undecidable: Basic Papers on Undecidable Propositions, Unsolvable Problems, and Computable Functions* (Hewlett, NY: Raven Press, 1965), 116–53.

—— (1939), 'Systems of Logic Based on Ordinals', in M. Davis (ed.), *The Undecidable: Basic Papers on Undecidable Propositions, Unsolvable Problems, and Computable Functions* (Hewlett, NY: Raven Press, 1965), 155–222; first pub. in *Proceedings of the London Mathematical Society*, 45: 161–228.

—— (1940/2004), 'Bombe and Spider', in Copeland (2004: 314–35). This is ch. 6 of Turing, *Treatise on the Enigma*; the complete document is available at <http://www.AlanTuring.net/profs_book>.

—— (1946), 'Proposal for Development in the Mathematics Division of an Automatic Computing Engine (ACE)', Report prepared for the National Physical Laboratory, UK, in B. E. Carpenter and R. W. Doran (eds.), *A. M. Turing's ACE Report of 1946 and Other Papers* (Cambridge, Mass.: MIT Press, 1986), 20–105.

—— (1947a), 'Lecture to the London Mathematical Society on 20 February 1947', first pub. in B. E. Carpenter and R. W. Doran (eds.), *A. M. Turing's ACE Report of 1946 and Other Papers* (Cambridge, Mass.: MIT Press, 1986), 106–24.

—— (1947b), 'Intelligent Machinery', Report prepared for the National Physical Laboratory, UK, first pub. in Meltzer and Michie (1969b: 3–23).

—— (1950), 'Computing Machinery and Intelligence', *Mind*, 59: 433–60; repr. in Boden (1990b: 40–66).

—— (1952), 'The Chemical Basis of Morphogenesis', *Philosophical Transactions of the Royal Society*, ser. B, 237: 37–72.

—— (1953), 'Digital Computers Applied to Games', in Bowden (1953: 286–95).

—— WELCHMAN, W. G., ALEXANDER, C. H. O'D., and MILNER-BARRY, P. S. (1941), 'Letter to Winston Churchill', in Copeland (2004: 338–40).

TURING, S. S. (1959), *Alan M. Turing* (Cambridge: W. Heffer).

TURK, G. (1991), 'Generating Textures on Arbitrary Surfaces Using Reaction-Diffusion', *Computer Graphics*, 25: 289–98.

TURKLE, S. (1984), *The Second Self: Computers and the Human Spirit* (New York: Simon & Schuster).

—— (1995), *Life on the Screen: Identity in the Age of the Internet* (New York: Simon & Schuster).

—— and PAPERT, S. (1990), 'Epistemological Pluralism: Styles and Voices Within the Computer Culture', *Signs: Journal of Women in Culture and Society*, 16/1, 128–57.

TURNER, K. J. (1974), 'Computer Perception of Curved Objects', *Proceedings of the AISB Summer Conference*, Sussex, 238–46.

TURNER, M. (1991), *Reading Minds: The Study of Literature in an Age of Cognitive Science* (Princeton: Princeton University Press).

TURNER, S. R. (1994), *The Creative Process: A Computer Model of Storytelling and Creativity* (Hillsdale, NJ: Lawrence Erlbaum).

TURVEY, M. T., and SHAW, R. E. (1995), 'Toward an Ecological Physics and a Physical Psychology', in R. Solso and D. Massaro (eds.), *The Science of the Mind: 2001 and Beyond* (Oxford: Oxford University Press), 144–69.

—— —— REED, E. S., and MACE, W. M. (1981), 'Ecological Laws of Perceiving and Acting: In Reply to Fodor and Pylyshyn', *Cognition*, 9: 237–304.

TVERSKY, A., and KAHNEMAN, D. (1973), 'Availability: A Heuristic for Judging Frequency and Probability', *Cognitive Psychology*, 5: 207–32.

—— —— (1974), 'Judgment Under Uncertainty: Heuristics and Biases', *Science*, 125: 1124–31.

—— —— (1981), 'The Framing of Decisions and the Psychology of Choice', *Science*, 211: 453–8.

—— —— (1982), 'Evidential Impact of Base Rates', in Kahneman *et al.* (1982: 153–60).

TWAIN, M. (1897), 'Horrors of the German Language', speech to the Vienna Press Club, 21 Nov. 1897, in German with literal translation, available at <http://www.esp.org> (and elsewhere on the web).

TYLER, S. A. (ed.) (1967), *Cognitive Anthropology: Readings* (London: Holt, Rinehart & Winston; reissued 1969).

TYLOR, E. B. (1871), *Primitive Culture: Researches into the Development of Mythology, Philosophy, Religion, Art, and Custom* (London: John Murray).

UHR, L., and VOSSLER, C. (1961*a*/1963), 'A Pattern Recognition Program that Generates, Evaluates, and Adjusts its Own Operators', in Feigenbaum and Feldman (1963: 251–68); a shorter version pub. in *Proceedings of the Western Joint Computer Conference*, 19 (1961), 555–70.

—— —— (1961*b*), 'Recognition of Speech by a Computer Program that Was Written to Simulate a Model for Human Visual Pattern Perception', *Journal of the Acoustical Society of America*, 33: 1426.

ULAM, S. M. (1958), 'John von Neumann 1903–1957', *Bulletin of the American Mathematical Society*, 64/3 (May), 1–49.

—— (1962), 'On Some Mathematical Problems Connected with Patterns of Growth of Figures', *Proceedings of Symposia in Applied Mathematics*, 14: 215–24; repr. in Burks (1970: 219–31).

—— (1989), 'Vita: Extracts from *Adventures of a Mathematician*', in N. G. Cooper (ed.), *From Cardinals to Chaos: Reflections on the Life and Legacy of Stanislaw Ulam* (Cambridge: Cambridge University Press), 8–22.

ULLMAN, S. (1979), *The Interpretation of Visual Motion* (Cambridge, Mass.: MIT Press).

—— (1980), 'Against Direct Perception', *Behavioral and Brain Sciences*, 3: 373–415.

ULRICH, R. S. (1983), 'Aesthetic and Affective Response to Natural Environment', in I. Altman and J. F. Wohlwill (eds.), *Behavior and the Natural Environment* (New York: Plenum Press), 85–126.

—— (1984), 'View Through a Window May Influence Recovery from Surgery', *Science*, 224: 420–1.

—— (1993), 'Biophilia, Biophobia, and Natural Landscapes', in S. R. Kellert and E. O. Wilson (eds.), *The Biophilia Hypothesis* (Washington: Island Press), 74–137.

UMEREZ, J. (2001), 'Howard Pattee's Theoretical Biology—A Radical Epistemological Stance to Approach Life, Evolution, and Complexity', *BioSystems*, 60: 159–77.

URMSON, J. O. (1956), *Philosophical Analysis: Its Development Between the Two World Wars* (Oxford: Clarendon Press).

USSELMANN, R. (2003), 'The Dilemma of Media Art: Cybernetic Serendipity at the ICA London', *Leonardo*, 36: 389–96.

UTTAL, W. R. (2001), *The New Phrenology: The Limits of Localizing Cognitive Processes in the Brain* (Cambridge, Mass.: MIT Press).

UTTLEY, A. M. (1951), 'Les Machines à calculer et la pensée humaine', *Colloque Internationale*, CNRS, Paris, 465–73.

—— (1954), 'The Classification of Signals in the Nervous System', *EEG and Clinical Neurophysiology*, 6: 479–94.

—— (1956), 'Conditional Probability Machines and Conditioned Reflexes', in Shannon and McCarthy (1956: 253–75).

—— (1959*a*), 'Conditional Probability Computing in a Nervous System', in Blake and Uttley (1959: i. 119–47).

—— (1959*b*), 'The Design of Conditional Probability Computers', *Information and Control*, 2: 1–24.

—— (1975), 'The Informon in Classical Conditioning', *Journal of Theoretical Biology*, 49: 355–76.

—— (1979), *Information Transmission in the Nervous System* (London: Academic Press).

UTTLEY, A. M. (1982), *Brain, Mind, and Spirit* (Brighton: University of Sussex, privately published).

VAINA, L. (ed.) (1991), *From the Retina to the Neocortex: Selected Papers of David Marr* (Boston: Birkhauser).

VAISH, A., and STRIANO, T. (2004), 'Is Visual Reference Necessary? Contributions of Facial Versus Vocal Cues in 12-month-olds' Social Referencing Behavior', *Developmental Science*, 7: 261–9.

VALENTINE, E. R. (1989), 'Neural Nets: From Hartley to Hebb and Hinton', *Journal of Mathematical Psychology*, 33: 348–57.

VALLAR, G., and SHALLICE, T. (eds.) (1990), *Neuropsychological Impairments of Short-Term Memory* (Cambridge: Cambridge University Press).

VAN CANEGHEM, M., and WARREN, D. H. (eds.) (1986), *Logic Programming and its Applications* (Norwood, NJ: Ablex).

VAN DER SMAGT, P. (1998), 'Cerebellar Control of Robot Arms', *Connection Science*, 10: 301–20.

VAN DIJK, T. A. (1972), *Some Aspects of Text-Grammars* (The Hague: Mouton).

VAN FRAASSEN, B. C. (1980), *The Scientific Image* (Oxford: Oxford University Press).

VAN GELDER, T. J. (1992/1995), 'What Might Cognition Be, If Not Computation?', *Journal of Philosophy*, 92 (1995), 345–81; first circulated as Cognitive Sciences Technical Report 75, Indiana University (1992).

———— (1997), 'Dynamics and Cognition', in Haugeland (1997: 421–50).

———— (1998), 'The Dynamical Hypothesis in Cognitive Science', *Behavioral and Brain Sciences*, 21: 615–28, 654–66.

———— (2005), 'Enhancing and Augmenting Human Reasoning', in Zilhao (2005: 162–81).

———— and PORT, R. F. (1995), 'It's About Time: An Overview of the Dynamical Approach to Cognition', in Port and van Gelder (1995: 1–44).

VARELA, F. J. (1987), 'Laying Down a Path in Walking', in W. I. Thompson (ed.), *Gaia: A Way of Knowing* (Great Barrington, Mass.: Lindisfarne Press), 48–64.

———— THOMPSON, E., and ROSCH, E. (1991), *The Embodied Mind: Cognitive Science and Human Experience* (Cambridge, Mass.: MIT Press).

VARTANIAN, A. (1960), *L'Homme machine: A Study in the Origins of an Idea*, by Julien Offray de La Mettrie, critical edn. with introductory monograph and notes (Princeton: Princeton University Press).

VAUCANSON, J. DE (1738/1742), *An Account of the Mechanism of an Automaton or Image Playing on the German Flute*, trans. from French, with letter to the Abbé de Fontaine, J. T. Desaguliers (London: Parker); first pub. (Paris: Guerin, 1738); facs. repr. of both (Buren: Fritz Knuf, 1979).

VELMANS, M. (ed.) (1996), *The Science of Consciousness: Psychological, Neuropsychological and Clinical Reviews* (London: Routledge).

———— (2000), *Understanding Consciousness* (London: Routledge/Psychology Press).

———— (2002), 'How Could Conscious Experiences Affect Brains?', *Journal of Consciousness Studies*, 9/11: 3–29.

VERA, A. H., and SIMON, H. A. (1993), 'Situated Action: A Symbolic Interpretation', *Cognitive Science*, 17: 7–48.

VIDAL, F. (1993), 'Psychology in the Eighteenth Century: A View from Encyclopaedias', *History of the Human Sciences*, 6: 89–120.

VON BERTALANFFY, L. (1933/1962), *Modern Theories of Development: An Introduction to General Systems Theory* (New York: Harper); first pub. in German (1933).

———— (1950), 'An Outline of General Systems Theory', *British Journal for the Philosophy of Science*, 1: 134–65.

VON DER MALSBURG, C. (1973), 'Self-Organization of Orientation Sensitive Cells in the Striate Cortex', *Kybernetik*, 14: 85–100; repr. in J. A. Anderson and Rosenfeld (1998: 212–27).

———— (1979), 'Development of Ocularity Domains and Growth Behavior of Axon Terminals', *Biological Cybernetics*, 32: 49–62.

———— and SCHNEIDER, W. (1986), 'A Neural Cocktail-Party Processor', *Biological Cybernetics*, 54: 29–40.

VON DOMARUS, E. (1944), 'The Specific Laws of Logic in Schizophrenia', in J. S. Kasanin (ed.), *Language and Thought in Schizophrenia* (Berkeley: University of California Press, 1944), 135–7.

VON FOERSTER, H. (1950–5), *Cybernetics, Circular Causal, and Feedback Mechanisms in Biological and Social Systems: Published Transactions of the Sixth, Seventh, Eighth, Ninth, and Tenth Conferences*, 5 vols. (New York: Josiah Macy Foundation).

VON NEUMANN, J. (1945), 'First Draft of a Report on the EDVAC', Report prepared for the US Army Ordnance Department; repr. in W. Aspray and A. W. Burks (eds.), *Papers of John von Neumann on Computing and Computer Science* (Cambridge, Mass.: MIT Press, 1987), 17–82.

____ (1951), 'The General and Logical Theory of Automata', in Jeffress (1951: 1–13).

____ (1956), 'Probabilistic Logics and the Synthesis of Reliable Organisms from Unreliable Components', in Shannon and McCarthy (1956: 43–98).

____ (1958), *The Computer and the Brain* (New Haven: Yale University Press).

____ and MORGENSTERN, O. (1944), *Theory of Games and Economic Behavior* (Princeton: Princeton University Press).

VON UEXKÜLL, J. (1926), *Theoretical Biology*, trans. D. L. Mackinnon, International Library of Psychology, Philosophy, and Scientific Method (London: Kegan Paul, Trubner).

____ (1934), *A Stroll Through the Worlds of Animals and Men: A Picture Book of Invisible Worlds*, trans. from German in Schiller (1957: 5–82).

VYGOTSKY, L. S. (1962), *Thought and Language*, ed. and trans. E. Hanfmann and G. Vakar (Cambridge, Mass.: MIT Press); first pub. in Russian (1934), but suppressed 1936–56.

WADDINGTON, C. H. (ed.) (1966–72), *Toward a Theoretical Biology*, 4 vols. (Edinburgh: Edinburgh University Press).

WAGSTAFF, G. F. (1981), *Hypnosis, Compliance, and Belief* (Brighton: Wayland).

WAISMANN, F. (1955), 'Verifiability', in A. G. N. Flew (ed.), *Logic and Language (First Series)* (Oxford: Blackwell), 117–44.

WALD, A. (1950), *Statistical Decision Functions* (New York: Wiley).

WALDE, P., WICK, R., FRESTA, M., MANGONE, A., and LUISI, P. L. (1994), 'Autopoietic Self-Reproduction of Fatty Acid Vesicles', *Journal of the American Chemical Society*, 116: 11649–54.

WALDEN, D. C., and NICKERSON, R. S. (eds.) (2005), *Computing at Bolt, Beranek and Newman: The First 40 Years, Part I, IEEE Annals of the History of Computing*, special issue, 27/2: 4–78 (Apr.–June 2005); pt. II is forthcoming.

WALDROP, M. M. (1992), *Complexity: The Emerging Science at the Edge of Order and Chaos* (New York: Simon & Schuster).

____ (2001), *The Dream Machine: J. C. R. Licklider and the Revolution that Made Computing Personal* (Harmondsworth: Viking Press).

WALKER, E. H. (2000), *The Physics of Consciousness: The Quantum Mind and the Meaning of Life* (Cambridge, Mass.: Perseus).

WALL, P. D. (1974), ' "My Foot Hurts Me", An Analysis of a Sentence', in R. Bellairs and E. G. Gray (eds.), *Essays on the Nervous System* (Oxford: Clarendon Press), 391–406.

____ (1979), 'On the Relation of Injury to Pain', *Pain*, 6: 253–64.

WALLACE, A. F. C. (1965), 'Driving to Work', in M. E. Spiro (ed.), *Context and Meaning in Cultural Anthropology* (London: Collier-Macmillan), 277–96.

WALLACE, I., KLAHR, D., and BLUFF, K. (1987), 'A Self-Modifying Production System Model of Cognitive Development', in Klahr, Langley, and Neches 1987: 359–436.

WALLIN, N. L., MERKER, B., and BROWN, S. (eds.) (2000), *The Origins of Music* (Cambridge, Mass.: MIT Press).

WALLMAN, J. (1992), *Aping Language* (Cambridge: Cambridge University Press).

WALKER, C. C., and ASHBY, W. R. (1966), 'On the Temporal Characteristics of Behavior in Certain Complex Systems', *Kybernetik*, 3: 100–8.

WALTER, G. (1953), *Caesar*, trans. E. Craufurd, 2 vols. (London: Cassell).

WALTER, N. (1990), 'Personal Notes for the *Dictionary of National Biography*', copy given to M.A.B. by Owen Holland.

WALTON, A. (1930), 'Demonstrational and Experimental Devices', *American Journal of Psychology*, 42: 109–15.

WALTON, K. L. (1990), *Mimesis as Make-Believe: On the Foundations of the Representational Arts* (Cambridge, Mass.: Harvard University Press).

WALTZ, D. L. (1975), 'Understanding Line Drawings of Scenes with Shadows', in Winston (1975: 19–92).

____ (1999), 'The Importance of Importance', presidential address to AAAI-98, *AI Magazine* (Fall), 19–35.

WANG, H. (1960), 'Towards Mechanical Mathematics', *IBM Journal of Research Development*, 4: 224–68; reprinted in K. M. Sayre and F. J. Crosson (eds.), *The Modeling of Mind: Computers and Intelligence* (Notre Dame, Ind.: University of Notre Dame Press), 91–120.

WANG, P., BECKER, A. A., JONES, I. A., GLOVER, A. T., BENFORD, S. D., GREENHALGH, C. M., and VLOEBERGHS, M. (forthcoming), 'Virtual Reality Simulation of Surgery with Haptic Feedback Based on the Boundary Element Method', submitted to B. H. V. Topping and C. A. Mota Soares (eds.), *Computers and Structures:* This is a fuller version of a paper published in *Proceedings of the Seventh International Conference on Computational Structures Technology*, Lisbon, Sept. 2004 (Stirling: Civil-Comp Press).

WANNER, E., and MARATSOS, M. P. (1978), 'An ATN Approach to Comprehension', in M. Halle, J. W. Bresnan, and G. A. Miller (eds.), *Linguistic Theory and Psychological Reality* (Cambridge, Mass.: MIT Press), 119–61.

WARD, J. (1902), 'Psychology', in *Encyclopaedia Britannica*, 10th edn.

WARREN, N. E. (ed.) (1977), *Advances in Cross-Cultural Psychology*, 2 vols. (London: Academic Press).

WARRINGTON, E. K. (1975), 'The Selective Impairment of Semantic Memory', *Quarterly Journal of Experimental Psychology*, 27: 635–57.

WASON, P. C. (1966), 'Reasoning', in B. M. Foss (ed.), *New Horizons in Psychology*, i (Harmondsworth: Penguin), 131–51.

——and JOHNSON-LAIRD, P. N. (1972), *Psychology of Reasoning: Structure and Content* (London: Batsford).

WATERMAN, D. A. (1970), 'Generalization Learning Techniques for Automating the Learning of Heuristics', *Artificial Intelligence*, 1: 121–70.

——(1985), *A Guide to Expert Systems* (New York: Addison-Wesley).

WATERS, K., and LEVERGOOD, T. (1995), 'DECface: A System for Synthetic Face Applications', *Multimedia Tools and Applications*, 1: 349–66.

WATKINS, J. W. N. (1997), 'Karl Raimund Popper 1902–1994', *Proceedings of the British Academy*, 94: 645–84.

WATSON, J. B. (1913), 'Psychology as the Behaviorist Views It', *Psychological Review*, 20: 158–77.

——(1924), *Psychology from the Standpoint of a Behaviorist*, 2nd edn. (London: J. P. Lippincott; 1st edn., 1919).

——(1925), *Behaviorism* (London: Kegan Paul, Trench, & Trubner).

——and RAYNER, R. (1920), 'Conditioned Emotional Reactions', *Journal of Experimental Psychology*, 3: 1–14.

WATSON, J. D. (1968), *The Double Helix: A Personal Account of the Discovery of the Structure of DNA* (London: Weidenfeld & Nicolson).

WATSON, P. J., and THORNHILL, R. (1994), 'Fluctuating Asymmetry and Sexual Selection', *Trends in Ecology and Evolution*, 9: 21–5.

WEAVER, W. (1949), 'On Translation', internal memorandum of the Rockefeller Foundation; repr. in W. N. Locke and Booth (1955: 15–23).

WEBB, B. (1994), 'Robotic Experiments in Cricket Phonotaxis', in D. Cliff, P. Husbands, J.-A. Meyer, and S. W. Wilson (eds.), *From Animals to Animats 3*, Proceedings of the Third International Conference on Simulation of Adaptive Behavior (Cambridge, Mass.: MIT Press), 45–54.

——(1996), 'A Cricket Robot', *Scientific American*, 275/6: 94–9.

——(2001a), 'View from the Boundary', *Biological Bulletin*, 200: 184–9.

——(2001b), 'Can Robots Make Good Models of Biological Behaviour?', with peer commentary, *Behavioral and Brain Sciences*, 24: 1033–94.

——(2001c), 'A Spiking Neuron Controller for Robot Phonotaxis', in B. Webb and T. R. Consi (eds.), *Biorobotics: Methods and Applications* (Cambridge, Mass.: MIT Press), 3–20.

——and HARRISON, R. R. (2000) 'Integrating Sensorimotor Systems in a Robot Model of Cricket Behavior', *Sensor Fusion and Decentralised Control in Robotic Systems III*, Society of Photo-Optical Instrumentation Engineers, Boston, 113–24.

——and SCUTT, T. (2000), 'A Simple Latency-Dependent Spiking-Neuron Model of Cricket Phonotaxis', *Biological Cybernetics*, 82: 247–69.

WEBER, A., and VARELA, F. (2002), 'Life After Kant: Natural Purpose and the Autopoietic Foundations of Biological Individuality', *Phenomenology and Cognitive Science*, 1: 97–125.

WEBSTER, G., and GOODWIN, B. C. (1996), *Form and Transformation: Generative and Relational Principles in Biology* (Cambridge: Cambridge University Press).

WEIMER, W. (1986), 'Interview with Walter Weimer', in Baars (1986: 299–315).

WEIR, S. (1974), 'Action Perception', *Proceedings of the Society for Artificial Intelligence and Simulation of Behaviour*, 1: 247–56.

_____ (1987), *Cultivating Minds: A LOGO Casebook* (New York: Harper & Row).

_____ and EMANUEL, R. (1976), *Using LOGO to Catalyse Communication in an Autistic Child*, Research Report 15, Jan. (Edinburgh: Edinburgh University AI Department).

_____ ADLER, M. R., and McLENNAN, M. (1975), *Final Report on Action Perception Project* (Edinburgh: Edinburgh University AI Dept).

WEISER, M. (1991), 'The Computer for the 21st Century', *Scientific American*, 265/3 (Sept.), 94–104.

_____ (1994), 'The World Is Not a Desktop', *Interactions* (New York: ACM Press), 1/1: 7–8.

WEISKRANTZ, L. (1972), 'Behavioural Analysis of the Monkey's Visual Nervous System', *Proceedings of the Royal Society*, ser. B, 182: 427–55.

_____ (1986), *Blindsight: A Case Study and Implications* (Oxford: Oxford University Press).

_____ (1988), 'Some Contributions of Neuropsychology of Vision and Memory to the Problem of Consciousness', in Marcel and Bisiach (1988: 183–99).

_____ (1997), *Consciousness Lost and Found: A Neuropsychological Exploration* (Oxford: Oxford University Press).

_____ WARRINGTON, E. K., SANDERS, M. D., and MARSHALL, J. (1974), 'Visual Capacity in the Hemianopic Visual Field Following a Restricted Occipital Ablation', *Brain*, 97: 709–28.

_____ (1986), *Blindsight: A Case Study and Implications* (Oxford: Oxford University Press).

WEISS, A. P. (1925), 'One Set of Postulates for a Behavioristic Psychology', *Psychological Review*, 32: 83–7.

WEISS, G. (ed.) (1999), *Multiagent Systems: A Modern Approach to Distributed Artificial Intelligence* (Cambridge, Mass.: MIT Press).

WEITZENFELD, A., ARBIB, M. A., and ALEXANDER, A. (2002), *The Neural Simulation Language: A System for Brain Modeling* (Cambridge, Mass.: MIT Press).

WEIZENBAUM, J. (1966), 'ELIZA—A Computer Program for the Study of Natural Language Communication Between Man and Machine', *Communications of the Association for Computing Machinery*, 9: 36–45.

_____ (1972), 'On the Impact of the Computer on Society: How Does One Insult a Machine?', *Science*, 176: 609–14.

_____ (1976), *Computer Power and Human Reason: From Judgment to Calculation* (San Francisco: W. H. Freeman).

WELD, D. S., and DE KLEER, J. (eds.) (1990), *Readings in Qualitative Reasoning About Physical Systems* (Los Altos, Calif.: Morgan Kaufmann).

WELLMAN, H. M. (1990), *The Child's Theory of Mind* (Cambridge, Mass.: MIT Press).

WELLS, J. C. (1982), *Accents of English* (Cambridge: Cambridge University Press).

WERGER, B. B. (1999), 'Cooperation Without Deliberation: A Minimal Behavior-Based Approach to Multi-Robot Teams', *Artificial Intelligence*, 110: 293–320.

WERNALD, J. D., and EDMONDS, E. A. (1985), 'A Three-Dimensional Computer Graphics Workstation', in S. A. R. Scrivener (ed.), *Professional Workstations: State of the Art Report* (Oxford: Pergamon Press), 71–9.

WERTH, P. (1999), *Text Worlds: Representing Conceptual Space in Discourse* (Harlow: Longman).

WERTHEIMER, M. (1945), *Productive Thinking*, ed. S. E. Asch, W. Kohler, and C. W. Mayer, based on publications in German, 1920/1925 (New York: Harper).

WHEELER, M. W. (1996), 'From Robots to Rothko: The Bringing Forth of Worlds', in Boden (1996: 209–36).

_____ (2005), *Reconstructing the Cognitive World: The Next Step* (Cambridge, Mass.: MIT Press).

_____ ZIMAN, J. M., and BODEN, M. A. (eds.) (2002), *The Evolution of Cultural Entities*, Proceedings of the British Academy, 112 (Oxford: Oxford University Press).

_____ HUSBANDS, P., and HOLLAND, O. (eds.) (forthcoming), *The Mechanization of Mind in History* (London: MIT Press).

WHEELER, P. J. (1987), 'SYSTRAN', in M. King (ed.), *Machine Translation Today: The State of the Art* (Edinburgh: Edinburgh University Press), 192–208.

WHEWELL, W. (1833), *Astronomy and General Physics Considered with Reference to Natural Theology*, The First Bridgwater Treatise (London: William Pickering).

WHITBY, B. (1988), *Artificial Intelligence: A Handbook of Professionalism* (Chichester: Ellis Horwood).

_____ (1996a), 'The Turing Test: AI's Biggest Blind Alley?', in Millican and Clark (1996: 53–62); repr. in Whitby (1996b: 35–43).

_____ (1996b), *Reflections on Artificial Intelligence: The Legal, Moral, and Ethical Dimensions* (Exeter: Intellect Books).

WHITBY, B. (2002*a*), 'Futurologists—Don't You Just Love Them?', *AISB Quarterly*, no. 8: 2.

———(2002*b*), 'Let's Stop Throwing Stones', *AISB Quarterly*, no. 109: 1.

———(2003), *The Myth of AI Failure*, CSRP 568 (Brighton: University of Sussex); forthcoming in C. MacAllister (ed.), *Critical Issues in AI* (Amsterdam: Rodopi).

———and GRAND, S. (2001), 'Robots in the Mist—Steve Grand and the "Lucy" Project', *AISB Quarterly*, no. 105 (Spring), 13–23.

WHITE, L. A. (1959*a*), 'The Concept of Evolution in Cultural Anthropology', in B. J. Meggers (ed.), *Evolution and Anthropology: A Centennial Appraisal* (Washington: Anthropological Society of Washington), 106–25.

———(1959*b*), *The Evolution of Culture: The Development of Civilization to the Fall of Rome* (New York: McGraw-Hill).

WHITEHEAD, A. N. (1898), *A Treatise on Universal Algebra: With Applications*, i (Cambridge: Cambridge University Press).

WHITELAW, M. (2004), *Metacreation: Art and Artificial Life* (London: MIT Press).

WHITEN, A. (ed.) (1991), *Natural Theories of Mind: The Evolution, Development and Simulation of Everyday Mindreading*, Symposium on 'The Emergence of Mindreading: Evolution, Development and Second-Order Representations' (Oxford: Blackwell).

WIDROW, B. (1990), 'Thirty Years of Adaptive Neural Networks: Perceptron, Madaline, and Backpropagation', *Proceedings of the Institute of Electronics and Electrical Engineers*, 78: 1415–42.

———and HOFF, M. E. (1960), 'Adaptive Switching Circuits', *Institute of Radio Engineers WESCON Convention Record* (New York: IRE), 96–104; repr. in J. A. Anderson and Rosenfeld (1988: 126–34).

WIENER, N. (1947), Letter to Grey Walter, 12 June, copy given to M.A.B. by Owen Holland.

———(1948), *Cybernetics: or Control and Communication in the Animal and the Machine* (Cambridge, Mass.: MIT Press); printed offset from the 1948 French edn. (Paris: Hermann).

———(1950), *The Human Use of Human Beings: Cybernetics and Society* (Boston: Houghton Mifflin).

———(1961), 'Introduction to the Second Edition', and 'Supplementary Chapters', in N. Wiener, *Cybernetics: or Control and Communication in the Animal and the Machine*, 2nd edn., enlarged (Cambridge, Mass.: MIT Press).

WILCOX, S. K. (1998), *Web-Developer.Com: Guide to 3D Avatars* (Chichester: Wiley).

WILKES, K. V. (1988), '———, Yishi, Duh, Um, and Consciousness', in Marcel and Bisiach (1988: 16–41).

WILKES, M. V. (1953), 'Calculating Machine Development at Cambridge', in Bowden (1953: 130–4).

———(1967), 'Computers Then and Now', in *ACM Turing Award Lectures: The First Twenty Years* (New York: ACM Press), 197–205.

———(1971), 'Babbage as a Computer Pioneer', *Report of Proceedings, Babbage Memorial Meeting, London, 18th October 1971* (London: British Computer Society), 1–18; also pub. in *Historia Mathematica*, 4 (1977), 415–40.

———(1982), 'J. G. Brainerd on the ENIAC', *Annals of the History of Computing*, 4: 53–5.

———(1991), 'Babbage's Expectations for his Engines', *Annals of the History of Computing*, 13/2: 141–5.

WILKINS, D. E. (1988), *Practical Planning: Extending the Classical AI Planning Paradigm* (San Mateo, Calif.: Morgan Kaufmann).

WILKINS, J. (1668), *An Essay Towards a Real Character, and a Philosophical Language* (Menston: Scolar Press, 1968).

WILKS, Y. A. (1972), *Grammar, Meaning, and the Machine Analysis of Natural Language* (London: Routledge & Kegan Paul).

———(1975), 'A Preferential, Pattern-Seeking, Semantics for Natural Inference', *Artificial Intelligence*, 6: 53–74.

———(1976), 'Dreyfus's Disproofs', *British Journal for the Philosophy of Science*, 27: 177–85.

———(1978), 'Good and Bad Arguments for Semantic Primitives', *Communication and Cognition*, 10: 181–221.

———(ed.) (forthcoming), *Language, Cohesion and Form: Margaret Masterman* (Cambridge: Cambridge University Press).

———(in preparation), *Virtual Persons* (provisional title).

WILLIAMS, B. A. O. (1978), *Descartes: The Project of Pure Enquiry* (Hassocks: Harvester Press).

WILLIAMS, T. I. (1982), *A Short History of Twentieth-Century Technology (c.1900-c.1950)* (Oxford: Clarendon Press).

WILLSHAW, D. J. (1981), 'Holography, Associative Memory, and Inductive Generalization', in Hinton and Anderson (1981: 83–104).

—— and BUCKINGHAM, J. T. (1990), 'An Assessment of Marr's Theory of the Hippocampus as a Temporary Memory Store', *Philosophical Transactions of the Royal Society*, ser. B, 329: 205–15.

—— and LONGUET-HIGGINS, C. H. (1969a), 'The Holophone—Recent Developments', in Meltzer and Michie (1969a: 349–57).

—— —— (1969b), 'Associative Memory Models', in Meltzer and Michie (1969b: 351–9).

—— and VON DER MALSBURG, C. (1976), 'How Patterned Neural Connections Can Be Set Up by Self-Organization', *Proceedings of the Royal Society*, ser. B, 194: 431–45.

—— BUNEMAN, O. P., and LONGUET-HIGGINS, C. H. (1969), 'Non-Holographic Associative Memory', *Nature*, 222: 960–2.

WILSON, A. M. (1972), *Diderot* (New York: Oxford University Press).

WILSON, E. A. (1998), *Neural Geographies: Feminism and the Microstructure of Cognition* (London: Routledge).

WILSON, E. O. (1984), *Biophilia* (Cambridge, Mass.: Harvard University Press).

—— (1993), 'Biophilia and the Conservation Ethic', in S. R. Kellert and E. O. Wilson (eds.), *The Biophilia Hypothesis* (Washington: Island Press), 31–41.

WILSON, R. A., and KEIL, F. C. (eds.) (1999), *MIT Encyclopedia of the Cognitive Sciences* (Cambridge, Mass.: MIT Press).

WILTSHIRE, J. (1991), *Samuel Johnson in the Medical World: The Doctor and the Patient* (Cambridge: Cambridge University Press).

WIMMER, H., and PERNER, H. (1983), 'Beliefs About Beliefs: Representations and Constraining Function of Wrong Beliefs in Young Children's Understanding of Deception', *Cognition*, 13: 103–28.

WINCH, P. G. (1958), *The Idea of a Social Science: And its Relation to Philosophy* (London: Routledge & Kegan Paul).

WINOGRAD, T. (1972), *Understanding Natural Language* (Edinburgh: Edinburgh University Press); rev. version of *Procedures as a Representation for Data in a Computer Program for Understanding Natural Language*, AI-TR-17 (Cambridge, Mass.: MIT AI Lab, 1971).

—— (1973), 'A Procedural Model of Language Understanding', in Schank and Colby (1973: 152–86).

—— (1974), *Five Lectures on Artificial Intelligence*, paper presented to the Electrotechnical Laboratory in Tokyo, 18–23 Mar. 1974, Stanford AI Lab Memo AIM-246 (Stanford, Calif.: Stanford University Computer Science Department).

—— (1975), 'Frame Representations and the Declarative/Procedural Controversy', in Bobrow and Collins (1975: 185–210).

—— (1976), 'Artificial Intelligence and Language Comprehension', in T. Winograd, I. Goldstein, S. Papert, M. L. Minsky, and Y. A. Wilks, *Artificial Intelligence and Language Comprehension* (Washington: National Institute of Education), 1–26.

—— (1977), 'On Some Contested Suppositions of Generative Linguistics About the Scientific Study of Language', *Cognition*, 5: 151–79.

—— (1980a), 'What Does It Mean to Understand Language?', *Cognitive Science*, 4: 209–42.

—— (1980b), 'Extended Inference Modes in Reasoning by Computer Systems', *Artificial Intelligence*, 13: 5–26.

—— (1983), *Language as a Cognitive Process*, i: *Syntax* (Reading, Mass.: Addison-Wesley).

—— (1991), 'Strategic Computing Research and the Universities', in C. Dunlop and R. Kling (eds.), *Computerisation and Controversy: Value Conflicts and Social Choices* (Boston: Academic Press), 704–16.

—— and FLORES, F. (1986), *Understanding Computers and Cognition: A New Foundation for Design* (Norwood, NJ: Ablex).

WINSTON, P. H. (1970a), *Learning Structural Descriptions from Examples*, MIT AI Lab Memo AI-TR-231 (Cambridge, Mass.: MIT); rev. version in Winston (1975: 157–210).

—— (1970b), *Holes*, MIT AI Lab Memo 163 (Cambridge, Mass.: MIT).

—— (1972), 'The MIT Robot', in Meltzer and Michie (1972: 431–63).

—— (ed.) (1975), *The Psychology of Computer Vision* (New York: McGraw-Hill).

—— (1977), *Artificial Intelligence* (Reading, Mass.: Addison-Wesley).

—— and BRADY, M. (1979), 'Series Forward', in P. H. Winston and R. H. Brown (eds.), *Artificial Intelligence: An MIT Perspective*, i (Cambridge, Mass.: MIT Press), pp. ix–x.

WINSTON, P. H. and PRENDERGAST, K. A. (eds.) (1984), *The AI Business: Commercial Uses of Artificial Intelligence* (Cambridge, Mass.: MIT Press).

WINTERSTEIN, D., BUNDY, A., and GURR, C. A. (2004), 'Dr Doodle: A Diagrammatic Theorem Prover', in D. Basin and M. Rusinowitch (eds.), *Automated Reasoning* (Heidelberg: Springer-Verlag), 331–5.

WISDOM, A. J. T. D. (1944), 'Gods', *Proceedings of the Aristotelian Society*, 45: 185–206.

——— (1952), 'Ludwig Wittgenstein 1934–1937', *Mind*, 61: 258–60.

WISDOM, J. O. (1951), 'The Hypothesis of Cybernetics', *British Journal for the Philosophy of Science*, 2: 1–24.

WITHINGTON, P. T. (1991), *The LISP Machine: Noble Experiment or Fabulous Failure?*, draft paper, 11 July, <http://pt.withy.org.publications>.

WITTGENSTEIN, L. (1922), *Tractatus Logico-Philosophicus*, trans. C. K. Ogden and F. Ramsey (London: Routledge).

——— (1933), Letter to the Editor, *Mind*, 42: 415–16.

——— (1953), *Philosophical Investigations*, trans. G. E. M. Anscombe (Oxford: Blackwell).

WOLFRAM, S. (1983), 'Statistical Mechanics of Cellular Automata', *Review of Modern Physics*, 55: 601–44.

——— (1984a), 'Cellular Automata as Models of Complexity', *Nature*, 311: 419–24.

——— (1984b), 'Universality and Complexity in Cellular Automata', *Physica D*, 10: 1–35.

——— (1986), *Theory and Applications of Cellular Automata* (Singapore: World Scientific).

WOLPERT, D. M., and GHAHRAMANI, Z. (2000), 'Computational Principles of Movement Neuroscience', *Nature Neuroscience Supplement*, 3: 1212–17.

——— ——— and JORDAN, M. I. (1995), 'An Internal Model for Sensorimotor Integration', *Science*, 269: 1880–2.

——— ——— and FLANAGAN, J. R. (2001), 'Perspectives and Problems in Motor Learning', *Trends in Cognitive Sciences*, 5: 487–94.

WOLPERT, L. (1992), *The Unnatural Nature of Science* (London: Faber & Faber).

WOOD, G. (2002), *Living Dolls: A Magical History of the Quest for Mechanical Life* (London: Faber).

WOOD, J., and GARTHWAITE, J. (1994), 'Model of the Diffusional Spread of Nitric Oxide: Implications for Neural Nitric Oxide Signalling and its Pharmacological Properties', *Neuropharmacology*, 33: 1235–44.

WOODGER, J. H. (1929), *Biological Principles: A Critical Study* (London: Routledge).

——— (1937), *The Axiomatic Method in Biology* (Cambridge: Cambridge University Press).

WOODS, S., DAUTENHAHN, K., and SCHULZ, J. (2004), 'The Design Space of Robots: Investigating Children's Views', *Proceedings of IEEE Ro-Man*, Kurashiki, Okayama, Japan (New York: IEEE Press), 47–52.

——— ——— ——— (2005), 'Child and Adults' Perspectives on Robot Appearance', *Proceedings of the AISB-05 Symposium on Robot Companions: Hard Problems and Open Challenges in Robot–Human Interaction*, Hatfield, Herts., 126–32.

WOODS, W. A. (1970), 'Transition Network Grammars for Natural Language Analysis', *Communications of the Association for Computing Machinery*, 13: 591–606.

——— (1973), 'An Experimental Parsing System for Transition Network Grammars', in R. Rustin (ed.), *Natural Language Processing* (New York: Algorithmics Press), 111–54.

——— (1975), 'What's in a Link: Foundations for Semantic Networks', in Bobrow and Collins (1975: 35–82).

——— (1978), 'Semantics and Quantification in Natural Language Question Answering', in M. Yovits (ed.), *Advances in Computers*, xvii (New York: Academic Press), 2–64.

WOOLDRIDGE, M. J. (1999), 'Intelligent Agents', in G. Weiss (1999: 27–78).

——— (2001), *An Introduction to Multiagent Systems* (Chichester: Wiley).

——— and JENNINGS, N. R. (1995), 'Intelligent Agents: Theory and Practice', *Knowledge Engineering Review*, 10: 115–52.

WOS, L., and VEROFF, R. (1987), 'Resolution, Binary: Its Nature, History, and Impact on the Use of Computers', in S. C. Shapiro (ed.), *Encyclopedia of Artificial Intelligence*, ii (Chichester: Wiley), 892–902.

——— CARSON, D., and ROBINSON, G. A. (1964), 'The Unit Preference Strategy in Theorem Proving', *Proceedings of the Fall Joint Computer Conference, 1964* (New York: Thompson Book Co.), 615–21.

——— ——— ——— (1965), 'Efficiency and Completeness of the Set of Support Strategy in Theorem Proving', *Journal of the Association for Computing Machinery*, 12: 536–41.

WRIGHT, I. P. (1997), *Emotional Agents*, Ph.D. thesis, School of Computer Science, University of Birmingham, available at <http://www.cs.bham.ac.uk/research/cogaff/>.

_____ and SLOMAN, A. (1997), *MINDER1: An Implementation of a Protoemotional Agent Architecture*, Technical Report CSRP-97-1, University of Birmingham, School of Computer Science, available at <ftp://ftp.cs.bham.ac.uk/pub/tech-reports/1997/CSRP-97-01. ps.gz>.

_____ _____ and BEAUDOIN, L. P. (1996), 'Towards a Design-Based Analysis of Emotional Episodes', *Philosophy, Psychiatry, and Psychology*, 3: 101–37.

WU, C. H. (1984), 'Electric Fish and the Discovery of Animal Electricity', *American Scientist*, 72: 598–607.

WUENSCHE, A. (1997), *Attractor Basins of Discrete Networks: Implications on Self-Organisation and Memory*, D.Phil. thesis, Cognitive Science Research Paper 461 (Brighton: University of Sussex).

_____ (1999), 'Classifying Cellular Automata Automatically: Finding Gliders, Filtering, and Relating Space-Time Patterns, Attractor Basins, and the Z Parameter', *Complexity*, 4/3: 47–66.

_____ and LESSER, M. J. (1992), *The Global Dynamics of Cellular Automata*, Santa Fe Institute Studies in the Sciences of Complexity (Reading, Mass.: Addison-Wesley).

WUNDT, W. (1916), *Elements of Folk Psychology*, trans. from German E. L. Schaub (London: Allen & Unwin; first pub. 1912).

YAKIMOVSKY, Y., and FELDMAN, J. A. (1973), 'A Semantics-Based Decision Theory Region Analyser', *Proceedings of the Third International Joint Conference on Artificial Intelligence*, Los Angeles, 580–8.

YARBUS, A. L. (1967), *Eye Movements and Vision*, trans. from Russian Basil Haigh (New York: Plenum Press; first pub. 1965).

YATES, F. A. (1982), *Lull and Bruno: Collected Essays*, i, ed. J. N. Hillgarth and J. B. Trapp (London: Routledge & Kegan Paul).

YEDIDIA, J., FREEMAN, W., and WEISS, Y. (2000), 'Generalized Belief Propagation', *Advances in Neural Information Processing Systems*, 13: 689–95.

YNGVE, V. H. (1955), 'Syntax and the Problem of Multiple Meaning', in W. N. Locke and Booth (1955: 208–26).

_____ (1958), 'A Programming Language for Mechanical Translation', *Mechanical Translation*, 5: 25–41.

_____ (1960), 'A Model and an Hypothesis for Language Structure', *Proceedings of the American Philosophical Society*, 104: 444–66.

_____ (1961), 'Random Generation of English Sentences', *Proceedings of the International Conference on Mechanical Translation and Applied Language Analysis*, NLP, Teddington, Middx. (London: HMSO), 1: 65–80.

_____ (1962*a*), *An Introduction to COMIT Programming* (Cambridge, Mass.: MIT Press).

_____ (1962*b*), *COMIT Programmers' Reference Manual* (Cambridge, Mass.: MIT Press).

_____ (1962*c*), 'Computer Programs for Translation', *Scientific American*, 206 (June), 68–76.

_____ (1967), 'MT at M.I.T. 1965', in Booth (1967: 451–523).

_____ (1972), *Computer Programming with COMIT II* (Cambridge, Mass.: MIT Press).

_____ (1986), *Linguistics as a Science* (Bloomington: Indiana University Press).

YOUNG, A. W., and BURTON, A. M. (1999), 'Simulating Face Recognition: Implications for Modeling Cognition', *Cognitive Neuropsychology*, 16: 1–48.

_____ PHILLIPS, M. L., SENIOR, C., BRAMMER, M., ANDREW, C., CALDER, A. J., BULLMORE, E. T., PERRETT, D. I., ROWLAND, D., WILLIAMS, S. C. R., GRAY, J. A., and DAVID, A. S. (1997), 'A Specific Neural Substrate for Perceiving Facial Expressions of Disgust', *Nature*, 389: 494–8.

YOUNG, J. Z. (1964), *A Model of the Brain* (Oxford: Clarendon Press).

_____ (1978), *Programs of the Brain*, The Gifford Lectures, 1975–7 (Oxford: Oxford University Press).

YOUNG, M. P. (ed.) (2000), *Brain-Structure–Function Relationships: Advances from Neuroinformatics* (London: Royal Society), *Philosophical Transactions of the Royal Society*, ser. B, special issue, 355: 1–161.

_____ HILGETAG, C-C., and SCANNELL, J. W. (2000), 'On Imputing Function to Structure from the Behavioral Effects of Brain Lesions', *Philosophical Transactions of the Royal Society*, ser. B, 355: 147–61.

YOUNG, RICHARD M. (1974), 'Production Systems as Models of Cognitive Development', *Proceedings of the First AISB Summer Conference*, University of Edinburgh, July 1976, 284–95.

_____ (1976), *Seriation by Children: An Artificial Intelligence Analysis of a Piagetian Task* (Basel: Birkhauser).

_____ PLOTKIN, G. D., and LINZ, R. F. (1977), 'Analysis of an Extended Concept-Learning Task', *Proceedings of the Fifth International Joint Conference on Artificial Intelligence*, Cambridge, Mass., 285.

YOUNG, ROBERT M. (1967), 'The Development of Herbert Spencer's Concept of Evolution', *Actes du XIe Congrès international d'histoire des sciences* (Warsaw: Ossolineum), 2: 273–8.

_____ (1970), *Mind, Brain, and Adaptation in the Nineteenth Century: Cerebral Localization and its Biological Context from Gall to Ferrier* (Oxford: Clarendon Press).

Yovits, M. C., Jacobi, G. T., and Goldstein, G. D. (eds.) (1962), *Self-Organizing Systems* (Washington: Spartan Books).

Yu, V. L., Fagan, L. M., Wraith, S. M., Clancey, W. J., Scott, A. C., Hannigan, J. F., Blum, R. L., Buchanan, B. G., and Cohen, S. N. (1979), 'Antimicrobial Selection by a Computer—A Blinded Evaluation by Infectious Disease Experts', *Journal of the American Medical Association*, 242: 1279–82.

Zadeh, L. A. (1965), 'Fuzzy Sets', *Information and Control*, 8: 338–53.

—— (1972), 'Fuzzy Languages and their Relation to Human and Machine Intelligence', *Proceedings of the International Conference on Man and Computer*, Bordeaux (Basel: S. Karger), 130–65.

—— (1975), 'Fuzzy Logic and Approximate Reasoning', *Synthese*, 30: 407–28.

Zawidzki, T. W. (1998), 'Competing Models of Stability in Complex Evolving Systems', *Biology and Philosophy*, 13: 541–54.

Zangwill, O. L. (1980), 'Kenneth Craik: The Man and his Work', *British Journal of Psychology*, 71: 1–16.

Zeki, S. M. (1993), *A Vision of the Brain* (Oxford: Blackwell).

Zelený, M. (1977), 'Self-Organization of Living Systems: A Formal Model of Autopoiesis', *International Journal of General Systems*, 4: 13–28.

—— (ed.) (1981), *Autopoiesis: A Theory of Living Organization* (New York: Elsevier).

—— Klir, G. J., and Hufford, K. D. (1989), 'Precipitation Membranes, Osmotic Growths, and Synthetic Biology', in Langton (1989: 125–39).

Ziff, P. (1959), 'The Feelings of Robots', *Analysis*, 19: 64–8.

Zilhão, A. J. T. (ed.), *Cognition, Evolution, and Rationality: A Cognitive Science for the Twenty-First Century* (London: Routledge).

Ziman, J. M. (2000a), *Real Science: What It Is, and What It Means* (Cambridge: Cambridge University Press).

—— (2000b), 'Evolutionary Models for Technological Change', in Ziman (2000c: 3–12).

—— (ed.) (2000c), *Technological Innovation as an Evolutionary Process* (Cambridge: Cambridge University Press).

—— Wheeler, M. W., and Boden, M. A. (eds.) (2002), *The Evolution of Cultural Entities*, Proceedings of the British Academy (Oxford: Oxford University Press).

Zinn, M., Khatib, O., Roth, B., and Salisbury, J. K. (2002), 'A New Actuation Approach for Human Friendly Robot Design', *International Journal of Robotics Research*, 23: 379–98; first pub. in *Proceedings of the International Symposium on Experimental Robotics*, Italy, July 2002.

Zivanovic, A. (2005), 'The Development of a Cybernetic Sculptor: Edward Ihnatowicz and the Senster', in Candy (2005: 102–8).

Zubek, J. P. (ed.) (1969), *Sensory Deprivation: Fifteen Years of Research* (New York: Appleton-Century-Crofts).

Zuse, K. (1969/1970), *Calculating Space*, MIT Technical Translation AZT-70-164-GEMIT (Cambridge, Mass.: MIT, 1970); first pub. in German (1969).

—— (1993), *The Computer—My Life*, trans. from German P. McKenna and J. A. Ross (London: Springer); first pub. as *Computer, mein Lebenswerk*, autobiography (Munich: Verlag Moderne Industrie, 1970; rev. 1984).

Zwaan, R. A. (1999), 'Embodied Cognition, Perceptual Symbols, and Situation Models', *Discourse Processes*, 28: 81–8.

Zytkow, J. (ed.) (1997), *Machine Discovery* (London: Kluwer Academic); also available as a special issue of *Foundations of Science*, 1/2 (1995–6).

LIST OF ABBREVIATIONS

AAA	American Anthropological Association
AAAI	American Association for Artificial Intelligence
ABC	Agent-Based Computing
ABSET	Aberdeen set-theory programming language
ABSYS	Aberdeen compiler system
ACE	Automatic Computing Engine
ACT	adaptive control of thought
ACTF	a version of ACT, in a series named by successive letters of the alphabet
Adaline	Adaptive Linear Element
ADIOS	Automatic Distillation of Structure
AI	artificial intelligence
AISB	Society for the Study of Artificial Intelligence and Simulation of Behaviour (UK/Europe-based)
ALICE	Algorithmically Integrated Composing Environment
ALPAC	Automatic Language Processing Advisory Committee (of the US government)
AM	Automated Mathematician
AMG	Architecture Machine Group (the MIT group that preceded the Media Lab)
AN	auditory neurone
APA	American Psychological Association
APU	Applied Psychology Unit (University of Cambridge)
ARPA	Advanced Research Programs Agency (of the DOD)
ASL	American Sign Language
ATF	axon transfer factor
ATN	augmented transition network
BBC	British Broadcasting Corporation
BBN	Bolt, Beranek & Newman, now called BBC Technologies
BBS	*Behavioral and Brain Sciences*
BGA	Bruner, Goodnow, and Austin (1956)
BNA	British Nationality Act
BPS	British Psychological Society
BR	belief representation
BVSR	Blind Variation and Selective Retention
C3	command, control, and communications
CA	cellular automaton
CAD/CAM	computer-aided design/manufacture
CAS	Center for Adaptive Systems (Boston University)
CASTE	Course Assembly System and Tutorial Environment
CBC	Computer Based Consultant
CD	conceptual dependency
CDU	cathode-display unit
CF	context-free
CFPS	context-free phrase-structure
CFPSG	context-free phrase-structure grammar
CHIP	Center for Human Information Processing (UCSD; now renamed Center for Brain Cognition)
CIA	Central Intelligence Agency (USA)
CLRU	Cambridge Language Research Unit
CLS	Concept Learning System
CMAC	Cerebellar Model Arithmetic Computer

CMU	Carnegie Mellon University
CNE	computational neuro-ethology
CNM	Connectionist Navigational Map
CNS	central nervous system
COGS	Cognitive Studies Programme/School of Cognitive and Computing Studies (University of Sussex)
CORA	Conditioned Reflex Analogue
CPSR	Computer Professionals for Social Responsibility
CQG	complete quantum gravity
CREA	Centre de recherche en épistémologie appliquée (Paris)
CRT	cathode-ray tube
CS	context-sensitive
CSLI	Center for the Study of Language and Information (Stanford, Calif.)
CSPSG	context-sensitive phrase-structure grammar
CTR(N)N	continuous-time recurrent (neural) network
CUNY	City University of New York
DARPA	Defense Advanced Research Programs Agency (USA) (formerly ARPA)
DCB	Direct-Conflict-Brother
DEC	Digital Equipment Corporation
det (*or* Det)	determiner
DNA	deoxyribonucleic acid
DOD	(US) Department of Defense
DOG	Difference of Gaussians
DPA	Darwinian psychological anthropology
DTF	dendrite transfer factor
EAM	Elementary Adaptive Mechanism
EDSAC	Electronic Delay Storage Automatic Calculator
EDVAC	Electronic Discrete Variable Automatic Computer
EE	evolutionary epistemology
EEG	electroencephalograph
ELMER	Electro-Mechanical Robot
ELSIE	Electro-Mechanical Robot, Light-Sensitive with Internal and External Stability
EM	external memory
EMCSR	European Meeting on Cybernetics and Systems Research
EMI	Experiments in Musical Intelligence (now Emmy)
ENIAC	Electronic Numerical Integrator and Computer
EPAM	Elementary Perceiver and Memorizer
EPSRC	Engineering and Physical Sciences Research Council (UK)
ESRI	elementary self-replicating instructions
FAHQT	fully automatic high-quality translation
FANs	folding architecture networks
FAP	fixed action pattern
FAQ	frequently asked question
FARG	Fluid Analogies Research Group
FFA	fixed functional architecture
fMRI	functional magnetic resonance imaging
FORTRAN	Formula Translation programming language
FRAME	Fund for the Replacement of Animals in Medical Experiments
FREDDY	Friendly Robot for Education, Discussion and Entertainment, the Retrieval of (FREDERICK) Information, and the Collation of Knowledge
FRS	Fellow of the Royal Society
FSG	finite-state grammar
GA	genetic algorithm
GB	government-binding
GEB	Hofstadter, *Gödel, Escher, Bach: An Eternal Golden Braid* (1979)
GNP	gross national product

GOFAI	good old-fashioned AI
GOFAIR	good old-fashioned AI and robotics
GPS	General Problem Solver
GSIA	Graduate School of Industrial Administration (Carnegie Mellon University)
GSOH	good sense of humour
GSPG	generalized phrase-structure grammar
HAM	human associative memory
H&H	Hunt and Hovland
HCI	human–computer interaction
HOT	higher-order thought
HPSG	head-driven phrase-structure grammar
HRR	holographic reduced representations
IAAI	'Innovative Applications of AI'
IAS	Institute of Advanced Study (Princeton)
IBE	inference to the best explanation
ICA	Institute of Contemporary Arts (London)
IEEE	Institute of Electronics and Electrical Engineers (international)
IFIP	International Federation for Information Processing
IFIPS	International Federation of Information Processing Societies
IJCAI	International Joint Conference on Artificial Intelligence
IJCNN	International Joint Conference on Neural Networks
IKBS	Intelligent Knowledge-Based Systems
IMHO	in my humble opinion
INNS	International Neural Network Society
IPL	Information Processing Language
IR	information retrieval
IRCAM	Institut de recherche et coordination acoustique/musique
IRE	Institute of Radio Engineers
IRM	Innate Releasing Mechanism
IRMA	Innate Releasing Mechanism Analogue
ISA	ideal sentential automaton
IT	information technology
JASSS	*Journal of Artificial Societies and Social Simulation*
K&T	Kahneman and Tversky
KIPS	knowledge information processing systems
KMB	Kilmer, McCulloch, and Blum
KR	knowledge representation
LAC	Lattice Artificial Chemistry
LBP	loopy belief propagation
Leabra	Local, Error-driven and Associative, Biologically Realistic Algorithm
LFG	lexical-functional grammar
LGN	lateral geniculate nucleus
LISP	list-processing language
LMS	least mean square
LOT	language of thought
LRAAM	labelling recursive auto-associative memories
LRB	*London Review of Books*
LT	Logic Theory Machine, or Logic Theorist
LTM	long-term memory
MAC	Multi-Access Computing (time sharing); Man and Computers; Machine-Aided Cognition
MACT	*Mathematical Anthropology and Culture Theory*
Madaline	a group of Many Adalines
MADM	Manchester Automatic Digital Machine
M&V	Maturana and Varela
MARIAL	Center for Myth and Ritual in American Life

MCC	Microelectronics and Computer Technology Corporation
MENACE	Matchbox Educable Noughts And Crosses Engine
MGP	Miller, Galanter, and Pribram (1960)
MIT	Massachusetts Institute of Technology
MITECS	*MIT Encyclopedia of the Cognitive Sciences*
MM	massive modularity
MMM	moderately massive modularity
MMORPG	Massively Multi-Player Online Role-Playing Game
MMPI	Minnesota Multiphasic Personality Inventory
MN	motor neurone
MOMA	Museum of Modern Art (New York)
MOP	memory organization packet
MRC	Medical Research Council (UK)
MT	machine translation
MTCLA	Machine That Can Learn Anything
MUD	Multi-User Dungeon
NASA	National Aeronautics and Space Administration (USA)
NCST	National Centre for Software Technology (Bombay)
NE	neuro-ethology
NEA	National Endowment for the Arts (USA)
NEH	National Endowment for the Humanities (USA)
NERISSA	Nerve Excitation, Inhibition, and Synaptic Analogue
NewFAI	newfangled AI
NIH	National Institutes of Health (USA)
NIPS	Neural Information Processing Systems
NK	network with N units, each with K inputs
NLP	natural language processing
NLS	online system
NOAH	Nets of Action Hierarchies
NOMAD	Neurally Organized Multiply/Mobile Adaptive Device
NORAD	North Atlantic Defence
NP	noun phrase
NPL	National Physical Laboratory (London)
NSF	National Science Foundation (USA)
NSL	Neural Simulation Language
NYU	New York University
OED	*Oxford English Dictionary*
ONR	Office of Naval Research (USA)
OPEC	Organization of Petroleum Exporting Countries
OPS	Official Production-System language
P&W	Premack and Woodruff
PARC	Palo Alto Research Center
PC	personal computer
PCB	Prerequisite-Conflict-Brothers
PCBG	Prerequisite-Clobbers-Brother-Goal
PDP	parallel distributed processing
PET	positron emission tomography
PM	Prerequisite-Missing
PNA	polyamide-linked nucleic acid
PROLOG	Programming in Logic
PS	production systems
PSG	control structure for production systems: see Newell (1973*a*) (The "G" is alphabetical: the 7th, and last, in the series)
PSS	Physical Symbol System
PURR-PUSS	Purposeful Unprimed Real-world Robot with Predictors Using Short Segments
RAAM	recursive auto-associative memory

RAF	Royal Air Force
RAND	Research And Defence Corporation (USA)
REM	rapid eye movement
RLE	Research Lab of Electronics (MIT)
RN	residual normality
RNA	ribonucleic acid
RR	representational redescription
RRE	Royal Radar Establishment
RT	run and twiddle
RTF	reticular formation
SAGA	Species Adaptation GA
SAGE	Semi-Automatic Ground Environment
SAIL	Stanford AI Laboratory
SAM	Sound Activated Mobile
SARA	Simple Analytic Recombinant Algorithm
SASci	Society for Anthropological Sciences
SCB	Strategy-Clobbers-Brother
SCI	Strategic Computing Initiative
SDI	Strategic Defense Initiative
SERC	Science and Engineering Research Council (UK)
SETI	Search for Extra-Terrestrial Intelligence
SFI	Santa Fe Institute
SFN	Society for Neuroscience
SIGART	Special Interest Group on Artificial Intelligence (of the ACM)
SIP	semantic information processing
SNARC	Stochastic Neural-Analog Reinforcement Calculator
SOAR	Success Oriented Achievement Realized
SPA	Society for Psychological Anthropology
S–R	stimulus–response
SRC	Science Research Council (UK)
SRI	Stanford Research Institute
STeLLA	Standard Telecommunications Laboratories Learning Automaton
STEVE	SOAR Training Expert for Virtual Environments
STM	short-term memory
STRIPS	Stanford Research Institute Problem Solver
SUNY	State University of New York
SYCO	*Syritta computatrix* (simulated hoverfly)
SYNICS	Syntax and Semantics programming language
TAU	thematic abstraction unit
TG	transformational grammar
THES	*Times Higher Education Supplement*
TINLAP	Theoretical Issues in Natural Language Processing
ToM	Theory of Mind
TOP	thematic organization point
TOTE	Test–Operate–Test–Exit
TPE	temporal propositional expression
TREAC	Telecommunications Research Establishment Automatic Computer (UK)
TT	Turing Test
TtB	Take the Best (Ignore the Rest)
TTT	Total Turing Test
TTTT	Total Total Turing Test
UC	University of California
*UCI	unit of cultural instruction
UCLA	University of California at Los Angeles
UCSD	University of California at San Diego
Umass Amherst	University of Massachusetts, Amherst

US	United States
UWE	University of the West of England
VDU	visual-display unit
VLSI	Very Large-Scale Integration
VP	verb phrase
VR	virtual reality
WHO	World Health Organization
XOR	exclusive or (i.e. p or q, but not both)

SUBJECT INDEX

NAME INDEX

Aarsleff, H. 596, 617
Abelson, H. 820
Abelson, R. P. 357, 376, 377–80, 381, 415, 690, 691, 692, 1009; *see also* Rosenberg et al 376
Ackley, D. H. 951
Adams, F. 667
Addison, J. 65, 91
Adrian, E. D. 116, 118, 216, 223
Agre, P. E. 1032–4, 1036, 1039, 1422–3
Aiken, H. 155, 219–20, 825, 1447
Albertus Magnus 57
Albus, J. S. 1151
Aleksander, I. 1220
Allen, J. F. 747
Allport, D. A. 512
Allport, F. H. 304
Allport, G. W. 246, 298, 307, 416, 424
Amarel, S. 712, 759, 1067
Amari, S.-I. 48, 939, 961, 980
Ambler, P. 868
Ames, J.: *see* Psujek et al 1330–1, 1332
Ampère, A. 200
Anderson, J. A. 930, 931, 935, 938, 940, 946, 947, 952, 960, 1166, 1206, 1208
Anderson, J. R. 427, 435–8, 455, 503, 509–10, 1000
Andreae, J. H. 926–8
Anscombe, G. E. M. 303, 1364, 1388–9
Apostel, L. 364
Apter, J. T. 889
Aquinas, T. 57, 69, 575, 577, 582
Arak, A. 553
Arbib, M. A. 190, 193, 363–4, 434, 980, 1140, 1166, 1167, 1211–12, 1215, 1271
    and language, evolution of 481, 667, 1176
    neuropsychology 396
    Rana computatrix 45, 1170–5, 1286, 1287
    schema theory 386, 979, 1173–7
    see also Guazelli et al 1175
Archimedes 54, 823
Aristotle 53, 60, 75, 1240
Armer, P. 68, 715–16, 826–7, 839, 842–3, 846
Armstrong, D. M. 1344
Arnauld, A. 71–2, 602–4
Arnold, W. J.: *see* Eysenck et al 244
Aronson, G., *see* Colby et al 372
Arrow, K. 1317
Arthur, B. 1317
Asch, S. E. 300–1
Ascott, R. 1088, 1089, 1091, 1099
Ashby, W. R. xli, 17, 222, 228–32, 235, 897, 1250, 1310, 1333, 1429
Asher 595–6

Atkinson, A. P. 486
Atran, S. 524, 579
Attneave, F. 285
Austin, G. 304–8
Austin, J. L. 389–90, 695
Averroes 56
Ayer, A. J. 1339
Azzopardi, P. 1225

Babbage, C. B. 19, 20, 52, 122, 131–5, 136–47, 151, 157, 162–6, 167
    Analytical Engine 121, 137, 142–6, 163, 164, 165
    and British Association for the Advancement of Science 134
    and calculating machines 131–2, 136–46
    Difference Engine 138–42
    and religion 135, 136, 137
Bach-y-Rita, P. 60–1
Bacon, F. 20, 22, 25, 61–2
Bacon, R. 602
Baddeley, A. 290, 292, 513, 925
Baker, J. K. 71, 698
Baldwin, J. M. 253
Bales, R. F.: *see* Stone et al 690
Ballonoff, P. A. 530
Banks, J. 134
Bar-Hillel, Y. 669, 672, 673, 674, 678–9, 681, 718, 1020
Barker, E. 570
Barker, R. G. 380
Barkow, J. H. 540–2, 543; *see also* Cosmides et al 542
Barlow, H. B. 40, 897–8, 934, 1122, 1203–4, 1209, 1218
    grandmother cells 1205, 1207–8
    on perception 1126–8, 1206–8
    and Ratio Club 1126
Baron, R. J. 1137
Baron-Cohen, S. 488, 490, 491, 500
Barr, A. 738
Barrett, K.: *see* Bertenthal et al 468
Barrow, H. G. 791, 961, 1192–3
Barsalou, L. W. 997
Barth, F. G. 537
Barthes, R. 25, 33
Bartlett, F. C. 211, 216, 250–1, 301–2, 335, 366, 589, 1346–7
Barto, A. G. 928
Bartrip, J., *see* Karmiloff-Smith 498
Bastin, E. W. xxxix, xli
Bates, E. 481; *see also* Elman et al 494, 993
Bates, J. 222